Understanding Chemistry

for Advanced Level

Ted Lister and Janet Renshaw

Trinity School, Leamington Spa

Stanley Thornes (Publishers) Ltd

First published in 1991 by:
Stanley Thornes (Publishers) Ltd
Ellenborough House
Wellington Street
CHELTENHAM GL50 1YD
England

Reprinted 1992, 1993, 1994

British Library Cataloguing in Publication Data

Lister, Ted
Understanding chemistry for advanced level.
1. Chemistry
I. Title II. Renshaw, Janet
540.3

ISBN 0-7487-0216-4

Typeset by Thomson Press (India) Ltd
Printed and bound in Hong Kong

ACKNOWLEDGEMENTS

The authors would like to thank a number of people for their help in producing this book. Many sixth formers at Trinity School, Leamington Spa read and commented on draft material. Professor P. J. Derrick and Dr D. W. Hutchinson of Warwick University Chemistry Department helped to locate mass and IR spectra. Thanks are also due to Malcolm Tomlin, the staff at Stanley Thornes (Publishers) Ltd and especially to Ruth Holmes for her meticulous editing. In addition, the authors would like to thank Colin Foster for the points he brought to their attention.

The following Examinations Boards granted permission to use questions from recent examination papers: Associated Examining Board, Joint Matriculation Board, University of Cambridge Local Examinations Syndicate, University of London School Examinations Board, University of Oxford Delegacy of Local Examinations. Please note that the numerical answers supplied on page 590 are the sole responsibility of the authors and have not been supplied or approved by the examining boards.

Data and technical information were obtained from *Revised Nuffield Advanced Science Book of Data* (Longman, 1984), and *Wiley/NBS Registry of Mass Spectral Data* (© Fred W. McLafferty and Douglas B. Stauffer 1989. Reprinted by permission of John Wiley & Sons, Inc.) All IR spectra are reproduced by courtesy of Aldrich Chemical Company, Inc.

The authors and publisher are grateful to all those who granted permission to reproduce photographs.
Claire Starkey pages 1, 11, 23, 25, 27, 28, 41, 56 (middle), 83, 84, 102, 108, 118, 144, 145(lower), 159, 161, 172, 174, 175, 215, 221, 224, 253, 255, 295(lower left), 296(lower), 317(lower), 318, 320(upper), 322, 328, 330, 344, 345, 346, 356, 368(lower), 382, 386, 391, 399, 401, 405, 406, 409, 415, 416, 421, 422, 425, 432(lower), 442, 444, 453, 465, 469(bottom), 481, 483, 486(top, middle), 494, 499, 503, 505, 506, 540, 546, 561, 563(upper), 565; Alex Renshaw pages 2, 468(top); Mary Evans Picture Library pages 13, 22, 46, 55, 231, 251, 252, 270, 280(upper); Education Development Center pages 39, 40; Hulton-Deutsch Collection pages 47, 141, 439; Ann Ronan Picture Library pages 49, 432(upper); Science Photo Library pages 56(bottom), 60(upper); 293, 316(lower right), 371(upper), 379, 452; U.K. Atomic Energy Authority pages 57(upper), 296(upper); Central Electricity Generating Board pages 57(lower), 233; Camera Press pages 60(lower), 115, 372(bottom); Ford page 73; Rex Features page 116; Philip Harris page 145(upper); The Associated Press Ltd page 171; Volkswagen Press page 184; Barnaby's Picture Library pages 195, 320(lower), 486(bottom), 487; John Payne/ICCE page 201; British Alcan Aluminium plc page 234; Perkin-Elmer Limited page 257; National Rivers Authority page 258; Royal Society page 270; ICI pages 294, 329, 388; Indusfoto page 295(upper); Kinetico page 295(lower right); Novosti Press Agency page 302; British Aerospace page 304; British Rail page 305; Mauri Rautkari/ICCE page 308; British Telecom page 316(upper); Imperial War Museum page 321; M. G. Duff Marine Ltd page 368(upper); Silva page 371(lower); British Steel pages 372(middle), 373; Shell Photographic Service page 413; Mavis Ronson/Barnaby's Picture Library page 427; Wellcome Institute Library, London page 469(middle); Peter Lumley page 474(upper); David Waterman/Camera Press page 474(lower); Biophoto Associates page 495; Alan Thomas page 504; Studio 70 page 508; Metropolitan Police page 509; Vauxhall Motors Ltd. page 547(upper); West Midlands Fire Brigade page 547(lower).

For Les and Roy

Contents

Colour plates (between pages 410 and 411, and pages 442 and 443)

1 Introduction

1.1 The importance of chemistry

Chemistry is the branch of science which is concerned with materials of every description. It is often called the central science as it overlaps with both biology and physics. On the one hand, chemists unravel the chemical reactions which are responsible for life, and on the other, they investigate new materials with exciting and potentially useful properties such as superconductors and electrically conducting plastics. Chemists are interested in the properties of substances—such as whether they are gas, liquid or solid, how hard, strong or brittle they are, whether they conduct electricity, and so on. They are also concerned with how to change one substance into another. Indeed chemistry evolved from the work of the early alchemists who tried to turn so-called base metals into gold. Although they failed they learned a lot of chemistry on the way.

Products of crude oil

Modern chemists are concerned with equally dramatic changes, turning, for example, crude oil into a whole range of useful and diverse products, such as nylon, aspirins, paint, adhesives and petrol. Other spectacular transformations include sand into glass and silicon chips, and nitrogen from the air into fertilizers and explosives. In fact we use very few materials which have not been changed in some way by a chemist. Even wood is likely to be treated by fungicides to prevent rot, and then painted or varnished.

All this means that chemistry is big business. The UK chemical business is the nation's fourth largest industry and it is the fifth largest chemical industry in the western world. It is our largest export earner with overseas sales of almost £10 billion in 1986. Two hundred major chemical companies employ 6 per cent of the UK work force, producing many diverse products, **Figure 1.1.**

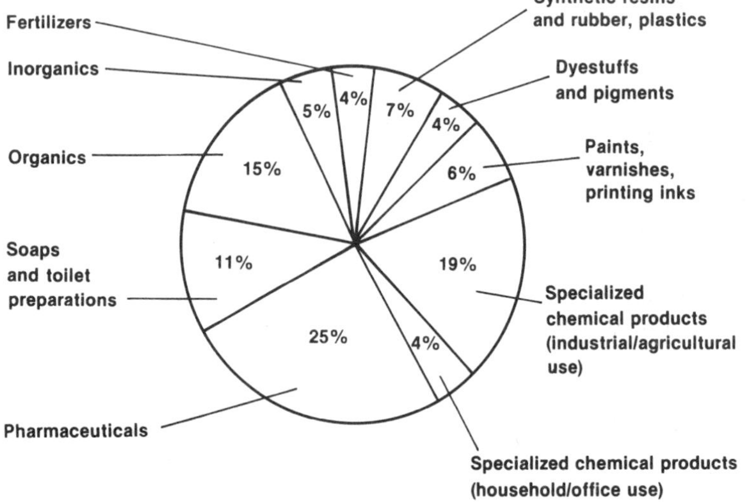

Figure 1.1 Products of the UK chemical industry

Disused quarries need not be just an eyesore

Chemistry has also created some serious problems:

- Quarrying for limestone and other raw materials for industry has produced what most people consider to be blots on the landscape.
- Accidents in industry, such as Flixborough in 1974 where a cyclohexane plant blew up, Seveso in 1976 where dioxin escaped, and Bhopal in 1984 where a cloud of methyl isocyanate poisoned the surroundings, have hit the headlines.
- Pollution problems such as acid rain, depletion of the ozone layer and the greenhouse effect are issues which must be tackled.
- Many people are worried about the long-term effects on health of food additives.

Chemical research may be 'targetted', where it is dedicated to making new substances which are more effective, cheaper and with fewer disadvantages than those used at present, or 'fundamental', where it is designed to further our understanding of how and why things are as they are. One of many current growth areas is biotechnology which involves tailoring enzymes, the super-efficient biological catalysts, to bring about useful reactions faster and with less drastic conditions. Recently it has become possible to synthesize 'factor 8', the clotting component of blood, by biotechnological means, instead of extracting it from blood. This eliminates the risk of haemophiliacs contracting AIDS from contaminated blood. What is undeniable is that chemistry affects the lives of every one of us.

1.2 Careers, higher and further education in chemistry

What use is an A-level or A/S-level qualification in chemistry? Broadly speaking, there are three possibilities.

You may go into a career in the field of chemistry itself where your A level will be directly relevant. You may go in a direction where chemistry is not your main interest but acts as a useful or even essential back-up. Medicine would be a good example. Thirdly, your future career may lie in a direction where your chemistry is not relevant but the fact that you have successfully studied the subject shows your intelligence, application and ability to learn. **Figure 1.2** shows just some of the many career options in chemistry itself. Some of these may be entered directly after leaving

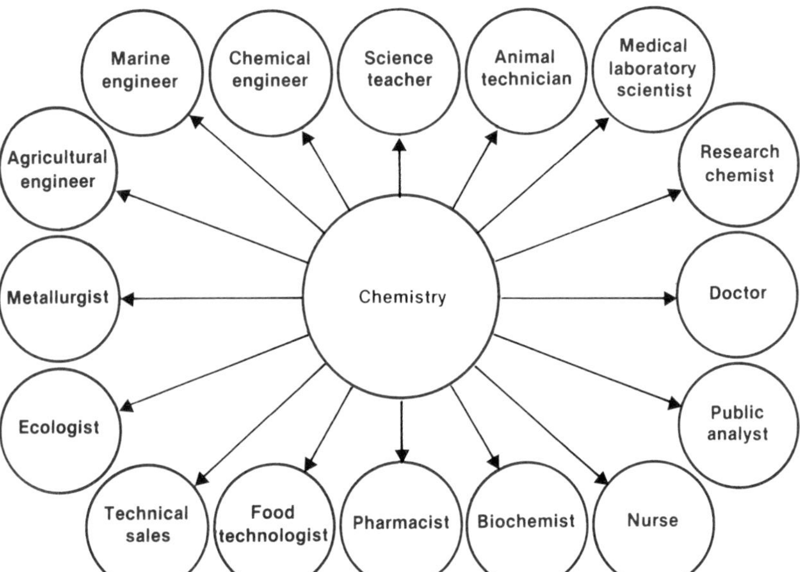

Figure 1.2 Some career directions which are possible with a qualification in chemistry

school with A levels, probably with on-the-job training or day release to gain further qualifications. Others may best be entered after further study, such as a degree or further education in chemistry or a related subject.

Figure 1.3 shows some of the further/higher education paths. Careers in which chemistry acts as a supporting subject offer an even wider choice—literally from agriculture to zoology. Again there is a choice of entry into employment either direct from school or after further or higher education. **Table 1.1** (below) shows a range of courses which could follow A- or AS-level chemistry. Careers following higher education offer the widest choice of all. Depending on your other A-level subjects very few career choices are barred. In our experience, students with combinations of chemistry and art and chemistry and foreign languages have been successful.

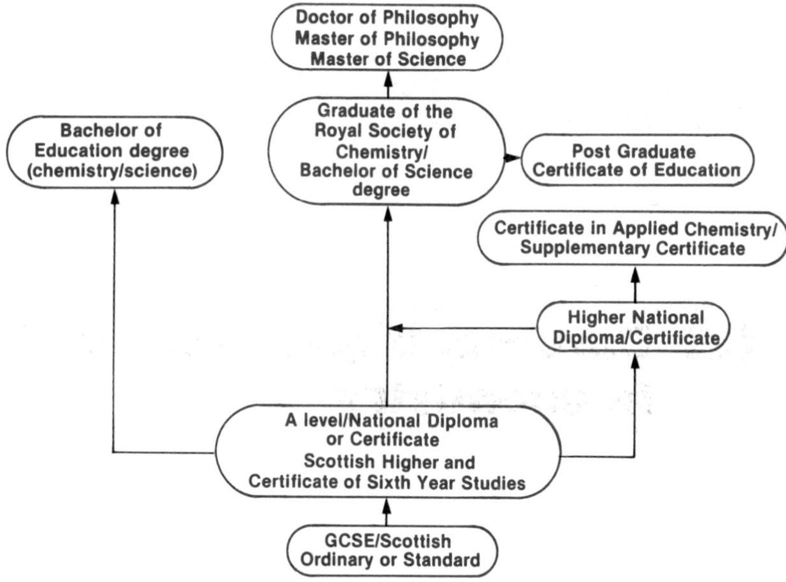

Figure 1.3 Further qualifications in chemistry

Table 1.1 Courses which could follow advanced study in chemistry

Agriculture	Cell biology	Genetics	Petrology
Agricultural sciences	Ceramics	Geochemistry	Pharmaceutical chemistry
Agricultural bacteriology	Chemical education	Geology	Pharmacology
Agricultural biochemistry	Chemical engineering	Glass technology	Pharmacy
Agricultural botany	Chemical physics	Human biology	Physical chemistry
Agricultural chemistry	Chemical process	Human ecology	Physical sciences
Agricultural zoology	Chemical technology	Human sciences	Physiology
Analytical chemistry	Chemistry	Inorganic chemistry	Plant sciences
Animal biology	Colour chemistry	Materials science	Polymer chemistry
Animal nutrition	Colour technology	Material technology	Polymer and colour science
Animal physiology	Crystallography	Marine biology	Polymer engineering
Animal sciences	Dentistry	Marine botany	Polymer sciences and technology
Applied biochemistry	Development physiology	Marine zoology	Science (general)
Applied biology	Dietetics	Medical biochemistry	Science (history and philosophy)
Applied electrochemistry	Dyeing and dyestuffs technology	Medicine	Soil science
Bacteriology	Earth sciences	Metallurgy	Textile technology
Biochemical engineering	Environmental sciences	Microbiology	Theoretical chemistry
Biochemistry	Exploration science	Mineral exploitation	Veterinary science
Biological chemistry	Farm animals (physiology and biochemistry)	Mineral sciences and technology	Virology
Biology	Fibre science	Minerology	Wood science
Biological sciences	Food science and technology	Molecular sciences	Zoology
Biomedical electronics	Forestry	Natural sciences	
Biomedical engineering	Fuel science	Neurobiology	
Botany		Nutrition	
Brewing		Oil technology	
		Paper science	
		Pathology	

1.3 Study skills for A- and A/S-level chemistry and how to use this book

This book has been written as an aid to studying, understanding and passing A-level and A/S-level chemistry. It is not a 'teach yourself' book and no book can be a substitute for the work you will do in the class with your teacher. In particular, no book could be a substitute for practical work, for chemistry is essentially a practical subject.

Many A-level students make the mistake of taking things easy during the lower sixth. This is unwise as the pace of sixth form work is such that it is very difficult to catch up. A good habit to cultivate would be to read through the notes you take *after each lesson* and relate them to the relevant section of this book. Then check that you have understood the work by trying the relevant questions at the end of the chapter. If you get stuck, read the chapter again. The summaries at the end of each chapter will also be useful—you can use these to do some quick revision, and again go back to the chapter content if you need help.

Understanding Chemistry for Advanced Level has been written after a thorough analysis of all the A and A/S syllabuses. While there is a common core of content for all the syllabuses, each one differs slightly so that it would be a good idea to find out which syllabus you are following and get a copy of it (or better still ask your teacher) to check which topics you should be studying. This will apply even more to A/S levels whose content is around half that of an A level.

Most chapters contain boxes featuring extension material, reference material or applications. It should be possible to follow the chapter without reading the boxed material and you may wish to do so at a first reading. **Extension boxes** (marked E) contain material which is harder than that in the main chapter and may go beyond the strict bounds of A level. For example, there might be the derivation of an equation which you would be happy to simply accept. **Applications boxes** (marked A) might best be read when you have grasped the main gist of the chapter or section, but you should not ignore them. The social, economic, environmental and technological applications of chemistry are increasing in importance in A- and A/S-level syllabuses and exams. **Reference boxes** (marked R) contain data, equations and definitions which may be needed throughout the chapter.

Chapters 2–7 we have called 'foundations', and they are just that. They are an attempt to bridge the gap between GCSE (either chemistry or double award science) and A or A/S level. This is basic material that you should thoroughly understand before starting your A-level course. It can also be used to revise the basics at any time during your course.

The rest of the book is divided into parts on physical chemistry, inorganic chemistry and organic chemistry. We have arranged them in what we believe is a logical sequence. There is no reason to tackle them in this order although the first chapter of both the inorganic and the organic parts sets the scene for the rest of the part.

1.4 Examinations in chemistry

If you have understood and revised your work you should have little to fear from the final examination. However we shall give you a few points to help you get the maximum marks from what you know.

One very important point is good preparation. Make sure you know well in advance the dates of your chemistry exams. This will help you plan your revision. Most A- and A/S-level exams have more than one paper—so you need to know when they are, how long they last and the format of the paper, i.e. the number, type and choice of questions. You should be able to get this information from your teacher but it will also be in the up-to-date syllabus of your course.

You should make a point of studying past papers to see the format as well as to practise questions. Once you know the format of the papers you can devise a plan as to how much time to spend on different sections of the paper. Where there is a choice of questions, allow time to read all the alternatives carefully. It is only too easy to make a hasty choice which you may regret later. For example, question 7 in Chapter 14 could easily be taken to be about organic chemistry, while it is actually about reaction rates. Even if you are pressed for time, *always* attempt the correct number of questions. Exam questions are almost always written so that the first few marks are easy, the next few a little harder and so on, until the last few marks will be gained by almost no one. So it makes sense to try and pick up the easy marks on all the questions rather than to try and complete answers to just a few. Do not attempt more than the maximum number of questions—the examiner will not be allowed to pick the best; he or she will usually have instructions to mark your answers in the order they are written, and ignore questions after the maximum asked for.

It should go without saying that you should make sure that you arrive at the exam in good time and fully equipped with pen, pencil, calculator, etc. including spares. Finally, some chemistry exams allow you to use data books, data sheets and even specified text books (*note*: only in one paper in the Nuffield exams from 1990). Make sure you know what, if anything, you are allowed to use and make sure you are familiar with the sheet, book or books.

1.4.1 Types of exam questions

There are four main types of questions used in A- and A/S-level chemistry exams. They are illustrated in the examples below.

Example 1

Which of the following analytical techniques makes use of vibration within molecules?

A) Infra-red absorption spectroscopy
B) Measurement of dipole moments
C) Mass spectroscopy
D) Nuclear magnetic resonance
E) X-ray diffraction (Nuffield 1988)

Multiple-choice questions such as Example 1 are normally answered on a machine-read answer sheet. Try to see a copy of the type you will actually be using so that you are familiar with it. The usual time allowed is something like one hour for forty questions. Usually you have to select the best response from five. There are different types of multiple-choice questions such as ones where you have to select which *combination* of statements is correct. Again, find out what types you should expect.

Example 2

Br

reaction A

fuming H_2SO_4

compound C

benzene

concentrated HNO_3 + concentrated H_2SO_4

compound B

A) State the reagent and conditions required for reaction A. (2 marks)
B) Give the name and structural formula for compound B. (2 marks)
C) Give the name and structural formula for compound C. (2 marks)
(Nuffield 1983)

Short-answer questions such as Example 2 are normally answered on the paper itself. The amount of space allowed for the answer is a guide to how much you are expected to write. The marks available for each part of the question are usually given. You are guided through most calculations.

Example 3

An organic liquid, A, when refluxed with 100% phosphoric acid, produced a low boiling liquid, B, containing carbon and hydrogen only. When the organic liquid, A, was oxidized with nitric acid a solid, C, was formed which was a dibasic acid.

0.240 g of A gave 73.5 cm^3 of vapour at 100 °C and 1 atmosphere pressure.

0.372 g of B on combustion produced 1.20 g of carbon dioxide and 0.410 g of water.

0.150 g of C when titrated with 0.1 M sodium hydroxide solution gave an end point at 20.5 cm^3.

Use the data and information provided to calculate the empirical and molecular formulae of A, B, and C. Draw diagrams of their molecular structures and write equations for the formation of B and C from A.

(Nuffield 1985)

Example 4

Describe the main reactions of alcohols, quoting formulae and equations wherever appropriate.

The behaviour of the hydroxyl group in phenol is often significantly different from its behaviour in ethanol. Give examples of such differences in behaviour and explain them. (Nuffield 1985)

Figure 1.4 Scaling graphs

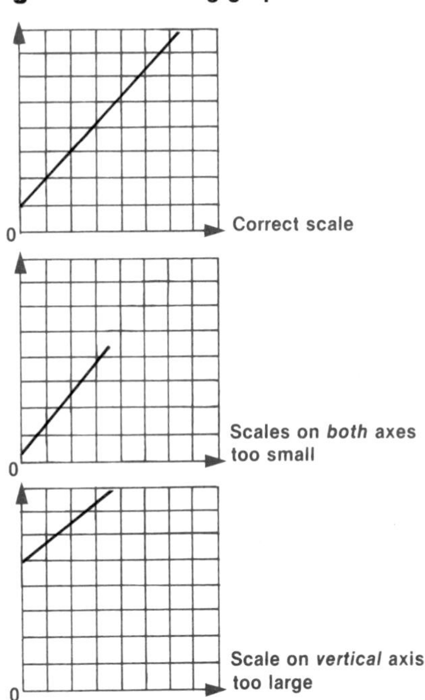

Free-response questions such as Examples 3 and 4 are longer. They may require an extended calculation, an extended essay-type answer or something in between. Either way, once you have chosen the question, some planning of the answer is required, particularly for the essay-type questions. Note that although the term 'essay-type' is often used, the examiners do not require a structured essay unless they specifically ask for it. An answer set out using headings will almost always be preferable and easier to do in a short time. Whenever you refer to a chemical reaction, write a chemical equation if you can. Ideally it should be a balanced symbol equation including state symbols. Long answers may include tables of data, graphs and diagrams. Make sure that tables of data have headings with units (if appropriate) and that graphs have titles, labels (with units) on axes, an appropriate scale (see examples in **Figure 1.4**) and that points are clearly plotted and a straight line or smooth curve (as appropriate) is drawn.

Numerical answers to problems should always have units.

1.5 Practical skills and techniques

Chemistry is essentially a practical subject, so anyone qualified as a chemist must be able to show that they can work safely in a laboratory and that they have mastered some practical skills. Thus all A-level examinations assess practical skills in some way—either by a practical exam or by continuous assessment during the course, carried out by your teacher.

If you are to be assessed by an exam, find out the details of the exam's format, what you will be expected to do and what books or notes you will be allowed. Most practical exams allow you to take in books or notes as it is your practical skills that are being tested.

If your practical work is to be assessed by your teacher, he or she will explain the system to you.

Safety and careful, methodical working are the keys to good practical work. Always wear eye protection when doing practical work, and ideally a lab coat as well. Take seriously any safety warnings given in instructions or on chemical bottles. Carefully check the names on bottles and above all make sure you know what you are doing before starting. Try to obtain a clear working area of bench and keep it clear with only the chemicals and equipment you need around. Good practical skills can be gained only by practice but a few tips may be useful.

- When reading scales you should aim for an accuracy of half the smallest scale reading. For example when reading a burette, where the smallest marked division is $0.1\,cm^3$, you should estimate the reading to $0.05\,cm^3$, i.e. decide whether the meniscus is exactly on the division, in which case the reading should be given as x.00, or in between divisions in which case the reading should be given as x.05.
- Try to estimate the likely errors in your experimental work. These may be of two types—systematic (to do with the design of the experiment) or measuring (to do with how accurately a measuring instrument can be read).

Figure 1.5 A simple apparatus for measuring the enthalpy (heat) change of combustion of an alcohol

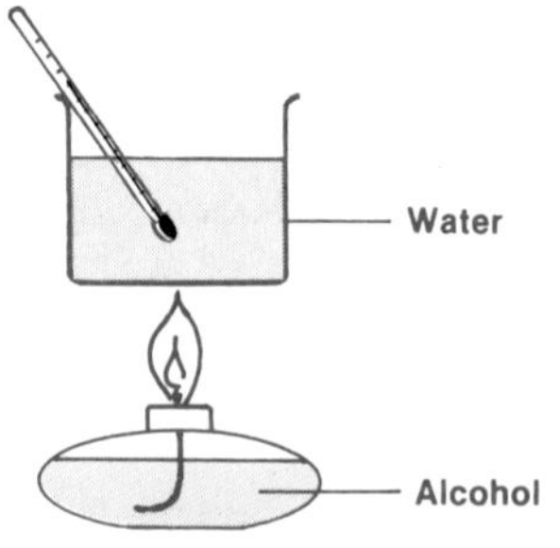

For example: suppose we have to measure the enthalpy (heat) change of combustion of an alcohol using simple apparatus, as shown in **Figure 1.5** (see **section 9.3.1**). Systematic errors revolve around the fact that much of the heat fails to enter the water in the beaker, and that some of that will be lost from the sides and top. Also, incomplete combustion may occur. Measurement errors occur in reading the thermometer, weighing the alcohol and weighing the water.

Try to develop the idea of looking critically at experiments. Point out any systematic errors and suggest ways of overcoming them—for example the use of a lid and insulation for the sides of the beaker in the above experiment. Try to estimate likely measuring errors, for example, 'the initial temperature of the water was $21.5\,°C \pm 0.5\,°C$'. If the temperature rose by $10\,°C$ in the experiment this gives an uncertainty of $\pm 1.0\,°C$ on a measurement of $10\,°C$ ($\pm 0.5\,°C$ for two readings). This is a possible error of

$$\frac{1.0 \times 100}{10} = 10\%$$

It would be foolish to quote the enthalpy change of combustion as $-2056.3712\,kJ\,mol^{-1}$ with this sort of experiment. $-2060\,kJ\,mol^{-1}$ might be appropriate, so think carefully about errors and their sizes.

If possible, repeat measurements to get a check on their accuracy. In titrations, for example, you should aim to get two titres within $0.1\,cm^3$ and then use the average of these two in your calculations.

Part A Foundations

2 The kinetic particulate theory of matter

2.1 Introduction

One of the key ideas of science is that matter, which is anything with mass, is made of small particles, that is, it is *particulate*. These particles are in motion, and so the whole idea is called 'the *kinetic* particulate theory of matter' (kinetic meaning moving). This is important because it gives us a picture of matter (a model) which we can use to explain things like rates of reactions or properties of gases.

2.2 The three states of matter

Table 2.1 sets out the model for the three states of matter. Some of the experimental evidence is described later and this is marked*. It is very rare for a single piece of evidence to be conclusive.

Figure 2.1 Compressing liquids and gases

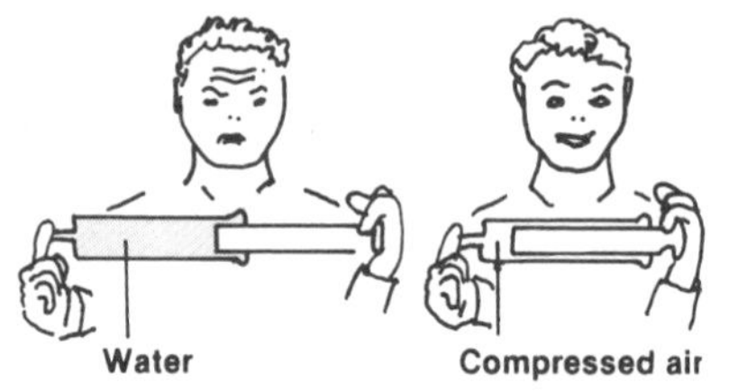

Water in a syringe cannot be compressed

Air in a syringe can be compressed easily

Figure 2.2 A drop of ethanol liquid takes up 500 times more space when made into a gas

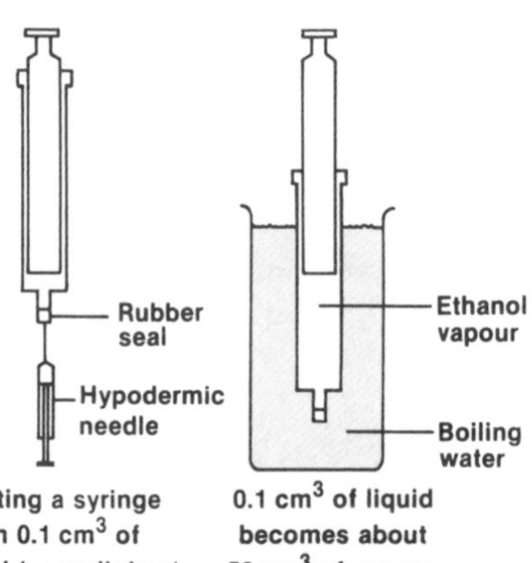

Injecting a syringe with 0.1 cm^3 of ethanol (a small drop)

0.1 cm^3 of liquid becomes about 50 cm^3 of vapour

Table 2.1 The three states of matter

	Solid	Liquid	Gas
Arrangement particles	Regular	Random	Random
Evidence	Crystal shapes have straight edges. Solids have definite shapes.	None direct but a liquid changes shape to fill the bottom of its container.	None direct but a gas will fill its container.
Spacing	Close	Close	Far apart
Evidence	Solids are not easily compressed.	Liquids are not easily compressed. (See **Figures 2.1 and 2.2.)**	Gases are easily compressed.
Movement	Vibrating about a point	Rapid 'jostling'	Rapid
Evidence	*Diffusion is *very* slow. Solids expand on heating.	*Diffusion is slow. Liquids evaporate.	Diffusion is rapid. *Gases exert pressure. *Brownian motion
Models	**Vibration** Particles vibrate about a point.	heat → / ← cool **Evaporation** Particles move but are too close to travel far except at the surface. Melting temperature T_m	heat → / ← cool Particles are free and have rapid random motion. Boiling temperature T_b

2.3 Evidence for particles

There is very little *direct* evidence that we can observe in the school laboratory for the particulate nature of matter, and indeed most of the behaviour of matter could be explained using a model of matter as being continuous, perhaps like a stretchy piece of gum. (This could be why the idea of matter being made of particles was rejected by the ancient Greek philosophers.) The results of the following simple experiment are not easily explained, except by using the idea of particles.

2.3.1 Mixing liquids

If 50 cm^3 of ethanol is mixed with 50 cm^3 of water, the final volume of the mixture is around 97 cm^3. Mixing water and water or ethanol and ethanol gives the expected 100 cm^3, so spillage or evaporation cannot be the cause.

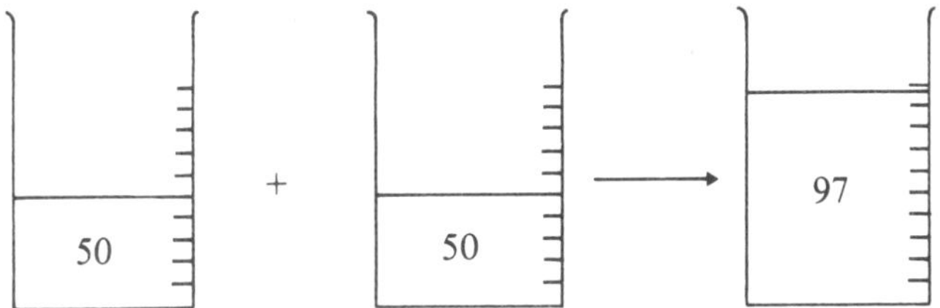

However, the same kind of loss of volume would happen if peas and sand were mixed together.

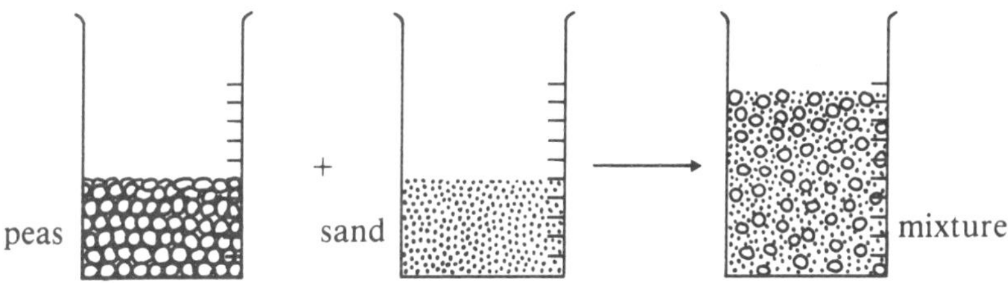

This suggests that the two liquids could be made of particles of different sizes. On mixing, the smaller particles can fit into the gaps between the larger ones, thus reducing the total volume of the mixture.

2.4 Evidence for movement

2.4.1 Diffusion

The tendency for substances to spread out and mix due to the motion of their particles is called **diffusion**.

Gases diffuse rapidly, which is why we can smell the gas from a distant gas tap a few *seconds* after it is turned on.

Liquids also diffuse, but more slowly. We can see this if we pour ethanol very carefully on to a solution of copper sulphate, so that two layers form. It will take a few *weeks* for the two layers to become a uniform colour, see **Plate 1.**

If two different solid metals, such as silver and gold, are clamped together, some small evidence of diffusion may be seen after several *years*. The particles, which are vibrating, can gradually intermingle at the face where they meet.

The oil can demonstration

2.4.2 Brownian motion

Specks of pollen in water, or specks of smoke in air, can be seen to be 'jigging about' when observed under a microscope. The explanation for this is that the specks are being bombarded on all sides by particles of water or air which themselves are too small to be seen. This experiment suggests that particles exist, *and* that they can move rapidly. Also, since the smoke or pollen specks move at random, it seems reasonable to suggest that the particles bombarding them also move at random.

2.4.3 Pressure

Gases exert a pressure on the walls of their containers. This is not always obvious as most everyday objects (including ourselves) have air both inside and out, thus cancelling the effect of air pressure. However, pumping out the air from, for example, a gallon oil can will cause the can to collapse spectacularly, see photograph and **Figure 2.3.**

Figure 2.3 Collapsing a can by air pressure

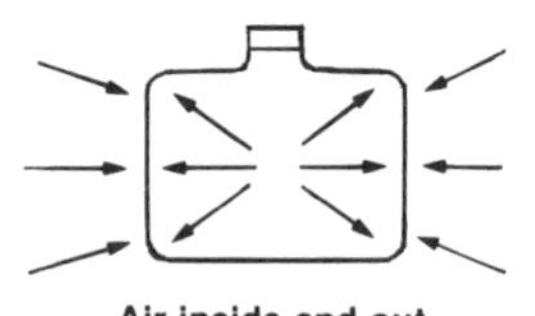

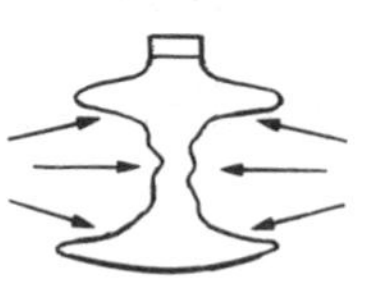

This collapse is caused by pressure from the fast-moving particles of air bombarding the walls of the can on one side only.

Figure 2.4 Measuring the effect of temperature change on pressure of a gas

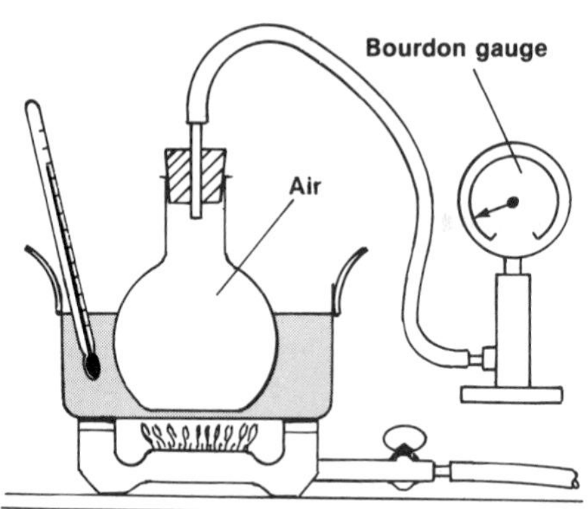

The effect of temperature on pressure

This can be investigated using the apparatus shown in **Figure 2.4.** It is found that increasing the temperature increases the gas pressure. This can be explained if the particles of the gas move more quickly at high temperatures, thus colliding more often with the sides of the container and hitting them harder.

2.5 The size of particles

2.5.1 The oil drop experiment

This is a simple way of getting some idea of the size of a particle. More sophisticated and accurate methods are now available, see **section 19.2.1.** The basic assumption is that, when a liquid floats on top of water, it spreads out (if there is enough room) into a layer that is only one particle thick—a monolayer. This means that if we can measure the thickness of the monolayer, we are measuring the diameter of a particle.

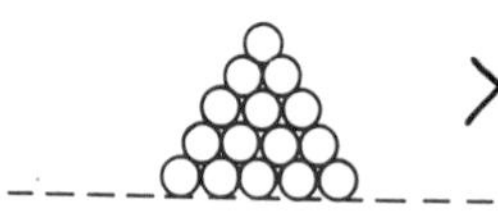

A monolayer rather than a clump

The thickness of the layer can be calculated if we know the *volume* of the liquid and the *area* covered by the oil slick, since:

$$\text{volume} = \text{area} \times \text{height}$$

so

$$\text{height} = \frac{\text{volume}, V}{\text{area}, A}$$

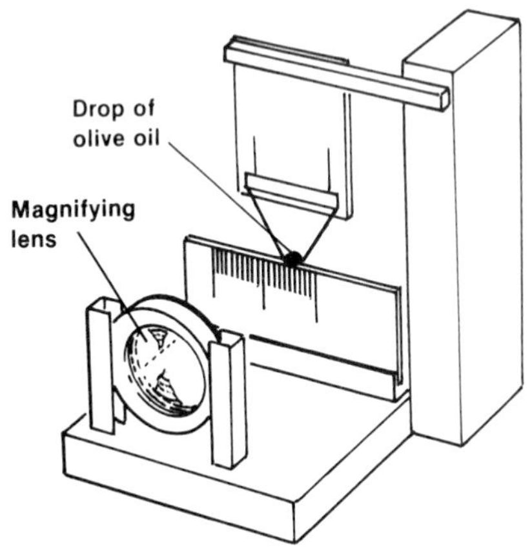

Figure 2.5 Apparatus for measuring the diameter of an oil drop

The experiment is done in the laboratory by putting a tiny drop of oil on to a clean powdered water surface. The drop must be tiny, otherwise it would not have room to spread, and the water surface is powdered lightly so that the area of the oil drop can be seen and measured.

The volume of the drop can be estimated in many ways, but the most direct method is to measure the diameter of a drop in a loop of wire using the apparatus shown in **Figure 2.5.**

The volume can be found from the formula for the volume of a sphere:

$$V = \frac{4}{3}\pi r^3$$

If this drop is put on to the water, the area of the oil slick can then be found by measuring its diameter (assuming a perfect circle is formed). The area can be found from the formula for the area of a circle:

$$A = \pi r^2$$

then

$$\text{thickness} = \frac{V}{A}$$

This experiment will give an answer for the thickness of a particle of olive oil of about

$$\frac{3}{10\,000\,000}\ \text{cm} = \frac{3}{1\,000\,000\,000}\ \text{m} = 3 \times 10^{-9}\ \text{m} = 3\ \text{nanometres (nm)}$$

This would imply that there are 300 000 oil particles in a space the size of a pin head. Olive oil molecules are actually quite large (0.1 nanometres is a typical diameter for an atom).

3 Atomic structure and bonding

3.1 Inside the atom

The quest to find out what atoms were made of started when radioactive elements (whose atoms naturally break up revealing their contents) were discovered. This search is not finished and new subatomic particles are still being discovered. However, we can work with quite a simple picture which can explain the patterns of the Periodic Table (see Chapter 4) and why elements react with each other.

3.1.1 The structure of the atom

The most surprising feature is that the atom is mostly space. The mass is concentrated in the middle, in the **nucleus**, which is positively charged, and orbiting round this nucleus, but a long way from it, are very light, negatively charged particles called **electrons**. It is the attraction between positive charge and negative charge that holds the electrons in place round the atom. The nucleus is composed of two sorts of subatomic particles, **protons**, which each have a positive charge, and **neutrons** which have the same mass as protons but are neutral, **Figure 3.1.**

Table 3.1 Main subatomic particles

	Proton	Neutron	Electron
Position	Nucleus	Nucleus	Orbiting nucleus
Mass (atomic mass units)	1	1	$\frac{1}{1840}$
Charge (relative)	1+	0	1−

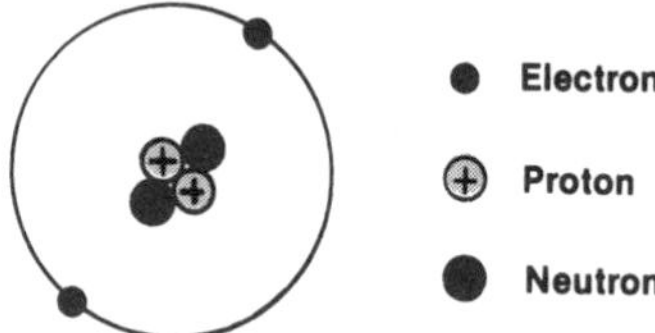

Figure 3.1 The arrangement of subatomic particles in the atom (not to scale)

The atomic number, *Z*

The number of protons in any atom is equal to the number of electrons and thus the atom is neutral. This number of protons or electrons is called the **atomic number** of the element and it is unique for the element. It is usually given the symbol *Z*. If you look at the Periodic Table you will see that each element has two numbers, for example

lithium $\;\;{}^{7}_{3}\text{Li}$ — 7 ← relative atomic mass; 3 ← atomic number

The smaller number is the atomic number and in most versions of the table it is written at the bottom. The Periodic Table is arranged in order of the number of protons and there are no numbers missing, starting with hydrogen which has one proton and ending with unnilhexium which has 106.

In 1869, when Mendeleev arranged the elements which were known at the time into the Periodic Table, he had no knowledge of protons, so his achievement was really astonishing.

Dmitri Mendeleev – father of the Periodic Table

The relative atomic mass, A_r

The larger number, the **relative atomic mass**, A_r, tells us the total number of protons and neutrons. The number of neutrons in an atom can be found by taking the atomic number from the relative atomic mass, for example:

lithium $^{7}_{3}Li$ 7 protons + neutrons / 3 protons

has $7 - 3 = 4$ neutrons in its nucleus

Isotopes

The only complication is that the relative atomic mass is the *average* mass of an atom of the element. Some elements have atoms with different numbers of neutrons, which means that some atoms of that element will be slightly heavier than others. These slightly different atoms of the same element are called **isotopes**. The number of protons in each atom of any given element is always the same, so the atomic number, and not the relative atomic mass, actually defines the element. A more accurate way of representing an atom is to use the **mass number**, because this is not an average. The mass number is the total number of protons and neutrons in a given atom, for example:

	mass numbers		relative atomic mass
chlorine	$^{35}_{17}Cl$	$^{37}_{17}Cl$	$^{35.5}_{17}Cl$

three of this to every one of this

Chlorine has two isotopes in a ratio of about three to one, so that the average mass is 35.5, as shown below.

$$\text{mass of four atoms} = 35 + 35 + 35 + 37 = 142$$
$$\text{average mass} = 142/4 = 35.5$$

Figure 3.2 Electron shells

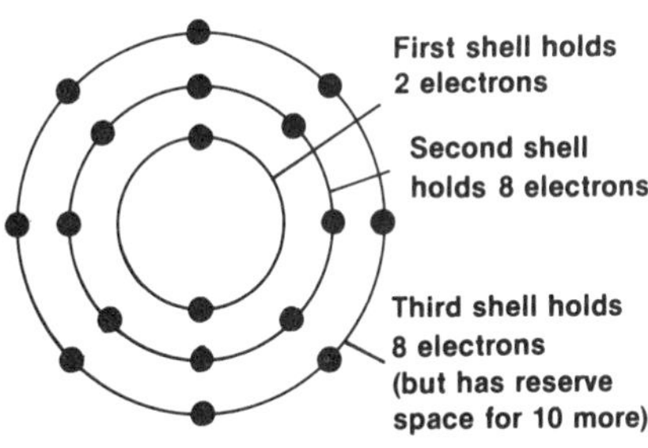

The arrangement of the electrons

Again we can use a simple picture that will work well enough at this stage. The electrons orbit the nucleus in a series of 'shells' which are further and further from the nucleus and so become bigger and bigger.

Each shell can hold only a given number of electrons, **Figure 3.2.** The first shell, which is always filled first, is closest to the nucleus and can hold only two electrons.

The second shell is full when it has eight electrons in it.

The third shell will also hold eight, but has reserve room for a further ten and fills up when we reach the transition elements. For our simple picture this is far enough.

We can start with hydrogen and go through to calcium with the electron arrangement of each atom, as in **Table 3.2.**

Table 3.2 Electron arrangements of the first 20 elements

	Shell			
Element	First	Second	Third	Fourth
$_1H$	1			
$_2He$	2			
$_3Li$	2	1		
$_4Be$	2	2		
$_5B$	2	3		
$_6C$	2	4		
$_7N$	2	5		
$_8O$	2	6		
$_9F$	2	7		
$_{10}Ne$	2	8		
$_{11}Na$	2	8	1	
$_{12}Mg$	2	8	2	
$_{13}Al$	2	8	3	
$_{14}Si$	2	8	4	
$_{15}P$	2	8	5	
$_{16}S$	2	8	6	
$_{17}Cl$	2	8	7	
$_{18}Ar$	2	8	8	
$_{19}K$	2	8	8	1
$_{20}Ca$	2	8	8	2

Looking for a pattern

If we now write the elements in the same groups together, we can see a pattern:

Group I		Group II		Group III		Group 0	
H	1					He	2
Li	2, 1	Be	2, 2	B	2, 3	Ne	2, 8
Na	2, 8, 1	Mg	2, 8, 2	Al	2, 8, 3	Ar	2, 8, 8
K	2, 8, 8, 1	Ca	2, 8, 8, 2				

We can see that all Group I elements have one electron in the outer shell, Group II elements have two electrons in the outer shell, Group III have three, and so on, until we get to the inert gases in Group 0. These all have a full outer shell.

From these patterns we can infer that:

1. It is having the same number of electrons in the *outer* shell that makes elements have similar chemical properties.
2. A full outer shell of eight electrons (except for the first shell which is full with two electrons) makes an atom chemically inert.
3. The electrons are all-important in the chemical behaviour of atoms. We might expect this anyway, since electrons are on the outside of the atom.

Figure 3.3 The electron arrangement of sodium

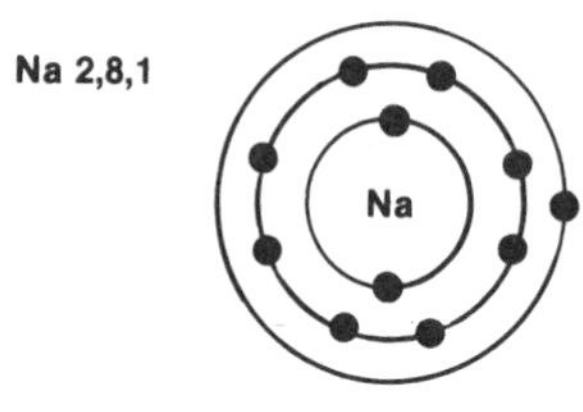

Electron diagrams

Atoms can be represented as flat circles. This is not at all accurate, but gives a simple picture of the electrons, which is very helpful for showing how atoms bond together, for example, **Figure 3.3.**
In these diagrams, it helps to show electrons as having four positions on the circle. These four positions are first filled singly, then filled up as pairs, for example, **Figure 3.4.**

Figure 3.4 (right) The electron arrangements of carbon, nitrogen and oxygen

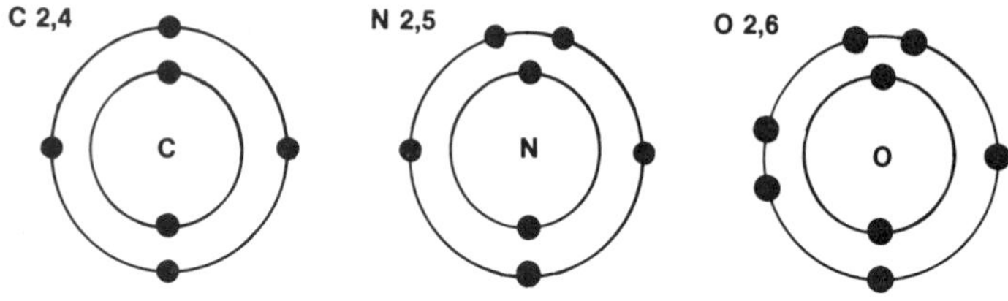

3.2 Bonding

Atoms bond together to become more stable, which means they are less reactive when bonded. Since inert gases are very unreactive (and so must be stable), it is logical to suppose that a full outer shell leads to stability. We can thus say that atoms bond together to get a full outer shell. There are three types of strong chemical bonds: **ionic**, **covalent** and **metallic**.

3.2.1 Ionic bonding

This occurs between metals and non-metals, for example, in sodium fluoride, **Figure 3.5.** The outer electron of the sodium atom moves into the outer shell of the fluorine atom. An electron is *transferred*, **Figure 3.6.** Each outer shell is now full, and both sodium and fluorine have an inert gas electron structure.

Figure 3.5 A dot cross diagram to illustrate the different electron arrangements of sodium and fluorine (remember that electrons are all identical whether shown by a dot or a cross)

Na

F

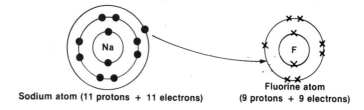

Sodium atom (11 protons + 11 electrons)

Fluorine atom (9 protons + 9 electrons)

Figure 3.6 Transfer of electron from a sodium atom to a fluorine atom

Na^+ Sodium ion (11 protons + 10 electrons)

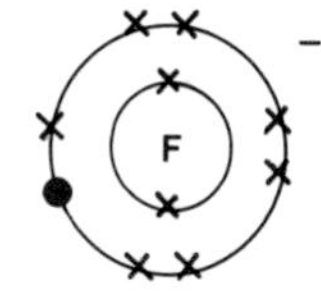

F^- Fluorine ion (called fluoride) (9 protons + 10 electrons)

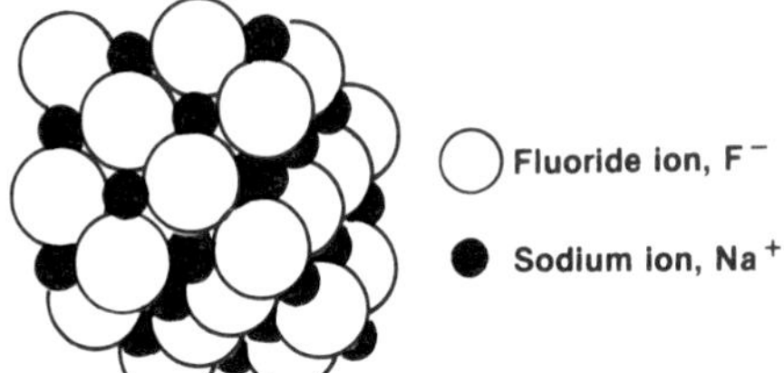

Figure 3.7 The sodium fluoride structure. This is an example of a giant structure – the bonding extends throughout the compound, and because of this the compound will be difficult to melt

This means that the two particles left are no longer neutral. The sodium will be positively charged, because it has lost a negative electron. The fluorine will be negatively charged, because it has gained an extra electron. These charged particles are called **ions**. They are attracted to each other and to other ions in the compound and it is this attraction of positive for negative which makes the bond between atoms. The attraction extends throughout the compound, **Figure 3.7.** We can also say that the formula of sodium fluoride is NaF, because we know that one sodium atom is required for each fluorine atom. We call this smallest repeating unit an **entity**.

For example:

1. Magnesium oxide:

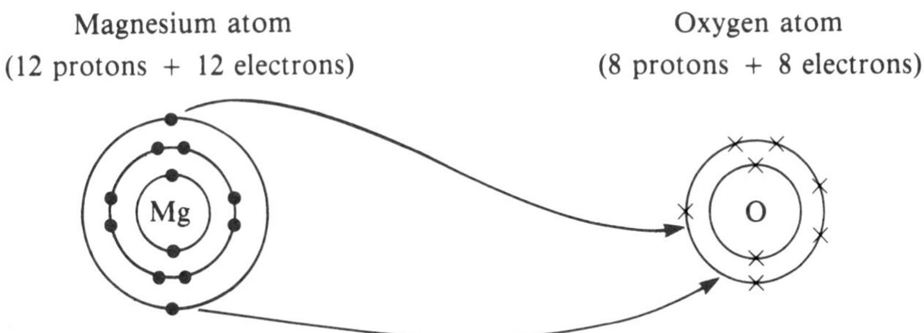

This time *two* electrons from the magnesium atom move to the outer shell of the oxygen atom.

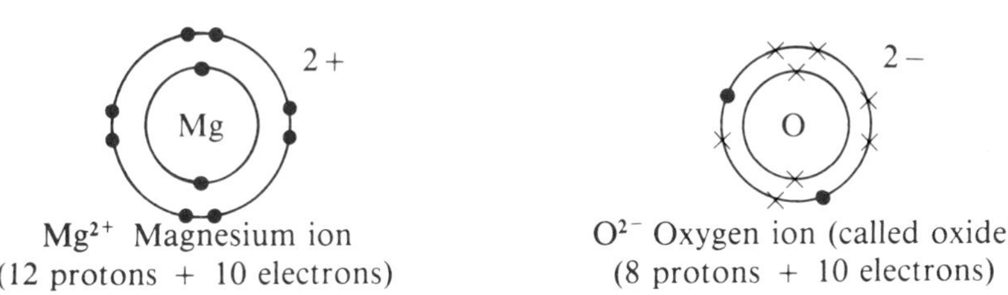

Mg^{2+} Magnesium ion
(12 protons + 10 electrons)

O^{2-} Oxygen ion (called oxide)
(8 protons + 10 electrons)

The magnesium atom has lost two electrons and has a 2^+ charge. The oxygen atom has gained two electrons and has a 2^- charge.
The formula of magnesium oxide is MgO.

2. Calcium chloride:

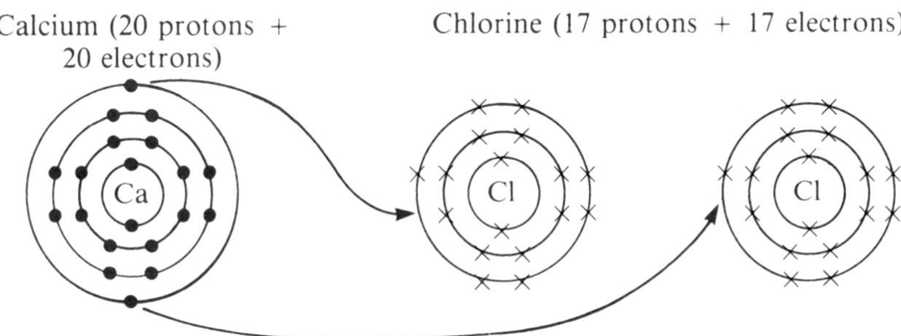

This time we need *two* chlorine atoms for each calcium atom because the calcium atom has two electrons in its outer shell to give away, but each chlorine atom can only receive one electron.

2+ Ca

Cl − Cl −

Cl^- Cl^-

Ca^{2+} Calcium ion
(20 protons + 18 electrons)

Chloride ions
(17 protons + 18 electrons)

The formula of calcium chloride is $CaCl_2$.

Properties of compounds with ionic bonding

- Ionic compounds are solids which have high melting temperatures.
- They will not conduct electricity when solid. (The bonding is too strong for the ions to separate.)
- If they are melted, i.e. liquid, they will conduct electricity. (The charged ions are then free to move and carry a current.)
- Those which dissolve in water produce solutions which will conduct electricity. (Again the ions are free to move.)

3.2.2 Covalent bonding

This occurs between non-metals, for example fluorine gas, F_2:

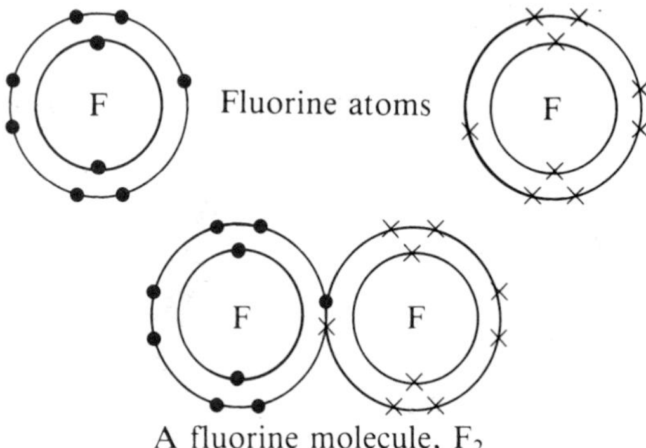

A fluorine molecule, F_2

In this sort of bonding the atoms move close together and *share* a pair of electrons. Each atom then has a full outer shell.

You can test this by covering each atom in turn, as in **Figure 3.8.** This small group of covalently bonded atoms is called a **molecule**. The formula in this case is F_2.

Figure 3.8 Each fluorine atom has eight electrons

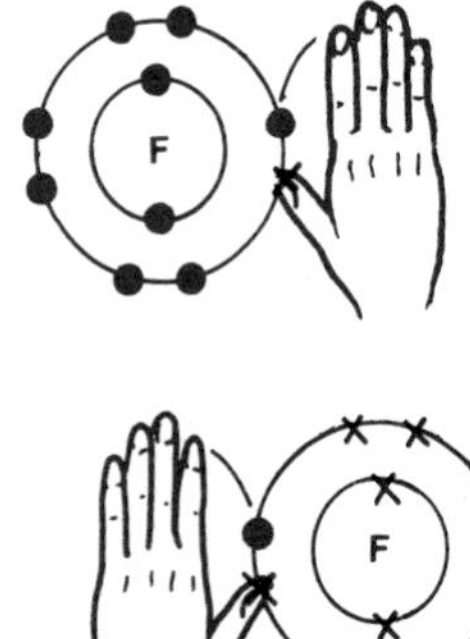

Other examples:

1. Water:

O

Oxygen

H H

Hydrogen atoms

O H H

A water molecule, formula H_2O

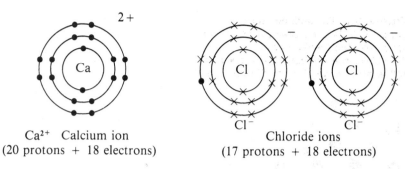

Note that hydrogen is completing the *first* shell with two electrons, so that it has the same structure as the inert gas helium.

2. Oxygen, O_2:

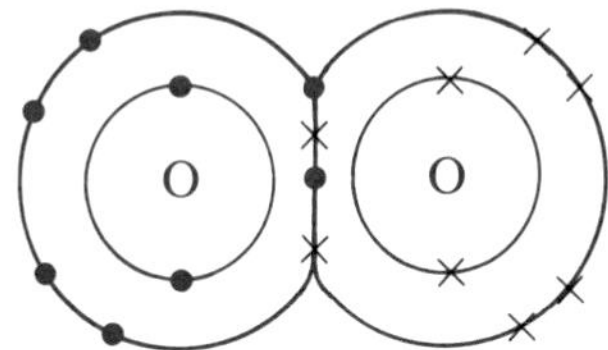

Here we have *two* pairs of shared electrons—a double bond.

3. Nitrogen, N_2:

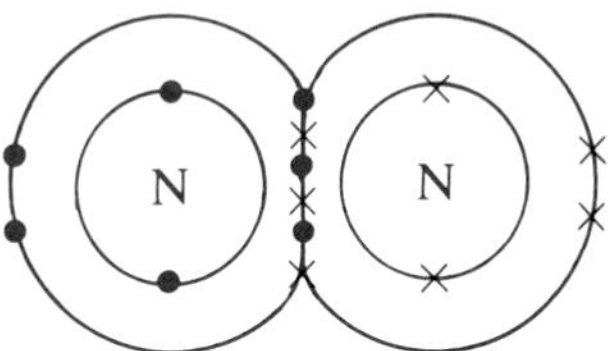

Here we have *three* pairs of shared electrons—a triple bond.

Notice that in the examples above we have formed neutral molecules, which are not strongly attracted to each other. These are examples of **molecular covalent structures**.

Giant covalent compounds

Another example of covalent bonding occurs in, for example, sand. Sand is silicon dioxide, SiO_2, and it is a covalently bonded compound with a very high melting temperature. This is because, instead of forming small molecules, silicon dioxide is a giant structure in which a network of bonds extends throughout the compound, **Figure 3.9.** The simplest entity is SiO_2.

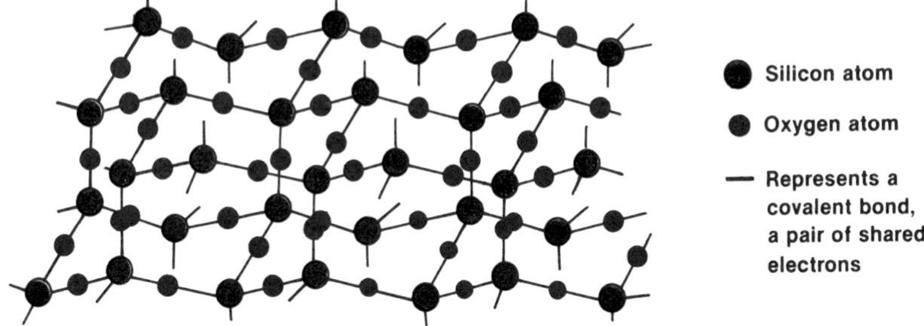

Figure 3.9 The structure of silicon dioxide

Properties of compounds with covalent bonding

1. Molecular covalent

- Molecular covalent compounds are gases, liquids or solids with low melting temperatures. (The strong bonds are between the atoms and do not extend throughout the compound so that the molecules can easily be given the energy to move apart.)
- They are poor conductors of electricity (there are no charged particles to carry the current).
- If they dissolve in water, and often they do not, the solution is a poor conductor of electricity. (Again there are no charged particles.)

2. Giant covalent

- Giant covalent compounds have high melting temperatures (they have a giant structure).
- They are poor conductors of electricity.
- A solution in water is a poor conductor of electricity, though, as before, covalent compounds often do not dissolve in water.

3.2.3 Metallic bonding

This bonding is present in metals or alloys of metals, for example, magnesium metal, Mg:

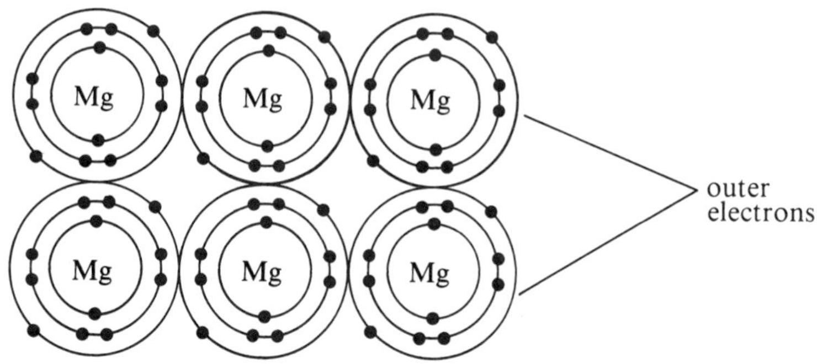

The two electrons in the outer shell of each magnesium atom are free to move not only around the atom to which they 'belong', but around the atoms closest to them. Thus the outer electrons form a general pool. The atoms remain neutral, but the electrons are free to move.

This is often referred to as a 'sea of electrons', and it is this 'sea' that gives the metals their unique properties.

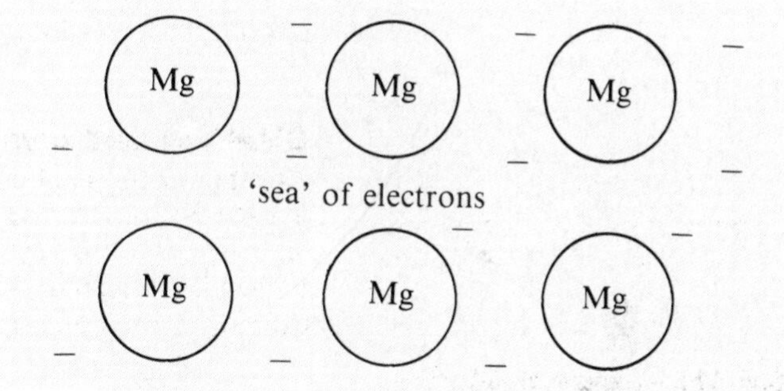

Properties of compounds with metallic bonding

- They have high melting temperatures because they are giant structures.
- They conduct electricity as solids or liquids.
- They are shiny, malleable (can be beaten into shape) and ductile (can be pulled into thin wires).

3.2.4 Summary of properties

Table 3.3 Properties of elements and compounds according to type of bonding

Bond	Conductivity: Solid	Conductivity: Liquid	Conductivity: Aqueous solution	Structure	Melting temperature T_m
Ionic	No	Yes	Yes	Giant	High
Covalent (giant)	No	No	No	Giant	High
Covalent (molecular)	No	No	No	Molecular	Low
Metallic	Yes	Yes	—	Giant	High

4 The Periodic Table

4.1 The arrangement of the Periodic Table

The Periodic Table is a list of all the elements in order of their atomic numbers, see Chapter 3. A copy of the Periodic Table is found on page 602.

4.1.1 Groups

The Periodic Table is arranged so that similar elements fall into vertical columns forming chemical 'families', called **groups**.

So H, Li, Na, K, Cs, Rb is Group I; Be, Mg, Ca, Sr, Ba, Ra is Group II, and so on.

4.1.2 Periods

Horizontal rows of elements are called **periods**.

So H, He is Period 1; Li, Be, B, C, N, O, F, Ne is Period 2, and so on. Some areas of the Periodic Table are given names which are shown in **Figure 4.1**.

Figure 4.1 Named areas of the Periodic Table

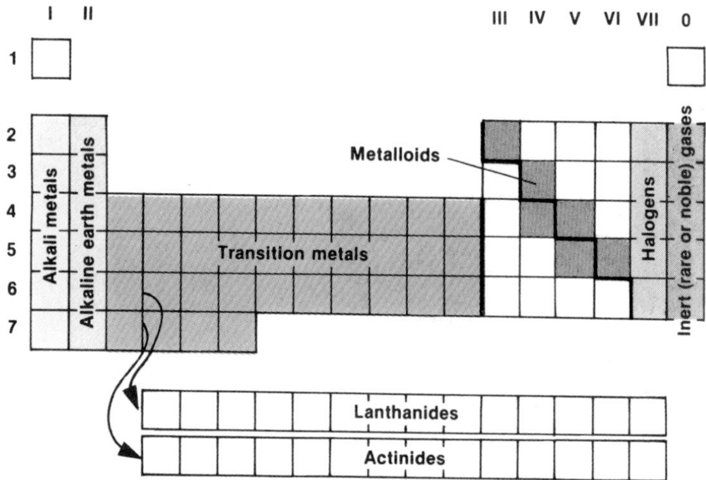

4.1.3 Semi-metals

The heavy line in **Figure 4.1** divides metals (on its left) from non-metals (on its right). Elements close to this line, like silicon, have properties intermediate between those of metals and non-metals. These elements are called metalloids or semi-metals. Silicon, in particular, has become very important. When 'doped' with other elements its electrical conduction can be altered. This has formed the basis of microelectronics and table-top computers.

Notice that the form of the Periodic Table should really look like **Figure 4.2** (overleaf). The conventional way of presenting it simply saves paper.

Figure 4.2 The full form of the Periodic Table

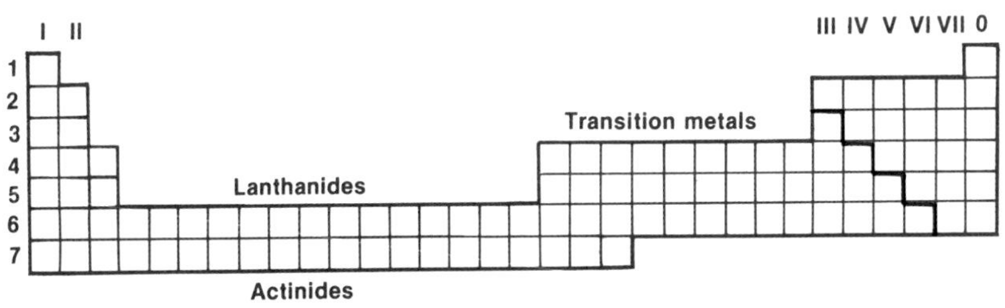

4.2 Using the Periodic Table

The Periodic Table can be used to predict how any element would behave by looking at its position in the table. The overall pattern of the Periodic Table can be seen by considering just a few groups.

4.2.1 Group I: The alkali metals

Although hydrogen is usually (but not always) included in Group I, it is not a typical member of the group and is not included as an alkali metal, though recent experiments have shown that when solid it has the appearance of a shiny reddish metal.

The alkali metals are soft and shiny in appearance. Some physical properties are set out in **Table 4.1.**

Table 4.1 Properties of the alkali metals

Atomic number	Metal	Electron arrangement	Density/ $g\,cm^{-3}$ (increase ↓)	Melting temperature/K (decrease ↓)	Boiling temperature/K (decrease ↓)
3	Li	2, 1	0.53	454	1615
11	Na	2, 8, 1	0.97	371	1156
19	K	2, 8, 8, 1	0.86	336	1033
37	Rb	2, 8, 18, 8, 1	1.53	312	959
55	Cs	2, 8, 18, 18, 8, 1	1.88	302	942
87	*Fr	2, 8, 18, 18, 32, 8, 1	?	?	?

Humphry Davy discovered sodium and potassium during electrolysis experiments on moistened salts

The densities generally increase as we go down the group, while the melting and boiling temperatures decrease. Thus, caesium is much denser than water (density $1\,g\,cm^{-3}$), but would be a liquid on a very hot day (temperature 303 K).

*Francium is a very rare radioactive metal and not enough of it has been isolated for long enough to know its properties for certain, but from the table there are trends which would allow us to guess the value of the gaps. The graph in **Figure 4.3** gives a better prediction.

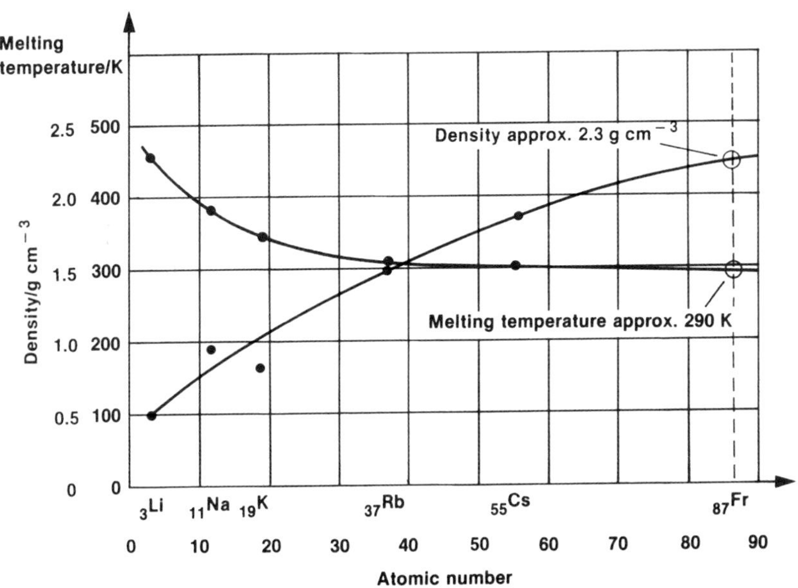

Figure 4.3 Graph to predict the melting temperature and density of francium

Chemical properties and trends

These metals are very reactive indeed and, for example, have to be stored carefully to prevent them from reacting with air.

REACTION IN AIR

They all react immediately with air to form oxides. The oxides then react with the water in the air to form strongly *alkaline* hydroxides, for example:

$$\underset{\text{lithium}}{4Li(s)} + \underset{\text{oxygen}}{O_2(g)} \longrightarrow \underset{\text{lithium oxide}}{2Li_2O(s)}$$

$$\underset{\text{lithium oxide}}{Li_2O(s)} + \underset{\text{water}}{H_2O(l)} \longrightarrow \underset{\text{lithium hydroxide}}{2LiOH(aq)}$$

REACTION WITH WATER

They all react with water to form *alkaline* solutions of the hydroxides and hydrogen is given off, for example:

$$\underset{\text{lithium}}{2Li(s)} + \underset{\text{water}}{2H_2O(l)} \longrightarrow \underset{\text{lithium hydroxide}}{2LiOH(aq)} + \underset{\text{hydrogen}}{H_2(g)}$$

The reactions of lithium, sodium and potassium (left to right) in water

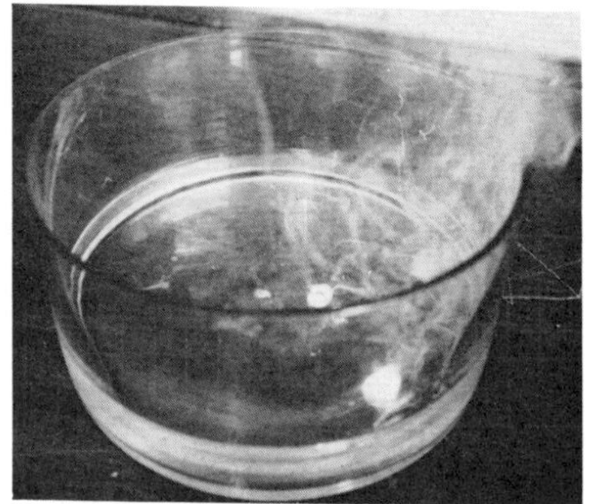

TRENDS

The reactivity of the elements *increases* as we go down the group. A good example is the reaction with water. Lithium fizzes gently, sodium less gently, potassium catches fire and from then on the reaction becomes more and more explosive.

If we look at the electron arrangements of the elements, we can try to explain this trend.

Group I metals all react by giving away their single outer electron to form a singly positively charged ion, which has a full outer shell of eight electrons:

$$\text{metal} \longrightarrow \text{metal}^{+} + \text{electron}^{-}$$

The attraction for the positively charged nucleus is what holds the negatively charged electrons to it.

As we go down the group, the outer electron gets further away from the nucleus, **Figure 4.4**, so it can more easily leave the atom, which means that the elements will be more reactive as we go down the group.

Figure 4.4 Loss of the outer electron becomes easier as we go down the group in Group I metals

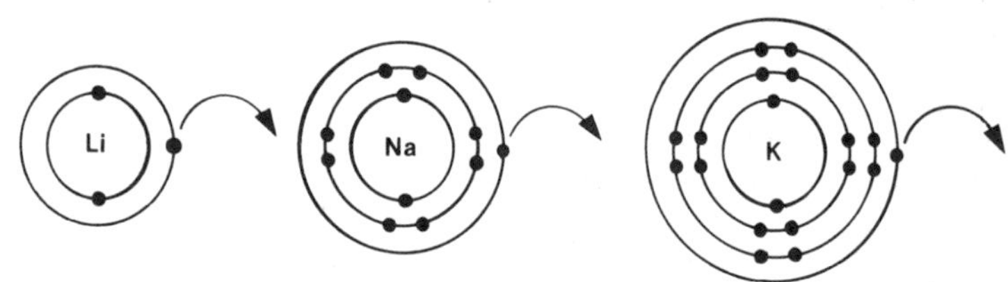

4.2.2 Group II metals: the alkaline earth metals

These are similar in many ways to Group I metals, though they are less reactive. This time, the atoms all have two electrons in their outer shells and *both* these are given away when the elements react to form a doubly charged metal ion:

$$\text{metal} \longrightarrow \text{metal}^{2+} + 2\ \text{electrons}^{-}$$

Loss of two electrons is more difficult than loss of one, so any Group II metal is less reactive than the Group I metal in the same period.
The reactivity of the elements increases as we go down the group in the same way as before, and for the same reason.

4.2.3 Group VII non-metals: the halogens

Again there is a pattern of general similarity of the elements with trends in both physical properties and chemical reactivity (**Table 4.2**).

Table 4.2 Properties of the halogens

Atomic number	Element	Electron arrangement	Appearance	Melting temperature/K (increase ↓)	Boiling temperature/K (increase ↓)
9	F	2, 7	Colourless gas	53	85
17	Cl	2, 8, 7	Yellow/green gas	172	238
35	Br	2, 8, 18, 7	Brown liquid	266	332
53	I	2, 8, 18, 18, 7	Black solid	387	457
85	At	2, 8, 18, 32, 18, 7	?	?	?

As before, we can plot a graph and predict the missing properties of the rare element astatine (**Figure 4.5**):

Figure 4.5 The boiling and melting temperatures of the halogens

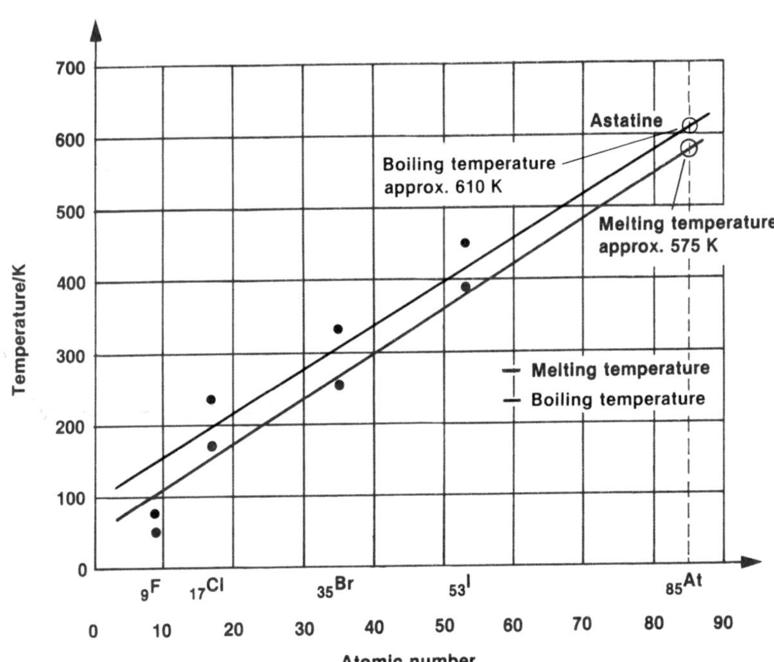

While the appearances of the elements are not similar, their smells are. All have a characteristic 'swimming pool' smell. Chlorine is used to sterilize the water in swimming pools, though the amounts are carefully controlled, since chlorine, like the other halogens, is harmful to us.

Chemical properties and trends

The elements exist as covalently bonded molecules X_2 where X represents any halogen atom, and all are reactive. They react with metals by gaining one electron and so form ions with one negative charge in these compounds:

X_2	+ 2 electrons$^-$	⟶	$2X^-$
halogen	(from the metal)		halogen ions (halide ions)

For example:

$2Na(s)$	+ $Cl_2(g)$	⟶	$2NaCl(s)$
sodium	chlorine		sodium chloride (sodium chloride exists as Na^+ and Cl^-)

They all react with sodium hydroxide to form bleaches, for example:

$2NaOH(aq)$	+ $Cl_2(g)$	⟶	$NaOCl(aq)$	+ $NaCl(aq)$	+ $H_2O(l)$
sodium hydroxide	chlorine		sodium chlorate(I)	sodium chloride	water

Sodium chlorate(I), which used to be called sodium hypochlorite, is household bleach.

TRENDS

The halogens get more reactive as we go *up* the group, so that fluorine is a remarkably reactive gas and will attack (combine with) almost all of the elements, even including some inert gases. Many of the reactions take place at room temperature.

We can again try to explain this trend in terms of electron structures, **Figure 4.6.** All are one electron short of having a full outer shell and this time, unlike the Group I and Group II metals, the *nearer* the outer shell is to the nucleus, the more readily it can *attract* an electron.

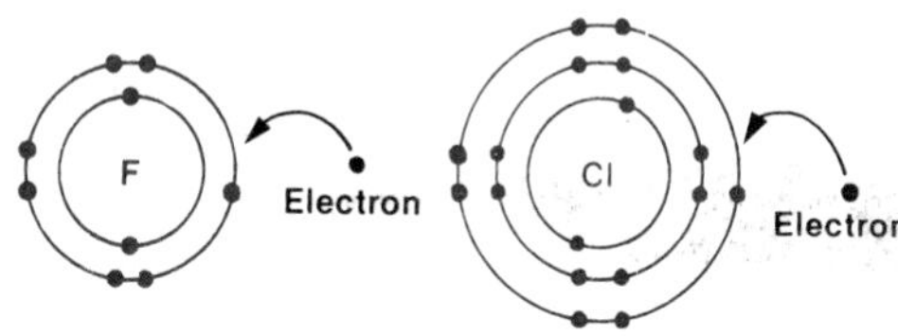

Figure 4.6 Electron gain by halogen atoms

4.2.4 The transition metals

This block of elements is neither a group nor a period. It contains most of the everyday metals that we are familiar with, and many have been known to chemists for a long time. Their position in the table seems to destroy the simple pattern and may have been one of the reasons that it took so long for chemists to work out the overall shape that finally became the Periodic Table.

Uses of transition metals

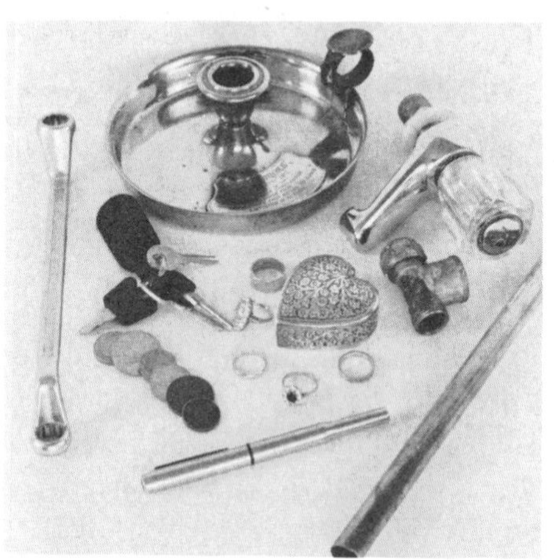

Properties of transition elements

These are hard, dense, rather similar metals, which are good conductors of electricity.

Compared with the metals in Groups I and II, they are relatively unreactive. These properties make them useful. They often form coloured compounds such as blue copper sulphate or green iron(II) sulphate. The reason for their similarities is again due to their electron structure. In these elements, the atoms usually have two electrons in their outer shell, but the inner shells are not complete. This is dealt with in Chapter 25.

4.3 Trends across a period

A number of trends occur as we go across a period, for example from Na to Ar. The most obvious is the gradual change in properties from metallic to non-metallic, but there are other trends that we can notice and they are shown in **Table 4.3.**

Table 4.3 Trends across a period

Group	I	II	III	IV	V	VI	VII	0
Element	Na	Mg	Al	Si	P	S	Cl	Ar
	Metals			Semi-metal	Non-metals			(Inert gas)
	less reactive →				← less reactive			
Structure of element	Giant metallic			Giant covalent	Molecular			(Atomic)
Ion	Na^+	Mg^{2+}	Al^{3+}	None	P^{3-}	S^{2-}	Cl^-	None
Oxide	Na_2O	MgO	Al_2O_3	SiO_2	P_4O_{10} (P_4O_6)	SO_2 (SO_3)	Cl_2O (Cl_2O_7 etc.)	none
	Strongly basic		Amphoteric	Acidic				
Structure of oxide	Giant ionic			Giant covalent	Molecular			None

Points to notice

- From left to right we move from metals, through semi-metals, to non-metals.
- Metals form positive ions by loss of one, two or three electrons.
- At silicon the loss of four electrons is too difficult and it forms covalent bonds only.
- The other non-metals can form negative ions by gain of three, two, or one electron.
- Argon has no tendency to gain, lose or share electrons.
- The oxides go from strongly basic to strongly acidic as we move from left to right. Aluminium oxide is amphoteric. It can display both acidic and basic properties (basic compounds neutralize acids).
- There is a general trend in both elements and compounds from giant structures on the left to molecular structures on the right (except for the inert gas argon which is atomic).

4.4 Predicting formulae of simple ionic compounds

If we look at **Table 4.3**, we can see the charges on the ions for the period Na to Ar:

$$Na^{+}\ Mg^{2+}\ Al^{3+}\ Si\ P^{3-}\ S^{2-}\ Cl^{-}\ Ar$$

These charges will tend to be the same for all the elements in the same group, for example Na forms an ion with a single positive ion and so do all the rest of the Group I metals.

Compounds are always electrically *neutral* so we can predict the formulae of simple ionic compounds, for example:

1. Magnesium fluoride:

   Mg is in Group II and will form Mg^{2+} ions

   F is in Group VII and will form F^{-} ions

   To form a neutral compound, two F^{-} ions are needed to balance the charge on the Mg^{2+}. So the formula will be MgF_2 ($Mg^{2+} + 2F^{-}$).

2. Aluminium oxide:

   Al in Group III forms Al^{3+}

   O in Group VI forms O^{2-}

   The simplest formula for a neutral compound is Al_2O_3 ($2Al^{3+} + 3O^{2-}$).

Note

This method of balancing the charges can be extended to find the formulae of other ionic compounds if the charge on the anion (the atom or group of atoms that make up the negatively charged part) is known.

Some common anions

Name	Formula and charge
sulphate	SO_4^{2-}
carbonate	CO_3^{2-}
nitrate	NO_3^{-}
phosphate	PO_4^{3-}
hydroxide	OH^{-}

For example, we might want to find the formula of magnesium phosphate with the ions Mg^{2+} and PO_4^{3-}. To get the charges balanced, we need:

$$Mg^{2+}\ Mg^{2+}\ Mg^{2+}\quad PO_4^{3-}\ PO_4^{3-}$$

So the formula must be

$$Mg_3(PO_4)_2\quad (3Mg^{2+} + 2PO_4^{3-}).$$

5 Moles

5.1 Introduction

An atom is too small to be seen by even the most powerful microscope, and impossible to weigh individually. When chemists need to know how many atoms are involved they must count by weighing large numbers. This is the normal method used in a bank, when large amounts of money are weighed.

5.2 Relative atomic mass, A_r

A 1p coin weighs 3.5 g and a 2p coin 7 g

The relative atomic mass, A_r, was first used to show how heavy an atom of an element is compared to an atom of hydrogen. We can find A_r from the Periodic Table, see Chapter 4.

For example, an atom of helium, He, which has $A_r = 4$, is *four* times as heavy as an atom of hydrogen, H, and an atom of lithium, Li, which has $A_r = 7$, is *seven* times as heavy as an atom of hydrogen.

$A_r \rightarrow {}^{1}_{1}H$ •	$A_r \rightarrow {}^{4}_{2}He$ ●	$A_r \rightarrow {}^{7}_{3}Li$ ⬤
Hydrogen	Helium 4 × as heavy as H atom	Lithium 7 × as heavy as H atom

Coins can be counted by weighing. Each stack has the same number of coins

Suppose we could weigh out 1 g of hydrogen atoms. Then, to get *the same number* of atoms of helium, we would have to weigh out 4 g of helium, because each atom of helium is four times as heavy as each atom of hydrogen. In the same way, to get the same number of lithium atoms we would have to weigh out 7 g of lithium, because each atom of lithium is seven times as heavy as each atom of hydrogen.

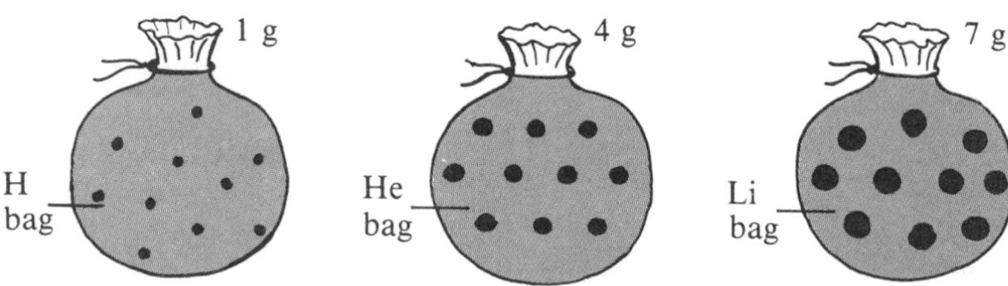

Each bag has the *same* number of atoms

5.3 Relative molecular mass, M_r

The relative molecular mass, M_r, is the mass of a molecule (small group of atoms covalently bonded together) compared with the mass of an *atom* of hydrogen. It is found from the formula, by adding together the relative atomic masses of the elements present. The term relative molecular mass is used whatever the bonding, as for calcium nitrate shown below, which is not molecular, but forms a giant ionic structure. In this case M_r refers to the simplest formula unit, which is called an **entity**. This term is useful because it includes all types of structure. For example:

1. Water, H_2O:

Element	A_r		Number present			Total
H	1	×	2		=	2
O	16	×	1		=	16
				M_r	=	18

2. Calcium nitrate, $Ca(NO_3)_2$:

Element	A_r		Number present			Total
Ca	40	×	1		=	40
N	14	×	2		=	28
O	16	×	6		=	96
				M_r	=	164

So one entity of water is 18 times as heavy as an *atom* of hydrogen, and one entity of calcium nitrate is 164 times as heavy as an *atom* of hydrogen. To get the *same number* of entities as there are in 1 g of hydrogen, we need to weigh out 18 g of water or 164 g of calcium nitrate.

5.4 The mole

The number of atoms or other entities we get when we weigh out the relative atomic mass or the relative molecular mass is enormous. It is called the Avogadro constant and is equal to 6.023×10^{23}. The amount of substance that contains this number of atoms or entities is called a **mole**.

Whenever we weigh out the relative atomic mass of an element in grams we have a mole of atoms. Similarly, the relative molecular mass in grams gives a mole of entities.

A mole of carbon (left) and a mole of copper (right). Each has the same number of atoms

Although the mole is defined in terms of an *amount* of substance, it is often simpler to treat it as though it were a *number* with a special name, just as a dozen or a score is a number with a special name. For example:

a mole of carbon atoms C has a mass of 12 g
a mole of copper atoms Cu has a mass of 63.5 g
a mole of hydrogen atoms H has a mass of 1 g

but note: a mole of hydrogen *molecules* H_2 has a mass of 2 g
a mole of calcium nitrate $Ca(NO_3)_2$ entities has a mass of 164 g

To avoid ambiguity, it is best to give the formula of the entity you are referring to. (The symbol for mole is mol.)

5.4.1 Number of moles

We can always find out how many moles of substance are present in a given mass of the substance if we know its formula.

$$\text{number of moles of substance} = \frac{\text{mass in g}}{\text{mass of 1 mole in g}}$$

For example, in 0.09 g of water H_2O ($M_r = 18$, so 1 mole has a mass of 18 g):

$$\text{number of moles} = \frac{0.09}{18} = 0.005$$

Self check

Try the following. The answers are at the end of Chapter 6.

1. What is the mass of one mole of:
   a) oxygen atoms, O **b)** oxygen molecules, O_2 **c)** sulphuric acid, H_2SO_4 **d)** aluminium sulphate $Al_2(SO_4)_3$? A_r for O = 16, S = 32, Al = 27, H = 1.

2. How many moles of atoms or entities (state which) are there in:
   a) 23 g of sodium **b)** 2 g of hydrogen **c)** 1 g of calcium carbonate $CaCO_3$? A_r for Na = 23, H = 1, Ca = 40, C = 12, O = 16.

5.4.2 Moles and solutions

Concentrations of solutions are measured in moles per cubic decimetre ($mol\,dm^{-3}$).
(One *cubic* decimetre = 1000 cm^3, because 1 decimetre = 10 cm.)
1 $mol\,dm^{-3}$ means there is one mole per cubic decimetre of solution.
2 $mol\,dm^{-3}$ means there are two moles per cubic decimetre of solution, and so on.
Until recently these were written 1 M and 2 M respectively.

Number of moles in a given volume of solution

It is often necessary to work out how many moles of solute are present in a solution, for example:

How many moles are there in 20 cm^3 of a solution of concentration 0.5 $mol\,dm^{-3}$? From the definition,

1000 cm^3 of a solution of 1 $mol\,dm^{-3}$ contains 1 mole
so 1000 cm^3 of a solution of 0.5 $mol\,dm^{-3}$ contains 0.5 mole

so 20 cm^3 of a solution of 0.5 $mol\,dm^{-3}$ contains $\frac{0.5 \times 20}{1000} = 0.01$ mole

Notice that the answer is the same whatever solute is used.
In general, in a solution of concentration M $mol\,dm^{-3}$ and volume V cm^3:

$$\text{number of moles} = \frac{M \times V}{1000}$$

Self check

3. **a)** How many moles of sodium hydroxide are in 25 cm^3 of a 0.2 $mol\,dm^{-3}$ solution?
   b) How many grams of copper sulphate ($CuSO_4{\cdot}5H_2O$) would be dissolved in 100 cm^3 of a 2 $mol\,dm^{-3}$ solution?

5.4.3 Moles and gases

A mole of *any* gas has the same volume under the same conditions of temperature and pressure. At room temperature and pressure this volume is approximately 24 000 cm^3 (24 dm^3). Thus a mole of carbon dioxide gas, CO_2 (mass 44 g), has the same volume as a mole of hydrogen gas, H_2 (mass 2 g).

This may seem unlikely at first, but a gas particle (even quite a heavy one) is extremely small compared with the space in between the particles. Think of several people running around a large hall, and imagine the space between them. It would not make any measurable difference to this space if the people were fat or thin. In the same way, the space between the gas particles is what accounts for the volume of a gas. The size of the gas molecules themselves is negligible compared to this.

For example, how many moles of methane gas (CH_4) are present in a volume of 240 cm^3 at room temperature and pressure? What is the mass of this volume of gas? A_r for C = 12, H = 1.
One mole of CH_4 has a volume of 24 000 cm^3, so

$$\text{number of moles of methane in } 240\,\text{cm}^3 = \frac{240}{24\,000} = 0.01$$

One mole of CH_4 has a mass of 16 g, so 0.01 moles has a mass of $16 \times 0.01 = 0.16$ g.

5.4.4 Moles of liquids

It often more convenient to measure a mole of liquid by volume, rather than by weighing. This is easy to do if we know the density ρ (called rho) of the liquid, because we can use the formula:

$$\text{density, } \rho = \text{mass/volume}$$

(If you think of the units of density, g cm^{-3}, it is easy to remember this formula.)

For example, what is the volume of 0.1 moles of ethanol C_2H_5OH, density 0.8 g cm^{-3}?

$$\text{Mass of 1 mole of ethanol} = 46\,\text{g, so mass of 0.1 mole} = 4.6\,\text{g}$$

$$\begin{aligned} \text{density} &= \text{mass/volume} \\ \text{volume} &= \text{mass/density} \\ \text{volume} &= 4.6/0.8\,\text{cm}^3 \\ &= 5.75\,\text{cm}^3 \end{aligned}$$

Self check

4. **a)** How many moles of hydrogen molecules, H_2, are present in a volume of 600 cm^3 at room temperature and pressure? What is the mass of this volume of gas? (A_rH = 1)
   b) What would be the volume at room temperature and pressure of 22 g of carbon dioxide gas, CO_2? (A_rC = 12, O = 16)
   c) What is the volume of 0.5 moles of mercury, density 13.6 g cm^{-3}, A_r 201?

5.5 Moles and formulae

Suppose we needed to find the formula of magnesium oxide. We could weigh some magnesium, burn it and weigh the magnesium oxide formed.

For example, 0.24 g of magnesium ribbon burn to form 0.40 g of magnesium oxide.

A_r for Mg = 24, O = 16.

Mass of magnesium = 0.24 g.
Number of moles of magnesium = 0.24/24 = 0.01.
Mass of oxygen combined with the magnesium = 0.40 − 0.24 = 0.16 g.
Number of moles of oxygen = 0.16/16 = 0.01.
So the ratio of magnesium moles to oxygen moles = 0.01:0.01
= 1:1.

The simplest formula of magnesium oxide is therefore MgO.

Thus we can find the simplest formula of any compound if we know the amounts of elements present in it.

For example, 9.8 g of an oily liquid was found to contain 0.2 g of hydrogen, 3.2 g of sulphur and 6.4 g of oxygen.
A_r for H = 1, S = 32, O = 16.

Number of moles of hydrogen = 0.2/1 = 0.2
Number of moles of sulphur = 3.2/32 = 0.1
Number of moles of oxygen = 6.4/16 = 0.4

Ratio of hydrogen : sulphur : oxygen moles
= 0.2 : 0.1 : 0.4
= 2 : 1 : 4
The formula is therefore H_2SO_4.

5.5.1 Points to remember

The final ratio of atoms is normally written in whole numbers, so that 0.5:1 is 1:2. Use the simplest ratio, so that 2:4 is 1:2.

Self check

5. Find the simplest formula of the following:
   a) 2.39 g of lead sulphide containing 2.07 g of lead
   b) 22 g of a gas containing 6 g of carbon and 16 g of oxygen
   c) 5 g of calcium carbonate containing 2 g of calcium, 0.6 g of carbon and the rest oxygen.
   A_r for Pb = 207, S = 32, C = 12, O = 16, Ca = 40.

5.6 Summary

- Number of moles of substance = $\dfrac{\text{mass in grams}}{\text{mass of 1 mole in grams}}$.
- Number of moles of solute *in a solution* = $\dfrac{M \times V}{1000}$, where

M = concentration in mol dm^{-3} and V = volume of solution in cm^3.
- A mole of any gas has a volume of approximately 24 dm^3 (24 000 cm^3) at room temperature and pressure.

6 Equations

6.1 Introduction

When a chemical change takes place, chemical bonds are broken and atoms are rearranged to form new substances.

The substances we start with are called the *reactants* and these rearrange to form the *products*.

6.2 Word equations

Equations are used as a simple way to represent what is happening in a chemical reaction, and the simplest is the word equation.
For example when magnesium is burnt in air the word equation would be:

$$\text{magnesium} + \text{oxygen} \longrightarrow \text{magnesium oxide}$$

Equations should always be based on experimental evidence.

6.2.1 State symbols

Equations do not say how fast a reaction happens. Conditions (heat, pressure, etc.) may be shown on arrows, for example $\xrightarrow{\text{heat}}$. Letters in brackets can be added to say what state the reactants and products are in: (s) means solid; (l) means liquid; (g) means gas; (aq) means in aqueous solution (dissolved in water).

6.3 Balanced symbol equations

A balanced symbol equation is useful because it tells us not only what happens in a reaction but also the *quantities* that react together, and how much of the products we should get from a given amount of reactants. 'Balanced' means that there are the same number of each atom on both sides of the ⟶ sign, because atoms cannot be made or destroyed, just rearranged. It is very important to start with the correct formulae.
For example:

$$\begin{array}{lclcl} \text{magnesium} & + & \text{copper sulphate} & \longrightarrow & \text{magnesium sulphate} + \text{copper} \\ Mg(s) & + & CuSO_4(aq) & \longrightarrow & MgSO_4(aq) + Cu(s) \end{array}$$

This is a balanced symbol equation because the number of atoms of each element is the same on both sides of the ⟶ sign. (It also has state symbols to show you how the symbols are used.)

If we now take the first example and write in the correct formulae:

$$\begin{array}{lclcl} \text{magnesium} & + & \text{oxygen} & \longrightarrow & \text{magnesium oxide} \\ Mg(s) & + & O_2(g) & \longrightarrow & MgO(s) \end{array}$$

This is *not* balanced, because there are two oxygens on the reactants side (left-hand side) but only one on the products side (right-hand side).

It is very tempting to change the formula of magnesium oxide to MgO_2 at this stage, which would indeed balance the equation, or to write O for the formula of oxygen, but you must *never change the correct formulae*. These have been established by experiment and cannot be changed at your convenience!

Instead, you must add another MgO to get two oxygens, and this is written:

$$Mg(s) + O_2(g) \longrightarrow 2MgO(s)$$

This is still not balanced because now we have ended up with two magnesium atoms on the products side and only one on the reactants side. If we now put a 2 in front of the Mg, the equation will be balanced:

$$2Mg(s) + O_2(g) \longrightarrow 2MgO(s)$$

We can say from this that two atoms of magnesium react with one molecule of oxygen to produce two entities of magnesium oxide, so that two moles of magnesium atoms react with one mole of oxygen molecules to give two moles of magnesium oxide entities. Remember the word entity is used to describe the simplest formula unit. Magnesium oxide has a giant structure and does not exist as molecules.

At this point we can also work out the masses that will react together according to our equation:

$2Mg(s)$	+	$O_2(g)$	$\longrightarrow$	$2MgO(s)$	
2 moles		1 mole		2 moles	(A_r for Mg = 24, O = 16.)
48 g		32 g		80 g	

Note that the total mass is the same on both sides, which is a good way of checking whether the equation is balanced.
An analogy to balancing an equation might be the task of buying batteries for torches. Suppose torches each take three batteries, and batteries are sold in twos. The task is to buy enough torches and enough batteries to fit them, with none left over.

torch	+	batteries	$\longrightarrow$	working torch
2T	+	$3B_2$	$\longrightarrow$	$2TB_3$ (where B = 1 battery, T = 1 torch)

So there are three steps to writing a balanced symbol equation:

1. Write a word equation.

2. Put in the correct formulae.

3. Add large numbers in front of the formulae where necessary to get the same number of atoms of each element on both sides. *Do not change formulae*! The number of moles can then be put in and the masses that react worked out.

Another example: write a balanced symbol equation for the reaction when aluminium burns in air.

1. Word equation
   aluminium + oxygen $\longrightarrow$ aluminium oxide

2. Correct formulae
   $Al + O_2 \longrightarrow Al_2O_3$

3. Balancing: **a)** Tackling Al

$$2Al + O_2 \longrightarrow Al_2O_3$$

b) Now the O

$$2Al + 3O_2 \longrightarrow 2Al_2O_3$$

c) Back to Al

$4Al$	$+ 3O_2$	$\longrightarrow 2Al_2O_3$
4 moles	3 moles	2 moles
108 g	96 g	204 g

Note the similarity between this and the batteries and torches problem on the opposite page.

Self check

6. Write a balanced symbol equation for the reaction between lithium and oxygen and hence find how many grams of lithium oxide would be formed from 0.7 g of lithium. The formula of lithium oxide is Li_2O. A_r for Li = 7, O = 16.

6.4 Using the balanced symbol equation

6.4.1 Predicting quantities

As we have seen, we can predict the quantities produced during a reaction if we have a balanced symbol equation. For example, the reaction between calcium carbonate and hydrochloric acid is used to produce carbon dioxide gas (see **Figure 6.1**).

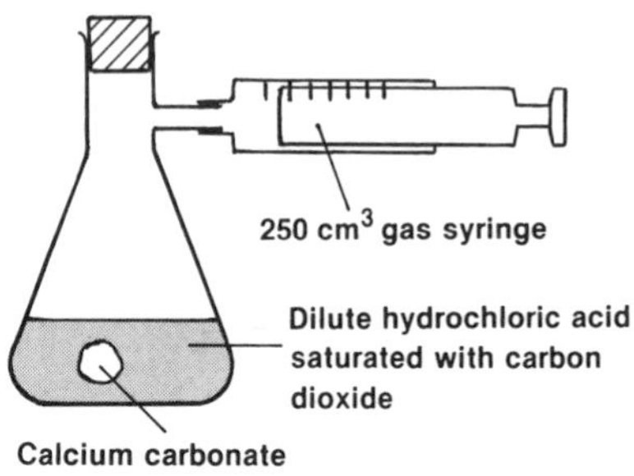

Figure 6.1 Apparatus for collecting carbon dioxide gas

1. How much gas would be produced by 1 g of calcium carbonate and excess acid?

calcium carbonate	+ hydrochloric acid	$\longrightarrow$ calcium chloride	+ water	+ carbon dioxide
$CaCO_3(s)$	$+ 2HCl(aq)$	$\longrightarrow CaCl_2(aq)$	$+ H_2O(l)$	$+ CO_2(g)$
1 mole	2 moles	1 mole	1 mole	1 mole

So one mole of $CaCO_3$ (100 g) produces one mole of CO_2 (and one mole of any gas has a volume of 24 000 cm^3 at room temperature and pressure).
So 1 g of $CaCO_3$ = 0.01 mole and produces 0.01 mole CO_2 which has a volume of 240 cm^3.
Answer is 240 cm^3.

2. How much gas would be produced by a large lump of calcium carbonate in 25 cm^3 of 2 mol dm^{-3} HCl?

From the above balanced symbol equation we can see that two moles of HCl will produce one mole of CO_2.

$$\text{number of moles in solution} = \frac{M \times V}{1000}$$

where M is concentration in mol dm^{-3} and V is volume in cm^3

so $$\text{number of moles of HCl} = \frac{2 \times 25}{1000} = 0.05 \text{ moles}$$

0.05 moles of HCl will produce 0.025 moles of CO_2.
Volume of 0.025 moles of gas = 24 000 × 0.025 cm^3 = 600 cm^3.
Answer is 600 cm^3 (and the syringe would not be large enough for this amount of gas).

Self check

7. By writing a balanced symbol equation for the reaction between magnesium and hydrochloric acid, find out if the following quantities would be suitable to use in the apparatus in **Figure 6.1:**
a) 0.12 g of magnesium ribbon in excess acid
b) 25 cm^3 of 0.5 mol dm^{-3} HCl with excess magnesium ribbon.

6.4.2 Finding concentrations

The technique of finding the concentration of a solution is important. The simplest case is for the reaction between an acid and an alkali when
a) the equation for the reaction is known
b) the concentration of one of the reactants is known.

Usually a **titration** is done using the apparatus in **Figure 6.2.** An accurately measured amount of alkali is added to the conical flask with a few drops of a suitable indicator, and acid is run in from the burette until the colour just changes, showing that the solution in the conical flask is now neutral.

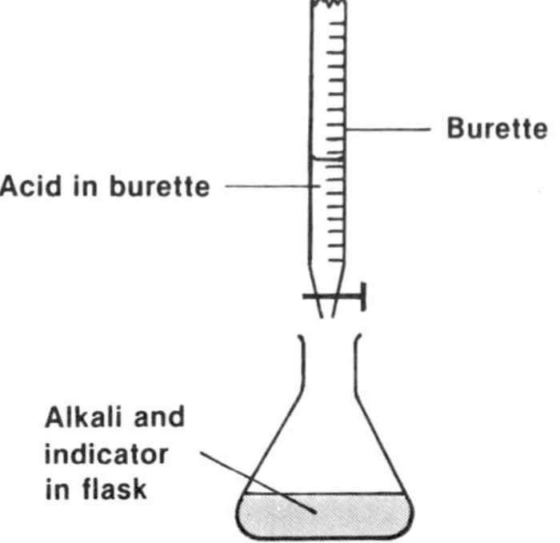

Figure 6.2 Apparatus for a titration

For example, 10 cm^3 of a 2 mol dm^{-3} solution of sodium hydroxide were neutralized by 25 cm^3 of hydrochloric acid. What is the concentration of the acid?

First write a balanced symbol equation:

NaOH(aq)	+	HCl(aq)	⟶	NaCl(aq)	+	H_2O(l)
sodium hydroxide	+	hydrochloric acid	⟶	sodium chloride	+	water

So one mole of NaOH reacts with one mole of HCl to give one mole of NaCl and one mole of water. In other words, equal numbers of moles of acid and alkali are needed.

$$\text{number of moles of solute in a solution} = \frac{M \times V}{1000}$$

so $$\text{number of moles of NaOH} = \frac{2 \times 10}{1000}$$

Let the concentration of acid be $Y \text{ mol dm}^{-3}$.

$$\text{number of moles of HCl} = \frac{Y \times 25}{1000}$$

Since we know there must be an equal number of moles for neutralization, we can say:

$$\text{number of moles of NaOH} = \text{number of moles of HCl}$$

so

$$\frac{2 \times 10}{1000} = \frac{Y \times 25}{1000}$$

From this $Y = 0.8$, and the concentration of the acid is 0.8 mol dm^{-3}.

Self check

8. What is the concentration of NaOH, given that 25 cm^3 of it was neutralized by 10 cm^3 of HCl of concentration 0.15 mol dm^{-3}?

Another example: 25 cm^3 of sodium hydroxide were neutralized by 10 cm^3 of sulphuric acid of concentration 0.1 mol dm^{-3}. What is the concentration of the NaOH?

$$\begin{array}{lclclcl} 2NaOH(aq) & + & H_2SO_4(aq) & \longrightarrow & Na_2SO_4(aq) & + & 2H_2O(l) \\ \text{2 moles} & + & \text{1 mole} & \longrightarrow & \text{1 mole} & + & \text{2 moles} \end{array}$$

From the balanced equation we can see that two moles of NaOH are required to neutralize every mole of sulphuric acid.

$$\text{number of moles of acid} = \frac{0.1 \times 10}{1000} = 0.001$$

From the equation, the number of moles of sodium hydroxide must be twice this, i.e. 0.002 mol. Let Y be the concentration of NaOH in mol dm^{-3}.

$$\text{number of moles of NaOH} = \frac{Y \times 25}{1000} \text{ and this must be equal to } 0.002$$

so

$$0.002 = \frac{Y \times 25}{1000}$$

and from this $Y = 0.08$, so the concentration of the sodium hydroxide is 0.08 mol dm^{-3}.

Self check

9. 20 cm^3 of 0.1 mol dm^{-3} NaOH were neutralized by 25 cm^3 of H_2SO_4. What is the concentration in mol dm^{-3} of the acid?

Answers to self check questions (Chapters 5 and 6)

1. a) 16 g **b)** 32 g **c)** 98 g **d)** 342 g

2. a) 1 mole of atoms **b)** 1 mole of molecules or 2 moles of atoms **c)** 0.01 mole of entities

3. a) 0.005 mol **b)** 50 g

4. a) 0.025 mol, 0.05 g **b)** $12\,000 \text{ cm}^3$ **c)** 7.38 g

5. **a)** PbS **b)** CO_2 **c)** $CaCO_3$

6. $4Li + O_2 \longrightarrow 2Li_2O$, 1.5 g

7. $Mg + 2HCl \longrightarrow MgCl_2 + H_2$ **a)** 120 cm^3, yes
b) 150 cm^3, yes

8. 0.06 mol dm^{-3}

9. 0.04 mol dm^{-3}

7 Physics and mathematics background

7.1 Introduction

There are a number of important ideas from physics which you should be familiar with in order to fully understand chemistry. Different sciences are interdependent and progress in one area often sparks off new developments in another. Indeed, there are scientists who refer to themselves as physical chemists or chemical physicists, so close are the two disciplines. There are also some mathematical techniques which you will find useful.

7.2 Waves

Note: Waves occur again in Chapters 8 (*The electromagnetic spectrum*) and 19 (*X-ray diffraction*)

Waves are disturbances in a **medium** which carry energy through the medium without the medium itself moving from place to place. A good example is dropping a stone into a pond, **Figure 7.1.** Energy is transferred from the stone to the boat (which moves up and down) via the water which is the medium. The water does not move from place to place, it merely moves up and down as the wave passes. Waves have two important measurements associated with them:

Figure 7.1 Water waves

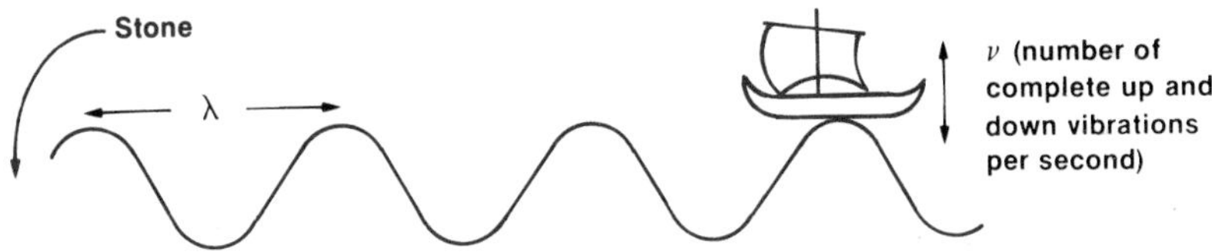

Diffraction in a ripple tank

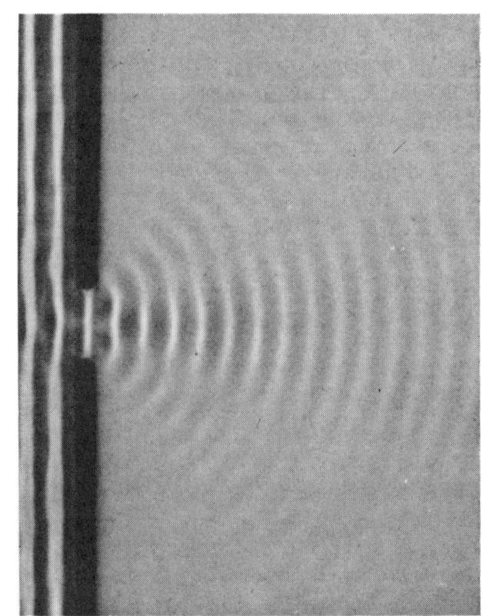

- **Wavelength** is the distance between two neighbouring peaks (or troughs). This is given the Greek letter λ (lambda) as its symbol.
- **Frequency** is the number of complete up and down vibrations per second. This is also given a Greek symbol ν (nu).

These two are related by the equation:

$$V = \nu\lambda$$

where V is the velocity with which the wave travels along. Velocity is speed in a given direction.

Waves have two important and interesting properties—**interference** and **diffraction.**

7.2.1 Diffraction

When waves pass through a gap comparable in size with their wavelength, they spread out as shown in **Figure 7.2(A)** overleaf, rather than as you might expect as in **Figure 7.2(B).** This is called diffraction. It is the reason that you can hear someone in another room talking even though you cannot see them. The sound waves are diffracted as they pass through the door, so someone standing at C would hear sound from A, but would be quite safe from a stream of bullets, which would not be diffracted.

Figure 7.2 Diffraction

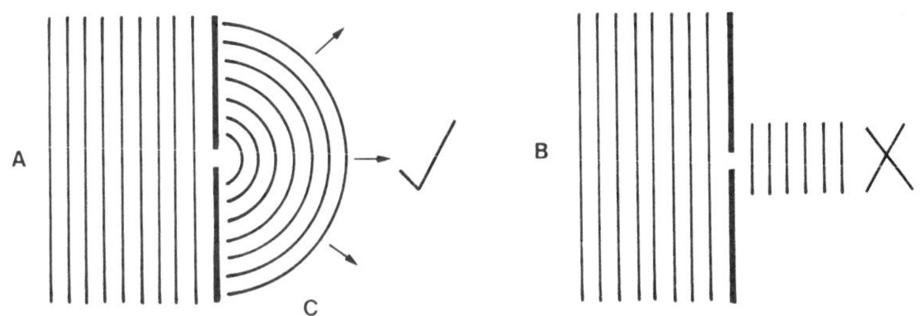

7.2.2 Interference

This occurs when two waves meet. Think about water waves. If two peaks coincide, an extra large peak will result. Similarly, if two troughs coincide, an extra deep trough will occur. This is called constructive interference. However, if a trough and a peak coincide, the two will cancel out and undisturbed water will result. This is called destructive interference. See **Figure 7.3.** The two effects can be illustrated in the experiment shown in the photograph (below left).

Figure 7.3 (right) Interference

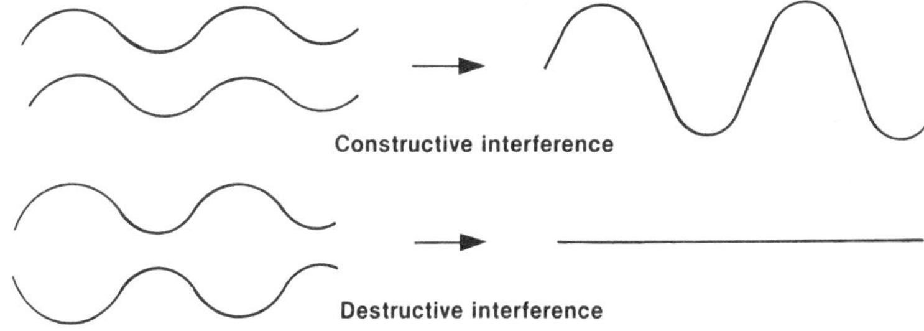

Interference of water waves produced by two vibrators in a ripple tank

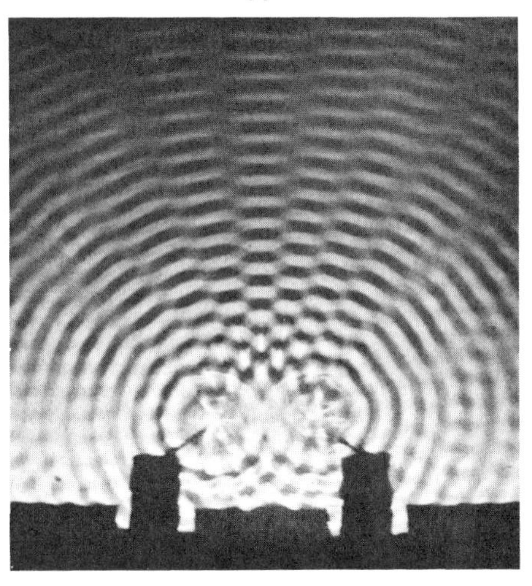

7.2.3 Electromagnetic radiation

Electromagnetic radiation shows an exception to the rule that waves travel through a medium. Electromagnetic radiations are able to travel through a vacuum as, for example, sunlight reaching the earth from the sun. Different types of electromagnetic radiation have different wavelengths and frequencies but they all travel through a vacuum at the same speed, 'the speed of light', 300 000 000 m s^{-1}. Examples include radiowaves, infra-red (radiant heat), visible light, ultraviolet (which causes tanning) and X-rays.

Since $V = v\lambda$, waves with short wavelengths have high frequencies (this is associated with high energy) and waves with long wavelengths have low frequencies.

7.3 Electricity

7.3.1 Electric charge and current

Experiments show that there are two types of electric charge: + and −. Like charges repel one another while opposite charges attract.

The *size* of an electric charge is given the symbol Q and is measured in units called **coulombs**, symbol C. An electric current, symbol I, is a flow of electric charge, usually through a wire. The **current** is the *rate* at which the charge flows and is measured in units called **amperes** (usually abbreviated to amps), symbol A. One amp is a rate of flow of one coulomb per second. So the number of coulombs of charge which flow through a circuit in a given time is current × time.

$$\text{charge } Q = I \times t \quad (t \text{ is in seconds})$$

Figure 7.4 Electric current

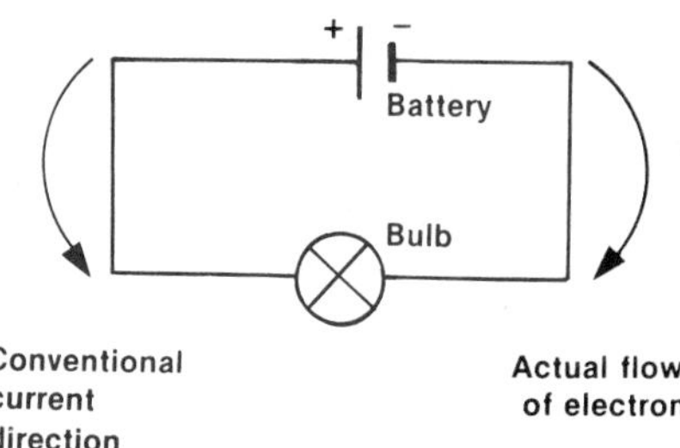

Usually an electric current is visualized as a flow of positive charge from the positive to the negative terminal, although we now know that a current is actually a flow of negatively charged electrons from negative to positive, **Figure 7.4.** The coulomb is a rather large unit of charge. The electron has a charge of only 1.6×10^{-19} coulomb.

7.3.2 Voltage

The force which drives the charge through the circuit is provided by the voltage, V (sometimes called potential difference) of the battery. The voltage of the supply is measured in **volts**. Voltage can also be interpreted as the **amount of energy** given to each unit of charge by the supply. One volt means one joule per coulomb—each coulomb of charge is given an energy of one joule by a 1 V supply. This energy will be given up by the charge on its journey through the circuit. It may be changed into light energy by a bulb, heat energy by a heater, moving (or kinetic) energy by a motor, and so on.

7.4 Force, pressure and momentum

The effect of a force applied over a small area is greater than when it is applied over a large area

7.4.1 Force

A **force** is a push or pull which moves or tends to move something. Force is measured in units called **newtons**, symbol N, after Isaac Newton. One newton is approximately the force needed to lift a mass of 100 g on the earth's surface—roughly the force needed to lift an average sized apple.

7.4.2 Pressure

In chemistry we more often come across the idea of **pressure**. This is the force per unit area.

$$\text{pressure } P = \text{force/area}$$
$$= F/A$$

The units of pressure are newtons per square metre, N m^{-2}. These are also called pascals, Pa.

Forces often have more effect when applied over a small area—a stiletto heel can make a hole in a lawn while the same person wearing flat shoes would leave no impression, see photograph.

In chemistry, pressure is usually encountered in connection with gases. Gases exert a pressure on the walls of their container because their molecules constantly bombard the walls. This is what keeps a balloon inflated and holds up the plunger of the syringe, **Figure 7.5.**

Figure 7.5 Some effects of pressure

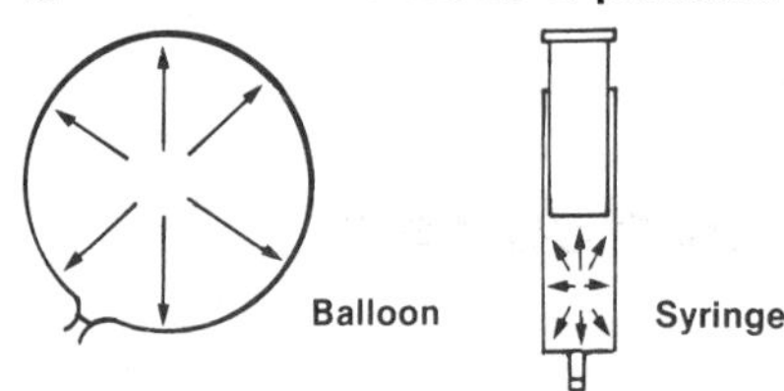

7.4.3 Momentum

The impact of a particle in a collision depends on its **momentum**, the product of its mass, m, and velocity, v.

$$\text{momentum} = mv$$

The units are therefore kg m s^{-1}. The more massive the particle, the harder it hits, and the faster the particle, the harder it hits. Newton's second law of motion says that the force exerted in a collision is equal to the rate of change of momentum.

Imagine a tennis ball colliding with a wall, **Figure 7.6.** The ball will rebound with the same speed as before *but in the opposite direction* so its velocity has changed from $+v$ to $-v$. So the change of momentum is from $+mv$ (in one direction to) $-mv$ (in the opposite direction), a change of $2mv$.

Figure 7.6 Tennis ball colliding with a wall

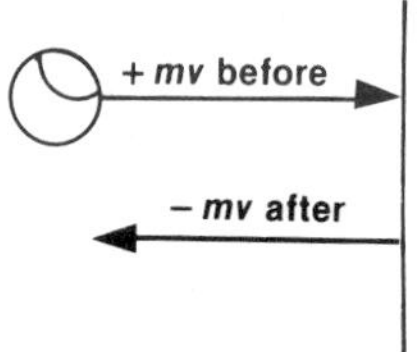

7.5 Mathematics

7.5.1 Indices

The number 100 is ten squared, 10^2. 1000 can be written 10^3, 10 000 can be written 10^4, and so on.
The power to which ten is raised, i.e. the 2 in 10^2 and 3 in 10^3, is called the **index** (plural indices). When *multiplying* numbers, we *add* the indices, for example:

$$100 \times 1000 = 100\,000$$
$$10^2 \times 10^3 = 10^{2+3} = 10^5$$

On *dividing* we subtract the indices, for example:

$$1000 \div 10 = 100$$
$$10^3 \div 10^1 = 10^{3-1} = 10^2$$

In this system, 1 is written 10^0. Inverse numbers can be represented by negative indices, for example:

$$\frac{1}{100} = 1 \div 100 = 10^0 \div 10^2 = 10^{0-2} = 10^{-2}$$

7.5.2 Standard form

This provides a convenient way of writing very large or very small numbers, for example:

$$387\,000.0 = 3.87 \times 100\,000 = 3.87 \times 10^5$$

To get 3.87 from 387 000.0 we had to move the decimal point five places, so the index is 5.

Similarly, 0.003 87 could be written 3.87×10^{-3}.

These ideas can be applied to units too, so an area measured in square metres can be written in m^2. A pressure in newtons per square metre can be written in $N\,m^{-2}$. It is sometimes useful to know that 'per' any value means divide by it:

pressure = force ÷ area
newtons per square metre
(units of force) ÷ (units of area)
units are $N\,m^{-2}$

Other examples include:

Quantity	Unit	
Speed	m/s	$m\,s^{-1}$
Voltage	J/C	$J\,C^{-1}$
ΔH	J/mol	$J\,mol^{-1}$

7.5.3 Manipulating units

When quantities with units are multiplied or divided, then the units are also multiplied or divided, to give the units of the resulting quantity. An example of this is the units of equilibrium constants, see Chapter 11, which can vary.

For the Haber process K, the equilibrium constant, is defined:

$$K = \frac{[NH_3]^2}{[N_2][H_2]^3}$$

The brackets [ ] are an abbreviation for concentration in $mol\,dm^{-3}$.
Thus the units for K will be:

$$\frac{(mol\,dm^{-3})^2}{(mol\,dm^{-3})(mol\,dm^{-3})^3} = \frac{1}{(mol\,dm^{-3})^2} = dm^6\,mol^{-2}$$

7.5.4 Logarithms

The logarithm of a number is the power to which some other number (the base) must be raised to give the original number. One common base is ten. Such logarithms are written $\log_{10}$ or just log or sometimes lg. So for example, what is log 1000?

$$1000 \text{ is } 10^3$$

so

$$\log 1000 = 3$$

similarly

$$\log 100\,000 = 5$$

When we want to *multiply* numbers, we *add* their logs:

$$1000 \times 100\,000 = 100\,000\,000$$
$$\log 1000 + \log 100\,000 = 3 + 5 = 8$$
$$8 \text{ is the log of } 100\,000\,000$$

We say that the **antilog** or **inverse log** of 8 is 100 000 000. Antilogs can be found using the INV LOG functions on a calculator.

So multiplication of two numbers is done by adding their logs, and division by subtracting their logs, followed by taking the antilog of the result.

It is easy to see what the log is of numbers like 10, 100, 1000, etc. Other numbers are more difficult, but they can be worked out. Tables are available and calculators have a log function. For example, check these using your calculator:

$$\log 3.142 = 0.4972$$
$$\log 483.1 = 2.6840$$

One use of logs is to help scaling on graphs. The numbers 10, 1000 and 100 000 would be difficult to fit on the same scale on a graph, but their logs 1, 3 and 5 will fit easily.

Natural logarithms

An alternative base for logs is the number 2.718, called 'e'. Such logs are called natural logs and the symbol ln is used. They have certain mathematical advantages and are more often used than logs to the base ten. The rules for their manipulation are the same as for logs to the base ten, however.

To multiply two numbers add their logs;
to divide two numbers subtract their logs;
to raise a number to the power n, multiply its log by n;
followed by taking the antilog in each case.

7.5.5 Use of a calculator

Efficient use of a calculator will be vital both during your course and in exams. You will need log, ln, cos, sin, tan, $\sqrt{\ }$, square functions and brackets as well as $+$, $-$, $\times$ and $\div$. Learn how to use your calculator effectively. Two common errors are noted below:

1. When entering numbers like 3.8×10^8, enter:

3.8 EXP 8

Not 3.8 × 10 EXP 8

The EXP function means '× 10 to the power'—the extra '× 10' is not needed.

2. When calculating a fraction like $\dfrac{8 \times 4}{3 \times 16}$ enter:

$8 \times 4 \div 3 \div 16$ or $8 \times 4 \div (3 \times 16)$

Not $8 \times 4 \div 3 \times 16$ which would give $\dfrac{(8 \times 4)}{3} \times 16$

Part B Physical Chemistry

8 Atomic and nuclear structure

8.1 Introduction

Atoms are the basic building bricks of chemistry. In order to understand how they bond together we must have an idea of their structure. This chapter will build on the simple picture of the atom in Chapter 3 to give a more detailed and sophisticated model. Some of the key experiments which led to the working out of the atom's structure are described but the whole story, while fascinating, is too long for this book. The unravelling of the structure of the atom, which indeed is still going on, is rather like doing a jigsaw puzzle—as each piece fits into place, it confirms that all the others were in the right place and the whole picture is slowly revealed.

8.2 Subatomic particles

8.2.1 The electron

The idea of matter being made up of atoms was first proposed in relatively modern times by John Dalton in the early 1800s, but it was not until the end of the nineteenth century that it became clear that particles *smaller* than atoms existed.

The first to be identified was the electron—a negatively charged particle which seemed to exist in all atoms. Electrons, then called 'cathode rays', were first detected in discharge tubes used to investigate the conduction of electricity by gases at low pressure.

8.2.2 The proton

Shortly after the discovery of cathode rays, 'positive rays' were discovered in the same type of experiment. These turned out to be charged particles with a variety of masses, the smallest of which we now know as protons—the fundamental positively charged particles. Protons were found to be about 2000 times more massive than electrons. They were later detected directly by Rutherford after bombarding nuclei with alpha particles—positively charged particles with the mass of a helium atom, see **section 8.4.**

8.2.3 Models of the atom

These discoveries led to speculation about how these particles might be arranged to form neutral atoms. One theory was the so-called 'plum pudding' atom, where electrons were seen as being dotted about like plums in a positively charged pudding. This is shown in **Figure 8.1.**

Figure 8.1 The 'plum pudding' model of the atom – electrons dotted about within a sphere of positive charge

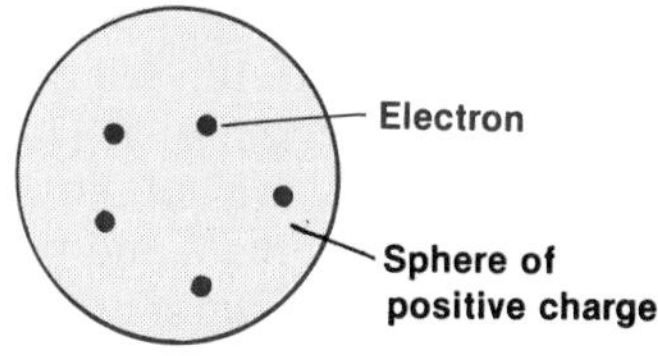

E

THE DISCOVERY OF THE ELECTRON

By the last two decades of the nineteenth century it was known that there was some connection between atoms and electricity, because of the formation of positive and negative ions which moved during electrolysis.

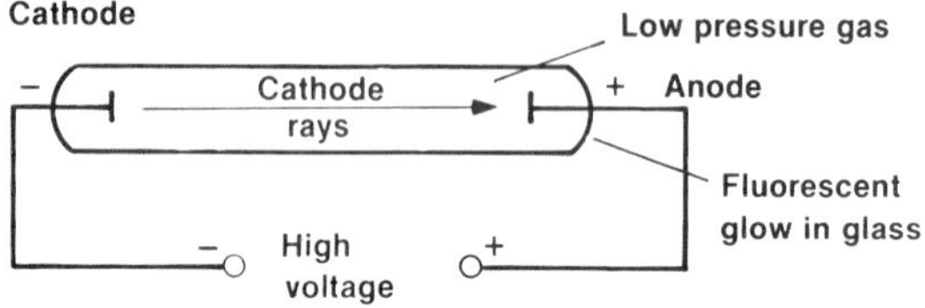

During experiments on the conduction of electricity by gases at low pressure, in gas discharge tubes like that above, a glow was noted at the anode end of the tube. Metal objects placed in the tube caused a 'shadow' and the glow seemed to be caused by rays of some sort coming from the cathode, hence the name 'cathode rays'. The rays could be deflected towards the positive plate in an electric field showing that

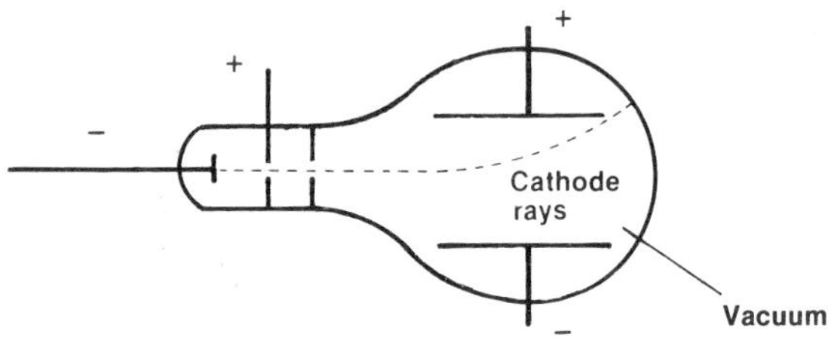

they carried a negative charge. Cathode rays were investigated by the British scientist J.J. Thomson who deflected them with magnetic and electric fields and demonstrated that they were charged particles by measuring their charge/mass ratio. The deflection depends both on the mass and the charge, so only the ratio could be determined. Thomson found that cathode rays always had the same charge/mass ratio whatever the material of the cathode (or if any other details were changed). He had discovered a fundamental constituent of all matter—the first subatomic particle, called the electron.

J. J. Thomson – discoverer of the electron

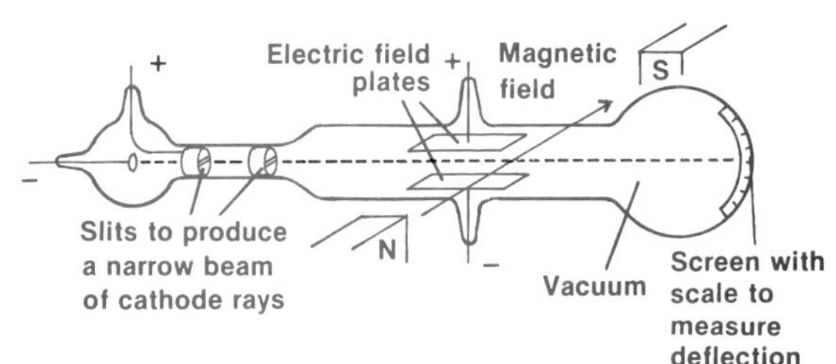

J. J. Thomson's apparatus for measuring the charge/mass ratio of cathode rays. The charge of the electron was measured by Millikan some years later

E

THE DISCOVERY OF THE PROTON

The discovery of the electron left the question of what constituted the positive part of the atom, as atoms are electrically neutral. It also left a question about the mass of the atom, as electrons were known to have a mass of around 1/2000 of the mass of the lightest atom, hydrogen. Further experiments with gas discharge tubes led to the discovery of positive rays travelling in the opposite direction to cathode rays. These had a variety of masses, depending on the gas in the tube, but the smallest had the same mass as a hydrogen atom (assuming they carried one unit of electric charge). These were named protons. Protons were observed directly in 1914 by Rutherford who produced them by bombarding various nuclei with alpha particles.

8.2.4 Rutherford's nuclear atom

The 'plum pudding' idea was disproved by Rutherford, Geiger and Marsden's famous alpha scattering experiment carried out in 1911. They used fast-moving alpha particles as 'bullets' which they fired at thin sheets of gold atoms. Some bounced back, indicating that the mass and positive charge of the gold atom must be concentrated in a dense core or nucleus.

Mathematical analysis of the detailed scattering patterns enabled them to estimate that the diameter of the nucleus was less than 1/10 000 of that of the whole atom. Scaled up, this corresponds to a nucleus the size of a pinhead in an atom the size of a house. The 'house' is occupied only by the electrons, which are very light. This means that the nucleus is exceptionally dense—approximately 6×10^{10} kg cm^{-3} so that a matchbox full of nuclei would weigh around 600 million tonnes!

Exactly what was in the nucleus was a puzzle for a number of years until a neutral particle of the same mass as a proton was discovered by Chadwick in 1934. It was named the neutron. Being neutral, it was more difficult to detect than charged particles which was why it had escaped detection for so long. This led to a model of the atom consisting of protons and neutrons packed closely together in the nucleus surrounded by electrons orbiting at a considerable distance.

E

THE NUCLEUS

How were the positive and negative bits of the atom arranged? By 1906, a new tool existed for probing the structure of the atom. It had been discovered that some radioactive elements gave off alpha particles. New Zealander Ernest Rutherford, one of the greatest experimental scientists ever, and his collaborators Geiger and Marsden, used alpha particles as a sort of nuclear artillery. They fired them at gold foil just a few atoms thick and investigated how they were scattered by observing scintillations (flashes of light) as the particles hit a zinc sulphide screen. The scattering pattern showed that the vast majority of alpha particles passed almost undeflected through the foil. This would be expected for the plum pudding model. However, a few were deflected through large angles, and a very few even bounced back. This could not be explained by the plum pudding model. These results suggested that the gold atoms were mostly empty space occupied by the electrons, too light to deflect an alpha particle, with most of the mass and positive charge concentrated in a small core called the nucleus.

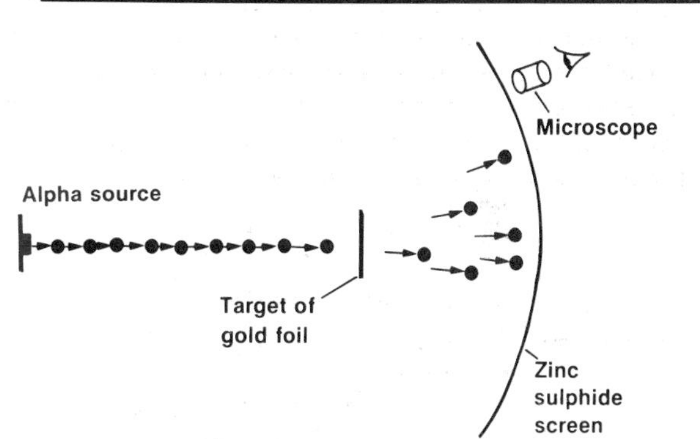

Schematic diagram of the alpha scattering experiment

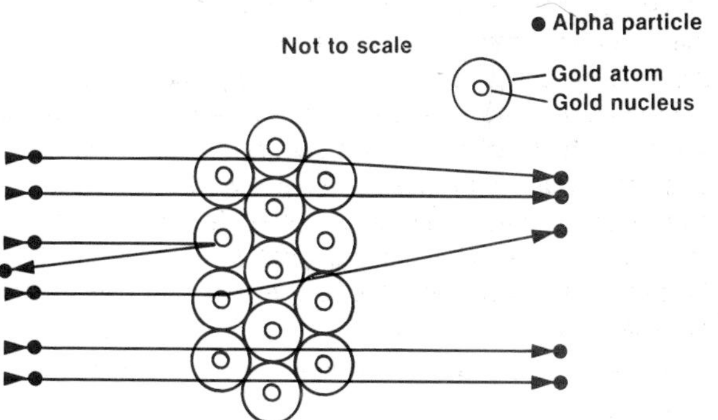

The process of alpha scattering

Ernest Rutherford. His nuclear model of the atom forms the basis of current ideas on atomic structure

Note that a collision between an alpha particle and a gold nucleus involves repulsion between the two positively charged entities. Mathematical analysis of the scattering pattern allowed the charge and size of the gold nucleus to be estimated.

The experiment is rather like finding and measuring the size of a cannonball hidden in a bale of hay by firing machine gun bullets at the bale.

E

THE NEUTRON

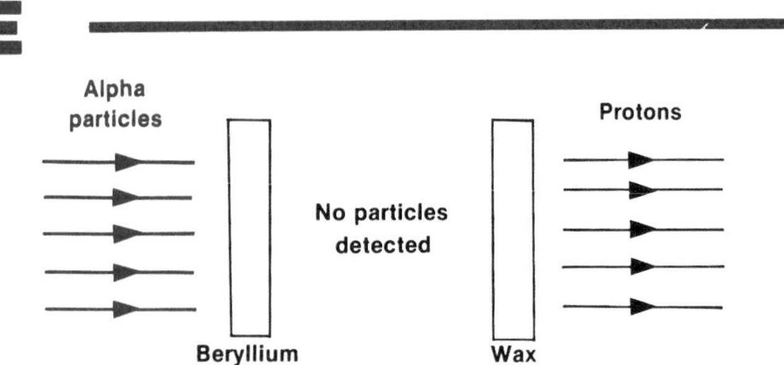

Chadwick's discovery of the neutron

While the proton and electron were the only subatomic particles known, in order to account for the known charge and mass of nuclei it was suggested that nuclei contained both protons and electrons. For example He ($A_r = 4$, $Z = 2$) was thought to contain four protons and two electrons in the nucleus, protons having a mass of one unit each and electrons negligible mass. The nucleus was orbited by two electrons. It was not till 1934 that Chadwick found evidence for neutral particles within the nucleus, although Rutherford had proposed this some years earlier. Chadwick bombarded beryllium with alpha particles and detected no resulting particles. However, on placing a wax screen close to the beryllium, protons were detected. He explained this result by supposing that the alpha particles ejected neutral particles (neutrons) from the beryllium which in turn knocked protons out of the wax. Neutrons, being uncharged, were not detectable by methods then in use.

This led to a model for the nucleus made up of protons and neutrons both of relative mass 1 unit. Both contributed to the mass, but only the protons to the charge. So a helium atom would be made up of two protons and two neutrons in the nucleus, surrounded by two electrons. Protons and neutrons are both referred to as nucleons.

8.2.5 The strong nuclear force

Clearly nuclei must be held together by a very powerful force to hold many protons close together against their electrostatic repulsion. The force which does this is called the **strong nuclear force**. It is much stronger than the electrostatic forces which hold electrons and protons together in the atom, but it operates only at very short range—within the nucleus. It is this force which gives rise to the large amounts of energy which are released in nuclear reactions such as in the atom bomb and nuclear reactors.

The idea of a small positive nucleus led to the model of electrons orbiting the nucleus like planets in the solar system, their orbital motion counteracting the electrostatic attraction of the nucleus, and preventing them from being dragged inwards.

8.2.6 Atomic number

The atomic number was originally merely the number of an element's position in the Periodic Table and was not thought to be of fundamental significance. However, in 1914 Henry Moseley discovered that on bombarding metallic elements with electrons, X-rays were produced, and that the frequency of the X-rays produced was related to the atomic number of the bombarded element. This suggested that the atomic number represented a fundamental property of the atom. It is in fact the number of protons in an atom, and so the atomic number is now often called the proton number. It is usually given the symbol Z.

Thus by the Second World War, atomic structure was well enough understood to allow nuclear reactions to be brought about, resulting in the atomic bombs used at Hiroshima and Nagasaki and, postwar, the controlled use of nuclear energy. Atoms were by then known to consist of a small dense nucleus containing protons and neutrons held together by the strong nuclear force. The diameter of the nucleus is of the order of 10^{-14} m compared with that of the atom of 10^{-10} m. The nucleus was visualized as being surrounded by orbiting electrons whose arrangement will be considered next.

8.2.7 Summary of subatomic particles

This is given in **Table 8.1**.

Table 8.1 Summary of subatomic particles

Name	Mass number	Mass/kg	Relative charge	Charge/coulomb
Proton	1	1.673×10^{-27}	+1	$+1.6 \times ^{-19}$
Neutron	1	1.675×10^{-27}	0	0
Electron	0	0.911×10^{-31}	−1	-1.6×10^{-19}

E

ATOMIC NUMBER

When metals are bombarded with a stream of electrons (cathode rays), with the apparatus shown below, X-rays (which are a form of high-energy electromagnetic radiation) are given off. In 1913 Moseley measured the frequency of the X-rays emitted by different metals. There seemed to be a relationship between the square root of the frequency and

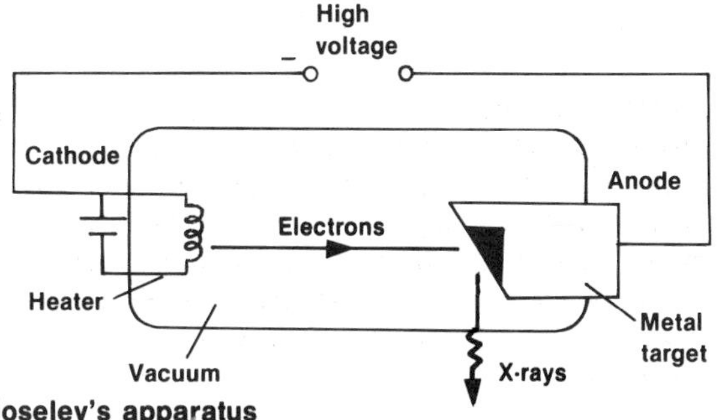

Moseley's apparatus

Henry Moseley. His work demonstrated the significance of atomic number as the number of protons in the nucleus. He was killed fighting in the First World War

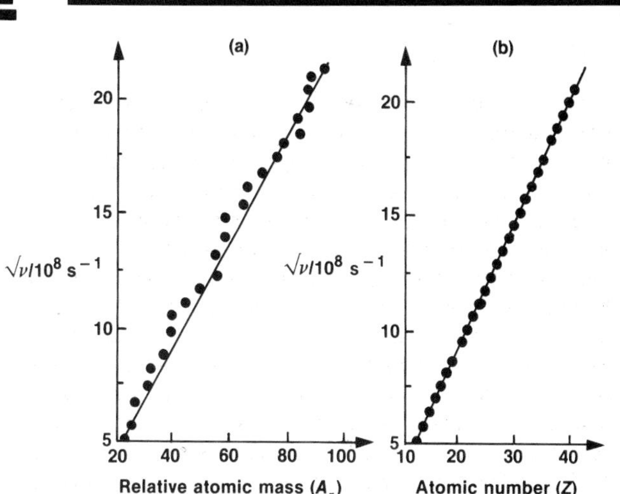

Graphs of $\sqrt{\nu}$ for characteristic X-rays of various elements. Note the much better straight line when plotted against Z rather than A_r

the relative atomic mass (see graph (a) above). However, an even better straight line is obtained if $\sqrt{\nu}$ is plotted against the atomic number—the order in which the elements appear in the Periodic Table (see graph (b) above). Thus the atomic number represents something more fundamental than just the element's position in an artificial list. Moseley proposed that the atomic number represented the number of protons in the nucleus. This link between a real physical quantity and position in the Periodic Table made it possible to know with certainty whether any elements were missing from the Periodic Table. This technique also made possible certain identification of any element. The mechanism for the emission of characteristic X-rays is similar to that for the production of line spectra.

Note that the accurate masses of the proton and neutron are not the same. The charges on the proton and the electron are *exactly* the same. The **mass number** is simply the number of nucleons (protons + neutrons) in the particle.

8.3 The arrangement of electrons in atoms

The evidence for the way electrons are arranged in atoms comes largely from the study of light and other types of electromagnetic radiation given out by atoms.

8.3.1 The electromagnetic spectrum

Electromagnetic radiation is energy which can travel through a vacuum. The most familiar form is light, but other types include heat (infra-red), radio waves and X-rays. Electromagnetic radiation can be thought of as a wave and therefore like all waves has a wavelength (λ), the distance between two successive peaks, and a frequency (ν), the number of vibrations per second. See **section 7.2.**

Wavelength and frequency are linked by the equation:

$$c = \nu\lambda$$

where c is the velocity of the electromagnetic radiation (the velocity of light, approximately $3 \times 10^8\ m\ s^{-1}$).

Figure 8.2 The electromagnetic spectrum

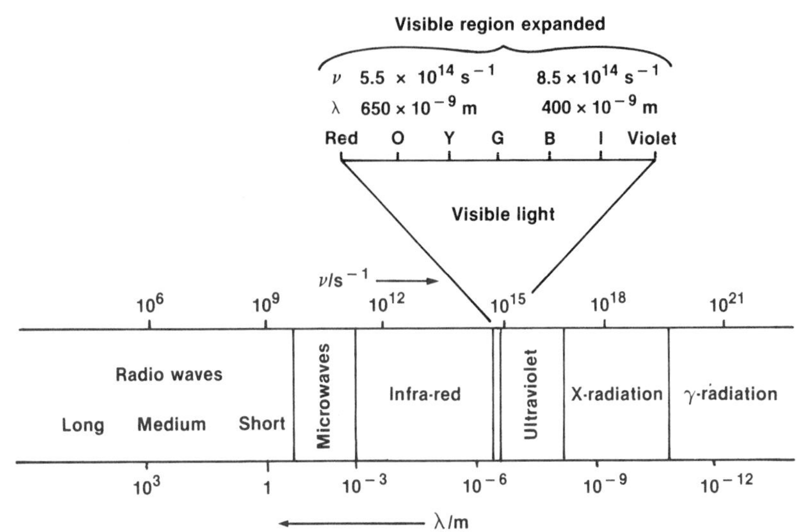

The different types of electromagnetic radiation have different frequencies and wavelengths **Figure 8.2.** Notice that high frequency radiation has a short wavelength and vice versa. The boundaries between different types are approximate. Electromagnetic radiation may also be thought of as coming in individual 'packets', rather like particles, called **quanta** (often called photons, especially quanta of visible light). Each quantum carries energy, depending on its frequency, given by the equation

$$E = h\nu$$

where h is a constant called Planck's constant, having the value 6.6×10^{-34} J s. So a photon of yellow visible light of frequency $6 \times 10^{14}\ s^{-1}$ has energy $6.6 \times 10^{-34} \times 6 \times 10^{14}\ J = 3.96 \times 10^{-19}$ J per quantum. This idea of electromagnetic radiation having some properties of both waves and particles is sometimes called **wave–particle duality**. It was once regarded as a contradiction in terms by scientists, but now the two views are seen as complementary.

8.3.2 Line spectra

When atoms are **excited** (given extra energy) they give out light and other forms of electromagnetic energy. Excitation may be done in a variety of ways. One of the simplest is to heat the substance in a bunsen flame; another is to pass an electric discharge through the vapour using the apparatus shown in **Figure 8.3.** This will produce a glow.

Figure 8.3 A discharge lamp

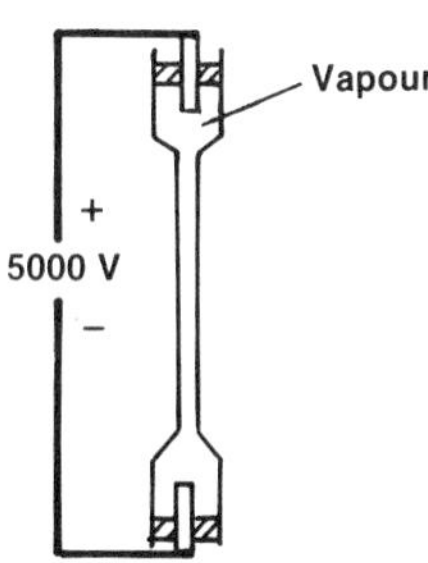

If the light is observed through a spectroscope (a device which splits light into its component wavelengths or frequencies) each element will be seen to produce a pattern of **lines**. Part of the spectrum of hydrogen is shown in **Figure 8.4.**

Notice how the lines are in groups and that the lines in each group get closer together or **converge** towards the high frequency (high energy) end of the spectrum. Each element produces a completely individual pattern of lines which can be used to identify it. When a previously unknown set

Figure 8.4 Part of the line spectrum of hydrogen showing two of the groups of lines

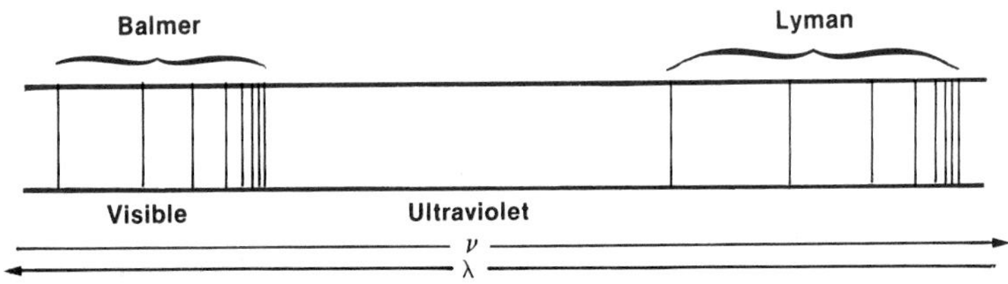

of lines was found in the spectrum of the sun by the French astronomer Janssen in 1868, the British scientist Lockyer was able to predict that a then unknown element, subsequently named helium (helios means sun in Greek) had been identified.

Spectral lines

All elements are found to have several series of lines, each fitting a formula similar to the one below for hydrogen:

$$\nu = cR\left(\frac{1}{n_1{}^2} - \frac{1}{n_2{}^2}\right)$$

ν = frequency of line
c = velocity of light
R = constant (Rydberg constant)
n_1 and n_2 are whole numbers, $n_2 > n_1$

For hydrogen the groups of lines have been given the names of their discoverers. They are listed in **Table 8.2.**

Table 8.2 Named series of lines in the hydrogen spectrum

Lyman	$n_1 = 1$, $n_2 = 2, 3, 4$, etc.	found in the ultraviolet region
Balmer	$n_1 = 2$, $n_2 = 3, 4, 5$, etc.	found in the visible region
Paschen	$n_1 = 3$, $n_2 = 4, 5, 6$, etc.	found in the infra-red region
Brackett	$n_1 = 4$, $n_2 = 5, 6, 7$, etc.	found in the infra-red region
Pfund	$n_1 = 5$, $n_2 = 6, 7, 8$, etc.	found in the infra-red region

An interpretation of these line spectra was proposed by the Dane Niels Bohr in 1913. He suggested that electrons could occupy only certain fixed energy levels around the nucleus which can be visualized as orbits of increasing radius, see **Figure 8.5.** Electrons in orbits of larger radius have higher energy. Even in hydrogen, with only one electron, an infinite number of levels exists, though only one is occupied at any one time. The situation is rather like a set of shelves with only one object on them. The object can be placed only on a shelf (not in the gaps between) and the shelves still exist even if they have no object on them. When an electron moves from one orbit to another, it absorbs (if going to a higher level) or emits (if dropping to a lower level) a **single quantum** of energy equal in size to the energy gap (ΔE) between the two levels. So the line spectrum of hydrogen is explained as follows. When hydrogen is excited in a discharge tube, the electron is promoted from the lowest level (the **ground state**), which it normally occupies, to a higher level. It then drops back to the lowest level either directly or via a series of steps. At each drop in level, it gives out a quantum of energy equal to the difference in levels ΔE. Because of the relationship $E = h\nu$, each quantum has a different frequency and therefore contributes one line to the line spectrum, as shown in **Figure 8.6.** Electrons dropping into level 1 give rise to the Lyman series of lines, those dropping into level 2 the Balmer series (of overall lower energy), and so on.

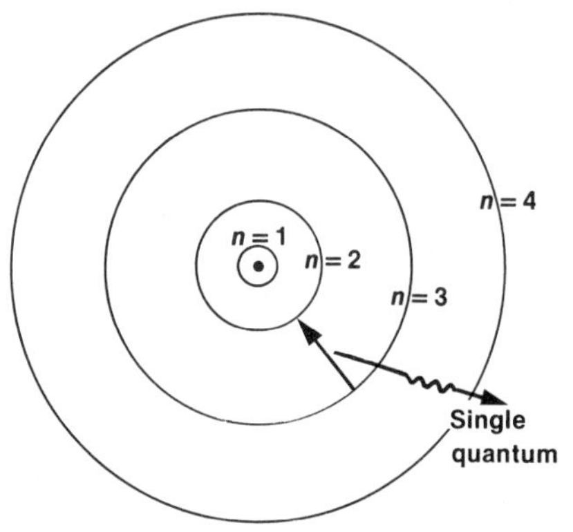

Figure 8.5 The first four Bohr orbits of hydrogen showing a transition between level 3 and level 2. The quantum of radiation would produce a line in the Balmer series

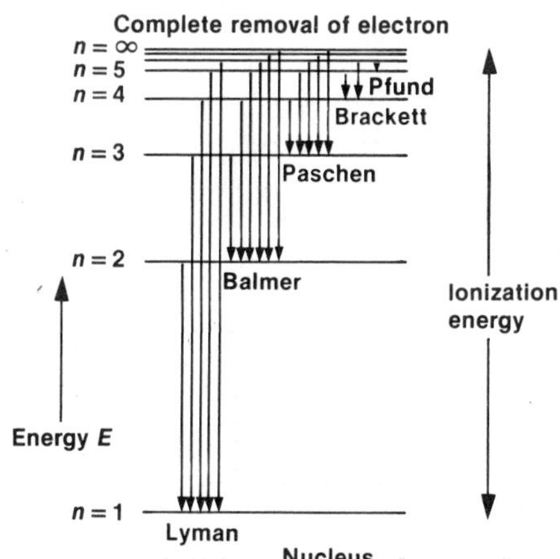

Figure 8.6 Energy levels in hydrogen. Note how this fits in with Figure 8.5. In Figure 8.5, the *radii* of the levels are drawn to scale. Here, the *energies* are drawn to scale. Also, the energy levels are drawn as straight lines rather than as circles

8.3.3 Ionization energy

As the energy levels get further from the nucleus, they get closer together. Eventually they are effectively 'touching' and they merge together. This corresponds to complete removal of the electron from the nucleus, or **ionization**. The amount of energy required to do this (for level 1) is called the **ionization energy** (strictly enthalpy of ionization, see **section 9.2.1**). A value of 1312 kJ mol^{-1} is found for hydrogen. The method described in the box on page 53 can be used to find the ionization energy of any atom. Also, the spacing between the electronic energy levels can be found. This can be done by measuring ionization energies from levels 1, 2, 3, etc. and finding the differences. For example, the ionization energy of hydrogen from level 1 is 1312 kJ mol^{-1} and from level 2 is 362 kJ mol^{-1} so the gap between levels 1 and 2 must be $1312 - 326 = 986$ kJ mol^{-1}.

FLAME EMISSION SPECTROPHOTOMETRY

This powerful analytical technique has become very important in recent years for the qualitative (what's there?) and quantitative (how much is there?) analysis of trace amounts of metals. It is used in pharmaceutical, metallurgical, medical, geological and forensic work. It depends on the fact that every element produces a unique set of spectral lines. A solution containing the sample is sprayed into a flame and the spectral lines identified using a spectrometer. Quantities are measured by comparing the intensity of the lines with a standard. Modern instruments are computer controlled and can produce a printout of results in seconds—vital in quality control work where stopping a production line to await analysis results could be very costly. A further advantage of the method is that only very small amounts of sample are required. The method is a much more sophisticated version of flame testing which you may have carried out to identify metals (sodium yellow, potassium lilac, etc.).

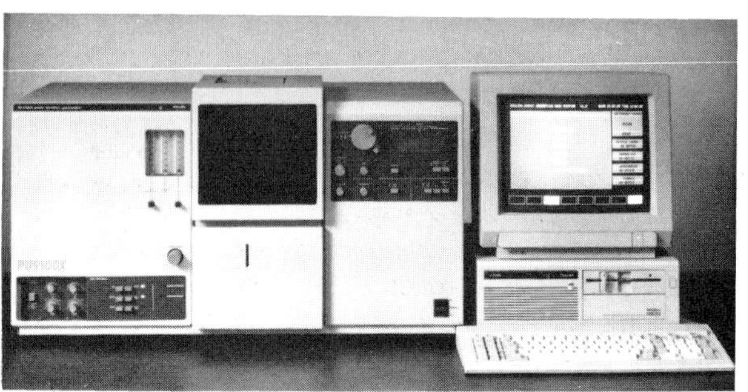

Flame emission spectrophotometer

8.3.4 Measuring ionization energies by electron bombardment

Figure 8.7 Apparatus for measuring the ionization energy of argon by electron bombardment

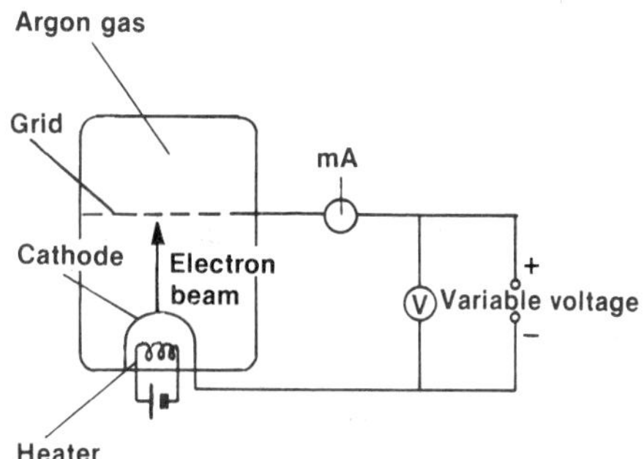

An alternative method of measuring ionization energy is by using a beam of fast-moving electrons as 'bullets' to knock electrons out of the atom using the apparatus shown in **Figure 8.7** for argon gas. Electrons are attracted from the heated cathode by the positively charged grid. A current flows and is detected by the milliammeter. As the voltage increases, so does the current. Eventually, a voltage is reached at which the electrons have enough energy to knock out an electron from an argon atom if they collide with it:

$$Ar(g) + e^- \longrightarrow Ar^+(g) + 2e^-$$

so three charged particles replace the original one, the electron. The current thus jumps. The ionization energy of argon can be calculated from the voltage at which the jump in current starts, see box opposite.

8.3.5 The number of electrons in different orbits

The energies required to remove the electrons one by one from an atom with many electrons can be measured. These are usually called successive ionization energies (IEs). Each electron requires more energy than the previous one, as the first electron is being removed from a neutral atom, the second from a 1+ ion, the second from a 2+ ion, and so on.

For example sodium:

$$Na(g) \longrightarrow Na^+(g) + e^- \quad \text{first IE}$$
$$Na^+(g) \longrightarrow Na^{2+}(g) + e^- \quad \text{second IE}$$
$$Na^{2+}(g) \longrightarrow Na^{3+}(g) + e^- \quad \text{third IE, and so on}$$

Table 8.3 Successive IEs of sodium/kJ mol^{-1}

1	496
2	4563
3	6913
4	9544
5	13352
6	16611
7	20115
8	25491
9	28934
10	141367
11	159079

MEASURING THE IONIZATION ENERGY OF HYDROGEN FROM ITS LINE SPECTRUM

It is possible to determine the frequency at which the spectral lines 'merge' (which corresponds to ionization) graphically. A graph is plotted of the *difference* in frequency of successive lines ($\Delta\nu$) of one series against the frequency and extrapolated to find the frequency when $\Delta\nu$ is zero. This frequency is called the **convergence limit** of the series of lines. When this is done for the Lyman lines of hydrogen, a convergence limit of $3.313 \times 10^{15}\,s^{-1}$ is found. Using

$$E = h\nu$$
$$E = 6.6 \times 10^{-34} \times 3.313 \times 10^{15}\,J$$
$$E = 2.186 \times 10^{-18}\,J \text{ per atom}$$

For one mole of atoms

$$E = 2.186 \times 10^{-18} \times 6 \times 10^{23}\,J\,mol^{-1}$$
$$E = 1\,312\,000\,J\,mol^{-1}$$

In more familiar units, $E = 1312\,kJ\,mol^{-1}$. This is the energy required to remove an electron from the lowest energy level of hydrogen, the **ground state**, and is called the **ionization energy**.

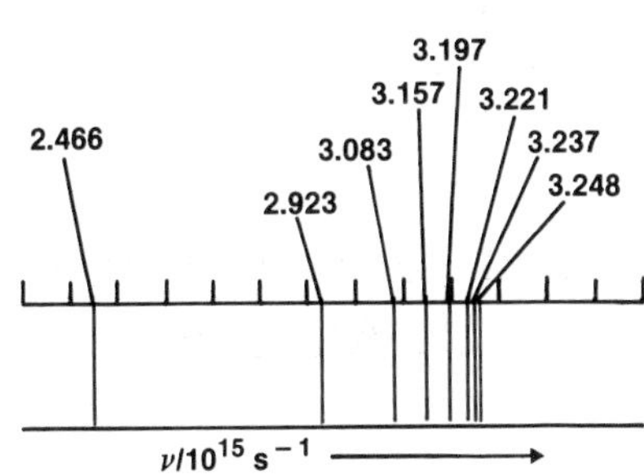

The Lyman lines in the hydrogen spectrum

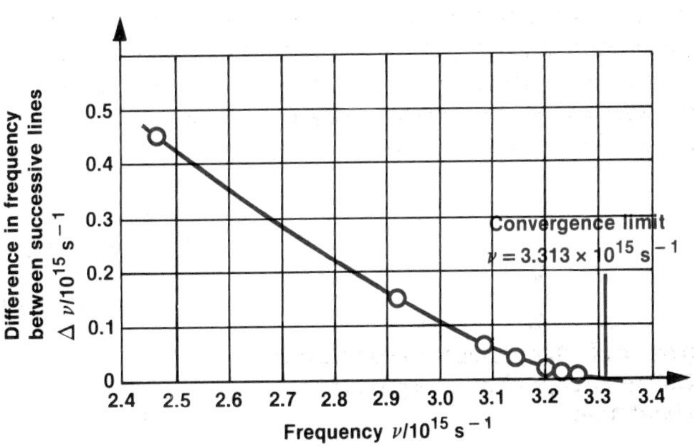

CALCULATING IONIZATION ENERGY FROM ELECTRON BOMBARDMENT EXPERIMENTS

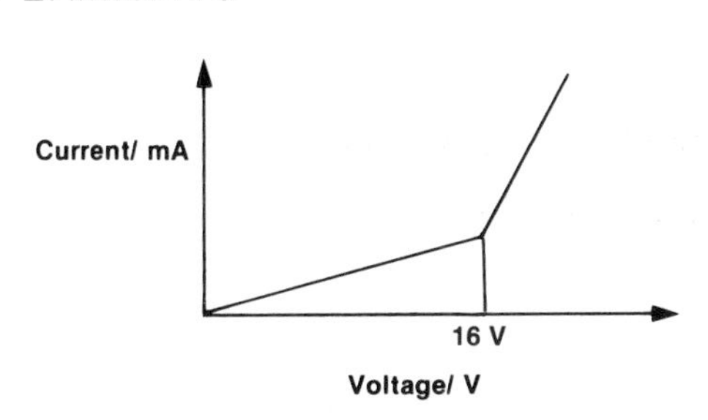

The voltage is equal to the energy of the bombarding electrons in joules per coulomb.

Number of joules per electron $= 16 \times$ charge on electron
$= 16 \times 1.6 \times 10^{-19}\,J$
$= 2.56 \times 10^{-18}\,J$

Ionization energies are usually quoted per mole, so

$$IE = 2.56 \times 10^{-18} \times 6 \times 10^{23}$$
$$= 1\,536\,000\,J\,mol^{-1}$$
$$= 1536\,kJ\,mol^{-1}$$

Remember that the first ionization energy is ΔH for the process $E(g) \longrightarrow E^{+}(g) + e^{-}$.

Figure 8.8 Successive ionization energies of sodium

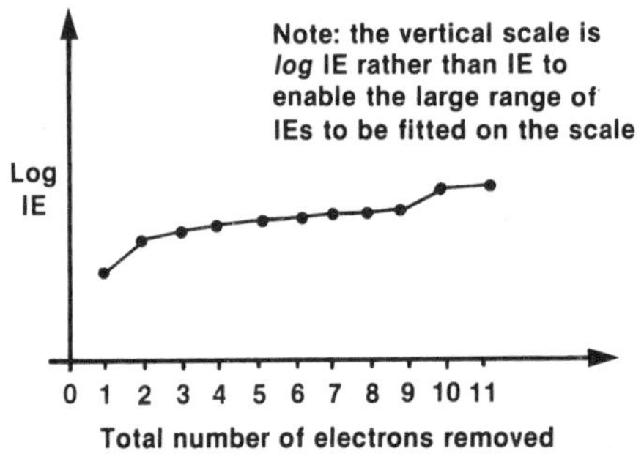

Notice that the second IE is *not* the energy change for

$$Na(g) \longrightarrow Na^{2+}(g) + 2e^{-}$$

The energy change for this process would be the first IE + the second IE.

If a graph is plotted of the figures in **Table 8.3** we get **Figure 8.8.** Notice that one electron is relatively easy to remove, then comes a group of eight rather more difficult to remove, and then two very difficult to remove. This suggests that sodium has two electrons very close to the nucleus (which are the most difficult to remove), eight further out and one further away still. So it is possible to work out the number of electrons in each shell or orbit. A closer look at the figures would show that the group of eight electrons is in fact split into a group of two and a group of six, although the difference in energy is small and is not noticeable in **Figure 8.8.**

Orbitals

For a more complete description of electrons in atoms we must turn to a theory called **quantum mechanics**. This considers the electron as a wave (wave–particle duality again) and describes the atom with an equation (the Schrödinger equation). The solutions to this equation give the *probability* of finding an electron in a given volume of space, rather than its exact location at any time. For our purposes, it may be best to think of the electron as a cloud of negative charge which may fill a volume in space called its **orbital**. The solutions tell us the distance from the nucleus and hence the energy of an electron, and these solutions correspond to the shell numbers 1, 2, 3,... The shapes of the orbitals are described by the letters s, p, d and f. These shapes are shown in **Figure 8.9.**

s orbitals can hold up to two electrons.

p orbitals can hold up to two electrons each but always come in groups of three of the same energy to give a total of six electrons.

d orbitals can hold up to two electrons each but come in groups of five of the same energy to give a total of 10 electrons.

f orbitals can hold up to two electrons each but come in groups of seven of the same energy to give a total of 14 electrons.

The energy level diagram in **Figure 8.10** shows the orbitals for the first few elements of the Periodic Table. Notice that 4s is actually of slightly lower energy than 3d. Also notice that level 1 has only an s orbital, level 2 has s and p, level 3 has s, p and d, and so on.

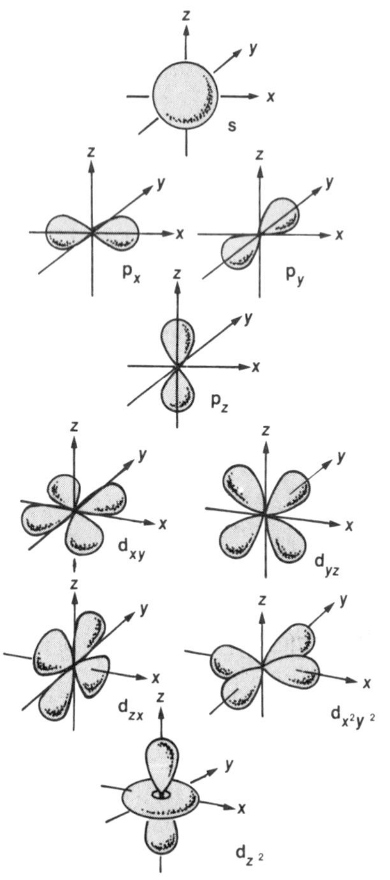

Figure 8.9 Shapes of orbitals. f orbitals have a more complex shape still. These shapes represent a volume of space in which there is a 95% probability of finding the electron

Figure 8.10 (above right) The energy levels of the first few atomic orbitals

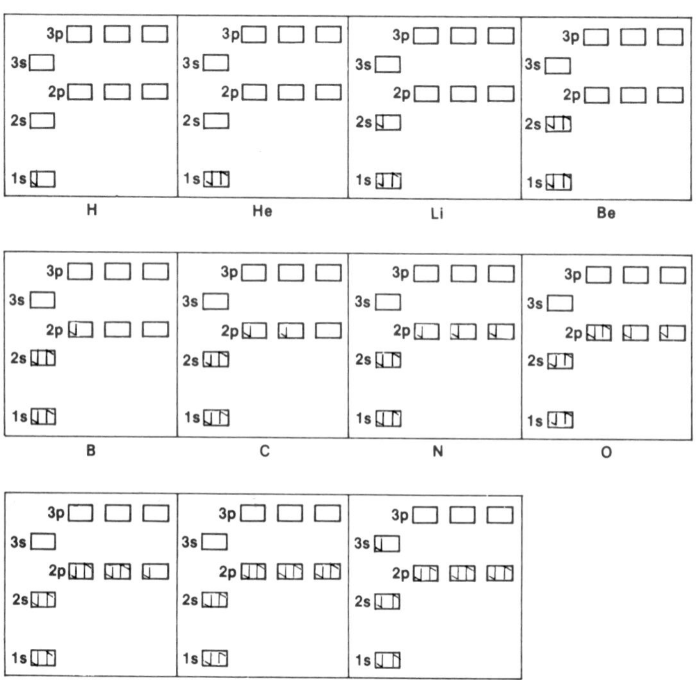

Figure 8.11 The electron arrangements for the elements hydrogen to sodium

Each 'box' in **Figure 8.10** represents an orbital of the appropriate shape which can hold up to two electrons. Electrons also have the property of **spin**. Two electrons in the same orbital must have opposite spins. This is usually represented by arrows pointing up or down to represent the different directions of spin. The rules for allocating electrons to orbitals state that the orbitals of lower energy are filled first. Some energy is needed to pair electrons, so orbitals of the same energy fill singly before pairing starts. The electron diagrams for the elements hydrogen to sodium are shown in **Figure 8.11.** A shorthand way of writing electronic structures is as follows, for example for sodium:

$$\begin{array}{ccc} 1s^2 & 2s^2 2p^6 & 3s^1 \\ 2 & 8 & 1 \end{array}$$

Note how this corresponds with the simpler 2, 8, 1.
Calcium would be:

$$1s^2\, 2s^2\, 2p^6\, 3s^2\, 3p^6\, 4s^2$$

and argon would be:

$$1s^2\, 2s^2\, 2p^6\, 3s^2\, 3p^6$$

Sometimes it simplifies things to refer back to the previous inert gas. So the electron arrangement of calcium, Ca, could be written (Ar)$4s^2$ as a shorthand for $1s^2\, 2s^2\, 2p^6\, 3s^2\, 3p^6\, 4s^2$, as $1s^2\, 2s^2\, 2p^6\, 3s^2\, 3p^6$ is the electron arrangement of argon.

8.4 Radioactivity—nuclear reactions

Marie Curie, whose work on radioactive elements earned her the Nobel Prize for Chemistry in 1911. She died a victim of the radiation she studied

The nuclei of some atoms are unstable. They can break down forming new elements and giving out radiation. Such atoms are called **radioactive**. The main radioactive processes are:

1. **Alpha decay**—the nucleus emits an alpha particle (α-particle), a group of two protons and two neutrons.
2. **Beta decay**—an electron, called a beta particle (β-particle) in this context, is emitted from the *nucleus.*
3. **Gamma emission**—high-energy electromagnetic radiation (γ-radiation) is emitted from the nucleus. This only occurs along with alpha or beta decay.
4. **Fission**—the nucleus splits into two to give two new nuclei, each of smaller mass.

In the first three cases, the radiations given off can knock out electrons from atoms to produce ions. They are thus called **ionizing radiations**.

If radiation passes through the human body, ions are formed in the cells and these can interfere with the complex reactions which occur there. We are all exposed to a small level of natural radioactivity from cosmic rays and from rocks, for example. We call this **background radiation**. Our bodies can deal with a certain amount of ionization, but it makes sense to cut down exposure from other sources to a minimum.

In experimental work, the background radiation must be measured and subtracted from each reading to give the reading produced by the source which we are interested in.

8.4.1 Nuclides and isotopes

Any combination of protons and neutrons forming the nucleus of an atom is called a **nuclide**. Nuclides with the same number of protons but different numbers of neutrons are called **isotopes**. The chemical identity of an atom depends on its number of protons which governs the number of electrons and thus its chemical reactivity. Different isotopes *of the same element* will react the same as one another chemically, but they may have different

Film badge which monitors exposure to radiation

Figure 8.12 The Geiger–Müller tube

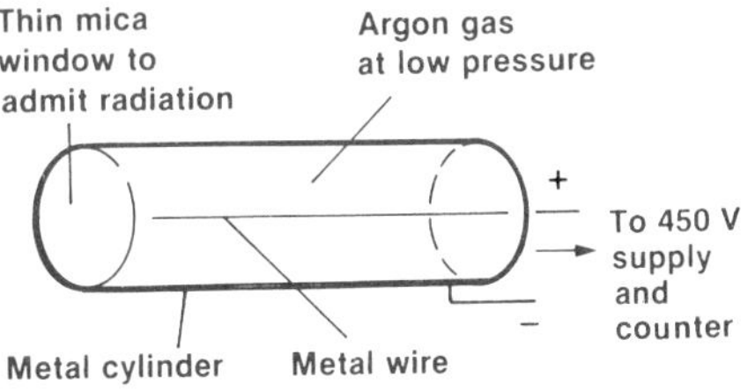

Geiger–Müller tube (front) and counter (behind)

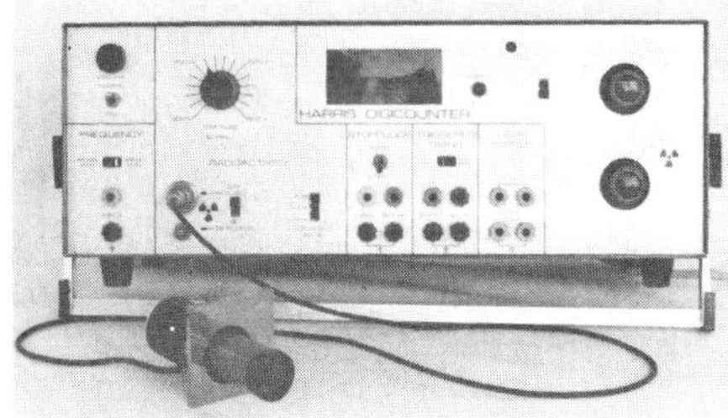

Tracks in a bubble chamber

radioactive properties. For example, carbon has three common isotopes. All have six protons and each has either six, seven or eight neutrons.

Nuclides can be written with the following shorthand:

$$^{\text{mass number}}_{\text{atomic number}}\text{chemical symbol}$$

So the isotopes of carbon are:

$$^{12}_{6}C \quad ^{13}_{6}C \quad ^{14}_{6}C$$

The first two are stable but $^{14}_{6}C$ (pronounced carbon-14) is a beta emitter.

8.4.2 Detection of radiation

Photographic film

Henri Becquerel first discovered the radiation from uranium in 1896 when he noticed that a photographic plate wrapped in black paper had become fogged when it was placed close to a mineral sample. The sample contained uranium. This method of detection is still used today—people who work with ionizing radiation wear a film badge to monitor their total exposure over a period of time, see photograph (left).

Geiger–Müller tube

See **Figure 8.12** and photograph. When radiation enters the tube through the window (which is thin so that it does not absorb much radiation) it forms ions from argon atoms:

$$Ar(g) \longrightarrow Ar^{+}(g) + e^{-}$$

The ions momentarily make the gas conduct electricity, and the pulse of electricity which flows is detected by the counter.

Cloud chamber/bubble chamber

In the cloud chamber, a vapour is cooled so that it is supersaturated. When radiation passes through it, liquid droplets form along the radiation's track (like the vapour trail of an aeroplane). This is useful because the actual path of the radiation can be followed, see photograph (below left).

The bubble chamber is similar except that a liquid, kept above its boiling point by pressure, is used instead of the supersaturated vapour. Just before observing the radiation, the pressure is reduced. As the radiation passes through the liquid, a trail of vapour bubbles is formed along its track. The inventor is said to have had the idea from contemplating the bubbles forming in a glass of beer.

The cloud chamber has in fact been a powerful tool in the discovery of sub-nuclear particles. Every nuclear breakdown produces a set of tracks which can help to identify the sub-nuclear particles which have made them. A 'strange' track may be the first sign of the presence of an unknown species.

8.4.3 Nuclear decay processes

Alpha decay

An α-particle is the same as the nucleus of a helium atom (written $^{4}_{2}He$)—two protons and two neutrons. When this is emitted from the nucleus, the atomic number goes down by two, and the mass number by four. This can be represented by a **nuclear equation**, for example:

$$\underset{\text{thorium}}{^{229}_{90}Th} \longrightarrow {^{4}_{2}He} + \underset{\text{radium}}{^{225}_{88}Ra}$$

Notice that *a new element is formed*—thorium changes into radium in this example. This is quite unlike chemical reactions which merely involve rearrangement of existing atoms. Note also that both the total mass

Nuclear fission has both military and peaceful applications
(a) Atomic explosion

(b) Fuelling at Sizewell nuclear power station

Figure 8.13 **A nuclear chain reaction**

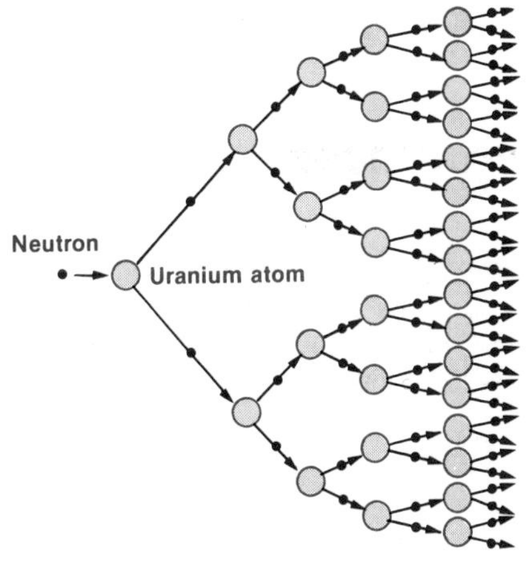

numbers and total atomic numbers are *conserved*—the total mass number of the products is equal to the mass number of the original element, and similarly for the atomic number.

α-particles are relatively large with a mass number of four, and they have a positive charge of two. They are therefore easily absorbed, being stopped by a sheet of paper for example. They become helium atoms by picking up two electrons.

Beta decay

A β-particle (an electron, written $_{-1}^{0}e$) is emitted from the nucleus, *not* from the electron shells. For this to happen, a neutron turns into a proton and an electron. This may be represented:

$$_{0}^{1}n \longrightarrow {}_{1}^{1}p + {}_{-1}^{0}e$$

Notice again the conservation of total mass number and total atomic number, if the negative charge on the electron is counted as an atomic number of -1.

The proton remains in the nucleus while the electron is ejected. So again a new element is created, this time with the same mass number and an atomic number that is one greater than the original element. For example:

$$\underset{\text{polonium}}{^{216}_{84}\text{Po}} \longrightarrow \underset{\text{astatine}}{^{216}_{85}\text{At}} + {}_{-1}^{0}e$$

β-particles travel further than α-particles and it takes, for example, aluminium which is a few centimetres thick to absorb them.

Gamma radiation

Following either α or β decay, the new nucleus may be formed in a high energy state. In order to lose this energy it may give off a quantum of γ-radiation—a high energy form of electromagnetic radiation. γ-radiation therefore accompanies either α or β decay. No particle is emitted and therefore the identity of the nucleus does not change. γ-radiation is the most penetrating of the three types of radiation, and is only stopped by several centimetres of lead.

Nuclear fission

Large nuclei with many protons are relatively unstable due to the mutual repulsion of the protons. Such nuclei can split or undergo **fission** to form two new elements of smaller mass. This can happen spontaneously, but more often it is induced by a collision with a neutron. For example, if an atom of uranium captures a neutron, the following nuclear reaction occurs:

$$\underset{\text{uranium}}{^{235}_{92}\text{U}} + {}_{0}^{1}n \longrightarrow \underset{\text{barium}}{^{144}_{56}\text{Ba}} + \underset{\text{krypton}}{^{90}_{36}\text{Kr}} + 2{}_{0}^{1}n$$

The significant thing is that *one* neutron is captured but *two* are produced. If the lump of uranium is large enough so that these neutrons do not escape from the surface, each one can initiate a further fission which in turn releases two more neutrons. The result is an escalating process—a **chain reaction, Figure 8.13.** The mass of uranium which is just large enough to support a chain reaction is called the **critical mass**. Lumps smaller than this lose too many neutrons from their surface for a chain reaction to continue.

Although the mass numbers are conserved in the above reaction, if the *exact* masses of starting materials and products are measured, those of the products total slightly less than those of the starting materials. This 'lost' mass is converted to energy, the amount of which can be calculated using Einstein's famous equation

$$E = mc^2$$

E = energy produced
m = mass destroyed
c = velocity of light ($3 \times 10^8\ \text{m s}^{-1}$)

which tells us the amount of energy produced when mass is destroyed.

Since the term c^2 is so large, a minute amount of mass produces a great deal of energy. An uncontrolled chain reaction is the source of the energy produced in atomic bombs like that used at Hiroshima. Such a reaction can be started by bringing together two pieces of uranium, each slightly smaller than the critical mass, to produce a lump which is larger than the critical mass. Chain reactions are used in nuclear reactors, using substances like graphite as moderators to absorb some of the neutrons and control the process.

A

ENERGY FROM NUCLEAR FISSION

In the nuclear reaction:

$$^{235}_{92}U + ^{1}_{0}n \longrightarrow ^{144}_{56}Ba + ^{90}_{36}Kr + 2^{1}_{0}n$$

the total mass numbers are conserved, i.e. the total mass number before reaction (235 + 1 = 236) is the same as the total mass number after reaction (144 + 90 + 2 = 236).

However, if we add up the *exact* masses of each particle, some mass is apparently lost. Exact masses:

U = 235.043 93 u
Ba = 143.923 u
Kr = 89.9197 u
n = 1.008 665 u

Total mass before reaction = 235.043 93 + 1.008 665
= 236.052 595 u
Total mass after reaction = 143.923 + 89.9197 + (2 × 1.008 665) = 235.860 03 u
Loss of mass = 0.192 565 u

This lost mass is converted into energy. The amount of energy is given by $E = mc^2$, where c is the speed of light (3×10^8 m s^{-1}).

So 253 kg of uranium, when fissioned, converts 0.192 565 kg of mass into energy.

$$E = mc^2$$
$$E = 0.192\,565 \times (3 \times 10^8)^2$$
$$E = 1.73 \times 10^{16}\text{ J} \quad \text{or} \quad 1.73 \times 10^{13}\text{ kJ}$$

This is the same as the amount of energy produced by burning 300 000 tonnes of petrol!

THE SAFETY OF IONIZING RADIATIONS

All three types of radiation can cause cell damage. The more cells which are damaged, the greater the danger to the body. The number of cells damaged depends, of course, on the intensity of the radiation, but it also depends on the type. γ-radiation is so penetrating that in small doses most of it passes right through the body, causing little harm. β-radiation is less penetrating. That is, more is absorbed in cells, making it more damaging. α-radiation is the least penetrating and if it gets into the body, it causes a great deal of damage. However, because it is so easily absorbed, most of it is stopped by the layer of dead cells which form the outer surface of our skin. So external α sources are relatively safe. α sources *within* the body are a different story. With no protective layer of dead skin to protect the body, α particles cause intense local cell damage. So, for example, inhaled α emitters such as radon gas or particles of plutonium dust are very dangerous.

8.4.4 Radioactive decay—half-life

Different radioactive elements decay at widely different rates. It is always found that the rate of decay is proportional to the amount of radioactive substance remaining. This is called a first-order process (see **section 14.4**).

Mathematically this means that the time taken for the radioactivity to decay to half the original amount is always the same, irrespective of the original amount. This time is called the **half-life**, $t_{\frac{1}{2}}$. It can be found from a graph of count rate (or mass of radioactive substance remaining, since this is directly proportional to count rate) against time, **Figure 8.14**. Half-lives of radioisotopes vary from fractions of a second to millions of years, for example:

^{235}U	7.13×10^8 years
^{212}Po	3×10^{-7} seconds
^{90}Sr	28 years

The lengths of the half-lives of radioisotopes present in radioactive waste,

Figure 8.14 **Graph showing the decay of a radioactive element. $t_{\frac{1}{2}}$ is the time taken for the radioactivity to drop from *any* initial value to half that value**

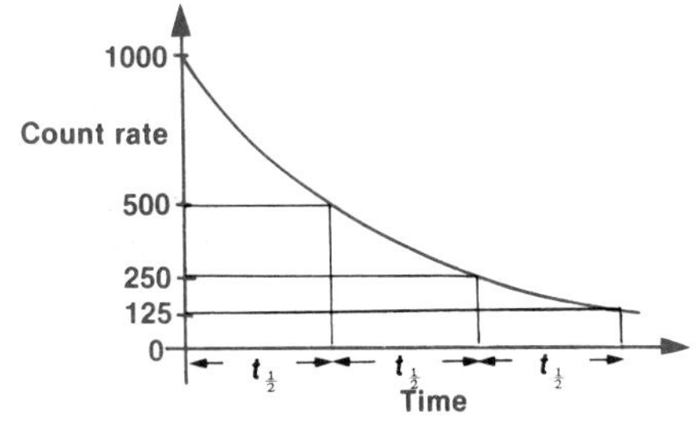

from nuclear power stations for example, have an important bearing on the pollution problems they present.

For example, if radioactive waste has a half-life of 10 years, after 50 years it will have dropped in activity to:

$$\frac{1}{2} \times \frac{1}{2} \times \frac{1}{2} \times \frac{1}{2} \times \frac{1}{2} = \frac{1}{32} \text{ of its original activity}$$

whereas waste with a half-life of 1000 years will barely have dropped in activity at all.

Radioactive decay processes are unaffected by temperature, catalysts or anything else. So the only way of dealing with such waste is to store it until it decays naturally. Present methods involve storing the waste in steel drums or chemically treating it to make it into a glass-like substance—virtrification. However, with half-lives of hundreds of years, radioactive waste is an unpleasant legacy to future generations.

8.4.5 Uses of radioisotopes

The term 'radioisotope' is a contraction of 'radioactive isotope'. Because they can be detected at a distance by the methods described in **section 8.4.2,** radioisotopes have a large number of industrial and medical applications.

Leak tracing

A small amount of a radioisotope, placed in an inaccessible system of pipework, can be detected by a Geiger–Müller tube when it leaks out of the pipe. This pinpoints the leak without the need to expose the whole system. On the same principle, it is possible to track the direction of currents in tidal rivers, and find where silting up is occurring, so that the mud can be dredged from the river at this point.

Figure 8.15 Using a radioactive source to monitor the thickness of sheets during production

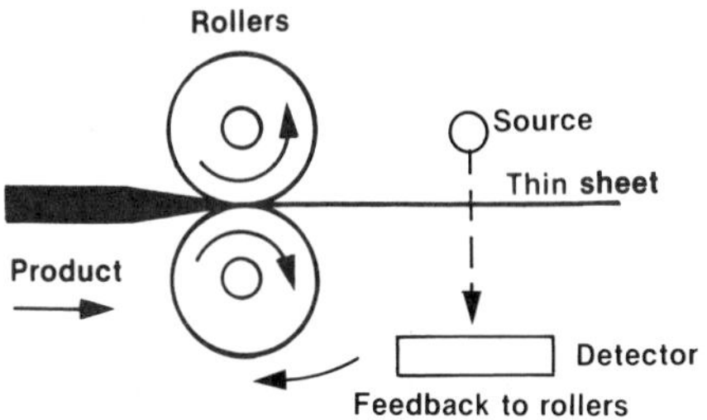

Thickness gauging

Sheets of paper, aluminium foil or rubber are manufactured by squeezing the product between a pair of rollers to the required thickness. The thickness can be measured by beaming a radioactive source at the sheet and detecting it on the other side. If the sheet is too thick or too thin the reading will vary and the pressure of the rollers can be automatically changed, **Figure 8.15.**

Figure 8.16 Using a radioactive source to monitor fill levels

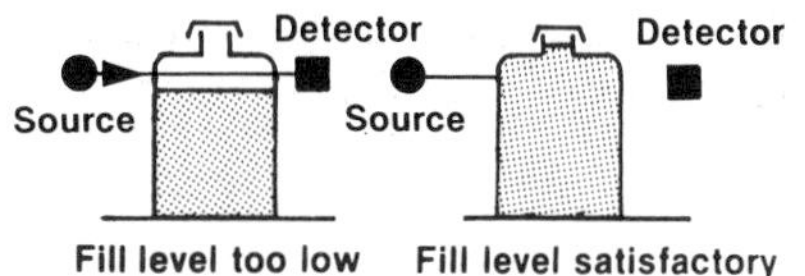

Measuring fill levels

In the canned food or drinks industries, a radioactive source may be used to check that bottles or cans have been filled to the correct level, **Figure 8.16.** The filled bottles on a conveyor belt pass a source and detector as shown. Unfilled bottles absorb less radiation than full ones and can be automatically rejected.

A

A DEADLY GAS UNDER THE FLOORBOARDS

This headline was how *New Scientist* in February 1987 reported a recently discovered radioactive health hazard. The gas concerned is radon-222, a chemically inert but radioactive gas, an α-emitter of half-life 3.8 days. High levels of this gas have recently been found in a number of homes, particularly those built on igneous rocks such as are found in Devon and Cornwall. The radon is formed by a series of nuclear reactions beginning with $^{238}_{92}U$ which is present in many types of granite. A particular danger is presented by radon as it is a gas and can 'bubble up' through rock faults and accumulate under floorboards in houses, from where it can seep into the house and be breathed in by the occupants.

The reactions involved are:

$$^{238}_{92}U \xrightarrow{\alpha} {}^{234}_{90}Th \xrightarrow{\beta} {}^{234}_{91}Pa \xrightarrow{\beta} {}^{234}_{92}U$$

$$\xrightarrow{\alpha} {}^{230}_{90}Th \xrightarrow{\alpha} {}^{226}_{88}Ra \xrightarrow{\alpha} {}^{222}_{86}Rn$$

Note the two different isotopes of both U and Th and the shorthand representation of α and β decay.

The National Radiological Protection Board is currently monitoring the problem and methods to reduce it, which include sealing floors and providing better ventilation, are being investigated.

Food irradiation

Because of its ability to kill living things, radiation can be used to treat foods. Depending on the type of food and the dose, this can prevent sprouting (of, say, potatoes), kill infesting insects, or completely sterilize the food. Such sterile food can be fed to hospital patients who are especially susceptible to infection. The technique of food irradiation is still controversial at present as the radiation inevitably causes chemical changes in the food which may affect its flavour and even safety. Irradiation does not make the food itself radioactive. γ-radiation from ^{60}Co is used to treat food. Because of the large doses of radiation used, the food is carried past the source on a conveyor belt inside a shielded building.

Patient being treated with radiotherapy

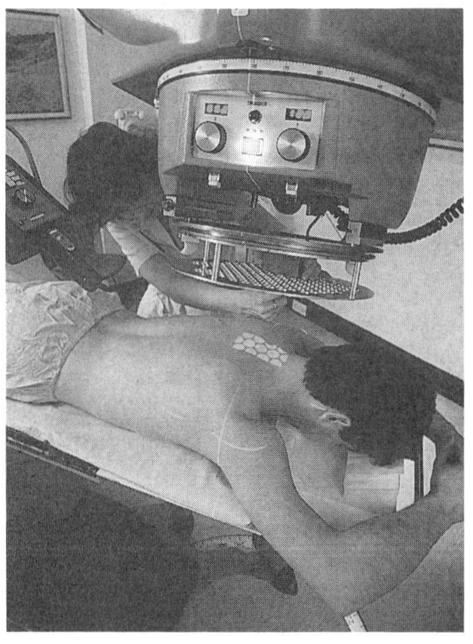

Medical uses

Radioisotopes can be used both in treatment and diagnosis. Radiation can be used to kill cancer cells, but as it also kills normal cells, the cancer must be relatively localized and the rest of the body well shielded, see photograph (left). In diagnosis, isotopes of iodine can be injected into the patient to test the activity of the thyroid—a gland in the neck which concentrates iodine. The radioactive iodine behaves chemically in the same way as non-radioactive iodine but the radiation can be detected outside the body to allow an estimate of how much iodine is being absorbed by the gland.

Technetium is absorbed by bones, so radioactive technetium is used to monitor bone growth. Areas of rapid bone deposition, such as a fracture healing, concentrate technetium and such areas can be detected by the concentration of radiation. In both cases, the doses of radioactivity are small. The patient rapidly excretes the isotopes used and as isotopes of short half-life are used, the radioactivity decays anyway.

Another medical use is in heart pacemakers, implanted in patients to control their heart rates by giving the heart a series of small electric shocks to 'pace' it. These can be powered by the energy given off by decay of a small amount of a radioisotope. Isotopes with a relatively long half-life must be used to avoid the need for frequent replacement operations. α sources are used to minimize the shielding problem.

The Turin shroud. Radiocarbon dating showed it to date from the thirteenth century

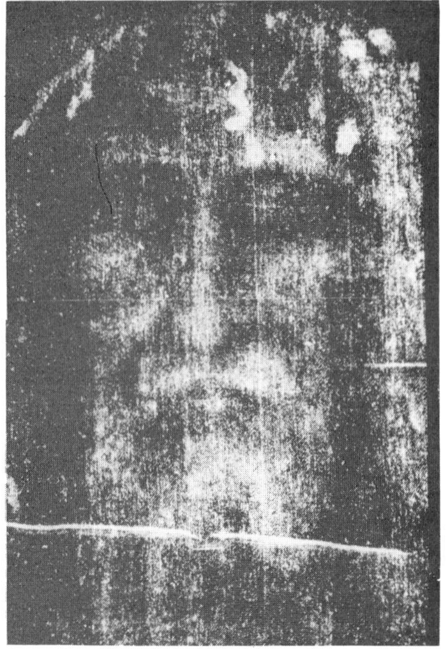

Radiocarbon dating

The radioactive isotope ^{14}C is continually being generated in the upper atmosphere by nuclear reactions brought about by cosmic ray bombardment. Thus atmospheric carbon dioxide contains a small known proportion of $^{14}CO_2$ which is incorporated into the tissues of all living things while they are alive. On death, no more ^{14}C is added, and that existing decays with a half-life of 5568 years. This enables archaeological specimens which contain any once-living material, such as wood or natural fibres, to be dated by measuring the proportion of ^{14}C remaining and calculating back. For example, if a sample contains $\frac{1}{4}$ of the normal ^{14}C proportion it must have stopped living 2 half-lives of ^{14}C ago, i.e. 11 136 years ago. This method has recently been used to date the Turin shroud, see photograph opposite.

Chemical studies

Radioisotopes can be used to help decide which atom in a compound is involved in a particular reaction, for example:

$$\underset{\text{ethanoic acid}}{CH_3C(=O)OH} + \underset{\text{ethanol}}{C_2H_5OH} \longrightarrow \underset{\text{ethyl ethanoate}}{CH_3C(=O)O-C_2H_5} + \underset{\text{water}}{H_2O}$$

Suppose we wish to find out which oxygen atom is the one that appears in the water. If ethanol with a radioactive isotope of oxygen could be prepared, it would be easy to find out if the radioactivity appears in the water or the ethyl ethanoate. The same procedure could be carried out with non-radioactive isotopes if they could be detected. This can be done with the mass spectrometer.

8.5 The mass spectrometer

An early form of this instrument was first used by Francis Aston in 1920 to detect the existence of isotopes. Now it is one of the most useful instruments in the chemist's 'armoury' for accurate measurements of relative atomic and molecular masses and for structure determination, especially in organic chemistry. A form of the instrument is shown in **Figure 8.17**. The sample to be investigated is injected directly, if a gas or a volatile liquid, or else vaporized first in an oven.

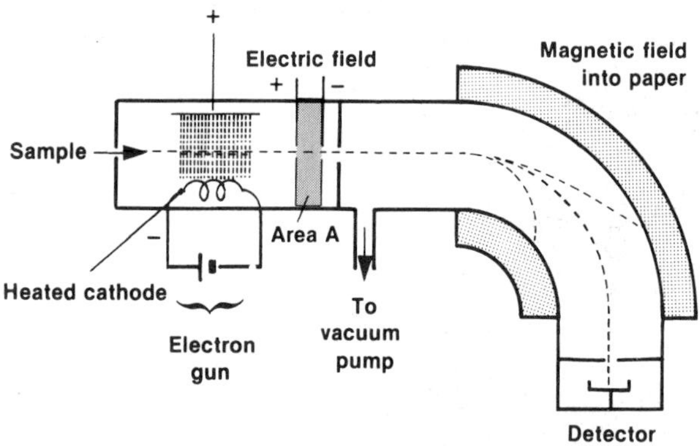

Figure 8.17 Schematic diagram of a mass spectrometer

A beam of electrons from an 'electron gun' then knocks out electrons from molecules of the sample to form positive ions. These then drift into area A where they are accelerated towards a plate at a negative potential.

Some ions pass through a pair of slits which form them into a beam. The speed they reach depends on their mass, lighter ions going faster.

The beam then moves into a magnetic field at right angles to its direction of travel. This deflects the ions into an arc of a circle. The deflection depends on mass, heavier ions being deflected less than lighter ones. The magnetic field is gradually increased so that ions of increasing mass enter the detector, which produces a signal proportional to the number of ions reaching it.

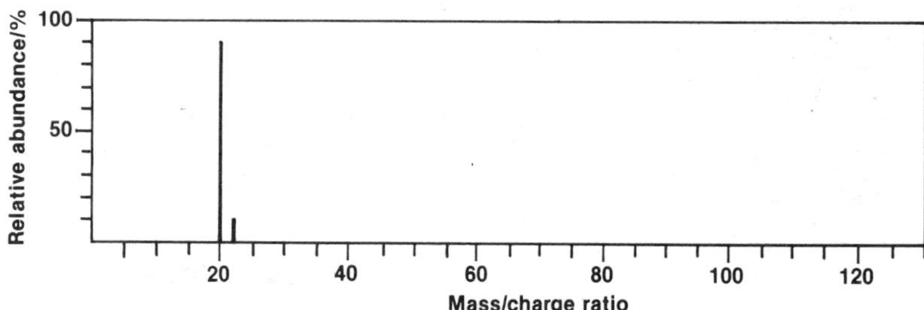

Figure 8.18 The mass spectrum of neon

Notice that the instrument detects individual ions so that isotopes of different mass are detected separately. The output is normally presented as a graph of abundance of ions present against mass number (strictly the mass/charge ratio, but since virtually all the ions have one unit of positive charge, the two are effectively the same). For example, neon would give the trace shown in **Figure 8.18.** This shows that neon has two isotopes of mass numbers 20 and 22, present in the approximate ratio 9:1. Hence neon has an average relative atomic mass of

$$[(9 \times 20) + (1 \times 22)]/10 = 20.2$$

Note that the peak height gives the abundance of each isotope and the horizontal scale gives the relative atomic mass. They are often confused.

Mass spectrometry is the main method for determining relative molecular masses of compounds. In the case of compounds, during their time of flight through the mass spectrometer, some of the ions break up or **fragment**, the bonds having been weakened by loss of an electron during ionization. The fragments may also appear in the mass spectrum and give useful clues to the structure of the compound. Structure determination is covered in more detail in Chapters 19 and 37.

A

THE ELECTRON GUN

The electron gun is a vital part of a number of pieces of apparatus. Its function is to produce a beam of electrons. This beam may be used to ionize a sample as in the mass spectrometer or in the apparatus used to measure ionization energies (see **section 8.3.4**), or to bombard a target as in an X-ray tube (see box *Atomic number* in this chapter).

An anode and cathode are sealed into an apparatus from which the air has been removed. This is necessary as collisions with air molecules would stop the electron beam. The cathode is heated by a very low voltage heater. This gives the mobile electrons in the metal extra kinetic energy to enable them to escape from the metal surface. This is often referred to as 'boiling off' electrons. These electrons are attracted by the positive charge of the anode producing a high-speed beam.

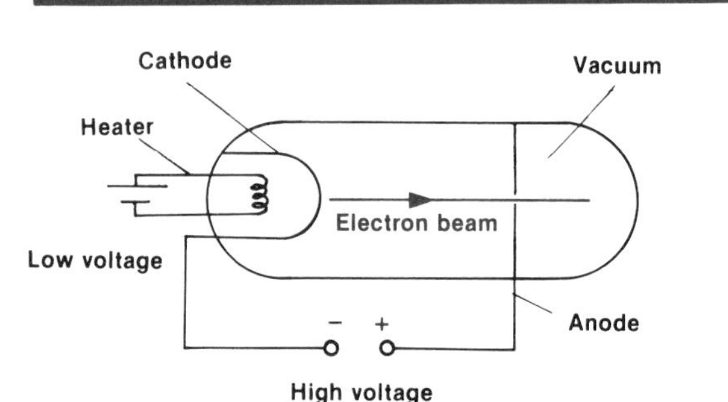

8.6 The ^{12}C scale of atomic masses

You are probably used to using relative atomic and molecular masses on the scale based on the mass of a hydrogen atom as 1. Since 1961 relative atomic and therefore molecular masses have been based on a scale such that the mass of one atom of ^{12}C is *exactly* 12.

^{12}C is the only atom with a relative atomic mass which is *exactly* a whole number. For carbon itself $A_r = 12.0111$, as this is an average of several isotopes of different relative atomic masses and abundance.

On the ^{12}C scale neither the relative mass of the proton nor that of the neutron is exactly 1, see **Table 8.4.**

Table 8.4 Relative masses of subatomic particles

Particle	A_r on ^{12}C scale
Proton	1.0072
Neutron	1.0086
Electron	0.000 548

Thus the relative masses of other atoms are not exact whole numbers. However the difference between the H = 1 and the $^{12}C = 12$ scales is small and for many purposes relative atomic masses rounded to the nearest whole number will suffice for calculations.

8.7 Summary

- Atoms consist of a dense nucleus containing protons and neutrons (nucleons), of approximate diameter 10^{-14} m compared with the diameter of the atom of 10^{-10} m.
- Any species of nucleus with a given number of protons and neutrons is called a nuclide and is written:

$$^{\text{mass number}}_{\text{atomic number}}\text{chemical symbol} \qquad \text{e.g. } ^{23}_{11}\text{Na}$$

- The mass number is the total number of nucleons.
- The atomic number is the total number of protons.
- The atomic number governs the chemical identity of the atom.
- The nucleus is surrounded by light electrons which may be thought of as clouds of negative charge.
- Electrons exist in orbitals which define the volume of space in which the electron is likely to be found.
- Orbitals are described by a number and a letter. The number 1, 2, 3, etc. gives the distance from the nucleus and hence the energy. The letter s, p, d, f defines the shape of the orbital. The possibilities are:

1	s only
2	s and p
3	s, p and d
4	s, p, d and f

- s orbitals come singly, p orbitals in threes, d orbitals in fives and f orbitals in groups of seven. Each orbital can hold up to two electrons, one with spin 'up' and one with spin 'down'.
- An electron can jump from one orbital to another and in doing so emits or absorbs a single quantum of electromagnetic radiation, equal in energy to the gap between the levels. This gives rise to line spectra. The frequency of the lines is given by the equation:

$$\Delta E = h\nu$$

- From the **convergence limit** of a line spectrum the ionization energy of the atom may be found (the energy required to completely remove an electron from the atom).
- Successive ionization energies are the energy changes (ΔH) for the processes:

$$\mathrm{E(g)} \longrightarrow \mathrm{E^+(g)} + \mathrm{e^-} \quad \text{first IE}$$
$$\mathrm{E^+(g)} \longrightarrow \mathrm{E^{2+}(g)} + \mathrm{e^-} \quad \text{second IE,} \quad \text{etc.}$$

where the symbol E represents any element. They are all endothermic.
- Line spectra tell us about the existence of energy levels; successive ionization energies tell us how many electrons are in each energy level.
- Nuclei of atoms can decay in ways which include:
  α decay—loss of $^4_2$He from the nucleus
  β decay—loss of $^{\ 0}_{-1}$e from the nucleus
  fission—the nucleus splitting into fragments.
  These processes may be accompanied by emission of γ-radiation.
- If mass is destroyed during these processes, a large amount of energy is produced:

$$E = mc^2$$

where c is the velocity of light.
- Nuclear processes can be represented by nuclear equations in which both total mass number and total atomic number are conserved and new elements are formed, for example:

$$\underset{\text{lead}}{^{211}_{82}\mathrm{Pb}} \longrightarrow \underset{\text{bismuth}}{^{211}_{83}\mathrm{Bi}} + {^{\ 0}_{-1}\mathrm{e}}$$

- Radioactive decay occurs in such a way that the time taken for half the source to decay (the half-life, $t_{\frac{1}{2}}$) is constant.
- The modern instrument for measuring relative atomic masses, A_r, and relative molecular masses, M_r, is the mass spectrometer. This can measure the masses of individual isotopes.
- Relative atomic and molecular masses are now measured on the ^{12}C scale on which the relative mass of an atom of the ^{12}C isotope is defined as exactly 12.

8.8 Questions

Use the following data as necessary:

$$h = 6.6 \times 10^{-34}\,\text{J s}$$
$$c = 3 \times 10^{8}\,\text{m s}^{-1}$$
$$\text{Avogadro constant} = 6 \times 10^{23}\,\text{mol}^{-1}.$$

1 The electron arrangement of fluorine may be written $1s^2 2s^2 2p^5$, or more simply 2, 7. Using each of these shorthands write the arrangements of the elements B, C, O and K.

2 $^{40}_{20}V$, $^{13}_{6}W$, $^{42}_{20}X$, $^{13}_{7}Y$, $^{19}_{9}Z$

A) How many protons, neutrons and electrons do each of the nuclides above have? (V, W, X, Y, Z are not their real symbols.)

B) Which two are a pair of isotopes?

3 The mass spectrum of copper is given. Calculate the average relative atomic mass, A_r, of copper.

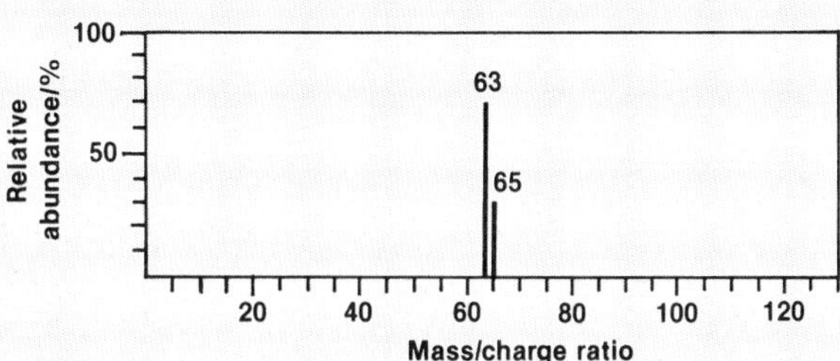

4 A) Calculate the energy of a quantum of the following types of electromagnetic radiation.

visible light $\lambda = 10^{-6}\,\text{m}$
ultraviolet $\nu = 10^{16}\,\text{s}^{-1}$
infra-red $\nu = 10^{14}\,\text{s}^{-1}$
γ-radiation $\lambda = 10^{-14}\,\text{m}$

B) Calculate the energy of a mole of each type of photon.

5 The table gives the frequencies for lines in the Balmer series in the spectrum of hydrogen.

Line number	Frequency $\nu/10^{14}\,\text{s}^{-1}$
1	4.568
2	6.167
3	6.907
4	7.309
5	7.551
6	7.709
7	7.817
8	7.894

$\Delta\nu$ for lines 1 and 2 is $1.599 \times 10^{14}\,\text{s}^{-1}$.

Calculate $\Delta\nu$ for all the other pairs of lines, i.e. 2 and 3, 3 and 4, etc.

Plot a graph of $\Delta\nu$ against the smaller frequency of the pair of lines.

Use the graph to find the frequency of the convergence limit for this series of lines (i.e. the frequency when $\Delta\nu = 0$). Using the equation $E = h\nu$, what ionization energy does this correspond to:

A) in joules per atom
B) in kJ mol^{-1}?

In the Balmer series, electrons fall back to level 2. The ionization energy from level 1 is 1312 kJ mol^{-1}. What is the energy difference between levels 1 and 2 in the hydrogen atom?

6 Complete the following nuclear equations.

A) $^{235}_{92}U \longrightarrow ? + ^{4}_{2}He$
B) $^{227}_{89}Ac \longrightarrow ? + ^{0}_{-1}e$
C) $^{215}_{84}Po \longrightarrow ^{211}_{82}Pb + ?$
D) $^{1}_{0}n + ^{10}_{5}B \longrightarrow ? + ^{4}_{2}He$
E) $^{4}_{2}He + ^{9}_{4}Be \longrightarrow ^{1}_{0}n + ?$
F) $^{235}_{92}U + ^{1}_{0}n \longrightarrow ^{90}_{38}Sr + ^{144}_{54}Xe + ?$

Could **F)** support a chain reaction? Explain your answer.

7 In the nuclear reaction:

$$^{235}_{92}U + ^{1}_{0}n \longrightarrow ^{144}_{56}Ba + ^{90}_{36}Kr + 2^{1}_{0}n$$

the total mass of the products is approximately 0.2 atomic mass units less than that of the products. One atomic mass unit is approximately 1.6×10^{-27} kg. Using the relationship $E = mc^2$, calculate the amount of energy produced. This is the amount of energy produced from the fission of one atom of uranium. How much energy is produced when 1 mole (235 g) of uranium fissions?

One tonne of coal produces approximately 30 MJ on burning. How much uranium must be fissioned to produce the same amount of energy?

8 The successive ionization energies (IEs) of sodium, Na, and magnesium, Mg, are given in kJ mol^{-1}

Na

496, 4563, 6913, 9544, 13 352, 16 611, 20 115, 25 491, 28 934, 141 367, 159 079

Mg

738, 1451, 7733, 10 541, 13 629, 17 995, 21 704, 25 657, 31 644, 35 463, 169 996, 189 371

A) What is the total energy required to form a Mg^{3+} ion from Mg?

B) Use the figures to explain why Mg usually forms Mg^{2+} and Na usually forms Na^{+}.

C) Why is each successive IE greater than the previous one?

D) Why is the last IE of Mg higher than the last IE of Na?

9 The first five IEs of an element are given in $kJ\,mol^{-1}$:

578, 1817, 2745, 11 578, 14 831

In which group of the Periodic Table would you expect it to be?

Explain your answer.

10 Look at the box about radon gas on page 59. Radon decays by emitting an α-particle to produce a 'daughter' nuclide, which is itself an α emitter. Identify the 'daughter' and 'granddaughter' nuclides.

11 Use the equation $\nu = cR\,(1/n_1^2 - 1/n_2^2)$ where $R = 1.09 \times 10^7\,m^{-1}$ to calculate the frequency of the spectral line produced when an electron drops from level 4 to level 3 in the hydrogen atom. In which of the named series of lines does this appear?

12 Use the following data to plot a graph and calculate the half-life of ^{198}Au.

Time/days	**Count rate**/min^{-1} (corrected for background)
Start	43 200
1	35 400
2	27 800
3	21 600
4	17 700
5	13 900
6	10 800
7	8900

How would the count rate have been 'corrected for background'? If the background count was $40\,min^{-1}$, what would have been the actual reading taken at day 7?

13 The radius of a ^{238}U nucleus is $74 \times 10^{-15}\,m$. What is the volume of this nucleus? Remembering that it contains 238 nucleons (each of mass approximately $1.6 \times 10^{-27}\,kg$), what is the density of the uranium nucleus in $kg\,m^{-3}$? Compare this with the density of uranium metal ($1.905 \times 10^4\,kg\,m^{-3}$).

14 A gamma source of half-life 1 hour is manufactured at Harwell for use in a hospital that is a 3-hour road journey away. When manufactured its count rate is $16\,000\,min^{-1}$.

A) What will be its count rate on arrival at the hospital?

B) The source can then only be used for treatment while its count rate is greater than $500\,min^{-1}$. For how long can it be used to treat patients?

C) A hospital administrator suggested that the lifetime of the source could be extended by cooling it, as he recalled that chemical reactions could be slowed down by cooling. Comment on this idea.

15 Chlorine has two isotopes, ^{35}Cl and ^{37}Cl.

A) What three values of M_r are possible for the molecule Cl_2?

B) Bearing in mind that the molecule might fragment, give the mass numbers of five peaks that might be expected in the mass spectrum of chlorine.

16 A quantum of electromagnetic radiation has an energy of $10^{-20}\,J$. Calculate

A) its frequency

B) its wavelength.

Use **Figure 8.2** on page 50 to decide what type of electromagnetic radiation it represents.

17 The radioactive count rate in living matter due to the decay of ^{14}C is approximately $16\,min^{-1}$ per gram of carbon. A wooden carving was dug up and found to have a count rate of approximately $4\,min^{-1}$ per gram of carbon. How old is the wood? Take the half-life of ^{14}C as 5600 years.

18 Research/debate. What methods are presently being proposed for the storage of nuclear waste? What are the problems and advantages of storing in steel drums, underground storage, and dumping at sea? Prepare a short talk either to reassure a community near a proposed underground storage site, or to stir them into action against the proposal.

19 Research/debate. Discuss the issues involved in deciding whether food irradiation should be allowed.

20 **A)** Explain what is meant by each of the following terms:

i electron
ii proton
iii neutron
iv isotope.

B) The atomic number provides three pieces of information about an element. What are they?

C) The radioactive atom $^{224}_{88}Ra$ decays by α emission with a half-life of 3.64 days.

i What is meant by a half-life of 3.64 days?

ii Referring to the product of the decay, what will be its mass number and its atomic number?

iii Radium is in Group 2 of the Periodic Table. In what group will the decay product be?

D) Explain briefly the principles underlying

i the use of radioactive isotopes as tracers

ii The dating of dead organic matter using radiocarbon ^{14}C. (London 1980)

9 Energetics

9.1 Introduction

When a chemical reaction takes place, chemical bonds break and new ones are formed. It takes energy to break bonds and energy is released when bonds are formed, so most chemical reactions involve some sort of energy change. The energy may be given out from the system to the surroundings, or it may need to be put in to the system in order for the reaction to take place.

The energy involved may be in different forms—light, electrical or most usually heat.

Studying heat changes is called *thermochemistry* and is important for both practical and theoretical reasons. Practical applications include measuring the energy values of fuels (both those now in use and proposed new ones), and determining energy requirements of industrial processes, etc. Theoretical considerations include the apportioning of amounts of energy to particular bonds, calculating energy changes for hypothetical reactions and attempting to predict the feasibility of reactions.

9.2 What is energy?

Energy is defined as 'the ability to do work', work being the moving of a force through a distance, for example lifting a load. This is the basis of the unit of energy (and work), the joule.

One joule is the energy required to move a force of one newton through one metre.

To get a feel for the size of this unit, consider that it is about the energy required to lift a medium-sized apple one metre. To boil the water for a cup of tea requires about 80 000 joules, or 80 kJ (kilojoules) in the units more usually used by chemists. Anything which can directly or indirectly move a force is a form of energy. Heat, for example, could be used to produce steam which will drive a turbine and lift a load, electricity could turn a motor to lift a load, and so on.

9.2.1 Enthalpy

The type of energy of most importance in chemistry is heat. The precise amount of heat given out or taken in by a reaction varies with the conditions—temperature, pressure, concentration of solutions, etc. Therefore if different experimental results are to be compared, there must be agreement on the conditions under which measurements are made. For example, chemists normally measure heat changes at constant pressure. This makes practical sense as it means that flasks open to the atmosphere are used, which is normal practice. To illustrate how pressure affects heat change, think of a reaction in which a gas is given off, such as:

$CaCO_3(s)$	+	$2HCl(aq)$	$\longrightarrow$	$CaCl_2(aq)$	+	$H_2O(l)$	+	$CO_2(g)$
calcium carbonate		hydrochloric acid		calcium chloride		water		carbon dioxide

A

THE CALORIE

Until fairly recently, chemists measured energy in calories rather than joules.

One calorie is the amount of heat energy needed to raise the temperature of 1 g of water by 1 °C (1 K). So 1 calorie = 4.2 J.

Nutritionists still measure the energy content of food in calories, or kilocalories for which they use the abbreviation Cal, with a capital C.

$$1\ \text{Cal} = 1\ \text{kcal} = 10^3\ \text{cal}$$

Energy values of foods are measured by burning them in a flame calorimeter (see section 9.3.2) although the actual energy available to the body is slightly less than this because the food is not completely absorbed or completely oxidized in the body.

The energy values of the foods we eat must equal the energy expenditure of our daily activity. If we eat more than this the body accumulates fat as a reserve of available energy.

Typical energy expenditures are given below.

Activity	Energy expenditure per hour /Cal	/kJ
Sleeping	70	293
School work	120	503
Housework	210	880
Cycling	360	1909
Coal mining	480	2012

The energy values of some foods are also given.

Food	Energy value per 100 g /Cal	/kJ
Potatoes, boiled	76	342
Potato crisps	559	2348
White bread	251	1068
Beef	226	940
Lettuce	8	36

A calorie-controlled diet involves eating food whose total energy value is less than your energy expenditure, thus forcing the body to use up some of its stored fat to make up the difference.

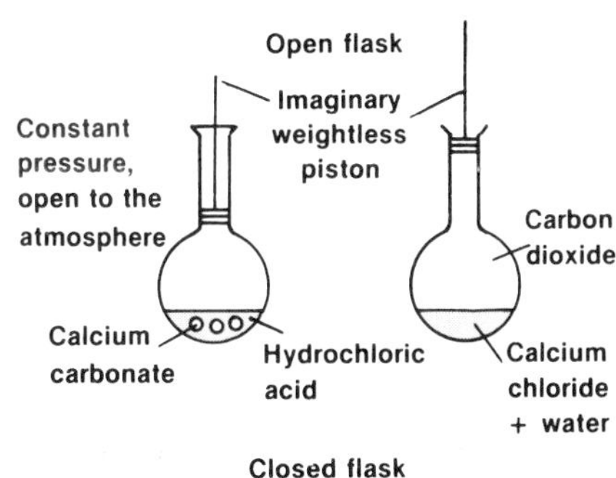

Figure 9.1 Comparison between a reaction taking place at constant pressure and at constant volume

If the reaction is carried out in an open flask, the gas which is evolved has to push back the air of the atmosphere, thus doing work and using energy. This becomes clear if we imagine a weightless piston placed across the neck of the flask as shown in **Figure 9.1.** In the closed flask, the pressure inside the vessel increases but there is no work done to push back the atmosphere. The total heat output, called the internal energy change, will therefore be greater in this case, since no energy is lost in pushing back the atmosphere. However, even when a whole mole of gas is evolved, the energy difference is only 2.4 kJ, which is very small compared with the total energy evolved by the reaction at constant pressure in the open flask (166 kJ).

A heat change measured at constant pressure is called an enthalpy change. Enthalpy changes are given the symbol ΔH. The Greek letter Δ (delta) is used to indicate a change in any quantity.

The standard conditions for measuring enthalpies are:

pressure of 10^2 kPa (approximately normal atmospheric pressure)
temperature of 298 K (around normal room temperature, 25 °C)
(0 °C = 273 K)

The symbol $\Delta H^{\ominus}_m$ (298 K) is used to represent an enthalpy change per mole measured under these conditions. Substances which appear in the equations will be in the states which they adopt at standard temperature (298 K) and pressure (10^2 kPa). For example, in the equation:

$$2H_2(g) + O_2(g) \longrightarrow 2H_2O(l)$$
hydrogen oxygen water

hydrogen and oxygen are gases and water is liquid. Occasionally allotropes (different forms of the same element, like graphite and diamond—allotropes of carbon) will be encountered. The most stable of these is the standard state. The subscript 'm' meaning molar (i.e. per mole) is often dropped, as the units of ΔH are generally stated as kJ mol^{-1}. The 298 K in brackets is also frequently omitted. The symbol $\ominus$ (called 'standard') is taken to imply a pressure of 10^2 kPa *and* a temperature of 298 K, unless otherwise stated. It is important when stating enthalpy changes that the equation to which the measurement refers is given. For example, the statement 'the enthalpy change of atomization of hydrogen is 218 kJ mol^{-1}' is ambiguous as it could refer to:

Figure 9.2 A reaction giving out heat at 298 K

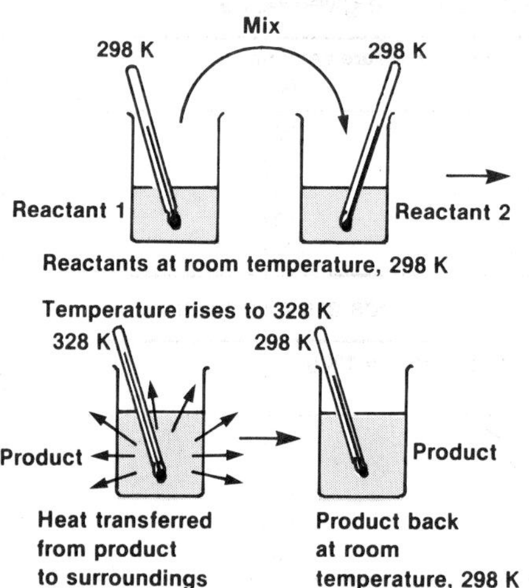

Figure 9.3 Enthalpy diagrams

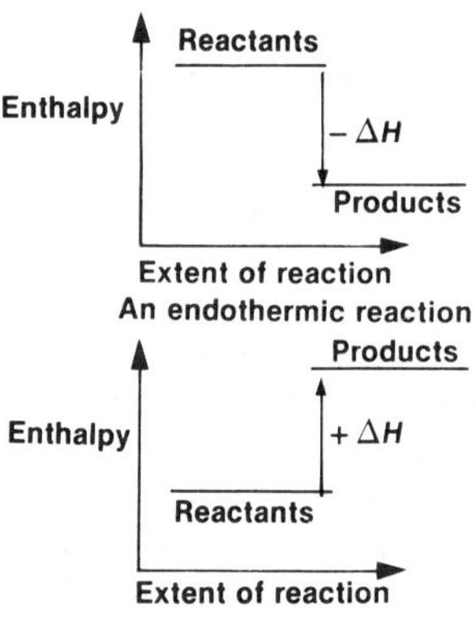

$$H_2(g) \longrightarrow 2H(g)$$

or

$$\tfrac{1}{2}H_2(g) \longrightarrow H(g)$$

The figure given actually represents the latter. The former would be 436 kJ mol^{-1}. So the equation needs to be given.

It may seem odd to refer to heat changes taking place at a constant temperature, when a heat change inevitably leads to a temperature change. The best way to regard this is to imagine the reactants starting at 298 K, see **Figure 9.2.** In this rection, heat is given out.

The reaction is not considered to be complete *until the products have cooled back to 298 K*. The heat transferred to the surroundings while cooling is the enthalpy change for the reaction. This sort of reaction is called **exothermic**, and the convention is that **ΔH is given a negative sign.**

If energy needs to be absorbed in order for a reaction to take place, the reaction is called **endothermic**, and **ΔH is given a positive sign.**

Endothermic reactions which take place in aqueous solution can absorb heat from the water and cool it down quite dramatically. As in the exothermic example, the reaction is not considered to be complete until the products have returned to the temperature at which they started, but in this case the solution has to absorb heat from the surroundings to do this. Unless this is remembered, it can seem strange that a reaction that is absorbing heat can get cold.

exothermic reaction: heat given out: ΔH negative
endothermic reaction: heat taken in: ΔH positive

9.2.2 Enthalpy diagrams (energy level diagrams)

Enthalpy changes are often presented on enthalpy diagrams, sometimes more loosely called energy level diagrams. The vertical axis represents enthalpy while the horizontal axis represents the extent of reaction—the percentage change from reactants to products. Often we are interested only in the extremes—100% reactants and 100% products (0% reactants) so the axis is left unlabelled.

A

PHOTOSYNTHESIS

This process is the origin of the energy used by living systems.

In green plants the following reaction takes place:

$$6CO_2(g) + 6H_2O(l) \longrightarrow \underset{\text{glucose}}{C_6H_{12}O_6(s)} + 6O_2(g)$$

$$\Delta_R H^\ominus = +2820\ \text{kJ mol}^{-1}$$

This is an endothermic reaction and energy is absorbed. The energy for this endothermic reaction comes not from heat but from light. The green pigment in plants, chlorophyll, absorbs the light which provides the energy. Glucose acts as an energy store, later releasing energy by the reverse of the reaction shown (left). Some of the released energy appears in a molecule called adenosine triphosphate (ATP) which fuels many reactions in living cells including protein synthesis and muscular movement.

Photosynthesis is the original source of the energy in fossil fuels.

9.3 Measurement of enthalpy changes—calorimetry

Before tackling this, we must be clear about the difference between heat and temperature. These words are often used interchangeably in daily conversation but as scientific terms they are quite distinct.

Essentially, temperature represents the average kinetic energy of the molecules in the system—it is independent of the quantity of the substance.

R

DEFINITION OF TERMS

$\Delta_R H^\ominus$, the enthalpy change of reaction, is the general name for the enthalpy change for any reaction. Some commonly used enthalpy changes are given names, though it is always better to give an equation for the reaction to avoid ambiguity. Some enthalpy changes that have particular definitions follow. Note that the word 'molar' now means 'per mole', rather than 'moles per dm^3'. All values are corrected to standard conditions (298 K, 10^2 kPa).

1. Standard molar enthalpy change of formation

$$\Delta_f H^\ominus$$

The enthalpy change for the formation of one mole of a compound from its elements, for example:

$$H_2(g) + \tfrac{1}{2}O_2(g) \longrightarrow H_2O(l) \quad \Delta_f H^\ominus = -286\ \text{kJ mol}^{-1}$$

Note that the elements and compounds are all in the state (gas, liquid, solid) that they are in under standard conditions (298 K, 10^2 kPa) even though the reaction may not actually take place under these conditions.

2. Standard molar enthalpy change of combustion

$$\Delta_c H^\ominus$$

The enthalpy change when one mole of substance is completely burned in oxygen, for example:

$$CH_4(g) + 2O_2(g) \longrightarrow CO_2(g) + 2H_2O(l)$$
$$\Delta_c H^\ominus = -890\ \text{kJ mol}^{-1}$$

3. Standard molar enthalpy change of atomization, often called atomization energy

$$\Delta_{at} H^\ominus$$

The enthalpy change when one mole of a substance decomposes completely into gaseous atoms, for example:

$$\tfrac{1}{2}Cl_2(g) \longrightarrow Cl(g) \quad \Delta_{at} H^\ominus = +121\ \text{kJ mol}^{-1}$$

Note that this is given per mole of chlorine atoms and not per mole of chlorine molecules. It is particularly important to have an equation to refer to for this type of enthalpy change.

4. Standard molar enthalpy change of lattice formation, often called lattice energy or LE

$$\Delta_L H^\ominus$$

The enthalpy change when one mole of ionic compound is formed from its gaseous ions, for example

$$Na^+(g) + Cl^-(g) \longrightarrow NaCl(s) \quad \Delta_L H^\ominus = -780\ \text{kJ mol}^{-1}$$

Some books refer to the process of lattice breaking, the reverse process, for which $\Delta_L H^\ominus = +780\ \text{kJ mol}^{-1}$ in this example.

5. Standard molar enthalpy change of first ionization (first ionization energy, first IE).

The enthalpy change when one mole of gaseous atoms is converted into singly positively charged ions, for example:

$$Na(g) \longrightarrow Na^+(g) + e^- \quad \Delta H^\ominus = +496\ \text{kJ mol}^{-1}$$

Note: The molar enthalpy change of second ionization (second ionization energy, second IE) refers to the loss of a mole of electrons from a mole of singly positively charged ions, for example:

$$Na^+(g) \longrightarrow Na^{2+}(g) + e^- \quad \Delta H^\ominus = +4563\ \text{kJ mol}^{-1}$$

6. Standard molar enthalpy change of electron gain (electron affinity or first electron affinity).

The enthalpy change when a mole of gaseous atoms is converted to a mole of singly negatively charged ions, for example:

$$O(g) + e^- \longrightarrow O^-(g) \quad \Delta H^\ominus = -141.1\ \text{kJ mol}^{-1}$$

Note that the second electron affinity refers to the addition of a mole of electrons to a mole of singly negatively charged ions, for example:

$$O^-(g) + e^- \longrightarrow O^{2-}(g) \quad \Delta H^\ominus = +798\ \text{kJ mol}^{-1}$$

7. Standard molar enthalpy change of bond dissociation (bond dissociation energy) refers to a specific bond in a specific compound, in molar amounts as specified by an equation, for example:

$$CH_4(g) \longrightarrow CH_3(g) + H(g) \quad \Delta H^\ominus = +423\ \text{kJ mol}^{-1}$$

Mean standard molar enthalpy change of bond dissociation (mean molar bond dissociation energy) refers to the average molar enthalpy change of bond dissociation of all the same bonds in the same molecule, for example:

$$\tfrac{1}{4}(CH_4(g) \longrightarrow C(g) + 4H(g)) \quad \Delta H^\ominus = +416\ \text{kJ mol}^{-1}$$

Average bond enthalpies in tables refer to an average of the mean molar bond dissociation enthalpy of several compounds for the bond under consideration. These are referred to as 'tabulated bond enthalpies' or 'tabulated bond energies'.

Heat is a measure of the total energy in a given amount of substance. It does depend on how much of the substance is present.

For example, two buckets of hot water at 50 °C have twice as much *heat* as one bucketful although both have the *same temperature*. Temperature is measured with a thermometer which takes no account of the amount of substance present. It would give the same reading in a thimbleful or a bucketful of the same water.

Heat always flows from high to low temperature, so heat will flow from a red-hot nail into a bucketful of lukewarm water, even though the total heat content of the water is greater than that of the nail.

To measure heat we must take into account the temperature, the mass

Figure 9.4 The water in the bucket has the *same temperature* as that in the thimble but the water in the bucket has *more heat*

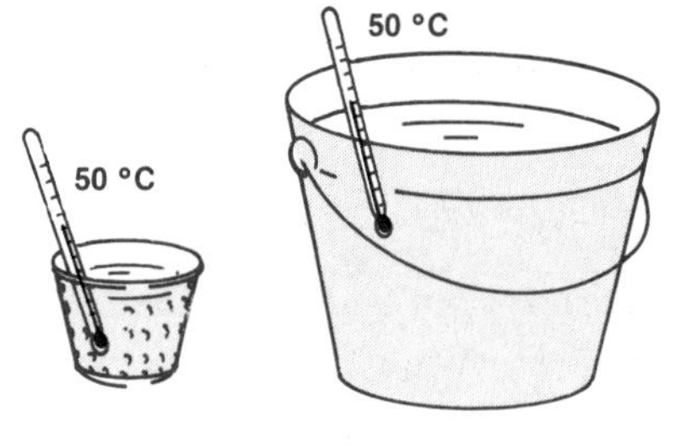

of the substance and the type of substance itself, as some substances take more heat to raise their temperatures than others.

The amount of heat needed to raise the temperature of 1 g of substance by 1 K is called the **specific heat capacity**, *c*.

For water the specific heat capacity is $4.2\,J\,g^{-1}\,K^{-1}$, an unusually high figure. Incidentally this makes water an excellent liquid for use in hot-water bottles. For a given temperature rise above room temperature, it stores almost twice as much heat as, for example, ethanol whose specific heat capacity is only $2.4\,J\,g^{-1}\,K^{-1}$.

There is no instrument which measures heat directly. We have to arrange for the heat we wish to measure to be transferred into a known mass of a substance (usually water) and then we must measure the temperature rise. The heat is then given by the expression:

$$\text{heat} = \text{mass of water} \times \text{specific heat capacity} \times \text{temperature rise}$$
$$\text{heat} = m \times c \times \Delta T$$

The apparatus used is called a **calorimeter** (from the Latin 'calor' meaning heat).

9.3.1 The simple calorimeter

Figure 9.5 A simple calorimeter

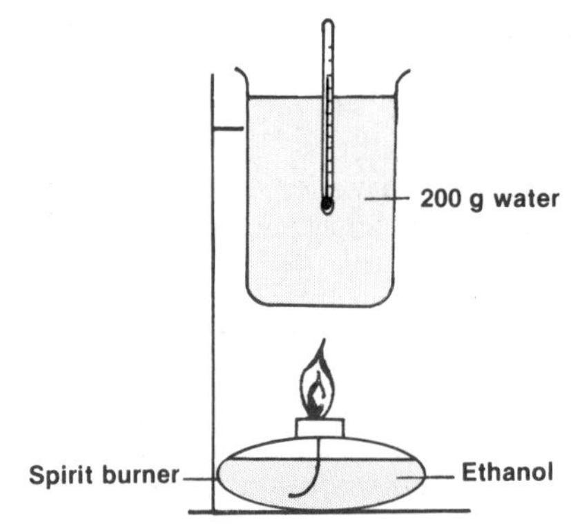

Figure 9.5 shows the simplest form of calorimeter being used to measure the enthalpy change of combustion of ethanol:

$$C_2H_5OH(l) + 3O_2(g) \longrightarrow 2CO_2(g) + 3H_2O(l)$$

If 0.46 g (0.01 mol) of ethanol is burned, a temperature rise of the water of 8 K is found.

$$\text{heat} = m \times c \times \Delta T$$
$$= 200 \times 4.2 \times 8 = 6720\,J$$
$$0.01\text{ mol gives } 6720\,J$$
$$1\text{ mol would give } 672\,000\,J \text{ or } 672\,kJ$$
$$\Delta_c H = 672\,kJ\,mol^{-1}$$

As the experiment was not necessarily done under standard conditions, the symbol $\ominus$ should not strictly speaking be used, but the difference in values will be small.

This value is about half the 'accepted' value for $\Delta_c H$. It is such a poor approximation because of:

1. heat which never enters the calorimeter, because of draughts, for example
2. heat lost from the top and sides of the calorimeter
3. heat which goes into the material of the beaker rather than the water
4. incomplete combustion—in a restricted supply of oxygen, some of the ethanol will not burn completely, giving carbon and carbon monoxide rather than carbon dioxide. Soot (carbon) on the bottom of the beaker may indicate this.

Despite these disadvantages, the simple calorimeter may be satisfactory for comparing the $\Delta_c H$ values of a series of similar compounds, as the errors will be comparable.

Improvements may be made to this simple calorimeter by insulating the sides of the beaker, using a lid and providing a draught screen as shown in **Figure 9.6.**

Figure 9.6 An improved calorimeter

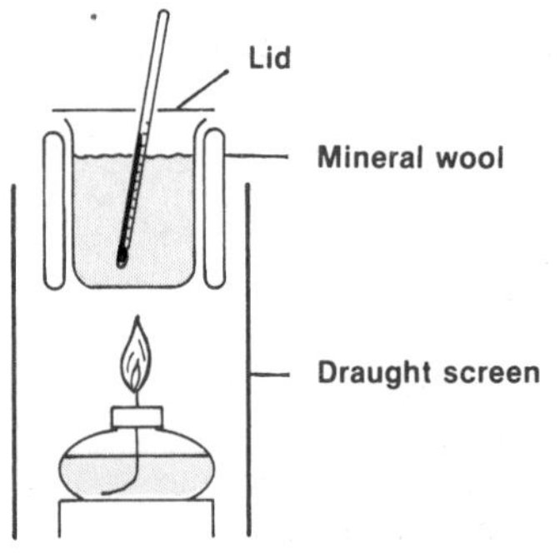

9.3.2 The flame calorimeter

This is a more sophisticated version of the simple calorimeter. It is shown in **Figure 9.7** (overleaf). The chimney is made of copper, to conduct heat from the combustion products to the water. The flame is enclosed, to reduce heat loss. Oxygen is supplied to ensure complete combustion.

Figure 9.7 A flame calorimeter

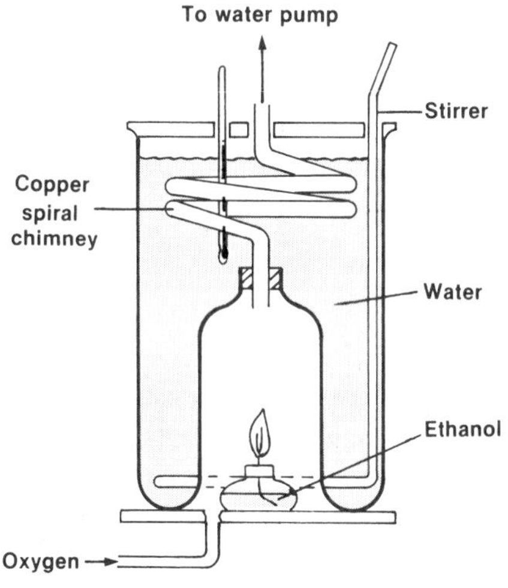

Figure 9.8 Electrical methods for determining enthalpy changes

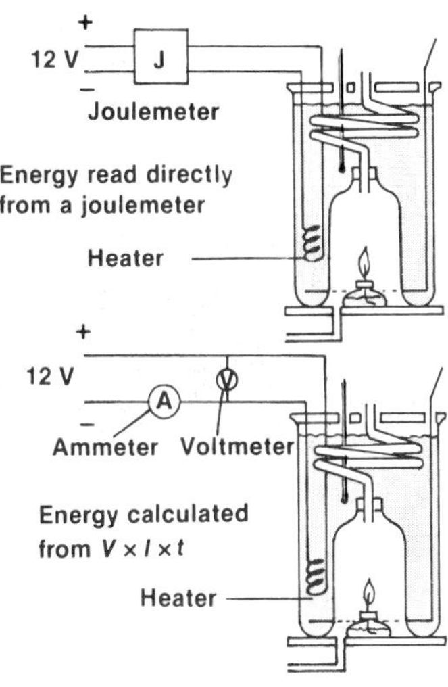

The heat output may be calculated in one of three ways:

1. As before (heat $= m \times c \times \Delta T$), allowing for the heat capacity of the materials (e.g. copper, glass, rubber) of the calorimeter.
2. The **heat capacity** of the whole apparatus may be measured by burning a substance of known molar enthalpy change of combustion $\Delta_c H^\ominus$ in the apparatus. The heat capacity of the apparatus is the heat needed to raise the temperature of the whole apparatus by 1 K, so heat capacity = heat change/temperature change. Since heat losses will be similar for both experiments these will tend to cancel out and be eliminated.

   Benzenecarboxylic acid ($M_r = 122$) is often used to calibrate calorimeters as its molar enthalpy change of combustion $\Delta_c H^\ominus$ is accurately known and the compound can be obtained in a very pure state.

   The following example shows how to determine the heat capacity of a calorimeter, and then use it for further enthalpy change determinations.

   1.22 g (0.01 mol) of benzenecarboxylic acid was burned in a flame calorimeter and gave a temperature rise of 10 K. $\Delta_c H^\ominus$ for benzenecarboxylic acid is $-3230\,\text{kJ mol}^{-1}$. So burning 0.01 mol benzenecarboxylic acid would produce 32.3 kJ of heat. This raises the temperature of the calorimeter by 10 K.

   But as we have seen above, the heat capacity of the calorimeter = heat change/temperature change, so the heat capacity is $32.3/10.0\,\text{kJ K}^{-1} = 3.23\,\text{kJ K}^{-1}$.

   0.46 g (0.01 mol) of ethanol was burned in the same calorimeter producing a temperature rise of 4 K.

$$\begin{aligned}\text{heat produced by the ethanol} &= \text{heat capacity} \times \text{temperature change}\\ &= 4\,\text{K} \times 3.23\,\text{kJ K}^{-1}\\ &= 12.9\,\text{kJ}\\ \text{0.01 mol ethanol produces } &12.9\,\text{kJ}\\ \text{1 mol ethanol produces } &1290\,\text{kJ}\\ \Delta_c H^\ominus \text{ of ethanol} &= -1290\,\text{kJ mol}^{-1}\end{aligned}$$

3. An electrical method may be used. A small electric immersion heater is placed in the water and run until the same temperature rise is produced as was produced by burning the ethanol. Again the heat losses in both experiments will cancel out. The heat produced in the heater is either measured by an electrical joulemeter, which records the energy directly, or determined by measuring the voltage, V, and current, I, of the heater and the time in seconds, t, for which it ran and using the expression:

$$\text{energy} = V \times I \times t$$

Both setups are shown in **Figure 9.8.**

9.3.3 Measuring enthalpy changes of reactions in solution

The methods already described have been concerned with combustion reactions. The situation is simpler with reactions which take place in solution. The heat is generated in the solutions themselves and merely has to be kept in. Experiments are often carried out in expanded polystyrene beakers which are good insulators and have a low heat capacity, and so absorb little heat themselves. For simple work the specific heat capacity of dilute solutions is usually taken to be the same as that of water, $4.2\,\text{J g}^{-1}\,\text{K}^{-1}$.

For example, to find the standard molar enthalpy change of reaction $\Delta_R H^\ominus$ for

$$HCl(aq) + NaOH(aq) \longrightarrow NaCl(aq) + H_2O(l)$$

$50\,\text{cm}^3$ of $1\,\text{mol dm}^{-3}$ hydrochloric acid and $50\,\text{cm}^3$ of $1\,\text{mol dm}^{-3}$ sodium hydroxide solution were mixed in an expanded polystyrene beaker. The temperature rose by 6.6 K.

A

ALTERNATIVE FUELS

Fuels are vital for our present-day lifestyle, for transport, heating, cooking and generation of electrical energy. Fuels are really a means of storing energy which can be released when required. For many purposes the **energy density** of the fuel is important, that is, the amount of energy which can be stored in a given mass or volume of fuel. This is obvious in the case of motor vehicles and aircraft. The reason that chemical fuels are used is because of their high energy density. For example, petrol stores approximately 42 MJ kg^{-1}, while a lead–acid car battery such as those used on milk floats stores only 0.13 MJ kg^{-1}, which is why petrol and similar hydrocarbon fuels like kerosene, aviation spirit and diesel are overwhelmingly used for transport purposes.

These fuels are derived mainly from crude oil, although some countries with no access to crude oil produce synthetic petrol. Crude oil is a fossil fuel derived from the remains of long-dead sea creatures—essentially it stores the energy of the sun in chemical bonds first built by photosynthesis.

We know that, at the present rate of use, known reserves of oil will run out early in the next century. New reserves may be found, new technology and the increasing price of oil may make some presently uneconomic sources, such as oil bearing shales, worth developing. However, the resources are finite. The search is therefore on for alternative fuels to those based on petroleum.

Two candidates which have been proposed are ethanol and hydrogen.

Ethanol

This has a slightly poorer energy density than petrol, around 30 MJ kg^{-1}. It has been used as a motor fuel since as long ago as the 1930s. Presently most interest is being shown in Brazil where sugar cane is fermented to produce the ethanol.

This makes ethanol a renewable fuel, unlike petrol whose resources are finite. Current petrol engines can easily be modified to run on ethanol and mixtures of ethanol and conventional petrol can be used. Ethanol is more expensive to produce than petrol—the Brazilians have no crude oil of their own and save on foreign exchange by their use of ethanol. But as the price of crude oil goes up and as improvements are made to the technology of ethanol production, so the difference in costs between them is likely to decrease.

Hydrogen

At first sight, this is an excellent fuel—its energy density is around 140 MJ kg^{-1} and it burns to form water as the only product, unlike carbon-containing fuels where the production of poisonous carbon monoxide is always a problem. However, as it is a gas at room temperature it is difficult to store, being prone to leakage. It is also extremely flammable so safe storage and use could be difficult. Rather than store the gas under pressure in cylinders, transport as metal hydrides has been suggested (see section 21.3.1). Either way, the storage system would be rather heavy. At present most hydrogen is made from hydrocarbons. This would defeat the object, but one possible alternative is the electrolysis of water. Currently research is being carried out into the use of sunlight to decompose water, either with catalysts or using living things such as algae. A hydrogen-fuelled transport aircraft is currently being test-flown in the USSR.

An ethanol-powered car

The total volume of the mixture is 100 cm^3 which has a mass of approximately 100 g.

$$\text{heat produced} = m \times c \times \Delta T$$
$$= 100 \times 4.2 \times 6.6$$
$$= 2772\ \text{J}$$
$$\text{this is produced by } \frac{50 \times 1}{1000} = 0.05\ \text{mol}$$
$$1\ \text{mol would give } \frac{2772}{0.05}\ \text{J mol}^{-1}$$
$$= 55\,440 = 55.4\ \text{kJ mol}^{-1}$$
$$\Delta_R H^{\ominus} = -55.4\ \text{kJ mol}^{-1}\ \text{(negative as heat is given out)}$$

Note that the volume on mixing 50 cm^3 and 50 cm^3 is 100 cm^3. A common mistake is to use 50 cm^3.

For more accurate work a vacuum flask can replace the polystyrene beaker and the electrical compensation method as described on page 72 can be used. This has the advantage that the specific heat capacity of the solution does not need to be known.

For example, to find the standard molar enthalpy change of:

$$\underset{\text{nitric acid}}{HNO_3(aq)} + \underset{\text{sodium hydroxide}}{NaOH(aq)} \longrightarrow \underset{\text{sodium nitrate}}{NaNO_3(aq)} + \underset{\text{water}}{H_2O(l)}$$

$100\,cm^3$ of $1\,mol\,dm^{-3}$ nitric acid solution and $100\,cm^3$ of $1\,mol\,dm^{-3}$ sodium hydroxide solution were mixed in a vacuum flask. A temperature rise of 6.8 K was noted. The solutions were allowed to cool back to room temperature and next heated with an electrical immersion heater connected to a joulemeter. 5600 J were required to produce the same temperature rise as before.

$$5600\,\text{J were produced by } \frac{100 \times 1}{1000}\,\text{mol} = \frac{1}{10}\,\text{mol}$$

$$\text{so 1 mol would produce } 56\,000\,\text{J}$$

$$= 56\,kJ\,mol^{-1}$$

$$\text{so } \Delta_R H^{\ominus} = -56\,kJ\,mol^{-1} \text{ (negative as heat is given out)}$$

The close agreement between the results of this experiment and the previous one is no coincidence. If we look at the reactions ionically, we see that the Cl^-, NO_3^- and Na^+ ions take no part in the reaction—they are **spectator ions**. Eliminating these from the reactions shows that both reactions are the same:

1. $H^+(aq) + \cancel{Cl^-}(aq) + \cancel{Na^+}(aq) + OH^-(aq)$
   $\longrightarrow H_2O(l) + \cancel{Cl^-}(aq) + \cancel{Na^+}(aq)$
   which leaves $H^+(aq) + OH^-(aq) \longrightarrow H_2O(l)$

2. $H^+(aq) + \cancel{NO_3}^-(aq) + \cancel{Na^+}(aq) + OH^-(aq)$
   $\longrightarrow H_2O(l) + \cancel{NO_3}^-(aq) + \cancel{Na^+}(aq)$
   which leaves $H^+(aq) + OH^-(aq) \longrightarrow H_2O(l)$ once more

In fact *all* neutralization reactions involving a strong acid and a strong base (see section 12.4.1) have the same standard molar enthalpy change.

9.4 Hess's law

Hess's law states that the total energy (or enthalpy) change for a chemical reaction is the same whatever route is taken. It is the outcome of a vast number of measurements of energy changes. It is also a consequence of a more general physical law—the law of conservation of energy, sometimes called the first law of thermodynamics—which states that energy can never be created or destroyed. To see exactly what Hess's law means, let us take a concrete example.

Ethyne reacts with two moles of hydrogen molecules to give ethane:

$$\underset{\text{ethyne}}{H{-}C\equiv C{-}H\ (g)} + 2H_2(g) \longrightarrow \underset{\text{ethane}}{\begin{array}{ccccccc} & & H & & H & & \\ & & | & & | & & \\ H & - & C & - & C & - & H \\ & & | & & | & & \\ & & H & & H & & \end{array}\ (g)}$$

This reaction could take place directly *or* ethyne could react first with one mole of hydrogen molecules to give ethene, which could then react with a second mole of hydrogen to give ethane.

$$H{-}C\equiv C{-}H\ (g) + H_2(g) \longrightarrow \underset{\text{ethene}}{\begin{array}{ccccc} H & & & & H \\ & \diagdown & & \diagup & \\ & & C{=}C & & \\ & \diagup & & \diagdown & \\ H & & & & H \end{array}}\ (g)$$

$$\begin{array}{ccccc} H & & & & H \\ & \diagdown & & \diagup & \\ & & C{=}C & & \\ & \diagup & & \diagdown & \\ H & & & & H \end{array}\ (g) + H_2(g) \longrightarrow \underset{\text{ethane}}{\begin{array}{ccccccc} & & H & & H & & \\ & & | & & | & & \\ H & - & C & - & C & - & H \\ & & | & & | & & \\ & & H & & H & & \end{array}\ (g)}$$

Hess's law states that the total energy change is the same whichever route is taken, direct or via ethene (or, in fact, any other route).

$$H-C\equiv C-H\,(g) + 2H_2(g) \xrightarrow[1.]{\Delta H_1(?)} H_3C-CH_3\,(g)$$

2. ΔH_2(–176 kJ mol^{-1}) → $H_2C=CH_2$ (g) + H_2(g) → 3. ΔH_3 (–137 kJ mol^{-1})

Hess's law tells us that:

$$\Delta H_1 = \Delta H_2 + \Delta H_3$$

The actual figures are:

$$\Delta H_2 = -176\,\text{kJ mol}^{-1}$$
$$\Delta H_3 = -137\,\text{kJ mol}^{-1}$$
$$\text{so } \Delta H_1 = -176 + -137 = -313\,\text{kJ mol}^{-1}$$

Another alternative route might be via the elements carbon and hydrogen, i.e. ethyne is converted to its elements carbon and hydrogen which then can react to form ethane.

$$H-C\equiv C-H\,(g) + 2H_2(g) \xrightarrow[1.]{\Delta H_1(?)} H_3C-CH_3\,(g)$$

4. ΔH_4 → $2C(gr) + 3H_2(g)$ → 5. ΔH_5

E

THE BOMB CALORIMETER

This is the most accurate method for measuring enthalpies of combustion and it is how accurate 'data book' values are determined.

The actual 'bomb' is a thick stainless steel pressure vessel into which the sample is placed in a crucible. A wick leads from the sample to an electrical ignition coil. The bomb is filled with oxygen at a pressure of several atmospheres, sealed and placed in a calorimeter full of water which itself sits in a water-filled tank. The sample is ignited electrically and as the heat passes into the calorimeter, a thermostat and heater ensure that the water in the outer tank is kept at the same temperature as the water in the calorimeter. This eliminates heat loss from the calorimeter as its surroundings are always at its own temperature. The heat capacity of the calorimeter is calibrated by the combustion of benzene-carboxylic acid, as in the flame calorimeter. The mass of the sample is determined to 0.0001 g and the temperature measured to three decimal places using a platinum resistance thermometer.

Since the reaction takes place in a sealed container, the energy change measured is not in fact the enthalpy change. However, there is a simple relationship between the measured energy change and the enthalpy change.

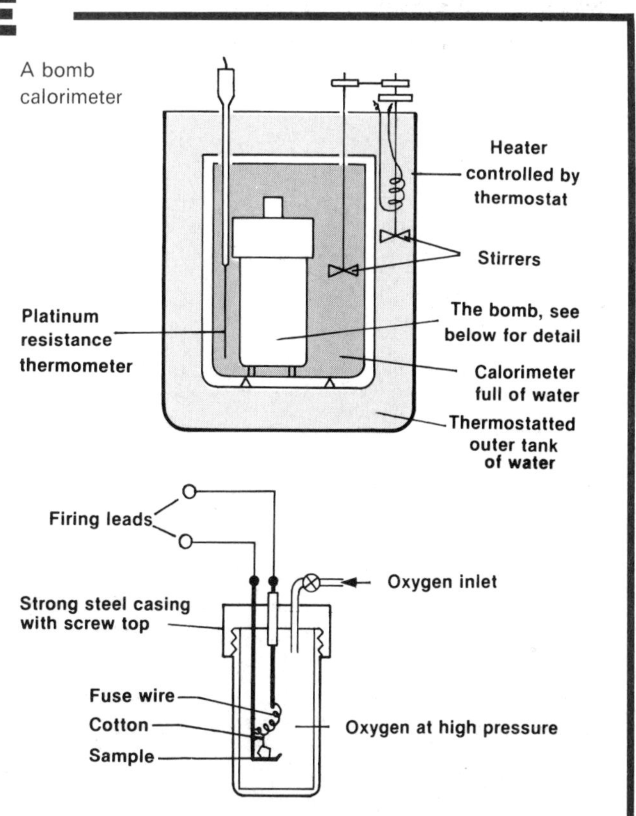

Note the special state symbol (gr) for graphite. Carbon (solid under standard conditions) has two **allotropes**, graphite and diamond. The more stable allotrope is the standard state—this is graphite.

$$\text{Hess's law tells us that } \Delta H_1 = \Delta H_4 + \Delta H_5.$$

ΔH_5 is the enthalpy of formation of ethane, $\Delta_f H^{\ominus}$, while reaction 4 is the *reverse* of the formation of ethyne. Another consequence of Hess's law is that the **reverse of a reaction has the negative of its ΔH value.**

The values we need are:

$$\Delta_f H^{\ominus}\ C_2H_2 = +228\ \text{kJ mol}^{-1}$$

and

$$\Delta_f H^{\ominus}\ C_2H_6 = -85\ \text{kJ mol}^{-1}$$

so

$$\Delta H_4 = -228\ \text{kJ mol}^{-1} \quad \text{(remember to change the sign)}$$
$$\Delta H_5 = -85\ \text{kJ mol}^{-1}$$

thus

$$\Delta H_1 = -228 + -85 = -313\ \text{kJ mol}^{-1}$$

which was the result we got from the previous method, as we should expect from Hess's law.

You may have noticed that in reaction 4 there are two moles of hydrogen 'spare', as only one of the three moles of hydrogen is involved.

$$H{-}C{\equiv}C{-}H(g) \longrightarrow 2C(gr) + H_2(g)$$

is the reaction we are considering, but we have:

$$H{-}C{\equiv}C{-}H(g) + 2H_2(g) \longrightarrow 2C(gr) + 3H_2(g)$$

However, this makes no difference. The 'extra' hydrogen *is not involved in the reaction* and there is no ΔH to consider. We could have added any amount of any other substance which is not involved and the same would be true. This point is worth remembering in the next example. This shows a third way of calculating the enthalpy change for the original reaction, and is a case where the hydrogen does need to be considered, though the oxygen does not. This time we will go via enthalpy changes of combustion of the starting material and product. This is of particular importance since all three substances involved, ethyne, hydrogen and ethane, burn readily, which means their enthalpy changes of combustion can be easily measured. Indeed the combustion of ethyne is of great practical importance. The non-systematic name for ethyne is acetylene and its combustion provides the heat for oxyacetylene welding and cutting torches.

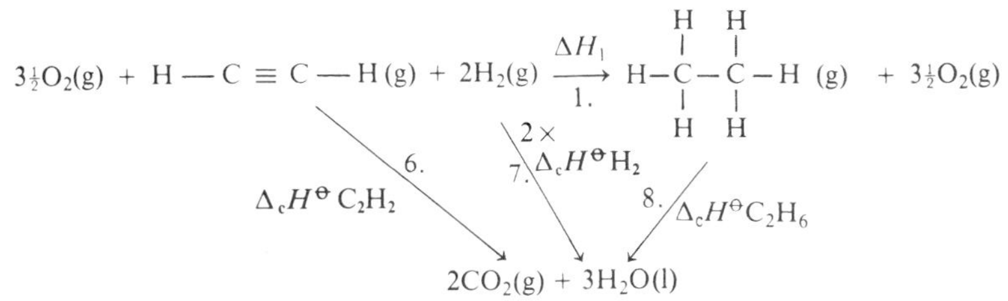

$$3\tfrac{1}{2}O_2(g) + H{-}C{\equiv}C{-}H(g) + 2H_2(g) \xrightarrow[1.]{\Delta H_1} H_3C{-}CH_3(g) + 3\tfrac{1}{2}O_2(g)$$

6. $\Delta_c H^{\ominus}\ C_2H_2$

7. $2\times\ \Delta_c H^{\ominus}\ H_2$

8. $\Delta_c H^{\ominus}\ C_2H_6$

$$2CO_2(g) + 3H_2O(l)$$

Note

1. Both reactions 6 *and* 7 have to occur to get from the starting materials to the combustion products. Do not forget the hydrogen.
2. In this case there are two moles of hydrogen molecules so we need *twice* the value of $\Delta_c H^{\ominus}$ which refers to one mole of hydrogen molecules.

$$\Delta_c H^{\ominus} C_2H_2 = -1301\ \text{kJ mol}^{-1}$$
$$\Delta_c H^{\ominus} H_2 = -286\ \text{kJ mol}^{-1}$$
$$\Delta_c H^{\ominus} C_2H_6 = -1560\ \text{kJ mol}^{-1}$$

Putting in the figures:

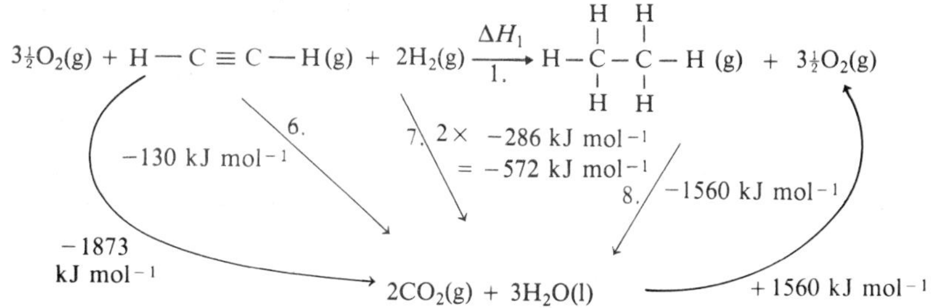

If we want the enthalpy change for reaction 1 we must go round the cycle in the direction of the brown arrows. We are dealing with the reverse of reaction 8 and so its sign must change.

So

$$\Delta H_1 = -1873 + 1560\,\text{kJ mol}^{-1}$$
$$\Delta H_1 = -313\,\text{kJ mol}^{-1}$$

again, the same answer as before.

The above instances are examples of **thermochemical cycles** and illustrate the use we can make of Hess's law to calculate ΔHs which cannot be measured directly, because the reaction either does not take place at all, does not take place under the required conditions (298 K, 10^2 kPa) or cannot easily or safely be carried out in a calorimeter. A good example is the measurement of lattice energies—reactions which cannot be brought about at all. So these have to be calculated using Hess's law (see section 10.2.1).

9.4.1 Enthalpy diagrams

It is often useful to put thermochemical cycles on enthalpy diagrams (see overleaf). One advantage is that these show more clearly the relative energy levels of different species, and this allows us to compare their relative energetic stabilities. If a species is of lower energy than another, it is said to be energetically more stable.

So far, all the data we have considered have been enthalpy *changes*, not absolute values. When drawing enthalpy diagrams it is useful to have a zero to work from.

By convention, the enthalpies of elements in their standard states (i.e. at 298 K and 10^2 kPa) are taken as zero.

Note that this means that the standard state of say, oxygen, is $O_2(g)$, not separate O atoms.

Of the two allotropes of carbon, graphite and diamond, graphite is lower in enthalpy by 2 kJ mol^{-1}. This means that diamond is actually less stable than graphite, which may surprise those with diamond jewellery!

Here are some further examples of thermochemical cycles presented both as cycles and as enthalpy diagrams.

Example 1

What is $\Delta_R H^\ominus$ for the change from methoxymethane to ethanol? (The compounds are a pair of isomers—they have the same formula but different structures.)

The standard molar enthalpy changes of formation of the two compounds are:

methoxymethane	$CH_3OCH_3(g)$	$\Delta_f H^\ominus = -184\,\text{kJ mol}^{-1}$
ethanol	$C_2H_5OH(l)$	$\Delta_f H^\ominus = -277\,\text{kJ mol}^{-1}$

USING A THERMOCHEMICAL CYCLE

1. Write down the equation for the required reaction.
2. Write down the elements in the two compounds with the correct quantities of each.

1.

$CH_3OCH_3(g) \xrightarrow{\Delta_R H^\ominus} C_2H_5OH(l)$

methoxymethane ethanol

4. 3. $\Delta_f H^\ominus$ -184 kJ mol^{-1} 3. $\Delta_f H^\ominus$ -277 kJ mol^{-1} 4.

5. $+184$ kJ mol^{-1} -277 kJ mol^{-1}

$2C(gr) + 3H_2(g) + \frac{1}{2}O_2(g)$

2. elements

6. $\Delta_R H^\ominus = +184 - 277$
$= -93$ kJ/mol^{-1}

3. The $\Delta_f H^\ominus$s of the two compounds represent the change from the elements to the compounds. Put arrows on the diagram to show the direction of change (the blue arrows), and insert $\Delta H^\ominus$ values.

4. Put in the arrows to go from methoxymethane to ethanol via the elements (the brown arrows).

5. Reverse the sign of $\Delta H^\ominus$ if the brown arrow is in the reverse direction to the blue arrow.

6. Go round the cycle in the direction of the brown arrows and add up the $\Delta H^\ominus$ values as you go.

Hess's law tells us that this is the same as $\Delta_R H^\ominus$ for the direct reaction.

USING AN ENTHALPY DIAGRAM

1. Draw a line at level 0 to represent the elements.

2. Look up the values of $\Delta_f H^\ominus$ for each compound and enter these on the enthalpy diagrams, taking account of the signs—negative values are below 0, positive values above.

The enthalpy diagram makes it clear that ethanol is energetically more stable than methoxymethane

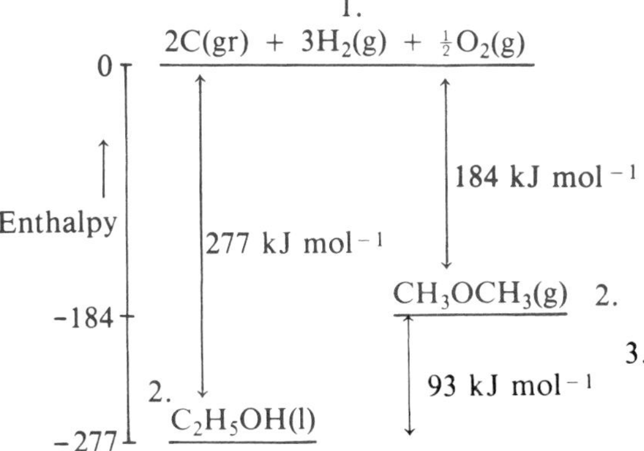

3. Find the difference in levels between the two compounds which represents the difference in their enthalpies.

4. $\Delta_R H^\ominus$ is the difference in levels taking account of the direction of change. Up is positive and down negative. From methoxymethane to ethanol is *down*, so the sign is negative.

$$\Delta_R H^\ominus = -93 \text{ kJ mol}^{-1}$$

Example 2

To find $\Delta_R H^\ominus$ for the reaction

$$NH_3(g) + HBr(g) \longrightarrow NH_4Br(s)$$

The standard molar enthalpy changes of formation of the compounds are

NH_3 $\Delta_f H^\ominus = -46$ kJ mol^{-1}
HBr $\Delta_f H^\ominus = -36$ kJ mol^{-1}
NH_4Br $\Delta_f H^\ominus = -271$ kJ mol^{-1}

USING A THERMOCHEMICAL CYCLE

1. Write an equation for the reaction.
2. Write down the elements.
3. Put in the $\Delta_f H^\ominus$ values with arrows showing the direction, i.e. from elements to compounds.
4. Put in the arrows to go from the starting materials to products via the elements (the brown arrows).
5. Reverse the sign of $\Delta H^\ominus$ if the brown arrow is in the opposite direction to the blue arrow.

1. $NH_3(g) + HBr(g) \xrightarrow{\Delta_R H^\ominus} NH_4Br(s)$

4. 3. $\Delta_f H^\ominus$ -46 kJ mol^{-1} 3. $\Delta_f H^\ominus$ -36 kJ mol^{-1} 3. $\Delta_f H^\ominus$ -271 kJ mol^{-1} 4.

2. $\frac{1}{2}N_2(g) + 2H_2(g) + \frac{1}{2}Br_2(g)$

$+82$ kJ mol^{-1} -271 kJ mol^{-1}

$$\Delta_R H^\ominus = +82 + -271 \text{ kJ mol}^{-1}$$
$$\Delta_R H^\ominus = -189 \text{ kJ mol}^{-1}$$

6. Go round the cycle in the direction of the brown arrows and add up the values of $\Delta H^\ominus$ as you go.

USING AN ENTHALPY DIAGRAM

1. Draw a line at level 0 to represent the elements.
2. Draw a line for ammonium bromide 271 kJ mol^{-1} *below* this, as $\Delta_f H^\ominus$ is negative.
3. Draw a line representing ammonia 46 kJ mol^{-1} below the level of the elements. There is still $\frac{1}{2}H_2$ and $\frac{1}{2}Br_2$ left unused.
4. Draw a line 36 kJ mol^{-1} below ammonia. This represents the formation of hydrogen bromide.

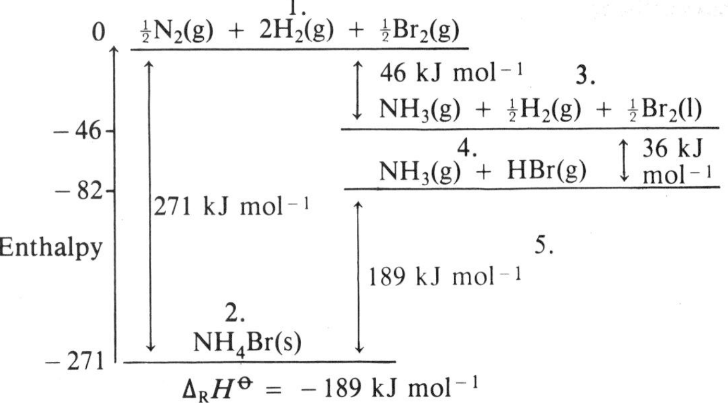

5. Find the difference in levels between the $NH_3 + HBr$ line and the NH_4Br one. This represents $\Delta_R H^\ominus$ for the reaction. As the change from $NH_3 + HBr$ to NH_4Br is 'downhill', $\Delta_R H^\ominus$ must be negative.

9.5 Bond enthalpies

The idea of bond enthalpies, often called bond energies, is to apportion an amount of enthalpy to each covalent bond in a molecule. This enthalpy has to be put in to break the bond and is given out when the bond is formed.

The **standard molar enthalpy change of bond dissociation (bond dissociation energy)** refers to a specific bond in a specific molecule. So methane, CH_4, has four different bond dissociation enthalpies.

$$CH_4(g) \longrightarrow CH_3(g) + H(g) \quad \Delta H^\ominus = +425\,kJ\,mol^{-1}$$
$$CH_3(g) \longrightarrow CH_2(g) + H(g) \quad \Delta H^\ominus = +480\,kJ\,mol^{-1}$$
$$CH_2(g) \longrightarrow CH(g) + H(g) \quad \Delta H^\ominus = +425\,kJ\,mol^{-1}$$
$$CH(g) \longrightarrow C(g) + H(g) \quad \Delta H^\ominus = +335\,kJ\,mol^{-1}$$

The average of these, corresponding to

$$\tfrac{1}{4}(CH_4(g) \longrightarrow C(g) + 4H(g)) \quad \Delta H^\ominus = +416\,kJ\,mol^{-1}$$

is called the **mean molar bond dissociation enthalpy** and is an average for the bonds in the specific compound methane.

The mean molar bond dissociation enthalpy will vary slightly for C—H bonds in other compounds and the average of these over several compounds is often used—the 'tabulated' bond enthalpy or just bond enthalpy.

As they are only averages, calculations done by applying tabulated bond enthalpies to specific compounds can only give approximate answers. As they are only approximate the distinction between bond enthalpy and bond energy is relatively unimportant and many chemists refer to them as simply tabulated bond energies. Nevertheless, they are useful and are often quick and easy to use.

A thermochemical cycle can be used to calculate the mean bond dissociation enthalpy.

$CH_4(g) \xrightarrow{\Delta_R H^\ominus} C(g) + 4H(g)$

$\Delta_f H^\ominus$ -75 kJ mol^{-1} $\Delta_{at} H^\ominus$ $+717$ kJ mol^{-1} $4 \times \Delta_{at} H^\ominus$ $4 \times +218$ kJ mol^{-1} $= +1589$ kJ mol^{-1}

$C(gr) + 2H_2(g)$

$+75$ kJ mol^{-1} $+1589$ kJ mol^{-1}

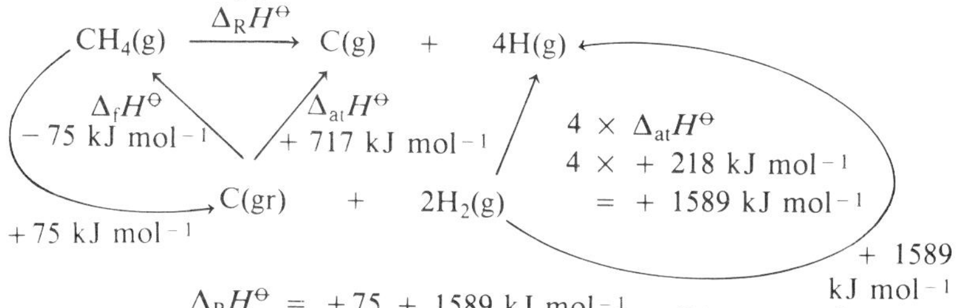

$$\Delta_R H^\ominus = +75 + 1589\ kJ\,mol^{-1}$$
$$\Delta_R H^\ominus = +1664\ kJ\,mol^{-1}$$

(*Note*: $\Delta_{at} H^\ominus$: the subscript $_{at}$ stands for *atomization*)

Mean bond dissociation enthalpy $= 1664/4 = 416\,kJ\,mol^{-1}$.

The usefulness of bond enthalpies is shown by the following example, calculating the enthalpy change of the following reaction:

$$C_2H_6(g) + Cl_2(g) \longrightarrow C_2H_5Cl(g) + HCl(g)$$

1. First draw out the molecules showing all the bonds:

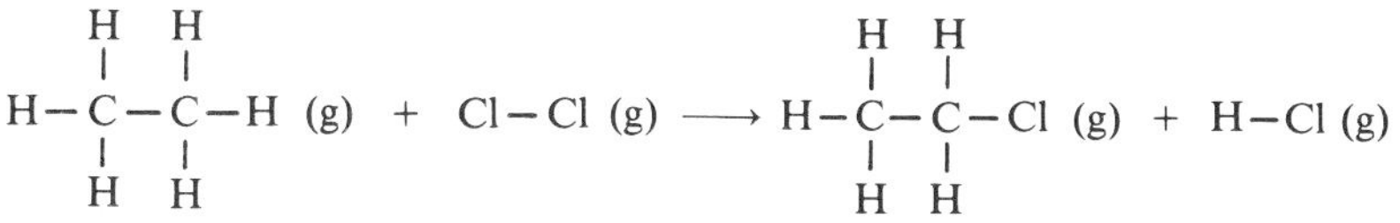

```
   H  H                                H  H
   |  |                                |  |
H--C--C--H (g)  +  Cl--Cl (g)  --->  H--C--C--Cl (g)  +  H--Cl (g)
   |  |                                |  |
   H  H                                H  H
```

Imagine all the bonds in the reactants break, leaving separate atoms.

2. Add the bond enthalpies up to give the total energy which must be *put in* to do this.

3. Now imagine the separate atoms reassemble to give the products. Add the bond enthalpies of the bonds which form to find the total enthalpy *given out* by the bonds forming.

The difference is the approximate enthalpy change of the reaction. The tabulated bond enthalpies required for the above examples are given in **Table 9.1.**

Table 9.1. Tabulated bond enthalpies

Bond	Bond enthalpy
C—H	413 kJ mol^{-1}
C—C	347 kJ mol^{-1}
Cl—Cl	243 kJ mol^{-1}
C—Cl	346 kJ mol^{-1}
Cl—H	432 kJ mol^{-1}

Note: Not exactly the same as C—H in methane, as the tabulated bond enthalpy is an average over many compounds.

We need to *break* these bonds:

	kJ mol^{-1}
6 × C—H	6 × 413 = 2478
1 × C—C	1 × 347 = 347
1 × Cl—Cl	1 × 243 = 243
	3068

So 3068 kJ mol^{-1} must be *put in*.

We need to *make* these bonds:

	kJ mol^{-1}
5 × C—H	5 × 413 = 2065
1 × C—C	1 × 347 = 347
1 × C—Cl	1 × 346 = 346
1 × Cl—H	1 × 432 = 432
	3190

So 3190 kJ mol^{-1} is *given out*.
The difference is 3190 − 3068 = 122 kJ mol^{-1}.
More enthalpy is given out than put in so the reaction is exothermic, so:

$$\Delta_R H^\ominus = -122\,\text{kJ mol}^{-1}$$

We could in fact have shortened this calculation:

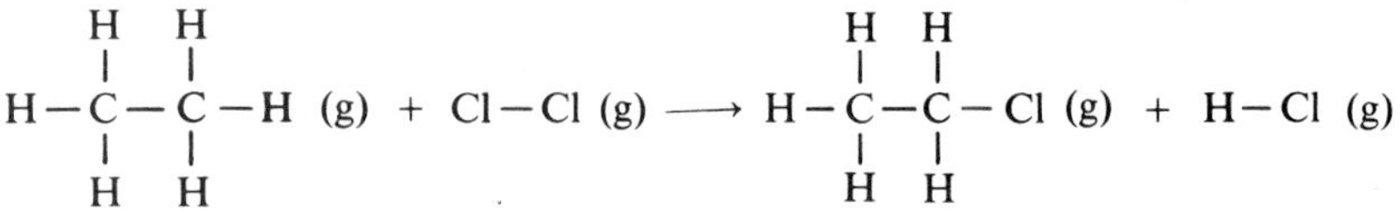

$$H_3C\text{—}CH_2\text{—}H\,(g) + Cl\text{—}Cl\,(g) \longrightarrow H_3C\text{—}CH_2\text{—}Cl\,(g) + H\text{—}Cl\,(g)$$

Only the bonds drawn in brown make or break during the reaction, so we need to break:

	kJ mol^{-1}
1 × C—H	= 413
1 × Cl—Cl	= 243
	656

We need to make

	kJ mol^{-1}
1 × C—Cl	= 346
1 × H—Cl	= 432
	778

The difference is 778 − 656 = 122 kJ mol^{-1}.

More enthalpy is given out than taken in so:

$$\Delta_R H^\ominus = -122\,\text{kJ mol}^{-1} \text{ (as before)}$$

Either method is acceptable. The first is longer, but there is less risk of forgetting any bonds.

This is only an approximate value because the tabulated bond enthalpies do not refer to the specific compounds used. An accurate value can be found using a thermochemical cycle, see below.

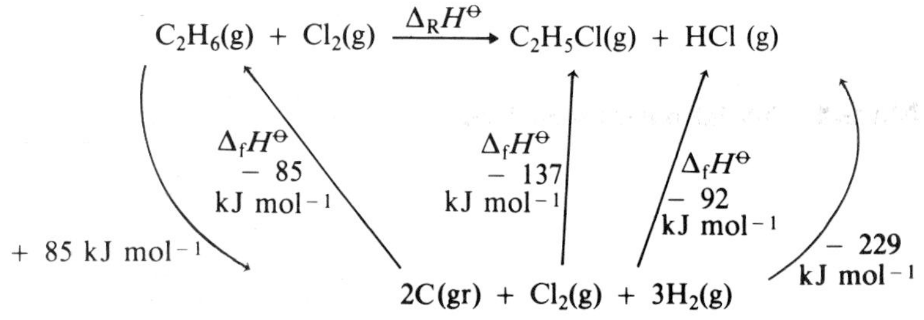

$$C_2H_6(g) + Cl_2(g) \xrightarrow{\Delta_R H^\ominus} C_2H_5Cl(g) + HCl\,(g)$$

$\Delta_f H^\ominus$ − 85 kJ mol^{-1} $\Delta_f H^\ominus$ − 137 kJ mol^{-1} $\Delta_f H^\ominus$ − 92 kJ mol^{-1}

+ 85 kJ mol^{-1} − 229 kJ mol^{-1}

$$2C(gr) + Cl_2(g) + 3H_2(g)$$

Remember $Cl_2(g)$ is an element so no $\Delta_f H^\ominus$ value is needed.

$$\Delta_R H^\ominus = 85 - 229\ \text{kJ mol}^{-1}$$
$$\Delta_R H^\ominus = -144\ \text{kJ mol}^{-1}$$

compared with $-122\ \text{kJ mol}^{-1}$ calculated from tabulated bond enthalpies. This discrepancy is typical of what may be expected using tabulated bond enthalpies. The answer obtained from the thermochemical cycle is the 'right' one as all the $\Delta_f H^\ominus$ values have been obtained from the actual compounds involved.

Another use of bond enthalpies is to predict which bonds might be expected to break during reactions. For example, the compound diethyl peroxide:

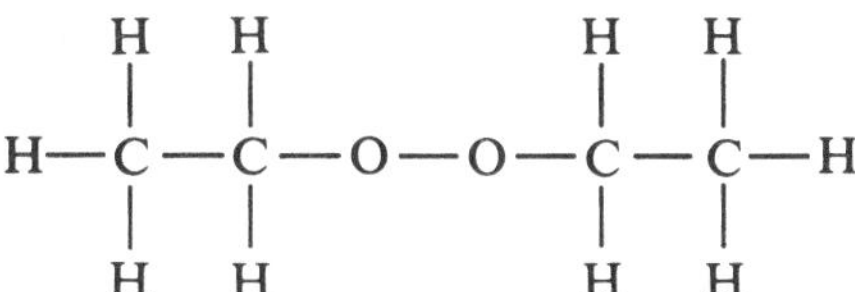

The tabulated bond enthalpies are:

	kJ mol^{-1}
C—H	413
C—O	358
O—O	144
C—C	347

The O—O bond is by far the weakest (requires the least energy to break it). We could predict that this molecule might react by cleavage of the O—O bond. This actually occurs, and similar compounds are used as initiators for chain reactions in fibreglass resins, such as are used in making canoes and repairing car bodies. The O—O bond breaks to give two highly reactive fragments called **radicals**, which initiate the setting reaction.

9.6 Predicting the feasibilities of reactions

9.6.1 Randomness or entropy

Chemists use the terms **feasible** or **spontaneous** to describe reactions which could take place of their own accord, that is, without an external supply of energy. The terms take no account of the rate of the reaction, which could be so slow as to be unmeasurable at room temperature.

You may have noticed that many reactions that occur of their own accord are exothermic. It is tempting to think that all exothermic reactions are spontaneous, and that all spontaneous reactions are exothermic. However this is not the case, as a number of endothermic reactions are spontaneous. One common factor in spontaneous endothermic reactions is that they tend to involve 'spreading out' in some way, for example solids dissolving to form solutions, or solids and liquids reacting to give off gases.

Both the following reactions which occur spontaneously are endothermic:

$$C_6H_8O_7(aq) + 3NaHCO_3(aq) \longrightarrow Na_3C_6H_5O_7(aq) + 3H_2O(l) + 3CO_2(g)$$

citric acid; sodium hydrogen carbonate; sodium citrate; water; carbon dioxide

$$NH_4NO_3(s) + (aq) \longrightarrow NH_4NO_3(aq)$$

ammonium nitrate; aqueous ammonium nitrate

A

PORTABLE HEAT AND COLD

Ice packs have long been used as a simple treatment for minor strains and sprains. Other injuries may be more effectively treated by gentle warmth. Nowadays hot and cold packs are available which produce almost instant heat or cold when needed, but which can be kept for months in a first-aid kit until they are needed. How do they work?

The cold pack is the simplest. It consists of a stout plastic bag inside which is ammonium nitrate (a white powder) and a second thinner plastic bag full of water. When cold is required, the user simply gives the bag a sharp blow to burst the inner bag of water. The ammonium nitrate then dissolves in the water—a strongly endothermic reaction. The actual

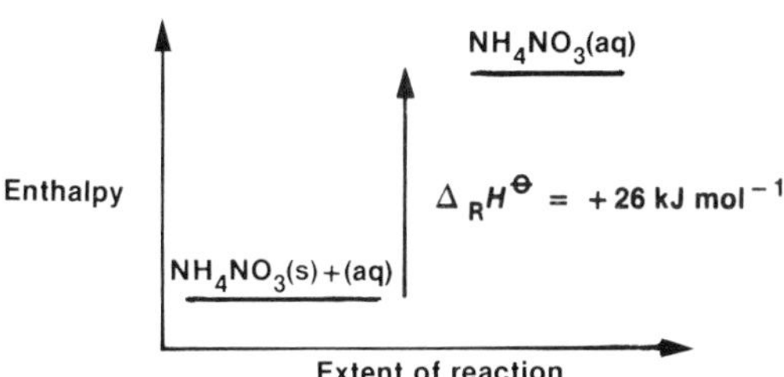

temperature drop depends on the quantity of ammonium nitrate and water, for example for 120 g of ammonium nitrate ($M_r = 80$) in 500 cm³ (just under a pint) of water:

120 g NH_4NO_3 is 1.5 moles

total heat absorbed = 1.5 × 26 kJ because $\Delta_R H^\ominus = +26\ kJ\ mol^{-1}$

$$\text{heat absorbed} = 39\ kJ = 39\,000\ J$$
$$\text{heat} = m \times c \times \Delta T$$
$$= 500 \times 4.2 \times \Delta T$$
$$39\,000 = 500 \times 4.2 \times \Delta T$$
$$\Delta T = \frac{39\,000}{500 \times 4.2}$$
$$\Delta T = 18.6\ K$$

This temperature drop is enough to cool the water almost to freezing point even on quite a warm day.

Unfortunately the cold pack is not reusable as we can not easily 'undissolve' the ammonium nitrate. However, ammonium nitrate is quite cheap and the only other ingredients are water and two plastic bags.

There are two types of hot pack. One is reusable and the other is not. The reusable type contains molten sodium thiosulphate ($Na_2S_2O_3$)—familiar to photographers as 'hypo'. The sodium thiosulphate stays liquid below its freezing point (48 °C, 321 K)– a phenomenon known as supercooling. By squeezing a compartment in the corner of the pack a 'seed crystal' of solid sodium thiosulphate is released. The liquid then freezes, triggered by the seed crystal, and gives out its enthalpy change of freezing. The change takes place at the freezing point of 'thio', a pleasantly warm 48 °C, 321 K. The pack can be restarted by placing it in boiling water until the 'thio' melts again. It can be used over and over again until the supply of seed crystals is used up.

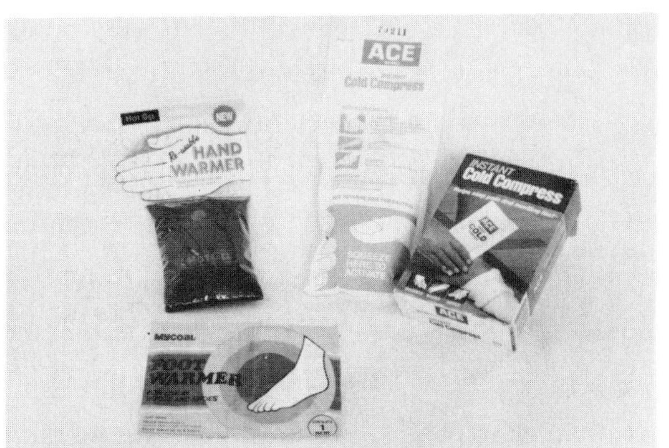

Ammonium nitrite cold packs, a 'thio' heat pack (handwarmer) and an iron heat pack (footwarmer)

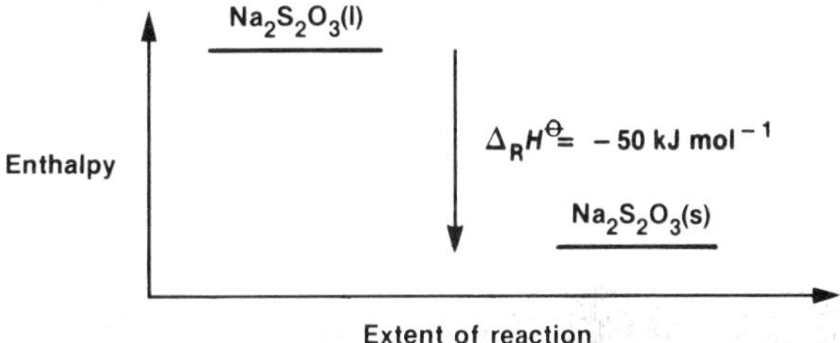

The second type of hot pack relies on the reaction:

$$4Fe(s) + 3O_2(g) \longrightarrow 2Fe_2O_3(s)$$
$$\Delta_R H^\ominus = -1648\ kJ\ mol^{-1}$$

A mixture of moist iron powder and sodium chloride is enclosed in a perforated bag, not unlike a huge teabag. This is sealed inside a plastic bag until required. It is then taken out of the plastic bag and shaken. This allows oxygen in, the above reaction occurs and heat is given out. The water and sodium chloride act as a catalyst in the same way as in rusting, which is the same reaction. Normally the rusting reaction takes place so slowly that we do not notice the heat. Here the finely powdered iron powder has such a large surface area that the reaction goes much faster.

A little thought will show that 'spreading out' is a feature of many spontaneous processes, both chemical and physical, whether they are exothermic or endothermic. See **Figure 9.9.** Indeed the reverse of the last process would be quite extraordinary. We should be most surprised if the oxygen and the nitrogen in the air were to unmix and give, say, a room with all the oxygen on one side and all the nitrogen on the other!

This gives us a clue to the second factor which governs which reactions occur of their own accord and which do not.

Processes which involve 'spreading out', randomizing or disordering the

Figure 9.9 Spontaneous processes

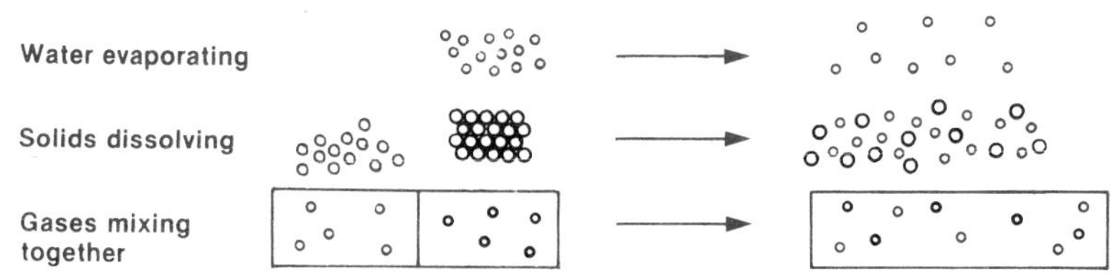

arrangement of molecules tend to happen of their own accord because they happen by chance alone.

The randomness of a system, expressed mathematically, is called the **entropy** of the system and is given the symbol S.

So *two* factors govern the feasibility of a chemical reaction:

the **enthalpy change** ΔH which should be **negative**
the **entropy change** ΔS which should be **positive**

We can now classify four types of reactions.

1. $+\Delta H$ cannot $-\Delta S$ 'go'		**2.** $-\Delta H$ might $-\Delta S$ 'go'
3. $+\Delta H$ might $+\Delta S$ 'go'		**4.** $-\Delta H$ must $+\Delta S$ 'go'

Reactions corresponding to area 4 with a negative ΔH and an increase in entropy or randomness would be sure to be spontaneous.

An example is the decomposition of solid ammonium nitrate:

$$NH_4NO_3(s) \longrightarrow N_2O(g) + 2H_2O(l) \quad \Delta_R H^{\ominus} = -124\,kJ\,mol^{-1}$$

This is a solid with an ordered crystal lattice becoming a gas and a liquid—much less ordered, so the reaction involves an increase in entropy. Its enthalpy change is negative.

An example of an area 3 reaction is:

$$(NH_4)_2CO_3(s) \longrightarrow 2NH_3(g) + CO_2(g) + H_2O(l)$$
$$\Delta_R H^{\ominus} = +68\,kJ\,mol^{-1}$$

The entropy change is positive and as the reaction does occur this must outweigh the positive ΔH value.

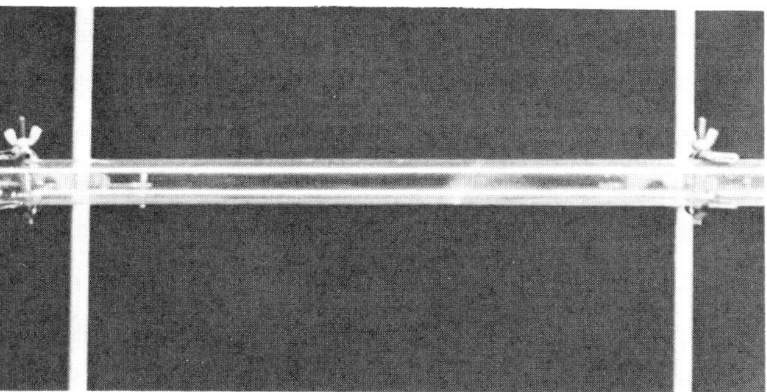

The white ring produced from the reaction of ammonia and hydrogen chloride

An example of an area 2 reaction is:

$$NH_3(g) + HCl(g) \longrightarrow NH_4Cl(s) \quad \Delta_R H^{\ominus} = -176\,kJ\,mol^{-1}$$

This is the reaction which produces a white ring in diffusion experiments. The entropy change is negative in going from gas to solid, but the fairly large negative enthalpy change outweighs this.

An area 1 reaction will not happen spontaneously.

9.6.2 Kinetic factors

However, there are still complications. Firstly, how do we measure randomness and thus give a value to an entropy change, and how do we decide on the relative importance of the entropy and enthalpy changes when they predict opposite outcomes? These problems will be tackled in Chapter 18.

Secondly, neither enthalpy changes nor entropy changes tell us anything about how quickly or slowly a reaction is likely to go. So, we might predict that a certain reaction should occur spontaneously because of enthalpy and entropy changes but the reaction might take place so slowly that for practical purposes it does not occur at all.

Carbon gives an interesting example:

$$C(gr) + O_2(g) \longrightarrow CO_2(g) \quad \Delta_R H^\ominus = -394\,\text{kJ mol}^{-1}$$

The entropy increases as we go from an ordered solid to a disordered gas, and the enthalpy change is negative, so we would expect the reaction to 'go' on both counts. However, experience with graphite (the 'lead' in pencils) tells us that the reaction does not take place at room temperature—although it will take place at higher temperatures. At room temperature the reaction is too slow.

Since the branch of chemistry dealing with enthalpy and entropy changes is called **thermodynamics**, and that dealing with rates is called **kinetics**, graphite is said to be thermodynamically unstable but kinetically stable.

9.7 Summary

- Energy is the ability to do work. It is measured in joules.
- The enthalpy change ΔH of a chemical reaction is the energy given out or taken in measured under conditions of constant pressure (10^2 kPa) and temperature (298 K). To avoid ambiguity, the equation of the reaction should always be given.
- Enthalpy changes are often represented on enthalpy diagrams (energy level diagrams, **Figure 9.10**).

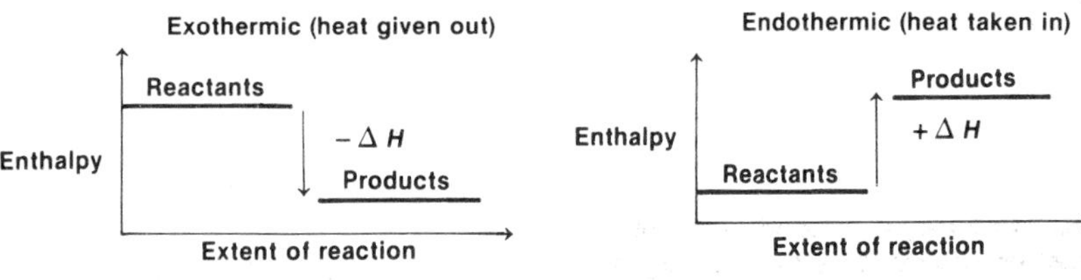

Figure 9.10 Enthalpy diagrams

A

ENERGETICS IN INDUSTRIAL CHEMISTRY

The efficient use of energy in industrial processes is vital for their economic success. Energy considerations include the following:

- Reactions may take in or give out heat.
- Reactions may need to be heated or cooled in order to give suitable reaction rates.
- Reactions may need to be heated or cooled to give a suitable equilibrium position.
- Energy may be needed to compress gases and it may need to be removed from compressed gases (compressing gases heats them up).
- Many purification processes require energy—for example, distillation.
- Electrolysis reactions require large amounts of electrical energy.
- Energy is used by ancillary equipment—pumps, lighting, etc.

Efficient industrial plants are designed to make use of energy given out at different stages of the process.

For example in the manufacture of sulphuric acid from sulphur, the two key steps are:

$$\underset{\text{sulphur}}{S} + \underset{\text{oxygen}}{O_2} \longrightarrow \underset{\text{sulphur dioxide}}{SO_2} \quad \Delta_R H^\ominus = -297\,\text{kJ mol}^{-1}$$

$$\underset{\text{sulphur dioxide}}{SO_2} + \underset{\text{oxygen}}{\tfrac{1}{2}O_2} \longrightarrow \underset{\text{sulphur trioxide}}{SO_3} \quad \Delta_R H^\ominus = -98\,\text{kJ mol}^{-1}$$

The maximum energy given out of 395 kJ mol^{-1} is the equivalent of approximately 4 GJ per tonne of sulphuric acid (1 GJ = 10^9 J or 10^6 kJ). Most of this is recovered and after supplying the plant's energy requirements, some 3 GJ per tonne is left for sale. The income from the sale of this energy is enough to cover all the costs of the plant (including maintenance and the salaries of the operators) except the raw material sulphur. This explains the very low price of around £25 per tonne of sulphuric acid.

• Enthalpy changes are measured by transferring the heat into a container of water (a calorimeter).

heat change = mass of water × specific heat capacity of water × temperature change

• Sometimes the heat capacity of the whole apparatus is found by a separate experiment or electrical calibration. In this case:

heat change = heat capacity × temperature change

• The most accurate determinations are made with a bomb calorimeter.
• Hess's law states that the overall heat change of a chemical reaction is independent of the route by which the reaction occurs. This enables heat changes of reactions to be calculated using thermochemical cycles. Two examples are given in **Figure 9.11.**

Figure 9.11 Enthalpy cycles

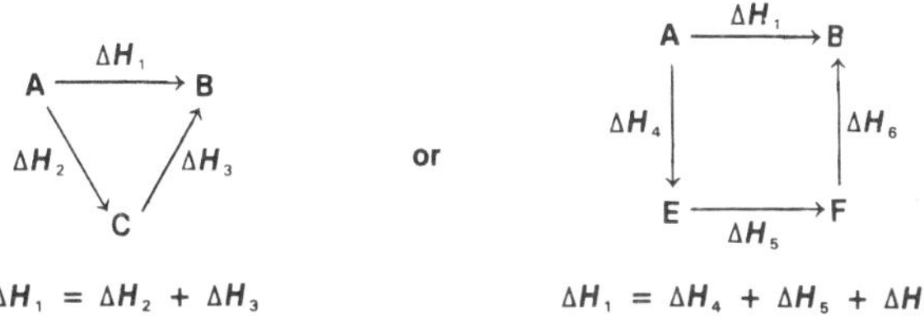

• In every case due regard must be taken of the sign of ΔH. Reversing a reaction means the sign of ΔH is reversed.
• Changes may also be represented on enthalpy diagrams using the convention that elements in their standard states have an enthalpy of zero. **Figure 9.12** gives some examples.

Figure 9.12 Enthalpy diagrams

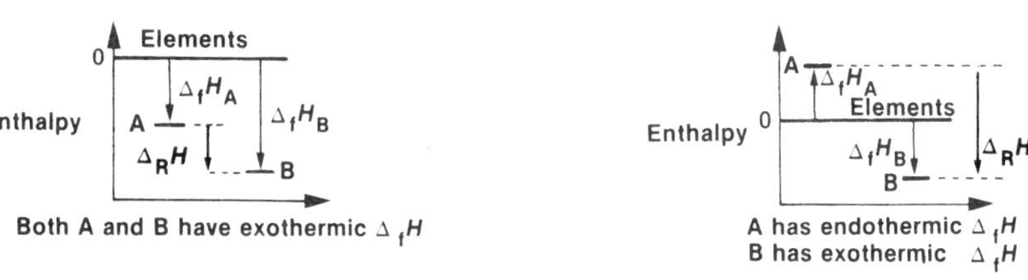

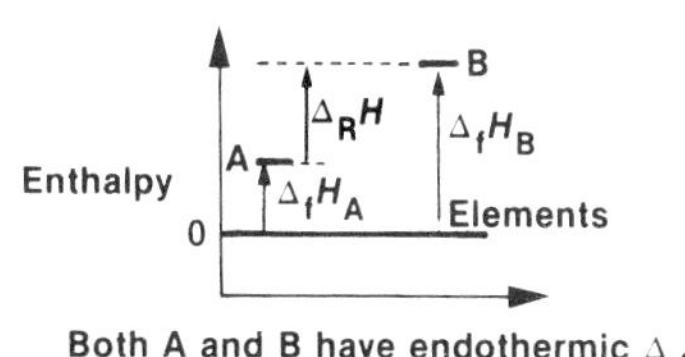

• Tabulated bond enthalpy is the enthalpy required to break a particular bond averaged over many compounds.
• Tabulated bond enthalpies can be used to calculate approximate enthalpy changes for reactions, by working out the difference between the enthalpies to break all the bonds in the reactants and make all the bonds in the products.
• Tabulated bond enthalpies may also suggest which bonds might break in a given molecule.
• Entropy is a measure of the degree of disorder or randomness of a system.
• Two factors govern the feasibility of reactions: the enthalpy change and the entropy change.
• Negative enthalpy change ($-\Delta H$) which denotes an exothermic reaction, and positive entropy change ($+\Delta S$) which means an increase in disorder, favour reactions taking place.

9.8 Questions

1 $50\,cm^3$ of $2\,mol\,dm^{-3}$ sodium hydroxide and $50\,cm^3$ of $2\,mol\,dm^{-3}$ hydrochloric acid were mixed in an expanded polystyrene beaker. The temperature rose by 11 K.

A) Calculate $\Delta H^\ominus$ for the reaction in $kJ\,mol^{-1}$.

B) How will this value compare with the accepted value for this reaction?

Explain your answer.

C) What value of $\Delta H^\ominus$ would you expect if $50\,cm^3$ of $2\,mol\,dm^{-3}$ potassium hydroxide and $50\,cm^3$ of $2\,mol\,dm^{-3}$ nitric acid were mixed in the same apparatus?

Explain your answer.

2 $100\,cm^3$ of $0.1\,mol\,dm^{-3}$ copper sulphate was placed in an electrical compensation calorimeter and 6.5 g of powdered zinc added. A 12 V heater taking a current of 0.2 A was switched on for 15 minutes to produce the same temperature rise.

A) Write an equation for the reaction.

B) Which reactant was in excess?

C) How many joules of heat were supplied? (Use energy $= VIt$.)

D) Calculate $\Delta H^\ominus$ for the reaction in $kJ\,mol^{-1}$.

E) Represent the reaction on an enthalpy diagram.

F) Why was it not necessary to know the actual value of the temperature rise in order to calculate $\Delta H^\ominus$?

G) What two errors are eliminated by use of the electrical heater to produce the same temperature rise, compared with the method of question 1?

3 A bomb calorimeter was calibrated by burning a pellet of benzenecarboxylic acid ($C_7H_6O_2$) of mass 0.7934 g. The temperature rise was 2.037 K. $\Delta_cH^\ominus$ for benzenecarboxylic acid is $-3227.0\,kJ\,mol^{-1}$.

A) What fraction of a mole of benzenecarboxylic acid was used?

B) Calculate the heat capacity of the calorimeter using the relationship

heat given out = heat capacity × temperature rise

The calorimeter was then used to calculate $\Delta_cH^\ominus$ for other compounds and the following results were obtained.

Compound	M_r	Mass used/g	Temperature rise/K
Propan-1-ol	60	0.7563	2.445
Butan-1-ol	74	0.8233	2.860
Pentan-1-ol	88	0.8378	3.059

C) Calculate $\Delta_cH^\ominus$ for each alcohol using the following steps

i What fraction of a mole was used?

ii How much heat has been produced?

iii How much heat would be produced by a mole?

4 This question refers to the ammonium nitrate 'cold pack' described on page 83.

A) Explain why ammonium nitrate should dissolve in the water even though the reaction is endothermic.

B) Calculate the temperature drop if only 80 g of ammonium nitrate had been used.

C) What mass of ammonium nitrate would be used if a temperature drop of 20 K was desired?

5 Use the following tabulated bond energies to calculate the standard enthalpy change of atomization of methanol.

$$CH_3OH(g) \longrightarrow C(g) + 4H(g) + O(g)$$

$$C\!-\!H = 413\,kJ\,mol^{-1},\quad C\!-\!O = 358\,kJ\,mol^{-1},\quad O\!-\!H = 464\,kJ\,mol^{-1}$$

6 Use a thermochemical cycle to calculate $\Delta_RH^\ominus$ for the following reactions using the values of $\Delta_fH^\ominus$ given below. Represent each reaction on an enthalpy diagram.

A) $CH_3COCH_3(l) + H_2(g) \longrightarrow CH_3\!-\!CH(OH)\!-\!CH_3(l)$

B) $H_2C\!=\!CH_2(g) + Cl_2(g) \longrightarrow H\!-\!CHCl\!-\!CHCl\!-\!H(l)$

C) $H_2C\!=\!CH_2 + HCl(g) \longrightarrow H\!-\!CH_2\!-\!CH_2\!-\!Cl(g)$

D) $Zn(s) + CuO(s) \longrightarrow ZnO(s) + Cu(s)$

E) $Pb(NO_3)_2(s) \longrightarrow PbO(s) + 2NO_2(g) + \tfrac{1}{2}O_2(g)$

Compound	$\Delta_fH^\ominus/kJ\,mol^{-1}$
$CH_3COCH_3(l)$	−248
$CH_3CH(OH)CH_3(l)$	−318
$C_2H_4(g)$	+52
$C_2H_4Cl_2(l)$	−165
$HCl(g)$	−92
$CH_3CH_2Cl(g)$	−137
$CuO(s)$	−157
$ZnO(s)$	−348
$Pb(NO_3)_2(s)$	−452
$PbO(s)$	−217
$NO_2(g)$	+33

7 For the following reactions say whether you would expect an increase or a decrease in entropy.

A) $NaNO_3(s) \longrightarrow NaNO_2(s) + \tfrac{1}{2}O_2(g)$

B) $H_2O(l) \longrightarrow H_2O(g)$

C) $Na(s) + \tfrac{1}{2}Cl_2(g) \longrightarrow NaCl(s)$

D) $Ca^{2+}(aq) + CO_3^{2-}(aq) \longrightarrow CaCO_3(s)$

8 For the reaction

$$\mathrm{H_2C{=}CH_2(g) + HCl(g) \longrightarrow H{-}CH_2{-}CHCl{-}H(g)}$$

calculate $\Delta_R H^\ominus$ by using tabulated bond energies.

C—H $413\,kJ\,mol^{-1}$, C=C $612\,kJ\,mol^{-1}$,
H—Cl $432\,kJ\,mol^{-1}$, C—C $347\,kJ\,mol^{-1}$,
C—Cl $346\,kJ\,mol^{-1}$

9

$$\mathrm{CH_3CHO(l) + H_2(g) \longrightarrow CH_3CH_2OH(l)}$$

Calculate $\Delta_R H^\ominus$ for the reaction by thermochemical cycles

A) via $\Delta_f H^\ominus$ values
B) via $\Delta_c H^\ominus$ values.

Comment on the two answers.

	$\Delta_f H^\ominus$/kJ mol^{-1}	$\Delta_c H^\ominus$/kJ mol^{-1}
CH_3CHO	−192	−1167
H_2	—	−286
CH_3CH_2OH	−277	−1367

10 Research/debate

Make a list of qualities (chemical and otherwise) of a good fuel for a motor vehicle. How do the following match up with your list?

petrol, alcohol, hydrogen, lead–acid batteries, nuclear fuel

11 The apparatus shown in the diagram (above right) was used to find the enthalpy change of combustion of propanone, CH_3COCH_3.

A) Why would an inadequate supply of air lead to error in the results?

B) The following information was obtained during the experiment:

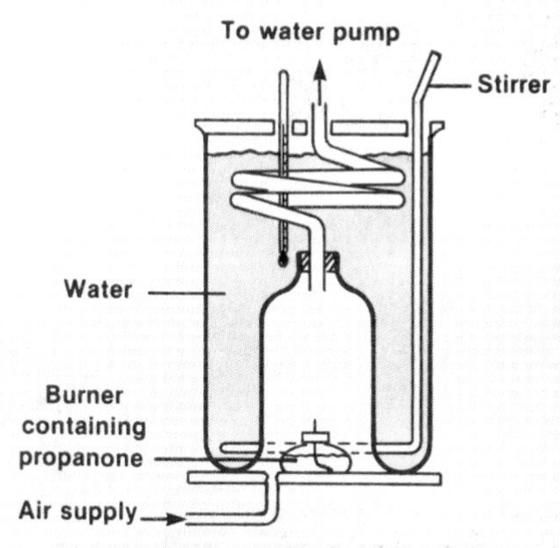

heat capacity of the apparatus = $3.34\,kJ\,K^{-1}$
loss of mass of burner = 2.90 g
temperature rise = 25.3 °C

i Calculate the heat (in kJ) produced in the experiment.

ii Calculate the enthalpy change of combustion of propanone ($M_r = 58$).

C) Construct a thermochemical cycle to determine the enthalpy change of atomization $\Delta_{at}H^\ominus$ of propanone using the appropriate data. (The enthalpy change of atomization refers to the formation of one mole of gaseous atoms of the element concerned.)

D) Use tabulated bond energies to calculate another value for the enthalpy change of atomization of propanone.

E) Comment on the agreement, or disagreement, between the two values calculated in C) and D).

Tabulated bond enthalpies/kJ mol^{-1}:

C—C +346
C—H +413
C=O +749
$\Delta_f H^\ominus$ propanone (l) $-216\,kJ\,mol^{-1}$
$\Delta_{at} H^\ominus$ carbon (gr) $+715\,kJ\,mol^{-1}$
$\Delta_{at} H^\ominus$ hydrogen (g) $+218\,kJ\,mol^{-1}$ (per mole of H atoms)
$\Delta_{at} H^\ominus$ oxygen (g) $+249\,kJ\,mol^{-1}$ (per mole of O atoms)

(Nuffield 1984)

10 Bonding and structure

10.1 Introduction

When nylon first came on the market in 1928, the public was told it was made of 'coal, air and water'. This is true in that the elements in nylon—carbon, oxygen, hydrogen and nitrogen—do indeed come from these materials. However, it is the way the atoms are held together (bonding) and how they are arranged in space (structure) that accounts for the quite distinct properties of these different substances.

Chapter 3 gave a simple account of three types of bonding—ionic, covalent and metallic—which involve respectively transfer, sharing and pooling of electrons between atoms; and two types of structure—giant and molecular. In this chapter we shall see that these types of bonding are extremes—the character of many bonds is somewhere between these. We shall also use some of the ideas of atomic structure from Chapter 8 to give a better picture of bonding, in particular the idea of energy being involved in bonding. Finally, we shall see how bonding ideas can explain the shapes of molecules and the geometries of some giant structures.

10.2 Bonding

10.2.1 Ionic bonding

In Chapter 3 we looked at a simple model of ionic bonding where we saw that electrons were transferred from metal atoms to non-metal atoms so that the resulting charged ions each gained a full outer shell of electrons. However, this is a considerable oversimplification. In Chapter 8 we saw that the removal of electrons from alkali metals required energy to be put in. Why then do ionic compounds form so readily and exothermically? To answer this question we need to look much more carefully at the energy changes that take place when ionic compounds are formed.

Energy changes on forming ionic compounds—the Born–Haber cycle

Note: All energies in this chapter are measured at or converted to 298 K and 10^2 kPa and are therefore enthalpies. However, most chemists still refer to ionization *energies* and lattice *energies* and we shall use these terms.

If a cleaned piece of solid sodium is placed in a gas jar containing chlorine gas, a rapid exothermic reaction takes place, forming solid sodium chloride.

$$Na(s) + \tfrac{1}{2}Cl_2(g) \longrightarrow (Na^+ + Cl^-)(s) \quad \Delta_f H^\ominus = -411\,kJ\,mol^{-1}$$

Let us examine the enthalpy changes which take place.

At some stage the sodium atom must give up an electron to form Na^+. We know from Chapter 9 that the energy change for the process:

$$Na(g) \longrightarrow Na^+(g) + e^-$$

is the **first ionization energy** of sodium and is $+486\,kJ\,mol^{-1}$, i.e. energy must be *put in* for this process to occur.

At some stage a chlorine atom must gain an electron:

$$Cl(g) + e^- \longrightarrow Cl^-(g)$$

The energy change for this process of electron *gain* is called the **first electron affinity** of chlorine and should not be confused with the first ionization energy, which refers to electron *loss*. The first electron affinity for the chlorine atom is $-349\,kJ\,mol^{-1}$, i.e. energy is given out when this process occurs.

However, this is not the whole story. The reaction we are considering involves *solid* sodium, not gaseous, and chlorine *molecules*, not separate atoms, so we must consider the energy changes for the processes below:

$$Na(s) \longrightarrow Na(g) \quad \Delta_{at}H^\ominus = +108\,kJ\,mol^{-1}$$
$$\tfrac{1}{2}Cl_2(g) \longrightarrow Cl(g) \quad \Delta_{at}H^\ominus = +122\,kJ\,mol^{-1}$$

These energy values are called **atomization energies**. Notice that energy has to be *put in* to 'pull apart' the atoms (ΔH is positive).

There is a further energy change to be considered. At room temperature, sodium chloride exists as a solid lattice of alternating positive and negative ions and not as separate gaseous ions. If oppositely charged ions are allowed to come together into a solid lattice, energy is given out due to the attraction of the ions. This is called the **lattice energy** and it refers to the process:

$$Na^+(g) + Cl^-(g) \longrightarrow (Na^+ + Cl^-)(s) \quad LE = -788\,kJ\,mol^{-1}$$

So we have five processes which lead to the formation of NaCl(s) from its elements. These are listed below:

atomization of Na

$$Na(s) \longrightarrow Na(g) \qquad \Delta_{at}H^\ominus = +108\,kJ\,mol^{-1}$$

atomization of Cl

$$\tfrac{1}{2}Cl(g) \longrightarrow Cl(g) \qquad \Delta_{at}H^\ominus = +122\,kJ\,mol^{-1}$$

ionization (e^- loss) of Na

$$Na(g) \longrightarrow Na^+(g) + e^- \qquad \text{first IE} = +496\,kJ\,mol^{-1}$$

gain of e^- by Cl

$$Cl(g) + e^- \longrightarrow Cl^-(g) \qquad \text{first EA} = -349\,kJ\,mol^{-1}$$

formation of lattice

$$Na^+(g) + Cl^-(g) \longrightarrow (Na^+ + Cl^-)(s) \quad LE = -788\,kJ\,mol^{-1}$$

formation of NaCl(s)
from its elements

$$Na(s) + \tfrac{1}{2}Cl_2(g) \longrightarrow (Na^+ + Cl^-)(s) \quad \Delta_f H^\ominus = -411\,kJ\,mol^{-1}$$

Hess's law (see **section 9.4**) tells us that the sum of the first five energy changes is equal to the enthalpy change of formation of sodium chloride. If we put all the changes together in a logical sequence, we get a cycle called a **Born–Haber cycle** (the same Haber as in the Haber process). We can calculate any of the quantities, provided all the others are known.

The cycle is constructed by starting with the elements in their standard states, and taking this as the energy zero.

Then, each step is added as shown in **Figure 10.1,** positive changes going upwards and negative one downwards. When drawing Born–Haber cycles pay attention to the following points.

- Devise a rough scale, say, one line on lined paper = $100\,kJ\,mol^{-1}$. This will enable you to see which energy changes are the most significant and give you the correct relative levels.
- Allow plenty of space and plan it out roughly first to avoid going off the top or bottom of the paper.
- It is better not to put signs in on the diagram because if you do you

Figure 10.1 **Stages in the construction of the Born–Haber cycle for sodium chloride, NaCl. All energies in kJ mol^{-1}**

Na(s) + ½Cl_2(g)

1 Elements in their standard states. This is the energy zero of the diagram

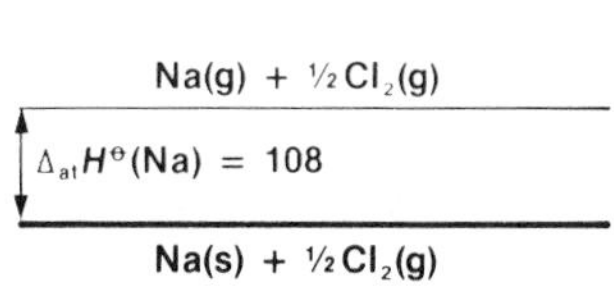

2 Add in the atomization of sodium. This is positive so it is drawn 'uphill'

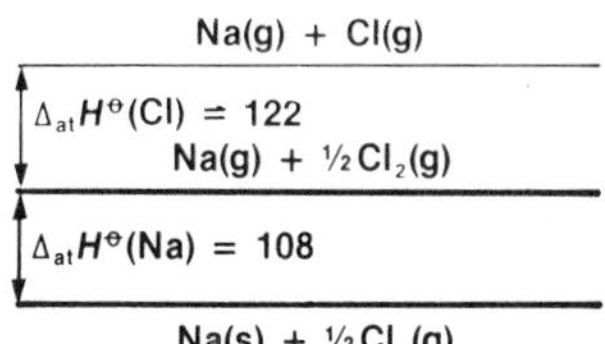

3 Add in the atomization of chlorine. This too is positive and so drawn 'uphill'

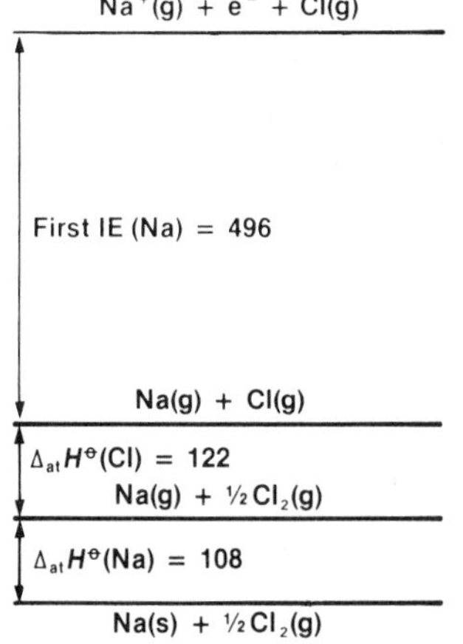

4 Add in the ionization of sodium, also positive and so drawn 'uphill'

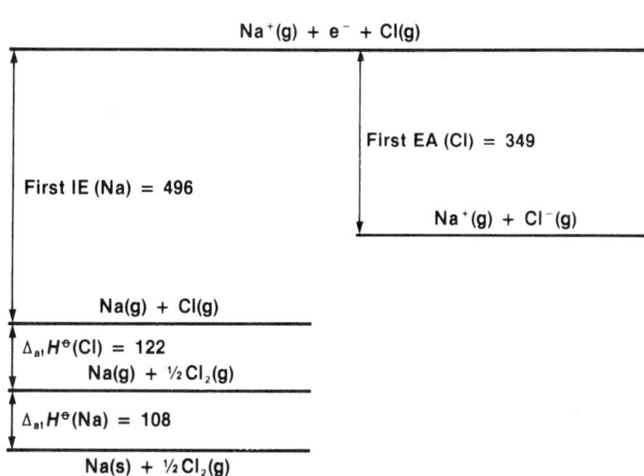

5 Add in the electron affinity of chlorine. This is a negative energy change and so is drawn 'downhill'

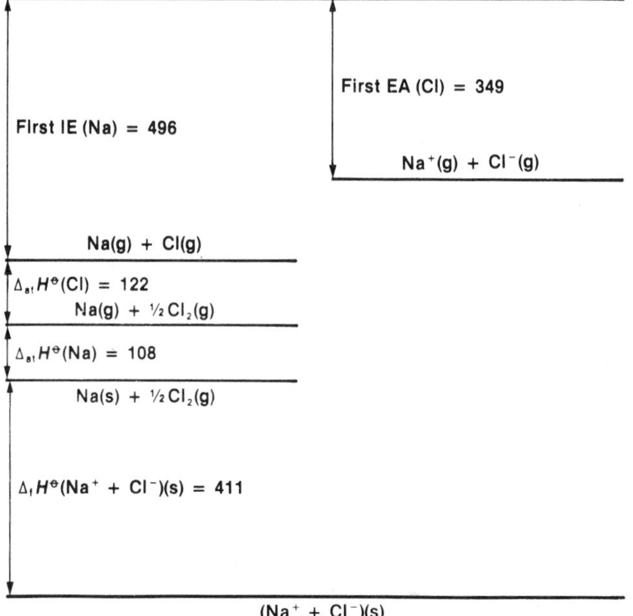

6 Add in the enthalpy of formation of sodium chloride, also negative and drawn 'downhill'

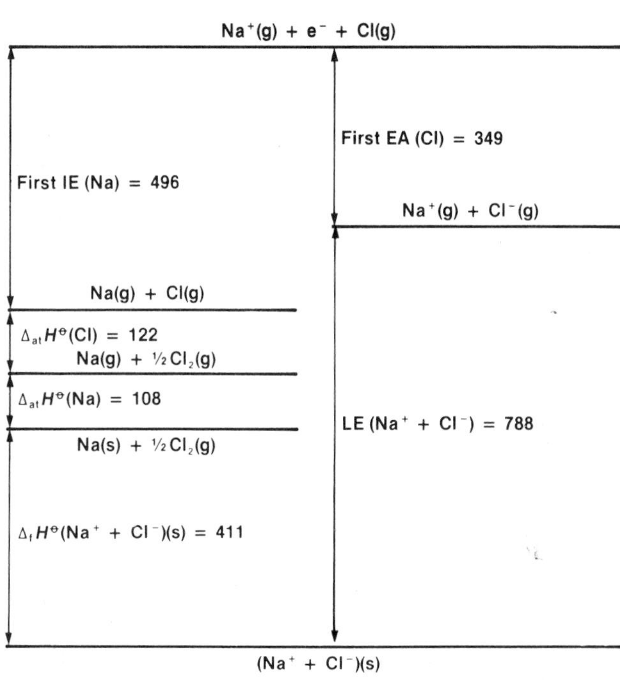

7 The final unknown quantity is the lattice energy of sodium chloride. The size of this is 788 kJ mol^{-1} from the diagram. The definition of lattice energy is the change from separate ions to solid lattice and we must therefore go 'downhill', so LE (Na^+ + Cl^-)(s) = −788 kJ mol^{-1}

The complete Born–Haber cycle for sodium chloride NaCl

IONIZATION ENERGIES AND ELECTRON AFFINITIES

It is interesting to see why the ionization energy of sodium is positive—energy put in, while the electron affinity of chlorine is negative—energy given out.

If we look at a sodium atom with eleven protons and eleven electrons:

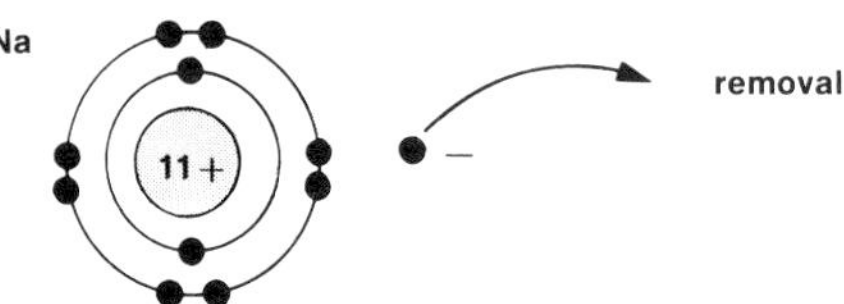

The electron which is being removed is attracted by the positive charge of the nucleus and therefore energy must be put in to pull it away. It does not 'feel' the full 11+ charge of the nucleus because there are ten electrons in inner shells which **shield** the nuclear charge. It therefore 'feels' 11+ less 10, i.e. an effective nuclear charge of 1+.

In the case of chlorine (17 protons and 17 electrons):

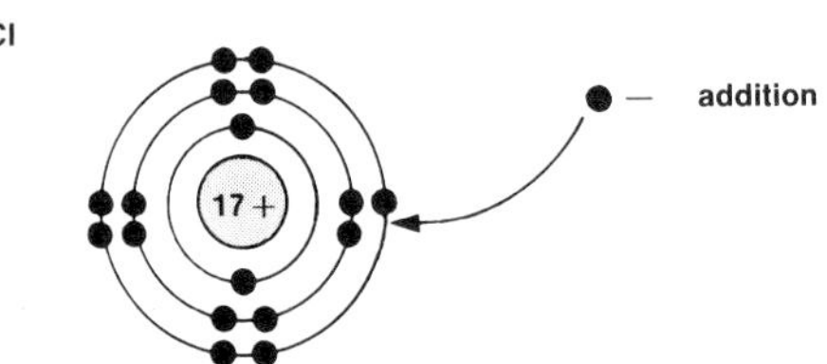

The electron which is being added 'feels' the nuclear charge of 17+ shielded by 10 electrons in inner shells, i.e. an effective nuclear charge of 7+. It is therefore attracted by the nucleus and energy is given out by the process.

Addition of a second electron would require energy to be put in, as the original ion would now have an overall negative charge. Forcing the two negatively charged entities together requires energy to be put in. For example, adding the first electron in the formation of O^{2-}:

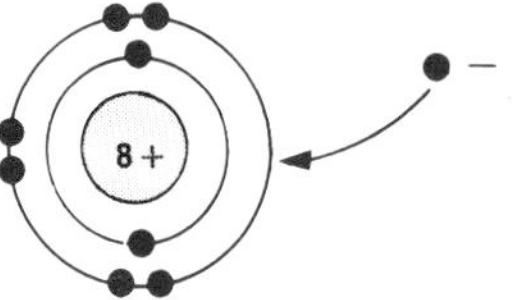

The added electron is attracted by an effective charge of 8+ − 2 = 6+.

first electron affinity = −141 kJ mol^{-1}

Adding the second electron:

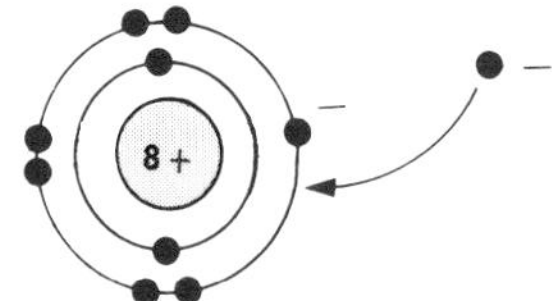

Although the added electron still feels a shielded charge of 6+, the O^- ion has an overall negative charge and repels the electron.

second electron affinity = +798 kJ mol^{-1}

The overall process is the sum of these two:

$$O(g) + e^- \longrightarrow O^-(g) \quad \Delta H = -141\ \text{kJ mol}^{-1}$$
$$O^-(g) + e^- \longrightarrow O^{2-}(g) \quad \Delta H = +798\ \text{kJ mol}^{-1}$$

$$O(g) + 2e^- \longrightarrow O^{2-}(g) \quad \Delta H = +657\ \text{kJ mol}^{-1}$$

can easily count them twice. If you go 'uphill', from a lower line to a higher line, the energy change is positive (endothermic—energy put in), 'downhill', from a higher line to a lower, the energy change is negative (exothermic—energy given out) so the quantities can be given signs when you have calculated them. For example, the lattice energy refers to the change *from* separate gaseous ions *to* crystalline solid ('downhill') so the lattice energy of sodium chloride is −788 kJ mol^{-1}.

Using a Born–Haber cycle we can see that the formation of an ionic compound from its elements is an exothermic process, principally because the large amount of energy given out when the lattice is formed outweighs the energy which has to be put in to form the positive ions.

Further examples of Born–Haber cycles are shown in **Figure 10.2** together with notes on their construction.

Magnesium chloride: since magnesium forms Mg^{2+} ions, we must include the first and second ionization energies of magnesium (*not* 2 × first ionization energy).

Since *two* chlorines are involved, all the quantities related to chlorine are doubled, i.e. $2 \times \Delta_{at}H^{\ominus}$, 2 × first electron affinity (*not* first + second electron affinities).

Magnesium oxide: here the sum of the first and second electron affinities of oxygen is used (*not* 2 × first electron affinity), as oxygen forms O^{2-}. Remember that the first electron affinity is negative and the second is positive, hence the 'dip' at the top of the cycle. The sum of the first and second IEs of magnesium is used as this forms Mg^{2+}.

Sodium oxide: the sum of the first and second electron affinities of oxygen, and 2 × first IE of sodium, are used.

Figure 10.2a The Born–Haber cycle for magnesium chloride, $MgCl_2$. All energies in kJ mol^{-1}

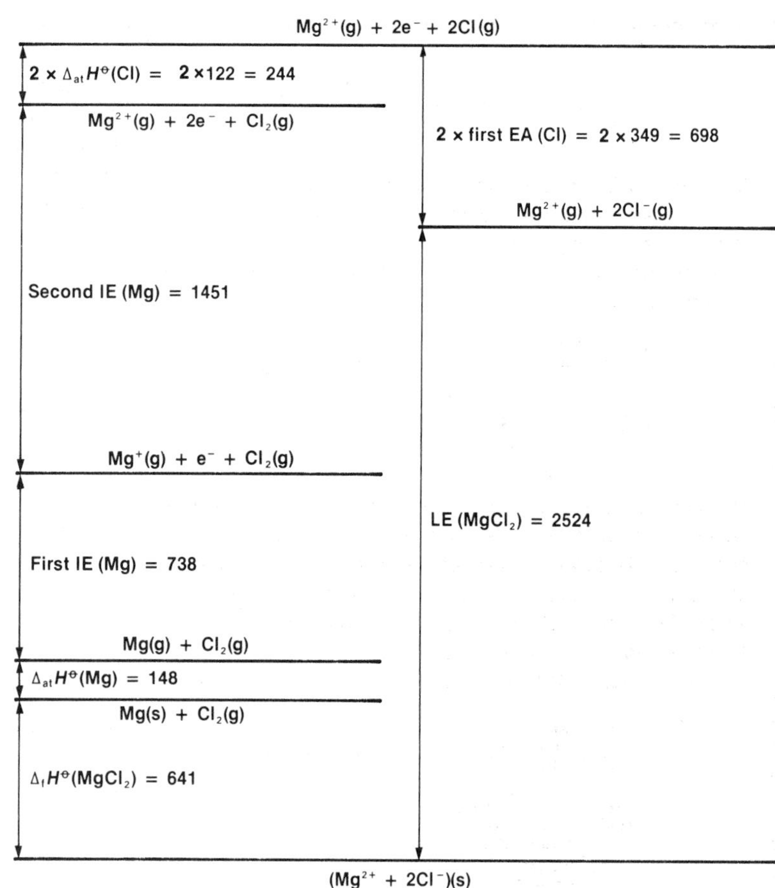

Figure 10.2b The Born–Haber cycle for magnesium oxide, MgO. All energies in kJ mol^{-1}

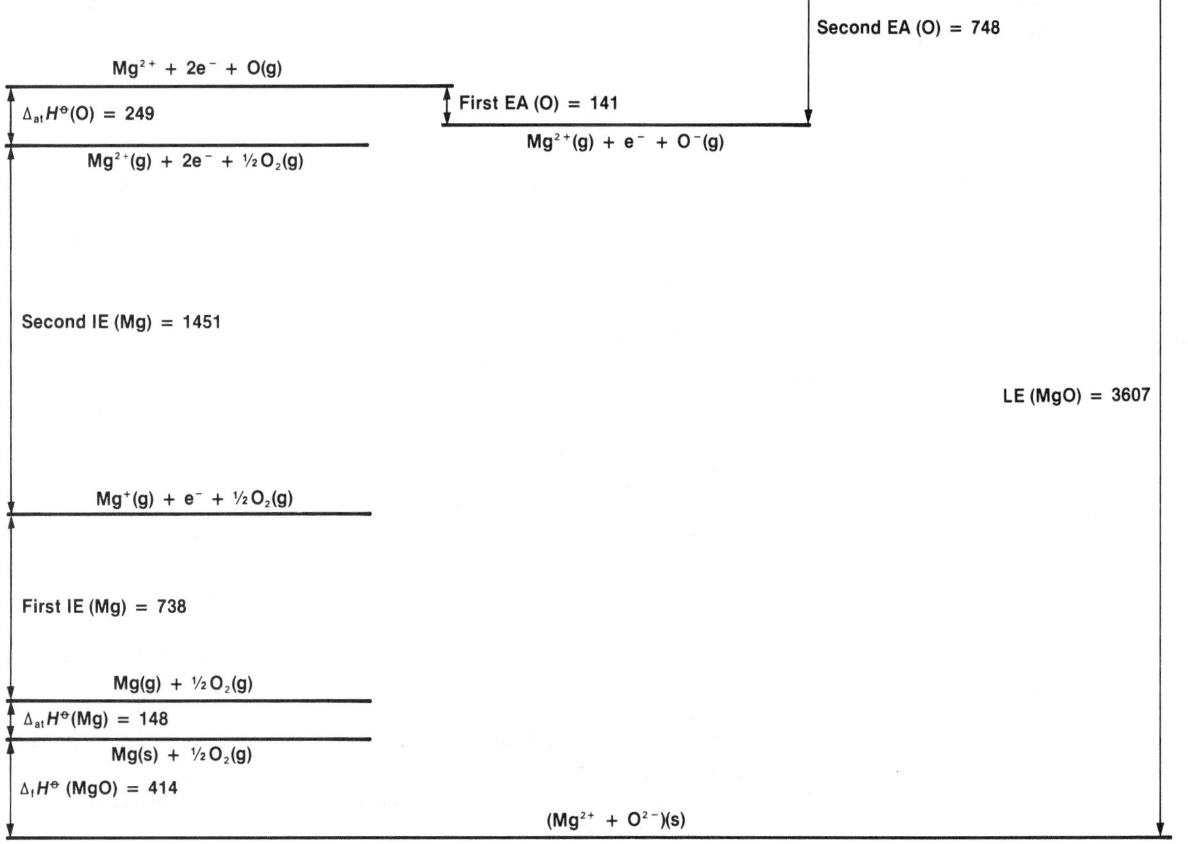

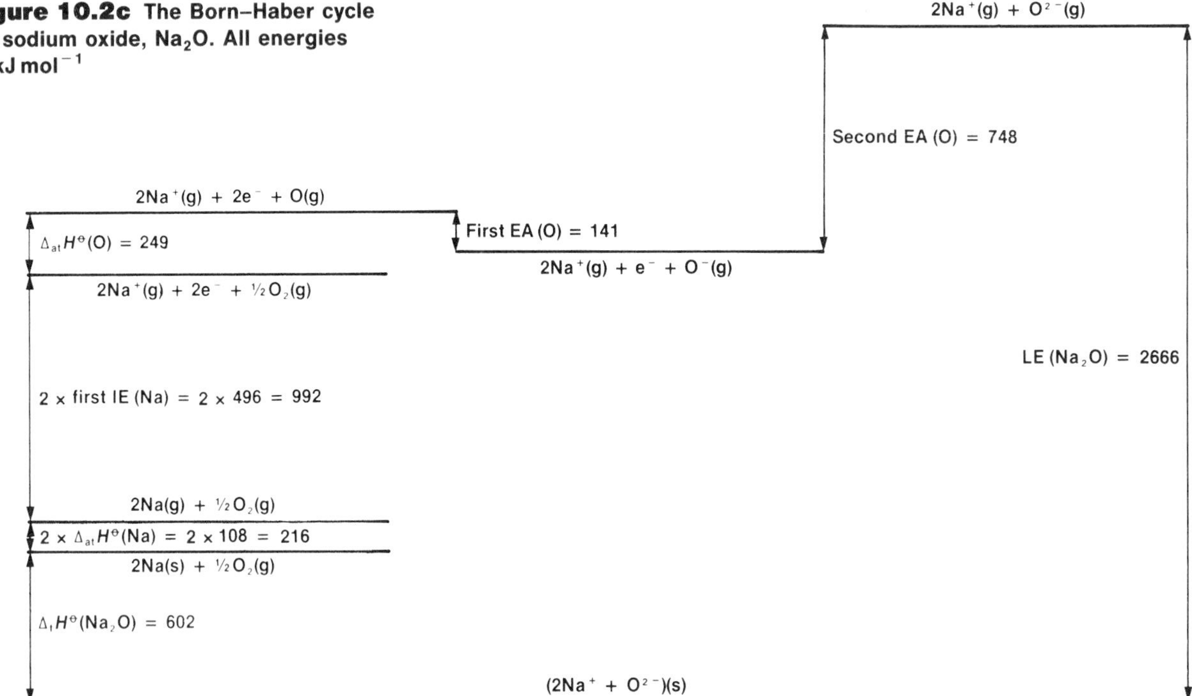

Figure 10.2c The Born–Haber cycle for sodium oxide, Na_2O. All energies in kJ mol^{-1}

Table 10.1 Lattice energies in kJ mol^{-1} for M^+X^- (all values are negative)

		Larger negative ions (anions) →			
		F^-	Cl^-	Br^-	I^-
Larger positive ions (cations) ↓	Li^+	1031	848	803	759
	Na^+	918	780	742	705
	K^+	817	711	679	651
	Rb^+	783	685	656	628
	Cs^+	747	661	635	613

Table 10.2 Lattice energies in kJ mol^{-1} for $M^{2+}X^{2-}$ (all values are negative)

		Larger anions →	
		O^{2-}	S^{2-}
Larger cations ↓	Be^{2+}	4443	3832
	Mg^{2+}	3791	3299
	Ca^{2+}	3401	3013
	Sr^{2+}	3223	2843
	Ba^{2+}	3054	2725

Notice how in each case the two cycles are dominated by the ionization energy and the lattice energy terms as these are the largest.

There are trends in lattice energies related to the charge and size of the ions involved in the compound. These are illustrated in Tables 10.1 and 10.2.

Two relationships are apparent.

1. Larger ions have smaller lattice energies—this is because large ions cannot approach as closely as small ones, so less energy is given out when the lattice forms.
2. More highly charged ions produce *much* larger lattice energies than singly charged ones of approximately the same size. Compare Li^+F^- and $Be^{2+}O^{2-}$. Doubling both charges increases the lattice energy around fourfold.

Both these relationships illustrate the **Born–Mayer equation** (see box)—a formula for calculating the lattice energy from the size and charge of the ions and a knowledge of the geometry of the lattice which is formed. This formula has been derived assuming pure ionic bonding, i.e. complete transfer of electrons from metal to non-metal.

HYPOTHETICAL COMPOUNDS

Born–Haber cycles can be used to investigate the likely $\Delta_f H^\ominus$s of hypothetical compounds to see if they might be expected to exist. The cycles in **Figure 10.3** are for CaF ($Ca^+ + F^-$), CaF_2($Ca^{2+} + 2F^-$), and CaF_3($Ca^{3+} + 3F^-$). Only CaF_2 actually exists, so intelligent guesses have to be made about the likely geometries of the lattices of CaF and CaF_3 in order to make an estimate of their lattice energies via the Born–Mayer equation.

Look at the $\Delta_f H^\ominus$s. A large amount of energy would have to be put in to form CaF_3. CaF's formation would give out energy, but not as as much as for CaF_2. This explains why only CaF_2 has ever been prepared as a stable compound. This is a much more satisfactory explanation than that 'calcium only needs to lose two electrons for a full outer shell'.

Figure 10.3 The Born–Haber cycle for CaF_2 and the hypothetical compounds CaF ($Ca^+ + F^-$) and CaF_3 ($Ca^{3+} + 3F^-$). All energies in kJ mol^{-1}

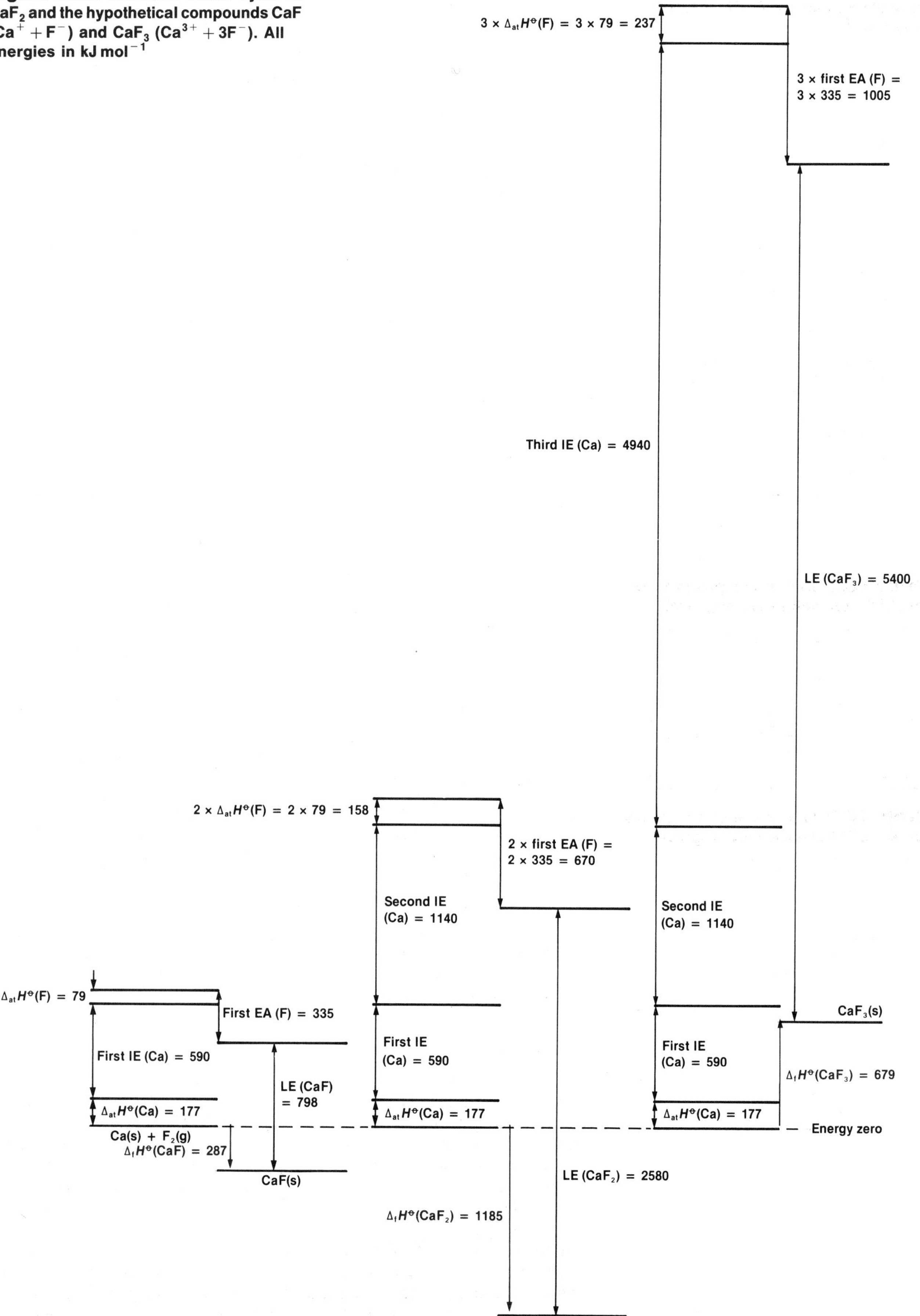

THE BORN–MAYER EQUATION

The Born–Mayer equation states that the lattice energy is proportional to the product of the charges on each ion, and is inversely proportional to the distance between the centres of the ions. It also includes a factor related to the geometry of the crystal lattice. The equation is:

$$\text{lattice energy} = \frac{LMz^+z^-e^2}{4\pi\varepsilon_0 r}\left(1 - \frac{\rho}{r}\right)$$

where

L is the Avogadro constant

M is the Madelung constant which depends on the geometry of the crystal lattice

z^+ is the number of charges on each positive ion

z^- is the number of charges on each negative ion

e is the charge on the electron

r is the distance between the centres of the ions

ρ is a constant which takes account of the repulsion between the electron shells of adjacent ions

ε_0 is a constant called the permittivity of space which governs the strength of electrostatic forces

For many ionic compounds, the lattice energy determined from experimental values via a Born–Haber cycle agrees extremely well with that calculated by the Born–Mayer expression, giving further confirmation of the correctness of the ionic model.

Polarization

For many ionic compounds there is excellent agreement—within 1%—between experimental values of the lattice energies and those calculated by the Born–Mayer expression which assumes pure ionic bonding. For some compounds however, the discrepancy is large. Zinc selenide ($Zn^{2+} + Se^{2-}$) has an experimental lattice energy of $-3611\ \text{kJ mol}^{-1}$. The value of the lattice energy calculated by the Born–Mayer equation is $-3305\ \text{kJ mol}^{-1}$, a difference of almost 10%. As the experimental lattice energy is greater than the theoretical one, this implies some extra bonding is present. What has happened is this. The Zn^{2+} ion is relatively small and has a high positive charge, while the Se^{2-} is relatively large and has a high negative charge. The small Zn^{2+} can approach closely to the electron clouds of the Se^{2-} and distort them by attracting them towards it. The Se^{2-} is fairly easy to disort because its large size means the electrons are far from the nucleus and its double charge means that there is plenty of negative charge to distort. This distortion results in there being more electrons than expected concentrated *between* the Zn and Se nuclei and represents a degree of electron *sharing* or covalency which accounts for the lattice energy discrepancy. The Se^{2-} ion is said to be **polarized**. This is shown in **Figure 10.4.**

The factors which increase polarization are:

- positive ion (cation): small size, high charge } Fajans' rules
- negative ion (anion): large size, high charge. }

So pure ionic bonding can be seen as one extreme of a continuum from pure ionic to pure covalent.

Figure 10.4 Polarization in zinc selenide

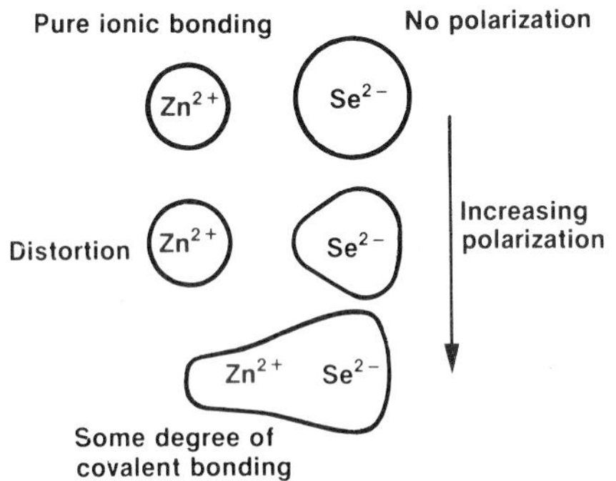

Properties of ionically bonded compounds

- The main property of ionic compounds is that they conduct electricity when molten but not when solid. This is because the ions which carry the current are not free to move in the solid state, but are in the liquid state.
- When a molten ionic compound does conduct electricity, it is decomposed at the electrodes. Aqueous solutions of ionic compounds also conduct electricity.
- To melt an ionic compound, the lattice energy (which is high) must be supplied and therefore ionic compounds have high melting points and high enthalpy changes of melting (fusion).
- Many ionic compounds dissolve in water. The energy required to break up the lattice is supplied from hydration by water molecules (see **section 16.3.4**). The water molecules also get between the ions and reduce the strength of their electrostatic attraction.

Figure 10.5 An ionic lattice

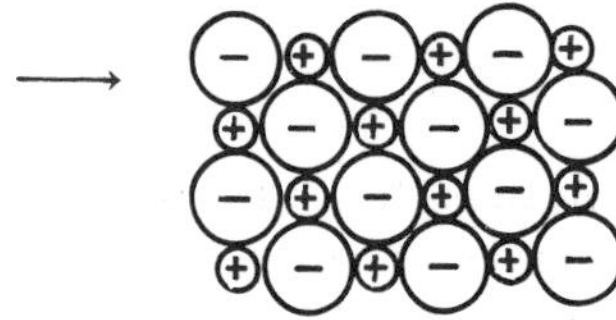

• Ionic compounds tend to be brittle, i.e. they shatter easily when given a firm blow. This is because they form a lattice of alternating positive and negative ions, **Figure 10.5.** A blow in the direction shown may move the ions sufficiently to produce contact between like charged ions, **Figure 10.6.** The repulsion between these ions makes the crystal fly apart, **Figure 10.7.**

10.2.2 Covalent bonding

Figure 10.6 A small displacement causes contact between like charged ions...

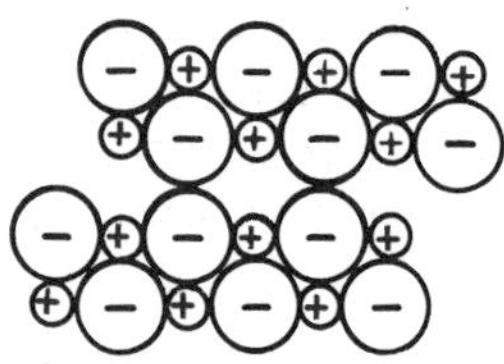

A simple picture of covalent bonding was developed in Chapter 3, which involved atoms sharing a pair or pairs of electrons in such a way that the atoms obtained a full outer shell. It is most simply described by 'dot cross' diagrams, for example water:

H × O × H

(outer electrons only drawn, as is usual).

This simple picture does not at first sight explain how the bond holds the atoms together. It also suggests that the electrons are always equally shared between the atoms and gives the impression that molecules are flat and two-dimensional. Neither of these is always the case. In this section we shall develop a more sophisticated model of covalent bonding which will enable us to understand the distribution of electrons in molecules and also to explain their shapes. We shall also look at the strength of covalent bonds in terms of the energy needed to pull atoms apart.

Figure 10.7 ...and the structure shatters

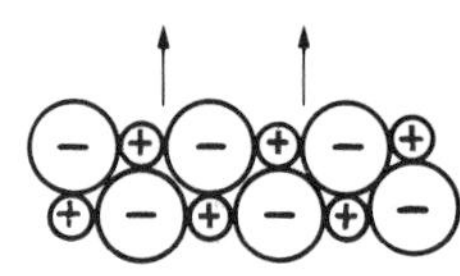

Shatters

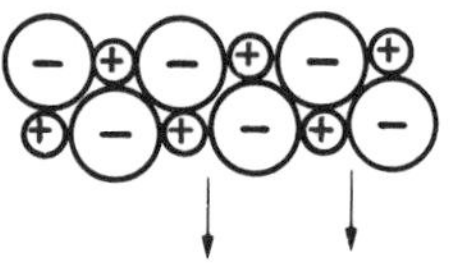

How are atoms held together in molecules?

Atoms are held together by electrostatic attractions within the molecules. The simplest example is hydrogen. The hydrogen molecule consists of two protons held together by a pair of electrons. The electrostatic forces are shown in **Figure 10.8.** The attractive forces are in blue, and the repulsive forces in brown. These forces just balance at a particular distance apart called the **bond length**.

Figure 10.8 Electrostatic forces within the hydrogen molecule

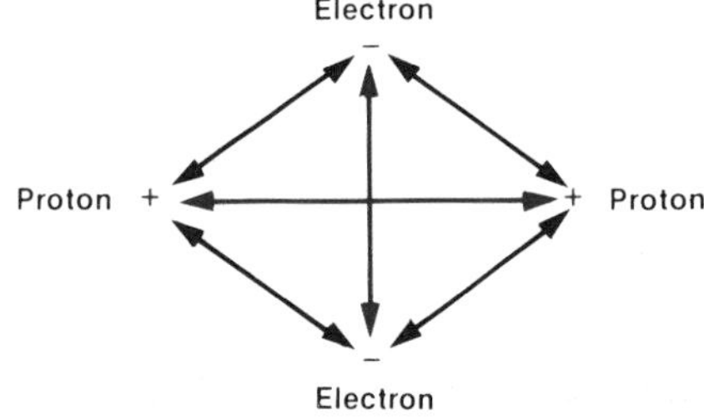

Dative covalent bonding

We have used the idea that atoms share electrons in order to attain a full outer shell. However, this is not always possible.

For example, the compounds beryllium chloride, $BeCl_2$ and boron trifluoride, BF_3, are shown below:

The beryllium in beryllium chloride has only four electrons in its outer shell and the boron in boron trifluoride, six. These are sometimes called **electron deficient compounds**.

One way in which species like this can obtain a full outer shell is by accepting a pair of electrons from a species with an unshared pair (or lone pair) of electrons. Ammonia, NH_3, is a good example, see margin.

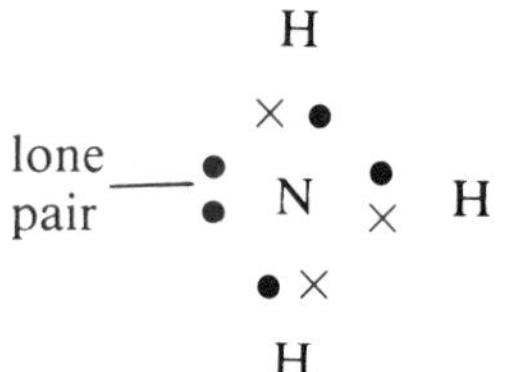

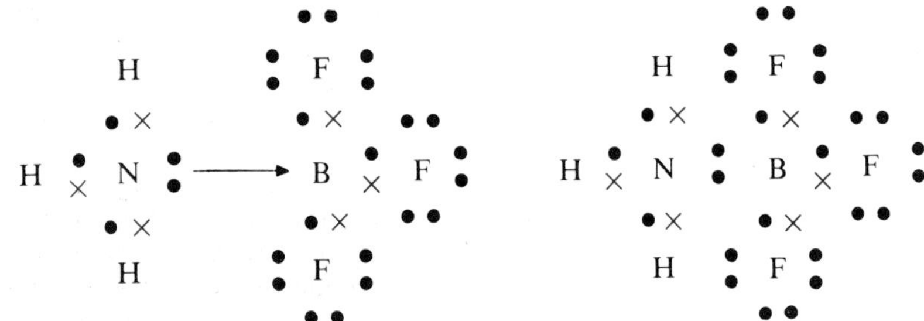

Figure 10.9 Representing single, double and triple bonds

—	•×	Single, e.g.	F — F
=	•×•×	Double, e.g.	O = O
≡	•×•×•×	Triple, e.g.	N ≡ N

This type of bonding, shown on the previous page, is called **dative** or **coordinate bonding**. 'Ordinary' covalent bonds are represented by a short line as a shorthand for an electron pair, **Figure 10.9.** Dative bonds are represented by an arrow showing the direction of donation.

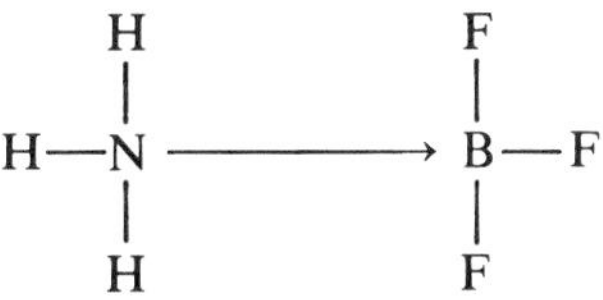

However, this is simply book keeping. Once established, a dative bond is identical in all respects to ordinary bonds between the same two atoms. For example in the ammonium ion, the nitrogen uses its lone pair to form a dative bond to an H^+ ion ('bare' proton with no electrons at all):

Ammonium ion

The ammonium ion is completely symmetrical and all the bonds are identical in strength, and length. The ammonium ion is also an example of a species with *covalently* bonded atoms, but with an overall charge, and is therefore an *ion.* Such species are often called **complex ions.** The OH^- ion is another example, see margin.

The electron • has come from the positive ion.

Dative bonds can also form part of a multiple bond, as in carbon monoxide:

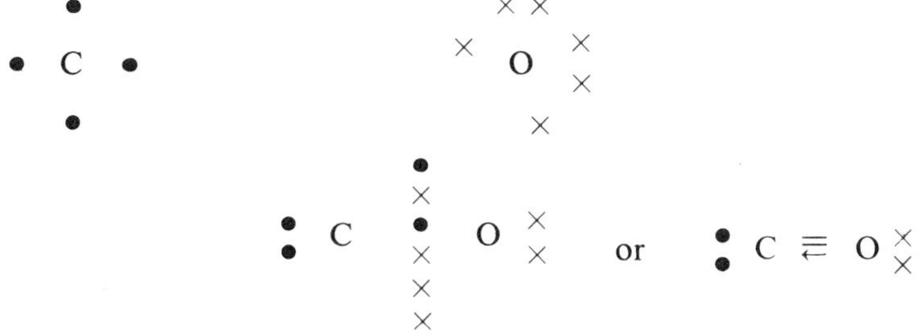

Because they are important in the chemistry of many species, lone pairs are often drawn on the formula as above.

Further examples of dot cross diagrams

Although as we will see, there are more sophisticated descriptions of bonding, dot cross diagrams are still very useful. Some examples are given below which include dative and multiple bonding. Only outer electrons are shown. This is acceptable as inner electrons are not involved in bonding.

ethanol

unshared pairs

Each atom has a full outer shell (hydrogen only needs 2 electrons)

ethene

double bond (4 electrons shared)

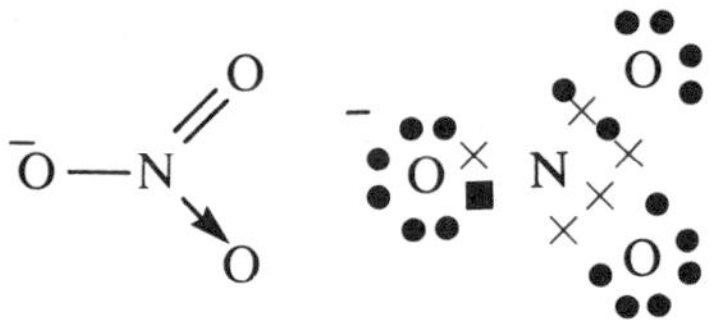

The electron ■ is the negative charge on the ion and came originally from a positive ion like Na^+. Since all electrons are identical it could appear anywhere, not just on this oxygen.

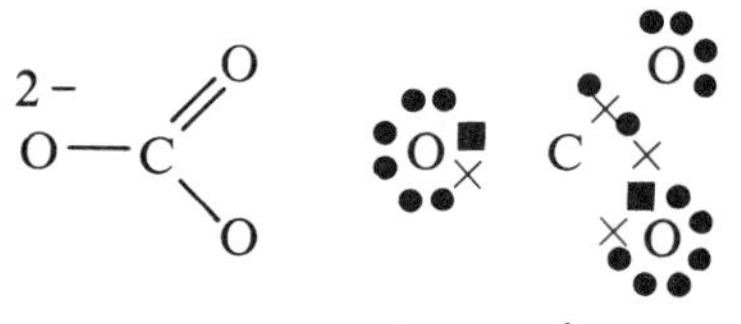

Both electrons ■ are from the positive ion

Figure 10.10 A molecular orbital picture of the hydrogen molecule, H_2

Figure 10.11 The shape of a σ orbital is unaffected by rotation about the axis shown

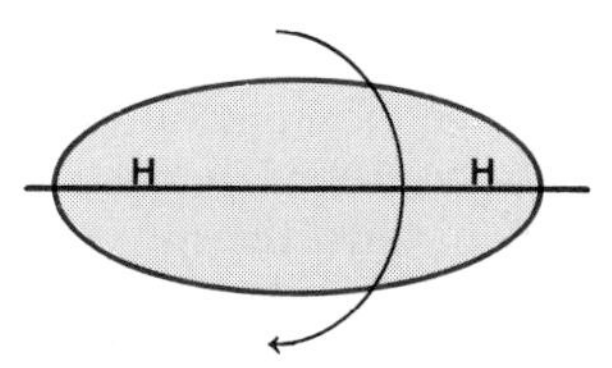

Figure 10.12 A molecular orbital picture of the bonding in hydrogen fluoride, HF

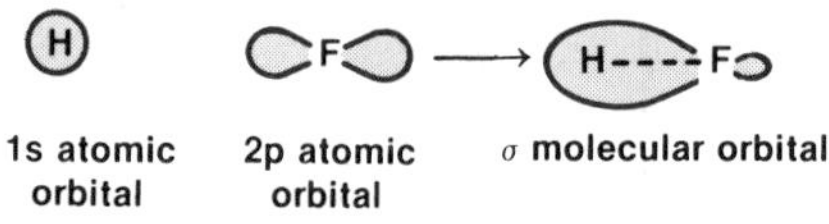

Figure 10.13 A molecular orbital picture of the bonding in fluorine, F_2

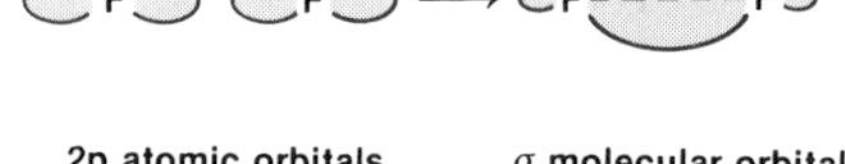

Figure 10.14 Formation of a π molecular orbital

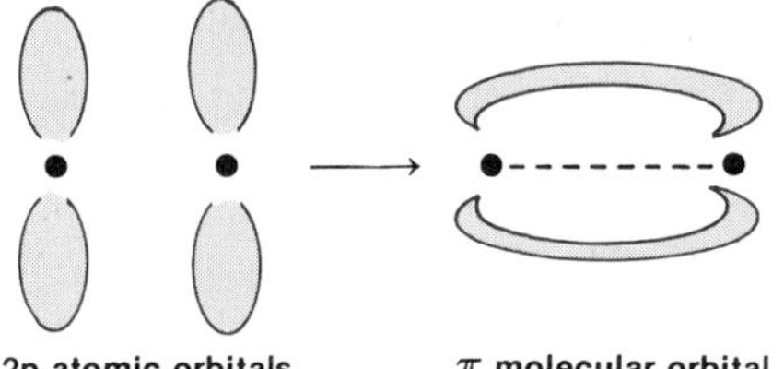

Figure 10.15 σ and π molecular orbitals in oxygen, O_2

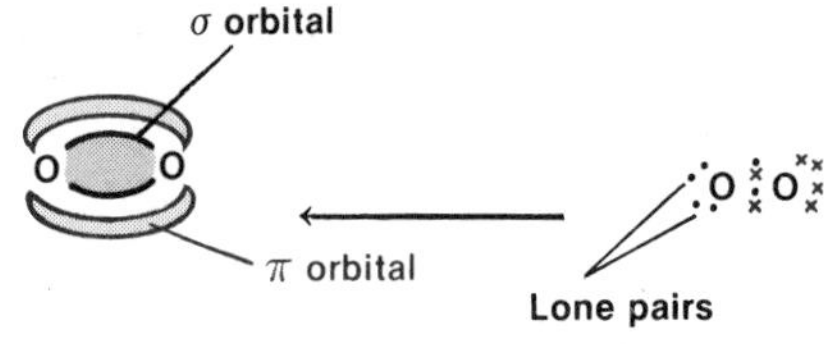

Molecular orbitals

A more sophisticated picture of covalent bonding considers not electrons but **orbitals**. Remember that an orbital is a volume of space where an electron charge is likely to be found. This model of covalent bonding uses the idea that an orbital from one atom overlaps with one from the other atom to form a new **molecular orbital**.

σ (SIGMA) ORBITALS

For example in hydrogen, H_2, the two 1s orbitals can overlap, as in **Figure 10.10.** The new molecular orbital shows a significant electron density *between* the nuclei.

Notice that the molecular orbital has rotational symmetry about a line joining the nuclei, **Figure 10.11.** Such molecular orbitals are called σ (sigma) orbitals. They may also be formed by overlap of an s and a p orbital as in hydrogen fluoride, HF, **Figure 10.12.** A pair of p orbitals can also overlap to give a σ orbital as in fluorine, F_2 **Figure 10.13.**

π (PI) ORBITALS

A pair of p orbitals can also overlap in a different way, **Figure 10.14.** Here the new orbital is called a π (pi) orbital. π orbitals produce electron density above and below a line joining the nuclei. They are less effective at holding the atoms together than σ orbitals. They only form where there is already a σ bond, but the $\sigma + \pi$ combination is usually less than twice as strong as the σ alone. Double bonds are composed of σ and π bonds, as in oxygen, O_2, **Figure 10.15.**

It is possible to form a triple bond—a σ and two π orbitals—as in nitrogen, N_2, **Figure 10.16.** Notice that there are electron pairs that take no part in the bonding. They are called **lone** or **unshared pairs** and exist in non-bonding orbitals.

Hybridization

This description of bonding starts by considering that on any atom, atomic orbitals of similar energy, e.g. 2s and 2p, can 'mix together' to produce

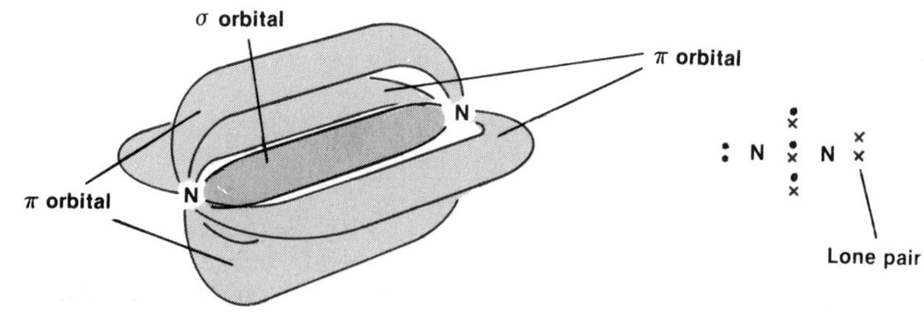

Figure 10.16 (right) Nitrogen, N_2, has one σ and two π molecular orbitals

the *same number* of new orbitals of a different shape. These **hybrid orbitals** can then form bonds by overlapping with orbitals on other atoms. This model helps to explain the shapes of the molecules that result.

For example, in carbon where the orbitals involved in bonding are the 2s and the three 2p orbitals, they may mix in three ways.

1. $s + 3 \times p \longrightarrow 2 \times sp + 2 \times p$
2. $s + 3 \times p \longrightarrow 3 \times sp^2 + 1 \times p$
3. $s + 3 \times p \longrightarrow 4 \times sp^3$

} Notice that the original four orbitals always give rise to four new ones

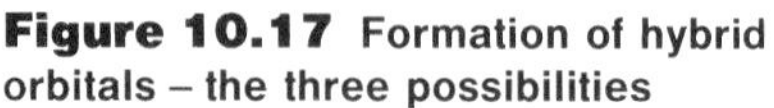
Figure 10.17 Formation of hybrid orbitals – the three possibilities

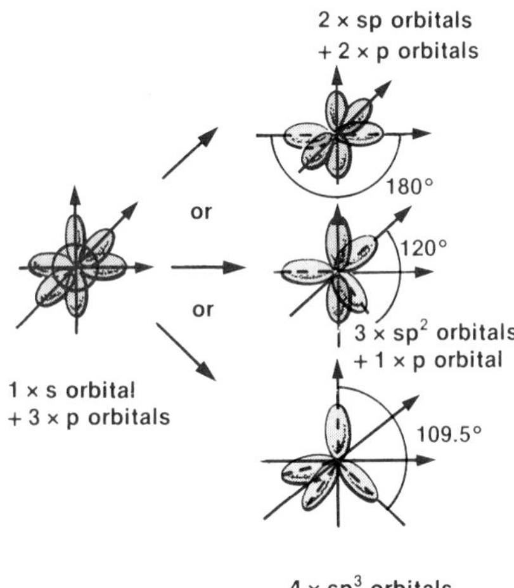

The shapes are illustrated in **Figure 10.17.** The new hybrid orbitals are identified by the symbols sp, sp^2 and sp^3, which indicate the proportions of s and p orbitals from which they were formed. The angles between the hybridized orbitals are important as they help to explain the shapes of molecules. The sp orbitals are at 180° to one another, the sp^2 at 120° and the sp^3 point to the corners of a tetrahedron (109.5°). This is particularly important in organic chemistry, see **section 26.3.**

These three explanations of bonding—dot cross diagrams, molecular orbitals and hybridization—are not contradictory. They represent different levels of sophistication. Chemists tend to use the simplest model which is adequate to deal with a particular situation. Thus we shall frequently use the dot cross method.

Bond polarity

In a molecule like fluorine, F_2, where both atoms are the same (a 'homonuclear diatomic' molecule), the electrons in the bond *must* be shared equally:

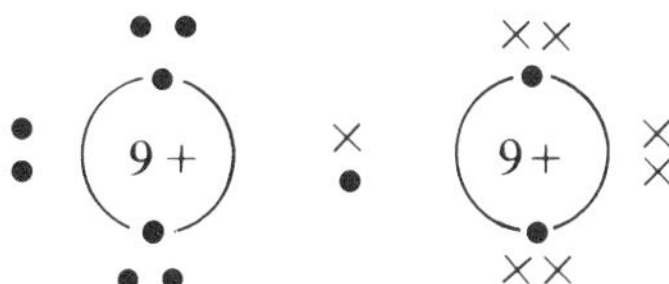

Both nuclei have a charge of 9+ but the inner electrons tend to shield or screen the nuclear charge so that the **effective nuclear charge** which is felt by the shared electrons is 9+ −2 = 7+.

So both 'feel' an effective nuclear charge of 7+ (9+ −2 inner electrons). However, in a diatomic molecule with *different* atoms, the sharing will not be equal. In hydrogen fluoride, HF, the shared electrons 'feel' one positive charge from the hydrogen (no shielding) and seven from the fluorine.

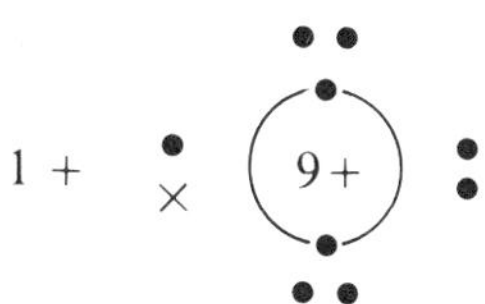

The electrons will be attracted closer to the fluorine than to the hydrogen, or to express it another way, the electron charge cloud or molecular orbital will be distorted towards the fluorine. This makes the fluorine end of the molecule relatively negative and the hydrogen end relatively positive.

This may be represented:

H F

showing the charge cloud distorted towards the fluorine, or more usually:

$$^{\delta+}H\text{—}F^{\delta}$$

+ or − represent one 'electron's worth' of charge. δ^+ and δ^- represent a small charge of less than one 'electron's worth'. Molecules like this are said to be **polar** or to have a **dipole.** Another way of looking at this is to say that the H—F bond is covalent, but has some ionic character. Compare this with the situation described earlier of ionic bonds with some covalent character.

H—F has been calculated to have 45% ionic character.

Electronegativity

Because of its greater electron attracting ability, fluorine is said to be more **electronegative** than hydrogen. Electronegative atoms are those whose outer electrons 'feel' a large effective nuclear charge after allowing for the shielding effect of the inner electrons. They therefore attract electrons toward themselves. As we go across a period in the Periodic Table, the effective nuclear charge increases from Group I to Group VII, see **Table 10.3.**

Table 10.3 Effective nuclear charge increases across a period

	Li	Be	B	C	N	O	F	
Nuclear charge	3	4	5	6	7	8	9	All have a shielding factor of 2
Effective nuclear charge	1	2	3	4	5	6	7	(2 electrons in the inner shell)
	—increasing electronegativity ⟶							

On descending a group in the Periodic Table the effective nuclear charge remains the same but the outer electrons get further away from the nucleus, so electronegativity decreases.

Thus the most electronegative atoms are found at the top right-hand corner of the Periodic Table (ignoring the inert gases which form few compounds). The most electronegative atoms are fluorine, oxygen, nitrogen followed by chlorine.

A scale of electronegativity values has been proposed by the US chemist Linus Pauling. The greater the number, the more electronegative the atom. Some examples are shown in **Table 10.4.** It is the *difference* in electronegativities of two atoms which tells us how polar the bond between them is. A difference of zero corresponds to pure covalent bonding. A difference of more than around 2.1 is usually considered to be ionic bonding.

Table 10.4 Trends in electronegativity

Increasing electronegativity ⟶							
Li	Be	B	C	N	O	F ↑	
1.0	1.5	2.0	2.5	3.0	3.5	4.0	
						Cl	Increasing
						3.0	electronegativity
						Br	
						2.8	
(H = 2.1)							

For example in lithium fluoride, Li^+F^-, the difference is $4.0 - 1.0 = 3.0$.

Between 0 and 2.1, the bonds are considered to be polar covalent bonds, i.e. covalent with some degree of ionic character.

For example in hydrogen fluoride the difference is $4.0 - 2.1 = 1.9$, a highly polar bond, but in a H—C bond the difference is $2.5 - 2.1 = 0.4$, almost non-polar.

Figure 10.18 The effect of an electric field on molecules with dipoles

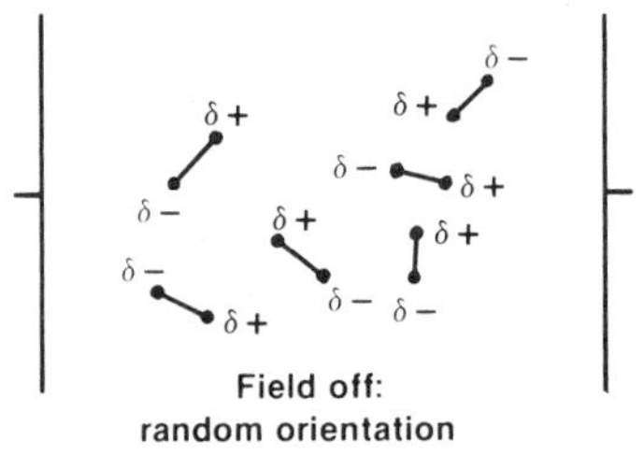

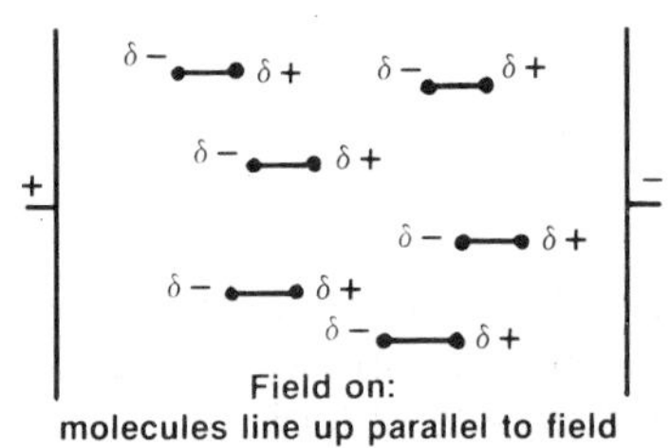

Dipole moment

Polarity is the property of a particular bond. Molecules as a whole may have a **dipole moment**. This means that if the molecules are placed in an electric field, they will tend to flip, as shown in **Figure 10.18**, until they lie with their negative ends towards the positive plate and their positive ends towards the negative plate. In molecules with more than one polar bond, the effects of each bond may cancel, to leave a molecule with no dipole moment.

For example, carbon dioxide:

$${}^{\delta-}O = C^{\delta+} = O^{\delta-} \qquad \text{(linear) the dipoles cancel}$$

tetrachloromethane:

$$\begin{array}{c} Cl^{\delta-} \\ | \\ C^{\delta+} \\ {}^{\delta-}Cl \quad Cl^{\delta-} \quad Cl^{\delta-} \end{array}$$

(tetrahedral) the dipoles cancel

water:

$$\begin{array}{c} O^{\delta-} \\ {}^{\delta+}H \diagup \quad \diagdown H^{\delta+} \end{array}$$

(angular) the dipoles do *not* cancel.

Figure 10.19 A water molecule is attracted by a charged rod

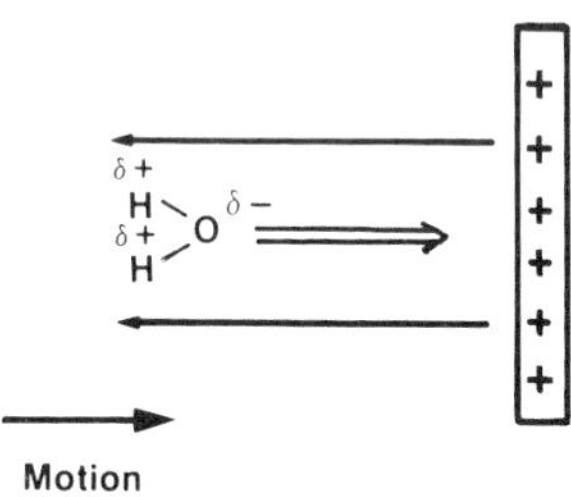

A stream of water is deflected by a charged plastic rod

The dipole moment of water explains why a stream of water bends when a plastic rod, charged by being rubbed, is brought near it, **Figure 10.19** and photograph. The rod is charged positively by friction. The water molecules flip, so that their negative ends are towards the rod. Now since the negative ends which are attracted are closer to the rod than the the positive ends which repel, the attraction outweighs the repulsion, and the molecules move towards the rod. Would you observe anything different if a negatively charged rod were used?

Bond lengths

It is not easy to measure the size of atoms. Electron clouds are so diffuse (spread out) that it is difficult to say where the atom stops. Furthermore, isolated atoms are not easy to come by. What can be measured with some accuracy is the distance apart of two nuclei—the centres of the atoms—which are covalently bonded together. This distance is usually called the **bond length.** These are usually in the range 0.1–0.2 nm. When two identical atoms are covalently bonded together, half the bond length is called the **covalent radius**, shown in **Figure 10.20.**

Figure 10.20 Bond length and covalent radius

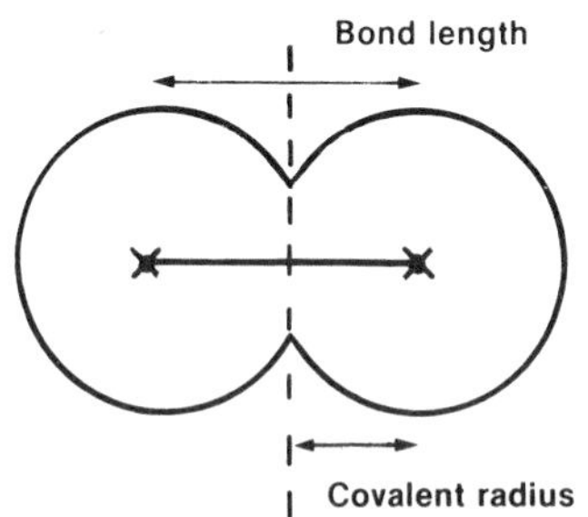

BOND LENGTH AND BOND ORDER

The **order** of a bond is whether it is single, double or triple. In the case of atoms which can form multiple bonds, a relationship between the bond length and bond order can be seen for bonds between the same two atoms. This is shown in **Table 10.5** for carbon–carbon bonds:

Table 10.5 Relationship between bond length and bond order

Bond order		Bond length/nm
1	C—C	0.154
2	C═C	0.134
3	C≡C	0.120

This shows that multiple bonds are shorter—the extra attraction caused by more electrons being shared pulls the atoms closer before the repulsion of the electron clouds intervenes. This relationship allows bond orders to be determined experimentally.

Delocalization

If we attempt to draw a dot cross diagram of the bonding of nitric acid, HNO_3, whose *skeleton* is shown below (the dashed lines showing simply which atom is bonded to which):

H - - - O - - - N (- - - O) (- - - O)

we get this result:

[dot cross diagram of HNO_3] i.e. using the usual conventions H—O—N(→O)(═O)

However, we could just as easily have drawn:

```
          ××        ××
   ××    •× O ×
H •× O •× N •××
   ××    •× O ×
          ××
```

i.e. H—O—N(=O)→O

Studies of the nitric acid molecule have measured the bond lengths shown:

H- - -O- - -N (0.141 nm) ; N- - -O 0.122 nm ; N- - -O 0.122 nm

The usual N—O bond length is 0.136 nm, and that for N=O is 0.114 nm. Thus the N—OH bond is approximately the expected length for a single bond, but neither of the other N—O bonds seems to be double. In fact, both bonds have a length midway between that expected for a double and a single bond and both are identical in length and in bond order.

This can be explained if we think about the orbitals. The nitrogen atom is hybridized sp^2, producing a p orbital and three sp^2 orbitals, as shown in **Figure 10.21.**

Figure 10.21 A nitrogen atom: hybridized sp^2

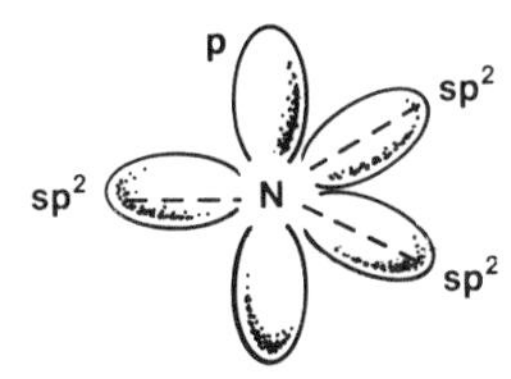

The sp^2 orbitals can overlap with p orbitals on the oxygen to form σ bonds. This leaves p orbitals on each oxygen able to overlap with the p orbital on the nitrogen, to form a π orbital. Since both oxygens are identical, there is no reason for overlap to occur with one rather than the other and overlap occurs with *both.* This produces an orbital which extends across all three atoms. This is called a **delocalized** π orbital, shown in **Figure 10.22.** Although it covers three atoms, there are only two electrons in this orbital. Thus the two N—O bonds are neither double nor single, but come somewhere in between, as the bond lengths indicate. This type of delocalization is shown on structural formulae as in **Figure 10.23.** Whenever delocalization occurs, the resulting molecule is always more stable (of lower energy) than either of the alternative structures with a double and a single bond.

Figure 10.22 The delocalized π orbital in nitric acid

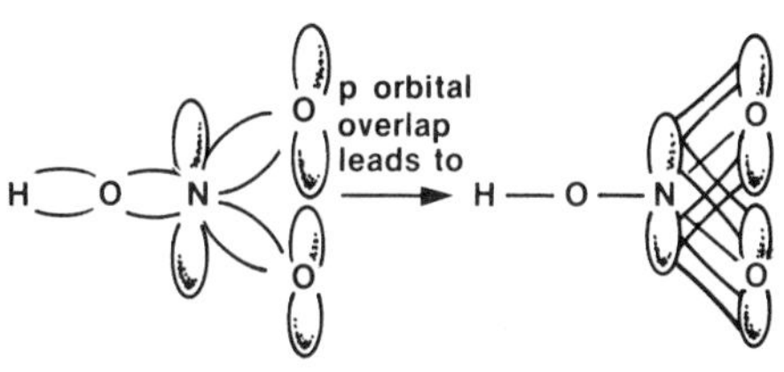

Another example where delocalization is a feature is the benzene ring, C_6H_6, which is dealt with in detail in **section 29.2.**

Figure 10.23 Representing delocalized π bonding on the structural formula of nitric acid

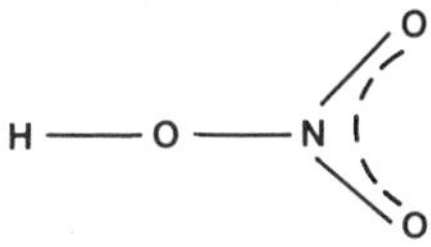

Bond energies

How strong are covalent bonds? As we have seen in section 9.5, their strengths are usually measured by the amount of energy which must be put in to completely separate the two atoms concerned. For example:

$$H_2(g) \longrightarrow 2H(g) \quad \Delta H^{\ominus} = +436\,\text{kJ mol}^{-1}$$

The bond energies given in tables of data are averages taken over a number of compounds which may have slightly different energies. Thus C—H is usually given as $413\,\text{kJ mol}^{-1}$.

As you might expect, there is a relationship between bond energy and bond order in bonds between the same two atoms. This is shown in **Table 10.6** for carbon–carbon bonds.

Table 10.6 Bond order, bond length and bond energy

Bond	Order	Length/nm	Energy/kJ mol^{-1}
C—C	1	0.154	+347
C=C	2	0.134	+612
C≡C	3	0.120	+838

Notice:

1. The relationship between length and energy. Shorter bonds are stronger.
2. A double bond is not twice as strong as a single bond. A single bond is a σ bond, a double bond is a $\sigma + \pi$ bond so the difference in energy represents the π bond energy: $612 - 347 = 265\,\text{kJ mol}^{-1}$, significantly less than the σ bond energy.

It is important to note that it is easier to break one of the bonds of a double bond than it is to break a single bond, though more difficult to break both bonds of a double bond than to break a single bond.

Properties of covalently bonded compounds

Covalently bonded compounds do not conduct electricity in general, whatever their state. They tend not to dissolve in or mix with water unless they have relatively polar bonds, as in sugars for example.

10.2.3 Metallic bonding

Only a simple picture of metallic bonding will be given here. Metals are elements which can easily lose up to three electrons therefore forming positive ions, for example:

Na $1s^2 2s^2 2p^6 3s^1$ or 2, 8, 1 Al $1s^2 2s^2 2p^6 3s^2 3p^1$ or 2, 8, 3

Figure 10.24 Metallic bonding

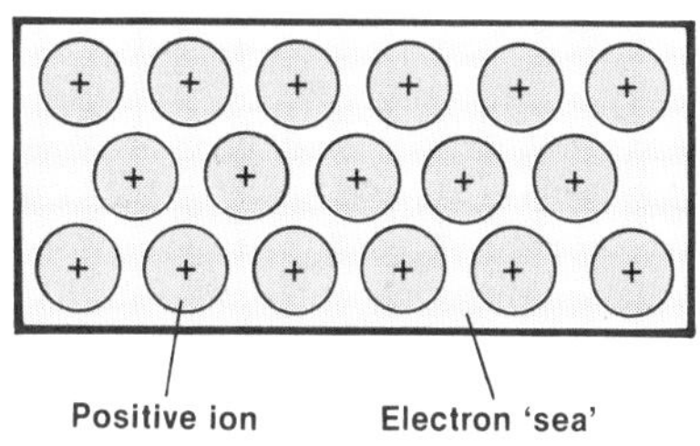

Transition metals have more complex electronic structures but still tend to give away electrons easily.

Metals exist as a lattice of positive ions in a freely moving 'sea' or 'cloud' of electrons. This is shown in **Figure 10.24.** The number of electrons in the 'sea' depends on how many electrons have been lost by each metal atom. This 'sea' may be thought of as a giant orbital, delocalized over the whole metal structure. The 'sea' of mobile electrons is responsible for the good electrical conduction of metals: an electron from the negative terminal of the supply may join the electron cloud at one end of a metal wire and at the same time a different electron leaves at the positive terminal.

Figure 10.25 The valence and conduction bands in conductors, semiconductors and insulators

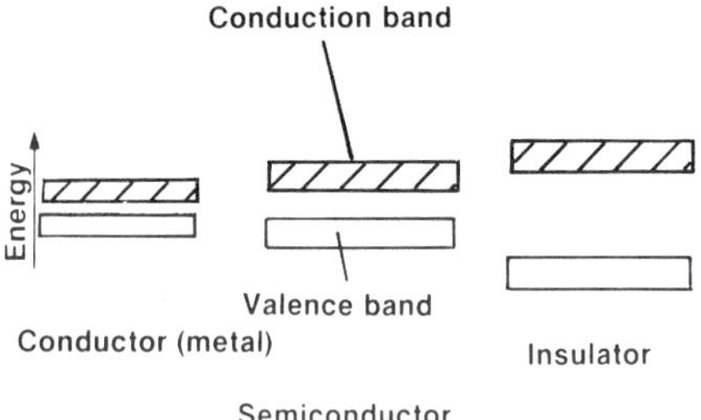

Conductors, semiconductors and insulators

A molecular orbital picture of solids leads to the idea that there are two groups of energy levels or orbitals. The lower is called the **valence band** and the higher the **conduction band**, see **Figure 10.25.** If the conduction band is partly filled with electrons, the electrons can move throughout the crystal (this is the electron 'sea' in metals). The difference between conductors, semiconductors and insulators depends on the energy gap between the two bands. In conductors, the two bands overlap, so the conduction band is always partly filled. In semiconductors, the bands are close together so that a few high-energy electrons can jump into the conduction band. In insulators the gap between the bands is large and no electrons exist in the conduction band. Increasing the temperature of a semiconductor improves its conductivity as more electrons can gain enough energy to attain the conduction band.

Figure 10.26 The malleability and ductility of metals

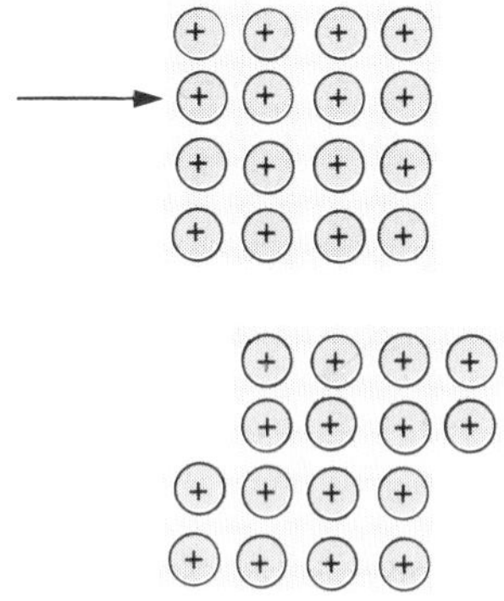

Metallic radius

The distance between metal atoms in a solid metallic lattice can be measured accurately by X-ray diffraction, see **section 19.2.1.** Half of this distance is called the **metallic radius**—a measure of the size of the metal atoms.

Properties of metals

Metals are good conductors of electricity due to their 'sea' of electrons. These electrons are also responsible for the high thermal conductivities of metals—they conduct heat well.

The metallic lattice also explains the strength, ductility (ability to be drawn out into wire) and malleability (the ability to be dented) of metals. The last two are illustrated in **Figure 10.26.** After a small distortion, each metal ion is still in exactly the same environment as before so the new shape is retained. Contrast this with the brittleness of ionic compounds (see page 97).

10.3 Structure

This section is concerned with the geometries of the arrangements of atoms in covalent molecules, giant covalent structures, giant ionic structures, and metals.

Note: when drawing diagrams of structures, it is difficult to represent three-dimensional structures in two-dimensional drawings. Two main

Figure 10.27 Close packing of spheres in two dimensions

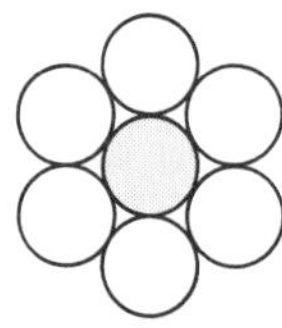

Figure 10.28 Adding another close-packed layer, below the first layer

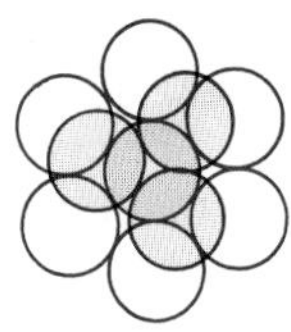

Figure 10.29 The two close-packed arrangements

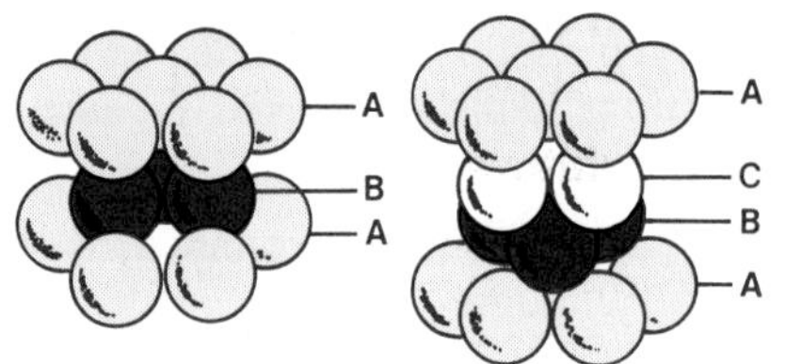

Figure 10.30 The two close-packed arrangements showing hexagonal and cubic shapes

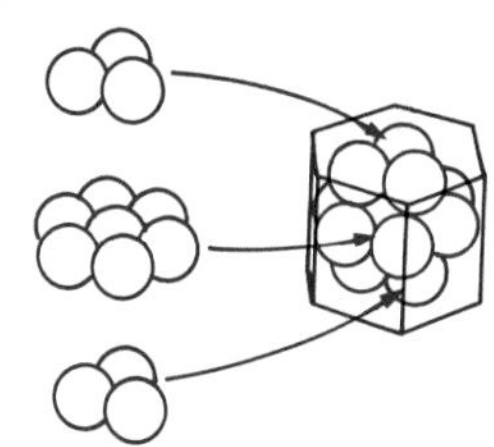

(a) ABA - hexagonal close packing

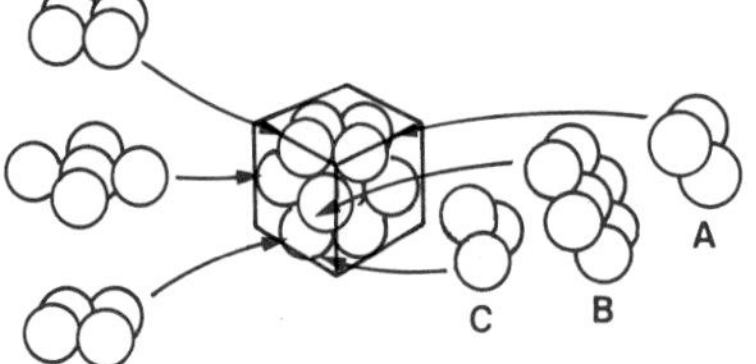

(b) ABC - cubic close packing

types of diagrams are used: 'space-filling' as in **Figure 10.31a** and 'expanded' as in **Figure 10.31b** (overleaf).

The space-filling diagrams are better representations but expanded diagrams often make it easier to see geometries. In expanded diagrams the lines are to make the geometry clearer and do *not* necessarily represent covalent bonds.

10.3.1 The structures of metals

You may find it helpful to have some spherical objects available while reading this section—a bag of marbles would be suitable.

In a metal, all the units which make up the lattice are the same size which simplifies the geometries considerably. X-ray diffraction, see **section 19.2.1**, shows that the ions are spherical.

Close packing

In many metals, the ions pack as closely together as they possibly can. This of course still leaves some gaps, as you will realize if you think about a box of tennis balls.

In two dimensions a sphere can be surrounded by a maximum of six spheres which touch it as shown in **Figure 10.27.** Extending to three dimensions, a further layer can be placed *below* this one as shown in brown in **Figure 10.28.**

There are now *two* ways of placing a further layer *above* the original one:

a) directly above the second one, called an ABA arrangement
b) twisted through 60° from position (a), called an ABC arrangement.

Both are shown in **Figure 10.29.** In both these arrangements the central sphere is surrounded by 12 nearest neighbours which touch it. This atom is said to have a **coordination number** of 12. In both cases the spheres occupy 74% of the space.

Arrangement (a), ABA, is called **hexagonal close packing** and arrangement (b), ABC, is called **face-centred cubic close packing**, or just **cubic close packing**. The hexagonal and cubic aspects of the symmetry can be seen in **Figure 10.30** but inspection of models would be most useful. Around 50 metals crystallize in one or other of these arrangements.

The body-centered cubic structure

This is the other common packing geometry for metals. It is not a close-packed arrangement, only 68% of the space being filled. It is best visualized by thinking of a sphere situated in the centre of a cube surrounded by eight other spheres, one at each corner of the cube, **Figure 10.31** (overleaf). The coordination number is 8.

Unit cells

The unit cell of a structural arrangement is the smallest unit of it which contains enough information to generate the whole structure by repeating itself. **Figure 10.32** (overleaf) shows some unit cells.

A

INTERSTITIAL COMPOUNDS

In a close-packed metal structure there are many 'holes'. Some of these holes are of the right size to trap small molecules such as hydrogen. The holes in the lattice are called interstices (pronounced in-ter-stiss-ees), and the resulting substances are called interstitial hydrides. They are not compounds in the usual sense of the word, as their formulae can vary from MH_0 to MH_2 depending on the hydrogen pressure. Heating the hydrides can make them release their hydrogen. Such compounds are under investigation for possible use for storage of hydrogen. One application would be as fuel storage for hydrogen-powered vehicles of the future. Such a system seems likely to be safer than storage under pressure in cylinders.

Figure 10.31 The body-centred cubic structure shown (a) space-filling and (b) expanded

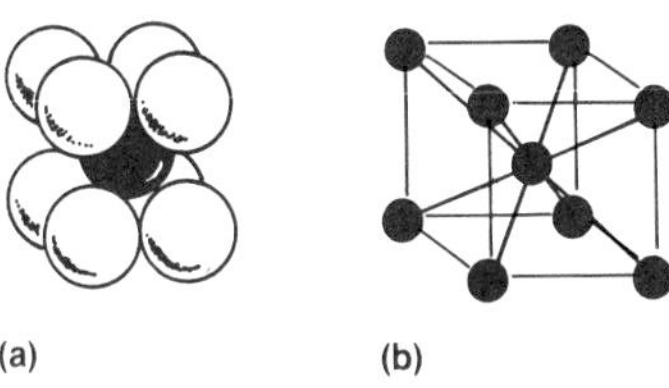

Figure 10.32 Some unit cells

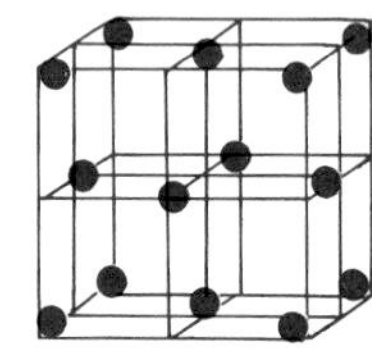

(a) Face-centred cubic close-packed unit cell

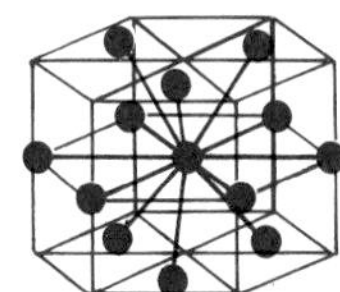

(b) Hexagonal close-packed unit cell

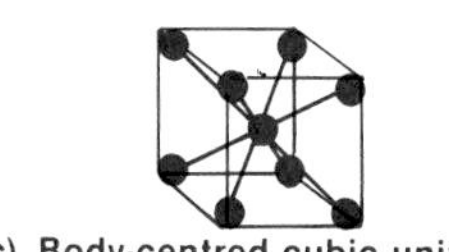

(c) Body-centred cubic unit cell

Density

The density (mass per unit volume) of a metal depends on three factors: the atomic mass; the size of the atom, i.e. the metallic radius; and the type of packing. Smaller atoms lead to a higher density, heavier atoms lead to a higher density and closer packing leads to a higher density.

10.3.2 The structures of ionic compounds

The situation here is more complex than in metals as in any ionic compound there are at least two different ions of opposite charge and of different size. These may be in the ratio 1:1, 1:2, 2:1, etc. Only 1:1 compounds will be considered here in any detail.

The particular geometry which is adopted depends on the relative sizes or **radius ratio** of the cations (+) to anions (−). The essential point to grasp is that positive ions will be surrounded by as many negative ions as possible, but that ions of the same charge cannot touch. It is worth remembering that in general positive ions, which have *lost* electrons, are smaller than negative ions, which have *gained* them. 1:1 ionic compounds adopt one of three structures which are referred to by the names of representative compounds. These are:

1. the **caesium chloride** structure where each cation is surrounded by eight anions and each anion by eight cations. This is referred to as 8:8 coordination
2. the **sodium chloride** structure where each cation is surrounded by six anions and vice versa. This is 6:6 coordination
3. the **zinc blende** structure—4:4 coordination.

The caesium chloride structure

This is illustrated in **Figure 10.33.** Notice the similarity between this and the body-centred cubic structure. Think of the Cs^+ ion as being at the centre of a cube. This is surrounded by eight Cl^- ions situated at the corners of that cube. The expanded diagram shows the geometry, but the space-filling one is more realistic as it shows the positive and negative ions touching.

Figure 10.33 (right) The caesium chloride structure shown (a) space-filling and (b) expanded

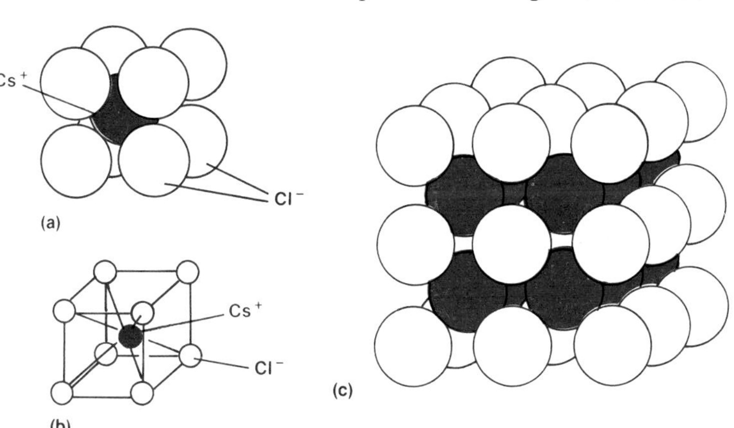

This structure is quite stable as the negative ions do not touch each other. This is because of the large size of the Cs^+ ion. However, imagine the positive ion shrinking. The structure would remain stable until the negative ions started to touch. If the positive ion then shrunk further, fewer than eight negative ions could fit round the positive ion and a different structure would have to be adopted.

Figure 10.34 The sodium chloride structure

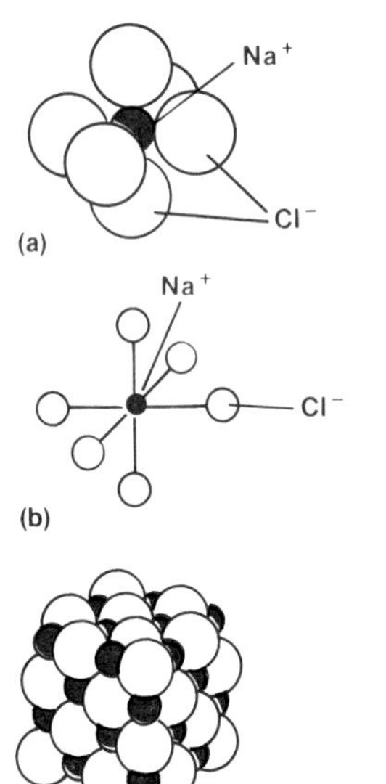

The sodium chloride structure

Here each cation is surrounded by six anions and vice versa, as shown in **Figure 10.34.** The negative ions form an octahedron with the positive ion at the centre. Again, imagine the positive ion shrinking until the negative ions touch, at which point the structure is no longer stable.

Figure 10.35 The zinc blende structure

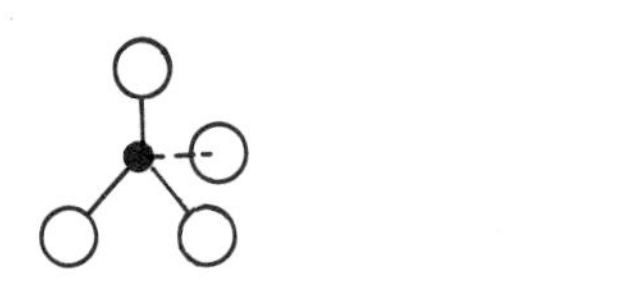

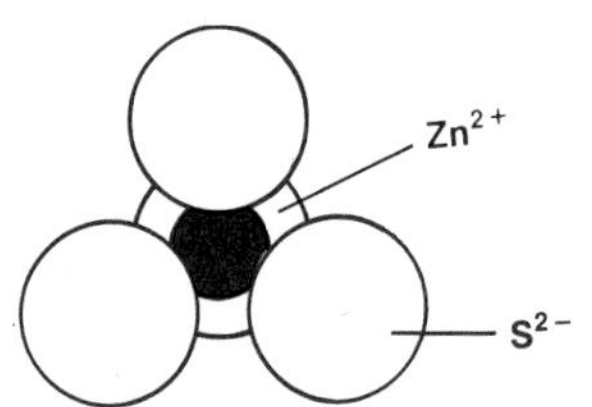

The zinc blende structure

Zinc blende is a form of zinc sulphide, $Zn^{2+} + S^{2-}$. Here the anion is at the centre of a tetrahedron of cations and vice versa, as shown in **Figure 10.35.**

The fluorite structure

Fluorite is calcium fluoride, $Ca^{2+} + 2F^-$. This is a 1:2 structure—the only one we shall mention. This structure can be thought of as based on the caesium chloride structure but with half of the positive ions missing, see **Figure 10.36.** Look at the central F^- ion. You should be able to see four nearest neighbour Ca^{2+} ions arranged tetrahedrally around it. Each Ca^{2+} is the centre of a cube with an F^- ion at each corner (totalling eight nearest neighbours), i.e. in exactly the same environment as a Cs^+ in Cs^+Cl^-. The structure is therefore sometimes referred to as 8:4 coordination.

Figure 10.36 The fluorite structure

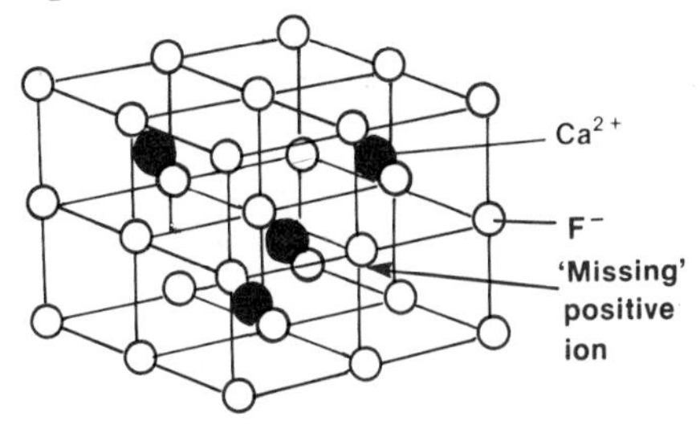

10.3.3 Shapes of covalent molecules

Shapes of simple molecules can be predicted by the Sidgwick–Powell or electron pair repulsion theory. The essential idea is that groups of electrons around an atom (such as a pair in the same orbital) will repel each other and therefore take up positions in space as far apart as possible. A group of electrons may be a shared pair forming a single bond, a group of four in a double bond, a group of six forming a triple bond or an unshared pair. If there are two groups of electrons, as for example in beryllium chloride, they will be 180° apart forming a **linear** molecule:

Two groups of electrons

Cl × Be × Cl   180°   Cl — Be — Cl

linear

Three groups of electrons, for example in boron trifluoride, will arrange themselves at 120° to each other. The resulting molecule is flat and described as **trigonal planar** in shape:

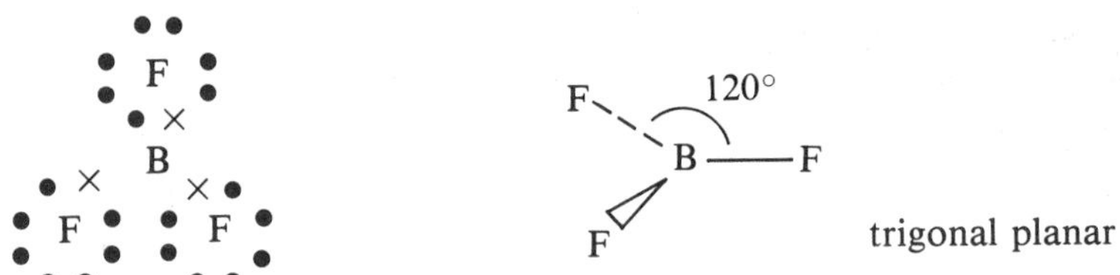

Four groups of electrons, as for example in methane, will arrange themselves **tetrahedrally** (pointing towards the four corners of a tetrahedron). The angle here is 109.5°. Remember this is a three-dimensional, not flat, arrangement.

H C H H H   109.5°   tetrahedral

Note: In three-dimensional representations of molecules ▷ is used to represent a bond coming out of the paper and a dashed line to represent one going into the paper, away from the reader.

The above three shapes represent the angles of sp, sp^2 and sp^3 hybrid orbitals respectively, see page 100. For elements in Period 3, there may be more than eight electrons (four pairs) in the outer shell.

Five groups of electrons can lead to two shapes—a **trigonal bipyramid** or a **square-based pyramid, Figure 10.37**. Six groups of electrons lead to an **octahedral** shape where all angles are 90°, **Figure 10.38**. Notice that an octahedron is so named because of its eight *faces*. We are interested in the fact that it has six *points*.

Figure 10.37 The two possible shapes for five coordination

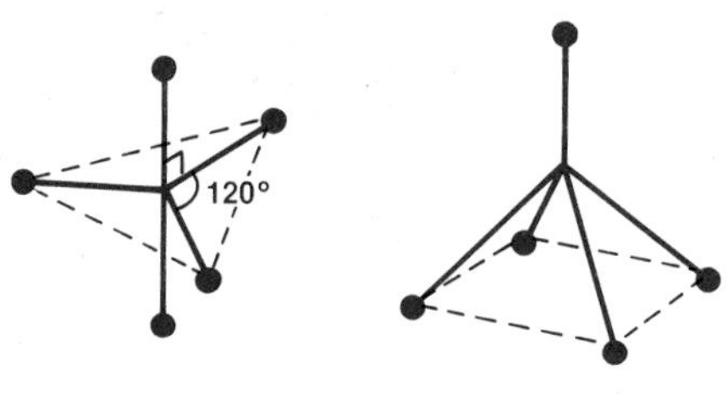

Trigonal pyramid   Square-based pyramid

Figure 10.38 Six coordination gives an octahedral shape

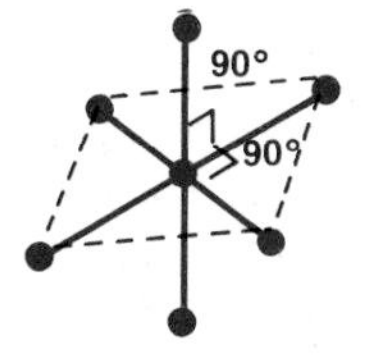

Lone pairs

The previous examples are fairly straightforward but considerable confusion can be caused by molecules with unshared pairs of electrons, for example water:

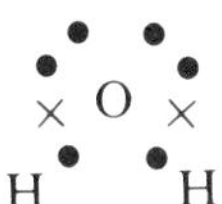

There are four groups of electrons so the shape is based on a tetrahedron, but two of the 'arms' of the tetrahedron are lone pairs and the resulting molecular shape is angular:

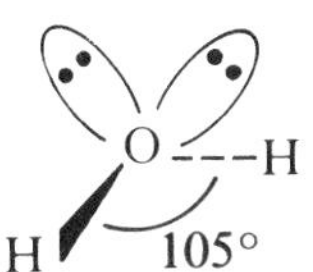

If the lone pair orbitals are drawn in, they are less likely to be forgotten. The angle of a perfect tetrahedron is 109.5° but lone pairs are closer to the oxygen than shared pairs (which are also attracted by the other nucleus). Therefore lone pairs repel more effectively than shared pairs, and 'squeeze' the H—O—H angle together. An approximate rule of thumb is a 2° reduction per lone pair, so the bond angle in water is approximately 105°.

In ammonia:

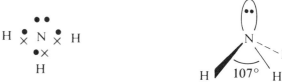

four groups of electrons mean the shape is based on a tetrahedron but with only three 'arms'. The bond angle is approximately 107°. The shape is often described as pyramidal.

Multiple bonds

These can also cause confusion. For example in methanal:

H C O   H C=O H

three groups of electrons mean that the shape is based on trigonal planar (the double bond counts as *one* group). However the group of four electrons in the double bond repels more than the groups of two in the single bonds, thus squeezing down the H—C—H angle to less than the 120° of a perfect trigonal shape.

In carbon dioxide, there are two groups of electrons therefore the molecule is linear:

More complex molecules

This treatment can be extended to molecules with more than one centre by treating each centre separately. For example, in methanol:

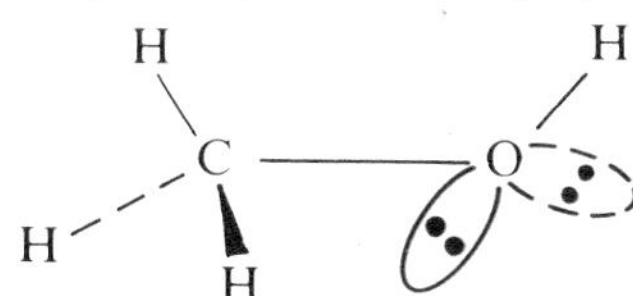

the carbon, which has four groups of electrons, is tetrahedral. There are also four groups of electrons around the oxygen atom but as two of these are lone pairs, the shape is angular with the C—O—H angle rather less than 109.5°.

Model of methanol molecule

Giant covalent structures

Giant covalent structures differ from giant ionic and metallic ones in that covalent bonds are directional and act between specific atoms, while the electrostatic forces that hold together the other structures act equally in all directions.

Only three detailed structures will be mentioned—examples of a three-dimensional giant structure, a two-dimensional one and a chain.

Figure 10.39 The structure of diamond

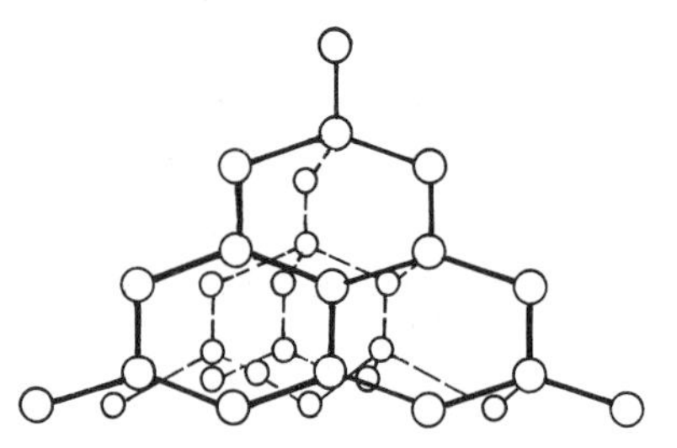

DIAMOND

This is a three-dimensional structure. Diamond is a form of pure carbon, in which the carbon atoms form four covalent bonds with other carbon atoms through four sp^3 hybrid orbitals. The resulting three-dimensional structure, **Figure 10.39**, is very rigid leading to diamond being one of the hardest substances known. Diamond also has a very high melting point of 4000 K. Both its hardness and its high melting point are caused by the need to break many covalent bonds to break the structure. As well as the well-known use in jewellery, diamonds are used industrially to tip drills because of their great hardness. Synthetic industrial diamonds can be made from graphite (see below) using temperatures of 1500 K and pressures of 10^7 kPa (see box *Artificial diamonds*, **section 24.3**).

Figure 10.40 The structure of graphite

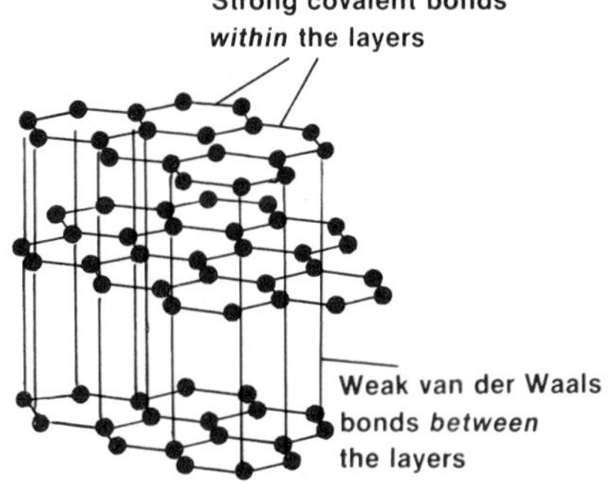

GRAPHITE

This is a two-dimensional structure. Graphite is also a form of pure carbon, differing from diamond only in the arrangement of atoms. The two are **allotropes**. The contrasting properties of diamond and graphite provide a dramatic illustration of the effects of structure on physical properties. The carbon atoms in graphite form bonds through sp^2 hybridized orbitals, giving three sigma bonds in the same plane. p orbitals also provide a delocalized π bonding system. Electrons can move through this so that graphite conducts electricity to some extent. This leads to two-dimensional layers of atoms, **Figure 10.40.** The bonding between the layers is weak (van der Waals—see **section 16.2.2**), allowing the layers to slide easily across one another. This leads to the use of graphite as a non-oily lubricant. It is also familiar as pencil 'lead', the layers flaking off onto the paper.

Figure 10.41 The chain structure of solid beryllium chloride

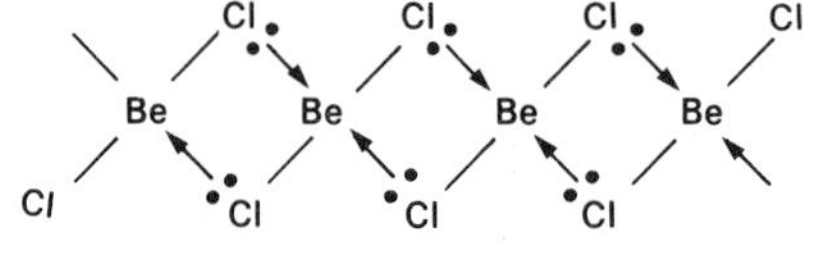

BERYLLIUM CHLORIDE

This exists as $BeCl_2$ molecules in the gas phase, but in the solid forms chains in which lone pairs on the chlorine atoms form dative covalent bonds to beryllium. Thus the beryllium ends up with a full outer shell of electrons and a chain structure results, **Figure 10.41.**

THE STRUCTURE AND PROPERTIES OF PYROGRAPHITE

Graphite's structure consists of layers of carbon atoms arranged in hexagons joined by covalent bonding. The layers are held together by much weaker van der Waals' bonds. It is not surprising therefore that the properties of graphite vary considerably depending on whether they are measured along the planes of the layers or at right angles to them. For example, graphite conducts heat 80 times and electricity 1000 times better along the planes of the layers than across them. It is also four times stronger along the planes than across them.

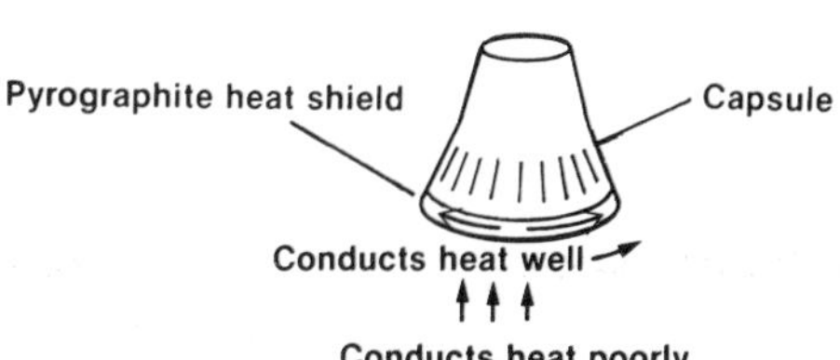

The difference in thermal conductivity with direction has led to the use of pyrographite (a crystalline form of graphite where all the crystals are oriented in the same direction) for the heat shields of spacecraft re-entry modules. Re-entry to the atmosphere generates a great deal of heat by friction. The pyrographite is oriented so that it conducts heat poorly into the capsule, thus protecting the occupants. The heat is conducted away from the hotter zones along the graphite planes.

10.4 Summary

- Born–Haber cycles summarize the energy terms involved in the formation of ionic compounds from their elements. The most significant terms are the ionization energy:

$$M(g) \longrightarrow M^+(g) + e^- \quad \Delta H^\ominus = IE$$

and lattice energy:

$$M^+(g) + X^-(g) \longrightarrow (M^+ + X^-)(s) \quad \Delta H^\ominus = LE$$

- Lattice energies are greater for:

a) smaller ions
b) more highly charged ions.

- Polarization causes some ionic compounds to have a degree of covalency. This is increased by small highly charged positive ions and large highly charged negative ions (Fajans' rules).
- Covalent bonding may be described in terms of dot cross diagrams or by overlap of atomic orbitals to form molecular orbitals.
- σ molecular orbitals have electron density concentrated along the line joining the atoms. π orbitals have electron density above and below this line.
- π bonds only occur where a σ bond already exists.
- In dative bonds, both shared electrons come from one atom.
- Where a bond is between two different atoms, it will be polar if the atoms differ in electronegativity (electron-attracting power).
- Bond energy is the energy required to separate the two atoms in the bond:

$$X_2 \longrightarrow 2X \quad \Delta H^\ominus = \text{bond energy}$$

- Delocalization involves a system of π bonds spread over more than two atoms. It results in greater stability for the molecule than would be predicted without it.
- Metallic bonding involves metal ions held together by a delocalized 'sea' of electrons, which also allows the structure to conduct both electricity and heat well.
- Metallic structures may have a hexagonal close-packed (ABA) or face-centred cubic close-packed (ABC) structure (both of coordination number 12) or body-centred cubic lattice (of coordination number 8).
- 1:1 ionic compounds adopt the CsCl (8:8), NaCl (6:6) or ZnS (4:4) structure depending on the relative sizes of the cations and anions.
- The shapes of covalent molecules may be predicted by considering that the groups of electrons around a central atom will adopt positions as far away from each other as possible due to their repulsion.
- This can lead to the following shapes—linear, angular, trigonal planar, pyramidal, tetrahedral, trigonal bipyramidal, square-based pyramidal or octahedral.

10.5 Questions

1 Draw dot cross diagrams for the following compounds:

Ionic	Covalent
LiCl	H_2S
CaO	$C{\equiv}N^-$ ion
CaF_2	CF_4
Li_2O	$H_2C{=}CH_2$

2 Predict the shapes of the following molecules or ions. You will need to draw dot cross diagrams first. Remember the lone pairs.

F_2O, $H{-}C{\equiv}N$, NH_3, NH_4^+, NH_2^-

PF_5, PH_3, SF_6, CF_4, $H_2C{=}CH_2$, $H_2N{-}NH_2$, $H_3C{-}CH_3$

3 Draw Born–Haber, cycles for the following, using the data in the tables below.

$NaBr(Na^+ + Br^-)$, $CaCl_2(Ca^{2+} + 2Cl^-)$, $MgO(Mg^{2+} + O^{2-})$, $Li_2O(2Li^+ + O^{2-})$

Calculate the lattice energy in each case.

	$\Delta_{at}H^\ominus$	First IE	Second IE	First EA	Second EA
Li	159.4	520	7298		
Na	107.3	496	4563		
Ca	178.2	590	1145		
Mg	147.7	738	1451		
Br	111.9			−342.6	
Cl	121.7			−348.8	
O	249.2			−141.1	798

All data in/kJ mol^{-1}

	$\Delta_f H^\ominus$
NaBr	−361.1
$CaCl_2$	−795.8
MgO	−601.7
Li_2O	−597.9

4 Indicate the polarity of the following molecules by drawing δ^+ and δ^- on the appropriate atoms.

H—Cl, C≡O, F—Cl, CH_2Cl_2 (H—C—Cl with H above and H below C),

O=C=O, $H_2C{=}O$ (two H on C, C=O), CCl_4 (Cl—C—Cl with Cl above and Cl below C)

Which two will have no overall dipole moment? Explain your answer.

5 **A)** For each of the following atoms, calculate the effective nuclear charge felt by one of the outer electrons:

Mg, Ca, F, Cl, N, O, H

B) Use your answer to explain why HF is a polar molecule.

6 Why is fluorine the most electronegative atom? What do you expect to be the most electropositive atom (the one which will give away electron(s) most readily)?

7 Draw a Born–Haber cycle for the hypothetical compound XeF $(Xe^+ + F^-)$ using the value of the lattice energy of CsF (−747 kJ mol^{-1}). Calculate $\Delta_f H^\ominus$ for XeF.

$$\Delta_{at}H^\ominus(F) = +79.0\,\text{kJ mol}^{-1}$$
$$\text{first IE (Xe)} = +1170\,\text{kJ mol}^{-1}$$
$$\text{first EA (F)} = -328.0\,\text{kJ mol}^{-1}$$

Why is $\Delta_{at}H^\ominus$ of Xe not given (or needed)? Explain why XeF does not exist. Comment on the validity of using the lattice energy of CsF.

8 The carbonate ion CO_3^{2-} is trigonal planar with the carbon atom in the centre. All bond angles are exactly 120° and all the bond lengths are 0.129 nm. The bond length C—O is 0.143 nm and that for C=O is 0.116 nm. What does this suggest about the bonding? Draw a dot cross diagram of the bonding and sketch the orbitals involved.

9 Explain these trends in bond energies (in kJ mol^{-1}) in terms of nuclear charge and sizes.

Increase →

N—H	O—H	F—H
391	464	568

↑ Increase (from bottom to top):

F—H 568
Cl—H 432
Br—H 366
I—H 298

10 Which of the following compounds would you expect to show the greatest degree of covalency? Explain.

BeF_2,	LiF,	NaF
$BeCl_2$,	LiCl,	NaCl
BeI_2,	LiI,	NaI

11 **A)** The bond lengths (in nm) and also the dipole moments (in units called Debyes) of the various hydrogen halides are shown below.

	Bond length	Dipole moment
HF	0.092	1.91
HCl	0.127	1.05
HBr	0.141	0.80
HI	0.161	0.42

i What is meant by 'dipole moment'?
ii Comment briefly on the reasons for the decrease in dipole moment from HF to HI.

B) i State whether the following molecules would have a dipole moment or not, giving reasons.

CH_3Cl, CCl_4, BCl_3, NH_3

ii CO_2 has no dipole moment, while SO_2 has quite a large one. What difference in structure does this suggest? (London 1979)

11 Equilibrium

11.1 Introduction

Equilibrium is a key idea in chemistry, although many A-level students find it difficult. It is also an idea which is applicable in other fields like biology, physics and even economics. An understanding of the principles of equilibrium is essential to the appreciation of acid–base and redox chemistry, many industrial processes, and procedures such as solvent extraction and fractional distillation.

11.2 The characteristics of equilibrium

It is best to start with a simple example. Imagine a puddle of water. The water molecules attract one another, but some of them move fast enough (i.e. have enough kinetic energy) to escape from the liquid and evaporate. If the puddle of water is outside, it is unlikely that many of the water molecules will ever return to it, so the process will continue until the water is gone.

However, if some water is placed in a closed container the situation is quite different, see **Figure 11.1.**

Figure 11.1
Water will evaporate into an empty container. Eventually the rates of evaporation and condensation will be equal. Equilibrium is set up

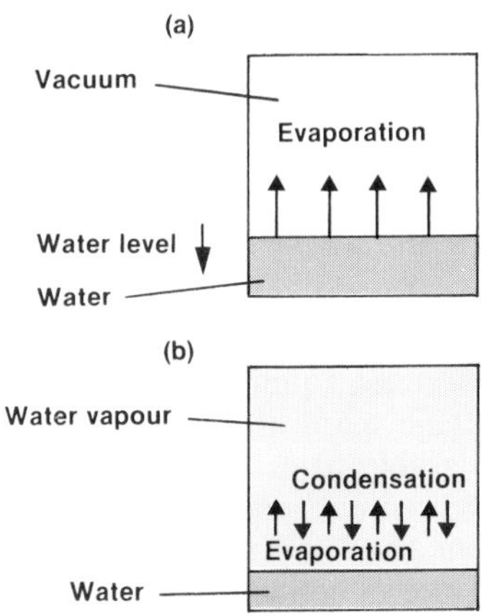

Evaporation will take place as before, but now the water molecules in the vapour are unable to escape and it is certain that some will collide with the water surface and return to liquid, i.e. condense. At first, the rate of condensation will be small, as there will be few molecules in the vapour. Evaporation will continue at a greater rate than condensation, the volume of liquid water will shrink and the number of molecules in the vapour will increase. This makes it more likely that vapour molecules will collide with, and thus rejoin, the liquid. Eventually, the rate of evaporation and the rate of condensation will become equal and the level of the water will stay the same, as will the number of molecules in the vapour and hence its pressure—the **vapour pressure**. The properties of the system will now remain constant but the evaporation and condensation are still going on *at the same rate*. This situation is called a **dynamic equilibrium** and is one of the key ideas of this topic.

It is important to realize that equilibrium can only be set up in a **closed system**. If the container had a leak which allowed water vapour to escape, the equilibrium would not be set up. Eventually all the water would escape. A closed system means a system in which nothing is added or taken away, and does not have to be *literally* closed. For example, for an equilibrium between a solid and its solution, an open beaker would provide a closed system, assuming the solvent did not evaporate.

Notice that we used the fact that the properties of the system were not changing, i.e. the water level and vapour pressure stayed the same, to show that an equilibrium existed. We could have measured the pressure exerted by the vapour by using a pressure gauge. It is also important to realize that we could have started by injecting the same mass of steam into the empty container and allowing some of it to condense. We would have reached the same equilibrium position. This is a further characteristic of equilibrium—it can be approached from either 'reactants' (in this case liquid water) or 'products' (in this case water vapour) and the *same* equilibrium mixture results. This particular example of equilibrium is called a **phase equilibrium** because it involves two **phases** (gas and liquid) of the same substance. The term phase refers to parts of a system which are physically separated from each other, for example solid, liquid, gas (or vapour), solution. Although the system we have used is a very simple one, we can discern four conditions which are applicable to all equilibria:

1. Equilibria can only be set up in a closed system (one where no components can escape).
2. Equilibrium has been reached when the properties of the system do not change with time or more strictly, when properties which do not depend on the total quantity of matter (**intensive properties**) do not change. Such properties include density, concentration, colour and pressure.
3. Equilibrium can be approached from either direction (in the above example liquid or vapour) and the final equilibrium position will be the same.
4. Equilibrium is a dynamic process—it occurs when the rates of two opposing processes (in the above case evaporation and condensation) are the same.

Equilibrium is denoted by the symbol $\rightleftharpoons$, for example:

$$\text{liquid water} \rightleftharpoons \text{water vapour}$$

or

$$H_2O(l) \rightleftharpoons H_2O(g)$$

E

DYNAMIC EQUILIBRIUM

How can we tell whether the equilibrium between liquid water and steam is dynamic or not? Surely the equilibrium could just as well be static, with all the liquid molecules remaining in the liquid and the vapour ones in the vapour, as no changes in physical properties occur? Since all the water molecules are identical how can we trace their movement? The answer is to use water molecules 'labelled' with a radioactive isotope such as ^3H (tritium). Allow the equilibrium to establish by checking that the pressure of the water vapour stays constant, then replace some of the liquid water with radioactively labelled water. The radioactivity is soon found to be spread between the liquid and the vapour phases. The same thing happens if some of the water vapour is replaced by labelled molecules.

11.3 Phase equilibria

As the name implies, these are equilibria between different phases of the same system. We will continue to consider the same example—water and its vapour. External conditions may affect the equilibrium. For example, if we increase the temperature, the average speed of the liquid molecules will increase and more of them will be able to escape from the liquid. For a moment, evaporation will be proceeding faster than condensation and the system will not be at equilibrium. However, this will lead to more molecules in the vapour and thus the rate of condensation will increase until the two are again equal. The new equilibrium position will have a higher vapour pressure. The pressure of the vapour when in contact with the liquid phase is called the **saturated vapour pressure** (SVP). If we plot this against temperature we get the graph in **Figure 11.2.** This is called a **phase diagram** for water and shows that the vapour pressure increases with temperature. The line shows the set of values of temperature and pressure at which water and water vapour can exist at equilibrium. To the left of the line (high pressure, low temperature) liquid is the stable phase, while to the right (high temperature, low pressure) vapour exists.

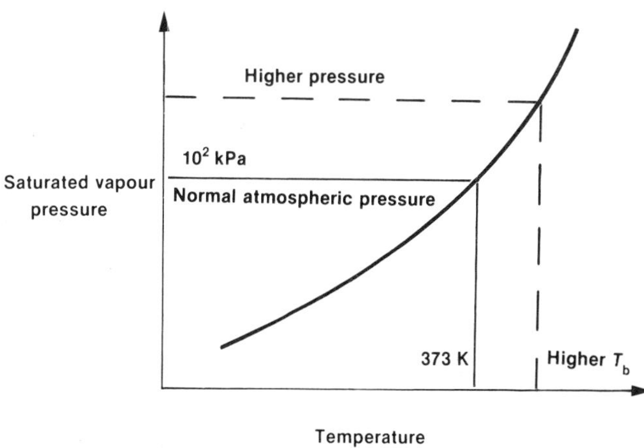

Figure 11.2 The saturated vapour pressure/temperature curve for water

E

MEASURING SATURATED VAPOUR PRESSURE

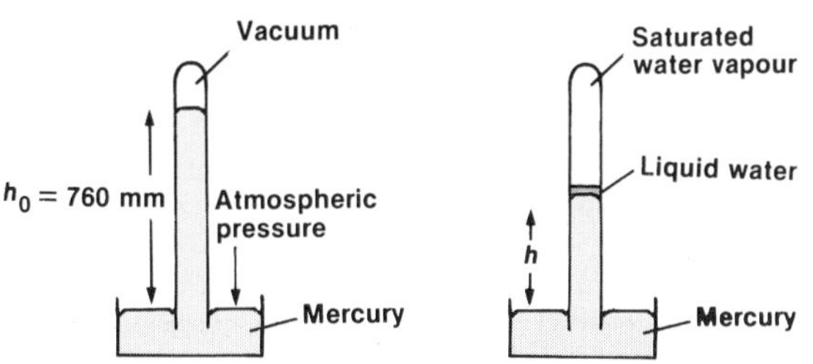

This measurement can be made as shown. A barometer is set up. Atmospheric pressure supports the column of mercury. Standard atmospheric pressure (10^2 kPa) can support a column about 760 mm high. The space above is a vacuum: it is not filled with air. If water is now put into this space, it will evaporate and its vapour pressure will force the level of mercury down. If enough water is added to produce a saturated vapour, i.e. some liquid water is left, the difference $h_0 - h$ represents the saturated vapour pressure of water. Because of this method of measuring, pressures are sometimes measured in units of mm Hg. Atmospheric pressure, 10^2 kPa, is approximately 760 mm Hg. *Note*: Strictly, the space above the mercury column contains a small amount of mercury vapour and is not a perfect vacuum, but the pressure is very low because mercury is not volatile. In practice, saturated vapour pressures are measured by more sophisticated means.

Note: At the summit of Mount Everest (8848 m) the boiling temperature of water is approximately 71 °C (344 K). At the summit of Mont Blanc (4807 m) it is 85 °C (358 K) and on top of Ben Nevis (1392 m) 95 °C (368 K)

11.3.1 Boiling

Liquids can evaporate at all temperatures. Whatever the temperature of the liquid there will always be a few molecules with enough energy to escape. The boiling temperature, T_b, is the temperature at which the vapour pressure is equal to atmospheric pressure. This means the vapour pressure is high enough to form bubbles in the body of the liquid. The normal boiling temperature is when the saturated vapour pressure is equal to standard atmospheric pressure, 10^2 kPa. This is marked on **Figure 11.2.** The graph shows that if the pressure increases, the boiling temperature will also increase, and the reverse. This lowering of the boiling temperature at low pressures explains why it is hard to make a good cup of tea high on a mountainside where the air pressure is lower than at sea level, and water therefore boils at a lower temperature.

11.3.2 The phase diagram

Figure 11.2 refers to two phases, liquid and vapour. Solids also have vapour pressures. This is obvious in the case of a solid like naphthalene (moth balls) which has a distinctive smell, showing that molecules must be escaping into the air as vapour. Solid water (ice), too, has a vapour pressure and the line showing how this varies with temperature can be added to **Figure 11.2.** The change directly from solid to vapour is called **sublimation.** Finally, if the variation of melting temperature with pressure is added, we get the complete phase diagram for water, **Figure 11.3** (overleaf).

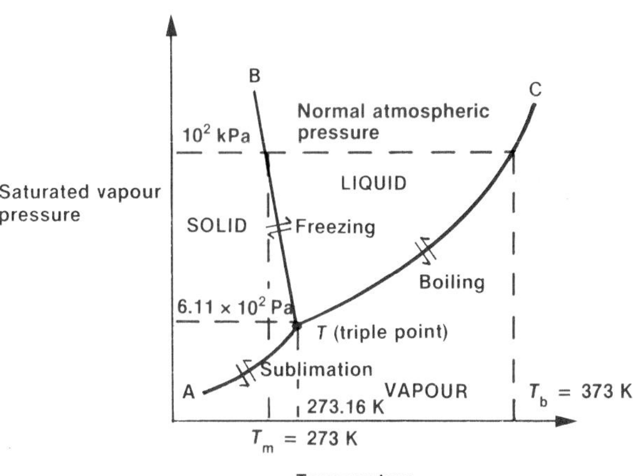

Figure 11.3 The phase diagram for water (not to scale)

T–C represents the variation of boiling temperature with pressure.
T–B represents the variation of melting temperature with pressure.
T–A represents the variation of sublimation temperature with pressure.

Each line represents the set of pressures and temperatures at which the two phases on either side of it can exist in equilibrium.

T is a unique point, the only temperature and pressure where all three phases are in equilibrium. It is called the **triple point** and for water is 273.16 K and 6.11×10^2 Pa. Moving along a horizontal line (increasing the temperature at constant pressure) at a pressure above the triple point we pass in succession from solid to liquid to vapour, but below the triple point we pass directly from solid to vapour, i.e. under these conditions ice will sublime.

The phase diagram for water is not typical in that the melting temperature line, T–B, slopes to the *left*, i.e. the melting temperature decreases with pressure. This is connected with the fact that ice is less dense than water, while most solids are denser than their liquids. This is a result of hydrogen bonding in water and ice (see **section 16.2.4**).

The phase diagram of carbon dioxide **Figure 11.4**, is more typical showing a *rightward* sloping melting temperature line. (Solid carbon dioxide is denser than liquid carbon dioxide.) Notice that the triple point is above atmospheric pressure so that at atmospheric pressure carbon dioxide sublimes. So 'dry ice', which is solid carbon dioxide, does not melt, but changes directly from solid to gas.

Carbon dioxide sublimes at atmospheric pressure. The low temperature causes water droplets to form in the air causing a mist

Figure 11.4 (right) The phase diagram for carbon dioxide (not to scale)

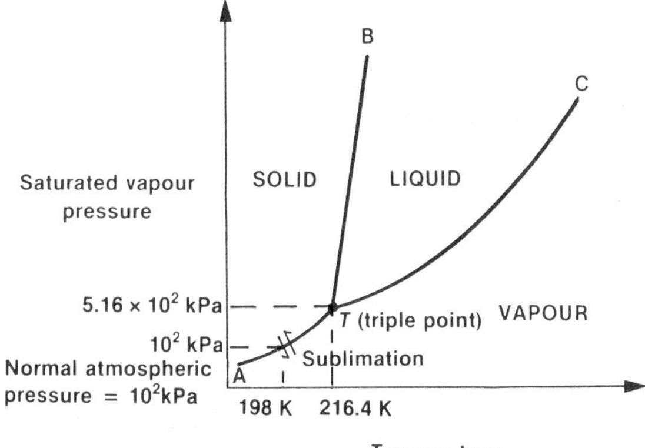

11.3.3 Partition

Let us now look at a different example of equilibrium. If we take two liquids which do not mix (are immiscible) such as water and 1,1,1-trichloroethane, CH_3CCl_3, water, which is the less dense, will float on top, so that they will form two layers. If we now add a solute like iodine, I_2, and shake the mixture, some iodine will dissolve in each layer.

(We actually use potassium iodide solution instead of water, so that the iodine will dissolve, but this is a detail.) Iodine can be seen in each layer because of its colour, pink in 1, 1, 1-trichloroethane, and yellow in potassium iodide solution. After shaking several times, the colours of the layers will not change, showing that an equilibrium has been set up. See **Figure 11.5.**

Figure 11.5 Partition of iodine between water and trichloroethane

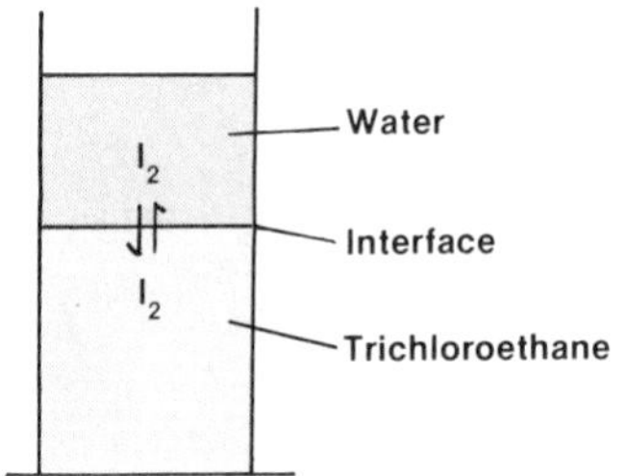

At the interface (the point where the two layers meet) some solute moves 'up' and some 'down', but at equal rates—a dynamic equilibrium exists. If we measure the concentration in mol dm^{-3} of the solute in each layer, we find that there is a constant ratio between them, irrespective of the total amount of solute used (provided that there is not enough to saturate either layer) or the volume of solvent in each layer, provided the temperature remains constant. For example, for iodine:

$$\frac{[I_2(CH_3CCl_3)]_{eqm}}{[I_2(aq)]_{eqm}} = \text{constant}$$

Note that the square brackets represent the concentration in mol dm^{-3} of the species inside. The subscript $_{eqm}$ shows that we have allowed the system to come to equilibrium before measuring the concentrations.

The constant is called the **partition coefficient** or **distribution ratio**, and is a particular example of a more general constant, the **equilibrium constant**. In this case it has no units as the units cancel out (mol dm^{-3}/mol dm^{-3}). Its value (around 80 at room temperature) shows that iodine has a much greater tendency to dissolve in trichloroethane than in water. In other words, the position of the equilibrium

$$I_2(aq) \rightleftharpoons I_2(CH_3CCl_3)$$

is well over to the right.

Practical use of partition is made in the techniques of solvent extraction and chromatography.

Solvent extraction

This is a useful method of separating one component from a mixture, if we can find a solvent in which the component we want is much more soluble than any of the others. This situation often occurs after an organic preparation. The organic substance we want is dissolved in water which also contains one or more inorganic impurities. Ethoxyethane dissolves many organic compounds well but is a poor solvent for inorganic ones. The mixture is shaken with ethoxyethane which is separated off with a separating funnel. If the partition coefficient for our component is 20 and we use equal volumes of ethoxyethane and water, 20/21 of the original substance will be extracted into the ethoxyethane, which may then be distilled off.

$$\frac{[X(\text{ethoxyethane})]_{eqm}}{[X(aq)]_{eqm}} = \frac{20}{1}$$

so if the total amount of X is 21 units, 20 of these will be in the ethoxyethane and one in the water.

The technique is even more efficient if the ethoxyethane is used in several small portions rather than all at once (see box, overleaf). Solvent extraction can be used to extract the stimulant caffeine from a solution of coffee, tea or cola drinks using dichloroethane as the solvent.

Chromatography

Chromatography describes a whole family of analytical techniques, all of which depend on the principle of partition.

PAPER CHROMATOGRAPHY

Figure 11.6 Paper chromatography

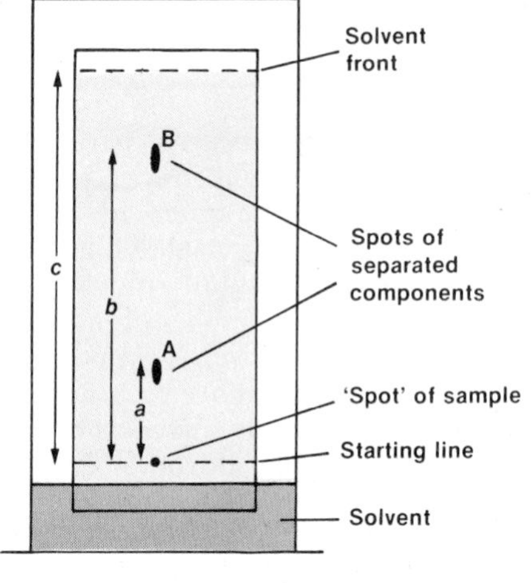

The most familiar chromatographic technique is paper chromatography (**Figure 11.6**), often used for separating dye mixtures, in which a solvent moves up a piece of filter paper by capillary action. The cellulose of which the paper is made holds many trapped water molecules. These form the **stationary phase**. As the solvent, called the **mobile phase** or **eluant**, moves

A

SOLVENT EXTRACTION

Imagine we have 5 g of organic compound X dissolved in 100 cm^3 water along with other inorganic impurities. This can be extracted with ethoxyethane. The partition coefficient for X is 20.

$$\frac{[\text{X (ethoxyethane)}]_{\text{eqm}}}{[\text{X(aq)}]_{\text{eqm}}} = 20$$

and the inorganic impurities are insoluble in ethoxyethane.

1. If we extract with 100 cm^3 of ethoxyethane and x g of X dissolves in the ethoxyethane leaving $(5 - x)$ g in 100 cm^3 of water, then:

$$\frac{x/100}{(5-x)/100} = 20$$

$$x = 20(5 - x)$$
$$x = 100 - 20x$$
$$21x = 100$$
$$x = 100/21 = 4.76\text{ g}$$

So 4.76 g of X are recovered and 0.24 g is left in the water.

2. However, if we use the 100 cm^3 of ethoxyethane in two portions of 50 cm^3 each, the first time y g of X dissolves in the ethoxyethane, leaving $(5 - y)$ g in 100 cm^3 of water.

So:

$$\frac{y/50}{(5-y)/100} = 20$$

$$2y = 20(5 - y)$$
$$2y = 100 - 20y$$
$$22y = 100$$
$$y = 100/22 = 4.54\text{ g}$$

So 0.46 g of X remain in the water.

For the second extraction when the other 50 cm^3 of ethoxyethane are added to the water, z g of X are extracted into the ethoxyethane.

So:

$$\frac{z/50}{(0.46-z)/100} = 20$$

$$2z = (0.46 - z) \times 20$$
$$2z = 9.2 - 20z$$
$$22z = 9.2$$
$$z = 9.2/22 = 0.42\text{ g}$$

So in total $4.54 + 0.42 = 4.96$ g of X is extracted into the ethoxyethane this time—an improvement of $0.2/4.76 \times 100\% = 4.2\%$, although no more ether has been used. In general, the more extractions the better. The same principle applies to cleaning paintbrushes. Many small rinses are more efficient than one large rinse using the same amount of white spirit.

A

CAFFEINE

Caffeine is a drug present in coffee, tea, cocoa, and 'cola' drinks, as well as in over-the-counter 'tonics' and 'hangover cures'. Caffeine stimulates the central nervous system, reduces fatigue and is also a diuretic (stimulates urine production)—all effects known to coffee drinkers! The average cup of coffee contains 100–200 mg of caffeine, tea 50–100 mg and cocoa 50–200 mg depending on strength, while a bottle of cola has around 35–55 mg. The dose required to produce stimulation is 100–300 mg, while doses of over 1000 mg produce effects like sleeplessness and rapid heart beat. It seems likely that the majority of people are actually addicted to caffeine.

Decaffeinated coffee is produced by a solvent extraction process. Originally, chlorinated solvents like dichloromethane were used, but these were dropped, following concern about their toxicity. Nowadays the caffeine is extracted using liquid carbon dioxide. Look at the phase diagram (**Figure 11.4**) to work out what conditions of temperature and pressure are needed for liquid carbon dioxide to exist.

The caffeine molecule

O, CH_3, C, CH_3, N, C, N, C, C, C, O, N, N, H, CH_3

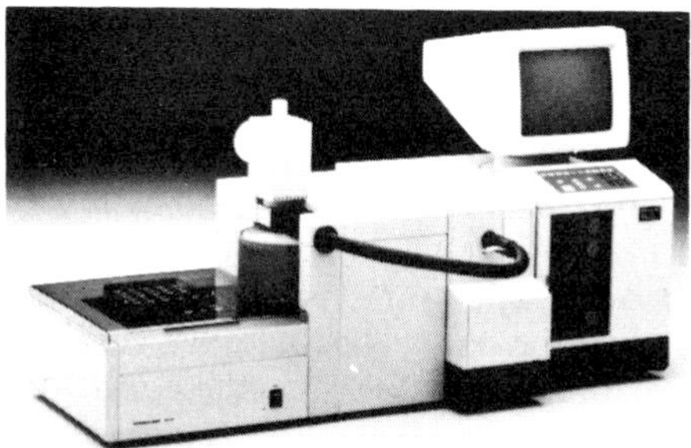

Gas–liquid chromatography apparatus (middle left) **being used for research on photosynthesis**

past the sample spot, the substances in the sample become partitioned between the mobile and stationary phases (solvent and trapped water). Dyes which are more soluble in the eluant than in water are carried along rapidly with the mobile phase, while those more soluble in water tend to remain where they are. Fresh eluant is constantly moving past the spot and the situation is rather like an enormous number of successive solvent extractions. The speed at which each dye moves is proportional to its partition coefficient. When the solvent front has moved a convenient distance, it is stopped by removing the paper from the solvent. For each spot, the R_f value (retardation factor) can be calculated from:

$$R_f = \frac{\text{distance moved by spot}}{\text{distance moved by solvent}}$$

Notice that the R_f value must always be less than one, as the spot cannot move further than the solvent. The R_f value of any substance will differ for different solvents, but two spots with identical R_f values, with the same solvent, may well be the same substance. An elaboration of this technique is two-way chromatography.

TWO-WAY CHROMATOGRAPHY

See **Figure 11.7.** A square of paper is used and a spot placed in one corner.

Figure 11.7 Two-way paper chromatography

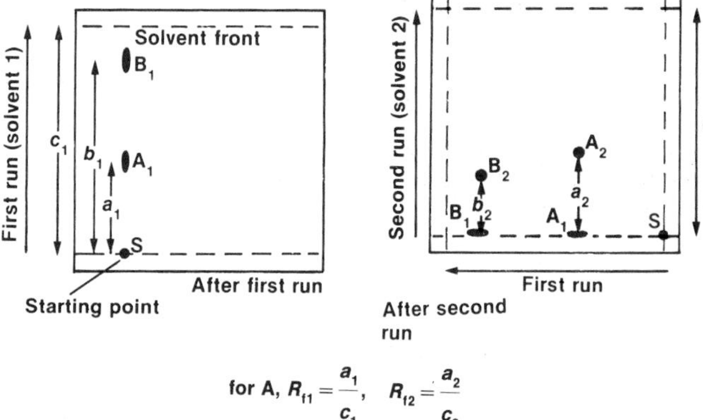

The chromatogram is run as normal, vertically, using one solvent, then the paper is turned through 90° and the chromatogram run again in another solvent. Spots not clearly separated by the first solvent may be separated by the second. Two different R_f values are obtained for each component. If both R_fs of a spot match with those of a known compound, then the two are likely to be the same compound.

THIN-LAYER CHROMATOGRAPHY (TLC)

This is a variation of paper chromatography in which the paper is replaced by a thin layer of powder such as silica or alumina spread on a glass or plastic backing. It is regularly used to determine which amino acids are in blood samples and for analysis of food dyes. See **Plate 3.**

Figure 11.8 Column chromatography

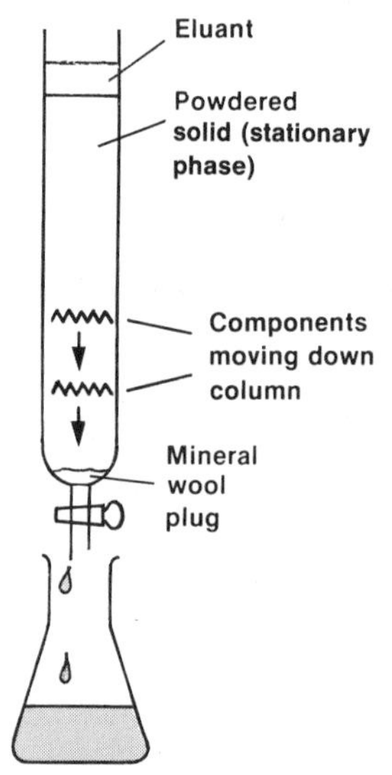

COLUMN CHROMATOGRAPHY

This is essentially a scaled-up version of thin-layer chromatography. The mixture to be separated is placed at the top of a column. Solvent is added at the top. Each component is carried down the column at a different rate and can be collected separately in a flask at the bottom, **Figure 11.8.** This method has the advantage that fairly large amounts can be separated and therefore it can be used for preparation of compounds rather than just analysis.

Figure 11.9 Gas–liquid chromatography

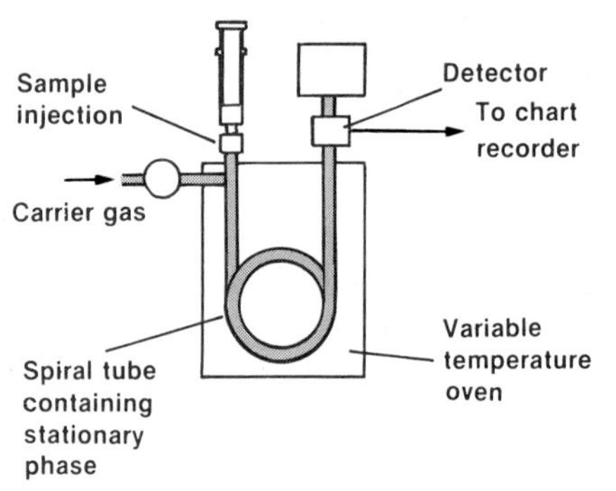

GAS–LIQUID CHROMATOGRAPHY (GLC)

This is one of the most important modern analytical techniques. The apparatus is shown in **Figure 11.9** and in the photo on the opposite page. The stationary phase is a powder coated with oil while the mobile phase is an unreactive gas, often nitrogen. The sample is partitioned between the oil and the gas, and the components leave the column at different times (retention times) after injection. Various types of detectors are used, including those which measure the thermal conductivity of the emerging gas. The results are usually presented as a graph, **Figure 11.10**, the area under each peak being proportional to the amount of that component. In some machines the emerging components are fed directly into a mass spectrometer for identification, see **section 19.3.4.**

As an analytical method for separating mixtures, GLC is extremely sensitive. It can separate minute traces of substances in foodstuffs, and even link crude oil pollution found on beaches with its tanker of origin by comparing oil samples. Among its best known uses are analysing the alcohol content of blood samples taken from suspected drunken drivers and testing athletes for drug taking.

Figure 11.10 Typical trace obtained from gas–liquid chromatography of a sample mixture containing three components

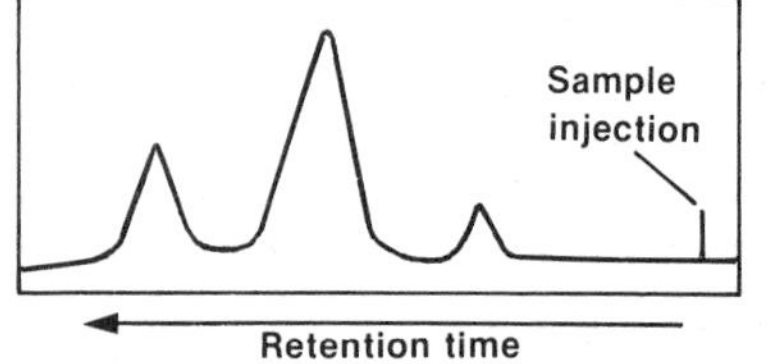

11.4 Chemical equilibria

Many chemical reactions can be reversed by suitably changing the conditions. For example ethanol, C_2H_5OH, will react with ethanoic acid, CH_3CO_2H, to produce the ester ethyl ethanoate, $CH_3CO_2C_2H_5$ and water. However, mixing the ester and water will produce ethanol and ethanoic acid, i.e. the reaction is reversible:

$C_2H_5OH(l)$	+	$CH_3CO_2H(l)$	$\longrightarrow$	$CH_3CO_2C_2H_5(l)$	+	$H_2O(l)$
ethanol		ethanoic acid		ethyl ethanoate (an ester)		water

If ethanol and ethanoic acid are mixed in a flask (stoppered to prevent evaporation) and left, a mixture is eventually obtained in which *all four* substances are present. (A strong acid catalyst is required if this is to occur within a reasonable length of time.) The system has reached equilibrium, so we can write:

$C_2H_5OH(l)$	+	$CH_3CO_2H(l)$	$\rightleftharpoons$	$CH_3CO_2C_2H_5(l)$	+	$H_2O(l)$
ethanol		ethanoic acid		ethyl ethanoate		water

The mixture may be analysed by titrating the ethanoic acid with standard alkali (making allowance for the catalyst). This gives the number of moles of ethanoic acid, from which the number of moles of the other components can be worked out, and hence their concentrations, if the total volume of the mixture is known (see box opposite).

If several experiments are done with different quantities of starting materials in aqueous solution, it is always found that the ratio:

$$\frac{[CH_3CO_2C_2H_5(aq)]_{eqm}[H_2O(l)]_{eqm}}{[CH_3CO_2H(aq)]_{eqm}[C_2H_5OH(aq)]_{eqm}}$$

has a constant value, provided the experiments are done at the same temperature. The value obtained using the figures in the box is 3.53 (no units—they cancel out). This ratio is called the **equilibrium constant** and is given the symbol K_c. The subscript $_c$ indicates that it is a ratio of concentrations.

For the reaction below, which takes place in ethanol solution:

$$CH_3COCH_3 + HCN \rightleftharpoons CH_3\text{—}C(OH)(CN)\text{—}CH_3$$

propanone, hydrogen cyanide, 2-hydroxy-2-methylpropanenitrile

we find that the expression:

$$\frac{[CH_3C(CN)(OH)CH_3]_{eqm}}{[CH_3COCH_3]_{eqm}[HCN]_{eqm}}$$

is constant at constant temperature.

11.4.1 The equilibrium law

Both the above expressions are examples of a general law—the **equilibrium law**. This is expressed as follows. For a reaction:

$$aA + bB + cC \rightleftharpoons xX + yY + zZ$$

the expression

$$\frac{[X]_{eqm}^{x}[Y]_{eqm}^{y}[Z]_{eqm}^{z}}{[A]_{eqm}^{a}[B]_{eqm}^{b}[C]_{eqm}^{c}}$$

has a constant value, provided the temperature is constant. The constant

E

EQUILIBRIUM CALCULATIONS EXAMPLE

4.6 g (0.1 mol) of ethanol and 12.0 g (0.2 mol) of ethanoic acid were mixed in a flask with 20 cm^3 of 1 mol dm^{-3} hydrochloric acid as catalyst at room temperature (20° C, 293 K). The contents of the flask were left for a week to reach equilibrium, and then titrated with 1.0 mol dm^{-3} sodium hydroxide.

137 cm^3 were required, but 20 cm^3 of these were used to react with the hydrochloric acid catalyst:

$$HCl(aq) + NaOH(aq) \longrightarrow NaCl(aq) + H_2O(l)$$

The rest (117 cm^3) reacted 1:1 with the ethanoic acid which remained at equilibrium.

$$NaOH(aq) + CH_3CO_2H(aq) \longrightarrow CH_3CO_2Na(aq) + H_2O(l)$$

so number of moles NaOH = number of moles CH_3CO_2H

number of moles = $M \times V/1000$

= 1 × 117/1000

= 0.117 mol

= number of moles ethanoic acid present at equilibrium

We can now calculate the number of moles of all the other components, from the equation and the number of moles of reactants that we started with.
Always set out problems of this type as follows:

1. Write the equation.

2. Below it write the number of moles at the start.

3. Below this write the number of moles known to be present at equilibrium.

4. Work out from 3 the number of moles of the other components of the equation.

	CH_3CO_2H +	$C_2H_5OH \rightleftharpoons$	$CH_3CO_2C_2H_5$ +	H_2O
1. / **2.** start	0.2 mol	0.1 mol	0 mol	0 mol
3. at equilibrium	0.117 mol	?	?	?

Therefore 0.2 − 0.117 = 0.083 mol of ethanoic acid has been used.

The equation tells us that each mole of ethanoic acid which reacts also takes with it 1 mole of ethanol and produces 1 mole of each of the products.

So 0.083 mol of ethanol is used and (0.1 − 0.083) = 0.017 mol is left.

0.083 mol each of ethyl ethanoate and water are formed. So we can now fill in the ? above.

	CH_3CO_2H +	$C_2H_5OH \rightleftharpoons$	$CH_3CO_2C_2H_5$ +	H_2O
4. at equilibrium	0.117 mol	0.017 mol	0.083 mol	0.083 mol

If we know the volume of the whole mixture at equilibrium we can work out the concentrations at equilibrium. The volume was 37.3 cm^3 or 0.0373 dm^3 including the catalyst which makes it an aqueous solution.
So

$$[CH_3CO_2H(aq)]_{eqm} = \frac{0.117\ mol}{0.0373\ dm^3} = 3.1\ mol\ dm^{-3}$$

$$[C_2H_5OH(aq)]_{eqm} = \frac{0.017\ mol}{0.0373\ dm^3} = 0.45\ mol\ dm^{-3}$$

$$[CH_3CO_2C_2H_5(aq)]_{eqm} = \frac{0.083\ mol}{0.0373\ dm^3} = 2.22\ mol\ dm^{-3}$$

$$[H_2O(aq)]_{eqm} = \frac{0.083\ mol}{0.0373\ dm^3} = 2.22\ mol\ dm^{-3}$$

$$K_c = \frac{[CH_3CO_2C_2H_5(aq)]_{eqm}[H_2O(aq)]_{eqm}}{[CH_3CO_2H(aq)]_{eqm}[C_2H_5OH(aq)]_{eqm}}$$

$$= \frac{2.22 \times 2.22}{3.1 \times 0.45} = 3.53 \text{ (no units)}$$

is called the **equilibrium constant,** K_c. The units of K_c vary, and you must work them out for each example by cancelling out the units of each term, for example:

$$A + B \rightleftharpoons C \qquad K_c = \frac{[C]^1}{[A]^1[B]^1}$$

units are
$$\frac{mol\ dm^{-3}}{mol\ dm^{-3} \times mol\ dm^{-3}} = dm^3\ mol^{-1}$$

It is worth noting that the size of the equilibrium constant gives an indication of the position of the equilibrium. Since the equilibrium law expression is always of the general form products/reactants, then if the equilibrium constant is much greater than 1, products predominate over reactants. We usually say that the equilibrium is over to the right. If the equilibrium constant is much less than 1, reactants predominate, and the equilibrium position is over to the left. Reactions where the equilibrium constant is greater than 10^{10} are usually regarded as going to completion, while those with an equilibrium constant of less than 10^{-10} are regarded as not taking place at all.

11.4.2 Gaseous equilibrium

Equilibrium reactions which take place in the gas phase also obey the equilibrium law, although their concentrations are usually expressed in a different way, using **partial pressures**.

Partial pressure

In a mixture of gases, each gas can be thought of as contributing to the total pressure. This contribution is called its partial pressure. The sum of all the partial pressures of the gases in the mixture is equal to the total pressure. This is called Dalton's law, see **section 15.4.1.** The partial pressure of one gas in a mixture is the pressure it would produce if it occupied the container on its own. The symbol p is used for partial pressure. For example, air is a mixture of one-fifth oxygen and four-fifths nitrogen. If the total pressure is 10^2 kPa:

partial pressure of oxygen, $pO_2 = 1/5 \times 10^2\ \text{kPa}$
partial pressure of nitrogen, $pN_2 = 4/5 \times 10^2\ \text{kPa}$

which when added together give the total pressure 10^2 kPa.

More precisely, the partial pressure of any gas in a mixture is given by its mole fraction × total pressure.

$$\text{mole fraction of a gas A} = \frac{\text{number of moles of gas A}}{\text{total number of moles of gases in the mixture}}$$

Notice that since pressure depends on the number of molecules in a given volume, it is really measuring the same thing as concentration. When equilibrium constants are expressed in terms of partial pressures, they are given the symbol K_p.

Applying the equilibrium law to gaseous equilibria

For the equilibrium:

$$H_2(g) + I_2(g) \rightleftharpoons 2HI(g)$$

$$K_p = \frac{p^2HI(g)_{eqm}}{pH_2(g)_{eqm}\,pI_2(g)_{eqm}}$$

Notice how this fits the equilibrium law expression given earlier. In this case, K_p has no units as they cancel. However, the equation could equally well have been written:

$$2HI(g) \rightleftharpoons H_2(g) + I_2(g)$$

in which case:

$$K_p = \frac{pH_2(g)_{eqm}\,pI_2(g)_{eqm}}{p^2HI(g)_{eqm}}$$

K_p will have a different value, so it is *vital* when writing expressions for equilibrium constants that the equation is written.

Furthermore, the same equation could also have been written:

$$HI(g) \rightleftharpoons \tfrac{1}{2}H_2(g) + \tfrac{1}{2}I_2(g)$$

for which:

$$K_p = \frac{p^{1/2}H_2(g)_{eqm}\,p^{1/2}I_2(g)_{eqm}}{pHI(g)_{eqm}}$$

$$= \frac{\sqrt{pH_2(g)_{eqm}}\sqrt{pI_2(g)_{eqm}}}{pHI(g)_{eqm}}$$

A further example is the following reaction, a key step in the Haber process for the manufacture of ammonia for use in the fertilizer, dyestuffs, explosives and fabric industries:

$$3H_2(g) + N_2(g) \rightleftharpoons 2NH_3(g)$$

$$K_p = \frac{p^2NH_3(g)_{eqm}}{p^3H_2(g)_{eqm}\,pN_2(g)_{eqm}}$$

Equilibrium constants may or may not have units. Notice that this equilibrium constant has units:

$$\frac{Pa^2}{Pa^3 Pa} = \frac{Pa^2}{Pa^4} = \frac{1}{Pa^2} = Pa^{-2}$$

Can you see at least two other ways of writing the equation and the appropriate expression for K_p? Work the units out by cancelling the units of all the terms, as above.

11.4.3 Calculations using equilibrium law expressions

What use is the equilibrium law? It enables us to calculate the composition of a reaction mixture which has reached equilibrium and is therefore useful if we are preparing a 'target compound' via a reversible reaction.

Example 1

Let us go back to the reaction of ethanol and ethanoic acid:

$$C_2H_5OH(aq) + CH_3CO_2H(aq) \rightleftharpoons CH_3CO_2C_2H_5(aq) + H_2O(aq)$$

$$K_c = \frac{[CH_3CO_2C_2H_5(aq)]_{eqm}[H_2O(aq)]_{eqm}}{[C_2H_5OH(aq)]_{eqm}[CH_3CO_2H(aq)]_{eqm}}$$

Imagine that $K_c = 4$ at the temperature of our experiment. Suppose we want to know how much ethyl ethanoate we could produce by mixing 1 mol of ethanol and 1 mol of ethanoic acid. Set out the information as shown below:

	$C_2H_5OH(aq)$ +	$CH_3CO_2H(aq)$ $\rightleftharpoons$	$CH_3CO_2C_2H_5(aq)$ +	$H_2O(aq)$
Start	1 mol	1 mol	0 mol	0 mol
At equilibrium	$(1-x)$ mol	$(1-x)$ mol	x mol	x mol

We do not know how many moles of ethyl ethanoate will be produced, so we call this x. The equation tells us that x mol of water will also be produced and in doing so x mol of both ethanol and ethanoic acid will be used up. So the amount of each of these remaining at equilibrium is $(1-x)$ mol.

These figures are in moles, but we need concentrations in mol dm^{-3} to substitute in the equilibrium law expression. Suppose the volume of the system at equilibrium was V dm^3. Then:

$$[C_2H_5OH(aq)]_{eqm} = (1-x)/V \text{ mol dm}^{-3}$$
$$[CH_3CO_2H(aq)]_{eqm} = (1-x)/V \text{ mol dm}^{-3}$$
$$[CH_3CO_2C_2H_5(aq)]_{eqm} = x/V \text{ mol dm}^{-3}$$
$$[H_2O(aq)]_{eqm} = x/V \text{ mol dm}^{-3}$$

These figures may now be put into the expression for K_c:

$$K_c = \frac{x/V \quad x/V}{(1-x)/V\,(1-x)/V}$$

The Vs cancel, so we do not need to know the actual volume of the system. (This will happen for all systems with equal numbers of moles of product and reactant, so V is sometimes omitted. It is always better to include V, so you will not forget it for systems where the Vs do not cancel out.)

$$4 = \frac{x \times x}{(1-x)(1-x)}$$

$$4 = \frac{x^2}{(1-x)^2}$$

Taking the square root of both sides, we get:

$$2 = \frac{x}{(1-x)}$$
$$2(1-x) = x$$
$$2 - 2x = x$$
$$2 = 3x$$
$$x = \tfrac{2}{3}$$

So $\frac{2}{3}$ mol of ethyl ethanoate and $\frac{2}{3}$ mol of water are produced if the reaction reaches equilibrium, and the composition of the equilibrium mixture would be:

ethanol $\frac{1}{3}$ mol, ethanoic acid $\frac{1}{3}$ mol, ethyl ethanoate $\frac{2}{3}$ mol, water $\frac{2}{3}$ mol

Alternatively, we can use K_c to find the amount of a reactant that will give a required amount of product.

Example 2

For the following reaction in ethanol solution, $K_c = 30\,dm^3\,mol^{-1}$:

$$CH_3COCH_3 + HCN \rightleftharpoons CH_3-C(OH)(CN)-CH_3$$

propanone   hydrogen cyanide   2-hydroxy-2-methylpropanenitrile

$$K_c = \frac{[CH_3C(CN)(OH)CH_3]_{eqm}}{[CH_3COCH_3]_{eqm}[HCN]_{eqm}} = 30\,dm^3\,mol^{-1}$$

Note the units. Can you see how they were obtained? If not, turn back to **section 11.4.1.** Suppose we are carrying out this reaction in $2\,dm^3$ of ethanol. How much hydrogen cyanide is required to produce 1 mol of product if we start with 4 mol of propanone?

Let x be the number of moles of hydrogen cyanide required.

	CH_3COCH_3 +	HCN	$\rightleftharpoons$	$CH_3C(CN)(OH)CH_3$
Start	4 mol	x mol		0 mol

At equilibrium, we want 1 mol of product, so:

At equilibrium	(4 − 1) mol	(x − 1) mol	1 mol

We next divide through by 2, because the volume of the solution is $2\,dm^3$ and we need the concentration in $mol\,dm^{-3}$:

$\frac{3}{2}\,mol\,dm^{-3}$ $(x-1)/2\,mol\,dm^{-3}$ $\frac{1}{2}\,mol\,dm^{-3}$

Putting the figures into the equilibrium law expression:

$$30\,dm^3\,mol^{-1} = \frac{\frac{1}{2}\,mol\,dm^{-3}}{\frac{3}{2}\,mol\,dm^{-3}(x-1)/2\,mol\,dm^{-3}}$$
$$30(\tfrac{3}{2}(x-1)/2) = \tfrac{1}{2}$$
$$45(x-1) = 1$$
$$45x = 46$$
$$x = \tfrac{46}{45} = 1.022 \quad \text{(the units cancel out)}$$

So, to obtain 1 mol of product we must start with 1.022 mol hydrogen cyanide. In this example the volume of the system *does* make a difference. Try reworking the problem with a volume of $1\,dm^3$ of ethanol. (You should get $x = 1.01$ mol.)

Example 3

K_p is 0.02 at 700 K for the reaction:

$$2HI(g) \rightleftharpoons H_2(g) + I_2(g)$$

If the initial pressure of hydrogen iodide was 10^2 kPa, what will be the partial pressure of hydrogen when equilibrium is reached?

Set out as before:

Let the partial pressure of hydrogen at equilibrium ($p\mathrm{H}_{2\mathrm{eqm}}$) be x kPa.

	$2\mathrm{HI(g)}$	$\rightleftharpoons$ $\mathrm{H_2(g)}$	+ $\mathrm{I_2(g)}$
Start	10^2 kPa	0 kPa	0 kPa
At equilibrium	$10^2 - 2x$ kPa	x kPa	x kPa

The equation tells us:

a) that there will be the same number of moles of both hydrogen and iodine at equilibrium, therefore if $p\mathrm{H}_{2\mathrm{eqm}} = x$, then $p\mathrm{I}_{2\mathrm{eqm}} = x$ as well;
b) that for each mole of hydrogen (or iodine) produced, 2 moles of hydrogen iodide are used up, therefore if $p\mathrm{H}_{2\mathrm{eqm}}$ is x, $p\mathrm{HI}_{\mathrm{eqm}}$ must be $10^2 - 2x$.

$$K_p = \frac{p\mathrm{H_2(g)_{eqm}} \times p\mathrm{I_2(g)_{eqm}}}{p^2\mathrm{HI(g)_{eqm}}}$$

Putting in the figures gives:

$$0.02 = \frac{x \times x}{(10^2 - 2x)^2}$$
$$0.02(10^4 + 4x^2 - 4 \times 10^2 x) = x^2$$
$$-0.92x^2 - 8x + 2 \times 10^2 = 0$$

This equation has only one unknown, x, so it must have a solution. Equations of the general form ($\mathrm{a}x^2 + \mathrm{b}x + \mathrm{c}$) are called quadratic and if no short cut is obvious they can be solved by using the formula:

$$x = \frac{-b \pm \sqrt{b^2 - 4ac}}{2a}$$

Using this formula gives a value of $x = 11$ kPa. So $p\mathrm{H}_{2\mathrm{eqm}} = 11$ kPa.

You are most unlikely to be given a problem at A-level requiring this method of solution. We can make a simpler approximation to check that the above answer is of the right order.

Since $K_p = 0.02$ is fairly small, x will be relatively small (the equilibrium will be to the left) and *very* approximately $10^2 - 2x \simeq 10^2$. This gives:

$$0.02 \simeq \frac{x^2}{(10^2)^2}$$
$$2 \times 10^2 \simeq x^2$$
$$x \simeq 14\ \mathrm{kPa}$$

$p\mathrm{H_2(g)_{eqm}} \simeq 14$ kPa which is of the same order of size as the accurately calculated answer above.

11.4.4 Heterogeneous equilibria

In all the examples of equilibrium that we have considered so far, each of the reactants and products have been in the same phase or **homogeneous**, for example all gases or all aqueous solutions. Often there will be more than one phase involved in an equilibrium, such as solid and aqueous solution or solid and gas. Such situations are called **heterogeneous equilibria**.

Solubility product

When excess of the sparingly soluble salt silver chloride is placed in water the following equilibrium is set up:

$$\mathrm{AgCl(s)} \rightleftharpoons \mathrm{Ag^+(aq)} + \mathrm{Cl^-(aq)}$$

The equilibrium law expression is

$$K_c = \frac{[\mathrm{Ag^+(aq)}]_{\mathrm{eqm}}[\mathrm{Cl^-(aq)}]_{\mathrm{eqm}}}{[\mathrm{AgCl(s)}]_{\mathrm{eqm}}}$$

However, the term [AgCl(s)] is constant as it is not possible to change the concentration of a solid, unlike a solution where more solute can be added to the same amount of solvent, or a gas where more molecules can be squeezed into the same volume. The concentration of a solid is governed by its density and can be worked out from it. To calculate the concentration we need to know the number of moles in $1\,dm^3$. For silver chloride, $M_r = 143.3$ and density $5.56\,g\,cm^{-3}$.

$$\text{density}, \rho = \text{mass/volume}$$
$$\text{mass} = \text{volume} \times \rho$$

$1\,dm^3$ of AgCl has a mass of $1000 \times 5.56 = 5560\,g$.
This is $5560/143.3\,mol = 38.8\,mol$.
So $[AgCl(s)] = 38.8\,mol\,dm^{-3}$ and cannot be varied from this value. Changing the *amount* of AgCl(s) has no effect at all on the equilibrium as long as some is present.

If we rearrange the equilibrium law expression:

$$K_c[AgCl(s)]_{eqm} = [Ag^+(aq)]_{eqm}[Cl^-(aq)]_{eqm}$$

and define a modified equilibrium constant $K_c' = K_c[AgCl(s)]_{eqm}$, then

$$K_c' = [Ag^+(aq)]_{eqm}[Cl^-(aq)]_{eqm}$$

and the units of K_c' are $mol^2\,dm^{-6}$

THE CONCENTRATION OF WATER

Ask any chemist the question 'what is the concentration of water?' and you may puzzle him or her. You may get answers like 'one', 'it hasn't got one', 'zero' or '100%'. Its concentration is actually $55.55\,mol\,dm^{-3}$. See if you can confirm this using a calculation like the one for silver chloride in **section 11.4.4.** Use the density of water $= 1\,g\,cm^{-3}$ and M_r for $H_2O = 18$.

Equilibrium law expressions are always treated like this, so that they incorporate all constant terms into a modified equilibrium constant. Constant terms arise whenever the equilibrium involves a solid, and sometimes when there is a pure liquid. Some examples are given in the box on page 128. To avoid confusion, it is important to write down the expression which you are using.

In the particular case of an equilibrium between a sparingly soluble salt and its ions in solution, as in the example above, the modified equilibrium constant K_c' is called the **solubility product** and given the symbol K_{sp}.

Another example:

$$PbI_2(s) \rightleftharpoons Pb^{2+}(aq) + 2I^-(aq)$$
$$K_{sp} = [Pb^{2+}(aq)]_{eqm}[I^-(aq)]_{eqm}{}^2 \quad \text{units } mol^3\,dm^{-9}$$

The value of K_{sp} in this case is $7 \times 10^{-9}\,mol^3\,dm^{-9}$, a very small value, indicating that lead iodide is a virtually insoluble salt. We can use the solubility product to work out the conditions needed for lead iodide to precipitate. In any solution where the value of $[Pb^{2+}(aq)] \times [I^-(aq)]^2$ is greater than 7×10^{-9}, lead iodide solid (which is bright yellow) will precipitate. This is often used as a test for lead ions—add potassium iodide solution and look for a yellow precipitate.

For example, if potassium iodide solution of $2\,mol\,dm^{-3}$ concentration is used, what concentration of lead will just form a precipitate?

$$[Pb^{2+}(aq)]_{eqm}[I^-(aq)]_{eqm}{}^2 = 7 \times 10^{-9}$$
$$[Pb^{2+}(aq)]_{eqm} \times 2^2 = 7 \times 10^{-9}$$
$$[Pb^{2+}(aq)]_{eqm} = \frac{7 \times 10^{-9}}{4}$$
$$[Pb^{2+}(aq)]_{eqm} = 1.75 \times 10^{-9}\,mol\,dm^{-3}$$

So this is the limit of sensitivity of the test. Solutions containing fewer lead ions than this will not form a precipitate with $2\,mol\,dm^{-3}$ potassium iodide while more concentrated solutions will.

SOLUBILITY PRODUCT AND SOLUBILITY

These two terms are often confused. Solubility product is defined on the opposite page. **Solubility** is the molarity of a saturated solution formed from the substance in question. The two quantities are linked, e.g. for silver iodide, AgI, $K_{sp} = 8 \times 10^{-17}\,mol^2\,dm^{-6}$.

$$AgI(s) \rightleftharpoons Ag^+(aq) + I^-(aq)$$
$$K_{sp} = [Ag^+(aq)]_{eqm}[I^-(aq)]_{eqm} = 8 \times 10^{-17}\,mol^2\,dm^{-6}$$

In a saturated solution:

$$[Ag^+(aq)] = [I^-(aq)], \text{ let this equal } x$$

so $x^2 = 8 \times 10^{-17}\,mol^2\,dm^{-6}$

and $x = 8.9 \times 10^{-9}\,mol\,dm^{-3}$

So the concentration of a saturated solution of silver iodide would be $8.9 \times 10^{-9}\,mol\,dm^{-3}$, as 1 mol of Ag^+ is produced by 1 mol of AgI. This is the solubility of silver iodide. Another example is calcium fluoride, CaF_2, $K_{sp} = 4 \times 10^{-11}\,mol^3\,dm^{-9}$. This is a little more complex:

$$CaF_2(s) \rightleftharpoons Ca^{2+}(aq) + 2F^-(aq)$$
$$K_{sp} = [Ca^{2+}(aq)]_{eqm}[F^-(aq)]_{eqm}{}^2 = 4 \times 10^{-11}$$

In a saturated solution, $[F^-(aq)]_{eqm} = 2[Ca^{2+}(aq)]_{eqm}$ as the chemical equation tells us that each mol of CaF_2 which dissociates produces 1 mol of $Ca^{2+}(aq)$ and 2 mol of $F^-(aq)$.

Let $[Ca^{2+}(aq)]_{eqm} = x$

then $[F^-(aq)]_{eqm} = 2x$

Substituting in the above expression for K_{sp} gives:

$$x \cdot (2x)^2 = 4 \times 10^{-11}$$
$$4x^3 = 4 \times 10^{-11}$$
$$x^3 = 10^{-11}$$
$$x = 2.15 \times 10^{-4}\,mol\,dm^{-3}$$
$$[Ca^{2+}(aq)]_{eqm} = 2.15 \times 10^{-4}\,mol\,dm^{-3}$$

Since 1 mol of $Ca^{2+}(aq)$ is produced by 1 mol of CaF_2, this is the solubility of CaF_2.

The ionization of water

This is another example of a heterogeneous equilibrium:

$$H_2O(l) \rightleftharpoons H^+(aq) + OH^-(aq)$$
$$K_c = \frac{[H^+(aq)]_{eqm}[OH^-(aq)]_{eqm}}{[H_2O(l)]_{eqm}}$$

As with a solid, the concentration of a *pure* liquid cannot be varied and so $[H_2O(l)]_{eqm}$ is incorporated into a modified equilibrium constant:

$$K_c' = [H^+(aq)]_{eqm}[OH^-(aq)]_{eqm}$$

This is given the special symbol K_w and named the **ionic product** of water. Its value is $10^{-14}\,mol^2\,dm^{-6}$ at 298 K.

Dissociation of solids to gases

As a final example of heterogeneous equilibrium, consider the reaction:

$$CaCO_3(s) \rightleftharpoons CaO(s) + CO_2(g)$$

We could write
$$K_p = \frac{pCaO(s)_{eqm}\,pCO_2(g)_{eqm}}{pCaCO_3(s)_{eqm}}$$

but as the partial pressures produced by solids are constant (at constant temperature) these are incorporated into a modified equilibrium constant, sometimes called the **dissociation pressure** of $CaCO_3$. So:

$$K_p' = pCO_2(g)_{eqm}$$

R

EXAMPLES OF EQUILIBRIUM CONSTANT EXPRESSIONS

1. $Fe^{3+}(aq) + SCN^-(aq) \rightleftharpoons FeSCN^{2+}(aq)$

$$K_c = \frac{[FeSCN^{2+}(aq)]_{eqm}}{[Fe^{3+}(aq)]_{eqm}[SCN^-(aq)]_{eqm}}$$

Units are $dm^3\,mol^{-1}$.

2. $BiCl_3(aq) + H_2O(l) \rightleftharpoons BiOCl(s) + 2HCl(aq)$

$$K_c = \frac{[HCl(aq)]_{eqm}^2}{[BiCl_3(aq)]_{eqm}}$$

Units are $mol\,dm^{-3}$.

[BiOCl(s)] does not appear, as the value of its concentration has been incorporated into the value of K_c. $[H_2O(l)]$ does not appear either. As the reaction is carried out in aqueous solution $[H_2O(l)]$ will not vary at all. Therefore in dilute aqueous solutions it is incorporated into the value of K_c.

3. $Ag_2CrO_4(s) \rightleftharpoons 2Ag^+(aq) + CrO_4^{2-}(aq)$

$$K_{sp} = [Ag^+(aq)]_{eqm}^2[CrO_4^{2-}(aq)]_{eqm}$$

Units are $mol^3\,dm^{-9}$.

$[Ag_2CrO_4(s)]$ is constant and is therefore incorporated into the value of the equilibrium constant which is called K_{sp} as this is a case of a sparingly soluble salt in equilibrium with its own ions.

4. $CH_3CO_2H(aq) + H_2O(l) \rightleftharpoons CH_3CO_2^-(aq) + H_3O^+(aq)$

$$K_a = \frac{[CH_3CO_2^-(aq)]_{eqm}[H_3O^+(aq)]_{eqm}}{[CH_3CO_2H(aq)]_{eqm}}$$

Units are $mol\,dm^{-3}$.

$[H_2O(l)]_{eqm}$ is effectively constant for the same reasons as in the bismuth chloride equilibrium above. It is therefore incorporated into the value of the equilibrium constant. The equilibrium constant is called K_a as this is the dissociation of a weak acid, see **section 12.4.1.**

5. $2H_2O(l) \rightleftharpoons H_3O^+(aq) + OH^-(aq)$

$$K_w = [H_3O^+(aq)]_{eqm}[OH^-(aq)]_{eqm}$$

Units are $mol^2\,dm^{-6}$.

The concentration of water is effectively constant so it is incorporated into the value of K_w. The equilibrium constant is called K_w as it represents the dissociation of water, see **section 12.3.**

6. $PCl_5(g) \rightleftharpoons PCl_3(g) + Cl_2(g)$

$$K_p = \frac{pPCl_3(g)_{eqm}\,pCl_2(g)_{eqm}}{pPCl_5(g)_{eqm}}$$

Units are kPa.

As this is a gaseous reaction we normally use K_p and use the value of the partial pressures of the reactants and the products. It is quite possible to use K_c but the units will be different.

$$K_c = \frac{[PCl_3(g)]_{eqm}[Cl_2(g)]_{eqm}}{[PCl_5(g)]_{eqm}}$$

Units are $mol\,dm^{-3}$.

7. $3Fe(s) + 4H_2O(g) \rightleftharpoons Fe_3O_4(s) + 4H_2(g)$

$$K_p = \frac{p^4H_2(g)_{eqm}}{p^4H_2O(g)_{eqm}}$$

No units.

Like all solids Fe and Fe_3O_4 have a constant partial pressure (at a fixed temperature) and so the values of these have been incorporated into K_p. We could also write:

$$K_c = \frac{[H_2(g)]_{eqm}^4}{[H_2O(g)]_{eqm}^4}$$

No units.

Note that in this example, water is in the gaseous state (steam) so its pressure and concentration can be changed.

11.4.5 The effect of changing conditions on equilibria

Once an equilibrium has been established, the composition of the equilibrium mixture can be changed by varying the temperature, the concentration of species involved and the pressure (in the case of reactions involving gases).

The effect of changing temperature

All the above discussions of equilibria have assumed that the temperature remains constant. *Changing the temperature results in a new value of the equilibrium constant.* This is shown by the data for the following reactions, **Table 11.1.**

Table 11.1 Values of K_p at different temperatures for two reactions

$N_2(g) + 3H_2(g) \rightleftharpoons 2NH_3(g)$ $\Delta_R H^\ominus = -92\,kJ\,mol^{-1}$

T/K	$K_p/10^{-10}Pa^{-2}$	
298	6.76×10^{5}	decrease
500	3.55×10^{-2}	
700	7.76×10^{-5}	
900	1.00×10^{-6}	

$H_2(g) + CO_2(g) \rightleftharpoons H_2O(g) + CO(g)$ $\Delta_R H^\ominus = +41\,kJ\,mol^{-1}$

T/K	K_p (no units)	
298	1.00×10^{-5}	increase
500	7.76×10^{-3}	
700	1.23×10^{-1}	
900	6.01×10^{-1}	

Note: Reversible reactions that are exothermic in one direction are endothermic in the other direction

$$+\Delta H \quad A + B \underset{\text{endothermic}}{\overset{\text{exothermic}}{\rightleftharpoons}} C + D \quad -\Delta H$$

The rule is that for an exothermic reaction (ΔH is negative), increasing the temperature decreases the equilibrium constant, while for an endothermic reaction (ΔH is positive), increasing the temperature increases the equilibrium constant. This is a general rule. So **for an exothermic reaction, increasing the temperature will move the equilibrium to the left** and **for an endothermic reaction, increasing the temperature will move the equilibrium to the right**. (The larger is K_p, the more the equilibrium has moved to the right to give more products and fewer reactants.)

We can show this experimentally for the following reaction:

$$N_2O_4(g) \rightleftharpoons 2NO_2(g) \quad \Delta_R H^\ominus = +58\,kJ\,mol^{-1}$$

If a mixture of the two gases at equilibrium in a syringe is placed in a beaker of hot water (see **Figure 11.11**), the volume of the mixture increases. This would be expected anyway as all gases expand on heating, but the volume increase is greater than would be predicted, as shown on the graph (**Figure 11.12**). This shows that the equilibrium has in fact moved to the right (producing more moles of gas).

Figure 11.11 Apparatus for investigating the equilibrium $N_2O_4(g) \rightleftharpoons 2NO_2(g)$

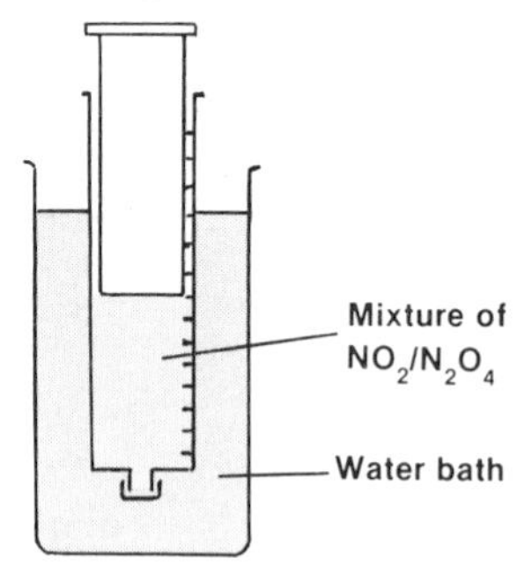

Figure 11.12 The effect of increasing the temperature of a mixture of NO_2 and N_2O_4

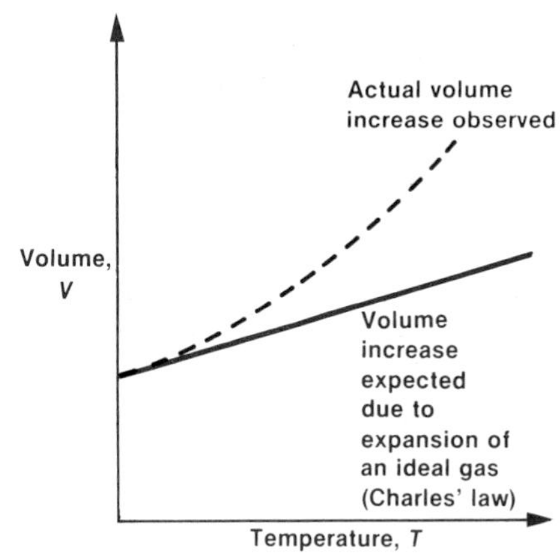

The effect of changing concentration

If, for example, potassium dichromate is dissolved in water at constant temperature, an equilibrium is set up between chromate ions and dichromate ions in aqueous solution:

$$\underset{\text{chromate ions (yellow)}}{2CrO_4^{2-}(aq)} + 2H^+(aq) \rightleftharpoons \underset{\text{dichromate ions (orange)}}{Cr_2O_7^{2-}(aq)} + H_2O(l)$$

$$K_c = \frac{[Cr_2O_7^{2-}(aq)]_{eqm}}{[CrO_4^{2-}(aq)]_{eqm}^2[H^+(aq)]_{eqm}^2} \text{ as } [H_2O(l)] \text{ is constant}$$

If an acid (H^+ ions) is added, the mixture turns orange, showing that more dichromate ions are being formed, i.e. the equilibrium position moves to the right. This is logical because increasing the concentration of a reactant will speed up a reaction, so the forward reaction by which dichromate ions *form* will go faster. The back reaction by which they are *destroyed* will not. Eventually a new equilibrium mixture is established which has a greater proportion of dichromate ions than before. The composition of the new equilibrium is such that K_c has the *same* value as before the addition of the acid. K_c does not change as the temperature has remained constant.

$$K_c = \frac{\overset{\text{this goes up}}{[Cr_2O_7^{2-}(aq)]_{eqm}}}{\underset{\text{this goes down}}{[CrO_4^{2-}(aq)]_{eqm}^2}\underset{\text{this goes up (it has been added)}}{[H^+(aq)]_{eqm}^2}}$$

The effect of changing pressure

Changing the pressure at constant temperature will only affect reactions involving gases and these will react faster in *both* directions under increased pressure. However, the equilibrium position will only be changed when there is a different total number of moles on each side of the equation.

For example:

$$\underbrace{H_2(g) + I_2(g)}_{\text{2 moles}} \rightleftharpoons \underset{\text{2 moles}}{2HI(g)}$$

The equilibrium position will not change in this reaction when the pressure is increased, so the proportions of H_2, I_2 and HI will stay the same. But in the equilibrium:

$N_2O_4(g)$	$\rightleftharpoons$	$2NO_2(g)$
dinitrogen tetroxide		nitrogen dioxide
(colourless)		(brown)
1 mole		2 moles

increasing the pressure will change the composition of the equilibrium mixture. We can use the colour of the mixture to roughly judge the equilibrium position. If an equilibrium mixture of gases is kept in a syringe and the plunger quickly pressed in to increase the pressure, the following is seen. First the mixture gets darker, as the concentration of brown NO_2 momentarily increases (the same amount is squeezed into a smaller space). Then the mixture goes paler than it was at the beginning, as the new equilibrium is established, which results in more N_2O_4 and less NO_2.

The contribution of pN_2O_4 to the total pressure increases and that of pNO_2 decreases in such a way that the value of

$$K_p = \frac{p^2NO_2(g)_{eqm}}{pN_2O_4(g)_{eqm}} \text{ remains constant.}$$

Le Chatelier's principle

This is a useful rule which makes it easier to predict what will happen when the conditions of an equilibrium mixture are changed. It states that **if a system at equilibrium is disturbed, the equilibrium moves in the direction which tends to reduce the disturbance.**

Let us apply it to the example above:

$$N_2O_4(g) \rightleftharpoons 2NO_2(g) \quad \Delta_R H^\ominus = +58\,\text{kJ mol}^{-1}$$

1. **Increasing pressure**: Le Chatelier's principle predicts that the equilibrium will move so as to try and decrease the pressure. The equilibrium will move left to do this (fewer molecules will exert less pressure). As we have seen, this is what happens.
2. **Increasing temperature**: Le Chatelier's principle predicts that the equilibrium moves in the direction which cools the system down. To do this it will move in the direction which absorbs heat (is endothermic), i.e. to the right.
3. **Increasing concentration**: Consider the example of a solution of potassium dichromate that we looked at earlier:

   $$2CrO_4^{2-}(aq) + 2H^+(aq) \rightleftharpoons Cr_2O_7^{2-}(aq) + H_2O(l)$$

   Le Chatelier's principle predicts that adding H^+ ions should move the equilibrium to the right where there will be fewer H^+ ions, which, as we have seen is what happens. Adding an alkali would remove H^+ ions and drive the equilibrium to the left.

Catalysts

Catalysts have no effect on the position of equilibrium. They work in such a way that they affect both the forward and reverse reactions equally (see **section 14.6**). Therefore they speed up the rate at which equilibrium is set up but not the composition of the equilibrium mixture.

11.4.6 Measuring equilibrium constants

In principle this is straightforward. We simply have to let the system we are interested in reach equilibrium and measure the concentrations of one or more species—enough to calculate the concentrations of the rest.

Any of the standard methods of measuring concentration may be used, with the one proviso that the method of measuring the concentration must not disturb the equilibrium. This might occur if a titration were used, for example for the equilibrium mixture:

$Ca(OH)_2(s)$	$\rightleftharpoons$	$Ca^{2+}(aq)$	$+$	$2OH^-(aq)$
calcium hydroxide		calcium ions		hydroxide ions

$[OH^-(aq)]_{eqm}$ could not be found by titration of the hydroxide ions with standard acid because, as the titration removed hydroxide ions, the equilibrium would move to the right (Le Chatelier's principle) and more calcium hydroxide would dissociate. If the titration continued, all the calcium hydroxide would dissociate. Titration can only be used if the rate at which the equilibrium is set up is very slow compared with that of the titration reaction. An example of this is the equilibrium

CH_3CO_2H	$+$	C_2H_5OH	$\rightleftharpoons$	$CH_3CO_2C_2H_5$	$+$	H_2O
ethanoic acid		ethanol		ethyl ethanoate		water

Here the acid could be titrated without disturbing the equilibrium as the titration reaction is more or less instantaneous while the equilibrium takes several days to establish, even with a catalyst. (See box on page 121.)

Another equilibrium which can be investigated by titration is:

$2HI(g)$	$\rightleftharpoons$	$H_2(g)$	$+$	$I_2(g)$
hydrogen iodide		hydrogen		iodine

The equilibrium can be set up at a high temperature (over 500 K) and the mixture rapidly cooled to slow down the reaction rate and effectively 'freeze' the system in its equilibrium position. Either the hydrogen iodide (dissolved in water and therefore acidic) can be titrated with standard alkali or the iodine can be titrated with sodium thiosulphate.

If possible, concentrations should be measured by methods which do not disturb the system, e.g. colorimetry or $E^{\ominus}$ values (see **section 14.3.2** and box $E^{\ominus}$ *and equilibrium constants*, **section 13.3**, page 167).

The average relative molecular mass method

This can be used for gaseous equilibria. If the gas mixture is kept in a syringe, its volume can be easily measured and used to calculate the relative molecular mass of the contents (see **section 15.3**). This will be the **weighted average** of the relative molecular masses of all the gases present. This can be used to calculate K_p as shown in the following example:

$$N_2O_4(g) \rightleftharpoons 2NO_2(g)$$

At 60 °C (333 K) and atmospheric pressure a mixture was found to have an average relative molecular mass of 60. What is K_p? The weighted average relative molecular mass takes account of the number of each species present. A simple way of working out the proportions of each gas present is as shown. The M_r can vary from 46 (NO_2 only) to 92 (N_2O_4 only).

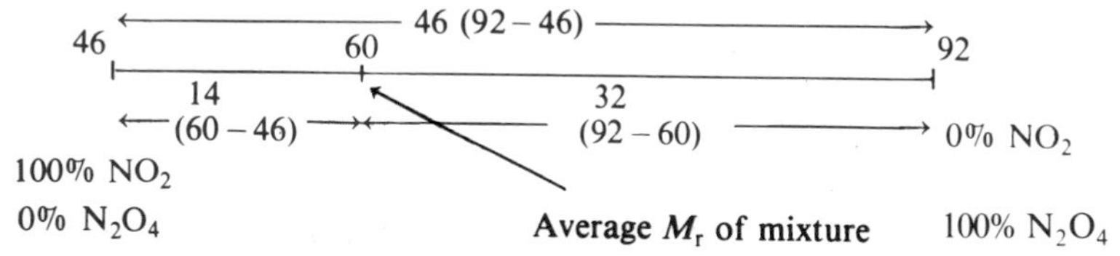

46 (92 − 46)
46 60 92
14 (60 − 46) 32 (92 − 60)
100% NO_2 0% N_2O_4
0% NO_2 100% N_2O_4
Average M_r of mixture

$M_r = 60$ represents a mixture comprising:

$$\tfrac{14}{46} \times 100\% N_2O_4 = 30.4\% N_2O_4$$
$$\tfrac{32}{46} \times 100\% NO_2 = 69.6\% NO_2$$

Since the total pressure is 10^2 kPa (atmospheric)

$$pN_2O_4 = 30.4\% \text{ of } 10^2 \text{ kPa} = 30\,400 \text{ Pa}$$
$$pNO_2 = 69.6\% \text{ of } 10^2 \text{ kPa} = 69\,600 \text{ Pa}$$

$$K_p = \frac{p^2NO_2(g)_{eqm}}{pN_2O_4(g)_{eqm}}$$

$$K_p = \frac{(69\,600)^2}{30\,400}\ \text{Pa}$$
$$= 1.59 \times 10^5\ \text{Pa}$$
$$= 1.59 \times 10^2\ \text{kPa} \quad \text{at } 333\ \text{K}$$

11.5 Industrial applications

11.5.1 The Haber process

Ammonia is a vital industrial chemical, current world production being in excess of 80 million tonnes annually. Around 80 per cent is used to make fertilizers like urea, ammonium sulphate and ammonium nitrate. The rest is used in the manufacture of synthetic fibres including nylon, dyes, explosives and plastics like polyurethane.

Virtually all ammonia produced today is made by the Haber process, the key step of which is direct reaction of nitrogen and hydrogen via a reversible reaction. The winner of the 1918 Nobel prize for chemistry, Fritz Haber, a German, devised the process in 1911 and therefore unwittingly lengthened the First World War. Prior to this process, the world's main source of nitrogen compounds, which are vital for making explosives, was the nitrate deposits in Chile. During the First World War, the Royal Navy's blockade denied Germany access to these. Without the Haber process it is unlikely that Germany could have sustained her munitions programme over four years of war. The process was scaled up for industrial use by another German, Karl Bosch, who also won a Nobel prize.

The key step of the Haber process is:

$$N_2(g) + 3H_2(g) \rightleftharpoons 2NH_3(g) \quad \Delta_R H^{\ominus} = -92\ \text{kJ mol}^{-1}$$

Application of Le Chatelier's principle to this equilibrium shows that high pressure and low temperature will both tend to move the equilibrium to the right and produce a large proportion of ammonia in the equilibrium mixture. This is illustrated by the table of the percentage of ammonia in equilibrium mixtures under different conditions, **Table 11.2.**

Table 11.2 Equilibrium % conversion of nitrogen and hydrogen to ammonia under different condition

Pressure/10^2 kPa	**Temperature**/K			
	473	573	673	773
10	51	15	4	1
100	82	53	25	11
200	89	67	39	18
300	93	71	47	24
400	94	80	55	32
600	95	84	65	42

Thus we might expect a low temperature and high pressure to be used. 600×10^2 kPa and 473 K would give close to 100% conversion. However, other economic and practical factors must also be considered such as:

- A high pressure plant is expensive, both to build and to maintain.
- A low temperature means that equilibrium is reached slowly.
- Catalysts can be used to speed up the rate of attainment of equilibrium, but they do not last indefinitely due to 'poisoning' by impurities, and lower temperatures prolong their life.
- Unreacted gases can be easily recycled through the reactor.

These factors lead to a 'compromise' set of conditions of around 200×10^2 kPa and 673 K in most plants. An iron-based catalyst is used to speed up the reaction to compensate for the relatively low temperature. Catalyst life is of the order of five years. These conditions would give an equilibrium conversion of about 40% (see **Table 11.2**) but in fact the gases are not allowed to spend long enough in the reaction vessel to reach equilibrium and the usual conversion is about 15%. The ammonia is removed by cooling, still under pressure, when it condenses to a liquid. The unconverted nitrogen and hydrogen are recycled back into the reaction vessel.

The raw materials for the process are air, to provide the nitrogen, and natural gas (methane, CH_4) to provide the hydrogen by the following reaction:

$$\underset{\text{methane}}{CH_4(g)} + \underset{\text{steam}}{H_2O(g)} \longrightarrow \underset{\text{carbon monoxide}}{CO(g)} + \underset{\text{hydrogen}}{3H_2(g)}$$

A

LE CHATELIER'S PRINCIPLE AND THE HABER PROCESS

heat given out
exothermic (surroundings get hot)

$N_2(g) + 3H_2(g) \rightleftharpoons 2NH_3(g) \quad \Delta H = -92\ kJ\ mol^{-1}$

heat absorbed
endothermic (surroundings are cooled)

1 mole 3 moles 2 moles
4 moles

Le Chatelier's principle would predict that increasing the temperature would move equilibrium to the left where heat would be absorbed, whereas decreasing the temperature would move the equilibrium to the right, because in this direction the reaction gives out heat. Increasing the pressure will move the reaction to the right where there are fewer moles of gas and therefore less pressure.

To increase the yield of ammonia we need the equilibrium to move to the right, so cooling the mixture and high pressure will increase the yield.

As supplies of natural gas are depleted, coal and water seem likely to replace the natural gas:

$C(s)$ +	$H_2O(g)$	$\longrightarrow$	$CO(g)$ +	$H_2(g)$
carbon (from coal)	steam		carbon monoxide	hydrogen

11.6 Summary

- An equilibrium may be set up only in a closed system.
- Equilibria have the following characteristics:

They can be approached from either reactants or products.
Chemical equilibria are dynamic, i.e. they occur when the rates of the forward and back reactions are equal.
Equilibrium mixtures have constant physical properties.

- Phase equilibria are equilibria between different phases (gas, liquid and solid) of the same substance. They are represented on phase diagrams—graphs of saturated vapour pressure against temperature on which areas represent phases and lines show the values of pressure and temperature where equilibrium occurs between phases. See **Figure 11.13.**

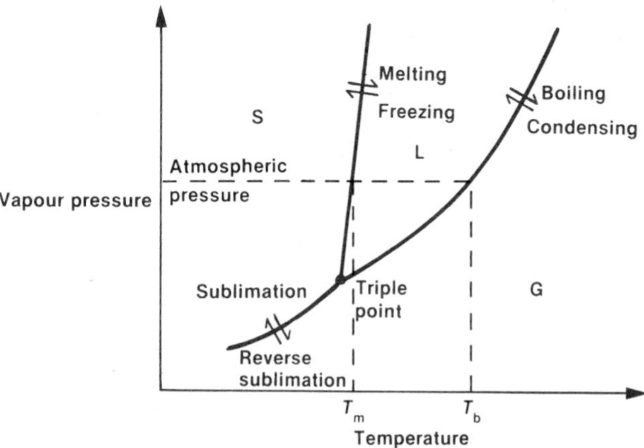

Figure 11.13 A typical phase diagram

- Partition: solutes partition themselves between two immiscible solvents and an equilibrium exists across the interface.

$$\text{partition coefficient} = \frac{\text{concentration of solute in solvent 1}}{\text{concentration of solute in solvent 2}}$$

- Solvent extraction: a solution is shaken with an immiscible solvent and solute may be extracted from it if it is more soluble in the second solvent. The extraction is more efficient if the available solvent is used in many small portions rather than in one large one.

- Chromatography is a family of separation techniques in which a moving solvent phase extracts components from a stationary phase at a rate depending on their relative solubilities in the two phases. Types include paper, thin-layer, column and gas–liquid. Components of mixtures are identified by their R_f values.
- A reversible reaction in a closed system reaches an equilibrium position. The equilibrium law states that for a reaction:

$$aA + bB \rightleftharpoons cC + dD$$

$$K_c = \frac{[C]_{eqm}{}^c[D]_{eqm}{}^d}{[A]_{eqm}{}^a[B]_{eqm}{}^b}$$

at constant temperature.

The concentrations may be expressed in mol dm^{-3} when the equilibrium constant is written K_c or by partial pressures when it is written K_p. These expressions allow the equilibrium concentrations (or partial pressures) of different species to be calculated.
- The concentrations (or partial pressures) of pure solids and liquids (not solutions) cannot vary at constant temperature. Conventionally these are incorporated in a modified equilibrium constant and do not appear in the expressions for K_c and K_p.
- Solubility products are equilibrium constants for equilibria between relatively insoluble salts and their ions in solution. If a solution is made by mixing two solutions such that the solubility product is exceeded, precipitation will occur until the concentrations are reduced to those 'allowed' by the solubility product.
- Le Chatelier's principle is a rule describing the effect on an equilibrium of changing conditions of concentration, pressure and temperature. In each case the position of equilibrium changes so as to reduce the disturbance.
- Measurement of equilibrium constants can be done by normal analytical methods provided that these do not disturb the equilibrium mixture or that the analysis is completed before the system has time to react.
- Many industrial processes involve reversible reactions and Le Chatelier's principle may be applied. However, often the conditions used in industry are not those which produce maximum conversion. The rate of the process is important and often a lower percentage conversion is accepted in order to increase the rate.

11.7 Questions

1 Write down the expression for K_c for the following equilibria in solution. State the units of K_c.

A) $Fe^{3+}(aq) + I^-(aq) \rightleftharpoons Fe^{2+}(aq) + \frac{1}{2}I_2(aq)$
B) $C_5H_{10} + CH_3CO_2H \rightleftharpoons CH_3CO_2C_5H_{11}$ (in ethanol)
C) $NH_3(aq) + H^+(aq) \rightleftharpoons NH_4{}^+(aq)$
D) $NH_4OH(aq) \rightleftharpoons NH_4{}^+(aq) + OH^-(aq)$
E) $2Fe^{3+}(aq) + 2I^- \rightleftharpoons 2Fe^{2+}(aq) + I_2(aq)$
F) $Cu^{2+}(aq) + 4NH_3(aq) \rightleftharpoons Cu(NH_3)_4{}^{2+}(aq)$
G) $Fe^{3+}(aq) + NCS^-(aq) \rightleftharpoons FeNCS^{2+}(aq)$
H) $Sn^{4+}(aq) + 2Fe^{2+}(aq) \rightleftharpoons Sn^{2+}(aq) + 2Fe^{3+}(aq)$

What mathematical relationship is there between K_c in **A)** and that in **E)**?

2 Write down the expression for K_p for the following gaseous equilibria. State the units of K_p.

A) $N_2(g) + 3H_2(g) \rightleftharpoons 2NH_3(g)$
B) $2SO_2(g) + O_2(g) \rightleftharpoons 2SO_3(g)$
C) $PCl_5(g) \rightleftharpoons PCl_3(g) + Cl_2(g)$
D) $CO_2(g) + H_2(g) \rightleftharpoons CO(g) + H_2O(g)$
E) $N_2(g) + O_2(g) \rightleftharpoons 2NO(g)$
F) $2NO(g) \rightleftharpoons N_2(g) + O_2(g)$
G) $COCl_2(g) \rightleftharpoons CO(g) + Cl_2(g)$

What mathematical relationship exists between K_p in **E)** and that in **F)**?

3 Write down the expression for the modified equilibrium constant for the following heterogeneous equilibria. Give K_c or K_p as appropriate and state the units.

A) $BiCl_3(aq) + H_2O(l) \rightleftharpoons BiOCl(s) + 2HCl(aq)$
B) $AgBr(s) \rightleftharpoons Ag^+(aq) + Br^-(aq)$
C) $PbCl_2(s) \rightleftharpoons Pb^{2+}(aq) + 2Cl^-(aq)$
D) $3Fe(s) + 4H_2O(g) \rightleftharpoons Fe_3O_4(s) + 4H_2(g)$
E) $H_2O(g) + C(s) \rightleftharpoons H_2(g) + CO(g)$
F) $Ag_2CO_3(s) \rightleftharpoons Ag_2O(s) + CO_2(g)$
G) $Ca^{2+}(aq) + CO_3{}^{2-}(aq) \rightleftharpoons CaCO_3(s)$
H) $H_2O(l) \rightleftharpoons H_2O(g)$

Which two of these are solubility products?

4 $H_2(g) + I_2(g) \rightleftharpoons 2HI(g)$

1.9 mol of H_2 and 1.9 mol of I_2 were allowed to reach equilibrium at 710 K. The equilibrium mixture contained 3.0 mol of HI. What is the equilibrium constant? What, if anything, would be the effect on the equilibrium position of increasing the total pressure? Explain your answer. Suggest a method of analysing the equilibrium mixture to show that 3.0 mol of HI was present.

5 $C_2H_5OH(aq) + CH_3CO_2H(aq) \rightleftharpoons CH_3CO_2C_2H_5(aq) + H_2O(aq)$

0.10 mol of ethanol was mixed with 0.10 mol of ethanoic acid and allowed to reach equilibrium. At equilibrium, the remaining ethanoic acid was titrated with 1 mol dm^{-3} sodium hydroxide. 33 cm^3 were required.

A) Use the titration figure to find how many moles of ethanoic acid remain at equilibrium.

B) What is the composition of the equilibrium mixture?

C) What is the value of K_c?

D) Explain why you do not have to know the volume of the mixture to calculate K_c.

6

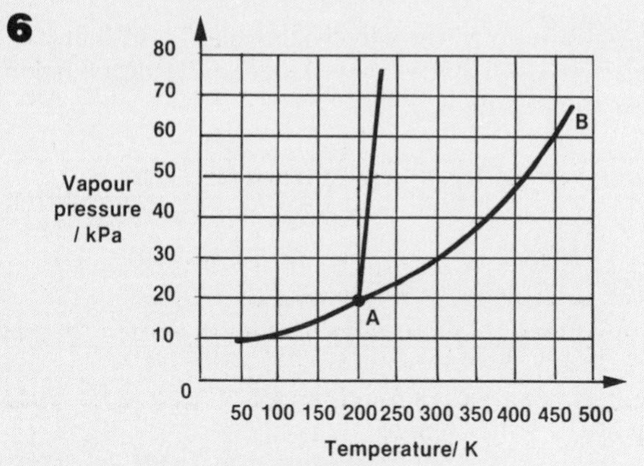

Look at the phase diagram.

A) Mark the areas representing vapour, liquid and solid.

B) What are the boiling temperatures and melting temperatures of this substance at 60 kPa pressure?

C) What does line AB represent?

D) Below what pressure does the substance sublime (turn directly from solid to vapour)?

7 The average M_r of an equilibrium mixture of NO_2 and N_2O_4 was 74 at 304 K. Calculate the value of K_p for

$$N_2O_4(g) \rightleftharpoons 2NO_2(g)$$

at atmospheric pressure (10^2 kPa). What could you say about the equilibrium position in a mixture whose average M_r was 92?

8 Predict the effect on the equilibria below of

A) increasing the temperature

B) increasing the pressure.

i $N_2O_4(g) \rightleftharpoons 2NO_2$ $\quad \Delta_R H^{\ominus} = +58\ \text{kJ mol}^{-1}$

ii $N_2(g) + 3H_2(g) \rightleftharpoons 2NH_3(g)$ $\quad \Delta_R H^{\ominus} = -92\ \text{kJ mol}^{-1}$

iii $H_2(g) + CO_2(g) \rightleftharpoons CO(g) + H_2O(g)$ $\quad \Delta_R H^{\ominus} = +40\ \text{kJ mol}^{-1}$

iv $KCl(s) \rightleftharpoons K^+(aq) + Cl^-(aq)$ $\quad \Delta_R H^{\ominus} = +19\ \text{kJ mol}^{-1}$

9 $CO_2(g) + H_2(g) \rightleftharpoons CO(g) + H_2O(g)$
$K_p = 0.72$ at 1273 K

If 1 mol of CO_2 and 1 mol of H_2 are mixed at this temperature, what will be the composition of the equilibrium mixture? What would be the composition of the equilibrium mixture if 10 mol of CO_2 were mixed with 1 mol of H_2? Does your result agree with the prediction of Le Chatelier's principle? If so, how? If not, why not?

This question requires the use of the formula:

$$x = \frac{-b \pm \sqrt{b^2 - 4ac}}{2a}$$

to solve a quadratic equation.

10 The partition coefficient of compound X between ethoxyethane and water is 10, i.e.

$$\frac{[\text{X in ethoxyethane}]_{eqm}}{[\text{X in water}]_{eqm}} = 10$$

1 g of X is dissolved in 100 cm^3 of water. How much X can be extracted using 100 cm^3 of ethoxyethane:

A) in one extraction, using all the ethoxyethane

B) in two extractions, using 50 cm^3 of ethoxyethane in each?

11

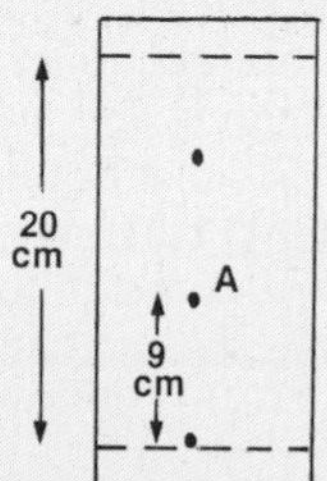

Calculate the R_f value of spot A.

12

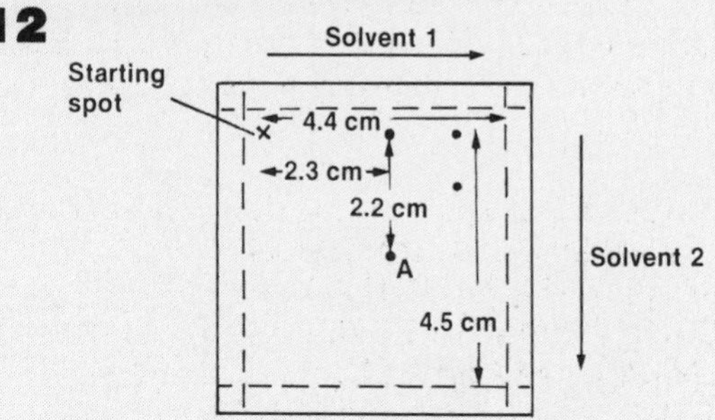

Calculate the R_f values of spot A in solvent 1 and in solvent 2 (page 135).

13 The trace from a gas–liquid chromatogram is shown. Calculate the approximate composition of the original mixture from the peak areas. Which substance has least affinity for the stationary phase?

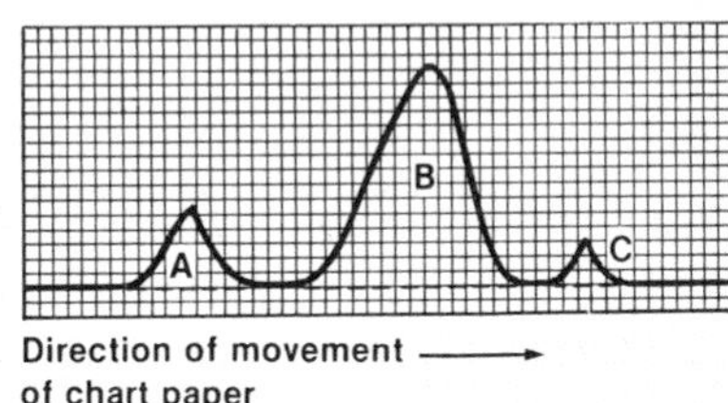

Direction of movement of chart paper →

14 Given the solubility products:

$PbSO_4$ $1.6 \times 10^{-8}\,mol^2\,dm^{-6}$
$MgCO_3$ $1 \times 10^{-5}\,mol^2\,dm^{-6}$

which of the following mixtures will produce a precipitate? Equal volumes of the two solutions were mixed in each case.

A) $10^{-3}\,mol\,dm^{-3}$ $Pb(NO_3)_2$ and $10^{-3}\,mol\,dm^{-3}$ Na_2SO_4
B) $1\,mol\,dm^{-3}$ $Mg(NO_3)_2$ and $10^{-4}\,mol\,dm^{-3}$ Na_2CO_3
C) $10^{-2}\,mol\,dm^{-3}$ $MgCl_2$ and $10^{-2}\,mol\,dm^{-3}$ Na_2CO_3
D) $10^{-5}\,mol\,dm^{-3}$ $Pb(NO_3)_2$ and $1\,mol\,dm^{-3}$ K_2SO_4

15
$$C_2H_5OH(aq) + CH_3CO_2H(aq) \rightleftharpoons CH_3CO_2C_2H_5(aq) + H_2O(l)$$
$K_c = 4$ at room temperature

A) If 4 mol of ethanol are mixed with 1 mol of ethanoic acid, what will be the composition of the mixture when equilibrium is reached?
B) If 2 mol of ethanol are available, how many moles of ethanoic acid must be added if there is to be a yield of 1 mol of ethyl ethanoate?
This question may need the solution of a quadratic equation $0 = ax^2 + bx + c$

via $$x = \frac{-b \pm \sqrt{b^2 - 4ac}}{2a}$$

16
$$H_2(g) + I_2(g) \rightleftharpoons 2HI(g)$$

In an equilibrium mixture the partial pressures were:

H_2 2.2×10^4 Pa
I_2 0.5×10^4 Pa
HI 7.3×10^4 Pa

What is the total pressure of the equilibrium mixture? What is the equilibrium constant at the temperature of the experiment?

17 Ethanoic acid will dissolve to some extent in trichloroethane. Is it possible to titrate the acid with an aqueous solution of alkali which does not mix with trichloroethane? Describe what would happen with regard to any equilibria which would occur.

18 **A)** Explain what is meant by the terms *exothermic*, *endothermic* and *equilibrium constant*, in relation to a chemical reaction.
B) For the reaction
$$N_2(g) + 3H_2(g) \rightleftharpoons 2NH_3(g)$$
$$\Delta H(298\,K) = -92.2\,kJ\,mol^{-1}$$
state how each of the three quantities listed below is affected by each of the three changes of condition listed below.

Quantity	Change of condition
Equilibrium constant	Increase in temperature
Yield of ammonia at equilibrium	Increase in pressure
Rate of attainment of of equilibrium	Introduction of a catalyst

(Your nine answers should be either "goes up", "goes down" or "little change".)
C) Discuss the importance of these considerations in the design of an industrial process for the manufacture of ammonia.

(COSSEC, A/S specimen)

12 Acids and bases

12.1 Introduction

***Acids* are among the best-known groups of compounds. They were originally identified by their sour taste but they are now recognized by the colour changes of dyes called *indicators* and by their reactions with metal oxides, hydroxides and carbonates and also with metals themselves. All these reactions produce ionic compounds called *salts*. Some examples are shown below.**

$$2HNO_3(aq) + CaO(s) \longrightarrow Ca(NO_3)_2(aq) + H_2O(l)$$

nitric acid	calcium oxide	calcium nitrate	water

$$H_2SO_4(aq) + 2NaOH(aq) \longrightarrow Na_2SO_4(aq) + 2H_2O(l)$$

sulphuric acid	sodium hydroxide	sodium sulphate	water

$$2CH_3CO_2H(aq) + K_2CO_3(aq)$$

ethanoic acid	potassium carbonate

$$\longrightarrow 2CH_3CO_2K(aq) + H_2O(l) + CO_2(g)$$

potassium ethanoate	water	carbon dioxide

$$2HCl(aq) + Zn(s) \longrightarrow ZnCl_2(aq) + H_2(g)$$

hydrochloric acid	zinc	zinc chloride	hydrogen

***Bases* were originally identified by their slimy feel. Now they too are recognized by their effect on indicators, and by the fact that they react with or neutralize acids to gave salts. The term *alkali* is used to refer to water-soluble bases.**

12.2 Theories of acidity

Many theories of acidity have been proposed, including those of:

1. **Lavoisier** (1777) who proposed that all acids contain oxygen
2. **Davy** (1816) who suggested that all contain hydrogen
3. **Liebig** (1838) who defined acids as substances containing hydrogen which could be replaced by a metal
4. **Arrhenius** (1887) who thought of acids as producing hydrogen ions (H^+).

You might like to consider how far each of these theories can go in explaining the properties of acids as you know them.

We shall use the **Lowry–Bronsted theory** of acidity which defines an acid as a substance which can donate a proton (an H^+ ion) and a base as a substance which can accept a proton. *Note*: It is important to realize that an H^+ ion *is* a proton. Hydrogen has only one electron and if that is lost all that remains is a proton. Its small size and intense electric field cause it to have unusual properties compared with other positive ions. At this stage it is also worth pointing out that positive ions are often called **cations**, as they move towards the negatively charged cathode. Similarly, negative ions are called **anions**.

Another theory (the **Lewis theory**) is also used to describe acids. This theory regards acids as electron pair acceptors and bases as electron pair donors in the formation of dative covalent bonds (see **section 10.2.2**). For example:

$$\begin{array}{c}\text{F}\\|\\\text{F—B}\\|\\\text{F}\end{array} \quad + \quad \begin{array}{c}\text{H}\\|\\\text{:N—H}\\|\\\text{H}\end{array} \longrightarrow \begin{array}{c}\text{F}\quad\text{H}\\|\qquad|\\\text{F—B}\longleftarrow\text{N—H}\\|\qquad|\\\text{F}\quad\text{H}\end{array}$$

boron trifluoride   ammonia

Here boron trifluoride is acting as a Lewis acid (electron pair acceptor) and ammonia as a Lewis base (electron pair donor). The Lewis definition of acids is wider than the Lowry–Bronsted one. Boron trifluoride contains no hydrogen and so cannot be an acid under the Lowry–Bronsted definition. H^+ ions can *only* form bonds by accepting an electron pair. For example:

$$H^+ \quad + \quad {}^-\text{:OH} \longrightarrow H \longleftarrow O\text{—}H$$

So all Lowry–Bronsted acids are also Lewis acids.

12.2.1 Examples of proton transfer

$$\underset{\text{hydrogen chloride}}{HCl} + \underset{\text{ammonia}}{NH_3} \longrightarrow \underset{\text{chloride ion}}{Cl^-} + \underset{\text{ammonium ion}}{NH_4^+}$$

Here hydrogen chloride is acting as an acid by donating a proton to ammonia, which is acting as a base by accepting it. Notice that acids and bases can only react in pairs—one acid and one base. Notice also that the Cl^- ion left after HCl has donated a proton is itself capable of accepting a proton (to go back to HCl) and is therefore a base. It is called the **conjugate base** of HCl. In the same way, the NH_4^+ ion is an acid as it can donate a proton and so return to being NH_3. It is the **conjugate acid** of ammonia. So we have two conjugate acid–base pairs:

$$\underset{\text{acid}}{HCl} \quad \text{and} \quad \underset{\text{conjugate base}}{Cl^-}$$

and

$$\underset{\text{acid}}{NH_4^+} \quad \text{and} \quad \underset{\text{conjugate base}}{NH_3}$$

HCl is a much better acid (proton donor) than NH_4^+ as it is able to give away its proton and force NH_3 to accept it.

HCl can also donate a proton to water:

$$HCl + H_2O \longrightarrow H_3O^+ + Cl^-$$

↑ base because it accepts a proton (refers to H_2O)

Here water is acting as a base, H_3O^+ being its conjugate acid. H_3O^+ is called the **oxonium ion** but the names hydronium ion and hydroxonium ion are also used.

Water may also act as an acid, for example:

$$H_2O + NH_3 \longrightarrow OH^- + NH_4^+$$

↑ acid because it donates a proton (refers to H_2O)

Here OH^- is the conjugate base and H_2O the conjugate acid.

Acid and base reactions can be thought of as competition for protons, the better base 'winning' the proton, so in:

$$HCl + NH_3 \longrightarrow NH_4^+ + Cl^-$$

ammonia must be a better base than the chloride ion, as it 'wins' the proton.

12.3 The ionization of water

We have seen in the box *Examples of equilibrium constant expressions*, **section 11.4.4**, that liquid water is slightly ionized:

$$H_2O \rightleftharpoons H^+ + OH^-$$

or this may be written:

$$H_2O + H_2O \rightleftharpoons H_3O^+ + OH^-$$

to emphasize that this is an acid–base reaction in which one water molecule donates a proton to another. Throughout this chapter we shall represent a proton in an aqueous solution by $H^+(aq)$ rather than $H_3O^+(aq)$.

This equilibrium:

$$H_2O(l) \rightleftharpoons H^+(aq) + OH^-(aq)$$

is established in water and all aqueous solutions.

The equilibrium constant:

$$K_w = [H^+(aq)][OH^-(aq)]$$

is called the **ionic product** of water, and at 298 K it is equal to $1 \times 10^{-14}\,mol^2\,dm^{-6}$. In pure water, each water molecule which dissociates (splits up) gives rise to one H^+ ion and one OH^- ion, so at 298 K:

$$[OH^-(aq)] = [H^+(aq)]$$

so

$$10^{-14} = [H^+(aq)]^2$$

$$[H^+(aq)] = 10^{-7}\,mol\,dm^{-3} = [OH^-(aq)]$$

12.4 Measurement of acidity—the pH scale

The acidity of a solution depends on the concentration of $H^+(aq)$ and is measured on the pH scale. pH is defined as $-\log_{10}[H^+(aq)]$. Although more complicated than simply stating the concentration of $H^+(aq)$, it does do away with awkward numbers like 10^{-13}, etc. A difference of one pH number means a tenfold difference in acidity, so that pH2 is ten times as acidic as pH3.

The pH of pure water

In pure water:

$$[H^+(aq)] = 10^{-7}\,mol\,dm^{-3}, \text{ see above}$$

$$\log[H^+(aq)] = -7$$

$$-\log[H^+(aq)] = 7$$

so the pH of pure water = 7 at 298 K

The pH of hydrochloric acid solutions

In dilute solution hydrochloric acid dissociates completely in water to give H^+ ions and Cl^- ions, i.e. the reaction:

$$HCl(aq) \longrightarrow H^+(aq) + Cl^-(aq)$$

goes to completion.

$$\text{So in } 1\,\text{mol dm}^{-3}\,\text{HCl},\quad [\text{H}^+(\text{aq})] = 1\,\text{mol dm}^{-3}$$
$$\log 1 = 0$$
$$-\log 1 = 0$$
$$\text{so the pH of } 1\,\text{mol dm}^{-3}\,\text{HCl} = 0$$

In a 0.1 mol dm^{-3} solution of HCl,

$$[\text{H}^+(\text{aq})] = 0.1\,\text{mol dm}^{-3} = 10^{-1}\,\text{mol dm}^{-3}$$
$$\log[\text{H}^+(\text{aq})] = -1$$
$$\text{so the pH of } 0.1\,\text{mol dm}^{-3}\,\text{HCl} = 1$$

In a 2 mol dm^{-3} solution of HCl,

$$[\text{H}^+(\text{aq})] = 2\,\text{mol dm}^{-3}$$
$$\text{pH} = -\log[\text{H}^+(\text{aq})]$$
$$= -0.3$$

Although pH numbers of less than zero and more than 14 are possible, zero and 14 are the practical limits of the scale, as in very concentrated solutions acids and alkalis tend not to be fully dissociated.

DIPROTIC ACIDS

These are acids where one molecule dissociates to give *two* protons, for example sulphuric acid which also dissociates more or less completely in dilute solutions:

$$\text{H}_2\text{SO}_4(\text{aq}) \longrightarrow 2\text{H}^+(\text{aq}) + \text{SO}_4^{2-}(\text{aq})$$

In 0.1 mol dm^{-3} sulphuric acid:

$$[\text{H}^+(\text{aq})] = 0.2\,\text{mol dm}^{-3}$$
$$\log[\text{H}^+(\text{aq})] = -0.7$$
$$-\log[\text{H}^+(\text{aq})] = 0.7$$
$$\text{pH} = 0.7$$

The pH scale can be confusing until you are used to it. Neutral solutions are pH 7 and acids have pH values lower than this. Many people think instinctively that a *lower* pH means a *less* acidic solution. It does not, it means a *more* acidic one. Even experienced chemists sometimes trip up over this.

The pH of alkaline solutions

In alkaline solutions, the situation is slightly more complex. It is tempting to think that there are no H^+ ions in alkaline solutions, but this is not the case.

The relationship:

$$[\text{H}^+(\text{aq})][\text{OH}^-(\text{aq})] = 10^{-14}\,\text{mol}^2\,\text{dm}^{-6}$$

applies to *all* aqueous solutions, acid, alkaline or neutral at 298 K. To find the pH of an alkaline solution we must first calculate $[\text{OH}^-(\text{aq})]$ then use:

$$[\text{H}^+(\text{aq})][\text{OH}^-(\text{aq})] = 10^{-14}\,\text{mol}^2\,\text{dm}^{-6}$$

to calculate $[\text{H}^+(\text{aq})]$, from which the pH can be calculated.

For example, to find the pH of 1 mol dm^{-3} sodium hydroxide:

Sodium hydroxide is fully dissociated in dilute aqueous solution.

$$\text{NaOH}(\text{aq}) \longrightarrow \text{Na}^+(\text{aq}) + \text{OH}^-(\text{aq})$$

so $$[\text{OH}^-(\text{aq})] = 1\,\text{mol dm}^{-3}$$

but $$[\text{OH}^-(\text{aq})][\text{H}^+(\text{aq})] = 10^{-14}\,\text{mol}^2\,\text{dm}^{-6}$$
$$1 \times [\text{H}^+(\text{aq})] = 10^{-14}\,\text{mol dm}^{-3}$$
$$[\text{H}^+(\text{aq})] = 10^{-14}\,\text{mol dm}^{-3}$$

taking logs $$\log[\text{H}^+(\text{aq})] = -14$$
$$\text{pH} = 14$$

For example, to find the pH of 0.1 mol dm^{-3} sodium hydroxide:

$$[\text{OH}^-(\text{aq})] = 10^{-1}\,\text{mol dm}^{-3}$$
$$[\text{H}^+(\text{aq})][\text{OH}^-(\text{aq})] = 10^{-14}\,\text{mol}^2\,\text{dm}^{-6}$$

$$[H^+(aq)] \times 10^{-1} = 10^{-14}\,mol\,dm^{-3}$$
$$[H^+(aq)] = 10^{-13}\,mol\,dm^{-3}$$
taking logs $$\log[H^+(aq)] = -13$$
$$pH = 13$$

Try calculating the pH of 0.2 mol dm^{-3} sodium hydroxide. You should get pH = 13.3.

A

ACID RAIN

Everyone has become aware of acid rain recently, but it is by no means a new problem. In the early 1950s a series of 'smogs' (the word is a combination of 'smoke' and 'fog') brought traffic to a halt in London and caused many deaths through respiratory complaints. At the time, smoke was the most obvious problem, but now it is thought that sulphuric acid was the main cause of deaths. The pH of smog has been estimated at less than 2.

Smog obscures Piccadilly Circus, 1955

Almost a century before this, the Norwegian playwright Ibsen wrote in the play *Brand*:

> 'Dimmer visions, worse foreboding
> Glare upon me through the gloom
> Britain's smoke-cloud sinks corroding
> On the land in noisome fumes...'

Scandinavians are still complaining about acid pollution from Britain.

The precise causes of acid rain are the subject of fierce debate, but it is indisputable that the burning of sulphur-containing fuels leads to the formation of sulphur dioxide which reacts with air and water in the atmosphere to produce sulphuric acid. British power stations are among the culprits, and gradually flue gas desulphurization equipment is being fitted. This process involves passing the gaseous combustion products through a suspension of calcium hydroxide which reacts with any acidic gases to produce salts, mostly gypsum, calcium sulphate.

The effects of acid pollution are many; two of the most important are the deaths of fish and trees. A fall in pH of many streams and lakes has been observed in many areas. This kills fish both directly and also by leaching poisonous aluminium ions out of the soil into the water. Trees are being killed by absorption of the acid both through their roots and directly through their leaves.

12.4.1 Strong and weak acids and bases

The above examples have dealt with acids like hydrochloric and sulphuric acids which, when dissolved in water, dissociate *completely* into ions. Although in the gas phase hydrogen chloride is a covalent molecule, a dilute solution of it in water is wholly ionic. To all intents and purposes, there are no covalently bonded hydrogen chloride molecules remaining. Acids which completely dissociate into ions in aqueous solutions are called **strong acids**. The word strong refers *only* to the extent of dissociation and *not in any way* to the concentration.

So it is perfectly possible to have a very dilute solution of a strong acid. It is vital to realize that strength and concentration are completely independent. Careful use of the two words is most important.

Many acids are far from fully dissociated when dissolved in water. Ethanoic acid (the acid in vinegar, also known as acetic acid) is a typical example. In a 1 mol dm^{-3} solution of ethanoic acid, only about four in every thousand ethanoic acid molecules are dissociated into ions (so the **degree of dissociation** is 4/1000); the rest remain dissolved as wholly covalently bonded molecules. In fact an equilibrium is set up:

	$CH_3CO_2H(aq)$	$\rightleftharpoons$	$H^+(aq)$	+	$CH_3CO_2^-(aq)$
	ethanoic acid		hydrogen ions		ethanoate ions
before dissociation	1000		0		0
at equilibrium	996		4		4

Acids like this are called **weak acids**. Again, note that weak refers only to the degree of dissociation, not the concentration. In a 5 mol dm^{-3} solution, ethanoic acid is still a weak acid, while in a 10^{-4} mol dm^{-3} solution, hydrochloric acid is still a strong acid.

The same arguments apply to bases. Strong bases are completely dissociated into ions in aqueous solutions, while weak bases are only partially split up. For example, sodium hydroxide is a strong base:

$$NaOH(aq) \longrightarrow Na^+(aq) + OH^-(aq)$$

Ammonia NH_3, is a weak base in aqueous solution:

$$NH_3(aq) + H_2O(l) \rightleftharpoons NH_4^+(aq) + OH^-(aq)$$

12.4.2 Calculating pH of solutions of weak acids and bases

Weak acids

Imagine a weak acid, HA, which dissociates:

$$HA(aq) \rightleftharpoons H^+(aq) + A^-(aq)$$

The equilibrium constant is given by:

$$K_c = \frac{[H^+(aq)]_{eqm}[A^-(aq)]_{eqm}}{[HA(aq)]_{eqm}}$$

This could equally well be written:

$$HA(aq) + H_2O(l) \rightleftharpoons A^-(aq) + H_3O^+(aq)$$

$$K_c' = \frac{[H_3O^+(aq)]_{eqm}[A^-(aq)]_{eqm}}{[HA(aq)]_{eqm}[H_2O(l)]_{eqm}}$$

Since the concentration of $H_2O(l)$ is effectively constant, it is incorporated into the value of K_c so that as before:

$$K_c = \frac{[H_3O^+(aq)]_{eqm}[A^-(aq)]_{eqm}}{[HA(aq)]_{eqm}}$$

For a weak acid, K_c is usually given the symbol K_a and called the **acid dissociation constant**.

The larger the value of K_a, the stronger the acid. Acid dissociation constants for some acids are given in **Table 12.1**.

Table 12.1 Values of K_a for some weak acids

Acid	K_a/mol dm^{-3}
Chloroethanoic	1.3×10^{-3}
Benzenecarboxylic	6.3×10^{-5}
Ethanoic	1.7×10^{-5}
Hydrocyanic	4.9×10^{-10}

THE pH OF 1 MOL DM^{-3} ETHANOIC ACID

Using the same method as in Chapter 11 for equilibrium calculations:

	$CH_3CO_2H(aq)$	$\rightleftharpoons$ $CH_3CO_2^-(aq)$	+ $H^+(aq)$
before dissociation	1	0	0
at equilibrium	$1-[H^+(aq)]$	$[CH_3CO_2^-(aq)]$	$[H^+(aq)]$

$$K_a = \frac{[CH_3CO_2^-(aq)]\ [H^+(aq)]}{[CH_3CO_2H(aq)]}$$

But as each CH_3CO_2H molecule which dissociates produces one $CH_3CO_2^-$ ion and one H^+ ion, $[CH_3CO_2^-(aq)] = [H^+(aq)]$:

$$K_a = \frac{[H^+(aq)]^2}{1-[H^+(aq)]}$$

Since the degree of dissociation is so small, the concentration of $H^+(aq)$ is very small and to a good approximation $1-[H^+(aq)] \simeq 1$. So:

$$K_a = \frac{[H^+(aq)]^2}{1}$$

$$1.7 \times 10^{-5}\ mol\ dm^{-3} = [H^+(aq)]^2$$

$$[H^+(aq)] = \sqrt{1.7 \times 10^{-5}}$$
$$[H^+(aq)] = 4.12 \times 10^{-3}\,\text{mol dm}^{-3}$$

so
$$pH = 2.384$$

Incidentally, how good the approximation is can be seen by comparing the value of pH calculated above, 2.384, with that calculated using the quadratic formula (see **section 11.4.3**) which gives 2.386. With a weak acid, the approximation can virtually always be used with confidence.

THE pH OF 0.1 MOL DM^{-3} ETHANOIC ACID

By the same reasoning, we get:

$$K_a = \frac{[H^+(aq)]^2}{0.1 - [H^+(aq)]}$$

Again, $0.1 - [H^+(aq)] \simeq 0.1$ so:

$$1.7 \times 10^{-5} = \frac{[H^+(aq)]^2}{0.1}$$
$$1.7 \times 10^{-6} = [H^+(aq)]^2$$
$$[H^+(aq)] = 1.3 \times 10^{-3}\,\text{mol dm}^{-3}$$
$$pH = 2.88$$

Weak bases

The dissociation of a weak base, B, in aqueous solution may be represented:

$$B(aq) + H_2O(l) \rightleftharpoons BH^+(aq) + OH^-(aq)$$

$$K_c = \frac{[BH^+(aq)]_{eqm}[OH^-(aq)]_{eqm}}{[B(aq)]_{eqm}[H_2O(l)]_{eqm}}$$

The concentration of $H_2O(l)$ is effectively constant and is incorporated into the equilibrium constant to give K_b, the **base dissociation constant**:

$$K_b = \frac{[BH^+(aq)]_{eqm}[OH^-(aq)]_{eqm}}{[B(aq)]_{eqm}}$$

The larger the value of K_b, the stronger the base. **Table 12.2** lists K_bs for some weak bases.

Table 12.2 Values of K_b for some weak bases

Base	K_b/mol dm^{-3}
Ethylamine	5.6×10^{-4}
Ammonia	1.7×10^{-5}
Phenylamine	4.3×10^{-10}

Note: Some books do not list values of K_b. Instead they have values of K_a for the conjugate acid, e.g. ammonia, NH_3 ($K_b = 1.7 \times 10^{-5}$). For NH_4^+, the conjugate acid, $K_a = 5.8 \times 10^{-10}$.
We can use the relationship:

$$K_a \times K_b = 10^{-14}$$

to find K_b:

$$K_b = \frac{10^{-14}}{5.8 \times 10^{-10}}$$
$$= 1.7 \times 10^{-5}\,\text{mol dm}^{-3}$$

THE pH OF 1 MOL DM^{-3} AMMONIA

	$NH_3(aq)$ + $H_2O(l)$ ⇌	$NH_4^+(aq)$ +	$OH^-(aq)$
before dissociation	1	0	0
at equilibrium	$1 - [OH^-(aq)]$	$[NH_4^+(aq)]$	$[OH^-(aq)]$

$$K_b = \frac{[NH_4^+(aq)][OH^-(aq)]}{[NH_3(aq)]}$$

$[NH_4^+(aq)] = [OH^-(aq)]$, as each NH_3 involved produces one of each

$$1.7 \times 10^{-5} = \frac{[OH^-(aq)]^2}{1 - [OH^-(aq)]}$$

since $[OH^-(aq)]$ is small, $1 - [OH^-(aq)] \simeq 1$

$$1.7 \times 10^{-5} = [OH^-(aq)]^2$$
$$[OH^-(aq)] = 4.12 \times 10^{-3}$$

Now we must calculate $[H^+(aq)]$ using:

$$K_w = [H^+(aq)][OH^-(aq)] = 10^{-14}\,\text{mol}^2\,\text{dm}^{-6}$$
$$10^{-14} = [H^+(aq)] \times 4.12 \times 10^{-3}$$
$$[H^+(aq)] = \frac{10^{-14}}{4.12 \times 10^{-3}}$$
$$[H^+(aq)] = 2.42 \times 10^{-12}\,\text{mol dm}^{-3}$$

taking logs

$$\log[H^+(aq)] = -11.6$$
$$pH = 11.6$$

Try calculating the pH of 0.1 mol dm^{-3} ammonia in the same way. You should get 11.1.

A

HOUSEHOLD ACIDS AND BASES

A large number of common household substances contain acids or bases. Probably the most well-known is vinegar, a dilute solution of the weak acid ethanoic acid. It is also widely known that oranges, lemons and limes contain citric acid, a weak triprotic acid.

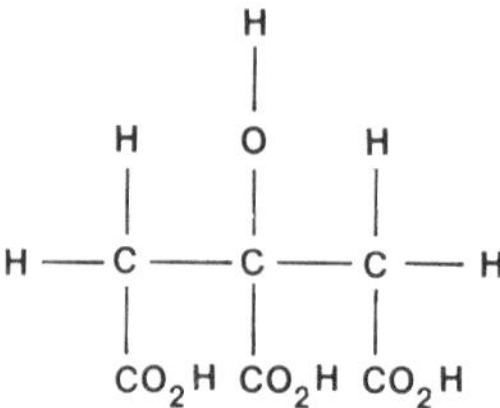

Citric acid (2-hydroxypropane-1,2,3-tricarboxylic acid). The acid hydrogens (those lost on dissociation) are marked in blue

The electrolyte in car batteries is a solution of the strong acid, sulphuric acid. Many powder types of lavatory cleaner contain sodium hydrogensulphate, $NaHSO_4$. This is called an acid salt and is produced when sulphuric acid, a diprotic acid, is half neutralized:

$$H_2SO_4(aq) + NaOH(aq) \longrightarrow NaHSO_4(aq) + H_2O\ (l)$$

The acid salt dissociates to give H^+ (aq):

$$NaHSO_4(aq) \longrightarrow H^+(aq) + Na^+(aq) + SO_4^{2-}(aq)$$

The purpose of this is to dissolve hard water scale, containing calcium carbonate from the lavatory bowl:

$$2H^+(aq) + CaCO_3(s) \longrightarrow H_2O(l) + CO_2(g) + Ca^{2+}(aq)$$

However, care must be taken that this type of cleaner does not come into contact with other brands of cleaner containing bleach.

Bleach contains a mixture of chloride ions, Cl^- and chlorate(I), ClO^- ions (often still called hypochlorite ions). In the presence of acid the following reaction occurs:

$$2H^+(aq) + ClO^-(aq) + Cl^-(aq) \longrightarrow H_2O(l) + Cl_2(g)$$

producing chlorine, an irritant gas which is toxic at high concentrations.

Also used for removing scale, this time in kettles, is sulphamic acid:

$$NH_2SO_3H$$

This acid is present in descaling products for kettles and coffee machines.

Bases to be found around the home include ammonia, which is present in many heavy duty cleaners. Bases are particularly good at dissolving grease. Some types of paint stripper contain sodium hydroxide and commercial paint strippers dip articles to be stripped in a large vat of sodium hydroxide solution which they refer to as 'caustic' after the common name for sodium hydroxide—caustic soda. In the garden, 'lime' or slaked lime is used to treat acid soils. Slaked lime is powdered calcium hydroxide. A final piece of household acid–base chemistry is provided by effervescent 'liver salts'. These contain dry citric acid and sodium hydrogencarbonate. These do not react together when dry and can be kept indefinitely in a closed tin or sealed sachet, but both dissociate in water and react to give off carbon dioxide which causes the effervescence.

The descaler (left) contains sulphamic acid; the paint stripper sodium hydroxide; and the liver salts citric acid and sodium hydrogencarbonate.

A

THE ACID–BASE CHEMISTRY OF SOAP

Soap is the sodium salt of a long-chain organic acid such as stearic acid:

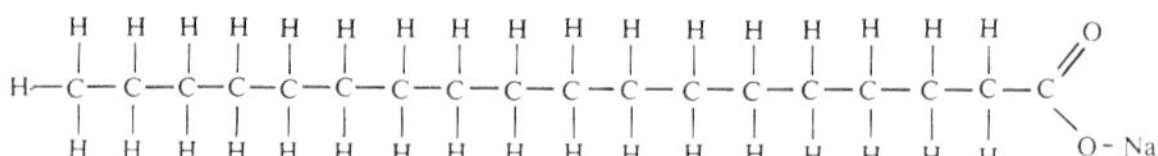

It works by the long hydrocarbon chain dissolving well in grease and the ionic end dissolving well in water, thus promoting the mixing of grease and water. Since soap is the salt of a weak acid and a strong base (stearic acid and sodium hydroxide), salt hydrolysis occurs, forming an alkaline solution:

$$C_{17}H_{35}C\begin{matrix}\diagup O \\ \diagdown O^-\end{matrix}(aq) + H_2O(l) \rightleftharpoons C_{17}H_{35}C\begin{matrix}\diagup O \\ \diagdown OH\end{matrix}(aq) + OH^-(aq)$$

The alkaline solution is potentially harmful to the skin, so soap has additional stearic acid added to displace the equilibrium to the left (Le Chatelier's principle) and produce a more neutral solution.

12.5 pH changes during acid–base titrations

Figure 12.1 Apparatus to investigate pH changes during a titration

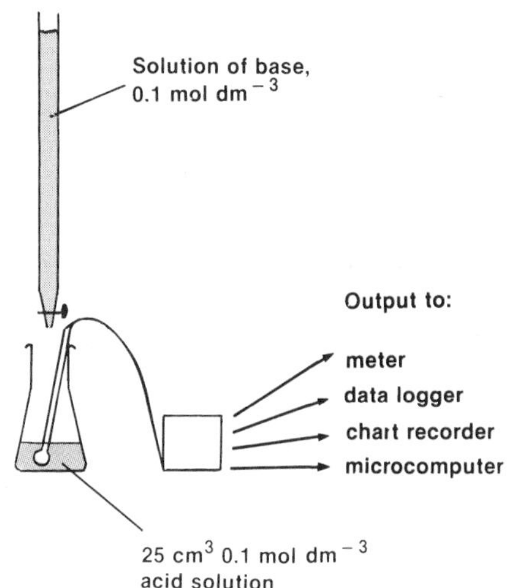

pH values of solutions can be accurately measured using a pH electrode which produces a voltage related to the concentration of $H^+(aq)$ in the solution. The principles of this device are discussed in the box *The pH meter*, **section 13.3.** The voltage may be measured on a meter calibrated to read pH directly or, to monitor changing pHs, recorded on a chart recorder, data logging device or microcomputer to produce a graphical output directly. This method can be used to follow the changes in pH when an acid reacts with a base. The graph in **Figure 12.2** shows the results obtained for four cases:

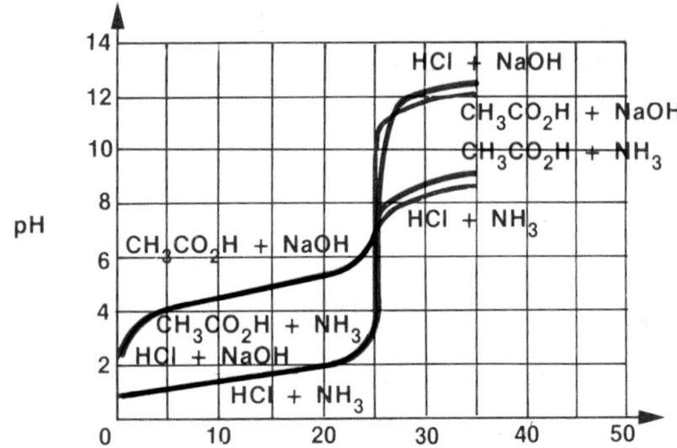

Figure 12.2 (right) Graphs of pH changes for titrations of different acids with different bases

In each case, we start with $25\,cm^3$ of the acid in the flask and add the base from a burette.

The first thing to notice about these curves is that pH does not change linearly as the base is added. Each curve has an almost horizontal section where a lot of base can be added without changing the pH much. There is also a very steep portion of each curve where a single drop of base changes the pH by several units.

In a titration, the **equivalence point** is the point at which exactly the same number of moles of hydroxide ions have been added as there are moles of hydrogen ions. In each of these titrations, the equivalence point is reached after $25\,cm^3$ of acid have been added. Notice that the pH at the equivalence point is not always seven. However, in each case, except the weak acid/weak base titration, there is a large and rapid change of pH at the equivalence point (i.e. the curve is almost vertical) even though this may not be centred on pH seven.

pH meter

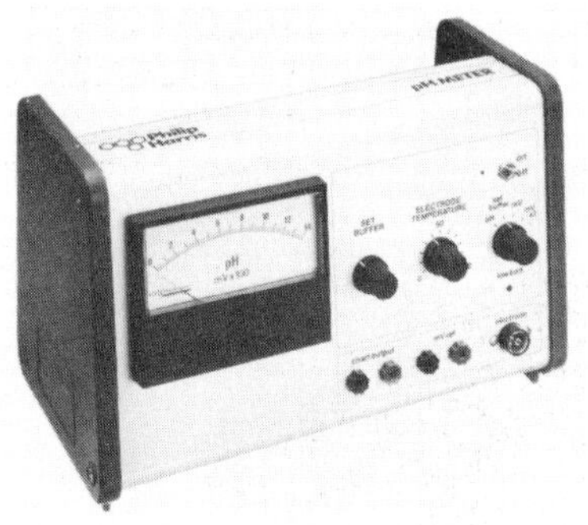

The set-up shown in Figure 12.1

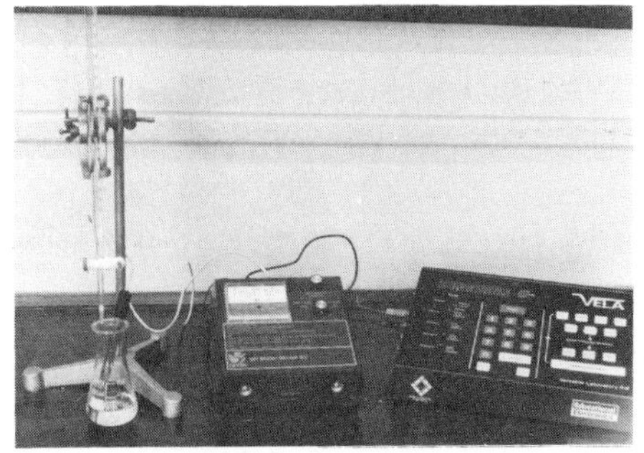

12.5.1 Choice of indicators for titrations

When we do a titration to find the concentration of an unknown solution of an acid or alkali, we use an indicator whose colour change tells us when the titration is complete—shows us the **end point**. It is important that:

1. The colour change is sharp rather than gradual, i.e. that no more than one drop of acid (or alkali) is needed to give a complete colour change. An indicator which changes colour gradually over several cubic centimetres would be difficult to use.

2. The end point of the titration given by the indicator must be the same as the equivalence point. Otherwise the titration will give us the wrong answer.

3. It is also desirable that an indicator gives a distinct colour change. For example, the colourless to red change of phenolphthalein is easier to see than the red to yellow of methyl orange, but this is much less important than the previous two points.

Some common indicators are given overleaf. Notice that the colour change takes place over a pH range of around 2 units.

Figure 12.3 Titration of a weak acid/strong base. Adding 0.1 mol dm^{-3} NaOH (aq) to 25 cm^3 of 0.1 mol dm^{-3} CH_3CO_2H(aq)

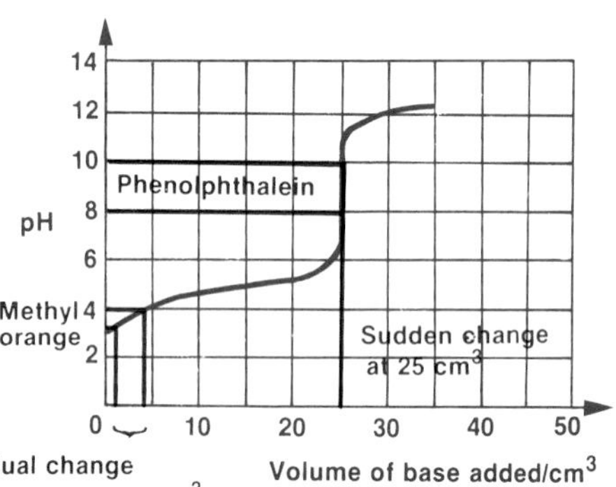

Figure 12.4 Titration of a strong acid/weak base. Adding 0.1 mol dm^{-3} NH_3(aq) to 25 cm^3 of 0.1 mol dm^{-3} HCl(aq)

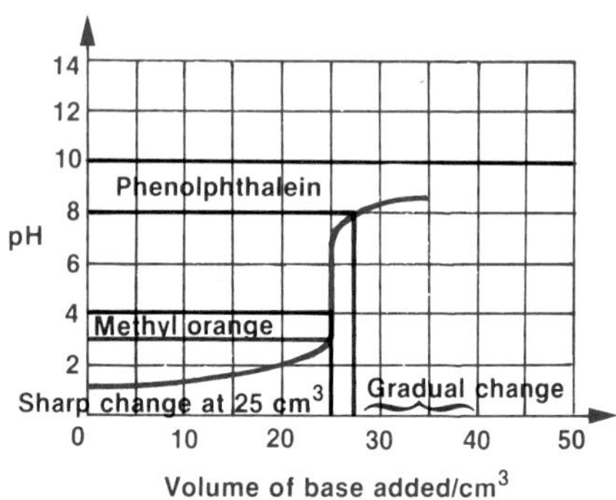

Figure 12.5 Titration of a strong acid/strong base. Adding 0.1 mol dm^{-3} NaOH(aq) to 25 cm^3 of 0.1 mol dm^{-3} HCl(aq)

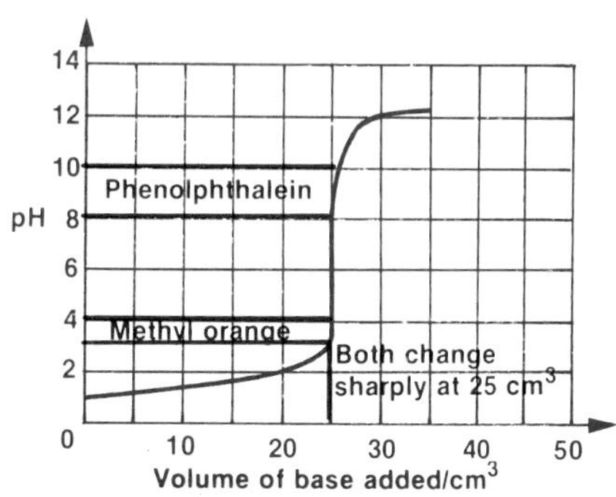

Figure 12.6 Titration of a weak acid/weak base. Adding 0.1 mol dm^{-3} NH_3(aq) to 25 cm^3 of 0.1 mol dm^{-3} CH_3CO_2H(aq)

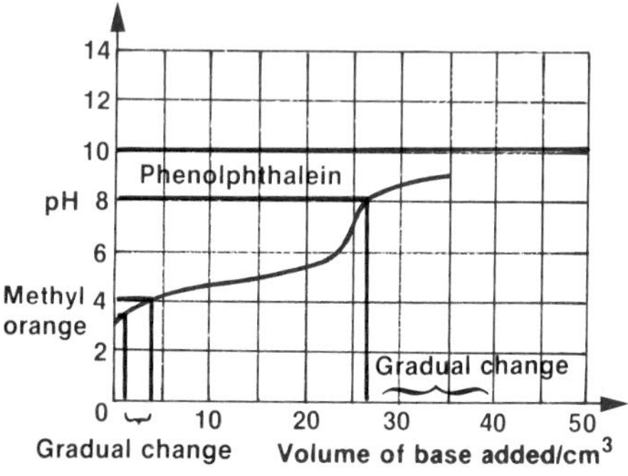

methyl orange	red 3.2	4.2 yellow		
litmus		red 5.0	8.0 blue	
bromothymol blue		yellow 6.0	7.0 blue	
phenolphthalein			colourless 8.2	10.0 red

Weak acid/strong base titration

Let us now examine the titration of a weak acid with a strong base and compare the suitability of phenolphthalein as an indicator with that of methyl orange, **Figure 12.3**. Methyl orange would change gradually at between 1 and 5 cm^3 of base compared with the equivalence point of 25 cm^3. Clearly this indicator is not suitable.

Phenolphthalein will change sharply at exactly 25 cm^3, the equivalence point, and would therefore be a good choice.

Strong acid/weak base titration

Let us try the same thing, **Figure 12.4.** Here methyl orange will change sharply at the equivalence point but phenolphthalein would be of no use.

Strong acid/strong base titration

See **Figure 12.5.** Here either indicator would be suitable but phenolphthalein is usually preferred because of its more easily seen colour change.

Weak acid/weak base titration

See **Figure 12.6.** Here neither indicator is suitable, both changing over a considerable range and well away from the equivalence point. In fact no indicator could be suitable, as an indicator requires a vertical portion of the curve of at least two pH units at the equivalence point to give a sharp change.

The pH value at the equivalence point

A close look at the titration curves for weak acid/strong base and strong acid/weak base shows that in neither curve is the pH at the equivalence point *exactly* equal to seven. See **Figure 12.7.**

Figure 12.7 pH values at equivalence point

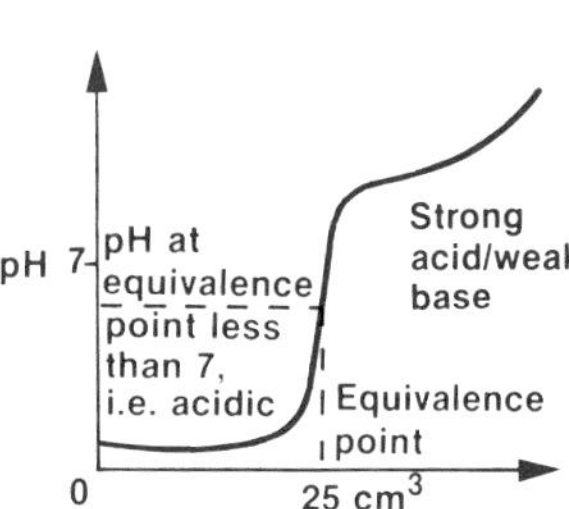

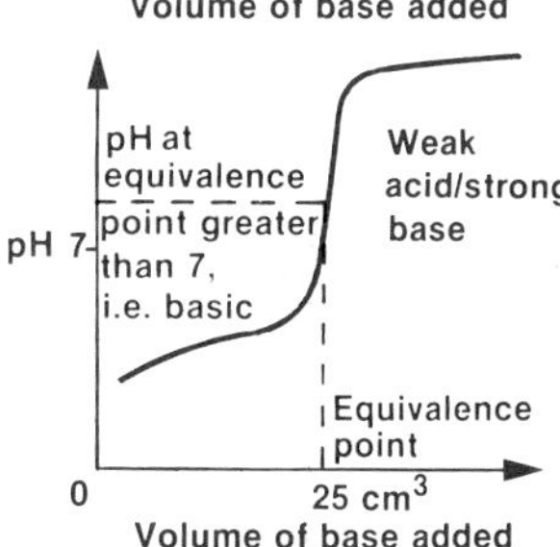

At the equivalence point:

$$NH_3 + HCl \longrightarrow NH_4{}^+Cl^- + H_2O$$

The only ionic compound left is $NH_4{}^+Cl^-$. But $NH_4{}^+$ is itself a weak acid, as it can dissociate:

$$NH_4{}^+ \rightleftharpoons NH_3 + H^+$$

This produces H^+ ions in solution and gives a pH of less than 7 (i.e. acidic).

At the equivalence point:

$$NaOH + CH_3CO_2H \longrightarrow CH_3CO_2{}^-Na^+ + H_2O$$

The only ionic compound is $CH_3CO_2{}^-Na^+$. But $CH_3CO_2{}^-$ is itself a base, as it can remove a proton from water:

$$CH_3CO_2{}^- + H_2O \rightleftharpoons CH_3CO_2H + OH^-$$

This produces OH^- ions in solution and gives a pH of more than 7 (i.e. alkaline).

This phenomenon is sometimes called **salt hydrolysis** (hydrolysis meaning reaction with water). Whenever we have a salt of a strong acid and weak base (for example ammonium chloride—a salt of ammonia and hydrochloric acid) or of a weak acid and strong base (for example sodium ethanoate—a salt of sodium hydroxide and ethanoic acid), the salt solution will not be neutral. A rule which is easy to remember is that the pH is governed by the 'strong' part of the salt so

strong acid/weak base ⟶ acidic salt
weak acid/strong base ⟶ alkaline salt

So ammonium chloride is acidic; sodium ethanoate is alkaline.

A salt of a weak acid and a weak base may be either acidic or alkaline depending on the relative strengths of the acid and base concerned.

Titration with diprotic acids

A diprotic acid such as maleic acid (*cis*-butenedioic acid) has two protons which can be successively replaced in what are essentially two separate acid–base reactions. So the titration curve has two 'kinks', see **Figure 12.8**. It may be possible to select indicators so that the two reactions may be titrated separately. With indicator 1, an end point of 25 cm^3 is observed and with indicator 2, an end point of 50 cm^3 is obtained. The first end point represents:

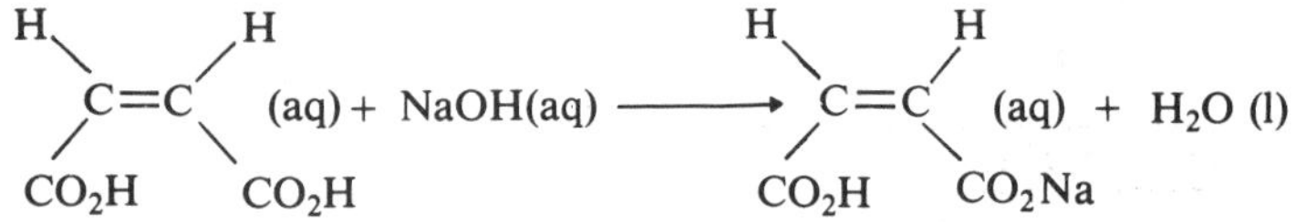

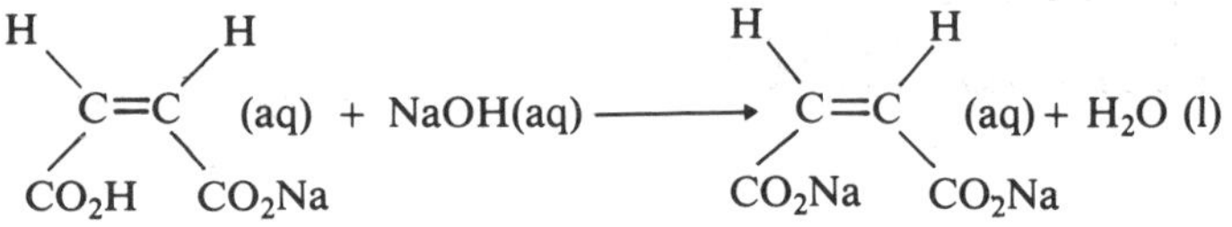

Figure 12.8 Titration of a diprotic acid. Adding 0.1 mol dm^{-3} NaOH(aq) to 25 cm^3 of 0.1 mol dm^{-3} acid

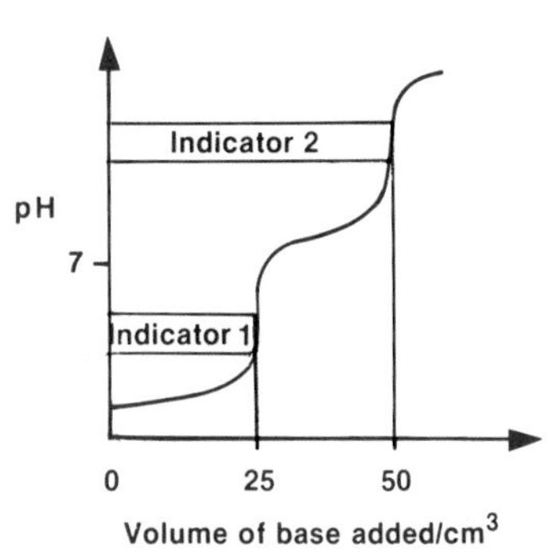

$$(H)(CO_2H)C{=}C(H)(CO_2H)\,(aq) + NaOH(aq) \longrightarrow (H)(CO_2H)C{=}C(H)(CO_2Na)\,(aq) + H_2O\,(l)$$

and the second:

$$(H)(CO_2H)C{=}C(H)(CO_2Na)\,(aq) + NaOH(aq) \longrightarrow (H)(CO_2Na)C{=}C(H)(CO_2Na)\,(aq) + H_2O\,(l)$$

12.5.2 Indicators

How do indicators work? Why do different indicators change colour over different pH ranges? Why do indicators change colour over a pH range of about two units?

The structures of some indicators are given in **Figure 12.9**. Indicators are weak acids or weak bases which can dissociate in aqueous solution. Imagine a weak acid HIn:

Figure 12.9 Structures of some indicators. The acidic hydrogen is marked in brown

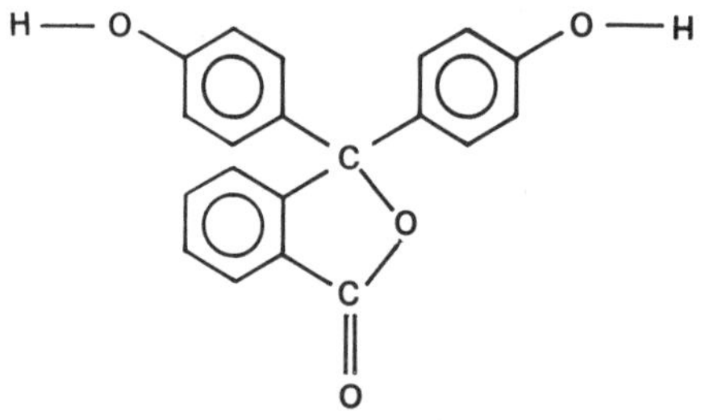

Phenolphthalein

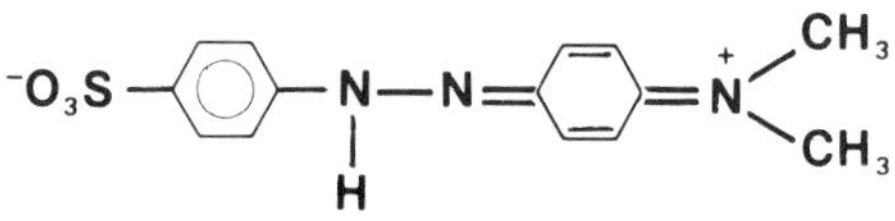

Methyl orange

$$\underset{\text{red}}{HIn(aq)} \rightleftharpoons H^+(aq) + \underset{\text{blue}}{In^-(aq)}$$

If HIn and In$^-$ are different colours, say red and blue respectively, then HIn will act as an indicator. In a solution with a high concentration of H$^+$ (an acid) the equilibrium will be forced to the left by Le Chatelier's principle, almost all the indicator will exist as HIn and the solution will be red. In an alkaline solution where the concentration of OH$^-$ is high, H$^+$ ions will be removed as H_2O:

$$HIn(aq) \rightleftharpoons H^+(aq) + In^-(aq)$$
$$\downarrow {+OH^-}$$
$$H_2O$$

Le Chatelier's principle predicts that the equilibrium will move to the right and most of the indicator will exist as In$^-$ (blue).

Quantitatively, we can use the expression:

$$K_{In} = \frac{[H^+(aq)][In^-(aq)]}{[HIn(aq)]}$$

where K_{In} is the acid dissociation constant for the indicator.

This can be rearranged:

$$K_{In} = [H^+(aq)] \times \frac{[In^-(aq)]}{[HIn(aq)]}$$

Taking logs and remembering that multiplication of numbers is achieved by adding logs:

$$\log K_{In} = \log [H^+(aq)] + \log\left(\frac{[In^-(aq)]}{[HIn(aq)]}\right)$$

Multiplying both sides by -1:

$$-\log K_{In} = -\log [H^+(aq)] - \log\left(\frac{[In^-(aq)]}{[HIn(aq)]}\right)$$

$-\log [H^+(aq)]$ is of course the pH of the solution and $-\log K_{In}$ can be called pK_{In}. So

$$pK_{In} = pH - \log\left(\frac{[In^-(aq)]}{[HIn(aq)]}\right)$$

or

$$pH = pK_{In} + \log\left(\frac{[In^-(aq)]}{[HIn(aq)]}\right)$$

or

$$pH = pK_{In} - \log\left(\frac{[HIn(aq)]}{[In^-(aq)]}\right)$$

which is equivalent.

This is called the **Henderson equation**. Values of pK_{In} are given in data books along with the values of K_{In} see **Table 12.3.** The Henderson equation lets us calculate the ratio $[HIn(aq)]/[In^-(aq)]$ in a solution of a given pH.

Table 12.3 pK_{In} **for some common indicators**

Indicator	pK_{In}
Methyl orange	3.7
Methyl red	5.1
Phenolphthalein	9.3

In a mixture of two colours of about the same intensity, the human eye can detect only the more concentrated if it is present in a ratio of more than 10:1. So if $[HIn(aq)]/[In^-(aq)] > 10$, the solution will appear red with no trace of blue. If $[HIn(aq)]/[In^-(aq)] < 1/10$, the eye will see blue only, with no trace of red. Between these limits, the eye will see a range of purple colours.

Putting this in the Henderson equation, the pH when the eye just sees red is given by:

$$pH = pK_{In} - \log\left(\frac{[HIn(aq)]}{[In^-(aq)]}\right)$$

$$pH = pK_{In} - \log\frac{10}{1}$$

as $\log 10 = 1$, $pH = pK_{In} - 1$.

Similarly the pH when the eye just sees blue is:

$$pH = pK_{In} - \log\frac{1}{10}$$

$$\text{as } \log 1/10 = -1, \quad pH = pK_{In} - (-1)$$

$$pH = pK_{In} + 1$$

acid pH — red : purple : blue — alkali

0 ... $pK_{In} - 1$ ... pK_{In} ... $pK_{In} + 1$ ... 7 ... 14

So the indicator will begin to change visibly from red to reddish-purple at $pK_{In} - 1$ and be completely blue by $pK_{In} + 1$—a range of 2 pH units centered on pK_{In}. This will only be approximate as the 10:1 ratio is only a rule of thumb based on the average human eye and it also assumes that both colours are equally intense.

Indicators should always be used in small amounts so that the H^+ ions they produce on dissociation do not significantly affect the pH of the solution they are placed in.

12.5.3 Buffer solutions

Buffers are solutions which have the ability to resist changes of acidity or alkalinity, their pHs remaining almost constant when small amounts of acid or alkali are added. One example of a system involving a buffer is blood, whose pH is buffered at approximately 7.4. A change of as little as 0.5 of a pH unit is likely to be fatal.

Buffers work via the dissociation of weak acids. Imagine a weak acid HA. It will dissociate in solution:

$$HA(aq) \rightleftharpoons H^+(aq) + A^-(aq)$$

As it is a weak acid $[H^+(aq)] = [A^-(aq)]$ and is small.

If a little alkali is added, the OH^- ions will react with H^+ ions to remove them as water molecules:

$$HA(aq) \rightleftharpoons H^+(aq) + A^-(aq)$$

$$\downarrow OH^-(aq)$$

$$H_2O$$

This disturbs the equilibrium and by Le Chatelier's principle more HA will dissociate to restore the situation so the pH tends to remain the same. If more H^+ is added, again the equilibrium shifts, this time to the left, H^+ ions combining with A^- ions to produce undissociated HA. However, since $[A^-]$ is small, the supply of A^- soon runs out and there is no A^- left to 'mop up' the added H^+. So we have half a buffer! However, we can add to the solution a supply of extra A^- by adding a soluble salt of HA such as Na^+A^-. This increases the supply of A^- so that more H^+ can be used up. So there is a way in which small amounts of both added H^+ and OH^- can be removed. A buffer therefore consists of a mixture of a weak acid and a soluble salt of that acid in dilute solution.

We can use the Henderson equation:

$$pH = pK_a - \log\left(\frac{[HA(aq)]}{[A^-(aq)]}\right)$$

For example, we could make a buffer from a mixture of 0.1 mol dm^{-3} ethanoic acid and 0.1 mol dm^{-3} sodium ethanoate. pK_a for ethanoic acid is 4.8.

$$\text{so the pH of the buffer} = 4.8 - \log\left(\frac{(0.1)}{(0.1)}\right)$$

$$= 4.8 - \log 1$$

$$= 4.8 - 0$$

$$pH = 4.8$$

Blood is an example of a solution which is buffered to a pH of 7.4

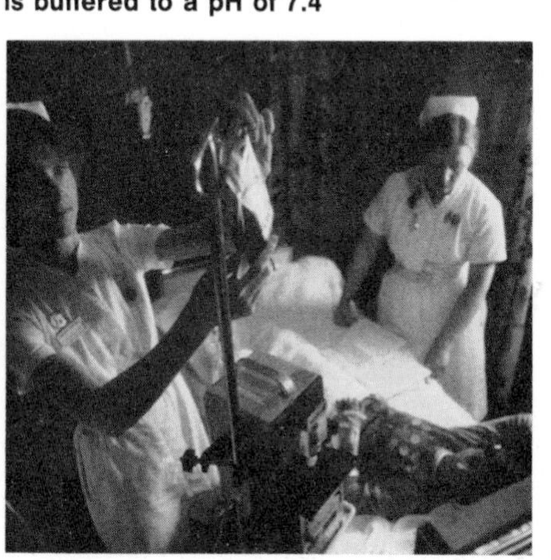

Suppose we require a buffer of pH = 5, again using ethanoic acid and sodium ethanoate. What concentrations are required?

$$pH = pK_a - \log\left(\frac{[HA(aq)]}{[A^-(aq)]}\right)$$

$$5 = 4.8 - \log\left(\frac{[HA(aq)]}{[A^-(aq)]}\right)$$

$$0.2 = -\log\left(\frac{[HA(aq)]}{[A^-(aq)]}\right)$$

Antilogging both sides of the equation:

$$0.63 = \frac{[HA(aq)]}{[A^-(aq)]}$$

So a solution which is 0.63 mol dm^{-3} in ethanoic acid (HA(aq)) and 1 mol dm^{-3} in sodium ethanoate (A^-(aq)) would give a buffer of pH 5.0.

When making a buffer to a specified pH, select a weak acid whose pK_a is approximately the same as the required pH, then calculate the required ratio of concentrations of HA and A^-. Notice that it is the *ratio* of HA to A^- which is important, not the actual concentrations. However, if more concentrated solutions are used, the buffer can 'mop up' more added acid or alkali before becoming saturated.

Buffers, however, are not perfect. Addition of acid or alkali will change the pH slightly, but by far less than the change caused by adding the same amount to a non-buffer. It is also possible to 'saturate' a buffer—to add so much acid or alkali that all of the available HA or A^- is used up.

Buffers correspond to the almost horizontal sections (see **Figure 12.2**) of titration curves. During any acid–base titration, a mixture of an acid and the salt is produced which therefore acts as a buffer. A useful rule is that a weak acid or base which is half neutralized (i.e. half as many moles of base added as there are moles of acid) by a strong base or acid forms a buffer of $pH = pK_a$.

12.6 Summary

- Acids are substances which donate protons (H^+ ions) and bases are substances which accept them.
- Water can act both as an acid and a base:

$$2H_2O(l) \rightleftharpoons H_3O^+(aq) + OH^-(aq) \quad K_w = 10^{-14}\,mol^2\,dm^{-6}$$

- Acidity and alkalinity are measured on the pH scale:

$$pH = -\log_{10}[H^+(aq)]$$

- Strong acids like hydrochloric acid are fully dissociated in dilute aqueous solution:

$$HCl(aq) \longrightarrow H^+(aq) + Cl^-(aq)$$

- Weak acids like ethanoic acid are only partially dissociated in aqueous solution:

$$CH_3CO_2H(aq) \rightleftharpoons CH_3CO_2^-(aq) + H^+(aq)$$

- Similarly, strong bases are fully dissociated in dilute aqueous solution and weak bases only partially:

$$NaOH(aq) \longrightarrow Na^+(aq) + OH^-(aq) \quad \text{a strong base}$$
$$NH_3(aq) + H_2O(l) \rightleftharpoons NH_4^+(aq) + OH^-(aq) \quad \text{a weak base}$$

- The acid dissociation constant, K_a, of a weak acid HA is defined by:

$$K_a = \frac{[H^+(aq)]_{eqm}[A^-(aq)]_{eqm}}{[HA(aq)]_{eqm}}$$

- The base dissociation constant, K_b, of a weak base B is:

$$K_b = \frac{[BH^+(aq)]_{eqm}[OH^-(aq)]_{eqm}}{[B(aq)]_{eqm}}$$

- The equivalence point of a titration is that point at which the correct amount of acid has been added to react with all the base or vice versa.
- The end point is the volume of acid or base added when the indicator changes.
- Correct choice of indicator makes the end point and the equivalence point the same.
- Indicators change over a pH range of about 2 units so there must be a vertical portion of the titration curve of at least 2 pH units for an indicator to be used.
- Indicators are weak acids where the acid and its conjugate base are different colours. The position of the equilibrium:

$$HIn(aq) \rightleftharpoons In^-(aq) + H^+(aq)$$

is governed by the concentration of $H^+(aq)$ in the solution.

• Buffers are mixtures of a weak acid and one of its salts which can absorb small quantities of added acid and base without significant pH change by movement of the position of the equilibrium:

$$HA(aq) \rightleftharpoons A^-(aq) + H^+(aq)$$

The Henderson equation can be applied to both buffers and indicators:

$$pH = pK_a - \log\left(\frac{[HA(aq)]}{([A^-(aq)]}\right)$$

12.7 Questions

1 Calculate the pH of the following:

A) hydrochloric acid, HCl (a strong acid) of:
- **i** 0.2 mol dm^{-3};
- **ii** 0.001 mol dm^{-3};
- **iii** 0.5 mol dm^{-3}.

B) sulphuric acid, H_2SO_4 (a diprotic strong acid) of:
- **i** 0.1 mol dm^{-3};
- **ii** 0.001 mol dm^{-3};
- **iii** 0.5 mol dm^{-3}.

C) potassium hydroxide, KOH (a strong base) of:
- **i** 0.01 mol dm^{-3};
- **ii** 0.001 mol dm^{-3};
- **iii** 0.5 mol dm^{-3}.

D) benzenecarboxylic acid (a weak acid, $K_a = 6.3 \times 10^{-5}$ mol dm^{-3}) of:
- **i** 1 mol dm^{-3};
- **ii** 0.1 mol dm^{-3};
- **iii** 0.01 mol dm^{-3}.

E) phenylamine (a weak base, $K_b = 4.3 \times 10^{-10}$ mol dm^{-3}) of:
- **i** 0.001 mol dm^{-3};
- **ii** 0.01 mol dm^{-3}.

2 A 0.1 mol dm^{-3} solution of a weak acid has a pH of 2.8. What is the K_a for the acid? What is its pK_a?

3 A buffer is made which is 0.1 mol dm^{-3} in propanoic acid of pK_a 4.9 and 0.05 mol dm^{-3} in sodium propanoate. What is its pH?

4 Methanoic acid has a pK_a of 3.8. What ratio of concentrations of methanoic acid and sodium methanoate would give a buffer of pH = 4?

5 Given the following values of pK_a:

methanoic acid	3.8
ethanoic acid	4.8
propanoic acid	4.9

how would you prepare a buffer of pH 4.6?

6 Which of the following indicators would be suitable for the titration of a strong acid with a weak base? (See **Figure 12.4.**)

methyl yellow	$pK_{In} = 3.5$
congo red	$pK_{In} = 4.0$
bromothymol blue	$pK_{In} = 7.0$

7 **A)** Define:
- **i** pH;
- **ii** p*K*;
- **iii** K_w.

B) The pK_a of ethanoic acid is 4.74. Calculate the approximate pH of an aqueous solution of ethanoic acid of concentration 0.1 mol dm^{-3}.

C) State whether the pH of 0.1 mol dm^{-3} sodium ethanoate solution would be less than 7, equal to 7, or greater than 7 and explain the reasons for your answer.

(Oxford 1987)

8 **A)** Define pH.

B) Calculate the pH of a 0.02 mol dm^{-3} aqueous solution of a weak base for which $K_b = 5.0 \times 10^{-4}$ mol dm^{-3} at the temperature of the solution. ($K_w = 1.0 \times 10^{-14}$ mol^2 dm^{-6}.)

C) Draw rough curves to show the change in pH as 25 cm^3 of 0.1 mol dm^{-3} solutions of:
- **i** hydrochloric acid;
- **ii** ethanoic (acetic) acid

are titrated with a 0.10 mol dm^{-3} solution of potassium hydroxide. (K_a for ethanoic acid is 1.8×10^{-5} mol dm^{-3} at room temperature.)

D) Describe qualitatively how the pH of an aqueous solution of ethanoic acid and sodium ethanoate is maintained almost constant when small amounts of hydrochloric acid are added.

(Oxford 1986, part question)

9 **A)** Write an expression for the dissociation constant, K_a, of ethanoic (acetic) acid.

B) If the pH of a 0.1 mol dm^{-3} solution of ethanoic (acetic) acid is 2.88 at 20 °C, calculate:
- **i** the hydrogen ion concentration of the solution at 20 °C;
- **ii** the dissociation constant, K_a, for ethanoic (acetic) acid at 20 °C.

(JMB 1987, part question)

13 Redox reactions

13.1 Oxidation and reduction

The word 'redox' is a contraction of 'reduction–oxidation'. Historically, oxidation was used to describe addition of oxygen. For example, in the reaction

$$Cu(s) + \tfrac{1}{2}O_2(g) \longrightarrow CuO(s)$$

the copper has been **oxidized** to copper oxide. Oxygen is the **oxidizing agent** or **oxidant**. The reverse process, such as removing oxygen from a metal oxide to leave the metal, was called reduction. For example in the reaction:

$$CuO(s) + H_2(g) \longrightarrow Cu(s) + H_2O(l)$$

copper has been **reduced** and hydrogen is the **reducing agent** or **reductant**. As hydrogen was often used as a reductant, addition of hydrogen was also considered to be a reduction, and removal of hydrogen an oxidation. So in the reaction:

$$Cl_2(g) + H_2(g) \longrightarrow 2HCl(g)$$

chlorine has been reduced.

A closer look at such oxidation and reduction reactions shows that when a species is oxidized it has lost electrons, and when reduced it has gained electrons.

loss of 2 electrons

$$Cu(s) + \tfrac{1}{2}O_2(g) \longrightarrow (Cu^{2+} + O^{2-})(s) \qquad \text{oxidation}$$

$$(Cu^{2+} + O^{2-})(s) + H_2(g) \longrightarrow Cu(s) + H_2O(l) \qquad \text{reduction}$$

gain of 2 electrons

This is the basis of the modern definition of oxidation and reduction. **Oxidation Is Loss** of electrons, **Reduction Is Gain** of electrons. The phrase **OIL RIG** makes this easy to remember.

So copper could equally well be oxidized by, for example, chlorine:

$$Cu(s) + Cl_2(g) \longrightarrow (Cu^{2+} + 2Cl^-)(s)$$

loss of 2 electrons

Look at the chlorine. It has gained electrons:

reduction
gain of electrons

$$Cu(s) + Cl_2(g) \longrightarrow (Cu^{2+} + 2Cl^-)(s)$$

loss of electrons
oxidation

So neither oxidation nor reduction can take place alone in a chemical reaction. If one species loses electrons (is oxidized), another must gain them (be reduced). An oxidizing agent oxidizes something else, but is itself reduced. In the same way, a reducing agent reduces, but is itself oxidized.

In the previous chapter we thought of acid–base reactions as competitions for protons. Redox reactions are competitions for electrons and are often called **electron transfer** reactions.

13.2 Oxidation number

Oxidation numbers, sometimes called **oxidation states**, are really a simple way of keeping track of redox reactions, so that it is easy to see which species has been oxidized and which reduced. They also provide a useful technique for constructing balanced equations for redox reactions and are used in the systematic naming system for inorganic compounds. They really form the basis of chemical 'book-keeping'. The abbreviation for oxidation number is Ox.

13.2.1 Ionic compounds

First think of ionic compounds. The oxidation number of each element in an ionic compound is the charge on the ion (including the sign). So in sodium chloride ($Na^+ + Cl^-$), the oxidation number of Na is +I and of Cl is −I. Note that oxidation numbers are usually written in Roman numerals.

Here are some further examples of oxidation numbers in ionic compounds:

calcium chloride $CaCl_2$ i.e. ($Ca^{2+} + 2Cl^-$)
oxidation number of Ca = +II, and of *each* Cl = −I

aluminium oxide Al_2O_3 i.e. ($2Al^{3+} + 3O^{2-}$)
oxidation number of *each* Al = +III, and of *each* O = −II

Since all compounds are electrically neutral, **the sum of all the oxidation numbers in a compound = 0.**

So for Al_2O_3, $(2 \times +\mathrm{III}) + (3 \times -\mathrm{II}) = 0$.

As uncombined *elements* have no charge, their **oxidation number = 0.**

Group I metals always form ions of charge +1, so they have a **fixed oxidation number of +I** in all compounds.

Group II metals always form ions of charge +2, so they have a **fixed oxidation number of +II** in all compounds.

13.2.2 Covalent compounds

Here the rule is to work as if the compound were ionic and predict the most likely charges of the ions. Some examples are given below.

BORON TRICHLORIDE, BCl_3

Chlorine often forms Cl^- ions as in sodium chloride ($Na^+ + Cl^-$), so in BCl_3: oxidation number of Cl = −I, which means that the oxidation number of B will be +III to fit the rule that the sum of the oxidation numbers in a compound = 0. Alternatively, we could reason that boron (electron arrangement 2, 3) would be expected to form B^{3+} and chlorine (electron arrangement 2, 8, 7) to form Cl^- which would give the same answer.

AMMONIA, NH_3

N (electron arrangement 2, 5) would be more likely to form N^{3-} than N^{5+}, so its oxidation number is likely to be −III, and H usually forms H^+ ions so its oxidation number is likely to be +I.

oxidation number N = −III
oxidation number H = +I
sum of oxidation numbers = 0

METHANE, CH_4

H usually forms H^+ ions so the oxidation number of H = +I.
If C (electron arrangement 2,4) formed ions, we could imagine C^{4+} or C^{4-}, but in this case to fit in with H being positive, the C must be negative, so:

oxidation number of H = +I
oxidation number of C = −IV
sum of oxidation numbers = 0

CARBON DIOXIDE, CO_2

The usual oxidation number of oxygen = −II so:

oxidation number of carbon must be = +IV
sum of oxidation numbers = 0

The more electronegative atom (see **section 10.2.2**) in a compound has the negative oxidation number.

13.2.3 Complex ions

The final case is for compounds with complex ions. These are ions which have two or more atoms bonded covalently, but where the whole group of atoms has a charge, e.g. ammonium chloride, NH_4Cl ($NH_4^+ + Cl^-$).

H usually forms H^+, so the oxidation number of H = +I
N would be likely to form N^{3-}, so the oxidation number of N = −III
Cl often forms Cl^-, so the oxidation number of Cl = −I

This fits the rule that the sum of the oxidation numbers = 0 (−III + IV − I = 0). We could think of the NH_4^+ ion itself. Here the sum of the oxidation numbers = charge on the ion (−III + IV = +I).

13.2.4 Rules for finding oxidation numbers

It is useful to remember the following rules for calculating oxidation numbers:

- The oxidation number of an uncombined element = 0.
- The sum of the oxidation numbers of a neutral compound = 0.
- The sum of the oxidation numbers of a complex ion = the charge on the ion.
- Some elements always have the same oxidation numbers in their compounds:

a) Group I metals *always* +I.
b) Group II metals *always* +II.
c) Al *always* +III.
d) H +I *except* in compounds with Group I or II metals (metal hydrides) where it is −I.
e) F *always* −I.
f) O −II *except* in peroxides where it is −I and compounds with F where it is positive.
g) Cl −I *except* in compounds with F and O where it has positive values.

These rules should help you to work out the oxidation number of almost any element in any compound. It is often useful to remember the charges on some common complex ions as a useful short cut:

OH^-	hydroxide	NO_3^-	nitrate
SO_4^{2-}	sulphate	CO_3^{2-}	carbonate
NH_4^+	ammonium		

For example: What is the oxidation number of Cr in $Cr_2(SO_4)_3$? Since each SO_4^{2-} contributes −II, the total negative oxidation number must be −VI. The total positive oxidation number must be +VI, which shared between two Crs gives the oxidation number of Cr = +III.

Note that some oxidation numbers are so large that they could not possibly correspond to 'real' ionic charges, e.g. potassium dichromate, $K_2Cr_2O_7$. What is the oxidation number of Cr? Using the rules:

O is always $-$II, so the Os contribute $7 \times -\text{II} = -14$
K is always $+$I, so the Ks contribute $2 \times +\text{I} = +2$

To make the sum $= 0$, the two Crs must contribute a total of $+12$, i.e. $+6$ each. So the oxidation number of Cr $= +$VI.

Strict application of these rules can produce some oddities. For example, in the tetrathionate ion, $S_4O_6^{2-}$, the rules give each S atom an oxidation number of $+\text{II}\frac{1}{2}$. In fact there are two different types of S atom with oxidation numbers of 0 and $+$V respectively.

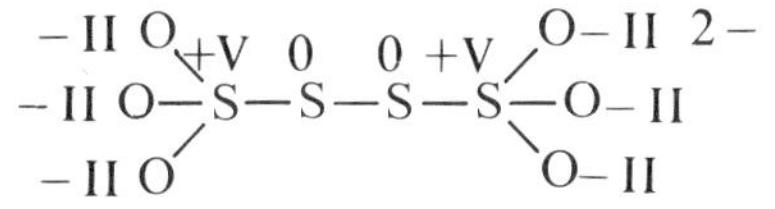

$$\begin{array}{c} -\text{II}\ \text{O}\quad\ +\text{V}\quad 0\quad\ \ 0\quad +\text{V}\quad \text{O}-\text{II}\ \ 2- \\ -\text{II}\ \text{O}-\text{S}-\text{S}-\text{S}-\text{S}-\text{O}-\text{II} \\ -\text{II}\ \text{O}\qquad\qquad\qquad\qquad\quad \text{O}-\text{II} \end{array}$$

The average oxidation number of sulphur is $+\text{II}\frac{1}{2}$.

Practice is the key to working with oxidation numbers.

13.2.5 Uses of oxidation numbers

Naming inorganic compounds

The naming of inorganic compounds containing only two elements follows the simple rule: metal name first, followed by non-metal name with the end changed to -ide. However, this would not distinguish between, for example, CuO and Cu_2O. This is done by placing the oxidation number in brackets after the metal name:

CuO	copper(II) oxide
Cu_2O	copper(I) oxide
$PbCl_2$	lead(II) chloride
$PbCl_4$	lead(IV) chloride

Compounds with three elements, one of which is oxygen, have the ending -ate, to indicate the presence of oxygen. For example, sodium sulphate contains sodium, sulphur and oxygen. However, there are two compounds which contain these elements—Na_2SO_4 and Na_2SO_3. In the past the ending -ite was given to the compound with the lower amount of oxygen, so that Na_2SO_3 used to be called sodium sulphite, but they are now distinguished by the use of oxidation numbers.

So Na_2SO_4 is sodium sulphate(VI) because here S has oxidation number $+$VI, and Na_2SO_3 is sodium sulphate(IV) because S has oxidation number $+$IV.

There is no need to include the oxidation numbers of elements whose oxidation numbers rarely change, like Na (which is always $+$I) and O (which is $-$II except in peroxides).

The full name therefore for $CuSO_4$ is copper(II) sulphate(VI), but it is often abbreviated to copper sulphate when no confusion is likely.

PCl_3 and PCl_5 are still usually called phosphorus trichloride and phosphorus pentachloride respectively and this applies to other molecular compounds.

The main thing is to be able to work from the name to the formula, rather than to be able to correctly name a particular compound.
Some further examples are:

$KMnO_4$	potassium manganate(VII)
K_2MnO_4	potassium manganate(VI)
$CoCl_2$	cobalt(II) chloride
$CoCl_3$	cobalt(III) chloride

Keeping track of redox reactions

The use of oxidation numbers helps us to decide which element has been oxidized and which reduced in a redox reaction. By remembering that oxidation is loss of electrons (OIL), we can see that if an atom is oxidized, its oxidation number will go up:

increase in oxidation number

$$\overset{0}{Cu(s)} + \tfrac{1}{2}O_2(s) \longrightarrow (\overset{+II}{Cu^{2+}} + O^{2-}(s)$$

loss of electrons

Similarly, when a substance is reduced, its oxidation number goes down. This includes an oxidation number becoming more negative (RIG). If we put in the oxidation numbers, it is easy to decide what has been oxidized and what reduced, even in fairly complicated equations:

oxidized, Ox goes up

$$\overset{+III}{Fe^{3+}} + \overset{-I}{I^-} \longrightarrow \overset{+II}{Fe^{2+}} + \tfrac{1}{2}\overset{0}{I_2}$$

reduced, Ox goes down

oxidized, Ox goes up

$$\overset{+V\,-II}{2IO_3^-} + \overset{+I+IV-II}{5HSO_3^-} \longrightarrow \overset{0}{I_2} + \overset{+VI\,-II}{5SO_4^{2-}} + \overset{+I}{3H^+} + \overset{+I\,-II}{H_2O}$$

reduced, Ox goes down

All the other species have unchanged oxidation numbers.

DISPROPORTIONATION

This happens when some atoms *of the same element* are oxidized and some reduced in the same reaction:

oxidized, Ox goes up

e.g.

$$\overset{+I\,-I}{2CuCl} \longrightarrow \overset{0}{Cu} + \overset{+II\,-I}{CuCl_2}$$

reduced, Ox goes down

oxidized, Ox goes up

or

$$\overset{0}{Cl_2} + \overset{+I\,-II\,+I}{NaOH} \longrightarrow \overset{+I\,-I}{NaCl} + \overset{+I\,-II\,+I}{NaOCl}$$

reduced, Ox goes down

This reaction is used in the manufacture of household bleach.

Balancing redox equations

If we know the starting materials and the products of a redox reaction, we can work out a balanced symbol equation for the reaction. Oxidation numbers can help us to balance redox reactions but they can only help us balance the redox change. Common sense and chemical intuition may allow us to do the rest. The key idea is that the *total* increase in oxidation number in one species must be balanced by the same *total* decrease in some other species. It is very important that you consider the *total* increase and decrease, not the increase and decrease per atom. For example, the equation:

$$Zn + HCl \longrightarrow ZnCl_2 + H_2$$

HOUSEHOLD BLEACH

Bleaching is a redox reaction. In household bleach, the sodium chlorate(I) (NaOCl) acts as the oxidizing agent, and oxidizes coloured dyes to colourless products. In hair bleach, the oxidizing agent is hydrogen peroxide H_2O_2, in which the oxidation number of the oxygens is −I. This is converted into water in which the oxidation number of oxygen is −II, and in the process, pigments are oxidized to colourless products.

is not balanced. How can we balance it? Note that all the formulae are correct. First write the oxidation numbers of all the atoms.

up 2 (total)

$$\overset{0}{Zn} + \overset{+I\,-I}{HCl} \longrightarrow \overset{+II\,-I}{ZnCl_2} + \overset{0}{H_2}$$

down 1

Then identify those which change (marked in brown above).

The total upward change in oxidation number must be balanced by the same total downward change, so *two* Hs must be involved.

$$Zn + 2HCl \longrightarrow ZnCl_2 + H_2$$

This was a straightforward example, but here is a less familiar one where manganate(VII) ions oxidize iron(II) ions to iron(III) ions:

$$\overset{+VII\,-II}{MnO_4^-} + \overset{+II}{Fe^{2+}} + \overset{+I}{H^+} \longrightarrow \overset{+II}{Mn^{2+}} + \overset{+III}{Fe^{3+}} + \overset{+I\,-II}{H_2O}$$

Again, the oxidation number of each atom is written down, with those that change in brown.

$$\overset{+VII}{MnO_4^-} + \overset{+II}{Fe^{2+}} + H^+ \longrightarrow \overset{+II}{Mn^{2+}} + \overset{+III}{Fe^{3+}} + H_2O$$

down 5

The total downward change of the Mn must be balanced by the same upward change of the Fe. Each Fe changes by 1, so five Fes must be involved.

up 1 each × 5 = 5

$$MnO_4^- + \overset{+II}{5Fe^{2+}} + H^+ \longrightarrow Mn^{2+} + \overset{+III}{5Fe^{3+}} + H_2O$$

Note that the process tells us nothing about the hydrogen or oxygen as their oxidation numbers do not change. However, to 'use up' all four oxygens of the MnO_4^-, 8 Hs are needed and this will produce $4H_2O$s.

So the balanced equation is:

$$MnO_4^- + 5Fe^{2+} + 8H^+ \longrightarrow Mn^{2+} + 5Fe^{3+} + 4H_2O$$

In this example sulphur dioxide (in solution) reduces aqueous bromine molecules to bromide ions:

up 2

$$\overset{+IV\,-II}{SO_2} + \overset{0}{Br_2} + \overset{+I\,-II}{H_2O} \longrightarrow \overset{+I}{H^+} + \overset{+VI\,-II}{SO_4^{2-}} + \overset{-I}{Br^-}$$

The total upward change of the sulphur is 2, so there must be a total downward change of 2. Each Br *atom* goes down by 1 so two of these are involved, i.e. one Br_2 molecule (not two molecules).

down 1 each × 2 = down 2

$$SO_2 + \overset{0}{Br_2} + H_2O \longrightarrow H^+ + SO_4^{2-} + \overset{-I}{2Br^-}$$

This is the redox balancing done. Again we are left with the task of 'sorting out' the Os and Hs. Two extra Os are required to go from SO_2 to SO_4^{2-} so there must be two H_2O molecules on the left and this means 4 Hs on the right.

$$SO_2 + Br_2 + 2H_2O \longrightarrow 4H^+ + SO_4^{2-} + 2Br^-$$

Whenever you believe you have balanced an equation, check that:

1. the chemical balance is right, i.e. that there are the same number of atoms of each element left and right of the arrow
2. that the charges balance, i.e. that the total charge of all the ions on the left is equal to that on the right.

If either of these does not balance, you have made a mistake and will need to check back. Remember redox balancing can only help with those species whose oxidation numbers change.

E

REDOX TITRATIONS

If we have a solution of an acid or base of unknown concentration we can find its concentration by titrating it with a base (or acid) of known concentration with a suitable indicator to show when the reaction is complete. The same principle can be used with oxidizing and reducing agents. One example is the analysis of iron(II) sulphate tablets, commonly called simply 'iron tablets', which may be taken by people with anaemia to ensure they have sufficient iron to produce the haemoglobin they need in their blood.

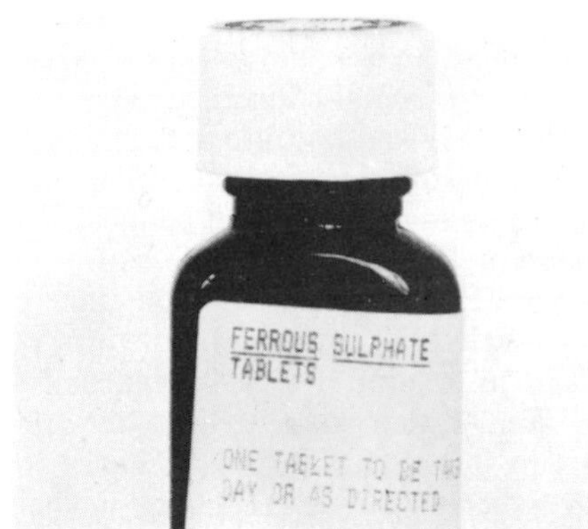

'Iron' tablets

The analysis depends on the reaction between iron(II) ions and manganate(VII) ions:

$$MnO_4^-(aq) + 8H^+(aq) + 5Fe^{2+}(aq) \longrightarrow Mn^{2+}(aq) + 4H_2O(l) + 5Fe^{3+}(aq)$$

MnO_4^- (aq)

Fe^{2+}(aq)/H^+(aq)

Apparatus for a redox titration

The oxidizing agent MnO_4^- can be titrated against the reducing agent Fe^{2+}, as they react together in the ratio 1:5. An acid solution is required to provide the H^+ ions. No indicator is required in this titration as MnO_4^- ions are a dark purple colour. All the other ions are very pale in colour—Fe^{2+} is green, Fe^{3+} brown, Mn^{2+} pink, so the manganate(VII) solution is added until a permanent purple colour just remains. Redox indicators are available for use when there is no convenient colour change.

A

SWIMMING POOL CHEMISTRY

The water in swimming pools is kept sterile by the addition of oxidizing agents, chlorine or chlorine compounds, which kill microorganisms by oxidation. The active agent is usually chloric(I) acid (HOCl). This may be formed by direct chlorination of the water:

$$Cl_2(aq) + H_2O(l) \rightleftharpoons HOCl(aq) + Cl^-(aq) + H^+(aq)$$

or by addition of sodium chlorate(I):

$$NaOCl(s) + H_2O(l) \rightleftharpoons Na^+(aq) + OH^-(aq) + HOCl(aq)$$

One way of analysing the amount of HOCl in the pool water is as follows. An excess of iodide ions and acid is added. The HOCl oxidizes the iodide ions to iodine:

Ox down 2

$$\overset{+I}{HOCl}(aq) + 2\overset{-I}{I^-}(aq) + H^+(aq) \longrightarrow \overset{-I}{Cl^-}(aq) + \overset{0}{I_2}(aq) + H_2O(l)$$

Ox up 1 × 2

The released iodine can then be titrated with a solution of sodium thiosulphate:

Ox down 1 × 2

$$\overset{0}{I_2}(aq) + 2Na_2\overset{+II}{S_2}O_3(aq) \longrightarrow 2Na\overset{-I}{I}(aq) + Na_2\overset{+II\frac{1}{2}}{S_4}O_6(aq)$$

Ox up $\frac{1}{2}$ × 4

The explanation of the odd oxidation number of S in $Na_2S_4O_6$ is given on page 156.

Aqueous iodine is brown, so thiosulphate solution is added until the colour disappears. The final stages are difficult to see, so it is usual to add starch at the stage when the colour of the solution is approximately that of lager. Starch forms an intense blue–black colour with iodine. This makes the final disappearance of iodine easy to judge visually.

13.3 Electrochemical cells

One redox reaction which is of considerable practical and theoretical interest is that of a metal dissolving in water and forming its own aqueous ions, for example:

$$Zn(s) + (aq) \longrightarrow Zn^{2+}(aq) + 2e^-$$

The zinc is oxidized, as its oxidation number has increased and it has lost electrons. Notice this is *not* the same as the ionization of gaseous zinc atoms to form gaseous ions ($Zn(g) \longrightarrow Zn^{2+}(g)$).

Such reactions are of interest because they are the basis of electrical batteries and because they enable us to predict which reactions might occur (are feasible) and which cannot take place, as we shall see.

13.3.1 Half cells

If we dip a rod of metal, say zinc, into a solution of its own ions, for example, zinc sulphate solution, then an equilibrium is set up:

$$Zn(s) + (aq) \rightleftharpoons Zn^{2+}(aq) + 2e^-$$

Figure 13.1 Solid zinc in equilibrium with its aqueous ions

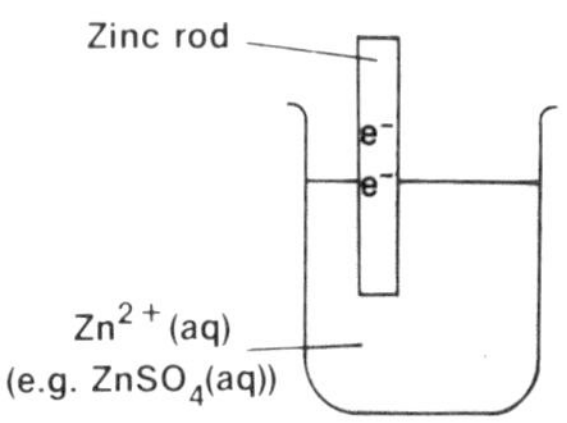

If some zinc dissolves, i.e. the reaction above goes from left to right, then the electrons left behind on the rod build up a negative electrical charge. We say the rod gains a negative potential. This arrangement is called a **half cell** see **Figure 13.1**.

If we could determine this potential, it would give us a measure of how readily electrons are released by the metal concerned. This would tell us how good a reducing agent the metal was, because reducing agents release electrons. However we cannot measure electrical potential, only potential *difference*, which is the same as voltage. So we can connect together two different half cells and measure the **potential difference** between them with a voltmeter, as shown in **Figure 13.2** for copper and zinc half cells.

Figure 13.2 Two half cells connected together

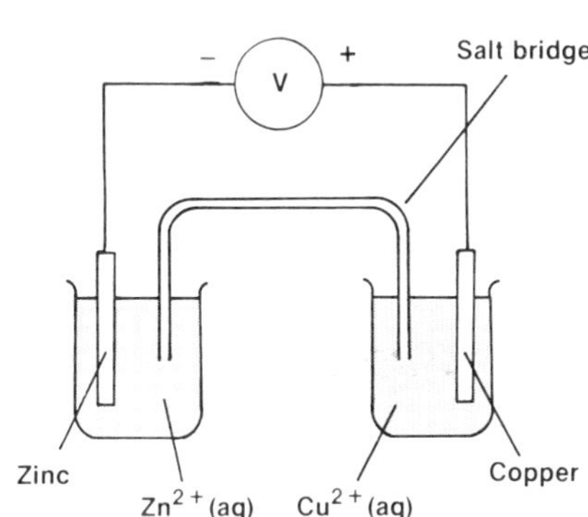

The circuit is completed by a **salt bridge**, the simplest form of which is a piece of filter paper soaked in a solution of a salt (usually saturated potassium nitrate). This is used, rather than a piece of wire, to avoid further metal/ion potentials in the circuit. The salt chosen for the salt bridge must not react with either of the salt solutions in the half cells. Note that a perfect voltmeter does not allow any current to flow—it merely measures the electrical 'push' or pressure which *tends* to make current flow.

If we connect the two half cells shown, we get a potential difference or voltage of 1.1 V with the zinc half cell the more negative (if the solutions are $1\ mol\ dm^{-3}$ and the temperature 298 K). The fact that the zinc half cell is the negative one tells us that zinc loses its electrons more readily than copper. If the voltmeter were removed and electrons allowed to flow, they would do so from zinc to copper. The following changes would take place:

1. Zinc would dissolve to form $Zn^{2+}(aq)$, increasing the concentration of $Zn^{2+}(aq)$.
2. The electrons would flow through the wire to the copper rod where they would combine with $Cu^{2+}(aq)$ ions (from the copper sulphate solution) so depositing fresh copper on the rod and decreasing the concentration of $Cu^{2+}(aq)$.

   That is, the following two half reactions would take place:

$$Zn(s) \longrightarrow Zn^{2+}(aq) + 2e^-$$

and

$$Cu^{2+}(aq) + 2e^- \longrightarrow Cu(s)$$

adding:

$$Zn(s) + Cu^{2+}(aq) + 2e^- \longrightarrow Zn^{2+}(aq) + Cu(s) + 2e^-$$

Figure 13.3 A Daniell cell lighting a bulb. The porous pot acts like a salt bridge

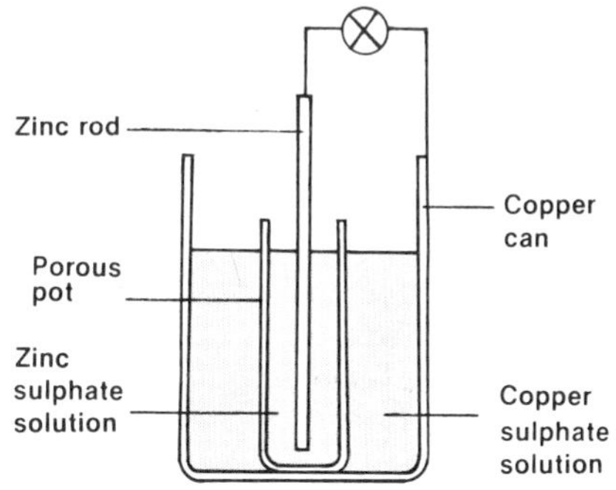

Adding the two half reactions, the electrons cancel out and we get the overall reaction shown, which is exactly the reaction we get on putting zinc into a solution of copper ions. It is a redox reaction with zinc

Figure 13.4 The hydrogen electrode

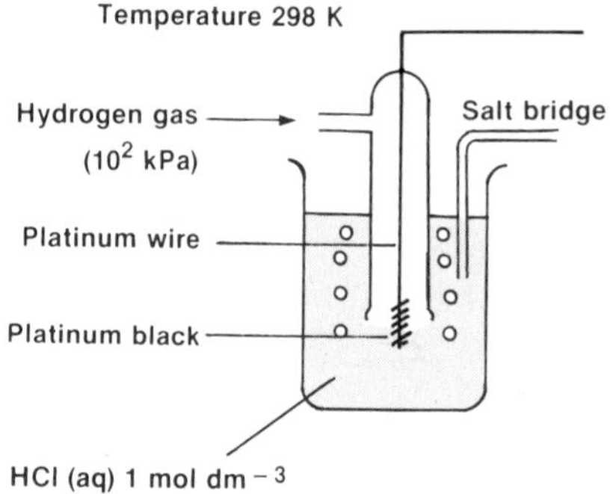

A hydrogen electrode

being oxidized and copper ions reduced. The system could be used to generate electricity and is, in fact, the basis of an electrical cell called the Daniell cell, see **Figure 13.3**.

13.3.2 The hydrogen electrode

If we want to compare the tendency of different metals to release electrons, we must agree on a standard half cell to which any other half cell can be connected for comparison. The half cell chosen is called the hydrogen electrode, see **Figure 13.4** and photograph (below left).

This might be expected to consist of a rod of hydrogen dipping into a solution of $H^+(aq)$ ions, but as hydrogen is a gas, we bubble hydrogen into the solution. As hydrogen does not conduct, we make electrical contact via a piece of the unreactive metal platinum which is coated with finely divided platinum called platinum black (to increase the surface area, and allow any reaction to proceed rapidly). The electrode is used under standard conditions of $[H^+(aq)] = 1\ mol\ dm^{-3}$ (in practice a $1\ mol\ dm^{-3}$ solution of hydrochloric acid), a pressure of 10^2 kPa and a temperature of 298 K.

The potential of the standard hydrogen electrode is *defined* as zero, so if it is connected to another half cell, the measured voltage, called the electromotive force (e.m.f., E) is the electrode potential of that cell. If the second cell is at standard conditions ($[\text{ions}] = 1\ mol\ dm^{-3}$, temperature = 298 K), then this voltage is given the symbol $E^\ominus$ (298 K), usually abbreviated to $E^\ominus$ and called E standard. Half cells with negative values of $E^\ominus$ are better at releasing electrons (better reducing agents) than hydrogen.

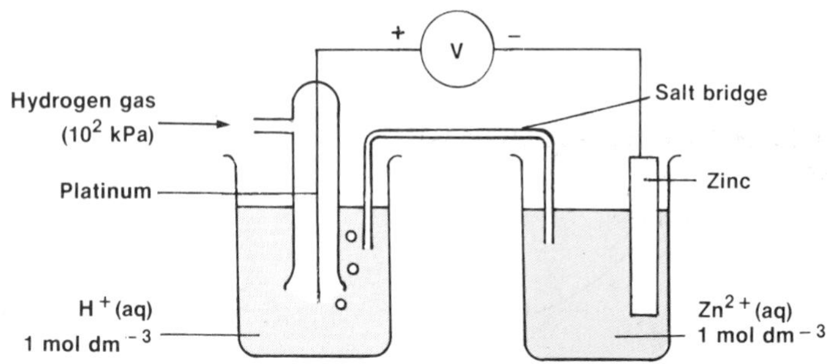

Figure 13.5 (right) Measuring $E^\ominus$ for a zinc half cell

Table 13.1 $E^\ominus$ values
Good reducing agents appear top right, e.g. Li(s)
Good oxidizing agents appear bottom left, e.g. $Ag^+(aq)$

Half reaction	$E^\ominus$/V
$Li^+(aq) + e^- \rightleftharpoons Li(s)$	−3.03
$Ca^{2+}(aq) + 2e^- \rightleftharpoons Ca(s)$	−2.87
$Al^{3+}(aq) + 3e^- \rightleftharpoons Al(s)$	−1.66
$Zn^{2+}(aq) + 2e^- \rightleftharpoons Zn(s)$	−0.76
$Pb^{2+}(aq) + 2e^- \rightleftharpoons Pb(s)$	−0.13
$2H^+(aq) + 2e^- \rightleftharpoons H_2(g)\|Pt$	0.00
$Cu^{2+}(aq) + 2e^- \rightleftharpoons Cu(s)$	+0.34
$Ag^+(aq) + e^- \rightleftharpoons Ag(s)$	+0.80

13.3.3 The electrochemical series

A list of some $E^\ominus$ values for metal/metal ion half cells is given in **Table 13.1**. Notice that the equilibria are written with the electrons on the left, i.e. as a reduction, so these are called **reduction potentials** (remember RIG). Arranged in this order with the most negative values at the top this list is called the **electrochemical series**. Notice that the number of electrons involved has no effect on the value of $E^\ominus$.

The voltage obtained by connecting two half cells together is found by the difference between the two $E^\ominus$ values. So connecting an $Al^{3+}(aq)/Al(s)$ half cell to a $Cu^{2+}(aq)/Cu(s)$ half cell would give a voltage of 2.00 V. See **Figure 13.6** (overleaf).

It is well worth sketching diagrams like the one shown in **Figure 13.6** (overleaf). It will prevent you getting muddled with signs. Again note the convention of listing the more negative values at the top.

13.3.4 Cell diagrams

These are a shorthand for writing down the cell formed by connecting two half cells. The usual apparatus is shown in **Figure 13.7** (overleaf), and the cell diagram is written as follows:

Figure 13.6 Calculating the value of the voltage when two half cells are connected

If we connect an $Al^{3+}(aq)/Al(s)$ half cell to a $Cu^{2+}(aq)/Cu(s)$ half cell the e.m.f. will be 2.00 V and the $Al^{3+}(aq)/Al(s)$ will be negative

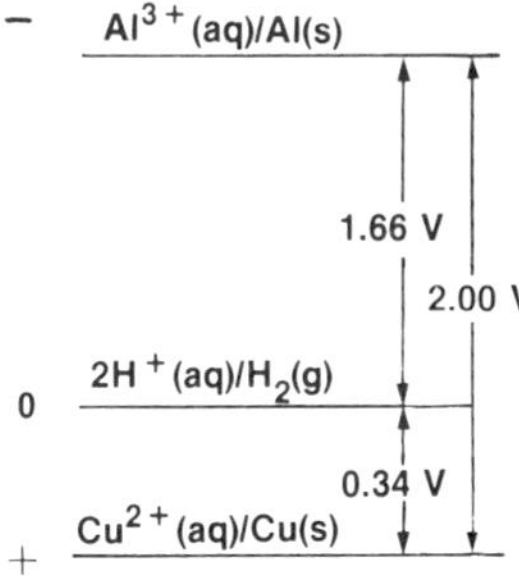

If we connect an $Al^{3+}(aq)/Al(s)$ half cell to a $Pb^{2+}(aq)/Pb(s)$ half cell, the e.m.f. will be 1.53 V and the $Al^{3+}(aq)/Al(s)$ will be negative

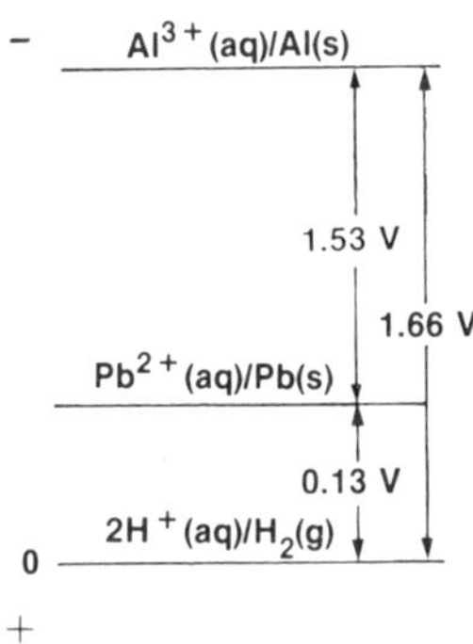

Figure 13.7 A pair of half cells

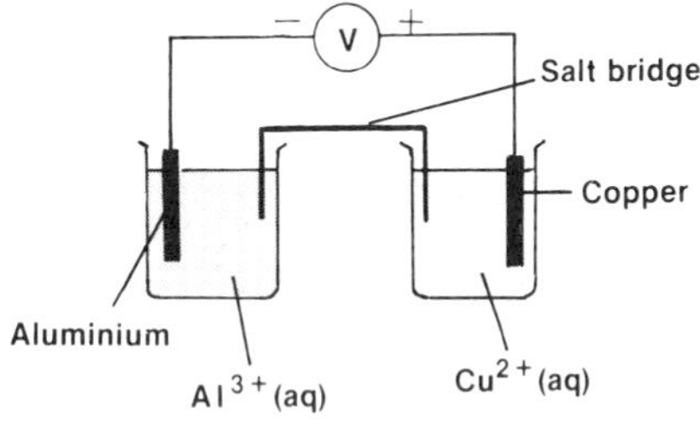

Figure 13.8 Measuring $E^\ominus$ for the $Fe^{3+}(aq)/Fe^{2+}(aq)$ system

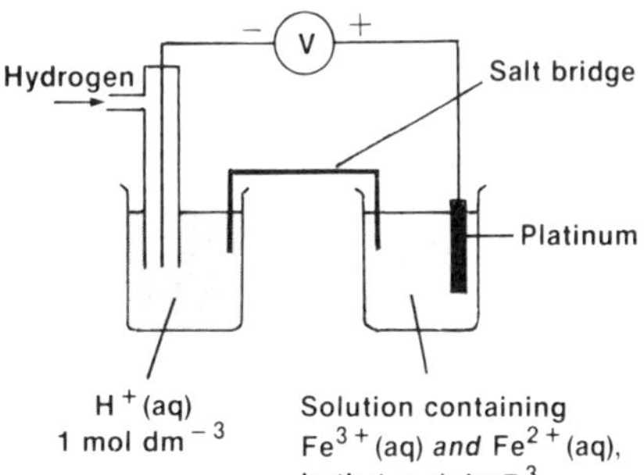

$$Al(s)|Al^{3+}(aq)\,\vdots\vdots\,Cu^{2+}(aq)|Cu(s) \quad E^\ominus = +2.00\,V$$

A vertical solid line indicates a **phase boundary**, e.g. between a solid and a solution, and a double vertical dotted line a **salt bridge**.

By convention, we give the polarity (whether it is positive or negative) of the right-hand electrode, as the cell diagram is written. In this case the copper half cell is positive and, if allowed to flow, electrons would go from aluminium to copper.

We could equally well have written the cell:

$$Cu(s)|Cu^{2+}(aq)\,\vdots\vdots\,Al^{3+}(aq)|Al(s) \quad E^\ominus = -2.00\,V$$

as we always give the polarity of the right-hand electrode.

The cell diagram for a silver electrode connected to a $Pb^{2+}(aq)/Pb(s)$ half cell would be:

$$Ag(s)|Ag^+(aq)\,\vdots\vdots\,Pb^{2+}(aq)|Pb(s) \quad E^\ominus = -0.93\,V$$

So you should be able to predict the value of $E^\ominus$ for a cell made up of any pair of half cells.

13.3.5 Extension to systems other than metal/metal ion

Cells can also be set up to measure $E^\ominus$ values for redox reactions which are not metal/metal ion systems. The problem is tackled in the same way as the hydrogen electrode. We ensure that both the species that we are interested in are present and make electrical contact via a piece of platinum, **Figure 13.8**, e.g.

$$Fe^{3+}(aq)/Fe^{2+}(aq) \quad (Fe^{3+}(aq) + e^- \rightleftharpoons Fe^{2+}(aq))$$

$$Pt|H_2(g)|2H^+(aq)\,\vdots\vdots\,Fe^{3+}(aq), Fe^{2+}(aq)|Pt \quad E^\ominus = +0.77\,V$$

Notice that in the cell diagram, the two aqueous species are separated by a comma and the more reduced form of the two ions (the lower oxidation number) goes nearest the platinum electrode.

The same rule applies to the system:

$$Br_2(aq)/2Br^-(aq) \quad (Br_2(aq) + 2e^- \rightleftharpoons 2Br^-(aq))$$

$$Pt|H_2(g)|2H^+(aq)\,\vdots\vdots\,Br_2(aq), 2Br^-(aq)|Pt \quad E^\ominus = +1.09\,V$$

Note the more reduced form ($Br^-(aq)$) closest to the platinum electrode.

More complicated systems are possible. For example, $MnO_4^-(aq)$ can be reduced to $Mn^{2+}(aq)$ but only in the presence of $H^+(aq)$ ions:

$$MnO_4^-(aq) + 8H^+(aq) + 5e^- \rightleftharpoons Mn^{2+}(aq) + 4H_2O(l)$$

The $E^\ominus$ values for this half reaction can be measured, provided all four species: $MnO_4^-(aq)$, $H^+(aq)$, $Mn^{2+}(aq)$, $H_2O(l)$ are present. Again, a platinum electrode is used.

In the cell diagram, the two pairs of reactants $[MnO_4^-(aq) + 8H^+(aq)]$ and $[Mn^{2+}(aq) + 4H_2O(l)]$, are bracketed together and separated by a comma. The rule that the more reduced species is written nearest the electrode still applies.

$$Pt|H_2(g)|2H^+(aq)\,\vdots\vdots\,[MnO_4^-(aq) + 8H^+(aq)], [Mn^{2+}(aq) + 4H_2O(l)]|Pt \quad E^\ominus = +1.51\,V$$

Table 13.2 (page 164) is an extension of **Table 13.1**, page 161, by including some systems other than metal/metal ion.

13.3.6 Predicting whether a redox reaction could take place (the feasibility of the reaction)

It is possible to use half cell e.m.f.s to decide on the feasibility of a reaction. When we connect a pair of half cells, the electrons will flow from the

A

BATTERIES

Batteries for torches, radios, calculators, etc. are the most familiar applications of electrochemical cells. Each man, woman and child buys on average five batteries per year. Nowadays there is a bewildering variety of types and brands advertised with slogans like 'long life' and 'high power'

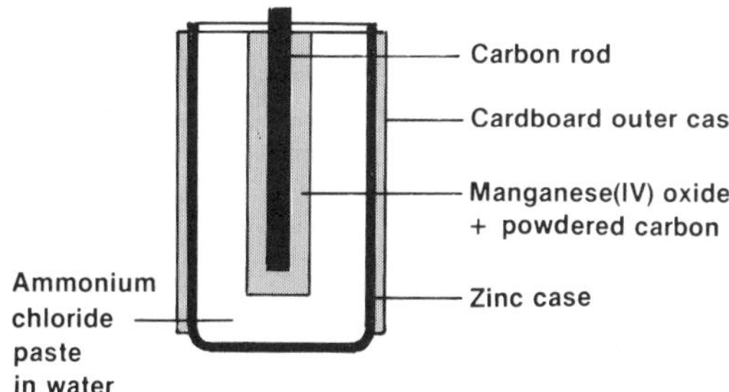

Leclanché cell

The majority of batteries for household use are based on the zinc/carbon system with a variety of different electrolytes. The most common is the Leclanché cell, named after its inventor. It consists of a zinc canister filled with a paste of ammonium chloride, NH_4Cl, and water—the electrolyte. In the centre is a carbon rod which performs the same function as platinum, an inert electrode, surrounded by a mixture of manganese(IV) oxide and powdered carbon to make it conduct. The reactions are:

$$Zn(s) \longrightarrow Zn^{2+}(aq) + 2e^-$$

and at the carbon rod:

$$2NH_4^+(aq) + 2e^- \longrightarrow 2NH_3(g) + H_2(g)$$

The hydrogen gas is oxidized to water by the manganese(IV) oxide, preventing a build-up of pressure, while the ammonia dissolves in the water of the paste. The cell diagram is

$$Zn(s)|Zn^{2+}(aq) \vdots\vdots 2NH_4^+(aq), [2NH_3(aq) + H_2(g)]|C(gr) \qquad E = 1.5\,V$$

As the cell discharges, the zinc is used up and the walls of the zinc canister become thin and prone to leakage. The ammonium chloride electrolyte is acidic (being the salt of a strong acid and a weak base, see **section 12.5.1**), and can be corrosive. That is why you should remove spent batteries from equipment. This cell is ideal for, say, doorbells, which need a small current intermittently.

Types of batteries

A variant of this cell is the zinc chloride cell. It is similar to the Leclanché but uses zinc chloride as the electrolyte. Such cells are better at supplying high currents than the Leclanché cell and are marketed as 'extra life' batteries for radios, torches and shavers. Long-life alkaline batteries are also based on the zinc/carbon/manganese(IV) oxide electrode system with an electrolyte of potassium hydroxide. Powdered zinc is used, whose greater surface area allows the battery to supply high currents. The cell is enclosed in a steel container to prevent leakage. These cells are suitable for equipment taking continuous high currents such as personal stereos. In this sort of situation they can last up to six times as long as ordinary zinc/carbon batteries, but they are more expensive.

Many other electrode systems are in use, especially for miniature batteries such as those used in watches, hearing aids, cameras and electronic equipment. These include zinc/air, mercury(II) oxide/zinc, silver oxide/zinc, and lithium/manganese(IV) oxide. Which is used for a particular application depends on the precise requirements of voltage, current, size and cost.

RECHARGEABLE BATTERIES

These are now available in standard sizes to replace traditional zinc/carbon batteries. Although more expensive to buy, they can be recharged up to 500 times, reducing the effective cost to around 1 p per battery (not including the price of the charger or the very small amount of electricity it uses). These cells are called nickel/cadmium cells and have an alkaline electrolyte. The reaction that occurs is:

$$2NiOOH + Cd + 2H_2O \rightleftharpoons 2Ni(OH)_2 + Cd(OH)_2$$

The reaction goes from left to right on discharge (electrons flowing from Cd to Ni) and right to left on charging.

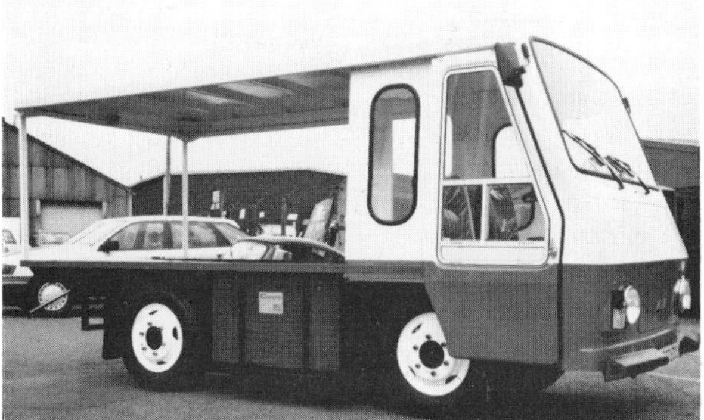

Lead–acid batteries on a milk float

Considerable research on batteries, both rechargeable and single use, is taking place. One challenge is to produce a rechargeable battery system which is suitable for powering electric vehicles. One promising condidate is the sodium sulphur battery which can store up to six times the energy of a conventional lead–acid battery for the same weight. Present-day electric vehicles, like milk floats, have to carry half their own weight in batteries.

Table 13.2 $E^\ominus$ **Values**

Half reaction	$E^\ominus$/V
$Li^+(aq) + e^- \rightleftharpoons Li(s)$	−3.03
$Ca^{2+}(aq) + 2e^- \rightleftharpoons Ca(s)$	−2.87
$Al^{3+}(aq) + 3e^- \rightleftharpoons Al(s)$	−1.66
$Zn^{2+}(aq) + 2e^- \rightleftharpoons Zn(s)$	−0.76
$Cr^{3+}(aq) + e^- \rightleftharpoons Cr^{2+}(aq)\mid Pt$	−0.41
$Pb^{2+}(aq) + 2e^- \rightleftharpoons Pb(s)$	−0.13
$2H^+(aq) + 2e^- \rightleftharpoons H_2(g)\mid Pt$	0.00
$Cu^{2+}(aq) + e^- \rightleftharpoons Cu^+(aq)\mid Pt$	+0.15
$Cu^{2+}(aq) + 2e^- \rightleftharpoons Cu(s)$	+0.34
$I_2(aq) + 2e^- \rightleftharpoons 2I^-(aq)\mid Pt$	+0.54
$Fe^{3+}(aq) + e^- \rightleftharpoons Fe^{2+}(aq)\mid Pt$	+0.77
$Ag^+(aq) + e^- \rightleftharpoons Ag(s)$	+0.80
$Br_2(aq) + 2e^- \rightleftharpoons 2Br^-(aq)\mid Pt$	+1.09
$Cl_2(aq) + 2e^- \rightleftharpoons 2Cl^-(aq)\mid Pt$	+1.36
$[MnO_4^-(aq) + 8H^+(aq)] + 5e^- \rightleftharpoons [Mn^{2+}(aq) + 4H_2O(l)]\mid Pt$	+1.51
$Ce^{4+}(aq) + e^- \rightleftharpoons Ce^{3+}(aq)\mid Pt$	+1.70

Figure 13.9 **Connecting $Zn^{2+}(aq)/Zn(s)$ and $Cu^{2+}(aq)/Cu(s)$ half cells**

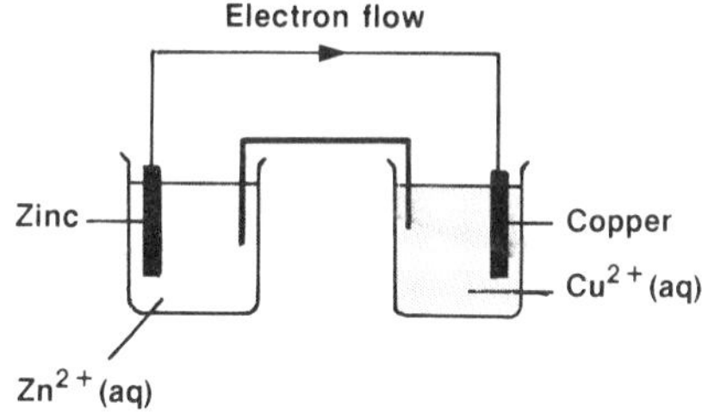

Figure 13.10 $E^\ominus$**–Ox graph. Note that the graph is plotted with the most negative $E^\ominus$ value at the top, as in the electrochemical series, and the values of oxidation number are most positive on the left**

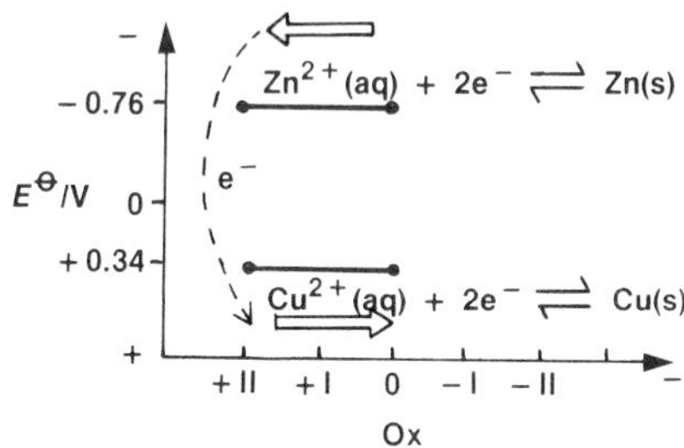

more negative to the more positive and not in the opposite direction.

Think of $Zn^{2+}(aq)/Zn$, $E^\ominus = -0.76$ V and $Cu^{2+}(aq)/Cu$, $E^\ominus = +0.34$ V half cells, **Figure 13.9**.

So the equilibrium:

$$Zn^{2+}(aq) + 2e^- \overset{\leftarrow}{\rightleftharpoons} Zn(s)$$

actually moves to the left (Zn(s) releases electrons) and the equilibrium:

$$Cu^{2+}(aq) + 2e^- \overset{\rightarrow}{\rightleftharpoons} Cu(s)$$

moves to the right ($Cu^{2+}(aq)$ accepts electrons).

The overall effect is:

$$Zn(s) \longrightarrow Zn^{2+}(aq) + 2e^-$$

adding:
$$Cu^{2+}(aq) + 2e^- \longrightarrow Cu(s)$$

$$Cu^{2+}(aq) + Zn(s) + 2\cancel{e}^- \longrightarrow Cu(s) + Zn^{2+}(aq) + 2\cancel{e}^-$$

which is the reaction that actually happens, either by connecting the two half cells or by adding Zn to Cu^{2+} ions in a test tube. The reverse reaction:

$$Cu(s) + Zn^{2+}(aq) \longrightarrow Zn(s) + Cu^{2+}(aq)$$

does not occur.

We could go through this mental process each time we wished to predict the outcome of a redox reaction. However, there are two short cuts which we can use, one diagrammatic and the other arithmetical.

$E^\ominus$–oxidation number diagrams—the anti-clockwise rule

We plot each half reaction on a graph of $E^\ominus$ against the oxidation number of the oxidized and reduced form in each half cell, **Figure 13.10.** On such a diagram, electrons will be transferred from the more negative to the more positive half cell, i.e. they will move downwards. So electrons will flow from $Zn^{2+}(aq)/Zn(s)$ to $Cu^{2+}(aq)/Cu(s)$. For this to happen, the two equilibria must move in the directions shown by the brown arrows so that $Zn^{2+}(aq)/Zn(s)$ releases electrons and $Cu^{2+}(aq)/Cu(s)$ accepts them, i.e. the top equilibrium moves from *right* to *left* and the bottom one from *left* to *right*, hence the term 'anti-clockwise rule'.

$$Zn(s) \longrightarrow Zn^{2+}(aq) + 2e^-$$

$$Cu^{2+}(aq) + 2e^- \longrightarrow Cu(s)$$

adding:
$$Zn(s) + Cu^{2+}(aq) + 2\cancel{e}^- \longrightarrow Zn^{2+}(aq) + Cu(s) + 2\cancel{e}^-$$

Calculation of $E^\ominus_{reaction}$

Taking the same example of the $Zn^{2+}(aq)/Zn(s)$ and $Cu^{2+}(aq)/Cu(s)$ half cells:

$$Zn^{2+}(aq) + 2e^- \rightleftharpoons Zn(s) \qquad E^\ominus = -0.76\text{ V}$$

$$Cu^{2+}(aq) + 2e^- \rightleftharpoons Cu(s) \qquad E^\ominus = +0.34\text{ V}$$

so
$$Zn^{2+}(aq) + 2e^- \longrightarrow Zn(s) \qquad E^\ominus = -0.76\text{ V}$$

but
$$Zn(s) \longrightarrow Zn(aq) + 2e^- \qquad E^\ominus = +0.76\text{ V}$$

as this is the reverse of the half reaction above.

and
$$Cu^{2+}(aq) + 2e^- \longrightarrow Cu(s) \qquad E^\ominus = +0.34\text{ V}$$

but
$$Cu(s) \longrightarrow Cu^{2+}(aq) + 2e^- \qquad E^\ominus = -0.34\text{ V}$$

as this is the reverse of the above reaction.

There are two possibilities for combining the half equations:

1.
$$Zn^{2+}(aq) + 2e^- \longrightarrow Zn(s) \qquad E^\ominus = -0.76\text{ V}$$

$$Cu(s) \longrightarrow Cu^{2+}(aq) + 2e^- \qquad E^\ominus = -0.34\text{ V}$$

adding:
$$Zn^{2+}(aq) + Cu(s) + 2\cancel{e}^- \longrightarrow Zn^{2+}(aq) + Cu^{2+}(s) + 2\cancel{e}^- \qquad E^\ominus_{reaction} = -1.10\text{ V}$$

2.

$$Zn(s) \longrightarrow Zn^{2+}(aq) + 2e^- \qquad E^\ominus = +0.76\,V$$
$$Cu^{2+}(aq) + 2e^- \longrightarrow Cu(s) \qquad E^\ominus = +0.34\,V$$

adding: $Zn(s) + Cu^{2+}(aq) + 2\not{e}^- \longrightarrow Zn^{2+}(aq) + Cu(s) + 2\not{e}^- \quad E^\ominus_{reaction} = +1.10\,V$

We know that in fact the second reaction takes place. It has a positive value of $E^\ominus_{reaction}$. This is a general rule. **Reactions which have a positive value of $E^\ominus_{reaction}$ are feasible**. Those with negative values are not feasible.

Any of these three techniques will of course give the same answer. Note that a prediction of **feasibility** does not necessarily mean that a reaction *will* occur. It could be that under the conditions of the experiment the reaction is too slow. So the only safe predictions are ones that say a reaction *cannot* occur. They also refer only to standard conditions—i.e. 1 mol dm^{-3} solutions, 298 K and 10^2 kPa. This technique is only useful for electron transfer (redox) reactions.

Further examples of predicting feasibility

$$Fe^{3+}(aq) + I^-(aq) \longrightarrow Fe^{2+}(aq) + \tfrac{1}{2}I_2(aq)$$

Will the reaction above occur or not?

1. VIA THE ANTI-CLOCKWISE RULE

Look up the values of $E^\ominus$ in **Table 13.2** for

$$Fe^{3+}(aq) + e^- \rightleftharpoons Fe^{2+}(aq) \quad \text{and} \quad \tfrac{1}{2}I_2(aq) + e \rightleftharpoons I^-(aq)$$

and place on the diagram, see **Figure 13.11**. (Remember that all reactions in the table are written with the electrons on the left.)

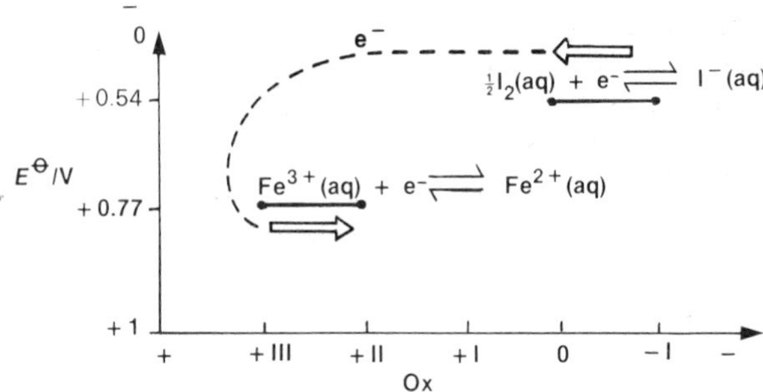

Figure 13.11 $E^\ominus$–Ox graph. Note that the number of electrons makes no difference to $E^\ominus$. So we can write $I_2(aq) + 2e^- \rightleftharpoons 2I^-(aq)$ or $\frac{1}{2}I_2(aq) + e^- \rightleftharpoons I^-(aq)$. Both have the same $E^\ominus$ value

The anti-clockwise rule tells us that

$$I^-(aq) \longrightarrow \tfrac{1}{2}I_2(aq) + e^-$$

and will provide an electron for:

$$Fe^{3+}(aq) + e^- \longrightarrow Fe^{2+}(aq)$$

adding: $I^-(aq) + Fe^{3+}(aq) \longrightarrow Fe^{2+}(aq) + \tfrac{1}{2}I_2(aq)$

The reaction is feasible.

2. VIA THE CALCULATION OF $E^\ominus_{reaction}$

To construct the required reaction we need to add:

$$Fe^{3+}(aq) + e^- \longrightarrow Fe^{2+}(aq) \quad E^\ominus = +0.77\,V \text{ (from the table)}$$

and $I^- \longrightarrow \tfrac{1}{2}I_2 + e^- \quad E^\ominus = -0.54\,V$

This is the reverse of the reaction in the table, so we change the sign of $E^\ominus$.

$$Fe^{3+}(aq) + I^-(aq) \longrightarrow Fe^{2+}(aq) + \tfrac{1}{2}I_2(aq) \quad E^\ominus_{reaction} = +0.23\,V$$

$E^\ominus_{reaction}$ is positive, so the reaction is feasible.

Another example:

$$Fe^{3+}(aq) + Br^-(aq) \longrightarrow Fe^{2+}(aq) + \tfrac{1}{2}Br_2(aq)$$

1. VIA THE ANTI-CLOCKWISE RULE

See **Figure 13.12**.

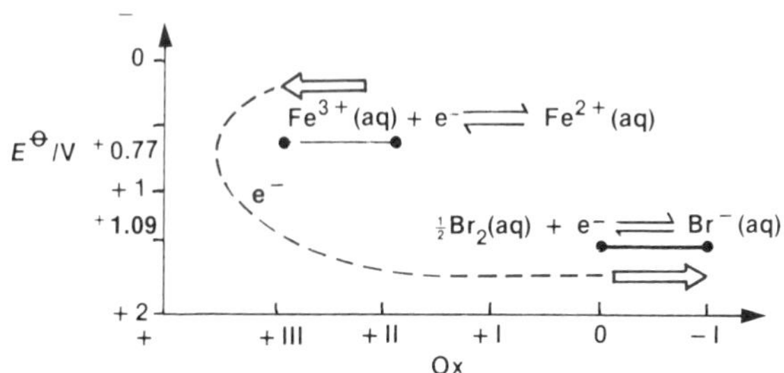

Figure 13.12 $E^\ominus$-Ox graph

Here the anti-clockwise rule predicts that:

$$Fe^{2+}(aq) \longrightarrow Fe^{3+}(aq) + e^-$$

will provide an electron for

$$\tfrac{1}{2}Br_2(aq) + e^- \longrightarrow Br^-(aq)$$

adding: $$Fe^{2+}(aq) + \tfrac{1}{2}Br_2(aq) \longrightarrow Fe^{3+}(aq) + Br^-(aq)$$

This is the *reverse* reaction to the one we required so our reaction is not feasible.

2. VIA THE CALCULATION OF $E^\ominus_{reaction}$

To construct the required reaction we must add:

$$Fe^{3+}(aq) + e^- \longrightarrow Fe^{2+}(aq) \qquad E^\ominus = +0.77\text{ V (from the table)}$$

$$Br^-(aq) \longrightarrow \tfrac{1}{2}Br_2(aq) + e^- \qquad E^\ominus = -1.09\text{ V}$$

(the reverse of the reaction in the table, therefore we change the sign of $E^\ominus$).

adding: $Fe^{3+}(aq) + Br^-(aq) \longrightarrow Fe^{2+}(aq) + \tfrac{1}{2}Br_2(aq) \quad E^\ominus_{reaction} = -0.32$ V

Note: Summary of anticlockwise rule

Overall

$$B + C \longrightarrow A + D$$

(B oxidized to A; C reduced to D)

1. Find the right system from the data

$$A \rightleftharpoons B$$
$$C \rightleftharpoons D$$

2. Write them down on the graph.
3. Put in the arrows, anti-clockwise from top reaction to bottom reaction.
4. Work out the overall equation.

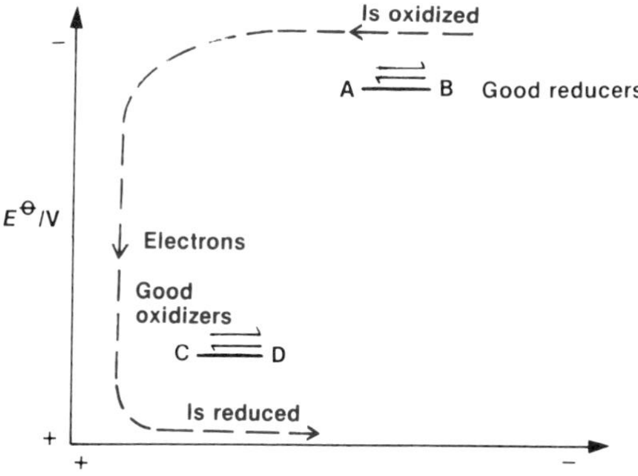

So the reaction is not feasible.

So, Fe^{3+}(aq) is a good enough oxidizing agent to oxidize $I^-(aq) \longrightarrow \tfrac{1}{2}I_2(aq)$, but not good enough to oxidize $Br^-(aq) \longrightarrow \tfrac{1}{2}Br_2(aq)$.

E

$E^\ominus$ AND EQUILIBRIUM CONSTANTS

We can be rather more precise than just saying that a reaction is feasible or not, because $E^\ominus_{reaction}$ is actually related to the equilibrium constant, K_c, for the reaction. As a rule of thumb, if $E^\ominus_{reaction}$ is greater than +0.6 V, the reaction goes to completion. If $E^\ominus_{reaction}$ is less than −0.6 V the reaction does not 'go' at all. Between these values, we would usually regard the system as being in equilibrium.

(The actual relationship between $E^\ominus_{reaction}$ and K_c is:

$$E^\ominus_{reaction} = \frac{RT}{zF} \ln K_c$$

where R is the gas constant (8.31 J K^{-1} mol^{-1})
T is the temperature (in K)
z is the number of electrons transferred in the half-cell reaction
F is the Faraday constant 96 484 C mol^{-1})

E

THE EFFECT OF CONCENTRATION ON $E^\ominus$ VALUES—THE NERNST EQUATION

How is the e.m.f. of a half cell affected by changing the concentration of the solution? This can easily be investigated as follows:

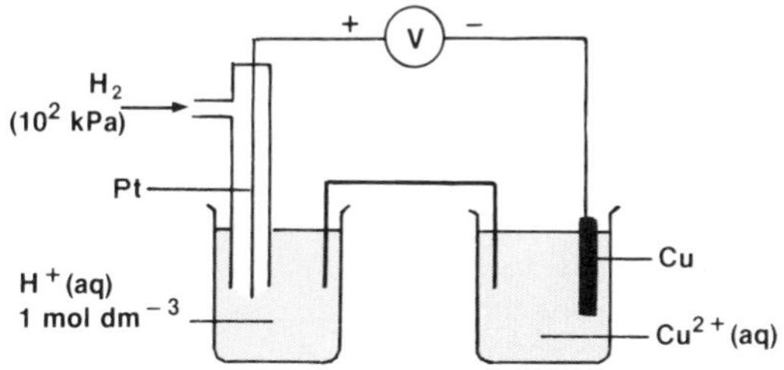

A Cu^{2+}(aq)/Cu(s) half cell is connected to a hydrogen electrode and the voltage recorded for various different concentrations of Cu^{2+}(aq) ions. The results show that the value of the e.m.f. is proportional to ln [Cu^{2+}(aq)] as shown by the graph.

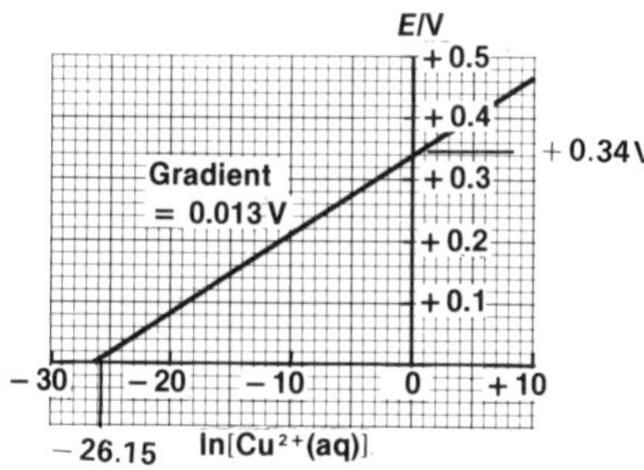

Graph to show the variation of *E* for Cu^{2+} (aq)/Cu(s) with [Cu^{2+}(aq)]

We can find $E^\ominus$ from this graph. The concentration of Cu^{2+}(aq) is 1 mol dm^{-3} for the standard half cell, i.e. [Cu^{2+}(aq)] = 1. Therefore ln [Cu^{2+}(aq)] = 0. From the graph, when ln [Cu^{2+}(aq)] = 0, $E^\ominus = 0.34$ V. This agrees with the value of $E^\ominus$ from **Table 13.1** on page 161. The gradient of the graph is 0.013 V. The equation of this graph is therefore:

$$E = E^\ominus + 0.013 \ln [Cu^{2+}(aq)]$$

which is an example of a more general expression which may be written:

$$E = E^\ominus + \frac{0.026}{z} \ln [\text{ion}]$$

or

$$E = E^\ominus \frac{RT}{zF} \ln [\text{ion}]$$

where R is the gas constant (8.3 J K^{-1} mol^{-1})
T is the temperature in K (298 K)
F is the value of the Faraday constant (96 484 C mol^{-1})
z is the number of electrons transferred in the half cell reaction (in this case, two)

This expression is called the **Nernst equation**. It is also worth noting that the results on the graph are consistent with the predictions of Le Chatelier's principle.

In the Cu^{2+}(aq)/Cu(s) half cell we have the equilibrium:

$$Cu^{2+}(aq) + 2e^- \rightleftharpoons Cu(s)$$

If the concentration of Cu^{2+}(aq) is reduced, then Le Chatelier's principle would predict that the equilibrium would move to the left giving more Cu^{2+}(aq) and more electrons. In other words, the $E^\ominus$ value of the half cell would become more negative (less positive), which is what occurs (see the graph).

USES OF THE NERNST EQUATION

1. Determination of equilibrium constants, K_c

The graph can be used to find K_c for the reaction

$$Cu(s) + 2H^+(aq) \rightleftharpoons Cu^{2+}(aq) + H_2(g)$$

For this reaction:

$$K_c = \frac{[Cu^{2+}(aq)]_{eqm}[H_2(g)]_{eqm}}{[H^+(aq)]_{eqm}^2}$$

Remember, pure solids do not appear in the expression for K_c (see **section 11.4.4**). When $E = 0$, there is no tendency for electrons to flow either way, so the two half cells must be at equilibrium at this value of E. From the graph, when $E = 0$, ln [Cu^{2+}(aq)] = −26.15. So [Cu^{2+}(aq)]$_{eqm}$ = 4.39 × 10^{-12} mol dm^{-3} (the antilog of −26.15 = 4.39 × 10^{-12}). [H^+(aq)] = 1 mol dm^{-3} in a standard hydrogen half cell. For H_2 at 10^2 kPa and 298 K, 1 mole has a volume of 24.46 dm^3, so the number of moles in 1 dm^3 is 1/24.46 = 0.041.

$$[H_2(g)]_{eqm} = 0.041 \text{ mol dm}^{-3}$$

Using these values, we get:

$$K_c = \frac{4.39 \times 10^{-12} \times 0.041}{1^2}$$

$$K_c = 1.8 \times 10^{-13}$$

Since K_c is much less than 1, the equilibrium is well over to the left, which we would expect as we know that copper does not react with dilute acids.

2. Measurements of concentrations

If we have a solution containing Cu^{2+}(aq) ions, we can measure the concentration by using it as the solution of Cu^{2+}(aq) in the apparatus (top left). We measure the value of E and calculate [Cu^{2+}(aq)] by extrapolating from the graph, above left, or substituting the value of E into the Nernst equation. This is especially useful for measuring small concentrations.

E

The pH METER

In principle the hydrogen electrode (connected to any other electrode) could be used to measure [$H^+(aq)$]. The Nernst equation gives:

$$E = E^{\ominus} + \frac{0.026}{z} \ln [H^+(aq)]$$

For the hydrogen electrode $E^{\ominus} = 0$ and $z = 1$, so

$$E = 0.026 \ln [H^+(aq)]$$

To convert ln into log we multiply by 2.303:

$$E = 0.026 \times 2.303 \log [H^+(aq)]$$
$$E = -0.59\,\text{pH}$$

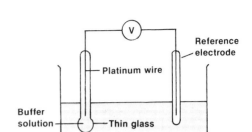

However, a hydrogen electrode is both cumbersome and in some situations potentially dangerous because it produces explosive hydrogen gas. Another more convenient electrode system whose potential depends on the pH is the glass electrode. This consists of a platinum wire sealed into a very thin walled glass tube which contains a buffer solution (a solution of fixed hydrogen ion concentration). The e.m.f. of this electrode depends on the pH of the solution in which it is placed. This electrode, coupled with a reference electrode, comprises a pH electrode. The voltage difference is measured on a high resistance voltmeter calibrated directly in pH units. The main drawback is that the thin wall of the glass electrode is rather delicate.

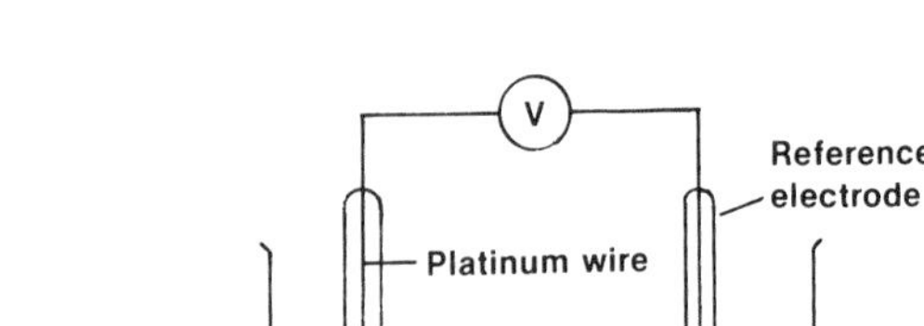

The pH meter

13.4 Summary

13.4.1 Redox reactions

- Redox reactions involve electron transfer.
- a species has been oxidized if it has lost electrons and reduced if it has gained them (OIL RIG).
- Alternatively, oxidation involves an increase in oxidation number and reduction a decrease in oxidation number.
- Oxidation number, abbreviated Ox, is the charge on each ion in an ionic compound or, more loosely, in a covalent compound, the charge that each atom would have if the compound is imagined to be ionic.
- Elements have Ox = 0.
- The sum of the Ox of all the atoms in a compound = 0.
- The sum of the Ox of all the atoms in a complex ion = the charge of the ionic species.

13.4.2 Electrochemical cells

- A half cell consists of a metal in contact with a solution of its own ions. Half cells develop electrical potential due to the tendency of all metals to give away electrons.
- If two half cells are connected by a voltmeter and salt bridge their potential difference (e.m.f.) can be measured.
- If the other half cell is a standard hydrogen electrode, the measured e.m.f. is called $E^{\ominus}$ (E standard), provided standard conditions are used—1 mol dm^{-3} solution, 10^2 kPa pressure and 298 K temperature.
- $E^{\ominus}$ values for half cells are listed in order, with the most negative (the best electron releasers) at the top, to give the electrochemical series.
- The electrochemical series can be extended to systems other than metal/metal ion by making electrical contact with a platinum electrode.
- Half cells high in the electrochemical series can transfer electrons to a half cell lower in the list. This leads to a method of predicting the feasibility of reactions by the 'anti-clockwise rule'.
- An alternative way of doing this is by calculating $E^{\ominus}_{\text{reaction}}$. This is done by adding together two half reactions and their $E^{\ominus}$ values. If $E^{\ominus}_{\text{reaction}}$ is positive, the reaction is feasible.

- Electrochemical cells made by connecting two half cells can be represented in diagrammatic form, for example:

$$Cu(s)|Cu^{2+}(aq) \,\vdots\vdots\, Zn^{2+}(aq)|Zn(s) \quad E^{\ominus} = -1.1\,V$$

| represents a phase boundary and ⁞⁞ a salt bridge, and the polarity of the right-hand electrode (as written down) is given.

13.5 Questions

1 Say whether the underlined atom is oxidized, is reduced, disproportionates or remains unchanged in the following reactions:

A) $\underline{Pb}Cl_2(s) + Cl_2(g) \longrightarrow PbCl_4(s)$
B) $\underline{Na}OH(aq) + HCl(aq) \longrightarrow NaCl(aq) + H_2O(l)$
C) $2\underline{I}O_3^-(aq) + 5HSO_3^-(aq) \longrightarrow I_2(aq) + 5SO_4^{2-}(aq) + 3H^+(aq) + H_2O(l)$
D) $Cu(s) + \frac{1}{2}\underline{O}_2(g) \longrightarrow CuO(s)$
E) $3\underline{Cl}_2(aq) + 6NaOH(aq) \longrightarrow 5NaCl(aq) + NaClO_3(aq) + 3H_2O(l)$

2 What is the oxidation number of the underlined atom in the following?

A) $\underline{Co}Cl_3$
B) $\underline{N}O_2$
C) $\underline{N}_2O_4$
D) $\underline{U}F_6$
E) $Ca\underline{C}O_3$
F) $\underline{Pb}$
G) $\underline{Cl}^-$
H) $\underline{N}H_4NO_3$
I) $NH_4\underline{N}O_3$
J) $\underline{P}Cl_3$
K) $\underline{P}Cl_5$
L) $H_2\underline{S}O_4$

3 Name the following compounds using oxidation numbers where necessary:

A) $CoCl_3$
B) H_2SO_4
C) H_2SO_3
D) $NaNO_3$
E) PbO
F) PbO_2
G) $NaClO_4$
H) $NaClO_3$
I) $NaClO$

4 Use values of $E^{\ominus}$ on page 164 to calculate $E^{\ominus}_{reaction}$ for the following:

A) $Ce^{4+}(aq) + Fe^{2+}(aq) \longrightarrow Ce^{3+}(aq) + Fe^{3+}(aq)$
B) $I_2(aq) + 2Br^-(aq) \longrightarrow Br_2(aq) + 2I^-(aq)$
C) $MnO_4^-(aq) + 8H^+(aq) + 5I^-(aq) \longrightarrow Mn^{2+}(aq) + 4H_2O(l) + 2\frac{1}{2}I_2(aq)$
D) $2H^+(aq) + Pb(s) \longrightarrow Pb^{2+}(aq) + H_2(g)$

5 A) Represent the following on conventional cell diagrams:

i)

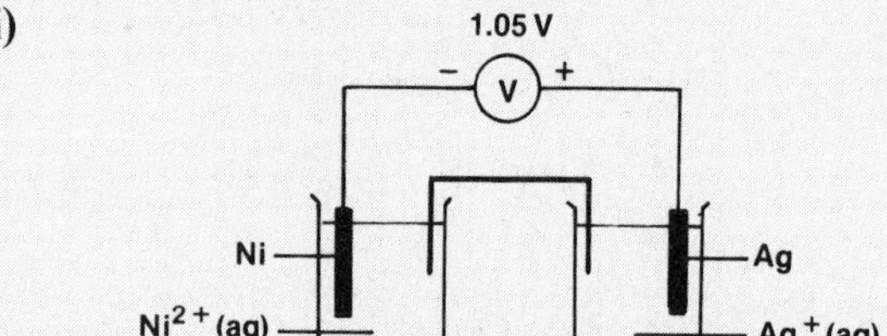

ii)

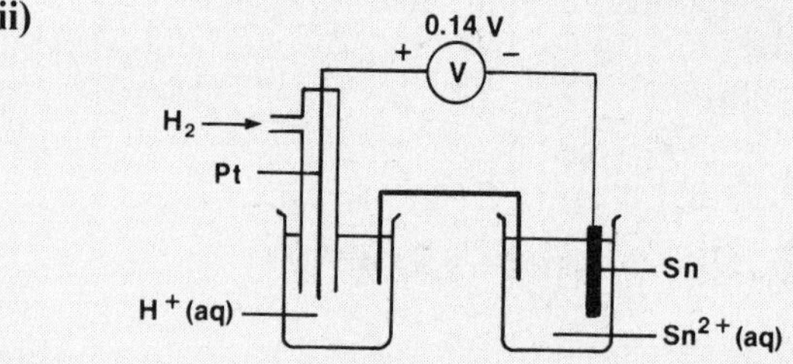

iii)

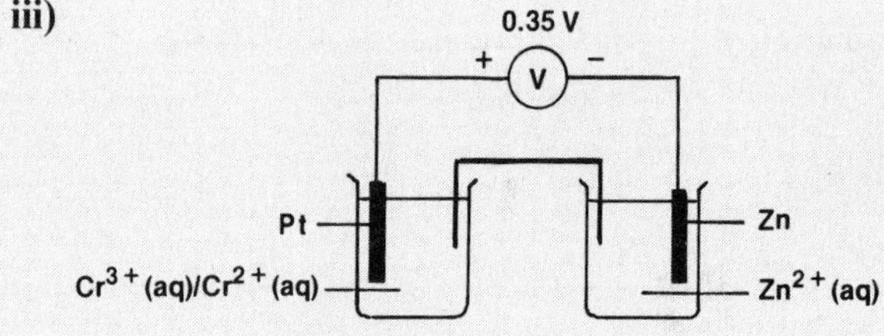

iv)

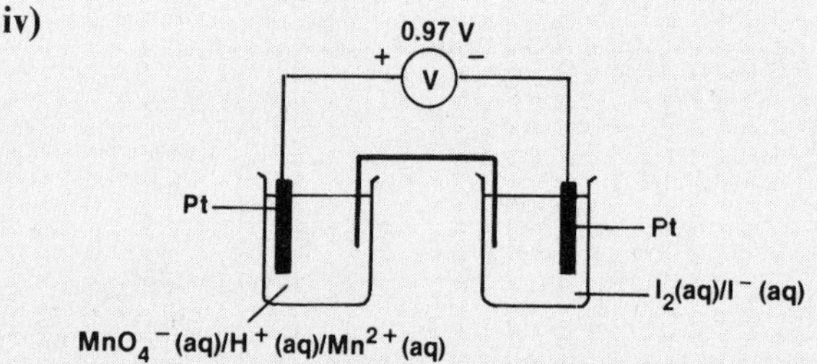

B) What reactions would take place if each of these cells was short-circuited (i.e. the voltmeter was replaced with a piece of wire so that electrons could flow)?

6 A) Using the data in **Table 13.2** on page 164, insert the following half cells on a diagram of $E^{\ominus}$ against Ox:

$$I_2(aq)/2I^-(aq),\ Cl_2(aq)/2Cl^-(aq),$$
$$Br_2(aq)/2Br^-(aq),\ Ag^+(aq)/Ag(s)$$

B) Which of the halogens could possibly oxidize Ag(s) to $Ag^+(aq)$ ions?

C) Is the reaction:

$$Br_2(aq) + 2Cl^-(aq) \longrightarrow Cl_2(aq) + 2Br^-(aq)$$

feasible?

7

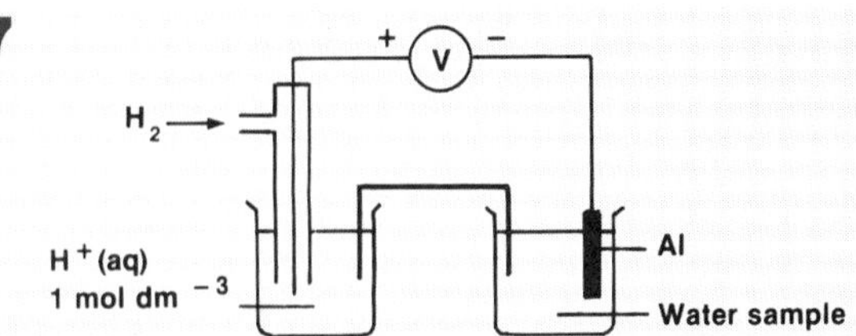

The apparatus above was set up to measure small concentrations of Al^{3+}(aq) ions in water samples where it was suspected that aluminium ions were being leached out of the soil by acid rain. Using the equation:

$$E = E^{\ominus} + \frac{0.026}{z} \ln [\text{ion}]$$

A) What values should be put in the equation for $E^{\ominus}$ and for z in this case?

B) With a solution containing Al^{3+}(aq) ions, the voltmeter reading obtained was 1.70 V. What concentration of Al^{3+}(aq) does this give?

C) In order to use the apparatus for field work, it was suggested that the hydrogen electrode be replaced by a Cu^{2+}(aq)/Cu(s) one. How, if at all, would that affect the value of:

i $E^{\ominus}$

ii z in the equation?

8 An electrochemical cell was set up to investigate the variation with concentration of the electrode potential of the nickel/aqueous nickel(II) ion system. A standard copper/aqueous copper(II) ion electrode was used as a reference electrode. From the results, the nickel electrode potential was deduced and plotted against the corresponding nickel(II) ion concentration.

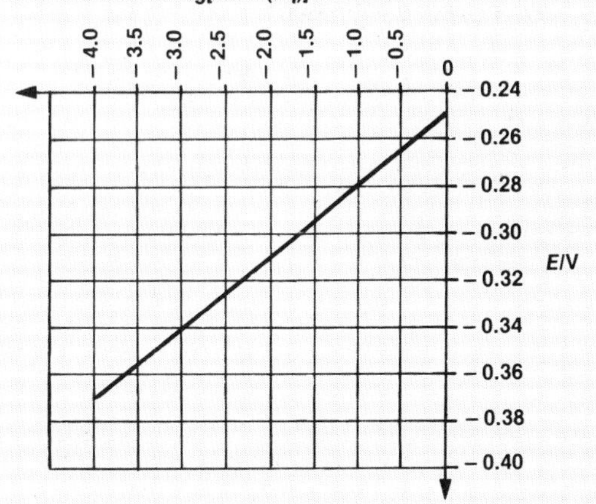

A) Write down the conventional cell diagram for the cell used.

B) From the graph deduce the standard electrode potential for the nickel electrode system.

C) If the standard electrode potential of the copper reference electrode system is +0.34 volts, what would be the e.m.f. of the cell composed of a standard copper electrode and a standard nickel electrode?

(Nuffield 1983, part question)

9 Aqueous hydrogen peroxide is a useful oxidizing agent, but its use in quantitative analysis is limited by its slow decomposition during storage.

$$2H_2O_2(aq) \longrightarrow 2H_2O(l) + O_2(g)$$

The table below contains data which may be required in answering the questions about hydrogen peroxide which follow.

Half cell	E/V
$[SO_4^{2-}(aq) + 4H^+(aq)], [H_2SO_3(aq) + H_2O(l)]\|Pt$	+0.17
$[O_2(g) + 2H^+(aq)], [H_2O_2(aq)]\|Pt$	+0.68
$[MnO_4^-(aq) + 8H^+(aq)], [Mn^{2+}(aq) + 4H_2O(l)]\|Pt$	+1.51
$[H_2O_2(aq) + 2H^+(aq)], [2H_2O(l)]\|Pt$	+1.77

A) Using oxidation numbers, explain why the decomposition of hydrogen peroxide solution is said to be a disproportionation reaction.

B) Put the half cell potentials on a diagram of $E^{\ominus}$ against oxidation number.

C) A common method for determining the concentration of hydrogen peroxide solutions involves a redox titration with a standard potassium manganate(VII) solution.

i Use the data in the diagram in **B)** to construct a balanced equation for the overall titration.

ii How would the end-point of the titration be indicated?

iii When 25.0 cm^3 of a hydrogen peroxide solution was titrated with 0.10 mol dm^{-3} acidified potassium manganate(VII), 24.0 cm^3 of the latter were required at the end-point. Calculate the concentration of the hydrogen peroxide solution.

(Nuffield 1985, part question)

14 Kinetics

14.1 Introduction

Bhopal, 1984 – victims queuing for treatment

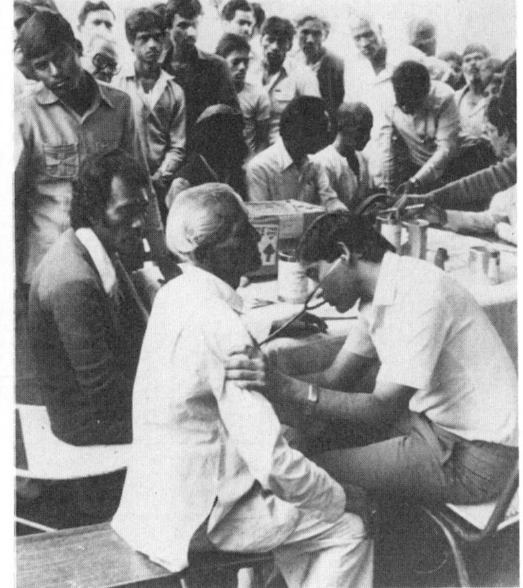

Chemical kinetics is the study of the rates of reactions and the factors which influence them. You will know from your own experience that there is a very large variation in reaction rates, from 'popping' a test tube full of hydrogen which is over in a fraction of a second, to the complete rusting away of an iron nail which could take several years. There are obvious practical advantages in being able to control the rates of reactions: it was a 'runaway' reaction in a pesticide factory that caused the poison gas tragedy in Bhopal, India in 1984. A more down-to-earth example is the use of low temperatures in a refrigerator to slow down spoilage reactions in food.

In terms of chemical theory, kinetic evidence is often used to piece together the details of the series of simple steps by which most reactions occur—the reaction mechanism.

14.1.1 Factors which affect the rates of chemical reactions

There are six factors which can affect the rates of chemical reactions.

Temperature

A rough rule of thumb which applies to many reactions at around room temperature is that a 10 K (10 °C) increase in temperature approximately doubles the rate of a reaction.

REFRIGERATION

Refrigeration (cooling down to 0 °C, 273 K) and freezing (below 0 °C) are used for food storage because spoilage reactions are slowed down and multiplication of micro-organisms considerably reduced. However, even at freezer temperatures, reactions do continue. Any warming up, such as during transfer from one freezer to another, speeds up deterioration. The combination of the duration and temperature of the warming up governs the shortening of storage life caused. One way to monitor this is a device called a TTM (time temperature monitor) attached to the food. This is based on an enzyme reaction which goes at the same rate as the reactions in the food. After the food has been warmed up to some preset combination of time and temperature an indicator changes colour. This colour change is not reversible. Food whose TTM has changed should be rejected. These devices have been developed in Sweden and are not yet in general use in the UK.

Concentration

Increasing the concentration of certain reactants increases the rate of reactions in solution.

Pressure

Increasing the pressure of a gas phase reaction increases the rate. Increasing either pressure or concentration increases the number of molecules, atoms or ions per cubic centimetre.

Surface area

Increasing the surface area of solid reactants increases the rate. For example, powdered zinc will react faster with acids than zinc granules. The smaller the size of the zinc granules, the greater the total surface area.

POWDER EXPLOSIONS

Increasing the surface area of solids can have a dramatic effect on the rate of chemical reactions. This can cause a considerable safety hazard in industries which deal with powdered substances. There are several instances of explosions in mines caused by the extremely rapid burning of coal dust in the air. This is hard to believe when you think how difficult it can be to set light to lumps of coal in a grate. Even substances as apparently innocuous as flour or custard powder pose a significant explosion risk due to their large surface areas. Workers in these industries have to wear soft shoes to guard against sparks and special precautions have to be taken with electrical equipment which might cause sparks. Even floor sanders which produce clouds of very finely divided sawdust have to be used with care.

Adhesive activated by sunlight

Light

Light affects the rates of certain reactions. For example, the reaction of bromine with alkanes proceeds quite quickly under a photoflood lamp but much more slowly in ordinary light. Some recently introduced adhesives for sticking glass make use of this principle. The adhesive does not begin to react until exposed to sunlight.

Catalysts

Catalysts are substances which can change the rate of a chemical reaction without being chemically changed themselves. For example, addition of manganese(IV) oxide dramatically increases the rate of decomposition of hydrogen peroxide to oxygen and water. Catalysts are of great economic importance in industrial processes.

14.2 Measuring reaction rates

Note: Remember the use of square brackets to mean concentration in mol dm^{-3}.

Figure 14.1 **Changes of concentration with time**

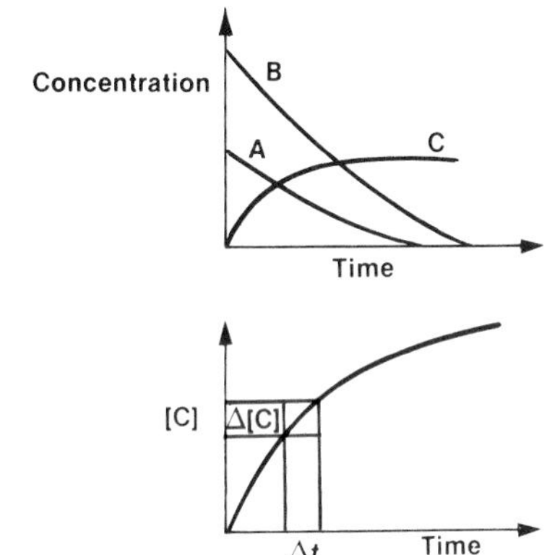

In a reaction, say:

$$A + 2B \longrightarrow C$$

the concentrations of the reactants A and B decrease with time and that of the product C increases, as shown in **Figure 14.1.** The average rate of the reaction with respect to C, during a period of time Δt, is the change in concentration of C, $\Delta[C]$, divided by Δt.

$$\text{rate of change concentration of C} = \frac{\Delta[C]}{\Delta t}$$

so if in 10 seconds [C] changed from 1.0 mol dm^{-3} to 1.1 mol dm^{-3},

$$\text{average rate of change of [C]} = 1.1 - 1.0\ \text{mol dm}^{-3}/10\ \text{s}$$
$$= 0.1/10 = 10^{-2}\ \text{mol dm}^{-3}\ \text{s}^{-1}$$

We are often interested in the rate of change of [C] at a particular instant in time rather than the average rate of change over a period of time. This rate of change at a particular instant is found from the gradient (slope) of the tangent to the curve at that time, **Figure 14.2.** The mathematical notation for rate of change of [C] with time is $d[C]/dt$ (pronounced dee cee by dee tee). So $d[C]/dt = a/b$. 'd' is used rather than Δ, to show that the change is taking place over a vanishingly small period of time. We shall use this notation as it is shorter than writing out 'rate of change of [C] with time'.

The rate of change of concentration of different species may be different. As [C] is increasing, [A] is decreasing. As one molecule of A disappears, one of C appears, so:

Figure 14.2 The rate of change of [C] at time *t* is the gradient of the concentration–time graph

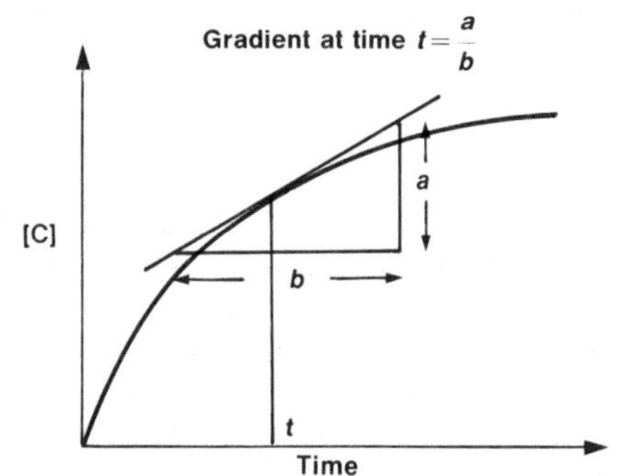

$$\frac{d[C]}{dt} = -\frac{d[A]}{dt}$$

As the equation tells us that two molecules of B are used up for every one of A, then:

$$\frac{d[B]}{dt} = \frac{2d[A]}{dt}$$

So when talking about the rate of a reaction, we must say which reactant or product we are referring to.

14.3 Experimental methods for measuring reaction rates

To measure reaction rates, we need a method of measuring the concentration of one of the reactants or products as time goes on. The method chosen depends on the substance whose concentration is being measured and also on the speed of the reaction. A method such as a titration which takes a few minutes to complete would be perfectly satisfactory for a reaction which takes some hours to go to completion, but not for a reaction which is over in a few seconds.

14.3.1 Sampling and titration

The reaction mixture is sampled by pipette at suitable time intervals and the concentration of one of the reactants or products found by a suitable titration. If the reaction is relatively fast, it may be possible to stop it or slow it down in some way after sampling, so that the titration can be carried out at leisure. For example, a sample could be taken from a reaction with an acid catalyst and run into an alkali to remove the catalyst and slow the reaction down. Alternatively, the reaction mixture can be cooled down rapidly which will also slow down the reaction to give more time for titration. Either method is referred to as 'freezing' or 'quenching' the reaction. A reaction for which this would be suitable is:

$$I_2(aq) + CH_3COCH_3(aq) \xrightarrow[\text{catalyst}]{H^+} CH_2ICOCH_3(aq) + HI(aq)$$

iodine propanone iodopropanone hydrogen iodide

The reaction is complete in about 30 minutes, depending on the concentrations. The reaction mixture can be sampled every five minutes or so and the sample run into a flask containing excess sodium hydrogencarbonate. This neutralizes the acid catalyst, slowing the reaction considerably. The iodine remaining at this time can be titrated with sodium thiosulphate solution using starch at the end point (see box *Swimming pool chemistry*, **section 13.2**) according to the equation below:

$$I_2(aq) + 2Na_2S_2O_3(aq) \longrightarrow 2NaI(aq) + Na_2S_4O_6(aq)$$

14.3.2 Using a colorimeter

This is a more satisfactory method than titration for two reasons. Firstly, no sampling is needed, and secondly, a reading can be taken almost instantaneously so that quite rapid reactions can be followed, especially if the colorimeter is interfaced to a data logger or microcomputer which would enable a graph of concentration versus time to be plotted as the reaction proceeds (on-line). The method is only useful, of course, if one of the reactants or products is coloured.

A colorimeter consists of a light bulb, and filters to select a suitable colour (or set of wavelengths) of light which is absorbed by the sample. (For example, a red coloured sample transmits red light and therefore absorbs most of the other colours of light—see box below.) The light passes through the sample onto a detector whose output goes to a meter or a recording device, see **Figure 14.3**. This method could be used as an alternative method for the investigation of the reaction on the previous page, as iodine (brown) is the only coloured species involved in the reaction. The colorimeter will usually need to be calibrated to establish the relationship between its reading and the concentration of the species being observed.

Colorimeter interfaced to VELA data logger. The sample is in the tube

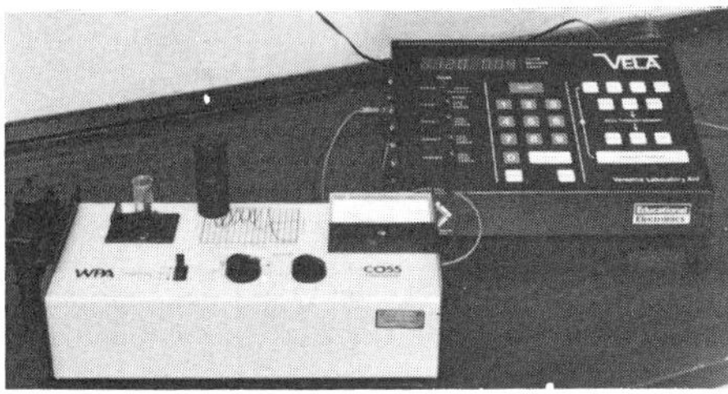

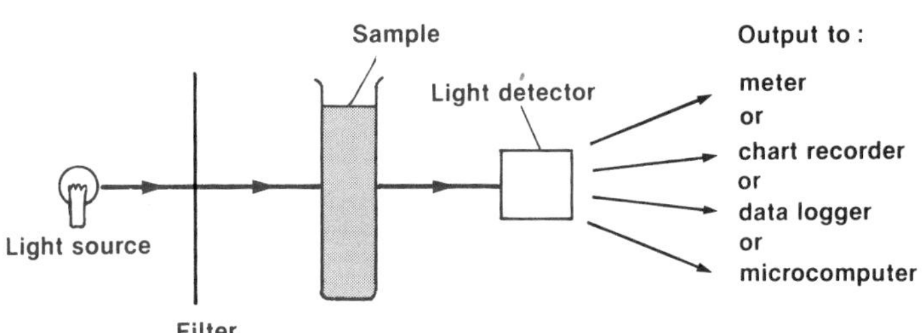

Figure 14.3 (right) Using a colorimeter to measure reaction rates

E

CHOICE OF FILTERS FOR A COLORIMETER

This is a very simplified account considering just the three primary colours: red R, green G, and blue B.

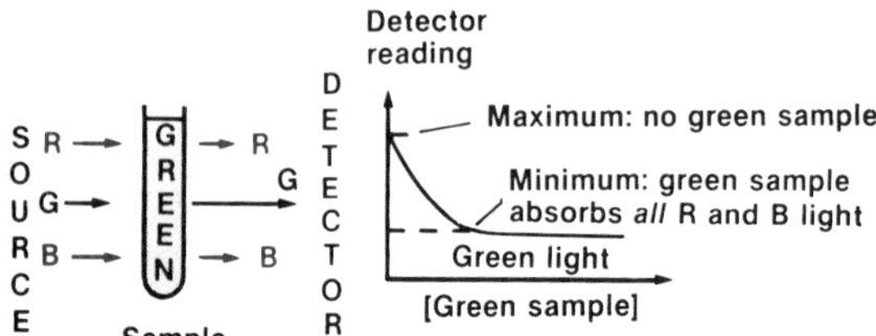

NO FILTER

A green sample lets green light through and absorbs some of the rest, the amount absorbed depending on the concentration of the green sample. So the meter reading can never drop to zero as there will always be green light passing through the sample unabsorbed, as well as some red and blue.

WITH A SUITABLE FILTER

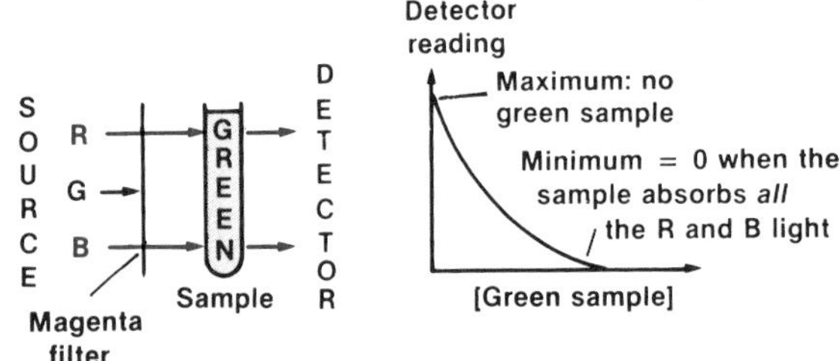

A suitable filter absorbs green (the filter colour is called magenta—a purplish colour) just allowing red and blue through. Now if the green sample is concentrated enough, the meter reading can drop to *zero* as there is now none of the green light which always gets through without a filter. In practice, this means that we can use the meter on a more sensitive setting to measure how well the green solution absorbs red and blue.

Figure 14.4 Measuring gas volume to monitor reaction rates

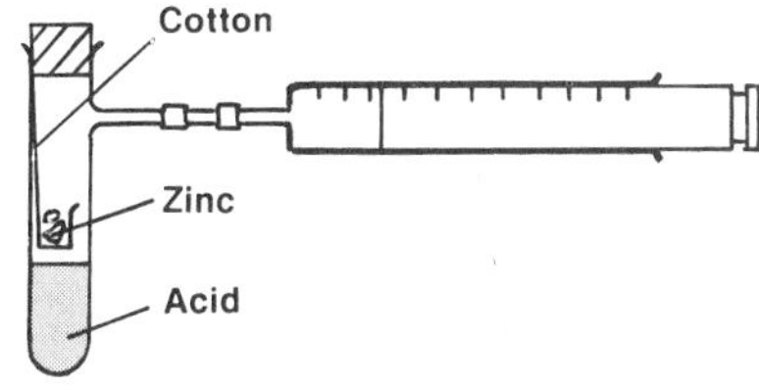

14.3.3 Measuring the volume of gas given off

In the reaction:

$$Zn(s) + H_2SO_4(aq) \longrightarrow ZnSO_4(aq) + H_2(g)$$

zinc + sulphuric acid ⟶ zinc sulphate + hydrogen

we can follow the reaction by measuring the volume of hydrogen evolved at any one time. The apparatus can be assembled and the reaction started by shaking the zinc into the acid.

Figure 14.5 A conductivity meter can be used to measure concentrations of ionic species

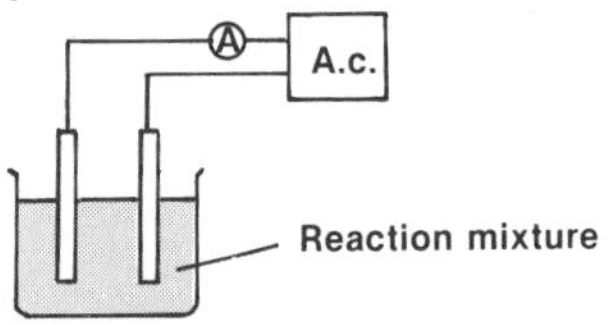

The apparatus in Figure 14.4 interfaced to a data logger

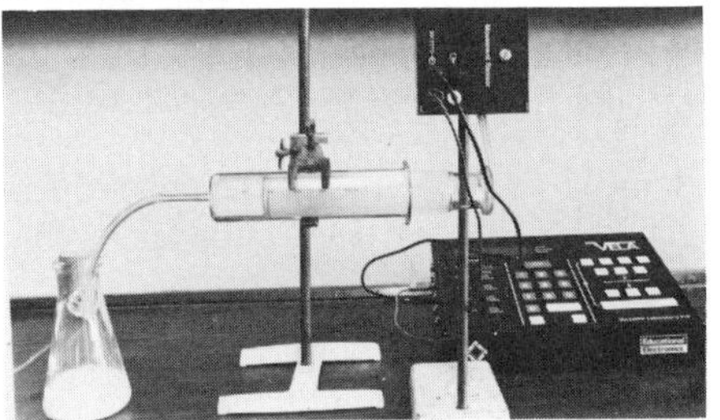

Figure 14.6 A dilatometer – used for measuring small volume changes during reactions

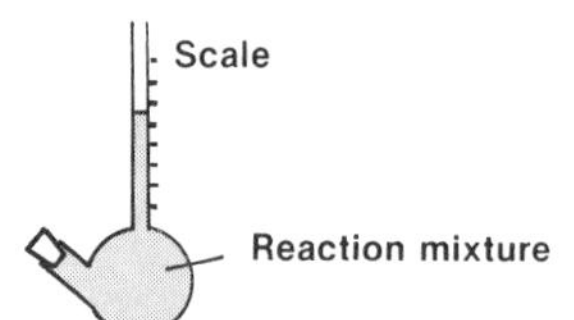

14.3.4 Measuring the conductivity of a solution

This method, **Figure 14.5** is suitable for reactions where the number of ions in the solution changes. The more ions, the greater the conductivity and the bigger the ammeter reading. Alternating current must be used to avoid electrolysing the solution. In the reaction below, neither of the reactants is ionic but one of the products is, so the conductivity should increase as the reaction proceeds.

C_4H_9Br	+ H_2O	$\longrightarrow$ C_4H_9OH	+ H^+	+ Br^-
bromobutane	water	butanol	hydrogen ion	bromide ion

14.3.5 Using a dilatometer

Most reactions in solution involve small volume changes. These can be measured with a dilatometer, **Figure 14.6.** The narrow vertical tube means that small volume changes produce measurable changes in the liquid level.

14.3.6 Other methods

A variety of other methods can be used to measure reaction rates. In fact, any property of the reaction mixture which can be related to the extent of reaction can be used. Examples include pressure changes for gas reactions, and the rotation of the plane of polarization for optically active compounds (see **section 26.7.2**).

14.4 The rate expression

The outcome of an experimental kinetic investigation of a reaction is a **rate expression**. This is an equation which describes how the rate of the reaction depends on the concentration of various species involved in the reaction. For example, the reaction:

$$X + Y \longrightarrow Z$$

might have the rate expression:

$$\frac{d[Z]}{dt} \propto [X][Y]^2$$

which can be written:

$$\frac{d[Z]}{dt} = k[X][Y]^2$$

k is called the **rate constant** for the reaction.

The **order** of the reaction with respect to X is the power to which [X] is raised in the rate expression—in this case it is one, so that doubling the concentration of X would double the rate.

In the same way the order with respect to Y is two, so that doubling the concentration of Y would quadruple the rate.

The **overall order** of the reaction is the sum of the orders of all the species which appear in the rate expression. In this case the overall order is three. So this reaction is said to be first order with respect to X, second order with respect to Y and third order overall.

The rate expression is entirely derived from the experimental evidence and there is no way that it can be predicted from the chemical equation for the reaction. Therefore it is quite unlike the equilibrium law expression which it seems to resemble at first sight.

Species which do not appear in the chemical equation *may* appear in the rate expression and species which appear in the chemical equation do not necessarily appear in the rate equation.

For example, in the reaction:

$$\underset{\text{propanone}}{CH_3COCH_3(aq)} + \underset{\text{iodine}}{I_2(aq)} \xrightarrow{H^+ \text{ catalyst}} \underset{\text{iodopropanone}}{CH_2ICOCH_3(aq)} + \underset{\text{hydrogen iodide}}{HI(aq)}$$

the rate expression is:

$$\frac{-d[I_2]}{dt} = k[CH_3COCH_3(aq)][H^+(aq)]$$

Note that as I_2 is a reactant, its concentration decreases as the reaction proceeds and there is a minus sign in the expression above. So the reaction is first order with respect to propanone, first order with respect to H^+ ions and second order overall.

The rate does not depend on $[I_2(aq)]$ so we can say the reaction is zero order with respect to iodine. The H^+ ions act as a catalyst in this reaction.

For the reaction:

$$\underset{\text{bromate ions}}{BrO_3^-(aq)} + \underset{\text{bromide ions}}{5Br^-(aq)} + \underset{\text{hydrogen ions}}{6H^+(aq)}$$

$$\longrightarrow \underset{\text{bromine}}{3Br_2(aq)} + \underset{\text{water}}{3H_2O(l)}$$

$$\frac{-d[BrO_3^-(aq)]}{dt} = k[BrO_3^-(aq)][Br^-(aq)][H^+(aq)]^2$$

The reaction is: first order with respect to $BrO_3^-(aq)$
first order with respect to $Br^-(aq)$
second order with respect to $H^+(aq)$
fourth order overall.

Notice that there is *no relationship* between the coefficients in the chemical equation and the powers in the rate expression.

Note that since $BrO_3^-(aq)$ is a reactant, $d[BrO_3^-(aq)]/dt$ is negative, hence the negative sign in the rate expression, as all the other terms are positive numbers.

14.4.1 The rate constant, *k*

The units of the rate constant vary depending on the overall order of the reaction. For a first-order reaction where:

$$\frac{d[A]}{dt} = k[A]$$

the units of $d[A]/dt$ are mol dm^{-3} s^{-1} and the units of [A] are mol dm^{-3}, so the units of k are s^{-1} obtained by cancelling:

$$\cancel{\text{mol}\,\text{dm}}^{-3}\,\text{s}^{-1} = k\,\cancel{\text{mol}\,\text{dm}}^{-3}$$

For a second-order reaction where:

$$\frac{d[B]}{dt} = k[B][C]$$

the units of $d[B]/dt$ are mol dm^{-3} s^{-1} and the units of both [A] and [B] are mol dm^{-3}, so by cancelling:

$$\cancel{\text{mol}\,\text{dm}}^{-3}\,\text{s}^{-1} = k\,\cancel{\text{mol}\,\text{dm}}^{-3} \times \text{mol}\,\text{dm}^{-3}$$

this gives the units of k as dm^3 mol^{-1} s^{-1}. It is best to work out the units for a particular situation rather than trying to remember them.

It is perfectly possible to have orders of reaction which are not whole numbers, or are less than one, although you are most unlikely to come across these in an A-level course.

Rate constants can be used to compare the rates of different reactions, as they represent the rate of a reaction when the concentrations of all relevant species are 1 mol dm^{-3}.

Rate constants vary with temperature so they must be stated with the temperature at which they were obtained.

Figure 14.7 A concentration–time graph

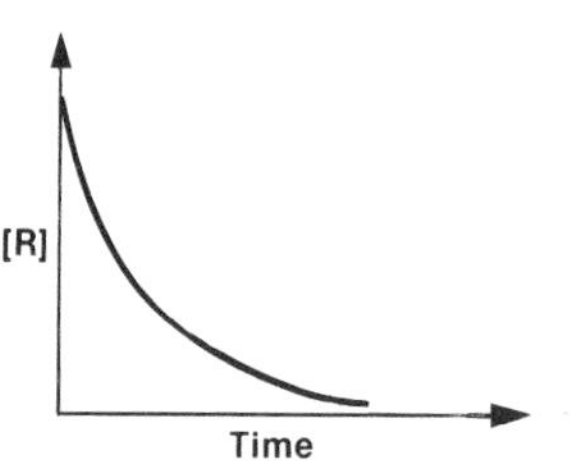

14.4.2 Determination of the order of a reaction from experimental data

The usual outcome from an experiment on reaction rates is a series of readings of concentration at different times. The first step is usually to plot these on a concentration–time graph.

To help interpretation, it is usual to make sure that reactants other than the one being investigated do not change in concentration. This can be done by making their concentrations very large compared with that of the reactant being investigated, for example:

$$\underset{\text{reactant R}}{R} + \underset{\text{reactant B}}{B} \longrightarrow \text{products}$$

if $[R] = 0.01\ \text{mol dm}^{-3}$ at the start
and $[B] = 1.00\ \text{mol dm}^{-3}$ at the start
then at the end of the reaction, $[R] = 0$ (all used up)
and $[B] = 1.00 - 0.01 = 0.99\ \text{mol dm}^{-3}$

So during the whole reaction [B] is practically constant.
A graph could then be plotted of [R] against time, **Figure 14.7**.

Sometimes it is possible to tell the order with respect to the species measured simply by looking at the graph. A reaction which is zero order with respect to reactant R will give a straight line, as the rate (the gradient of the graph) is unchanged by varying the [R]. (Zero order with respect to R means the rate is unaffected by changes in [R].) First- and second-order reactions both give curves, but the second-order curve is deeper than the first-order one, **Figure 14.8**. A zero-order graph will be obvious, but it will not be easy to distinguish first- and second-order curves just by eye.

Figure 14.8 Concentration–time graphs for zero-, first- and second-order reactions

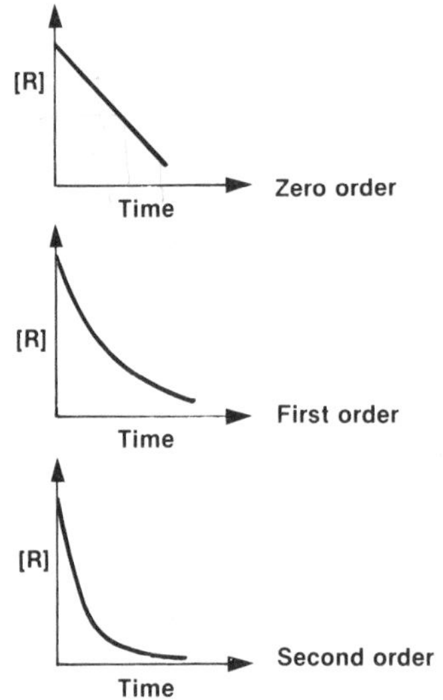

Successive half-lives

First- and second-order curves can be distinguished on the basis of their **half-lives**. The half-life of a reaction, $t_{1/2}$, is the time taken for [R] to fall from any chosen value to half that value. It is most commonly encountered in the context of radioactive decay, see **section 8.4.4**. The first half-life is taken from the start of the reaction, the second from the end of the first half-life, and so on, see **Figure 14.9.**

In first-order reactions, all the half-lives will be the same (within experimental error). A second-order curve will have half-lives which get successively larger. This should make identification easy. Do not worry about why this is the case—it arises from mathematical manipulation of the rate expression.

Figure 14.9 Successive half-lives for first- and second-order reactions

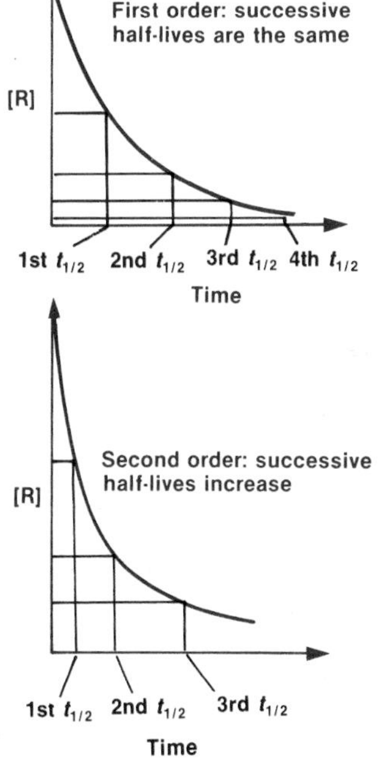

Rate–concentration graphs

A second method of determining the order of a reaction with respect to a particular species is by plotting a graph of **rate** against concentration (rather than concentration against time). On the original graph of concentration against time, draw tangents at different values of [R]. The gradients of these tangents are the rates of change of [R] (the reaction rate $d[R]/dt$) at different concentrations. These rates can be used to construct a second graph of rate ($d[R]/dt$) against concentration, see **Figure 14.10** (overleaf). If this graph is a straight line then rate $\propto [R]^1$ so the order is 1. If this graph is not a straight line try plotting rate against $[R]^2$. If this is a straight line then the order is 2, and so on.

The initial rate method

The method above only enables you to find the order with respect to the *one* reactant whose concentration has been measured. The initial rate

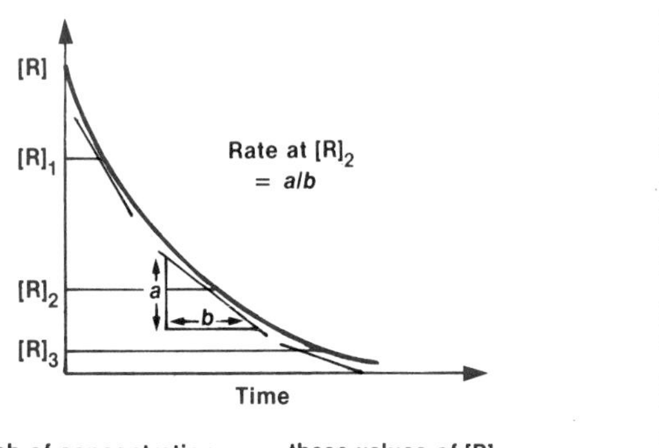

Original graph of concentration against time. The gradients of the tangents drawn at $[R]_1$, $[R]_2$, $[R]_3$, etc. give the rate of the reaction at these values of [R]. Now you can plot a graph of rate against [R]

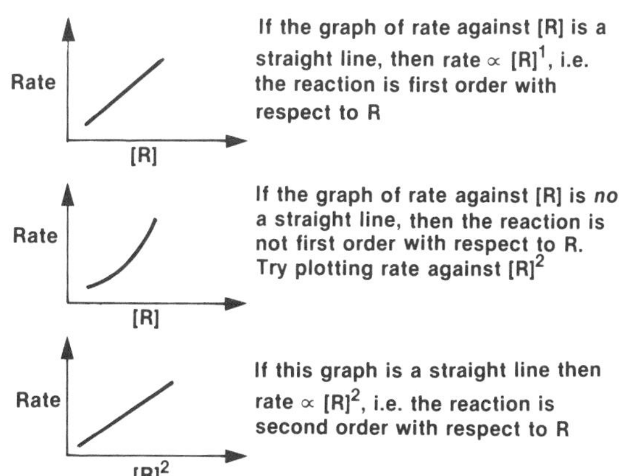

Rate
[R]
If the graph of rate against [R] is a straight line, then rate ∝ [R]1, i.e. the reaction is first order with respect to R
Rate
[R]
If the graph of rate against [R] is not a straight line, then the reaction is not first order with respect to R. Try plotting rate against [R]2
Rate
[R]2
If this graph is a straight line then rate ∝ [R]2, i.e. the reaction is second order with respect to R

Figure 14.10 Constructing graphs of rate against concentration

method allows us to find the order with respect to any species in the reaction mixture. A series of experiments is done with different initial concentrations of reactants, catalysts, etc. The concentration of one reactant is followed and a concentration–time graph plotted as before, **Figure 14.11.** The gradient of the tangent drawn in at $t = 0$ is the **initial rate**. The importance of measuring the *initial* rate is that the concentrations of all substances in the reaction are known *exactly* at this time.

Figure 14.11 Finding the initial rate of a reaction

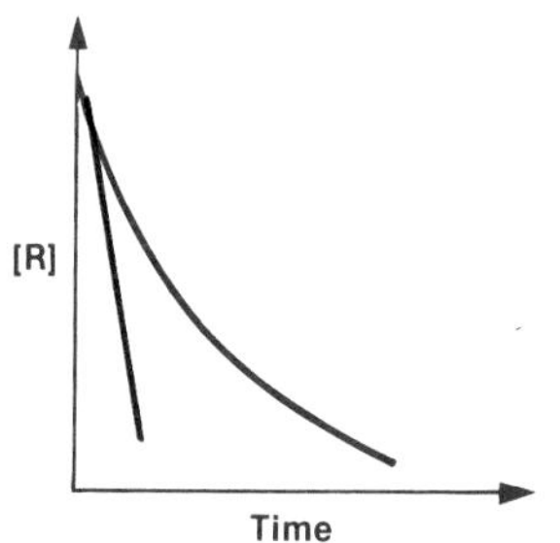

Comparing the initial concentrations and the initial rates for pairs of experiments should then enable the order with respect to each reactant to be found. For example, for the reaction:

$$2NO(g) + O_2(g) \longrightarrow 2NO_2(g)$$

nitrogen monoxide, oxygen, nitrogen dioxide

the initial rates found are shown in **Table 14.1.**

Table 14.1 Experimental results obtained for the reaction $2NO(g) + O_2(g) \rightarrow 2NO_2(g)$

Experiment number	Initial [NO]/10^{-3} mol dm^{-3}	Initial $[O_2]$/10^{-3} mol dm^{-3}	Initial rate $-d[NO]/dt$/10^{-4} mol dm^{-3} s^{-1}
1	1	1	7
2	2	1	28
3	3	1	63
4	2	2	56
5	3	3	189

On going from mixture 1 to mixture 2 [NO] is doubled while $[O_2]$ stays the same. The rate quadruples (from 7 to 28) suggesting rate $\propto [NO]^2$. This is confirmed by considering mixtures 1 and 3 where [NO] is trebled while $[O_2]$ stays the same. Here the rate is increased ninefold as would be expected if rate $\propto [NO]^2$. So the order with respect to NO is two.

Now take mixtures 2 and 4. Here [NO] is constant but $[O_2]$ doubles from mixture 2 to mixture 4. The rate doubles so it looks as if rate $\propto [O_2]$. This is confirmed by considering mixtures 3 and 5. Again [NO] is constant, but $[O_2]$ triples. The rate triples too, confirming that the order with respect to O_2 is one.

So: rate $\propto [NO]^2$

rate $\propto [O_2]^1$,

i.e. rate $\propto [NO]^2[O_2]^1$

The overall order is three and:

$$\text{rate} = k[NO]^2[O_2]^1$$

To find the rate constant, k, we then substitute any set of values of rate, [NO] and $[O_2]$ in the equation.

Taking the values for experiment 2:

$$28 \times 10^{-4}\,\text{mol}\,\text{dm}^{-3}\,\text{s}^{-1} = k(2 \times 10^{-3})^2(\text{mol}\,\text{dm}^{-3})^2 \times 1 \times 10^{-3}\,\text{mol}\,\text{dm}^{-3}$$
$$28 \times 10^{-4}\,\text{mol}\,\text{dm}^{-3}\,\text{s}^{-1} = k(4 \times 10^{-6}) \times 1 \times 10^{-3}\,\text{mol}^3\,\text{dm}^{-9}$$
$$28 \times 10^{-4} = k \times 4 \times 10^{-9}\,\text{mol}^2\,\text{dm}^{-6}\,\text{s}$$
$$k = \tfrac{28}{4} \times 10^5\,\text{dm}^6\,\text{mol}^{-2}\,\text{s}^{-1}$$
$$k = 7 \times 10^5\,\text{dm}^6\,\text{mol}^{-2}\,\text{s}^{-1}$$

You should get the same value by substituting the set of results from any of the experiments. Try it.

The discussion above has assumed that the reaction order is a whole number. Mathematically more sophisticated methods are available to deal with other cases but you will not need them at A level.

You will probably find the techniques easier to apply to problems than to read about.

14.5 Theories of reaction rates

Any theory of kinetics must be able to explain why rates increase with increasing temperature and pressure and also with increasing concentration or surface area of reactants. There are two theories, each stressing a different aspect of the process of a reaction. They are **collision theory** and **transition state theory**.

14.5.1 Collision theory

For a reaction to take place between, for example, two molecules A and B, the molecules must collide. This explains why reaction rates increase with increased concentration or pressure. If there are more molecules per cubic centimetre, collisions will occur more frequently. Also, increased surface area of a solid leads to more molecules colliding with the solid.

As molecules move faster when the temperature increases, there will be more collisions per second and the molecules will also hit each other harder (with more energy).

We can calculate the number of collisions per second from the known speeds of the molecules at a given temperature and the concentrations of the reactants. When this is done, it is found that the actual rates of many reactions are less than the collision rates by factors of around 10^{11}, so only a very small proportion of collisions actually result in a reaction. This is for two reasons.

1. For a collision to result in a reaction, the molecules must approach each other in the right orientation. This is sometimes called the **steric factor**.

   For example, OH^- ions react with bromobutane and replace the Br atom:

```
   H  H  H  H                           H  H  H  H
   |  |  |  |                           |  |  |  |
H—C—C—C—C—Br  +  OH⁻  ——→  H—C—C—C—C—OH  +  Br⁻
   |  |  |  |                           |  |  |  |
   H  H  H  H                           H  H  H  H
```

   A collision of an OH^- ion with a bromobutane molecule is unlikely to result in a reaction if it hits the end of the molecule away from the Br. So only a small proportion of collisions result in a reaction for this reason.

2. If a collision is to result in a reaction, the molecules must have a certain minimum energy for bond breaking, if they are to be able to react. So again, many collisions will not lead to a reaction. This leads on to the second theory.

14.5.2 Transition state theory

This considers the details of the actual collision. When two molecules approach closely on a collision course, their electron clouds begin to repel.

Unless they are moving very fast (i.e. have a lot of kinetic energy), this repulsion will push them apart before they get close enough for new bonds to form. If they get sufficiently close, rearrangement of the electrons in the outer shells can take place so that new bonds form and old ones break. The kinetic energy of the collision is converted into potential energy. The highly energetic and unstable species which exists briefly at the point of maximum potential energy is called the **transition state** or **activated complex**. In this species, bonds are in the process of making and breaking, for example:

$$\text{A—B} + \text{C} \longrightarrow \text{A} + \text{B—C}$$

The transition state could be represented as:

$$\text{A} - - - - - \text{B} - - - - - \text{C}$$

where a dotted line represents bonds in the process of breaking and forming.

We can represent the reaction on an energy (enthalpy) diagram, sometimes called the **reaction profile**, **Figure 14.12.**

Figure 14.12 Reaction profile for an exothermic reaction

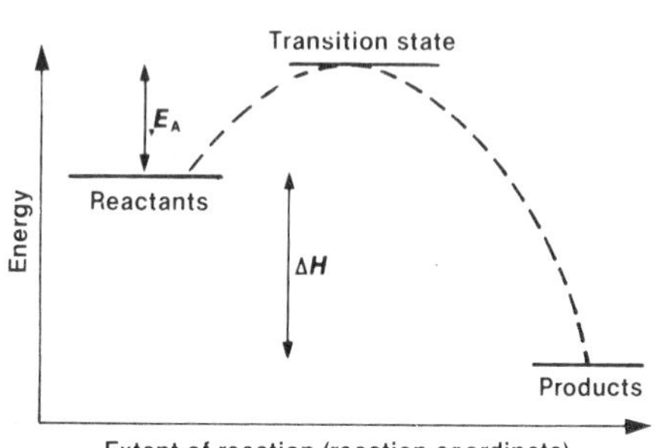

The transition state is the highest point in this reaction profile. The energy gap between the reactants and the transition state is called the **activation energy**, E_A, for the reaction. If a pair of molecules collide with less energy than E_A, they cannot react. The situation is rather like a ball on a mountainside, **Figure 14.13.** A small amount of energy has to be supplied to lift the ball over the lip, even though energy is given out when the ball rolls down into the 'valley'.

Figure 14.13 Ball on a mountainside model

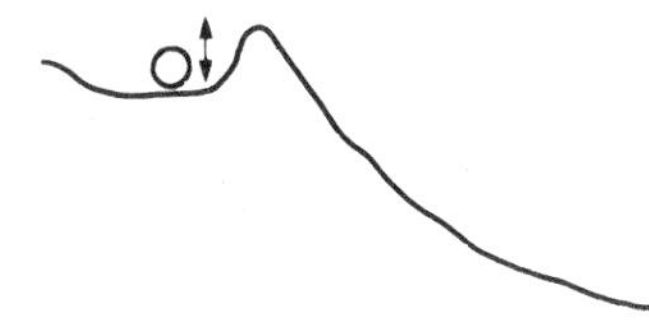

The Arrhenius equation

Molecules in a gas (or in a liquid) do not all have the same speed. Their speeds and therefore their energies are distributed according to the Maxwell–Boltzmann distribution—a few having low speeds, a few having high speeds and most somewhere in the middle, **Figure 14.14.** This distribution is such that the fraction of molecules with energies greater than E_A is given by $e^{-E_A/RT}$ where R is the gas constant and T the temperature in kelvin.

This type of relationship is called an exponential relationship and means that a small rise in temperature results in a large increase in the fraction of molecules with energy greater than E_A. This is shown by the shaded areas on **Figure 14.14.** You can see that the fraction of molecules with energy greater than E_A increases rapidly with temperature.

Figure 14.14 The Maxwell–Boltzmann distribution of molecular energies at three temperatures. Notice at higher temperatures the peak moves towards a higher energy value and the curve broadens out

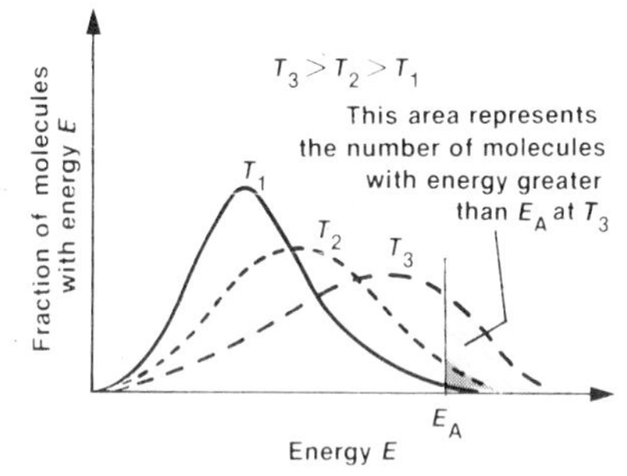

This graph can be compared with, for example, a graph of the percentage of people who take a certain shoe size against shoe size. The peak of this gives the most common shoe size.

According to the **Arrhenius equation**, the reaction rate is the product of three factors: the **collision rate** or number of collisions per unit volume per second is multiplied by the **steric factor** (the fraction of molecules which collide in the right orientation) multiplied by the **fraction of collisions with sufficient energy to react** (E_A or greater). It is usually written:

$$k = pz\,e^{-E_A/RT}$$

k: rate constant (which is proportional to the actual rate); p: steric factor; z: collision rate; $e^{-E_A/RT}$: fraction of molecules with energy to react

$p \times z$ is often given the symbol A and called the **pre-exponential factor**, so:

$$k = A\,e^{-E_A/RT}$$

The mathematics are easier if we take logs of both sides:

$$\ln k = \ln A - E_A/RT$$

(Taking log to the base e, i.e. ln, of $e^{-E_A/RT}$ leaves $-E_A/RT$, as the log of a number is the power to which the base must be raised to obtain the number. Taking logs of the product of two numbers means we must *add* their logs.) Rearranging:

$$\ln k = \ln A - (E_A/R) \times 1/T$$

Figure 14.15 An Arrhenius plot

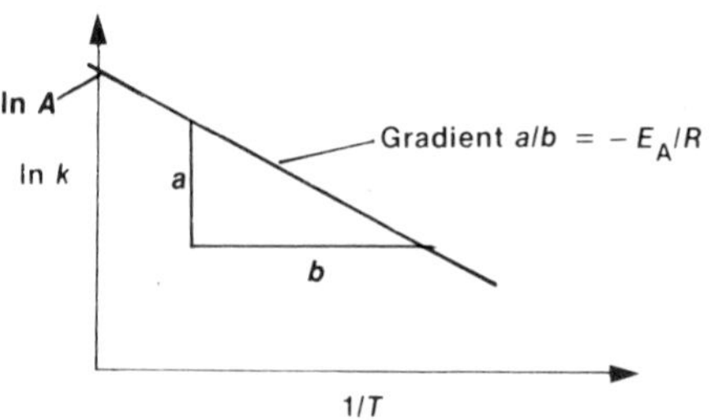

This is of the form:

$$y = c + mx$$

so a graph of $\ln k$ against $1/T$ will give a straight line of gradient $-E_A/R$ enabling E_A to be found, see **Figure 14.15.**

The intercept on the $\ln k$ axis is $\ln A$ (the pre-exponential factor), i.e. the log of the number of collisions of correct orientation. Note that if you enter R in $J\,K^{-1}\,mol^{-1}$ the value of E_A will be in joules, not in kilojoules. *It is a very common mistake to forget this factor.*

A

USING THE ARRHENIUS EQUATION

Imagine a reaction with an activation energy of 50 kJ mol^{-1}, a fairly typical value.

$$50\ \text{kJ mol}^{-1} = 50\,000\ \text{J mol}^{-1}$$

At a temperature of 300 K (a warm room):

fraction of molecules with energy > 50 kJ mol^{-1}

$$= e^{-\frac{50\,000}{8.3 \times 300}}$$
$$= e^{-20.08}$$
$$= 1.90 \times 10^{-9} = \frac{19}{10^{10}}$$

So only 19 out of every 10^{10} molecules have enough energy to react.

At 310 K:

fraction of molecules with energy > 50 kJ mol^{-1}

$$= e^{-\frac{50\,000}{8.3 \times 310}}$$
$$= e^{-19.43}$$
$$= 3.63 \times 10^{-9} = \frac{36.3}{10^{10}}$$

Now 36 out of every 10^{10} molecules can react. Notice how the 10 K rise in temperature has almost doubled the number of molecules able to react and therefore almost doubled the rate. This is the rule of thumb mentioned earlier. Beware, though, it is only applicable for reactions with $E_A \simeq 50$ kJ mol^{-1} and at about room temperature.

Don't be frightened of the expression $e^{-E_A/RT}$—just press the right buttons on your calculator.

Table 14.2 Values of the rate constant, *k*, at different temperatures for the reaction

$$2HI(g) \longrightarrow I_2(g) + H_2(g)$$

T/K	$1/T$/10^{-3} K^{-1}	k/dm^3 mol^{-1} s^{-1}	$\ln k$
633	1.579	1.78×10^{-5}	−10.936
666	1.501	1.07×10^{-4}	−9.142
697	1.434	5.01×10^{-4}	−7.599
715	1.398	1.05×10^{-3}	−6.858
781	1.280	1.51×10^{-2}	−4.193

Example

For the reaction:

$$2HI(g) \longrightarrow I_2(g) + H_2(g)$$

the values of k at different temperatures shown in **Table 14.2** were obtained.

From these we calculate $\ln k$ and $1/T$ (in brown). Next, we plot a graph and calculate the gradient (**Figure 14.16**). Activation energies are usually of approximately the same size as bond energies, as bond breaking occurs in the transition state. Typical values are between around 40 and 400 kJ mol^{-1}.

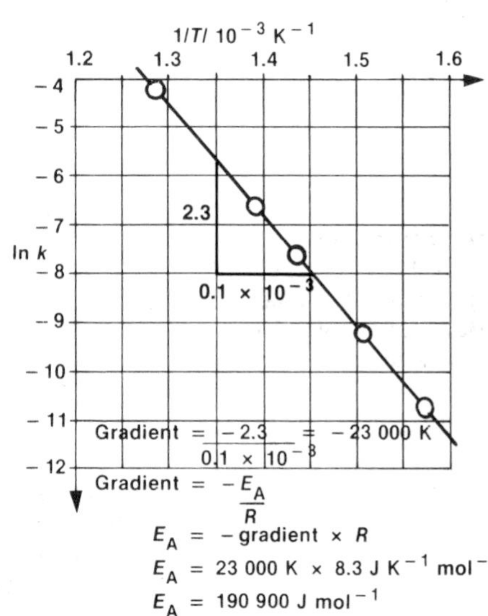

Figure 14.16 An Arrhenius plot for the reaction $2HI(g) \rightarrow H_2(g) + I_2(g)$

14.6 Catalysis

Catalysts are substances which affect the rate of a chemical reaction without being chemically changed themselves. Normally we use the term catalyst to refer to substances which speed up reactions. The terms **negative catalyst** or **inhibitor** are used for those which slow down reactions. Inhibitors are added to the water in central heating systems and car radiators to slow down corrosion reactions.

Positive catalysts work by reducing the activation energy of the reaction, either by making it easier to reach the transition state or by allowing the reaction to go by a different pathway or mechanism with a lower activation energy, **Figure 14.17.**

Figure 14.17 Reaction profiles for a reaction, catalysed and uncatalysed

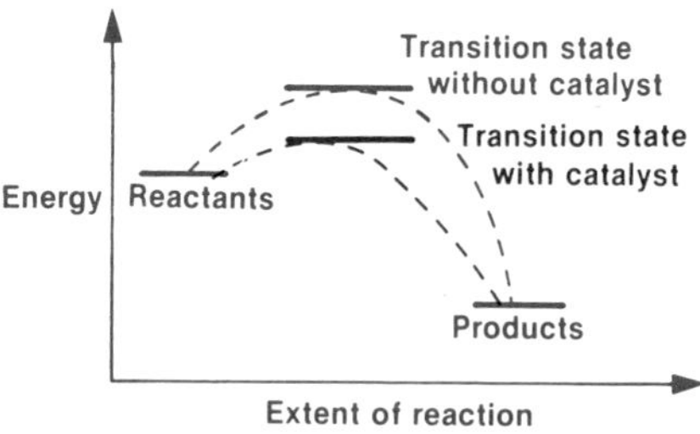

For example, for:

$$2N_2O(g) \longrightarrow 2N_2(g) + O_2(g)$$
$$E_A = 240\,\text{kJ mol}^{-1} \text{ (uncatalysed)}$$
$$E_A = 120\,\text{kJ mol}^{-1} \text{ (with gold catalyst)}$$

There is no single theory to explain catalysis, and many catalysts were discovered simply by trial and error. They are usually divided into two groups:

1. **Heterogeneous catalysts** (where the catalyst is in a different phase to the reactants—usually solid catalysts and liquid or gaseous reactants); and

2. **Homogeneous catalysts** (where catalyst and reactants are in the same phase).

Some examples are given in **Table 14.3.**

Table 14.3 Examples of catalysis

Reaction	Catalyst	Type
$N_2(g) + 3H_2(g) \longrightarrow 2NH_3(g)$	Many metals, including Fe	Heterogeneous
$SO_2(g) + \frac{1}{2}O_2(g) \longrightarrow SO_3(g)$	V_2O_5	Heterogeneous
$H_2O_2(aq) \longrightarrow H_2O(l) + \frac{1}{2}O_2(g)$	MnO_2 + other metal oxides and enzymes	Heterogeneous
$CH_3CO_2H(l) + CH_3OH(l) \longrightarrow CH_3CO_2CH_3(aq) + H_2O(l)$	H^+ or OH^-	Homogeneous
$CH_3COCH_3(aq) + I_2(aq) \longrightarrow CH_2ICOCH_3(aq) + HI(aq)$	H^+	Homogeneous

14.6.1 Heterogeneous catalysis

This frequently involves transition metals catalysing gas phase reactions. Transition metals have unfilled d orbitals (see **section 25.1**) and can use these to form new bonds. The gases are **adsorbed** on the surface of the metal; that is, they form weak bonds with the metal atoms. This may catalyse the reaction in one or both of two ways:

1. The formation of bonds with the metal may use some of the electrons in bonds within the gas molecule, thus weakening these bonds and allowing them to break more easily, **Figure 14.18.**

Figure 14.18 Adsorption of hydrogen on to a metal surface

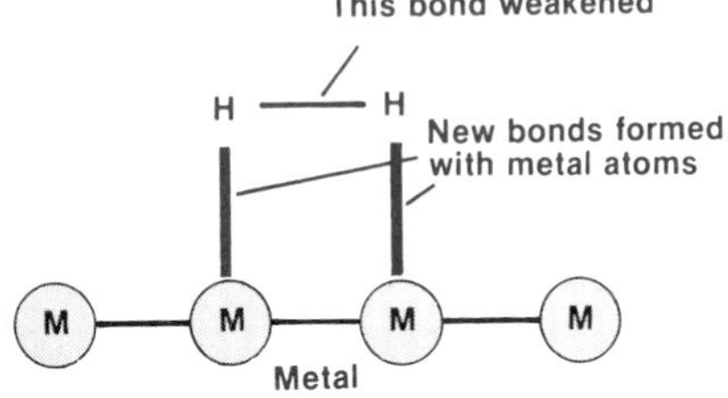

2. The adsorbed gases may be held on the metal surface in just the right orientation for reaction to occur, thus essentially increasing the number of favourable collisions, **Figure 14.19.**

Figure 14.19 Ethene and hydrogen adsorbed on to a nickel catalyst in the right orientation for new bonds (in brown) to form to make ethane

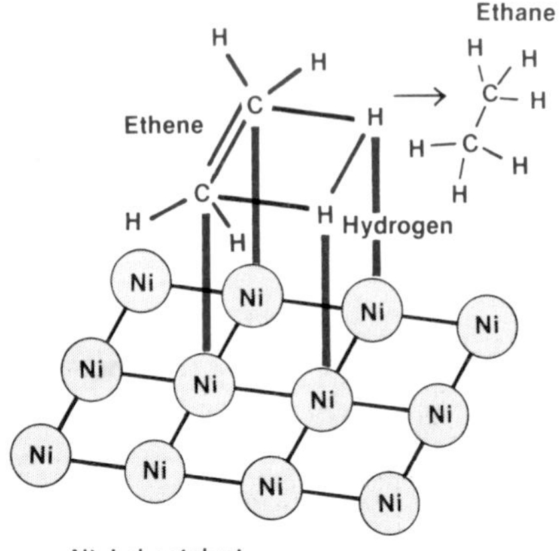

The strength of the adsorption of gases on the surface is critical—if it is too weak, little adsorption will occur; if it is too strong, the product molecules will tend to remain on the surface and block further catalysis. Impurities in the reaction mixture often adsorb more strongly than the reactants and 'poison' the catalyst. This is of critical importance in industrial processes so the reaction mixtures need to be of a high purity, as closing down a plant to change the catalyst can be very expensive in terms of lost production.

The surface area of heterogeneous catalysts is important; pea-sized lumps of iron are used in the Haber process for this reason.

14.6.2 Homogeneous catalysis

A large number of mechanisms is possible. We shall look at just two.

Acid catalysed esterification

This is the reaction between a carboxylic acid like ethanoic acid and an alcohol to form an ester:

$$\underset{\text{ethanoic acid}}{CH_3-\overset{\overset{O}{||}}{C}-OH} + \underset{\text{methanol}}{CH_3OH} \xrightarrow{H^+ \text{ catalyst}} \underset{\text{methyl ethanoate}}{CH_3-\overset{\overset{O}{||}}{C}-O-CH_3} + \underset{\text{water}}{H_2O}$$

Both ethanoic acid and methanol have dipoles (see Chapter 10):

$$CH_3\overset{\delta+}{C}(=O^{\delta-})OH \quad \text{and} \quad {}^{\delta+}H_3C-\overset{\delta-}{O}H$$

and the reaction occurs when the $O^{\delta-}$ on the methanol attacks the $C^{\delta+}$ on the acid to form a new bond using a lone pair of electrons, followed by loss of water from the new species formed.

$$CH_3C^{\delta+}(={}^{\delta-}O)(OH) + {}^{\delta-}{:}O(H)({}^{\delta+}CH_3) \xrightarrow{\text{step 1}} CH_3-C(-O^-)(OH)-\overset{+}{O}(H)(CH_3) \xrightarrow{\text{step 2}} CH_3-C(=O)-O-CH_3 + H_2O$$

The first step is made easier if the acid first accepts a H^+ ion (is protonated) from the catalyst:

$$CH_3C^{\delta+}(=O^{\delta-})(OH) + \mathbf{H^+} \xrightarrow{\text{step 1}} CH_3C^+(\mathbf{OH})(OH) + {}^{\delta-}{:}O(H)({}^{\delta+}CH_3) \xrightarrow{\text{step 2}} CH_3-C(\mathbf{OH})(OH)-\overset{+}{O}(H)(CH_3) \xrightarrow{\text{step 3}} CH_3-C(=O)-OCH_3 + H_2O + \mathbf{H^+}$$

The protonated acid has more positive charge and is more easily attacked by the $O^{\delta-}$ on the alcohol. Note how the H^+ catalyst is regenerated.

Similar mechanisms to this occur often in organic chemistry.

Redox reactions catalysed by transition metal ions

Peroxodisulphate(VI) ions oxidize iodide ions to iodine. This reaction is catalysed by Fe^{2+} ions:

$$S_2O_8^{\ 2-}(aq) + 2I^-(aq) \xrightarrow{Fe^{2+} \text{ catalyst}} 2SO_4^{\ 2-}(aq) + I_2(aq)$$

It is believed that the catalysed reaction takes place in two steps. First the peroxodisulphate ions oxidize iron(II) to iron(III):

$$S_2O_8^{\ 2-}(aq) + 2Fe^{2+}(aq) \longrightarrow 2SO_4^{\ 2-}(aq) + 2Fe^{3+}(aq)$$

The Fe^{3+} then oxidizes the I^- to I_2, regenerating the Fe^{2+} ions so that none are used up in the reaction:

$$2Fe^{3+}(aq) + 2I^-(aq) \longrightarrow 2Fe^{2+}(aq) + I_2(aq)$$

The reaction profile would look like **Figure 14.20**.

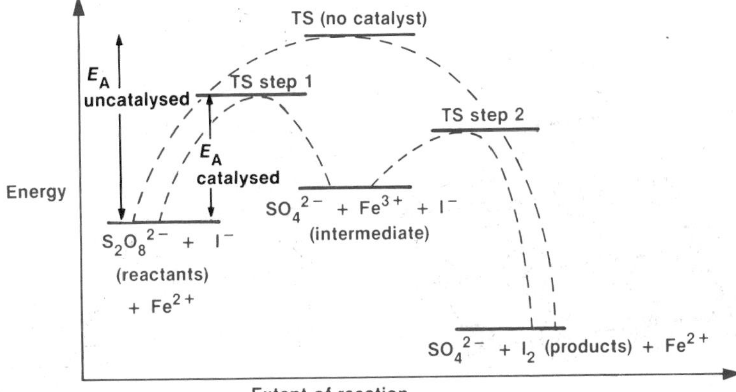

Figure 14.20 Possible reaction profile for the iodine/peroxodisulphate reaction. Note E_A for the catalysed reaction is the energy gap between the reactants and the *higher* of the two transition states

Although there are two steps in the catalysed reaction, the overall activation energy is lower than that for the uncatalysed reaction. Part of the reason for this may be that the uncatalysed reaction takes place between two ions of the same charge (both negative). Both steps of the catalysed reaction involve reaction between pairs of oppositely charged ions. Transition metals and their compounds frequently catalyse redox reactions in this sort of way because they have variable oxidation numbers (see **section 25.3.1**) and can act as a temporary 'warehouse' for electrons. In this case the iron first gives an electron to the peroxodisulphate and later takes one back from the iodide ions. Incidentally, Fe^{3+} ions also catalyse this reaction. Can you work out how?

Autocatalysis

An interesting example of catalysis is found where one of the products of the reaction is a catalyst for the reaction. Such a reaction proceeds slowly at first, at the uncatalysed rate, until a significant concentration of the product (which is also the catalyst) is established. Then the reaction speeds up to the catalysed rate and from then on behaves like a normal reaction. This leads to an odd-looking rate curve, **Figure 14.21**.

Figure 14.21 A concentration–time graph for an autocatalytic reaction

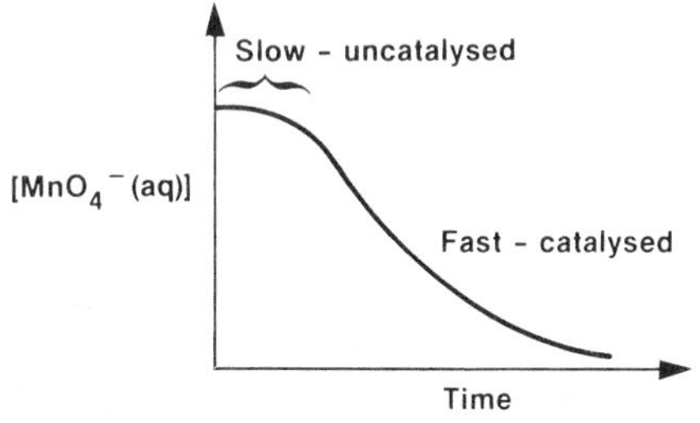

The curve shown in **Figure 14.21** is for the reaction:

$$2MnO_4^-(aq) + 16H^+(aq) + 5C_2O_4^{2-}(aq)$$
manganate(VII) ions, hydrogen ions, ethanedioate ions

$$\longrightarrow 2Mn^{2+}(aq) + 8H_2O(l) + 10CO_2(aq)$$
manganese(II) ions, water, carbon dioxide

The catalyst is Mn^{2+} ions. The reaction can easily be followed using a colorimeter to measure the concentration of $MnO_4^-(aq)$ which is purple. Can you suggest another method of following the reaction? Can you devise an experiment to show that the autocatalyst is Mn^{2+} rather than CO_2?

A

CATALYTIC CONVERTERS FOR CAR EXHAUSTS

Motor vehicle exhausts are a major source of air pollution, being responsible for significant amounts of carbon monoxide, hydrocarbons and nitrogen oxides (NO and NO_2—often abbreviated to NOx), as well as lead oxides from anti-knock petrol additives. Various methods have been proposed to control this including reducing maximum speed limits (less pollution is emitted at low speeds) and improved engine design (the so-called lean burn engine which is more efficient).

A catalytic converter

In the US car exhausts are fitted with a catalytic converter. This speeds up the reactions below, neatly removing the pollutants by getting them to react with one another to form harmless products.

carbon monoxide + nitrogen oxides ⟶ carbon dioxide + nitrogen

hydrocarbons + nitrogen oxides ⟶ carbon dioxide + nitrogen + water

The catalyst system consists of a honeycomb shape coated with platinum and rhodium.

It has a number of drawbacks, however. A converter costs around £500, mostly due to the cost of the platinum and rhodium, but only has a lifetime of around 50 000 miles.

The car must run on lead-free petrol or the catalyst will be 'poisoned' by the lead. The catalyst system is only effective at temperatures over 400 °C, 673 K. With stop-start usage, which is typical of British motoring, the converter would be inactive for much of the time as it would not reach the operating temperature.

For these reasons, as well as political considerations, converters are as yet little used in the UK and much research has gone into the lean-burn engine.

A

ENZYMES AND BIOTECHNOLOGY

Enzymes are protein-based molecules found in living things which are extremely effective catalysts. They are extremely specific—one enzyme normally catalyses one reaction of one molecule, which is known as its **substrate**. Many enzymes convert their substrate so fast that the rate-determining step is the diffusion of the substrate towards the catalyst, so reaction rates can be 10^3 molecules of substrate *per second per molecule of enzyme*!

Protein molecules are extremely delicate and their shape, which governs their catalytic activity, is easily changed by relatively small changes in temperature or pH. These changes can **denature** the molecule. Thus enzymes have an optimum temperature, often around body temperature (37 °C, 310 K). Below this temperature the reaction rate increases with temperature in the usual way; above this temperature the rate decreases as the enzyme is denatured. People have used enzymes for their own purposes for thousands of years—to ferment sugar into alcohol and turn milk into cheese and yoghurt.

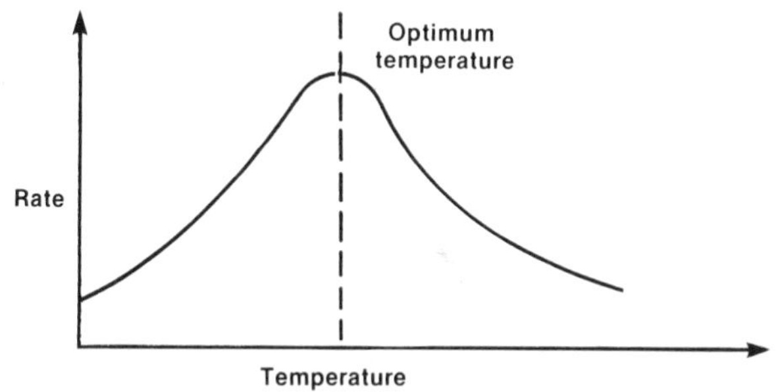

However, in the last 20 years vast strides have been made in this field, so much so that a new name has been coined—**biotechnology**. This includes the 'tailoring' of enzymes to catalyse particular processes with their usual super-efficiency. It is possible to 'fix' enzyme molecules on a solid so that the enzyme can be retained for re-use, rather than being mixed with the product and lost after each batch is produced. The enzymes become effectively heterogeneous rather than homogeneous catalysts.

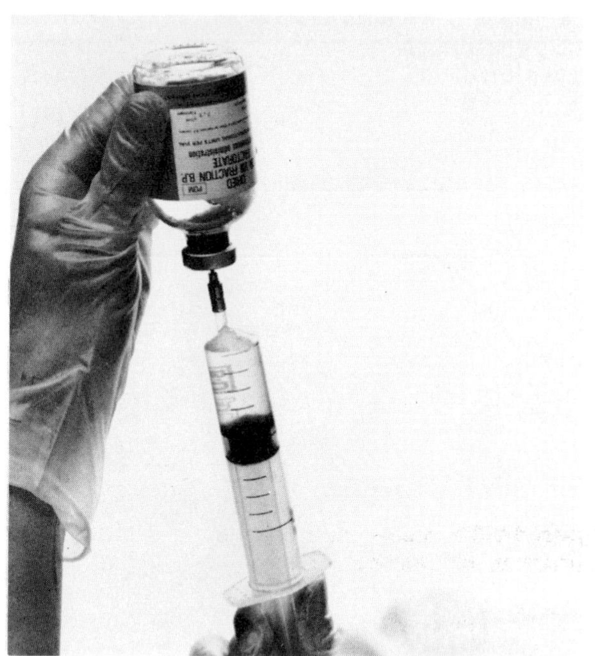

Synthetic factor VIII, made using enzymes, avoids the possibility of haemophiliacs contracting AIDS from contaminated factor VIII obtained from human blood

Many drugs, especially, are now synthesized by biotechnological processes including insulin, the hormone required by diabetics, and the new 'wonder drug' interferon which is being tested for anti-viral and anti-cancer effects.

In the home, an enzyme called alkaline protease is an ingredient of many biological washing powders. It is used for its ability to digest protein-based stains, such as blood, at low temperatures and in the slightly alkaline conditions of washing powder solutions. One drawback is that a few people appear to be allergic to the enzyme.

The enzyme industry has grown because enzymes are so efficient. Not only can small quantities of enzymes convert large quantities of chemicals, but they can do so at low temperatures and normal pressures. The idea of working with enzymes on a large scale was started by two British chemists Malcolm Lilly and Peter Dunnill working in a basement at University College London.

A

ZIEGLER-NATTA CATALYSTS

Catalysts are of enormous industrial importance, as illustrated by the polymerization of ethene to give poly(ethene), usually called polythene. When this was first manufactured by ICI in the early 1950s, a process was used requiring a pressure of 2000 atmospheres (2×10^8 Pa) and a temperature of 300 °C (573 K). The discovery of a catalyst for this process by Karl Ziegler in the late 1950s enables the process to be run now at less than 50 atmospheres (5×10^6 Pa) and at temperatures of between 310 and 360 K. The catalyst is a mixture of titanium(IV) chloride and triethylaluminium, the precise details of which are secret.

The Italian Giulio Natta developed a similar catalyst for the manufacture of poly(propene). As well as accelerating the polymerization, this catalyst made it possible to produce stereoregular poly(propene), which has a regular rather than random arrangement of the methyl side chains. By controlling the precise proportions of the different types of polymers, it is possible to tailor the properties of the final plastic (such as density, tensile strength and elasticity) to a particular specification. Ziegler and Natta received the 1963 Nobel prize for their work on polymerization catalysts.

Ziegler–Natta catalysts are discussed further in the box *Rubber*, **section 28.8,** and **section 35.3.1.**

14.7 Reaction mechanisms

We have seen that reactions can involve more than one step. Indeed, except for the very simplest processes, reactions invariably have more than one step. The separate steps which lead from reactants to products are collectively called the **reaction mechanism**.

Mechanisms usually have to be pieced together by detective work using several pieces of evidence, as it is rare that the intermediate species can be isolated and identified. For example:

$$BrO_3^-(aq) + 6H^+(aq) + 5Br^-(aq) \longrightarrow 3Br_2(aq) + 3H_2O(l)$$

This reaction *must* involve several steps, as a simultaneous collision of twelve ions is immensely improbable.

14.7.1 Rate determining step

In a multi-step reaction, the steps are usually sequential, that is the product(s) of one step is/are the starting material for the next. Therefore the rate of the slowest step governs the rate of the whole process. The slowest step forms a 'bottleneck'. This can easily be seen by imagining a canteen where meat, vegetables and potatoes are served by three different people. The rate of getting your meal will be governed by the rate of the slowest server, no matter how rapidly the other two work.

So, in a chemical reaction, any step which occurs *after* the rate determining step will not affect the rate, and species which are involved in the mechanism after the rate determining step will not appear in the rate expression. For example, the reaction:

$$A + B \longrightarrow I + J$$

might occur in the following steps:

1. $$A + B \xrightarrow{\text{fast}} C + D$$

2. $$C + E \xrightarrow{\text{slow}} F + G$$

3. $$G + H \xrightarrow{\text{fast}} I + J$$

Step two is the slowest step and so determines the rate. As soon as some G is produced, it is rapidly converted to I and J. However, the concentration of C will depend on the rate of step 1 and so the rate of step 1 could contribute to the overall rate. Thus any species involved in or before the rate determining step could affect the rate and hence appear in the rate expression.

The reaction between iodine and propanone illustrates this, although you would not be expected to recall details of this for an A-level examination. The overall reaction is:

$$CH_3-\overset{\overset{\displaystyle O}{\|}}{C}-CH_3(aq) + I_2(aq) \xrightarrow[\text{catalyst}]{H^+} CH_2I\overset{\overset{\displaystyle O}{\|}}{C}CH_3(aq) + HI(aq)$$

and the rate expression is:

$$\frac{-d[I_2]}{dt} = k[CH_3OCH_3][H^+]$$

The mechanism is:

1. $$H-\underset{\underset{\displaystyle H}{|}}{\overset{\overset{\displaystyle H}{|}}{C}}-\overset{\overset{\displaystyle O}{\|}}{C}-CH_3 + H^+ \xrightarrow{\text{slow}} H-\underset{\underset{\displaystyle H}{|}}{\overset{\overset{\displaystyle H}{|}}{C}}-\underset{+}{\overset{\overset{\displaystyle O-H}{|}}{C}}-CH_3$$

2. $$H{-}CH_2{-}\overset{+}{C}(OH){-}CH_3 \xrightarrow{\text{fast}} H_2C{=}C(OH){-}CH_3 + H^+$$

3. $$H_2C{=}C(OH){-}CH_3 + I_2 \xrightarrow{\text{fast}} H{-}CHI{-}CI(OH){-}CH_3$$

4. $$H{-}CHI{-}CI(OH){-}CH_3 \xrightarrow{\text{fast}} H{-}CHI{-}C({=}O){-}CH_3 + H^+ + I^-$$

The rate determining step is the first one, so I_2 does not appear in the rate expression.

14.7.2 Kinetic evidence for reaction mechanisms

It is rarely possible to deduce a reaction mechanism from a single piece of evidence, but kinetic evidence can be useful. The reaction of bromobutanes with alkalis is a simple example. The reaction is:

$$C_4H_9Br + OH^- \longrightarrow C_4H_9OH + Br^-$$

Two mechanisms are possible.

a) A two-step mechanism:

1. $$C_4H_9Br \xrightarrow{\text{slow}} C_4H_9{}^+ + Br^-$$

2. $$C_4H_9{}^+ + OH^- \xrightarrow{\text{fast}} C_4H_9OH$$

The slow step involves breaking the C—Br bond while the second (fast) step is a reaction between oppositely charged ions.

b) A one-step mechanism:

1. $$C_4H_9Br + OH^- \longrightarrow C_4H_9OH + Br^-$$

The C—Br bond breaks at the same time as the C—OH bond is forming.

There are three compounds of formula C_4H_9Br:

$$H{-}CH_2{-}CH_2{-}CH_2{-}CH_2{-}Br$$ 1-bromobutane

$$H{-}CH_2{-}CH_2{-}CHBr{-}CH_2{-}H$$ 2-bromobutane

$$H{-}CH_2{-}CBr(CH_3){-}CH_2{-}H$$ 2-bromo-2-methylpropane

These are a set of isomers—they have the same formula but different structures, see **section 26.7.** The first of these reacts by a second-order mechanism—the rate depends on the concentration of *both* reactants, suggesting mechanism b. The last of these reacts by a first-order mechanism, suggesting mechanism a. Further discussion of these reactions appears in **section 30.7.1.**

14.7.3 Photochemical reactions

A mixture of hydrogen and chlorine gases is stable in the dark but reacts explosively to form hydrogen chloride if exposed to sunlight.

$$H_2(g) + Cl_2(g) \xrightarrow{\text{sunlight}} 2HCl(g)$$

The reaction starts when a chlorine molecule absorbs a quantum of light which breaks the Cl—Cl bond, forming two Cl atoms or radicals. These are very reactive and start a **chain reaction**. Chain reactions are discussed more fully in **section 27.7.3.**

14.8 Summary

Figure 14.22 Concentration–time graphs showing reactions of different orders

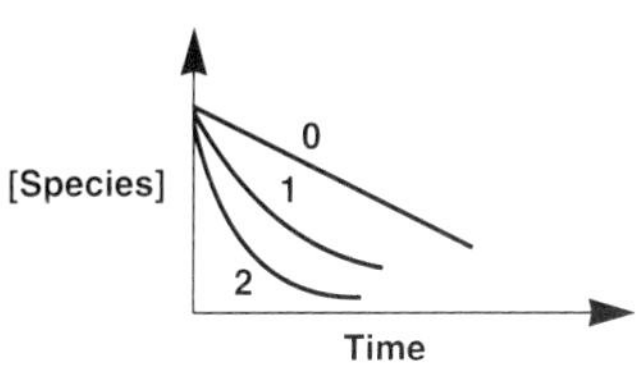

Figure 14.23 Rate–concentration graphs to determine reaction order

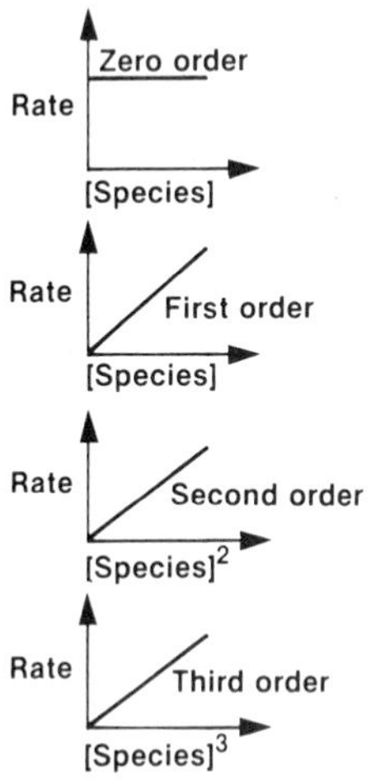

Figure 14.24 Reaction profile for an exothermic reaction

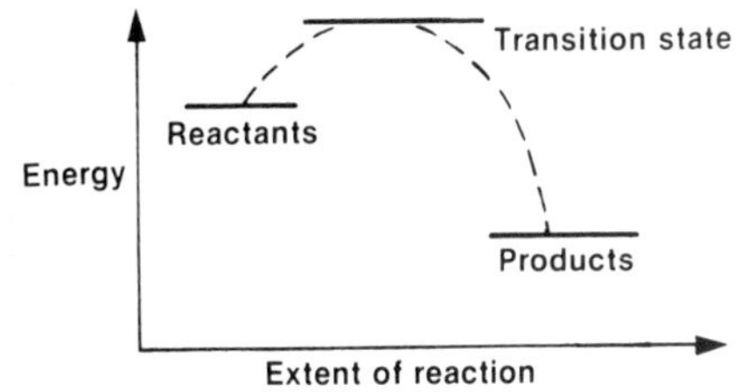

- Rates of chemical reactions can be affected by six factors: concentration of reactants, pressure (for gas phase reactions), temperature, surface area of solids, light and catalysts.
- Rates may be measured by monitoring any property of the reaction mixture which depends on the concentration of a reactant or product. Common methods include quenching followed by a titration, colorimetry, conductivity measurement and measurement of volumes of gases involved.
- The rate expression is an equation which shows how the reaction rate depends on the concentration of species present in the reaction mixture. It *cannot* be predicted from the chemical equation.
- The order of a reaction with respect to a particular species is the power to which its concentration is raised in the rate expression.
- The overall order of a reaction is the sum of the orders with respect to each species.
- The rate constant is the constant of proportionality in the rate expression.
- Order with respect to a particular species may be determined by:

1. The shape of a graph of species concentration against time, **Figure 14.22.** Zero-order processes give a straight line. First-order processes have constant half-lives. For processes of order two and above, successive half-lives increase.

2. Graph of rate against species concentration raised to some power, **Figure 14.23.**

3. The initial rate method. Several experiments are carried out in which the initial concentration of each species is varied. The change in initial rate when the species concentration is doubled is monitored. If the rate doubles, the order is one; if it increases fourfold (2^2), the order is two; if it increases eightfold (2^3), the order is three, and so on.

- Collision theory says that reactions occur if two molecules collide with the correct orientation and with at least a set minimum energy.
- Transition state theory identifies this energy as the activation energy E_A, that is required to reach a high-energy species called the transition state, which is the peak of the reaction profile, **Figure 14.24.**

These are summarized by the Arrhenius equation:

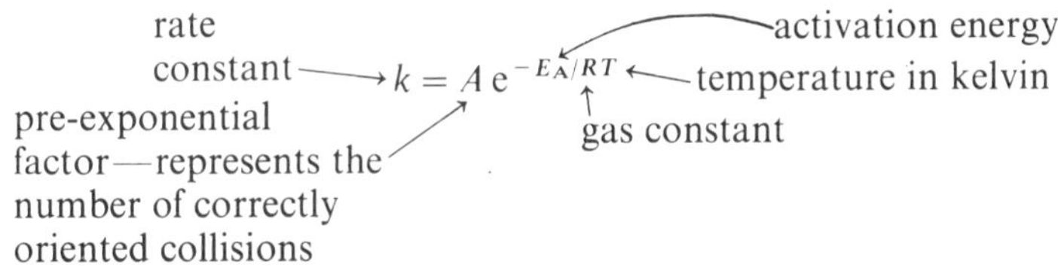

or

$$\ln k = \ln A - E_A/RT$$

Hence E_A can be found from the gradient of a graph of $\ln k$ against $1/T$.

• Catalysts change the rate of a reaction without themselves being chemically changed. They are classified as homogeneous (in the same phase as the reactants) or heterogeneous (different phase from the reactants—usually solid catalyst and liquid or gaseous reactants). They work by lowering the activation energy of the reaction in some way. Many reactions are catalysed by acids, bases or transition metal compounds. Heterogeneous catalysts often work by first adsorbing the reactants on to their surface, hence their surface area is important.

• Reaction mechanism: all but the simplest reactions take place via a sequential set of simple steps. The overall rate is governed by that of the slowest or rate determining step. Only species which take part in the mechanism in or before the rate determining step appear in the rate expression.

• Photochemical reactions: these start by the adsorption of one quantum of light which usually breaks a bond, followed by a chain reaction.

14.9 Questions

1 Devise a method for measuring the rate of the following reactions:

A) $CaCO_3(s) + 2HCl(aq) \longrightarrow CaCl_2(aq) + H_2O(l) + CO_2(g)$

B) $CuCl_2(aq) + Zn(s) \longrightarrow ZnCl_2(aq) + Cu(s)$

C) $2NO(g) + O_2(g) \longrightarrow 2NO_2(g)$

D) $CH_3{-}C({=}O){-}CH_3(aq) + H^+(aq) + C{\equiv}N^-(aq) \longrightarrow CH_3{-}C(OH)(C{\equiv}N){-}CH_3(aq)$

E) $BrO_3^-(aq) + 5Br^-(aq) + 6H^+(aq) \longrightarrow 3Br_2(aq) + 3H_2O(l)$

F) $C_2H_5Br + NaOH \longrightarrow C_2H_5OH + NaBr$

2 In the reaction in E) above, if the rate with respect to bromate ions:

$$\frac{d[BrO_3^-]}{dt} = -10^{-3}\,mol\,dm^{-3}\,s^{-1}$$

what will be:

A) the rate with respect to Br^- ions $d[Br^-]/dt$

B) the rate with respect to Br_2 molecules $d[Br_2]/dt$?

3 For the reaction:

$$2H_2(g) + 2NO(g) \longrightarrow 2H_2O(g) + N_2(g)$$

the following initial rates were determined:

Initial [NO] /10^{-3} mol dm^{-3}	**Initial [H_2]** /10^{-3} mol dm^{-3}	**Initial rate** /10^{-3} mol dm^{-3} s^{-1}
6	1	3
6	2	6
6	3	9
1	6	0.5
2	6	2
3	6	4.5

A) What is the order of the reaction:

i with respect to NO;

ii with respect to H_2;

iii overall?

Explain your reasoning for each answer.

B) Write the rate expression for the reaction.

C) Using values from the table above, determine the rate constant (with appropriate units).

D) What would the rate of the reaction be if $[H_2]$ was 10^{-3} mol dm^{-3} and [NO] was 10^{-3} mol dm^{-3}?

4 $Br_2(aq) + HCO_2H(aq) \xrightarrow[\text{catalyst}]{H^+} 2Br^-(aq) + 2H^+(aq) + CO_2(g)$

Time/min	**[Br_2]**/mol dm^{-3}
0.0	0.0100
0.5	0.0900
1.0	0.0081
1.5	0.0073
2.0	0.0066
3.0	0.0053
4.0	0.0044
6.0	0.0028
8.0	0.0020
10.0	0.0013
12.0	0.0007

A) These figures were obtained using a colorimeter to measure the $[Br_2]$. Give two other ways by which the rate of reaction could have been followed.

B) Plot a graph of $[Br_2]$ vertically versus time. Determine the order of the reaction with respect to Br_2 by the following procedures:

i determine the first three half-lives of the reaction;

ii find the rate of the reaction (gradient of the graph) at at least five different values of $[Br_2]$ and plot a graph of rate against $[Br_2]$.

C) In this experiment the $[HCO_2H]$ was kept constant. How could this be achieved in practice?

D) Is it possible to determine the rate with respect to HCO_2H from the above figures? Explain your answer.

5 In the reaction:

$$A + B \longrightarrow C$$

the following data were obtained:

Initial [A]	Initial [B]	Initial rate
1	1	3
1	2	12
2	2	24

A) What is the order of the reaction with respect to: **i)** A; **ii)** B?

B) What is the overall order?

6 This question concerns the following reaction:

$C_6H_5-N(Cl)-C(=O)-CH_3$ (compound A) $\xrightarrow{HCl}$ $Cl-C_6H_4-N(H)-C(=O)-CH_3$ (compound B)

The progress of this reaction can be followed by using the fact that compound A reacts with acidified potassium iodide, liberating iodine, whereas compound B does not.

A series of experiments was carried out to determine the initial rate of reaction for various initial concentrations of compound A. The following data were obtained:

Concentration of compound A/mol dm^{-3}	Initial rate/ mol dm^{-3} s^{-1}
0.060	3.12×10^{-4}
0.120	6.23×10^{-4}
0.180	9.38×10^{-4}
0.240	12.50×10^{-4}

A) Outline an experimental procedure for the determination of the concentration of the reactant (compound A) at intervals as the reaction proceeds.

B) From the data in the table, deduce the order of reaction with respect to compound A. Explain your reasoning.

(Nuffield 1983, part question)

7 Given the reaction $2X \longrightarrow A + B$, the results of an experiment which measured the concentration of X with time are shown on the following graph.

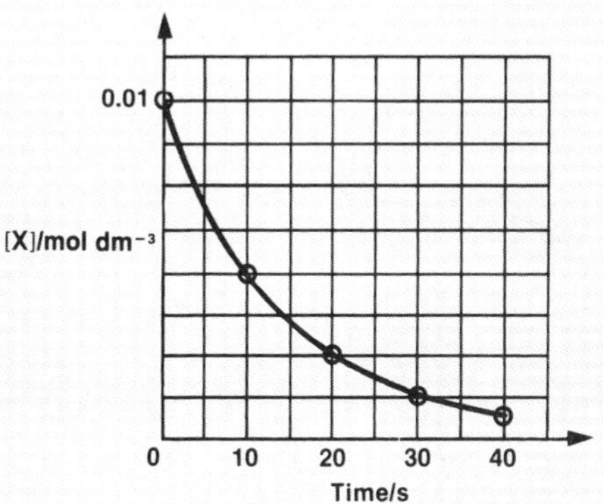

Find the first three half-lives. What is the order of the reaction with respect to X? (AEB, specimen)

8 An experiment was carried out to investigate the rate of reaction of an organic chloride of molecular formula C_4H_9Cl with hydroxide ions. The reaction was carried out in solution in a mixture of propanone and water with the following results.

Time elapsed/s	Concentration of C_4H_9Cl/mol dm^{-3}	Concentration of hydroxide ions/ mol dm^{-3}
0	0.0100	0.0300
294	0.0050	0.0250
595	0.0025	0.0225

A) Write a balanced equation for the reaction of an organic chloride of formula C_4H_9Cl with hydroxide ions.

B) Suggest a reason why water alone is not used as the solvent in the experiment.

C) i From the results given deduce an order of reaction. Explain your answer.

ii Write a rate expression for the overall reaction.

iii Suggest a mechanism for the reaction that would be consistent with your rate expression.

D) Suggest a practical method by which it should be possible to follow the extent of the reaction.

(Nuffield 1980)

9 The following data were obtained for the reaction:

$$N_2O_5 \longrightarrow 2NO_2 + \tfrac{1}{2}O_2$$

in a solution of tetrachloromethane at 318 K.

Time/s	$[N_2O_5]$/mol dm^{-3}
0	2.33
184	2.08
319	1.91
526	1.67
867	1.36
1198	1.11
1877	0.72
2315	0.55
3144	0.34
3500	0.21

A) Plot a graph of $[N_2O_5]$ against time. Find the first three half-lives. What order does this give for the reaction with respect to N_2O_5?

B) Confirm your answer to (a) by drawing tangents to the graph at five different values of $[N_2O_5]$. Calculate the reaction rate (gradient of the tangent) for each value of $[N_2O_5]$. Plot a graph of rate against $[N_2O_5]$. Does this confirm the order of reaction with respect to $[N_2O_5]$?

C) Write the rate expression for this reaction.

D) Use one of your sets of values of rate and concentration to calculate the rate constant k.

E) If the NO_2 remains dissolved in the tetrachloromethane, but the oxygen does not, suggest a method of following the reaction.

10 The decomposition of N_2O_5 in the gas phase:

$$N_2O_5 \longrightarrow 2NO_2 + \tfrac{1}{2}O_2$$

was investigated at different temperatures with the following result.

Temperature/K	Rate constant $k/10^3\,s^{-1}$
338	48 700
328	15 000
318	4980
308	1350
298	346
273	7.87

A) Work out the values of $1/T$ and $\ln k$ and plot a graph of $\ln k$ versus $1/T$.

B) Use the gradient of the graph to calculate the activation energy of the reaction.

11 The first-order hydrolysis of a chloroalkane gave the following values of rate constant, k, at different temperatures:

Temperature/K	Rate constant, $k/10^{-4}\,s^{-1}$
273	0.106
298	3.19
308	9.86
318	29.20

Plot a suitable graph to find the activation energy.

12 The results (below) were found for the reaction:

$$X + Y \longrightarrow \text{products}$$

Initial concentrations/ mol dm^{-3}		
[X]	[Y]	Initial rate/mol dm^{-3} s^{-1}
1	1	1
1	2	4
1	3	9
2	3	18
4	3	36

What are the orders with respect to X and Y? Find the rate constant and give the correct units.

13 The example on page 181 calculates the activation energy for the reaction:

$$2HI \longrightarrow H_2 + I_2$$

without a catalyst. The reaction is catalysed by a number of metals. With one metal, the following data were recorded:

Temperature/ K	Rate constant, k/ $10^{-5}\,dm^3\,mol^{-1}\,s^{-1}$
625	2.27
667	7.56
714	24.87
769	91.18
833	334.59

A) Plot a graph of $\ln k$ against $1/T$ and find the activation energy with this catalyst.

B) Can you make any suggestions as to how the metal might operate as a catalyst?

15 Gases

15.1 Introduction

Gases are the least dense phase of matter. They fill the whole of any container in which they are placed and exert pressure on the walls of the container. Gases consist of well spaced-out particles moving rapidly at random, whose collisions with the walls of the container cause their pressure. The behaviour of gases at moderate temperature and pressure is described by several experimental laws and can be predicted by relatively simple theoretical principles.

This chapter describes the kinetic particulate theory of gases and develops two simple experimental methods for measuring relative molecular masses. Since many industrial reactions take place in the gas phase an understanding of the behaviour of gases is of considerable economic importance.

15.2 The gas laws

15.2.1 Gay-Lussac's law

Gases are not easy to handle experimentally, simply because they have to be in closed containers. Despite this, they were quite well studied even 200 years ago, and in the early 1800s Joseph Gay-Lussac put forward the law named after him. This said that **gases always react in volumes which are in a simple whole-number relationship to one another—both reactants and products**. The volumes had to be measured under the same conditions of temperature and pressure. This law was simply a generalization of many measurements made on gas reactions over many years.

Some examples of gas reactions which obey Gay-Lussac's law are:

hydrogen	+	iodine	⟶	hydrogen iodide
1 vol		1 vol		2 vols
hydrogen	+	oxygen	⟶	water
2 vols		1 vol		2 vols (if over 373 K, i.e. the water exists as a gas) or 0 vol (if below 373 K, i.e. the water is a liquid of negligible volume)
nitrogen	+	hydrogen	⟶	ammonia
1 vol		3 vols		2 vols

15.2.2 Avogadro's law

Gay-Lussac's law was explained by Avogadro, who proposed that **equal volumes of all gases, measured at the same temperature and pressure, contain the same number of particles (atoms or molecules)**. This implies that in

E

MEASURING THE VOLUMES OF GASES IN REACTIONS

Gas reactions can be most easily studied at atmospheric pressure using gas syringes. Their ground glass barrels and plungers make them both gas tight and free running. An example is the decomposition of ammonia. (You may be aware that this is a reversible reaction. Under the conditions of this experiment the decomposition is practically complete.)

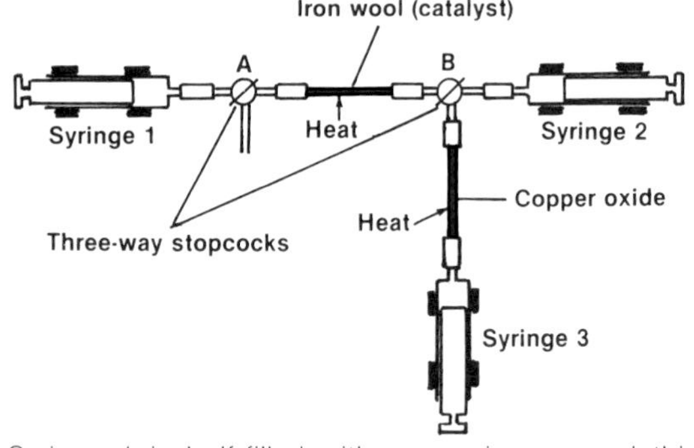

Syringe 1 is half-filled with ammonia gas and this is passed backward and forward over heated iron (a catalyst) from syringe 1 to syringe 2 by pushing the appropriate plungers. The gas volume increases. When no further increase occurs, the gas mixture (now nitrogen and hydrogen) is allowed to cool to room temperature and the volume measured. The hydrogen is now removed by passing the gas mixture to and fro from syringe 2 to syringe 3 so that it is passed over the heated copper oxide. This removes the hydrogen leaving water:

$$\text{hydrogen} + \text{copper oxide} \longrightarrow \text{copper} + \text{water}$$

which condenses in the cool syringe to leave a few droplets of water of negligible volume. The black copper oxide is seen to turn to pink copper.

When no further volume changes occur, the remaining gas (nitrogen) is allowed to cool to room temperature and its volume is measured. Typical results are:

$$\text{initial volume of ammonia} = 48\,\text{cm}^3$$
$$\text{volume of nitrogen and hydrogen} = 96\,\text{cm}^3$$
$$\text{volume of nitrogen} = 24\,\text{cm}^3$$
$$\therefore \text{volume of hydrogen} = 72\,\text{cm}^3$$

So the gases react in the ratios:

$$\text{ammonia} : \text{hydrogen} : \text{nitrogen} = 2:3:1$$

which fits Gay-Lussac's law.

any gas, the *spaces* between the gas molecules are so large, compared with the size of the molecules, that the volume the gas occupies is unaffected by the volume of the molecules themselves. The situation is shown in **Figure 15.1**. So if we had a second gas with molecules of double the volume of those of the first gas, only a tiny extra volume would result if the spacing remained the same.

Figure 15.1 The actual volume of gas molecules is very small compared with that of the gas

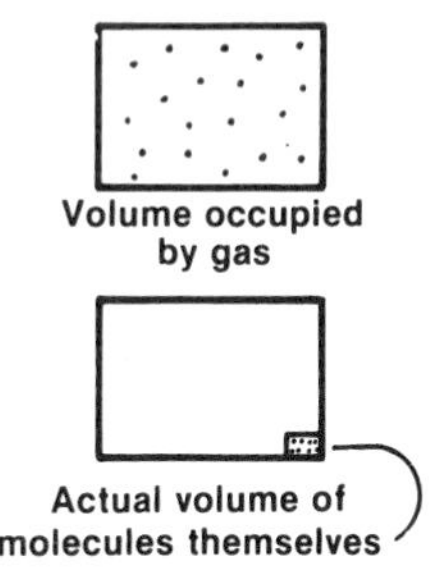

Note that the *masses* of equal volumes of different gases will be different.

Avogadro's law means that we can measure out equal numbers of different atoms or molecules simply by measuring out the same volumes of different gases. So a $100\,\text{cm}^3$ flask full of oxygen contains the same number of oxygen molecules as there are hydrogen molecules in $100\,\text{cm}^3$ of hydrogen. By weighing the gases we can compare the masses of the molecules directly.

In this example,

$100\,\text{cm}^3$ of oxygen has a mass of 0.133 g
$100\,\text{cm}^3$ of hydrogen has a mass of 0.0083 g

measured under the same conditions of temperature and pressure.

Since the number of molecules in each is the same according to Avogadro's law, an oxygen molecule must weigh $0.133/0.0083 = 16$ times as much as a hydrogen molecule. This forms the basis of the scale of relative atomic and molecular masses. Hydrogen is chosen to have $M_r = 2$ rather than 1, as there is chemical evidence that there are two atoms in the hydrogen molecule.

In order for volumes of gases to be comparable, they must be measured under the same conditions of temperature and pressure. This is inconvenient but a knowledge of how conditions affect gases may help. A number of simple relationships between pressure, temperature and volume of gases have been found.

Figure 15.2 Boyle's law. Doubling the pressure (by adding a load equal in weight to the plunger) halves the volume

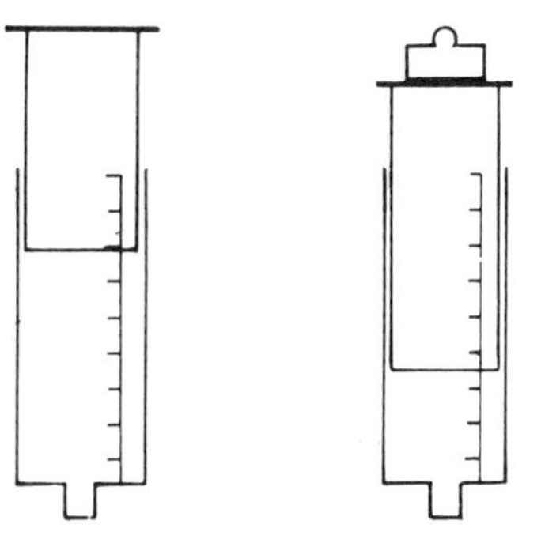

15.2.3 Boyle's law

Increasing the pressure on a fixed mass of gas causes its volume to decrease.

A

AIRSHIPS—GASES FOR LIFTING

You would find it quite hard to push a balloon below the surface of water in a bath. This is because you have to push away the water to make room for the balloon. The water trying to get back to where it was causes an upthrust on the balloon. Archimedes' principle tells us that the upthrust is equal to the weight of water displaced by the balloon. The same thing happens in the atmosphere.

A helium-filled balloon has an upthrust on it due to the air it displaces. Since helium molecules are lighter than air molecules (A_r for helium = 4, M_r for oxygen = 32, nitrogen = 28) and the same number of molecules of any gas have the same volume, the upthrust on a helium balloon is greater than the weight of the helium and it will rise and be able to lift a load.

Hydrogen is twice as good a lifting gas as helium (M_r for hydrogen = 2, A_r for helium = 4) and it was originally used in airships which were used for commercial aviation until the 1930s, when explosions destroyed the R101 and the Hindenberg and signalled the end of the airship era.

A modern airship

More recently, helium-filled airships have been making a comeback—for coastal anti-smuggling patrols and for airborne early warning over naval vessels. As they need no fuel to keep aloft, they have the advantage of long endurance and are cheap to run.

We find that the product of pressure and volume is a constant value, provided the temperature remains constant, **Figure 15.2** (opposite page).

$$\text{pressure} \times \text{volume} = \text{constant}$$

$$P \times V = \text{constant}$$

Alternatively,

$$V = \text{constant}/P$$

so

$$V \propto 1/P$$

and a graph of V against $1/P$ gives a straight line, **Figure 15.3**. The volume of a fixed mass of gas is inversely proportional to the pressure, at constant temperature.

Figure 15.3 Graphical illustrations of Boyle's law

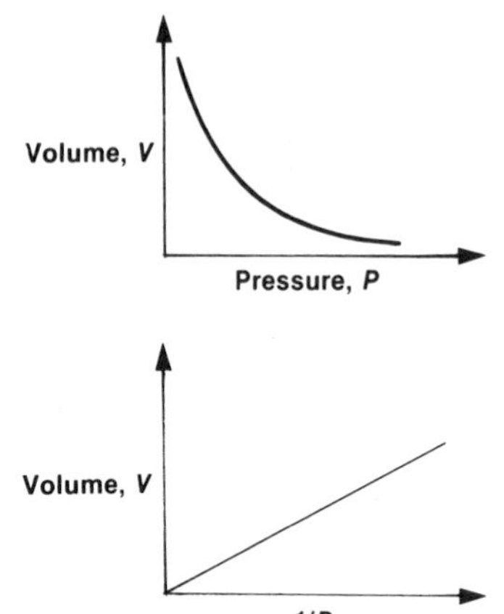

15.2.4 Charles' law

Increasing the temperature of a fixed mass of gas at constant pressure increases the volume. The volume is proportional to the temperature (measured in K) provided the pressure remains constant, **Figure 15.4.**

$$V \propto T$$

$$\frac{V}{T} = \text{constant}$$

Figure 15.4 Charles' law

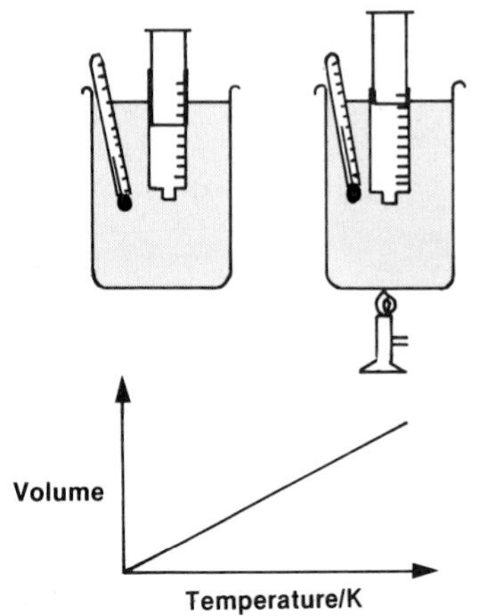

15.2.5 The 'constant volume' law

The pressure of a fixed mass of gas increases when the temperature is increased. The pressure is proportional to the temperature (measured in K) provided its volume remains constant. See **Figure 15.5** (overleaf).

$$P \propto T$$

$$\frac{P}{T} = \text{constant}$$

15.2.6 The ideal gas equation

Combining the three laws above leads to the equation:

Figure 15.5 The 'constant volume' law

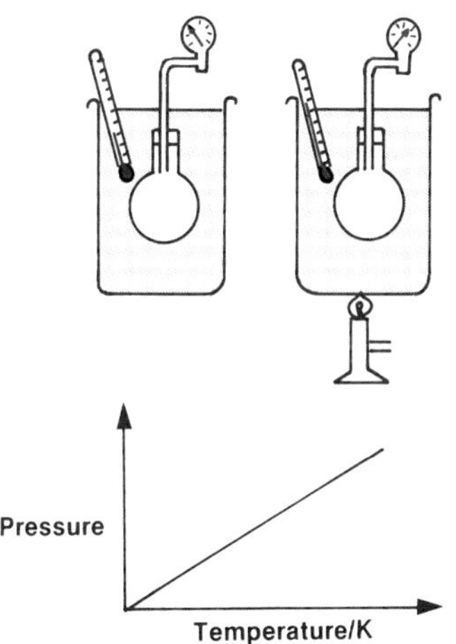

$$\frac{PV}{T} = \text{constant for a fixed mass of gas}$$

If we take one mole of gas, the constant is given the symbol R and is called the **gas constant**. For n moles of gas we have:

$$\frac{PV}{T} = nR$$

the value of R is $8.314\,\mathrm{J\,K^{-1}\,mol^{-1}}$

This is called the **ideal gas equation**. No gases obey it *exactly*, but for most gases at fairly low pressures and fairly high temperatures (say around normal room conditions) it holds quite well. It is often useful to imagine a gas which obeys this equation perfectly—an **ideal gas**.

We can use the ideal gas equation to calculate the volume a gas would occupy at temperatures and pressures other than those at which it was actually measured. If a gas has a volume V_1 measured at pressure P_1 and temperature T_1, then:

$$\frac{P_1 V_1}{T_1} = nR$$

If the same gas's volume V_2 is now measured at a new pressure P_2 and temperature T_2:

$$\frac{P_2 V_2}{T_2} = nR$$

so

$$\frac{P_1 V_1}{T_1} = \frac{P_2 V_2}{T_2}$$

(This could be rearranged to $V_2 = (P_1 V_1 T_2 / T_1 P_2)$ but you will almost certainly find it easier to remember in the form given above.)

The agreed set of conditions for comparison of gas volumes is called **standard temperature and pressure: 273 K and 10^2 kPa—usually called STP**. Don't confuse STP (273 K) with the standard conditions for enthalpy changes (298 K).

Conversion to STP

Example: A sample of gas had a volume of $50\,\mathrm{cm^3}$ measured at 20 °C and 101 315 Pa. What is its volume at STP?

$V_1 = 50\,\mathrm{cm^3}$ $\quad V_2 = ?$

$P_1 = 101\,315\,\mathrm{Pa}$ $\quad P_2 = 100\,000\,\mathrm{Pa}$ (remember to convert the temperatures to kelvin)

$T_1 = 20\,°\mathrm{C} = 293\,\mathrm{K}$ $\quad T_2 = 273\,\mathrm{K}$

$$\frac{P_1 V_1}{T_1} = \frac{P_2 V_2}{T_2}$$

so

$$\frac{101\,315 \times 50}{293} = \frac{100\,000 \times V_2}{273}$$

$$V_2 = \frac{101\,315 \times 50 \times 273}{293 \times 100\,000}$$

$$= 47.2\,\mathrm{cm^3}$$

It is easy to make arithmetic mistakes, so look to see if the answer is reasonable. Small changes in pressure and temperature will not make a very large difference to the volume. It is possible to convert to any set of conditions, not just STP.

Note on units of *P*, *V* and *T*

When using the ideal gas equation, consistent units must be used. **If you want to calculate *n***, the number of moles of gas, then:

P must be in Pa
V must be in m^3
T must be in K
R must be in $J\,K^{-1}\,mol^{-1}$

Other units are often encountered. Most volumes will be measured in cm^3 or dm^3:

$$1\,m^3 = 10^6\,cm^3 = 10^3\,dm^3$$

Industrial chemists often measure pressures in atmospheres:

$$1 \text{ atmosphere} \simeq 10^5\,Pa \text{ or } 10^2\,kPa$$

Many laboratory measurements of pressure are made using a mercury manometer which may give a reading in millimetres of mercury (mm Hg or torr):

$$760\,mm\,Hg \simeq 10^2\,kPa$$

Weather forecasters use millibars. 1000 millibars $\simeq 10^2$ kPa. When converting to STP (or any other temperature and pressure) you can use these units *provided they are used for both the initial and final conditions*, for example, converting 100 cm^3 at 740 mm Hg to volume in cm^3 at 760 mm Hg. But temperatures must *always* be in K, never °C or any other units.

Example

A sample of gas has a volume of 100 cm^3 measured at 25 °C and 760 mm Hg. What is its volume at 30 °C and 740 mm Hg?

$P_1 = 760\,mm\,Hg \qquad P_2 = 740\,mm\,Hg$
$V_1 = 100\,cm^3 \qquad V_2 = ?$
$T_1 = 25\,°C = 298\,K \quad T_2 = 30\,°C = 303\,K$

$$\frac{P_1V_1}{T_1} = \frac{P_2V_2}{T_2}$$

$$\frac{760 \times 100}{298} = \frac{740 \times V_2}{303}$$

$$V_2 = \frac{760 \times 100 \times 303}{298 \times 740}\,cm^3$$

$$V_2 = 104.4\,cm^3$$

15.2.7 The volume of a mole of gas

Avogadro's theory states that the same volume of all gases contains the same number of molecules. Thus one mole of *any* gas should have the same volume. This is confirmed by the ideal gas equation from which the volume of one mole of gas can be calculated at STP.

$n = 1$
$R = 8.3\,J\,K^{-1}\,mol^{-1}$
$T = 273\,K$
$P = 100\,000\,Pa$

$$PV = nRT$$

$$V = \frac{nRT}{P}$$

$$\text{so } V = \frac{1 \times 8.3 \times 273}{100\,00}$$

$$= 0.0226\,m^3 \text{ (note the units)}$$

$1\,m^3 = 10^3\,dm^3 \qquad \text{so } V = 22.6\,dm^3$

The volume of one mole of any gas is 22 600 cm^3 at STP. A useful rule of thumb is that one mole of any gas has a volume of 24 dm^3 or 24 000 cm^3 at room temperature and pressure.

15.3 Measurement of relative molecular mass

15.3.1 The relative molecular mass of a gas

The ideal gas equation allows us to measure easily the relative molecular masses of gases:

$$PV = nRT$$

$$n = \frac{PV}{RT}$$

so we can calculate the number of moles in a sample of gas if we can measure its pressure, volume and temperature. If we can find the mass of the gas (in grams), we know the mass of a known number of moles. We can therefore calculate the mass of one mole. This is numerically equal to the relative molecular mass.

The experimental method varies with the method of production of the gas and the apparatus used.

Gas in a pressurized container

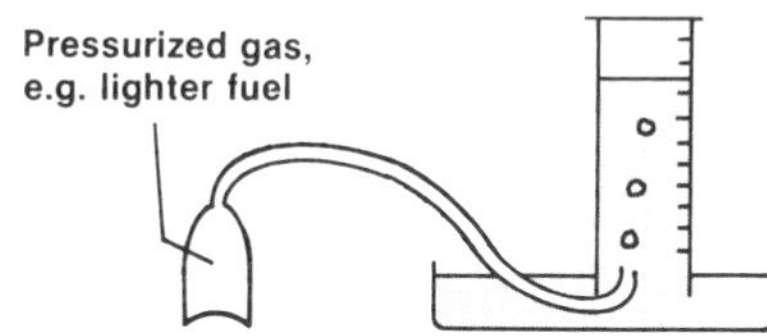

Figure 15.6 Measuring M_r for lighter fuel

1. The apparatus is shown in **Figure 15.6**. The gas container is weighed and then, say, 1 dm^3 of gas dispensed into a measuring cylinder keeping the final levels of water inside and outside the measuring cylinder the same to ensure that the pressure of the gas is the same as atmospheric pressure. The pressure can then be measured with a barometer, and room temperature measured. The gas container is then reweighed.

   For example, a sample of gas sold for filling gas lighters was used.

   loss of weight of the can = 2.29 g
   temperature = 14 °C (287 K)
   volume of gas = 1 dm^3 (10^{-3} m^3)
   atmospheric pressure = 100 100 Pa (measured with a barometer)

$$n = \frac{PV}{RT}$$

$$n = \frac{100\,100 \times 10^{-3}}{287 \times 8.3}$$

$$n = 0.042 \text{ mol}$$

$$2.29 \text{ g} = 0.042 \text{ mol}$$

$$1 \text{ mol} = \frac{2.29}{0.042} = 54.5 \text{ g}$$

$$M_r = 54.5$$

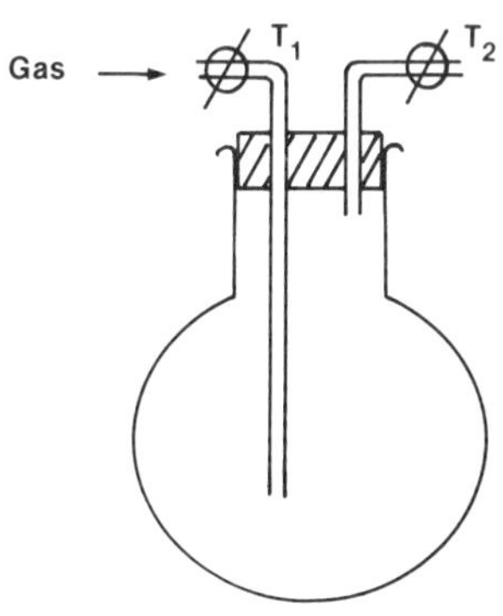

Figure 15.7 Apparatus for weighing a known volume of gas

2. A variation of this method uses a glass flask from which the air can be pumped with a vacuum pump as shown in **Figure 15.7.** The flask is evacuated using a vacuum pump and weighed. The gas (from any source) is allowed to flow through the container for a few seconds. Tap T_1 is then closed *followed by* tap T_2 to ensure that the gas in the flask is at atmospheric pressure. Atmospheric pressure and the temperature are then found. The volume of the container can be found (by filling it with water for example). Substitution in $n = PV/RT$ gives the number of moles of gas in the weighed amount.

Example

A flask of 250 cm^3 volume was evacuated and then filled with carbon dioxide at atmospheric pressure of 100 000 Pa and 20 °C (293 K). The flask increased in mass by 0.44 g.

$V = 250\,cm^3\ (250 \times 10^{-6}\,m^3)$
$R = 8.3\,J\,K^{-1}\,mol^{-1}$
$P = 100\,000\,Pa$
$T = 293\,K$

$$n = \frac{PV}{RT}$$

$$n = \frac{100\,000 \times 250 \times 10^{-6}}{8.3 \times 293}$$

$$n = 0.01\,mol$$

now $0.01\,mol = 0.44\,g$
so 1 mol has a mass of 44 g
and $M_r = 44$

If no vacuum pump is available, an ordinary flask can be weighed full of air, then filled with gas as shown in **Figure 15.8** and reweighed. To find the mass of air we use the relationship

$$\text{density, } \rho = \frac{\text{mass } (m)}{\text{volume } (V)}$$

or $$m = V\rho$$

This is subtracted from the mass of flask + air to give the mass of an empty flask.

Figure 15.8 Methods of filling flasks with gases

(a) Filling a flask with a gas which is denser than air
(b) Filling a flask with a gas which is less dense than air

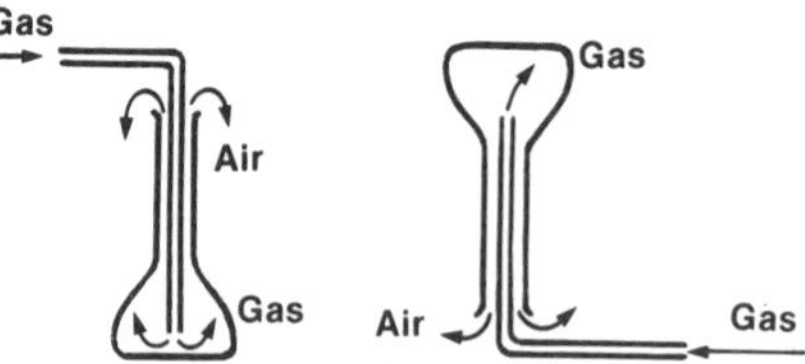

Example

A flask was weighed full of air and then full of methane (methane is less dense than air therefore method (b) was used for filling).

volume of flask $= 115.0\,cm^3$
mass of flask + air $= 51.139\,g$
mass of flask + methane $= 51.076\,g$
temperature $= 20\,°C\ (= 293\,K)$
pressure $= 100\,050\,Pa$
density of air $= 1.293 \times 10^{-3}\,g\,cm^{-3}$ at STP

First convert the volume of air in the flask to STP:

$$\frac{P_1 V_1}{T_1} = \frac{P_2 V_2}{T_2}$$

$P_1 = 100\,050\,Pa$ $\quad P_2 = 100\,000\,Pa$
$V_1 = 115.0\,cm^3$ $\quad V_2 = ?$
$T_1 = 293\,K$ $\quad T_2 = 273\,K$

so $$\frac{100\,050 \times 115}{293} = \frac{V_2 \times 100\,000}{273}$$

$$V_2 = \frac{100\,050 \times 115 \times 273}{293 \times 100\,000}$$

$$= 107.2\,cm^3$$

mass of air in the flask $= V \times \rho$
$= 107.2 \times 1.293 \times 10^{-3}$
$= 0.139\,g$

so mass of the empty flask $= 51.139 - 0.139\,g$
$= 51.000\,g$

mass of methane in the flask $= 51.076 - 51.000\,g$
$= 0.076\,g$

The volume of methane at STP is $107.2\,cm^3 = 107.2 \times 10^{-6}\,m^3$.

$$PV = nRT$$

$$n = \frac{PV}{RT}$$

$P = 100\,000\,Pa$
$V = 107.2\,cm^3\ (= 107.2 \times 10^{-6}\,m^3)$
$R = 8.3\,J\,K^{-1}\,mol^{-1}$
$T = 273\,K$

$$= \frac{100\,000 \times 107.2 \times 10^{-6}}{8.3 \times 273}$$

so $n = 4.73 \times 10^{-3}$ mol

4.73×10^{-3} mol has a mass of 0.076 g

so 1 mol has a mass of 16.0 g

and M_r of methane = 16

15.3.2 The relative molecular mass of a volatile liquid

The same type of method can be applied to volatile liquids simply by vaporizing them and measuring the volume of the vapour as shown in **Figure 15.9**. The method uses a steam jacket and can be used with liquids of boiling point up to about 80 °C (353 K). A 100 cm^3 gas syringe is heated to approximately 100 °C (373 K) using a steam jacket. The syringe is fitted with a self-sealing rubber cap and about 5 cm^3 of air are left in the syringe.

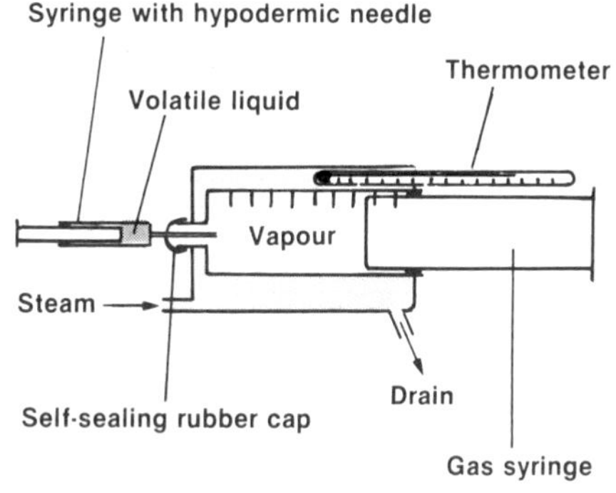

Figure 15.9 Apparatus for measuring M_r for a volatile liquid

When the temperature is steady, the initial reading of the gas syringe is taken and about 0.2 cm^3 of the volatile liquid is injected into the air space left in the gas syringe, using a syringe with a hypodermic needle to pierce the rubber cap.

The liquid vaporizes and when the gas syringe plunger has stopped moving, a final volume reading is taken.

The temperature inside the steam jacket and atmospheric pressure are taken. The mass of liquid injected is found by weighing the hypodermic syringe before and after injection.

Example: Methanol

change in mass of the hypodermic syringe = 0.083 g

increase in volume of the gas syringe = 80 cm^3 (= 80×10^{-6} m^3)

temperature of the steam jacket = 100 °C (= 373 K)

atmospheric pressure = 99 950 Pa

$$n = \frac{PV}{RT}$$

$$n = \frac{99\,950 \times 80 \times 10^{-6}}{8.3 \times 373}$$

$$n = 2.58 \times 10^{-3} \text{ mol}$$

so 2.58×10^{-3} mol has a mass of 0.08 g

so 1 mol has a mass of 32.0 g

and M_r of methanol = 32

15.4 More gas laws

15.4.1 Dalton's law of partial pressures

This law, already encountered in **section 11.4.2**, is about mixtures of gases. The **partial pressure** of a gas in a mixture is the pressure that the gas would exert if it was in the container alone. Dalton's law states that the total pressure of a mixture of gases is the sum of the partial pressures of all the components. This is simply saying that the gases do not affect each other's behaviour so their pressures add up.

A consequence of Dalton's law is that the partial pressure of a gas is the *total* pressure multiplied by the *mole fraction* of that gas, where the mole fraction is the number of moles of that gas divided by the total number of moles of all the gases in the mixture.

In a mixture of gases A, B and C:

$$\text{the partial pressure of gas A, } p_A = \frac{n_A}{(n_A + n_B + n_C)} \times P_{total}$$

A

ATMOSPHERIC CHEMISTRY

1. THE GREENHOUSE EFFECT

Greenhouses get hot inside because glass is transparent to visible light which can thus enter the greenhouse. However, glass does not allow infra-red (heat) radiation through—it remains trapped inside. Carbon dioxide gas molecules in the atmosphere behave in the same way as glass, letting in light from the sun but not letting heat out. Burning fossil fuels produces large quantities of carbon dioxide and there is concern that increasing concentrations of this gas in the atmosphere could lead to a rise in the temperature of the earth. Even a small rise in temperature could have far-reaching effects, for example, melting of part of the polar ice caps which would lead to a rise in sea level and flooding of low-lying areas.

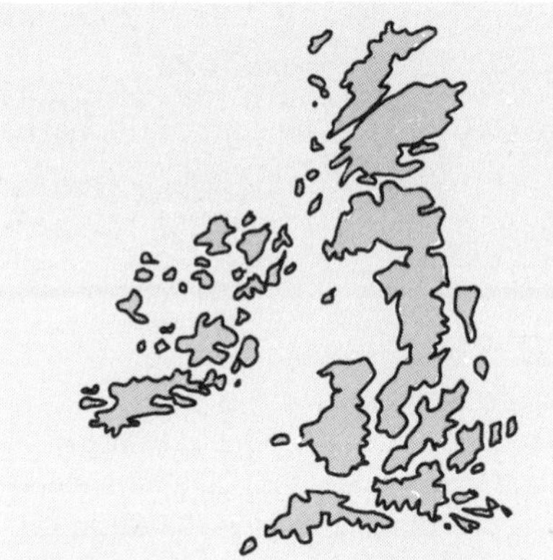

Areas of the UK which would be flooded after melting of the polar ice caps

There are mechanisms for removing carbon dioxide from the atmosphere: plants use it in photosynthesis and large amounts dissolve in the sea. However, there is evidence of raised carbon dioxide levels and of a gradual raising of the earth's temperature.

2. OZONE

The earth is bathed in high-energy (short-wavelength) ultraviolet (UV) light from the sun. Most of this is filtered out before it reaches the earth's surface. It is absorbed by the small concentration (0.25 parts per million) of ozone in the earth's stratosphere, the layer of the atmosphere about 25 km above the earth's surface. Ozone (trioxygen) is an allotrope of oxygen. It has the formula O_3 and it is formed by photochemical reactions in the atmosphere:

$$\underset{\text{oxygen molecule}}{O_2(g)} \xrightarrow{UV} \underset{\text{oxygen atoms}}{2O(g)}$$

$$\underset{\text{oxygen atom}}{O(g)} + \underset{\text{oxygen molecule}}{O_2(g)} \longrightarrow \underset{\text{ozone}}{O_3(g)}$$

Ozone is removed by the following reactions with dinitrogen oxide, a gas which is present in the atmosphere in small amounts, due to the action of bacteria on nitrogen in the soil:

Burning the Amazon rainforest removes many trees, increasing carbon dioxide levels due to both the burning and to the resulting loss of photosynthesis

1. $O(g) + N_2O(g) \longrightarrow 2NO(g)$

2. $O_3(g) + NO(g) \longrightarrow NO_2(g) + O_2(g)$

3. $O(g) + NO_2(g) \longrightarrow NO(g) + O_2(g)$

The overall effect of reactions 2 and 3 is:

$$O_3(g) + O(g) \longrightarrow 2O_2$$

and the NO is regenerated thus acting as a catalyst.

Normally the generation of ozone balances its destruction and the concentration of the gas remains steady. However, in recent years, the amount of NO (nitrogen monoxide) in the atmosphere has increased due to its production in motor vehicle exhausts. Chlorine atoms originating in chlorofluorocarbons (used as aerosol propellents among other things) can also act in the same way as NO in the ozone destruction reactions and their concentration in the atmosphere has also increased.

$$O_3(g) + Cl(g) \longrightarrow ClO(g) + O_2(g)$$
$$O(g) + ClO(g) \longrightarrow Cl(g) + O_2(g)$$

overall

$$O_3(g) + O(g) \longrightarrow 2O_2(g)$$

and the chlorine is regenerated.

The net result of this is that the rate of loss of ozone is greater than its rate of production, see **Plate 4.** The UV filtering effect is reduced and more high-energy UV is reaching the earth's surface. This may well result in increased incidence of skin cancer as well as affecting marine plankton. This latter effect could have far-reaching consequences as plankton are at the base of the marine food chain.

or

$$p_A = \frac{n_A}{n_{total}} \times P_{total} \quad n_{total} = n_A + n_B + n_C$$

p_A = partial pressure of A
n_A = number of moles of A
n_B = number of moles of B
n_C = number of moles of C
n_{total} = total number of moles
P_{total} = total pressure

Example

A mixture of two moles of hydrogen and one mole of oxygen has a total pressure of 10^2 kPa.

$$\text{partial pressure of oxygen} = \frac{1}{1+2} \times 10^2 \text{ kPa} = 33.3 \text{ kPa}$$

$$\text{partial pressure of hydrogen} = \frac{2}{1+2} \times 10^2 \text{ kPa} = 66.7 \text{ kPa}$$

Notice that the total pressure (100 kPa) is the sum of the partial pressures (33.3 + 66.7 kPa).

A

SOLUBILITY OF GASES—FROM SOFT DRINKS TO DIVING

Gases dissolve in liquids in proportion to their partial pressure. This is called Henry's law. Carbon dioxide is dissolved in fizzy drinks under pressure during manufacture. When we open the bottle or can, the pressure is reduced and the gas comes out of solution and the drink fizzes.

Henry's law is very important to divers who breathe air at high pressure when working at depths down to about 50 metres. The high pressure causes nitrogen to dissolve in the blood to a larger extent than at atmospheric pressure. This causes two problems. Firstly, the large quantities of nitrogen cause a condition called nitrogen narcosis. The symptoms are rather like drunkenness—not the best conditions for intricate work in a dangerous situation! The second problem occurs when the diver ascends back to atmospheric pressure. The dissolved nitrogen comes out of the blood and bubbles can collect in the joints causing excruciating pain called 'the bends'. The bends can be avoided by ascending slowly to allow the nitrogen to come out of solution gradually. Divers at depths of greater than 50 metres breathe mixtures of helium and oxygen to avoid nitrogen narcosis. Currently, experiments are being carried out with mixtures containing hydrogen to try to find gas mixtures suitable for even greater diving depths.

15.4.2 Graham's law

This applies to **diffusion** (the tendency of two gases to mix together) and **effusion** (the tendency of a gas to escape through a small hole). Graham's law states that the rate of both of these processes is inversely proportional to the square root of the density of the gas.

$$\text{rate of effusion (or diffusion)} \propto \frac{1}{\sqrt{\rho}}$$

$$\text{rate of effusion (or diffusion)} = \frac{\text{constant}}{\sqrt{\rho}}$$

Since ρ = mass/volume and for one mole of any gas the volume is the same (22 600 cm^3 at STP)

$$\text{rate of diffusion} \propto 1 \bigg/ \sqrt{\frac{\text{mass of 1 mol}}{22\,600 \text{ cm}^3}}$$

$$\text{rate of diffusion} \propto \frac{\sqrt{22\,600\ \text{cm}^3}}{\sqrt{\text{mass of 1 mole}}}$$

Since the mass of 1 mole in grams is M_r:

$$\text{rate of diffusion} \propto \frac{\sqrt{22\,600}}{\sqrt{M_r}}$$

$$\text{rate of diffusion} = \frac{\text{constant}}{\sqrt{M_r}}$$

Effusion is easier to measure than diffusion and it can be used as the basis of a method of measuring relative molecular masses. The apparatus shown in **Figure 15.10** can be used to measure rates of effusion.

The nozzle of the gas syringe is blocked by metal foil with a pinhole in it and the weight of the plunger produces a forced effusion of the gas through the hole.

Figure 15.10 Apparatus for measuring the rate of effusion of a gas

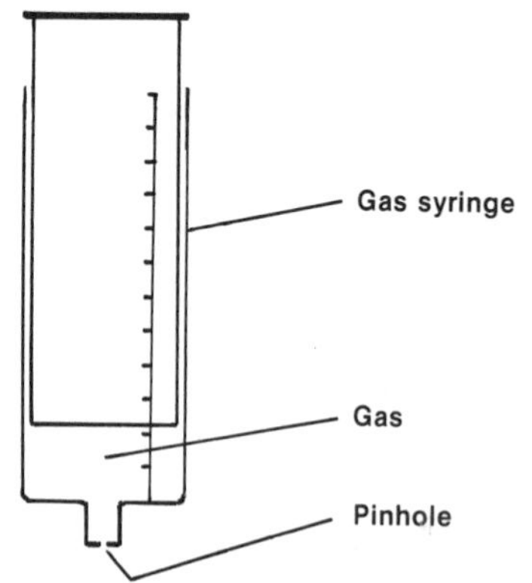

Example

$100\ \text{cm}^3$ of carbon dioxide, $M_r = 44$, diffused in 60 seconds. Nitrogen diffused at a rate of $100\ \text{cm}^3$ in 48 seconds. What is the relative molecular mass of nitrogen?

For carbon dioxide:

$$\text{rate of diffusion} = \frac{\text{constant}}{\sqrt{M_r}}$$

$$\text{But rate of diffusion} = \frac{100\ \text{cm}^3}{60\ \text{s}} = \frac{100}{60}\ \text{cm}^3\ \text{s}^{-1}$$

$$\frac{100}{60}\ \text{cm}^3\ \text{s}^{-1} = \frac{\text{constant}}{\sqrt{44}}$$

$$1.67\ \text{cm}^3\ \text{s}^{-1} = \frac{\text{constant}}{6.63}$$

$$\text{constant} = 1.67 \times 6.63\ \text{cm}^3\ \text{s}^{-1}$$
$$= 11.0\ \text{cm}^3\ \text{s}^{-1} \text{ for this apparatus}$$

so $$\text{rate of diffusion} = \frac{11.0}{\sqrt{M_r}} \text{ for this apparatus}$$

For nitrogen:

$$\text{rate of diffusion} = \frac{100}{48} = 2.08\ \text{cm}^3\ \text{s}^{-1}$$

$$2.08 = \frac{11.0}{\sqrt{M_r}}$$

$$\sqrt{M_r} = \frac{11.0}{2.08} = 5.29$$

$$M_r = 28$$

So M_r of nitrogen is 28.

A

SEPARATION OF ISOTOPES BY DIFFUSION

Graham's law has an important industrial application as it enables uranium isotopes to be separated.

Natural uranium is a mixture of two isotopes: 0.7% ^{235}U and 99.3% ^{238}U. Only ^{235}U is suitable for fuel for nuclear power stations.

The isotopes can be separated by making uranium hexafluoride, UF_6 (boiling point 329 K). Gaseous $^{235}UF_6$ ($M_r = 349$) diffuses through a porous membrane faster than $^{238}UF_6$ ($M_r = 352$) and this forms the basis for the method of separating them.

Use Graham's law to find out how much faster $^{235}UF_6$ diffuses than $^{238}UF_6$. (You should get an answer of 1.004!)

15.5 The kinetic theory

Kinetic theory is an attempt to explain the observed properties of gases on the basis that gases are made up of small particles, well spaced out, moving rapidly at random. More formally, kinetic theory makes five assumptions about gas molecules:

1. The particles are moving at random.
2. We can neglect the volume of the particles themselves in comparison with the total volume of the gas.
3. The particles do not attract one another.
4. The kinetic energy of the particles is proportional to the temperature of the gas.
5. No energy is lost in collisions between particles.

Without going into the mathematics, these assumptions explain the properties of gases.

Bombardment of the walls of the container explains pressure. The fact that kinetic energy increases with increasing temperature means that molecules will bombard the walls harder and more often at increased temperature, thus increasing the pressure.

The random motion of well spaced-out molecules explains diffusion.

Using these assumptions quantitatively, it is possible to derive the equation:

$$PV = \tfrac{1}{3}Nmc^2$$

where P = pressure, V = volume, N = number of molecules, m = mass of each particle and c = average velocity of the particles. See box below.

E

DERIVATION OF $PV = \frac{1}{3}Nmc^2$

(You should be able to follow this if you have done some physics.) This is a simplified version and you may not be happy about some of the assumptions. However, more detailed treatments do give the same answer. Imagine the particles, all moving at speed c, are in a cube of side l.

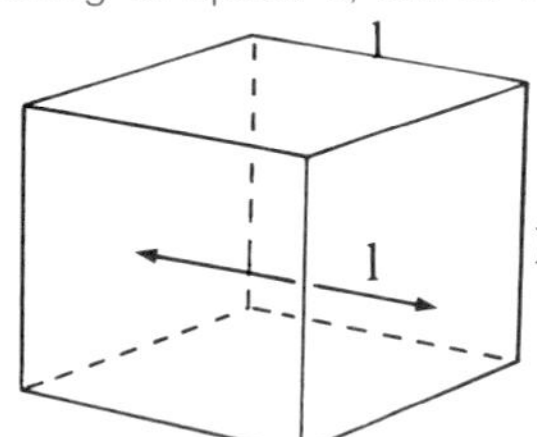

Imagine that one-third of the particles move along each axis of the box, i.e. up and down, left to right and front to back. This means that each particle would repeatedly bounce back and forth from one face to the opposite one. It travels a distance of $2l$ between each collision with the same face. Travelling at a speed of c, the time between collisions is given by:

$$\text{speed} = \frac{\text{distance}}{\text{time}} \qquad \text{time} = \frac{\text{distance}}{\text{speed}} \qquad \text{time} = \frac{2l}{c}$$

The **momentum** of each particle is mc. So the change of momentum on each collision is from mc to $-mc$ (moving in the opposite direction), i.e. $2mc$.

The force exerted on the face in each collision is given by **the rate of change of momentum** (Newton's second law of motion). Each particle collides with a face of the cube once every $2l/c$ seconds. So it makes $c/2l$ collisions per second (i.e. if it collided once every quarter second it would make four collisions per second).

So the rate of change of momentum for each particle is:

$$2mc \times \frac{c}{2l} = \frac{mc^2}{l}$$

so

$$\text{force per particle} = \frac{mc^2}{l}$$

But one-third of the N particles are moving along each axis. So the total force on each face is:

$$\frac{Nmc^2}{3l}$$

$$\text{pressure} = \frac{\text{force}}{\text{area}} \text{ and the area of a face is } l^2$$

$$\text{pressure} = \frac{Nmc^2}{3l} \Big/ l^2$$

$$P = \frac{Nmc^2}{3l^3}$$

but l^3 is the volume V of the box, so:

$$P = \frac{Nmc^2}{3V}$$

or

$$PV = \tfrac{1}{3}Nmc^2$$

15.5.1 Deriving the gas laws

Using the equation on page 204 we can predict the gas laws.

Boyle's law

Boyle's law refers to a fixed mass of gas at a constant temperature, so the kinetic energy ($\frac{1}{2}mc^2$) is constant, as is N for a fixed mass of gas.
So:

$$PV = \frac{1}{3}N(mc^2)$$ ← constant at constant temperature

↑ (N) constant for a fixed mass of gas

so $$PV = \text{constant}$$

Charles' law

This refers to a fixed mass of gas at a constant pressure. The kinetic theory assumes temperature to be proportional to kinetic energy ($\frac{1}{2}mc^2$). So:

$$PV = \frac{1}{3}N(mc^2)$$ ← proportional to temperature, T

↑ (P) constant; ↖ (N) constant for a fixed mass of gas

so $$V = \text{constant} \times T$$

'Constant volume' law

You should be able to see that:

$$P = \text{constant} \times T$$

in the same way as for Charles' law above.

Graham's law

The rate of effusion depends on how often a particle 'hits' the pinhole. This depends on the speed of the particles, c. The faster the particles move, the more often they will 'hit' the pinhole and go through it:

Note: If molecules were the size of people, a pinhole would be about 1000 km across

$$PV = \frac{1}{3}Nmc^2$$

$$c^2 = \frac{3PV}{Nm}$$

$$c = \sqrt{\frac{3PV}{Nm}}$$

So at constant pressure and volume:

$$c \propto \frac{1}{\sqrt{m}}$$

and m, the mass of the particle, is obviously related to the relative molecular mass, M_r.

Figure 15.11 The distribution of speeds of gas molecules at temperatures T_1, T_2 and T_3

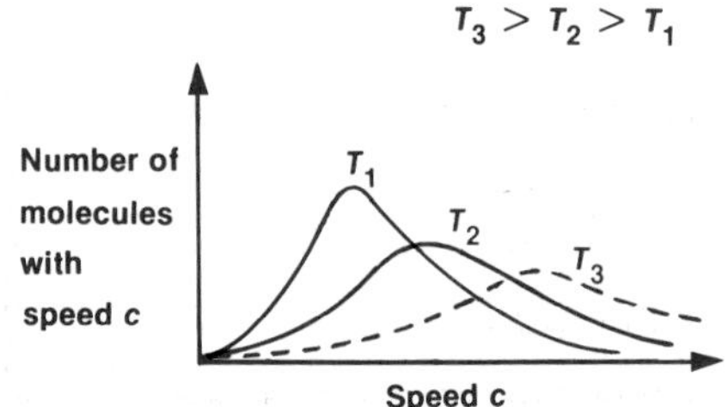

15.5.2 The distribution of molecular velocities

The above simple treatment of the kinetic theory has assumed that all the particles in a gas have the same (average) speed. This is not actually the case. There is a distribution of speeds, some molecules moving slowly, some fast and the majority having speeds in between. This can be shown on a graph, **Figure 15.11.** Notice that the distribution is not symmetrical. At higher temperatures, the peak of the graph, which represents the most probable speed, moves to the right (towards higher speed) but the height of the peak gets less. This means that the most probable speed gets faster

A

EVAPORATION—THE FRIDGE

The distribution of molecular velocities in a liquid is similar to that in a gas and this explains why liquids evaporate at any temperature, even well below their boiling points.

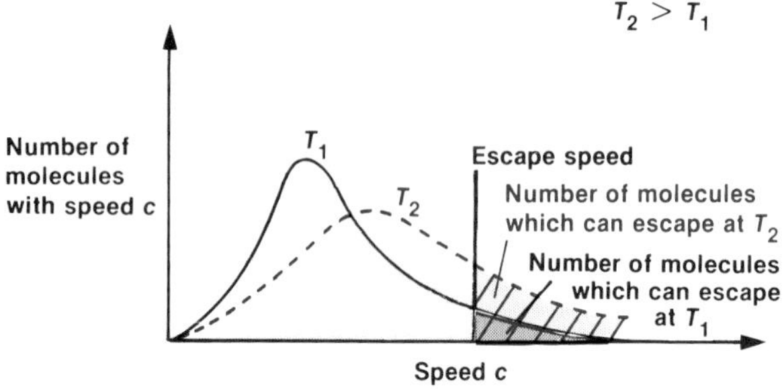

Because of the shape of the distribution curve, there will always be a few molecules with sufficient speed (and energy) to escape the attraction of the liquid (provided they are moving towards the surface, not away from it). At a high temperature, more molecules have the speed to escape.

EVAPORATIVE COOLING

The molecules which escape from a liquid are the faster (more energetic) ones so the average energy of the remaining molecules falls, i.e. the temperature drops. The liquid will then absorb heat from its surroundings, thus acting as a refrigerant, and continue to evaporate. A beaker of the volatile liquid ethoxyethane will cool so much on evaporation that it can freeze a pool of water in which it stands. This is the principle of the fridge. A volatile liquid, either ammonia or one of the freons (see **sections 30.1** and **30.12**) evaporates inside the cooling coils within the fridge and this causes cooling. The vapour is then pumped into pipes outside the fridge, is compressed back to a liquid and recycled to the cooling coils. While being compressed, it gives out heat. The device can be thought of as a heat 'pump' pumping heat from inside the fridge to outside. You may have noticed that the coils on the outside of a fridge are warm.

and there is a wider spread of speeds. The curve always starts at the origin. The area under the whole curve is constant as it represents the total number of molecules. This distribution is referred to as the Maxwell–Boltzmann distribution, which we have already met in terms of energy distribution in **section 14.5.2.**

15.6 Real gases—deviations from the ideal gas equation

No real gas actually obeys the ideal gas equation in its entirety. All real gases become liquid when they are cooled and/or compressed enough. The volume of a liquid hardly changes when the pressure or temperature changes. The ideal gas equation predicts a gradual decrease in volume with decreasing temperature until the volume becomes zero at 0 K.

The reason for this failure to obey the ideal gas equation at high pressure and low temperature is that two of the assumptions of the kinetic theory are no longer valid under these conditions. Once the molecules get close together, as they do at high pressures and low temperatures, then we can no longer ignore the attraction between them. Attraction between molecules is discussed in Chapter 16. Eventually, the attraction pulls the molecules together and the liquid state results. Also, when the volume of gas is small, it is no longer possible to ignore the volume of the actual molecules in comparison with the volume of the whole gas.

15.7 Summary

- Gay-Lussac's law states that gases react in volumes which are in simple whole number ratios (measured under the same conditions of temperature and pressure).
- This is explained by Avogadro's law which states that equal volumes

(measured at the same temperature and pressure) of any gases contain the same numbers of particles. This gives a simple way of 'counting' molecules by measuring gas volumes.

- The ideal gas equation is:

$$PV = nRT$$

where $P =$ pressure
$V =$ volume
$n =$ number of moles
$T =$ temperature (K)
R is the gas constant ($8.3\,\mathrm{J\,K^{-1}\,mol^{-1}}$)

It is a combination of:

1. Boyle's law $V \propto 1/P$

2. Charles' law $V \propto T$

3. 'constant volume' law $P \propto T$

- The ideal gas equation is obeyed fairly closely by most gases at room temperature and pressure.

- The equation: $$\frac{P_1 V_1}{T_1} = \frac{P_2 V_2}{T_2}$$

allows volumes of gases to be 'corrected' to standard temperature and pressure (STP), 10^2 kPa and 273 K.
- One mole of an ideal gas has a volume of $22\,600\,\mathrm{cm^3}$ at STP (approximately $24\,000\,\mathrm{cm^3}$ at room temperature and pressure).
- The ideal gas equation allows the relative molecular mass, M_r, of a gas to be calculated if P, V, T and the mass of the gas can be measured. The method can be extended to volatile liquids if they can be heated to above their boiling points.
- The partial pressure of a gas in a mixture of gases is the pressure that the gas would exert if it alone occupied the container.
- Dalton's law states that the total pressure of the mixture is the sum of the partial pressures of all the components.
- Diffusion is the mixing of gases and effusion is the tendency of gases to escape through a small hole.
- Graham's law states that:

$$\text{rate of diffusion or effusion} \propto \frac{1}{\sqrt{\rho}} \propto \frac{1}{\sqrt{M_r}}$$

$\rho =$ density of the gas

This provides another method of measuring M_r.
- Kinetic theory assumes gases to be made of well spaced-out particles moving rapidly at random. The particles are assumed to have negligible volume and no mutual attraction. Using these assumptions, the equation:

$$PV = \tfrac{1}{3}Nmc^2$$

can be derived, where $N =$ number of molecules, $m =$ mass of one molecule and c represents the average speed of the molecules.
- The distribution of velocities among molecules in a gas can be shown graphically, **Figure 15.12.**
- Real gases do not obey the ideal gas equation exactly. At low temperatures and high pressures the volume of the gas is small and the volume of the molecules and their mutual attraction cannot be neglected. At sufficiently low temperatures and/or high pressures, gases can be liquefied.

Figure 15.12 Distribution of molecular speeds in a gas

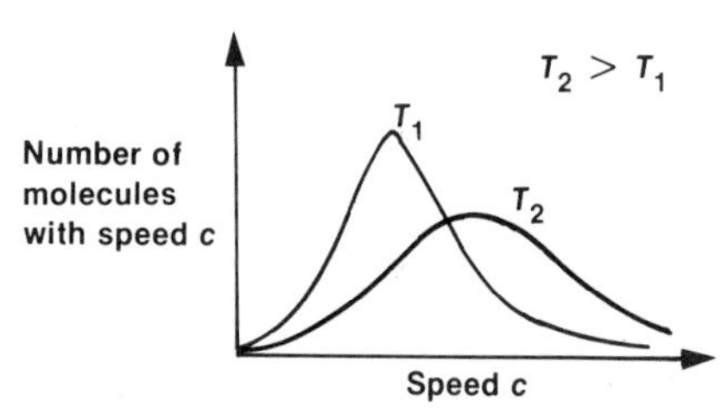

15.8 Questions

1 Convert the following gas volumes to STP (273 K and 100 000 Pa or 760 mm Hg and 1 atmosphere). Answer in the same volume units as the question.

A) 100 cm^3 at 290 K and 100 060 Pa;
B) 80 cm^3 at 373 K and 90 040 Pa;
C) 1 dm^3 at 200 K and 100 000 Pa;
D) 100 m^3 at 20 °C and 750 mm Hg;
E) 1 pint at 50 °C and 2 atmospheres.

2 The gas hydrogen chloride can be decomposed into its elements, hydrogen, H_2, and chlorine, Cl_2, by heating it over a suitable catalyst. 80 cm^3 of hydrogen chloride gives 80 cm^3 of the mixture of hydrogen and chlorine.

On removing the hydrogen, 40 cm^3 of the gas remained.

A) What is the formula of hydrogen chloride? (You probably know this already so make sure you can explain your reasoning. What law or laws did you assume to get the answer?)

B) Briefly describe how you might remove the hydrogen.

C) Can you think of a way of removing the chlorine instead?

D) Sketch a suitable arrangement of gas syringes for the experiment.

3 Find the relative molecular mass of a liquid, 0.15 g of which on vaporization gives 30 cm^3 of vapour at 373 K and 10^2 kPa.

(London 1979, part question)

4 In a particular experiment it was found that 1.013 g of a hydrocarbon occupied a volume of 227 cm^3 corrected to 0 °C and 1 atmosphere. What is the relative molecular mass of the hydrocarbon used? (Molar volume at 0 °C and 1 atmosphere = 22.4 dm^3 mol^{-1}.)

(Nuffield 1979, part question)

5 The compound $SbCl_5$ is a liquid which boils at 140 °C. A sample of $SbCl_5$ weighing 3.00 g was completely vaporized at its boiling point, and the volume of vapour obtained, converted to STP, was 285 cm^3. (Molar volume of a gas at STP = 22 400 cm^3 mol^{-1}.)

A) From these data, calculate a value for the relative molecular mass of $SbCl_5$ in the vapour state.

B) How does the value in (A) compare with the expected value? What explanation can you offer for any discrepancy? (Relative atomic masses: Sb = 122, Cl = 35.5.)

(Nuffield 1985, part question)

6 Which one of the following diagrams correctly represents the Boltzmann distribution of molecular speeds at two temperatures, T_1 and T_2 where $T_2 > T_1$?

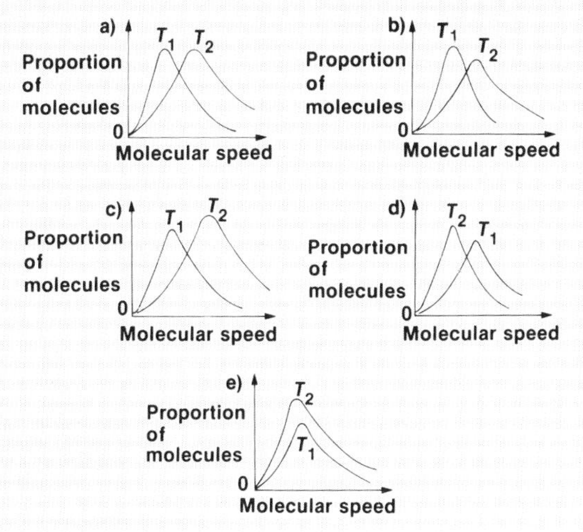

(Cambridge 1987)

7 In an effusion experiment using an apparatus like that on page 203, a sample of methane, $M_r = 16$, effused at a rate of 100 cm^3 in 50 seconds.

A) How long would 100 cm^3 of the following gases take to effuse:

i carbon dioxide $M_r = 44$;
ii hydrogen $M_r = 2$?

B) Which gas would be best for calibrating the apparatus? Explain.

C) A sample of nitrogen dioxide, NO_2, was investigated in the same apparatus.

100 cm^3 of the gas took 100 seconds to diffuse. What value of M_r does this give for the nitrogen dioxide? What value would you expect for nitrogen dioxide? Can you suggest an explanation for the difference?

16 Intermolecular forces: liquids and solutions

16.1 Introduction

Chapter 10 described the different types of bonds which hold atoms together: covalent, ionic and metallic. For example, the water molecule has two atoms of hydrogen covalently bonded to an atom of oxygen. However, there must be other forces which hold one water molecule to the next. Otherwise the water molecules would not 'stick together' to form a liquid and water would only exist as a gas. These intermolecular forces are the subject of this chapter. Do not confuse the intermolecular forces *between* molecules with the covalent bonds between atoms *within* molecules. Covalent bonds are much stronger than intermolecular forces—water needs to be heated to only 100 °C (373 K) to break the intermolecular forces and separate the molecules, but much higher temperatures are needed to break the bonds holding the hydrogen atoms to the oxygen atoms.

Intermolecular forces exist between all molecules and atoms. Even the separate atoms of the inert gases can form liquids. There are three distinct types of intermolecular forces called hydrogen bonding, dipole-dipole bonding and van der Waals bonding. The first two act only between certain types of molecules. van der Waals forces act between all atoms and molecules.

Intermolecular forces are responsible for the existence of the liquid state and also govern the processes of mixing, dissolving and boiling.

16.2 Intermolecular forces

16.2.1 Dipole–Dipole forces

In the hydrogen chloride molecule, **Figure 16.1**, the hydrogen and chlorine molecules differ in electronegativity. The shared electrons 'feel' one proton

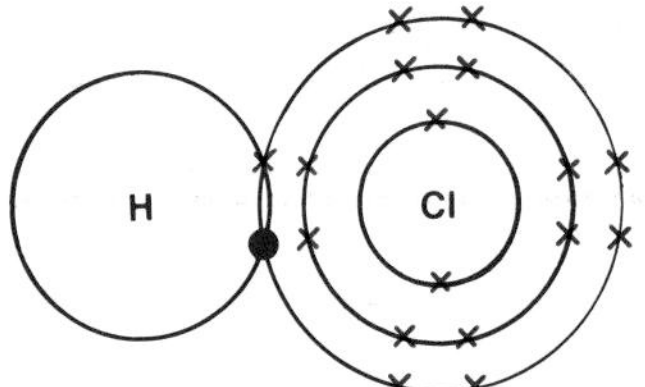

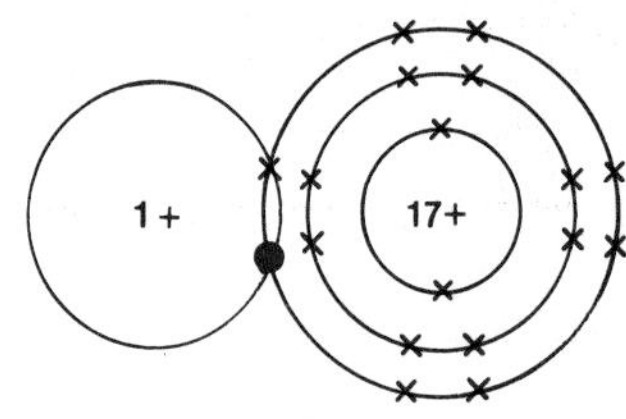

Figure 16.1 Dot cross diagram for hydrogen chloride

attracting them on the hydrogen and seven (after allowing for shielding) on the chlorine (see **section 10.2.2**). Thus the electrons are unequally shared, being pulled towards the chlorine atom rather than the hydrogen atom. The molecule is therefore written:

$$H^{\delta+}—Cl^{\delta-}$$

and is said to have a **dipole**.

Electronegative elements are found in the area of the Periodic Table

Figure 16.2 The most electronegative elements

N 3.0	O 3.5	F 4.0	
	S 2.5	Cl 3.0	
		Br 2.8	
		I 2.5	

marked in **Figure 16.2**. The electronegativity values are given in brown—the larger the number, the more electronegative is the element.

Two molecules which both have dipoles will always attract one another, see **Figure 16.3**. Whatever their original orientation, the molecules will 'flip' to give arrangement A, where the two molecules attract.

Figure 16.3 Two polar molecules will always attract one another

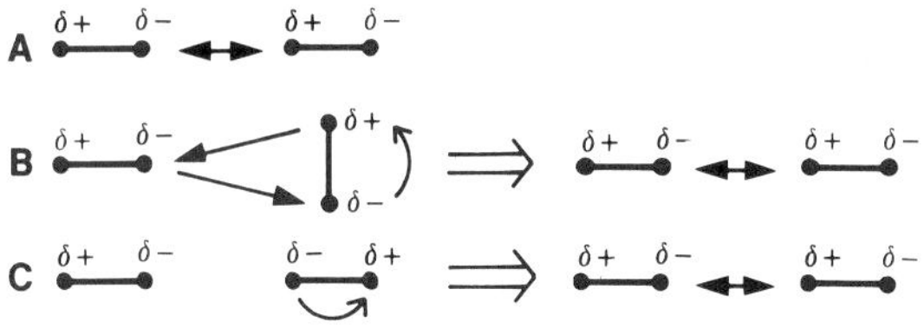

Examples of molecules which exhibit dipole–dipole attraction include:

$^{\delta+}H - Cl^{\delta-}$ hydrogen choloride

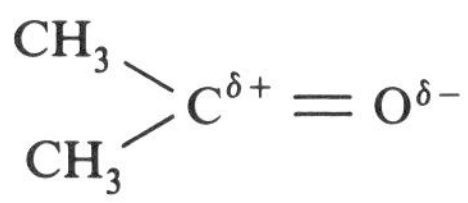

$(CH_3)_2C^{\delta+} = O^{\delta-}$ propanone

$(Cl^{\delta-})_3C^{\delta+} - H$ trichloromethane

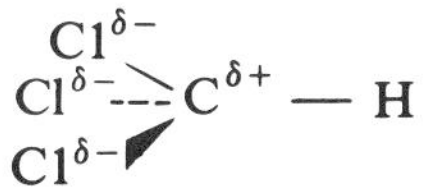

16.2.2 van der Waals forces

Imagine a helium atom which has two positive charges on its nucleus and two electrons. The atom is electrically neutral, but at any instant its charge distribution may not be symmetrical. Any of the arrangements in **Figure 16.4** could occur *at any instant*, or their equivalents with the electrons being thought of as 'charge clouds'. Any of the arrangements in **Figure 16.4** will result in the atom having a dipole at that instant. An instant later, the dipole may be in a different direction, or even not exist, but the overwhelming likelihood is that at any instant the atom will have a dipole. This instantaneous dipole can affect the electron distribution in nearby atoms, so that they too are distorted, **Figure 16.5**. The result of this

Figure 16.4 Possible arrangements of the electrons in helium

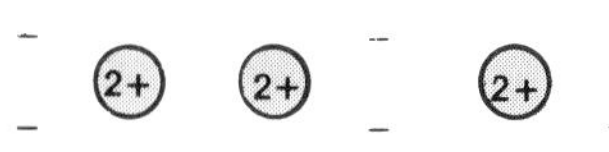
2+
2+
2+

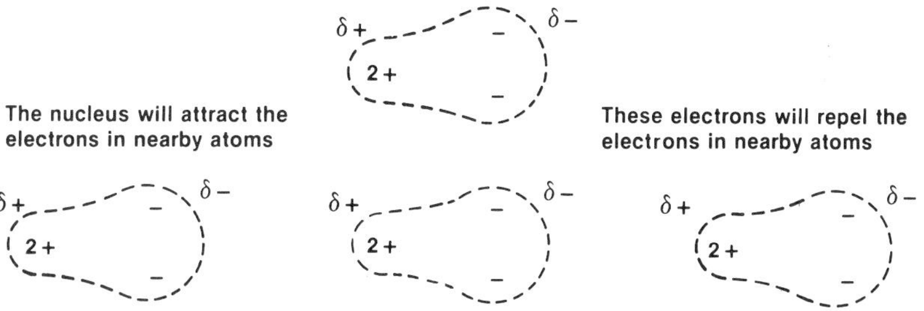
The nucleus will attract the electrons in nearby atoms
These electrons will repel the electrons in nearby atoms
δ+
δ−
2+

Figure 16.5 Instantaneous dipoles induce dipoles in nearby atoms

is to *induce* dipoles in nearby atoms which will be *attracted* to the original dipoles. As the electron distribution of the original molecule changes, it will induce new dipoles in the atoms around it, but they will always be attracted to the original atom. These forces are sometimes called **instantaneous dipole-induced dipole forces**, but this is rather a mouthful. The more usual name is van der Waals forces. These forces are very weak but act between all atoms or molecules, in addition to any other intermolecular forces. The ease of distortion of the electron cloud, and therefore the size of the van der Waals forces, increases with the number of electrons present. This is why the boiling temperatures of the inert gases increase with their atomic numbers, **Figure 16.6.** This is also the reason for the increase in boiling temperatures with increased chain length for hydrocarbons, where the greater number of electrons increases the van der Waals forces.

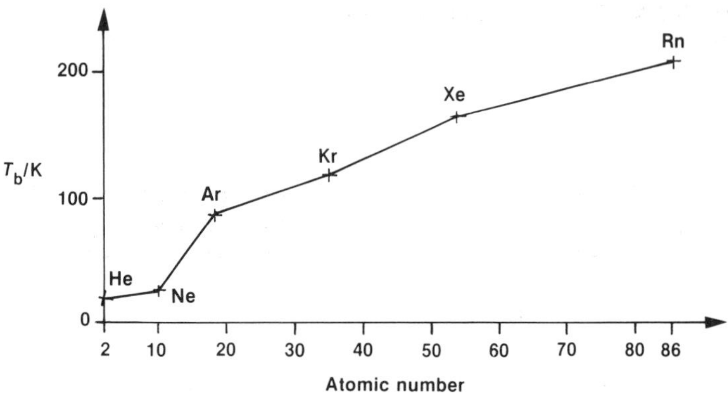

Figure 16.6 Boiling temperature against atomic number for the inert gases. The increased boiling temperatures are caused by the greater numbers of electrons

16.2.3 Hydrogen bonding

Think of the water molecule, H_2O. Oxygen is much more electronegative than hydrogen and so water is polar:

$\delta-$ O
$\delta+$ H H $\delta+$ dipole - dipole attraction
$O^{\delta-}$ — $H^{\delta+}$
$H^{\delta+}$

We should expect dipole–dipole attractions as shown. However, in this case, there is more to it than that, for two reasons:

1. the oxygen atoms have lone pairs;
2. the hydrogen atom has almost lost its shared electron to the oxygen to which it is covalently bonded, leaving it with none. The exposed proton which remains has a very strong electric field due to its small size. The other oxygen can begin to form a dative covalent bond (see **section 10.2.2**) with the hydrogen, using one of its lone pairs:

$O^{\delta-}$ — $H^{\delta+}$
$\delta+$ H $O^{\delta-}$
$\delta+$ H $H^{\delta+}$

The bond which is thus formed is considerably stronger than simple dipole–dipole interaction, although significantly weaker than a 'proper' covalent bond. Because of its small size, hydrogen is the only atom which can take part in this type of bonding—effectively acting as a bridge between two very electronegative atoms. **Hydrogen bonding only occurs between a very electronegative atom and a hydrogen covalently bonded to another very electronegative atom.** The only atoms electronegative enough are fluorine, oxygen and nitrogen, although chlorine is a borderline case.

One exception to the rule is the group:

$\delta-$ Cl
$\delta-$ Cl — $C^{\delta+}$ —
$\delta-$ Cl

The combined effect of the three chlorines, all quite electronegative, makes this group almost as electronegative as a nitrogen, oxygen, or fluorine atom, so that trichloromethane can form hydrogen bonds with, for example, propanone:

$Cl^{\delta-}$

$Cl^{\delta-}$ — C — $H^{\delta+}$ - - - $O^{\delta-}$ = $C^{\delta+}$(CH_3)(CH_3)

$Cl^{\delta-}$

hydrogen bond

A hydrogen bond could form between water, H_2O and ammonia, NH_3.

$^{\delta+}H$ — O^{δ} — $H^{\delta+}$ - - - $N^{\delta+}$($H^{\delta+}$)($H^{\delta+}$)($H^{\delta+}$)

hydrogen bond

Hydrogen bonds are usually represented by a dashed line, – – –. The hydrogen bonds formed with fluorine as the electronegative atom are the strongest, because fluorine is the most electronegative atom. That between two molecules of hydrogen fluoride, HF, has a bond energy of $125\,kJ\,mol^{-1}$. Average hydrogen bond energies are around $25\,kJ\,mol^{-1}$. This should be compared with $300–400\,kJ\,mol^{-1}$ for typical covalent bonds, see **Table 16.1**.

Table 16.1 Strengths of intermolecular forces compared with covalent bonds

Type of bond	Typical bond energy/$kJ\,mol^{-1}$	Acts between
Hydrogen bonding	Approx. 25 (max 125) ↑	Electronegative atom and hydrogen atom covalently bonded to electronegative atom
Dipole–dipole	between these ↓	Two molecules, both with permanent dipoles
van der Waals	Less than 1	All the above
(Covalent	Typically 300–400	Within molecules)

16.2.4 Examples of hydrogen bonding

The consequences of hydrogen bonding are considerable. On several counts, life as we know it would not exist but for hydrogen bonding.

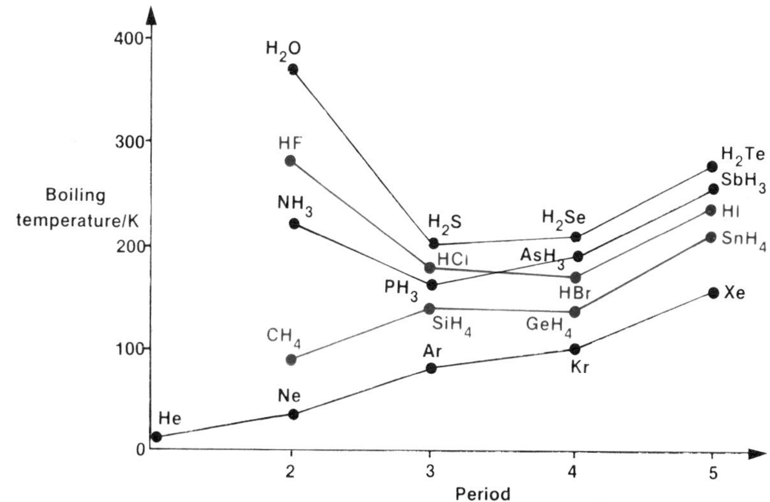

Figure 16.7 Graph of boiling temperatures against period number for some hydrides and the inert gases

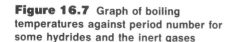

Boiling points of hydrides

The graph, **Figure 16.7**, shows the boiling points of the inert gases and the hydrides of elements of Groups IV, V, VI and VII plotted against the period number. The inert gases and the Group IV hydrides show a gradual decrease in boiling temperature going from Period 5 to Period 1, as we should expect. These substances have van der Waals forces only—no hydrogen bonding as there are no hydrides for the inert gases and no

Group IV element is electronegative enough to be involved in hydrogen bonding. The same trend is observed for most of the Group V, VI and VII hydrides but hydrogen fluoride, water and ammonia show a marked deviation from the trend. This is because these three hydrides can form hydrogen bonds, making them much more difficult to boil (boiling involves separating the molecules and thus breaking intermolecular bonds). Simple extrapolation of the graph predicts the boiling point of water to be about 200 K without hydrogen bonding, compared with the actual value of 373 K. But for hydrogen bonding, water would be a gas at room temperature. There are two other points of interest on the graph. Firstly, the boiling point of water seems to be more affected than that of hydrogen fluoride. This seems odd as fluorine is more electronegative than oxygen, so we should expect hydrogen bonding in hydrogen fluoride to be stronger than in water. The explanation is simple. To form a hydrogen bond, we need a hydrogen atom *and* a lone pair. It is the numbers of these that are significant:

HF gas, $T_b = 293\,K$:F—H (three lone pairs on F)

Hydrogen fluoride can form only one hydrogen bond per molecule as it has only one hydrogen atom.

H_2O liquid, $T_b = 373\,K$:O(H)H (two lone pairs on O)

Water can form two hydrogen bonds per molecule as it has two hydrogen atoms and two lone pairs.

NH_3 gas, $T_b = 240\,K$:N(H)(H)H (one lone pair on N)

Ammonia can form only one hydrogen bond per molecule as it has only one lone pair. So water can form more hydrogen bonds than hydrogen fluoride, leading to a three-dimensional network of water molecules.

Secondly, look at the graph for Group VII hydrides. Hydrogen chloride's boiling temperature is slightly greater than hydrogen bromide's. This suggests that chlorine is almost electronegative enough to form hydrogen bonds. In fact, we usually think of hydrogen chloride as having strong dipole–dipole bonds, but it is a borderline case.

The structure and density of ice

We have just seen that the oxygen atom in water can form *two* hydrogen bonds per molecule as well as its two covalent bonds. In the liquid state, the hydrogen bonds, being quite weak, break easily as the molecules are moving about. When water freezes, the water molecules are no longer free to move about and the hydrogen bonds form permanently. The resulting three-dimensional structure resembles the structure of diamond, see **Figure 16.8** and below:

H
|
—H···O—H
—H···

The molecules in this structure are slightly less closely packed than in liquid water so ice is less dense than water and forms on top of ponds rather than at the bottom—thus enabling fish to survive through the winter and life to continue in the relative warmth of the water under the ice during the Ice Ages.

Figure 16.8 The three-dimensional network of covalent bonds (blue) and hydrogen bonds (brown) in ice

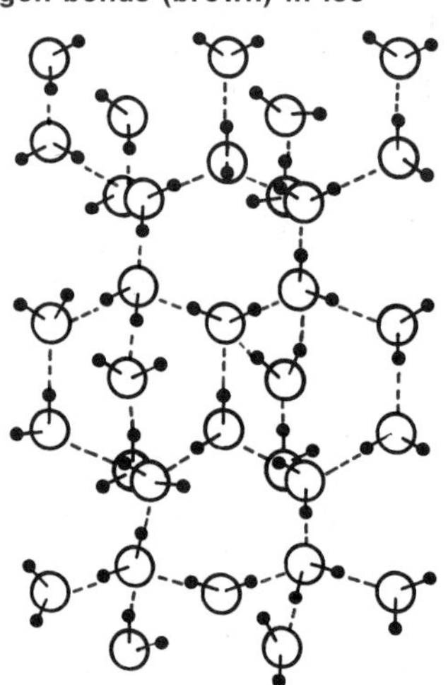

The viscosity of alcohols

Viscosity is a measure of the 'treacliness' of liquids—how resistant they are to flowing. When liquids flow, the intermolecular bonds between them are broken. The more hydrogen bonding between comparable molecules, the more viscous the liquid will be.

Propan-1-ol, C_3H_7OH, can form one hydrogen bond per molecule, as only the hydrogen bonded to the oxygen can hydrogen bond:

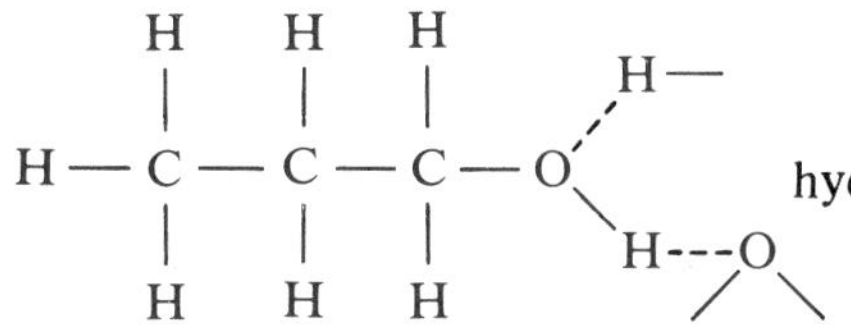

```
     H   H   H
     |   |   |       H—
                    .
H — C — C — C — O           hydrogen bond
                    \
     |   |   |       H---O
     H   H   H          / \
```

Propan-1,2-diol is more viscous as it can form two hydrogen bonds per molecule:

```
           H   H   H
           |   |   |
      H — C — C — C — H
           |   |   |
           H   O   O
  \           / :  : \         /
   O ---H      H   H   H---O
  /            |   |         \
```

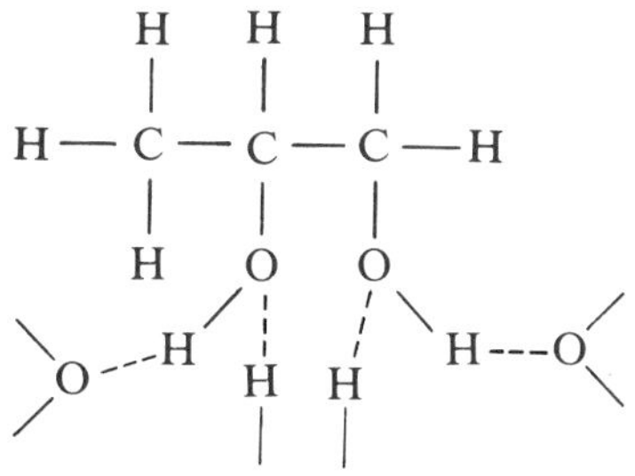

Propan-1,2,3-triol (glycerine) forms three hydrogen bonds per molecule and can therefore form three-dimensional hydrogen bonded networks of molecules. It is about as viscous as treacle.

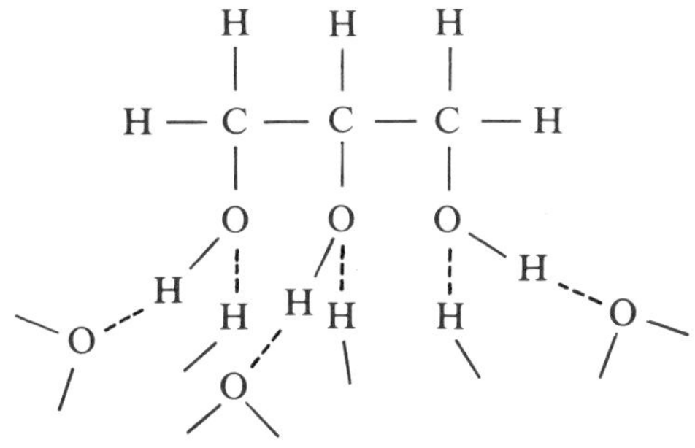

```
         H     H     H
         |     |     |
    H — C  —  C  —  C — H
         |     |     |
         O     O     O
        / :   / :   : \
      H   H  H  H   H   H
  \  .   / .    \     \   .  O —
   O      O                    /
  /      / \
```

Dimerization

If the relative molecular mass of ethanoic acid is measured in the gas phase, by the method described in **section 15.3.2**, a value of 120 is found, while a measurement carried out on the acid in aqueous solution (by, for example, titrating a weighed amount with standard sodium hydroxide) gives a value of 60. The molecules seem to have formed pairs or **dimers** in the gas phase. *Two* hydrogen bonds can form as shown:

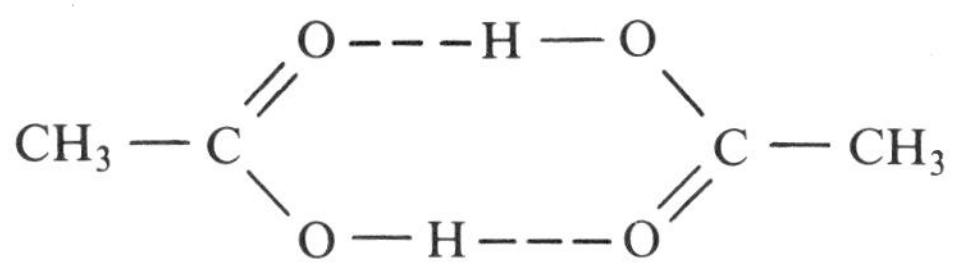

```
           O---H—O
          //      \
$CH_3$ — C          C — $CH_3$
          \        //
           O—H---O
```

Their combined effect holds the molecules together as pair.

Dimers are also formed when carboxylic acids are dissolved in solvents like hexane, which cannot form hydrogen bonds. When the acids are dissolved in water, the water molecules form hydrogen bonds to the acid as shown:

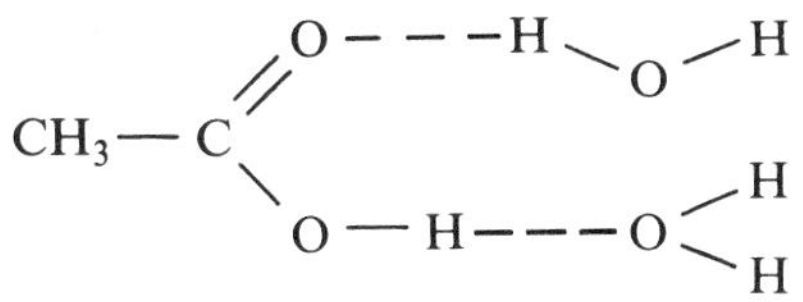

```
              O — —H       H
            //       \ O /
$CH_3$— C
            \                H
              O—H——O <
                             H
```

Since there are far more water than acid molecules, dimer formation is unlikely.

Figure 16.9 The double helix of DNA is held together by hydrogen bonds

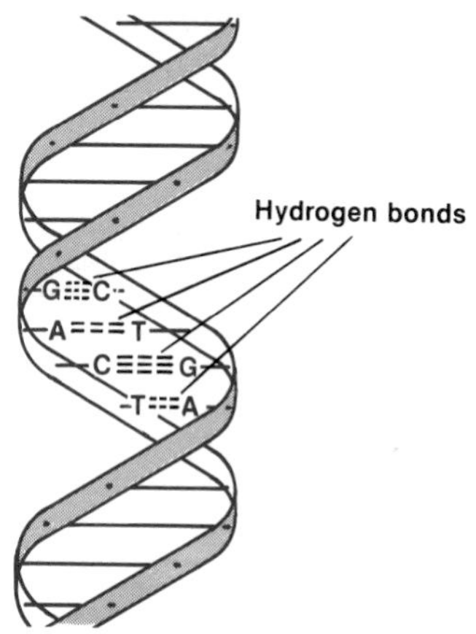

DNA and replication

DNA (deoxyribonucleic acid) is the molecule which stores the genetic information which makes offspring resemble their parents. The molecule exists as a double-stranded helix, the two spirals being linked by hydrogen bonds. When cells divide, or replicate, the two strands of the double helix break apart. Each strand then acts as a pattern for a new helix to build around it as an identical copy of the first. The weakness of hydrogen bonds relative to covalent bonds allows them to break easily in conditions which do not damage the rest of the molecule.

Surface tension

Surface tension is the 'skin' effect on the surface of liquids. The surface tension of water is unusually large, enabling a needle to be supported by the water's surface, if it is lowered carefully. Water's high surface tension is caused by hydrogen bonding, which holds together very strongly the molecules on the surface of the water. Liquids with high surface tensions tend to form droplets with a small surface area, rather than spreading out and wetting surfaces.

Surface tension allows a needle to 'float' on water

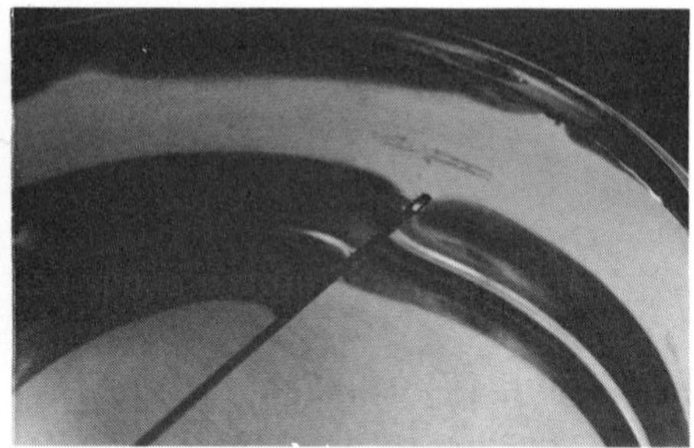

Intramolecular hydrogen bonds

Intra means 'within'. Intramolecular hydrogen bonds form between atoms *within the same* molecule. This can affect the properties of the molecule considerably. The molecules 4-nitrophenol and 2-nitrophenol are a good example, **Figure 16.10**. In 2-nitrophenol, an **intra**molecular hydrogen bond forms as shown. This means the hydrogen is not free to form hydrogen bonds with other 2-nitrophenol molecules. The O—H group in 4-nitrophenol is not close enough to the Os of the $—NO_2$ group to

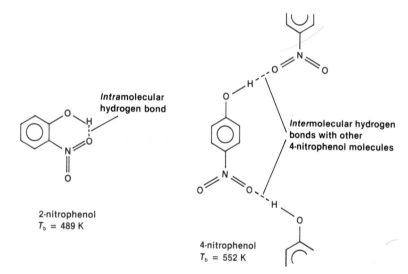

*Intra*molecular hydrogen bond

2-nitrophenol
T_b = 489 K

*Inter*molecular hydrogen bonds with other 4-nitrophenol molecules

4-nitrophenol
T_b = 552 K

Figure 16.10 (right) Intramolecular hydrogen bonding in 2-nitrophenol compared with intermolecular hydrogen bonding in 4-nitrophenol

hydrogen bond. In this molecule, **inter**molecular hydrogen bonds form, so 4-nitrophenol is more soluble than 2-nitrophenol—it can form hydrogen bonds with water while 2-nitrophenol cannot. Also, 2-nitrophenol, with fewer bonds between the molecules, melts at a lower temperature than 4-nitrophenol.

16.3 Mixtures and solutions

Since molecules are in constant random motion, substances tend to mix to form solutions by chance alone. There are far more molecular arrangements corresponding to mixing than to separate components. It is like a constantly shuffled pack of cards—random arrangements are vastly more probable than those where all the cards appear in their suits. So mixing involves a positive entropy change (see **section 9.6.1**).

16.3.1 Mixtures of liquids

A very general rule of mixing is 'like dissolves like'. This really refers to the type of intermolecular forces between the molecules. Liquids with similar intermolecular forces will mix. Ethanol and water, both of which form hydrogen bonds, will mix. Hexane and tetrachloromethane, which are held together by van der Waals forces, will mix. Water will mix with neither hexane nor tetrachloromethane. The explanation is as follows. In order to separate water molecules, we must supply energy to break fairly strong hydrogen bonds. If water and hexane mixed, no new bonds of comparable strength would form between the water and the hexane. In the case of ethanol and water, hydrogen bonds must be broken to separate the original molecules but bonds of comparable strength form between water and ethanol. With hexane and tetrachloromethane, we must break weak van der Waals bonds in the pure liquids but, on mixing, bonds of comparable strength form. Substances with dipole–dipole bonds tend to be intermediate and will often mix with both hydrogen-bonded and van der Waals-bonded substances. Propanone is a good example. It will mix quite well with both water and hexane. The size of a molecule may have an effect on how well it mixes. Methanol, CH_3OH, ethanol, C_2H_5OH and propanol, C_3H_7OH all mix completely with water because of hydrogen bonding between the —OH group and water molecules. Longer chain alcohols mix less well as their behaviour is dominated by the hydrocarbon chain, which cannot hydrogen bond. However, the addition of a second polar group may improve the solubility once more so, for example, glucose is very soluble, see **Figure 16.11**.

Figure 16.11 Glucose has five OH groups and is very soluble in water

```
        OH   H    OH   OH   OH
 O      |    |    |    |    |
  \\ C— C —  C —  C —  C —  C — H
 H /    |    |    |    |    |
        H    OH   H    H    H
```

Enthalpy changes on mixing

When a pair of liquids is mixed, there is often an enthalpy change, i.e. heat is given out or taken in. The observed enthalpy change is the difference between the enthalpy which has to be *put in* to break the intermolecular bonds in the separate liquids and the enthalpy *given out* when new intermolecular bonds are formed in the mixture.

E

THE ENTHALPY CHANGE ON MIXING TRICHLOROMETHANE AND ETHYL ETHANOATE

A trichloromethane molecule cannot form hydrogen bonds with other molecules of the same sort, nor can ethyl ethanoate. Each can form only dipole–dipole bonds:

```
        Clδ-                         Oδ-
        |                            ||
Clδ- —  C — Hδ+           CH3 — Cδ+ — O — C2H5
        |
        Clδ-
```

trichloromethane **ethyl ethanoate**

However, when mixed, hydrogen bonds can form between them:

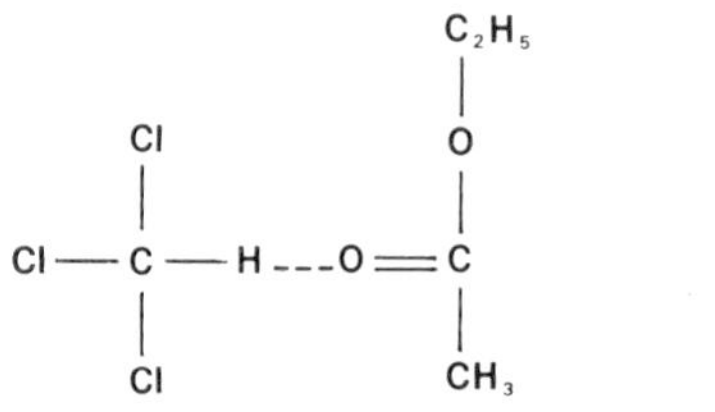

Remember that Cl_3C—can act as a pseudo-electronegative atom (see **section 16.2.3**). Since hydrogen bonds form and only weaker dipole–dipole bonds break, $\Delta H^\ominus$ for this reaction gives an approximate value for the hydrogen bond energy.

If 11.95 g (0.1 mol) of trichloromethane and 8.8 g (0.1 mol) of ethyl ethanoate are mixed, a temperature rise of 10 K is observed.

$$\text{heat} = m \times c \times T \quad (c\text{, the SHC of the mixture, is } 1.5\,\text{J g}^{-1}\text{K}^{-1})$$
$$= 20.75 \times 1.5 \times 10\,\text{J}$$
$$= 311.25\,\text{J}$$

so

0.1 mol gives 311.25 J

and

1 mol gives 3112.5 J or 3.1 kJ

The bond energy of the hydrogen bond is 3.1 kJ mol^{-1}.

This is a lower figure than most hydrogen bonds. Note—it is possible that not all the molecules form hydrogen bonds under these conditions.

A

DETERGENTS AND SOAPS

Detergent molecules are often described as 'tadpole shaped'. They have a long hydrocarbon 'tail' and an ionic or polar 'head'. The 'like dissolves like' rule indicates that the hydrocarbon end of the molecule should mix with other non-polar molecules such as grease, while the head should dissolve well in water.

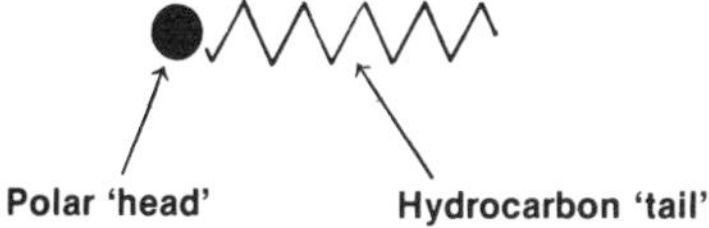

Detergent molecules work in two ways. Firstly, they tend to congregate at the surface of the water, with their heads in the water and tails out of it. This disrupts the

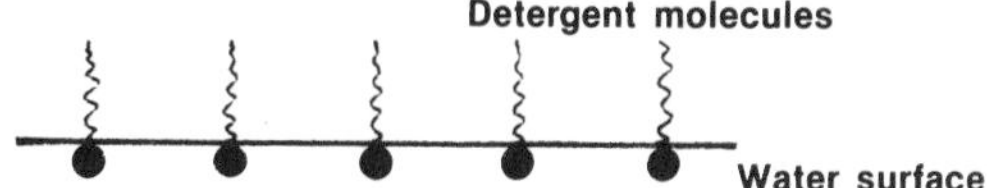

hydrogen bonding at the surface of the water, reducing the surface tension and allowing the water to wet the surface better. You can see this if you sprinkle water on the surface of a non-stick frying pan. The water collects in droplets rather than wetting the whole surface of the pan. Add a little washing up liquid and the water spreads out over the whole surface of the pan.

Secondly, detergent molecules enable grease to mix with water. Grease is what holds most dirt to surfaces like clothes and skin. The hydrocarbon end of the detergent molecule dissolves in the grease and the polar head pulls the grease into solution.

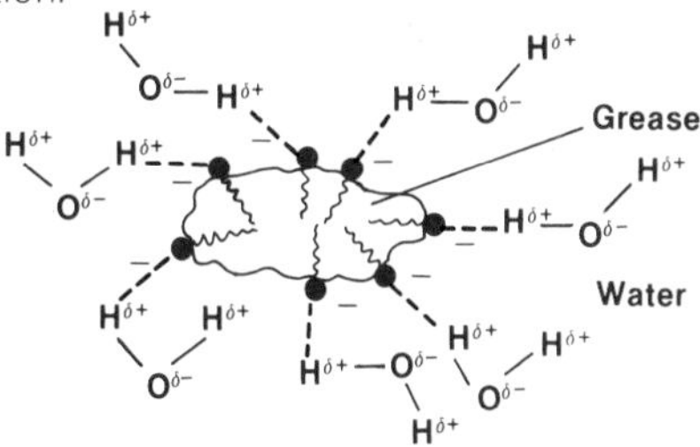

Detergents are of three types: anionic, cationic and non-ionic. Examples of anionic detergents include:

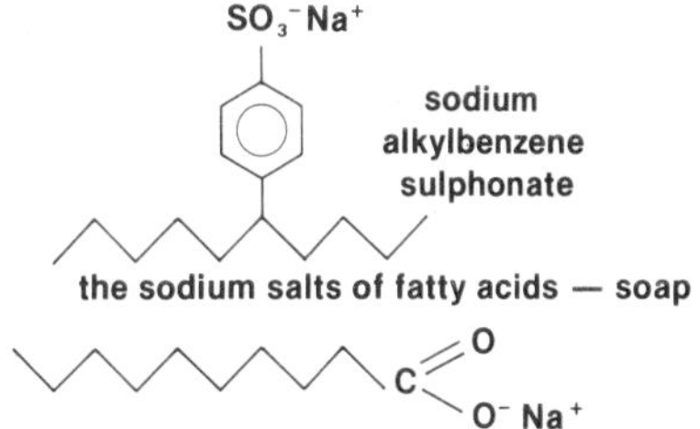

the sodium salts of fatty acids — soap

Cationic detergents are less common and cannot be mixed with anionic detergents as the two react together to form a scum. They are, however, used in fabric conditioners. An example of a cationic detergent is:

Non-ionic detergents are often combined with anionic detergents in, for example, washing powders. They do not produce foam and are unaffected by acids and alkalis. One example is the polyethene oxide shown below:

A typical washing powder contains anionic and non-ionic detergents as well as a number of other ingredients.

Sodium tripolyphosphate acts as a buffer to control the pH and also as a sequestering agent to remove from the washing water ions such as Mg^{2+} and Ca^{2+} which cause hardness. Sodium perborate reacts with water to form hydrogen peroxide—a bleach.

Other ingredients include a fluorescent or optical brightener, perfume, dye and sodium sulphate to make the powder flow freely.

16.3.2 Boiling points and vapour pressures of mixtures of liquids

Remember (from **section 11.2**) that all liquids have a vapour pressure, caused by molecules escaping at their surface. Liquids with low boiling temperatures have high vapour pressures and vice versa. The vapour pressure increases more and more rapidly with increasing temperature and when the vapour pressure reaches atmospheric pressure, the liquid boils.

Raoult's law

This states that each liquid in a mixture contributes to the total vapour pressure. It does so in proportion to **a)** its vapour pressure when pure and **b)** its mole fraction in the mixture.

For example, in a mixture of two liquids A and B, of vapour pressures p_{0A} and p_{0B} respectively, n_A moles of A are mixed with n_B moles of B.

The vapour pressure p_A of A above the mixture is given by:

A

STAIN REMOVAL AND DRY CLEANING

Removing stains from clothes is sometimes considered an art, and a folklore of tips and remedies exists advising how to remove particular stains from different materials. The chemistry of stain removal is straightforward. To dissolve out a stain, we need to find a solvent which forms stronger intermolecular forces with the stain than the stain does with the material. This is just an application of the 'like dissolves like' rule. So greasy stains are usually removable by treatment with a non-polar solvent like 1, 1, 1-trichloroethane and tar is removed by dabbing with petrol. Stains made by polar or ionic materials will be best removed by hydrogen bonding solvents such as water or methylated spirits (ethanol with added methanol and dye to make it unsuitable for drinking).

The type of cloth will also affect the ease of stain removal. Wool, for example, is a protein containing many NH and CO groups to which hydrogen bonds can form, so it stains fairly easily with polar stains (like other proteins). Other materials may have fewer polar groups, and for example, poly(propene) (from which carpets may be made) has only non-polar CH bonds. Some methods of stain removal involve chemical reactions such as bleaching with various oxidizing agents like hydrogen peroxide or sodium chlorate(I) (sodium hypochlorite—household bleach). Another interesting tip is to use salt on red wine stains or blood stains. This is thought to work by osmosis—water moving from the cloth to the salt and taking the stain with it.

Many materials shrink in water. For these, dry cleaning is used. Here chlorinated hydrocarbons such as 1, 1, 1-trichloroethane are used. They are much less polar than water and will dissolve grease directly without the use of a detergent. Small quantities of water are added to remove polar stains like salt and sugar, but the quantities used are too small to cause shrinkage of fabrics.

Commercial stain removers

$$p_A = \frac{n_A}{(n_A + n_B)} \times p_{0A} \text{ where } \frac{n_A}{(n_A + n_B)} \text{ is the \textbf{mole fraction} of A}$$

and the vapour pressure p_B of B above the mixture is given by:

$$p_B = \frac{n_B}{(n_A + n_B)} \times p_{0B} \text{ where } \frac{n_B}{(n_A + n_B)} \text{ is the \textbf{mole fraction} of B}$$

This means that, at any temperature, the total vapour pressure over the mixture is just the weighted average of the vapour pressures, at that temperature, of the pure liquids.

For example, in a mixture at a given temperature containing 2 moles of A of $p_{0A} = 60\,\text{kPa}$ and 1 mole of B of $p_{0B} = 30\,\text{kPa}$,

$$p_A = \frac{2}{2+1} \times 60 = 40\,\text{kPa} \quad \text{and} \quad p_B = \frac{1}{2+1} \times 30 = 10\,\text{kPa}$$

so the total vapour pressure

$$p_{TOT} = p_A + p_B = 50\,\text{kPa}$$

Figure 16.12 **A vapour pressure–composition graph for a pair of liquids A and B**

Vapour pressure of *pure* B
Vapour pressure of *pure* A
Vapour pressure
p_{0B} p_B p_A p_{0A}
A 0% B 100% 50 50 100% A 0% B
Composition/mole %

A vapour pressure–composition graph at any given temperature would be a straight line joining p_{0A} to p_{0B}, **Figure 16.12**. The dotted lines are the separate contributions of A and B to the total vapour pressure, and at any composition their values will add up to the total vapour pressure.

We could also plot a graph of boiling temperature against composition **Figure 16.13**, because as we have already established, boiling temperatures are inversely related to vapour pressures.

Note: This graph slopes the other way as a liquid with a *high* vapour pressure has a *low* boiling temperature. Also, the line is not quite straight. Boiling temperatures are more easily measured than vapour pressures and the apparatus in **Figure 16.14** is used. The boiling point is taken when the mixture is boiling steadily. The condenser prevents vapour escaping.

Raoult's law assumes that the two liquids behave, when mixed, in the same way as they do when pure. Many pairs of liquids obey Raoult's law, in which case the graphs of vapour pressure or boiling temperature against

Figure 16.13 **A boiling temperature–composition graph for a pair of liquids A and B**

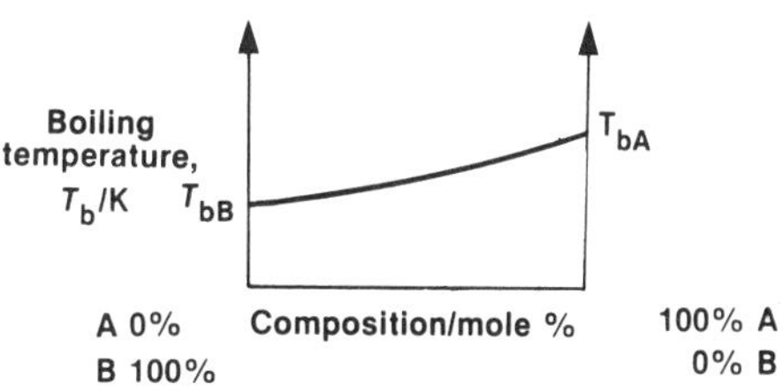

Figure 16.14 Apparatus for measuring boiling temperatures of mixtures

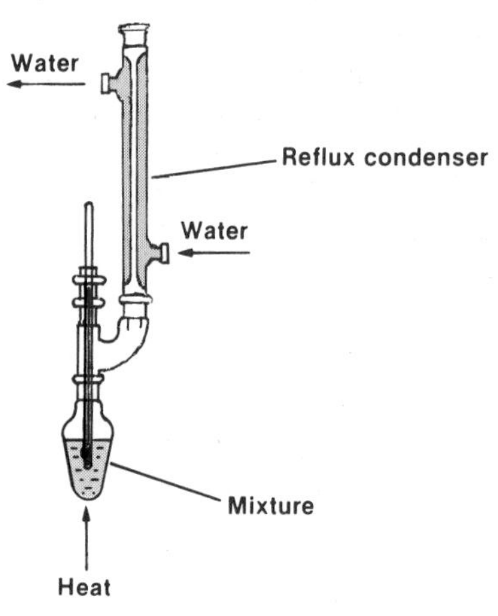

composition will be as shown in **Figures 16.12** and **16.13**. These pairs are those where the intermolecular forces are similar, e.g. propan-1-ol and propan-2-ol, both of which form hydrogen bonds, or hexane and heptane, both of which form van der Waals bonds only. Mixtures which obey Raoult's law are called **ideal mixtures**.

Deviations from Raoult's law

By no means all liquids obey Raoult's law. The boiling temperature–composition and vapour pressure–composition curves are shown in **Figure 16.15** for a mixture such as ethanol/hexane. Concentrate on the vapour pressure curves, as Raoult's law is expressed in terms of vapour pressure, not boiling point. The mixture shows a *higher* vapour pressure than Raoult's law predicts. This is called a *positive deviation* from Raoult's law. (This means that molecules escape from the liquid more easily than expected for an ideal mixture and so it has a lower boiling temperature.)

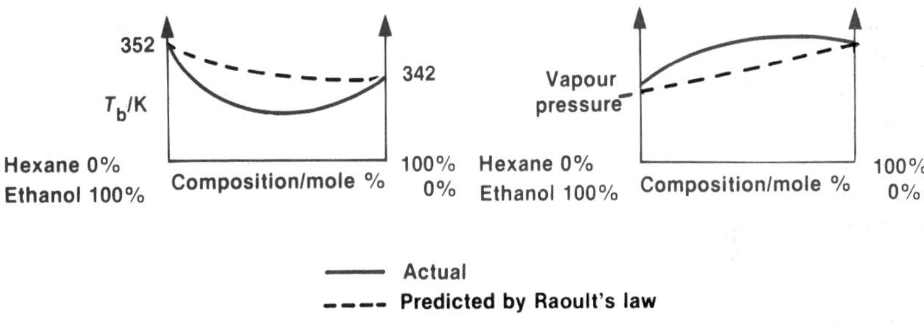

Figure 16.15 Boiling temperature–composition and vapour pressure–composition curves for mixtures of ethanol and hexane

This ease of escape into the vapour is caused by the hexane molecules, which cannot form hydrogen bonds, coming between the ethanol molecules and disrupting the hydrogen bonding, making it easier for ethanol molecules to escape into the vapour. The graphs in **Figure 16.16** are for a mixture such as trichloromethane/ethyl ethanoate. The mixture shows a *lower* vapour pressure than Raoult's law predicts. This is called a *negative deviation*. It means that it is harder for molecules to escape into the vapour than for an ideal mixture, so it has a *higher* boiling point. This is because hydrogen bonds form between ethyl ethanoate and trichloromethane, see box *The enthalpy change on mixing trichloromethane and ethyl ethanoate*, so molecules in the mixture are more strongly bonded than those in the pure liquids.

Figure 16.16 Boiling temperature–composition and vapour pressure–composition curves for mixtures of trichloromethane and ethylethanoate

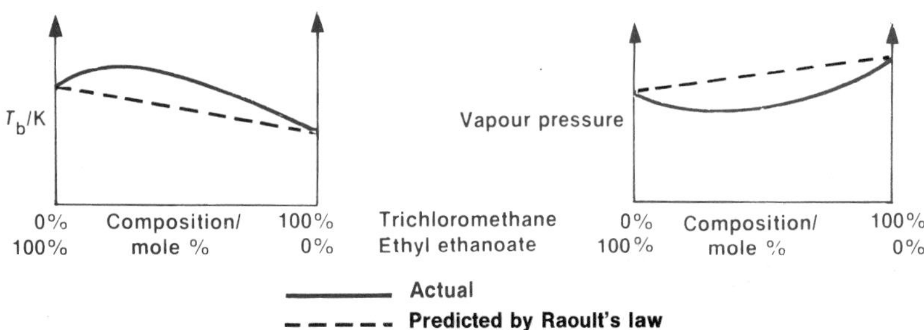

16.3.3 The separation of mixtures

Simple distillation can separate, for example, salt and water. The salt is completely involatile so the vapour contains only water, which can be condensed and thus separated. Fractional distillation allows the separation of two components differing only slightly in volatility.

Fractional distillation

For all mixtures of liquids (ideal or not), the vapour above the mixture will be richer in the more volatile component than the liquid. A composition–boiling temperature curve for two liquids A and B (B more volatile than A) is shown in **Figure 16.17** (overleaf). Notice that at any temperature the vapour mixture contains more B (the more volatile

Figure 16.17 Boiling temperature–composition curve for two liquids A and B

component) than the liquid mixture at the same temperature. This can be used as a method of separating mixtures, called **fractional distillation**.

Imagine starting with a liquid of composition W. This will boil at temperature T_W and give a vapour of composition X, which has more B and less A. If this vapour of composition X is condensed it will form a liquid which also has a composition of X. So, in one act of boiling, although we have not separated A and B, we have ended up with a liquid which is richer in B than the liquid with which we started. This new liquid will boil at temperature T_X (lower than T_W) to give a vapour of composition Y. This vapour could be condensed to give a liquid of composition Y with even more B and even less A. We could continue this process with more boiling/condensation cycles until we reach a vapour consisting entirely of pure B. A would all be liquid.

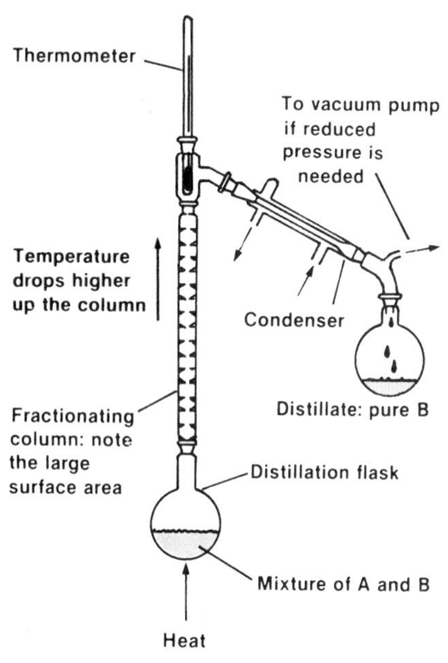

Figure 16.18 Separating two liquids by fractional distillation

To actually boil, condense, boil, condense, etc. would be very tiresome. Instead, a **fractionating column** is used which produces the effect of many successive boiling/condensation cycles automatically. The column sides have a large surface area, see **Figure 16.18**. The mixture of A and B boils in the flask and rises up the column. As the temperature drops, the vapour condenses, the liquid (richer in A) drops back into the boiling flask and the vapour (richer in B) moves up the column. The same thing happens many times as the vapour rises up the tube—effectively forming an enormous series of boiling/condensation cycles. A long enough column will result in a vapour of pure B at the top of the column and pure A left in the boiling flask.

Some liquids decompose before reaching their boiling temperatures. In this case, distillation can be done under reduced pressure. Since boiling

A

INDUSTRIAL FRACTIONATION OF CRUDE OIL

In the industrial fractional distillation of crude oil, the column is replaced by a series of trays. Bubble caps allow the vapour to rise up the column and overflow pipes let liquid descend. Vaporized crude oil is fed in continuously. Each tray contains a mixture of different composition corresponding to petrol, kerosene, etc. These are drawn off so that the process operates without stopping.

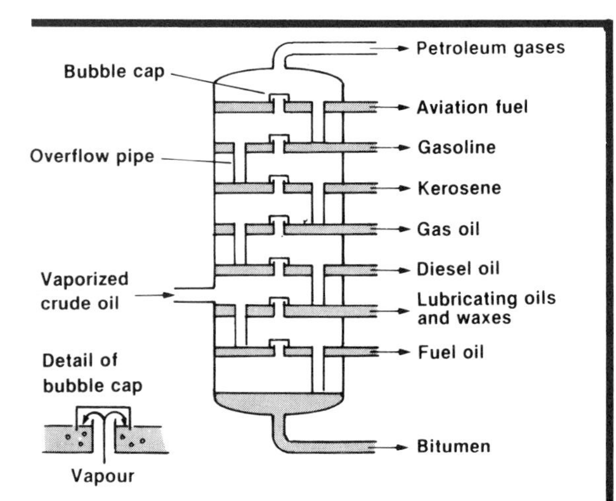

A

THE FRACTIONAL DISTILLATION OF AIR

Air consists of 78% nitrogen, 21% oxygen and 1% carbon dioxide, argon and other inert gases. In the UK 2 million tonnes of oxygen are produced annually, along with 700 000 tonnes of nitrogen (not including that used for ammonia production), 30 000 tonnes of argon, and smaller quantities of other inert gases.

Nitrogen is used as an inert atmosphere for various processes like food packaging and silicon chip manufacture. Argon is also used as an inert atmosphere for arc welding and casting, as well as the filling of electric light bulbs. Most oxygen is used in steel manufacture but also has uses in medicine and cutting torches. All these gases are produced by the fractional distillation of liquid air.

The air is filtered and compressed in several stages, being cooled between each stage. Water condenses out on cooling and carbon dioxide is removed by passing the air through sodium hydroxide. The air is then liquefied and fractionally distilled twice at different pressures. Nitrogen has the lower boiling temperature. Argon, which has a boiling temperature between those of oxygen and nitrogen, is separated at this stage too.

Crisp packets are filled with nitrogen to keep the crisps fresh

Figure 16.19 Steam distillation

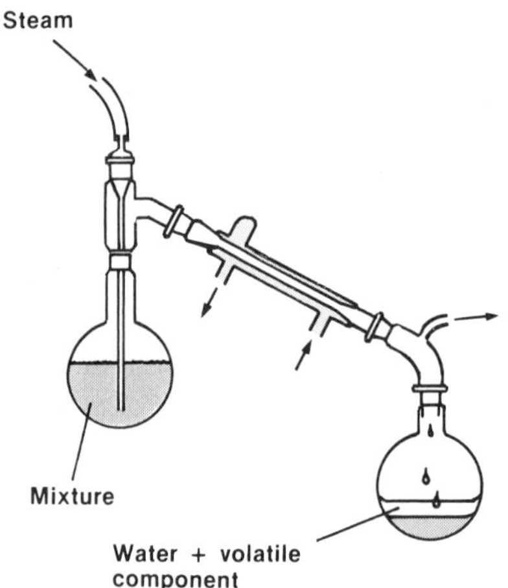

occurs when the vapour pressure of the liquid reaches the external pressure, boiling occurs at a lower temperature.

Another technique to solve the same problem is steam distillation.

Steam distillation

With two liquids which do not mix (are **immiscible**) both contribute their own vapour pressure to the mixture, provided they are stored so that both vapours can escape. The total vapour pressure of the mixture will be the sum of the vapour pressures of each component:

$$\text{total vapour pressure} = \text{vapour pressure A} + \text{vapour pressure B}$$

so the vapour pressure will be higher and the boiling temperature will be lower than that of either separate liquid.

Steam is passed into the mixture and heats it until it boils, **Figure 16.19**. Boiling occurs at a temperature below that of the boiling temperatures of either of the immiscible liquids. The vapour contains both steam and the vapour of the immiscible liquids. The vapour condenses to give a mixture of water and the volatile component. These can be separated with a separating funnel.

Steam distillation can be used to separate eucalyptus oil (boiling point 449 K) from eucalyptus leaves. The oil decomposes at its boiling point and so cannot be separated by ordinary distillation. Eucalyptus oil is used in cough mixtures, gargles and medicated soaps.

Figure 16.20 Solvation of cations, M^+ and anions, X^- by water molecules

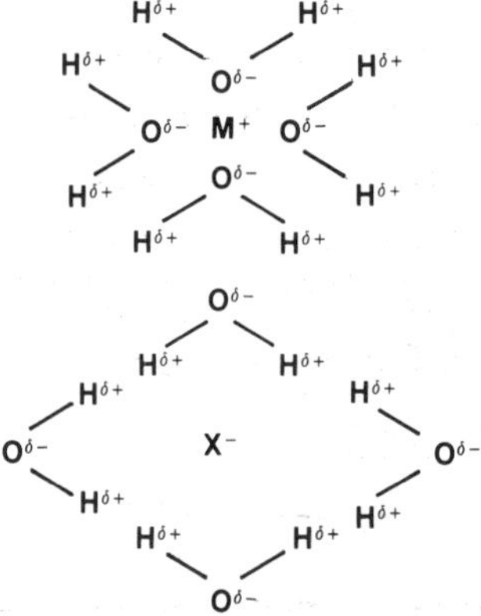

16.3.4 Solid/liquid mixtures—solutions

The 'like dissolves like' rule applies here too. Solids which form hydrogen bonds dissolve in hydrogen bonding solvents, so glucose dissolves in water. Iodine, I_2, molecules which are held together by van der Waals forces dissolve poorly in water but well in non-polar solvents like hexane.

Solubility of ionic compounds

Ionic compounds can only dissolve well in polar liquids. In order to dissolve an ionic compound, the lattice must be broken up. This requires the lattice energy (see **section 10.2.1**) to be put in. The separate ions are then **solvated** by solvent molecules. These cluster around the ions in such a way that positive ions are surrounded by the negative ends of the dipoles of the solvent molecules and the negative ions by the positive ends of the dipoles, **Figure 16.20**. This is called **hydration** when the solvent is water.

Solvation (or hydration) gives out energy, as opposite charges come together. This energy change is called the enthalpy change of hydration or solvation and given the symbol $\Delta_{\text{hydration}}H^{\ominus}$ or $\Delta_{\text{solvation}}H^{\ominus}$.

Enthalpy changes of hydration are approximately the same size as lattice energies for most ionic compounds and the process of dissolving can be represented on an enthalpy diagram, **Figure 16.21**. For example, for sodium chloride:

$$\text{lattice energy} = 771\ \text{kJ mol}^{-1}$$

and

$$\Delta_{\text{hydration}}H^{\ominus} = -770\ \text{kJ mol}^{-1}$$

Since $\Delta_{\text{hydration}}H^{\ominus}$ depends on the same factors as the lattice energy (i.e. both are larger for more highly charged ions), there is usually only a small enthalpy change of solution of ionic compounds. It may be either positive or negative. In the actual process of dissolving, the solvent molecules have two functions. Firstly, they solvate the ions as described above. Secondly, they come between the positive and negative ions and weaken the forces of attraction between them.

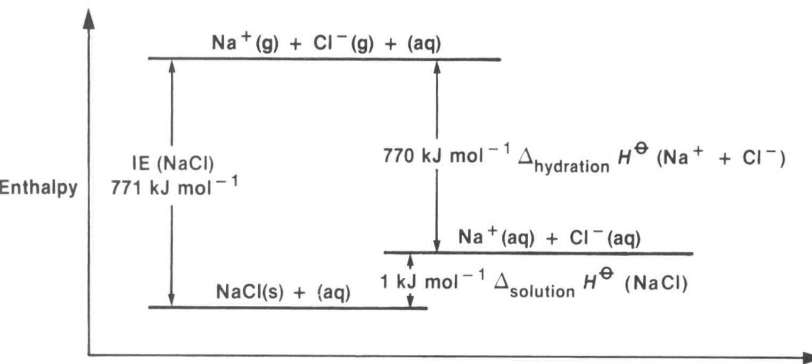

Figure 16.21 Enthalpy diagram for the dissociation of sodium chloride in water

16.3.5 Colligative properties of solutions

These are properties of solutions which depend on the *number* of solute particles in the solvent and do not depend on their nature, so a sodium ion, Na^+ has the same effect as a sucrose molecule, $C_{12}H_{22}O_{11}$. They are dealt with below.

Boiling temperature elevation and freezing temperature depression

One effect on dissolving a non-volatile (i.e. solid) solute in a liquid solvent is that the vapour pressure is reduced. This is because solute particles take up some of the surface of the solvent, and since they have no tendency to escape, they reduce the surface area available for the escape of solvent molecules. As they do not prevent molecules of solvent returning from the vapour, the vapour pressure is reduced. In **Figure 16.22**, the phase

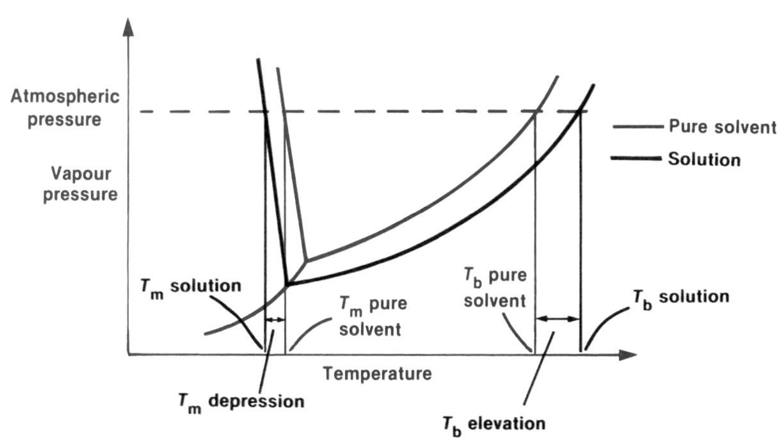

Figure 16.22 Phase diagram for a solution (brown) compared with that of pure solvent (blue)

Note that the freezing temperature and melting temperature are the *same*. The symbol T_m is used for both.

diagram (see **section 11.3**) of the solution is shown in brown compared with that of the pure solvent in blue. It can be seen that the presence of the solute both *increases* the boiling temperature and *decreases* the freezing temperature. Both these effects are found to be proportional to the *number* of solute particles present and are independent of the type of particle.

$$\text{boiling temperature elevation} \propto [\text{solute}]$$

or
$$\text{boiling temperature elevation} = k\ [\text{solute}]$$

where k is the boiling temperature constant (or **ebullioscopic constant**) for the *solvent*.

$$\text{freezing temperature depression} \propto [\text{solute}]$$

or
$$\text{freezing temperature depression} = k'\ [\text{solute}]$$

Note that both ebullioscopic and cryoscopic constants are properties of the *solvent*, not the solute.

where k' is the freezing temperature constant (or **cryoscopic constant**) for the *solvent*. Boiling temperature and freezing temperature constants are usually quoted in K kg mol^{-1}, so the concentrations of solutions must be measured in mol kg^{-1}. **Table 16.2** gives some examples.

Table 16.2 Cryoscopic and ebullioscopic constants for some solvents

Solvent	Boiling temperature constant/K kg mol^{-1}	Freezing temperature Constant/K kg mol^{-1}
Water	0.52	1.86
Benzene	2.53	3.90
Camphor	3.07	40.00

This means that a 1 mol kg^{-1} solution of *any* solute in water will increase the boiling temperature by 0.52 K. A 1 mol kg^{-1} solution of *any* solute in camphor will lower the freezing temperature by 40.00 K. Either method can be used to enable the number of moles of a solute to be found and hence the relative molecular mass.

For example, the freezing temperature of camphor is 452.5 K. A solution of 0.01 g of compound X is dissolved in 1 g of camphor. The solution's freezing temperature was 440.5 K. What is the relative molecular mass of X?

$$\begin{aligned}\text{freezing temperature depression} &= k' \times [\text{X}]\\ 12.0 &= 40 \times [\text{X}]\\ [\text{X}] &= 12/40\ \text{mol kg}^{-1}\\ &= 0.30\ \text{mol kg}^{-1}\\ &= 3 \times 10^{-4}\ \text{mol g}^{-1}\end{aligned}$$

$$\text{so } 3 \times 10^{-4}\ \text{mol is } 0.01\ \text{g of X}$$

$$1\ \text{mol X is contained in } \frac{0.01}{3 \times 10^{-4}} = 33.3\ \text{g}$$

$$M_r \text{ of X} = 33.3$$

Notice how the large freezing temperature constant of camphor gives a large freezing point depression which can be easily measured.

Figure 16.23 A semi-permeable membrane

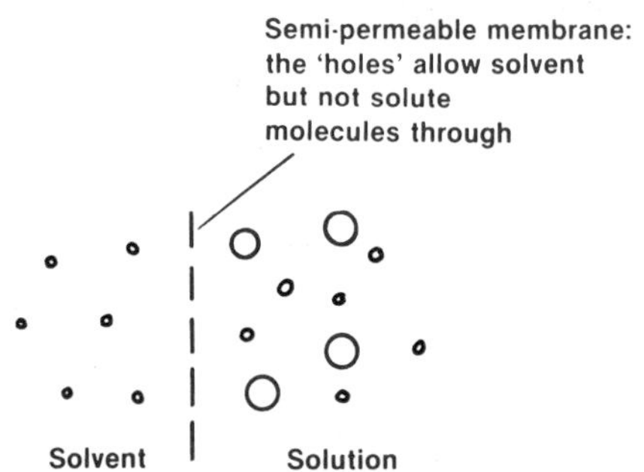

Figure 16.24 Osmosis

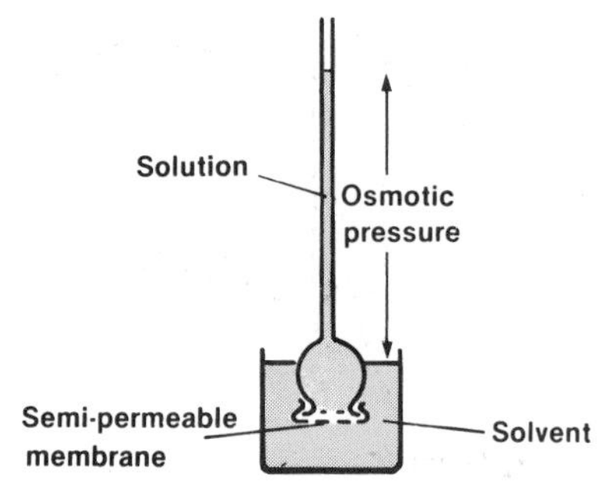

Osmotic pressure

A semi-permeable membrane is one through which solvent molecules can pass, but larger solute particles cannot, **Figure 16.23**. Animal and plant cell walls are semi-permeable membranes, and cellophane is a reasonably good one. If a pure solvent is separated from a solution by a semi-permeable membrane as in **Figure 16.24**, solvent moves through the semi-permeable membrane in an attempt to equalize the concentrations on either side of the membrane by diluting the solution. If a pressure is applied to the solution side of the membrane, osmosis can be stopped. This pressure is called the **osmotic pressure**, π. Osmotic pressure follows an equation similar to the ideal gas equation, see **section 15.2.6**.

$$\pi V = nRT$$

$$\pi = \frac{nRT}{V}$$

This allows n, the number of moles of solute, to be measured. The measurement of osmotic pressure can be used for substances of large relative molecular mass like polymers. These are too involatile to be measured by the gas syringe method described in **section 15.3.2**, they are often too insoluble to give easily measurable freezing temperature depressions and boiling temperature elevations and they frequently fragment in the mass spectrometer and give no molecular ion.

For example: the osmotic pressure at 298 K of a solution of protein (1.35 g in 100 cm^3) is 100 Pa. What is the relative molecular mass of the protein?

$\pi = 100\ \text{Pa}$, $V = 100\ \text{cm}^3 = 10^{-4}\ \text{m}^3$, $n = ?$, $R = 8.3\ \text{J K}^{-1}\ \text{mol}^{-1}$, $T = 298\ \text{K}$

$$\pi V = nRT$$

$$n = \frac{\pi V}{RT}$$

$$n = \frac{100 \times 10^{-4}}{8.3 \times 298}$$

$$n = 4.04 \times 10^{-6}\ \text{mol}$$

4.04×10^{-6} mol has a mass of 1.35 g

1 mol has a mass of 33 390 g

M_r of the protein is 33 390

E

COLLIGATIVE PROPERTIES

Freezing temperature depression, boiling temperature elevation and osmotic pressure all depend on the *number* of particles present in the solution. So substances which dissociate or associate will give an apparent M_r different from that expected, for example:

$$KCl \longrightarrow K^+ + Cl^-$$

gives twice as many particles as expected, therefore gives an apparent M_r of half that expected. The reaction:

$$MgCl_2 \longrightarrow Mg^{2+} + 2Cl^-$$

gives three particles, so the apparent M_r will be one-third that expected.

A

OSMOSIS

The walls of living cells are semi-permeable membranes and they thus allow water to be taken into cells. Cells contain a solution of salts so if a cell is placed in pure water, it will take in water by osmosis until it bursts. In a concentrated solution of salt, it will lose water and shrivel up. Red corpuscles in the blood do this, so any fluids injected into the blood must be of the same concentration as the contents of a red cell so that no osmosis occurs. Such solutions are called isotonic solutions. The phenomena of swelling and shrinking can be illustrated with an egg whose shell has been dissolved away by acid. The photograph shows eggs (without their shells) placed in distilled water (left) and concentrated salt solution (right).

The technique of reverse osmosis can be used to purify water, to make drinking water from sea water. Pressure is applied to the solute side of the membrane and forces pure water through from the solution. The solute particles are too large to pass through the semi-permeable membrane.

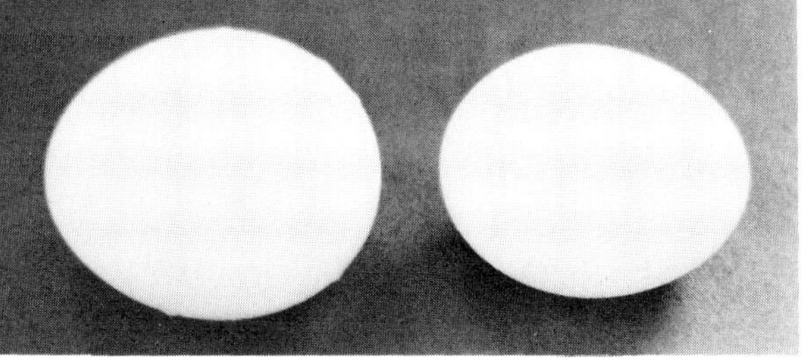

These two eggs started off the same size. The one on the left has been in distilled water, the one on the right in concentrated salt solution

16.4 Summary

- Intermolecular forces may be classified into three types:

1. Hydrogen bonding (between a hydrogen atom covalently bonded to a nitrogen, oxygen or fluorine atom and a nitrogen, oxygen or fluorine atom).
2. Dipole–dipole (between two molecules with permanent dipoles).
3. van der Waals forces (between all atoms—instantaneous dipole-induced dipole). Strength depends on the total number of electrons.

- These forces are responsible for the existence of the liquid state.
- Mixtures of liquids form if molecules of the two liquids are able to form bonds of approximately the same strength or greater strength than the bonds in the pure liquids.
- The general rule is 'like dissolves like'.
- Raoult's law states that the components of a mixture contribute to the total vapour pressure of the mixture in proportion to their mole fraction and to the vapour pressure of the pure liquids.

$$p_{\text{total}} = p_{0\text{A}}m_{\text{A}} + p_{0\text{B}}m_{\text{B}}$$ where p_0 is vapour pressure and m is mole fraction

- Raoult's law is obeyed by pairs of liquids with similar intermolecular forces.
- If there is *stronger* bonding in the mixture than in the pure liquids, a *negative* deviation will occur (*lower* vapour pressure than expected and *higher* boiling temperature).
- If there is *weaker* bonding in the mixture than in the pure liquids, a *positive* deviation (*higher* than expected vapour pressure and *lower* boiling temperature) will be observed.
- Liquids can be separated by fractional distillation which takes advantage of the fact that the vapour is richer in the more volatile component compared with the liquid.
- For immiscible liquids, the vapour pressure is the sum of the vapour pressures of each liquid provided each liquid has its surface exposed.
- This is the basis of steam distillation where a liquid component of a mixture may be distilled at below its boiling point.
- For solutions of solids in liquids the rule 'like dissolves like' also applies.
- Ionic compounds will dissolve in polar liquids. $\Delta_{\text{solution}}H^{\ominus}$ is small, being the difference between the lattice energy of the ionic crystal and $\Delta_{\text{solvation}}H^{\ominus}$, both of which are approximately the same size.
- Colligative properties depend on the number of solute particles, not their nature.
- They include freezing temperature depression and boiling temperature elevation, both of which are caused by the lowering of the vapour pressure of a solution which occurs due to the presence of solute particles near the surface.
- Both freezing temperature depression and boiling temperature elevation can be used to measure the relative molecular mass of the solute.
- Osmosis is the tendency of solvent to move through a semi-permeable membrane from pure solvent to solution. This tendency or osmotic pressure can also be used to measure relative molecular masses and is used for measuring relative molecular masses of polymers.

16.5 Questions

1 What type of intermolecular forces will predominate in the following liquids

A) ammonia, NH_3;
B) octane, C_8H_{18};
C) argon, Ar;
D) propanone, $CH_3\overset{\overset{\displaystyle O}{\|}}{C}CH_3$
E) methanol, CH_3OH

2 Propanone, CH_3COCH_3, propan-1-ol, $CH_3CH_2CH_2OH$ and butane, $CH_3CH_2CH_2CH_3$ have very similar relative molecular masses. List them in the expected order of increasing boiling point. Explain your answer.

3 Explain as fully as possible: Ethanol and propanoic acid are both miscible with water in all proportions and ethyl propanoate is only slightly miscible with water.

4 Many of the differences between water and hydrogen sulphide are ascribed to the fact that the former has hydrogen bonds but the latter has not.

A) Explain what is meant by 'hydrogen bonds'.
B) List three differences between water and hydrogen sulphide which can be ascribed to hydrogen bonding in the former.

(London 1979, part question)

5 Three liquids have the following properties:

	Water	Propanone	Pentane
Molecular formula	H_2O	C_3H_6O	C_5H_{12}
Relative molecular mass	18	58	72
Enthalpy change of vaporization/kJ mol^{-1}	41.1	31.9	27.7
Boiling point/°C	100	56	36

A) What type of intermolecular force predominates in each liquid:
i in water;
ii in propanone;
iii in pentane?
B) What can you deduce about the relative strength of these forces in the liquids? Justify your conclusions.
C) **i** If the liquids were shaken together in pairs, which pair would be unlikely to mix?
ii Explain this immiscibility in terms of the forces between the molecules.
D) Choose *one* of the pairs that mix and say whether the enthalpy change on mixing would be exothermic or endothermic. Justify your answer.

(Nuffield 1984)

6 This question is about hydrogen bonds and their effects on the properties of substances.

A) Which of the following hydrides can form hydrogen bonds?
i CH_4;
ii H_2O;
iii NH_3;
iv HF.
B) Describe how the physical properties of hydrides alter down a group in the Periodic Table because of the presence or absence of hydrogen bonds.
C) State two physical properties of water which can be attributed to hydrogen bonds.
D) Draw a diagram of an apparatus you could use to investigate the variations in boiling point with composition for a mixture of two liquids.
E) The diagram shows a typical plot of boiling point against volume composition for two liquids which form hydrogen bonds on mixing.

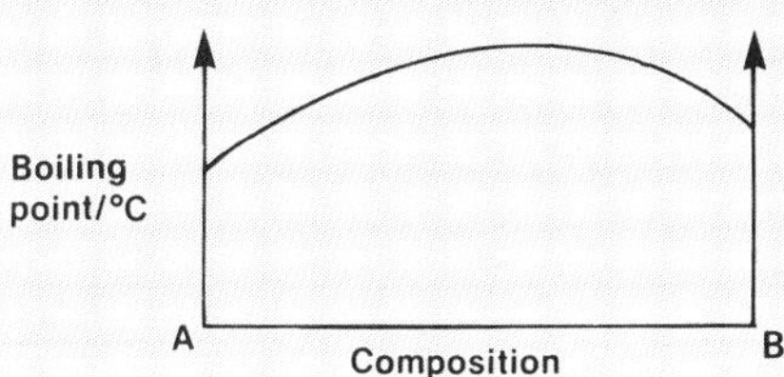

i In what way(s) does the graph differ from the graph that would be obtained from the ideal mixing of two liquids?
ii Why does the presence of hydrogen bonds in the mixture result in the boiling point differing from that expected for ideal mixing? (Nuffield 1982)

7 The graph shown below represents the partial vapour pressures of the miscible liquids dioxane and water above various mixtures of the two liquids at 35 °C.

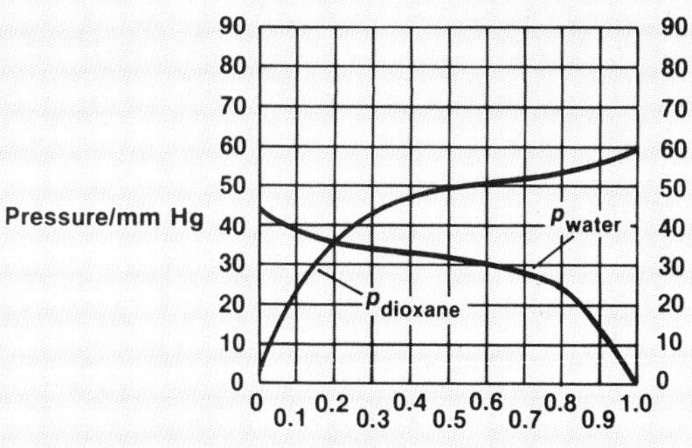

The structure of dioxane is:

$$\begin{array}{ccc} & O & \\ H_2C & & CH_2 \\ | & & | \\ H_2C & & CH_2 \\ & O & \end{array}$$

A) i Using the values on the graph, plot accurately on the same axes a line which would correspond to the variation in total vapour pressure with composition for any mixture of the two liquids. Label this line 'Line A'.

ii Also on the same axes, draw accurately a line to represent the variation of total vapour pressure with composition for any mixture of the two liquids, had the liquids behaved ideally and obeyed Raoult's law. Label this 'Line B'.

B) What would be the actual vapour pressure over a mixture containing a mole fraction of 0.8 of water and 0.2 of dioxane?

C) What would be the actual total vapour pressure over a mixture containing 9 g of water and 132 g of dioxane? (Relative atomic masses: H = 1, C = 12, O = 16.)

D) What type of interaction predominates between molecules in:

i pure water;
ii pure dioxane?

E) Explain, in terms of intermolecular forces, the deviation from Raoult's law that occurs in a mixture of water and dioxane. (Nuffield 1981)

8 A) i State Raoult's law;

ii Explain what is meant by an ideal solution of two liquids. Sketch vapour pressure/composition diagrams for mixtures of two liquids which correspond to (A) an ideal solution; (B) a non-ideal solution.

iii Indicate, giving your reasons, whether the following mixtures would show a large or small departure from ideal behaviour:

benzene and methylbenzene
benzene and methanol

B) In an experiment, 7.50 g of a solid A were dissolved in 100 g of benzene and the boiling temperature was found to rise from 80.10 °C to 80.70 °C. Explain why the boiling temperature rises and calculate the relative molecular mass of A given that the boiling temperature (ebullioscopic) constant for benzene is 2.63 K kg mol^{-1}. You may assume that A does not dissociate or associate in benzene.

Explain how the observed boiling point would change if A were a solid compound which associated completely to form A_2 molecules in benzene.

Suggest a compound which might behave in this way.

9 The graph shows boiling point composition curves for mixtures of propanone and ethanol. ADC represents the composition of the liquid and ABC the vapour.

A) If a mixture containing 10% propanone was boiled, what would be the composition of the vapour?

B) If the vapour was then condensed, and reboiled, what would be the composition of the vapour now?

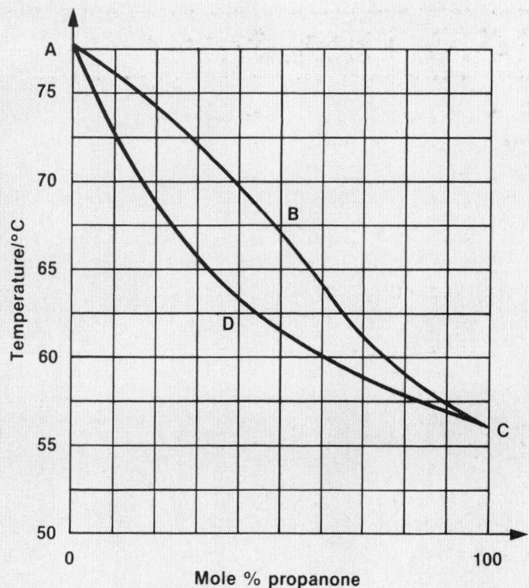

C) How many boiling/condensation cycles would be needed for the composition to reach 80% propanone?

10 A new antibiotic has been isolated of relative molecular mass approximately 10 000. It dissolves in water to the extent of 1 g per 100 g of water. Calculate the freezing temperature depression, boiling temperature elevation and osmotic pressure of this solution at 25 °C. Which method would give the most easily measured value for an accurate determination of relative molecular mass? (Use the data from question 12.)

11 A solution of 58.5 g of sodium chloride dissolved in 1 kg of aqueous solution. What value would you expect for:

A) freezing temperature depression;
B) boiling temperature elevation;
C) osmotic pressure?

(Remember sodium chloride dissociates into two ions in solution.)

12 1.2 g of ethanoic acid (CH_3CO_2H) dissolved in 100 g of benzene raised the boiling temperature by 0.253 K, and 1.2 g of ethanoic acid dissolved in 100 g of water depressed the freezing temperature of water by 0.372 K. What values of the relative molecular mass of ethanoic acid do these figures give? Can you explain the difference?

Freezing temperature constant for water
= 1.86 K kg mol^{-1}
Boiling temperature constant for water
= 0.52 K kg mol^{-1}
$R = 8.3$ J K^{-1} mol^{-1}
Boiling temperature constant for benzene
= 2.53 K kg mol^{-1}

17 Electrolysis

17.1 Introduction

It is not surprising that there is a close relationship between electricity and chemistry because atoms themselves are made up of electrically charged particles—protons and electrons. Electrochemistry has two aspects—the production of electricity by chemical reactions, dealt with in Chapter 13, and the chemical effects of electricity which this chapter will consider. Both are of considerable economic importance. The first includes the technology of batteries, while the second is used in the extraction of many elements from their compounds, including aluminium and chlorine.

17.2 The conduction of electricity by compounds

Figure 17.1 The electrolysis of molten lead bromide

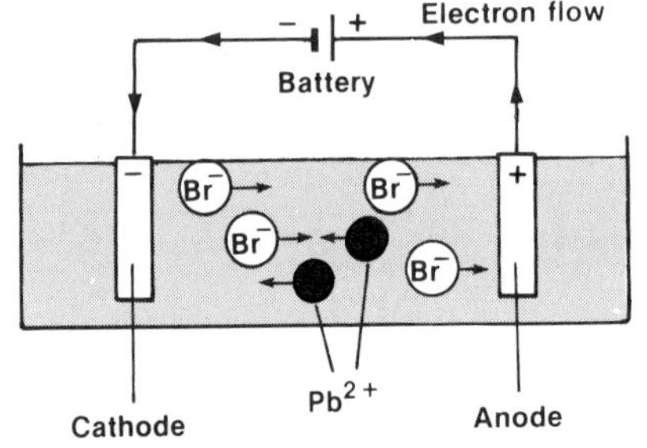

When metals conduct, the current is carried by the electrons in their delocalized 'sea' (see **section 10.2.3**). No chemical change occurs—the metal is left completely unaltered. Metals conduct in both the solid and liquid states. In contrast, ionic compounds conduct only when liquid (molten) or when in solution. The current is carried by the movement of both positive and negative ions through the liquid. In the solid state these ions are fixed in a lattice, but in the liquid state or in solution, they are free to move. Positive ions (**cations**) migrate towards the negative electrode (cathode) and negative ions (**anions**) migrate towards the positive electrode (anode). Chemical reactions occur at the electrodes so that after passage of electricity the liquid (called the **electrolyte**) is chemically changed.

For example, when a current is passed through molten lead bromide ($Pb^{2+} + 2Br^-$) (l) the Pb^{2+} ions migrate towards the cathode and Br^- ions migrate towards the anode, as shown in **Figure 17.1.**

E

THE MOVEMENT OF COLOURED IONS

The movement of coloured ions can be demonstrated easily. Copper chromate(VI) ($CuCrO_4$) is green due to the blue Cu^{2+} ions (as in copper sulphate) and yellow CrO_4^{2-} ions (as in potassium chromate(VI)). A drop of concentrated copper chromate(VI) is placed on a piece of filter paper (moistened so that it will conduct) as shown. After a few minutes a blue band is seen moving towards the cathode. This is the copper ions. A yellow band of chromate(VI) ions is seen moving towards the anode.

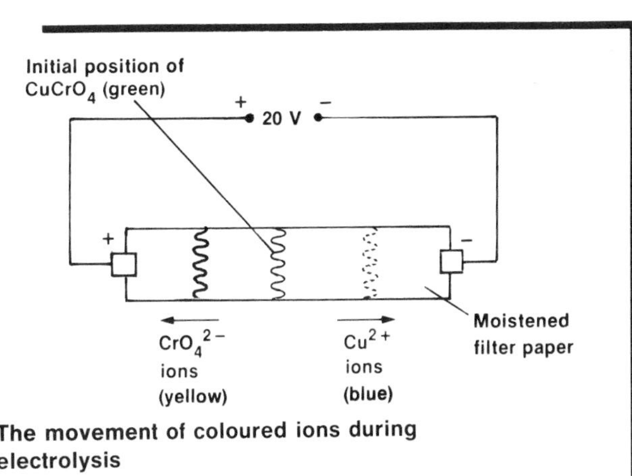

The movement of coloured ions during electrolysis

17.2.1 Reactions at the electrodes

The cathode releases electrons which are supplied by the battery. These are accepted by the cations (Pb^{2+} in this example) and the following reaction occurs, represented by a half equation:

$$Pb^{2+} + 2e^- \longrightarrow Pb$$

The oxidation number (see **section 13.2**) of the lead has *decreased* from +II to 0 so it has been *reduced*. Cathode reactions are *always* reductions.

The anode accepts electrons from the anions (Br^- in this example) and the following reaction occurs, represented by the half equation:

$$Br^- \longrightarrow Br + e^-$$

immediately followed by:

$$2Br \longrightarrow Br_2$$

These may be written:

$$2Br^- \longrightarrow Br_2 + 2e^-$$

The oxidation number of the bromine has *increased* from −I to 0 so it has been *oxidized*. *All* anode reactions are oxidations.

The overall reaction is found by adding the two half reactions:

$$\begin{aligned} Pb^{2+} + 2e^- &\longrightarrow Pb \\ 2Br^- &\longrightarrow Br_2 + 2e^- \\ \hline \cancel{2e^-} + Pb^{2+} + 2Br^- &\longrightarrow Pb + Br_2 + \cancel{2e^-} \end{aligned}$$

or

$$PbBr_2 \longrightarrow Pb + Br_2$$

The lead bromide has been decomposed into its elements.

Notice that the electrons cancel out. The battery simply moves them from the anode to the cathode through the external circuit.

17.3 Faraday's laws

17.3.1 Faraday's first law of electrolysis

This states that the mass of a substance produced at an electrode during electrolysis is proportional to the quantity of electricity passed. This can be illustrated using the apparatus shown in **Figure 17.2.** The quantity of electricity passed is measured in coulombs (C). One amp is a rate of flow of electricity (an electric current) of one coulomb per second. So:

$$\text{quantity of electricity in coulombs} = \text{current in amps} \times \text{time in seconds}$$

Figure 17.2 Apparatus for the electrolysis of silver nitrate solution

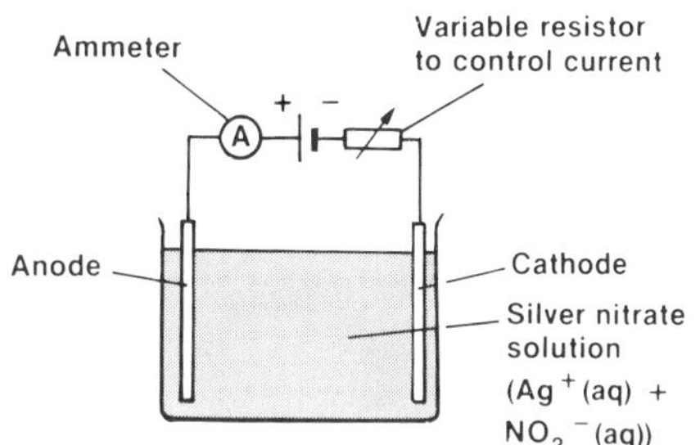

At the cathode the following reaction occurs:

$$Ag^+(aq) + e^- \longrightarrow Ag(s)$$

The quantity of silver deposited can be measured by weighing the cathode before and after the experiment. If several experiments are done with different currents and times, the results can be plotted on a graph, **Figure 17.3.** This is a straight line so shows Faraday's first law to be true. Extrapolation of the graph shows that approximately 96 000 coulombs are required to deposit one mole (108 g) of silver.

Figure 17.3 Graph illustrating Faraday's first law of electrolysis

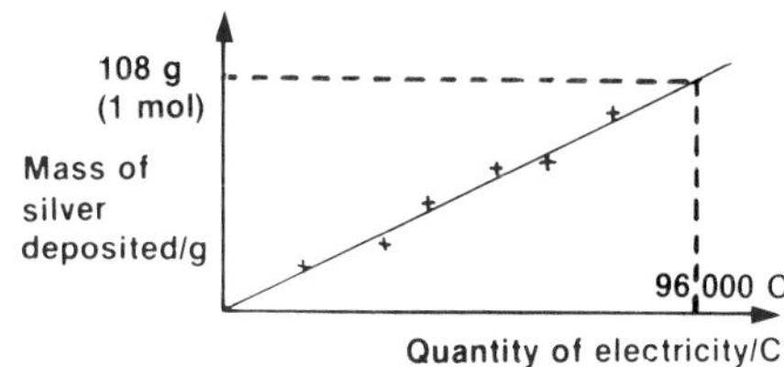

17.3.2 Faraday's second law of electrolysis

This states that the quantity of electricity required to produce a mole of a substance from its ions is proportional to the charge on the ions.

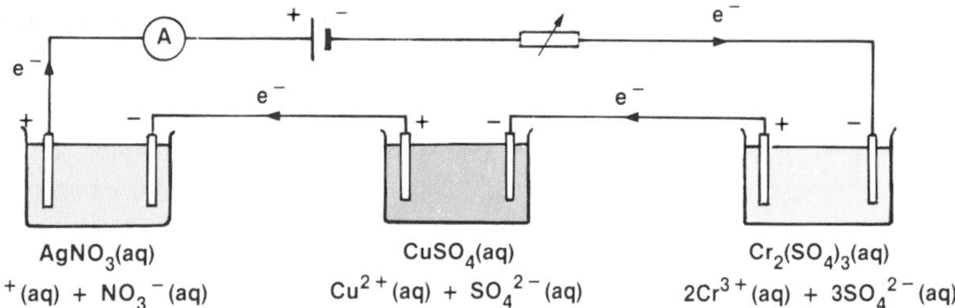

Figure 17.4 Apparatus illustrating Faraday's second law of electrolysis. Notice that the left-hand electrode in each beaker is the anode (+). Always trace the leads right back to the battery to decide on the polarity of electrodes

Michael Faraday

This can be illustrated experimentally with the apparatus in **Figure 17.4.** In this arrangement, the current flowing through each solution is the same. So in a given time the same quantity of electricity flows through each solution. In an experiment, a current of 0.1 amp flowing for 9600 seconds produced 1.08 g (1/100 mol) of silver, 0.32 g (1/200 mol) of copper and 0.173 g (1/300 mol) of chromium.

$$\begin{aligned} \text{quantity of electricity passed} &= \text{current} \times \text{time} \\ &= 0.1 \quad \times 9600\,\text{s} \\ &= 960\,\text{C} \end{aligned}$$

so to produce 1 mol of silver from Ag^+ requires 96 000 C
to produce 1 mol of copper from Cu^{2+} requires 192 000 C
to produce 1 mol of chromium from Cr^{3+} requires 288 000 C

17.3.3 The Faraday constant

The above example shows that a quantity of electricity of 96 000 C is required to produce one mole of substance from singly charged ions, twice this for a mole of doubly charged ions, and so on. The number of coulombs to discharge a mole of singly charged ions is called the **Faraday constant,** *F*. An accurate value is 96 485 C mol^{-1}, but it is often rounded off to 96 500 or 96 000 C mol^{-1} for rough calculations.

The equation below shows silver being discharged at a cathode:

Ag^+	+ e^-	⟶ Ag
1 ion	1 electron	1 atom
1 mole of ions	1 mole of electrons	1 mole of atoms

It shows that to produce a mole of atoms from singly charged ions requires a mole of electrons. So another way of looking at the Faraday constant is that it represents a mole of electrons. The charge of an electron is 1.6022×10^{-19} C and the Avogadro constant (the number of particles in a mole) is $6.022 \times 10^{23}\,mol^{-1}$. So a mole of electrons has a charge of:

$$1.6022 \times 10^{-19}\text{C} \times 6.022 \times 10^{23}\,\text{mol}^{-1} = 96\,485\,\text{C mol}^{-1}$$

17.4 Discharge of ions at electrodes

Table 17.1 The electrochemical series

Half reaction	Voltage/V
$Ca^{2+}(aq) + 2e^- \rightleftharpoons Ca(s)$	−2.89
$Na^+(aq) + e^- \rightleftharpoons Na(s)$	−2.71
$Mg^{2+}(aq) + 2e^- \rightleftharpoons Mg(s)$	−2.37
$Al^{3+}(aq) + 3e^- \rightleftharpoons Al(s)$	−1.66
$Zn^{2+}(aq) + 2e^- \rightleftharpoons Zn(s)$	−0.76
$Pb^{2+}(aq) + 2e^- \rightleftharpoons Pb(s)$	−0.13
$2H^+(aq) + 2e^- \rightleftharpoons H_2(g)$	0.00
$Cu^{2+}(aq) + 2e^- \rightleftharpoons Cu(s)$	+0.34
$O_2(g) + 2H_2O(l) + 4e^- \rightleftharpoons 4OH^-(aq)$	+0.4
$I_2(aq) + 2e^- \rightleftharpoons 2I^-(aq)$	+0.54
$Ag^+(aq) + e^- \rightleftharpoons Ag(s)$	+0.80
$Br_2(aq) + 2e^- \rightleftharpoons 2Br^-(aq)$	+1.09
$Cl_2(aq) + 2e^- \rightleftharpoons 2Cl^-(aq)$	+1.36

In a molten electrolyte there is usually only one type of positive and one type of negative ion, and therefore, there is no choice of ion to be discharged at each electrode. In an aqueous solution of an ionic compound there will be H^+ and OH^- ions from the dissociation of water as well as ions from the solute. So a solution of copper sulphate will contain Cu^{2+}(aq), SO_4^{2-}(aq), H^+(aq) and OH^-(aq) ions. Both Cu^{2+}(aq) and H^+(aq) will migrate to the cathode and both SO_4^{2-}(aq) and OH^-(aq) will migrate to the anode. Which ions will be discharged?

This depends on the position of the ions in the **electrochemical series** (see **section 13.3**). This is a list of ions in order of the voltage they produce in an electrochemical cell. Here the voltage represents the voltage required to discharge the ions at an electrode. The version of the electrochemical series in **Table 17.1** includes the half reactions most often met in electrolysis.

17.4.1 Reactions at the cathode

The cathode supplies electrons, so that the equilibria in **Table 17.1** will go from left to right at the cathode. The lower the half reaction is in the table, the more readily it will occur at the cathode.

So in an aqueous solution of copper ions,

$$Cu^{2+}(aq) + 2e^- \longrightarrow Cu(s)$$

will occur in preference to:

$$2H^+(aq) + 2e^- \longrightarrow H_2(g)$$

17.4.2 Reactions at the anode

The anode removes electrons so that the equilibria in the table will go from right to left at the anode. The higher the half reaction in the table, the more readily it will occur at the anode. So in an aqueous solution of iodide ions,

$$4OH^-(aq) \longrightarrow O_2(g) + 2H_2O(l) + 4e^-$$

will tend to occur in preference to:

$$2I^-(aq) \longrightarrow I_2(aq) + 2e^-$$

Note: Nitrate and sulphate ions are almost never discharged in electrolysis of aqueous solutions. Hydroxide ions are always discharged instead.

Two factors can affect these predictions. Firstly, concentration: if one ion is present in a much greater concentration than another, the more concentrated ion may be discharged, even though the electrochemical series predicts that the other ion would. For example, in the electrolysis of 1 mol dm^{-3} sodium chloride, $[Cl^-(aq)] = 1$ mol dm^{-3} and $[OH^-(aq)] = 10^{-7}$ mol dm^{-3} (see **section 12.3**). Although the electrochemical series predicts that hydroxide ions would be discharged, chloride ions are in fact discharged as chlorine due to their overwhelmingly larger concentration, though there will also be a little oxygen from the hydroxide ions. Secondly, the nature of the electrode may affect the reactions. In the industrial electrolysis of sodium chloride solution (see box, **section 23.4.4**), if a graphite electrode is used, hydrogen ions are discharged as predicted by the electrochemical series. However, if a mercury electrode is used, sodium ions are discharged.

17.5 Industrial applications of electrolysis

17.5.1 Extraction of metals

Most metals are found in nature as compounds rather than as pure metals. Metals exist as positive ions and must therefore be reduced to form the pure metal. Relatively unreactive metals can be obtained from their ores by chemical reduction, for example in the blast furnace iron is reduced by carbon monoxide:

$$Fe_2O_3(s) + 3CO(g) \longrightarrow 2Fe(s) + 3CO_2(g)$$

For more reactive metals it is hard to find powerful enough reducing agents and the reduction is done by electrolysis. Two metals extracted by electrolysis are aluminium and sodium.

Aluminium

Aluminium is one of the most versatile metals. Being light, strong and a good electrical conductor, it is used for everything from saucepans to airliners, cooking foil to electricity transmission lines. Aluminium is extracted by the Hall–Héroult process.

Aluminium is used for overhead cables

The raw material for aluminium manufacture is bauxite which contains aluminium oxide, Al_2O_3(alumina). The main inpurities are oxides, particularly iron(III) oxide and silica, so bauxite is first purified. Powdered bauxite is dissolved in hot sodium hydroxide solution and the insoluble impurities are filtered off. The solution is cooled and pure alumina precipitates.

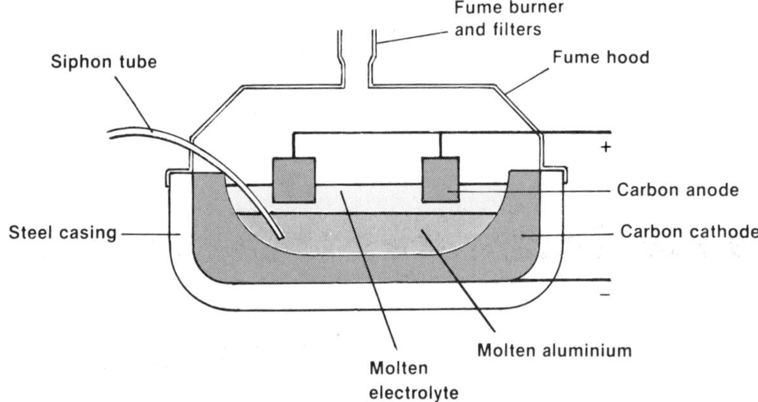

Figure 17.5 Electrolysis cell for aluminium extraction

The pure alumina is then dissolved in molten cryolite, Na_3AlF_6 at approximately 1200 K and electrolysed in a cell with carbon anode and cathode, **Figure 17.5.**

$$\text{At the cathode} \quad Al^{3+} + 3e^- \longrightarrow Al$$

$$\text{At the anode} \quad 2O^{2-} \longrightarrow O_2(g) + 4e^-$$

At the high temperature used, the carbon anode burns away in the oxygen and must be replaced frequently. Apart from replacement of the anode, the process is continuous, alumina being regularly added to the cell and the molten aluminium being siphoned off (the melting temperature of aluminium is 932 K). The electrical power requirements are enormous, both for electrolysis and for keeping the electrolyte molten. Each tonne of aluminium requires 18 megawatt hours of electricity—enough to keep 100 electric fires burning for a week. Plants are usually built close to a

Aluminium extraction cell

source of electrical energy. For example, the Anglesey aluminium plant near Holyhead is built close to the Wylfa nuclear power station.

Sodium

Sodium is used in the manufacture of titanium, indigo dye for dyeing blue jeans, and tetraalkyllead anti-knock petrol additives, as well as being used as a cooling medium for nuclear reactors.

Molten sodium chloride, obtained in this country from underground salt deposits in Cheshire, is electrolysed in the Down's cell, **Figure 17.6.**

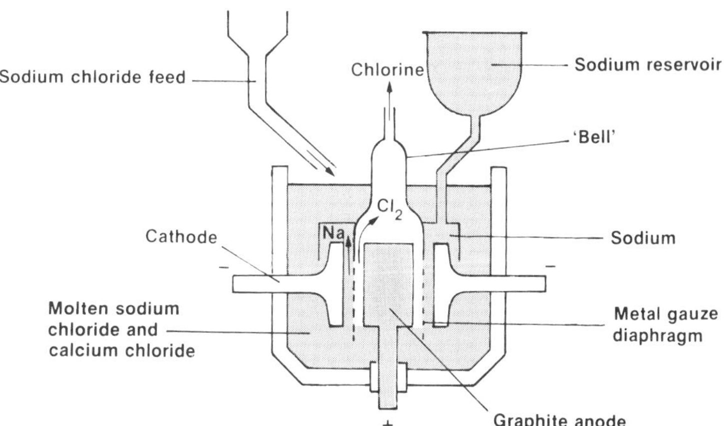

Figure 17.6 Down's cell for sodium manufacture

It is mixed with calcium chloride to form a mixture which melts at around 880 K, rather than 1074 K for pure sodium chloride. This brings about a significant saving in the energy costs of keeping the electrolyte molten, as well as making the process safer, by being below the boiling point of sodium—sodium vapour is explosive. At the cathode:

$$Na^+ + e^- \longrightarrow Na$$

and also

$$Ca^{2+} + 2e^- \longrightarrow Ca$$

as the discharge voltages of the two metals are very close.

A mixture of molten sodium and about 5% calcium is obtained. The calcium solidifies first as the mixture cools and it is replaced in the cell. At the anode:

$$2Cl^- \longrightarrow Cl_2 + 2e^-$$

A steel mesh or diaphragm separates the anode and cathode to keep the chlorine and sodium apart and prevent them reacting back to form sodium chloride. Because of the harsh environment to which it is exposed—chlorine at high temperature—the mesh requires fortnightly replacement. The process also produces chlorine, a saleable product.

Example

A Down's cell operates at a current of 30 000 amps. What mass of sodium is produced per day? What mass of chlorine is produced per day? (1 day is approximately 90 000 seconds.) A_rNa = 23, Cl = 35.5.

quantity of electricity = current × time
quantity of electricity used per day = 30 000 × 90 000 C = 27 × 10^8 C
to produce 1 mole of sodium from its singly charged ions takes 96 500 C
in 1 day we get 27 × 10^8 ÷ 96 500 = 27 980 mol of Na
= 27 980 × 23 g of Na
= 644 kg per day
for each mole of sodium atoms we get 1 mole of chlorine atoms
($\frac{1}{2}$ mole of Cl_2 molecules)
so the mass of chlorine = 27 980 × 35.5/1000 kg = 993 kg of chlorine per day

Purification of copper

In this process a lump of impure copper is the anode and a sheet of pure copper the cathode of an electrolysis cell. If you look at the electrochemical series you will see that the reaction $Cu(s) \longrightarrow Cu^{2+}(aq) + 2e^-$ will occur if copper is made the anode, i.e. the anode will dissolve. An electrolyte containing copper ions is used.

At the anode the copper anode itself dissolves:

$$Cu(s) \longrightarrow Cu^{2+}(aq) + 2e^-$$

while at the cathode, copper is deposited in the usual way:

$$Cu^{2+}(aq) + 2e^- \longrightarrow Cu(s)$$

As the copper anode dissolves, its impurities, which include valuable metals such as silver and gold, drop to the bottom of the container where they are recovered. The cathode deposit consists of very pure copper.

Figure 17.7 An unusual purification of a copper boulder on site

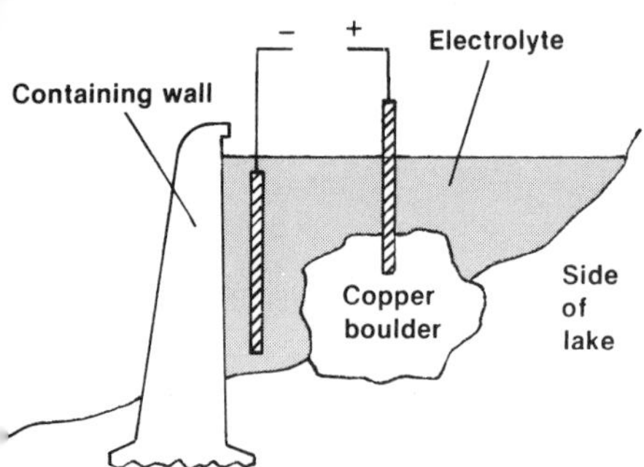

17.5.2 Electroplating

There are many everyday examples of electroplating. One of the most common is the chromium plating of bicycle parts and car bumpers. Chromium is less reactive than iron and prevents it rusting as well as improving its appearance. Pure chromium parts would be much more expensive than plated ones. Also, pure chromium is rather brittle for structural use.

17.6 Summary

- When electricity passes through ionic liquids, it is carried by migration of ions through the electrolyte.
- Cations (positive ions) go to the cathode where they are reduced by gaining electrons.
- Anions (negative ions) go to the anode where they are oxidized by electron loss.
- Faraday's laws state that a mole of electrons is required to discharge a mole of singly charged ions, two moles of electrons to discharge a mole of doubly charged ions, and so on.
- A mole of electrons is a quantity of electricity equivalent to approximately 96 500 C mol^{-1}—the Faraday constant.
- If a solution contains more than one type of ion of the same charge, the one which is most likely to be discharged can be predicted from the electrochemical series.
- Positive ions near the bottom of the electrochemical series are most easily discharged at the cathode.
- Negative ions near the top of the electrochemical series are preferentially discharged at the anode.
- The above predictions may fail if another ion is present in a much greater concentration than the predicted one, or because of electrode effects.

17.7 Questions

1 A mixture of zinc sulphate and copper sulphate is electrolysed in aqueous solution using carbon electrodes. What products would you predict at the anode and cathode? Explain your answer.

2 A solution containing nickel ions was electrolysed with a current of 0.1 amp for 160 minutes. 0.295 g of nickel was deposited at the cathode.

What is the charge on the nickel ion?

3 Give the name or formula of the principal ore from which aluminium is extracted. Indicate briefly how the ore is purified and how the metal is extracted from the purified ore. (London 1978, part question)

4 An aluminium plant produced 200 tonnes of aluminium per week. How many coulombs of electricity are required for 200 tonnes of aluminium? What current would be required if production is continuous?

1 tonne = 1000 kg. Take 1 week to be 600 000 seconds.

5 An industrial cell electrolyses an aqueous solution of sodium chloride using a current of 100 kiloamps for one day (90 000 seconds). How many kilograms of chlorine does it produce?

18 Thermodynamics—enthalpy, entropy and free energy

18.1 Introduction

Thermodynamics is the study of changes in heat energy and other related quantities that occur in chemical reactions. This chapter tackles the most basic problem in chemistry, which is: why do some reactions take place of their own accord while others do not? For example, gases mix and do not 'unmix'; zinc reacts with copper sulphate to give zinc sulphate and copper but the reverse of this does not happen; and so on.

In section 9.6 we discussed some clues to this and found that the enthalpy and entropy changes of a reaction seemed to be involved. The enthalpy change of a reaction is the heat change at constant pressure and temperature. Reactions which give out enthalpy (ΔH is negative) are more favoured. Entropy is a measure of randomness of a system and reactions with a positive entropy change (ΔS is positive) are more favoured. We built up a diagram:

1. $+\Delta H$ cannot $-\Delta S$ 'go'	2. $-\Delta H$ might $-\Delta S$ 'go'
3. $+\Delta H$ might $+\Delta S$ 'go'	4. $-\Delta H$ must $+\Delta S$ 'go'

('go' means 'be feasible')

We were left with two problems:

1. How do we measure or calculate entropies?
2. What are the relative importances of entropy and enthalpy changes when they are opposed?

18.2 Entropy

18.2.1 The arrangement of molecules

We believe that molecules move at random and cannot 'know where they should go.' This suggests that reactions happen *by chance alone*, i.e. the *most probable* thing happens. The most probable thing is disorder. This is the same as saying that on dealing a well-shuffled pack of cards you will very probably get a haphazard arrangement of cards rather than, say, all one suit. There are a vast number of haphazard arrangements of cards but only a very few corresponding to getting all the cards in one suit.

It may seem odd to suggest that chemical reactions, which have a predictable outcome, are governed by chance alone, but it is easy to show that this could be the case. If you tossed 1000 coins, you could bet with confidence that there would be between 450 and 550 heads. Such an outcome could be predicted with near certainty. (It would however be unwise to bet on *exactly* 500 heads.) The same thing is true of molecules. Imagine a gas in box 1. A similar sized box, box 2, is empty (a vacuum),

Figure 18.1 Diffusion

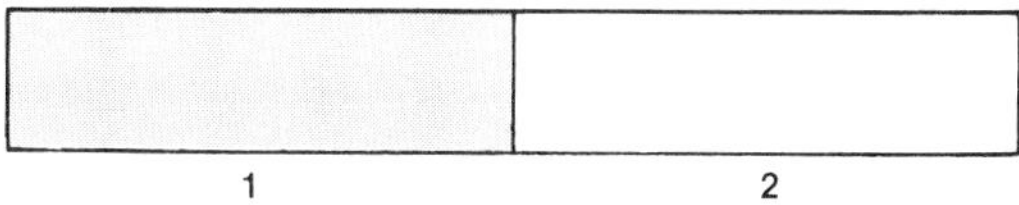

Figure 18.1. If we remove the partition, we know that a few seconds later the gas molecules will be distributed evenly between boxes 1 and 2. We call this **diffusion**. Note that the gas molecules cannot 'know' that there is a vacuum in the other container or that they are 'supposed' to spread out. They behave randomly. Could this happen by chance alone?

We shall start with a very simple example. Imagine there were only two molecules, A and B. These could be distributed as shown:

1	2
AB	0
0	AB
A	B
B	A

This gives 4 arrangements, i.e. 2^n (where n is the number of particles). There is a 2 in 4 chance of all the molecules being in either box 1 or box 2, and 2 out of 4 of the arrangements correspond to spreading out.

For three molecules, ABC:

1	2
ABC	0
AB	C
A	BC
BC	A
B	CA
CA	B
C	AB
0	ABC

There are 8 arrangements (2^3). There is a 2 in 8 chance of all the molecules being in either box 1 or box 2. Here 6 out of 8 arrangements correspond to spreading out.

For six molecules, we get 64 arrangements, and with this number it is very laborious to write the possibilities out in full, so we shall just consider the number of ways of getting each arrangement. For example, there is just one way of getting all six in container 1, and 6 ways of getting 5 in container 1 and 1 in container 2. These are ABCDE/F, ABCDF/E, ABCEF/D, ABDEF/C, ACDEF/B and BCDEF/A.

Container:	**1 2**	**1 2**	**1 2**	**1 2**	**1 2**	**1 2**	**1 2**
	6 0	5 1	4 2	3 3	2 4	1 5	0 6
Number of ways:	1	6	15	20	15	6	1

78% (15 + 20 + 15)

Total = 64, 50 of which (78%) correspond to considerable spreading out.

The more molecules we have, the more probable spreading out becomes. With 50 molecules there are 2^{50} (approx 10^{15}) arrangements, only two of which represent all the molecules in one box.

As we consider even greater numbers of molecules, so it becomes overwhelmingly likely that we will get an approximately even distribution of molecules, solely by chance.

So we have shown that the diffusion of gases can be accounted for just by probability.

If we double the volume of a gas, we allow many more ways for the molecules to arrange themselves. There were a vast number of possible arrangements in the single container. Now there is a large extra number because molecules can now be in container 1 *or* container 2. The system is now more random. There has been an increase in entropy and the change has happened of its own accord, by chance alone.

Entropy is related to the number of ways, *W*, in which the system can be arranged. The more ways, the greater the entropy. In fact the mathematical

Figure 18.2 Energy levels in a molecule. Translational levels are even more closely spaced than rotational ones and are not shown. Notice the vibrational levels are evenly spaced

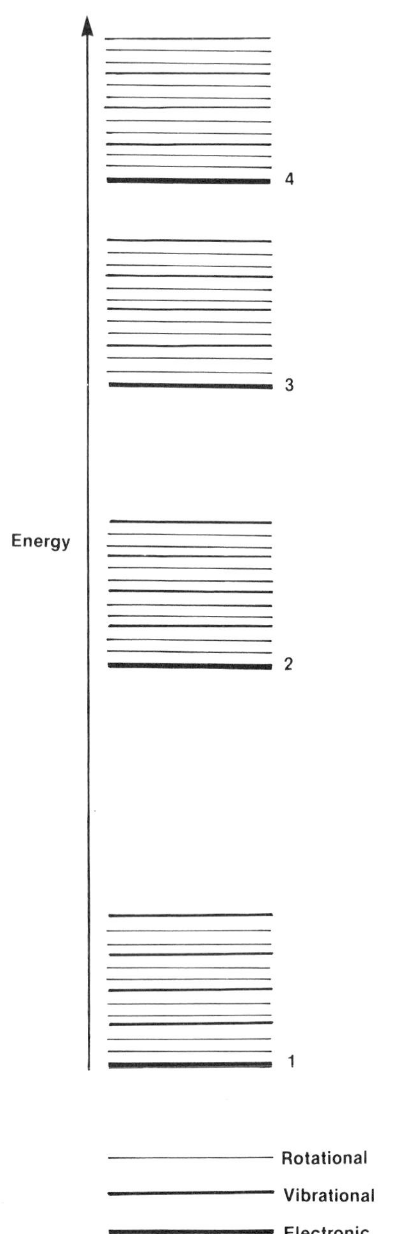

relationship is:

$S = k \ln W$ where S is entropy
W is the number of ways,
k is the Boltzmann constant ($1.38 \times 10^{-23}\,\mathrm{J\,K^{-1}}$)

and for any change in the system:

$\Delta S = k\Delta \ln W$ where ΔS is the entropy change
and $\Delta \ln W$ is the change in the ln of the number of ways of arranging the molecules in the system.

As $\ln W$ is a pure number with no units, the units of k ($\mathrm{J\,K^{-1}}$) are the units for entropy. The entropy change on doubling the volume of a *mole* of gas is $+6\,\mathrm{J\,K^{-1}\,mol^{-1}}$.

18.2.2 The arrangement of energy

Figure 18.3 Types of energy in a diatomic molecule

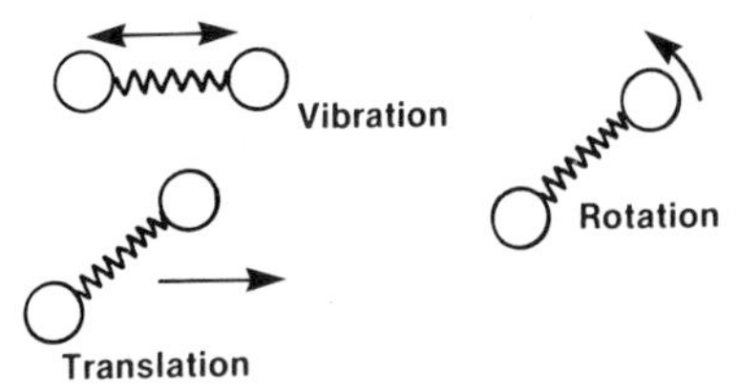

We have said that entropy and entropy changes are related to the number of ways of arranging the molecules (or atoms or ions) in the system. A further contribution comes from the number of ways that the **energy** can be arranged. We saw in **section 8.3.1** that atoms have discrete energy levels for their electrons and that energy comes in 'packets' of different sizes called quanta. A quantum of just the right size is needed to promote an electron from one level to the next. The same is true of electronic energy levels in molecules, and in fact quantization also applies to *all* the types of energy that a molecule can have. These include the energy that makes it vibrate, rotate and translate (move from place to place), **Figure 18.3.**

Figure 18.4 Vibrational energy levels in two molecules showing one possible arrangement of four quanta

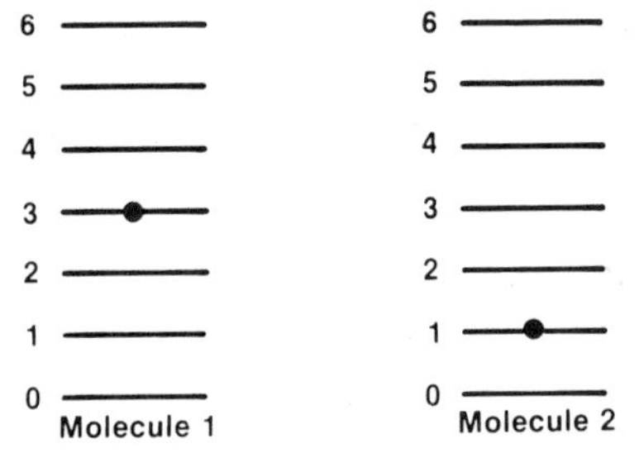

Energy levels in a molecule

These are shown schematically in **Figure 18.2** (opposite page).

Let us take a very simple example. Suppose we have just two molecules and four quanta of vibrational energy (which corresponds to heat energy) to share between them, **Figure 18.4.** In how many ways can this be done? One arrangement is all four quanta in molecule 1 and none in molecule 2. Another is three quanta in molecule 1 and one in molecule 2, and so on. **Table 18.1(a)** shows all the possibilities. There are five. With five quanta there are six arrangements, see **Table 18.1(b).** So, the more quanta of energy (i.e. the more heat), the more ways in which they can be arranged and the more random the system. This aspect of randomness has to be taken into account when calculating entropies of substances and entropy changes in reactions using:

$$S = k \ln W$$

or:

$$\Delta S = k\Delta \ln W$$

Table 18.1 Arrangements of quanta between energy levels. (a) 4 quanta—five arrangements, (b) 5 quanta—six arrangements

(a)

Molecule 1	Molecule 2
4	0
3	1
2	2
1	3
0	4

(b)

Molecule 1	Molecule 2
5	0
4	1
3	2
2	3
1	4
0	5

18.2.3 Entropy values

Entropies can be calculated for different substances. They differ with conditions such as temperature, and the same convention is used as for enthalpies. Standard conditions are as for enthalpies, 298 K and 10^2 kPa, so $S^{\ominus}(298\,\mathrm{K})$ is the entropy of 1 mole of substance in its standard state under standard conditions. As with enthalpies the (298 K) is often omitted, the symbol $^{\ominus}$ being taken to imply this.

Table 18.2 Some values of standard entropies

Substance		$S^{\ominus}$(298 K)/J K^{-1} mol^{-1}
Carbon (graphite)	(s)	5.7
Carbon (diamond)	(s)	2.4
Copper	(s)	33.2
Water	(l)	69.9
Water (steam)	(g)	188.7
Ethanol	(l)	160.7
Carbon dioxide	(g)	213.6
Chlorine	(g)	223.0

Some values are given in **Table 18.2.**
Note that the entropies of elements in their standard states are not zero. Because of the distribution of energy quanta, entropies could only be zero at 0 K when there are no quanta to be distributed.

You should notice that the entropies of solids are small, those of liquids larger and those of gases larger still. This reflects the randomness of these states. Notice also the entropy increase when water turns to steam. This is due to the increase in volume.

18.2.4 Entropy and enthalpy

We have seen that heating something up (giving it more quanta of heat energy) results in an increase in entropy.

Is there a relationship between the amount of heat energy or enthalpy given to it and the entropy increase?

Firstly, it is obvious that the more heat (more quanta) we transfer, the more entropy will be gained. This suggests that:

$$\Delta S \propto \Delta H$$

Secondly, if we give a certain number of quanta to a system which has very few (i.e. is at a low temperature) it will produce a large increase in entropy. If we give the same number of quanta to a system which already has many quanta (i.e. is at a high temperature) the difference will be less significant. The argument is rather like saying that a £50 wage increase is more important to someone earning £100 a week than to someone earning £1000 a week. This suggests that:

$$\Delta S \propto 1/T$$

so that, if we add a fixed amount of heat energy when T is large, ΔS is small and when T is small, ΔS is large. These relationships turn out to be the case and:

$$\Delta S = \frac{\Delta H}{T}$$

Trouton's rule

Can we find any evidence to back up the idea that:

$$\Delta S = \frac{\Delta H}{T}?$$

We need a situation where we are transferring heat to something and can measure the entropy increase produced.

Such a situation is found when liquids boil: their volumes increase causing an entropy increase. This volume increase is about the same for all liquids—the volumes of the liquids are all very small, especially when compared with the volumes of the gases that they turn into. The volume of the gases is approximately the same for all substances—about 24 dm^3, per mole (if they boil at around room temperature). Also, boiling occurs without a change in temperature, so there is no change of entropy due to change of temperature.

So all liquids which boil at close to room temperature should have approximately the same entropy change on boiling. (There are some exceptions, mostly liquids which form hydrogen bonds. It is possible to explain these, but this would lose the thread of the argument.)

So if:

$$\Delta S = \frac{\Delta H}{T},$$

then

$$\frac{\Delta_{vap} H}{T_b}$$

should be the same for all liquids. This is found to be the case, see **Table 18.3**, and is called **Trouton's rule**.

Table 18.3 Trouton's rule

Liquid	$\Delta_{vap}H$/kJ mol^{-1}	T_b/K	$\Delta_{vap}H/T_b$/J K^{-1} mol^{-1}
Carbon disulphide	27.2	319	85.2
Benzene	30.9	353	87.5
Octane	34.9	399	87.4
Trichloromethane	29.3	335	87.4

18.2.5 The system and the surroundings

If we accept the idea that reactions happen by chance and that this leads to disorder, we now seem to have the idea that reactions will go of their own accord if ΔS is positive, because chance alone makes it overwhelmingly likely. But what about a reaction like:

$$\underset{\text{ammonia}}{NH_3(g)} + \underset{\text{hydrogen chloride}}{HCl(g)} \longrightarrow \underset{\text{ammonium chloride}}{NH_4Cl(s)}$$

which is spontaneous, even though we are going from gases (disorder) to solid (order) so that the entropy change is clearly negative?

In fact, $\Delta S^{\ominus} = -284\,\text{J K}^{-1}\,\text{mol}^{-1}$

Tables show that:

$$S^{\ominus} \quad NH_3 = 192 \quad \text{J K}^{-1}\,\text{mol}^{-1}$$
$$S^{\ominus} \quad HCl = 187 \quad \text{J K}^{-1}\,\text{mol}^{-1}$$
$$S^{\ominus} \quad NH_4Cl = 95 \quad \text{J K}^{-1}\,\text{mol}^{-1}$$

(As expected, solid ammonium chloride has a lower value of $S^{\ominus}$ than either ammonia or hydrogen chloride gases.)

$$\begin{aligned}\Delta S &= S_{\text{products}} - S_{\text{reactants}}\\ &= 95 - (192 + 187)\\ &= -284\,\text{J K}^{-1}\,\text{mol}^{-1}\end{aligned}$$

The problem is that we have not considered *all* the entropy changes, only those of the reactants and products, which we can call the **system**. The reaction is exothermic, $\Delta H^{\ominus} = -176\,\text{kJ mol}^{-1}$. This energy is dumped into the surroundings and increases *its* entropy. Using the relation:

$$\Delta S = \Delta H/T$$

we can calculate ΔS.

The ΔS we are calculating is the entropy change of the *surroundings*, while the ΔH is that of the *system*. A negative value of ΔH means energy is being transferred to the surroundings and so the entropy *increase* of the surroundings is:

$$\Delta S_{\text{surroundings}} = -\Delta H_{\text{system}}/T = - -176/T$$

At room temperature, 298 K: $\Delta S_{\text{surroundings}} = +176/298$

$$\begin{aligned}&= 0.59\,\text{kJ K}^{-1}\,\text{mol}^{-1}\\ &= 590\,\text{J K}^{-1}\,\text{mol}^{-1}\end{aligned}$$

(Take care with units. Entropies are usually given in J K^{-1} mol^{-1}, enthalpies in kJ mol^{-1}.)

So the entropy *increase* of the *surroundings* more than compensates for the entropy *decrease* of the *system*.

The *total* entropy change: $\Delta S_{\text{tot}} = \Delta S_{\text{system}} + \Delta S_{\text{surroundings}}$

$$\Delta S_{\text{tot}} = -284 \quad + 590 = +306\,\text{J K}^{-1}\,\text{mol}^{-1}$$

We must always consider the *total* entropy change in a reaction including the system *and* the surroundings. Now we can understand the ammonia/hydrogen chloride reaction. It is spontaneous because the *total* entropy change is positive. We now have what we were looking for—an indication of reaction feasibility.

$$\text{if } \Delta S_{\text{tot}} > 0\text{, the reaction can 'go'}$$

This is an illustration of the **second law of thermodynamics**—the total entropy of the universe always increases in a natural change.

18.3 Free energy change, ΔG

We could write:

$$\Delta S_{tot} = \Delta S_{system} - \Delta H_{reaction}/T$$

Multiplying by $-T$ gives:

$$-T\Delta S_{tot} = -T\Delta S_{system} + \Delta H_{reaction}$$

Note that $-T\Delta S$ has units of energy per mole ($J\,K^{-1}\,mol^{-1} \times K = J\,mol^{-1}$). We call this energy change the **Gibbs free energy change**, ΔG.

$$\Delta G = \Delta H - T\Delta S$$

Since we have changed signs, by multiplying by $-T$, if ΔG is *negative* the reaction will go.

ΔG has the advantage that it can be calculated solely from measurements on the reacting system rather than the surroundings. ΔGs can be calculated using Hess's law cycles, just like ΔHs.

It is now possible to see why the idea that reactions with a negative value of ΔH are feasible is often correct:

$$\Delta G = \Delta H - T\Delta S$$

If T is small, the term $-T\Delta S$ is small, and $\Delta G \simeq \Delta H$. So at low temperature if ΔH is negative, ΔG will be too, unless the entropy change is very large. Since most of our chemical observations are made at room temperature, where T is quite small in absolute terms, ΔG is often approximately equal to ΔH.

18.3.1 Why do we use the term 'free energy'?

Consider the reaction:

$$2Ag^{+}(aq) + Cu(s) \longrightarrow Cu^{2+}(aq) + 2Ag(s) \quad \Delta H^{\ominus} = -147\,kJ\,mol^{-1}$$

We would predict that $\Delta S^{\ominus}_{system}$ will be negative, as there are two mobile ions on the left and only one on the right.

In fact this is so, as we can see from the following values of $S^{\ominus}$, taken from tables:

Note: Entropies of ions in solution are calculated relative to an arbitrary standard $S^{\ominus}H^{+}(aq) = 0$, hence negative values are possible.

$$S^{\ominus}Ag^{+}(aq) = 73\,J\,K^{-1}\,mol^{-1}$$
$$S^{\ominus}Cu(s) = 33\,J\,K^{-1}\,mol^{-1}$$
$$S^{\ominus}Cu^{2+}(aq) = -100\,J\,K^{-1}\,mol^{-1}$$
$$S^{\ominus}Ag(s) = 43\,J\,K^{-1}\,mol^{-1}$$

$$\Delta S^{\ominus}_{system} = \text{total } S^{\ominus}_{products} - \text{total } S^{\ominus}_{reactants}$$
$$\Delta S^{\ominus}_{system} = ((2 \times 43) + -100) - ((2 \times 73) + 33)$$
$$\Delta S^{\ominus}_{system} = -193\,J\,K^{-1}\,mol^{-1}$$

which is, as we predicted, negative and *inside the test tube* entropy has decreased. Now, the reaction can only become feasible if the *total* entropy change $\Delta S^{\ominus}_{tot} > 0$. So the entropy of the surroundings must increase by at least $193\,J\,K^{-1}\,mol^{-1}$, so that $\Delta S_{tot} > 0$.

Since in general: $\Delta S = \Delta H/T$
rearranged: $\Delta H = T\Delta S$

we can find the amount of heat ΔH which must be transferred in order to make $\Delta S_{tot} > 0$, at standard temperature, 298 K.

$$\Delta H^{\ominus} = 193 \times 298\,J\,mol^{-1}$$
$$= 57\,514\,J\,mol^{-1}$$
$$= 57.5\,kJ\,mol^{-1}$$

So of the $147\,kJ\,mol^{-1}$ (ΔH for the original reaction), $57.5\,kJ\,mol^{-1}$ *must* be transferred to the *surroundings* as heat in order for the total entropy change of system plus surroundings to be positive.

The rest ($147 - 57.5 = 89.5\,\text{kJ mol}^{-1}$) is *free* energy which is available in any form and for any purpose. It need not necessarily be given out as heat. The $57.5\,\text{kJ mol}^{-1}$ *must* be given out as heat, to provide an entropy *increase* of the surroundings to balance the entropy *decrease* of the system.

So
$$\Delta G^{\ominus} = \Delta H^{\ominus} - T\Delta S^{\ominus}$$
$$\Delta G^{\ominus} = -147 - \frac{(298 \times -193)}{1000}$$
$$\Delta G^{\ominus} = -147 + 57.5$$
$$\Delta G^{\ominus} = -89.5\,\text{kJ mol}^{-1}$$

18.3.2 $\Delta G^{\ominus}$ and $E^{\ominus}$

The free energy calculated above could, for example, be produced as electricity and used, say, to lift a load by powering a motor. This can be done if the reaction is carried out in an electrochemical cell, see **Figure 18.5.** $E^{\ominus}$ for this cell = 0.46 V. How much electrical energy can it produce?

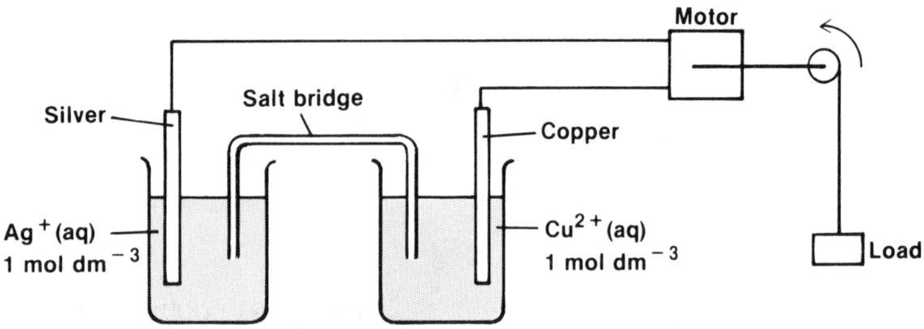

Figure 18.5 Using the free energy from a cell to lift a load

$E^{\ominus}$, the e.m.f., which is measured in volts, is equal to the number of joules of energy per coulomb of electric charge.

In other words: $E^{\ominus}$ = joules of energy/coulombs

Rearranging: joules of energy = $E^{\ominus}$ × coulombs

How do we find the number of coulombs?

$$2Ag^{+}(aq) + Cu(s) \longrightarrow 2Ag(s) + Cu^{2+}(aq)$$

(loss of $2e^-$: Cu → Cu^{2+}; gain of $2e^-$: Ag^+ → Ag)

For one mole of this reaction to take place, 2 moles of electrons must be transferred. 2 moles of electrons = 2 faradays = 2 × 96 500 coulombs (see **section 17.3.3**).

So: energy (in joules) = $E^{\ominus} \times 2 \times 96\,500$

and in general we can say:

$$\text{energy} = E^{\ominus} \times z \times F \text{ joules}$$

where z = number of moles of electrons transferred in the reaction

$$\text{energy} = 0.46 \times 2 \times 96\,500\,\text{J}$$
$$= 88\,800\,\text{J per mole of reaction}$$
$$= 88.8\,\text{kJ mol}^{-1}$$

Compare this with the value of $\Delta G^{\ominus}$ calculated above of $-89.5\,\text{kJ mol}^{-1}$. The small discrepancy is solely due to rounding errors in the values used in the calculations. This is the energy available (free) to do work, i.e. it is $\Delta G^{\ominus}$, and since the word 'available' implies it can be given out (as opposed to taken in), it has a *negative* value.

We can now write the following relationship:

$$\Delta G^{\ominus} = -zFE^{\ominus}$$

This fits in with what we found in Chapter 9, that if $E^{\ominus}$ is positive, the reaction is feasible. We now know that $\Delta G^{\ominus}$ must be negative for the reaction to be feasible and we can see from the equation that $E^\ominus$ must be positive for this to be so.

Example
What is the free energy change of the reaction:

$$Zn(s) + Cu^{2+}(aq) \longrightarrow Zn^{2+}(aq) + Cu(s)?$$

$E^\ominus$ for this reaction is $+1.1$ V.

$$\Delta G^\ominus = -zFE^\ominus$$

When one mole of reaction occurs, two electrons are transferred from zinc to copper, so $z = 2$.

$$\Delta G^\ominus = -2 \times 96\,500 \times 1.1 \text{ J mol}^{-1}$$
$$\Delta G^\ominus = -212\,300 \text{ J mol}^{-1}$$
$$\Delta G^\ominus = -212.3 \text{ kJ mol}^{-1}$$

18.3.3 $\Delta G^\ominus$ and the equilibrium constant

We can treat all reactions as reversible ones which reach equilibrium, even those that seem to go to completion. We can say that in this case the amounts of reactants remaining at equilibrium are so small that we can ignore them. $\Delta G^\ominus$ tells us whether a reaction is feasible or not. The equilibrium constant K tells us whether reactants or products predominate. Both are essentially telling us the same thing, because if $\Delta G^\ominus$ is negative, products will predominate over reactants and if $\Delta G^\ominus$ is positive, reactants will predominate over products. If $\Delta G^\ominus$ is zero we will have an equilibrium where products and reactants are equally likely. Is there a mathematical relationship between $\Delta G^\ominus$ and K? **Table 18.4** shows the situation.

Table 18.4 Comparison between equilibrium constant, *K*, and free energy change, ΔG

K	$\Delta G^\ominus$
Large	Negative
1	0
Small	Positive

The fact that when $\Delta G^\ominus = 0$, $K = 1$ suggests a logarithmic relationship ($\ln 1 = 0$). Negative values of $\Delta G^\ominus$ giving values of K suggests:

$$-\Delta G^\ominus \propto \ln K$$

Both $\Delta G^\ominus$ and K are temperature dependent. The relationship is:

$$\Delta G = -RT \ln K$$

where R is the gas constant, which frequently turns up in expressions involving molecular collisions.
Note: Units are consistent: R has units J K^{-1} mol^{-1}, T has units K, so $\Delta G^\ominus$ has units J K^{-1} mol^{-1} $\times$ K = J mol^{-1}.

We now have three parameters which tell us about reaction feasibility. They are $\Delta G^\ominus$, $E^\ominus$ and K, the equilibrium constant.

They are linked by these equations, which allow us to find values for $\Delta G^\ominus$:

$$\Delta G^\ominus = -RT \ln K$$
$$\Delta G^\ominus = -zFE$$

Table 18.5 compares them.

Table 18.5 Comparison between values of $\Delta G^\ominus$, $E^\ominus$ and K

	'Does not occur'		'Equilibrium' Reactants predominate		Products predominate	'Complete'	
$\Delta G^\ominus$/kJ mol^{-1}	+85.5	+57	+28.5	0	−28.5	−57	−85.5
$E^\ominus$/V	−0.885	−0.59	−0.295	0	+0.295	+0.59	+0.885
K/units depend on reaction	10^{-15}	10^{-10}	10^{-5}	1	10^{5}	10^{10}	10^{15}

As a rule of thumb, we often take it that reactions where $K > 10^{10}$ are complete and reactions where $K < 10^{-10}$ do not occur. Between these values we think of the reaction as an equilibrium between products and reactants. The values of $E^\ominus$ and $\Delta G^\ominus$ which correspond to $K = 10^{10}$ are

0.59 V and $-57\,\text{KJ mol}^{-1}$ respectively. Since this is only a rule of thumb, we often remember $K = 10^{10}$, $E^{\ominus} = 0.6\,\text{V}$ and $\Delta G^{\ominus} = -60\,\text{kJ mol}^{-1}$ as the borderlines.

18.4 Reaction feasibility and temperature

Since temperature appears in the equation:

$$\Delta G^{\ominus} = \Delta H^{\ominus} - T\Delta S^{\ominus}$$

some reactions may be feasible at some temperatures and not at others. Take the reaction:

$$CaCO_3(s) \longrightarrow CaO(s) + CO_2(g) \quad \Delta H^{\ominus} = +178\,\text{kJ mol}^{-1}$$

Here a solid produces a solid and a gas, so the entropy change of the system will be positive. The enthalpy change is positive. Will this reaction go or not? **Table 18.6** gives the values of the relevant standard entropies.

Table 18.6 Standard entropy values

Species	$S^{\ominus}/\text{J K}^{-1}\,\text{mol}^{-1}$
$CaCO_3(s)$	93
$CaO(s)$	40
$CO_2(g)$	214

$$\begin{aligned}\Delta S^{\ominus} &= S^{\ominus}_{\text{products}} - S^{\ominus}_{\text{reactants}}\\ &= (40 + 214) - 93\,\text{J K}^{-1}\,\text{mol}^{-1}\\ &= +161\,\text{J K}^{-1}\,\text{mol}^{-1}\end{aligned}$$

Neither $\Delta H^{\ominus}$ or $\Delta S^{\ominus}$ change a great deal with temperature, so the temperature itself is very significant. At 298 K:

$$\begin{aligned}\Delta G^{\ominus} &= \Delta H^{\ominus} - T\Delta S^{\ominus}\\ &= +178 - \frac{298 \times 161}{1000}\\ &= +178 - 48\,\text{kJ mol}^{-1}\\ \Delta G^{\ominus} &= +130\,\text{kJ mol}^{-1}, \quad \therefore \textit{not}\ \text{feasible}\end{aligned}$$

At 2000 K:

$$\begin{aligned}\Delta G &= +178 - \frac{2000 \times 161}{1000}\\ &= +178 - 322\\ &= -144\,\text{kJ mol}^{-1}, \quad \therefore \text{feasible}\end{aligned}$$

At what temperature does this reaction become feasible?

The borderline between feasible and not feasible is when $\Delta G = 0$. Substituting this into our equation:

$$\begin{aligned}0 &= +178 - \frac{T \times 161}{1000}\\ 0.161T &= 178\\ T &= 1106\,\text{K}\end{aligned}$$

The reaction is not feasible below 1106 K and is feasible above 1106 K. In fact the reaction does not 'jump' between being feasible and not feasible. An equilibrium will exist around this temperature.

18.4.1 Calculating $\Delta G^{\ominus}$ values from thermochemical cycles

It is possible to calculate $\Delta G^{\ominus}$ using thermochemical cycles, in just the same way as for enthalpy changes. For example, is the reaction:

$$Cu(s) + ZnO(s) \longrightarrow CuO(s) + Zn(s)$$

feasible under standard conditions?

$\Delta_f G^{\ominus}(ZnO) = -318\,\text{kJ mol}^{-1}$
$\Delta_f G^{\ominus}(CuO) = -128\,\text{kJ mol}^{-1}$

$\Delta_f G^{\ominus}$ of all elements in their standard states = 0

E

APPLICATIONS OF $\Delta G^\ominus$; ELLINGHAM DIAGRAMS

An important industrial problem is extracting metals from their ores. Ores are usually metal oxides or compounds like sulphides and carbonates which can easily be converted to oxides by heating them in air (roasting them). The problem is to find a suitable reducing agent which will convert the oxide to the metal. The reducing agent should ideally be cheap and carry out the reduction at as low a temperature as possible to reduce fuel costs. An Ellingham diagram allows us to predict which reducing agents are suitable for a particular application. It is a plot of $\Delta G^\ominus$ against T for a variety of relevant reactions. All the reactions are written so that they involve one mole of oxygen molecules.

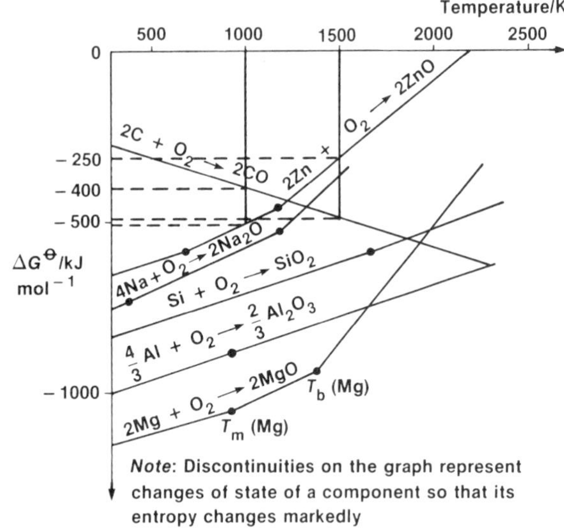

For example, can carbon be used to reduce zinc oxide to zinc at **a)** 1000 K **b)** 1500 K?

The reaction we require is:

$$2C + 2ZnO \longrightarrow 2CO + 2Zn$$

The graph shows that at 1000 K we have the following $\Delta G^\ominus$ values:

1. $2C + O_2 \longrightarrow 2CO \quad \Delta G^\ominus = -400\ \text{kJ mol}^{-1}$

2. $2Zn + O_2 \longrightarrow 2ZnO \quad \Delta G^\ominus = -550\ \text{kJ mol}^{-1}$

So, reversing reaction **2**,

3. $2ZnO \longrightarrow 2Zn + O_2 \quad \Delta G^\ominus = +550\ \text{kJ mol}^{-1}$

Reversing the reaction reverses the sign of $\Delta G^\ominus$.

Adding **1** and **3** we get:

$$2C + \not{O}_2 + 2ZnO \longrightarrow 2CO + 2Zn + \not{O}_2$$

and $\Delta G^\ominus = -400 + 550 = +150\ \text{kJ mol}^{-1}$

so at 1000 K the reaction is *not* feasible.

At 1500 K we have the following $\Delta G^\ominus$ values:

4. $2C + O_2 \longrightarrow 2CO \quad \Delta G^\ominus = -500\ \text{kJ mol}^{-1}$

5. $2Zn + O_2 \longrightarrow 2ZnO \quad \Delta G^\ominus = -250\ \text{kJ mol}^{-1}$

So:

6. $2ZnO \longrightarrow 2Zn + O_2 \quad \Delta G^\ominus = +250\ \text{kJ mol}^{-1}$

Adding **4** and **6**:

$$2ZnO + 2C \longrightarrow 2CO + 2Zn \quad \Delta G^\ominus = -250\ \text{kJ mol}^{-1}$$

At 1 500 K the reaction *is* feasible.

You can probably see from the graph that where the two lines cross, $\Delta G^\ominus = 0$. Thus the reaction above is feasible at temperatures above 1200 K.

The same procedure can be used to investigate the feasibility of other reductions with carbon and with other reducing agents.

It is possible to short-circuit the above procedure: at any temperature, the reaction represented by the *lower* line will go *forwards* and the reaction represented by the *upper* line will go *in reverse*. So that carbon will only reduce magnesium oxide at temperatures above 2000 K. Carbon, in the form of coke, is used as a reducing agent in a number of industrial processes, most notably in the production of iron, where iron oxide is reduced to iron in the blast furnace.

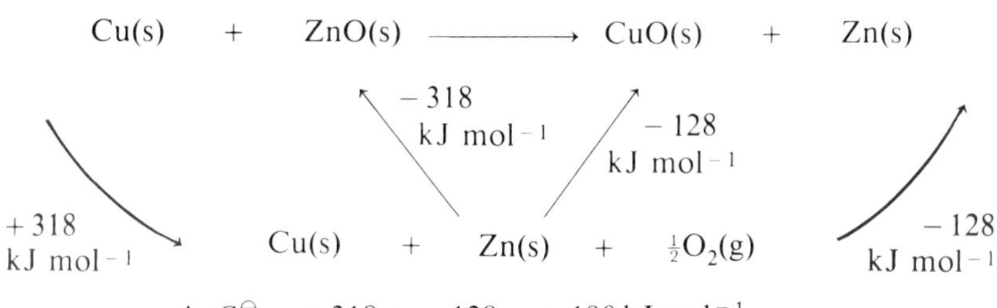

$$\Delta_R G^\ominus = +318 + -128 = +190\ \text{kJ mol}^{-1}$$

so the reaction is not feasible at 298 K.

18.4.2 Rates

We must always remember that if we predict that a reaction is feasible (i.e. that $\Delta G^\ominus$ is negative, $E^\ominus$ is positive or K is large), the reaction may not actually occur at a measurable rate. Thermodynamics tells us only about the initial and final states of a reaction, and nothing else. It may well be that there is a large activation energy barrier which prevents the reaction taking place at a reasonable speed, see **Figure 18.6.** $\Delta G^\ominus$ tells us *nothing* about the height of the activation energy 'barrier'. However, if $\Delta G^\ominus$ is negative but the reaction is too slow, we might look for a catalyst. If $\Delta G^\ominus$ is positive, there is no point in looking for a catalyst as the reaction is not feasible.

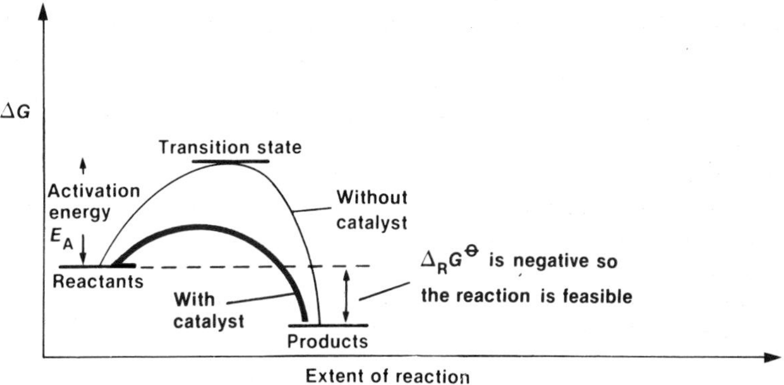

Figure 18.6 **A reaction profile for a reaction where ΔG is negative**

Increasing the temperature will always increase the rate of a reaction. Sometimes it makes $\Delta G^\ominus$ negative, which makes the reaction feasible. For example, as we saw earlier, calcium carbonate needs to be heated to above 1106 K before $\Delta G^\ominus$ is negative and the breakdown to calcium oxide and carbon dioxide is possible. At higher temperatures still the rate of breakdown will increase, but these two effects of temperature are quite separate.

18.5 Summary

- Since molecules move at random, reactions occur by chance alone, i.e. the most probable outcome occurs.
- Chance predicts that reactions will go in the direction of increased randomness.
- Entropy, S, is a measure of the randomness of a system—it is related to the number of ways (W) of arranging both the molecules and energy quanta in the system:

$$S = k \ln W \quad \text{where } k \text{ is the Boltzmann constant}$$

- Reactions can occur spontaneously when they involve an overall increase in entropy.
- This increase in entropy must be the total change of both the system (reactants and products) and the surroundings (everything else).
- If heat is transferred to the surroundings, the entropy of the surroundings will increase:

$$\Delta S_{\text{surroundings}} = \frac{-\Delta H_{\text{reaction}}}{T}$$

- For a spontaneous reaction, $\Delta S_{\text{total}} > 0$

$$\Delta S_{\text{total}} = \Delta S_{\text{system}} + \Delta S_{\text{surroundings}}$$

- The Gibbs free energy change $\Delta G^\ominus$ is the energy available from a reaction after enough enthalpy has been transferred to the surroundings to compensate for any entropy decrease of the system caused by the reaction.

$$\Delta G^\ominus_{\text{reaction}} = \Delta H^\ominus_{\text{reaction}} - T\Delta S^\ominus_{\text{system}}$$

This equation allows us to forget the surroundings. All the terms refer to the system only.

- Reactions are only spontaneous if $\Delta G^\ominus$ is negative, *but* they may proceed slowly: $\Delta G^\ominus$ tells us nothing about the rate.

- $\Delta G^\ominus$ is related to $E^\ominus$ for a cell reaction and to the equilibrium constant K (K_c or K_p):

$\Delta G^\ominus = -zFE^\ominus$ where z is the number of electrons transferred and F the Faraday constant

$\Delta G^\ominus = -RT\ln K$ where R is the gas constant and T the temperature

- $\Delta G^\ominus$ values can be calculated by thermochemical cycles in the same way as $\Delta H^\ominus$s.
- $\Delta S^\ominus$ values can be calculated by

$$\Delta S = S_{\text{products}} - S_{\text{reactants}}$$

18.6 Questions

1 **A)** How many ways are there of arranging four particles A, B, C and D in two boxes, 1 and 2? Write them down in a table beginning:

1.	**2.**
A B C D	0
A B C	D

B) What is the probability of finding all the particles in box 2?

C) What percentage of the arrangements are 3:1 or 2:2 arrangements?

2 **A)** Predict whether the entropy change for the following reactions will be significantly positive, significantly negative or approximately zero and explain your reasoning.

i $Mg(s) + ZnO(s) \longrightarrow MgO(s) + Zn(s)$

ii $2Pb(NO_3)_2(s) \longrightarrow 2PbO(s) + 4NO_2(g) + O_2(g)$

iii $MgO(s) + CO_2(g) \longrightarrow MgCO_3(s)$

iv $H_2O(l) \longrightarrow H_2O(g)$

B) Calculate $\Delta S^\ominus$ for each reaction using the data below. Comment on your answers.

	$S^\ominus$/J K^{-1} mol^{-1}
$Mg(s)$	32.7
$MgO(s)$	26.9
$MgCO_3(s)$	65.7
$Zn(s)$	41.6
$ZnO(s)$	43.6
$Pb(NO_3)_2(s)$	213.0
$PbO(s)$	68.7
$NO_2(g)$	240.0
$O_2(g)$	205.0
$CO_2(g)$	213.6
$H_2O(l)$	69.7
$H_2O(g)$	188.7

3 For the reaction:

$$MgO(s) \longrightarrow Mg(s) + \tfrac{1}{2}O_2(g)$$

$$\Delta H^\ominus = +602\ \text{kJ mol}^{-1}$$
$$\Delta S^\ominus = +109\ \text{J K}^{-1}\,\text{mol}^{-1}$$

A) Without doing a calculation, is it possible to predict if the reaction is feasible or not? Explain your reasoning.

B) Using the equation $\Delta G^\ominus = \Delta H^\ominus - T\Delta S^\ominus$, calculate $\Delta G^\ominus$ at

i 1000 K

ii 6000 K.

At which temperature is the reaction feasible?

C) Calculate the temperature when $\Delta G^\ominus = 0$.

4 For the following reactions calculate:

A) ΔS_{system}

B) $\Delta S_{\text{surroundings}}$ at **i)** 298 K **ii)** 2000 K

C) ΔS_{tot} at **i)** 298 K **ii)** 2000 K

A $N_2(g) + 3H_2(g) \longrightarrow 2NH_3(g)$
$\Delta H^\ominus = -92\ \text{kJ mol}^{-1}$

B $2C(gr) + SiO_2(s) \longrightarrow 2CO(g) + Si(s)$
$\Delta H^\ominus = +344\ \text{kJ mol}^{-1}$

	$S^\ominus$/J K^{-1} mol^{-1}
$N_2(g)$	191.6
$H_2(g)$	130.6
$NH_3(g)$	192.3
$C(gr)$	5.7
$SiO_2(s)$	41.8
$CO(g)$	197.6
$Si(s)$	18.8

5 Use a thermochemical cycle using $\Delta G^\ominus$ values to find $\Delta G^\ominus$ for the reaction:

$$Mg(s) + Cu^{2+}(aq) \longrightarrow Mg^{2+}(aq) + Cu(s)$$

$E^\ominus$ for the reaction is $+2.71$ V.

Calculate $\Delta G^\ominus$ using $\Delta G^\ominus = -zFE^\ominus$.

	$\Delta_f G^\ominus$/kJ mol^{-1}
Mg^{2+}(aq)	−455
Cu^{2+}(aq)	+ 66

6 For the reaction:

$$N_2O_4(g) \rightleftharpoons 2NO_2(g)$$

$\Delta G^\ominus$ at 500 K is −31 kJ mol^{-1}. Calculate the equilibrium constant at this temperature.

7 For the reaction:

$$H_2O(g) + C(gr) \longrightarrow H_2(g) + CO(g)$$

$\Delta G^\ominus$/kJ mol^{-1}	T/K
91.3	298
34.2	700
4.2	900
−26.0	1100
−57.6	1300

Plot a graph of $\Delta G^\ominus$ against T.

A) Does the graph support the relationship:

$$\Delta G^\ominus = -RT \ln K?$$

B) Find the temperature at which $\Delta G^\ominus = 0$.

C) In what range of temperature would the reaction be considered:

i not to take place at all
ii to be at equilibrium
iii to go to completion?

D) Calculate the equilibrium constant K_p at 700 K.

8 Use the Ellingham diagram on page 246 for this question.

A) Which substance could possibly reduce SiO_2 to Si at temperatures below 1700 K?

B) Above what temperature can carbon reduce SiO_2 to Si?

C) Which substances and what conditions would you recommend to reduce SiO_2 to Si, bearing in mind economic as well as chemical factors? Explain your choices.

9 A) Calculate $\Delta H^\ominus$ and $\Delta S^\ominus$ for the reaction:

$$4CuS(s) + 5O_2(g) \longrightarrow 2Cu_2O(s) + 4SO_2(g)$$

from the data below:

	$\Delta H_f^\ominus$/kJ mol^{-1}	$S^\ominus$/J K^{-1} mol^{-1}
CuS(s)	−48.5	+66.5
SO_2(g)	−296.9	+248.5
Cu_2O(s)	−166.7	+100.8
O_2(g)	0	+204.9

B) Use the expression $\Delta G^\ominus = \Delta H^\ominus - T\Delta S^\ominus$ to calculate $\Delta G^\ominus$ for this reaction.

C) For the reaction

$$2CuS(s) + 3O_2(g) \longrightarrow 2CuO(s) + 2SO_2(g)$$

$$\Delta H^\ominus = -807.2 \text{ kJ mol}^{-1}$$

$$\Delta S^\ominus = -163.7 \text{ J K}^{-1} \text{ mol}^{-1}$$

Calculate $\Delta G^\ominus$ for this reaction.

D) Which reaction is more likely to occur at 298 K?

E) Why can you not say for certain which reaction will actually occur at 298 K?

(Nuffield special paper, 1985)

10 A) If 80 molecules were placed in a box, what is the chance all 80 would be in the right-hand half of the box? (Use the expression: number of arrangements $= 2^n$ and remember the y^x button on your calculator.)

B) If a computer simulation could shuffle the molecules at a rate of 10^6 every second, how long would it take to try out all the combinations? How long is that in years?

C) i $S^\ominus$ of argon is 155 J K^{-1} mol^{-1}. 1 mole of argon has 6×10^{23} atoms. How many *extra* arrangements result if the volume is doubled? Imagine the gas was all in one jar and suddenly a second empty jar was added to the system. Don't try to work the number out.

ii Use $\Delta S = k\Delta \ln W$ to work out ΔS for the volume doubling, where $k = 1.38 \times 10^{-23}$ J K^{-1}.

iii What is the entropy of the mole of argon after doubling its volume?

19 Structure determination

19.1 Introduction

The structure of a substance is the way in which its atoms are arranged in space. Determining structure involves measuring the distances and the angles between the atoms—in fact obtaining enough information to build a scale model of the structure. It is important to realize that this is a three-dimensional problem.

One major difficulty is obvious—atoms are too small to be seen. If atoms were visible, structure determination would be easy. As it is, we have to use indirect methods, often several methods for one structure, and deduce the arrangement by detective work.

19.2 Structure determination of solids with giant structures

Figure 19.1 Water waves pass an object much smaller than their wavelength undisturbed

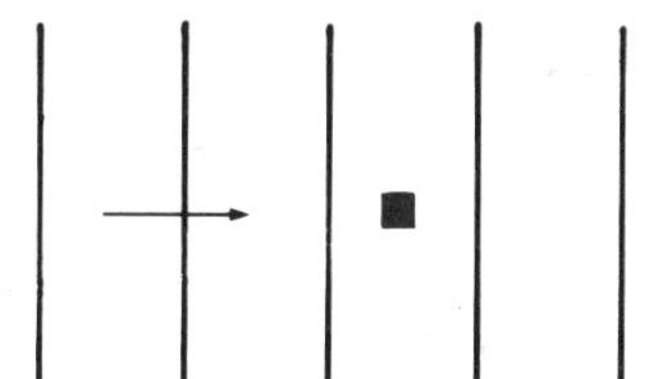

Max von Laue

The starting point for any structure determination is the formula of the substance concerned. This means we must know what elements are present and in what proportions. We may know this from the way we made or obtained the substance, or we may have to use a variety of analytical techniques. Some of these are described in Chapter 37.

19.2.1 X-ray diffraction

This is by far the most important method of determining the structure of solids. It depends on the fact that X-rays have wavelengths of about the same size as the distance between the nuclei of atoms—around 0.1 nm (10^{-10} m as 1 nm = 10^{-9} m). Atoms have no solid exteriors so we use the nuclei as the fixed points.

The reason we cannot see atoms is that the wavelength of visible light is too large. It is around 5×10^{-7} m = 500 nm, which is about 5000 times the size of an atom. So light waves are not affected by atoms, in the same way that the water waves in **Figure 19.1** are unaffected by the small obstacle. This is a fundamental limitation: the problem is not merely that microscopes that are powerful enough have not yet been built. It takes an object of around the same size as the wavelength to significantly affect the passage of waves.

So, to 'observe' atoms we need radiation of wavelength of approximately the same size as atoms—hence the use of X-rays. This was first realized by Max von Laue in 1912.

In **section 7.2** we saw how waves are diffracted when they pass through small gaps, of size comparable with their wavelengths and also how they can interfere constructively and destructively. The photograph in **section 7.2.2** shows a diffraction pattern produced by water waves

Figure 19.2 The Bragg condition

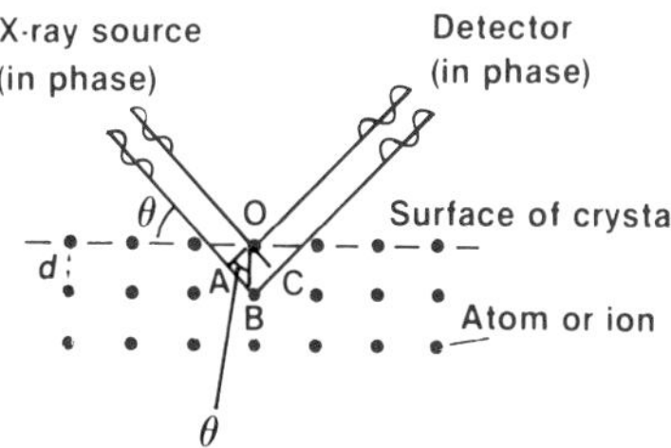

interfering after being diffracted through small gaps. A similar sort of pattern is observed when X-rays are directed at a crystal, which is a regularly spaced array of atoms or ions. Analysis of such a pattern enables the spacing between the atoms to be calculated. The basis of this calculation is shown in the next section.

The Bragg condition

Figure 19.2 shows what happens when a beam of X-rays is shone on to the surface of a crystal which is composed of a regular array of atoms or ions. The X-rays behave as through they are reflected by the atoms (or ions). They are actually absorbed and re-emitted by them. Those reflected

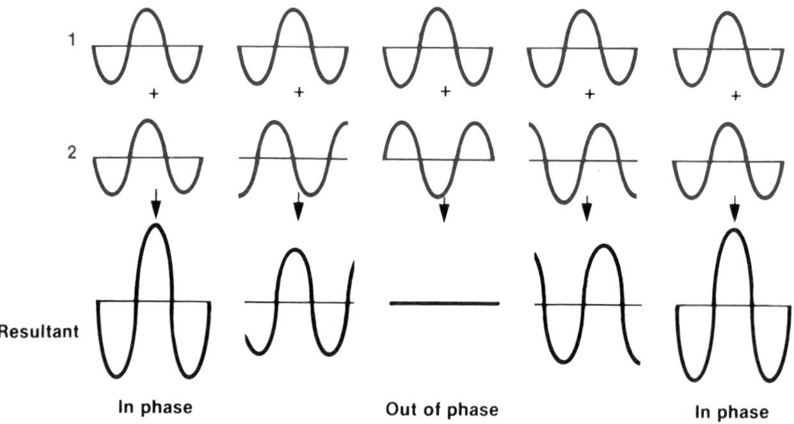

Figure 19.3 (right) Wave 2 lags further and further behind wave 1 until it is in phase again but one wavelength behind

by atoms in the second layer travel a distance ABC further than those reflected from the surface atoms, so they 'lag behind'. Unless they are in phase with the X-rays reflected from the first layer, they will cancel out (destructive interference) and no X-rays will be detected. Only if the second layer X-rays lag behind by a *whole number of wavelengths* does constructive interference occur and so X-rays can be detected, **Figure 19.3.**

So the X-rays are in phase when

$$ABC = n\lambda$$

where n is a whole number and λ is the wavelength of the X-rays.

In triangle OAB (**Figure 19.4**),

$$\frac{AB}{OB} = \sin\theta$$

so

$$AB = OB\sin\theta$$
$$AB = d\sin\theta$$

where d is the spacing between the atoms.

As triangle OBC is the same as triangle OAB:

$$\text{distance } ABC = 2d\sin\theta$$

and for the X-rays to be in phase:

$$n\lambda = 2d\sin\theta$$

William Bragg and his son Lawrence, who together won a Nobel Prize in 1915 for their work on X-rays and crystals

This relationship is called the **Bragg condition** after William and Lawrence Bragg, the father-and-son team who pioneered X-ray diffraction. It means that if we know the wavelength of the X-rays we are using and can measure the angle at which we get the maximum X-ray signal, we can calculate d, the distance between the atoms in the crystal. We may get more than one angle where constructive interference occurs, when the 'lag' is 1, 2 or 3 wavelengths, i.e. $n = 1$, 2 or 3.

The X-rays are produced by bombarding a metal target with electrons, as described in **section 8.2.6.** The detector may be a photographic film or a Geiger–Müller tube.

Figure 19.4 Deriving the Bragg condition

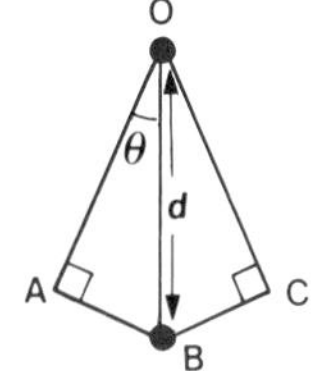

Figure 19.5 Apparatus for single crystal X-ray diffraction

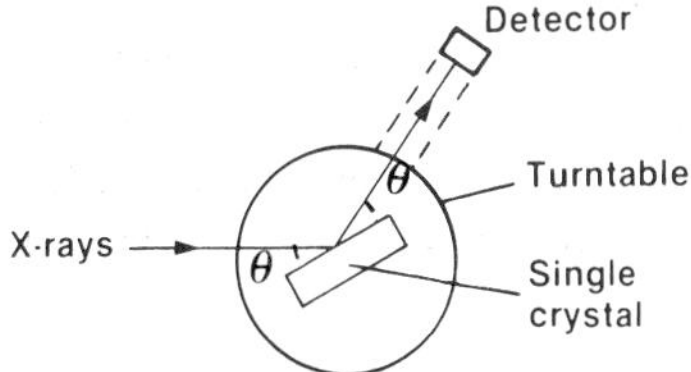

The single crystal method

In this method a single crystal is mounted on a rotating stand as shown in **Figure 19.5** and a beam of X-rays directed at its face. The crystal is

Figure 19.6 **Trace obtained from X-ray diffraction of sodium chloride**

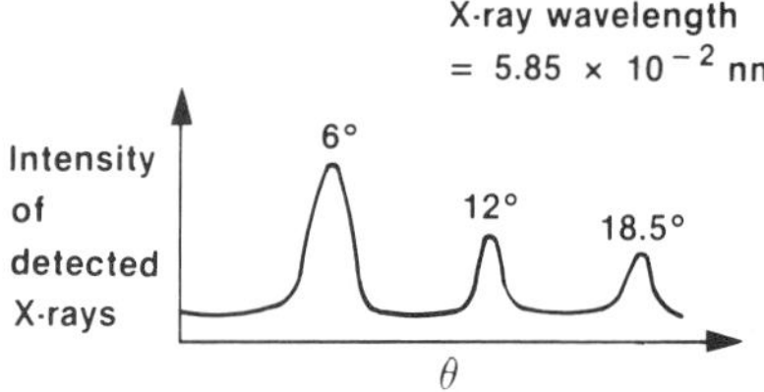

Apparatus for single crystal X-ray diffraction

Figure 19.7 **Electron density contour map (obtained from X-ray diffraction) of urea (carbamide) compared with its formula. The positions of the carbon, nitrogen and oxygen atoms can be clearly seen on the map. Oxygen, with eight electrons, has significantly more closely spaced contours than carbon with only six**

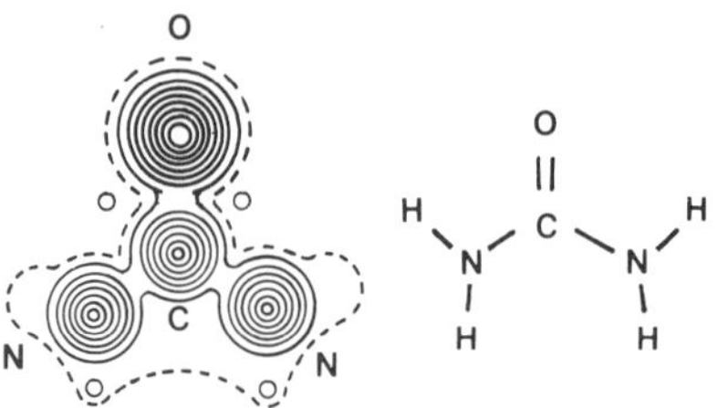

Figure 19.8 **Electron density contour map for solid iodine**

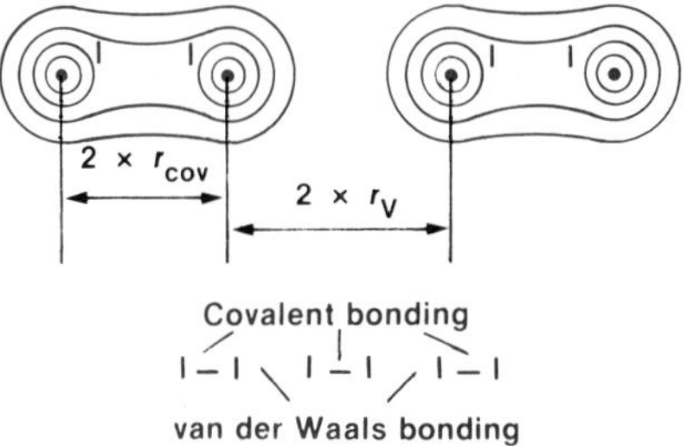

rotated along with the detector and the intensity of the reflected beam is measured at different angles.

A trace is obtained like the one shown in **Figure 19.6** for sodium chloride. The three peaks correspond to $n = 1$, $n = 2$ and $n = 3$.

So $$n\lambda = 2d \sin\theta$$

$$d = \frac{n\lambda}{2 \sin\theta}$$

For $n = 1$, $$d = \frac{1 \times 5.85 \times 10^{-2}}{2 \times 0.1045}\,\text{nm}$$

$$d = 0.280\,\text{nm}$$

Similarly for $n = 2$, we find:

$$d = 0.281\,\text{nm}$$

And for $n = 3$:

$$d = 0.277\,\text{nm}$$

The average of these gives the distance between the ions as 0.279 nm. This is the distance between the centres of the ions, their nuclei.

Electron density maps

The X-rays are actually diffracted by the *electrons* in the atoms. So one way of presenting the results is as an **electron density map**, where contours joining positions of equal electron density (electrons per cubic nanometre) represent the electron distribution. An example is shown in **Figure 19.7.**

19.2.2 Sizes of atoms and ions

X-ray diffraction can measure the distance between the centres of adjacent atoms in a structure. It does not measure from the centre of an atom to its edge, as it is not obvious where the atom stops. Look again at the electron density map of urea, **Figure 19.7.** It is clear where the centres of the atoms are, but not where their edges are. In this respect atoms are more like party balloons than billiard balls. They are squashy rather than hard. This has implications for the measurement of the radii of atoms and ions.

Metallic radius, r_m

The metallic radius is defined as half the distance between the centres of adjacent atoms in a metallic giant structure.

Covalent radius, r_{cov}

This is defined as half the distance between the centres of adjacent atoms of the same element which are held together by a single covalent bond.

van der Waals radius, r_v

In the electron density map of solid iodine (**Figure 19.8**), there are two I–I distances. The smaller is between a pair of covalently bonded atoms and is twice the covalent radius. The larger is between two iodine atoms in different molecules. These are not covalently bonded but held close

A

DNA

Since the Second World War, the structures of many complex substances have been worked out by X-ray diffraction. Perhaps the most significant was that of DNA, the molecule which carries the genetic information in cells. This structure was worked out in the 1950s by Francis Crick, James Watson, Maurice Wilkins and Rosalind Franklin at the Universities of Cambridge and London. The first three were awarded a Nobel Prize in 1962 for this work. Ms Franklin would almost certainly have joined them but for her death before the prize was awarded. The model is shown in **Plate 5.**

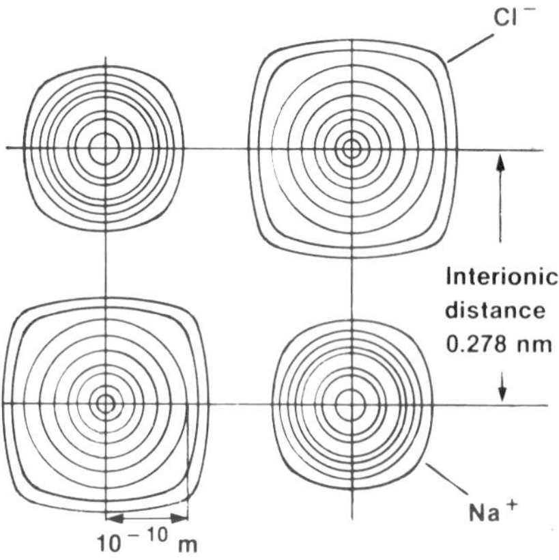

Figure 19.9 Electron density contour map for sodium chloride

together by van der Waals forces (see **section 16.2.2**). Half this distance is called the van der Waals radius. Since van der Waals bonding is weaker than covalent bonding, the atoms are less closely held together and the van der Waals radius is larger than the covalent radius, e.g. for iodine:

$$r_{cov} = 0.128\ \text{nm}$$
$$r_V = 0.177\ \text{nm}$$

Ionic radius, r_i

Here the situation is more tricky. Look at the electron density map for sodium chloride, **Figure 19.9.** It is easy to measure the interionic distance (from the centre of Na^+ to the centre of Cl^-), but it is not easy to see where the Na^+ stops and the Cl^- starts. Since the two ions are different we cannot just halve the interionic distance. The procedure is to make an intelligent guess at the radius of one ion, and then this effectively fixes all the other ionic radii. For example, if we give K^+ the value of 0.137 nm then, since the interionic distance in K^+Cl^- is 0.314 nm, the radius of Cl^- must be 0.177 nm. Then, since the interionic distance in Na^+Cl^- is 0.278 nm, the ionic radius of Na^+ must be 0.101 nm, and so on. Because of this element of guesswork there are often discrepancies between ionic radii given in different data books.

E

SOME VALUES OF ATOMIC AND IONIC SIZES

Note:

- Size increases going down a group.
- van der Waals radius is greater than covalent radius or metallic radius.
- Forming negative ions *increases* the size of the atom.
- Forming positive ions *decreases* the size of the atom.

(All radii in nm.)

	r_v	r_{cov}	r_{ionic}	$r_{metallic}$
I	0.195	0.133	0.215	—
Br	0.190	0.114	0.195	—
Cl	0.180	0.099	0.180	
Li	0.180	0.134	0.074	0.157
Na	0.230	0.154	0.102	0.191
K	0.280	0.196	0.138	0.235

FIXING THE IONIC RADIUS OF K^+

In the compound potassium chloride, K^+Cl^-, both ions have 18 electrons—they are **isoelectronic**. The interionic distance is 0.314 nm. How is this shared between K^+ and Cl^-? The nuclear charge on K^+ is 19 and on Cl^- 17.

K^+'s electronic structure is 2, 8, 8 so the outer electrons 'feel' an effective nuclear charge of 19 − 10 (for the inner electrons) = 9.

Cl^-'s electron structure is also 2, 8, 8. Here the outer electrons 'feel' an effective nuclear charge of 17 − 10 = 7. So Cl^- will be *bigger* than K^+ because the outer electrons are attracted less strongly by a factor of 7 to 9.

So radius of Cl^- = 9/16 × 0.314 = 0.177 nm

radius of K^+ = 7/16 × 0.314 = 0.137 nm

19.3 Structure determination of molecular substances

X-ray diffraction can be used to find the structure of molecular substances in the solid state. For example, we have seen the electron density maps of urea, $CO(NH_2)_2$, and iodine, I_2, in **section 19.2.** X-ray diffraction is the only method of finding for certain the shapes, bond lengths and bond angles of molecules. However, it is a specialized technique and by combining the evidence obtained from several simpler methods, we can often deduce a good deal about a molecule's structure without resorting to X-ray diffraction.

19.3.1 Formulae

First we need to know the formula of the substance under investigation. This requires **qualitative analysis** to find what elements are present and **quantitative analysis** to find the masses of each. This will give us the ratios of the number of moles of each element present. This is called the **empirical** (or simplest) **formula**. For example, ethene has an empirical formula of CH_2. However, this could represent several different molecules—CH_2, C_2H_4, C_3H_6, and so on. To decide which, we need to know the relative molecular mass, M_r. A variety of techniques are available for this, including gas volume measurement and Graham's law (**section 15.3**), freezing temperature depression, boiling temperature elevation and osmotic pressure (**section 16.3.5**), and mass spectrometry (**section 8.5**). The preferred technique is mass spectrometry as it is the most accurate, but this is beyond the resources of school laboratories. Ethene has a relative molecular mass of 28, twice the relative molecular mass of the empirical formula CH_2, so the molecule must contain two units, i.e. C_2H_4. This is called the **molecular formula**—it shows the number and type of atoms in the actual molecule. It is always a multiple of the empirical formula.

Figure 19.10 The three types of vibration of a water molecule

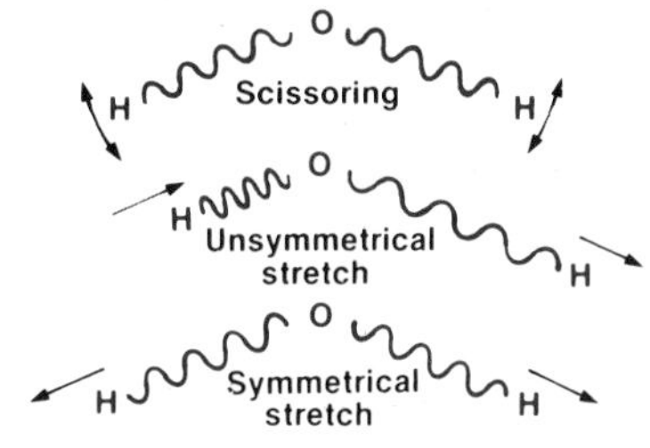

19.3.2 Chemical reactivity

This may give clues to the structure. For example, ethene decolorizes a solution of bromine. This test reveals that ethene has a carbon–carbon double bond (C=C). You may have used Benedict's test to identify a reducing sugar like glucose. In fact the test specifically indicates the present of an aldehyde group (C(=O)H) in the molecule (see **section 32.6.3**). There are many examples of specific reactions which show the presence of a particular feature in the molecule, many of which are described later in this book. Other simple clues include acidity, and also solubility in water which may indicate whether or not there are polar groups.

Figure 19.11 A mass vibrating on a spring

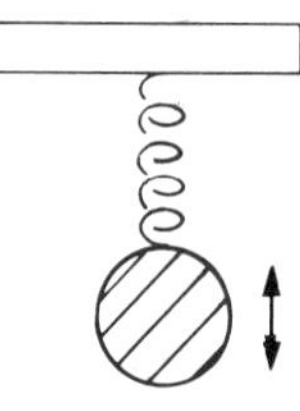

19.3.3 Infra-red (IR) spectroscopy

This is an instrumental method which can be used to identify particular groups of atoms. It depends on the fact that molecules or parts of molecules can vibrate—in fact, a molecule behaves as though its atoms were masses connected by springs (representing the bonds). For example, a water molecule can vibrate in three ways, **Figure 19.10.** Molecules with many atoms can have very complex vibrations. We shall concentrate on vibrations involving just two atoms. The frequency of vibration (number of vibrations per second) depends on the masses of the atoms and the strength of the bond. Imagine a ball of Plasticine vibrating on a spring, **Figure 19.11.** The more massive the Plasticine, the lower the frequency, and the weaker the spring, the lower the frequency. The same is true of atoms, stronger bonds behaving like stiffer springs, i.e. they vibrate at a higher frequency than weaker bonds. So a particular type of bond between a certain pair of atoms will have a characteristic frequency of vibration. Like all molecular energy, this vibration is **quantized** (its amplitude can only have discrete values). A particular bond can only absorb quanta of energy of its own characteristic frequency. When it has done so it vibrates with a greater amplitude (the atoms move more). This is like a child on a swing being pushed by its mother, see photograph. The child can only pick up energy and swing further if the mother times her pushes at exactly the frequency of the swing—always pushing as the swing is moving away from her.

Energy is transferred from mother to child only if she pushes at the correct frequency

The frequencies of molecular vibrations fall in the region of the electromagnetic spectrum called **infra-red**, which we feel as heat radiation. If we shine a beam of infra-red radiation at a substance, only those frequencies which correspond to the frequencies of bonds in the substance will be absorbed. The rest will pass through. The beam will then have

A

THE STRUCTURE AND ACTIVITY OF DRUGS

The first drugs were discovered more or less by chance and theories as to how they operated would be regarded as fanciful today. For example, aspirin was first extracted from willow bark. It was thought that the cure to a condition would be found close to the cause of the condition. So willow trees, which are found in damp places, were expected to hold a cure for rheumatism, a condition associated with cold and damp. In this case the theory worked—aspirin is good at relieving the pain and inflammation of rheumatism.

Nowadays some drugs are still discovered by chance, but increasingly we are able to relate their activity to their structure and thus carry out research much more systematically. A good example is provided by the sulphonamides—a group of antibacterial drugs. These are all related to the compound sulphanilamide.

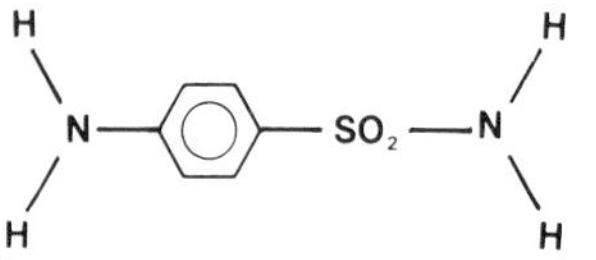

Sulphanilamide

The molecule para-aminobenzoic acid (PABA) is vital to most bacteria. They use it to synthesize folic acid, which is

Para-aminobenzoic acid (PABA)

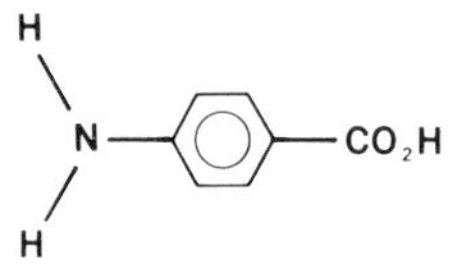

vital for their metabolism. Sulphanilamide has a very similar structure to PABA so it competes with PABA in the process of making folic acid. Obviously, it cannot form folic acid as it is the wrong molecule. The bacterial cells thus stop growing and eventually die. Folic acid is equally important to human metabolism, but we absorb it from our food, rather than synthesizing it, so sulphonamides do not harm us.

A number of variations on the basic sulphanilamide have been produced, for example:

Sulphathiazole

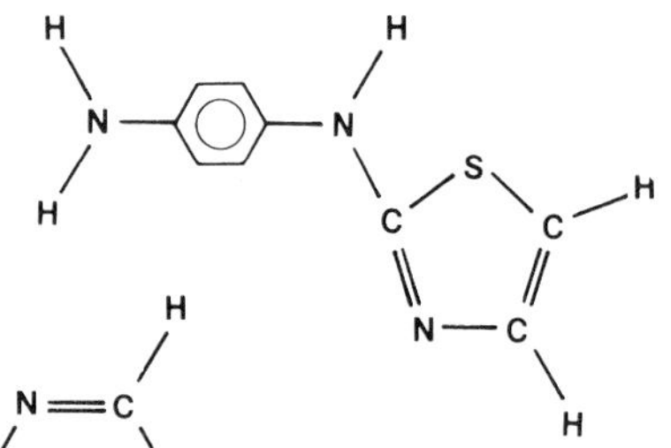

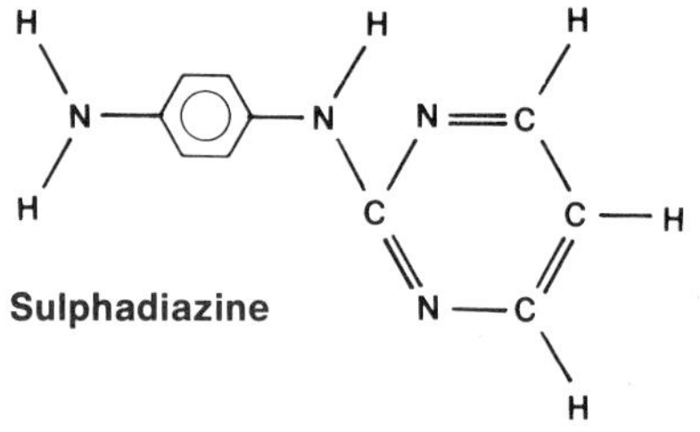

Sulphadiazine

These all work in the same way because of their similarity in structure to PABA, but have different rates of absorption and excretion by the body. Over 5400 chemical variations of sulphonamides have been synthesized and tested.

This method of synthesizing compounds which are closely related to a substance which is known to have a therapeutic effect is often used to discover new drugs. A good example is the group of narcotic pain relievers based on morphine. Morphine is probably the most effective pain killer known, but it is strongly addictive so it must be used with great care. However, replacing one of the marked H atoms by a CH_3 group produces codeine—a less potent drug, safe enough to be bought without prescription. Heroin, even more addictive than morphine, also retains the same basic structure.

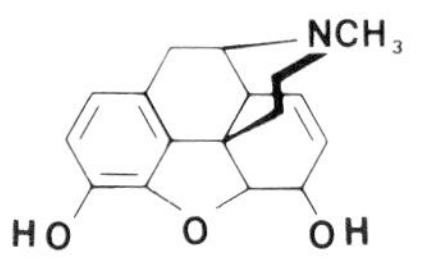

Morphine

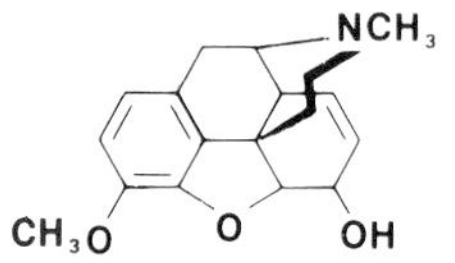

Codeine

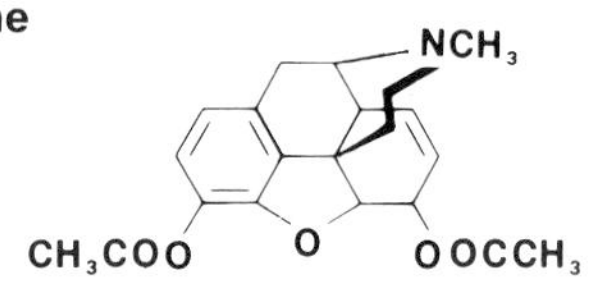

Heroin

Figure 19.12 Schematic diagram of an IR spectrophotometer

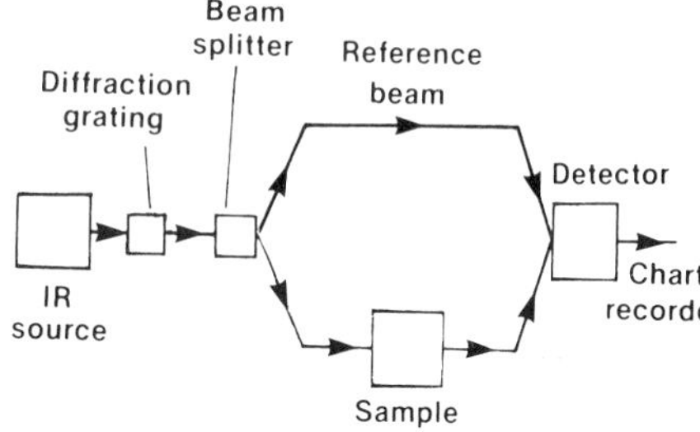

less energy, at those frequencies, than it started with. Only vibrations which result in a change in dipole moment will actually absorb infra-red energy.

This is therefore a way of deciding what bonds are present in a particular substance, by comparing the frequencies which are absorbed with the known frequencies of bonds. The principle of the infra-red spectrophotometer is shown in **Figure 19.12.** A source of infra-red radiation (rather like an electric bar heater) produces a spread of frequencies of infra-red. The diffraction grating splits this up into separate frequencies, just as a prism separates light into its separate colours. By rotating the grating infra-red of different frequencies passes into the beam splitter. This produces two beams, one of which passes through the sample, while the other is a reference beam. The detector compares the intensity of the two beams and sends a signal proportional to the difference to a chart recorder. This produces a graph of **transmission** (the amount of radiation which passes through the sample) against **frequency** of the radiation, although the axis is almost always labelled in cm^{-1} ($1/\lambda$, called the wave number

IR spectrophotometer

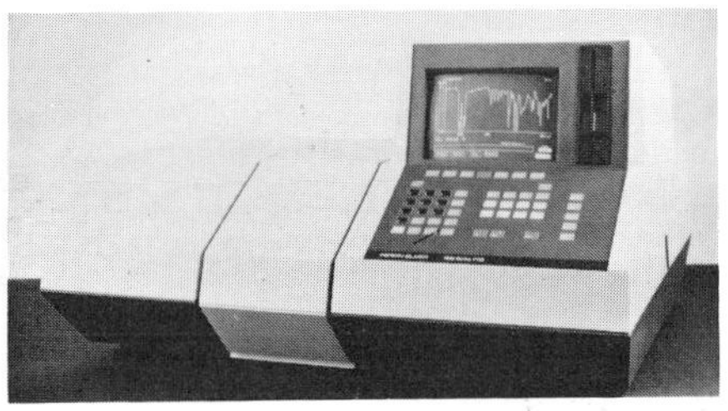

Since $V = \nu\lambda$, $\nu \propto 1/\lambda$). A typical infra-red spectrum is shown in **Figure 19.14.** High values of transmission (or transmittance) of, say, over 90% mean that the sample is absorbing little or no energy. *Dips* in the graph mean that energy is being absorbed. Confusingly these dips are often referred to as 'peaks'. The wave numbers at which particular bonds absorb are given in **Figure 19.13.** Notice that C—Cl absorbs at a lower

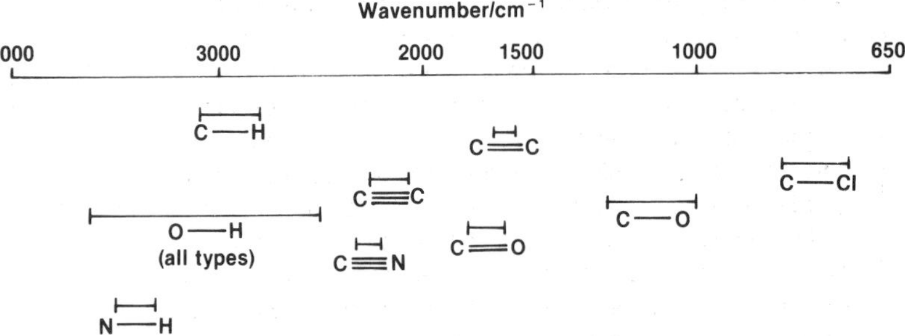

Figure 19.13 The wave numbers at which some bonds absorb infra-red. The exact frequency depends on a number of factors including the details of the rest of the molecule

frequency than C—O, as the chlorine atom is more massive than the oxygen atom, but C═O absorbs at a higher frequency than C—O as the bond is stiffer. The absorptions are given as bands of frequencies as the same bond will absorb at a slightly different frequency in different molecules. It is rarely possible to correlate every peak in an infra-red spectrum with a particular molecular vibration, but usually, specific bonds can be identified as shown in **Figure 19.14.**

If **Figure 19.14** was a spectrum of an unknown compound, we could say that it had C═O, C—O and O—H bonds present in it—valuable clues to its structure.

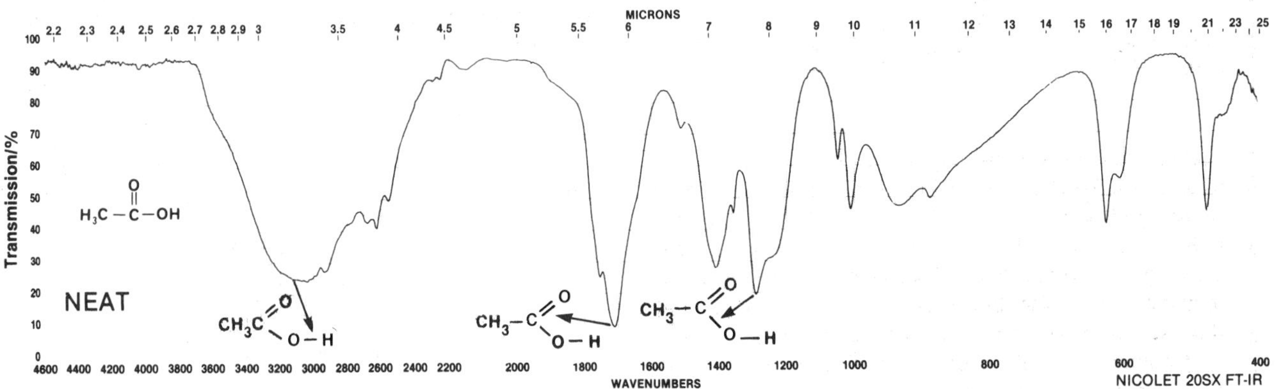

Figure 19.14 The infra-red spectrum of ethanoic acid

19.3.4 Mass spectrometry

The principles of this technique were described in **section 8.5**, where it was used to measure relative atomic masses of isotopes. When used with compounds, it can give much more information than just the relative molecular mass.

Fragmentation

Note: The molecular ion is the peak of highest mass, *not* the tallest peak.

During their time of flight through the mass spectrometer, many of the ionized molecules will break up, i.e. **fragment**, as their bonds have been weakened by the loss of an electron. These fragments give clues to the structure of the original molecules. Usually a few ions remain intact to give a peak at the relative molecular mass of the compound. This is called the **molecular ion** or **parent ion.**

An example of the use of mass spectra is illustrated by the spectra of butane and its isomer methylpropane (same molecular formula, C_4H_{10}, but different structure). See **Figure 19.15** (overleaf).

E

THE FINGERPRINT REGION

The area of an infra-red spectrum between about 1400 and 900 cm^{-1} usually has many peaks. These often correspond to complex vibrations of the whole molecule and cannot always be assigned to any particular vibration. However, the shape of this region is unique for any particular substance and can be used to identify it by comparison with a known spectrum, just as people can be identified by their fingerprints. This region is therefore called the fingerprint region.

E

MASS SPECTROMETRY AND WATER QUALITY

It is possible for many chemicals to pollute river water—diesel from pleasure boats, weedkillers and fertilizers from agriculture, and a whole variety of industrial effluents. Water boards routinely sample the water from the rivers in their areas to monitor pollutants. Pollutants are separated by gas–liquid chromatography and fed directly into a mass spectrometer. Each substance produces a unique pattern of fragments. The pollutant can be identified by comparing its mass spectrum with the mass spectrum of thousands of known compounds until a match is found. The matching can be done by computer—it would be an impossible task if done by hand. If unacceptable levels of certain pollutants are present in a particular river, water will not be taken from that river.

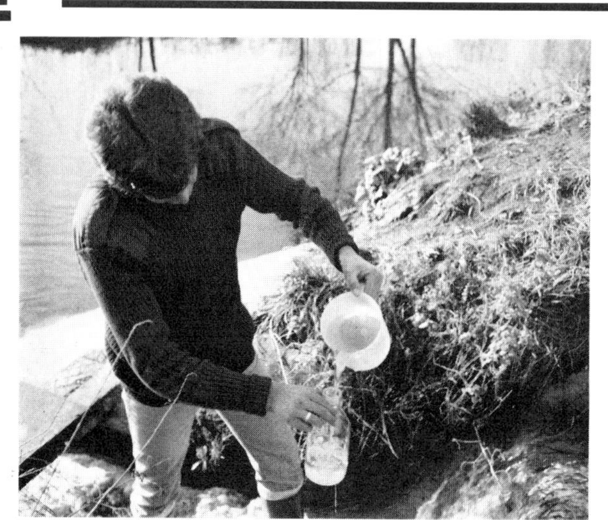

Water sampling

Figure 19.15 Mass spectra of butane and 2-methylpropane

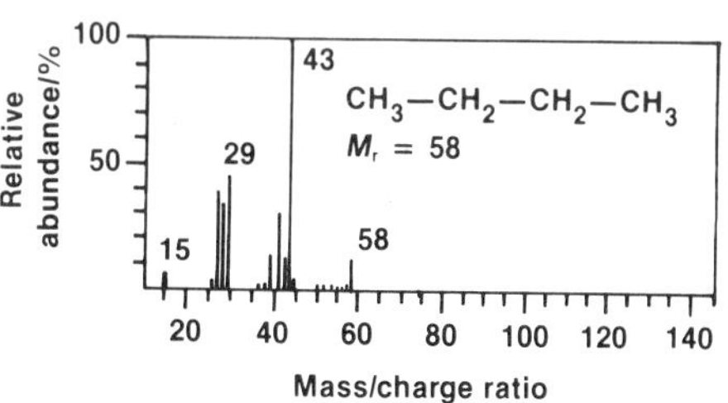

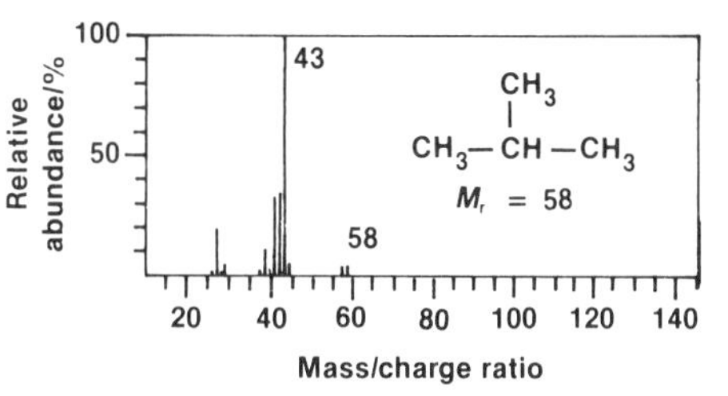

Butane, $H—\overset{\displaystyle H}{\underset{\displaystyle H}{C}}—\overset{\displaystyle H}{\underset{\displaystyle H}{C}}—\overset{\displaystyle H}{\underset{\displaystyle H}{C}}—\overset{\displaystyle H}{\underset{\displaystyle H}{C}}—H$, shows the following main peaks:

$M_r = 58$ molecular ion

$M_r = 43$ $CH_3CH_2CH_2^+$

the brown bond breaks

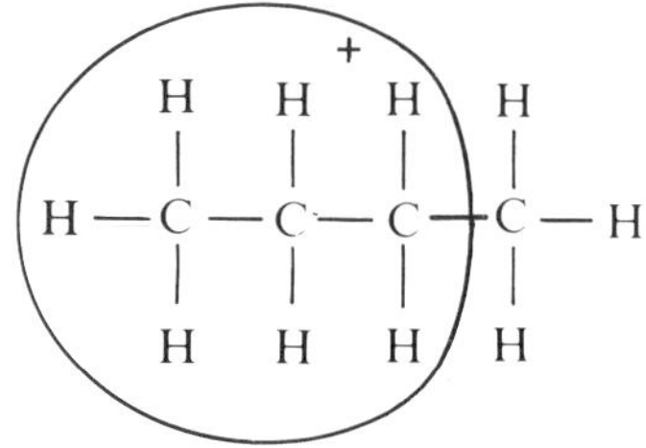

$M_r = 29$ $CH_3CH_2^+$

the brown bond breaks

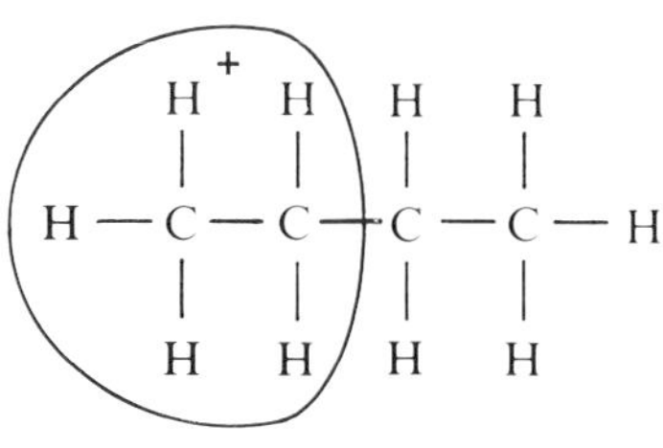

```
        H   H   H
        |   |   |
    H — C — C — C — H
        |   |   |
        H   |   H
        H — C — H
            |
            H
```

2-methylpropane, (structure above), shows the following main peaks:

$M_r = 58$ molecular ion

```
            H   H   H
            |   |   |  +
        H — C — C — C — H
            |   |   |
            H   |   H
            H — C — H
                |
                H
```

$M_r = 43$ $CH_3CHCH_3^+$

any one of the brown bonds breaks

But there is no significant peak at $M_r = 29$, as it is not possible to get a fragment of $M_r = 29$ by breaking just one bond in this molecule. So we can distinguish between two very similar molecules.

Note that it is often difficult to assign all the peaks in a mass spectrum.

Isotope peaks

The mass spectrometer is able to detect isotopes separately, so if an element with more than one common isotope is present, these will be detected. For example, chlorine has two isotopes, ^{35}Cl and ^{37}Cl, which are present in the ratio $^{35}Cl:^{37}Cl = 3:1$. So any molecule or fragment containing chlorine will give rise to *two* peaks, separated by two mass numbers and with heights in the ratio 3:1. If you spot this feature in a mass spectrum, it is almost certain that chlorine is present. The spectrum of chloroethane, **Figure 19.16**, illustrates this. The peaks at 64 and 66 are both molecular ion peaks—$CH_3CH_2\,^{35}Cl^+$ and $CH_3CH_2\,^{37}Cl^+$ respectively. The peaks at 49 and 51 are also in an approximate 3:1 ratio. These represent loss of a CH_3 group (mass = 15) leaving $CH_2\,^{35}Cl^+$ and $CH_2\,^{37}Cl^+$.

Figure 19.16 Mass spectrum of chloroethane

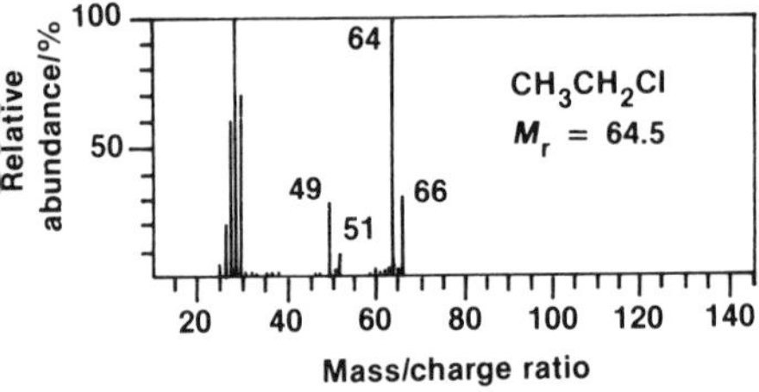

Note that most elements have more than one isotope, although often one is by far the most abundant. Carbon, for example, has ^{12}C (abundance 98.9%), ^{13}C (abundance 1%) and tiny amounts of others. Thus each mass spectrum peak for a carbon compound will have smaller peaks of higher mass close to it. These are often called **satellite peaks**. For example, ethane, C_2H_6 would have peaks at 30 corresponding to $^{12}C\,^{12}CH_6^+$, 31 corresponding to $^{12}C\,^{13}CH_6^+$ and even 32 corresponding to $^{13}C\,^{13}CH_6^+$. As there are also isotopes of hydrogen to consider, many combinations are possible. It is usually best to ignore these for simple interpretation of spectra.

High resolution mass spectra

The mass spectra we have seen so far have given masses to the nearest whole number. High resolution mass spectroscopy measures masses to four or five decimal places. This enables us to distinguish between different combinations of atoms which could make up the same approximate relative molecular mass. Accurate values of some relative atomic masses are:

$$^{12}C = 12.00000 \quad \text{(by definition)}$$
$$^{1}H = 1.00782$$
$$^{14}N = 14.00307$$
$$^{16}O = 15.99492$$

A compound whose $M_r = 123$ (low resolution, to the nearest whole number) could have the formula $C_6H_5NO_2$ *or* $C_7H_7O_2$, or a number of other possibilities. The accurate relative molecular masses are:

$$C_6H_5NO_2 = 123.0320$$
$$C_7H_7O_2 = 123.0446$$

High resolution mass spectroscopy could easily decide between these possibilities.

Tables of all combinations of elements adding up to a given whole-number relative molecular mass and giving the exact value of the relative molecular mass are available, enabling the formula to be found if the *exact* mass is known.

Figure 19.17 Orientations of bar magnets in a magnetic field

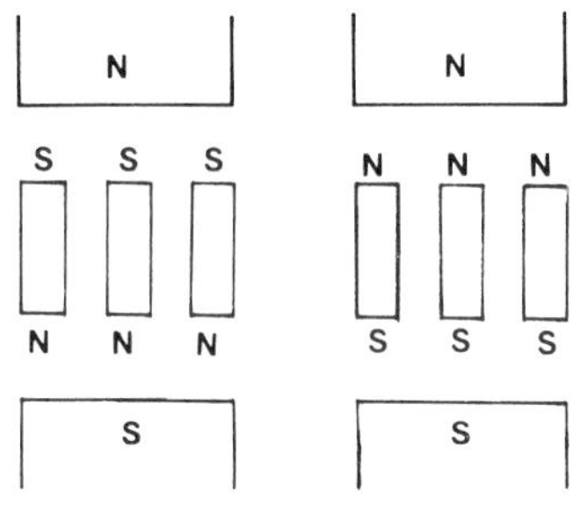

Figure 19.18 Energy level diagram of the two orientations of bar magnets in a magnetic field

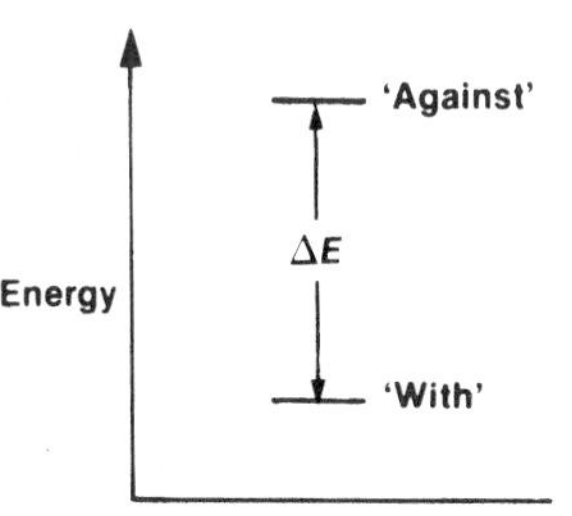

19.3.5 Nuclear magnetic resonance (NMR) spectroscopy

Certain nuclei, including hydrogen, have the property of *spin* (like electrons). This gives them a magnetic field like that of a bar magnet. If magnets are placed in a magnetic field, they will line up as shown in **Figure 19.17a.** It is also possible that they could line up as in **Figure 19.17b** but this has a higher energy, as the magnet has to be forced into that position against the repulsion of the field. It can only stay in that position if undisturbed. The same applies to nuclei with spin, such as hydrogen. They will line up in a magnetic field in the low energy position. However if energy quanta just equal to the difference between the two positions (ΔE in **Figure 19.18**) are supplied, a few nuclei will 'flip' into the higher energy position. The energy gap corresponds to electromagnetic radiation in the radio region of the electromagnetic spectrum for the usual values of magnetic field used. This is the basis of NMR spectroscopy.

The apparatus is shown in **Figure 19.19** and **Plate 6.** The sample containing hydrogen atoms is placed in a magnetic field, and radio waves directed at the sample. When the frequency of the radio waves corresponds to quanta of energy ΔE, the sample absorbs some of the radio waves and a smaller signal is received by a detector. (In practice the radio frequency is usually kept the same and the magnetic field changed, using an electromagnet.)

Figure 19.19 Schematic diagram of an NMR spectrometer

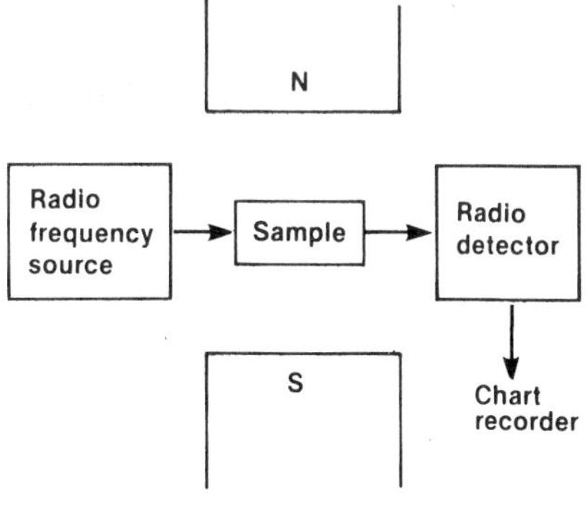

If all the hydrogen nuclei were identical, we would get only one position of absorption. However, nuclei are **shielded** from the external magnetic field by their electrons. Nuclei with more electrons around them are better shielded, so hydrogen atoms in different chemical environments will absorb the radio waves at slightly different frequencies. This is called the **chemical shift**. The greater the electron density around a hydrogen atom, the higher the field that is required for the nucleus to 'flip'. So, for methanol we get the NMR spectrum shown in **Figure 19.20.**

Figure 19.20 NMR spectrum of methanol (low resolution). Peak areas are in the ratio 1:3

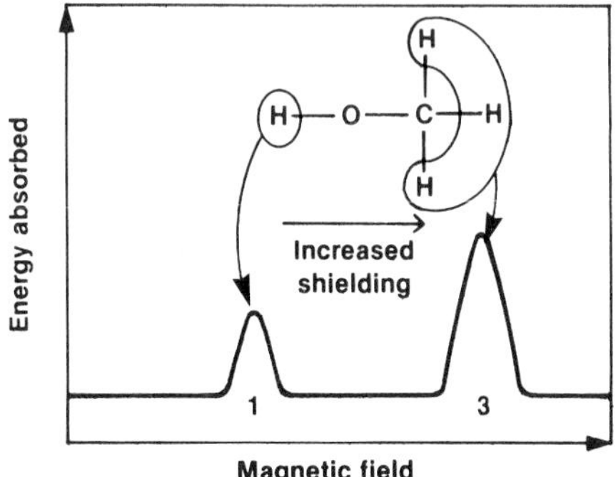

The single hydrogen attached to the oxygen has a low electron density, as it is attached directly to an electronegative oxygen atom which pulls electrons away from it. It has little shielding and therefore needs a relatively low field to 'flip'. The three hydrogens attached to the less electronegative carbon atom are shielded better by their electrons and need a higher field to 'flip'. The areas under the peaks are proportional to the number of hydrogen atoms of each type.

The NMR spectrum of ethanol is shown in **Figure 19.21.** This shows hydrogen atoms in *three* environments with different shielding. The single one with little shielding represents the O—H hydrogen, the pair with more shielding represent the CH_2 ones and the three with most shielding are the CH_3 hydrogens, which are furthest away from the electronegative oxygen. NMR therefore gives information about the numbers of different types of hydrogen atoms in the molecule and where they are with respect to electronegative atoms.

For example, we could predict that the NMR spectrum of propan-2-ol:

Figure 19.21 NMR spectrum of ethanol (low resolution)

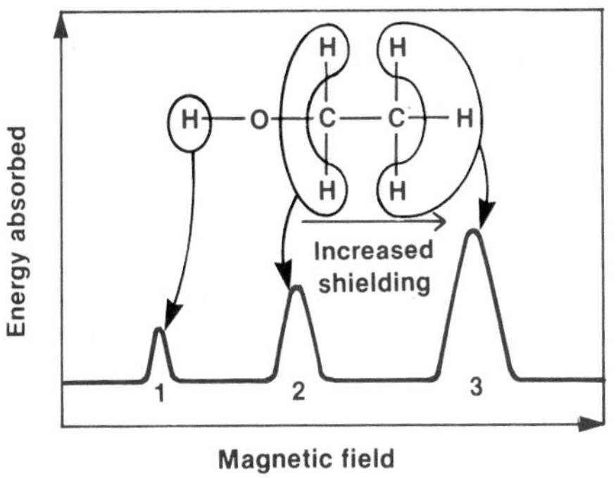

H(A)
(C)H O H(C)
(C)H—C—C—C—H(C)
H(C) H(B) H(C)

would have *three* peaks with areas in the ratios 1:1:6 representing the hydrogen atoms labelled A, B and C. Hydrogen A attached to the oxygen would have least shielding and therefore absorb energy at the lowest magnetic field, followed by hydrogen B, and then the six equivalent hydrogens labelled C, as they are further from the electronegative oxygen atom.

19.3.6 Dipole moments

Dipole moments (see **section 10.2.2**) can give information about the shapes of molecules. For example, carbon dioxide, CO_2, has no dipole moment but sulphur dioxide, SO_2, has. This tells us that CO_2 must be linear so that the dipoles of each bond cancel out, while SO_2 must be angular:

$\delta-$ $\delta+$ $\delta-$
O=C=O

carbon dioxide – linear, the dipoles of each bond cancel

$\delta+$
S
$\delta-$O O$\delta-$

sulphur dioxide, angular, the dipoles of each bond do not cancel

resultant dipole

Water, H_2O, has a dipole moment and so must be angular. Ammonia, NH_3, has a dipole moment and so cannot be a flat trigonal molecule or the dipoles of the individual bonds would cancel. It must therefore be pyramidal:

✗
$\delta+$ $\delta-$ $\delta+$ $\delta+$
H — N H H

If ammonia were flat, the dipoles of each bond would cancel

✓
$\delta-$ $\delta+$ $\delta+$ $\delta+$
N---H H

Ammonia is actually pyramidal so the bond dipoles do not cancel

resultant dipole

Measurement of the values of dipole moments is not easy but it is easy to find out whether a molecule has a resultant dipole. A stream of a liquid with a dipole will be deflected by a charged rod, see **section 10.2.2.**

19.4 Summary

- X-rays have approximately the same wavelength as the spacing between atoms. A beam of X-rays can be reflected from the atoms in a crystal.
- At certain angles of reflection the X-rays interfere constructively. The Bragg condition:

$$n\lambda = 2d \sin \theta$$

enables the spacing between the atoms, d, to be found if we know the angle of reflection, θ, and the wavelength, λ, of the X-rays. n is a whole number.
- More sophisticated analysis of X-ray scattering patterns enables the positions of all the atoms in the structure to be found.
- Atoms and ions do not have a definite radius. We refer to ionic,

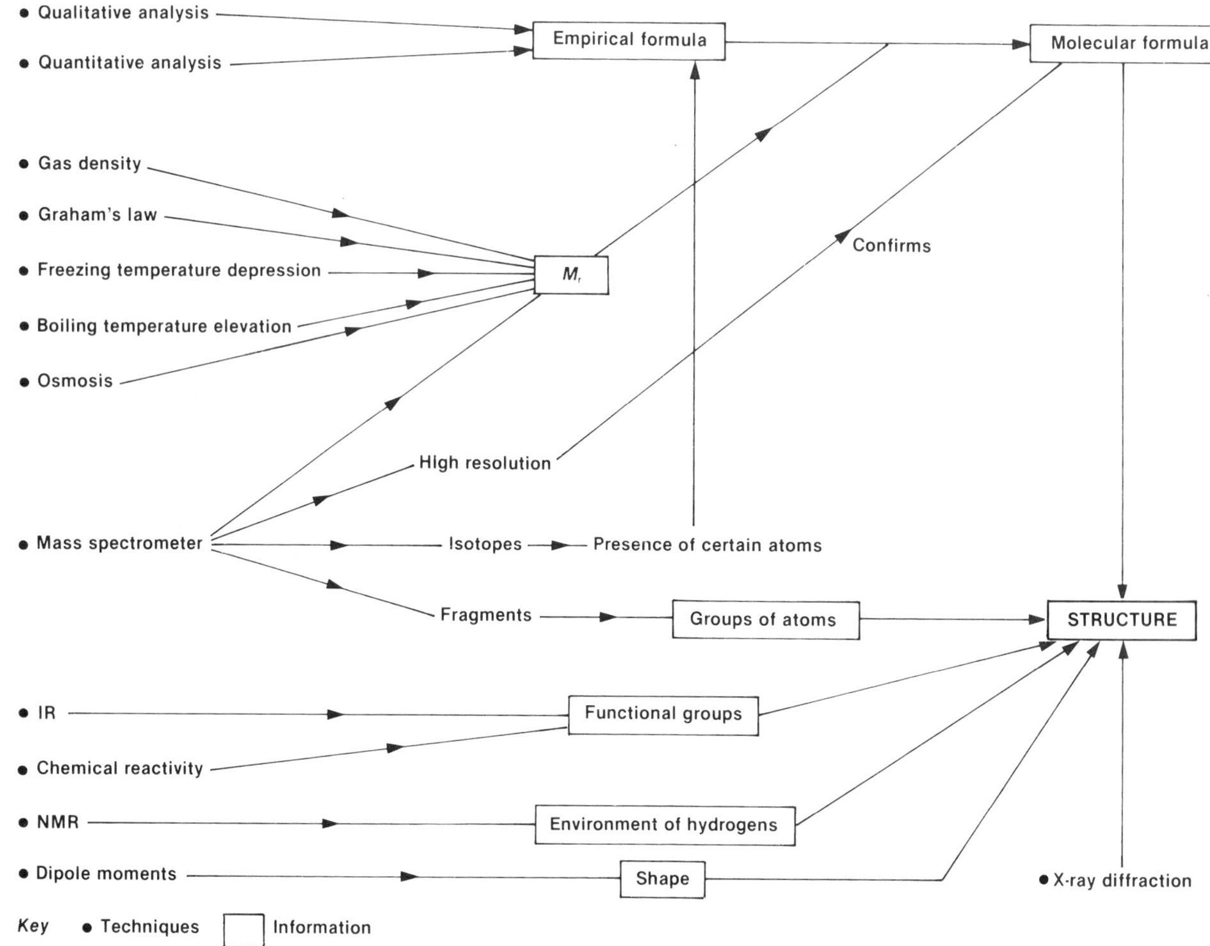

Figure 19.22 Summary of structure determination methods

metallic, covalent and van der Waals radii, depending on the way the atom is bonded.

• Structure determination of molecular compounds—**Figure 19.22** summarizes the information obtained from each technique and how it contributes to the determination of structure.

19.5 Questions

1 Using the single crystal method of X-ray diffraction with potassium chloride, a strong X-ray signal was found at any angle of 5.25° from the surface of the crystal. If the wavelength of the X-rays was 0.0586 nm, what is the interionic distance in potassium chloride? At what angle would you expect to find the next strong X-ray signal? (*Hint*: $n = 2$.)

2 A molecule X gives a molecular ion of $M_r = 60$ at low resolution. The following molecular formulae are possible:

$$C_3H_8O \quad C_2H_8N_2 \quad C_2H_4O_2$$

A) Can you find another molecular formula (using C, H, N and O only) which gives $M_r = 60$?

B) High resolution mass spectrometry gave the M_r of X as 60.057. Use the list of exact A_rs below to calculate the exact masses of the molecular formulae (below left) and then decide which one is X:

H = 1.00782
C = 12.00000
N = 14.00307
O = 15.99492

3 Explain the following in terms of molecular shape:

A) $BeCl_2$ has no dipole moment, but SCl_2 has

B) BCl_3 has no dipole moment, but PCl_3 has.

4 A compound Z contains carbon, hydrogen and oxygen only in the following percentages:

C 54.5%
H 9.1%
O 36.4%

It gives a brick-red precipitate when heated with Benedict's solution.

A) What is the empirical formula?

The mass, IR and NMR spectra are given below.

B) Look at the mass spectrum. What is the most likely mass of the molecular ion? What molecular formula does this give?

C) 0.1 g of Z produced 51.3 cm^3 of vapour (converted to STP). Does this confirm the value of M_r given by the mass spectrum?

D) Look at the IR spectrum. What does the peak at approximately 1700 cm^{-1} suggest? Is this confirmed by the result of the Benedict's test?

E) Write down the probable structural formula of Z from the information used so far.

F) How many types of protons does this structural formula have? Does this fit the observed NMR spectrum?

G) Would you expect Z to have a dipole moment?

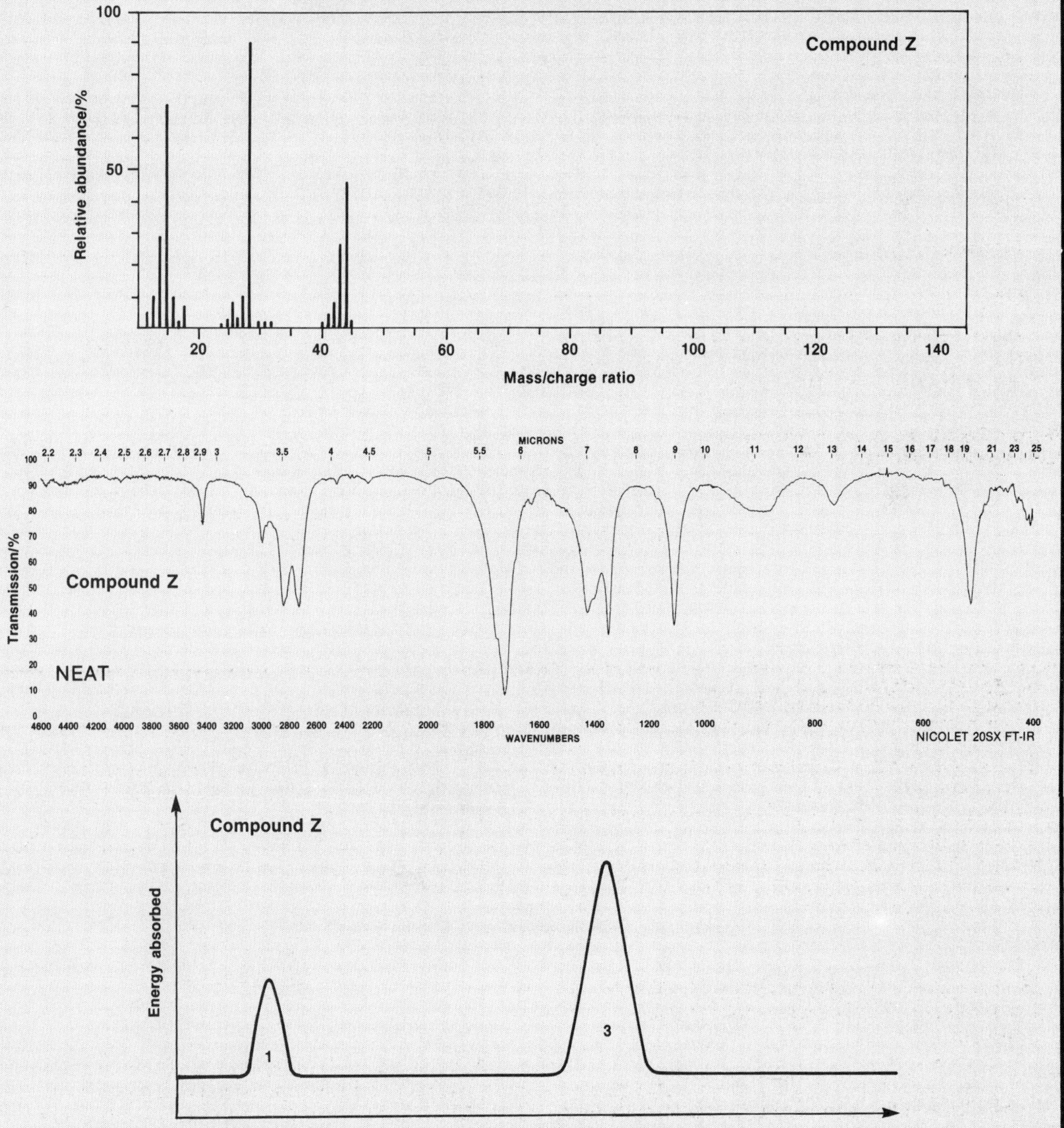

5 The NMR spectrum of a compound is given below:

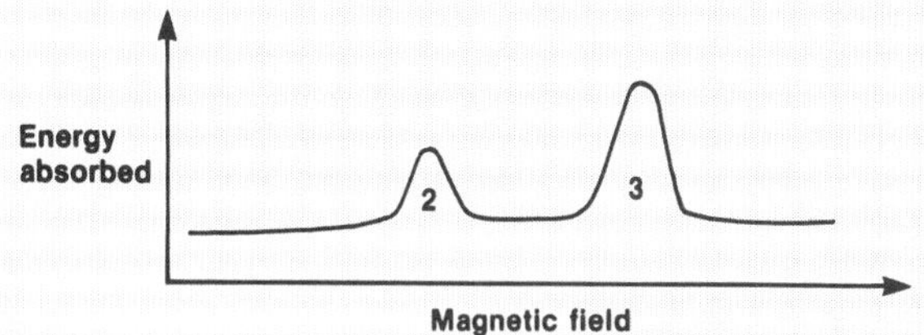

A) How many different types of hydrogen atoms are there?
B) Which group of hydrogen atoms is most likely to be closer to an electronegative atom?

6 The infra-red spectra of propane, propanal and propan-1-ol are given:

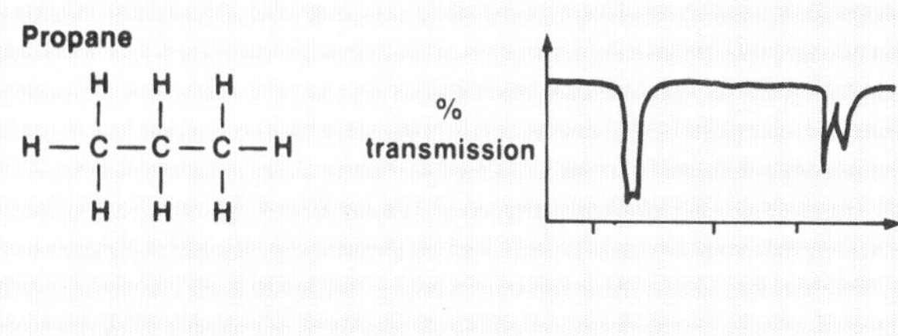

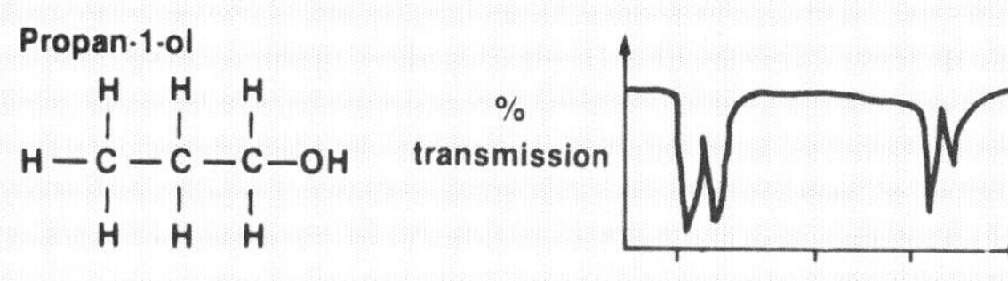

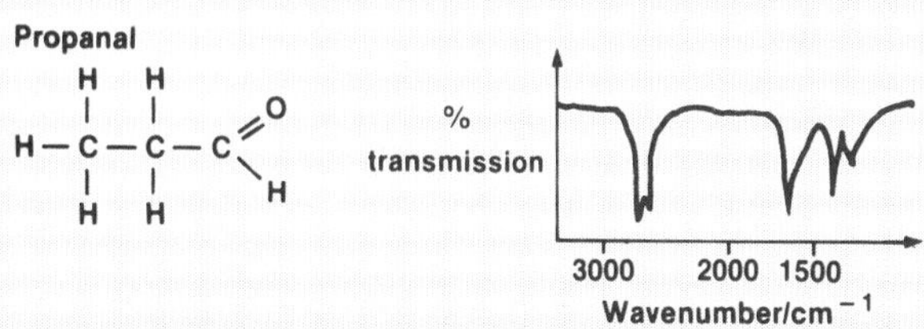

A) Using **Figure 19.13** on page 257 and the structures above, suggest which bonds are causing the peaks at around 3000 cm^{-1} and 1600 cm.$^{-1}$
B) Sketch the probable NMR spectrum of each of the compounds above.
C) Sketch the likely appearance of the IR spectrum of propanoic acid (top right).

7 The electron density map below was obtained from an X-ray diffraction experiment on a compound Y. The compound's molecular formula is $C_5H_6N_3Cl$. Which atom is the chlorine? How can you tell? Which atoms are not shown clearly on this map? Why? Can you distinguish clearly between carbons and nitrogens? Explain your answer. Attempt to draw Y's structural formula.

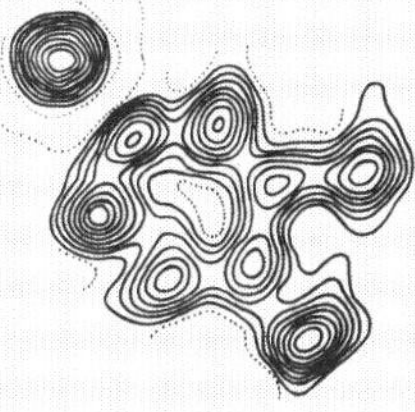

8 Below is part of the mass spectrum of a chlorine-containing compound:

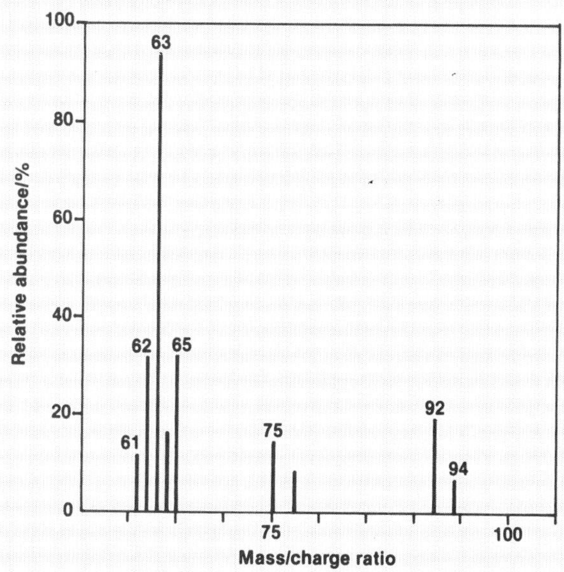

Two fragments certainly contain chlorine.
A) Identify them by their masses.
B) If the original molecule contains C, H and Cl only, suggest formulae for the two chlorine-containing fragments.

Part C Inorganic Chemistry

20 Periodicity

20.1 Introduction

The Periodic Table is of central importance to chemistry. It provides a logical framework for recognizing patterns in the properties of elements and their compounds. It also allows us to explain trends and similarities in properties in terms of the electronic structures of the elements. Without the Periodic Table, chemistry would be a hotchpotch of unrelated information about different substances. With it, we can see such data as part of a unified whole.

20.2 The structure of the Periodic Table

The Periodic Table has been written in many forms, including pyramids and spirals. The one shown on page 602 is the most usual and useful one.

Notes

- Areas of the table are labelled s-block, p-block, d-block and f-block. Elements in the s-block have their highest-energy electrons in s orbitals, e.g. Li ($1s^2\,2s^1$). Elements in the p-block have their highest energy electrons in p orbitals, e.g. C ($1s^2\,2s^2\,2p^2$), and so on. These are shown in **Figure 20.1**.

Figure 20.1 The Periodic Table – in most versions the f-block is placed below the d-block as in Figure 20.2

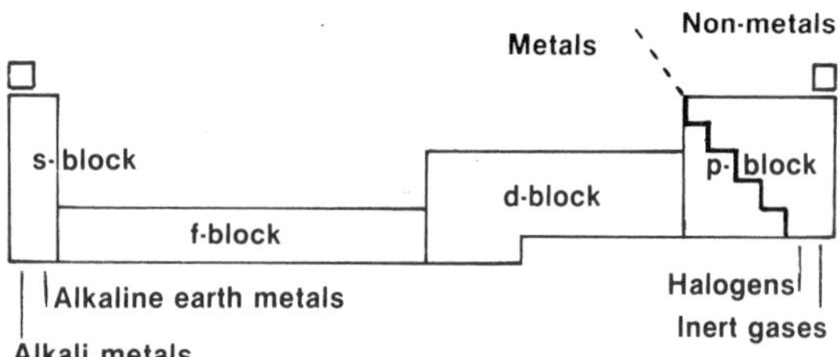

- The positions of hydrogen and helium vary in different versions of the table. Helium is usually placed above the inert gases (Group 0) because of its properties. However, it is not, of course, a p-block element, its electronic structure being $1s^2$.

  Hydrogen is placed above Group I but separated from it. It usually forms singly charged H^+ ions like the Group I elements but otherwise is not very similar to them. It can also form H^- ions and is placed above the halogens in some versions of the Periodic Table.
- The periods are numbered starting from Period 1 containing just hydrogen and helium. Period 2 contains the elements lithium to neon, and so on.
- The groups are usually numbered I to VII plus 0. However, it has recently been proposed that numbering from 1 to 18 across the d-block should be adopted, as shown in **Figure 20.2** (overleaf).
- Notice that if the table were drawn out and in full it would look as in **Figure 20.1.**
- There has been controversy over the discovery and naming of the

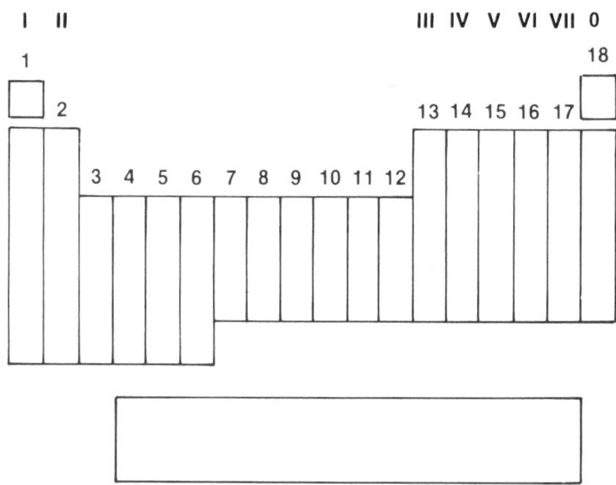

Figure 20.2 Recently proposed new numbering system for the groups of the Periodic Table. The more usual numbering is in brown

elements after number 103 (lawrencium). The Russians have named element number 104 kurchatovium, Ku, while the Americans have named it rutherfordium, Rd. It has now been agreed that new elements will be named unnilquadium, Unq (Latin for 104), unnilpentium, Unp (Latin for 105) and so on. This marks the introduction of three-letter symbols.

20.3 Periodicity of physical properties

Radii of atoms and ions

It is not possible to measure the radius of an isolated atom (see **section 19.2.2**) so we use half the distance between a pair of identical atoms. This differs depending on the type of bonding: covalent; metallic; van der Waals, etc. The covalent radius is most commonly used as a measure of the size of the atom. Since noble gases do not bond covalently with one another, they do not have covalent radii, and they often have to be left out of comparisons of atomic sizes.

The word 'periodic' means 'recurring regularly'. Most chemistry lessons are periodic as they occur regularly, on the same days of each week. One of the best ways of showing periodic behaviour of the properties of the elements is to plot graphs of the property against the atomic number.

20.3.1 Atomic sizes

One of the most obvious properties of an atom is its size. A plot of atomic volume (that is the volume of one mole of atoms of each element in the solid state) against atomic number for the first 60 elements is shown in **Figure 20.3**.

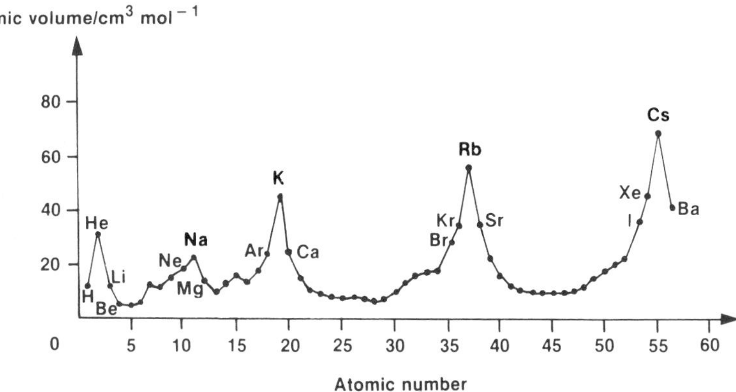

Figure 20.3 The periodicity of atomic volumes

Two features are worth noting:

1. Atomic volume is clearly a periodic property, the graph always peaking at a Group I metal (except for He);
2. The Group I metals get larger as we descend the group.

The same two features are found on a plot of covalent radius against atomic number, **Figure 20.4.** We can explain these trends by looking at the electronic structures of the elements in a period, for example, lithium to neon, as illustrated in **Figure 20.5.** As we go from lithium to neon we are adding electrons in the outer shell – the '2' shell. The charge on the nucleus is increasing from 3+ to 10+ (or 1+ to 8+ allowing for the

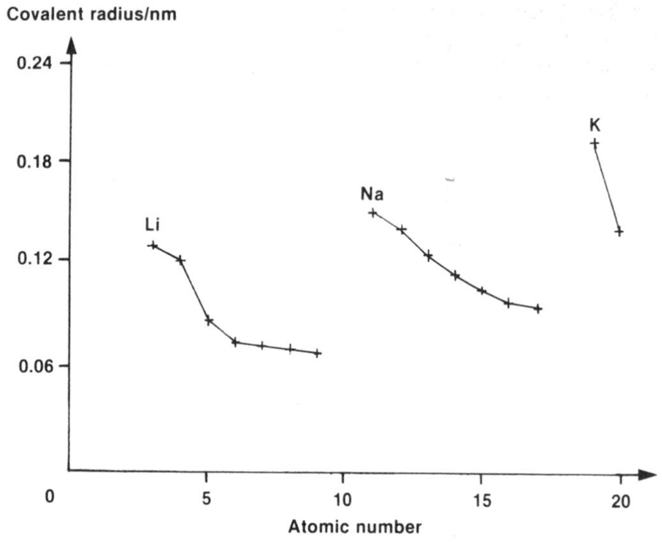

Figure 20.4 The periodicity of covalent radii. *Note*: The noble gases are not included on this graph as they do not form covalent bonds with one another. Even metals can form covalent molecules in the gas phase

Figure 20.5 The electronic structures of the elements lithium to sodium

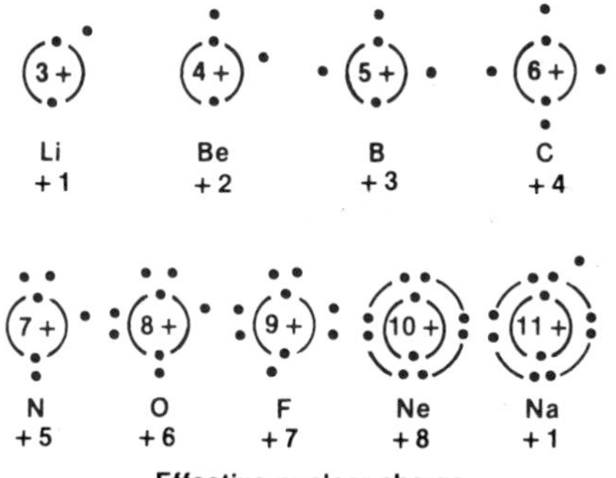

shielding of the inner shell), tending to pull the electrons in closer to the nucleus, so the size of the atom *decreases* as we go across the period. On moving from neon to sodium, the second shell is full and the extra electron goes into the '3' shell. The nuclear charge is 11+ but the shielding of the inner shells is now 10 and the outer electron feels an effective nuclear charge of just +1. Thus for two reasons we get a sudden jump in size—firstly the outer electron in sodium is one shell further out than those in neon, and secondly it feels an effective nuclear charge of 1+ rather than 8+.

20.3.2 The size of ions

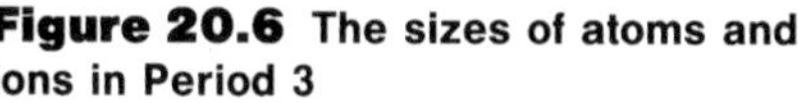

Figure 20.6 The sizes of atoms and ions in Period 3

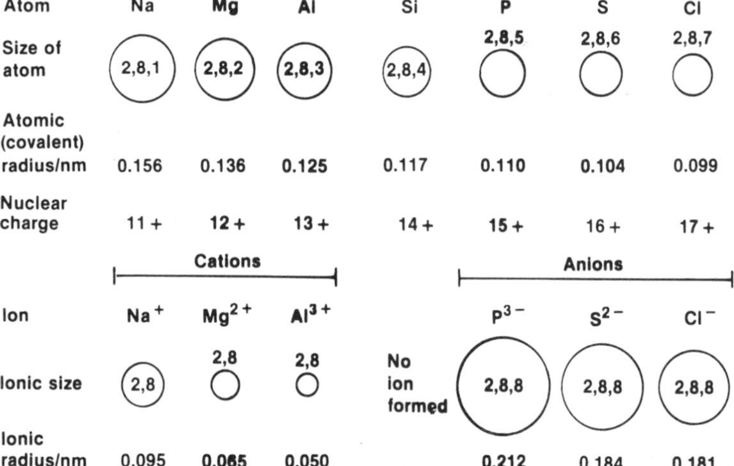

Atom
Na Mg Al Si P S Cl
Size of atom
2,8,1 2,8,2 2,8,3 2,8,4 2,8,5 2,8,6 2,8,7
Atomic (covalent) radius/nm
0.156 0.136 0.125 0.117 0.110 0.104 0.099
Nuclear charge
11+ 12+ 13+ 14+ 15+ 16+ 17+
Cations
Anions
Ion
Na^+ Mg^{2+} Al^{3+} P^{3-} S^{2-} Cl^-
Ionic size
2,8 2,8 2,8
No ion formed
2,8,8 2,8,8 2,8,8
Ionic radius/nm
0.095 0.065 0.050 0.212 0.184 0.181

Figure 20.6 shows to scale the sizes of the covalent radii of the elements sodium to chlorine (even the metals can form covalent bonds in the gas phase). Below these are the sizes of the ions they form. You will see that *positive* ions (cations) are *smaller* than the parent atom because:

1. the whole outer shell of electrons has been lost
2. the nuclear charge can pull the remaining electrons closer to the nucleus, as there is now less electron–electron repulsion.

The *negative* ions (anions) are *larger* than the parent atoms as they have gained electrons, resulting in more electron–electron repulsion.

The periodicity of atomic (covalent) and ionic radii is shown in **Figure 20.7** (page 271). Notice that atomic radii peak at the Group I metals,

E

THE HISTORY OF THE PERIODIC TABLE

Chemists have long searched for some order in their science.

New Names. English.	Latin.	Correſpondent old Names.
Light	- -	Light.
Caloric	Caloricum	Heat. Principle or element of heat, Fire, Igneous fluid, Matter of fire and of heat.
Oxygen	Oxygenum	Dephlogiſticated air, Empyreal air, Vital air, or Baſe of vital air.
Azot	Azotum	Phlogiſticated air or gas, Mephitis, or its baſe.
Hydrogen	Hydrogenum	Inflammable air or gas, or the baſe of inflammable air.

Oxydable and Acidifiable ſimple Subſtances not Metallic.

New Names.		Correſpondent old Names.
Sulphur	Sulphurum	The ſame names.
Phoſphorus	Phoſphorum	
Carbon	Carbonum	The ſimple element of charcoal.
Muriatic radical	Murium	Still unknown.
Fluoric radical	Fluorum	
Boracic radical	Boracum	

Oxydable and Acidifiable ſimple Metallic Bodies.

New Names.			Correſpondent old Names.
Antimony	Antimonium	Regulus of	Antimony.
Arſenic	Arſenicum		Arſenic.
Biſmuth	Biſmuthum		Biſmuth.
Cobalt	Cobaltum		Cobalt.
Copper	Cuprum		Copper.
Gold	Aurum		Gold.
Iron	Ferrum		Iron.
Lead	Plumbum		Lead.
Manganeſe	Manganum		Manganeſe.
Mercury	Mercurium		Mercury.
Molybdena	Molybdenum		Molybdena.
Nickel	Nickolum		Nickel.
Platina	Platinum		Platina.
Silver	Argentum		Silver.
Tin	Stannum		Tin.
Tungſtein	Tungſtenum		Tungſtein.
Zinc	Zincum		Zinc.

Q 3 Salifiabl

Salifiable ſimple Earthy Subſtances.

New Names English.	Latin.	Correſpondent old Names.
Lime	Calca	Chalk, calcareous earth, Quicklime.
Magneſia	Magneſia	Magneſia, baſe of Epſom ſalt, Calcined or cauſtic magneſia.
Barytes	Baryta	Barytes, or heavy earth.
Argill	Argilla	Clay, earth of alum.
Silex	Silica	Siliceous or vitrifiable earth.

Lavoisier's original list of 'elements'

One of the first to make real progress was Antoine Lavoisier who in 1789 put forward a list of what we would now call elements, which he divided into sets on the basis of chemical similarities. To modern eyes his list looks decidedly odd as he included light and heat, which we do not regard as chemical substances at all, and also several materials which we now know to be compounds such as 'alumina' (aluminium oxide) and 'magnesia' (magnesium oxide). The chemical techniques of the day were not advanced enough to decompose these very stable compounds.

In 1817 a German, Johann Dobereiner, noticed some sets of strongly similar elements: lithium, sodium and potassium; calcium, barium and strontium; and chlorine, bromine and iodine. Not only did these 'triads', as he called them, have chemical similarities, but the relative atomic mass of the middle one was approximately the average of the outer two. For example, A_r for Li = 7, K = 39, Na = 23 = (39 + 7)/2. Dobereiner had spotted sets of elements which we now recognize as members of the same chemical group.

By the 1860s more accurate methods of determining relative atomic masses had been developed and three chemists discovered patterns in the then known elements when they were arranged in ascending order of atomic mass. The English chemist John Newlands noticed that there were similarities between the first, eighth and fifteenth elements, the second, ninth and sixteenth, and so on. He called this the 'law of octaves' and compared it with a musical scale. Partly because of this musical analogy, but more seriously because he made no allowance for elements not then known, his ideas were ridiculed by some of his contemporaries.

H	Li	Be	B	C	N	O
F	Na	Mg	Al	Si	P	S
Cl	K	Ca	Cr	Ti	Mn	Fe

Newland's octaves. There is a striking similarity to the modern Periodic Table.

The inert gases had yet to be discovered in his day.

John Newlands

At around the same time, the German Lothar Meyer had plotted a graph of what we would now call the volume of a mole of atoms against relative atomic mass and noted a repeating pattern. A modern version of his graph is shown in **Figure 20.3.**

The credit for the Periodic Table, however, goes firmly to the Russian Dmitri Mendeleev (see photograph, **section 3.1.1**). His greatest achievement was to realize that there were probably elements remaining to be discovered (there were only around 60 then known). He not only left gaps in his table for them but predicted their properties by averaging the properties of the known elements above and below the gaps. The table gives Mendeleev's predictions made in 1871 for the element between silicon and tin (which he called eka-silicon) and the actual data for the element germanium, discovered in 1886.

Eka-silicon	Germanium
Grey metal	Grey-white metal
Density 5.5 g cm^{-3}	Density 5.47 g cm^{-3}
A_r 73.4	A_r 72.6
$T_m > 1073$ K	$T_m = 1231$ K
Formula of oxide XO_2	Formula of oxide GeO_2
Density of oxide 4.7 g cm^{-3}	Density of oxide 4.7 g cm^{-3}
Formula of chloride XCl_4	Formula of chloride $GeCl_4$
Density of chloride 1.9 g cm^{-3}	Density of chloride 1.84 g cm^{-3}

Mendeleev also made departures from atomic mass order, placing tellurium, A_r 127.6, *before* iodine, A_r 126.9, so that these two would fall into the groups which fitted their properties (iodine has clear similarities to bromine and chlorine). We now know that it is the number of protons, not the atomic mass, which governs the positions in the Periodic Table. The significance of this was unknown to Mendeleev as this was many years before the structure of atoms was known. Mendeleev has been very appropriately honoured by having an element—number 101—named after him.

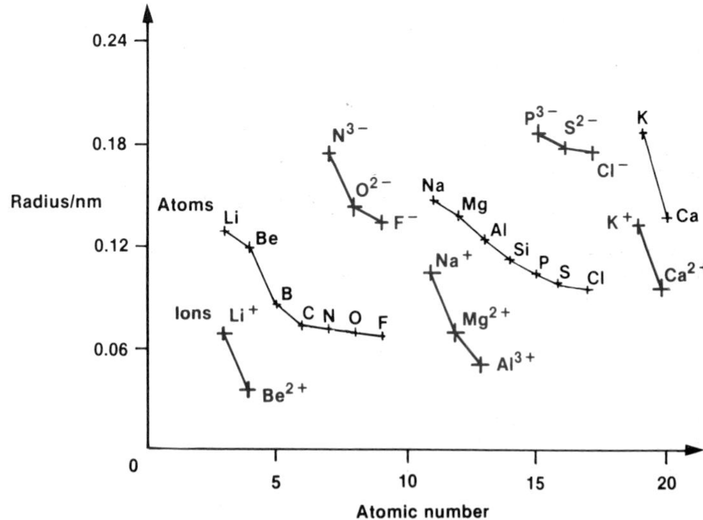

Figure 20.7 Radius against atomic number for the elements helium to calcium

where a new shell starts, while ionic radii peak at the most negatively charged ion. The noble gases are not included as they do not form covalent bonds with one another, nor do they form ions.

20.3.3 First ionization energy

This is the energy required to convert a mole of isolated, gaseous atoms into a mole of singly positively charged, gaseous ions, i.e. to remove one electron from each atom:

$$E(g) \longrightarrow E^+(g) + e^-$$

where E is any element. The periodicity of first ionization energies is shown in **Figure 20.8**.

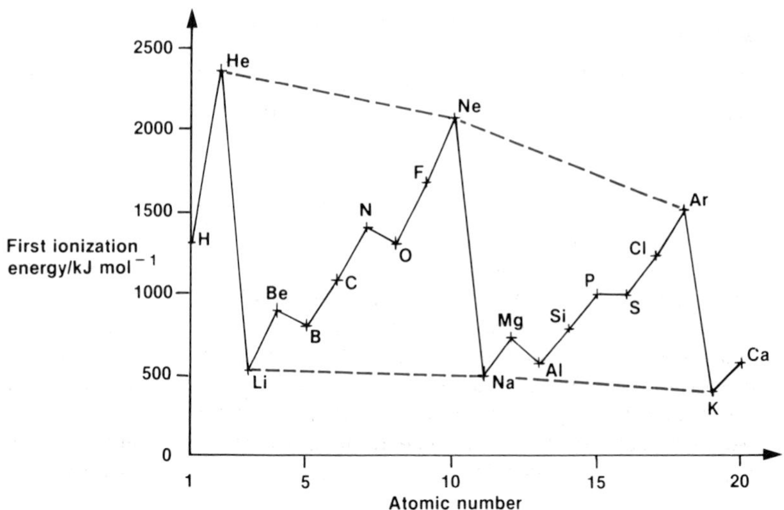

Figure 20.8 The periodicity of first ionization energies

A

USING THE PERIODIC TABLE: HIGH-TEMPERATURE SUPERCONDUCTORS

When metals are cooled, their electrical resistance drops—cooling reduces the vibrations of the atoms and makes it easier for electrons to travel through the lattice. At very low temperatures, close to 0 K, the resistance disappears entirely and a current will flow for ever without loss. This is called superconduction. Until 1986, the highest temperature superconductor was the substance Nb_3Ge which will superconduct up to 23 K which is called its transition temperature.

In 1986 a compound of lanthanum, barium, copper and oxygen was found to superconduct at 30 K and this signalled the start of a race to develop yet higher temperature superconductors, whose practical benefits would be enormous. The original compound had an even higher transition temperature under high pressure. This suggested that forcing the atoms closer together might help so it was decided to prepare a compound in which the barium was replaced by strontium—in the same group of the Periodic Table as barium and thus able to bond in the same way, but with smaller atoms.

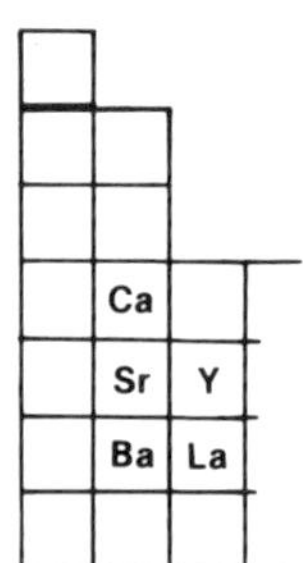

This worked—the transition temperature of the new compound was 10 K higher. Using the same reasoning a compound with yttrium instead of lanthanum was prepared and $YBa_2Cu_3O_7$ was found to superconduct at 95 K. The current 'record' is 125 K for thallium barium calcium copper oxide although this may well have been surpassed by the time you read this book, as research in this area is continuing apace with room temperature superconductors as the ultimate prize. One avenue is certain to be to try different combinations of elements from Groups II and III of the Periodic Table in place of the lanthanum and barium of the original compound.

The alkali metals are at troughs and the inert gases at peaks. There is also a trend of decreasing first ionization energy as we descend any group. Those for Groups I and 0 are shown dotted in brown on the graph.

This graph can also be explained by looking at electronic structures. As we go along a period, we are adding electrons to the same shell and the nuclear charge is increasing, so it gets increasingly difficult to pull the electron out of the atom, **Figure 20.9.** In neon, an outer electron feels an effective nuclear charge of 8+ and so neon has a high ionization energy.

3+ 4+ 5+ 6+ 7+ 8+ 9+ 10+

Li Be B C N O F Ne

Figure 20.9 (right) The electronic structures of the elements lithium to neon

On moving to sodium, **Figure 20.10**, we start a new shell. The nuclear charge felt by the outer electron is shielded by 10 (2 + 8) inner electrons and it becomes much easier to remove the outer electron. In addition, the outer electron is in shell 3 rather than shell 2 and is thus already further away from the nucleus. As we go down a group, the shielding means that the outer electron(s) in each atom all feel the same effective nuclear charge, **Figure 20.11.** The outer electron gets easier to remove because it is further away from the nucleus.

Figure 20.10 The electronic structure of sodium. The outer electron feels a nuclear charge of (11 + − 10) = 1+

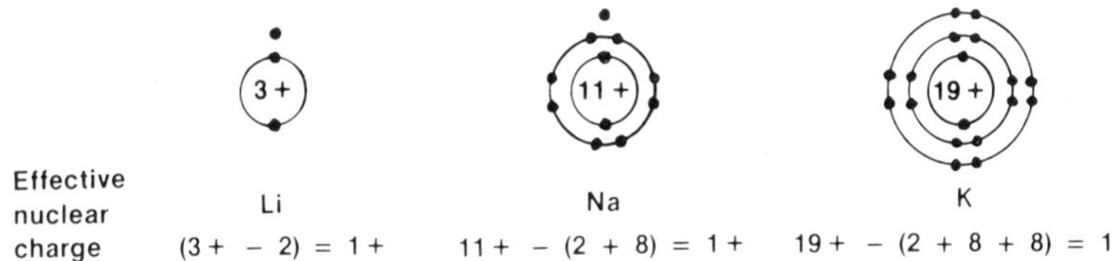

Figure 20.11 The electronic structures of Group I elements

A closer look at the graph of first ionization energy against atomic number for the lithium to neon and sodium to argon periods (**Figure 20.8**) is interesting. The rise in ionization energy is not regular and actually drops from beryllium to boron and from nitrogen to oxygen. The same thing happens with corresponding elements in Period 3. To explain this, we must remember that the '2' shell of electrons is actually subdivided

Figure 20.12 Electron arrangements of nitrogen and oxygen

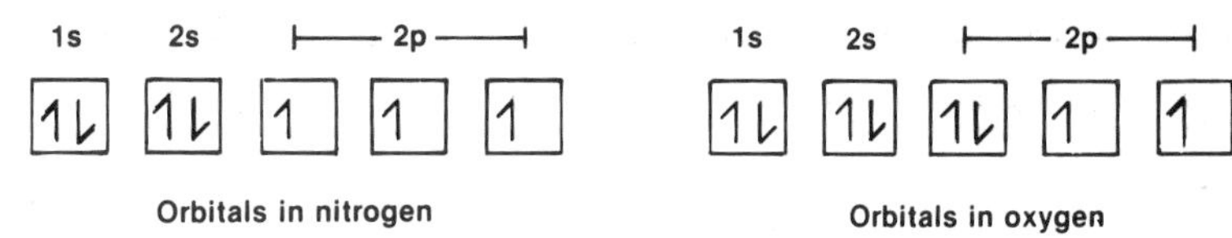

Figure 20.13 Energies of orbitals in beryllium and boron

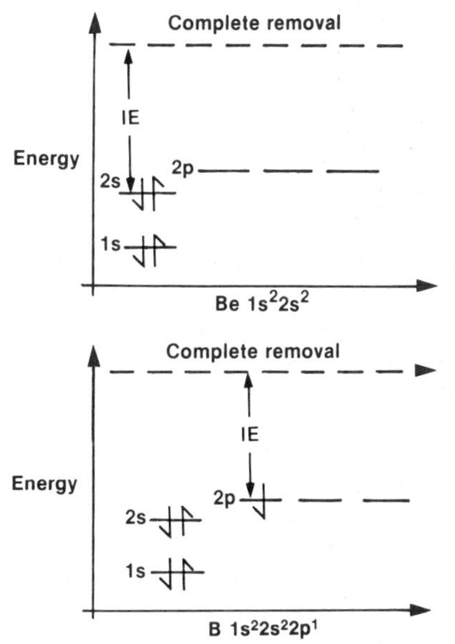

into 2s and 2p, **Figure 20.12.** The most easily removed electron in boron is that in the 2p orbital, higher in energy than the 2s electrons in beryllium. It therefore needs less energy for total removal than the 2s electron and this outweighs the effect of the increased nuclear charge of boron, which tends to make it harder to remove.

Comparing nitrogen and oxygen, the extra electron in oxygen has to be paired in one of the 2p orbitals while all three 2p electrons in nitrogen are unpaired, **Figure 20.13.**

An electron in a pair is easier to remove than one occupying an orbital alone due to the repulsion of the other electron in the orbital. This outweighs the effect of the greater nuclear charge of oxygen. The same situations occur with magnesium and aluminium and with phosphorus and sulphur in Period 3.

20.3.4 The transition elements

Changes on going across the transition elements are much less marked than on crossing the s- and p-blocks. This is because on crossing a transition series, the electrons are being added to an *inner* shell of electrons and therefore affect the properties less than electrons being added to the outer shell. The electron arangements of the first transition series from scandium to zinc are shown in **Table 20.1.**

The electron arrangement of scandium is $1s^2 2s^2 2p^6 3s^2 3p^6 3d^1 4s^2$ which may be written [Ne] $3s^2 3p^6 3d^1 4s^2$.

Table 20.1. Electronic structures of the elements of the first transition series

Sc	[Ne] $3s^2 3p^6 3d^1\ 4s^2$
Ti	$3d^2\ 4s^2$
V	$3d^3\ 4s^2$
Cr	$3d^5\ 4s^{1*}$
Mn	$3d^5\ 4s^2$
Fe	$3d^6\ 4s^2$
Co	$3d^7\ 4s^2$
Ni	$3d^8\ 4s^2$
Cu	$3d^{10} 4s^{1*}$
Zn	$3d^{10} 4s^{2*}$

Figure 20.14 The energies of the orbitals up to 4s. Note that the 3d energy levels are slightly *higher* than the 4s

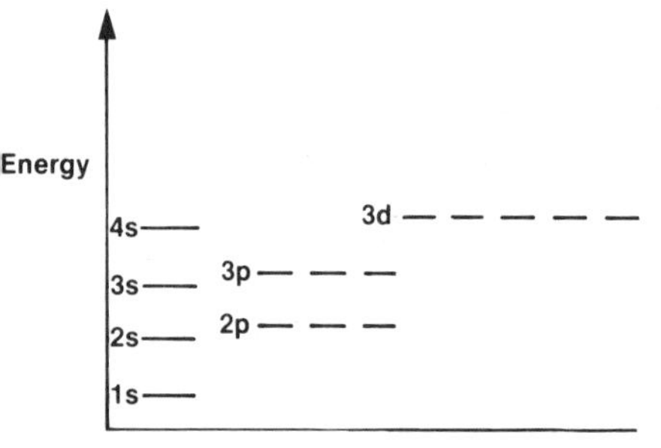

The electrons are being added to the 3d shell while the *outer* shell (4s) is almost unchanged, because the 3d shell is higher in energy than the 4s, **Figure 20.14.**

*Note that the $3d^5$ and $3d^{10}$ arrangements seem to be preferred—half-full and full shells seem to have extra stability which makes it worth transferring an electron from 4s. The 3d and 4s levels are very close in energy.

Many other physical properties show periodicity, for example, melting temperatures and boiling temperatures.

20.4 Periodicity of chemical properties

Detailed reactions of particular elements and compounds will be given in the relevant chapters. This section is devoted to more general trends in chemical properties. We have already seen in **section 4.3** two trends in chemical properties on moving left to right across a period:

1. a gradual change from metals to non-metals
2. a gradual change from basic oxides to acidic oxides.

We can add to these the following trends.

A

ANTI-KNOCK PETROL ADDITIVES

'Knocking' in petrol engines is a vibration caused by the fuel–air mixture exploding too early while the piston is still rising in the cylinder, rather than as it reaches the top. Knocking reduces the efficiency of the engine and can cause damage to the piston and cylinder head. In 1916 an American engineer, Thomas Midgeley, was working on ways to alleviate the problem. He had found by chance that iodine and its compounds when dissolved in the fuel reduced the knocking effect. Other halogens were cheaper but were much less effective. He then found a similar trend with Group VI compounds—an increasing anti-knock effect on descending the group, and the Group VI compounds were better than the corresponding compounds of Group VII elements.

increasing anti-knock effect ←

C	N	O	F
Si	P	S	Cl
Ge	As	Se	Br
Sn	Sb	Te	I
Pb	Bi	Po	At

increasing anti-knock effect ↓

Following these trends led to lead compounds being investigated. These indeed proved to be very effective and a petrol-soluble lead compound, tetraethyllead, has been added to petrol for many years, although it is now being phased out for environmental reasons.

Midgeley's investigations were not without their problems, however, and he suffered from tellurium poisoning, one symptom of which is a powerful garlic-like body odour!

20.4.1 Redox properties

A gradual change from the element being a good reducing agent (on the left) to a powerful oxidizing agent (on the right) is shown by the values of $E^\ominus$, see **section 13.3.** Values of $E^\ominus$ below refer to aqueous solutions.

	Na	Mg	Al	(Si P)	S	Cl
	$Na^+ + e^- \rightarrow Na$	$Mg^{2+} + 2e^- \rightarrow Mg$	$Al^{3+} + 3e^- \rightarrow Al$		$S + 2e^- \rightarrow S^{2-}$	$\frac{1}{2}Cl_2 + e^- \rightarrow Cl^-$
$E^\ominus$/V	−2.71	−2.37	−1.66		−0.48	+1.36

→ [element is a stronger oxidizing agent]

← [element is a stronger reducing agent]

element tends to give away electrons — element tends to accept electrons

20.4.2 Electronegativity

There is a gradual increase in electronegativity (electron-attracting power) as we go from left to right:

	Na	Mg	Al	Si	P	S	Cl
electronegativity	0.9	1.2	1.5	1.8	2.1	2.5	3.0

increasing electronegativity →

A graph of electronegativity against atomic number (**Figure 20.15**) shows both the *increase* across a period and the *decrease* on descending a group (e.g. fluorine, chlorine, bromine, iodine).

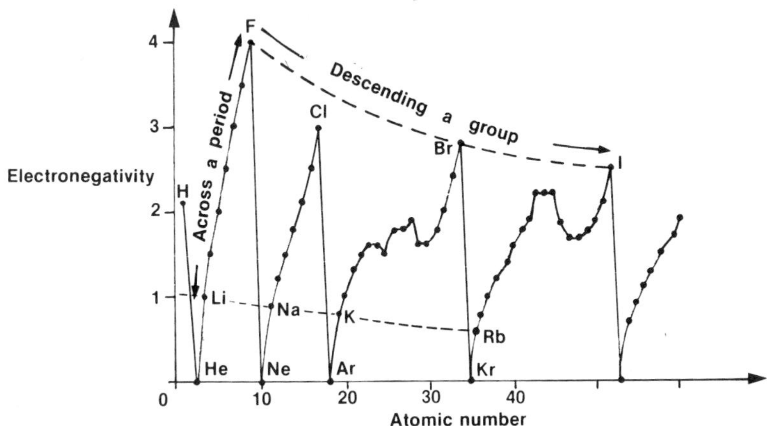

Figure 20.15 The periodicity of electronegativity values

20.4.3 Bond energies (enthalpies)

Table 20.2 shows element–hydrogen bond energies for the first two periods. (Typically, elements in Group I, II and III do not form covalent bonds with hydrogen.)

Table 20.2 E—H bond energies/kJ mol^{-1}

IV	V	VI	VII
C—H	N—H	O—H	F—H
435	391	464	568
Si—H	P—H	S—H	Cl—H
318	321	364	432

Notice the overall increase across each period. This is due to the increased nuclear charge on successive elements which hold the shared pair of electrons more strongly. For example, in methane, carbon has an effective nuclear charge of $6-2=4+$:

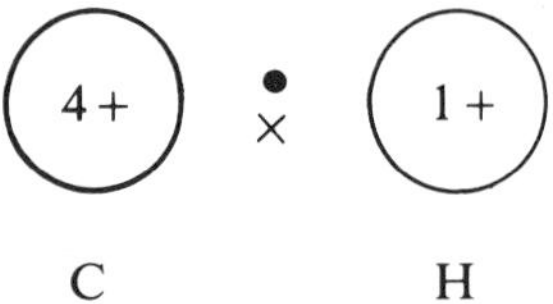

In hydrogen fluoride, fluorine has an effective nuclear charge of $9-2=7+$:

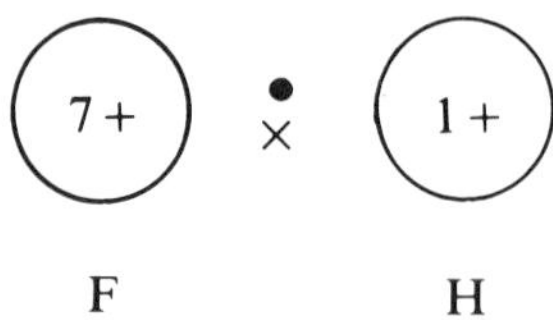

The decrease on descending the group is because the shared electrons become further from the nucleus.

20.4.4 Oxidation states

Figure 20.16 shows a plot of oxidation number in oxides, chlorides and hydrides against atomic number from lithium to argon. A certain periodicity is seen. The first three members of each group show only the positive oxidation number corresponding to loss of all their outer electrons. Both Group IV elements show oxidation numbers of +IV and −IV. The last three members of the period show negative oxidation numbers

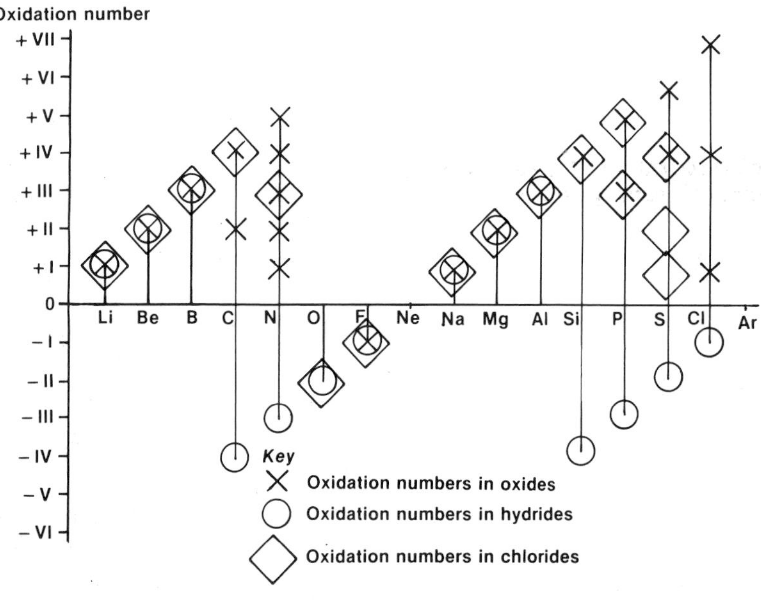

Figure 20.16 The variation of oxidation numbers with atomic number

corresponding to gaining sufficient electrons to form a full outer shell, although those in Period 3 (phosphorus, sulphur, chlorine) also show a number of positive oxidation numbers.

20.4.5 Periodicity of properties of chlorides of elements in Periods 2 and 3

These are summarized in **Table 20.3**. A number of trends are apparent:
- On moving from left to right there is a gradual change from giant ionic to molecular covalent structures.

Table 20.3 **Periodicity of chemical properties of the chlorides of elements of Periods 2 and 3.**

	Group	I	II	III	IV	V	VI	VII
Period 2	Formula	LiCl	$BeCl_2$	BCl_3	CCl_4	NCl_3	OCl_2	FCl
	Structure	Giant ionic	Covalent chain molecules	Molecular	Molecular	Molecular	Molecular	Molecular
	Effect of water	Dissolves to give ions $Li^+(aq)+$ $Cl^-(aq)$	Hydrolyses to give $Be(OH)_2(s)+$ $2H^+(aq)+$ $2Cl^-(aq)$	Hydrolyses to give $B(OH)_3(s)+$ $3H^+(aq)+$ $3Cl^-(aq)$	Does not mix or react	Hydrolyses to give $NH_3(aq)+$ $3HClO(aq)$	Hydrolyses to give HClO(aq)	Hydrolyses to give $H^+(ag)+F^-(aq)+$ HClO(aq)
Period 3	Formula	NaCl	$MgCl_2$	$AlCl_3$	$SiCl_4$	PCl_3	SCl_2	$ClCl(Cl_2)$
	Structure	Giant ionic	Ionic layer lattice	Mainly covalent layer lattice	Molecular	Molecular	Molecular	Molecular
	Effect of water	Dissolves to give ions $Na^+(aq)+$ $Cl^-(aq)$	Dissolves to give ions $Mg^{2+}(aq)+$ $2Cl^-(aq)$	Hydrolyses to give $Al(OH)_3(s)+$ $3H^+(aq)+$ $3Cl^-(aq)$	Hydrolyses to give $SiO_2(s)+$ $4H^+(aq)+$ $4Cl^-(aq)$	Hydrolyses to give $H_3PO_3(aq)+$ $3H^+(aq)+$ $3Cl^-(aq)$	Hydrolyses to give $S(s)+$ $H^+(aq)+$ $Cl^-(aq)$	Partially hydrolyses to give $HClO(aq)+$ $H^+(aq)+$ $Cl^-(aq)$

- The behaviour in water shows a trend from dissociation into ions on the left to reaction to give acidic solutions on the right.
- The periodicity of the formulae can be seen with the ratio of E:Cl, where E represents the other element, rising to a maximum in Group IV. Note though that other chlorides of phosphorus and sulphur exist—PCl_5, S_2Cl_2 and SCl_4. Other oxides of chlorine exist also.

20.4.6 Anomalous properties of the elements of the lithium to neon period

It is generally true that elements in the same group of the Periodic Table have similar properties. However, the elements in Period 2 (lithium to neon) show many properties which are not typical of their group. For example, the compounds of lithium and beryllium show a degree of covalency which is not found in compounds of other s-block elements; nitrogen and oxygen form diatomic molecules while the other members of their groups form polyatomic molecules or chains; and there are many other examples.

The reasons for these anomalies are:

1. These atoms are particularly small. This means that they have unusually high ionization energies because electrons must be removed from close to the nucleus. Their small size also makes them especially electronegative, i.e. good at attracting electrons (when they form covalent bonds, the electrons in their bonds approach very close to the nucleus). Their small size also means that their positive ions have a high **charge**

density and thus they can polarize negative ions very strongly—see section 10.2.1. These factors all lead to a greater than normal degree of covalency of the metal compounds.

2. The elements of Period 2 can form bonds only through their 2s and 2p orbitals. The next available orbital, 3s, is of a much higher energy and it is most unlikely that an electron would be promoted to it. The elements in Period 3 can more easily promote an electron as the 3d orbitals, fairly close in energy to 3s and 3p, are available. This allows them to form more covalent bonds. For example, nitrogen forms only NCl_3 while phosphorus forms both PCl_3 and PCl_5, as illustrated in **Figure 20.17**.

Figure 20.17 Nitrogen can form only three covalent bonds while phosphorus can form up to five

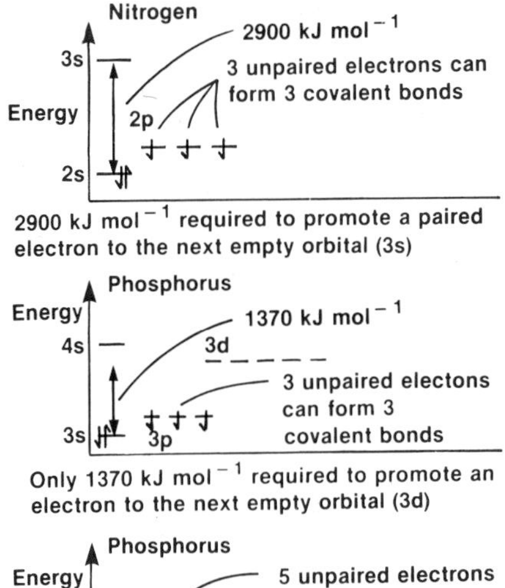

20.4.7 Diagonal relationships

The unusual electronegativity of lithium, beryllium and boron leads to what are called **diagonal relationships** – lithium is strikingly similar to magnesium, beryllium to aluminium and boron to silicon.

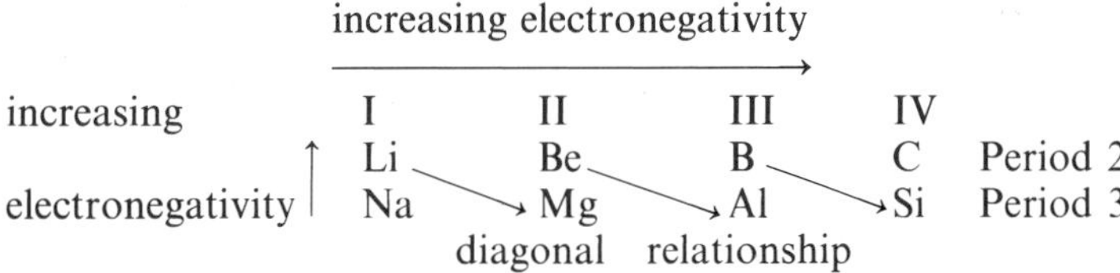

Electronegativity increases from left to right across a period. The untypically large electronegativities of Period 2 elements (caused by their small size) mean that they are in some ways more typical of elements one group to the right than elements of their own group.

Examples of the diagonal relationship of lithium and magnesium are:

- Lithium burns in air to form the 'normal' oxide Li_2O, the other alkali metals form peroxides and superoxides (M_2O_2 and MO_2). Magnesium and the Group II elements form 'normal' oxides.
- Lithium combines with nitrogen to form lithium nitride. The other Group I elements do not form nitrides. Magnesium does.
- Lithium nitrate decomposes on heating to form the oxide as do Group II nitrates. The other Group I nitrates decompose to form the metal nitrate(III) (nitrite).

Other examples will be found in Chapter 23.

20.5 Summary

- Areas of the Periodic Table are identified as s-block, p-block, d-block and f-block. This refers to the type of orbital which holds the highest energy electrons.
- Many physical properties are periodic, i.e. similar properties always recur in the same group of the Periodic Table. Examples include sizes of atoms and ions and first ionization energies.
- Many periodic properties can be explained in terms of electron structures and the effective nuclear charge felt by the outer electrons.
- Chemical properties of elements and their compounds are generally similar for elements in the same group.
- There are many trends in properties on crossing a period of s- and p-block elements, for example:

acidity of oxides →

element is better oxidizing agent →

increasing electronegativity of element →

metallic to non-metallic properties →

• d-block elements show fewer trends across a period and are similar to one another as d-electrons are being added to an *inner* shell.
• Elements in the lithium to neon period are often untypical of the rest of their group due to their small sizes.

20.6 Questions

1 For each element below, identify it as s-block, p-block, d-block or f-block.
A) Pb **B)** Cu **C)** Be **D)** U **E)** Ne

2 The table gives the melting temperatures and second ionization energies of the elements from hydrogen to calcium.
A) Plot a graph of property against atomic number for each. Label with the symbol of the element any points you consider to be significant.
B) To what extent is each property periodic?
C) Attempt to explain any periodicity you identify.

Element	T_m/K	Second ionization energy/kJ mol^{-1}
H	14	–
He	1	5251
Li	454	7928
Be	1551	1757
B	2573	2427
C	3925	2353
N	63	2856
O	55	3388
F	53	3374
Ne	25	3952
Na	371	4563
Mg	922	1451
Al	933	1817
Si	1683	1577
P	317	1903
S	386	2251
Cl	172	2297
Ar	84	2666
K	336	3051
Ca	1112	1145

3 A) Place the elements Cl, Br, I in order of the first ionization energy you would expect. Explain your answer.
B) Place the elements Si, P, S in order of the covalent radius you would expect. Explain your answer.

4 The table (top right) shows the atomic and ionic radii of the alkaline earth elements.

Element	Atomic radius /nm	Ionic radius /nm
Beryllium	0.112	0.030
Magnesium	0.160	0.065
Calcium	0.197	0.094
Strontium	0.215	0.110
Barium	0.221	0.134

A) Using magnesium as an example, explain carefully what is meant by:
i atomic radius;
ii ionic radius.
B) The ionic radius is, in each case, smaller than the atomic radius. Explain why this is so.
C) Explain why the atomic radius increases from Be to Ba.
D) The ions K^+ and Ca^{2+} have identical electronic configurations, yet the ionic radius of K^+ is larger than that of Ca^{2+}. Suggest why it is larger.
(London 1979)

5 The table gives some information about eight elements A to H, with consecutive atomic numbers; these letters are not the usual chemical symbols for the elements.

Element	Configuration of outer electrons	First ionization energy /kJ mol^{-1}	Formula of typical oxide	Formula of typical chloride
A	$3s^2 3p^1$	580		
B		790		
C		1010		
D		1000		
E		1260		
F		1520		
G		420		
H	$3s^2 3p^6 4s^2$	590		

A) Copy the table and fill in the blank spaces where appropriate.
B) Select from the elements A to H:
i a noble gas;
ii an alkaline earth metal;
iii a Group VI element.
C) Select from the elements A to H:
i one element which forms a basic oxide;
ii one which has an acidic oxide;
iii one which might form an amphoteric oxide.
(London 1979)

21 Hydrogen

21.1 Introduction

Hydrogen is the simplest atom, its most common isotope consisting of just one proton and one electron. It is the most abundant element in the universe. Estimates suggest that around 90% of the atoms in the universe are hydrogen. Hydrogen is vital to life on earth in many ways. It is nuclear fusion reactions, in which hydrogen is converted to helium, which provide the sun's energy. All life depends on the presence of water and it is the unique property of hydrogen—hydrogen bonding—which makes water liquid at normal temperatures. Virtually all of the complex organic compounds on which life depends contain hydrogen.

21.2 The element

21.2.1 Isotopes of hydrogen

Hydrogen has three isotopes:

$$^{1}_{1}H, \quad ^{2}_{1}H, \quad ^{3}_{1}H$$

$^{2}_{1}H$, which has one proton and one neutron, is often called deuterium (D), and $^{3}_{1}H$ (with one proton and two neutrons) is called tritium (T). Deuterium oxide, D_2O, is often called heavy water and is used as a moderator (it slows

E

FUSION REACTIONS IN THE SUN

In the sun, temperatures are so high that hydrogen nuclei collide fast enough to overcome the repulsion of their positive charges. If the nuclei get close enough, the **strong nuclear force** (see **section 8.2.5**) holds the protons together and the nuclei **fuse** (join together). Four atoms of hydrogen can fuse to form an atom of helium. The extra positive charge is lost by the emission of two positrons (particles of the same mass as electrons but with a positive charge). The reaction is:

$$4\,^{1}_{1}H \longrightarrow \,^{4}_{2}He + 2\,^{0}_{1}e^{+}$$

The mass numbers and the atomic numbers are conserved, but the exact masses are not. Exact masses:

$$H = 1.007825\ u$$
$$He = 4.002603\ u$$
$$e^{+} = 0.000549\ u$$

$$\text{total mass before reaction} = 4 \times 1.007825\ u = 4.0313\ u$$

$$\text{total mass after reaction} = 4.002603 + (2 \times 0.000549) = 4.0037\ u$$

$$\text{difference} = -0.0276\ u$$

This mass is converted to energy which fuels the sun. In the sun, 4 200 000 *tonnes* of mass are converted to energy every second! However, the sun is so massive that we estimate that it can continue shining for several billion years.

Similar nuclear fusion reactions are used in the hydrogen bomb, and research is proceeding towards a controlled nuclear fusion reaction for generating energy. The high temperature required (over 100 000 000 K) to get nuclei to fuse is the major stumbling block.

down neutrons) in nuclear power stations. Compounds in which deuterium replaces hydrogen can be made. These are useful for 'labelling' a particular hydrogen atom in a compound to trace its path during a reaction. Deuterated compounds can be detected by their mass using a mass spectrometer.

21.2.2 Occurrence

There is little uncombined hydrogen on the earth but it is abundant combined in compounds, particularly water and hydrocarbons (e.g. crude oil and coal).

21.2.3 Physical properties

An early airship using hydrogen

Hydrogen is a colourless odourless gas. Its boiling temperature is 20 K and its melting temperature 14 K. It has the lowest density of any gas ($0.09\,g\,dm^{-3}$), which led to its use as a lifting gas in airships in the early years of this century. It forms explosive mixtures with air which is the basis of the 'pop' test for hydrogen. However, hydrogen itself burns non-explosively and indeed was piped to British homes as a constituent of 'town gas', a mixture of hydrogen and carbon monoxide. Now, methane gas from the North Sea is used instead.

21.2.4 Industrial production and uses

1000 million m^3 of hydrogen are produced annually in the UK. Around 40% of this is used to manufacture ammonia via the Haber process; 30% is used in the treatment of petroleum and 10% is used in the manufacture of margarine. Hydrogen is usually manufactured where it is required, as it is not economic to transport it. It is stored under pressure in steel cylinders and, as the gas is so light, most of the transport costs are for moving the cylinders around.

Most hydrogen is produced from natural gas (methane) and water by the reaction:

$$\underset{\text{methane}}{CH_4(g)} + \underset{\text{steam}}{H_2O(g)} \xrightarrow[\text{catalyst}]{\text{metal}} \underset{\text{carbon monoxide}}{CO(g)} + \underset{\text{hydrogen}}{3H_2(g)}$$

More steam is added to convert the carbon monoxide to carbon dioxide, which can be removed by dissolving it in water:

$$\underset{\text{carbon monoxide}}{CO(g)} + \underset{\text{steam}}{H_2O(g)} \longrightarrow \underset{\text{carbon dioxide}}{CO_2(g)} + \underset{\text{hydrogen}}{H_2(g)}$$

A

MANUFACTURE OF MARGARINE

Margarine is an emulsion of water with oils and fats. Its also contains salt, vitamins, colouring, and emulsifiers—these prevent the oil and water separating. Both vegetable oils and animal fats are used, including unsaturated oils (those with carbon–carbon double bonds). Unsaturated oils tend to have low melting temperatures and deteriorate more easily than saturated fats. They are therefore partially saturated by hydrogenation, using hydrogen and a nickel catalyst. The reaction which occurs is:

$$\sim\!\sim\ \underset{H}{C}=\underset{H}{C}\ \sim\!\sim + H_2 \longrightarrow \sim\!\sim\ \underset{H}{\overset{H}{C}}-\underset{H}{\overset{H}{C}}\ \sim\!\sim$$

Different oils and fats are blended to give the correct consistency of the final product, for example to allow it to spread easily. Increasing the percentage of saturated fats makes the margarine more solid.

The percentage of unsaturated fat in margarine is controlled by hydrogenation

Carbon monoxide and hydrogen can react directly to give methanol, which is used as a solvent and also to make methanal for methanal-based plastics:

$CO(g)$	+	$2H_2(g)$	$\longrightarrow$	$CH_3OH(g)$
carbon monoxide		hydrogen		methanol

A catalyst of zinc and chromium oxides is used.

Some hydrogen is made by the action of steam on white-hot coke to produce a mixture of gases called water gas:

$C(s)$	+	$H_2O(g)$	$\longrightarrow$	$CO(g)$	+	$H_2(g)$
coke		steam		carbon monoxide		hydrogen
				water gas		

The carbon monoxide can be removed as before. Hydrogen is also a by-product of the electrolysis of brine (see boxes, pages 294 and 342) and the cracking of hydrocarbons.

21.3 Compounds of hydrogen

Hydrogen will combine with all elements apart from the noble gases. The more electropositive (less electronegative) s-block metals are able to form compounds with hydrogen in which the hydrogen accepts an electron to form the hydride ion, H^-. In the majority of cases hydrogen forms covalent bonds with the other elements. With more electronegative elements the E—H bond is strongly polar: $E^{\delta-}—H^{\delta+}$. The trend on crossing the sodium to argon period is shown:

NaH	MgH_2	AlH_3	SiH_4	PH_3	SH_2	ClH
ionic: $Na^+ + H^-$	intermediate between ionic and covalent	covalent polymeric	covalent—increasingly polar $E^{\delta-}—H^{\delta+}$ $\longrightarrow$			

21.3.1 Ionic hydrides

The ionic hydrides of the s-block metals conduct electricity when molten and produce hydrogen at the *anode*. These hydrides can be prepared by direct reaction of the metal with hydrogen at moderately high temperatures. They react with water to give the metal hydroxide and hydrogen, for example:

$NaH(s)$	+	$H_2O(l)$	$\longrightarrow$	$NaOH(aq)$	+	$H_2(g)$
sodium hydride		water		sodium hydroxide		hydrogen

They are good reducing agents, as are the complex hydrides $LiAlH_4$ (lithium tetrahydridoaluminate(III)) and $NaBH_4$ (sodium tetrahydridoborate(III)), both of which are much used in organic chemistry.

21.3.2 Covalent hydrides

These are generally molecular substances. Except for water they burn in air. The Group IV hydrides do not dissolve in water. The hydrides of Groups V, VI and VII dissolve in water to give solutions which increase in acidity on moving from Group V to Group VI to Group VII.

21.3.3 Transition metal hydrides

The structures of these are not well understood. Some of them are thought to be interstitial compounds with hydrogen molecules trapped in the holes in the metal lattice. Often hydrogen is given off when the compounds are heated. They have been investigated as a possible means of storing hydrogen for use as a vehicle fuel.

21.4 Acids

Hydrogen can form the H^+ ion, a bare proton, by loss of its electron. The ionization energy is high ($1312\,kJ\,mol^{-1}$). The bare proton is never found in chemical systems although it can be produced in a high-voltage discharge tube. The charge density of the proton is so high that it always attracts a pair of electrons to form a dative covalent bond with it, for example:

$$H^+ + H_2O \longrightarrow H_3O^+$$

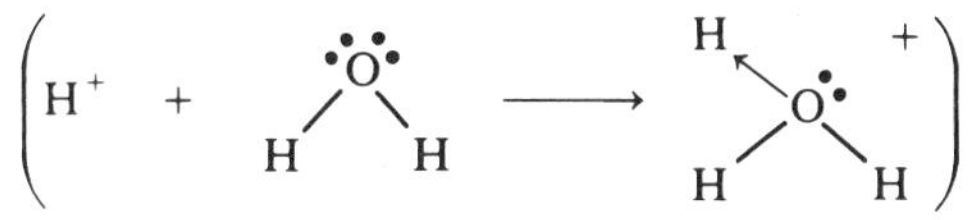

or

$$H^+ + NH_3 \longrightarrow NH_4{}^+$$

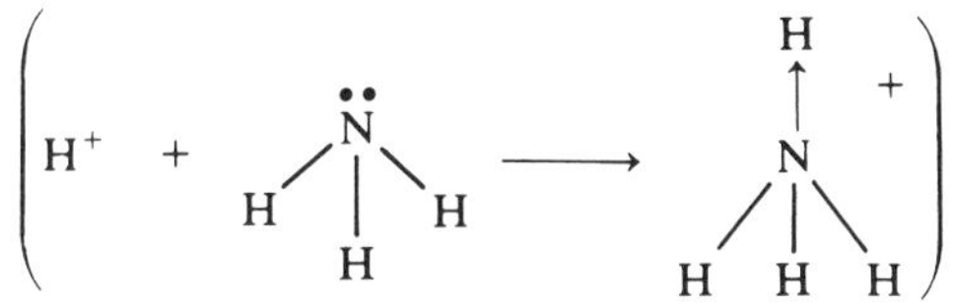

21.5 Summary

- Hydrogen is an element with unique properties consisting of one proton and one electron.
- It is an important industrial chemical for use in ammonia manufacture, petrol refining, margarine production and the synthesis of methanol.
- It is expensive to transport due to its low density so it is usually manufactured where it is required.
- Most hydrogen is made by the reaction of methane and steam.
- Hydrogen can form compounds with all elements except the noble gases. The trend is for ionic hydrides containing the H^- ion to form with elements on the left of the Periodic Table and covalent molecular hydrides to form with those on the right.
- The H^+ ion, a bare proton, cannot normally exist alone, rapidly accepting an electron pair to form dative covalent bonds.

21.6 Questions

1 Research/debate: Hydrogen has been proposed as a replacement for hydrocarbons as a fuel, particularly for vehicles and aircraft (see box *Alternative fuels*, **section 9.3**). Discuss the advantages and disadvantages of this proposal. Some topics you might consider are: source; renewability; safety; cost; storage; pollution; the environment. Do you think the advantages outweigh the disadvantages? Which problem(s) is (are) likely to be the most difficult to overcome?

2 **A)** Draw dot cross electron arrangement diagrams to show the bonding in:

i NaH (ionic)
ii CaH_2 (ionic)
iii NH_3 (covalent)
iv H_2O (covalent)

B) What changes (if any) would you have to make to (iv) if it referred to D_2O? Explain your answer.

3 Given a sample of deuterium oxide, D_2O, how would you make:

A) D_2
B) NaOD
C) DCl?

4 If H_2O and D_2O are mixed together, a mass spectrum of the sample shows three major peaks of masses 18, 19 and 20.

A) What species cause the peaks?
B) What must have happened to produce *three* peaks?

22 The noble gases

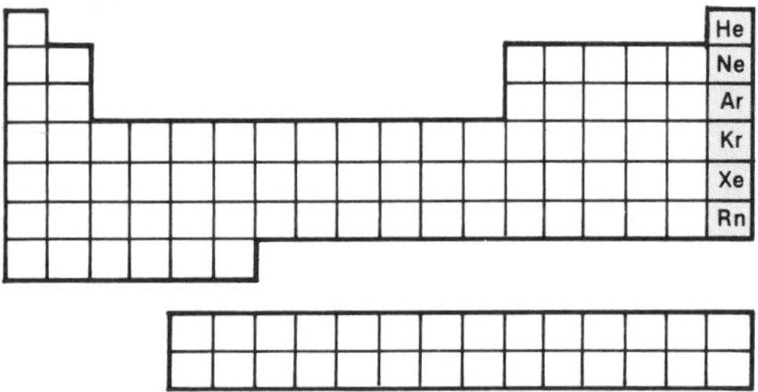

22.1 Introduction

The elements of Group 0, helium, neon, argon, krypton, xenon and radon, are often called the noble, inert or rare gases. They are the least reactive group of elements because they have a full outer shell of electrons. None of the names for the group is particularly appropriate. The names noble and inert imply a reluctance to form compounds. While this is generally true, over 60 compounds of these elements have been made, and only helium has no known compounds. Nor are the gases very rare. An average sized laboratory contains around 2500 dm^3 of argon, 5 dm^3 of neon, 1.5 dm^3 of helium, 300 cm^3 of krypton and 25 cm^3 of xenon.

22.2 The elements

22.2.1 Occurrence

All the gases except radon are present in the air in small amounts. Radon is formed in igneous rocks by the radioactive decay of uranium isotopes. Radon itself is radioactive with a short half-life. This has inhibited research into its properties.

22.2.2 Industrial preparation and uses

Apart from radon, the noble gases are extracted from the atmosphere by fractional distillation. Argon is obtainable as a by-product of the Haber process as it remains behind after the nitrogen has reacted with the hydrogen. Helium is often present in natural gas (it may comprise up to 50%).

Most of the uses of these gases depend on their inertness. The gas most in demand industrially is argon (30 000 tonnes per year are used in the UK) which is used to produce an inert atmosphere in metal manufacturing processes, such as welding or casting, where it prevents the formation of metal oxides by keeping air away from the hot metal. Electric light bulbs are filled with argon to prevent the white-hot filament burning away.

Neon-filled discharge lamps are used in advertising signs, giving a red light which is also ideal for airport lighting as it penetrates fog well. See **Plate 15.**

Because of its low density, helium is used as a lifting gas for weather balloons and airships. Deep-sea divers breathe a mixture of helium and oxygen, as helium is less soluble in the blood than nitrogen, and so prevents 'the bends' (see box *Solubility of gases*, **section 15.4**). Liquid helium, which boils at 4 K, is used as a coolant for research at low temperatures.

Krypton and xenon are both used as fillers for various types of lamps.

22.2.3 Physical properties

These are summarized in **Table 22.1.**

Table 22.1 Physical properties of the noble gases. The direction of increase in trends is shown in brown

	Atomic number, *Z*	Electron arrangement	First IE /kJ mol^{-1}	van der Waals radius	T_m/K	T_b/K	$\Delta_{vap}H^\ominus$ /kJ mol^{-1}	Abundance in air /p.p.m.
He Helium	2	2 $1s^2$	2372 ↑	0.18	1 (at 26 × 10² kPa)	4	0.08	5
Ne Neon	10	2, 8 $[He]2s^2 2p^6$	2081	0.16	25	27	1.8	18
Ar Argon	18	2, 8, 8 $[Ne]3s^2 3p^6$	1521	0.19	84	87	6.5	9300
Kr Krypton	36	2, 8, 18, 8 $[Ar]3d^{10} 4s^2 4p^6$	1351	0.20	116	121	9.0	1
Xe Xenon	54	2, 8, 18, 18, 8 $[Kr]4d^{10} 5s^2 5p^6$	1170	0.22	161	166	12.6	0.08
Rn Radon	86	2, 8, 18, 32, 18, 8 $[Xe]4f^{14} 5d^{10} 6s^2 6p^6$	1037	? ↓	202 ↓	211 ↓	? ↓	–

Points to note are:

- Increasing van der Waals radius on descending the group – each atom has one shell of electrons more than the previous one.
- Decreasing first ionization energy on descending the group – the outer electrons are further from the nucleus.
- The values of $\Delta_{vap}H^\ominus$, the enthalpy change on boiling, are all small but increase on descending the group. $\Delta_{vap}H^\ominus$ is a measure of the strength of van der Waals forces between the atoms. These are weak in general but rise with atomic number as the atoms have more electrons.

22.3 Compounds of the noble gases

No noble gas compounds were known until 1962 when Neil Bartlett, a British chemist working in Canada, found that the strong oxidizing agent platinum(VI) fluoride, PtF_6, would oxidize even oxygen, forming dioxygenylhexafluoroplatinate(V), $O_2^+ + (PtF_6)^-$. He then realized that the ionization energy of xenon was slightly less than that of the oxygen molecule, so that platinum(VI) fluoride should also react with xenon.

$$Xe(g) \longrightarrow Xe^+(g) + e^- \quad \Delta H^\ominus = +1170\,\text{kJ mol}^{-1}\ \text{(first IE of Xe)}$$
$$O_2(g) \longrightarrow O_2^+(g) + e^- \quad \Delta H^\ominus = +1183\,\text{kJ mol}^{-1}\ \text{(first IE of } O_2\text{)}$$

He therefore allowed xenon, a colourless gas, and platinum(VI) fluoride, a red vapour, to react, and an orange solid, xenon hexafluoroplatinate(V), $Xe^+ + (PtF_6)^-$ was formed—the first compound of a noble gas.

This development spurred on other chemists to try to prepare other noble gas compounds, and now only helium has no known compounds. In the majority of these compounds, the inert gas has a positive oxidation number and almost all the known compounds are formed with strongly electronegative elements, mainly oxygen or fluorine and the other halogens. Xenon, having the lowest first ionization energy, has the most known compounds. Because of its short half-life, radon has hardly been studied.

Since noble gas compounds involve positive oxidation states of the noble gas and, given the stability of the elements, we should expect the compounds to be good oxidizing agents, the noble gas atom readily dropping its own oxidation number and therefore increasing that of some other species. This is found to be the case, for example:

$$\overset{+\text{VI}}{Xe}O_3(aq) + 6\overset{-\text{I}}{I^-}(aq) + 6H^+(aq) \longrightarrow \overset{0}{Xe}(g) + 3\overset{0}{I_2}(aq) + 3H_2O(l)$$

The iodide ions are oxidized to iodine.

We shall not consider noble gas chemistry in any detail but we shall look at some applications of chemical principles to these compounds.

22.3.1 Energetics

Born–Haber cycles can be used to predict the enthalpy of formation of inert gas compounds and hence whether they are likely to exist, see also **section 10.2.1**. For example:

$$Xe^+ + (PtF_6)^-$$

$$\text{first ionization energy Xe} = +1170\,\text{kJ mol}^{-1}$$

$$\text{electron affinity } PtF_6 = -710\,\text{kJ mol}^{-1}$$

$$\text{lattice energy } Xe^+\,(PtF_6)^- = -460\,\text{kJ mol}^{-1}$$

Note: The value of $\Delta_{at}H^{\ominus}(Xe)$ is not required as xenon exists as separate gaseous atoms in its standard state, nor is $\Delta_{at}H^{\ominus}(PtF_6)$ which is also a gas. **Figure 22.1** shows the Born–Haber cycle for xenon hexafluoroplatinate(V).

Figure 22.1 The Born–Haber cycle for xenon hexafluoroplatinate (V). $\Delta_f H^{\ominus}$ for this compound is zero – this is unusual but entirely coincidental. No prediction can easily be made about the likely stability of the compound in this case

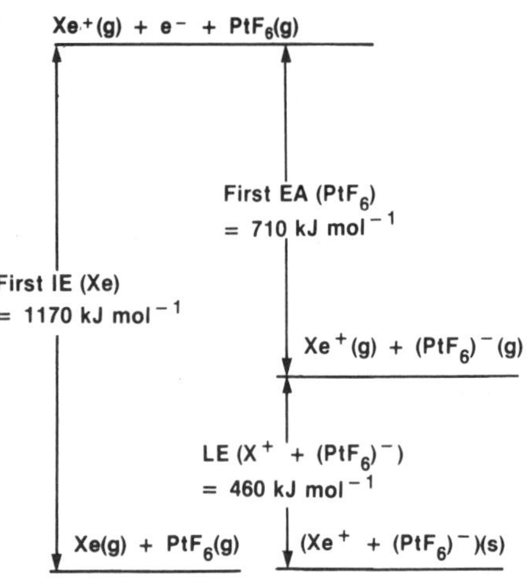

Bond energies

For example, we shall look at xenon(IV) fluoride, XeF_4, a covalent compound:

$$\text{bond energies: Xe—F} = 130\,\text{kJ mol}^{-1}$$

$$\text{F—F} = 158\,\text{kJ mol}^{-1}$$

What is $\Delta_f H^{\ominus}$ for:

$$Xe + 2F_2 \longrightarrow XeF_4?$$

$$\text{we must break two F—F bonds} = 2 \times 158 = 316\,\text{kJ mol}^{-1}$$

$$\text{we must make four Xe—F bonds} = 4 \times 130 = 520\,\text{kJ mol}^{-1}$$

The difference is 204 kJ mol^{-1}. More energy is given out than put in, so $\Delta_f H^{\ominus} = -204$ kJ mol^{-1}. Thus the compound has a good chance of being energetically stable.

22.3.2 Shapes of molecules

For example, we shall again look at the molecule xenon(IV) fluoride, XeF_4.

Xenon's outer electron structure is $5s^2 5p^6$. That is, xenon has eight electrons in its outer shell. The dot cross diagram must look like this:

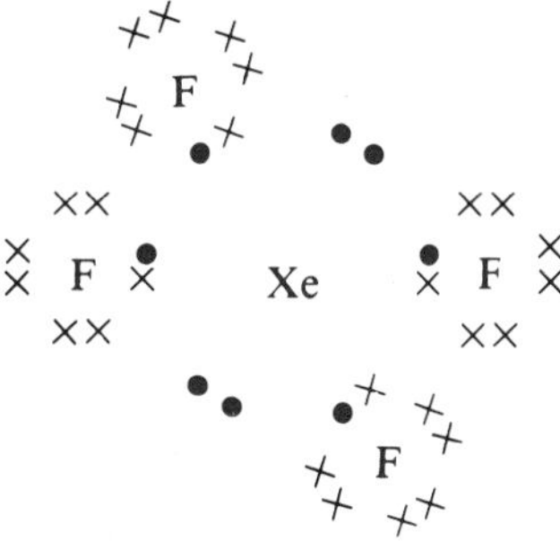

So the xenon atom has six pairs of electrons in its outer shell in this compound and the shape is based on an octahedron, with fluorine atoms at four points and lone pairs directed towards the other two. This is in fact the observed shape of XeF_4, see **Figure 22.2.**

Figure 22.2 The structure of the xenon(IV) fluoride molecule

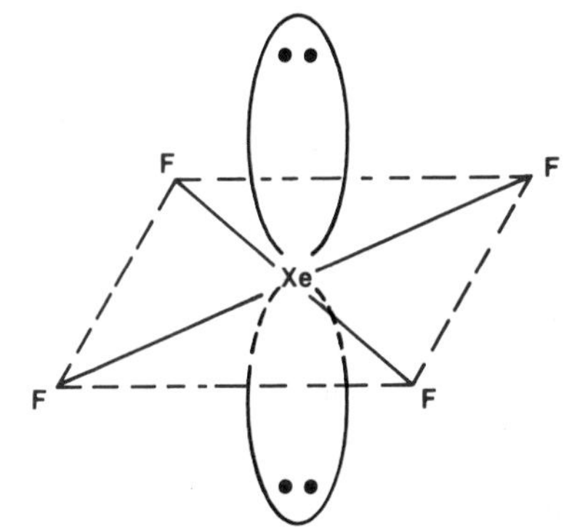

22.3.3 Redox potentials

We would expect noble gas compounds to be good oxidizing agents. If so they should have large positive values for $E^{\ominus}$ for their half reactions.

$$H_4XeO_6(aq) + 2H^+(aq) + 2e^- \rightleftharpoons XeO_3(g) + 3H_2O(l) \quad E^{\ominus} = +3.0\,\text{V}$$

Xenic(VIII) acid, H_4XeO_6, is therefore a better oxidizing agent than gaseous fluorine:

$$F_2(g) + 2e^- \rightleftharpoons 2F^-(aq) \quad E^{\ominus} = +2.87\,\text{V}$$

22.4 Summary

- The noble gases are chemically unreactive due to their full outer shells of electrons.
- Their main industrial use is as inert atmospheres.
- They are obtained by fractional distillation of liquid air.
- They form a few compounds in which they have positive oxidation states, especially xenon which has the lowest ionization energy except for radon.
- The compounds are almost invariably formed with fluorine or oxygen.
- The compounds are good oxidizing agents, often reacting to give the uncombined noble gas.

22.5 Questions

1 Use the bond energies Xe—F = $130\,kJ\,mol^{-1}$ and F—F = $158\,kJ\,mol^{-1}$ to calculate $\Delta_f H^\ominus$ for XeF_2 and XeF_6. Compare their likely thermodynamic stabilities with that of XeF_4 (see **section 22.3.1**).

2 Draw a dot cross diagram for XeF_2. Predict its likely shape.

3 Use the bond energies Xe—Cl = $85\,kJ\,mol^{-1}$, Kr—F = $50\,kJ\,mol^{-1}$, Cl—Cl = $243\,kJ\,mol^{-1}$, F – F = $158\,kJ\,mol^{-1}$ to calculate $\Delta_f H^\ominus$ for the molecules $XeCl_2$ and KrF_2. Both these molecules have been isolated at low temperatures. Sketch an enthalpy diagram for each, showing the compound and its elements. Why are the compounds stable only at low temperatures?

23 The s-block elements

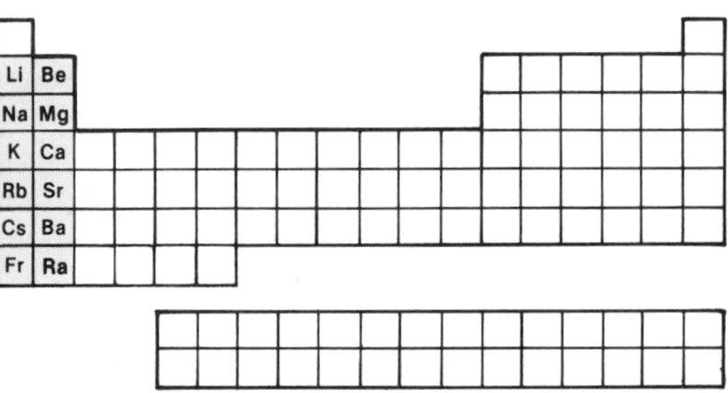

23.1 Introduction

The s-block elements are those in Groups I (the alkali metals) and II (the alkaline earth metals). As the name implies, their outer electrons are in s orbitals. The two groups have considerable similarities—both contain highly reactive metals of unusually low density. The elements are too reactive to be found uncombined and are normally extracted from their compounds by electrolysis. The elements are all good reducing agents. They almost invariably form ionic compounds, which are white unless a transition metal is also present.

23.2 Physical properties of the elements

A summary of some of the physical properties of the s-block elements is given in **Tables 23.1** and **23.2** (overleaf). Both francium and radium are radioactive elements and many of their properties are not known with precision. Trends in properties are indicated by brown arrows, showing the direction of increase.

A number of trends are clear. They are discussed below.

23.2.1 Atomic and ionic radii

Atomic (metallic) and ionic radii *increase* on descending each group, because each element has an extra full shell of electrons compared with its predecessor. Ionic radii are much smaller than the atomic radii for two reasons: firstly the loss of the outer electron(s); secondly this loss of an electron or electrons means there is less electron–electron repulsion and the nucleus can pull the remaining electrons in more closely towards it, **Figure 23.1.**

Figure 23.1 Electron–electron repulsion balances the nucleus–electron attraction

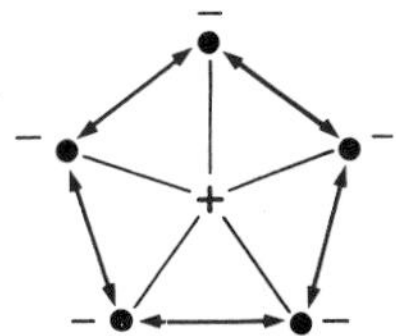

There is a *decrease* in both metallic and ionic radii from the Group I to the Group II element in the same period due to the increased nuclear charge.

23.2.2 Ionization energy

(Remember, this is the energy required for:

$$E(g) \longrightarrow E^+(g) + e^-$$

where E represents the element—the lower the ionization energy, the easier it is to lose the electron.) Ionization energies *decrease* on descending each group because the outer electron(s) get further from the nucleus, yet the outer electron(s) in each element feel the same effective nuclear charge. This partly explains why the metals get more reactive on going down the groups.

Table 23.1 Properties of Group I elements

	Atomic number, Z	Electron arrangement	Metallic radius /nm	Ionic radius M^{+}/nm	First IE /kJ mol^{-1}	Electro-negativity	$E^{\ominus}$ for $M^{+}(aq) + e^{-} \rightleftharpoons M(s)$ /V	$\Delta_{hyd}H^{\ominus}$ /kJ mol^{-1}	Oxidation number in compounds	T_m/K	T_b/K	Density, ρ/g cm^{-3}
Li lithium	3	[He]$2s^1$	0.157	0.074	520	1.0	−3.03*	−519	+I	454	1615	0.53
Na sodium	11	[Ne]$3s^1$	0.191	0.102	496	0.9	−2.71	−407	+I	371	1156	0.97
K potassium	19	[Ar]$4s^1$	0.235	0.138	419	0.8	−2.92	−322	+I	336	1033	0.86
Rb rubidium	37	[Kr]$5s^1$	0.250	0.149	403	0.8	−2.93	−301	+I	312	959	1.53
Cs caesium	55	[Xe]$6s^1$	0.272	0.170	376	0.7	−3.02	−276	+I	302	942	3.51
Fr francium	87	[Rn]$7s^1$	–	–	–	–	–	–	+I	–	–	–

*See **section 23.2.4**.

Table 23.2 Properties of Group II elements

	Atomic number, Z	Electron arrangement	Metallic radius /nm	Ionic radius M^{2+}/nm	First + second IEs /kJ mol^{-1}	Electro-negativity	$E^{\ominus}$ for $M^{2+}(aq) + e^{-} \rightleftharpoons M(s)$ /V	$\Delta_{hyd}H^{\ominus}$ /kJ mol^{-1}	Oxidation number in compounds	T_m/K	T_b/K	Density, ρ/g cm^{-3}
Be beryllium	4	[He]$2s^2$	0.112	0.027	900 + 1757 = 2657	1.5	−1.85	−2981	+II	1551	3243	1.85
Mg magnesium	12	[Ne]$3s^2$	0.160	0.072	738 + 1451 = 2189	1.2	−2.37	−2082	+II	922	1380	1.74
Ca calcium	20	[Ar]$4s^2$	0.197	0.100	590 + 1145 = 1735	1.0	−2.87	−1760	+II	1112	1757	1.54
Sr strontium	38	[Kr]$5s^2$	0.215	0.113	550 + 1064 = 1614	1.0	−2.89	−1600	+II	1042	1657	2.60
Ba barium	56	[Xe]$6s^2$	0.224	0.136	503 + 965 = 1468	0.9	−2.90	−1450	+II	998	1913	3.51
Ra radium	88	[Rn]$7s^2$	–	–	–	–	–	–	+II	–	–	–

The first ionization energy of a Group II metal is greater than that of the Group I element in the same period due to the increased nuclear charge. For Group II elements which invariably form M^{2+} ions, the sum of the first two ionization energies is the important factor.

23.2.3 Electronegativity

All the s-block elements have low electronegativity values, i.e. are very electropositive, which means they tend to lose their outer electrons relatively easily. As we go down the groups the elements become more electropositive, that is, the elements tend to lose electrons more readily. This is because the outer shell is further from the nucleus as we go down the groups.

Group II elements are more electronegative (less electropositive) than their Group I partners in the same period, as the nuclear charge (which attracts electrons) has increased by one.

23.2.4 Redox potential

$E^\ominus$ values for the half cell:

$$M^+(aq) + e^- \longrightarrow M(s) \text{ in Group I}$$

or: $$M^{2+}(aq) + 2e^- \longrightarrow M(s) \text{ in Group II}$$

are a measure of each metal's effectiveness as a reducing agent, i.e. as an electron donor. The more negative is $E^\ominus$, the more easily the metal will lose electrons (which means the better it is as a reducing agent). As we go down the groups, the elements get better as reducing agents. This would be expected, because as we have seen, the ionization energy decreases as we go down the group which means that the metals lose electrons more easily.

Lithium is a much better reducing agent than expected. This is due to the unusually large enthalpy change of hydration of the Li^+ ion (see next section).

23.2.5 Enthalpy change of hydration

By definition, this is the amount of energy given out when water molecules cluster round a lone metal ion:

$$M^+(g) + (aq) \longrightarrow M^+(aq)$$

See **Figure 23.2.**

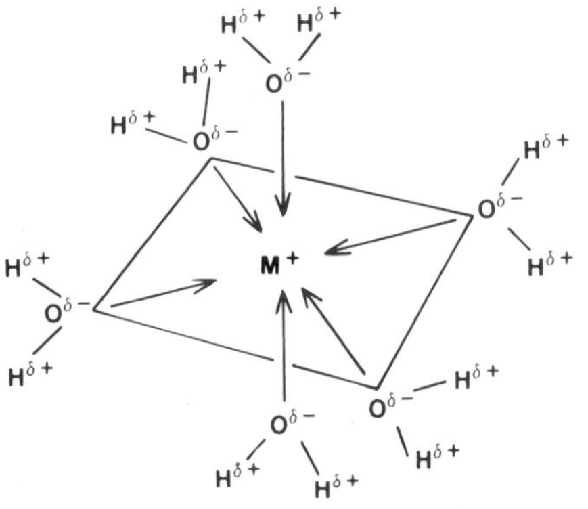

Figure 23.2 Water molecules clustering around a metal ion

The trend is for $\Delta_{hyd}H^\ominus$ to decrease on descending the groups. This is because the ions get larger and their charge density is less, so they attract the $O^{\delta-}$ end of the water less strongly. In Group I, lithium therefore has the highest $\Delta_{hyd}H^\ominus$. This is why it is the best reducing agent in the group. When lithium acts as a reducing agent:

$$Li(s) + (aq) \longrightarrow Li^+(aq) + e^-$$

the high negative $\Delta_{hyd}H^\ominus$ of lithium favours the formation of $Li^+(aq)$ and thus helps the release of the electrons for reduction.

Trends are also apparent in the melting and boiling points of Group I elements and the densities of all s-block metals. There are some irregularities in the densities, as these depend on three factors—atomic mass, metallic radius and the type of packing in the lattice.

23.2.6 Flame colours

We saw in **section 8.3.2** that each element produces a unique pattern of lines in its emission spectrum. This pattern is related to the electronic energy levels in the atom. Most of the s-block elements have spectra with lines in the visible region. This means that when the elements or their compounds are heated in a flame, they produce characteristic colours. This is the basis of atomic emission spectroscopy. However, for the s-block elements the colours themselves are often distinctive enough to identify the element. A list of flame colours is given in **Table 23.3.** See also **Plates 7–13.**

Table 23.3 Flame colours of s-block elements.

Li	Scarlet	Be	None
Na	Yellow	Mg	
K	Lilac	Ca	Brick-red
Rb	Red	Sr	Crimson
Cs	Blue	Ba	Apple green

23.3 Chemical reactions of the elements

23.3.1 Group I

These metals react very readily to form positive ions, as we could expect from their low ionization energies, low electronegativities and high negative values of $E^\ominus$.

Reaction with halogens

The Group I metals all react readily with the halogens to produce solid ionic halides, $M^+ + X^-$, where M represents any Group I metal, for example:

$$2Na(s) + Cl_2(g) \longrightarrow 2NaCl(s)$$

Reaction with water

All the Group I metals react with cold water to produce a solution of the metal hydroxide and hydrogen. The vigour of the reaction increases rapidly on descending the group, the reaction of caesium being explosive. For example:

$$2K(s) + 2H_2O(l) \longrightarrow 2KOH(aq) + H_2(g)$$

The solutions of the hydroxides are strongly alkaline, being completely dissociated into $M^+(aq)$ and $OH^-(aq)$ ions.

Reaction with oxygen

Here there is a difference in behaviour on descending the group. Lithium forms the 'normal' oxide—that which would be predicted using a simple ionic model—Li_2O ($2Li^+ + O^{2-}$):

$$4Li(s) + O_2(g) \longrightarrow \underset{\text{lithium oxide}}{2Li_2O(s)}$$

Sodium forms a mixture of the normal oxide, Na_2O ($2Na^+ + O^{2-}$), and the peroxide, Na_2O_2 ($2Na^+ + O_2{}^{2-}$)—the $O_2{}^{2-}$ ion is called the peroxide ion. The amounts of each depend on the amount of oxygen available. The peroxide is favoured in excess oxygen.

$$2Na(s) + O_2(g) \longrightarrow \underset{\text{sodium peroxide}}{Na_2O_2(s)}$$

The other alkali metals all form the superoxide, MO_2 ($M^+ + O_2{}^-$)—the $O_2{}^-$ ion is called the superoxide ion. For example:

$$K(s) + O_2(g) \longrightarrow \underset{\text{potassium superoxide}}{KO_2(s)}$$

Superoxides are exceptions for compounds of s-block metals. They are coloured. The reason for this difference in behaviour is found in the sizes of the ions: $O_2{}^-$ is larger than $O_2{}^{2-}$ which is larger than O^{2-}. Li^+ is too small for enough peroxide or superoxide ions to cluster round it to form a stable crystal lattice. All these reactions occur rapidly. When the metals are exposed to air they begin to tarnish after a few seconds, becoming coated with oxide. To prevent this, lithium, sodium and potassium are normally stored under oil. Rubidium and caesium are even more reactive and are normally stored in sealed containers.

Reaction with hydrogen

All the Group I metals react with hydrogen when heated to form ionic hydrides containing the H^- ion, $M^+ + H^-$, for example:

$$2Li(s) + H_2(g) \longrightarrow \underset{\text{lithium hydride}}{2LiH(s)}$$

23.3.2 Group II

Reaction with halogens

The halides MX_2 are formed. These are ionic, $M^{2+} + 2X^-$, except for beryllium chloride which is covalent.

Reaction with water

The Group II metals are less reactive than those in Group I. Beryllium does not react directly with water at all; magnesium reacts rapidly with steam and slowly with cold water to form the hydroxide and hydrogen. The rest react with increasing rapidity on descending the group, for example:

$$Ca(s) + 2H_2O(l) \longrightarrow Ca(OH)_2(aq) + H_2(g)$$

Reaction with oxygen

The 'normal' oxide, MO ($M^{2+} + O^{2-}$), is formed when the metals are heated in oxygen. Strontium and barium also form peroxides. As the M^{2+} ions are smaller than the M^{+} ions in Group I, peroxides become important lower down the group than in Group I.

Reaction with hydrogen

Hydrides of formula MH_2 are formed by heating the metal in hydrogen, except for beryllium. All are ionic ($M^{2+} + 2H^{-}$) except for those of beryllium and magnesium, the least electropositive Group II metals, which are covalent.

23.4 Compounds of the s-block elements

Generally the compounds are typically ionic. They have high melting and boiling temperatures and dissolve better in water than in non-polar solvents. The compounds of lithium and especially beryllium are exceptions and show a good deal of covalency. This occurs because the atoms are very small and the outer electrons very near to the nucleus and thus difficult to remove (high ionization energy). Their small size means that Li^{+} and Be^{2+} are highly polarizing (see **section 10.2.1**) which leads to a good deal of covalency in their compounds. Some magnesium compounds also show a tendency to covalency and there is a noticeable diagonal relationship, see **section 20.4.7**, between lithium and magnesium compounds.

23.4.1 Thermal stability of compounds

Thermal stability refers to decomposition of the compound on heating. Increased thermal stability means a higher temperature is needed to decompose the compound. When we say a compound is thermally stable, we usually mean that it is not decomposed at the temperature of a normal bunsen flame (approximately 1300 K). **Table 23.4** summarizes the patterns.

Table 23.4 **Effect of heat on compounds of s-block elements**

1+ → 2+ (Charge of cation)

Group I	Group II	
Carbonates (CO_3^{2-})		
Li ↓ Cs: $Li_2CO_3(s) \longrightarrow Li_2O(s) + CO_2(g)$; Rest stable	Mg ↓ Ba: $MgCO_3(s) \longrightarrow MgO(s) + CO_2(g)$; Successively higher temperatures needed to decompose the rest	Size of cation ↓
Hydroxides (OH^{-})		
Li ↓ Cs: $2LiOH(s) \longrightarrow Li_2O(s) + H_2O(g)$; Rest stable	Mg ↓ Ba: $Mg(OH)_2(s) \longrightarrow MgO(s) + H_2O(g)$; ↓ More stable	Size of cation ↓
Hydrogencarbonates (HCO_3^{-})		
All decompose at less than 373 K, e.g. $2NaHCO_3(s) \longrightarrow Na_2CO_3(s) + CO_2(g) + H_2O(g)$	Too unstable to exist as solids—exist only in solution	Size of cation ↓
Nitrates (NO_3^{-})		
$2LiNO_3(s) \longrightarrow Li_2O(s) + 2NO_2(g) + \frac{1}{2}O_2(g)$; $NaNO_3(s) \longrightarrow NaNO_2(s) + \frac{1}{2}O_2(g)$; Rest as sodium	$Mg(NO_3)_2(s) \longrightarrow MgO(s) + 2NO_2(g) + \frac{1}{2}O_2(g)$; Rest as magnesium	Size of cation ↓
Sulphates (SO_4^{2-}) and halides (X^{-})		
All thermally stable		

Beryllium compounds have not been included as they are quite covalent and not typical of the rest.

Some clear trends can be seen:

- Lithium often follows the pattern of Group II rather than Group I.

This is an example of the **diagonal relationship** and is indicated by the brown diagonal arrows.

• Nitrates decompose in two distinct ways: Group II metal (and lithium) nitrates give the metal oxide and a mixture of nitrogen dioxide (a brown gas) and oxygen. Group I metal nitrates decompose less (except for lithium) and give the nitrite and oxygen. (The systematic names of nitrate and nitrite are nitrate(V) and nitrate(III) respectively.)

• The overall pattern is that:

1. Compounds of Group I metals are as a whole more stable than those of Group II.

2. Stability increases on descending Group II.

These trends can be explained in terms of the **charge density** of the cations. This measures the concentration of charge on the cation. The smaller the ion the higher the charge density and the greater the charge on the ion the higher the charge density. Cations with a high charge density can **polarize** the negative ion (see **section 10.2.1**). This assists decomposition to the oxide, for example:

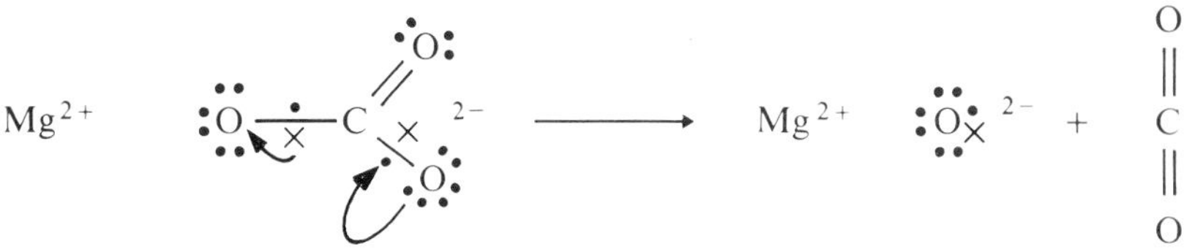

$$Mg^{2+} \quad CO_3^{2-} \longrightarrow Mg^{2+} \quad O^{2-} + O{=}C{=}O$$

('Curly' arrows show movement of pairs of electrons—see **section 26.9.5.**) Mg^{2+} pulls electrons towards it, helping the C—O bond to break, leaving O^{2-} and CO_2.

Group I cations, with only one positive charge, have a relatively lower polarizing power than Group II cations which have two positive charges. Li^+, which is very small, has a charge density more similar to Group II cations than Group I cations. The charge density *decreases* on descending the group as the ions get larger making the compounds of the metals at the bottom of the group more stable.

23.4.2 Solubilities of the salts of s-block metals

Virtually all salts of Group I metals are soluble in water except lithium fluoride, which has a very high lattice energy. Nitrates of Group II metals are all soluble. With anions of charge −2, there is a distinct pattern of solubility as shown in **Table 23.5**

Table 23.5 The solubility of some Group II metal compounds formed by mixing equal volumes of 0.2 mol dm^{-3} solutions of cations and anions. Solubilities of the salts, $M^{2+} + X^{2-}$/10^{-4} mol per 100 g, are given in brown, e.g. the solubility of $Mg^{2+}CO_3^{2-}$ is 1.5×10^{-4} mol per 100 g

0.2 mol dm^{-3} metal nitrate (increasing size of cation ↓) \ 0.2 mol dm^{-3} sodium salt	CO_3^{2-}	SO_4^{2-}	CrO_4^{2-}	$C_2O_4^{2-}$
Mg^{2+}	White precipitate **1.5**	No precipitate **1830**	No precipitate **8500**	No precipitate **5.7**
Ca^{2+}	White precipitate **0.13**	No precipitate **47**	No precipitate **870**	White precipitate **0.05**
Sr^{2+}	White precipitate **0.07**	White precipitate **0.71**	Yellow precipitate **5.9**	White precipitate **0.29**
Ba^{2+}	White precipitate **0.09**	White precipitate **0.009**	Yellow precipitate **0.01**	White precipitate **0.52**

Note:
CO_3^{2-} = carbonate
SO_4^{2-} = sulphate
CrO_4^{2-} = chromate (yellow)
$C_2O_4^{2-}$ = ethanedioate (oxalate)

This shows clearly that the *larger the cation*, the *less soluble the salt*. Such a clear-cut pattern of solubilities is unusual because the solubility of ionic salts is related to $\Delta_{soln}H^\ominus$, the enthalpy change of dissolution of the salt.

Figure 23.3 Enthalpy changes on dissolving an ionic solid

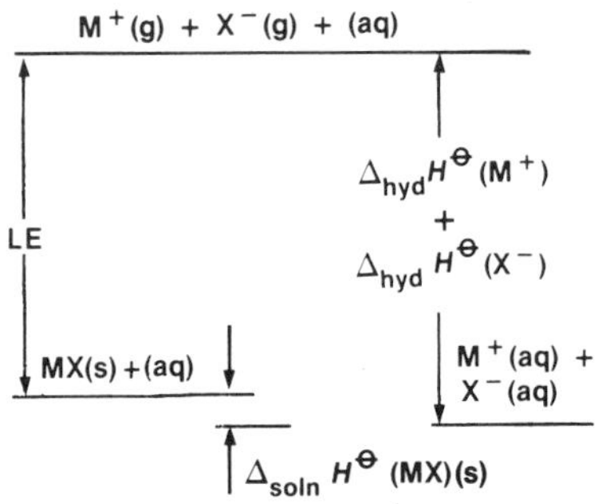

Figure 23.4 Processes which occur on dissolving an ionic solid

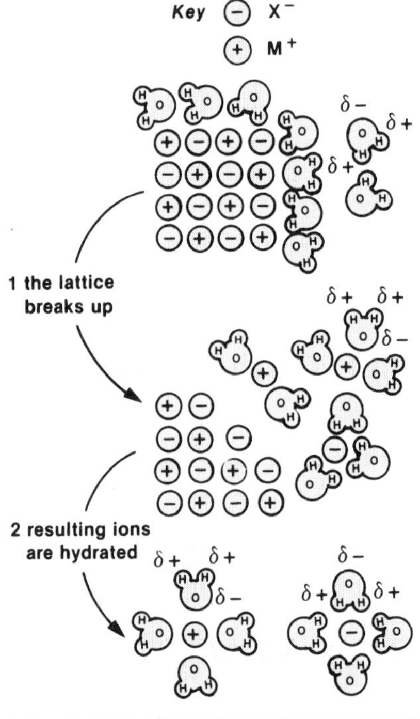

This in turn depends on the lattice energy of the salt and the enthalpy changes of hydration of the two ions, as shown in **Figure 23.3**, which make up the salt. This is not a simple relationship.

Figure 23.4 shows what happens when an ionic compound dissolves in water. The lattice breaks up, followed by hydration of the ions by water molecules.

Both the lattice energy and the enthalpy changes of hydration depend on the size and charge of the ions in the same sort of way.

Highly charged ions have high lattice energies *and* high $\Delta_{hyd}H^{\ominus}$s. Small ions have high lattice energies *and* high $\Delta_{hyd}H^{\ominus}$s so changes in size and charge tend to cancel out in their effects on $\Delta_{sol}H^{\ominus}$.

23.4.3 Reactions of the hydrides of s-block metals

These all react with water to give the metal hydroxide and hydrogen, for example

	$NaH(s)$	+	$H_2O(l)$	$\longrightarrow$	$NaOH(aq)$	+	$H_2(g)$
	sodium hydride		water		sodium hydroxide		hydrogen
and	$CaH_2(s)$	+	$2H_2O(l)$	$\longrightarrow$	$Ca(OH)_2(aq)$	+	$2H_2(g)$
	calcium hydride		water		calcium hydroxide		hydrogen

The hydrides are good *reducing agents*. They can be used to prepare the complex hydrides lithium tetrahydridoaluminate(III) (lithium aluminium hydride), $Li^+ + AlH_4^-$, and sodium tetrahydridoborate(III) (sodium borohydride), $Na^+ + BH_4^-$, which are used in organic chemistry as selective reducing agents which reduce C=O but not C=C (see **sections 32.6.3** and **36.4.2**).

23.4.4 Reactions of the oxides of s-block elements

These all react with water to form hydroxides. Normal oxides give the

A

THE BARIUM MEAL TEST

The use of X-rays is common in medical diagnosis. They are absorbed by atoms with large numbers of electrons, so bones with a high concentration of relatively heavy calcium atoms absorb X-rays better than soft tissue containing mostly lighter atoms like carbon, hydrogen, oxygen and nitrogen. Examinations of soft tissue can only be done indirectly. The digestive system can be examined by giving the patient a 'barium meal' containing barium sulphate. Barium absorbs X-rays very well indeed and an outline of the digestive tract can be obtained on an X-ray to help diagnose and locate, say, a stomach ulcer or intestinal blockage. Although barium compounds are poisonous, barium sulphate is so insoluble that it is not absorbed by the body and so it can be given safely.

Another use of heavy atoms in this way is in X-rays of the urinary system where an injection of an iodine-containing solution is given. This is soon extracted by the kidneys into the urine, enabling an X-ray of the kidneys, ureters and bladder to be obtained.

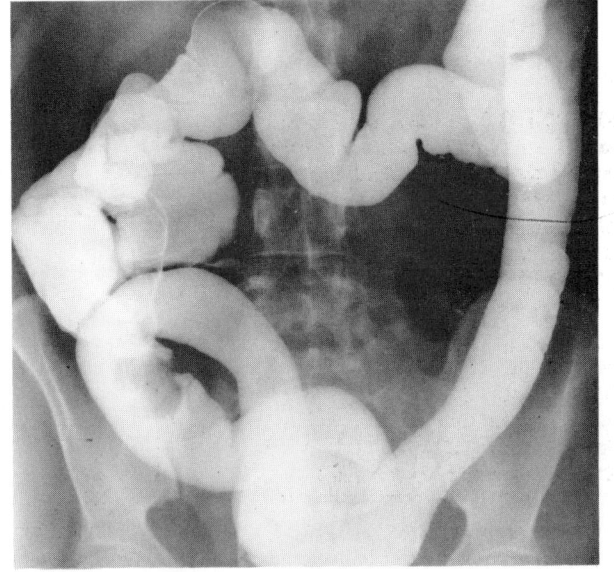

X-ray after taking a barium meal

metal hydroxide only, for example:

$$Li_2O(s) + H_2O(l) \longrightarrow 2LiOH(aq)$$

Peroxides give the metal hydroxide and hydrogen peroxide, for example:

$$Na_2O_2(s) + 2H_2O(l) \longrightarrow 2NaOH(aq) + H_2O_2(aq)$$

Superoxides give the metal hydroxide, hydrogen peroxide and oxygen, for example:

$$2KO_2(s) + 2H_2O(l) \longrightarrow 2KOH(aq) + H_2O_2(aq) + O_2(g)$$

Both peroxides and superoxides are oxidizing agents.

The metal hydroxides of both Group I and Group II dissolve in water to give $OH^-(aq)$ and are thus alkaline:

$$MOH(s) + (aq) \longrightarrow M^+(aq) + OH^-(aq)$$

The Group II hydroxides are less soluble than those of Group I and are therefore weaker alkalis.

Beryllium oxide, BeO, is an exception to the above in that it is amphoteric, and in solution can show either acidic or basic properties, depending on what it is reacting with. In this it resembles aluminium oxide—an example of the diagonal relationship between these elements.

Salt mining

23.4.5 Occurrence of the s-block elements

Both sodium and potassium are relatively abundant elements, each comprising around 2.5% of the earth's crust and occurring in deposits of their chlorides which have resulted from the evaporation of sea water. There are large underground deposits of sodium chloride in Cheshire, around which many chemical industries have grown up. The other Group I elements are less common.

A

MANUFACTURE OF SODIUM HYDROXIDE

Sodium hydroxide is manufactured from brine (sodium chloride solution) in two different types of electrolysis cell—the mercury cell described here and the diaphragm cell described in the box in **section 24.6.3.** Mercury cells cost more to build but produce sodium hydroxide which is both purer than that produced in a diaphragm cell and is also at the concentration required by the customers.

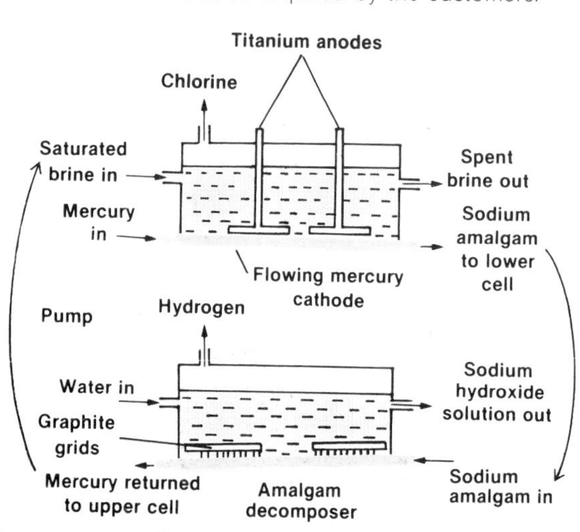

The mercury cell

The electrolysis takes place in a cell approximately 2 m x 15 m with a slightly sloping base like a swimming pool. Mercury flows down the floor of the cell, forming the cathode. The titanium anodes are adjusted so that they are no more than 2 mm away from the mercury. Saturated brine flows through the cell. At the anode, the electrochemical series would predict the discharge of OH^- ions but in fact chloride ions are discharged because of their high concentration, see **section 17.4.**

$$2Cl^-(aq) \longrightarrow Cl_2(g) + 2e^-$$

At the cathode, sodium is discharged at the mercury electrode, dissolves in it and is carried away before it can react with the water. The sodium/mercury mixture, called an amalgam, flows into a similar shaped cell—the decomposer—usually located below the electrolysis cell. The brine, now containing less salt, is recycled. In the decomposer, the sodium/mercury amalgam flows down the base against a current of water. The sodium reacts with the water forming sodium hydroxide solution and hydrogen. The reaction is catalysed by graphite grids which float on the mercury.

$$\underset{\text{(amalgam)}}{2Na} + 2H_2O(l) \longrightarrow 2NaOH(aq) + H_2(g)$$

The mercury is pumped back up to the electrolysis cell. All three products—sodium hydroxide, chlorine and hydrogen—are in demand.

A

HARDNESS OF WATER

While tap water has been treated to kill any bacteria and make it safe to drink, it is by no means pure. Water is taken from rivers, reservoirs or wells which it has reached after percolating through soil and rocks, etc. It will therefore contain substances which it has dissolved from the ground. Water with large amounts of dissolved solids may be hard water. How hard the water in a particular area is depends on the types of rock present. Some dissolved solids cause problems—particularly dissolved calcium and magnesium ions. These are present when rain water, which is slightly acidic due to dissolved carbon dioxide, has percolated through limestone rocks containing calcium and/or magnesium carbonates:

$$CaCO_3(s) + H_2O(l) + CO_2(g) \longrightarrow Ca(HCO_3)_2(aq)$$

calcium carbonate + water + carbon dioxide ⟶ calcium hydrogencarbonate

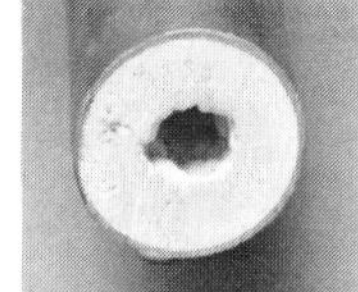

Hard water deposit in hot-water pipe

This reaction is reversed by heating and it leads to deposits of insoluble calcium (and magnesium) carbonates in hot water pipes, kettles, etc. which may cause blockages and reduce the efficiency of heating elements. The hardness caused by calcium and magnesium hydrogencarbonates is called temporary hardness because it is removed by heating. Hardness caused by calcium and magnesium sulphates or chlorides is unaffected by heat and is called permanent hardness.

Both types of hardness cause problems when soap is used. Soap is a mixture of salts such as sodium octadecanoate (sodium stearate), $C_{17}H_{35}CO_2Na$. This salt is soluble but the calcium and magnesium salts are not and these insoluble salts are the scum which form when soap is used in hard water:

$$2C_{17}H_{35}CO_2Na(aq) + Ca^{2+}(aq) \longrightarrow Ca(C_{17}H_{35}CO_2)_2(s) + 2Na^+(aq)$$

sodium stearate + calcium ions ⟶ calcium stearate + sodium ions

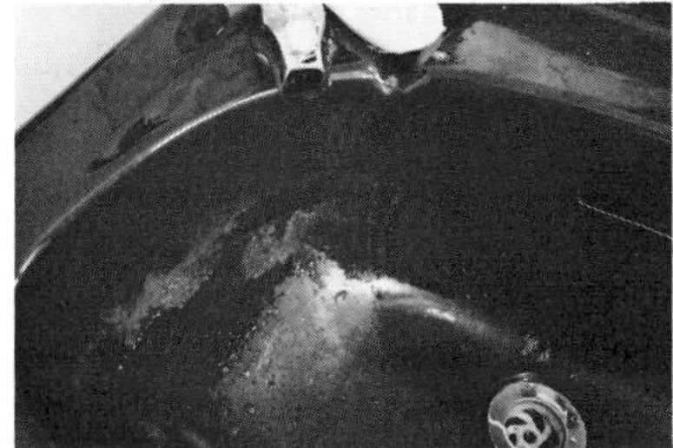

Hard water causes a scum with soap

Scum is unpleasant and can damage the fabrics in clothes. Its formation also means more soap is needed than in soft water.

A number of methods can be used for softening water. Distillation removes all types of hardness but is expensive in terms of fuel costs, as is boiling to remove temporary hardness. Calcium hydroxide (slaked lime) is cheap and can be added to precipitate out temporary hardness as calcium carbonate:

$$Ca(HCO_3)_2(aq) + Ca(OH)_2(s) \longrightarrow 2CaCO_3(s) + 2H_2O(l)$$

calcium hydrogencarbonate + calcium hydroxide ⟶ calcium carbonate + water

Exactly the right amount of slaked lime must be added or more hardness will be produced. Sodium carbonate may be added to precipitate out any calcium or magnesium ions:

$$Mg^{2+}(aq) + Na_2CO_3(aq) \longrightarrow MgCO_3(s) + 2Na^+(aq)$$

magnesium ions + sodium carbonate ⟶ magnesium carbonate + sodium ions

Sodium compounds do not cause hardness. This method is the principle on which bath salts depend. These are simply coloured and perfumed sodium carbonate (washing soda) which soften bath water.

Another method of water softening is the use of sodium tripolyphosphate (trade name Calgon) which forms complexes (see **section 25.3.2**) with the calcium and magnesium ions leaving them in solution but unable to react normally.

Some homes have water softeners which contain ion exchange-resins. These are plastic beads which contain sodium ions which are exchanged for calcium or magnesium ions in the hard water. The resin is regenerated each night by pumping sodium chloride through the resin to reverse the process. This is done automatically by a timer.

$$2R{-}Na + Ca^{2+} \underset{\text{regenerating}}{\overset{\text{softening}}{\rightleftharpoons}} 2R{-}Ca + 2Na^+$$

resin ⟶ resin

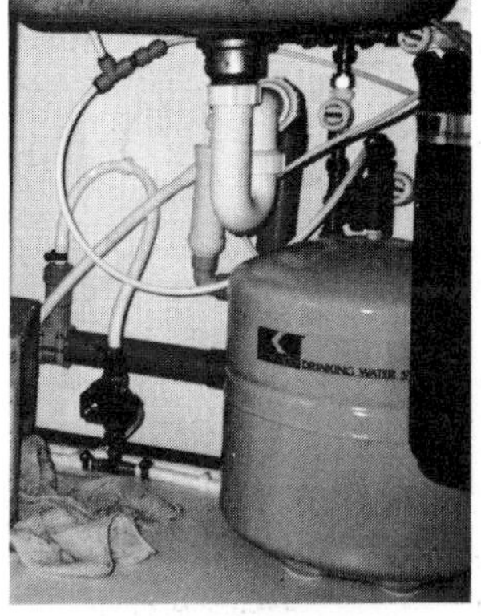

Domestic water softener

In Group II, calcium is found in limestone, $CaCO_3$, and magnesium in dolomite, $CaCO_3.MgCO_3$. Sea water is an important source of both sodium and magnesium salts, especially in hot countries where the heat of the sun can be used to evaporate the water.

23.4.6 Uses of s-block elements and compounds

Sodium is used in large quantities (100 000 tonnes per year) in the manufacture of tetraethyllead (a petrol additive), as a reducing agent in the manufacture of titanium and as a coolant for nuclear reactors. Sodium carbonate is used in the manufacture of glass, and sodium hydroxide for making paper, soap and man-made fibres.

Lithium salts are used to treat certain types of mental depression and the metal itself finds application in some alloys because of its lightness. Magnesium too is used in alloys and castings because it is cheap to cast. In large volumes it is not as reactive as in ribbon form, where it has a high surface area. Metal pencil sharpners are often made of magnesium. Calcium oxide is a constituent of cement.

Dounreay sodium-cooled nuclear reactor

23.4.7 Extraction of s-block elements

As they are themselves such powerful reducing agents, the metals cannot be extracted from their compounds by chemical reduction of the ore. So the s-block elements are normally produced from their ores by electrolysis. The extraction of sodium is described in **section 17.5.1.**

A

THE CHEMISTRY OF MORTAR

The mortar used by bricklayers is a mixture of calcium hydroxide (slaked lime, $Ca(OH)_2$), sand and water. It sets as the water dries. This is followed by a slower reaction in which the calcium hydroxide reacts with atmospheric carbon dioxide to form small crystals of calcium carbonate which bind the sand together.

Weathered mortar

$$Ca(OH)_2(s) + CO_2(g) \longrightarrow CaCO_3(s) + H_2O(l)$$

calcium hydroxide + carbon dioxide → calcium carbonate + water

A further simple reaction is responsible for the weathering of mortar which makes occasional repointing of brickwork necessary. Carbon dioxide dissolved in rainwater produces a weak acid called carbonic acid. This reacts with the calcium carbonate to form soluble calcium hydrogen-carbonate which is washed away by the rain until ultimately only the sand remains:

$$CaCO_3(s) + \underbrace{H_2O(l) + CO_2(g)}_{\text{carbonic acid}} \longrightarrow Ca(HCO_3)_2(aq)$$

calcium carbonate + carbonic acid → calcium hydrogen-carbonate

These reactions are exactly the same as those which occur in the limewater test for carbon dioxide. Carbon dioxide reacts with a solution of calcium hydroxide, $Ca(OH)_2(aq)$, to form a milky white precipitate of calcium carbonate. Excess carbon dioxide dissolves this, forming the soluble hydrogencarbonate and the milkiness clears.

23.5 Summary

23.5.1 Trends

- On descending a group: the properties *increase* in the direction of the arrows in **Table 23.6**, which refers to both Groups I and II.

Table 23.6 Trends in the properties of s-block metals

Solubilities of compounds	First ionization energy	Metallic radius	Ionic radius	Chemical reactivity	$E^\ominus$ for $M^{n+}(aq) + ne^- \rightleftharpoons M(s)$	Electronegativity	Thermal stability of compounds
↑	↑	↓	↓	↓	(not Li) ↓	↑	↓

- Moving from Group I to Group II, the properties increase in the direction of the arrows in **Table 23.7.**

Table 23.7 Trends on moving from Group I to Group II elements

	Group I	Group II
Atomic radius	←	
Ionic radius	←	
First ionization energy	→	
Electronegativity	→	
Thermal stability of compounds	←	
Solubility of compounds	←	

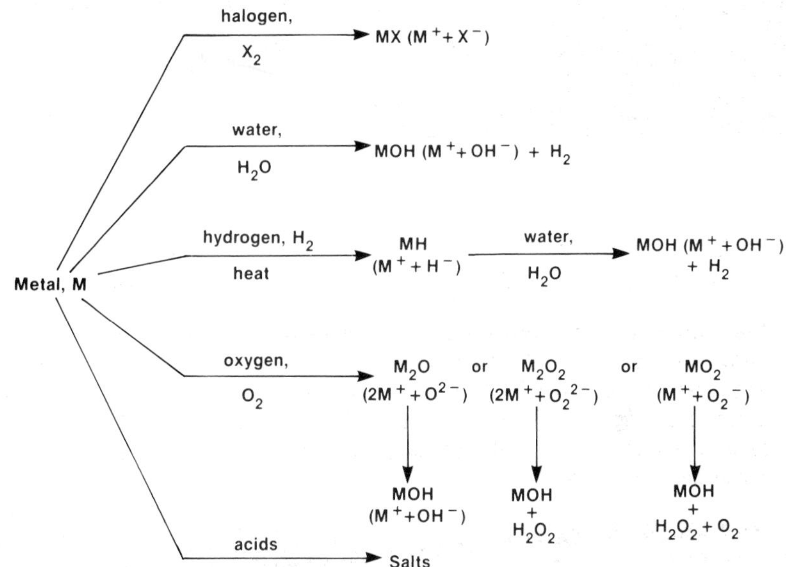

Figure 23.5 Reactions of Group I elements

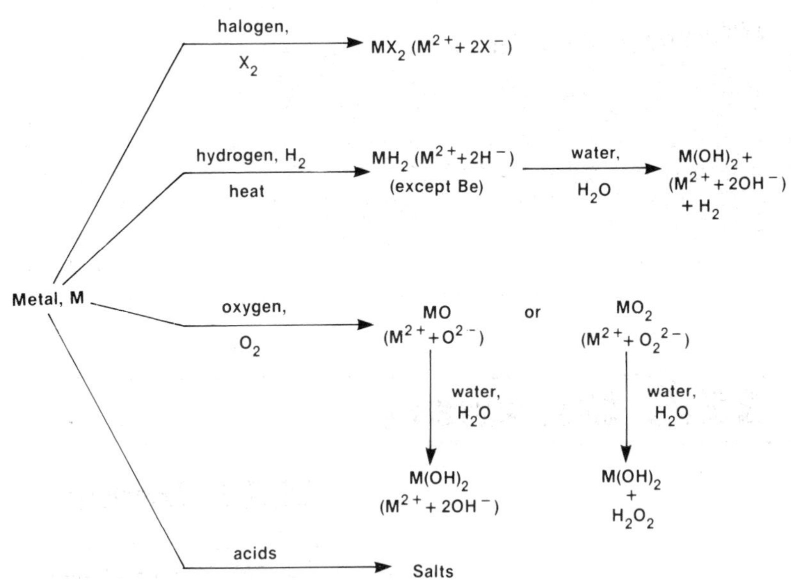

Figure 23.6 Reactions of Group II elements

23.5.2 Reactions

- Those of Group I are summarized in **Figure 23.5.**
- Those of Group II are summarized in **Figure 23.6.**

23.6 Questions

1 The following table gives the radii of four isoelectronic species:

Species	Radius/nm
S^{2-}	0.185
Cl^-	0.180
K^+	0.138
Ca^{2+}	0.100

A) Explain what is meant by the term isoelectronic.
B) Explain why the ionic radii differ in the way they do.
C) Write the electronic structures of these species in the form $1s^2$, etc.

2 Classify the following reactions as redox, acid/base, precipitation:

A) $Ba(NO_3)_2(aq) + Na_2SO_4(aq) \longrightarrow BaSO_4(s) + 2NaNO_3(aq)$

B) $2Na(s) + H_2(g) \longrightarrow 2NaH(s)$

C) $NaH(s) + H_2O(l) \longrightarrow NaOH(aq) + H_2(g)$

D) $2NaOH(aq) + H_2SO_4(aq) \longrightarrow Na_2SO_4(aq) + 2H_2O(l)$

E) $K_2O_2(s) + 2H_2O(l) \longrightarrow 2KOH(aq) + H_2O_2(aq)$

For the redox reactions, give the oxidation numbers of the atoms which are oxidized and reduced (before and after reaction).

3 A) The statements below are about a white solid compound containing an s-block element. For each statement say what conclusions can be drawn at that stage.
i The compound reacted with water to give a colourless gas and an alkaline solution;
ii The gas popped when tested with a lighted taper;
iii A flame test on the solution gave a lilac flame.
B) i If the original compound was melted and electrolysed, what products would you expect at each electrode?
ii Write an equation for the reactions at each electrode.

4 A) This question concerns a white solid compound of an s-block metal. For each statement below, say what conclusions you can draw at that stage:
i On heating it produced a brown gas and a white solid;
ii The resulting white solid was sparingly soluble in water to give an alkaline solution;
iii A flame test on the original solid gave an apple green colour.
B) Write equations for the reactions in (i) and (ii).

5 The s-block of the Periodic Table contains Group I, the alkali metals and Group II, the alkaline earths. Give an account of these two groups of elements and their compounds, paying particular attention to similarities and differences within the groups.

(London 1978)

6 A) Magnesium nitrate and barium nitrate decompose similarly on heating.
i Write a balanced equation for the decomposition of either of these nitrates.
ii There is a noticeable difference in the decomposition temperatures of these two nitrates. Which of these nitrates decomposes at a lower temperature? Explain this in terms of the ionic radii of the two cations.
iii Explain the relative numerical values of the lattice energies of the substances in each of these pairs:
A Magnesium nitrate and barium nitrate;
B magnesium nitrate and the residue formed after its thermal decomposition.
B) i Aqueous solutions of magnesium nitrate and barium nitrate can be distinguished by the addition of dilute sulphuric acid. Write an ionic equation, with state symbols, for the reaction which occurs with one of these nitrates.
ii Suggest reasons why one of these sulphates is insoluble and the other is soluble in water.

(Cambridge 1987)

7 What are the characteristics of a solvent which is capable of dissolving a wide range of ionic compounds? Discuss how these characteristic properties facilitate the solubility of such compounds.

Using the data below, calculate the heat of solution of each of the fluorides. Give a brief explanation for any trends which are visible in the tabulated and calculated values.

	LE/kJ mol^{-1}	$\Delta_{hyd}H^\ominus$/kJ mol^{-1}	
		M^+	F^-
LiF	1031	−499	−458
NaF	918	−390	−458
KF	817	−305	−458
RbF	783	−281	−458
CsF	747	−248	−458

Comment, in the light of the above, on the following solubility data:

formula of substance:	LiF	NaF	KF	RbF	CsF
solubility measured in grams per 100 g of water at 18 °C:	0.27	4.22	92	130	367

(Nuffield 1982)

8 This question concerns the chemistry of strontium (atomic number 38) and its compounds.

A) Write down the ground-stage electron configuration for an isolated strontium atom.

B) i Sketch a graph showing the pattern of the first five ionization energies for strontium.

ii Explain the shape of your graph.

C) Strontium metal can be obtained from strontium bromide ($SrBr_2.6H_2O$) by electrolysis of molten strontium bromide in the apparatus shown in the diagram.

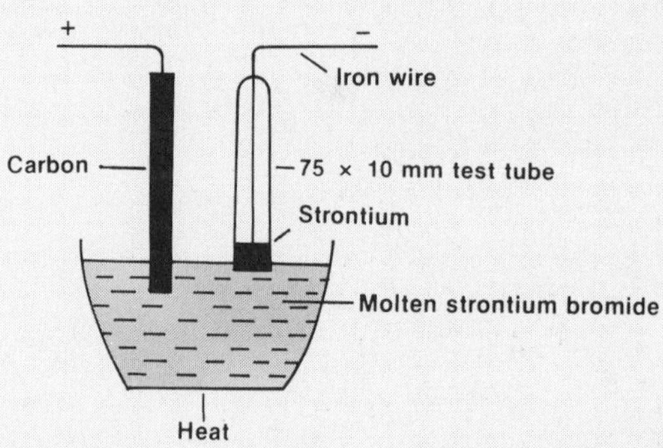

i What must be done to the strontium bromide before it can be used in this experiment?

ii Give two reasons for surrounding the cathode of iron wire with the glass test tube.

(Nuffield 1984)

9 Some of the alkaline earth metals are listed below:

Element	Atomic number
Magnesium	12
Calcium	20
Strontium	38
Barium	56

The solubilities in water of some salts of magnesium and barium are given in the following table (all values at 20 °C).

Salt	Formula	Solubility/mol dm^{-3}
Barium ethanedioate	BaC_2O_4	4.1×10^{-4}
Barium nitrate	$Ba(NO_3)_2$	3.3×10^{-1}
Magnesium ethanedioate	MgC_2O_4	9.3×10^{-3}
Magnesium nitrate	$Mg(NO_3)_2$	4.7

A) A flame test is carried out on the chlorides of each of the alkaline earth elements listed above.

i Which does *not* give a flame colour?

ii Which gives a pale green colour?

B) i Account for the production of such flame colours.

ii When the flame colours are viewed with a spectroscope, what type of spectra are seen?

C) i Complete the ground-state electron configuration of a barium atom:

$$1s^2 2s^2 2p^6 3s^2 3p^6 3d^{10} 4s^2 4p^2 4d^{10} 5s^2 \ldots$$

ii What is the oxidation number of barium in its compounds?

D) When 100 cm^3 of a saturated aqueous solution of magnesium ethanedioate is added to 100 cm^3 of a saturated aqueous solution of barium nitrate, a white precipitate is formed.

i What is the white precipitate?

ii Calculate the mass of the white precipitate formed when the two solutions are mixed at 20 °C. (Relative atomic masses: Ba = 137, C = 12, Mg = 24, N = 14, O = 16.)

(Nuffield 1981, part question)

24 The p-block elements

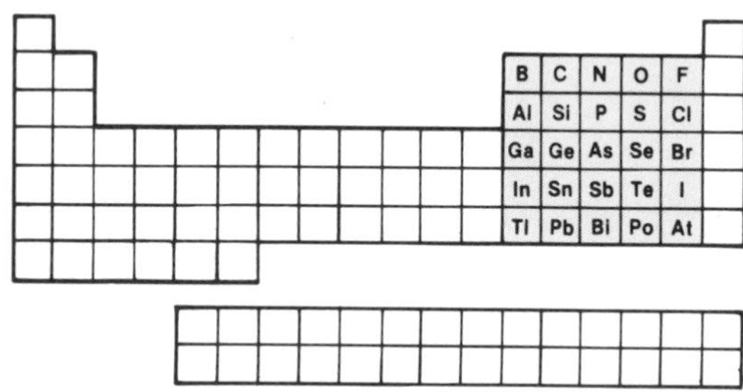

24.1 Introduction

The p-block elements have relatively little in common except that their outer electrons are in p orbitals, Figure 24.1. This contrasts with the considerable

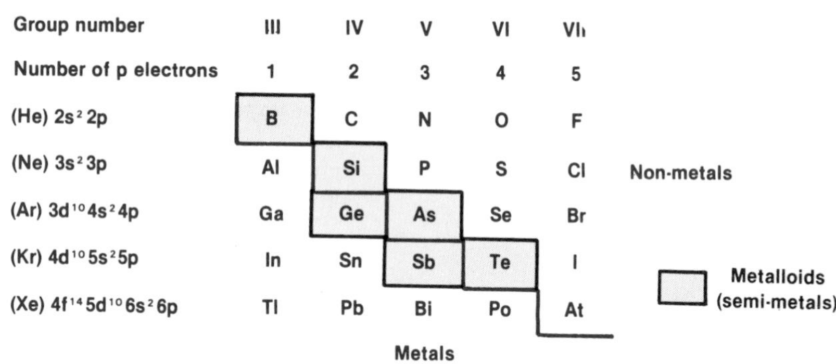

igure 24.1 The p-block elements howing electronic structures and the ositions of metals, metalloids and on-metals

similarities between the elements of the s-block and also between those of the d-block. The p-block is more notable for *trends*, which are broadly as shown in Figure 24.2. You may find it useful to relate the chemistry of each element to these trends. Each group will be treated separately, and within the groups we shall deal with each element following the overall plan: element;

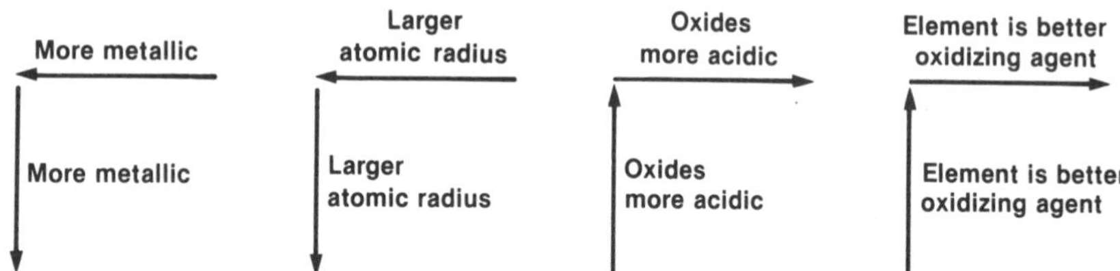

igure 24.2 Some trends in roperties within the p-block of the eriodic Table

hydride(s); oxide(s); hydroxide(s) or oxoacid(s); halides; industrial importance. We shall not attempt to cover the detailed chemistry of all the elements, only of the more significant ones.

The p-block contains several elements of great social and economic importance as well as chemical interest, for example aluminium's use as a structural material, the importance of silicon and germanium as semiconductors and the use of sulphur, phosphorus and nitrogen in fertilizers.

24.2 Group III

Only boron and aluminium, the first two members of the group, will be considered in any detail. The elements gallium, indium and thallium are relatively uncommon and have few uses at present. Their chemistry is broadly similar to that of aluminium. **Table 24.1** summarizes some important properties of the elements.

Table 24.1 Properties of elements in Group III

	Atomic number	Electron arrangement	Ionization energies/kJ mol^{-1} First	Second	Third	Electro-negativity	Atomic radius /nm	Ionic radius E^{3+}/nm	$E^{\ominus}$/V for $E^{3+}(aq) + 3e^- \rightleftharpoons E(s)$
B	5	$1s^2 2s^2 2p^1$	801	2427	3660	2.0	0.098	0.012 (est.)	–
Al	13	[Ne] $3s^2 3p^1$	578	1817	2745	1.5	0.143	0.053	−1.66
Ga	31	[Ar] $3d^{10} 4s^2 4p^1$	579	1979	2963	1.6	0.153	0.062	−0.52
In	49	[Kr] $4d^{10} 5s^2 5p^1$	558	1821	2705	–	–	0.081	−0.34
Tl	81	[Xe] $4f^{14} 5d^{10} 6s^2 6p^1$	589	1971	2878	–	–	0.095	+0.72

An important point to note is the large difference in the values of ionization energy and radii between boron and the other members of the group. These differences explain the noticeable difference between the chemistry of boron and that of the other elements.

24.2.1 Boron

The element

Pure boron is a very unreactive non-metal. It has several allotropes (different crystalline forms) and it is difficult to purify. 'Amorphous' boron is an impure form of the element which is much more reactive than pure boron. The element is a good absorber of neutrons and is used as a moderator in nuclear reactors. The Russians dumped quantities of it on the stricken Chernobyl reactor.

Boron being dumped on the Chernobyl reactor

Bonding in boron compounds

Boron has three electrons in its outer shell and might therefore be expected to form B^{3+} ions. However, the sum of the first three ionization energies is so large (6888 kJ mol^{-1}) that ions do not form. These large ionization energies arise because the atom is small and the electrons are being removed from close to the nucleus. Boron therefore forms only covalent compounds. If three covalent bonds are formed, boron still has only six electrons. Boron compounds will therefore readily accept a pair of electrons in order to gain an outer octet of electrons. They therefore act as **Lewis acids** (see **section 12.2**).

CERAMICS

In recent years, the technology of ceramic materials has been developed (ceramic means hardened by heat). Among these are many compounds of Group III and IV elements, including boron carbide and boron nitride. They have been used to make parts in internal combustion engines which are subjected to wear and high temperatures, for example, valves and cams. Automobile engineers are looking to ceramics to enable them to design and build engines with no need for lubrication or cooling. Boron nitride is isoelectronic with diamond (boron has one electron less and nitrogen one electron more)—it has the same structure and is almost as hard.

Boron compounds

BORON HYDRIDES

Boron forms a series of compounds with hydrogen called **boranes**. The simplest of these, diborane, has the molecular formula B_2H_6. The bonding in this compound (and other boranes) cannot be explained through normal bonding theories in which a covalent bond consists of a pair of electrons shared between two atoms.

Figure 24.3 The bonding in diborane

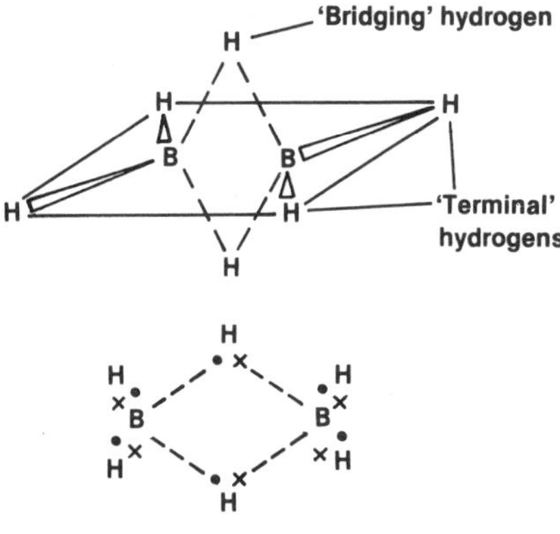

The structure is now known to be as shown in **Figure 24.3**. The four bonds in brown are normal bonds in which a pair of electrons is shared between a boron atom and a hydrogen atom. The bridging hydrogens are bonded to both borons via a delocalized bond in which just two electrons bond together three atoms. These are called 'two-electron three-centre bonds' or, because of their shape 'banana bonds'. This illustrates the central feature of boron chemistry. Boron has too few electrons to gain a full outer shell easily. Diborane is a gas which reacts immediately on exposure to air:

$$B_2H_6(g) + 3O_2(g) \longrightarrow \underset{\text{boron oxide}}{B_2O_3(s)} + 3H_2O(l)$$

Higher boranes (those with more boron in them) form a number of geometrically interesting structures held together by a combination of normal and two-electron three-centre bonds, **Figure 24.4.**

Figure 24.4 Structures of some boranes

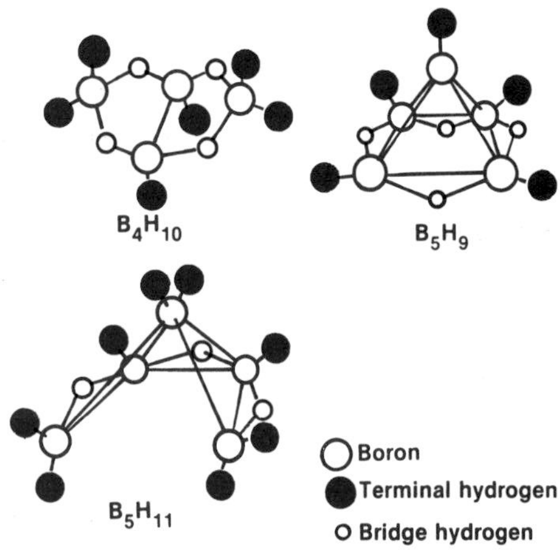

Figure 24.5 The structure of boron oxide

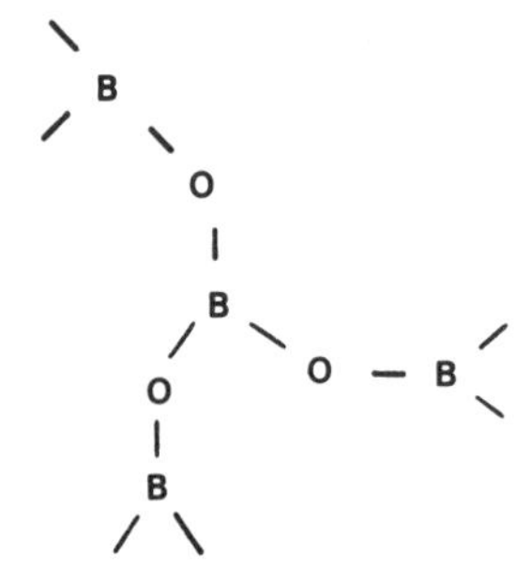

BORON OXIDE

The oxide, B_2O_3, can be produced by burning the element or by heating boric acid (trioxoboric(III) acid), H_3BO_3.

$$2H_3BO_3(s) \longrightarrow B_2O_3(s) + 3H_2O(l)$$

The oxide is usually formed as a glassy solid with a structure held together by B—O—B bonds, **Figure 24.5**.

BORIC ACID (TRIOXOBORIC(III) ACID)

The oxide is acidic, as would be expected for a non-metal. It dissolves in water to form boric acid, $B(OH)_3$.

$$B_2O_3(s) + 3H_2O(l) \longrightarrow 2B(OH)_3(aq)$$

Boric acid is a weak monoprotic acid:

$$B(OH)_3(s) + H_2O(l) \longrightarrow B(OH)_4^-(aq) + H^+(aq)$$

Strictly speaking it is acting as a Lewis acid, accepting an electron pair rather than donating a proton. Solid boric acid has a two-dimensional

$$(HO)_3B + :\ddot{O}H_2 \longrightarrow (HO)_3B \leftarrow :OH^- + H^+$$

layer structure with $B(OH)_3$ molecules joined by hydrogen bonds, **Figure 24.6.** The layers are held together by weak van der Waals bonding (rather like that in graphite) so the solid acid flakes easily.

Figure 24.6 Structure of solid trioxoboric(III) acid

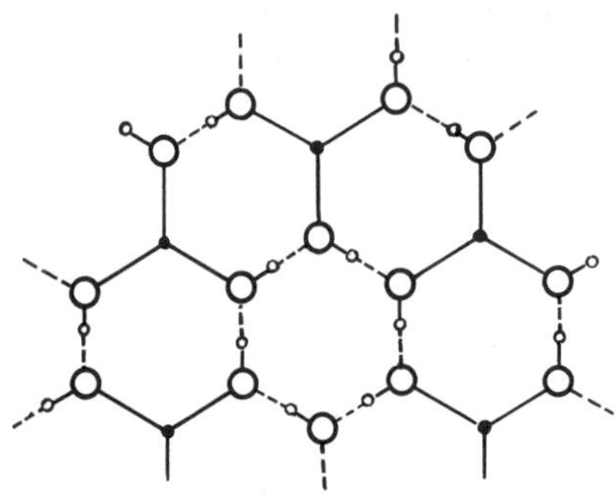

Figure 24.7 Structures of some borates

$B_3O_6^{3-}$ - a ring structure

$(B_2O_4^{2-})_n$ - polymeric

BORATES

Many boron–oxygen compounds are formed. These often form chains or rings with alternating boron and oxygen atoms. In this respect boron shows a marked similarity to silicon—another example of a diagonal relationship. The chains and rings are usually based on a trigonal planar BO_3 unit. Some examples are shown in **Figure 24.7.** The best known borate is borax, $Na_2B_4O_5(OH)_4$ ($2Na^+ + B_4O_5(OH)_4^{2-}$), deposits of which are the main source of boron compounds. Boric acid can be prepared by the action of acids on borax:

$$Na_2B_4O_5(OH)_4(aq) + 3H_2O(l) + 2H^+(aq) \longrightarrow 4B(OH)_3(aq) + 2Na^+$$

BORON HALIDES

These have the general formula BX_3 and exist as trigonal planar molecules. Boron trifluoride is a gas and the other trihalides become less volatile with increasing relative molecular mass. They are all electron deficient compounds and readily form addition compounds by accepting a pair of electrons, for example:

$$H_3N\!: \longrightarrow BCl_3$$

The arrangement of groups around both the nitrogen and the boron atom in the addition compound is tetrahedral.

The halides (except for boron trifluoride which gives fluoroboric acid, HBF_4) react rapidly with water to give boric acid:

$$BX_3 + 3H_2O \longrightarrow \underset{\text{boric acid}}{B(OH)_3} + 3HX$$

Industrial chemistry

The main sources of boron and its compounds are deposits of borax in the USA, Turkey and Tibet. Borax itself is used as a fertilizer and in large quantities for the manufacture of borosilicate glass, such as Pyrex which is used for ovenware and laboratory glassware because it is resistant to high temperatures.

Aluminium is widely used in aircraft construction

24.2.2 Aluminium

The element

Aluminium is a metal, familiar to most people from its use in saucepans and garage doors. Its low density ($2.7\,g\,cm^{-3}$) makes it an attractive constructional metal and although the pure metal is relatively weak, many of its alloys are very strong. It is a particularly good conductor of electricity. The value of $E^{\ominus}$ for the half reaction:

$$Al^{3+}(aq) + 3e^- \longrightarrow Al(s)$$

is -1.66 V. This suggests that aluminium should be a highly reactive metal and that it would be expected to corrode readily by reaction with air and water. However, it does not do so, and can be used unpainted while less reactive metals like iron cannot. This is because of the formation of a thin

A

ANODIZING ALUMINIUM

This is an electrolytic process for thickening the natural oxide layer on the surface of aluminium from around 15 nm (1.5×10^{-8} m) to 20 μm (2×10^{-5} m). The aluminium object is made the anode in an electrolytic cell where the electrolyte is dilute sulphuric acid. At the anode, the following reaction occurs:

$$2Al(s) + 3H_2O(l) \longrightarrow Al_2O_3(s) + 6H^+(aq) + 6e^-$$

It is possible to add dyes which are absorbed by the oxide layer to give the anodized aluminium an attractive finish.

(approximately 15 nm) insoluble layer of aluminium oxide which rapidly forms on the surface of the exposed metal and seals it off from further reaction. Unlike the oxides of most metals, it does not flake off or dissolve leaving the metal surface exposed. The true reactivity of aluminium can be demonstrated by dipping an aluminium milk botle top in mercury(II) chloride solution to remove the oxide layer. On exposure to air, a layer of white aluminium oxide forms very rapidly and the top becomes very hot due to the reaction with oxygen.

$$4Al(s) + 3O_2(g) \longrightarrow 2Al_2O_3(s)$$

Aluminium is **amphoteric**, reacting with both acids and alkalis. In hydrochloric acid:

$2Al(s)$	+	$6HCl(aq)$	$\longrightarrow$	$2AlCl_3(aq)$	+	$3H_2(g)$
aluminium		hydrochloric acid		aluminium chloride		hydrogen

In sodium hydroxide:

	$2Al(s)$	+	$2NaOH(aq)$	+	$6H_2O(l)$
	aluminium		sodium hydroxide		water
$\longrightarrow$	$2NaAl(OH)_4(aq)$	+	$3H_2(g)$		
	sodium tetrahydroxo-aluminate(III)		hydrogen		

Thus alkalis such as washing soda should not be used with aluminium saucepans.

Aluminium will react directly with the more reactive non-metals:

$4Al(s)$	+	$3O_2(g)$	$\longrightarrow$	$2Al_2O_3(s)$
aluminium		oxygen		aluminium oxide
$4Al(s)$	+	$6S(s)$	$\longrightarrow$	$2Al_2S_3(s)$
aluminium		sulphur		aluminium sulphide
$2Al(s)$	+	$N_2(g)$	$\longrightarrow$	$2AlN(s)$
aluminium		nitrogen		aluminium nitride
$2Al(s)$	+	$3F_2(g)$	$\longrightarrow$	$2AlF_3(s)$
aluminium		fluorine		aluminium fluoride

These compounds are covalent, except for the oxide and fluoride which are essentially ionic. The Al^{3+} ion is small and highly charged, making it very strongly polarizing (see **section 10.2.1**) so that anions, except for the very smallest like O^{2-} and F^-, are so distorted as to give a high degree of covalency.

Aluminium reacts with the oxides of less reactive metals, a spectacular example being the thermite reaction:

$2Al(s)$	+	$Fe_2O_3(s)$	$\longrightarrow$	$Al_2O_3(s)$	+	$2Fe(l)$
aluminium		iron oxide		aluminium oxide		iron

The thermite reaction is used to weld railway lines

The molten iron formed by this reaction has been used to weld together sections of railway line, see photograph.

Aluminium compounds

ALUMINIUM HYDRIDE

Alane (AlH_3), a polymer in the solid state, is prepared by the reaction of lithium hydride and aluminium chloride in ether solution. More significant are the complex hydrides such as lithium tetrahydridoaluminate(III), $LiAlH_4$, which is an important reducing agent in organic chemistry.

Although the similar tetrahydridoborate(III), BH_4^-, ion is stable in water, AlH_4^- is rapidly hydrolysed:

$$AlH_4^- + 4H_2O \longrightarrow Al(OH)_3 + 4H_2 + OH^-$$

and is therefore generally used in ether solution.

ALUMINIUM OXIDE

This is an ionic compound ($2Al^{3+} + 3O^{2-}$), in contrast with boron oxide which is covalent. It is amphoteric, unlike boron oxide, which is exclusively acidic.

$$\begin{array}{lllllll} Al_2O_3(s) & + & 6HCl(aq) & \longrightarrow & 2AlCl_3(aq) & + & 3H_2O(l) \\ \text{base} & + & \text{acid} & \longrightarrow & \text{salt} & + & \text{water} \end{array}$$

and

$$\begin{array}{lllllll} Al_2O_3(s) & + & 2KOH(aq) & + & 3H_2O(l) & \longrightarrow & 2Al(OH)_4K \\ & & & & & & \text{potassium tetrahydroxo-aluminate(III)} \\ \text{acid} & + & \text{base} & & & \longrightarrow & \text{salt} \end{array}$$

ALUMINIUM HYDROXIDE

Aluminium hydroxide, too, is amphoteric, contrasting with boric acid which is acidic. Notice that the formulae $B(OH)_3$ and $Al(OH)_3$ are comparable although the names make the two compounds sound quite different.

Whether a compound E—O—H is regarded as an oxoacid or as a hydroxide depends on which bond breaks. If the reaction:

$$E{-}O{-}H + (aq) \longrightarrow E^+(aq) + OH^-(aq)$$

takes place, then E—O—H is regarded as a hydroxide. If the reaction:

$$E{-}O{-}H + (aq) \longrightarrow EO^-(aq) + H^+(aq)$$

takes place, then E—O—H is regarded as an oxoacid.

The difference in behaviour can be related to the electronegativity of the element. Electropositive elements favour the first type of behaviour leading to positive ions and electronegative elements favour the second. Aluminium is intermediate in electronegativity, thus either can occur.

ALUMINIUM HALIDES

All four aluminium halides can be prepared by direct combination:

$$Al(s) + \tfrac{3}{2}X_2(g) \longrightarrow AlX_3(s)$$

The apparatus is shown in **Figure 24.8**. Aluminium fluoride is ionic ($Al^{3+} + 3F^-$); the other aluminium halides are covalently bonded. In the gas phase, measurements of relative molecular mass indicate that they exist as dimers, **Figure 24.9**, unlike the monomeric boron halides. The dative bonds complete aluminium's octet of electrons. Each aluminium atom is surrounded by chlorines roughly tetrahedrally, **Figure 24.10**.

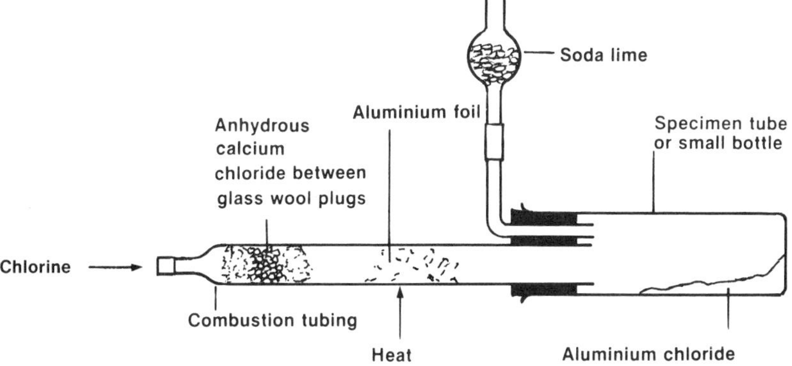

Figure 24.8 Apparatus for the preparation of aluminium chloride

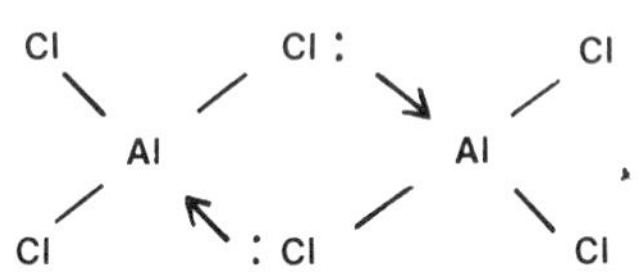

Figure 24.9 Bonding in the aluminium chloride dimer

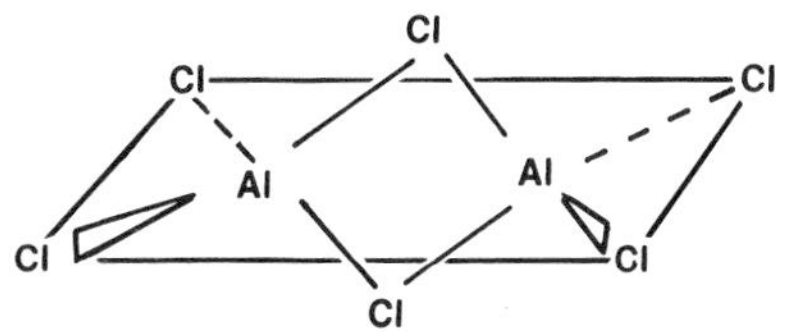

Figure 24.10 The shape of the aluminium chloride dimer

LAYER LATTICES

Layer lattices are composed of layers of two-dimensional giant structures stacked one on top of another like a pile of sheets of paper. Each sheet consists of a sandwich where the 'meat' is a layer of positive ions with layers of negative ions above and below it. The bonding within this sandwich is ionic, while the bonding between one sandwich and the next is van der Waals. Such crystals cleave easily between the van der Waals bonded layers as this bonding is much weaker than the ionic bonding.

Both magnesium chloride and aluminium chloride have this type of structure which is comparable with the structure of graphite.

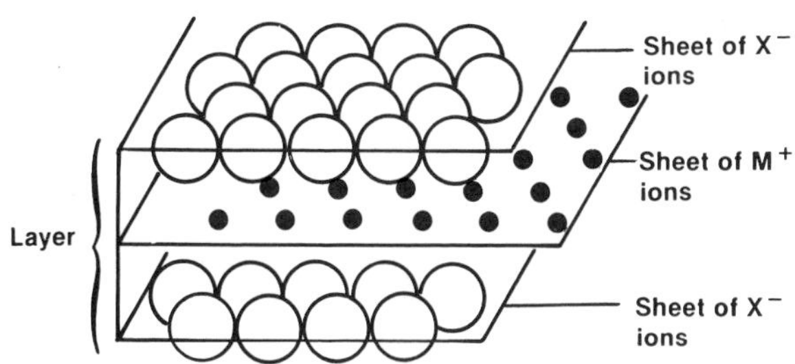

A single sandwich of a layer lattice, e.g. magnesium chloride. Another sandwich is placed above this and then another, and so on

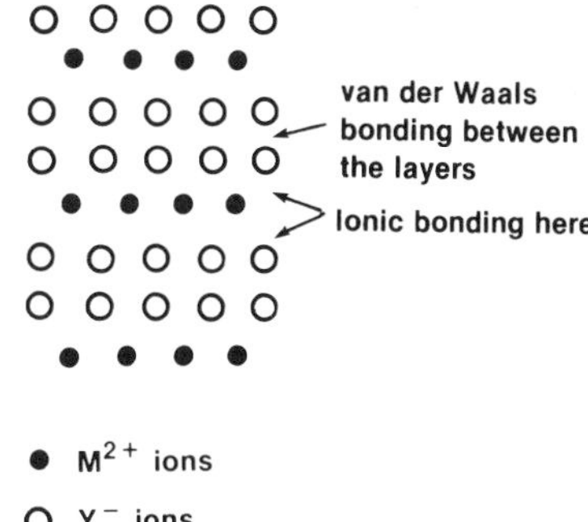

Bonding within a layer lattice

In the solid phase, they form layer lattices (see box) while at high temperatures in the gas phase, dissociation into monomers with a trigonal planar structure occurs. Aluminium chloride hydrolyses rapidly in water or even moist air:

$$AlCl_3(s) + 3H_2O(l) \longrightarrow Al(OH)_3(s) + 3HCl(g)$$

Fumes of hydrogen chloride are seen simply on opening the container—which should be done with care.

THE AQUEOUS Al^{3+} ION

Like other cations, aluminium ions in solution are hydrated, being surrounded by six water molecules at the corners of an octahedron, **Figure 24.11.** Each water molecule arranges itself around the Al^{3+} ion with the negative end of its dipole towards it. As the Al^{3+} ion is both small and highly charged, it is effective at attracting electrons away from the water molecules. This tends to weaken the O—H bonds which are already polarized $O^{\delta-}$—$H^{\delta+}$, making loss of a proton (H^+) more likely. So the hydrated $Al^{3+}(aq)$ ion is significantly acidic:

Figure 24.11 The hydrated $Al^{3+}(aq)$ ion

$$[Al(H_2O)_6]^{3+}(aq) \rightleftharpoons [Al(H_2O)_5OH]^{2+}(aq) + H^+(aq)$$

ALUMS

Alums are **double salts** of the general formula $M^+Al^{3+}(SO_4)_2.12H_2O$. M^+ may be Na^+, K^+, Rb^+ or NH_4^+. Li^+ is too small to fit into the crystal structure of the alums. The Al^{3+} may also be replaced by other ions of the same charge and similar size, e.g. Cr^{3+}, Mn^{3+}, Co^{3+}, Fe^{3+}.

Double salts are prepared by first dissolving aluminium in, say, potassium hydroxide:

$$2Al(s) + 2KOH(aq) + 2H_2O(l) \longrightarrow 2KAlO_2(aq) + 3H_2(g)$$

potassium dioxoaluminate(III) (potassium aluminate)

and then adding sulphuric acid:

$$KAlO_2 + 2H_2SO_4 \longrightarrow KAl(SO_4)_2 + 2H_2O$$

potassium aluminium sulphate

The industrial chemistry of aluminium

Aluminium is the most abundant metal on the earth, making up around 8% of the earth's crust. It is found in many clays. The ore bauxite (named after Baux in France) is the main source of aluminium. Known reserves of bauxite (impure hydrated aluminium oxide) are sufficient for hundreds of years at present rates of usage.

Aluminium is extracted from purified bauxite by electrolysis by the Hall–Héroult process which is described in **section 17.5.1**. World production of aluminium amounts to some 15 million tonnes annually. A large amount of this is used in alloys such as duralumin (95% aluminium, 4% copper, 1% magnesium) and magnalium (83% aluminium, 15% magnesium, 2% copper) which have structural strength and lightness making them ideal for transport applications such as the bodies of aircraft, buses and lorries. Motor engine parts are increasingly being made of aluminium alloys. Aluminium's good electrical conductivity and lightness make it ideal for overhead power lines. The average home may have aluminium window frames or garage doors, saucepans, cooking foil, drink cans and milk bottle tops. Many of these are disposable and can be recycled—indeed 30% of aluminium used in Europe is recycled. Recycled aluminium requires only 5% of the energy input needed to produce aluminium from the ore.

Aluminium compounds also have many varied uses. Aluminium triethyl is part of the catalyst system for the polymerization of ethene to polythene. Compounds containing aluminium oxide (alumina) and silicon oxide (silica) are used as furnace linings because of their resistance to high temperatures. Alumina is used as a catalyst support and as the stationary phase for chromatography. One form of alumina, corundum, is used as an abrasive. Semi-precious stones, such as sapphire and ruby, are forms

A

ACID RAIN—THE ROLE OF ALUMINIUM

Acid rain is now a well known environmental problem. Broadly it is caused when oxides of nitrogen and sulphur, produced during the burning of fuels, react with oxygen and water in the air so that rain becomes a mixture of dilute sulphuric and nitric acids. It now seems clear that this rain is responsible for the deaths of trees and fish all over Europe, but why? One culprit could be aluminium. Aluminium is present in most soils in the form of aluminosilicate minerals.

Aluminium kills fish by clogging their gills

Acidic rain can dissolve aluminium ions from the soil and carry them into lakes. Aluminium is toxic to fish. It kills them by clogging up their gills with mucus so that they eventually suffocate. High aluminium levels also seem to be involved in the death of trees.

Aluminium may also affect human health. Evidence is accumulating which suggests that it may be involved in Alzheimer's disease, a form of senility, following two recent discoveries. Firstly, it was found that inhabitants of islands with aluminium-rich soils suffer from a similar disease. Secondly, aluminium has been found in high concentrations in those parts of the brain associated with the disease. Could high aluminium levels in tap water be a factor? Ironically, aluminium compounds are used to treat drinking water. Aluminium sulphate is added to precipitate out anions as insoluble aluminium salts. However, a mix-up over deliveries at a treatment works in Camelford, Cornwall, led to high aluminium levels in the local tapwater for a short period in 1988. This resulted in some bizarre short-term effects such as oddly coloured hair. Some local doctors fear longer-term problems and the situation is being monitored.

of alumina with small amounts of transition metal ions as impurities which cause the colour. Ruby contains Cr^{3+}, sapphire Co^{2+}, Fe^{2+} and Ti^{4+}. Artificial gemstones for, say watch bearings, are made by fusing alumina.

Alum salts are used as mordants which help dyes to stick to materials and they are used in styptic pencils to coagulate blood and stop small cuts bleeding.

24.2.3 The chemistry of gallium, indium and thallium

The chemistry of these elements and their compounds generally resembles that of aluminium although there is a trend towards a more ionic character of the compounds on descending the group, and also increasing basicity. In other words the elements become more typically metallic. The three elements also show an increasing tendency to exhibit an oxidation state of +I as well as +III. The two s electrons tend not to participate in bonding. This is known as the **inert pair effect** and it also occurs in Group IV.

24.2.4 Summary (Group III)

- The elements of Group III have an outer electron arrangement of ns^2np^1.
- Due to its small size, boron's ionization energies are too high for B^{3+} ions to form.
- Aluminium forms an ionic fluoride and oxide but otherwise has covalent compounds.
- Lower down the group there is a greater tendency to form ionic compounds.
- Covalent Group III compounds have only six electrons in their outer shells and they tend to act as electron pair acceptors (Lewis acids).
- Boron is a fairly typical non-metal. It conducts electricity poorly and has acidic properties.
- Aluminium is a metal. It conducts electricity and heat well but it is amphoteric.
- Gallium, indium and thallium become increasingly typical metals. They do however show an increasing tendency to form E^+ ions rather than E^{3+} ions.
- Aluminium is important industrially, mainly for the high strength, low density and good electrical conduction of its alloys. It is extracted by the electrolysis of aluminium oxide dissolved in molten cryolite.

24.2.5 Questions (Group III)

1 **A)** The element aluminium is described as amphoteric. Explain this term. Give chemical evidence (including equations) which supports the fact that aluminium is amphoteric.

B) If aluminium is described as amphoteric, how would you describe boron and thallium (the lowest member of Group III)?

C) The boiling temperature of aluminium fluoride is 1564 K and that of aluminium bromide 536 K. Explain the difference. Estimate the boiling temperature of aluminium iodide.

2 The element gallium (symbol Ga) occurs in Group III of the Periodic Table, one row below aluminium and one row above indium. The existence of gallium was predicted by Mendeleev a few years before its discovery in 1875. Gallium has similar properties to aluminium, reacting with potassium hydroxide releasing hydrogen, forming an amphoteric hydroxide and burning in chlorine to form a chloride.

A) The value accepted in 1870 for the relative atomic mass of aluminium was 27 and that for indium was 113. Suggest, giving your reasons, a value for the relative atomic mass of gallium.

B) **i** What oxidation number would you expect gallium to have in its compounds?

ii Deduce a formula for gallium hydroxide.

iii What would be the product(s) of the reaction of gallium hydroxide with dilute aqueous nitric acid?

iv What would be the product(s) of the reaction of gallium hydroxide with dilute aqueous sodium hydroxide?

C) Suggest an experimental procedure by which gallium potassium sulphate, 'gallium alum', might be prepared from gallium metal.

D) i Gallium forms a chloride if the element is heated in a stream of chlorine gas. The chloride is a white solid melting at 78 °C and boiling at 202 °C. Complete the diagram to show how a sample of gallium chloride might be collected.

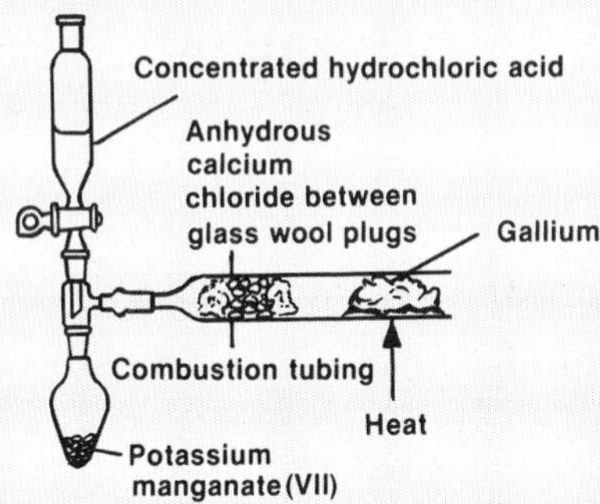

ii What is the purpose of the anhydrous calcium chloride in the apparatus?

E) A sample of gallium chloride weighing 0.1 g was vaporized and it was found to occupy 16 cm^3 at a temperature of 415 °C and at a pressure of 1 atmosphere.

i Calculate a value for the relative molecular mass of gallium chloride under the conditions of the experiment. (Molar volume of a gas at 0 °C and 1 atmosphere $= 22.4\,dm^3\,mol^{-1}$ or $R = 0.082$ atmos $dm^3\,K^{-1}\,mol^{-1}$)

ii Suggest a molecular formula for gallium chloride in the vapour state, indicating how you arrive at your answer. (Relative atomic mass: Cl = 35.5)

(Nuffield 1982)

3 Describe the extraction and manufacture of aluminium, starting from purified bauxite. Give two large scale uses of aluminium and show how they are related to the properties of the element.

Describe and explain what would happen in each of the following experiments.

A) Aluminium chloride is added to water containing a few drops of universal indicator solution.

B) A strongly alkaline solution is boiled in an aluminium saucepan.

(Cambridge 1987)

24.3 Group IV

The Group IV elements show very clearly the trend from non-metals through metalloids to metals on descending the group. The ionization energies are so large that M^{4+} ions hardly ever form (only in tin and lead fluorides) and the chemistry of the compounds is essentially covalent in the +IV oxidation state, although there are some Sn^{2+} and Pb^{2+} compounds—a consequence of the inert pair effect. Carbon, of course, has the most extensive chemistry of any element. The details of this will be found in Part D. **Table 24.2** summarizes some important properties.

A

ARTIFICIAL DIAMONDS

Both diamond and graphite are allotropes of carbon—the atoms are simply arranged differently. This raises the tantalizing possibility of converting cheap graphite into valuable diamond. The phase diagram for carbon shows the conditions of temperature and pressure under which the different allotropes are stable. The line AB shows the conditions under which it might be possible to convert graphite into diamond. This conversion was first brought about at around 2000 K and 55 000 atmos (5.5 × 10^9 Pa) in 1955 and industrial diamonds used for drill bits are now made under similar conditions using metal catalysts to speed up the conversion.

More recently Russian chemists have tried another approach—growing diamonds by deposition of carbon atoms from the gas phase or a solution of carbon in molten metals. A small seed crystal is used in the same way that, say, large crystals of copper sulphate can be grown from a saturated solution. It is now possible to coat objects with diamond film. The potential applications of this are enormous—virtually everlasting cutting surfaces on tools, totally scratch-resistant glass and wear-resistant bearings are just a few, if this technique becomes commercially viable.

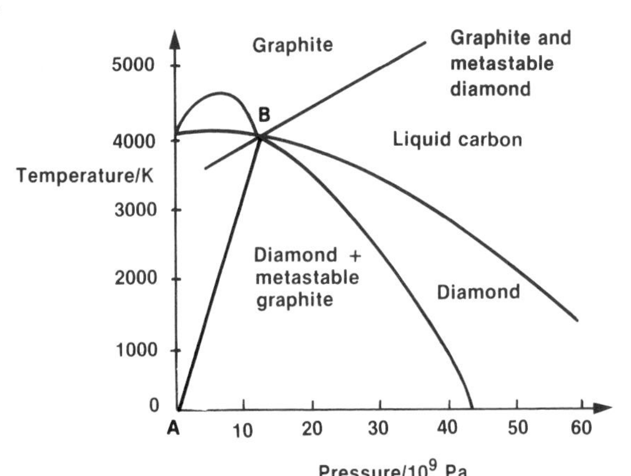

The phase diagram for carbon

Table 24.2 Properties of elements in Group IV

	Atomic number, Z	Electron arrangement	Electro-negativity	Atomic (covalent) radius/nm	Density /g cm^{-3}	T_m/K	T_b/K
C	6	$1s^22s^22p^2$	2.5	0.077	2.25(diamond) 3.51(graphite)	3930	(sublimes)
Si	14	[Ne] $3s^23p^2$	1.8	0.118	2.33	1683	2628
Ge	32	[Ar] $3d^{10}4s^24p^2$	1.8	0.122	5.35	1210	3103
Sn	50	[Kr] $4d^{10}5s^25p^2$	1.8	0.140	7.28(white) 5.79(grey)	505	2533
Pb	82	[Xe] $4f^{14}5d^{10}6s^26p^2$	1.8	0.154	11.34	601	2013

24.3.1 The physical properties of the elements

Carbon has two well-known allotropes—diamond and graphite. Diamond, silicon and germanium all have similar giant structures (see **section 10.3.3**). As the atoms get larger, the E—E bonds become weaker, making the structure less hard and giving a lower melting temperature. (The bond energies are C—C = 347 kJ mol^{-1}, Si—Si = 226 kJ mol^{-1}, Ge—Ge = 167 kJ mol^{-1}.) Both tin and lead have giant metallic structures. Diamond is an electrical insulator, silicon and germanium are semi-conductors while tin and lead are metallic conductors. Graphite's layer structure has been described in **section 10.3.3.** It conducts electricity to some extent.

E

CONDUCTION OF ELECTRICITY BY GRAPHITE

The layer structure of graphite has been described in **section 10.3.3.** Each carbon is hybridized to form three sp^2 orbitals, in a plane at 120° to one another, and a single p orbital. In graphite a σ-bonded framework is formed by overlap of the sp^2 orbitals. The π bonds overlap above and below this to form an orbital which spreads over the whole layer. Electrons can move freely within this orbital and so graphite conducts electricity well in the direction along the planes. Recently a number of polymers that conduct electricity have been made. One of the simplest is poly(ethyne). This, like graphite, has a π-orbital which spreads along the whole molecule so making it conduct. Here the orbital is a one-dimensional chain rather than a two-dimensional plane.

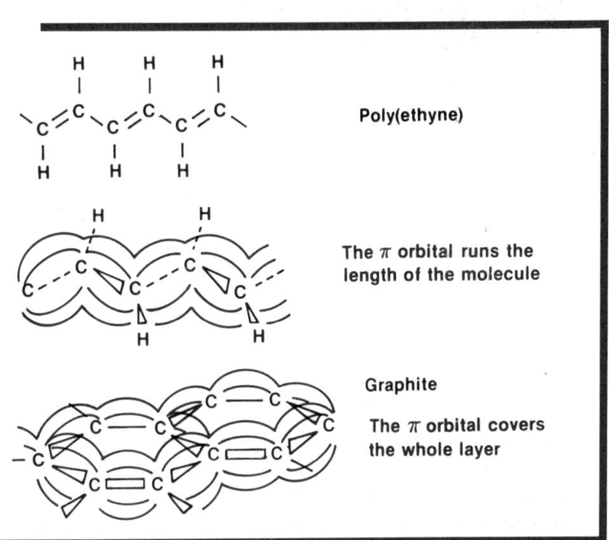

24.3.2 Chemical reactivity of the elements

In general the elements are not particularly reactive. Overall, reactivity increases on descending the group. All the elements react directly with oxygen to form EO_2. Lead also forms PbO. A similar pattern is found in the reaction with chlorine, ECl_4 being formed in every case except for lead which forms $PbCl_2$. These examples illustrate the general trend that the stability of the +IV oxidation state decreases on descending the group, while that of the +II state increases.

In carbon, the +II state is very unstable, carbon monoxide, $\overset{+II}{CO}$, reacting readily to form carbon dioxide, $\overset{+IV}{CO_2}$, for example:

$$3\overset{+II}{CO}(g) + \overset{+II}{Fe_2}O_3(s) \longrightarrow 2\overset{0}{Fe}(s) + 3\overset{+IV}{CO_2}(g)$$

so carbon monoxide is a good reducing agent. In the case of lead, PbO is more stable than PbO_2 so that PbO_2 is a good oxidizing agent, oxidizing Cl^- to Cl_2 for example:

$$\overset{+IV}{PbO_2}(s) + 4H\overset{-I}{Cl}(aq) \longrightarrow \overset{+II}{Pb}Cl_2(aq) + \overset{0}{Cl_2}(g) + 2H_2O(l)$$

The increased stability of the +II oxidation state lower down the group is a consequence of the inert pair effect. The two s electrons are less likely to be involved in bonding lower down the group. This effect also occurs in Groups III and V. The reason is that inner d and f orbitals, because of their shape, do not shield outer electrons as effectively as do s and p orbitals. Therefore in atoms with such inner electrons (those in Periods 4, 5 and 6) the outer s electrons are pulled more strongly towards the nucleus and are less available to take part in bonding, as shown in **Figure 24.12**.

Figure 24.12 The inert pair effect. Inner d orbitals do not shield the outer s electrons very effectively. The outer s electrons are attracted to the nucleus strongly and are less available to take part in bonding, thus the name 'inert pair'

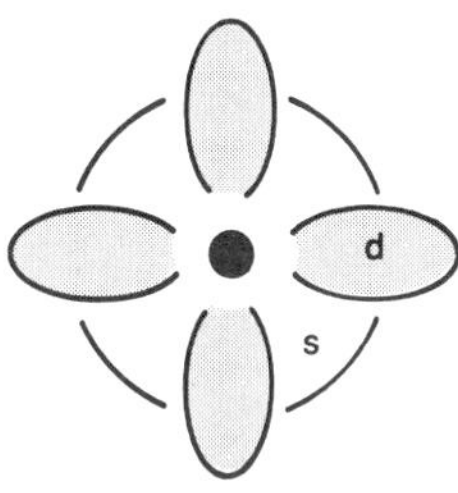

24.3.3 The uniqueness of carbon

Carbon has a unique ability to form compounds. At present over 6 million are known. The main reason for this is its ability to form stable chains and rings of carbon atoms, the simplest of which are hydrides. This property is sometimes called **catenation**. Carbon's ability to catenate is unparalleled, even among the other members of Group IV. Silicon and germanium form chains, including hydrides, but only to a very limited extent. Why? Part of the answer lies in bond energies, **Table 24.3.**

Table 24.3 Bond energies in compounds of carbon and silicon

Bond	energy/kJ mol^{-1}	Bond	energy/kJ mol^{-1}
C—C	347	Si—Si	226
C—H	413	Si—H	318
C—O	358	Si—O	466

Notice that the bond energies of C—C and C—O are of comparable value, while for silicon, the Si—Si bond is much weaker than the Si—O bond, so that thermodynamically Si—Si bonds will not be stable in the presence of oxygen and Si—O bonds will be more favoured than Si—Si bonds. Silicon forms many compounds with —Si—O—Si—O—chains, like silicon dioxide and silicone polymers.

The second factor is kinetic. Part of the reason for the kinetic stability of carbon compounds is that all the low energy orbitals of the carbon atoms are filled once the compound is formed, so that they are not available to form bonds with attacking molecules. The first step in the formation of a new compound is often the temporary filling of an unoccupied orbital. In carbon there is a large energy gap between the highest-energy filled orbital (2p) and the next unfilled orbital (3s) so this energy must be supplied if an attacking molecule is to form a bond. By contrast, in silicon, empty 3d orbitals are available at only a slightly higher energy than the filled orbitals (3p). So carbon compounds would have a higher activation energy than comparable silicon compounds for oxidation or reaction with, for example, water. This explains their kinetic stability. **Figure 24.13** shows the situation for CH_4 and SiH_4, but it is the same for more complex hydrides.

Thirdly, carbon forms many compounds with multiple bonds—C=C and C≡C. These involve overlap of p orbitals to form π bonds. Silicon atoms are larger than carbon and cannot approach closely enough to allow efficient overlap of p orbitals to form a π bond.

So carbon can form stable chains and rings held together by strong carbon–carbon bonds, both single and multiple. Such molecules are kinetically stable.

24.3.4 The hydrides

Carbon forms a vast number of hydrides with rings, chains of unlimited length and varying degrees of branching. They are called alkanes if they have no multiple bonds, alkenes with one or more double bonds and alkynes if they have a triple bond C≡C. They are considered in detail in Chapters 27 and 28. They are stable in air except at high temperatures and do not react with water, dilute acids or alkalis. They are gases, liquids or solids at room temperature, depending on the chain length.

Silicon forms a series of hydrides called silanes of chain length up to ten silicon atoms including some with branched chains. They are gases or volatile liquids. They burn on contact with air and are rapidly hydrolysed

Figure 24.13 Energy levels of orbitals in carbon, silicon, methane (CH_4) and silane (SiH_4)

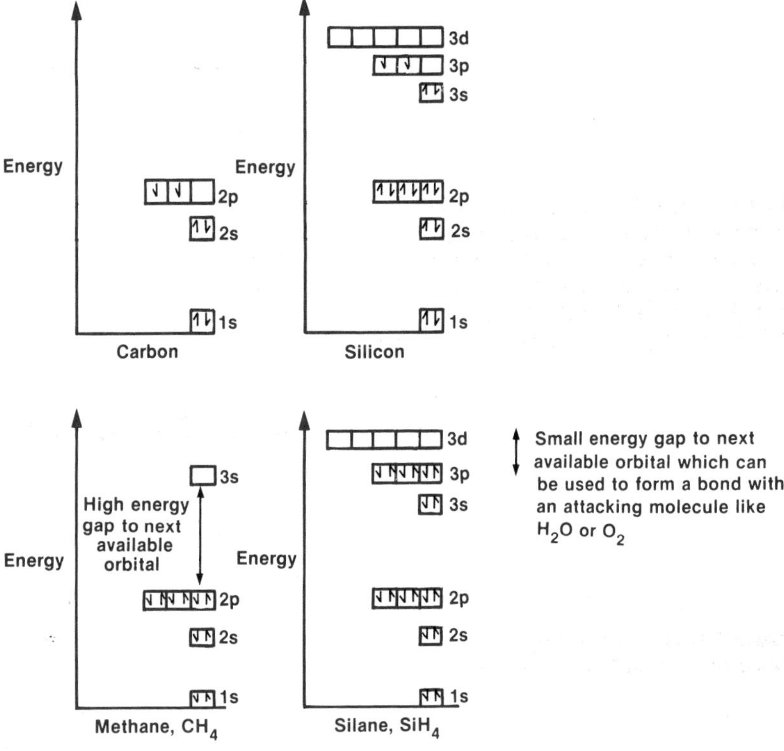

by water for the reasons discussed above, for example:

$$\underset{\text{monosilane}}{SiH_4(g)} + \underset{\text{oxygen}}{2O_2(g)} \longrightarrow \underset{\text{silicon dioxide}}{SiO_2(s)} + \underset{\text{water}}{2H_2O(l)}$$

$$\underset{\text{disilane}}{Si_2H_6(g)} + \underset{\text{water}}{4H_2O(l)} \longrightarrow \underset{\text{silicon dioxide}}{2SiO_2(s)} + \underset{\text{hydrogen}}{7H_2(g)}$$

Germanium forms hydrides called germanes of chain length up to six. They behave in the same way as the silanes. Tin and lead form only a single hydride each, stannane, SnH_4, and plumbane, PbH_4. Both are unstable. The tendency to shorter chains reflects the decreasing E—E bond strength on descending the group.

24.3.5 Halides

Carbon forms halides CX_4, with all the halogens and chain halides can also form. They are all stable in air and water, especially the fluorocarbons such as poly(tetrafluoroethene), ptfe, Teflon:

```
   F   F   F   F
   |   |   |   |
 — C — C — C — C —
   |   |   |   |
   F   F   F   F
```

whose inertness makes it useful for non-stick coatings on saucepans, etc.

Silicon forms halides SiX_4, as well as some chains. They are all rapidly hydrolysed, fumes of hydrogen halide being given off even in moist air, for example:

$$\underset{\text{silicon tetrachloride}}{SiCl_4(l)} + \underset{\text{water}}{2H_2O(l)} \longrightarrow \underset{\text{silicon dioxide}}{SiO_2(aq)} + \underset{\text{hydrogen chloride}}{4HCl(g)}$$

Germanium and tin form halides EX_4, and Ge_2Cl_6 is also known. These halides hydrolyse in the same way as the silicon compounds. $GeCl_2$ and $SnCl_2$ exist but are readily oxidized to the tetrachlorides. SnF_4 is ionic

$(Sn^{4+} + 4F^-)$. With lead, the +II oxidation state is more stable and lead(IV) chloride easily decomposes into lead(II) chloride and chlorine. The dihalides have an appreciable ionic character in their bonds and PbF_2 is definitely ionic.

The trends are clear. On descending the group, the halides have less tendency to form chains due to the decreasing strength of the E—E bond, and the dichloride becomes more stable due to the inert pair effect. All the tetrahalides hydrolyse easily except for those of carbon which has no available orbital to accept a lone pair from water as the first step of the reaction.

24.3.6 Oxygen compounds

Here again, there is a considerable contrast between carbon and the rest of the group. Carbon forms carbon dioxide, CO_2, a linear molecule with two carbon–oxygen double bonds. It also forms carbon monoxide, which is easily oxidized to the dioxide.

O═C═O

carbon dioxide

C≡O

carbon monoxide

Even so, it is unusual to find the +II oxidation state in the first member of the group. By complete contrast, silicon dioxide has a giant structure bonded by Si—O single bonds. The structure is based on silicon atoms surrounded by four oxygen atoms tetrahedrally. Each oxygen is shared by two silicon atoms. The geometry of the structure is similar to that of diamond and silicon but with an oxygen atom between each of the silicon atoms, see **Figure 24.14.** The resulting structure, quartz, is hard and has a high melting temperature. This difference between CO_2 and SiO_2 can be explained by looking at the bond energies, **Table 24.4.** The C═O bond energy is more than twice as great as that for C—O. Thus, more energy is evolved if carbon forms two double bonds (1610 kJ mol^{-1}) than four C—O bonds (1432 kJ mol). The Si═O bond is less than twice as strong as the Si—O bond so for silicon more energy is given out in forming four Si—O bonds (1864 kJ mol^{-1}) than two Si═O bonds (1276 kJ mol^{-1}).

Figure 24.14 The giant structure of silicon dioxide

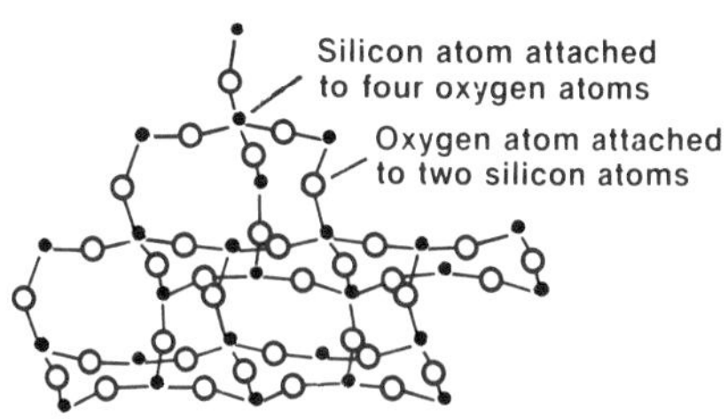

Table 24.4 Bond energies for carbon–oxygen and silicon–oxygen bonds

	Bond energy/kJ mol^{-1}	
C—O	358	
C═O	805	(in CO_2)
Si—O	466	
Si═O	638	

GeO_2, SnO_2 and PbO_2 all have giant structures, the last two being ionic. On descending the group, there is the expected trend from acid to base—CO_2 and SiO_2 are acidic, GeO_2 and SnO_2 amphoteric. SnO and PbO exist, lead(II) oxide being more stable than lead(IV) oxide. Pb_3O_4 is a mixed oxide which may be thought of as a mixture of PbO and PbO_2 in the ratio 2:1. Hence its name is dilead(II) lead(IV) oxide.

24.3.7 Carbonates and silicates

When carbon dioxide is dissolved in water the following equilibria occur:

$$CO_2(aq) + H_2O(l) \rightleftharpoons HCO_3^-(aq) + H^+(aq)$$

carbon dioxide, water, hydrogen-carbonate ion, hydrogen ion

$$\rightleftharpoons H_2CO_3 \rightleftharpoons 2H^+ + CO_3^{2-}$$

carbonic acid, hydrogen ions, carbonate ion

Carbonic acid itself cannot be isolated but metal carbonates and hydrogen-carbonates are common.

Silicon dioxide is also acidic. It reacts with molten bases to form silicates.

igure 24.15 **Structures of some ilicates**

Orthosilicate ion

Pyrosilicate ion

The simplest of these is the tetrahedral orthosilicate ion SiO_4^{4-}. Chains and rings can also form. See **Figure 24.15**.

24.3.8 Economic importance of Group IV elements and compounds

Carbon

dustrial diamond tools

Carbon compounds have enormous economic importance. Their industrial chemistry is discussed in Part D. Coal and crude oil are the most important sources of carbon compounds. Diamonds are used in cutting tools as well as gemstones, and graphite is used as a lubricant, in batteries, as the brushes of electric motors and baked with clay to form the 'lead' in pencils. Coke, an impure form of carbon obtained from coal, is used in the blast furnace both as a fuel and to form carbon monoxide which reduces iron ore to iron.

Silicon

Silicon forms almost 30% of the earth's crust. It is found in the form of silicates in many rocks and minerals including granite and clays and also as impure silicon dioxide in sand. It is fortunate that it is so abundant as its compounds have many uses.

Both silicon and germanium are semiconductors and are used to make

A

THE CARBON CYCLE

The atmosphere contains 0.05% carbon dioxide and this forms a vast reservoir of carbon which is continually being removed and returned by a variety of processes. These are summarized in the diagram (right). The system is normally at equilibrium, i.e. the rate of removal of atmospheric carbon dioxide is equal to its rate of return. At present, because of the large amounts of fossil fuels being burned, the amount of atmospheric carbon dioxide is increasing. The carbon dioxide acts like the glass in a greenhouse, keeping in heat. There are fears that undue amounts of carbon dioxide could cause the earth to heat up significantly and possibly cause some melting of the polar ice caps, see box, **section 15.4.**

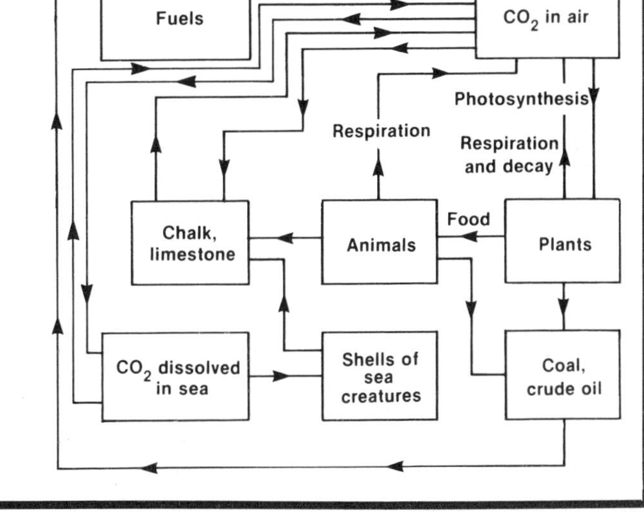

The carbon cycle

SEMICONDUCTORS

Silicon is itself a semiconductor. Its conduction properties can be modified by **doping** it with atoms of elements from the same period in Group III (Al) or Group V (P). Phosphorus has one electron more than silicon and the 'spare' electrons in phosphorus-doped silicon improve the conduction. This is called an *n*-type semiconductor (*n* for negative, the charge on the electron).

'spare' electron

Si ẋ Si ẋ Si
•× •× • •×
Si ẋ P ẋ Si
•× •× •×
Si ẋ Si ẋ Si

n-type semiconductor

'hole'

Si ẋ Si ẋ Si
•× • •×
Si ẋ Al ẋ Si
•× •× •×
Si ẋ Si ẋ Si

p-type semiconductor

Aluminium has one electron less than silicon. The missing electron produces a *hole* which acts like a positive charge. The holes can carry a current as electrons 'hop' into the hole from the next silicon atom. This is the same as the hole moving in the opposite direction. This is called a *p*-type semiconductor (*p* for positive, the 'charge' on the hole). At a junction between two types of semiconductor electrons can flow from *n* to *p* as they are moving from a negatively charged area to a positively charged one, but they will not move in the other direction. The junction acts like a turnstile, allowing the current to flow in one direction only. This is called a rectifier and can be used to convert alternating to direct current.

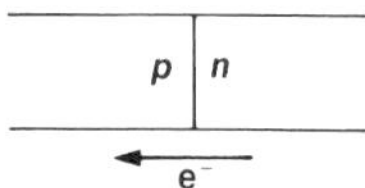

At a *p*, *n* junction, electrons can flow only from *n* to *p*

GLASS

A major use of silicon compounds is in making glass. The cheapest glass—soda glass—is made by melting together sand, SiO_2, sodium carbonate, Na_2CO_3, and limestone, $CaCO_3$, at 1800 K. A transparent mixture of sodium and calcium silicates is formed:

$$Na_2CO_3 + SiO_2 \longrightarrow CO_2 + Na_2SiO_3$$
$$CaCO_3 + SiO_2 \longrightarrow CO_2 + CaSiO_3$$

Addition of boron oxide gives borosilicate glasses which withstand high temperatures, while addition of small amounts of transition metal ions gives coloured glasses.

Floating glass on molten tin

Optically flat glass can be made by floating the glass on molten tin as it sets. Glass is not crystalline; the particles have very little order and it is described as a supercooled liquid. It still flows slowly at room temperature—old glass windows have been found to be significantly thicker at the bottom than at the top.

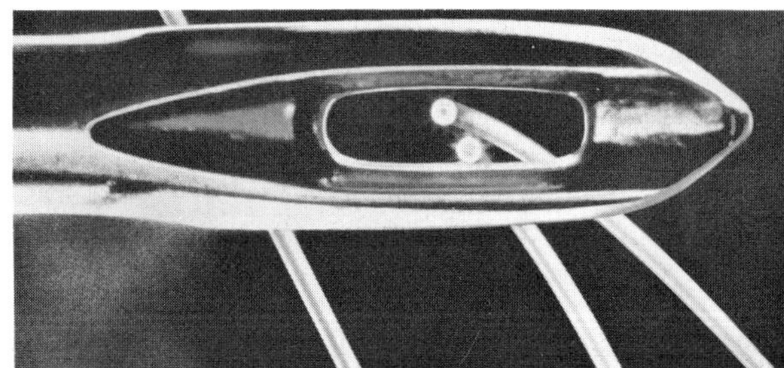

Fibre optics are replacing copper wires for communications

Glass has many conventional uses but the latest—fibre optics—may turn out to be as important as any. Light from a laser can be shone along glass fibre as thin as a human hair to carry information, say telephone conversations. These 'light pipes' have many advantages over conventional copper wires—in particular a single fibre can carry more information than a wire.

The Turner Special sports car has a glass fibre body

Glass fibres are also important for insulation and are embedded in resins for building canoes, repairing vehicle bodies, etc.

Figure 24.16 Zone refining

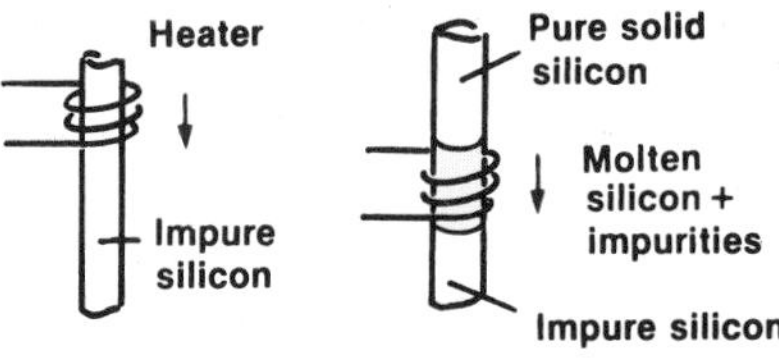

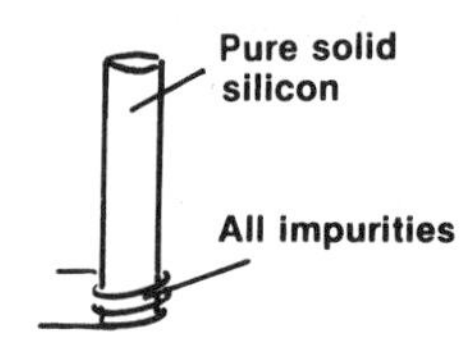

transistors and 'microchips' which are circuits containing thousands of transistors and other solid-state devices on a single wafer of silicon. Silicon for this purpose is made as follows.

Impure silica, SiO_2 (sand) is reduced to impure silicon using carbon. The impure silicon is reacted with chlorine to give liquid silicon tetrachloride. This can be purified by distillation and is then reduced back to silicon by reaction with hydrogen. The silicon is finally purified by zone refining. A heater is passed along a rod of impure silicon which melts and then resolidifies as the heater moves along. The impurities dissolve better in molten silicon than in the solid and so are swept along to the end of the rod, **Figure 24.16.**

Lead

Lead is used in car batteries

325 000 tonnes of lead are used each year in the UK. Over half goes into the manufacture of car batteries. Over 40% of this lead is recycled, which is why car batteries are traded in. The rest goes into pipes, cable sheathing, solder and in the manufacture of pigments and petrol additives. Lead ores are roasted in air to convert them into lead(II) oxide. This is reduced by coke in a blast furnace. Environmental lead is a major problem as lead is toxic and accumulates in the body, causing mental retardation, especially in children. Some historians have tried to blame the decline of the Roman Empire on lead poisoning, as the Romans used lead a great deal in plumbing and in drinking vessels. Certainly analysis of the bones of some Roman rulers has shown high levels of lead. Nowadays, 70% of environmental lead can be attributed to petrol additives. This should be reduced gradually as unleaded fuel becomes increasingly available and more cars have engines which can use it. Some older houses still have lead plumbing. This lead may dissolve slightly in the drinking water, especially in soft water areas. In such houses it is wise to run off a certain amount of water before drinking in the morning as it will have been standing in the pipe all night and may have had time to dissolve a significant quantity of lead.

24.3.9 Summary (Group IV)

- The Group IV elements have the outer electron arrangement ns^2np^2.
- Group IV shows clearly the trend from non-metal to metal on descending the group.

A

ZEOLITES

In the three-dimensional structure of silica (silicon dioxide) some of the silicon atoms may be replaced by aluminium. Essentially, Si^{4+} has been replaced by Al^{3+} and so other positive ions such as Na^+ are required to balance the charges. These aluminosilicates are found in many minerals. They have empirical formulae like $NaAlSi_2O_6$ and are called **zeolites**. The open mesh structures of zeolites contain many cavities—see **Plate 23.** They have two important uses.

Firstly they may exchange their sodium ions for calcium ions (two Na^+ for each Ca^{2+} to balance the charges) and can therefore be used as ion exchange resins to soften hard water (see box *Hardness of water*, **section 23.4.5**).

Secondly, small molecules may penetrate into the cavities and be trapped there, held by van der Waals forces. They can therefore be used to absorb selectively molecules of a certain size from a mixture. Used like this they are called molecular sieves. Artificial zeolites can be tailored with cavities of the right size to absorb only certain molecules. For example, straight-chain hydrocarbons may be absorbed while branched chains and rings are not. Silica gel, a drying agent, has a similar structure which absorbs water molecules very well. It is used in the laboratory in desiccators and in the home to prevent condensation in between the panes of double glazing. You may have come across silica gel in the packaging of cameras or binoculars to keep them dry. It is also sold as anti-damp crystals.

Silica gel is used as a drying agent in packaging

A

SILICONES

These are polymers based on chains:

```
     |         |         |
  — Si — O — Si — O — Si —
     |         |         |
```

They are made by polymerizing compounds of the formula:

```
          R
          |
   HO — Si — OH
          |
          R
```

were R represents an organic group like CH_3— or C_2H_5—.

During polymerization, water molecules are eliminated:

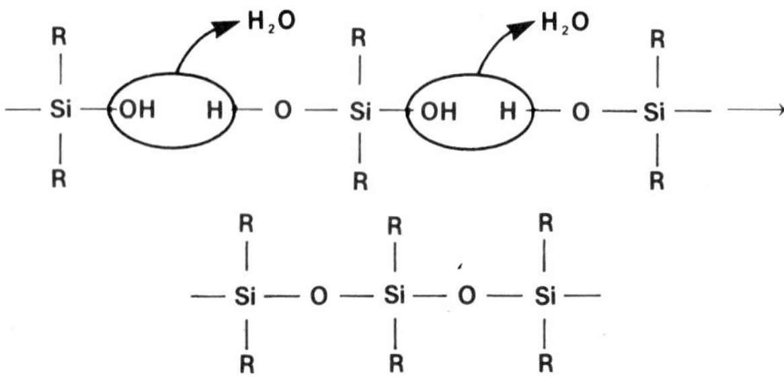

```
                  R           R
                  |           |
Small amounts of  R — Si — OH   are added to the mixture to
                  |
                  R
```

act as 'endstops' for the chain:

```
       R         R
       |         |
  R — Si — O — Si —
       |         |
       R         R
```

The more 'endstops' the shorter the average chain length. Molecules like:

```
         OH
         |
   R — Si — OH
         |
         OH
```

can be added to cross-link two chains:

```
     R         R         R
     |         |         |
  — Si — O — Si — O — Si —
     |         |         |
     R         O         R
     R         |         R
     |         |         |
  — Si — O — Si — O — Si —
     |         |         |
     R         R         R
```

Both chain length and cross-linking affect the properties of the polymer. The silicones are liquids; the longer the chains the thicker the liquid. They have a large variety of useful properties:

- lubricants, especially for high temperature applications
- they are added to polishes to reduce the elbow grease of polishing
- they are water repellant and are used to waterproof cloth for tents, anoraks, etc.
- they do not stick and are used to coat the backing papers of elastoplast, stick-on labels, etc.
- they have antifoaming properties—small traces added to stirred vats in industrial processes can reduce troublesome foaming.

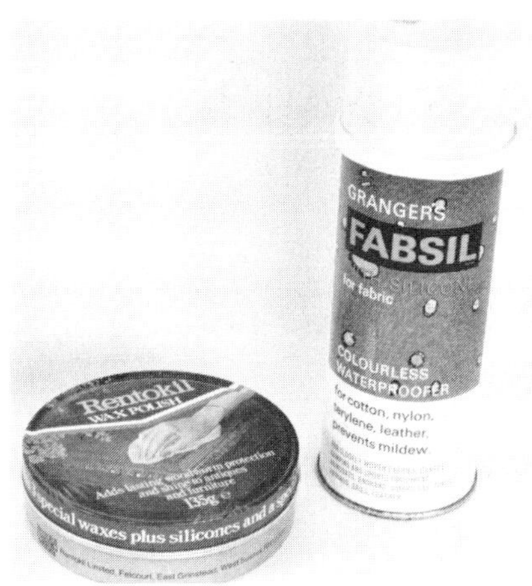

Household polish and waterproofer contain silicones

- The inert pair effect operates. The +II oxidation state gets more stable and the +IV state less stable as we descend the group.
- Except for the fluorides and oxides of tin and lead, all the compounds are covalent.
- Carbon compounds are very different from those of the rest of the group because of the greater relative strength of C—C bonds compared with E—E for the other elements, the unavailability of low energy d orbitals and carbon's small size which favours π bond formation.
- Carbon forms molecular oxides; the other elements form oxides with giant structures.
- Carbon tetrahalides (tetrahalogenomethanes) and methane are stable to both air and water at room temperature. The halides and hydrides of the other members of the group react with air and water.
- Carbon forms a vast number of compounds due to its ability to form chains (catenation).
- Catenation occurs much less as we descend the group, as the E—E

bonds get weaker.

- Silicon and germanium are both of commercial importance as semiconductors. Silicates are also important for making glass.
- Lead is of economic importance for making car batteries and solder and for petrol additives.

24.3.10 Questions (Group IV)

1 On page 317 there is a series of reactions for producing pure silicon from impure silica. Write balanced equations for each of the chemical steps.

2 Give a comparative account of the chemistry of the oxides and chlorides of carbon, silicon, germanium, tin and lead, paying particular attention to:

A) The formation, composition and stability of oxides;

B) The formation, hydrolytic behaviour (reaction with water) and stability of chlorides.

(London 1980)

3 State how the following vary on descending Group IV (C Si Ge Sn Pb):

A) Chain-forming ability (catenation)

B) Stability of the +IV oxidation state compared with the +II oxidation state

C) Electrical conductivity

D) Type of bonding in the chlorides.

Try to explain any trends you note.

4 Describe the main ways in which the chemistry of carbon compounds differs from that of their silicon analogues. Try to account for the differences.

5 What is meant by the 'diagonal relationship' as applied to the Periodic Table? Illustrate your answer by reference to the chemistry of boron and its compounds. What causes the diagonal relationship?

24.4 Group V

We shall look at only the top two members of this group—nitrogen and phosphorus—in any detail. These are both non-metals. Arsenic and antimony are metalloids while bismuth is metallic. We have seen in other groups that the first member is untypical, and this also occurs in Group V. Both nitrogen and phosphorus are important in fertilizers.

Some properties are given in **Table 24.5**.

Table 24.5 Properties of Group V elements

	Atomic number, Z	Electron arrangement	Electro-negativity	Atomic (covalent) radius/nm	T_m/K	T_b/K
N	7	$1s^2 2s^2 2p^3$	3.0 ↑	0.075	63	77
P (white)	15	$[Ne]3s^2 3p^3$	2.1	0.110	317	553
As (grey)	33	$[Ar]\,3d^{10} 4s^2 4p^3$	2.0	0.122	886 sublimes	
Sb	51	$[Kr]\,4d^{10} 5s^2 5p^3$	1.9	0.143 ↓	904	2023 ↓
Bi	83	$[Xe]\,4f^{14} 5d^{10} 6s^2 6p^3$	1.9		544 ↓	1833

Figure 24.17 Bonding in the nitrogen molecule, N_2

N≡N :N ⁝ N

24.4.1 The elements

Nitrogen is a gas comprising approximately 78% of the atmosphere. It exists as diatomic molecules, N_2, held together by a triple covalent bond, **Figure 24.17.** The bond energy is high, 945 kJ mol^{-1}, which goes some way to explaining nitrogen's lack of reactivity.

Phosphorus has three allotropes—white, black and red. White phosphorus exists as P_4 molecules, **Figure 24.18.** Arsenic and antimony also form this structure. To obtain a full outer shell of electrons, Group V elements would need to lose five electrons to form E^{5+} or gain three to form E^{3-}. No E^{5+} ions form, due to the large amounts of energy this would require for ionization, although Sb^{3+} and Bi^{3+} ions exist—the inert

Figure 24.18 The phosphorus molecule, P_4

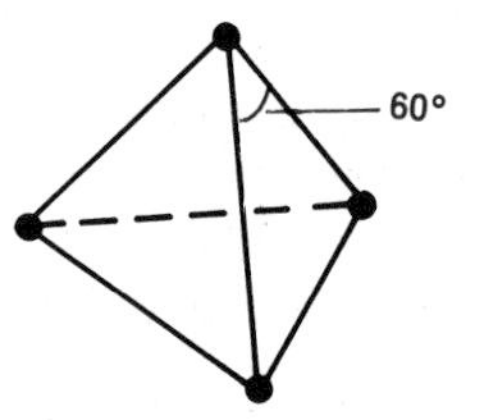

Figure 24.19 The corrugated layer structure of black phosphorus

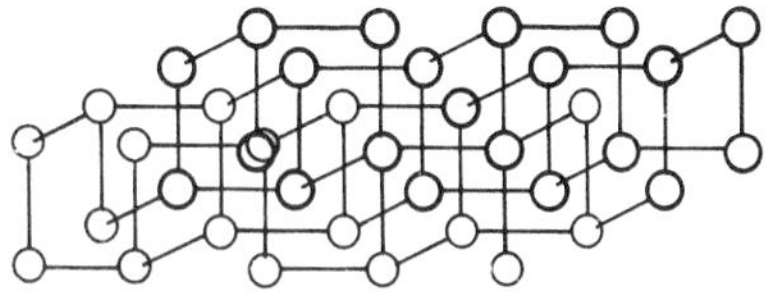

Ball and spring model of the P_4 molecule showing the bond strain

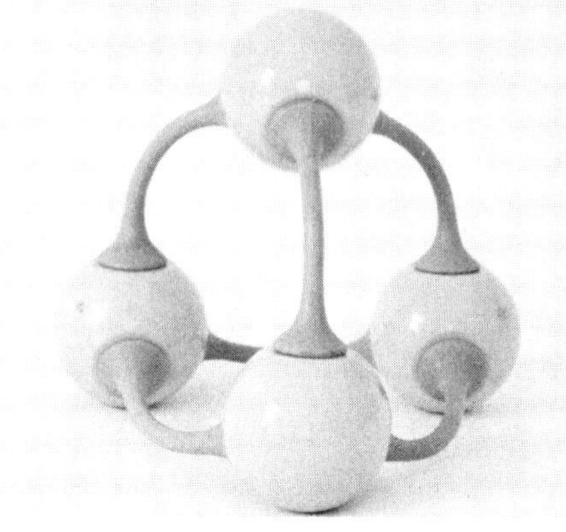

Nitrogen oxides are formed by lightning

pair effect operates again. N^{3-} exists in the nitrides of the s-block elements but otherwise the bonding is covalent.

In white phosphorus, each atom forms three covalent bonds giving it a full outer shell. P═P does not occur as, like silicon, the phosphorus atom is too large for p orbitals to overlap efficiently to give π bonds. Black phosphorus has a giant covalent structure, **Figure 24.19**, in which corrugated sheets are formed, each phosphorus atom again forming three covalent bonds. Bonding between the layers is weaker than within them leading to a flakiness not unlike graphite. Red phosphorus has a giant structure whose exact details are not known. White phosphorus is more reactive than the other two allotropes (it will ignite in air at just above room temperature) due to **bond strain**. If p orbitals are used to form the bonds the ideal P—P—P angle would be 90°. The shape of the molecule forces the angle to be 60°, and so the molecule has a tendency to 'spring apart', see photograph.

24.4.2 Chemical reactivity of the elements

Nitrogen undergoes few reactions. It will combine with s-block metals to form nitrides containing the N^{3-} ion, for example:

$$\underset{\text{magnesium}}{3Mg(s)} + \underset{\text{nitrogen}}{N_2(g)} \longrightarrow \underset{\text{magnesium nitride}}{\begin{matrix} Mg_3N_2(s) \\ (3Mg^{2+} + 2N^{3-})(s) \end{matrix}}$$

This reaction occurs when magnesium is burned in air. Magnesium nitride is responsible for the greyish colour of the product. Pure magnesium oxide is white.

At high temperatures nitrogen will combine with oxygen to give nitrogen oxides, for example:

$$\underset{\text{nitrogen}}{N_2(g)} + \underset{\text{oxygen}}{O_2(g)} \rightleftharpoons \underset{\text{nitrogen monoxide}}{2NO(g)}$$

Other oxides of nitrogen, dinitrogen oxide, N_2O, and nitrogen dioxide, NO_2, are also formed. This reaction occurs in car engines (where the resulting mixture of oxides is called NOx) and also lightning flashes.

Nitrogen will also combine with hydrogen to form ammonia:

$$N_2(g) + 3H_2(g) \rightleftharpoons 2NH_3(g)$$

This is basis of the Haber process.

Nitrogen can also be 'fixed' (i.e. chemically combined) by bacteria in the roots of leguminous plants such as clover peas and beans. They convert the nitrogen into proteins (see **section 35.2**).

Phosphorus is much more reactive, due to the bond strain in the molecule and the fact that P—P bonds (BE or bond energy = 298 kJ mol^{-1}) are much weaker than N≡N (BE = 945 kJ mol^{-1}). Phosphorus will react with metals to form phosphides although these are not ionic. It will burn easily in air to form a mixture of the phosphorus(III) oxide (phosphorus trioxide, P_2O_3) and phosphorus(V) oxide (phosphorus pentoxide, P_2O_5). With halogens, tri- or pentahalides, PX_3 and PX_5, are formed depending on which element is in excess. With alkalis, phosphine, PH_3, is formed.

$$\underset{\text{phosphorus}}{P_4(s)} + \underset{\text{sodium hydroxide}}{3NaOH(aq)} + \underset{\text{water}}{3H_2O(l)} \longrightarrow \underset{\text{phosphine}}{PH_3(g)} + \underset{\text{sodium phosphinate}}{3NaH_2PO_2(s)}$$

Concentrated nitric acid reacts with phosphorus to produce phosphoric(V) acid, H_3PO_4.

24.4.3 Compounds of Group V elements

Hydrides

All the Group V elements form hydrides of formula EH_3. These becomes less stable to heat on descending the group as the E—H bond gets weaker.

AMMONIA, NH_3

This is a gas at room temperature ($T_b = 240$ K) consisting of pyramid-shaped molecules of bond angle approximately 107°, see **section 10.3.3** and **Figure 24.20.** Because of its lone pair of electrons, it acts as a weak base by accepting a proton. Its solution in water is sometimes called ammonium hydroxide.

Figure 24.20 The ammonia molecule

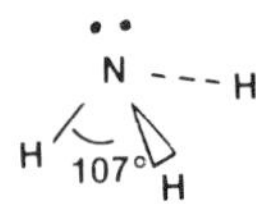

$$NH_3(g) + H_2O(l) \rightleftharpoons NH_4^+(aq) + OH^-(aq)$$

The high electronegativity of nitrogen means that ammonia can form hydrogen bonds and is therefore very soluble in water.

In the laboratory, ammonia is prepared by the action of strong bases on ammonium salts:

$$OH^-(aq) + NH_4^+(aq) \longrightarrow H_2O(l) + NH_3(g)$$

Industrially it is prepared in large quantities by direct reaction of nitrogen and hydrogen (see **section 11.5.1**).

$$N_2(g) + 3H_2(g) \rightleftharpoons 2NH_3(g)$$

Ammonia can be oxidized, for example:

$$\underset{\text{ammonia}}{2NH_3(g)} + \underset{\text{copper oxide}}{3CuO(s)} \longrightarrow \underset{\text{nitrogen}}{N_2(g)} + \underset{\text{water}}{3H_2O} + \underset{\text{copper}}{3Cu(s)}$$

It burns in oxygen:

$$\underset{\text{ammonia}}{4NH_3(g)} + \underset{\text{oxygen}}{3O_2(g)} \longrightarrow \underset{\text{nitrogen}}{2N_2(g)} + \underset{\text{water}}{6H_2O(l)}$$

With a catalyst such as platinum, an alternative oxidation is favoured:

$$\underset{\text{ammonia}}{4NH_3(g)} + \underset{\text{oxygen}}{5O_2(g)} \xrightarrow[\text{catalyst}]{\text{platinum}} \underset{\text{nitrogen monoxide}}{4NO(g)} + \underset{\text{water}}{6H_2O(l)}$$

This reaction is the basis of the industrial process for the manufacture of nitric acid, see box.

Another hydride of nitrogen exists: hydrazine, N_2H_4. It burns vigorously in air to form steam and nitrogen with the evolution of much heat. It has been used as a rocket fuel as in the German Second World War Messerschmitt 163 rocket fighter, see photograph.

Messerschmitt Me-163 which ran on hydrazine

$$N_2H_4(l) + O_2(g) \longrightarrow N_2(g) + 2H_2O(g)$$

E

LIQUID AMMONIA

Liquid ammonia is a good solvent. Among other things, it will dissolve the alkali metals to give blue solutions containing the metal ions and solvated electrons—electrons surrounded by ammonia molecules. These solutions readily donate their electrons and are therefore good reducing agents.

Acid–base chemistry can take place in liquid ammonia. Acids are species which produce ammonium ions, NH_4^+, and bases are species which produce amide ions, NH_2^-. These ions are analogous to H_3O^+ and OH^- in aqueous systems. Ammonia ionizes:

$$2NH_3 \rightleftharpoons NH_4^+ + NH_2^-$$

This is analogous to the ionization of water:

$$2H_2O \rightleftharpoons H_3O^+ + OH^-$$

A

NITRIC ACID MANUFACTURE (THE OSTWALD PROCESS)

The annual UK production of nitric acid is around 800 000 tonnes, around 90% of which is used in the manufacture of nitrate fertilizers such as ammonium nitrate, trade name 'Nitram'. Other uses include the manufacture of explosives for quarrying, silver nitrate for photography and the production of fibres like nylon and polyester.

The raw materials are ammonia (produced by the Haber process) and air. The air oxidizes the ammonia to nitrogen monoxide if a catalyst of platinum/rhodium gauze is used:

$$4NH_3 + 5O_2 \xrightarrow[\text{catalyst}]{\text{Pt/Rh}} 4NO + 6H_2O \quad \Delta H^{\ominus} = -909\,\text{kJ mol}^{-1}$$

The conditions used are typically 1100 K and a pressure of 4–10 atmospheres (4×10^2–10^3 kPa). This gives a conversion of about 96%. Too high a temperature results in the ammonia being oxidized to nitrogen rather than nitrogen monoxide.

The hot gas mixture is cooled to almost room temperature. This process has two purposes: firstly, the heat extracted is used to raise steam which can be used to generate electricity or power the compressors of the first stage. Secondly, the equilibria in the second stage move towards product at low temperature. These equilibria are:

$$2NO + O_2 \rightleftharpoons 2NO_2 \quad \Delta H^{\ominus} = -115\,\text{kJ mol}^{-1}$$

and

$$2NO_2 \rightleftharpoons N_2O_4 \quad \Delta H^{\ominus} = -58\,\text{kJ mol}^{-1}$$

so that nitrogen monoxide reacts with more air to form dinitrogen tetroxide. This gas then reacts with water in a tower where the rising stream of gas meets a downward flow of water:

$$3N_2O_4(g) + 2H_2O(l) \longrightarrow 4HNO_3(aq) + 2NO(g)$$

Nitric acid of approximately 60% concentration is produced. This is the concentration required for fertilizer manufacture. 99% nitric acid for explosives manufacture is obtained by using concentrated sulphuric acid as a drying agent to remove the water.

PHOSPHINE, PH_3

Phosphine is a similarly shaped molecule to ammonia. It is an evil-smelling poisonous gas which, despite having a higher relative molecular mass than ammonia, has a lower boiling temperature (185 K). This can be explained by the fact that phosphine does not form hydrogen bonds (see **section 16.2.3**). This also explains phosphine's low solubility in water. Like ammonia, phosphine can accept a proton to form phosphonium salts.

Oxygen compounds

Table 24.6 shows the oxides of nitrogen and phosphorus. The more important ones are marked*.

Table 24.6 The oxides of nitrogen and phosphorus

Oxidation number of Group V element	Nitrogen	Phosphorus
+I	*$N_2O(g)$	–
+II	*$NO(g)$	–
+III	$N_2O_3(g)$	*$P_4O_6(s)$
+IV	*$2NO_2(g) \rightleftharpoons N_2O_4(g)$	$PO_2(g)$
+V	$N_2O_5(s)$	$P_4O_{10}(s)$

Oxides of nitrogen

DINITROGEN OXIDE (NITROGEN(I) OXIDE, N_2O)

This is a sweet-smelling, colourless gas which is used as an anaesthetic, commonly known as 'laughing gas'. It is sometimes used as an aerosol propellant gas, especially in whipped cream.

It is slightly soluble in water, giving a neutral solution. Although it is relatively unreactive, it decomposes to its elements on heating. The heat of a glowing taper is sufficient to do this and the oxygen produced will relight the taper, so the gas can be mistaken for oxygen.

Whipped cream aerosols use dinitrogen oxide as a propellant

Dinitrogen oxide is prepared by heating a mixture of sodium nitrate and ammonium sulphate. These react to give ammonium nitrate which immediately decomposes to dinitrogen oxide and water:

$$(NH_4)_2SO_4(s) + 2NaNO_3(s) \longrightarrow Na_2SO_4(s) + 2NH_4NO_3(s)$$
$$NH_4NO_3(s) \longrightarrow N_2O(g) + 2H_2O(l)$$

Heating solid ammonium nitrate (as opposed to the mixture of two salts) is not advised as it can explode if the heating is not carefully controlled.

The molecule is asymmetrical, N—N—O rather than N—O—N, and linear, indicating that the central nitrogen has no lone pairs. Two structures can be written:

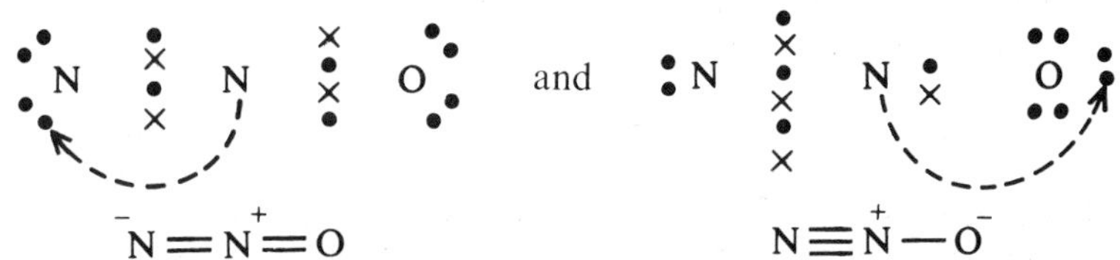

Both involve the central nitrogen giving an electron away, hence the charges. The actual molecule has an N—N bond length half-way between that expected for N=N and N≡N and an N—O bond length half-way between that for N=O and N—O. So it is often represented as N$\overset{...}{=}$N$\overset{...}{-}$O, a sort of average of the two structures. This illustrates that dot cross diagrams are only an aid to understanding bonding and should not be interpreted too literally.

NITROGEN MONOXIDE (NITROGEN(II) OXIDE, NO)

This is a colourless neutral gas which is insoluble in water. It can be prepared by the reaction of copper with a mixture of 50% concentrated nitric acid, 50% water:

$$\underset{\text{nitric acid}}{8HNO_3(aq)} + \underset{\text{copper}}{3Cu(s)}$$
$$\longrightarrow \underset{\text{copper nitrate}}{3Cu(NO_3)_2(aq)} + \underset{\text{water}}{4H_2O(l)} + \underset{\text{nitrogen monoxide}}{2NO(g)}$$

or by sodium nitrite (sodium nitrate(III)) and sodium iodide in sulphuric acid:

A

EXPLOSIVES

What is an explosion and why do things explode? An explosion is essentially a rapid exothermic reaction producing large quantities of gases. A good example of an explosive is nitroglycerine (propane-1,2,3,-triyl trinitrate). The large entropy increase associated with the production of 29 moles of gas coupled with the negative value of $\Delta H^{\ominus}$ means $\Delta G^{\ominus}$ is negative. The large amount of heat given off speeds up the reaction and also means the gases expand. Notice also that no oxygen is needed for the explosion of nitroglycerine—the molecule itself has more than enough. Most other explosives need oxygen from the air or oxidizing agents such as sodium chlorate(V), $NaClO_3$.

$$4\ H-\underset{ONO_2}{\overset{H}{C}}-\underset{ONO_2}{\overset{H}{C}}-\underset{ONO_2}{\overset{H}{C}}-H\ (s) \longrightarrow$$
$$12CO_2(g) + 10H_2O(g) + 6N_2(g) + O_2(g)$$
$$\Delta H^{\ominus} = 27\ 890\ kJ\ mol^{-1}$$

Nitroglycerine is so sensitive that a small shock will detonate it. This makes it too dangerous for routine use, but if it is absorbed in a porous form of silica called kieselguhr, it becomes much safer to handle. This was discovered in 1867 by Alfred Nobel who called the mixture dynamite. This was the basis of Nobel's fortune which enabled him to establish the Nobel prizes. With sad irony, shortly before his discovery of dynamite, Nobel's factory which manufactured nitroglycerine exploded, killing Nobel's younger brother.

Another common explosive is trinitrotoluene (methyl-2,4,6-trinitrobenzene) (TNT). Many nitrogen-containing compounds are explosives. One factor is the large bond energy of N≡N. A reaction which produces unreactive nitrogen gas gives out energy. It is an exothermic gas-producing reaction and likely to be explosive.

TNT (benzene ring with CH_3, O_2N, NO_2, NO_2)

$2NaNO_2(aq)$	+	$2NaI(aq)$	+	$4H_2SO_4(aq)$
sodium nitrite		sodium iodide		sulphuric acid

$\longrightarrow$	$I_2(aq)$	+	$4NaHSO_4(aq)$	+	$2H_2O(l)$	+	$2NO(g)$
	iodine		sodium hydrogensulphate		water		nitrogen monoxide

Nitrogen monoxide reacts immediately on exposure to air to give nitrogen dioxide:

$$2NO(g) + O_2(g) \longrightarrow 2NO_2(g)$$

The nitrogen monoxide molecule has an unpaired electron:

$\dot{N}{=}O$ or $N \rightleftarrows O$

Loss of this electron produces the NO^+ ion which has the same electron arrangement as nitrogen.

NITROGEN DIOXIDE (NITROGEN(IV) OXIDE, NO_2)

This is a brown gas. It is formed as part of a complex series of reactions in photochemical smogs and it is responsible for their colour.

Nitrogen dioxide exists in equilibrium with its colourless dimer dinitrogen tetroxide.

$$2NO_2(g) \rightleftharpoons N_2O_4(g) \quad \Delta H^{\ominus} = -58\,\text{kJ mol}^{-1}$$

Since the reaction is exothermic, lowering the temperature moves the equilibrium to the right.

Nitrogen dioxide can be prepared by heating lead nitrate:

$2Pb(NO_3)_2(s)$	$\longrightarrow$	$2PbO(s)$	+	$4NO_2(g)$	+	$O_2(g)$
lead nitrate		lead(II) oxide		nitrogen dioxide		oxygen

The nitrogen dioxide can be separated out by cooling the gases in an ice/salt freezing mixture. The nitrogen dioxide condenses to a liquid at 262 K (−11 °C).

An alternative preparation is from copper and concentrated nitric acid:

$Cu(s)$	+	$4HNO_3(aq)$	$\longrightarrow$	$Cu(NO_3)_2(aq)$	+	$2NO_2(g)$	+	$2H_2O(l)$
copper		nitric acid		copper nitrate		nitrogen dioxide		water

Nitrogen dioxide dissolves in water to give a mixture of nitric(V) acid (nitric acid) and nitric(III) acid (nitrous acid):

$2NO_2(g)$	+	$H_2O(l)$	$\longrightarrow$	$HNO_3(aq)$	+	$HNO_2(aq)$
nitrogen dioxide		water		nitric(V) acid		nitric(III) acid (nitrous acid)

The molecule is angular with bond angle 134°:

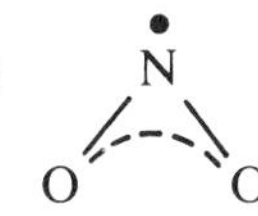

Again, as in N_2O, two structure can be written:

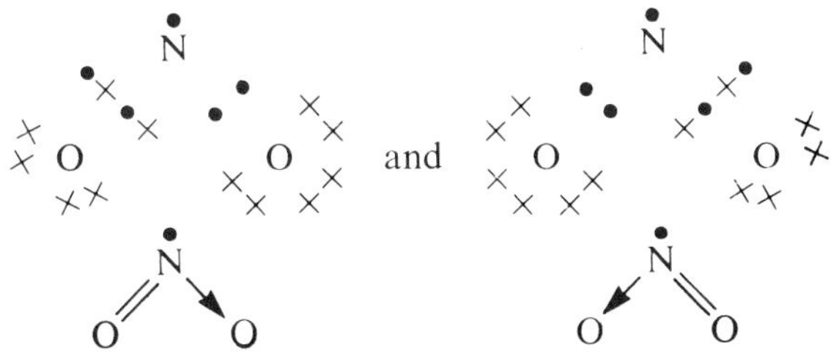

The nitrogen has a single non-bonding electron which repels the groups of bonded electrons less than a lone pair would. Thus the O—N—O angle is greater than 120°. An alternative description of the bonding is to imagine a delocalized π orbital spreading over all three atoms but containing just two electrons:

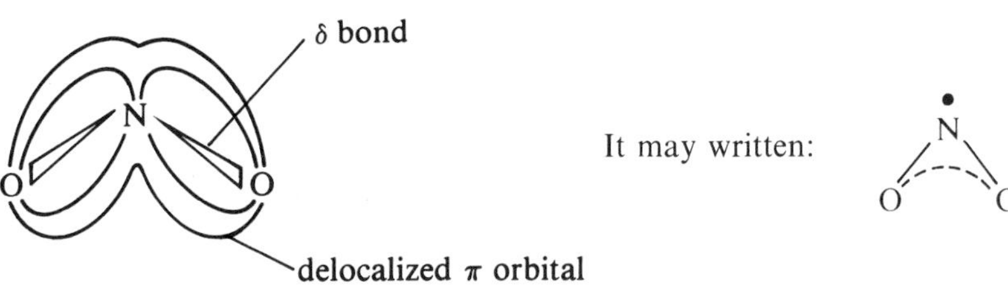

On dimerising to N_2O_4, the unpaired electrons on each NO_2 form the bond:

Oxoacids of nitrogen

Nitrogen has two important oxoacids, nitric(III) acid (nitrous acid, HNO_2) and nitric(V) acid (nitric acid, HNO_3).

NITRIC(III) ACID (NITROUS ACID, HNO_2)

The full systematic name is dioxonitric(III) acid. This is a weak acid, $pK_a = 3.3$.

$$\underset{\text{nitric(III) acid}}{HNO_2(aq)} \rightleftharpoons \underset{\text{hydrogen ion}}{H^+(aq)} + \underset{\text{nitrate (III) ion (nitrite ion)}}{NO_2^-(aq)}$$

It cannot be isolated pure from solution, as it disproportionates to nitric(V) acid and nitrogen monoxide:

$$\underset{\text{nitric(III) acid}}{3H\overset{+III}{N}O_2(aq)} \longrightarrow \underset{\text{nitric(V) acid}}{H\overset{+V}{N}O_3(aq)} + \underset{\text{nitrogen monoxide}}{2\overset{+II}{N}O(g)} + \underset{\text{water}}{H_2O(l)}$$

It is therefore prepared, as required, by the reaction of hydrochloric acid with sodium nitrate(III) (sodium nitrite).

$$\underset{\text{hydrochloric acid}}{HCl(aq)} + \underset{\text{sodium nitrate(III)}}{NaNO_2(aq)} \longrightarrow \underset{\text{nitric(III) acid}}{HNO_2(aq)} + \underset{\text{sodium chloride}}{NaCl(aq)}$$

Nitric(III) acid and the nitrate(III) (nitrite) ion NO_2^- are oxidizing agents, oxidizing iodide ions to iodine and Fe^{2+} ions to Fe^{3+} ions:

$$2\overset{-I}{I}(aq) + 4H^+(aq) + 2\overset{+III}{N}O_2^-(aq) \longrightarrow \overset{0}{I_2}(aq) + 2H_2O(l) + 2\overset{+II}{N}O(g)$$

$$\overset{+II}{Fe^{2+}}(aq) + 2H^+(aq) + \overset{+III}{N}O_2^-(aq) \longrightarrow \overset{+III}{Fe^{3+}}(aq) + H_2O(l) + \overset{+II}{N}O(g)$$

Nitric(III) acid is used to form diazonium salts for manufacturing dyes (see box **section 34.2.5**) and nitrate(III) salts (nitrites) are added to cooked meats as preservatives (sodium nitrite is E250) although some adverse effects have been reported. Nitrites can produce traces of nitrosamines, **Figure 24.21**, in the stomach and these have been shown to cause cancer in some animals, though not humans.

Figure 24.21 **A nitrosamine**

NITRIC(V) ACID (NITRIC ACID, HNO_3)

The full systematic name is trioxonitric(V) acid. This is a strong acid; the dissociation:

$$\underset{\text{nitric(V) acid}}{HNO_3(aq)} \longrightarrow \underset{\text{hydrogen ion}}{H^+(aq)} + \underset{\text{nitrate(V) ion}}{NO_3^-(aq)}$$

is practically complete in dilute aqueous solution. Nitric acid is also a good oxidizing agent. This factor complicates its reactions with metals. Dilute nitric acid reacts with magnesium to give a salt and hydrogen as would be expected for an acid:

$$\overset{0}{Mg(s)} + \overset{+I}{2HNO_3(aq)} \longrightarrow \overset{+II}{Mg(NO_3)_2(aq)} + \overset{0}{H_2(g)}$$

magnesium, nitric(V) acid, magnesium nitrate(V), hydrogen

Here the magnesium has been oxidized by the H^+ ions, and the oxidation number of the nitrogen is unchanged. In more concentrated acid and with less reactive metals, the nitrogen is reduced to one of its oxides, for example with 50% acid (approximately $10\,mol\,dm^{-3}$):

$$\overset{0}{3Cu(s)} + \overset{+V}{8HNO_3(aq)} \longrightarrow$$

copper, nitric(V) acid

$$\overset{+II\ +V}{3Cu(NO_3)_2(aq)} + 4H_2O(aq) + \overset{+II}{2NO(g)}$$

copper nitrate(V), water, nitrogen monoxide

With concentrated acid:

$$\overset{0}{Cu(s)} + \overset{+V}{4HNO_3} \longrightarrow \overset{+II\ +V}{Cu(NO_3)_2(aq)} + 2H_2O(l) + \overset{+IV}{2NO_2(g)}$$

copper, nitric(V) acid, copper nitrate(V), water, nitrogen dioxide

Concentrated nitric acid can also oxidize non-metal elements, for example:

$$\overset{0}{S} + \overset{+V}{6HNO_3} \longrightarrow$$

sulphur, nitric(V) acid

$$\overset{+VI}{H_2SO_4} + \overset{+IV}{6NO_2} + 2H_2O$$

sulphuric acid, nitrogen dioxide, water

$$\overset{0}{P} + \overset{+V}{5HNO_3} \longrightarrow$$

phosphorus, nitric(V) acid

$$\overset{+V}{H_3PO_4} + \overset{+IV}{5NO_2} + H_2O$$

phosphoric(V) acid, nitrogen dioxide, water

and also compounds, for example:

$$\overset{-II}{3H_2S} + \overset{+V}{2HNO_3} \longrightarrow 4H_2O + \overset{+II}{2NO} + \overset{0}{3S}$$

hydrogen sulphide, nitric(V) acid, water, nitrogen monoxide, sulphur

Metal nitrates are all soluble in water because they have low lattice energies due to the large size and single charge of the nitrate ion.

Two tests are available for the nitrate ion. Firstly, it is reduced to ammonia by **Devarda's alloy** (45% aluminium, 5% zinc, 50% copper) in alkaline solution:

$$4Zn(s) + NO_3^-(aq) + 7OH^-(aq) + 6H_2O(l) \longrightarrow 4Zn(OH)_4^{2-}(aq) + NH_3(g)$$

Secondly, the **brown ring test:** iron(II) sulphate is added followed by careful addition of sulphuric acid which forms a separate lower layer. At the junction, a brown ring appears caused by the ion $(Fe(H_2O)_5NO)^{2+}$

In the laboratory, nitric acid is made by the action of concentrated sulphuric acid on a nitrate salt:

$$NO_3^- + H_2SO_4 \longrightarrow HNO_3 + HSO_4^-$$

followed by distilling off the fairly volatile nitric acid ($T_b = 356$ K). Industrially the acid is made by oxidation of ammonia (see box, page 322).

Bonding in nitric acid and the nitrate ion

The bonding in nitric acid and the nitrate ion is discussed in **section 10.2.2.**

Oxygen compounds of phosphorus

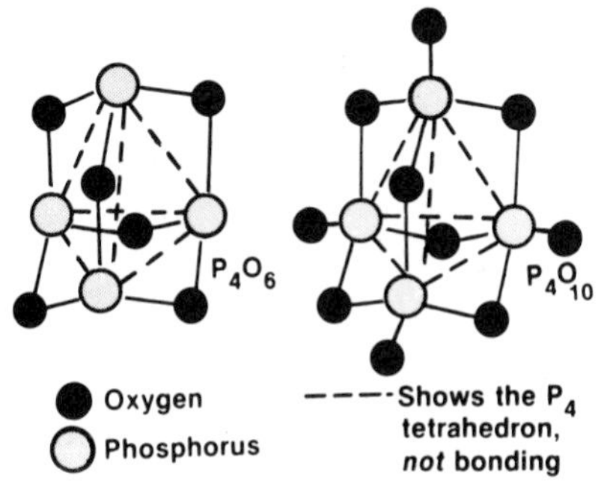

Figure 24.22 Structures of phosphorus(III) oxide and phosphorus(V) oxide

Phosphorus forms two important oxides—phosphorus(III) oxide and phosphorus(V) oxide. The names phosphorus trioxide and phosphorus pentoxide are not appropriate as the molecular formulae are P_4O_6 and P_4O_{10} respectively. The gas phase molecules have the structure shown in **Figure 24.22.** Each has a tetrahedron of phosphorus atoms, bridged by oxygen atoms. In P_4O_{10} in addition, each phosphorus atom is bonded to another, non-bridging oxygen atom. In P_4O_6 each phosphorus atom is forming three single covalent bonds. In P_4O_{10} each phosphorus atom forms three single covalent bonds and a double bond to the extra oxygen, i.e. it has 10 electrons in its outer shell. This is due to the presence of 3d orbitals which are only a little higher in energy than 3p orbitals. An electron can be promoted. This is accomplished by the promotion of one of the 3s electrons into a 3d orbital leaving five unpaired electrons for sharing with the other atoms to form bonds. The energy required for promotion is 1320 kJ mol^{-1}. In nitrogen, promotion of an electron to the lowest energy unoccupied orbital (in this case 3s) requires 2900 kJ mol^{-1} and is therefore much less likely, **Figure 24.23.** Nitrogen forms no compounds with five covalent bonds; phosphorus forms several.

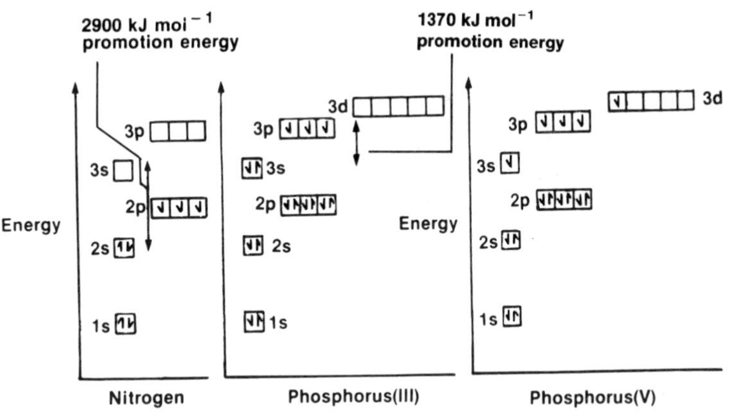

Figure 24.23 The electron arrangements in nitrogen and phosphorus in the +III and +V oxidation states

The oxides can be formed by burning phosphorus either in a limited supply of oxygen, when phosphorus (III) oxide is formed:

$$\underset{\text{phosphorus}}{P_4(s)} + \underset{\text{oxygen}}{3O_2(g)} \longrightarrow \underset{\text{phosphorus(III) oxide}}{P_4O_6(s)}$$

or in excess oxygen when phosphorus(V) oxide is formed:

$$P_4(s) + 5O_2(g) \longrightarrow \underset{\text{phosphorus(V) oxide}}{P_4O_{10}(s)}$$

The main use of phosphorus(V) oxide is as a drying agent. It will react with water to form one or more of the phosphoric acids depending on the amount of water. Not only will it remove water from mixtures but it will also remove water from compounds, dehydrating, for example, amides to nitriles (see **section 33.8.3**).

$$CH_3C(=O)NH_2 \xrightarrow[\text{[}P_4O_{10}\text{]}]{-H_2O} CH_3-C\equiv N$$

ethanamide → ethanenitrile

Table 24.7 Oxoacids of phosphorus

Phosphinic acid	H_3PO_2 monoprotic	O=P(H)(H)(HO)
Phosphonic acid	H_3PO_3 diprotic	O=P(H)(HO)(OH)
phosphoric(V) acid (orthophosphoric acid)	H_3PO_4 triprotic	O=P(OH)(HO)(OH)

Oxoacids of phosphorus

There are several of these, summarized in **Table 24.7.** The acidic hydrogen atoms are shown in brown.

Polymeric phosphoric acids are also formed, such as diphosphoric(V) acid, which may be thought of as two phosphoric(V) acid molecules joined via an oxygen atom after a molecule of water has been eliminated:

$$(HO)_2P(=O)-O-P(=O)(OH)_2$$

diphosphoric (V) acid

These are not unlike the silicates in structure. The most important oxoacid is phosphoric(V) acid, a crystalline solid which forms syrupy solutions in water due to the large amount of hydrogen bonding possible between the acid and water.

The acid is triprotic. The salt sodium dihydrogenphosphate(V) can be made with the same number of moles of sodium hydroxide as acid and disodium hydrogen phosphate(V) by reaction with two moles of sodium hydroxide with one mole of acid:

$$H_3PO_4(aq) + NaOH(aq) \longrightarrow NaH_2PO_4(s) + H_2O(l)$$

phosphoric(V) acid + sodium hydroxide → sodium dihydrogenphosphate(V) + water

$$H_3PO_4(aq) + 2NaOH(aq) \longrightarrow Na_2HPO_4(s) + 2H_2O(l)$$

disodium hydrogenphosphate(V)

Trisodium phosphate cannot be formed this way as the PO_4^{3-} ion is a strong base and is hydrolysed by water:

$$PO_4^{3-} + H_2O \longrightarrow HPO_4^{2-} + OH^-$$

phosphate ion + water → hydrogen-phosphate ion + hydroxide ion

Anti-rust jelly contains phosphoric acid

Phosphoric(V) acid can be made in the laboratory by the reaction of phosphorus(V) oxide with water:

$$P_4O_{10}(s) + 6H_2O(l) \longrightarrow 4H_3PO_4(s)$$

phosphorus (V) oxide + water → phosphoric(V) acid

Industrially it is made by the reaction of phosphate rock (calcium phosphate) with sulphuric acid:

$$Ca_3(PO_4)_2 + 3H_2SO_4 \longrightarrow 2H_3PO_4 + 3CaSO_4$$

90% of the phosphoric acid manufactured is used to make fertilizers like diammonium hydrogenphosphate(V):

$$2NH_3 + H_3PO_4 \longrightarrow (NH_4)_2HPO_4$$

ammonia + phosphoric acid → diammonium hydrogenphosphate(V)

Another use of phosphoric acid is the rustproofing of iron and steel—a coating of insoluble iron phosphate is formed. Phosphates, especially sodium tripolyphosphate (see **Figure 24.24**), are used in detergents to act as water softeners, removing calcium and magnesium ions from solution.

Figure 24.24 The tripolyphosphate ion

$$^-O-P(=O)(O^-)-O-P(=O)(O^-)-O-P(=O)(O^-)-O^-$$

Halides of Group V elements

Nitrogen forms halides of formula NX_3 only. The chloride, a liquid, hydrolyses rapidly to give ammonia and chloric(I) acid, by a mechanism which involves the lone pair of the nitrogen initially bonding with a hydrogen on water:

NCl_3	+	$3H_2O$	$\longrightarrow$	NH_3	+	$3HClO$
nitrogen trichloride		water		ammonia		chloric(I) acid

Phosphorus forms trihalides, PX_3, with a similar structure to the nitrogen halides. The most important, PCl_3, can be prepared by direct reaction of the elements:

$P_4(s)$	+	$6Cl_2(g)$	$\longrightarrow$	$4PCl_3(l)$
white phosphorus		chlorine		phosphorus trichloride

PCl_3, too, hydrolyses readily but the first step is the formation of a bond between phosphorus and oxygen using the oxygen's lone pair to form a bond using one of the d orbitals on phosphorus:

$$Cl_3\ddot{P} \longleftarrow :OH_2$$

Figure 24.25 The PCl_5 molecule

The products are different from those produced by nitrogen trichloride, NCl_3:

PCl_3	+	$3H_2O$	$\longrightarrow$	H_3PO_3	+	$3HCl$
phosphorus trichloride		water		phosphonic acid		hydrochloric acid

Phosphorus also forms pentahalides, PX_5, **Figure 24.25.** This is possible due to promotion of an electron from a 3s orbital into a 3d orbital (see **Figure 24.23**). Nitrogen pentachloride does not exist. Phosphorus pentachloride exists in the gas phase as trigonal bipyramidal molecules. The compound easily decomposes into the trichloride and chlorine:

$$PCl_5(g) \rightleftharpoons PCl_3(g) + Cl_2(g)$$

Both phosphorus trichloride and pentachloride are used in organic chemistry to replace —OH with —Cl, for example, carboxylic acids can be converted to acid chlorides:

$$R{-}C(=O){-}OH \xrightarrow[PCl_5]{PCl_3 \text{ or}} R{-}C(=O){-}Cl$$

and alcohols to chloroalkanes:

$$R{-}OH \xrightarrow[PCl_5]{PCl_3 \text{ or}} R{-}Cl$$

Nitrogen and phosphorus are used in fertilizers

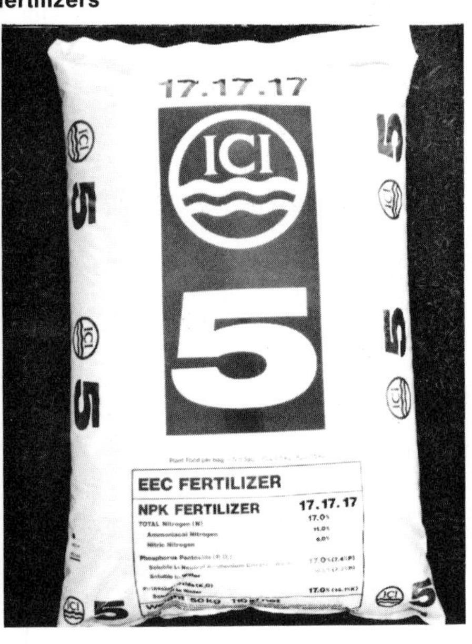

Table 24.8 Fertilizers containing nitrogen and/or phosphorus

Trade name	Formula
Nitram (ammonium nitrate)	NH_4NO_3
Superphosphate of lime	$Ca(H_2PO_4)_2 + CaSO_4$
Triple superphospate	$Ca(H_2PO_4)_2$
Nitrophos	$Ca(H_2PO_4)_2 + Ca(NO_3)_2$
Urea	$CO(NH_2)_2$
Ammonium sulphate	$(NH_4)_2SO_4$
Nitrochalk	$NH_4NO_3 + CaCO_3$

24.4.4 Economic importance of Group V elements and compounds

Both nitrogen and phosphorus are essential for plant growth. Fertilizer manufacture is the major use for compounds of these elements. Ammonia, nitric acid and nitrates are all used to manufacture fertilizers. Some of the more important are listed in **Table 24.8.**

- Nitric(V) acid is used in the manufacture of explosives such as

A

THE NITROGEN CYCLE

This is a summary of how nitrogen is fixed (converted into compounds) and subsequently returns to the atmosphere as molecular nitrogen. Nitrogen is fixed by bacteria in the root nodules of plants such as clover. It is also fixed by conversion to nitrogen oxides (and thence nitric acid) in lightning flashes and in combustion reactions in vehicle engines and power stations. The nitric acid then produces nitrates which may be converted back to nitrogen by soil bacteria or converted to organic nitrogen compounds directly by plants or plants and animals. Ammonia may be produced from these organic compounds by decay of their waste products or by decay of the animals and plants themselves after death. Other bacteria convert ammonia back to nitrogen or (via nitrites) to nitrates. The whole system is in equilibrium, the amount of atmospheric nitrogen remaining constant.

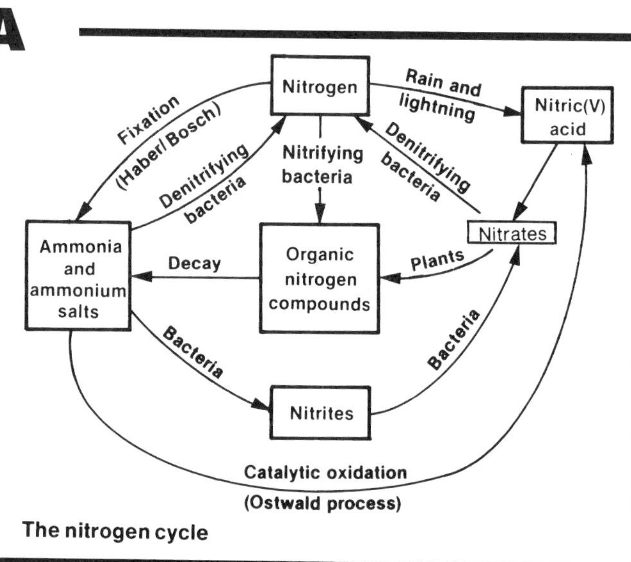

The nitrogen cycle

Phosphates are used in toothpaste

nitroglycerine, dynamite and TNT.

- Nitric(V) acid is also used to manufacture synthetic fibres, e.g. nylon, polyester.
- Phosphoric acid is used as an anti-rust treatment for iron and steel.
- Phosphates are used as 'builders' in washing powders. They buffer the detergent mixture to the correct pH, soften the water and prevent the redeposition of dirt.
- Phosphates are used in toothpastes as abrasives (calcium hydrogenphospate) and a source of fluoride (sodium monofluorophosphate).

24.4.5 Oxidation number charts

Both nitrogen and phosphorus display a variety of oxidation numbers in their compounds. This is in contrast to the s-block elements and those of Groups III and IV.

An oxidation number chart is shown in **Table 24.9** for nitrogen and phosphorus compounds. Only a few compounds are listed and the most important of these are shown in brown.

Table 24.9 Oxidation number chart for compounds of nitrogen and phosphorus

Nitrogen	Ox	Phosphorus
HNO_3, MNO_3	V	P_4O_{10}, H_3PO_4, PCl_5
N_2O_4, NO_2	IV	
HNO_2, MNO_2	III	P_4O_6, H_3PO_3, PCl_3
NO	II	
N_2O	I	
N_2	0	P_4
	−I	
N_2H_4	−II	
NH_3, MN	−III	PH_3, MP

(M is any metal)

24.4.6 Summary (Group V)

- The Group V elements have the outer electron arrangement ns^2np^3.
- There is a clear trend from non-metals to metals on descending the group.
- Nitrogen exists as N_2 molecules due to the great strength of the N≡N bond.
- The large N≡N bond energy makes the N_2 molecule unreactive.
- Phosphorus exists as P_4 white phosphorus molecules or giant structures (the red and black allotropes). Each phosphorus atom forms three single bonds.
- Phosphorus atoms are too large to form effective π bonds by overlap of p orbitals.
- Nitrogen can form three covalent bonds plus one dative covalent bond.
- Phosphorus can form five covalent bonds by promotion of an electron from a 3s orbital to a 3d orbital.
- Nitrogen forms ionic nitrides containing the N^{3-} ion with the more reactive s-block elements.
- There are no compounds containing the P^{3-} ion.
- The hydrides become less stable on descending the group because the E—H bond gets weaker.

• Both ammonia, NH_3, and phosphine, PH_3, are tetrahedral molecules, with a lone pair.
• Ammonia readily accepts a proton (i.e. it is a good base).
• Ammonia can be oxidized to nitrogen monoxide, NO.
• Nitrogen forms three important oxides, N_2O, NO and NO_2.
• Nitrogen dioxide can dimerize to form dinitrogen tetroxide:

$$2NO_2(g) \rightleftharpoons N_2O_4(g)$$

• Nitrogen forms two important oxoacids—nitric(III) acid (HNO_2) and nitric(V) acid (HNO_3).
• Nitric(V) acid is both a strong acid and an oxidizing agent.
• All metal nitrates are soluble.
• Phosphorus forms two oxides—phosphorus(III) oxide, P_4O_6, and phosphorus(V) oxide, P_4O_{10}—both are acidic.
• Phosphorus(V) oxide is used as a dehydrating agent.
• Phosphoric(V) acid, H_3PO_4, is a triprotic acid.
• Nitrogen forms halides NX_3 only, while phosphorus forms both PX_3 and PX_5.

24.4.7 Questions (Group V)

1 Nitrogen forms N_2 (N≡N) and phosphorus P_4 (tetrahedron).

A) How many P—P bonds are there in a phosphorus molecule?

B) Use bond energies to calculate $\Delta_R H^\ominus$ for:

$$2N_2 \longrightarrow N_4$$

Assume the hypothetical N_4 molecule has the same structure as P_4.

C) Does your answer explain why N_4 does not exist?

D) Use a similar strategy to explain why P_2 (P≡P) does not exist.

Bond energies/kJ mol^{-1}			
N≡N	945	N—N	158
P≡P	485	P—P	198

2 Name and give the formulae of three oxides of nitrogen. How can each be prepared?

3 Nitrogen forms only one chloride, NCl_3, while phosphorus forms two, PCl_3 and PCl_5. Describe the bonding in each of these compounds by means of dot cross diagrams and predict the shape of each molecule. Explain carefully why phosphorus can form PCl_5 while NCl_5 is unknown.

4 In compounds with the s-block elements, nitrides containing the N^{3-} ion exist. Why do compounds not form containing P^{3-} ions even though the sum of the first three ionization energies of phosphorus is less than the sum of the first three ionization energies for nitrogen?

5 **A)** Give the atomic number of nitrogen and the electron configuration of its isolated atom.

B) For nitrogen molecules, N_2, nitrate ions, NO_3^-, and ammonium ions, NH_4^+ respectively:

i State the oxidation number of the nitrogen;

ii Show the structural formula, including the outer electrons;

iii Describe, in words, or by a diagram, the shape of each ion.

(London 1980, part question)

6 The element antimony (atomic number = 51) forms compounds in two oxidation states, +III and +V. Instructions for the preparation of the antimony compound of composition SbI_3 are as follows:

1 Add 0.60 g of antimony powder to 40 cm^3 benzene in a 100 cm^3 round-bottom flask fitted with a reflux condenser.

2 Heat on a water-bath until the benzene is boiling and add 2.0 g of iodine in small portions.

3 Reflux for 20 minutes.

4 Allow to cool, when red crystals of the product will deposit from solution.

5 Filter the crystals and wash them with a little tetrachloromethane.

6 Allow to dry at room temperature, then weigh. The pure crystals melt at 170 °C.

A) Write down the ground-state electronic structure of an isolated antimony atom.

B) In what group of the Periodic Table is antimony?

C) Write down the formula for the oxide formed by antimony in its +V oxidation state.

D) In the preparation of SbI_3 described above, show clearly by means of a calculation that iodine was present in excess (relative atomic masses: Sb = 122, I = 127).

E) If the mass of pure SbI_3 crystals obtained was 2.06 g, what was the percentage yield?

F) What was the purpose of washing the crystals with a little tetrachloromethane (step 5, above)?

G) **i** The low melting temperature of the crystals

suggests that SbI_3 has a molecular structure. What other evidence is there in the description on the previous page to support this conclusion?

ii Which simple compound of antimony is most likely to contain Sb^{3+} ions? Give its systematic chemical name or its formula.

H The compound $SbCl_5$ is a liquid which boils at 140 °C. A sample of $SbCl_5$ weighing 3.00 g was completely vaporized at its boiling temperature and the volume of vapour obtained, converted to s.t.p., was 285 cm^3 (molar volume of a gas at s.t.p. = 22 400 cm^3 mol^{-1}).

i From these data, calculate a value for the relative molecular mass of $SbCl_5$ in the vapour state.

ii How does the value obtained in (h) (i) compare with the expected value? What explanation can you offer for any discrepancy?

(Relative atomic masses Sb = 122, Cl = 35.5)

(Nuffield 1985)

24.5 Group VI

We shall look at only the top two members of the group—oxygen and sulphur—in any detail. We can see the usual trend from non-metals to metals on descending the group. Oxygen and sulphur are clearly non-metals, selenium and tellurium show some properties of both metals and non-metals while polonium is clearly metallic. Typically, the elements either form E^{2-} ions or bond covalently, although there is some evidence for Te^{4+} and Po^{4+} (not 6^{+}, the inert pair effect operates).

Some physical properties are listed in **Table 24.10.**

Table 24.10 Properties of Group VI elements

	Atomic number, Z	Electron arrangement	Electro-negativity	Atomic (covalent) radius/nm	Ionic radius E^{2-}/nm	T_m/K	T_b/K
O	8	$1s^22s^22p^4$	3.5	0.073	0.140	55	90
S	16	[Ne] $3s^23p^4$	2.5	0.102	0.185	386(rh)	
						392 (mon)	718
Se	34	[Ar] $3d^{10}4s^24p^4$	2.4	0.117	0.195	490	958
Te	52	[Kr] $4d^{10}5s^25p^4$	2.1	0.135	0.220	723	1263
Po	84	[Xe] $4f^{14}5d^{10}6s^26p^4$				527	1235

rh = rhombic sulphur, mon = monoclinic sulphur } two allotropes

24.5.1 The elements

Oxygen (dioxygen) exists as a diatomic gas which comprises approximately 20% of the atmosphere. The atoms are held together by a double bond of bond energy 498 kJ mol^{-1}. Oxygen has an allotrope (trioxygen), O_3, usually called ozone, see box.

Sulphur, too, has allotropes described by their crystal shapes as rhombic and monoclinic (needles), **Figure 24.26.** Both allotropes are made up of different arrangements of S_8 molecules. These are eight-membered puckered rings with S—S single bonds and a bond angle of 105°, **Figure 24.27**—just what would be predicted, as each sulphur forms two covalent bonds and has two lone pairs. The comparison between O_2 and S_8 is similar to the situation in Group V where nitrogen forms triply bonded N_2 molecules and phosphorus singly bonded P_4 units. Rhombic sulphur is the more stable allotrope at below 369 K, monoclinic above this temperature.

Figure 24.26 Crystals of (a) rhombic and (b) monoclinic sulphur

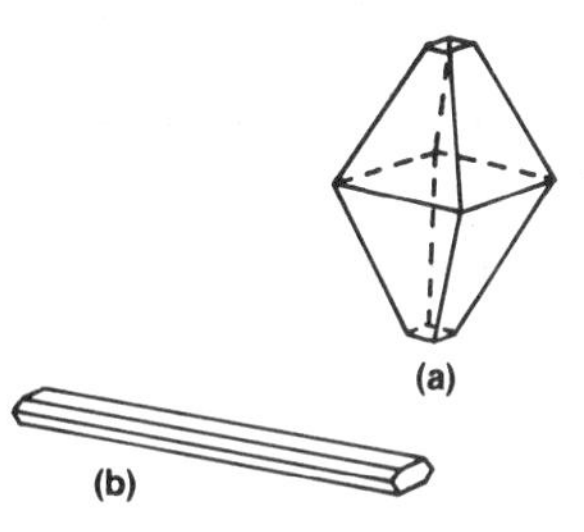

Figure 24.27 The structure of the S_8 molecule – a 'crown shaped' ring

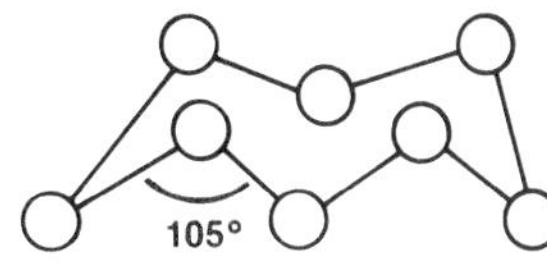

24.5.2 Chemical reactivity of the elements

Oxygen forms compounds with all elements except the noble gases and combines with most of them directly. Since it is the most electronegative element except for fluorine, it always has a negative oxidation number except in compounds with fluorine.

A

OZONE

Oxygen has an allotrope ozone (trioxygen), O_3. This is an angular molecule, bond angle 117°. It has delocalized π

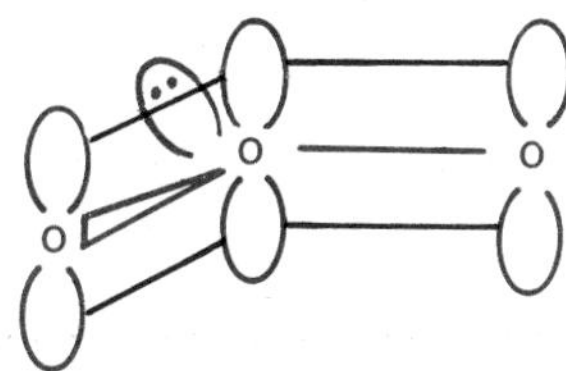

bonding. Ozone is unstable with respect to oxygen. It can be prepared by passing oxygen through an electric discharge and its characteristic smell can often be detected near high-voltage electrical apparatus such as some early models of photocopier. Contrary to the old-fashioned idea that ozone contributed to the bracing atmosphere of some seaside resorts, the gas is poisonous. Paradoxically, it is also vital to life on earth. A layer of the atmosphere at about 25 km contains ozone made by photochemical reactions. This absorbs damaging ultraviolet radiation and prevents it reaching the earth's surface. There is concern that pollutants are reducing the amount of ozone in this layer with far-reaching effects, see box *Atmospheric chemistry*, **section 15.4.**

A

SOURCES OF SULPHUR

Sulphur is found uncombined in underground deposits in various parts of the world, notably the USA and Poland. It is also found combined as sulphates and sulphides. A third important source is in sulphur compounds, particularly hydrogen sulphide, present in crude oil and natural gas. All these sources of sulphur are used commercially. 90% of the sulphur produced is used in the manufacture of sulphuric acid (see box, **section 9.6.2**). Vulcanization of rubber is another important use.

Uncombined sulphur deposits underground are 'mined' by the Frasch process. A hole is drilled through the earth into the sulphur deposit, typically 150 m down. A set of three concentric pipes is lowered into the hole. Water heated under pressure to around 420 K is pumped down the outer pipe. This melts the sulphur ($T_m = 393$ K). The molten sulphur is forced to the surface up the middle pipe under pressure of compressed air which is pumped down the central pipe. Nowadays the sulphur is often stored and transported as a liquid which is easier to handle than the solid.

Sulphur is recovered from crude oil and natural gas by first reacting the sulphur compounds with hydrogen to convert them into hydrogen sulphide. This is then removed by dissolving it in 2-aminoethanol. The hydrogen sulphide is then burned in a carefully controlled amount of air to produce sulphur:

$$2H_2S(g) + O_2(g) \longrightarrow 2H_2O(l) + 2S(s)$$

Some sulphur dioxide is produced but this can be reacted with hydrogen sulphide to recover the sulphur:

$$2H_2S(g) + SO_2(g) \longrightarrow 2H_2O(l) + 3S(s)$$

Removing the sulphur from natural gas and crude oil is important environmentally as well as economically. Sulphur compounds usually have unpleasant smells and are converted into sulphur oxides when the fuel is burned. Sulphur oxides are converted into sulphuric acid in the atmosphere and are a major cause of acid rain.

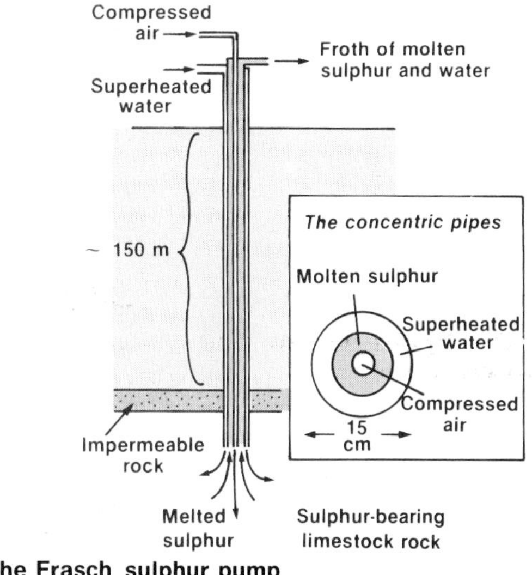

The Frasch sulphur pump

Normally, oxygen's oxidation number in compounds is −II, but in peroxides (containing O_2^{2-}) it has Ox = −I and in superoxides (containing O_2^-), it has Ox = $-\frac{1}{2}$. Oxygen forms ionic oxides containing O^{2-} as well as covalent oxides.

Sulphur combines directly with most metals to form ionic sulphides containing the S^{2-} ion. Sulphur will also combine with many non-metals and can form positive oxidation states up to +VI. This is made possible by the promotion of electrons from 3p to 3d and 3s to 3d orbitals, allowing a maximum of six covalent bonds to be formed, **Figure 24.28.**

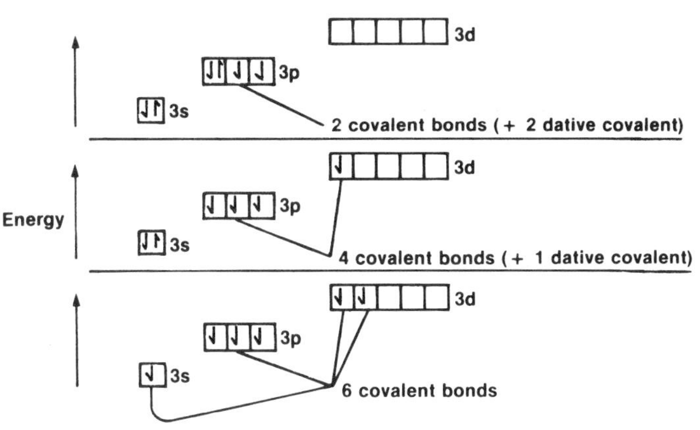

Figure 24.28 Promotion of electrons in sulphur allows up to six covalent bonds to form

24.5.3 *Compounds of Group VI elements*

Hydrides

Oxygen forms two hydrides—water, H_2O, and hydrogen peroxide, H_2O_2 Both are molecular with the structures shown in **Figure 24.29.** The angle are approximately as predicted by electron pair repulsion theory.

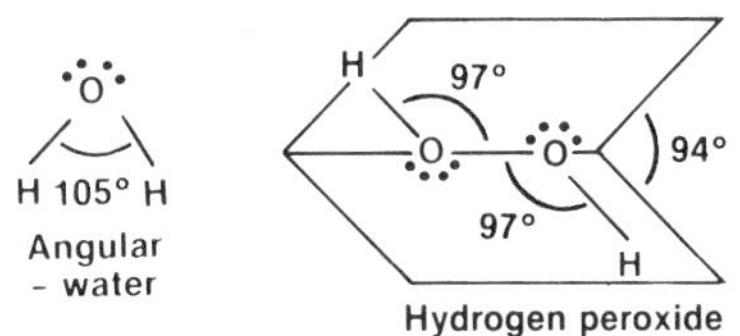

Figure 24.29 The structures of water and hydrogen peroxide

WATER

Many of the physical properties are unusual. Its melting and boilin temperatures are unusually high (273 K and 373 K respectively compare with 188 K and 212 K for H_2S which has a larger relative molecular mass These can be explained by hydrogen bonding resulting from the hig electronegativity of the oxygen atom, see **sections 16.2.3** and **16.2.4.** Wate is able either to donate or to accept a proton and thus show both acidi and basic properties so it may be called an **amphoteric** oxide. Wate donates a proton to ammonia and acts as an acid:

$$\underset{\text{acid}}{H_2O} + \underset{\text{base}}{NH_3} \longrightarrow OH^- + NH_4^+$$

Water accepts a proton from hydrogen chloride and therefore acts as base:

$$\underset{\text{base}}{H_2O} + \underset{\text{acid}}{HCl} \longrightarrow H_3O^+ + Cl^-$$

Water is a good solvent for ionic and polar compounds due to its polarit and ability to form hydrogen bonds:

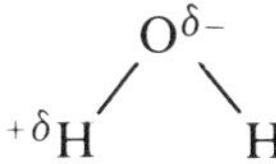

Water is able to hydrate ions, see **section 23.2.5.** This is an exothermi process which counterbalances the large endothermic lattice breakin process which must occur whenever ionic compounds dissolve.

HYDROGEN PEROXIDE

Hydrogen peroxide is weakly acidic. The pure compound is a pale blu liquid. It readily decomposes into water and oxygen:

$$2H_2O_2(l) \longrightarrow 2H_2O(l) + O_2(g)$$

The reaction is catalysed by many substances, including manganese(IV oxide, lead(IV) oxide, iron filings and enzymes present in blood, potat etc. This reaction occurs readily. The O—O bond is weak (144 kJ mol^{-1} the reaction is exothermic and as it involves the formation of a gas fro a liquid, it has a positive entropy change. The reaction is a disproportio nation.

VULCANIZATION OF RUBBER

Disulphur dichloride, S_2Cl_2, is used in the rubber industry in the process of **vulcanization**. Here sulphur atoms link hydrocarbon chains which are thus prevented from being pulled apart when the rubber is stretched. Vulcanized rubber is much harder and more elastic than untreated rubber and is suitable for applications such as car tyres. Vulcanization was discovered by chance in 1839 by Charles Goodyear, founder of the tyre firm.

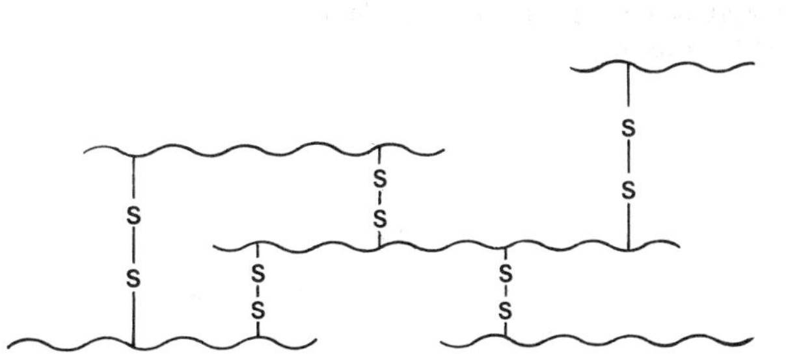

USES OF HYDROGEN PEROXIDE

Hydrogen peroxide is used as an oxidizing agent by hairdressers to bleach hair. Concentrated aqueous solutions are stored with a small amount of added acid which prevents them decomposing. Addition of a little ammonia removes this prior to use.

Washing powders contain sodium peroxoborate(III) (sodium perborate). This solid reacts with water giving hydrogen peroxide which bleaches the laundry.

Hydrogen peroxide has also been used in rocketry, either as an oxidant to provide oxygen in which a fuel is burned, allowing the rocket to operate beyond the atmosphere, or as a so-called 'mono propellant'. Here concentrated hydrogen peroxide called 'high test peroxide' is decomposed using a silver gauze catalyst. The resulting hot gases provide the thrust.

The concentration of hydrogen peroxide is usually measured in 'volumes'. 20 volume peroxide is a solution which will produce 20 cm^3 of oxygen gas when 1 cm^3 of solution is decomposed. The most concentrated solution available is 100 volume, which is not on sale over the counter. Hydrogen peroxide is a byproduct of many metabolic processes in both plants and animals. It is never found free because enzymes in the body or plant break it down to water and oxygen as it forms.

$$2H_2\overset{-I}{O}_2(l) \longrightarrow 2H_2\overset{-II}{O}(l) + \overset{0}{O}_2(g)$$

Hydrogen peroxide can be made in the laboratory by the reaction of barium peroxide with dilute sulphuric acid:

$BaO_2(s)$	+	$H_2SO_4(aq)$	$\longrightarrow$	$H_2O_2(aq)$	+	$BaSO_4(s)$
barium peroxide		sulphuric acid		hydrogen peroxide		barium sulphate

HYDROGEN SULPHIDE, H_2S

This is the only important hydride of sulphur. It is well known as a gas with a powerful smell of bad eggs. It is extremely toxic. It is a gas at room temperature while water with its smaller relative molecular mass is a liquid, because sulphur is not electronegative enough to form hydrogen bonds (see **section 16.2.3**). The hydrogen sulphide molecule has the same bonding and structure as water although the bond angle is smaller. It is prepared in the laboratory by the reaction of metal sulphides with acids, for example:

$FeS(s)$	+	$2HCl(aq)$	$\longrightarrow$	$H_2S(g)$	+	$FeCl_2(aq)$
iron(II) sulphide		hydrochloric acid		hydrogen sulphide		iron(II) chloride

In aqueous solution, hydrogen sulphide is weakly acidic:

$H_2S(aq)$	+	$H_2O(l)$	$\rightleftharpoons$	$H_3O^+(aq)$	+	$HS^-(aq)$	$pK_a = 7.1$
hydrogen sulphide		water		oxonium ion		hydrogen-sulphide ion	
$HS^-(aq)$	+	$H_2O(l)$	$\rightleftharpoons$	$H_3O^+(aq)$	+	$S^{2-}(aq)$	$pK_a = 12.9$
hydrogen sulphide		water		oxonium ion		sulphide ion	

Aqueous hydrogen sulphide forms precipitates of sulphides with meta ions, for example:

$Pb^{2+}(aq)$	+	$H_2S(aq)$	$\longrightarrow$	$PbS(s)$	+	$2H^+(aq)$
lead(II) ion		hydrogen sulphide		lead(II) sulphide		hydrogen ion

Hydrogen sulphide is a good reducing agent, for example:

$$\overset{-II}{H_2S}(aq) + \overset{0}{Br_2}(aq) \longrightarrow \overset{0}{S}(s) + 2H\overset{-I}{Br}(aq)$$

Oxides and oxoacids of sulphur

Sulphur forms two oxides: sulphur dioxide, SO_2, and sulphur trioxide SO_3.

SULPHUR DIOXIDE

Sulphur dioxide can be prepared in the laboratory by burning sulphur in air or oxygen:

$S(s)$	+	$O_2(g)$	$\longrightarrow$	$SO_2(g)$
sulphur		oxygen		sulphur dioxide

or by the action of dilute acids on sulphate(IV) (sulphite) salts:

$$SO_3{}^{2-}(aq) + 2H^+(aq) \longrightarrow SO_2(g) + H_2O(l)$$

The sulphur dioxide molecule is angular with a bond angle of 120°.

The S—O bonds are double so that the sulphur has ten electrons in its outer shell. This is possible due to promotion of electrons from the 3s and 3p orbitals into the empty 3d orbitals.

The gas dissolves in water when the following equilibria are set up:

$H_2O(l)$	+	$SO_2(g)$	$\rightleftharpoons$	$H_2SO_3(aq)$	$\rightleftharpoons$	$H^+(aq)$	+	$HSO_3{}^-(aq)$	$\rightleftharpoons$	$2H^+(aq)$	+	$SO_3{}^{2-}(aq)$
		sulphur dioxide		sulphuric(IV) acid				hydrogen-sulphate(IV) ions				sulphate(IV) ions

Sulphuric(IV) acid, the hydrogensulphate(IV) ion and the sulphate(IV) ion are still frequently referred to by their non-systematic names: sulphurous acid, the hydrogensulphite ion and the sulphite ion respectively. The sulphite ion is a good reducing agent, being oxidized to sulphate(VI) ions, for example:

$$\overset{0}{Cl_2}(aq) + \overset{+IV}{SO_3}{}^{2-}(aq) + H_2O(l) \longrightarrow 2\overset{-I}{Cl^-}(aq) + \overset{+VI}{SO_4}{}^{2-}(aq) + 2H^+(aq)$$

SULPHUR TRIOXIDE

Sulphur trioxide can be prepared by passing a dried mixture of sulphur dioxide and oxygen over a heated platinum catalyst:

$$SO_2(g) + \tfrac{1}{2}O_2(g) \xrightleftharpoons{\text{Pt catalyst}} SO_3(g)$$

The product can be collected as a solid in a receiver cooled in ice as its melting temperature is 290 K. Sulphur trioxide exists as a trigonal planar molecule in the gas phase. The bonding may be represented as:

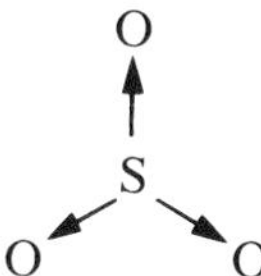

In the solid state, it exists in a number of polymeric forms. Sulphur trioxide reacts violently with water forming sulphuric acid:

$$SO_3(g) + H_2O(l) \longrightarrow H_2SO_4(l)$$

Figure 24.30 The sulphuric acid molecule

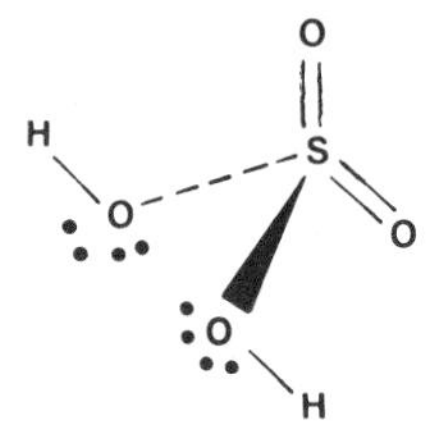

Figure 24.31 Hydrogen bonding in sulphuric acid

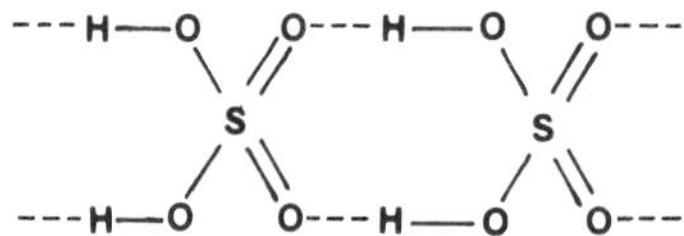

which reacts with more sulphur trioxide to give oleum (fuming sulphuric acid):

$SO_3(g)$	+	$H_2SO_4(l)$	$\longrightarrow$	$H_2S_2O_7(l)$
sulphur trioxide		sulphuric acid		oleum

SULPHURIC ACID

Sulphuric acid is a vital industrial chemical and it has been claimed that it is involved at some stage in the production of virtually all manufactured goods. The manufacture of sulphuric acid has been described—see box, **section 9.7.** Its importance is in part due to the fact that sulphuric acid can behave chemically in three distinct ways—as an acid, as an oxidizing agent and as a dehydrating agent.

The sulphuric acid molecule is approximately tetrahedral, **Figure 24.30.** It is a viscous liquid due to hydrogen bonding, **Figure 24.31**, which also accounts for its relatively high boiling temperature (611 K).

SULPHURIC ACID AS AN ACID

It is a strong acid; in dilute solution the ionization

$$H_2SO_4(l) + H_2O(l) \longrightarrow H_3O^+(aq) + HSO_4^-(aq)$$

goes to completion, but the second proton is lost less easily as H^+ is being removed from a negative ion:

$$HSO_4^-(aq) + H_2O(l) \longrightarrow H_3O^+(aq) + SO_4^{2-}(aq) \quad pK_a = 2$$

The acid can form two types of salts. Reaction with sodium hydroxide in a 1:1 molar ratio produces the salt sodium hydrogensulphate(VI):

$$NaOH(aq) + H_2SO_4(aq) \longrightarrow NaHSO_4(aq) + H_2O(l)$$

This salt contains the hydrogensulphate ion HSO_4^- and is therefore acidic. Reaction of sulphuric acid and sodium hydroxide in a 1:2 molar ratio produces the salt sodium sulphate which is neutral, being the salt of a strong acid and a strong base:

$$2NaOH(aq) + H_2SO_4(aq) \longrightarrow Na_2SO_4(aq) + 2H_2O(l)$$

Sulphuric acid undergoes the normal reactions of an acid with metals, metal oxides and carbonates.

SULPHURIC ACID AS A DEHYDRATING AGENT

Concentrated (almost pure) sulphuric acid reacts violently with water. It can therefore be used as a drying agent, removing traces of water from gases which are bubbled through it. It will also remove water from compounds. For example, it will remove the water of crystallization from hydrated copper(II) sulphate:

$$\underset{\text{blue}}{CuSO_4\cdot 5H_2O(s)} \xrightarrow[(H_2SO_4)]{-5H_2O} \underset{\text{white}}{CuSO_4(s)}$$

and it will also remove the elements of water from sucrose:

$$C_{12}H_{22}O_{11}(s) \xrightarrow[(H_2SO_4)]{-11H_2O} 12C(s)$$

It has important uses in organic chemistry as a dehydrating agent. For example, it can dehydrate ethanol, forming ethene if the acid is in excess or ethoxyethane if the ethanol is in excess:

$$C_2H_5OH \xrightarrow[(H_2SO_4)]{-H_2O} \underset{\text{ethene}}{CH_2CH_2}$$

$$2C_2H_5OH \xrightarrow[(H_2SO_4)]{-H_2O} \underset{\text{ethoxyethane}}{C_2H_5OC_2H_5}$$

It will also dehydrate skin and eye tissue and should be treated with great care.

SULPHURIC ACID AS AN OXIDIZING AGENT

Concentrated sulphuric acid acts as an oxidizing agent. The reactions with halide ions are interesting examples. Hydrogen iodide is easily oxidized:

$$8H\overset{-I}{I}(g) + H_2\overset{+VI}{S}O_4(l) \longrightarrow 4H_2O(l) + H_2\overset{-II}{S}(g) + 4\overset{0}{I_2}(s)$$

The oxidation number of the sulphur atom drops from +VI to −II enabling eight I^- ions to be oxidized to iodine.

Hydrogen bromide is less easily oxidized:

$$2H\overset{-I}{Br}(g) + H_2\overset{+VI}{S}O_4(l) \longrightarrow 2H_2O(l) + \overset{+IV}{S}O_2(g) + \overset{0}{Br_2}(l)$$

Here the oxidation number of the sulphur atoms only drops from +VI to +IV allowing only two Br^- ions to be oxidized to bromine.

Hydrogen chloride and hydrogen fluoride cannot be oxidized by sulphuric acid.

24.5.4 Oxidation number charts

You will have noticed that sulphur especially, and also oxygen, both display several oxidation states in their compounds. This is illustrated on the oxidation number chart in **Table 24.11.**

Notice that oxygen has a positive oxidation number only in compounds with fluorine, the only atom more electronegative than oxygen. Sulphur forms higher oxidation states than oxygen because it can use all six of its outer electrons in bonding by promoting some of them into its 3d orbitals. This does not occur in oxygen as its next empty orbital is 3s which is of considerably higher energy than the highest occupied orbital, 2p.

Table 24.11 **Oxidation number chart for compounds of sulphur and oxygen**

Sulphur	Ox	Oxygen
SO_3, SO_4^{2-}, SF_6	+VI	
	+V	
SO_2, SO_3^{2-}, SF_4	+IV	
	+III	
SCl_2	+II	OF_2
S_2Cl_2	+I	
S_8	0	O_2
	$-\frac{1}{2}$	KO_2 and other superoxides
	−I	H_2O_2 and other peroxides
H_2S, Na_2S and other metallic sulphides	−II	H_2O, Na_2O and other metallic oxides

24.5.5 Summary (Group VI)

- The Group VI elements have the outer electron arrangement ns^2np^4.
- The group shows a clear trend from non-metals at the top to metals at the bottom.
- Oxygen exists as O_2 due to the strength of the O=O bond while sulphur forms S_8 ring-shaped molecules with S—S single bonds.
- Oxygen has an allotrope O_3 called ozone (trioxygen)
- Sulphur has two allotropes, rhombic and monoclinic sulphur.
- Oxygen has Ox = −II in the vast majority of compounds, exceptions being peroxides (Ox(O) = −I), superoxides (Ox(O) = $-\frac{1}{2}$) and oxygen fluoride (Ox(O) = +II).
- Sulphur forms sulphides (Ox(S) = −II) but also forms compounds in which it has positive oxidation numbers up to +VI.
- Sulphur can form this increased variety of compounds by promotion of electrons into the 3d orbitals.
- Water can act both as a proton donor and a proton acceptor.
- Hydrogen bonding is responsible for many anomalous properties of water—high melting and boiling temperatures, low density of ice as well as its ability to dissolve ionic compounds.
- Hydrogen sulphide does not hydrogen bond.
- Hydrogen sulphide is weakly acidic.
- Hydrogen sulphide is a reducing agent.
- Sulphur dioxide is an acidic oxide forming sulphurous acid (sulphuric(IV) acid) in water.
- Sulphur trioxide is also an acidic oxide forming sulphuric acid in water.
- Sulphuric acid can act as a strong acid, a dehydrating agent and an oxidizing agent.

24.5.6 Questions (Group VI)

1 Sulphuric acid is manufactured from sulphur in a three-stage process. Sulphur is burnt in air:

$$S(s) + O_2(g) \longrightarrow SO_2(g) \quad \Delta H^\ominus = -296\,kJ\,mol^{-1}$$

The sulphur dioxide is reacted with further oxygen from the air to form the trioxide:

$$2SO_2(g) + O_2(g) \rightleftharpoons 2SO_3(g) \quad \Delta H^\ominus = -197\,kJ\,mol^{-1}$$

In the final stage, the sulphur trioxide is converted into sulphuric acid.

A) **i** The atomic number of sulphur is 16. Write down the ground-state electron configuration for an isolated sulphur atom.

ii What is the oxidation number of sulphur in each of the following?

$$SO_2; \qquad SO_3; \qquad H_2SO_4$$

iii Draw a diagram to show the electronic structure ('dot-and-cross' diagram) of sulphuric acid (only the outer-shell electrons need be shown).

B) Write the expression for the equilibrium constant, K_p, for the formation of sulphur trioxide.

C) Calculate the molar enthalpy change of formation of sulphur trioxide in the gas state.

D) The manufacture of sulphur trioxide for commercial sulphuric acid takes place at about 750 K and 1.3 atmospheres over a catalyst of vanadium(V) oxide.

i What would you expect to happen to the yield of sulphur trioxide if the temperature of the catalyst rose to 950 K? Explain your answer.

ii What would you expect to happen to the yield of sulphur trioxide if the manufacturing plant were to be operated at a pressure of 5 atmospheres? Explain your answer. (Nuffield 1986)

2 **A)** Give the atomic number of sulphur and the electronic configuration of its isolated atom.

B) For sulphur dioxide (SO_2) and hydrogen sulphide (H_2S) molecules and sulphate ions (SO_4^{2-}) respectively:

i state the oxidation number of the sulphur;

ii show the structural formula, including the outer electrons;

iii describe in words, or by a diagram, the shape of each of these.

C) Describe briefly, giving reagents, reaction conditions and equations, how substances containing sulphur of oxidation number +IV, +VI and 0 respectively could be obtained from sulphur dioxide.

i substance with oxidation number +IV;

ii substance with oxidation number +VI;

iii substance with oxidation number 0.

D) In the reaction:

$$Na_2H_{10}S_2O_8 + 4Br_2 \longrightarrow 2H_2SO_4 + 2NaBr + 6HBr$$

only the oxidation numbers of the sulphur and bromine alter. Calculate from the equation the oxidation number of the sulphur in $Na_2H_{10}S_2O_8$. (London 1980, part question)

3 **A)** What is the electron configuration (in terms of s, p and d orbitals) of

i an isolated oxygen atom;

ii an isolated sulphur atom?

B) For each of the following sulphur compounds give the oxidation number of the sulphur atom, a dot cross diagram of the bonding and describe the shape of the molecule in words or with a diagram:

i) H_2S; **ii)** SO_2; **iii)** SO_3.

4 In the reaction:

$$2H_2S(aq) + SO_2(aq) \longrightarrow 3S(s) + 2H_2O(l)$$

A) What are the oxidation numbers of the sulphur species in the equation?

B) Name the oxidizing agent and the reducing agent.

C) What use is made of this reaction industrially?

24.6 Group VII—the halogens

The members of this group are all reactive non-metals. All tend to react by forming ions with one negative charge as they are each one electron short of a full outer shell. There is a clear trend in reactivity which decreases as we descend the group. Astatine is both rare and radioactive and its chemistry has been little studied.

Table 24.12 gives some physical properties of the elements.

Table 24.12 Some properties of the halogens

	Atomic number, Z	Electron arrangement	Electro-negativity	Atomic (covalent) radius/nm	Ionic radius E^-/nm	T_m/K	T_b/K	E—E bond energy/ kJ mol^{-1}	$E^\ominus$ for $\frac{1}{2}X_2(aq) + e^- \rightleftharpoons X^-(aq)$
F	9	$1s^22s^22p^5$	4.0	0.071	0.133	53	85	158*	+2.87
Cl	17	$[Ne]\,3s^23p^5$	3.0	0.099	0.180	172	238	243	+1.36
Br	35	$[Ar]\,3d^{10}4s^24p^5$	2.8	0.114	0.195	266	332	193	+1.07
I	53	$[Kr]\,4d^{10}5s^25p^5$	2.5	0.133	0.215	387	457	151	+0.54
At	85	$[Xe]\,4f^{14}5d^{10}6s^26p^5$	–	–	–	575	610	–	–

*see text

†The value for $\frac{1}{2}F_2(g) + e^- \longrightarrow F^-(aq)$

24.6.1 The elements

All the elements exist as diatomic molecules with an E—E single bond. At room temperature (298 K) fluorine is a pale yellow gas, chlorine a greenish gas, bromine a volatile red–brown liquid and iodine a shiny black solid which gives off a purple vapour on gentle heating. The volatility of all the elements is due to the fact that the intermolecular forces are weak van der Waals forces. There is a clear reactivity trend in the group. Reactivity becomes less on descending the group. This is principally due to the increasing size of the atoms. The electron which is added to form the E^- ion goes into a shell further away from and thus less strongly attracted to the nucleus, i.e. the electron affinity generally becomes less negative on descending the group.

The elements are too reactive to be found uncombined. The main sources of the elements are halide salts. All the elements have a characteristic 'bleachy' smell. There are considerable similarities in the chemistry of the halogens but fluorine, like other elements in the Li–Ne period, is untypical in many ways. Its chemistry is described separately.

24.6.2 Fluorine compounds

Fluorine is the most electronegative element. Thus in all its compounds, both ionic and covalent, fluorine has an oxidation number of −I. **Table 24.12** shows that the fluorine molecule has an unexpectedly low bond energy compared with the other halogens. This is due to the small size of the atom, which leads to a short internuclear distance in F_2. Consequently there is repulsion between non-bonding electrons and this weakens the bond:

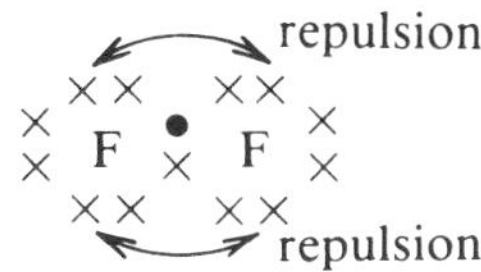

Since the E—F bond energies of fluorine with other elements are generally high (see **Table 24.13**) this means that the formation of covalent fluorides (involving breaking of F—F and making of E—F bonds) is usually strongly exothermic.

Similar considerations apply to the formation of ionic fluorides. Fluorine has a low value of $\Delta_{at}H^{\ominus}$ due to the weak F—F bond. Ionic fluorides have high lattice energies due to the small size of the F^- ion (see **Table 24.14.** Both factors lead to large exothermic values for $\Delta_f H^{\ominus}$ of ionic fluorides. See **Figure 24.32** which gives a Born–Haber cycle for sodium fluoride.

Table 24.13 Comparison of E—X tabulated bond energies for the halogens/kJ mol^{-1}

Element E \ Halogen X	F	Cl	Br	I
H	568	432	366	298
C	467	346	290	228

Table 24.14 Comparison of lattice energies for some ionic halides ($M^+ + X^-$)/kJ mol^{-1}

Metal M \ Halogen X	F	Cl	Br	I
Li	1031	848	803	759
Na	918	780	742	705
K	817	711	679	651

Figure 24.32 Born–Haber cycle for sodium fluoride. All values in kJ mol^{-1}

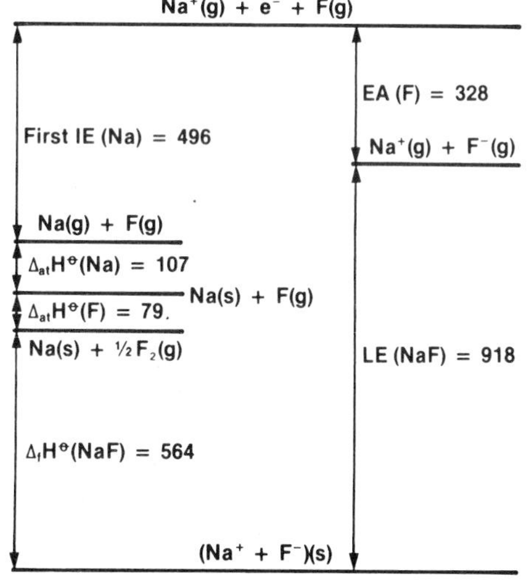

Extraction of fluorine

No chemical oxidizing agent is strong enough to easily oxidize F^- to F_2, so the element is produced electrolytically from a mixture of potassium fluoride and hydrogen fluoride. At the anode:

$$F^- \longrightarrow \tfrac{1}{2}F_2 + e^-$$

Fluorine as an oxidizing agent

Fluorine is an excellent oxidizing agent and will oxidize other elements to their highest oxidation states. For example, sulphur forms sulphur hexafluoride, SF_6 (Ox(S) = +VI) and oxygen forms oxygen difluoride, OF_2 (Ox(O) = +II). Fluorine will oxidize water to oxygen:

$$2F_2(g) + 2H_2O(l) \longrightarrow 4HF(aq) + O_2(g)$$

Hydrogen fluoride, HF

This is a liquid, $T_b = 293$ K. Its unusually high boiling temperature compared with the other hydrogen halides, which are gases at room temperature despite higher relative molecular masses, is due to the formation of strong hydrogen bonds. In fact the F—H---F—H *hydrogen bond* is the strongest known, having a bond energy of 125 kJ mol^{-1} which approaches the strength of some covalent bonds.

Unlike the other hydrogen halides, hydrogen fluoride dissolved in water is a *weak* acid:

$$HF(aq) + H_2O(l) \rightleftharpoons F^-(aq) + H_3O^+(aq)$$

This is due to the very strong H—F bond. Solutions of hydrofluoric acid will dissolve glass (they are used to etch glass) and are therefore stored in polythene containers.

A

FLUORIDATION OF TAP WATER

Both bones and teeth contain traces of fluorine compounds—in particular, fluorides appear to harden tooth enamel (the biting surface of the tooth) by combining with calcium phosphate in the enamel. Research has shown that lack of fluoride can result in excessive tooth decay. Natural water usually contains small amounts of fluorine-containing compounds but in some areas where the fluoride content is low, tap water is treated by adding sodium fluoride, NaF, sodium silicofluoride, Na_2SiF_6, or hydrofluorosilic acid to bring the total combined fluorine content up to 1 part per million (1 p.p.m.). Amounts of fluoride greater than 1.5 p.p.m. can cause mottling of the teeth. Many studies have proved beyond doubt that fluoridated water reduces tooth decay (especially in children) and is quite harmless, despite the fact that, in high concentrations, sodium fluoride is a poison.

Fluoridation has been strongly opposed by many people who are opposed in principle to 'tampering with nature' and the fact that once the water is fluoridated there is no choice but to drink it. Alternative ways of taking fluoride such as by use of fluoridated toothpaste have been shown to be less effective than water treatment. Nevertheless, many toothpastes contain fluoride either as sodium monofluorophosphate or sodium fluoride. Less than 10% of the population of the UK have fluoridated drinking water.

Fluorides

Fluorine is so reactive that it will combine directly with all the other elements (except helium, neon, krypton and nitrogen) to form fluorides. With the exception of beryllium, the fluorides of the s-block elements and aluminium are ionic. Some transition metals in lower oxidation states form ionic fluorides but those in higher oxidation states are covalent, for example, uranium hexafluoride. Such fluorides may be quite volatile—uranium hexafluoride boils at 329 K.

Figure 24.33 An $E^\ominus$–Ox diagram

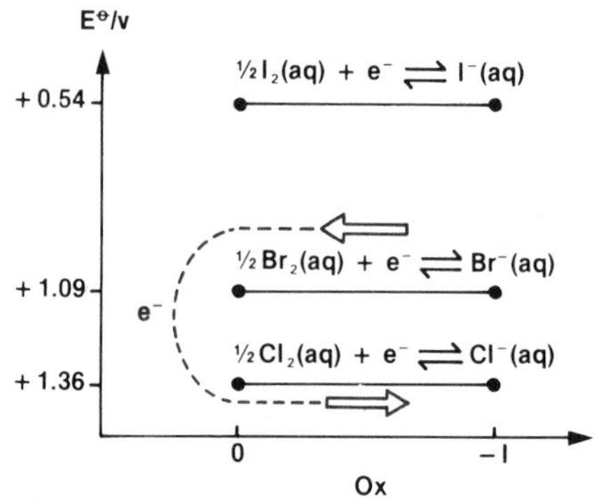

24.6.3 Extraction of the other halogens (chlorine, bromine and iodine)

Chlorine is extracted from sodium chloride by electrolysis (see box). Bromine is obtained from sea water, which contains bromide ions, by oxidation with chlorine. That this reaction is feasible is shown by the anticlockwise rule (see **section 13.3.6**). An $E^\ominus$–Ox diagram is shown in **Figure 24.33.**

$$Br^-(aq) \longrightarrow \tfrac{1}{2}Br_2(aq) + e^-$$

$$\tfrac{1}{2}Cl_2(aq) + e^- \longrightarrow Cl^-(aq)$$

Adding, $\tfrac{1}{2}Cl_2(aq) + Br^-(aq) + \cancel{e^-} \longrightarrow \tfrac{1}{2}Br_2(aq) + Cl^- + \cancel{e^-}$

This is an example of a displacement reaction—halogens can displace from their salts halide ions from lower down the group in the Periodic Table. The diagram also indicates that bromine will oxidize $I^-(aq)$ to $I_2(aq)$, as will chlorine. The latter reaction can be used to extract iodine from sea water which also contains iodide ions. A more important source of iodine is from deposits in Chile containing sodium iodate(V) ($NaIO_3$) from which it is extracted by reduction with sulphur dioxide. This reaction can also be predicted by the anti-clockwise rule, **Figure 24.34.**

Figure 24.34 An $E^\ominus$–Ox diagram

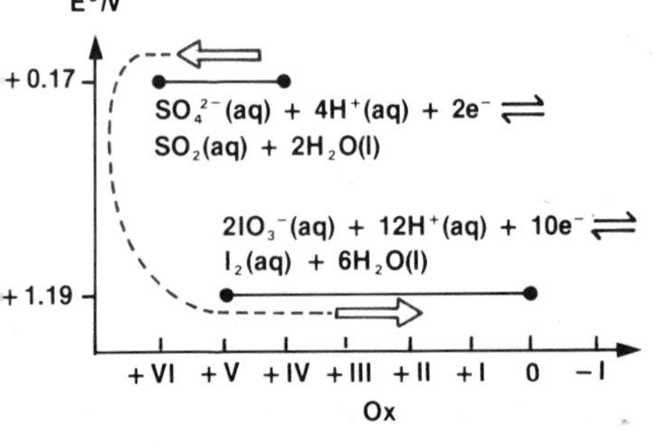

E

THE MANUFACTURE OF CHLORINE BY THE DIAPHRAGM CELL

We have already seen that chlorine is manufactured by electrolysis of molten sodium chloride by the Down's process, see **section 17.5.1**, and from sodium chloride solution (brine) in the flowing mercury cathode cell, see box, **section 23.4.5.** The diaphragm cell is another method for the electrolysis of brine which competes with the mercury cell.

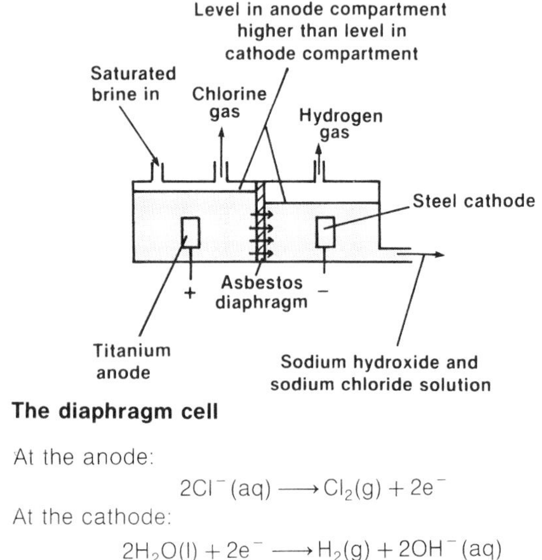

The diaphragm cell

At the anode:

$$2Cl^-(aq) \longrightarrow Cl_2(g) + 2e^-$$

At the cathode:

$$2H_2O(l) + 2e^- \longrightarrow H_2(g) + 2OH^-(aq)$$

Brine enters the cathode compartment where Cl^- ions are discharged as chlorine. The resulting solution, rich in Na^+ ions, seeps through the diaphragm into the anode compartment where hydrogen is discharged (as it is lower in the electrochemical series than sodium), resulting in a solution containing both sodium hydroxide and unchanged sodium chloride. The electrolyte level in the anode compartment is kept higher than in the cathode compartment which prevents $OH^-(aq)$ ions moving into the anode compartment where they would be discharged in preference to $Cl^-(aq)$, as $OH^-(aq)$ is above $Cl^-(aq)$ in the electrochemical series (see **section 17.4**). The solution from the cathode cell is evaporated to about one-fifth of its original volume when most of the less soluble sodium chloride crystallizes out. The resulting solution contains 50% sodium hydroxide and 1% sodium chloride, which is sufficiently pure for many purposes. Usually several diaphragm cells are run together connected in series.

An improvement on this cell is the membrane cell, where a fluorocarbon membrane replaces the asbestos diaphragm. This membrane is permeable to cations but not anions, thus allowing $Na^+(aq)$ through but not Cl^- nor $OH^-(aq)$. The resulting sodium hydroxide contains no sodium chloride impurity. Diaphragm cells are cheaper to build than mercury cells (which give the same product) but produce less pure sodium hydroxide. They do not use toxic mercury. The running costs of the two processes are similar.

$$5SO_2(aq) + 10H_2O(l) \longrightarrow 5SO_4^{2-}(aq) + 20H^+(aq) + 10e^-$$

$$2IO_3^-(aq) + 12H^+(aq) + 10e^- \longrightarrow I_2(aq) + 6H_2O(l)$$

Adding. $5SO_2(aq) + 10H_2O(l) + 2IO_3^- + 12H^+(aq) + \cancel{10e^-}$

$$\longrightarrow 5SO_4^{2-}(aq) + 20H^+(aq) + I_2(aq) + 6H_2O(l) + \cancel{10e^-}$$

$$5SO_2(aq) + 4H_2O(l) + 2IO_3^-(aq) \longrightarrow 5SO_4^{2-}(aq) + 8H^+(aq) + I_2(aq)$$

24.6.4 Laboratory preparations of the halogens

Chlorine is prepared by oxidation of concentrated hydrochloric acid with potassium manganate(VII). The apparatus is shown in **Figure 24.35.**

$2KMnO_4(s)$	+	$16HCl(aq)$
potassium manganate(VII)		hydrochloric acid

⟶	$2KCl(aq)$	+	$2MnCl_2(aq)$	+	$8H_2O(l)$	+	$5Cl_2(g)$
	potassium chloride		manganese(II) chloride		water		chlorine

The gas is bubbled through water to remove any unreacted hydrochloric acid and dried by bubbling through concentrated sulphuric acid. Being denser than air, it can be collected by downward delivery.

Bromine and iodine can be prepared by the reaction of a halide salt with sulphuric acid producing the hydrogen halide followed by oxidation of the hydrogen halide to the halogen by manganese(IV) oxide.

Figure 24.35 Preparation of chlorine

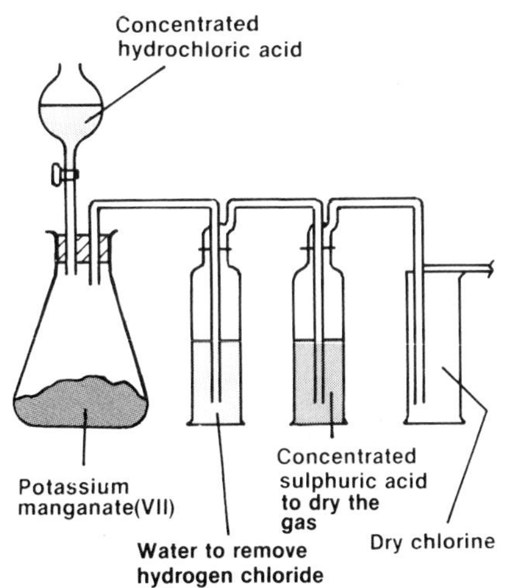

Reactions of the halogens as oxidizing agents

As the values of $E^\ominus$ show, there is a clear trend of decreasing oxidizing power as we descend the group. While fluorine oxidizes water, chlorine slowly disproportionates forming hydrochloric acid and chloric(I) acid:

$$\overset{0}{Cl_2}(aq) + H_2O(l) \longrightarrow \overset{+I}{HClO}(aq) + \overset{-I}{HCl}(aq)$$

chloric(I) acid hydrochloric acid

24.6.5 Compounds of chlorine, bromine and iodine

Chlorine, bromine and iodine all form halides with Ox = −I but also form compounds with positive oxidation states up to +VII, the odd numbers being the most important.

Hydrogen halides

Hydrogen chloride, HCl, hydrogen bromide, HBr, and hydrogen iodide, HI, are all gases at room temperature although hydrogen fluoride, HF, is a liquid due to its strong hydrogen bonding. They are very soluble in water, but unlike hydrogen fluoride, form *strong* acids: hydrochloric, hydrobromic and hydroiodic acids.

The hydrogen halides can be prepared in the laboratory by the action of phosphoric(V) acid on alkali metal halides:

$NaX(s)$	+	$H_3PO_4(s)$	$\longrightarrow$	$HX(g)$	+	$NaH_2PO_4(s)$
sodium halide		phosphoric(V) acid		hydrogen halide		sodium dihydrogen-phosphate

Hydrogen chloride can be made by the similar reaction of concentrated sulphuric acid on sodium chloride but bromides and iodides are oxidized to bromine and iodine respectively by sulphuric acid, so they cannot be prepared from it. The ease of oxidation of X^- is shown by the $E^\ominus$ values for the half reactions in **Table 24.15.**

$$\tfrac{1}{2}X_2(aq) + e^- \longrightarrow X^-(aq)$$

Table 24.15 Ease of oxidation of halide ions

	Half reaction	$E^\ominus$/V
X^- more easily oxidized ↑	$\frac{1}{2}I_2 + e^- \rightleftharpoons I^-$	+0.54
	$\frac{1}{2}Br_2 + e^- \rightleftharpoons Br^-$	+1.09
	$\frac{1}{2}Cl_2 + e^- \rightleftharpoons Cl^-$	+1.36

The hydrogen halides show a trend in thermal stability:

$$2HX(g) \longrightarrow H_2(g) + X_2(g, l, s)$$

Hydrogen iodide is easily decomposed into its elements by plunging a red-hot wire into a test-tube of the gas. Hydrogen bromide may or may not decompose depending on the exact temperature of the wire. Hydrogen chloride is not decomposed. This trend in the ease of decomposition reflects the strength of the H—X bond, **Table 24.16.**

Table 24.16 H—X bond energies for the halogens

	Bond	Tabulated bond energy /kJ mol^{-1}
stronger bond—less easily decomposed ↑	H—F	568
	H—Cl	432
	H—Br	366
	H—I	298

Metal halides

All the halogens form metal halides. Because the size of the halide ions increases from F^- to Cl^- to Br^- to I^-, the ions become more easily polarized and their properties less typically ionic. The aluminium halides are a good example. AlF_3 is ionic, $AlCl_3$ exists in the solid state as a layer lattice with some ionic and some covalent character while $AlBr_3$ and AlI_3 exist as covalent dimers.

Tests for halide ions

The presence of Cl^-, Br^- and I^- in aqueous solution can be confirmed by the formation of a precipitate with silver nitrate solution:

$AgNO_3(aq)$	+	$X^-(aq)$	$\longrightarrow$	$AgX(s)$	+	$NO_3^-(aq)$
silver nitrate		halide ion		silver halide		nitrate ion

Silver chloride is white, silver bromide cream and silver iodide yellow. The last two can be hard to distinguish by eye, but their solubility in aqueous ammonia helps. Silver bromide is soluble in concentrated aqueous ammonia but silver iodide is not. Fluorides do not form a precipitate with silver nitrate as silver fluoride is soluble in water—a reflection of its ionic character.

A

PHOTOGRAPHY

It has long been known that silver salts are darkened on exposure to light. That is why silver nitrate solution is normally stored in brown glass bottles.

In photography, silver bromide or silver iodide is used in an emulsion with gelatine to form the film. When the film is exposed, the halide ion actually absorbs the light, releasing an electron which is picked up by the silver ion, reducing it to metallic silver when the film is exposed:

$$Br^- \longrightarrow \tfrac{1}{2}Br_2 + e^-$$

$$Ag^+ + e^- \longrightarrow Ag$$

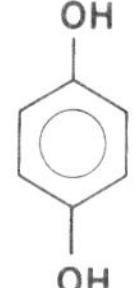

Benzene-1, 4-diol

These silver atoms act as a catalyst for the reduction of further silver ions around those which were originally produced by the light. To do this, a mild reducing agent like benzene-1, 4-diol (more familiarly called hydroquinone or quinol) is used. These silver grains produce the image.

This reduction stage is called developing. The remaining silver halide must be removed before the film can be safely exposed to light. This is done with a solution of sodium thiosulphate in which the silver ions dissolve as the complex $[Ag(S_2O_3)_2]^{3-}$. Photographers call sodium thiosulphate 'hypo' and this stage 'fixing'.

Photographic chemicals

At this stage the film is a 'negative'—black silver is deposited where the light was strongest. To produce a positive photograph, light is shone through the negative on to a further piece of film which is then processed as before.

E

QUANTITATIVE ANALYSIS OF HALIDES

The reaction:

$$X^-(aq) + AgNO_3(aq) \longrightarrow AgX(s) + NO_3^-(aq)$$

provides the basis of a method for analysis of solutions containing halide ions by titration. Portions of solution containing halide ions can be titrated with silver nitrate of known concentration. To indicate when the reaction is complete a few drops of yellow potassium chromate(VI), K_2CrO_4, solution are added. This reacts with Ag^+ ions to give a red precipitate of silver chromate(VI) Ag_2CrO_4, but less readily than halide ions react with silver ions. Only when all the halide ions have been used up will the chromate ions react with silver ions to give the red precipitate. When this appears, the titration is complete.

EXAMPLE

What is the formula of magnesium chloride given A_r for magnesium = 24 and for chlorine = 35.5? 0.95 g of magnesium chloride was dissolved in 1 dm³ of water. 25 cm³ of this solution was titrated with 0.01 mol dm⁻³ silver nitrate solution. 50 cm³ silver nitrate was added before the potassium chromate indicator gave a red precipitate.

50 cm³ of 0.01 mol dm^{-3} silver nitrate solution contains:

$$\frac{50 \times 0.01}{1000} = 5 \times 10^{-4} \text{ mol } AgNO_3$$

$$AgNO_3 + Cl^- \longrightarrow AgCl + NO_3^-$$

1 mol of $AgNO_3$ reacts with 1 mol Cl^-, so 25 cm³ magnesium chloride solution contains 5×10^{-4} mol Cl^-.

So 1 dm³ of this solution contains

$$\frac{5 \times 10^{-4}}{25} \times 1000 \text{ mol } Cl^- = 0.02 \text{ mol } Cl^-$$

so 1 dm³ contains 0.02 × 35.5 g of Cl^- = 0.71 g Cl^-

so 1 dm³ contains 0.95 − 0.71 g of Mg = 0.24 g Mg

so 0.24 g Mg combines with 0.71 g Cl^-

$$\frac{0.24}{24} = 0.01 \text{ mol Mg}$$

combines with $$\frac{0.71}{35.5} = 0.02 \text{ mol Cl}$$

So magnesium chloride has the formula $MgCl_2$.

24.6.6 Oxygen compounds of the halogens

Oxygen has an oxidation number of −II in its compounds so forcing halogen atoms to have positive oxidation states (except in the case of fluorine). Chlorate(I) ions are formed by the reaction of chlorine with water (see page 343) or with cold dilute alkalis. Chlorine disproportionates in this reaction:

$$\underset{\text{chlorine}}{\overset{0}{Cl_2(aq)}} + \underset{\text{hydroxide ion}}{2OH^-(aq)} \xrightarrow[\text{alkali}]{\text{cold dil.}} \underset{\text{chloride ion}}{\overset{-I}{Cl^-(aq)}} + \underset{\text{chlorate(I) ion}}{\overset{+I}{ClO^-(aq)}} + \underset{\text{water}}{H_2O(l)}$$

With hot concentrated alkali, chlorate(V) ions are formed. This is also a disproportionation reaction:

$$\underset{\text{chlorine}}{\overset{0}{3Cl_2(aq)}} + \underset{\text{hydroxide ion}}{6OH^-(aq)} \xrightarrow{\text{hot conc. alkali}} \underset{\text{chlorate(V) ion}}{\overset{+V}{ClO_3^-(aq)}} + \underset{\text{chloride ion}}{\overset{-I}{5Cl^-(aq)}} + \underset{\text{water}}{3H_2O(l)}$$

Chlorate(V) ions also disproportionate. When potassium or sodium chlorate(V) is heated at its melting temperature the chlorate(VII) forms:

$$\underset{\text{potassium chlorate(V)}}{\overset{V}{4KClO_3(s)}} \longrightarrow \underset{\text{potassium chlorate(VII)}}{\overset{+VII}{3KClO_4(s)}} + \underset{\text{potassium chloride}}{\overset{-I}{KCl(s)}}$$

Chlorate(I) ions are stable only in solution. Solutions of chlorate(I) ions are used commercially as a household bleach and as disinfectant. Sodium chlorate(V) is used in weedkillers but as the chlorate(V) ion is a powerful oxidizing agent it assists burning, and these weedkillers are being used less because of the fire risk. Sodium chlorate(V) decomposes on heating to give oxygen:

$$\underset{\text{sodium chlorate(V)}}{NaClO_3(s)} \longrightarrow \underset{\text{sodium chloride}}{NaCl(s)} + \underset{\text{oxygen}}{1\tfrac{1}{2}O_2(g)}$$

The reaction is catalysed by manganese(IV) oxide and a mixture of this and sodium chlorate(V) used to be called 'oxygen mixture'.

Chlorate(VII) salts are extremely powerful oxidants and are used in explosive detonators and as oxidizing agents for rocket fuels. Similar series of compounds are formed by bromine and iodine except that BrO_4^- is unknown.

The bonding in the chlorate(I), chlorate(V) and chlorate(VII) ions is shown in **Figure 24.36.** Chlorate(I), chlorate(V) and chlorate(VII) were previously known as hypochlorite, chlorate, and perchlorate respectively.

Figure 24.36 (right) Bonding in chlorate ions. The electron in brown comes from the cation

Chlorate(I) ClO^-	Cl O	Cl—O $^-$	Linear
Chlorate(V) ClO_3^-	O Cl O O	Cl O O O $^-$	Pyramidal
Chlorate(VII) ClO_4^-	O O Cl O O	O Cl O O O $^-$	Tetrahedral

Sodium hypochlorite is the non-systematic name for sodium chlorate(I)

The parent acids, chloric(I) acid, HClO, chloric(V) acid, $HClO_3$, and chloric(VII) acid, $HClO_4$ show a trend of increasing acidity:

$$HClO_4 > HClO_3 > HClO$$

The halogens form a number of oxides which are thermally unstable—

many of them explosively so. They are covalently bonded acidic oxides. Those of formulae Cl_2O, ClO_2, Cl_2O_6 and Cl_2O_7 are known. Chlorine dioxide, ClO_2, is a useful industrial oxidizing agent. Among its uses is as a flour additive which both bleaches flour and improves its properties for bread making. Because of its unstable nature, it is made where and when it is required by the reaction:

$$2NaClO_3(aq) + SO_2(aq) + H_2SO_4(aq) \longrightarrow 2ClO_2(g) + 2NaHSO_4(aq)$$

24.6.7 Interhalogen compounds

Table 24.17 Binary interhalogen compounds

ClF(s)	ClF_3(s)		
BrF(g)	BrF_3(s)	BrF_5(l)	
BrCl(g)			
ICl(l)	ICl_3(s)	IF_5(l)	IF_7(g)
IBr(s)			

These are molecular compounds containing different halogens. The known binary ones (i.e. with two halogens) are listed in **Table 24.17.**

They can be prepared by direct combination of the elements. Notice that as we descend the group, the maximum oxidation state increases, i.e. in ClF_3, Ox(Cl) = +III, in BrF_5, Ox(Br) = +V, in IF, Ox(I) = +VII.

Notice also that it is the fluoride compound which always has the maximum oxidation number. The molecules have dipoles with the more electronegative atom(s) forming the δ^- end.

24.6.8 Economic importance of Group VII elements and compounds

Fluorine is used in the manufacture of fluorocarbons (see Chapter 30) which have many uses including non-stick coatings for saucepans, aerosol propellants, refrigerator coolants and artificial blood. The element is also used to manufacture uranium hexafluoride, one of the few volatile uranium compounds, which is used to separate uranium isotopes by gaseous diffusion (see box, **section 15.4.2**). Uranium enriched in ^{235}U is used in nuclear reactors.

Chlorine is largely used in the manufacture of organochlorine compounds used as solvents for dry-cleaning and degreasing, plastics (for example, PVC) and insecticides (such as DDT). See Chapter 30 for more details. It is also used for treatment of water for drinking and swimming baths and in household bleach.

A

CHEMICAL WARFARE

Chlorine is a severe irritant to the eyes and lungs, as it reacts with the water in moist parts of the body to form a mixture of hydrochloric and chloric(I) acids. Prolonged exposure can cause permanent respiratory damage or blindness and high concentrations of the gas can cause death.

Victims of a First World War gas attack

Chlorine was the first gas to be used in warfare, by the Germans at Ypres in April 1915. The Allies were caught unprepared and troops had to resort to breathing through urine soaked handkerchiefs until proper gas masks were provided.

Cl—C(=O)—Cl (Cl, Cl, C=O) **Phosgene**

Cl—CH$_2$—CH$_2$—S—CH$_2$—CH$_2$—Cl **Mustard gas**

The Germans' use of gas rebounded on them, however, as the Allies soon retaliated in kind and the prevailing winds on the Western front blew from west to east towards the German-held territory. Later in the First World War, even more deadly gases were developed including phosgene and mustard gas. The 1925 Washington agreement outlawed gas warfare and neither side resorted to its use in the Second World War, although both sides stockpiled it in case the other should use it first. However, research has continued and highly potent 'nerve' gases have been developed. It is to be hoped that these are never used, but there have been allegations about their use in the Iraq–Iran war.

Over half the 30 000 tonnes of bromine manufactured annually in the UK is used to make 1,2-dibromoethane. This is added to petrol and on combustion combines with the lead in antiknock additives to form lead bromide which is fairly volatile at car engine temperatures. Without the dibromoethane, the lead would end up as involatile lead oxide which would build up in the engine, reducing its performance. Bromine is also used to make pesticides, flame retardants and photographic chemicals.

No iodine is made in the UK. However, 10 000 tonnes of it are extracted annually world-wide. It has a number of uses, the most familiar being in the manufacture of photographic film.

24.6.9 Summary (Group VII)

- All the halogens have the outer electron configuration ns^2np^5.
- They all form X^- ions.
- Except for fluorine, they all form compounds with positive oxidation states also.
- Fluorine is an extremely good oxidizing agent.
- Oxidizing power of the element decreases as we descend the group.
- Fluorine's high reactivity is enhanced by the weak F—F bond caused by repulsion of the non-bonding electrons.
- Fluorine forms strong covalent bonds with most other non-metals.
- Ionic fluorides have high lattice energies.
- Hydrogen fluoride is a weak acid due to the strength of the H—F bond.
- The other hydrogen halides are strong acids in aqueous solution.
- Hydrogen fluoride is a liquid at room temperature due to strong hydrogen bonding. The other hydrogen halides are gases.
- Halide ions can be identified by the formation of precipitates with silver nitrate.
- The halogens (except for fluorine) form a series of oxoacids and oxoanions in which the halogens have positive oxidation numbers.

24.6.10 Questions (Group VII)

1 Research/debate: Does your area have fluoridated water? How would you find out? The arguments *for* fluoridation are fairly clear. What arguments are there against it? Are these arguments scientific, emotional, political (or other)? To what extent are these arguments based on lack of understanding of the case for fluoridation?

2 This question is about interhalogen compounds which are compounds of two or more halogen elements only. Examples are ClF, ClF_3, ICl, BrF, BrCl, BrF_5 and IF_7. Several others are known. Like the halogens themselves, interhalogen compounds are usually quite reactive.

A) Draw diagrams to show the electronic structure ('dot-and-cross' diagrams) of the following interhalogens (only the outer-shell electrons need be shown):

i iodine monochloride, ICl;

ii bromine pentafluoride, BrF_5;

B) Draw a diagram to show the shape you would expect for a molecule of bromine pentafluoride, BrF_5;

C) Most of the interhalogen compounds known are fluorides. Suggest a reason for this.

(Nuffield 1981, part question)

3 Iodine dissolves in hot concentrated solutions of sodium hydroxide according to the equation:

$$3I_2(s) + 6NaOH(aq) \longrightarrow NaIO_3(aq) + 5NaI(aq) + 3H_2O(l)$$

In one experiment 3.81 g of iodine was dissolved in 4 M sodium hydroxide solution. (Relative atomic masses: H = 1, O = 16, Na = 23, I = 127)

A) i How many moles of iodine were used?

ii What volume of 4 M sodium hydroxide solution would be just sufficient to react with the iodine?

B) What reagent(s) would you add to solid sodium iodide to produce pure hydrogen iodide gas?

C) Iodine also reacts with aqueous sodium thiosulphate solution producing sodium iodide solution.

i Write a balanced equation, with state symbols, for the reaction.

ii Suggest a procedure by which solid sodium iodide might be obtained from the reaction mixture.

(Nuffield 1980)

4 This question is about the reaction of iodine with potassium hydroxide, which forms potassium iodate(V) and potassium iodide.

$$3I_2(s) + 6KOH(aq) \longrightarrow KIO_3(aq) + 5KI(aq) + 3H_2O(l)$$

The stoichiometric amount of iodine was added to 100 cm^3 of 4.0 M potassium hydroxide solution. The reaction mixture was warmed until reaction was complete and was then cooled to 20 °C. As the reaction mixture cooled, white crystals were precipitated. The solubility curves of potassium iodate(V) and potassium iodide are given in the diagram.

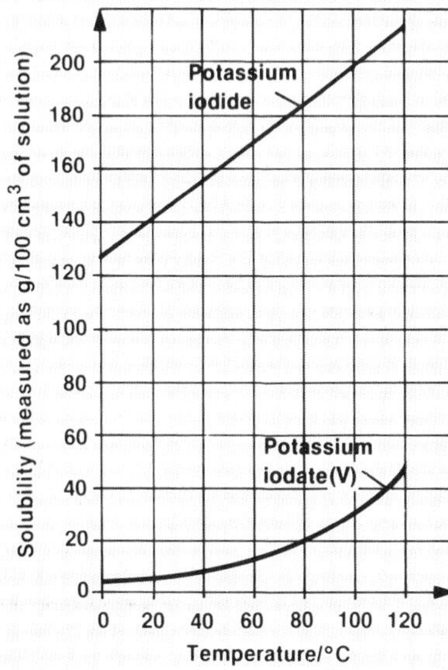

(Relative atomic masses: H = 1, O = 16, K = 39, I = 127)

A) What mass of solid iodine should be added to 100 cm^3 of 4.0 M potassium hydroxide?

B) How could you tell when the reaction was complete?

C) What mass of potassium iodate(V) and potassium iodide would be formed by the reaction?

i KIO_3;

ii KI.

D) At what temperature would the white crystals start to appear as the reaction mixture is cooled? Assume no water is lost during the reaction.

E) What would be the composition of the white crystals when the reaction mixture had been cooled to 20 °C?

(Nuffield 1984)

5 This question concerns the halogens chlorine, bromine and iodine and their compounds.

A) i By reference to the reaction between chlorine and bromide ions, explain what is meant by the terms *oxidation* and *reduction*.

ii Explain briefly why chlorine is a stronger oxidizing agent than bromine.

B) Explain why, although hydrogen chloride is conveniently prepared by warming sodium chloride with concentrated sulphuric acid, a similar method cannot be used for the preparation of hydrogen iodide.

C) On reacting chlorine with hot aqueous sodium hydroxide, *disproportionation* occurs.

i What is meant by the term *disproportionation*?

ii Write an equation for the reaction and give the oxidation states of chlorine in the products.

(London 1979, part question)

6 This question concerns the elements chlorine, bromine and iodine.

A) Give the electron configurations of the chlorine atom and the bromide ion.

B) Sketch graphs to indicate approximately the variation of:

i bond dissociation energy with atomic number;

ii ionic radius with atomic number.

C) Compare the reactions, if any, of chlorine and of iodine with water.

D) The following reaction takes place on warming:

$$3Br_2(l) + 6OH^-(aq) \longrightarrow 5Br^-(aq) + BrO_3^-(aq) + 3H_2O(l)$$

i Give the oxidation number of bromine in BrO_3^- and explain your reasoning.

ii What is the name given to this type of reaction?

iii Suggest the shape of BrO_3^- and explain your reasoning.

E) Which of the acids HClO and $HClO_3$ is the stronger? Briefly explain.

F) Use the following data to determine the enthalpy change for the reaction:

$$\tfrac{1}{2}I_2(s) \longrightarrow I^-(aq)$$

All steps in your calculation must be shown.

$$I_2(s) \longrightarrow I_2(g) \quad \Delta H = 30\,kJ\,mol^{-1}$$

$$\tfrac{1}{2}I_2(g) \longrightarrow I(g) \quad \Delta H = 76\,kJ\,mol^{-1}$$

$$I(g) \longrightarrow I^-(g) \quad \Delta H = -297\,kJ\,mol^{-1}$$

$$I^-(g) \longrightarrow I^-(aq) \quad \Delta H = -305\,kJ\,mol^{-1}$$

(London 1978)

24.7 Questions on s- and p-block elements in general

1 Write an account of the physical and chemical properties of the chlorides (where formed) of the elements of the third period (sodium to argon) showing how these properties depend on the position of the element in the period.

How do the chlorides of the elements of the second period (lithium to neon) differ from those of the elements of the third period in each group?

(Nuffield 1984)

2 **A)** Illustrate how the chemistry of the elements and compounds changes across a short period of the Periodic Table by writing a comparative account of the chemistry of:

i the elements sodium to chlorine;
ii their hydrides.

B) Discuss briefly the so-called diagonal relationship between either lithium and magnesium or beryllium and aluminium.

C) Explain why $SiCl_4$ is readily hydrolysed by cold water whereas CCl_4 is not. From this, suggest what may happen to NCl_3 and PCl_3 when they are brought into contact with water.

(Oxford 1986)

3 Give an account of the chlorides of the elements from sodium to chlorine in the Periodic Table, paying particular attention to:

A) methods of preparation including brief experimental details;
B) the nature of the bonding;
C) reactions with water.

0.800 g of a chloride of sulphur was hydrolysed with water and the solution diluted to 100 cm^3. 25.00 cm^3 of this solution was titrated with 0.100 M silver nitrate solution of which 29.60 cm^3 were required. Determine the empirical formula of the chloride.
[S = 32, Cl = 35.5] (London 1979)

25 The d-block elements

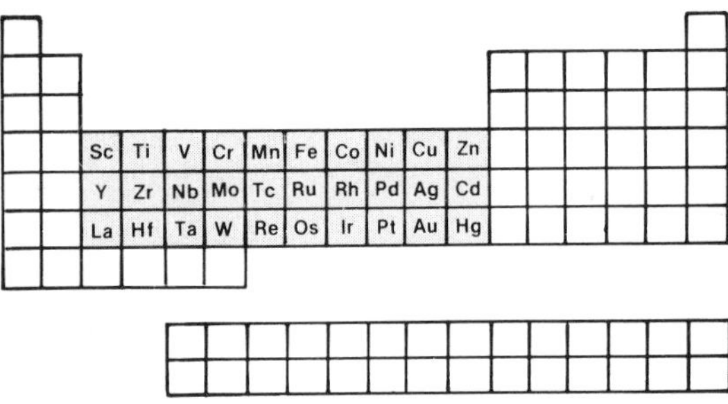

25.1 Introduction

The d-block elements are sometimes called the transition elements. We have seen that elements in the s- and p-blocks tend to have strong similarities to other members of their group, but that there are considerable trends as we move along a period. In the d-block, the elements show considerable similarities to one another even as we move across a period. This is shown in Figures 25.1 and 25.2. These remarkable similarities can be explained by looking at electronic structures. Figure 25.3 shows the energy levels of the first few orbitals drawn so as to show their distance from the nucleus. The electron arrangement of calcium is shown. The next level to be filled is 3d, which is where the d-block starts. Notice that although the 3d orbital

Figure 25.1 Atomic (covalent) radius against atomic number (van der Waals radius is used for the inert gases)

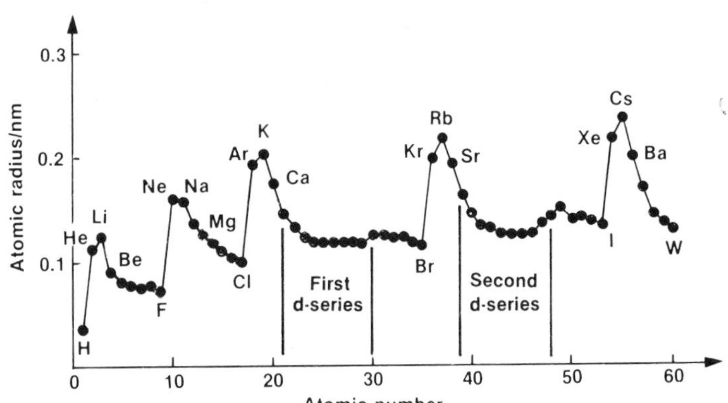

Figure 25.2 First ionization energy against atomic number

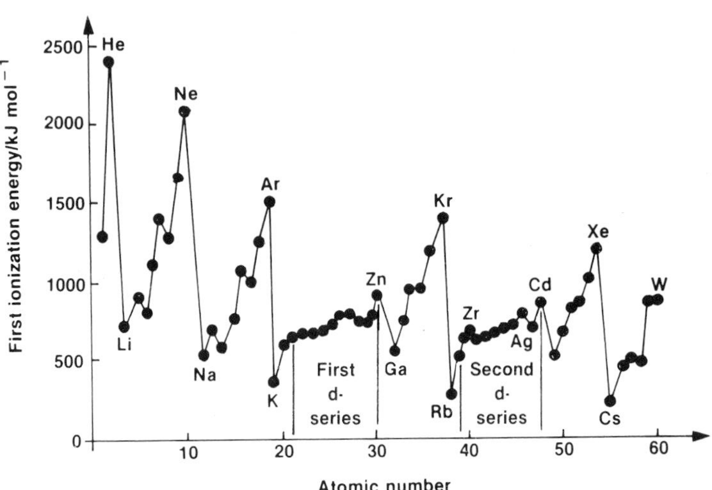

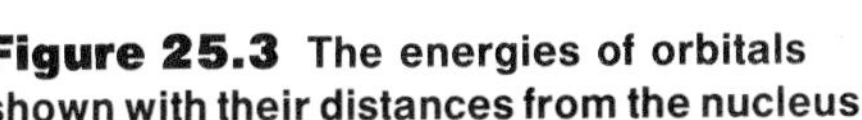
Figure 25.3 The energies of orbitals shown with their distances from the nucleus

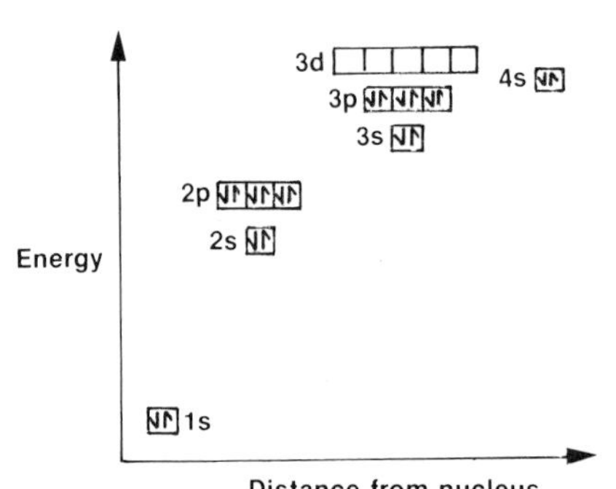

is slightly higher in energy, it is closer to the nucleus than the 4s orbital. So the next ten electrons will go into the 3d orbital, but the *outer* electrons will still be in the 4s orbital. This explains the similarities of the d-block elements. They all have the outer electron arrangement $4s^2$. (This is not quite true, as we shall see later, but it is approximately true.)

So the first d-block element (scandium) has the arrangement [Ar] 3d $4s^2$, the next (titanium) is [Ar] $3d^2 4s^2$, the next (vanadium) [Ar] $3d^3 4s^2$ and so on. Each atom has one more nuclear charge than the one before but the extra 3d electron, being in an inner shell, shields the nucleus quite effectively and the outer electrons feel almost the same effective nuclear charge. So their sizes are quite similar and so are many other properties.

The electronic structure of the d-block elements in Period 4 (the first transition series) are shown in Figure 25.4.

*The arrangements of chromium and copper do not quite fit the pattern. 4s and 3d are very close in energy and electrons can easily move from one to another. Chromium's and copper's arrangements show an extra stability which is associated with full ($3d^{10}$) or half full ($3d^5$) d orbitals.

The formal definition of a transition element is that it is one which forms *at least one compound* with a partially full d shell of electrons. Since scandium forms Sc^{3+} ($3d^0$) in all its compounds, and zinc forms Zn^{2+} ($3d^{10}$) in all its compounds they are not strictly transition elements, and their properties are much more comparable with those of the s-block elements.

Figure 25.4 Electronic structures of the elements of the first d-series

			3d	4s
Sc	[Ar]	$3d^1 4s^2$	↑	↑↓
Ti	[Ar]	$3d^2 4s^2$	↑ ↑	↑↓
V	[Ar]	$3d^3 4s^2$	↑ ↑ ↑	↑↓
Cr	[Ar]	$3d^5 4s^1$*	↑ ↑ ↑ ↑ ↑	↑
Mn	[Ar]	$3d^5 4s^2$	↑ ↑ ↑ ↑ ↑	↑↓
Fe	[Ar]	$3d^6 4s^2$	↑↓ ↑ ↑ ↑ ↑	↑↓
Co	[Ar]	$3d^7 4s^2$	↑↓ ↑↓ ↑ ↑ ↑	↑↓
Ni	[Ar]	$3d^8 4s^2$	↑↓ ↑↓ ↑↓ ↑ ↑	↑↓
Cu	[Ar]	$3d^{10} 4s^1$*	↑↓ ↑↓ ↑↓ ↑↓ ↑↓	↑
Zn	[Ar]	$3d^{10} 4s^2$	↑↓ ↑↓ ↑↓ ↑↓ ↑↓	↑↓

25.2 Physical properties of the elements

We have seen that physical properties of d-block metals, for example the first ionization energy and metallic radius, vary relatively little across a period. The same applies to many other properties like melting temperature, boiling temperature and hardness. The d-block metals can be said to be typical metals, being good conductors of heat and electricity, hard, strong, shiny, and having high melting and boiling temperatures. One notable exception is mercury, which is liquid at room temperature ($T_m = 234$ K). No simple explanation exists for this oddity. These physical properties, together with fairly low chemical reactivity, make transition metals extremely useful. Examples include iron (and its alloy steel) for vehicle bodies and to reinforce concrete, copper for water pipes and titanium for jet engine parts which encounter high temperatures. Melting and boiling temperatures are shown in **Figure 25.5**. The 'dips' at calcium ($3d^0$), manganese ($3d^5$) and zinc ($3d^{10}$) are caused by the extra stability of empty, half-full and full d shells. These electron arrangements make electrons less available for contribution to the 'pool' for metallic bonding thus weakening the metallic bonds and giving low melting and boiling temperatures.

Figure 25.5 Melting and boiling temperatures for the elements of the first d-series

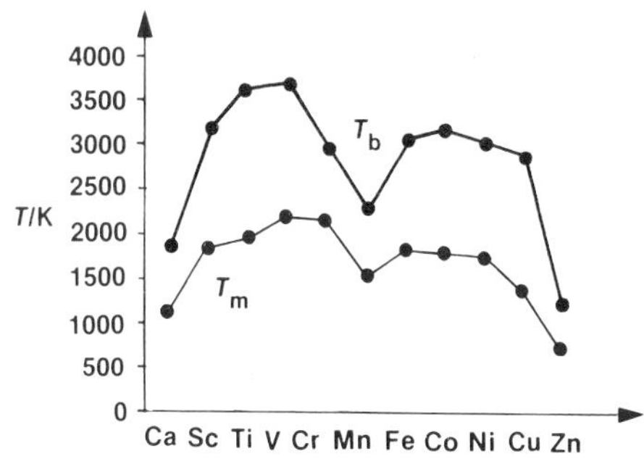

25.3 Chemical properties of the elements

The chemistry of d-block metals has four main features which are common to all the elements, except zinc and scandium which are not typical.

1. Variable oxidation states: typically, the d-block metals show more than one oxidation state in their compounds, as well as Ox = 0 in the element (see **Table 25.1**). This is in contrast with the s-block metals which have a fixed oxidation number in all their compounds (Ox = + I for Group I and Ox = + II for Group II).

2. Complex formation: d-block elements from **complexes**. A complex is a compound in which molecules or ions called **ligands** form dative covalent bonds with a metal atom or, more usually, a metal ion.

3. Colour: the majority of d-block metal ions are coloured. Some examples you may be familiar with are: hydrated copper(II) sulphate, blue;

nickel(II) carbonate, green; iron(III) chloride, brown. Contrast this with the s-block metals whose ions are colourless, e.g. sodium chloride, potassium nitrate and calcium oxide are all white, as are the compounds of zinc and scandium.

4. Catalysis. Many d-block metals and their compounds show catalytic activity. For example, iron in the Haber process, vanadium(V) oxide in the contact process and manganese(IV) oxide for the decomposition of hydrogen peroxide.

25.3.1 Variable oxidation states

Most d-block elements form compounds with different oxidation states. The variety is shown in **Table 25.1** for the first transition series.

Table 25.1 **Oxidation numbers exhibited by the elements of the first d series in their compounds**

Sc	Ti	V	Cr	Mn	Fe	Co	Ni	Cu	Zn
	+I	+I	+I	+I	+I	+I	+I	+I	
	+II	+II	+II	+II	+II	+II	+II	+II	+II
+III	+III	+III	+III	+III	+III	+III	+III	+III	
	+IV	+IV	+IV	+IV	+IV	+IV	+IV		
		+V	+V	+V	+V	+V			
			+VI	+VI	+VI				
				+VII					

The more commonly encountered states are shown in brown, though not all are stable. Some patterns emerge:

1. Except for scandium and zinc (which are not strictly transition metals) all the elements show the +I and +II oxidation states which correspond to forming bonds using only the 4s electrons.

2. There is an increase in the maximum oxidation number from scandium to manganese. For these metals, the maximum oxidation number corresponds to both 4s electrons and all the 3d electrons being used in bond formation.

For example, manganese ($3d^5 4s^2$) forms manganese(VII). From manganese to zinc, the maximum oxidation number decreases. This suggests that both 4s electrons and the *unpaired* d electrons can be used to form bonds, for example cobalt ($3d^7 4s^2$) has three unpaired d electrons so its maximum oxidation number in compounds is +V (two 4s electrons and three unpaired d electrons).

Note that only the lower oxidation states can actually exist as free ions, so that, for example, Mn^{2+} ions exist but no Mn^{7+} ions exist. In all manganese(VII) compounds, the manganese is covalently bonded, **Figure 25.6**.

Figure 25.6 **Bonding in the MnO_4^- ion**

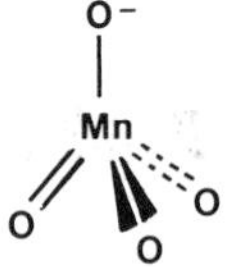

The anti-clockwise rule

Because the oxidation numbers of the d-block metals can vary in their compounds, they take part in many redox (electron transfer) reactions. The anti-clockwise rule (see **section 13.3.6**) can help to predict what happens in these reactions.

For example, the redox chemistry of iron is fairly straightforward as the only important oxidation numbers are 0 (Fe), +II (Fe^{2+}) and +III (Fe^{3+}).

We can look up the $E^\ominus$ values for the half reactions:

$$Fe^{2+}(aq) + 2e^- \rightleftharpoons Fe(s) \qquad E^\ominus = -0.44\,V$$

and

$$Fe^{3+}(aq) + e^- \rightleftharpoons Fe^{2+}(aq) \quad E^\ominus = +0.77\,V$$

and insert them on an $E^\ominus$–oxidation number diagram, **Figure 25.7** (overleaf), with any other half reactions that we are interested in, for example chlorine/chloride ions:

$$\tfrac{1}{2}Cl_2(aq) + e^- \rightleftharpoons Cl^-(aq) \quad E^\ominus = +1.36\,V$$

Figure 25.7 $E^\ominus$**–Ox diagram for iron compounds**

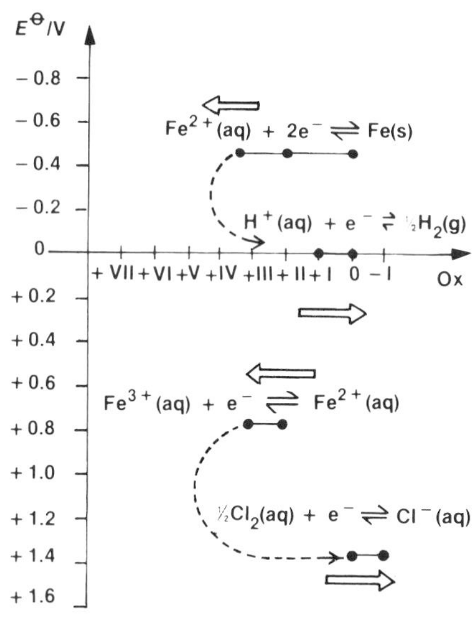

and hydrogen/hydrogen ions:

$$H^+(aq) + e^- \rightleftharpoons \tfrac{1}{2}H_2(g) \quad E^\ominus = 0.00\ V$$

The anti-clockwise rule tells us that half reactions at the top of the diagram will go from right to left and drive any reactions below them from left to right. So if we look at the reaction between Fe(s) and H^+(aq) we can predict that:

$$Fe^{2+}(aq) + 2e^- \rightleftharpoons Fe(s) \quad E^\ominus = -0.44\ V$$

will go from right to left so we change the sign of $E^\ominus$, and:

$$H^+(aq) + e^- \longrightarrow \tfrac{1}{2}H_2(g) \quad E^\ominus = 0.00\ V$$

will go from left to right.

Adding: $Fe(s) + 2H^+(aq) \longrightarrow Fe^{2+}(aq) + H_2(g) \quad E^\ominus = +0.44\ V$

In the same way we can predict from the diagram that chlorine will oxidize Fe^{2+} to Fe^{3+}:

$$Fe^{2+}(aq) + \tfrac{1}{2}Cl_2(aq) \longrightarrow Fe^{3+}(aq) + Cl^-(aq) \quad E^\ominus = +0.59\ V$$

Also, since $E^\ominus$ for this reaction is positive and almost 0.6 V, the reaction will essentially go to completion.

Both the above reactions are known to occur as predicted but you should remember that the anti-clockwise rule can only predict that a reaction *could* happen, not that it *must*. The reaction could be so slow as not to occur at all over a reasonable time scale.

25.3.2 Complex formation

A **complex** is formed when a d-block metal ion is surrounded by other molecules or ions which form dative covalent bonds with the d-block metal. The molecules or ions which form the dative bonds are called **ligands**. Ligands must therefore have a lone pair of electrons with which to form the dative bond. Ligands are negative ions or molecules with a dipole, the negative end of which forms the dative bond. Ligands which are commonly found include:

$$:Cl^-, \quad :NH_3, \quad :OH^-, \quad :OH_2$$

The number of ligands which surround the d-block metal ion is usually four or six. Typical complex ions include $[Co(NH_3)_6]^{3+}$ and $[CoCl_4]^{2-}$:

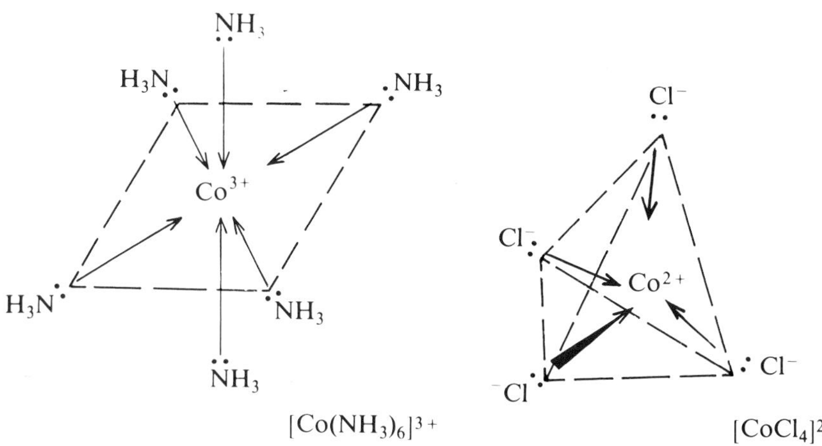

In $[Co(NH_3)_6]^{3+}$, since the metal ion has a charge of +3 and the ligands are all neutral, the complex ion has an overall charge of +3. The geometry of the ion is octahedral. In $[CoCl_4]^{2-}$, the metal ion has a charge of +2 and each of the four ligands has a charge of −1. The complex ion has an overall charge of −2. The geometry is tetrahedral.

The number of ligands surrounding the metal ion is called the **coordination number**. Coordination numbers of four and six are the most common. The usual geometries are tetrahedral and octahedral respectively. A few complexes of coordination number four adopt a square planar geometry, for example:

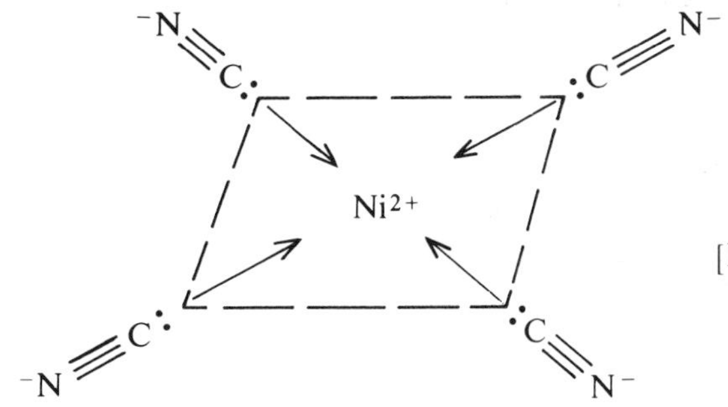

$[Ni(CN)_4]^{2-}$

Notice that complex ions may have a positive charge, negative charge or, less commonly, neutral complexes may be formed.

Bonding in complex ions

s-block metal ions may be surrounded by polar molecules with the negative ends of their dipoles towards the positive metal ion. This is called **solvation** (or hydration), see **section 16.3.4**. However, the bonding here is just the electrostatic attraction between the positive ion and the dipoles. In the case of transition metals, the dative covalent bonds are formed by lone pairs on the ligands donating lone pair electrons into the partly empty d-orbitals of the d-block ion.

Polydentate ligands—chelation

Some molecules have more than one site which can act as a ligand—these are called **polydentate ligands**. The word **dentate** comes from the Latin for tooth—they can 'bite' the metal ion more than once.

Bidentate ligands include:

1. Ethane-1, 2-diamine, sometimes called 1, 2-diaminoethane or ethylene diamine:

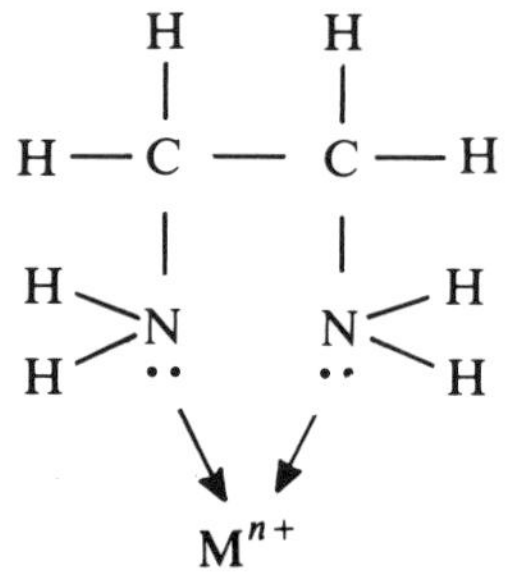

   M^{n+}

   Each nitrogen has a lone pair which can form a dative bond to the metal ion. The name of this ligand is often abbreviated to 'en', e.g. $[Cr(en)_3]^{3+}$.

2. Benzene-1, 2-diol, sometimes called 1, 2-dihydroxybenzene:

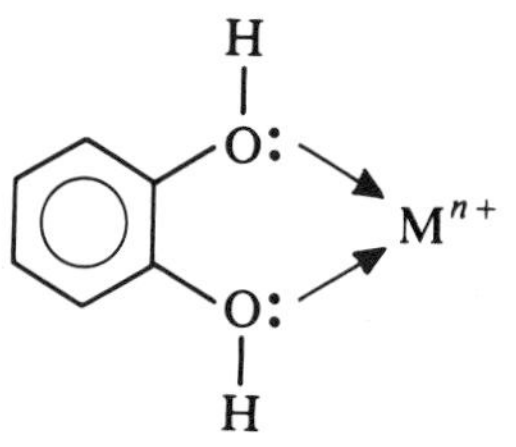

   M^{n+}

Both ligand sites of bidentate ions usually bond to the same metal forming a ring. However, they can act as bridges between two metal ions.

An important polydentate ligand is 1, 2-bis[bis(carboxymethyl)amino]-ethane, often still called by its non-systematic name ethylenediamine-tetraacetate or **edta**:

Model of edta

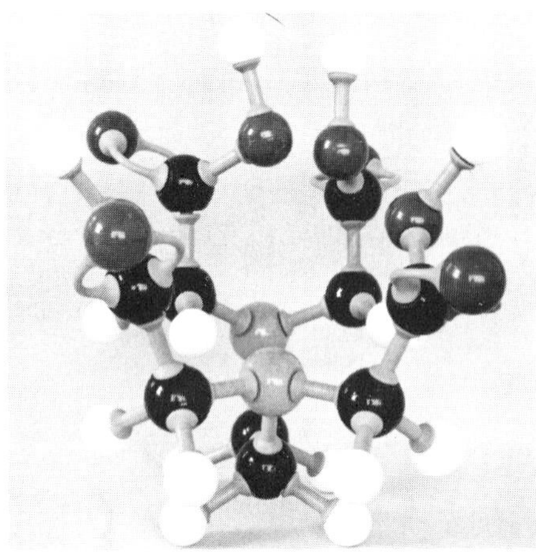

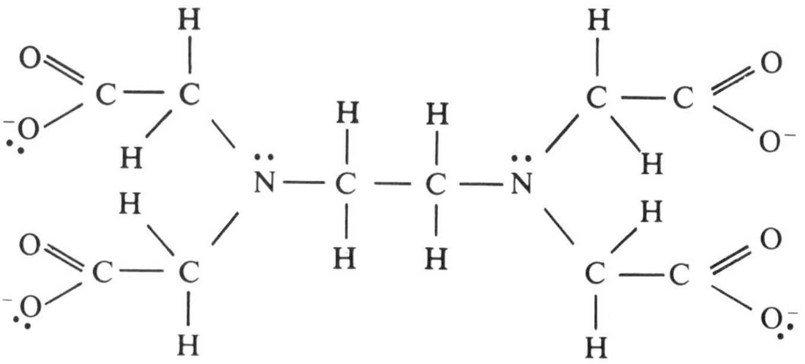

This can act as a hexadentate ligand using lone pairs on each of the four O^- ions and both the nitrogens. Complex ions with polydentate ligands are called **chelates** (pronounced key-lates) from the Greek word for claw. Chelates can be used to, in effect, remove free d-block metal ions from solution. For example, edta can be taken (under medical supervision) as an antidote to poisoning by d-block metal ions. The chelated ions have different properties from the uncomplexed ion.

Table 25.2 Names of some ligands

Ligand	Formula	Name
Ammonia	$:NH_3$	Ammine-
Carbon monoxide	$:C{\equiv}O$	Carbonyl-
Chloride ion	$:Cl^-$	Chloro-
Cyanide ion	$:C{\equiv}N^-$	Cyano-
Ethane-1,2-diamine	$H_2NCH_2CH_2NH_2$	Ethane-1,2-diamine-
Edta	See above	Edta-
Hydroxide ion	$:OH^-$	Hydroxo-
Water	$:OH_2$	Aqua-

Naming d-block metal complexes

Metal complexes are named by placing the name(s) of the ligand(s) (in alphabetical order if there are more than one, and ignoring prefixes like di, tri, etc. in deciding the alphabetical order) before the name of the metal atom. If the overall complex has a negative charge this is indicated by the suffix -ate. When the suffix -ate is used, the Latin version of the metal name is used so cuprate, ferrate, rather than copperate, ironate. The oxidation number of the metal is given in brackets at the end. The names used for ligands are given in **Table 25.2.**

Models of optical isomers of $CrCl_2en_2$

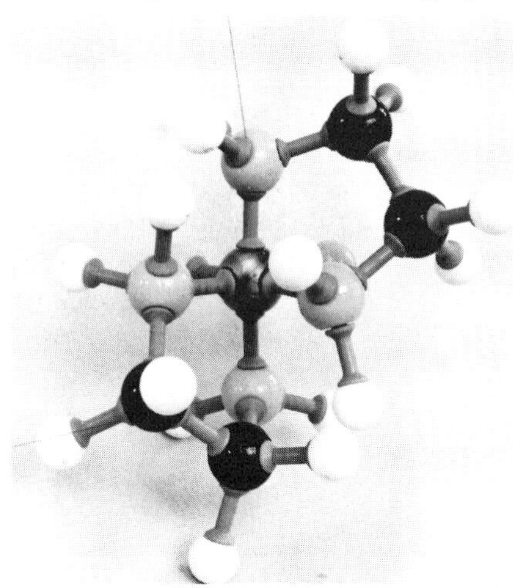

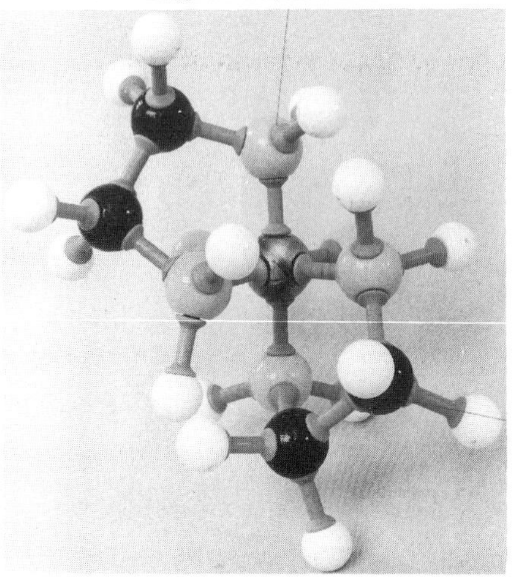

For example:

1. $[Cu(H_2O)_4]^{2+}$ tetraaquacopper(II) ion;
2. $[Fe(CN)_6]^{3-}$ hexacyanoferrate(III) ion (-ate is used as the ion is negatively charged, and called -ferrate not 'ironate');
3. $Ni(CO)_4$ tetracarbonylnickel(0);
4. $[CoCl_2(NH_3)_4]^+$ tetraamminedichlorocobalt(III) ion.

Note that the oxidation number of the metal is found by subtracting the total charge of the ligands (including sign) from the charge on the ion:

1. $2+ -0 \quad = +\text{II}$ because H_2O has no charge;
2. $3- -(-6) = +\text{III}$ because CN^- has charge -1;
3. $0 \quad -0 \quad = \quad 0$ because CO has no charge;
4. $1+ -(-2) = +\text{III}$ because NH_3 has no charge and Cl^- has charge -1

It is more important at this level to be able to work from the name to the correct formula rather than correctly name a given compound. The names of any other ions are given in the usual way so we might have a solution of hexaaquacobalt(II) chloride or potassium hexacyanoferrate(III).

Isomerism in complexes

Isomerism occurs where two or more compounds have the same molecular formula but different arrangements of their atoms in space. Three different types occur in d-block metal complexes:

IONIZATION ISOMERISM

Here ligands vary in their bonding. Three compounds exist with the overall formula $CrCl_3.6H_2O$. They are:

1. $[Cr(H_2O)_6]^{3+} + 3Cl^-$ violet hexaaquachromium(III) chloride;

2. $[CrCl(H_2O)_5]^{2+} + 2Cl^- + H_2O$ light green pentaaquachloro-chromium(III) chloride;

3. $[CrCl_2(H_2O)_4]^+ + Cl^- + 2H_2O$ dark green tetraaquadichloro-chromium(III) chloride.

As well as by their colours, these complexes can be distinguished by addition of silver nitrate. Only the free Cl^- ions will react to form a precipitate with silver nitrate, those bonded to the chromium will not. So 1 mole of compound (1) will react with 3 moles of silver nitrate, 1 mole of compound (2) with 2 moles of silver nitrate and 1 mole of compound (3) with 1 mole of silver nitrate.

GEOMETRICAL (*CIS–TRANS*) ISOMERISM

Here ligands differ in their position in space relative to one another. Compound (3) above also illustrates geometrical isomerism. The complex is octahedral and two arrangements of the ligands in space are possible, **Figure 25.8.** In (a) the chloride ligands are next to one another—this is called the *cis*-form. In (b) the chlorides are on opposite sides of the chromium atom—this is called the *trans*-form. So the full names of the compounds are:

Figure 25.8 *Cis–trans* **isomerism**

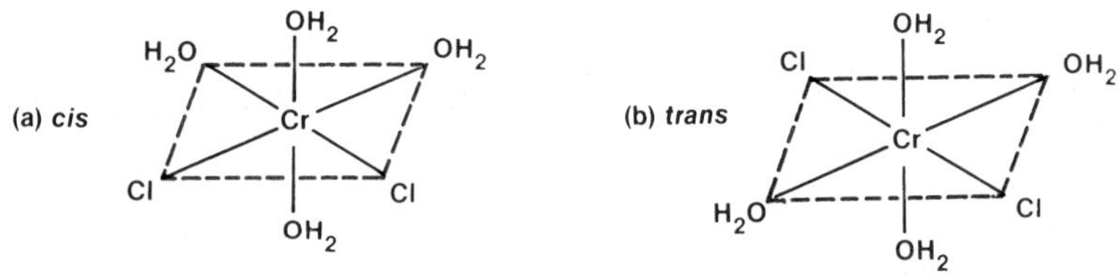

(a) *cis* (b) *trans*

a) *cis*-tetraaquadichlorochromium(III) chloride;
b) *trans*-tetraaquadichlorochromium(III) chloride.
You may need to look at models to convince yourself that there are only two isomers.

OPTICAL ISOMERISM

Here two isomers are non-identical mirror images of one another. This only occurs with two or more bidentate ligands. See **Figure 25.9.** You may need to look very carefully at **Figure 25.9** and the photographs on the previous page to see that the two molecules are not identical. Imagine rotating one of them round its vertical axis. If you are still unsure, you may need to look at models.

Optical isomers are said to be **chiral** which means 'handed'. They have identical chemical properties except that two solutions of equal concentrations, one of each isomer, rotate the plane of polarization of polarized light equally but in opposite directions. Optical isomerism is encountered more often in organic chemistry, see **section 26.7.2.**

Figure 25.9 Optical isomerism in $CrCl_2en_2$

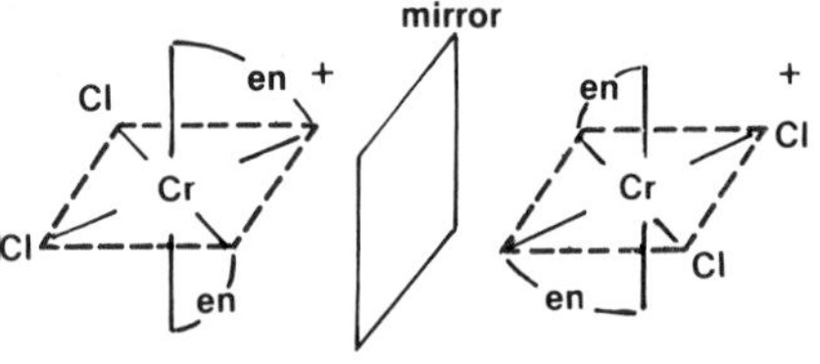

en is an abbreviation for ethane-1,2-diamine

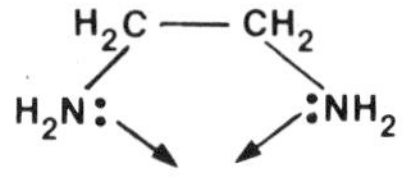

Competition between ligands—displacement reactions and stability constants

Some ligands may form stronger bonds than other ligands with a particular metal ion. In this case, the better ligand will displace the poorer one.

For example, copper(II) sulphate dissolved in water forms the tetraaquacopper(II) ion $[Cu(H_2O)_4]^{2+}$, which is responsible for the pale blue colour of the solution. Addition of chloride ions in high concentrations (e.g. concentrated hydrochloric acid) leads to the stepwise replacement of H_2O ligands by Cl^- ligands:

1. $Cu(H_2O)_4{}^{2+}(aq) + Cl^-(aq) \longrightarrow Cu(H_2O)_3Cl^+(aq) + H_2O(l)$ pale blue
2. $Cu(H_2O)_3Cl^+(aq) + Cl^-(aq) \longrightarrow Cu(H_2O)_2Cl_2(aq) + H_2O(l)$
3. $Cu(H_2O)_2Cl_2(aq) + Cl^-(aq) \longrightarrow Cu(H_2O)Cl_3{}^-(aq) + H_2O(l)$
4. $Cu(H_2O)Cl_3{}^-(aq) + Cl^-(aq) \longrightarrow CuCl_4{}^{2-}(aq) + H_2O(l)$ yellow

E

STEPWISE AND OVERALL STABILITY CONSTANTS

Each of the ligand replacement steps 1,2,3 and 4 in the formation of $CuCl_4^{2-}$ from $Cu(H_2O)_4^{2+}$ has a stability constant K_1, K_2, K_3, K_4 respectively.

$$K_1 = \frac{[Cu(H_2O)_3Cl^+(aq)]_{eqm}}{[Cu(H_2O)_4^{2+}(aq)]_{eqm}[Cl^-(aq)]_{eqm}}$$

$$K_2 = \frac{[Cu(H_2O)_2Cl_2(aq)]_{eqm}}{[Cu(H_2O)_3Cl^+(aq)]_{eqm}[Cl^-(aq)]_{eqm}}$$

$$K_3 = \frac{[Cu(H_2O)Cl_3^-(aq)]_{eqm}}{[Cu(H_2O)_2Cl_2(aq)]_{eqm}[Cl^-(aq)]_{eqm}}$$

$$K_4 = \frac{[CuCl_4^{2-}(aq)]_{eqm}}{[Cu(H_2O)Cl_3^-(aq)]_{eqm}[Cl^-(aq)]_{eqm}}$$

Now,

$$K_1 \times K_2 \times K_3 \times K_4$$

$$= \frac{\cancel{[Cu(H_2O)_3Cl^+(aq)]_{eqm}}}{[Cu(H_2O)_4^{2+}(aq)]_{eqm}[Cl^-(aq)]_{eqm}}$$

$$\times \frac{\cancel{[Cu(H_2O)_2Cl_2(aq)]_{eqm}}}{\cancel{[Cu(H_2O)_3Cl^+(aq)]_{eqm}}[Cl^-(aq)]_{eqm}}$$

$$\times \frac{\cancel{[Cu(H_2O)Cl_3^-(aq)]_{eqm}}}{\cancel{[Cu(H_2O)_2Cl_2(aq)]_{eqm}}[Cl^-(aq)]_{eqm}}$$

$$\times \frac{[CuCl_4^{2-}(aq)]_{eqm}}{\cancel{[Cu(H_2O)Cl_3^-(aq)]_{eqm}}[Cl^-(aq)]_{eqm}}$$

$$= \frac{[CuCl_4^{2-}(aq)]_{eqm}}{[Cu(H_2O)_4^{2+}(aq)]_{eqm}[Cl^-(aq)]^4_{eqm}}$$

But this expression is just K_c—the overall stability constant (see main text).

So:

$$K_c = K_1 \times K_2 \times K_3 \times K_4$$

Taking logs:

$$\log K_c = \log K_1 + \log K_2 + \log K_3 + \log K_4$$

This is a general result for any set of stepwise ligand replacement reactions. The values in this example are:

$$\log K_c = 5.62$$
$$\log K_1 = 2.80$$
$$\log K_2 = 1.60$$
$$\log K_3 = 0.49$$
$$\log K_4 = 0.73$$

Remember if $K > 1$, the equilibrium is over to the right. This is equivalent to $\log K$ being positive.

The overall reaction is:

5. $Cu(H_2O)_4^{2+}(aq) + 4Cl^-(aq) \longrightarrow CuCl_4^{2-}(aq) + 4H_2O(l)$

Each of the equilibria 1–4 lies somewhat to the right, implying that Cl^- is a better ligand than H_2O.

The equilibrium law expression is:

$$K_c = \frac{[CuCl_4^{2-}(aq)]_{eqm}}{[Cu(H_2O)_4^{2+}(aq)]_{eqm}[Cl^-(aq)]_{eqm}^4} = 3.9 \times 10^5\ dm^{12}\ mol^{-4}$$

(Remember: pure liquids do not appear in K_c expressions, see **section 11.4.4.**) The overall equilibrium is well over to the right and $CuCl_4^{2-}(aq)$ predominates over $Cu(H_2O)_4^{2+}(aq)$. If chloride ions, say in concentrated hydrochloric acid, are added to copper sulphate solution the pale blue colour of $Cu(H_2O)_4^{2+}(aq)$ is replaced by the yellow–green of $CuCl_4^{2-}(aq)$. K_c is called the **overall stability constant** for $CuCl_4^{2-}(aq)$. The larger K_c is, the more stable the complex.

Ammonia is an even better ligand than Cl^-. It will displace both water from $Cu(H_2O)_4^{2+}(aq)$ and Cl^- from $CuCl_4^{2-}(aq)$ forming the dark blue $Cu(NH_3)_4^{2+}(aq)$:

$$Cu(H_2O)_4^{2+}(aq) + 4NH_3(aq) \longrightarrow Cu(NH_3)_4^{2+}(aq) + 4H_2O(l)$$

The equilibrium constant for the overall replacement of water by ammonia

$$K_c = \frac{[Cu(NH_3)_4^{2+}(aq)]_{eqm}}{[Cu(H_2O)_4^{2+}(aq)]_{eqm}[NH_3(aq)]_{eqm}^4} = 1.5 \times 10^{13}\ dm^{12}\ mol^{-4}$$

Thus the overall stability constant for $Cu(NH_3)_4^{2+}(aq)$ is larger than that for $CuCl_4^{2-}(aq)$, showing that NH_3 is a better ligand than either Cl^- or H_2O.

It is usual to refer to the log (i.e. $\log_{10}$) of stability constants. Those for some Cu^{2+} complexes are given in **Table 25.3.**

able 25.3 Stability constants for some opper(II) complexes

	Ligand	Complex	$\log K_c$
Monodentate	Cl^-	$CuCl_4^{2-}$	5.6
	NH_3	$Cu(NH_3)_4^{2+}$	13.2
Bidentate	OH, C, O⁻, O	Cu (OH, C, O, O) 2	16.9
	OH, OH	Cu (OH, OH) 2, 2+	25
Tetradentate— in this situation— potentially hexadentate	edta	$Cu(edta)^{2+}$	18.9

The larger $\log K_c$, the more stable the complex.

Notice that the polydentate ligands have noticeably larger values of $\log K_c$ than monodentate ones. (Since these are logs, a value of 16.9 represents a K_c of about 5×10^3 times larger than does a value of 13.2.) This increased stability of complexes with polydentate ligands is mainly due to the entropy change of the reaction. Compare:

$$Cu(H_2O)^{2+}(aq) + 4Cl^-(aq) \longrightarrow CuCl_4^{2-}(aq) + 4H_2O(l)$$
$$\text{5 entities} \longrightarrow \text{5 entities}$$

with:

$$Cu(H_2O)_4^{2+}(aq) + edta \longrightarrow Cu(edta)^{2+}(aq) + 4H_2O(l)$$
$$\text{2 entities} \longrightarrow \text{5 entities}$$

The larger number of entities on the right in the second reaction means that there is an entropy increase as the reaction goes from left to right thus favouring the formation of chelates (complexes with polydentate ligands).

25.3.3 Colour of d-block metal compounds

Colours and colour changes are among the most striking aspects of d-block metal chemistry. A good example is the reaction of zinc with a solution containing a vanadium(V) compound. The solution gradually changes from the yellow of vanadium(V) to blue vanadium(IV) to green vanadium(III) to the mauve of vanadium(II) as the zinc gradually reduces the vanadium compounds, see **Plates 18–21.** We have seen in **section 8.3** that when electrons in a substance move from one energy level to a higher one, they absorb a quantum of electromagnetic energy equal in energy to the gap between the two levels. The frequency v of this is given by the equation $E = hv$ where E is the energy gap and h Planck's constant. If v is in the visible region of the spectrum, this will result in the substance being coloured. If a substance absorbs, say, green light, it will let through red and blue and thus appear purple. See **Figure 25.10.** In an isolated atom, all five d orbitals have exactly the same energy. However, when ligands approach the metal atom or ion, they affect the energy levels of different d orbitals differently, raising some slightly and lowering some slightly, see box on the next page.

igure 25.10 The d-block metal ompounds absorb light because of lectrons moving from one d-orbital to nother

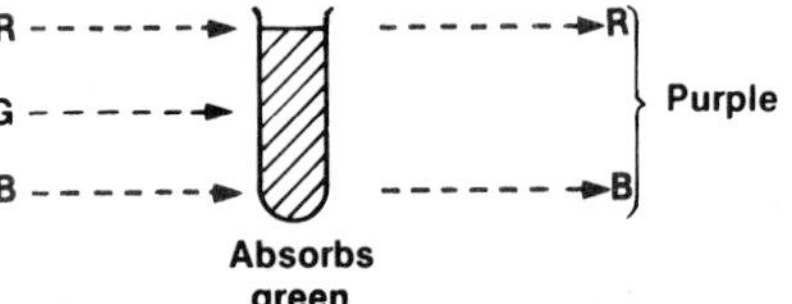

The resulting energy gap between the d orbitals turns out to correspond to frequencies of electromagnetic radiation in the visible region of the spectrum so most transition metal compounds are coloured. Chemical changes such as replacing one ligand by another will slightly affect the energies of the d orbitals and produce a compound with a different colour. Electrons can only move from one d orbital to another if there are spaces in the d orbitals, so Zn^{2+} compounds which have the electron arrangement $3d^{10}$ (i.e. the d orbitals are full) are not coloured. Sc^{3+} compounds ($3d^0$) have no d electrons and are therefore also colourless as, of course, are compounds of s-block metals. Compounds of s-block elements which also contain a transition metal like, for example, potassium manganate(VII), $KMnO_4$, may, of course, be coloured due to the presence of the transition metal.

COLOUR OF TRANSITION METAL IONS

The shapes of the five d orbitals are shown on an x, y, z axis system:

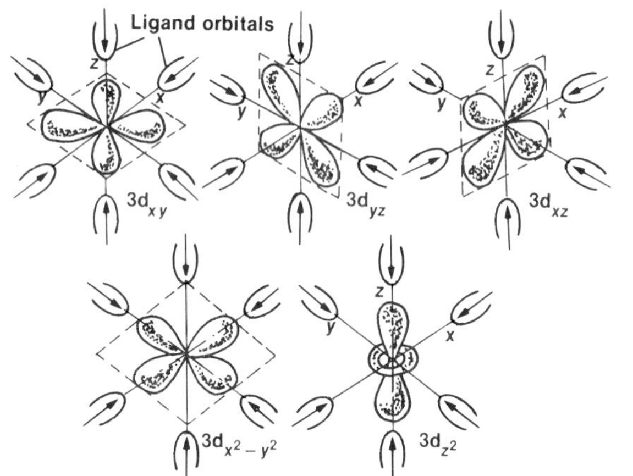

The shapes of the five 3d orbitals

The labels $3d_{xy}$, $3d_{yz}$, etc. simply distinguish the different d orbitals.

Think of an octahedral complex in which the six ligands approach the d-block metal ion along the x, y, and z axes. The orbitals on the ligands repel the d orbitals which point along the axes more than those which do not. So the energies of the $3d_{x^2-y^2}$ and $3d_{z^2}$ orbitals are raised compared with the other three orbitals. The resulting energy gap means that transitions between d orbitals absorb quanta of electromagnetic energy and cause colour.

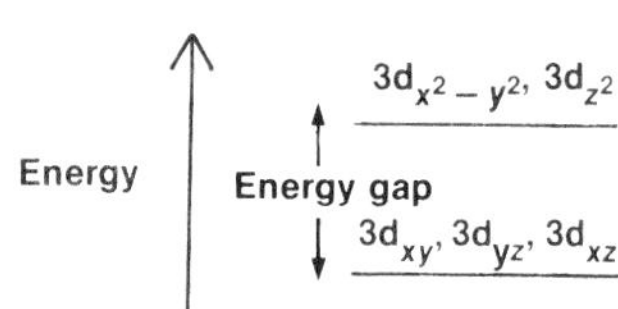

FINDING THE FORMULAE OF COMPLEXES USING A COLORIMETER

Because of their colours, transition metal complexes can often be investigated using a colorimeter. The intense blood-red complex formed between iron(III) ions and thiocyanate ions is a good example.

We make two dilute solutions of the same molarity, one of, say, iron(III) sulphate and the other of potassium thiocyanate. Both these solutions are virtually colourless. Eleven mixtures are then made in the proportions shown below and their colours measured on a colorimeter.

Mixture number	1	2	3	4	5	6	7	8	9	10	11
Volume of Fe^{3+}(aq) solution/cm^3	0	1	2	3	4	5	6	7	8	9	10
Volume of SCN^-(aq) solution/cm^3	10	9	8	7	6	5	4	3	2	1	0

Mixture number 1 has no colour as there is no Fe^{3+}(aq) and therefore no complex. Mixture 11 has no colour as there is no SCN^-(aq) and therefore no complex.

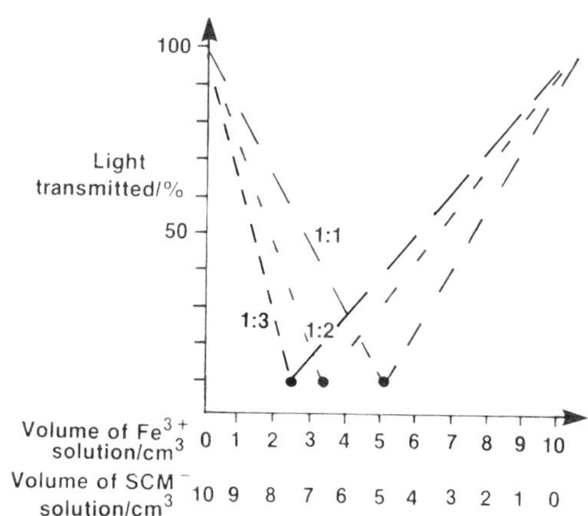

The maximum colour and therefore the minimum light transmitted is produced when the Fe^{3+}(aq) and SCN^-(aq) are mixed in the same ratio as they are in the complex. Graphs of the expected results for complexes with ratios of $[Fe^{3+}]:[SCN^-]$ of 1:1, 1:2 and 1:3 are shown. The iron(III)/thiocyanate complex is actually 1:1, the formula being $[FeSCN(H_2O)_5]^{2+}$.

25.3.4 Catalytic activity

Transition metals as catalysts may be divided into two distinct groups—**homogeneous catalysts** which involve transition metal compounds and **heterogeneous catalysts** where metals or metal compounds may be involved.

Homogeneous catalysts

We have already discussed in **section 14.6.2** the reaction of peroxodisulphate(VI) ions which oxidize iodide ions to iodine and explained how it is catalysed by *either* Fe^{2+} or Fe^{3+} ions:

$$S_2O_8^{2-}(aq) + 2I^-(aq) \xrightarrow[\text{catalyst}]{Fe^{2+}\text{ or }Fe^{3+}} I_2(aq) + 2SO_4^{2-}(aq)$$

$$\overset{-II}{^-O}-\underset{\underset{-II}{O}}{\overset{\overset{-II}{O}}{S}}-\overset{-I}{O}-\overset{-I}{O}-\underset{\underset{-II}{O}}{\overset{\overset{-II}{O}}{S}}-\overset{-II}{O^-}\ (aq) + \overset{-I}{2I^-}(aq) \xrightarrow[\text{catalyst}]{Fe^{2+}\text{ or }Fe^{3+}} \overset{0}{I_2}\ (aq) + 2\,\overset{-II}{^-O}-\underset{\underset{-II}{O}}{\overset{\overset{-II}{O}}{S}}-\overset{-II}{O^-}(aq)$$

by peroxodisulphate ions oxidizing Fe^{2+} to Fe^{3+} which then oxidizes I^- to I_2. Alternatively, I^- reduces Fe^{3+} to Fe^{2+}, which then reduces peroxodisulphate ions. Note that while the uncatalysed reaction occurs between two negatively charged ions (peroxodisulphate and iodide), each of the steps in the catalysed reaction occurs between oppositely charged ions ($S_2O_8^{2-}$ and Fe^{2+} in step 1 and I^- and Fe^{3+} in step 2). This may explain the increased rate.

An $E^\ominus$–Ox diagram is useful in explaining this, **Figure 25.11.** You may be able to see that *any* half cell with $E^\ominus$ between +0.54 V and +2.01 V is a possible catalyst for the peroxodisulphate/iodide reaction, while those outside this range are not.

Examples of possible catalysts include:

$$Co^{3+}(aq)/Co^{2+}(aq)$$
$$MnO_4^-(aq) + 8H^+(aq)/Mn^{2+}(aq) + 4H_2O(l)$$
$$CrO_4^{2-}(aq) + 8H^+(aq)/Cr^{3+}(aq) + 4H_2O(l)$$

These have been entered on the diagram.

Figure 25.11 $E^\ominus$–Ox diagram for the $S_2O_8^{2-}/I^-$ reaction catalysed by iron ions

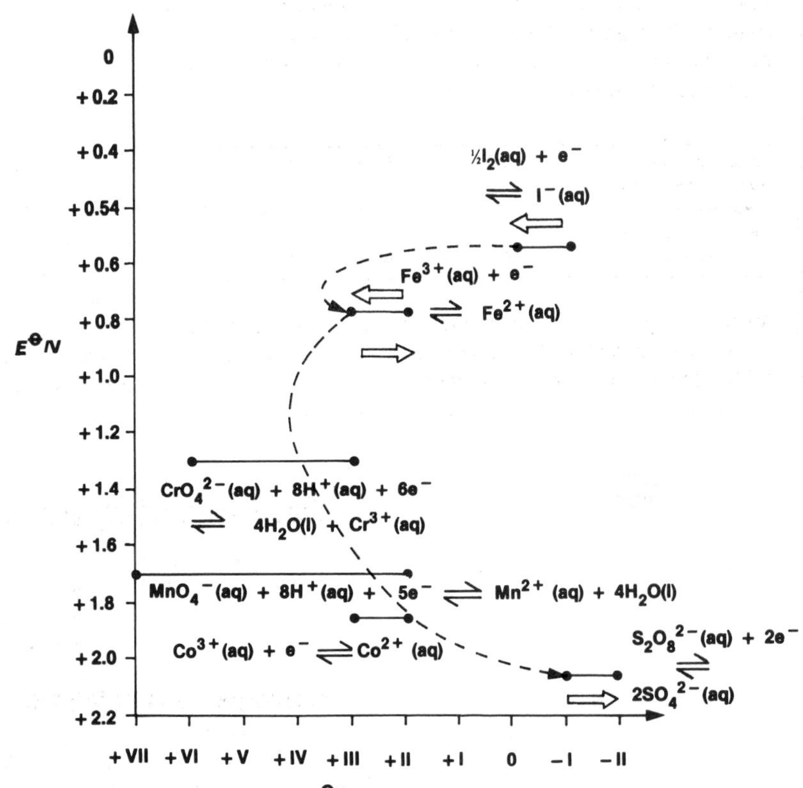

E

USING $E^\ominus$ VALUES TO PREDICT FEASIBLE REACTIONS—THE EFFECT OF CONDITIONS

It is important to realize that any predictions about reaction feasibility made using $E^\ominus$ values apply only to standard conditions (1 mol dm^{-3} solutions, 298 K and 10^2 kPa pressure). Changing the conditions slightly can change values of $E^\ominus$ and affect the predictions. We can illustrate this with an example from the chemistry of iron.

In acidic solution, the value of $E^\ominus$ for:

$$Fe^{3+}(aq) + e^- \longrightarrow Fe^{2+}(aq)$$

is +0.77 V, so Fe^{3+} will oxidize I^- to I_2. ($E^\ominus$ for $\frac{1}{2}I_2(aq) + e^- \rightleftharpoons I^-$ = +0.54 V.) However, if CN^-(aq) is added, the potential for the Fe(III)/Fe(II) system changes. The iron ions now exist as $Fe(CN)_6^{3-}$(aq) and $Fe(CN)_6^{4-}$(aq). The potential for:

$$Fe(CN)_6^{3-}(aq) + e^- \longrightarrow Fe(CN)_6^{4-}(aq)$$

is +0.36 V and iodine can oxidize Fe(II) to Fe(III) in the presence of CN^- ions.

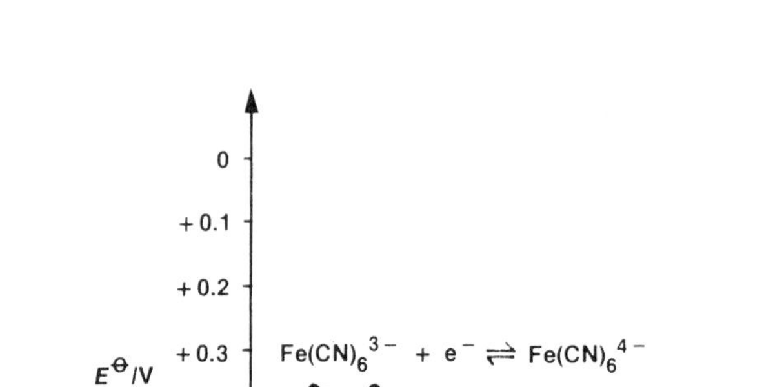

$E^\ominus$–Ox diagram for some iron species

E

REDOX TITRATIONS AND REDOX INDICATORS

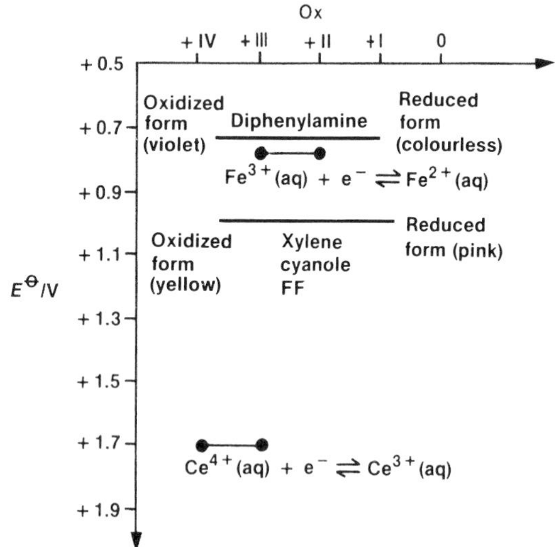

Note: **It is not easy to assign oxidation numbers to the oxidized and reduced forms of the indicators. This does not affect the argument.**

$E^\ominus$–Ox diagram for some redox indicators

It is possible to titrate a reducing agent against an oxidizing agent if a suitable indicator can be found. For example, the reducing agent Fe^{2+}(aq) can be titrated with the oxidizing agent Ce^{4+}(aq):

$$Ce^{4+}(aq) + Fe^{2+}(aq) \longrightarrow Ce^{3+}(aq) + Fe^{3+}(aq)$$

$$E^\ominus_{reaction} = +0.93\ V$$

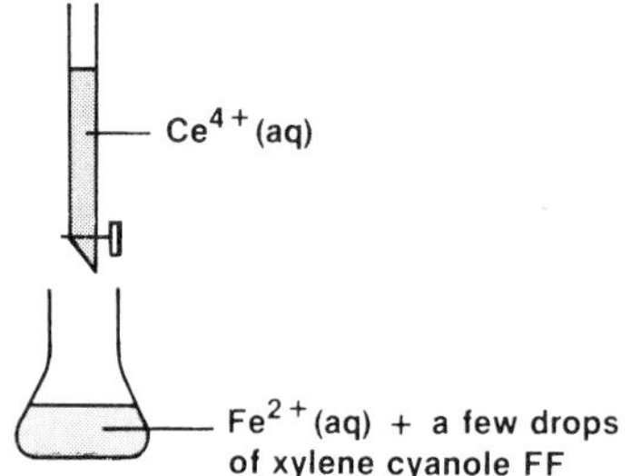

Titrating Fe^{2+}(aq)

The $E^\ominus$ value shows that this reaction goes to completion. If a few drops of the indicator xylene cyanole FF are added to Fe^{2+}(aq), the $E^\ominus$ values show that it will be reduced to the pink reduced form. As Ce^{4+}(aq) is added from the burette, it will react immediately with Fe^{2+}(aq). As soon as enough Ce^{4+}(aq) has been added to react with *all* the Fe^{2+}(aq), Ce^{4+}(aq) will be in excess. The $E^\ominus$ values show that Ce^{4+}(aq) will oxidize the indicator to the yellow oxidized form. So the indicator is pink just before the end point and yellow just after it. Can you explain why diphenylamine is not a suitable indicator for the reaction?

In practice, only Mn(VII) actually catalyses the reaction, a useful reminder that predictions made by the anti-clockwise rule can only suggest that a reaction is possible.

Heterogeneous catalysts

Section 14.6.1 describes how heterogeneous catalysts work. They adsorb reactants on their surfaces by forming weak chemical bonds with them. This has two effects—weakening bonds within the reactant and holding the reactants close together on the metal surface in the correct orientation for reaction. d-block metals have partly full d orbitals which can be used to form bonds with adsorbed reactants. Thus they are effective heterogeneous catalysts.

Examples of d-block elements and compounds used as heterogeneous catalysts for industrial processes are shown in **Table 25.4.**

able 25.4 Some examples of eterogeneous catalysis involving d-block ıetals or their compounds

Process	Product	Catalyst
Haber	Ammonia	Iron
Contact	Sulphuric acid	Vanadium(V) oxide
Hydrogenation of oils	Margarine	Nickel
Ostwald	Nitric acid	Platinum/rhodium
Oxidation of propan-2-ol	Propanone	Copper

25.4 Examples of the chemistry of some d-block elements

The rest of this section discusses the chemistry of selected d-block metals and their compounds.

25.4.1 Scandium

Scandium shows no typical transition metal properties. It forms only one oxidation state, +III, apart from Ox = 0, and the Sc^{3+} ion is colourless. This is because Sc^{3+} has no d electrons.

The metal is similar to calcium in its reactivity with water:

$$\underset{\text{scandium}}{Sc(s)} + \underset{\text{water}}{3H_2O(l)} \longrightarrow \underset{\text{scandium hydroxide}}{Sc(OH)_3(aq)} + \underset{\text{hydrogen}}{1\tfrac{1}{2}H_2(g)}$$

25.4.2 Vanadium

Redox chemistry of vanadium

Vanadium forms four important oxidation states: +II, +III, +IV and +V. All can be obtained in aqueous solution. **Table 25.5** summarizes the species present in acidic solution.

Table 25.5 Vanadium species in ıqueous solution

x	Species	Colour
-V	$VO_2^+(aq)$	Yellow
-IV	$VO^{2+}(aq)$	Blue
-III	$V^{3+}(aq)$	Green
-II	$V^{2+}(aq)$	Violet

The higher oxidation states of vanadium V^{4+} and V^{5+} are strongly polarizing due to their small size and high charge so that simple aqueous ions $V^{4+}(aq)$ and $V^{5+}(aq)$ cannot exist. Covalent bonds with oxygen atoms are formed and the vanadium(V) state is usually represented as $VO_2^+(aq)$ and the vanadium(IV) state as $VO^{2+}(aq)$. A similar situation occurs with high oxidation states of other transition metals. An $E^\ominus$–oxidation number diagram helps us to understand the aqueous redox chemistry of vanadium. The one in **Figure 25.12** shows the potentials in acid solution. Half reactions involving vanadium compounds are in brown.

Figure 25.12 $E^{\ominus}$**–Ox diagram for some vanadium oxidation states**

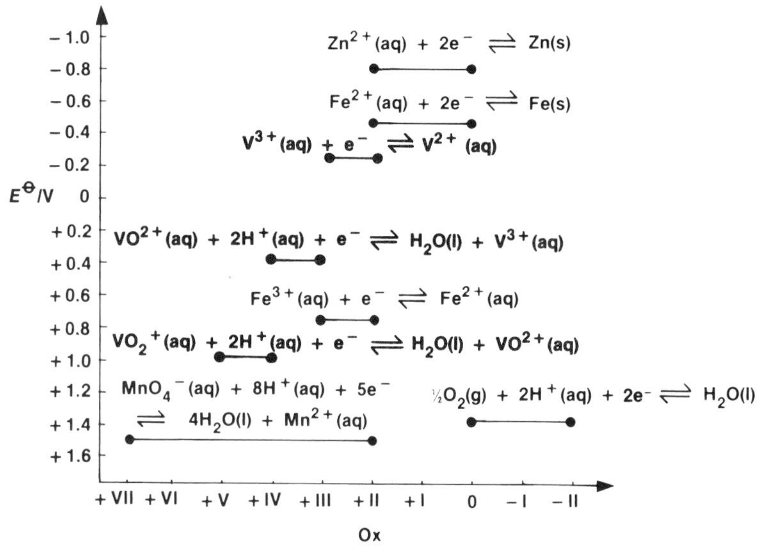

From this we can make a number of predictions:

- Zinc could reduce $\overset{+V}{VO_2^+}$(aq) to $\overset{+IV}{VO^{2+}}$(aq), $\overset{+IV}{VO^{2+}}$(aq) to $\overset{+III}{V^{3+}}$(aq), and $\overset{+III}{V^{3+}}$(aq) to $\overset{+II}{V^{2+}}$(aq)—a series of reactions that actually occurs.
- Oxygen will oxidize any vanadium species to $\overset{+V}{VO_2^+}$. In fact, $\overset{+II}{V^{2+}}$ is oxidized to $\overset{+III}{V^{3+}}$ and $\overset{+III}{V^{3+}}$ to $\overset{+IV}{VO^{2+}}$ by oxygen but not $\overset{+IV}{VO^{2+}}$ to $\overset{+V}{VO_2^+}$ for kinetic reasons. $\overset{+IV}{VO^{2+}}$ is the most stable state in aqueous solution.
- Fe^{2+} will reduce $\overset{+V}{VO_2^+}$ to $\overset{+IV}{VO^{2+}}$, another reaction which is observed.

These reactions can be followed by observing the colours of the differen vanadium species formed.

Oxides of vanadium

The oxides vanadium(II) oxide, VO, vanadium(III) oxide, V_2O_3 vanadium(IV) oxide, VO_2 and vanadium(V) oxide, V_2O_5 are known. Thei properties illustrate a general trend in the d-block of acidity of oxide increasing with oxidation number of the metal. Vanadium(II) oxide an vanadium(III) oxide are basic, vanadium(IV) oxide amphoteric an vanadium(V) oxide acidic.

25.4.3 Chromium

Chromium metal is bluish-white. It has a high melting temperatur (2130 K) and is quite resistant to chemical attack hence its most familia use as a shiny plating on car bumpers, bicycle handlebars, etc.

Redox chemistry of chromium

Chromium has an extensive redox chemistry, the most important oxidatio states being Cr(II) (blue), Cr(III) (green) and Cr(VI) (yellow/orange). I aqueous solution chromium(VI) exists in two forms, the chromate(VI) ion CrO_4^{2-}, and the dimeric dichromate(VI) ion, $Cr_2O_7^{2-}$.

The orange dichromate ion is stable in acid solution and the yellow chromate ion in alkalis, as shown by the equilibrium:

$$2CrO_4^{2-}(aq) + 2H^+(aq) \rightleftharpoons Cr_2O_7^{2-}(aq) + H_2O(l)$$

Figure 25.13 The structures of the chromate(VI) and dichromate(VI) ions

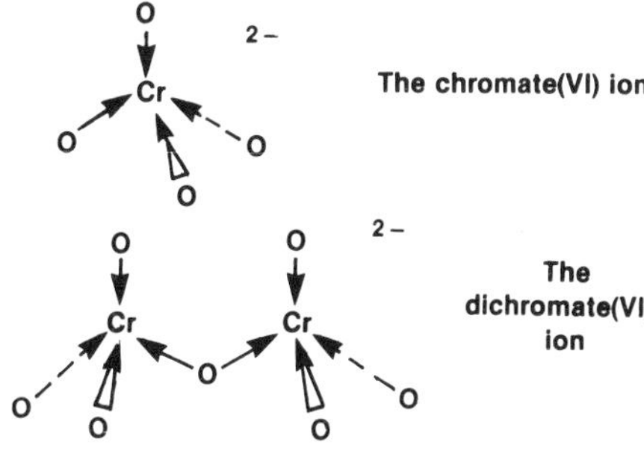

The chromate ion is tetrahedral having a similar structure to the sulphat ion, **Figure 25.13**, while the dichromate ion consists of two tetrahedr linked by a bridging oxygen.

Chromium(III) is the most stable oxidation state in aqueous solution

Chromium(II) is easily oxidized to chromium(III) and can only be prepared in the absence of air. If Cr(II) is prepared by reduction of chromium(VI) by zinc in acid solution, the hydrogen produced can be used to exclude air. The electrode potentials in **Figure 25.14** show that this reaction is feasible.

igure 25.14 $E^{\ominus}$–Ox diagram for me oxidation states of chromium

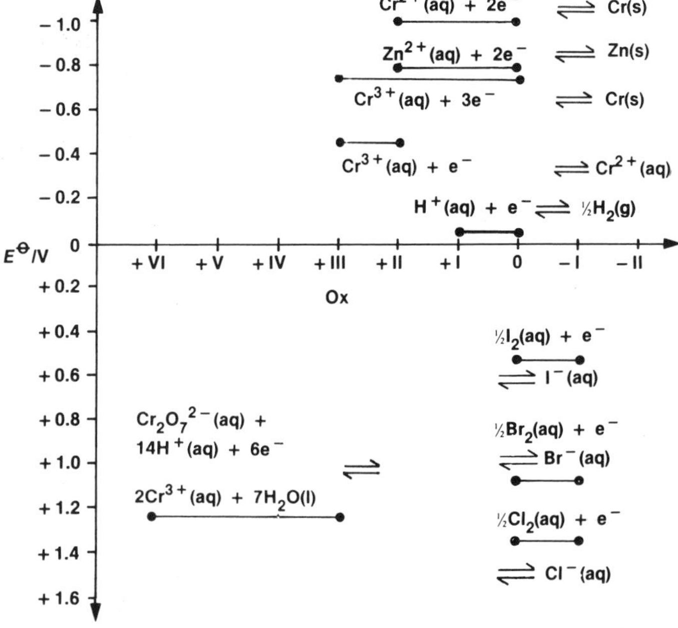

The chromium(VI) state is powerfully oxidizing and $Cr_2O_7^{2-}/H^+$ is much used as an oxidizing agent in organic chemistry. In the breathalyser, ethanol is oxidized by orange dichromate ions which are reduced to green chromium(III).

You should be able to see from the $E^{\ominus}$–Ox diagram that $Cr_2O_7^{2-}(aq)$ can oxidize I^- to I_2 and Br^- to Br_2, but not Cl^- to Cl_2. Also, the diagram shows that chromium(II) will reduce all the halogens to halide ions.

Oxides of chromium

The two oxides of chromium, chromium(III) oxide, Cr_2O_3, and chromium(VI) oxide, CrO_3, are typical in that the oxide with the higher oxidation number is more acidic. CrO_3 dissolves in water to give chromic(VI) acid, H_2CrO_4, which, as both an acid and an oxidizing agent, is useful for cleaning laboratory glassware. Chromium(III) oxide is amphoteric.

Chromium(III) oxide contains the Cr^{3+} ion while chromium(VI) oxide is covalently bonded.

25.4.4 Manganese

Manganese metal is brittle and fairly reactive, and thus has relatively few uses. It is, however, important as a constituent of many alloys. In its compounds, manganese displays more oxidation states than any other d-block metal in the first row. All the positive oxidation states up to VII are known, VII, IV and II being the most important.

Redox chemistry of manganese compounds

Much of the redox chemistry of manganese compounds can be related to **Figure 25.15.** $E^{\ominus}$ values show that manganese should react with acids to give manganese(II) salts. The reaction is observed.

Manganese(II) in the form of pale pink hexaaquamanganese(II) ions, $Mn(H_2O)_6^{2+}$, is the most stable oxidation state of manganese in aqueous

Figure 25.15 $E^\ominus$–Ox diagram for some oxidation states of manganese

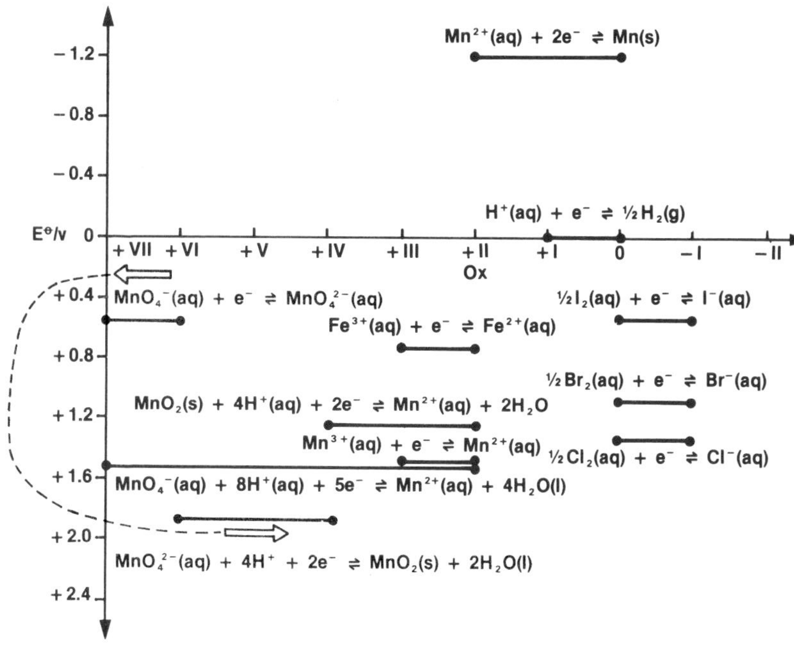

solution, due partly to its $3d^5$ electron arrangement. It is precipitated ou of alkaline solutions as insoluble manganese(II) hydroxide, $Mn(OH)_2$.

Manganese(IV) oxide is a black, insoluble, ionic solid. It is a goc oxidizing agent, $E^\ominus$ values predicting that it will oxidize $\overset{+II}{Fe^{2+}}$ to $\overset{+III}{Fe^{3+}}$, $\overset{-I}{I^-}$ to $\overset{0}{\tfrac{1}{2}I_2}$ and $\overset{-I}{Br^-}$ to $\overset{0}{\tfrac{1}{2}Br_2}$.

Manganese(VII) is an even more powerfully oxidizing state. It is usuall encountered as the dark purple manganate(VII) ion. $E^\ominus$ values show should oxidize $\overset{-I}{Cl^-}$ to $\overset{0}{\tfrac{1}{2}Cl_2}$, and this reaction is often used as a laborator preparation for chlorine:

$$MnO_4^-(aq) + 8H^+(aq) + 5Cl^-(aq) \longrightarrow Mn^{2+}(aq) + 4H_2O(l) + 2\tfrac{1}{2}Cl_2(aq)$$
$$E^\ominus = +0.15\ V$$

Potassium manganate(VII) can be used in titrations to measure th concentration of reducing agents. It needs no indicator as the end poin is shown by the disappearance of its purple colour.

The manganese(VI) state is represented in solution by the gree manganate(VI) ion MnO_4^{2-}. In acid solution it disproportionates. Th reaction is indicated in brown on **Figure 25.15.** The overall reaction is:

$$3\overset{+VI}{MnO_4^{2-}}(aq) + 4H^+(aq) \longrightarrow \overset{+IV}{MnO_2}(s) + 2\overset{+VII}{MnO_4^-}(aq) + 2H_2O(l)$$
$$E^\ominus = +0.9\ V$$

This reaction does not occur in alkaline solution.

Manganese(III) in acid solution also disproportionates to manganese(I and manganese(IV).

Oxides of manganese

These are summarized in **Table 25.6.**

Table 25.6 Oxides of manganese

Formula	Name	Properties
MnO	Manganese(II) oxide	Grey–green ionic solid, basic
Mn_2O_3	Manganese(III) oxide	Brown ionic solid, basic
MnO_2	Manganese(IV) oxide	Black solid, ionic, amphoteric
Mn_2O_7	Manganese(VII) oxide	Explosive dark coloured liquid, covalently bonded, dissolves in water to give manganic(VII) acid, $HMnO_4$

The usual trends of increasing acidity and increasing covalency as th oxidation number of the metal increases are seen.

Trimanganese tetroxide, Mn_3O_4, is a mixed oxide containing Mn^{2+} and Mn^{3+} ions in the ratio 1:2, i.e. $\overset{+II}{Mn}\overset{+III}{Mn_2}O_4$, so its systematic name is manganese(II) dimanganese(III) oxide.

25.4.5 Iron

Iron is without doubt the most important metal in present-day society. Around 25 000 tonnes are produced *daily* in the UK—most being used to produce steels, alloys of iron with up to 30% of other transition metals added to improve its properties. In many ways, iron is not an ideal constructional metal, especially as it corrodes quite readily. However, this disadvantage is outweighed by its cheapness, the result of its abundance (it is the second most abundant metal in the earth's crust) and relative ease of extraction. Also, alloying and heat treatment produce steels with a remarkable range of useful properties.

igure 25.16 $E^{\ominus}$**–Ox diagram for ome oxidation states of iron**

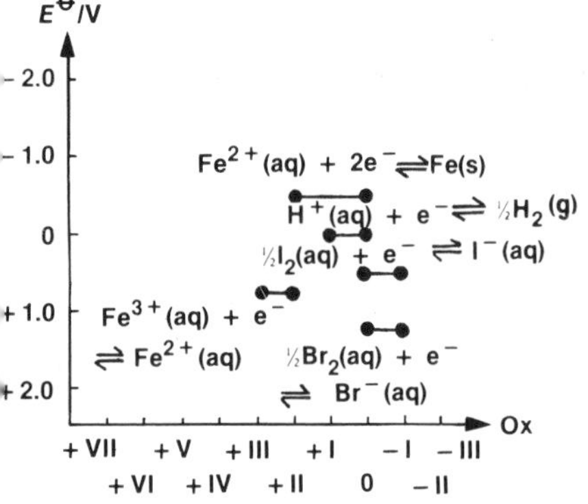

Redox chemistry of iron

This is relatively simple, as iron shows only two common oxidation states, iron(II) and iron(III). An $E^{\ominus}$–Ox diagram is shown in **Figure 25.16.** Of the two states, Fe^{3+} is the more stable as it has a $3d^5$ electron arrangement. Solutions containing Fe^{2+} are oxidized by air to Fe^{3+}.

The $E^{\ominus}$–Ox diagram shows that metallic iron could react with acids to form $Fe^{2+}(aq)$ ions and hydrogen, and this reaction is observed. Bromine will oxidize $Fe^{2+}(aq)$ to $Fe^{3+}(aq)$, but $Fe^{3+}(aq)$ will oxidize $I^-(aq)$ to iodine.

RUSTING

This is probably the most important redox reaction of iron. Metallic iron is converted in the presence of oxygen and water to rust (hydrated iron (III) oxide, $Fe_2O_3.xH_2O$).

The rusting or corrosion of iron is an electrochemical process in which an electrochemical cell is set up within the metal, **Figure 25.17.** In any

igure 25.17 The electrochemical usting of iron

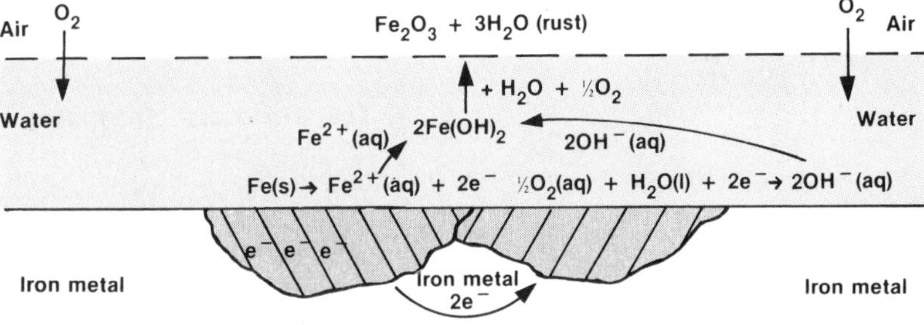

piece of iron, some areas will tend to accept electrons and others to give them away. Such areas may be caused by impurities in the iron—a more reactive impurity like zinc would release electrons better than iron, and be negative, while a less reactive impurity like copper would tend to form a positive area. Most iron has impurities but even in very pure iron, positive and negative areas can be produced by less obvious factors such as stress caused by bending the metal and by uneven oxidation of the surface.

If the iron surface is covered with water, which contains dissolved oxygen from the air, the following reactions can take place:

$$Fe(s) \longrightarrow Fe^{2+}(aq) + 2e^-$$

$$\tfrac{1}{2}O_2(aq) + H_2O + 2e^- \longrightarrow 2OH^-(aq)$$

The electrons produced by the first reaction flow through the iron (which, of course, is a good conductor) to take part in the second reaction. The $Fe^{2+}(aq)$ and $OH^-(aq)$ ions diffuse away through the solution and where they meet, they react to give iron(II) hydroxide:

$$Fe^{2+}(aq) + 2OH^-(aq) \longrightarrow Fe(OH)_2(s)$$

A

PREVENTION OF CORROSION

To prevent corrosion, iron may be coated with grease, paint or another metal to keep away the water and oxygen which are essential for the reaction. Alternatively, blocks of a more reactive metal than iron can be bolted to the iron surface. Magnesium is often used. In this case, the reaction:

$$Mg(s) \longrightarrow Mg^{2+}(aq) + 2e^-$$

occurs more readily than:

$$Fe(s) \longrightarrow Fe^{2+}(aq) + 2e^-$$

and the magnesium will dissolve away leaving the iron intact. This is called **sacrificial protection**. The magnesium lumps, of course, need regular replacement. This method is used on ships' hulls and pipelines where a paint surface would be unlikely to stay undamaged for long. Zinc plating or galvanizing has the same effect, zinc also being more reactive than iron. Once the coating is scratched, a cell is set up, but the more reactive zinc dissolves rather than the iron. On tin plated iron, as used in food cans, the reverse occurs. Tin is less reactive than iron and the iron dissolves leaving behind the tin. You may have noticed how rapidly food cans rust once they are scratched. On an unscratched can, the tin's function is to keep air and water away from the iron.

Sacrificial protection – the magnesium protects the ship's hull

A tin can rusts soon after being opened

This is then oxidized by dissolved oxygen to iron(III) hydroxide, whic forms hydrated iron(III) oxide (rust).

$$2Fe(OH)_2 + H_2O + \tfrac{1}{2}O_2 \longrightarrow Fe_2O_3 + 3H_2O$$

So rusting takes place only in the presence of air and water. It is accelerate by dissolved ionic salts which make the water a better electrical conducto

The aqueous chemistry of iron ions

TESTS FOR IRON IONS

Both Fe^{2+} and Fe^{3+} ions exist in aqueous solution as octahedral hexaaqu ions. $Fe(H_2O)_6{}^{2+}$ is pale green and $Fe(H_2O)_6{}^{3+}$ is pale brown. A simpl test to distinguish the two is to add dilute alkali which precipitates th hydroxides which are more obviously different coloured:

$$Fe(H_2O)_6{}^{2+}(aq) + 2OH^-(aq) \longrightarrow \underset{\text{iron(II) hydroxide (green)}}{Fe(OH)_2(s)} + 6H_2O(l)$$

$$Fe(H_2O)_6{}^{3+}(aq) + 3OH^-(aq) \longrightarrow \underset{\text{iron(III) hydroxide (brown)}}{Fe(OH)_3(s)} + 6H_2O(l)$$

Two other colour reactions provide tests to distinguish Fe^{2+} from Fe^{3} ions. Addition of potassium thiocyanate ($K^+ + SCN^-$) to $Fe(H_2O)_6{}^{3}$ produces the blood-red pentaaquathiocyanatoiron(III) ion:

$$Fe(H_2O)_6{}^{3+}(aq) + SCN^-(aq) \longrightarrow \underset{\text{blood-red}}{Fe(H_2O)_5SCN^{2+}(aq)} + H_2O(l)$$

The colour of the ion is very intense and can be used to detect very sma concentrations of $Fe^{3+}(aq)$.

The test for $Fe^{2+}(aq)$ is the formation of a deep blue colour, Prussia blue, on addition of potassium hexacyanoferrate(III).

$$Fe(H_2O)_6{}^{2+}(aq) + Fe(CN)_6{}^{3-}(aq) \longrightarrow \underset{\text{Prussian blue}}{Fe[Fe(CN)_6]^-(aq)} + 6H_2O(l)$$

A

HAEMOGLOBIN

Haemoglobin is the red pigment in blood. It is responsible for carrying oxygen from the lungs to the cells of the body. The molecule consists of an Fe^{2+} ion with a coordination number of six. Four of the coordination sites are taken up by a ring system called a porphyrin which acts as a tetradentate ligand. Below the plane of this ring is a fifth nitrogen atom, acting as a ligand. This atom is part of a complex protein called globin. The sixth site can accept an oxygen molecule as a ligand. The $Fe^{2+}—O_2$ bond is weak, allowing the oxygen molecule to be easily given up to cells.

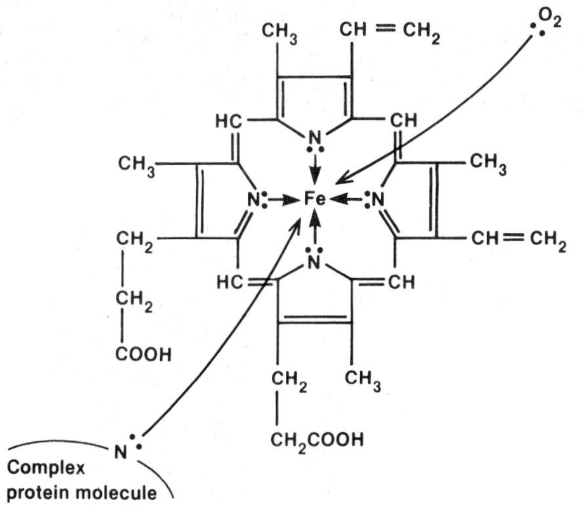

Better ligands than oxygen can bond irreversibly to the iron and thus destroy haemoglobin's oxygen-carrying capacity. This explains the poisonous effect of carbon monoxide and cyanide ions, both of which are very good ligands.

Anaemia is a condition caused by a shortage of haemoglobin. The body suffers from a shortage of oxygen and the symptoms include fatigue, breathlessness and a pale skin colour. The causes may be loss of blood or deficiency of iron in the diet. The latter may be treated by taking 'iron tablets'—tablets containing iron(II) sulphate.

THE ACIDITY OF AQUEOUS IRON IONS

Solutions of $Fe^{2+}(aq)$ are not appreciably acidic, while a solution of $Fe^{3+}(aq)$ is a stronger acid than ethanoic acid ($pK_a = 2.2$). This is because the Fe^{3+} ion is both smaller and more highly charged than the Fe^{2+} ion making it more strongly polarizing. So in the $Fe(H_2O)_6{}^{3+}(aq)$ ion the iron strongly attracts electrons from the oxygen atoms of the water ligands, thus weakening the O—H bonds in these water molecules. These will then readily donate H^+ ions, making the solution acidic, **Figure 25.18.** This is comparable to the situation with $Al^{3+}(aq)$, see **section 24.2.2.**

$$Fe(H_2O)_6{}^{3+}(aq) \longrightarrow Fe(H_2O)_5OH^{2+}(aq) + H^+(aq)$$

Figure 25.18 **The acidity of Fe^{3+}(aq) ions**

25.4.6 Copper

Copper is an extensively used metal. World production is some 9 million tonnes per year, most of which is extracted from sulphide ores. Its lack of chemical reactivity and ease of working makes it suitable for plumbing pipes and fittings. Its main use is as an electrical conductor. It is a constituent of many alloys such as brass, gunmetal and bronze. See **Figure 25.19.**

Redox chemistry of copper

There are only two stable oxidation states of copper in addition to the uncombined element, copper(I) and copper(II). See **Figure 25.20.** Copper(I) has a $3d^{10}$ electron arrangement and therefore it is not strictly transitional. As would be expected, most copper(I) compounds are white, although the best known compound copper(I) oxide, Cu_2O, is red–brown. Copper(II)

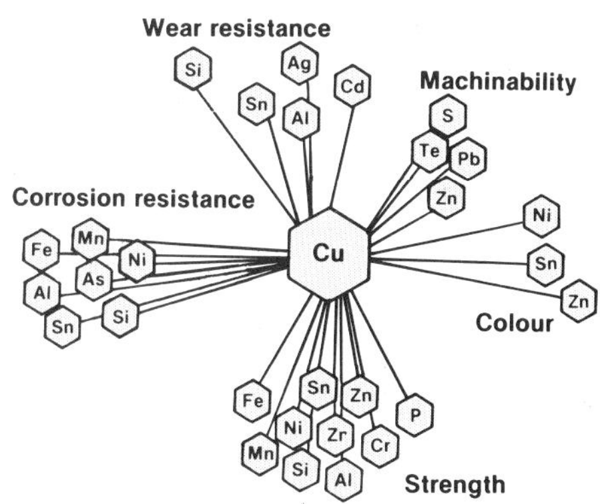

Figure 25.19 Copper can be alloyed with a variety of metals to produce the properties shown

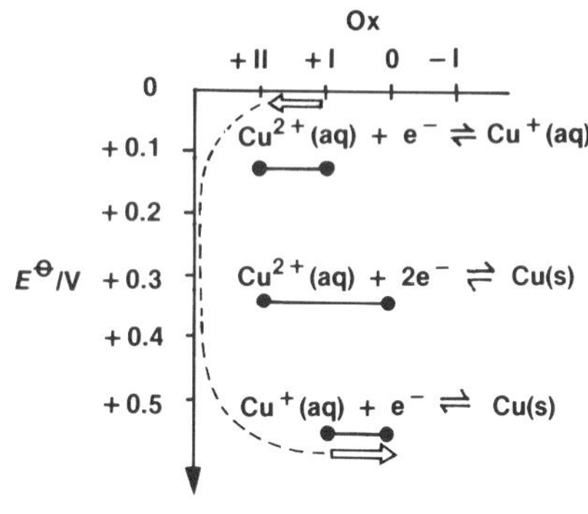

Figure 25.20 $E^\ominus$–Ox diagram for some copper compounds

($3d^9$) is transitional and most of its compounds are blue in aqueous solution due to the hydrated copper(II) ion.

In aqueous solution $Cu^+(aq)$ will disproportionate to $Cu^{2+}(aq)$ and Cu(s) as shown by the anti-clockwise rule:

$$\overset{+\text{I}}{2Cu^+(aq)} \longrightarrow \overset{0}{Cu(s)} + \overset{+\text{II}}{Cu^{2+}(aq)} \quad E^\ominus = +0.37\ \text{V}$$

Aqueous chemistry of copper ions

Some ligand displacement reactions of copper have been described above. That with ammonia is complicated by the precipitation of insoluble, pale blue copper hydroxide. An aqueous solution of ammonia contains $OH^-(aq)$ ions due to the equilibrium:

$$NH_3(aq) + H_2O(l) \longrightarrow NH_4{}^+(aq) + OH^-(aq)$$

These precipitate copper(II) hydroxide:

$$Cu(H_2O)_4{}^{2+}(aq) + 2OH^-(aq) \longrightarrow Cu(OH)_2(s) + 4H_2O(l)$$

On adding excess ammonia, the copper dissolves due to the tetraamine copper(II) ion which is a very dark blue:

$$Cu(OH)_2(s) + 4NH_3(aq) \longrightarrow Cu(NH_3)_4{}^{2+}(aq) + 2OH^-(aq)$$

25.4.7 Zinc

Zinc metal has the outer electron arrangement $3d^{10}4s^2$. It forms only one oxidation state, zinc(II) ($3d^{10}$). The metal is therefore non-transitional, its compounds typically being white. Zinc does, however, form complexes such as the tetrahedral $Zn(OH)_4{}^{2-}(aq)$, $Zn(NH_3)_4{}^{2+}(aq)$ and $ZnCl_4{}^{2-}(aq)$.

25.5 Economic importance of the d-block elements

25.5.1 Extraction of d-block metals

A few of the least reactive d-block metals such as copper, silver and gold (the coinage metals) are found uncombined. The rest are found in ores, usually as compounds of oxygen or sulphur. The general method is to convert sulphides into oxides by roasting them (heating them in air). Sulphur dioxide is a by-product at this stage. It may be used for sulphuric acid manufacture. The oxide must then be reduced to the metal by a suitable reducing agent. A suitable reducing agent must be able to reduce the oxide to the element, and also be cheap. Ellingham diagrams (see box, **section 18.4**) can be used to help select an appropriate reductant and a minimum temperature. Coke or carbon monoxide are used in many cases, notably in the blast furnace for making iron. Alternatively, a more reactive metal may be used. **Table 25.7** summarizes the industrial extraction of the first row d-block metals.

Table 25.7 Extraction of some transition metals from their ores

Element	Typical ore	Substance to be reduced	Reductant	Notes
Sc	Not extracted on an industrial scale			
Ti	Rutile, TiO_2	$TiCl_4$	Mg or Na	An inert atmosphere is needed
V	Carnolite, an ore of K, V, U and O	V_2O_5	Al	
Cr	Chromite, $FeCr_2O_4$	$Na_2Cr_2O_7$	C, then Al	
Mn	Pyrolusite, MnO_2, and hausmannite, Mn_3O_4	Mn_3O_4	Al	
Fe	haematite, Fe_2O_3, and magnetite, Fe_3O_4	Fe_2O_3	C(CO)	Blast furnace
Co	Smaltite, $CoAs_2$, and cobaltite, CoAsS	Co_3O_4	Al	
Ni	Millerite, NiS	NiO	C	
Cu	Copper pyrites, $CuFeS_2$, and copper glance, CuS	Cu_2S	S*	
Zn	Zinc blende, ZnS, and calomine, $ZnCO_3$	ZnO	C(CO)	Blast furnace

*The reaction:

$$\overset{+I\ \ -II}{Cu_2\ S}(s) + \overset{0}{O_2} \longrightarrow \overset{0}{2Cu} + \overset{+IV\ -II}{S\ O_2}$$

is used. Since Ox(S) goes up, it is the reducing agent.

Titanium replacement hip joint

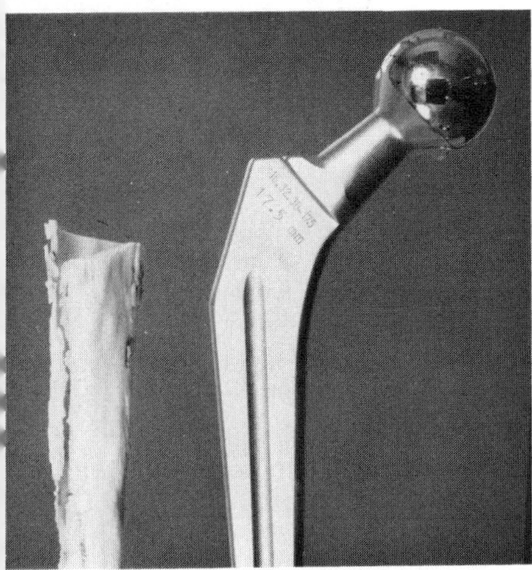

A

THE PURIFICATION OF NICKEL

Nickel is produced by reducing nickel(II) oxide with carbon:

$$NiO(s) + C(s) \longrightarrow Ni(s) + CO(g)$$

Nickel of extremely high purity can be obtained by heating the impure nickel with carbon monoxide at 330 K. Carbon monoxide is a good ligand and the tetrahedral complex tetracarbonylnickel(0) is formed:

$$Ni(s) + 4CO(g) \longrightarrow Ni(CO)_4(g)$$

This is unusual as the nickel is in oxidation state zero, and a neutral complex is formed. The complex is also unusually volatile for a metal compound ($T_b = 316$ K) and it can be separated from impurities by distillation. Heating the purified tetracarbonylnickel(0) to 473 K decomposes it, leaving very pure nickel:

$$Ni(CO)_4(g) \longrightarrow Ni(s) + 4CO(g)$$

25.5.2 Uses of d-block elements and their compounds

These are summarized in **Table 25.8.**

Table 25.8 Some uses of d-block elements and their compounds

Metal	Uses
Sc	Virtually no significant uses
Ti	Constructional—high-speed aircraft, nuclear reactors, heart pacemakers. Replacement joints, e.g. hip hoints. Construction of chemical plant. TiO_2 is a white pigment in paints. $TiCl_4$ is a catalyst for making poly(alkenes)
V	Alloys—special steels
Cr	Chrome plating. Alloys—special steels. Chrome alum mordant in dyeing, also compounds as pigments
Mn	Alloys—special steels
Fe	Alloy—steel (see box *steelmaking and alloy steels*) car bodies, reinforcing concrete, ships, bridges, general construction, domestic appliances. Catalyst in Haber process
Co	Alloy—stellite, CoCrW—cutting tools. AlNiCo, constantan, nichrome (see below). ^{60}Co—radioactive source for medical and other uses. Compounds as pigments (blue)
Ni	Alloys—coinage. AlNiCo—magnets, construction. Nichrome—electrical heating elements. Industrial catalyst, hydrogenation of oils to margarine
Cu	Electrical wiring, plumbing pipes. Many alloys, e.g. bronze, brass, gunmetal. Industrial catalyst methanol → methanal. Coinage. Destroying fungi and algae
Zn	Galvanizing, alloys, e.g. brass, dry batteries. ZnO white pigment in paints.

A practical application of magnets

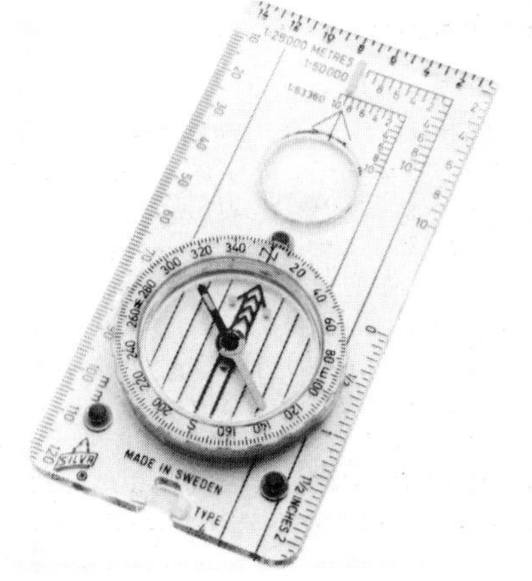

In addition, many d-block metal compounds are essential to the body's biochemistry, e.g. iron in haemoglobin in blood, cobalt in vitamin B, zinc and copper in several enzymes.

THE EXTRACTION OF IRON— THE BLAST FURNACE

Iron is the second most abundant metal in the earth's crust (5.8%, after aluminium, 8.0%). It is found as the ores haematite, Fe_2O_3, and magnetite, Fe_3O_4, which are obtained by quarrying or opencast mining. British deposits are of low quality and most iron ore is imported from as far afield as Scandinavia, America and the USSR. The major impurity is silica (SiO_2, sand).

Opencast iron ore mine

The principle of reducing iron ore to iron with carbon has been known for over 3000 years and the blast furnace has changed little except in size for 200 years. The furnace is shown below:

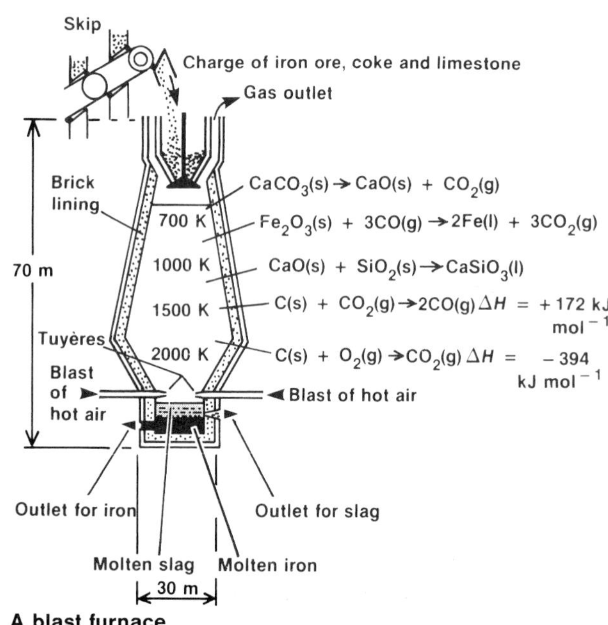

A blast furnace

It is made of steel, lined inside with firebricks. It is charged with a mixture of coke (carbon) and limestone (calcium carbonate). At the base of the furnace coke burns in a blast of preheated air (which gives the furnace its name). Heat is generated by this exothermic reaction resulting in a temperature of approximately 2000 K, which is required to melt the iron. The carbon dioxide produced reacts with more coke a little higher up to produce carbon monoxide, the reducing agent. This reaction is endothermic so the temperature drops. Higher up the furnace where the temperature is lower, the carbon monoxide reduces the ore to iron which is molten at the furnace temperature. Higher still, calcium carbonate is decomposed to calcium oxide which reacts with silicon dioxide impurities in the ore to form molten calcium silicate (slag). The molten iron and slag trickle to the base of the furnance. Slag floats on top of the molten iron and the two can be tapped off separately. Operation of the furnace is continuous, producing up to 3000 tonnes of iron per day. Every couple of years, the furnace must be shut down for replacement of the brick lining. The waste slag solidifies and must be disposed of. Many iron-making areas used to be surrounded by unsightly slag heaps. Nowadays slag can used be for road construction and cement manufacture and strains of grass are being found which will grow on slag heaps and reduce their environmental impact. The iron produced is 90–95% pure, containing silicon, sulphur and phosphorus impurities as well as around 4% carbon, which makes it brittle.

Blast furnace being tapped

Slag heaps – a blot on the landscape

A

STEELMAKING AND ALLOY STEELS

Iron straight from the blast furnace contains impurities, mainly non-metals, which make it too brittle for most purposes. In steelmaking, the impurities are removed by blowing a jet of pure oxygen into the molten steel and adding lime, calcium oxide. The impurities, being non-metals, produce acidic oxides which combine with the basic calcium oxide to form a slag containing calcium carbonate, calcium silicate, etc. This slag floats on the steel and can be tapped off. In the basic oxygen process, now the most common, a water cooled lance is used to supply the oxygen. A modern steel furnace can take a charge of 350 tonnes or iron and make it into steel in under an hour.

Basic oxygen steel converter

There are many kinds of steel tailored for particular purposes. All have less than 1.5% carbon. Other metals may be added to give alloy steels whose exact composition can be monitored by examining the emission spectrum of a sample while the steel is still in the converter. On-line computer controlled spectrometers can deliver a printout of the steel's composition to the furnace operator within minutes.

The table shows just a few of the thousands of alloy steels produced.

Name	Approximate composition	Special properties	Use
Manganese steel	86% Fe, 13% Mn, 1% C	Tough	Drill bits
Stainless steel	73% Fe, 8% Ni, 18% Cr, 1% C	Non-rusting	Cutlery, sinks
Cobalt steel	90% Fe, 9% Co, 1% C	Hard	Ball-bearings
Tungsten steel	81% Fe, 18% W, 1% C	Tough	Armour plate

Many alloy steels are made in an electric arc furnace where scrap steel is recycled. The scrap is melted by the heat of an electric arc struck from the steel to carbon electrodes. Precise quantities of other metals can be added to the molten scrap.

25.6 Summary

- The d-block elements are those in the series from scandium to zinc (the first row), yttrium to cadmium (the second row) and lanthanum to mercury (the third row).
- The definition of a transition metal is an element which forms at least one compound in which it has a part-filled d shell of electrons. For many purposes, d-block and transition metals are the same.
- The 3d shell is an inner shell and d electrons screen the nuclear charge well from the outer electrons in the 4s orbital.
- d-block elements tend to have fairly similar physical properties to one another.
- Generally, d-block metals are typical metals—hard, strong, shiny, good conductors of electricity and heat and with fairly high densities. Their chemical reactivities are moderate.
- Their chemistry is typified by: variable oxidation numbers; complex formation; coloured compounds; catalytic activity.
- Complexes are formed when ligands form dative covalent bonds with transition metal ions. They may be neutral, anionic or cationic.
- A ligand which can form more than one dative bond is called polydentate (many toothed).
- Complexes with polydentate ligands are called chelates.
- Isomers are compounds with the same molecular formula but different spatial arrangements of atoms.
- Three types of isomerism occur in d-block metal complexes: ionization isomerism; geometrical (*cis–trans*) isomerism; optical (mirror image) isomerism.
- Ligands may displace one another from complexes.

• The stability of a complex is measured by the log of its stability constant, log K. The greater log K, the more stable the complex.
• Chelates are usually very stable because their formation from complexes with monodentate ligands involves a large increase in entropy
• The colour of d-block metal compounds is caused by electrons moving from one d orbital to another whose energy is slightly different due to the effect of ligands.

25.7 Questions

1 This question is about an aqueous solution of copper(II) sulphate.

log K	Complex	
5.6	$CuCl_4^{2-}$(aq)	(yellow)
13.1	$Cu(NH_3)_4^{2+}$	(deep blue)
18.8	Cu edta^{2+}	(pale blue)

A) What colour is the solution?
B) What ion is responsible for this colour?
C) Describe and explain what happens when excess concentrated hydrochloric acid is added to the solution;
D) Describe and explain what would happen if excess ammonia solution is added to the solution remaining after (C);
E) Describe and explain what would happen if a solution of edta (aq) is added to the solution remaining after (D);
F) What would happen if ammonia solution is added directly to the original solution? Why is it different from what happened in (D)?
G)

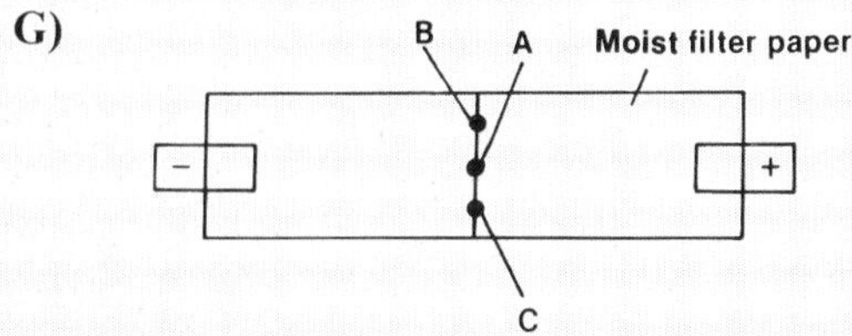

A drop of $CuSO_4$(aq) is placed at A, a drop containing $CuCl_4^{2-}$(aq) at B and a drop containing $Cu(NH_3)_4^{2+}$(aq) at C. What would happen when the current was switched on? Explain.

2 Debate/discussion. What would be the effect on present-day life if supplies of all transition metals stopped tomorrow? What replacements would be possible? Are any of these metals replaceable? What might be the scientific, political and economic effects?

3 Cobalt, a member of the d-block of the Periodic Table, exhibits many properties which are typical of d-block elements. Cobalt(II) sulphate-7-water contains the hydrated cobalt(II) ion, which is octahedral in shape. When a solution of this compound is treated with concentrated hydrochloric acid, the hydrated cobalt(II) ion is converted to the ion $[CoCl_4]^{x-}$ (where x is the charge on the ion).
A) What name is given to molecules (such as water) or ions which can bond to a central metal ion in this way?
B) How many water molecules surround the hydrated cobalt(II) ion present in cobalt(II) sulphate-7-water?
C) What is the value of x in the formula of the ion $[CoCl_4]^{x-}$ produced in the substitution reaction between hydrated cobalt(II) ions and chloride ions?
When a solution of cobalt(II) chloride is heated with a mixture of aqueous ammonia and ammonium chloride in the presence of a catalyst of charcoal, a solid of formula $Co(NH_3)_6Cl_3$ can be crystallized from solution. This compound has two isomers: $[Co(NH_3)_5Cl]Cl_2 \cdot NH_3$ and $[Co(NH_3)_4Cl_2]Cl \cdot 2NH_3$.
D) What type of reaction is involved in the formation of $Co(NH_3)_6Cl_3$ from cobalt(II) chloride? Explain your answer.
E) How many moles of silver chloride are formed from 1 mole of $[Co(NH_3)_5Cl]Cl_2 \cdot NH_3$ if it is reacted in aqueous solution with excess silver nitrate?
Another complex of cobalt is formed with ethanedioate ions,

$$\begin{array}{l} COO^- \\ | \\ COO^- \end{array}$$

and has the formula $[Co(C_2O_4)_3]^{3-}$. It is made by dissolving cobalt(II) carbonate in a solution of ethanedioic acid and potassium ethanedioate, and then adding lead(IV) oxide. The complex can then be isolated from the resulting solution.
F) What is the function of the lead(IV) oxide in this reaction?
G) Draw a diagram to illustrate the structure of this complex.
H) Would you expect this compound to have isomers? Explain your answer.

(Nuffield 1985)

4 Reactions involving some chromium compounds are shown below:

$$CrO_4^{2-}(aq) \overset{A}{\rightleftharpoons} Cr_2O_7^{2-}(aq) \rightarrow Cr^{3+}(aq) \overset{B}{\rightarrow} Cr(OH)_3(s)$$
$$\downarrow$$
$$[Cren_2Cl_2]^+(aq)$$

where 'en' is the bidentate ligand, 1,2-diaminoethane: $H_2N—CH_2—CH_2—NH_2$. Redox potentials for the conversion of dichromate(VI) ions to chromium(III) ions by iodide ions in acidified solution are:

$$I_2(aq) + 2e^- \rightleftharpoons 2I^-(aq) \quad E^\ominus = +0.54\,V$$

$Cr_2O_7^{2-}(aq) + 14H^+(aq) + 6e^- \rightleftharpoons 2Cr^{3+}(aq) + 7H_2O$
$E^\ominus = +1.33\ V$

A) i Write the equation, including state symbols, for the conversion of chromate(VI) ions to dichromate-(VI) ions in acidic solution (Reaction A).
ii Write the expression for the equilibrium constant, K_c, for this reaction.
B) i Write out the cell diagram for the reaction of iodide ions with acidified dichromate(VI) ions.
ii What would be the e.m.f of this cell?
iii Calculate the value of $\Delta G^\ominus$ for this reaction, using the relationship: $\Delta G^\ominus = -zFE^\ominus$ ($F = 96\,500$ C mol^{-1})
iv What can you deduce from your value for $\Delta G^\ominus$ about the equilibrium position of this reaction?
C) What reagent(s) could be used for reaction B?
D) Chromium forms three complex ions with the formula $[Cren_2Cl_2]^+$.
i Draw stereochemical diagrams of the three structures.
ii Which of the complex ions will exhibit optical activity? Justify your answer.
iii What structural feature of the ligand 'en' permits complexing?

(Nuffield 1984)

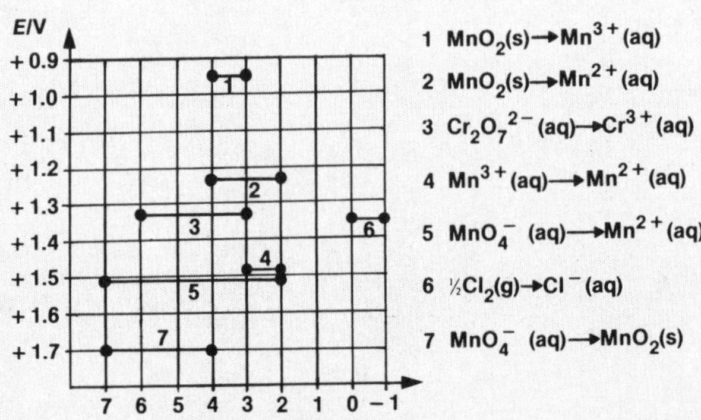

The chart displays the electrode potentials in acidic conditions for some possible reactions of manganese compounds or ions. You should answer the question on the basis of the information on the chart.
A) Which manganese ions or compound(s) would be expected to react with $Cr_2O_7^{2-}(aq)$?
B) Which manganese ions or compound(s) would be expected to react with $Cl^-(aq)$?
C) i Which manganese ions or compound(s) would be expected to disproportionate?
ii Write a balanced equation, with state symbols, for one of your answers to (C)(i).
D) For the cell:

$$Pt|[Cl_2(g) + 4H_2O(l)], [2Cl^-(aq) + 8H^+(aq)] \,\vdots\vdots\, [Cr_2O_7^{2-}(aq) + 14H^+(aq)], [2Cr^{3+}(aq) + 7H_2O(l)]|Pt$$

i What would be the e.m.f. of this cell?
ii Which would be the positive half cell?
iii Calculate a value for ΔG for the reaction between $Cl_2(aq)$ and $Cr^{3+}(aq)$. (The Faraday constant $F = 96\,500$ C mol^{-1}.)
iv In the reaction between $Cl_2(aq)$ and $Cr^{3+}(aq)$ would you expect reactants or products to predominate when the reaction reaches equilibrium? Justify your answer.
E) Why may a reaction expected to take place on the basis of the $E^\ominus$ value not actually take place?

(Nuffield 1980)

6 Chromium, manganese and iron are d-block elements and have atomic numbers 24, 25 and 26 respectively.
A) Explain what is meant by the term d-block element.
B) Give three properties characteristic of d-block elements, illustrating your answer by reference to the above elements.
C) Give the electronic configurations of Cr(III), Mn(VI) and Fe(O).
D) Give the name and structural formula of:
i a complex cation containing iron;
ii a complex anion containing iron.
E) Explain briefly why:
i the atomic radii of chromium, manganese and iron are very similar;
ii manganese can exist in various oxidation states, up to a maximum of seven.

(London 1980)

7 A) In the first long period of the Periodic Table the elements from Sc (Z = 21) to Zn (Z = 30) are said to belong to the d-block. Some but not all of these elements are 'transition elements'.
i What is meant by:
d-block element;
transition element?
ii which of the element(s) mentioned is/are in the d-block but not transition element(s)?
B) Copy the chart below and insert the electronic configurations of the atoms and ions shown.

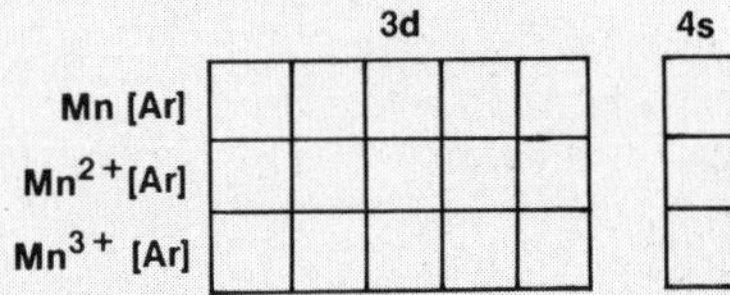

C) A solution of potassium manganate(VII), $KMnO_4$, can be standardized by titration under suitable conditions with a solution of arsenic(III) oxide, As_2O_3. If the arsenic(III) oxide is oxidized in the process to arsenic(V) oxide, As_2O_5, and 5 mol of arsenic(III) oxide are oxidized by 4 mol of manganate-(VII) ions, calculate the oxidation state to which the manganate(VII) is reduced.

(London 1979, part question)

8 A) Outline the structural requirements for the formation of a complex ion between a transition metal ion and ligands.

B) Describe, with the aid of one suitable example in each case, the commonly occurring shapes of complex ions formed by transition metals.

C) Describe and explain what happens in each of the following experiments.

i A concentrated aqueous solution of copper(II) chloride is gradually diluted and concentrated hydrochloric acid is then added until it is present in an excess.

ii Aqueous ammonia is added in an excess to aqueous copper(II) sulphate and an excess of an aqueous solution of edta,

$$(HO_2CCH_2)_2NCH_2CH_2N(CH_2CO_2H)_2$$

is then added to the resulting solution.

(Cambridge 1987)

9 A) Draw the electron arrangement of atomic chromium by copying the boxes below and inserting arrows representing the electrons:

1s	2s	2p	3s	3p	3d	4s
□	□	□□□	□	□□□	□□□□□	□

B) A, B and C are isomeric compounds of formula $CrCl_3 \cdot 6H_2O$. An aqueous solution containing 0.002 mol of A reacted with $50.0\,cm^3$ of $0.080\,mol\,dm^{-3}$ aqueous silver nitrate.

i How many moles of chloride ions are present in one mole of A?

ii Suggest a formula for A.

Aqueous solutions of both B and C, containing 0.002 mole of each, required $25.0\,cm^3$ of $0.080\,mol\,dm^{-3}$ aqueous silver nitrate.

iii How many moles of chloride ions are present in one mole of B or C?

iv Suggest two structural formulae for B and C and state what form of isomerism is shown by B and C.

(Cambridge 1987)

Part D Organic Chemistry

26 Fundamentals of organic chemistry

26.1 Introduction

Carbon is quite remarkable in the number and variety of compounds that it forms. While most elements have perhaps a few dozen compounds, carbon forms literally millions. All the complex and large molecules that make up life on our planet are carbon based. A recent estimate showed around six million compounds of carbon known, and there seems to be no limit to the number which can be made or discovered in nature. Many of these compounds have become vital to our present way of life. Plastics, fuels and lubricants, natural and man-made fibres, and many drugs are all carbon compounds.

A

ANAESTHETICS

Anaesthetics are substances which induce sleep so that pain cannot be felt. Many fairly unreactive gases have anaesthetic properties. They are believed to work by being absorbed in the fatty tissue which surrounds nerve endings and interfering with the transmission of sensations by the nerves. Early anaesthetics included dinitrogen oxide (nitrous oxide or 'laughing gas'), trichloromethane (chloroform) and ethoxyethane (ether). Each had a major disadvantage for surgical use. Dinitrogen oxide produces only light anaesthesia making it suitable for minor operations only—it is still used sometimes by dentists during tooth extractions. Trichloromethane is rather toxic and ethoxyethane is explosive. Chlorinated derivatives of methane show an interesting pattern in anaesthetic effectiveness and toxicity:

CH_2Cl_2 $CHCl_3$ CCl_4

— better anaesthetics ⟶

— more toxic ⟶

This led chemists in the 1950s to investigate other halogenated hydrocarbons as possible anaesthetics. The ideal product had to be potent, non-toxic, non-flammable and volatile enough to be administered to patients by the usual breathing apparatus used in operating theatres.

Eventually the compound 1-bromo-1-chloro-2, 2, 2-trifluoroethane was found to have an ideal combination of properties. Its trade name is Fluothane and it is now used in around three-quarters of the operations carried out in the UK.

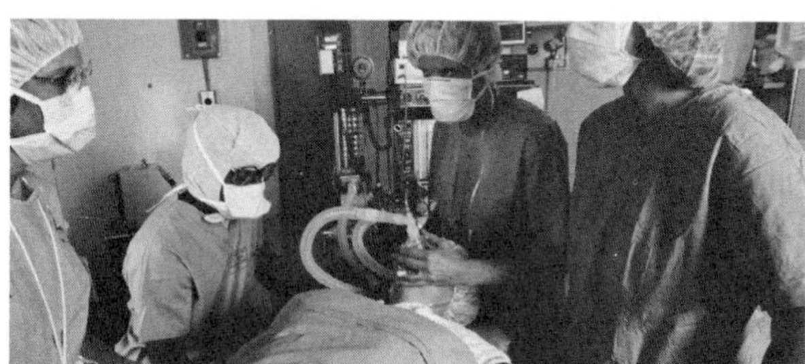

Anaesthetic gas administered through a face mask

```
      H    F
      |    |
Cl —— C —— C —— F
      |    |
      Br   F
```

Fluothane

The vast number of compounds of carbon is due to the ability of the element to form stable chains and rings of carbon atoms in which hydrogen is almost always also present. The study of compounds based on carbon chains is called organic chemistry. The name is derived from the time when it was thought that such compounds could be synthesized only by living organisms. The idea has long been disproved, but the name remains.

26.2 Reasons for the variety of carbon chemistry

Table 26.1 The bond energy of the carbon–carbon bond compared with other elements

Bond	Bond energy/kJ mol^{-1}
C—C	347
N—N	158
O—O	144
Si—Si	226

Figure 26.1 A hydrocarbon chain

```
  H   H   H   H
  |   |   |   |
—C—C—C—C—
  |   |   |   |
  H   H   H   H
```

Figure 26.2 A branched hydrocarbon chain

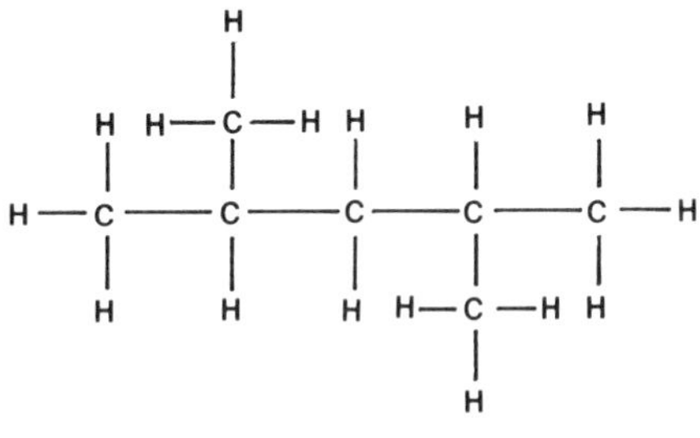

A carbon atom has the electron arrangement $1s^22s^22p^2$, i.e. it has fo electrons in its outer shell, allowing it to form four covalent bonds. forms particularly strong bonds with other carbon atoms (see **Table 26.** and thus readily forms chains and rings of carbon atoms. The C—H bo is also strong (435 kJ mol^{-1}) and hydrocarbon chains form the skelet of most organic compounds, **Figure 26.1.** Since we exist in an atmosphe of oxygen, we should consider the stability of organic compounds terms of their reaction with oxygen. Organic compounds are not *therm dynamically* stable in the presence of oxygen, i.e. oxidation reactions a exothermic, for example:

```
     H    H    H
     |    |    |
H — C — C — C — H (g) + 5O2(g) ——→ 3CO2(g)+4H2O(l)
     |    |    |            oxygen     carbon   water
     H    H    H                       dioxide
       propane                    ΔH⊖ = −2219 kJ mol
```

However, at room temperature the reaction with oxygen is so slow th organic compounds are *kinetically* stable. This is because of the lar activation energies for the reactions with oxygen. For comparison, silic chain molecules (called silanes) react immediately on exposure to air, do hydrides of the other Group IV elements. See **section 24.3.**

Another reason for the variety of carbon compounds is carbon's abili to form four covalent bonds, which means that it is possible to for branched chains, **Figure 26.2.** Sulphur also shows some ability to for chains, but since sulphur normally forms only two covalent bonds, the is no possibility of branching and therefore less variety.

26.3 Bonding in carbon compounds

Figure 26.3 Hybridized orbitals in carbon

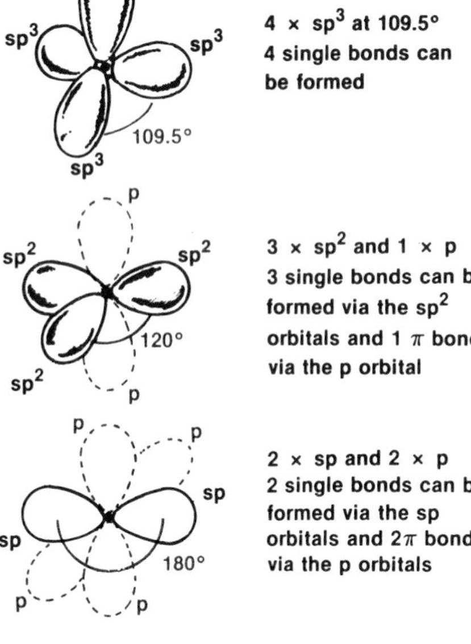

Carbon has an outer shell electron arrangement of $2s^22p^2$. However, 2s and 2p are close in energy, the one 2s and three 2p orbitals can hybridi (see **section 10.2.2**) to form either four sp^3 orbitals, three sp^2 and one orbital, or two sp orbitals and two p orbitals, **Figure 26.3.** Different typ of hybridization lead to the formation of different types of bond—sing double and triple, as shown in **Figure 26.4.** Figure 26.4 shows carb forming single (σ) bonds, double ($\sigma + \pi$) bonds and triple ($\sigma + 2\pi$) bon The shapes of the molecules are as predicted by electron pair repulsi theory, see **section 10.3.3.**

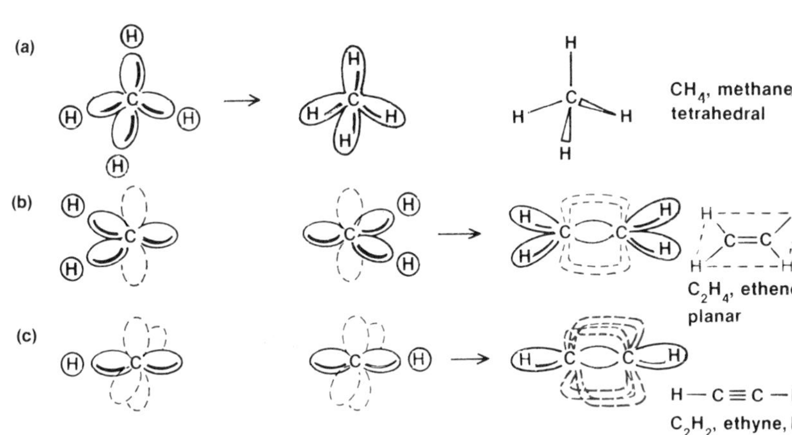

Figure 26.4 (right) Formation of (a) single, (b) double and (c) triple bonds

Often, however, a simpler picture of the bonding will suffice. Carbon has four electrons in its outer shell and can therefore form four single covalent bonds as in methane:

```
        H                        H
       •×                        |
   H •× C •× H               H — C — H
       •×                        |
        H                        H
```

or two single bonds and one double bond as in ethene:

```
  H              H          H         H
   ×•    •    •×             \       /
     C   ×   C                C == C
   •×    •    ×•             /       \
  H      ×       H          H         H
```

or one single bond and one triple bond as in ethyne:

```
           •
           ×
  H •× C •× C •× H            H—C≡C—H
           •
           ×
```

In all stable compounds, carbon forms the equivalent of four covalent bonds, i.e. it has eight electrons in its outer shell.

26.4 Formulae of carbon compounds

Because many organic compounds are more complex structurally than inorganic compounds, a number of different types of formulae are used depending on the amount of information required.

26.4.1 The empirical formula

This formula shows the simplest ratio of the atoms present. For example, ethanol:

> 6.56 g of ethanol is found to contain 3.42 g of carbon,
> 0.86 g of hydrogen and 2.8 g of oxygen
> 3.42 g of carbon is $3.42/12 = 0.285$ mol C
> 0.86 g of hydrogen is $0.86/1 = 0.86$ mol H
> 2.28 g of oxygen is $2.28/16 = 0.1425$ mol O
> so the ratio C:H:O is 0.285:0.86:0.1425
> dividing through by the smallest gives C:H:O as 2:6:1
> so the empirical formula of ethanol is C_2H_6O

26.4.2 The molecular formula

This shows the number of each type of atom in the molecule. To find the molecular formula we need to know the empirical formula *and* the relative molecular mass. For example, ethene has an empirical formula CH_2. This could represent formulae such as CH_2, C_2H_4, C_3H_6, etc., in fact any multiple of the empirical formula, $(CH_2)_n$. The relative molecular mass (M_r) of ethene is 28. The relative molecular mass of the empirical formula is $12 + (2 \times 1) = 14$ so the molecular formula must contain *two* empirical formula units, so it is $(CH_2)_2$ or C_2H_4.

To find the molecular formula, divide the relative molecular mass of the compound by the relative molecular mass of the empirical formula. This will tell you the number of empirical formula units in the molecular formula.

Example

The empirical formula of benzene is CH and the M_r is 78. The M_r of the empirical formula unit is $12 + 1 = 13$.

$78 \div 13 = 6$ so the molecular formula is $(CH)_6 = C_6H_6$

26.4.3 The structural formula

This gives information about the way the atoms are arranged within th molecules, as well as the numbers of each atom present. There are differen ways of doing this.

The displayed formula

This shows every atom and bond in the molecule. — represents a singl bond, = a double bound and ≡ a triple bond.

For example, for ethanol, C_2H_6O, the displayed formula is:

$$\begin{array}{ccccccc} & & H & & H & & \\ & & | & & | & & \\ H & - & C & - & C & - & O - H \\ & & | & & | & & \\ & & H & & H & & \end{array}$$

For ethene, C_2H_4, the displayed formula is:

$$\begin{array}{ccccc} H & & & & H \\ & \diagdown & & \diagup & \\ & & C = C & & \\ & \diagup & & \diagdown & \\ H & & & & H \end{array}$$

A shorthand form of structural formula is often used:

$$\begin{array}{ccccccc} & & H & & H & & \\ & & | & & | & & \\ H & - & C & - & C & - & O - H \\ & & | & & | & & \\ & & H & & H & & \end{array}$$ may be written CH_3CH_2OH

$$\begin{array}{ccccc} H & & & & H \\ & \diagdown & & \diagup & \\ & & C = C & & \\ & \diagup & & \diagdown & \\ H & & & & H \end{array}$$ may be written CH_2CH_2.

In this shorthand, branches in the carbon chains are indicated in brackets

$$\begin{array}{ccccccccc} & & H & & H & & H & & \\ & & | & & | & & | & & \\ H & - & C & - & C & - & C & - & H \\ & & | & & | & & | & & \\ & & H & & | & & H & & \\ & & & H - & C & - H & & & \\ & & & & | & & & & \\ & & & & H & & & & \end{array}$$ may be written $CH_3CH(CH_3)CH_3$

The three-dimensional structural formula

This attempts to show the three-dimensional structure of the molecule Bonds coming out of the paper are indicated ◢ , and those going into the paper by ⋰ . So ethanol could be written:

$$\begin{array}{ccccccc} & H & & H & & & \\ & \diagdown & & \diagup & & & H \\ H \cdots & C & - & C & \cdots & O & \diagup \\ & \blacktriangle & & \blacktriangle & & & \\ & H & & H & & & \end{array}$$

Model of ethanol

Skeletal notation

With more complex molecules, a normal structural formula becomes rathe cumbersome. In this notation the carbon atoms are not drawn at al Straight lines represent carbon—carbon bonds and carbon atoms ar assumed to be at each angle. Neither hydrogen atoms nor C—H bond are drawn. Each carbon is assumed to form enough C—H bonds to mak a total of four bonds.

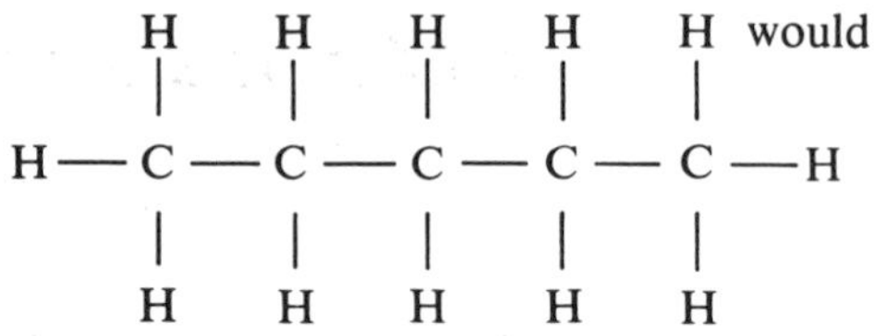

$H-CH_2-CH_2-CH_2-CH_2-CH_2-H$ (displayed) would be written (skeletal zig-zag)

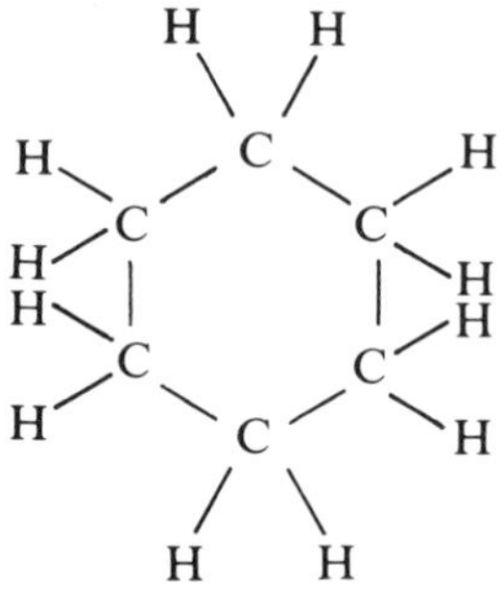

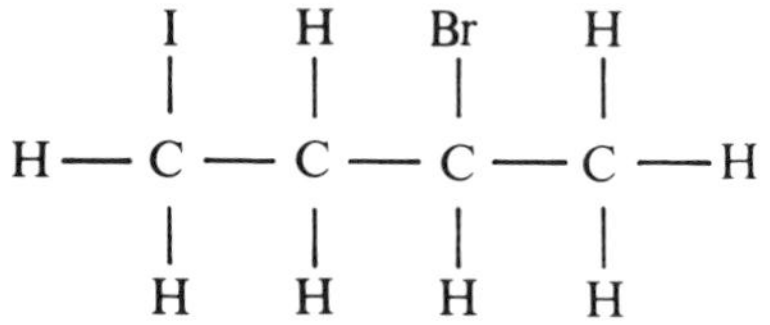

Cyclohexane (displayed) would be written (hexagon)

$H-CHI-CH_2-CHBr-CH_2-H$ (displayed) would be written (skeletal, with I and Br)

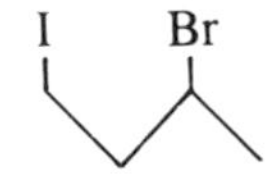

The choice of type of formula to use depends on the circumstances and the type of information you need to give. Notice that skeletal formulae give a rough idea of the bond angles. In an unbranched carbon chain these are 109.5°. Some examples are given in **Table 26.2.**

Table 26.2 **Examples of types of formulae used in organic chemistry**

Empirical formula	Molecular formula/Name	Shorthand	Displayed	Skeletal	Three-dimensional
CH_2	C_6H_{12} Hex-1-ene	$CH_2CHCH_2CH_2CH_2CH_3$	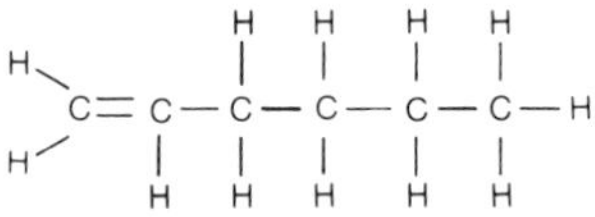		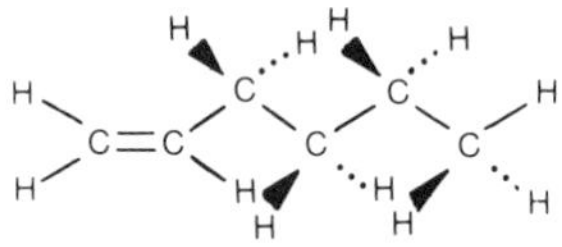
C_2H_6O	C_2H_6O Ethanol	CH_3CH_2OH (or C_2H_5OH)	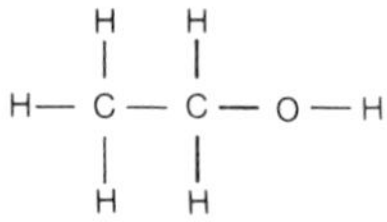	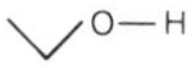	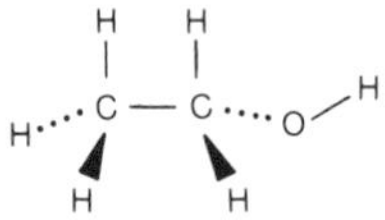
C_2H_6O	C_2H_6O Methoxymethane	CH_3OCH_3	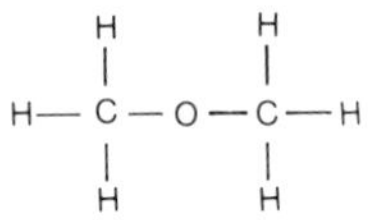	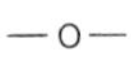	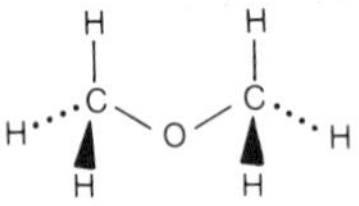
C_3H_7Cl	C_3H_7Cl 2-chloropropane	$CH_3CHClCH_3$	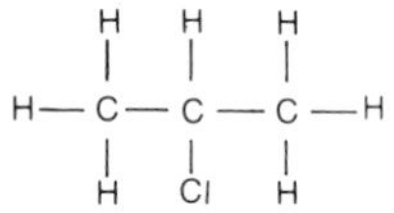		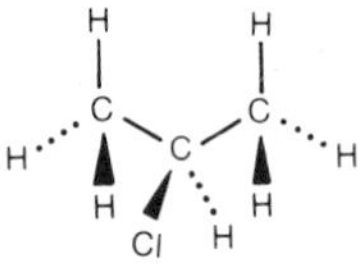
C_3H_6O	C_3H_6O Propanone	CH_3COCH_3	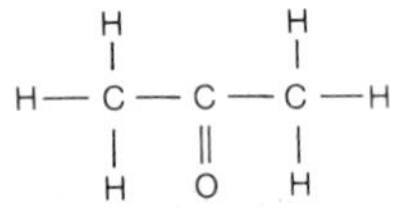		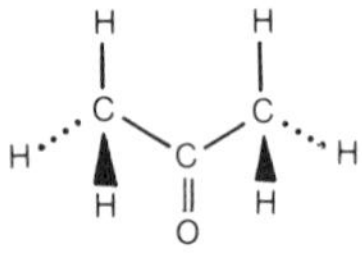

26.5 Naming organic compounds

Many organic compounds are known by **trivial names**—names which are not related to their structure. Examples include formic acid, first obtained from ants (Latin: formica = ants), citric acid obtained from citrus fruits, and Nylon—a trade name selected to trip easily off the tongue.

There is also a **systematic** scheme for naming compounds, developed by **IUPAC** (the International Union of Pure and Applied Chemistry). Systematic names are related to the structure of the compounds. The complete IUPAC rules are extremely complex and we can do no more than cover the basic principles here.

A systematic name consists of a **root**, which describes the overall geometry of the molecule by relating it to the longest unbranched hydrocarbon chain or ring. To this root are added one **suffix** (ending to a word) and as many **prefixes** (beginnings to a word) as necessary which describe the changes that have been made to the root molecule, i.e. hydrogen atoms replaced by other groups. A few simple roots are given in **Table 26.3.** Most of the roots are derived from simple hydrocarbons.

Table 26.3 Roots used in naming organic compounds

Unbranched chains	Rings	Other
Methane, CH_4		Benzene, C_6H_6
Ethane, C_2H_6		
Propane, C_3H_8	Cyclopropane, C_3H_6	
Butane, C_4H_{10}	Cyclobutane, C_4H_8	
Pentane, C_5H_{12}	etc.	
Hexane, C_6H_{14}		
Heptane, C_7H_{16}		
Octane, C_8H_{18}		
Nonane, C_9H_{20}		
Decane, $C_{10}H_{22}$		
Undecane, $C_{11}H_{24}$		
Dodecane, $C_{12}H_{26}$		

Side chains are indicated by a prefix, whose name is related to the number of carbons:

methyl, CH_3—
ethyl, C_2H_5—
propyl, C_3H_7—
cyclopropyl, C_3H_5—
cyclobutyl, C_4H_7—

but benzene as a side group is called phenyl, C_6H_5—

Other groups are indicated by a suffix or prefix, as shown in **Table 26.4.** The order in the list is important. If there is more than one group, the higher in the list is called the **principal group** and is named by a *suffix*, while other groups are named by *prefixes*.

```
                  NH2
                  |      //O
For example,  H — C — C           is called aminoethanoic acid,
                  |      \
                  H       OH
```

```
           //O
as the — C        is the principal group rather than —NH2.
           \
            OH
```

Note that some carbon-containing groups have two names, depending on whether the carbon is counted as part of the root or not.

```
      H   H
      |   |
So H— C — C — C≡N   is propanenitrile but   C6H5—C≡N (benzene ring)
      |   |
      H   H
```

is benzenecarbonitrile

Examples

```
               Cl  H
               |   |
Chloroethane: H— C — C —H
               |   |
               H   H
```

eth indicates that the molecule has a chain of two carbon atoms, **ane** that it has no multiple bonds and **chloro** that one of the H atoms of the ethane is replaced by a chlorine atom.

able 26.4 **The names of groups as ffixes and prefixes**

Formula	Suffix	Prefix
$[NH_4]^+$ (H–N(H)(H)–H)	-ammonium	
— COOH	-carboxylic acid (i.e. when hydrogen is replaced by—COOH)	carboxy-
—(C)(=O)OH	-oic acid	
— COO^-	-carboxylate (ion)	
—(C)(=O)O^-	...oate (ion)	
— COOR	alkyl...-carboxylate	alkoxycarbonyl-
—(C)(=O)OR	alkyl...-oate	
— CO — Hal	-carbonyl halide	halogenocarbonyl-
—(C)(=O)Hal	-oyl halide	
— $CONH_2$	-carboxamide	carbamoyl-
—(C)(=O)NH_2	-amide	
— CN	-carbonitrile	cyano-
—(C)≡N	-nitrile	
— CHO	-carbaldehyde	formyl- or methanoyl-
—(C)(=O)H	-al	oxo-
* —(C)(=O)—	-one	oxo-
—OH	-ol	hydroxy-
—SH	-thiol	
—NH_2	-amine	amino-
—NO_2		nitro-
—Hal		halo- (fluoro-, chloro-, bromo-, iodo-)

*When the C atom in brackets is counted in the carbon chain which forms the root. (Adapted from *Chemical Nomenclature, Symbols and Terminology for use in School Science*, ASE 1985)

Propene:

$$\begin{array}{c} \quad H \quad H \\ H-\overset{|}{\underset{|}{C}}-\overset{|}{C}=C\begin{smallmatrix} H \\ H \end{smallmatrix} \\ \quad H \end{array}$$

prop indicates a chain of three carbon atoms and **ene** that there is a C=C (double bond).

Methanol:

$$\begin{array}{c} H \\ | \\ H-C-OH \\ | \\ H \end{array}$$

meth indicates a single carbon and **ol** an OH group (an alcohol).

Ambiguities soon arise though. Bromopropane could indicate:

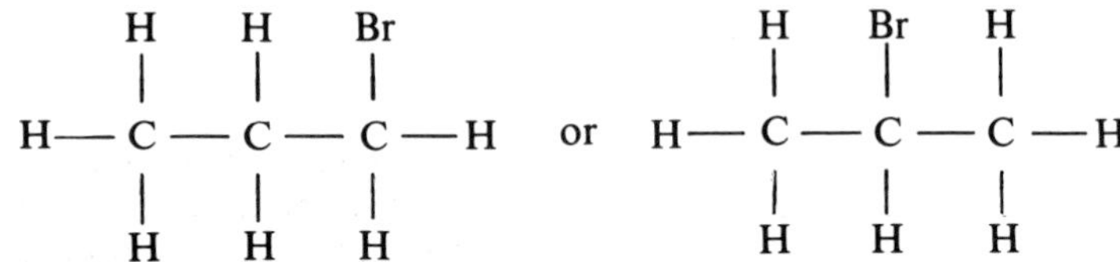

Model of 1-bromopropane

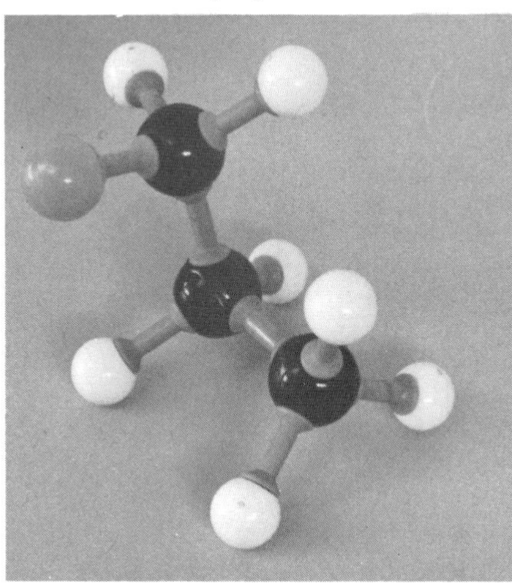

A number, called a **locant**, is used to distinguish these—the first exampl is 1-bromopropane the second 2-bromopropane. Note that:

$$\begin{array}{ccccccc} & & H & & H & & Br \\ & & | & & | & & | \\ H & — & C & — & C & — & C & — & H \\ & & | & & | & & | \\ & & H & & H & & H \end{array}$$

$$\begin{array}{ccccccc} & & H & & H & & H \\ & & | & & | & & | \\ H & — & C & — & C & — & C & — & Br \\ & & | & & | & & | \\ & & H & & H & & H \end{array}$$

$$\begin{array}{ccccccc} & & H & & H & & H \\ & & | & & | & & | \\ H & — & C & — & C & — & C & — & H \\ & & | & & | & & | \\ & & H & & H & & Br \end{array}$$

are *identical*, which is more obvious from the model (see photograph). Th apparent difference is due to the fact that we are trying to represent a three-dimensional molecule on flat paper. More than one locant may b needed:

$$\begin{array}{ccccccc} & & I & & Br & & H \\ & & | & & | & & | \\ H & — & C & — & C & — & C & — & H \\ & & | & & | & & | \\ & & H & & H & & H \end{array} \qquad \text{2-bromo-1-iodopropane}$$

Note that bromo is written before iodo, because the rule is **alphabetica order** of the substituting groups, rather than numerical order of the locants

$$\begin{array}{ccccccc} & & H & & H & & H \\ & & | & & | & & | \\ H & — & C & — & C & — & C & — & Br \\ & & | & & | & & | \\ & & H & & H & & H \end{array} \qquad \text{1-bromopropane}$$

This is not 3-bromopropane, as the rule is to use the smaller locant wher there is a choice. The extra prefixes di-, tri-, tetra- are used to indicat two, three and four respectively of the substituting group. Do not get thes mixed up with the locants. Di-, tri- and tetra- tell you *how man* substituents. 1, 2, 3, etc. tell you *where* they are located.

Further examples are given in **Table 26.5.**

Except for fairly simple molecules, working out the correct name can be a difficult task. Fortunately at A level it is much more important to be able to work out the structure of a molecule from its name, which is a much easier task than to correctly name a given molecule. Also, trivia names are still frequently used, especially in industry, simply because they are easier to say and remember. Terylene is less of a mouthful than poly(ethane-1, 2-dioylbenzene-1, 4-dicarboxylate)!

Further suffixes and prefixes will be introduced in the appropriate chapters, where examples will be given of their use. Naming organic compounds is best learned by practice—try to see how the name of each compound you come across is related to its structure.

Table 26.5 Examples of systematic naming of organic compounds

Structural formula	Name	Notes
H Br H H—C—C—C—H H Br H	2,2-dibromopropane	
H H H H H—C—C—C—C—OH H H Br H	2-bromobutan-1-ol	The suffix-ol defines the end of the chain we count from
H H OH H H—C—C—C—C—H H H H H	butan-2-ol	
H H H H\C=C—C—C—H H/ H H	but-1-ene	Not but-2-ene, as we use the smaller locant
H H H C H H C C H H C C H H C H H H	cyclohexane	Cyclo- is used to indicate a ring
H H H H H—C—C—C—C—H H H H H—C—H H	2-methylbutane	
H H H H H—C—C—C—C—H H H H H—C—H H—C—H H	3-methylpentane	This is not 2-ethylbutane. The rule is to base the name on the longest unbranched chain, in this case pentane (picked out in brown). Remember the bond angles are 109.5°, not 90°
H H H H H—C—C—C—C—H H H—C—H H H H—C—H H—C—H H	2,3-dimethylpentane	Again remember the root is based on the longest unbranched chain

26.6 Families of organic compounds—homologous series and functional groups

26.6.1 Functional groups

Many organic compounds consist of a relatively short hydrocarbon chain with one or more reactive groups attached. These reactive groups are called **functional groups**. Examples include alcohol, —OH, carboxylic acid

A

NON-SYSTEMATIC NAMES

Many non-systematic names are still in everyday use by chemists in industry and non-scientists. Most people know that vinegar is a solution of acetic acid (systematic name ethanoic acid) and in the pharmacist's you would probably get an odd look if you asked for a bottle of 2-ethanoyloxy-benzenecarboxylic acid tablets (aspirin)

Hazchem labels give vital information in an emergency

Many systematic names are simply too long to remember, or even say, easily and are not necessary when we just want to distinguish the contents of one drum from those of another. There may even be possibilities of confusion—ethanol and ethanal are easily mixed up both in speech and writing while their non-systematic names, ethyl alcohol and acetaldehyde, are quite distinct.

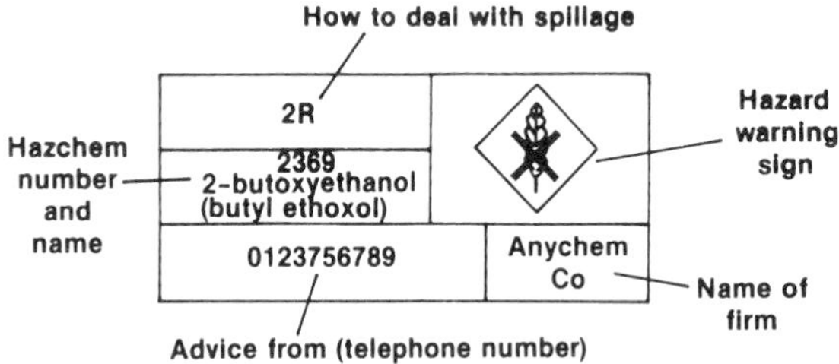

One naming system you might have seen in use on road tankers is the Hazchem system which is designed to give information about the hazards associated with a particular chemical and details of how the emergency services should deal with it in case of an accident. The Hazchem system identifies the chemical by a number as well as its usually used name (rather than a systematic name). A sequence of numbers and letters gives basic instructions about how to deal with a spillage and a symbol graphically illustrates the type of hazard it presents. In the example shown, 2-butoxyethanol, the hazard warning sign indicates that it is toxic but unlikely to be a serious acute health hazard. The emergency action code is 2R. The 2 advises that a fine spray of water should be used on spillages and the R that full body protective clothing should be used and the spillage diluted. See key.

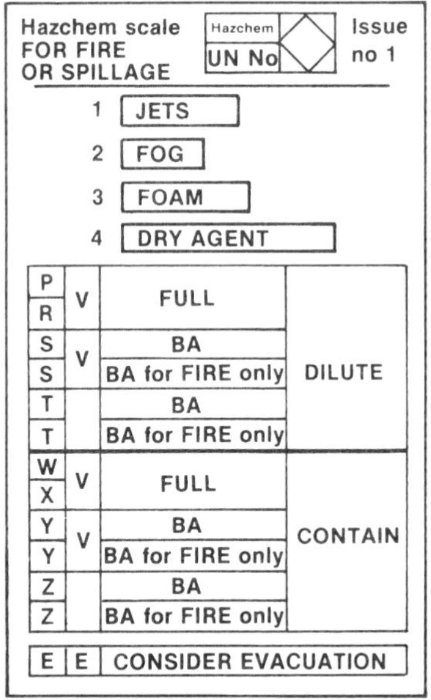

Key used by firefighters (BA means breathing apparatus)

It is interesting to note that both universities and industries do not always use the latest nomenclature. For example, propanone is more often called acetone in many scientific establishments and ethanoic acid is still called acetic acid on most food labels. Finally, some chemicals have quite amusing trivial names. The compound whose skeletal formula is shown below has been synthesized.

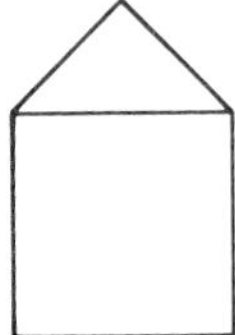

It is tricyclo(2, 1, 0)pentane but has been dubbed 'houseane'. Derivatives with a methyl group instead of one of the hydrogens have been made:

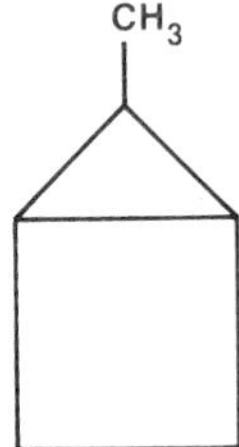

This has been dubbed 'roof-methyl houseane'. What do you think the structures of floor-methyl houseane and eave-methyl houseane are?

```
      O
    //
— C
    \
     OH
```

and amine, $—NH_2$. Often the functional group reacts in the same way, whatever the details of the rest of the molecule. This means that organic chemistry is usually divided up into chemical families with the same functional group. The reactions of the group are relatively unaffected by the details of the rest of the molecule to which it is attached. So, it is possible to learn the general reactions of, for example, alcohols and apply this knowledge to any alcohol.

26.6.2 Homologous series

An **homologous series** is a series of organic compounds with the same functional group, varying only in the length of the carbon chain to which the functional group is attached. Successive members of the series are called **homologues.** While the length of carbon chain has little effect on the chemical reactivity of the functional group, it does affect physical properties, like melting temperature, boiling temperature and solubility. For example, butane (four carbons) is a gas and hexane (six carbons) a liquid at room temperature. Ethanol (two carbons) is much more soluble in water than hexanol (six carbons). Because members of an homologous series react in the same way, we often want a way to indicate in equations any member of the series. This may be done by using the symbol R to indicate the rest of the molecule. For example R—OH could be used to indicate any alcohol.

26.7 Isomers

Isomers are compounds with the same molecular formula but a different arrangement of atoms in space. Organic chemistry provides many examples of isomerism.

26.7.1 Structural isomerism

This is where the structural formulae differ and the isomers either have different functional groups or the functional groups are attached to the main chain at different points.

Examples

The molecular formula C_2H_6O could represent:

```
    H   H                H       H
    |   |                |       |
H — C — C — O — H    H — C — O — C — H
    |   |                |       |
    H   H                H       H
```

ethanol (an alcohol) methoxymethane (an ether)

These isomers have different functional groups.

The molecular formula C_3H_8O could represent:

```
                                     H
                                     |
    H   H   H                    H   O   H
    |   |   |                    |   |   |
H — C — C — C — OH           H — C — C — C — H
    |   |   |                    |   |   |
    H   H   H                    H   H   H
```

propan-1-ol propan-2-ol

E

CIS- *AND* TRANS-*BUTENEDIOIC ACIDS*

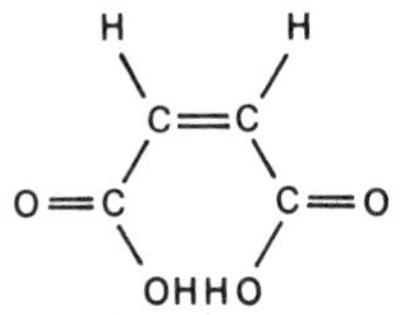

cis-butenedioic acid (maleic acid)

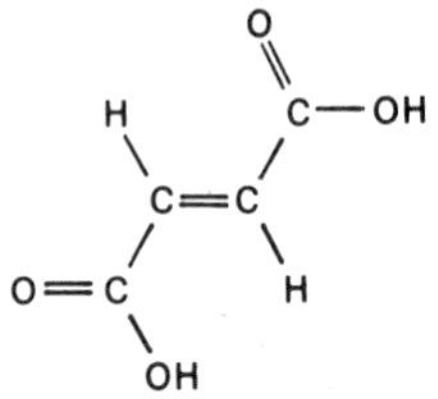

trans-butenedioic acid (fumaric acid)

Both these molecules are flat (planar). No rotation of the double bond is possible. On gentle heating (to 440 K) the *cis*-acid loses a molecule of water to form an anhydride.

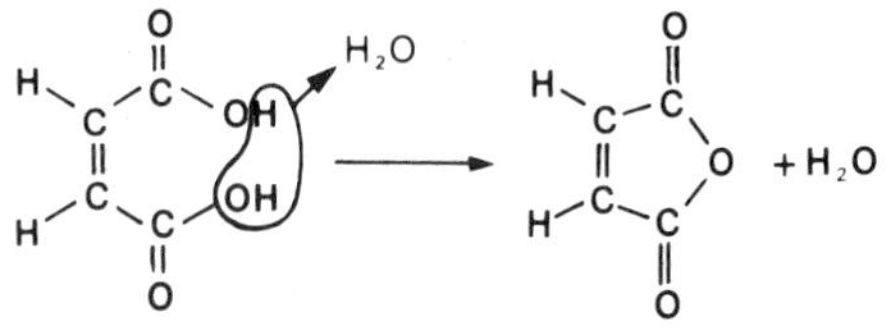

$\longrightarrow$ + H_2O

No such reaction is possible with the *trans*-acid, because the parts that would have to react are too far apart. The *trans*-acid has a much higher melting temperature than the *cis*-acid (559 K compared with 401 K). This is because the *trans*-acid molecules form hydrogen bonds with one another and this holds them together as a solid:

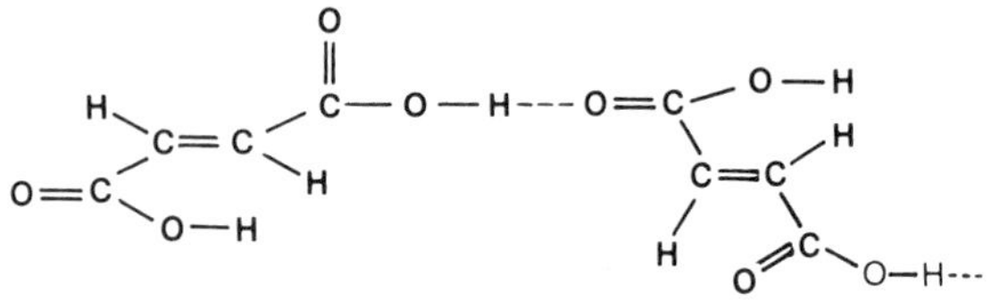

Intermolecular hydrogen bonds in *trans*-butenedioic acid

The *cis*-acid forms *intra*molecular (within the molecule) hydrogen bonds so the molecules are much less strongly held to one another.

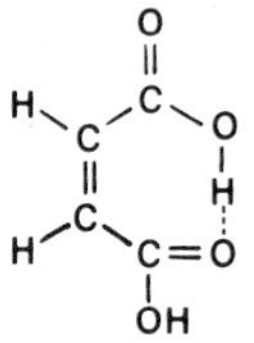

Intramolecular hydrogen bonds in *cis*-butenedioic acid

In these isomers, the functional group is attached to a different part o the carbon chain.

The molecular formula C_4H_{10} could represent:

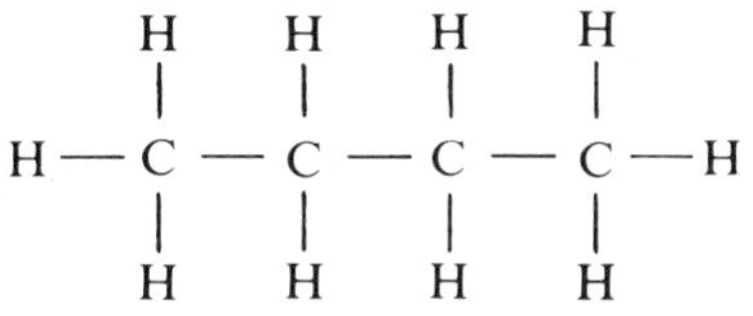

butane 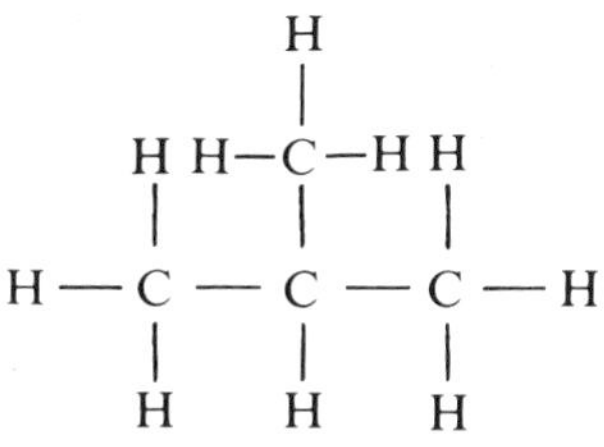 2-methylpropane

These isomers are called chain branching isomers.

Cis–trans isomerism

A particular case of structural isomerism is ***cis–trans* isomerism.** It involve two groups attached to the atoms at either side of a double bond. Thes may either be on the same side of the bond (*cis*) or opposite sides (*trans*)

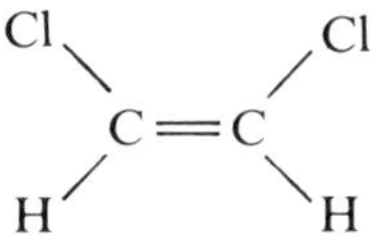

cis-1,2-dichloroethene 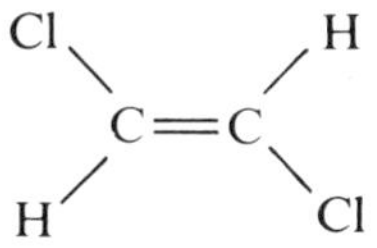 *trans*-1,2-dichloroethene

While molecules can rotate about single bonds, rotation about doubl bonds is not possible, as this would prevent the overlap of the p orbital forming the π part of the double bond. So there are two distinct compound and they are not easily interconvertible.

Figure 26.5 The optical isomers of bromochlorofluoromethane

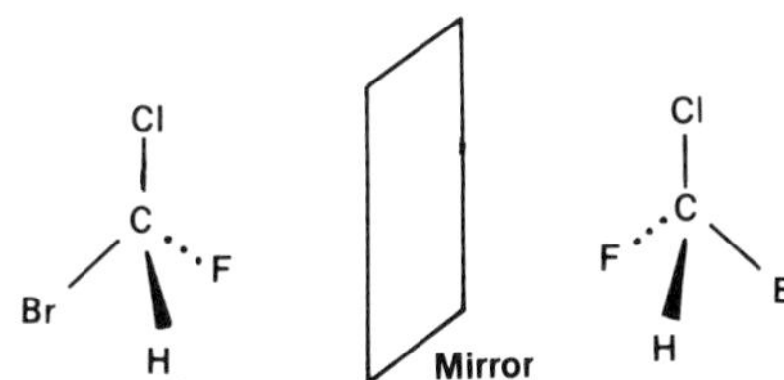

26.7.2 Optical isomerism

Here the two molecules are mirror images of one another. For example bromochlorofluoromethane exists as two mirror image forms, **Figure 26.5**

E

OPTICAL ACTIVITY

Light consists of vibrating electric and magnetic fields. We can think of it as vibrating waves like the skipping rope below, except that in the case of light, vibrations occur in *all* directions at right angles to the direction of motion of the light wave.

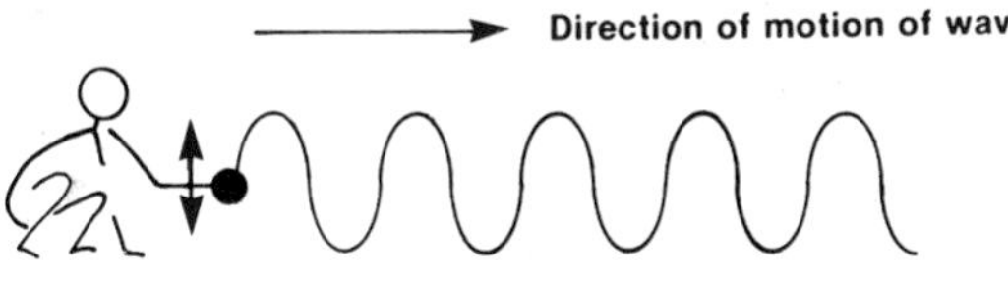

Vibration up and down

If the light passes through a special filter called a **polaroid** (as in polaroid sunglasses), all the vibrations are cut out except those in one plane, for example, the vertical plane.

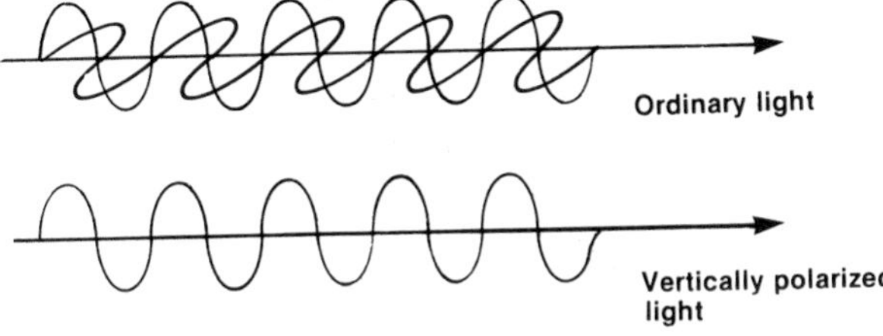

The light is now vertically polarized. If this polarized light passes through a solution containing one enantiomer of an optical isomer, the plane of polarization is rotated. So, vertically polarized light enters the solution and it emerges polarized at, say, 60° to the vertical plane. If one enantiomer rotates the plane of polarization *clockwise*, a solution of the other enantiomer (of the same concentration) would rotate it by the same angle in the *anti-clockwise* direction. The two enantiomers are labelled + and −, the +isomer rotating the plane of polarization clockwise (as viewed by an observer looking towards the light source) and the − isomer rotating it anti-clockwise.

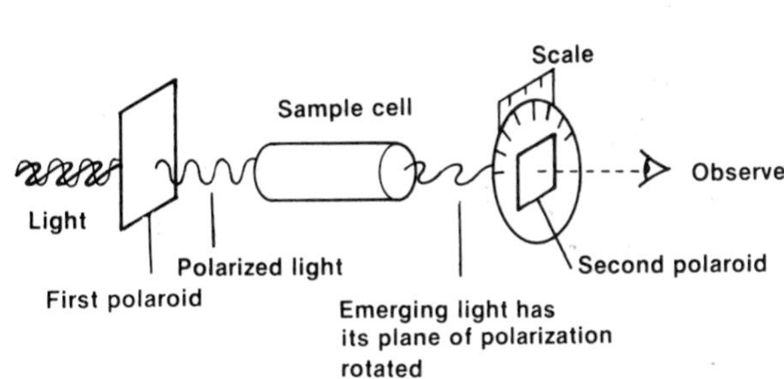

Optical rotation can be measured using a polarimeter. If there is no sample in the cell, and the second polaroid is rotated until it is at right-angles to the first, the observer will see no light, because the first polaroid will cut out all but the vertical vibrations and then the second polaroid will remove these too. (Many polaroid sunglasses are supplied with a small extra piece of polaroid to demonstrate this effect.) If the sample rotates the plane of polarization by, say, 30° clockwise, then the second polaroid will have to be rotated through this angle clockwise for the observer to see no light.

Molecules which rotate the plane of polarization of polarized light are called **optically active compounds.** They are often synthesized by methods which produce equal amounts of both + and − isomers. Such mixtures are not optically active and are called **racemates** (pronounced rass-em-ates) or **racemic mixtures**.

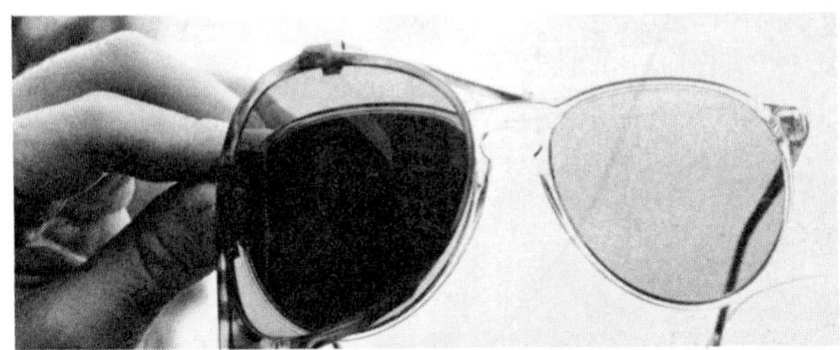

When two pieces of polaroid are at right angles, no light passes through

Models of the optical isomers of bromochlorofluoromethane

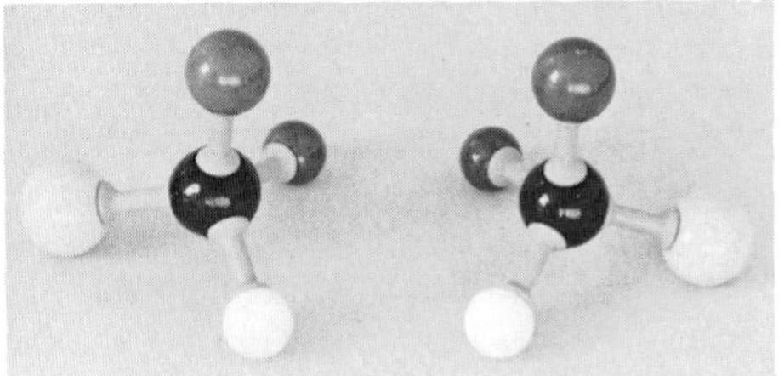

These two are not identical, as the photograph (left) shows. Such molecules have no plane of symmetry. They can easily be identified by the fact that the molecule contains at least one carbon atom to which four different groups are bonded. **Optical isomers** are said to be **chiral** (pronounced kiyral) meaning 'handed', as in left- and right-handed. The carbon bonded to the four different groups is called the **chiral centre** or the **asymmetric carbon atom** and is often indicated by *, and the two isomers are called a pair of **enantiomers.** They are also called **optical isomers** as they differ in the way they rotate the plane of plane of polarized light, see box.

Example

O–H on C*, with H_3C, H and CO_2H also bonded to C* 2-hydroxypropanoic acid

2-hydroxypropanoic acid (non-systematic name lactic acid) is chiral. Although the chiral carbon is bonded to two other carbon atoms, these carbons are part of different groups. Note that optical isomers result fro the three-dimensional structures of the molecules and isomers can onl be distinguished by three-dimensional formulae.

26.8 Physical properties of organic compounds

Physical properties include melting temperatures (T_m), boiling temper atures (T_b) and solubility in water and other solvents. These are determine largely by the intermolecular forces between the molecules (see **section 16.2**

26.8.1 Boiling temperatures

These are closely related to the *inter*molecular forces. Boiling occurs wher on average, molecules have enough energy to overcome the intermolecula forces. Two factors are important—the type of intermolecular force operating and the size of the molecule. Hydrogen bonding is stronger tha dipole–dipole forces which are stronger than van der Waals forces, sc other things being equal, hydrogen bonded liquids will have higher boilin temperatures than dipole–dipole bonded liquids, and these will have highe boiling temperatures than liquids where only van der Waals forces operate This can be illustrated by the liquids propan-1-ol, propanal and butane all of which have approximately the same relative molecular mass. Se **Table 26.6.**

Table 26.6 The effect of different types of intermolecular forces on boiling temperatures

Compound	Structure	Bonding and boiling temperature
Propan-1-ol $M_r = 60$	$CH_3CH_2CH_2$–O–H ··· O(H)C_3H_7	Hydrogen bonding $T_b = 371$ K
Propanal $M_r = 58$	$CH_3CH_2C^{\delta+}(=O^{\delta-})H$	Dipole–dipole bonding $T_b = 322$ K
Butane $M_r = 58$	H–C(H)(H)–C(H)(H)–C(H)(H)–C(H)(H)–H	van der Waals bonding $T_b = 135$ K

Increasing T_b ↑

In any series of similar compounds where the same type of inter molecular forces operate, the van der Waals forces will increase with the total number of electrons in the molecule and hence with increasing relative molecular mass. So in general a homologous series will show an increase of boiling temperature with relative molecular mass. See **Figure 26.6.** The substitution of a heavy atom like, say, chlorine ($A_r = 35.5$) for hydrogen will increase the boiling temperature, both because it makes the molecule more polar and more able to form dipole–dipole bonds, and also because it increases the relative molecular mass. For example methane $T_b = 109$ K and chloromethane $T_b = 249$ K.

Figure 26.6 The effect of relative molecular mass on boiling temperature for series of similar compounds

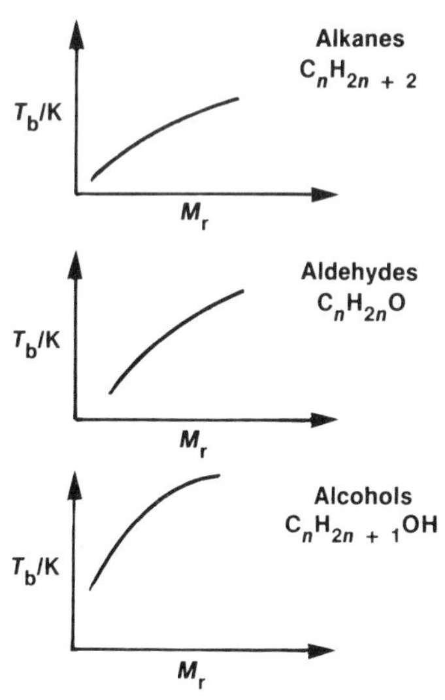

26.8.2 Melting temperatures

These are affected by the same factors as boiling temperatures. A furthe factor is also involved: the shape of the molecule, as this affects the way the molecules can pack together in the solid state. Those which can pack together well have higher melting temperatures than molecules which pack poorly, **Figure 26.7.** Of the two isomers butane and 2-methylpropane, the unbranched butane has the higher melting temperature as its molecules can pack together more closely and are therefore harder to separate.

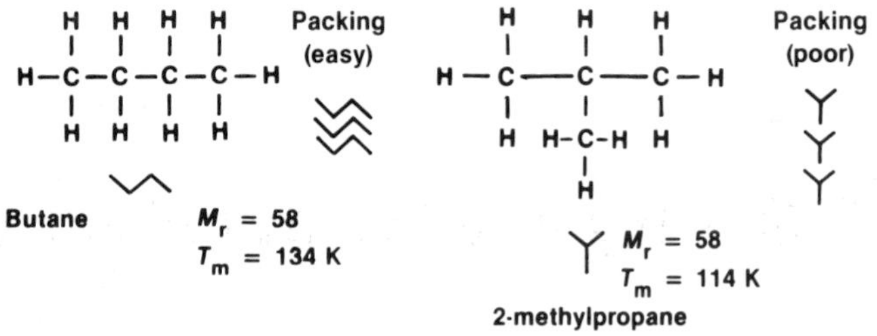

Figure 26.7 Butane molecules pack closer than those of 2-methylpropane

Butane $M_r = 58$ $T_m = 134$ K — Packing (easy)

2-methylpropane $M_r = 58$ $T_m = 114$ K — Packing (poor)

26.8.3 Solubility

The rule 'like dissolves like' introduced in **section 16.3.1** applies. In particular, molecules with a functional group which can form hydrogen bonds will dissolve well in water, while non-polar molecules like hydrocarbons will not. Examples include alcohols R—OH, amines R—NH_2 and carboxylic acids R—C(=O)OH. These will all dissolve well in water provided the hydrocarbon chain of R is less than about four carbon atoms long. With longer carbon chains, the compounds gradually become less soluble, as the insoluble hydrocarbon part begins to dominate the molecule's properties. However, a second hydrogen bonding group increases the solubility again.

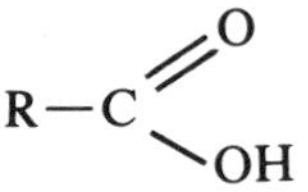

Polar, but non-hydrogen bonding groups can also confer water solubility, so that propanone (a ketone), $CH_3—C^{\delta+}(=O^{\delta-})—CH_3$ is very water soluble. As you might expect, ketones with longer hydrocarbon chains are less water soluble. Non-polar compounds will dissolve well in non-polar solvents like hydrocarbons. A liquid of intermediate properties can often be used to get two otherwise **immiscible** (un-mixable) substances to mix. For example, water and 1-iodobutane do not mix, but both water and 1-iodobutane will dissolve in ethanol. The ethanol is acting as a **co-solvent**.

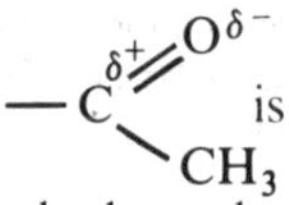

26.9 Reactivity of organic compounds

This section looks at some general principles which govern the reactivity of organic compounds. These include bond energy, bond polarity, the inductive effect and electron-rich areas in the molecule, as well as the type of reagent with which it is reacting. Details about the reactions of the different functional groups are given in the appropriate chapters.

26.9.1 Bond energies

Bond energies tell us the amount of energy required to break a particular bond—they represent the strength of that bond. If we examine the strengths of all the bonds in a molecule we might expect that the weakest would be the most likely to break. This is true up to a point and though there are a number of other factors which are involved in organic reactions, bond strength may give some pointers. For example, ethanol:

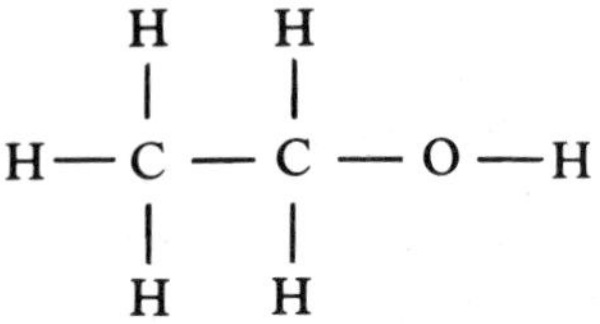

$$CH_3—CH_2—O—H$$

Tabulated bond energies/kJ mol^{-1}	
C—C	347
C—H	413
C—O	336
O—H	464

A

SURFACTANTS

Alcohols, R—OH, with chains of more than about five carbon atoms are not very soluble in water. In fact, the molecules tend to cluster at the surface with their —OH groups in the water, forming hydrogen bonds. The hydrocarbon 'tails' stick out. Thus they form a film on the surface of the water. Such films have been used to reduce evaporation from reservoirs in hot countries.

As the film is just one molecule thick, only a small amount of the alcohol need be added to cover a very large area.

The 'like dissolves like' rule explains the action of detergents. These molecules have a polar or hydrogen bonding 'head', described as **hydrophilic** (water loving), and a **hydrophobic** (water hating) hydrocarbon 'tail'. The hydrophobic tails dissolve in droplets of grease while the hydrophilic heads stick out into the water. As the heads are water soluble, enough of them can force the grease droplet to mix with the water.

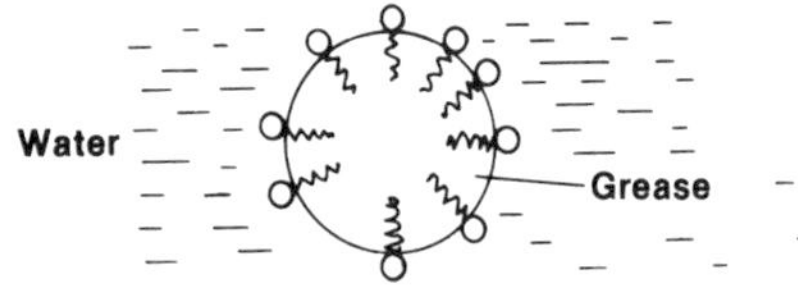

Some typical detergents

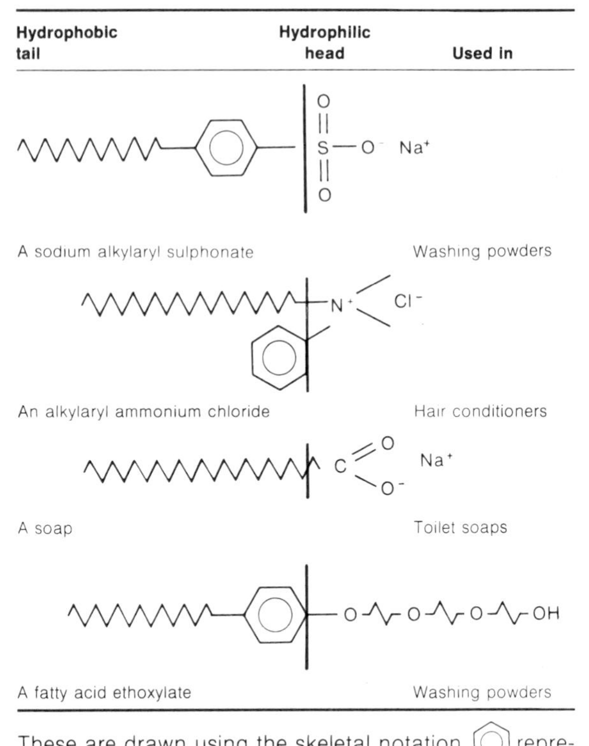

These are drawn using the skeletal notation. ⬡ represents a benzene ring (see Chapter 29).

Table 26.7 Tabulated bond energies of bonds encountered in organic chemistry

Bond	**Bond energy**/kJ mol^{-1}
C—H	413
C—C	347
C—O	358
O—H	464
C—F	467
C—Cl	346
C—Br	290
C—I	228
C=C	612
C≡C	838
C=O	743
C—N	286
C=N	615
C≡N	887
N—H	391
N—O	214
O—O	144

The C—O bond is the weakest and we might predict that it would be the most likely to break. This prediction is in fact upheld and the C—O bond does break in many of the reactions of ethanol, but there are reactions in which the O—H bond breaks, even though this is the strongest bond, so bond strength cannot be the only factor.

Again, a look at the bond energies in **Table 26.7** shows you that C—C and C—H bonds are among the strongest single bonds and so we might expect that the hydrocarbon skeleton of an organic molecule would tend to stay intact during reactions. This prediction, too, is borne out in practice, although again the bond energy factor is not the whole reason for this.

Multiple bonds

At first sight, multiple bonds seem to be stronger than single bonds, but we must remember that a double bond consists of a σ bond and a π bond, and a triple bond consists of a σ bond and two π bonds. If we take the C—C value (347 kJ mol^{-1}) to represent the σ bond, then this means that the π bond's contribution to C=C is $612 - 347 = 265$ kJ mol^{-1}, so the π bond is considerably weaker than the σ bond. Similarly, the energy of the *second* π bond in C≡C is given by $838 - 612 = 226$ kJ mol^{-1}, even weaker. So it may be possible to break the π component of a multiple bond and leave the σ part intact. This occurs often, for example, in the test for a double bond:

$$H_2C{=}CH_2 + Br_2 \longrightarrow H{-}\underset{Br}{\overset{H}{C}}{-}\underset{Br}{\overset{H}{C}}{-}H$$

able 26.8 Electronegativities of ements encountered in organic emistry

ement	Electronegativity
	2.5
	2.1
	3.5
	4.0
	3.0
	2.8
	2.5
	3.0

igure 26.8 Propanone, lectronegativities in brown

δ− O 3.5
δ+ C 2.5
H 2.1
C 2.5
H H H H H

able 26.9 Electronegativities and bond nergies for carbon–halogen bonds

lectronegativities	Bond energy/kJ mol^{-1}
5 4.0 C—F	467
5 3.0 C—C	347
5 2.8 C—Br	290
5 2.5 C—I	228

26.9.2 Polarity

This is the second factor which governs the reactivity of organic molecules. Except for bonds between two atoms of the same element, one of the atoms in a bond will always be more electronegative than the other, see **Table 26.8**, and will attract the shared electrons more strongly so that the bond will be polar (see **section 10.2.2**). This is indicated by the signs δ^+ and δ^-, showing the partial charges on each atom. For example in propanone, the C=O bond is far more polar than the C—H bonds, as the difference in electronegativities of C and O is much greater than that between C and H, **Figure 26.8.** So, we might expect negatively charged reagents (called **nucleophiles**, see page 396) to attack the carbon atom, and positively charged ones (called **electrophiles**, see page 397) to attack the oxygen atom. Either way we would get a reaction involving the C=O rather than the C—C or C—H bonds. This is observed in practice in most of the reactions of propanone.

Sometimes the bond energy and polarity approaches give conflicting results and so we cannot make a clear-cut prediction. For example, look at the data for the carbon–halogen bonds, **Table 26.9.**
The carbon–halogen bonds get weaker (and therefore *easier* to break) as we go from fluorine to iodine. However, the bonds also get less polar, which means that the carbon becomes *less easily* attacked by negatively charged reagents as we go from fluorine to iodine. So the two factors, bond energy and polarity, predict different outcomes. Bond energies predict C—F to be the least reactive, polarity predicts C—F to be the most reactive.

In fact, in the reaction with OH^- ions, CH_3I reacts the fastest and CH_3F not at all. This example illustrates the problems involved in making predictions without considering all the factors.

26.9.3 The inductive effect

Another way of representing the polarity of a bond with an electronegative atom is to draw an arrow on the bond in the direction in which the electrons are attracted. So a carbon–halogen bond would be represented:

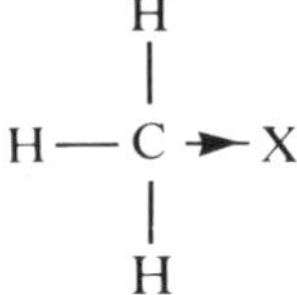

showing that the halogen draws electrons towards itself. This is sometimes called an **inductive effect**, and halogens are said to have a **negative inductive effect**.

Alkyl groups like methyl-, ethyl-, etc. have the opposite effect. They tend to *release* electrons and have a **positive inductive effect**, as shown by the direction of the arrow:

H H
H — C → C—
H H

The effect is increased if more than one alkyl group is attached to the same carbon:

CH_3
↓
H_3C → C —
↑
CH_3

The effect is relatively small but can be important. For example, a carbon atom carrying a positive charge is more stable if it is bonded to one or more alkyl groups rather than to hydrogens.

One prediction that is made by consideration of both bond energi and polarity is that hydrocarbon chains will be unreactive. Both C— and C—H bonds are relatively strong and relatively non-polar and bo these factors suggest low reactivity. This prediction is borne out in fac Most organic reactions involve the functional group rather than t carbon chain.

26.9.4 Electron-rich areas

In carbon–carbon double and triple bonds, there is a higher than norm concentration of electrons, four and six respectively per bond, compare with two in a single bond. These electrons are equally shared between t two carbon atoms. These areas of high electron density (**electron-rich area** make these bonds susceptible to attack by positively charged reagen (**electrophiles**) such as the hydrogen ion, H^+.

26.9.5 Types of reagent

Unless we try to classify organic reactions we shall end up with a mas of facts which simply have to be learnt. One way of classifying reaction is in terms of the type of reagent which attacks the organic molecule. Thes may be divided into **nucleophiles**, **electrophiles** and **radicals**.

Nucleophiles

Nucleophiles are ions or molecules which have electron-rich areas. The tend to attack positively charged areas of organic molecules. A nucleophil must have a lone pair of electrons with which it can form a dative bon with the positively charged atom in the organic molecule. Since they ten to denote electron pairs, nucleophiles are also Lewis bases (see **sectio 12.2**). Examples of nucleophiles include:

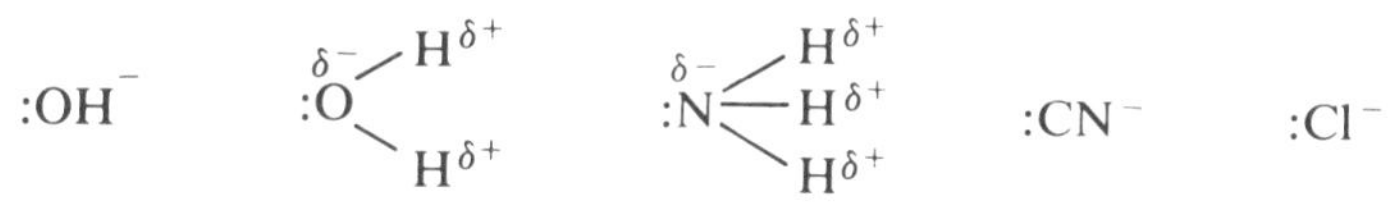

$:OH^-$	$:\overset{\delta-}{O}(H^{\delta+})_2$	$:\overset{\delta-}{N}(H^{\delta+})_3$	$:CN^-$	$:Cl^-$
hydroxide ion	water	ammonia	cyanide ion	chloride ion

The symbol :Nu or Nu^- is often used to indicate any nucleophile. Not that some are overall negatively charged while some are overall neutral.

The first step of a reaction with a nucleophile involves the nucleophil using its lone pair of electrons to form a bond with the electron-deficien (positively charged) carbon atom. Movement of electron pairs in organi reactions is usually shown by 'curly arrows'. Do not confuse them with a straight arrow representing a dative bond.

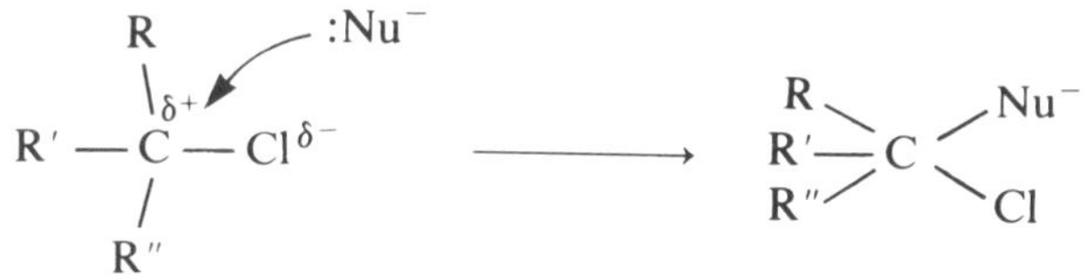

$$RR'R''C^{\delta+}—Cl^{\delta-} + :Nu^- \longrightarrow RR'R''C(Nu^-)(Cl)$$

starting material with $C^{\delta+}$ — intermediate

In the **intermediate**, the carbon is forming five bonds (has ten electrons in its outer shell) so needs to 'shed' a pair of electrons. One possibility is shown:

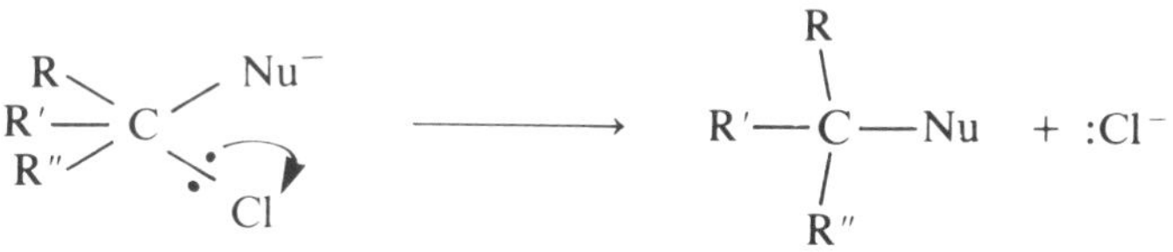

$$RR'R''C(Nu^-)(Cl) \longrightarrow RR'R''C—Nu + :Cl^-$$

The extra electrons are taken away by an atom or group of atoms called the **leaving group** (in this case a Cl^- ion). Remember that each single bond represents a pair of shared electrons.

Electrophiles

Electrophiles are ions or molecules which are electron deficient (positively charged) or have electron-deficient areas. They tend to attack electron-rich areas of organic molecules like double bonds.

Some examples of electrophiles include:

H^+ — hydrogen ion (proton)

NO_2^+

$\overset{\delta+}{S}(=O^{\delta-})_3$ — sulphur trioxide

The symbol El^+ or El is used to indicate any electrophile. Note that some are overall positively charged and some overall neutral.

Electrophiles can accept an electron pair and are therefore Lewis acids (see **section 12.2**).

The first step of the reaction between an electrophile and an organic compound involves the electrophile being attracted to an electron-rich area of the molecule, as for example a double bond, and then accepting an electron pair to form a bond with the carbon atom:

$$\mathrm{El^+} + \mathrm{RR'C{=}CR''R'''} \longrightarrow \mathrm{R{-}C(El)(R'){-}C^+(R''')}{-}\mathrm{R''}$$

In this case the electrophile might bond to either carbon atom although which is favoured might depend on what R, R′, R″ and R‴ are. Notice that the carbon not bonded to the electrophile is forming only three bonds (has only six electrons in its outer shell) and is thus positively charged. The resulting positive ion (sometimes called a **carbocation**) will then be attacked by a negative ion, for example:

$$\mathrm{R{-}C(El)(R'){-}C^+(R''){-}R'''} + \mathrm{X^-} \longrightarrow \mathrm{R{-}C(El)(R'){-}C(X)(R''){-}R'''}$$

Radicals

Radicals, sometimes called **free radicals**, are species with an unpaired electron. They can be formed by the breaking of a single bond in such a way that one of the shared electrons goes to each new species, for example:

$$\mathrm{Br \div Br \longrightarrow Br^{\cdot} + Br^{\cdot}}$$

This reaction can be brought about by ultraviolet light. Or

$$\mathrm{R{-}O \div O{-}R \longrightarrow RO^{\cdot} + RO^{\cdot}}$$

This reaction occurs with gentle heating.

Such bond breaking where one electron of the bond goes to each fragment is called **homolysis** ('equal breaking').

Most radicals are extremely reactive and will attack other molecules more or less indiscriminately in order to gain another electron:

$$\mathrm{Br^{\cdot} + CH_4 \longrightarrow \dot{C}H_3 + HBr}$$

Unless the reaction is between two radicals a new radical is always formed.

Another way in which a bond can break, which we note here simply for comparison, is called **heterolysis** and here, by contrast to homolysis, both shared electrons go to one species and none to the other, as in the example:

$$\mathrm{H \div Cl \longrightarrow H^+ + {:}Cl^-}$$

so there are no radicals formed.

26.9.6 Types of reaction

In principle, there are only three types of reaction and these are **addition**, **elimination** and **substitution**.

Examples

Addition:

$$\mathrm{H_2C{=}CH_2 + Br_2 \longrightarrow H{-}CHBr{-}CHBr{-}H}$$

Elimination:

$$\mathrm{H{-}CH_2{-}CH(OH){-}H \longrightarrow H_2C{=}CH_2 + H_2O}$$

Substitution:

$$\mathrm{H{-}CH_2{-}CH(OH){-}H + Cl^- \longrightarrow H{-}CH_2{-}CHCl{-}H + OH^-}$$

As the names imply, addition involves two molecules joining together, elimination a molecule being ejected, and substitution a replacement of one atom or group by another.

Combining this with the three types of reagent described above leads to **Table 26.10** which predicts a total of nine possible types of organic reaction (combinations of reagent type and reaction type), although not all need occur.

Table 26.10 Possible types of organic reaction

Reagent type \ Reaction type	Addition	Elimination	Substitution
Nucleophile	√	√*	√
Electrophile	√		√
Radical			√

*In nucleophilic elimination reactions, the nucleophile is often considered to be behaving as a base.

These reactions are usually referred to as a **nucleophilic addition**, **electrophilic substitution**, and so on. The types which are most important for A level are indicated with a √. Not all reactions fit easily into this scheme. Oxidation reactions, particularly, are often considered separately.

It is worth noting that substitution and elimination reactions are almost always possible but addition reactions are not possible in a compound in which all of the carbon atoms are already forming four single bonds. Such compounds are called **saturated** while compounds with multiple bonds where addition reactions are possible are called **unsaturated**. For example:

$$\mathrm{H{-}CH_2{-}CH_2{-}H}$$

ethane, saturated, so no addition reactions are possible

$$\mathrm{H_2C{=}CH_2 + H_2 \longrightarrow H{-}CH_2{-}CH_2{-}H}$$

ethene, unsaturated, addition is possible

$$\mathrm{H{-}C{\equiv}C{-}H + 2H_2 \longrightarrow H{-}CH_2{-}CH_2{-}H}$$

ethyne, unsaturated, addition is possible

26.10 Equations in organic chemistry

In inorganic chemistry it is usual to write balanced symbol equations for reactions. This is sometimes done in organic chemistry but reactions are often written in other forms. For example, it is common simply to write the organic starting material and organic product with the reagent and conditions along the arrow, as below, rather than a fully balanced equation. For example:

$$\underset{\text{bromoethane}}{CH_3CH_2Br} \xrightarrow[\text{reflux}]{\text{NaOH}} \underset{\text{ethene}}{H_2C{=}CH_2} \quad \text{(an elimination reaction)}$$

This is partly because organic chemists are usually interested mainly in the organic product and partly because organic reactions frequently give a mixture of products and so a balanced equation is not really appropriate. In the example above, some ethanol, CH_3CH_2OH, is also produced by a substitution reaction. By comparison, the majority of inorganic reactions give one product only.

26.11 Instrumental techniques in organic chemistry

26.11.1 *Infra-red spectroscopy*

Infra-red spectroscopy, see **section 19.3.3**, is one of the most versatile analytical techniques and is much used in organic chemistry. Particular bonds have specific frequencies, so infra-red spectroscopy allows us to identify the type of bonds present and hence the functional groups in a molecule. The technique is both quick and simple. For liquids, a drop is squeezed between a pair of sodium chloride plates to form a thin film. Solids are ground up with a liquid hydrocarbon called nujol to form a thin paste, called a **mull**, which is used to form the thin film. The spectrum can be scanned in less than a minute.

odium chloride plates used for infra-red pectroscopy

able 26.11 Wave numbers for bonds ommonly encountered in organic hemistry

ond	Wave number/cm^{-1}
—H	3700–3200
—H	3500–3100
—H	3200–2800
≡N	2400–2200
≡C	2300–2100
=O	1800–1650
=C	1700–1600
—O	1250–1000
—Cl	800–600

Figure 26.9 (overleaf) shows two examples of infra-red spectra while **Table 26.11** shows the wave numbers of some common bonds. The infra-red spectra of propan-2-ol and propanone are clearly different, the former showing a prominent peak at 3300 cm^{-1} due to an O—H stretch. This is absent in the latter which shows a peak at 1720 cm^{-1} due to a C=O stretch.

MICRONS

Transmission/%

NEAT

O—H stretch

WAVENUMBERS

NICOLET 20SX FT-IR

MICRONS

Transmission/%

CH_3COCH_3

NEAT

C = O stretch

WAVENUMBERS

NICOLET 20SX FT-IR

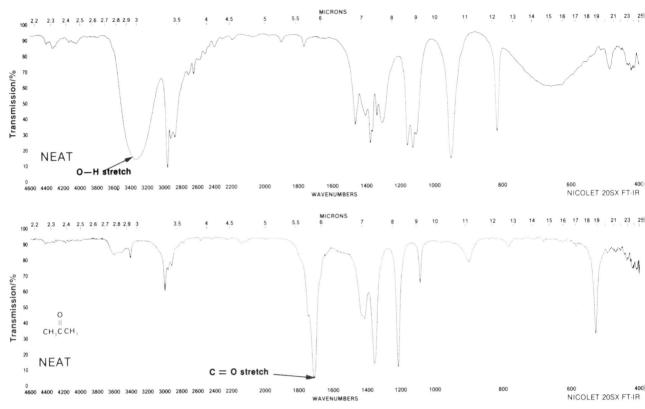

Figure 26.9 **The infra-red spectra of propan-2-ol (top) and propanone (bottom)**

26.11.2 Mass spectrometry

Low resolution mass spectrometry, see **section 19.3.4**, is used in organic chemistry both as a means of measuring relative molecular masses and to provide other structural information from isotope peaks and fragmentation patterns, **Figure 26.10**.

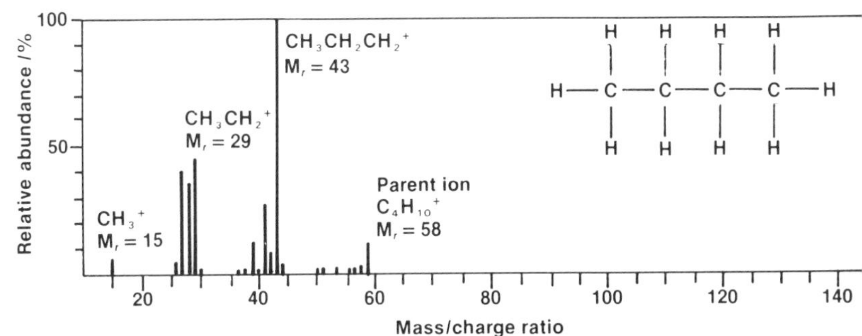

Figure 26.10 **The mass spectrum of butane. Minor peaks are caused by the loss of hydrogen atoms from the main peaks**

26.12 Summary

- Carbon can form four covalent bonds.
- It can form strong bonds with other carbon atoms and with hydrogen as well as with other elements.
- Carbon compounds are kinetically stable with respect to reactions with air and water at room temperature.
- Neither C—C nor C—H bonds are appreciably polar so they are not easily attacked by either positively or negatively charged reagents.
- The empirical formula is the simplest ratio of atoms present in the compound.
- The molecular formula shows the number and type of atoms present in one molecule of the compound.

- The number of empirical formula units in the molecular formula is found by dividing the relative molecular mass by the relative molecular mass of one empirical formula unit.
- Different types of structural formulae can be used:

$CH_3CH(CH_3)CH_2OH$

shorthand

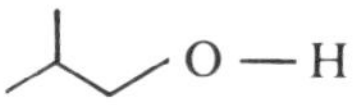

skeletal

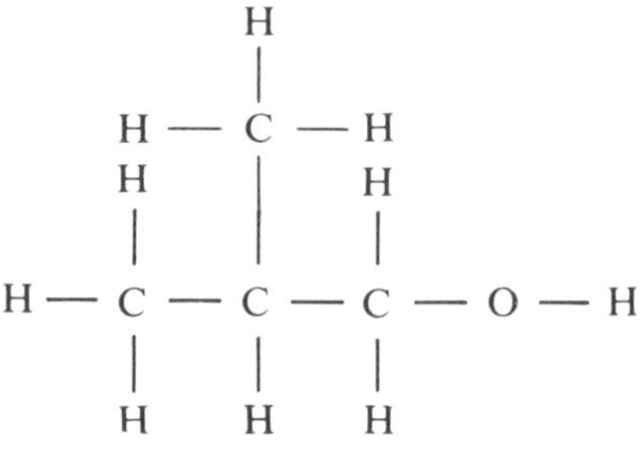

```
        H
        |
   H —  C — H
   H    |    H
   |    |    |
H— C —  C —  C — O — H
   |    |    |
   H    H    H
```

displayed

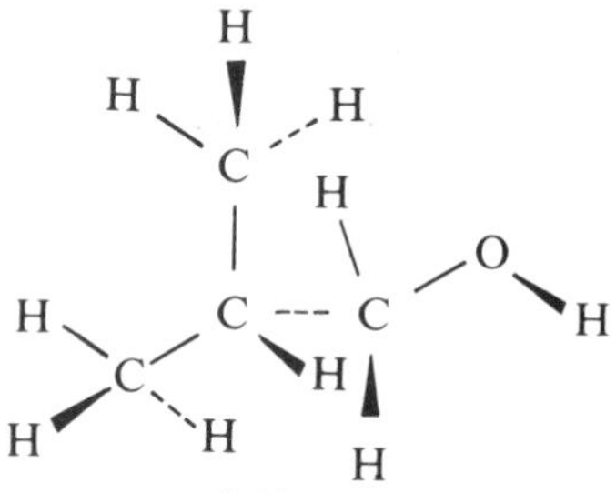

3-D

odel of 2-methylpropan-1-ol

All represent 2-methylpropan-1-ol, $C_4H_{10}O$.

- The IUPAC system for naming organic compounds is based on a root describing the longest carbon chain or ring with suffixes and prefixes to indicate groups attached to the main chain and locants to show where these are on the main chain.
- An homologous series is a family of organic compounds which vary only in the length of their carbon chain.
- Functional groups are the reactive groups in organic compounds.
- Isomers are molecules with the same molecular formula but different spatial arrangements of atoms.
- Melting and boiling temperatures of organic compounds increase with the polarity of the molecule. Molecules which form hydrogen bonds have the highest boiling temperatures.
- Solubility of organic compounds is governed by the 'like dissolves like' rule, compounds dissolving in solvents with a similar degree of polarity.
- Both bond energies and polarity can be used to help predict the likely course of organic reactions.
- Electronegative atoms bonded to carbon atoms produce a δ^+ charge on the *carbon.*
- Multiple bonds produce areas of high electron density in organic compounds.
- Carbon atoms with a δ^+ charge are attacked by nucleophiles.
- Areas of high electron density in organic compounds are attacked by electrophiles.
- Radicals are species with unpaired electrons, generated by homolytic breaking of a covalent bond. They are usually extremely reactive.
- Infra-red spectra are frequently used to help identify groups in organic chemistry.
- Mass spectrometry is used to determine relative molecular masses, as as well as give structural information.

26.13 Questions

1 Name the following compounds:

A)

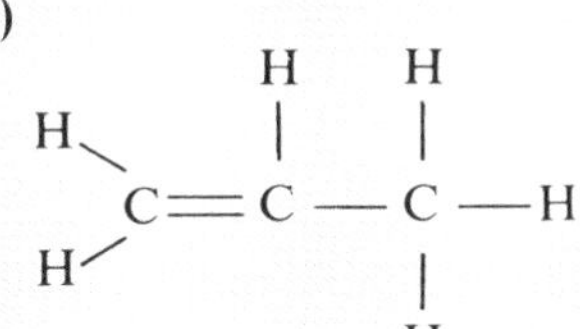

B)

```
    H   H   H   H
    |   |   |   |
H — C — C = C — C — H
    |           |
    H           H
```

C)

```
       H   H
H      |   |      H
  \    |   |     /
   C = C — C = C
  /              \
H                 H
```

D)

```
      H   H   OH  H
      |   |   |   |
HO —  C — C — C — C — H
      |   |   |   |
      H   H   H   H
```

E)

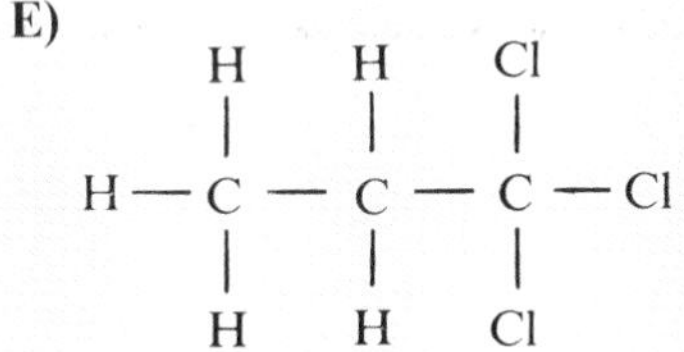

2 Draw the structural formulae of:
A) 2, 2-dichloropropane
B) butan-2-ol
C) 2-methylpropan-2-ol
D) cyclopropane
E) but-1-ene.
Which two form a pair of isomers?

3 There are five isomers with the molecular formula C_4H_9Br. Draw their displayed formulae. Identify the pair of optical isomers.

4 The substance:

```
       H
       |
CH3 —  C — CH2CH3
       |
       OH
```

displays optical isomerism. Indicate the chiral carbon atom with a*. Draw three-dimensional representations of the two enantiomers.

5 Copy the following formulae and insert δ^+ and δ^- signs to indicate the polarity of the molecules:

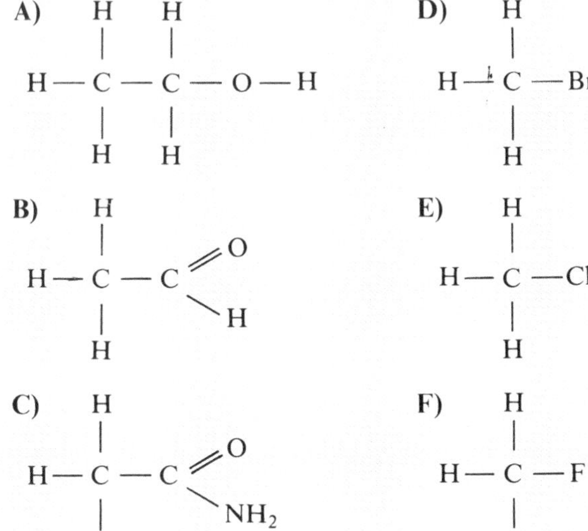

Put (D), (E) and (F) in order of δ^+ character of the carbon atoms (least δ^+ first).

6 Use the list of bond energies on page 394 to predict the weakest bond in the following compounds:

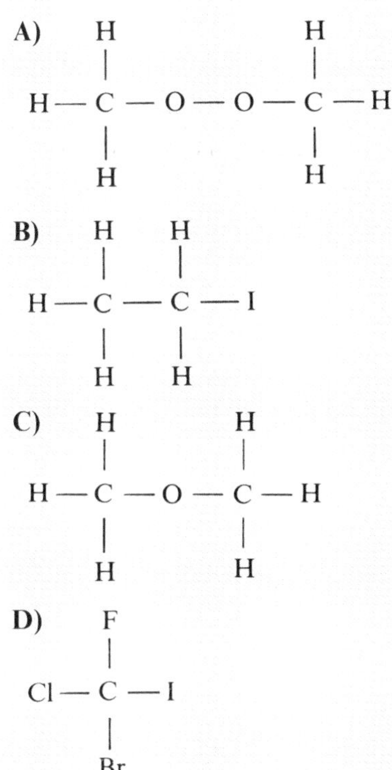

7 Classify the following reagents as nucleophiles, electrophiles or radicals:

H^+, H^-, CH_3NH_2, Cl^-, H_2O, $HSO_3{}^+$, $H^{\cdot}$

8 Draw a displayed formula of 'houseane' (see box *Non-systematic names*).

9 Classify the following reactions as addition, substitution or elimination:

A)

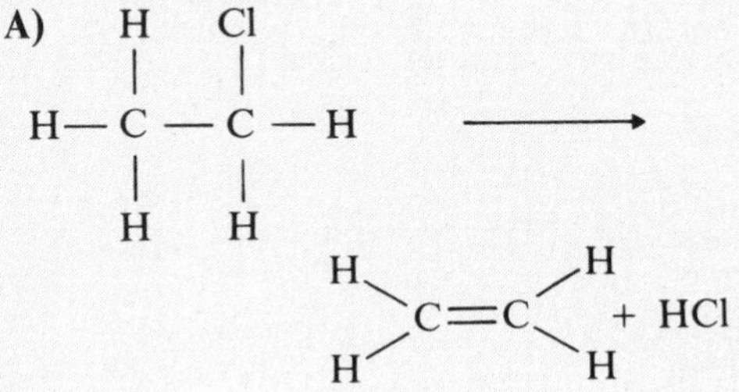

```
   H   Cl
   |   |
H— C — C — H   ——→
   |   |
   H   H
                H         H
                  \     /
                   C=C        + HCl
                  /     \
                H         H
```

B)

```
   H   H
   |   |
H— C — C —Cl + H2O   ——→
   |   |
   H   H
                H   H
                |   |
           H— C — C —OH + HCl
                |   |
                H   H
```

C)

```
H         H
  \     /
   C=C       + HCl   ——→
  /     \
H         H
                H   Cl
                |   |
           H— C — C — H
                |   |
                H   H
```

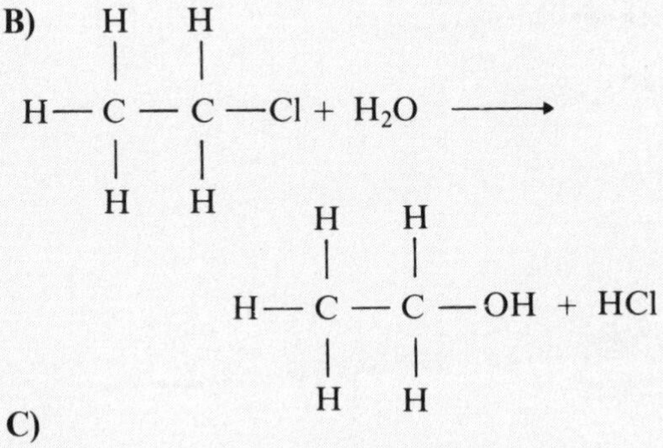

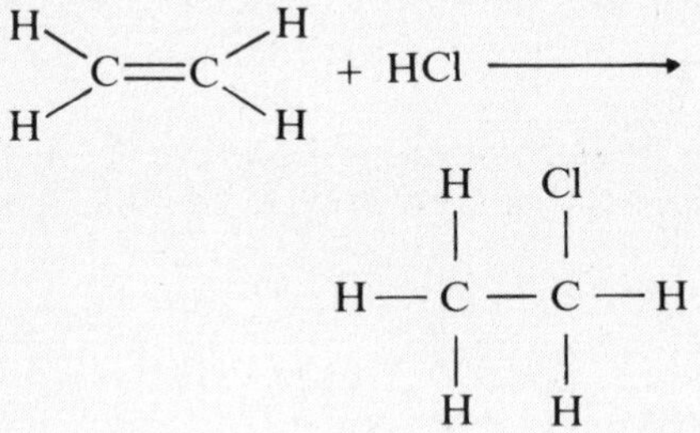

10 Which of the following compounds are members of the same homologous series?

A)

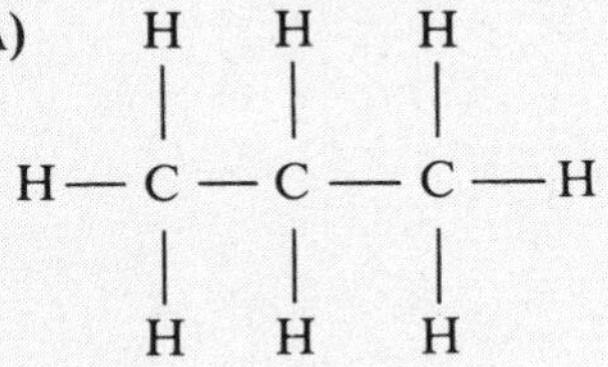

```
   H   H   H
   |   |   |
H— C — C — C — H
   |   |   |
   H   H   H
```

B)

```
   H   H
   |   |       H
   |   |     /
H— C — C = C
   |         \
   |           H
   H
```

C)

```
   H   H
   |   |
H— C — C — H
   |   |
H— C — C — H
   |   |
   H   H
```

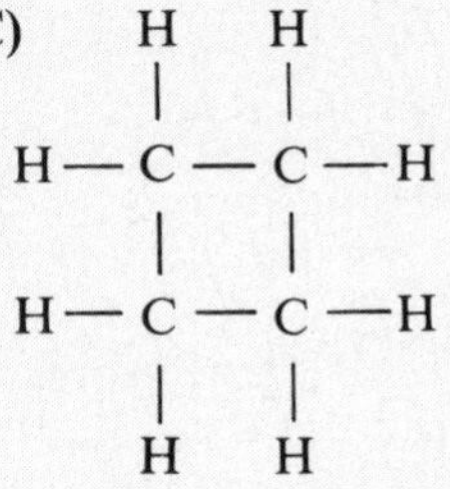

D)

```
   H   H   H   H
   |   |   |   |
H— C — C — C — C — H
   |   |   |   |
   H   H   H   H
```

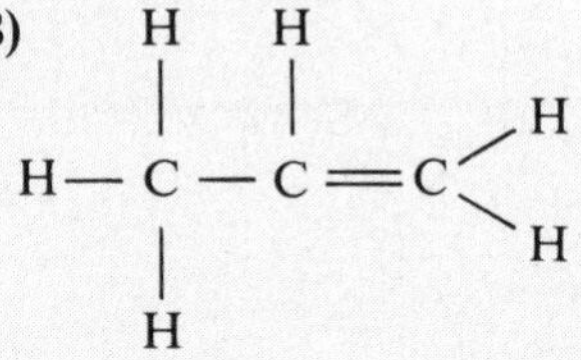

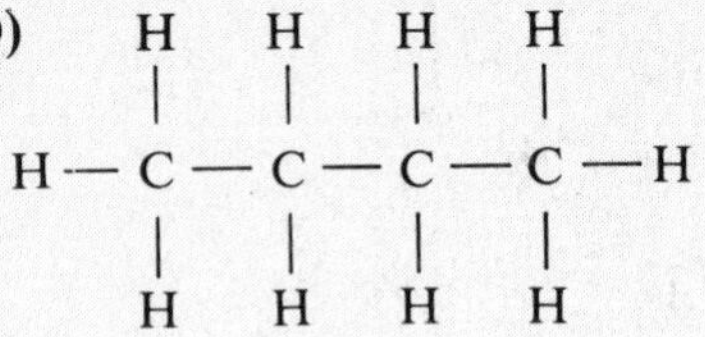

27 Alkanes

27.1 Introduction

Alkanes are saturated hydrocarbons, that is, they contain only carbon–hydrogen and carbon–carbon single bonds. They may occur as unbranched chains, branched chains and rings. They are among the least reactive organic compounds. Alkanes are the major constituents of both crude oil and natural gas and are thus compounds of enormous economic importance. They are used as fuels, lubricants and as starting materials for a vast range of other compounds. The fluctuations in the price of crude oil have significant economic and political impact. Crude oil is a non-renewable resource as it is derived from the remains of small marine plants and animals which lived millions of years ago. Current world consumption is of the order of 2×10^{10} barrels per year (1 barrel is approximately 160 dm^3 or almost 40 gallons). This rate of consumption would exhaust known reserves early in the next century. However, it is probable that the oil will last longer than this as:

1. **new reserves will be discovered,**
2. **the price will go up and this will decrease consumption,**
3. **as the price goes up, it will become economic to extract oil from lower quality deposits.**

Nevertheless, crude oil will run out and the issue of coping with this is one of the major challenges facing the human race today.

27.2 Formulae

Model of propane

Models of hexane – the molecule can twist

Alkanes exist as unbranched chains, branched chains and rings.

Unbranched chains

For example:

$$\begin{array}{ccccccc} & & H & & H & & H & & \\ & & | & & | & & | & & \\ H & — & C & — & C & — & C & — & H \\ & & | & & | & & | & & \\ & & H & & H & & H & & \end{array} \qquad CH_3CH_2CH_3$$

propane

Note: Unbranched chains are often called 'straight' chains—that is a misnomer; the $C{-}C{-}C$ angle is 109.5° so the chains are far from straight. Also, as there is free rotation about C—C bonds, the chains can twist.

Model of methylbutane

Model of cyclohexane

Branched chains

For example:

```
                  H
                  |
              H — C — H
      H     H     |     H
      |     |     |     |
  H — C  —  C  —  C  —  C — H      CH3CH2CH(CH3)CH3
      |     |     |     |
      H     H     H     H
```

$CH_3CH_2CH(CH_3)CH_3$

methylbutane

Rings

For example:

```
            H     H
             \   /
      H       C       H
       \    /   \    /
         C         C
      H /  |       |  \ H
      H \  |       |  / H
         C         C
       /    \   /    \
      H       C       H
             /  \
            H    H
```

C_6H_{12}

cyclohexane

The general molecular formula of chain alkanes (for n carbon atoms), both branched and unbranched, is C_nH_{2n+2}, as each carbon has two hydrogens except the two end ones which have three. Ring alkanes have the general molecular formula C_nH_{2n} as the 'end' hydrogens are not required.

27.3 Naming

Table 27.1 Names of the first twelve alkanes

Name	Formula
Methane	CH_4
Ethane	C_2H_6
Propane	C_3H_8
Butane	C_4H_{10}
Pentane	C_5H_{12}
Hexane	C_6H_{14}
Heptane	C_7H_{16}
Octane	C_8H_{18}
Nonane	C_9H_{20}
Decane	$C_{10}H_{22}$
Undecane	$C_{11}H_{24}$
Dodecane	$C_{12}H_{26}$

The names of the first twelve unbranched alkanes are given in **Table 27.1**. Their names are derived from a root, indicating the number of carbon atoms, and the suffix -ane, denoting an alkane.

When naming branched chains, first identify the longest unbranched chain. This gives the root name. Then the branches or **side chains** are named as prefixes according to their length, e.g. methyl-, ethyl-, propyl-, etc.

Examples

1.

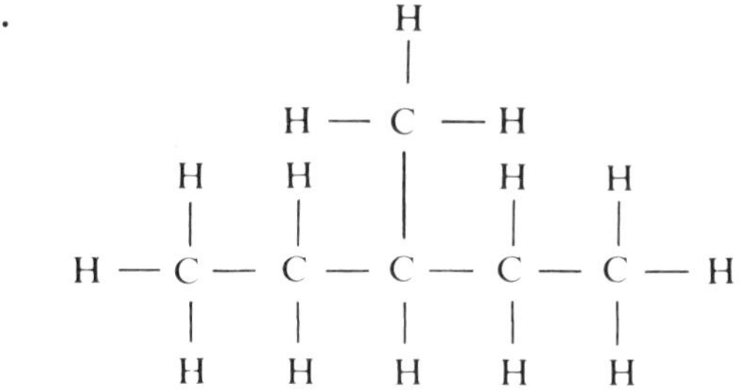

The longest unbranched chain (in brown) is five carbons so the molecule is named as a derivative of pentane. The one side chain has one carbon so it is methyl-. The side chain is attached at carbon number 3 so the full name is 3-methylpentane. The number 3 is the **locant**.

2.

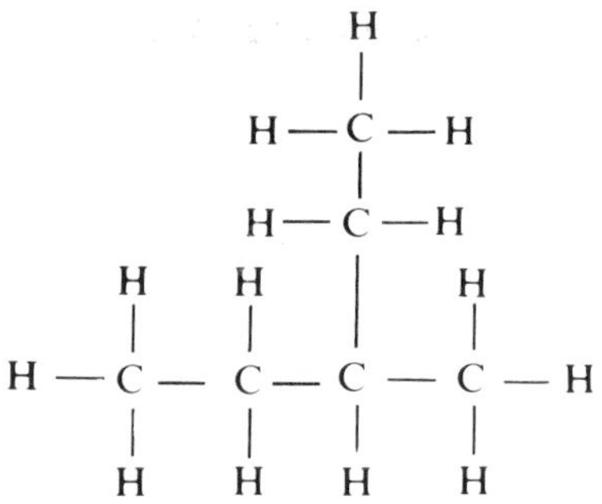

```
              H
              |
          H — C — H
              |
          H — C — H
    H     H   |     H
    |     |   |     |
H — C  —  C — C  —  C — H
    |     |   |     |
    H     H   H     H
```

At first sight, you might be tempted to call this molecule 2-ethylbutane, because of the way it has been drawn, but the longest unbranched chain (in brown) is five carbons so the root is pentane. A one-carbon chain is attached at the third carbon so it too is 3-methylpentane.

3.

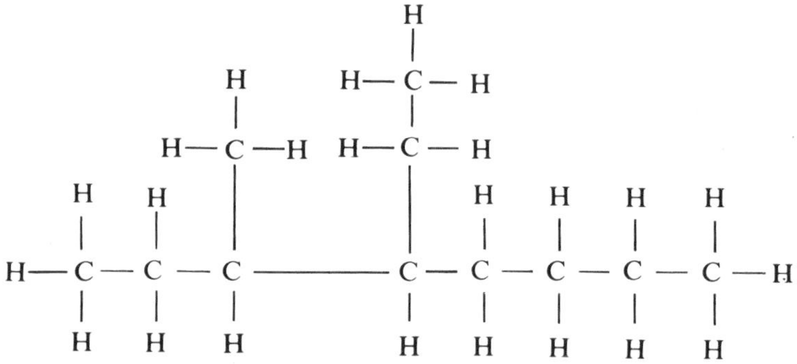

```
                          H
                          |
             H        H — C — H
             |            |
         H — C — H    H — C — H
   H    H    |            |    H    H    H    H
   |    |    |            |    |    |    |    |
H— C —  C  — C ————————— C — C — C — C — C — H
   |    |    |            |    |    |    |    |
   H    H    H            H    H    H    H    H
```

This is 4-ethyl-3-methyloctane as the side chains are listed in alphabetical order rather than in numerical order of their locants.

Rings are given the prefix cyclo- so:

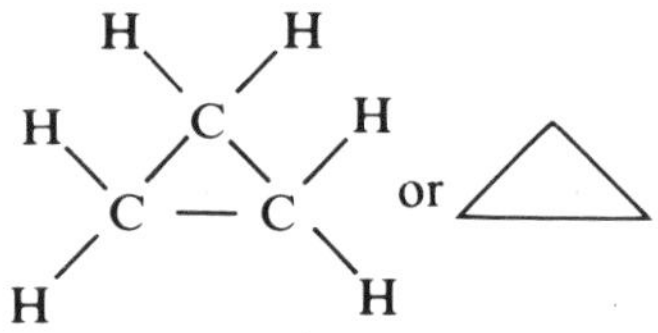

```
   H     H                      H   H
    \   /                       |   |
H    C    H                 H — C — C — H
 \  / \  /                      |   |        or  □
  C — C      or  △          H — C — C — H
 /     \                        |   |
H       H                       H   H
```

cyclopropane

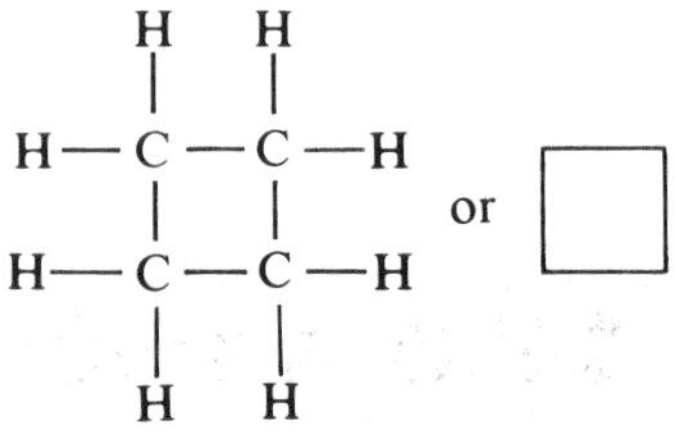

cyclobutane

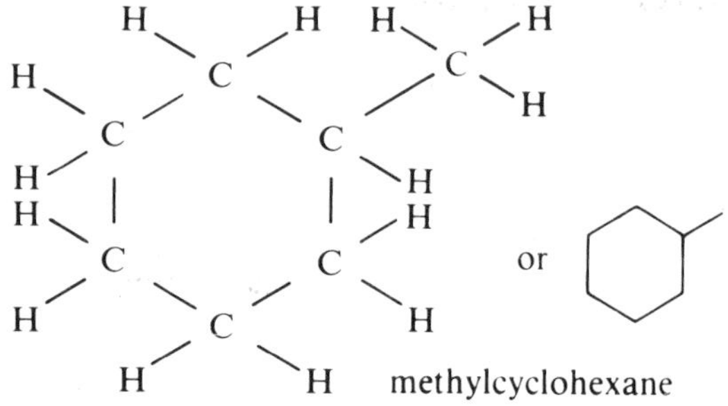

or

methylcyclohexane

or

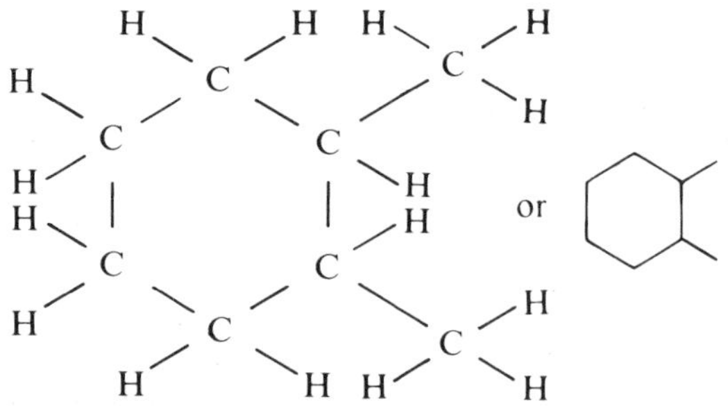

1,2-dimethylcyclohexane

27.4 Isomerism

Methane, ethane and propane have no isomers but butane and long chain alkanes have increasing numbers of isomers:

```
                                              H
                                              |
                                          H — C — H
     H    H    H    H                   H     |     H
     |    |    |    |                   |     |     |
 H — C —  C —  C —  C — H           H — C  —  C  —  C —
     |    |    |    |                   |     |     |
     H    H    H    H                   H     H     H
```

butane C_4H_{10} is isomeric with methylpropane C_4H_{10}

```
      H    H    H    H    H
      |    |    |    |    |
  H — C —  C —  C —  C —  C — H (pentane)
      |    |    |    |    |
      H    H    H    H    H
```

has two isomers

```
             H                                     H
             |                                     |
         H — C — H                             H — C — H
   H    H    |    H                         H      |      H
   |    |    |    |                         |      |      |
H— C —  C —  C —  C — H      and        H — C  —   C  —   C —
   |    |    |    |                         |      |      |
   H    H    H    H                         H      |      H
                                                   |
                                               H — C — H
                                                   |
                                                   H
```

2-methylbutane

2,2-dimethylpropane

Hexane has five isomers. Can you write their structures and name them?

27.5 Physical properties of alkanes

Alkanes are essentially non-polar so that the intermolecular forces betwee them are van der Waals forces only. Thus the shorter chains are gases room temperature. Pentane is a volatile liquid ($T_b = 309\,K$). At a chai length of around 18 carbons, the alkanes become solids at roo temperature. The solids have a waxy feel, candlewax being a mixture moderately long-chain alkanes, and polythene (despite its name) being very long chain *alkane*.

Alkanes are insoluble in water but mix with other relatively non-pola liquids.

27.6 Infra-red spectra of alkanes

The IR spectrum of hexane is shown in **Figure 27.1.** Apart from th fingerprint region, alkanes have little of interest in their IR spectra. Th strong peak at approximately $2950\,cm^{-1}$ is a C—H stretch, while tho

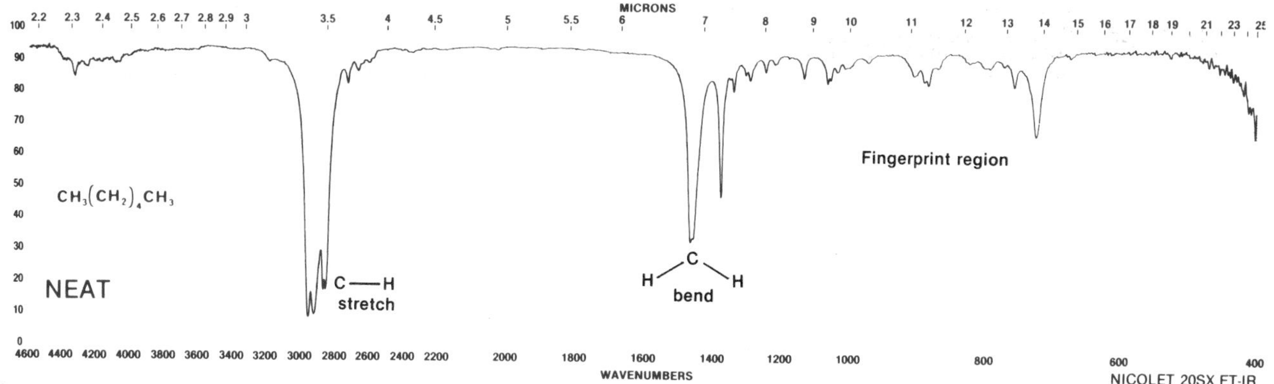

Figure 27.1 The infra-red spectrum of hexane (thin film)

at 1400–1500 cm^{-1} are H–C–H bending vibrations. These peaks are, of course, present in virtually all organic compounds.

Camping gas is a mixture of propane and butane. Polar expeditions use special gas mixtures with a higher proportion of propane as butane liquefies at 272 K (−1 °C)

The effect of increasing chain length (left to right) on the physical properties of alkanes

27.7 Reactions of alkanes

Alkanes are relatively inert, having no reaction with acids, bases, oxidizing or reducing agents, nucleophiles, electrophiles or polar reagents in general. In fact, alkanes have only three significant reactions.

27.7.1 Combustion

Alkanes will burn in a plentiful supply of oxygen to give carbon dioxide and water. For example, propane:

$$C_3H_8(g) + 5O_2(g) \longrightarrow 3CO_2(g) + 4H_2O(l) \quad \Delta H^\ominus = -2219\,kJ\,mol^{-1}$$

The enthalpies of combustion are important, as most of the shorter chain alkanes are used as fuels, e.g. methane (natural or 'North Sea' gas), propane ('camping' gas), butane ('Calor' gas), petrol (a mixture of hydrocarbons of approximate chain length C_8), paraffin (a mixture of hydrocarbons of chain lengths C_{10} to C_{18}).

In a limited supply of oxygen, carbon monoxide is formed, for example:

$$C_3H_8(g) + 3\tfrac{1}{2}O_2(g) \longrightarrow 3CO(g) + 4H_2O(l)$$

As carbon monoxide is poisonous, care must be taken in the design of gas burners to ensure complete combustion. For the same reason, care should also be taken to allow adequate ventilation in rooms where gas burners are used.

With even less oxygen available, carbon is deposited as soot. You wil notice this if you heat a test tube using a bunsen with its air hole closed Soot is also formed in the exhaust pipes of cars whose engines have bee running on an over-rich mixture, for example with the choke left out.

Although combustion reactions are important economically, they ar of no use for making new carbon compounds as they result in th destruction of the entire carbon chain.

27.7.2 Cracking

This reaction involves heating alkanes to a high temperature (often wit a catalyst—'cat' cracking). Carbon–carbon bonds are broken and two o more shorter chains are produced. As there are insufficient hydrogen atom to produce two alkanes, one of the new chains must have a carbon–carbo double bond, for example:

```
     H   H   H   H   H   H
     |   |   |   |   |   |          heat
H—C—C—C—C—C—C—H  ————→
     |   |   |   |   |   |        catalyst
     H   H   H   H   H   H
            hexane

          H   H   H              H   H
          |   |   |        H      |   |
     H—C—C—C—H  +     \    C=C—C—H
          |   |   |        /          |
          H   H   H      H            H
            propane                propene
```

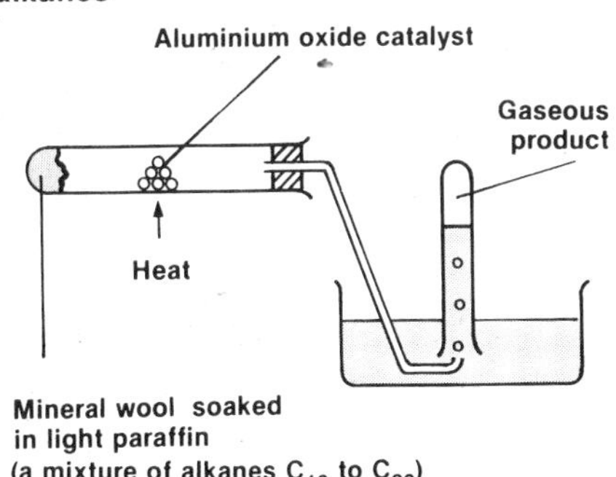

Figure 27.2 Laboratory cracking of alkanes

Any number of carbon–carbon bonds may break and the chain does no necessarily break in the middle. In the laboratory, cracking may be carrie out in the apparatus shown in **Figure 27.2**, using aluminium oxide as a catalyst. The products are mostly gases, showing that they have a chai length of less than C_5 and the mixture decolorizes bromine solution showing that it contains alkenes.

27.7.3 Reaction with halogens

Alkanes do not react with halogens in the dark at room temperature, bu react when irradiated with ultraviolet light. For example, a mixture o hexane and a little bromine liquid retains the colour of the bromine i the dark, but the bromine colour is rapidly lost if the mixture is expose to bright sunlight or a photoflood lamp. Fumes of hydrogen bromide ar produced suggesting that a substitution reaction has taken place—th hydrogen in the hydrogen bromide must have come from the alkane. Th overall reaction is:

$$C_6H_{14}(l) + Br_2(l) \longrightarrow C_6H_{13}Br(l) + HBr(g)$$

Chain reactions

The mechanism is of considerable interest and is an example of a **chai reaction**. We shall look at the similar reaction of bromine and metha for simplicity. The first step involves the absorption of a single quantu of ultraviolet light, of frequency approximately $10^{15}\,s^{-1}$, by a bromi molecule. The energy of such a quantum is greater than the Br—Br bon energy of 193 kJ mol^{-1} so the bond will break. As the two atoms in th bond are identical, it will break homolytically, i.e. one electron going t each bromine, resulting in two separate bromine atoms. These each hav an unpaired electron and are called radicals. They are written Br$^{\cdot}$ to stre the unpaired electron. The bond energy of C—H is 413 kJ mol^{-1}, whic is more than that available in a quantum of ultraviolet radiation, so th bond cannot break.

Plate 1 Layers of copper sulphate solution and ethanol diffuse into one another – see page 10

Plate 2 The reaction of chlorine with heated iron – see page 24

Plate 3 Thin-layer chromatography – see page 119

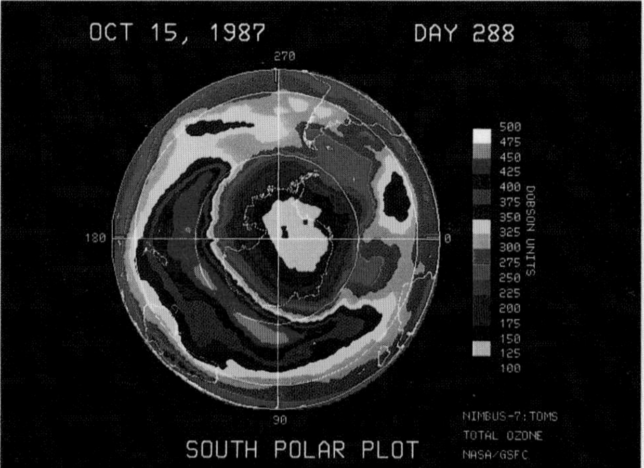

Plate 4 Computer simulation of the ozone 'hole' – see page 201

Plate 5 Model of the DNA molecule – see page 253

Plate 6 NMR spectrometer, used in structure determination – see page 260

Plates 7–13 s-block metal flame tests – see page 289

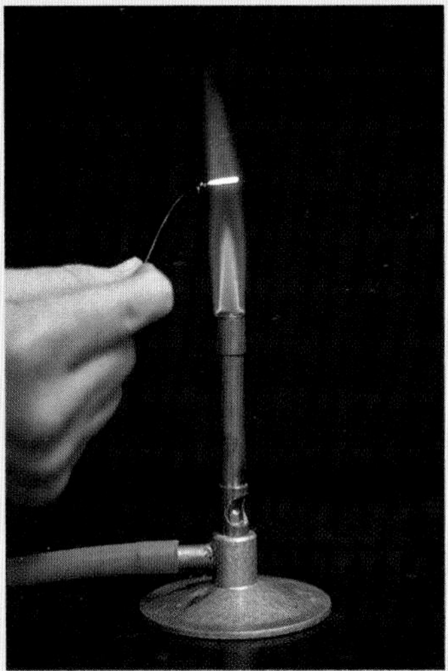

Plate 7 Lithium

Plate 8 Sodium

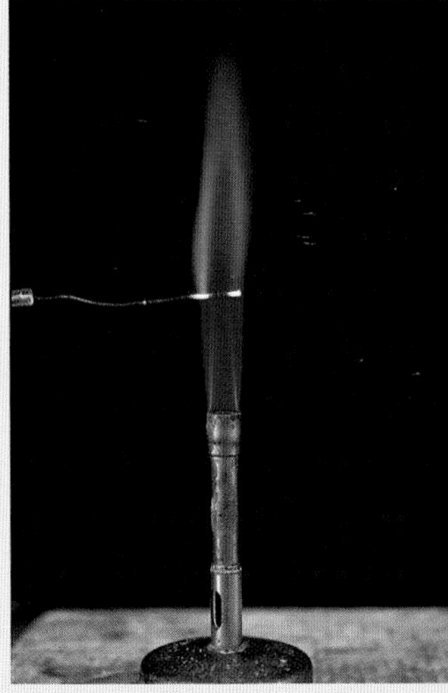

Plate 9 Potassium

Plate 10 Caesium

Plate 11 Calcium

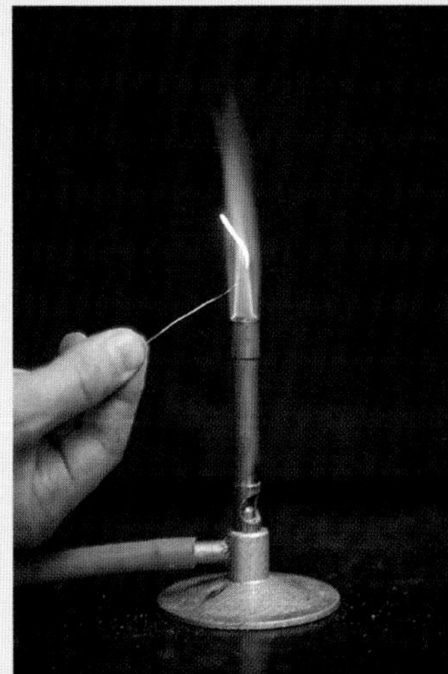

Plate 12 Strontium

Plate 13 Barium

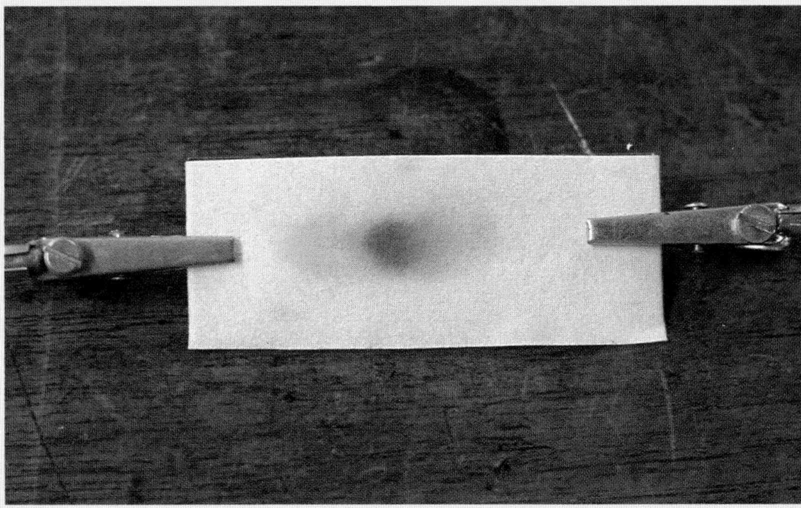

Plate 14 Coloured ions moving during electrolysis – see page 229

E

THE ENERGY OF LIGHT QUANTA

The energy of a quantum of electromagnetic radiation is given by:

$$E = h\nu$$

where h is the Planck constant, 6.6×10^{-34} J s, and ν the frequency of the radiation. So for UV light $\nu = 10^{15}\,s^{-1}$:

$$E = 6.6 \times 10^{-34} \times 10^{15} = 6.6 \times 10^{-19}\,J$$

For 1 mole of quanta the total energy is:

$$6.6 \times 10^{-19} \times 6 \times 10^{23} = 396\,000\,J\,mol^{-1}$$
$$= 396\,kJ\,mol^{-1}$$

For comparison, visible light has a typical frequency of $10^{14}\,s^{-1}$ and infra-red (heat) radiation $10^{13}\,s^{-1}$. So the energy of their quanta are respectively 39.6 and 3.96 kJ mol^{-1}. Typical covalent bond energies are around 300 kJ mol^{-1}, so visible and infra-red radiations are unable to break covalent bonds while ultraviolet light can.

The first or **initiation** step of the reaction is:

$$Br{-}Br \xrightarrow{\text{UV light}} 2Br^{\cdot}$$

Radicals (sometimes called free radicals) are highly reactive and can remove a hydrogen atom from methane forming hydrogen bromide and a methyl radical:

$$Br^{\cdot} + CH_4 \longrightarrow HBr + {}^{\cdot}CH_3$$

The resulting methyl radical is also reactive and can react with a bromine molecule:

$${}^{\cdot}CH_3 + Br_2 \longrightarrow CH_3Br + Br^{\cdot}$$

Each of the above steps results in a stable product and a reactive radical. After the two steps the original $Br^{\cdot}$ radical remains. The steps can be repeated over and over again and are therefore called **propagation** steps. It is believed that such steps take place thousands of times before the radicals are destroyed. This can happen in the following ways:

$$Br^{\cdot} + Br^{\cdot} \longrightarrow Br_2$$
$${}^{\cdot}CH_3 + {}^{\cdot}CH_3 \longrightarrow C_2H_6$$
$$Br^{\cdot} + {}^{\cdot}CH_3 \longrightarrow CH_3Br$$

These are all called **termination** steps.

Other products are possible as well as the main ones. The equations above show that some ethane is produced. Another possibility is:

$$Br^{\cdot} + CH_3Br \longrightarrow HBr + {}^{\cdot}CH_2Br$$
$${}^{\cdot}CH_2Br + Br_2 \longrightarrow CH_2Br_2 + Br^{\cdot}$$

With more complex alkanes such as hexane, many isomers may be produced as the $Br^{\cdot}$ radical is equally likely to remove any of the hydrogen atoms. Similar reactions occur without ultraviolet light at high temperatures.

27.8 Sources of alkanes

Although a number of methods exist for preparing alkanes in the laboratory, they are rarely used, as alkanes are readily available from the distillation of crude oil.

27.8.1 Crude oil and natural gas

Both these valuable sources of organic chemicals have been produced by the breakdown of plant and animal remains, brought about by high pressures and temperatures. In general, plant remains produce gas and animal (plankton) remains give oil.

Crude oil varies considerably in composition depending on its source. As well as alkanes, both unbranched and branched, it may contain aromatic compounds, see Chapter 29, and also oxygen-, nitrogen- and sulphur-containing compounds. The last mentioned are particularly troublesome, producing acidic sulphur dioxide when burned. Crude oil first separated into fractions by fractional distillation. The compositio of a typical North Sea crude oil is given in **Table 27.2.**

Table 27.2 The composition of a typical North Sea crude oil

Product	Gases	Petrol	Naphtha	Kerosene	Gas oil	Fuel oil and wax
Approximate boiling temperature/K	below 310	310–450	400–490	430–523	590–620	Above 620
Chain length	1–5	5–10	8–12	11–16	16–24	25+
Percentage present	2	8	10	14	21	45

27.9 Economic importance of alkanes

27.9.1 Fractional distillation of crude oil

Crude oil is separated into fractions by distillation in cylindrical tower typically 8 m in diameter and 40 m high. The process is described in th box *Industrial fractionation of crude oil*, **section 16.3.3.** The differen fractions are tapped off at suitable levels, higher boiling point (longe carbon chain) fractions being tapped off low in the column. A very hig boiling point fraction, called the **residue**, collects at the base of the columr while gases are taken from the very top. The residue can be used for roa tar, while the refinery gases are often used as fuel in the plant for heatin the furnance.

27.9.2 Industrial cracking

Crude oils from different sources contain different percentages of eac fraction. Unfortunately, the availability of each fraction rarely meets th demand. The major demand is for the petrol and naphtha fractions, whil longer-chain fractions are used less.

To rectify this mismatch, many of the longer-chain fractions are cracked which brings about two advantages. Firstly, shorter chains are produced and secondly, some of the products are alkenes, which are more reactiv and more suitable for use as **chemical feedstock** for conversion to othe compounds. Different processes can be used. In **steam cracking**, th naphtha or gas oil is mixed with steam and passed through a furnace a 1100 K for 0.2 seconds.

Catalytic cracking takes place at a lower temperature (approximatel 800 K) using a surface catalyst consisting of silicon dioxide and aluminiun oxide. A typical mixture obtained from the steam cracking of naphtha i shown in **Table 27.3.**

Table 27.3 A typical composition of the mixture obtained from steam cracking of naphtha

Component	Percentage
Hydrogen	1
Methane	15
Ethene	25
Propane	16
Butane	5
Buta-1,3-diene	5
Petrol	28
Fuel oil	4

27.9.3 Reforming

Straight-chain alkanes are relatively poor motor fuels. They caus 'knocking' in the engine by detonating rapidly rather than burning steadily Branched chains are much better in this respect. **Reforming** is a process where straight-chain alkanes are heated under pressure with a platinum catalyst. The chains break up and reform as branched chain molecules, for example:

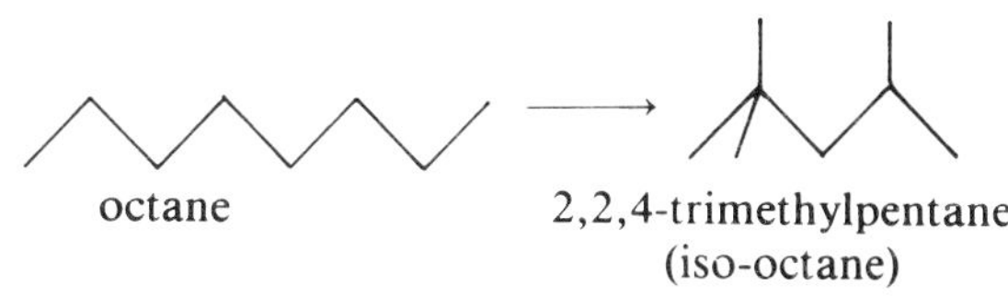

A

OCTANE RATING OF PETROL

Heptane causes considerable knocking in internal combustion engines while 2, 2, 4-trimethylpentane ('iso-octane') causes little. This is the basis of the **octane rating** of petrol. A petrol's tendency to knock is compared with mixtures of heptane and iso-octane. The petrol's octane number is the percentage of iso-octane in a mixture which has the same knocking characteristic, when run in a test engine, as the petrol. So 92 octane petrol knocks to the same extent as a mixture of 92% iso-octane, 8% heptane. Two-star petrol is approximately 92 octane, while four-star is 97 octane.

A number of additives can be used to improve the octane number of poor quality petrol, the most common being tetraethyllead. However, leaded petrol is being phased out on environmental grounds, as lead compounds are poisonous, especially to young children. Engines of all new cars are being designed to run on unleaded petrol.

High-octane petrols will continue to be made available by blending a higher proportion of branched chains or using alternative anti-knock additives such as methanol and 2-methylpropan-2-ol.

27.10 Summary

- Alkanes are saturated hydrocarbons of general formula C_nH_{2n+2} (chains) or C_nH_{2n} (rings).
- They may exist as unbranched chains, branched chains or rings.
- They are insoluble in water.
- At room temperature they are gases (C_1–C_4), liquids (C_5–C_{17}) or waxy solids (C_{18} upwards).
- They have only three significant reactions:
  combustion to give carbon dioxide and water (carbon monoxide or carbon and water if oxygen is not plentiful);
  cracking to give shorter chain alkanes and alkenes;
  radical substitution reactions with halogens on exposure to ultraviolet light or at high temperatures, to give halogenoalkanes.
- Radical substitution reactions have three stages:
  initiation where radicals are formed;
  propagation where products are formed and radicals are regenerated;
  termination where radicals are removed.
- Mixtures of alkanes are found in natural gas and crude oil.
- Crude oil is fractionally distilled to separate useful components.
- Longer-chain fractions are cracked to reduce the chain lengths and produce alkenes.
- Straight-chain alkanes may be reformed to produce branched chains which are more suitable for petrol.

27.11 Questions

1 Name the following:

A)

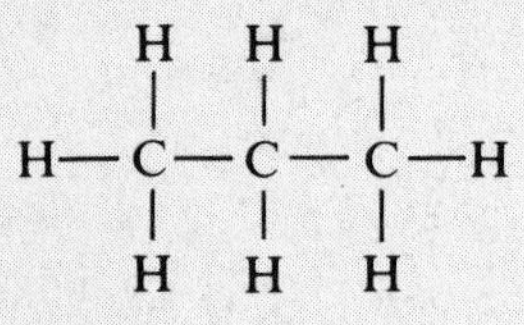

```
     H   H   H
     |   |   |
 H — C — C — C — H
     |   |   |
     H   H   H
```

B)

```
                     H
                     |
                 H — C — H
     H   H   H       |   H
     |   |   |       |   |
 H — C — C — C — C — C — H
     |   |   |   |   |
     H   |   H   H   H
     H — C — H
         |
         H
```

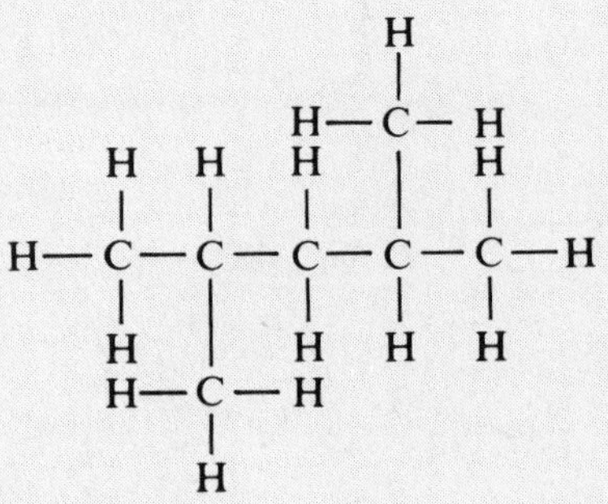

C) D)

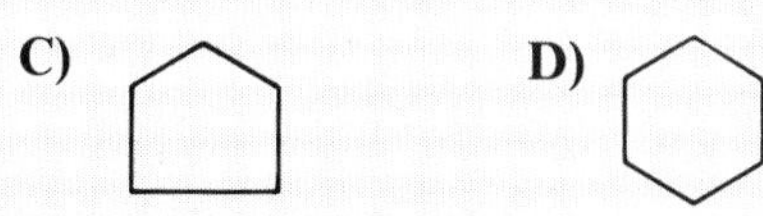

2 Draw the structural formula of:

A) 2-methylpentane
B) methylcyclohexane
C) 3-ethyl-4-methylheptane
D) cyclooctane

3 Write a balanced equation for the complete combustion of hexane.

4 Give three possible chlorine-containing products produced by the reaction of chlorine with hexane in ultraviolet light.

5 Calculate the lowest frequency of ultraviolet light which has enough energy to break a Cl—Cl bond (bond energy = 243 kJ mol^{-1}, $h = 6.6 \times 10^{-34}$ J s), the Avogadro constant = 6×10^{23} mol^{-1}).

6 A sample of hexane was cracked in the apparatus shown in **Figure 27.2.** 50 cm^3 of gas was produced. The gas decolorized 4 cm^3 of a 0.25 mol dm^{-3} solution of bromine. What percentage of alkene was produced? Take the volume of 1 mole as 24 000 cm^3.

7 A hydrocarbon contains 85.7% carbon and 14.3% hydrogen. The mass spectrum shows the peak of highest mass at $M_r = 84$.

A) What is the empirical formula?
B) What is the molecular formula?
C) The hydrocarbon does not decolorize bromine water. Give two possible structures.

28 Alkenes

28.1 Introduction

roducts originating from ethene

Alkenes are unsaturated hydrocarbons. They have one or more carbon–carbon double bond. This makes them more reactive than alkanes and makes their chemistry more interesting and varied. The reactivity of the alkenes and their availability from cracking makes them attractive to industrial chemists. Ethene, the simplest alkene, is the basic building block for a large range of products, including the polymers polythene, PVC and polystyrene, as well as products like antifreeze, paints and terylene fabrics.

28.2 Formulae, naming and structure

odels of *cis*- and *trans*-but-2-ene

Alkenes contain at least one carbon–carbon double bond. There is therefore no alkene with only one carbon atom.

The simplest is ethene:

$$\begin{array}{ccc} H & & H \\ & C = C & \\ H & & H \end{array}$$

followed by propene:

$$\begin{array}{ccccccc} & & H & & & & \\ & & | & & & & H \\ H & - & C & - & C & = & C \\ & & | & & | & & H \\ & & H & & H & & \end{array}$$

Longer chains than propene can form isomers with different positioning of the double bond, as well as chain-branching isomerism and *cis–trans*-isomerism (see **section 26.7.1**).

In:

$$\begin{array}{ccccccccc} & & & & H & & H & & H \\ & & & & | & & | & & | \\ H & & C & = & C & - & C & - & C & - & H \\ H & & & & & & | & & | \\ & & & & & & H & & H \end{array}$$

the double bond is between carbons 1 and 2. It is named but-1-ene rather than but-2-ene. The smaller number is always used.

$$\begin{array}{ccccccccccc} & & H & & H & & H & & H & & \\ & & | & & | & & | & & | & & \\ H & - & C & - & C & = & C & - & C & - & H \\ & & | & & & & & & | & & \\ & & H & & & & & & H & & \end{array}$$

is but-2-ene (*cis*- and *trans*-isomers exist).

Branched chain alkenes are named in the same way, so:

Model of ethene

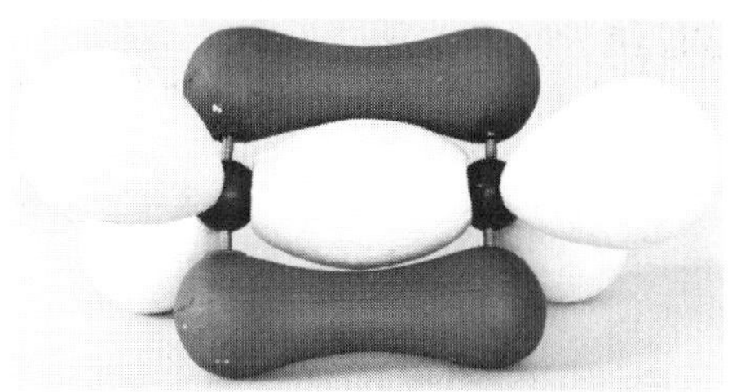

```
             H
             |
         H — C — H
    H        |    H     H
    |        |    |     |
H — C  —     C  = C  —  C — H
    |                   |
    H                   H
```

is 2-methylbut-2-ene (*cis* and *tra* isomers),

and

```
             H
             |
        H —  C — H
             |    H
             |    |
H\           |    |
  C = C  —   C  — C — H
H/    |      |    |
      H      H    H
```

is 3-methylbut-1-ene

Compounds with more than one double bond are named using di-, tr etc. to indicate the number of double bonds, as well as locants to sho where they are.

```
      H     H     H     H     H
      |     |     |     |     |
So H— C  =  C  —  C  =  C  —  C — H
      |                       |
      H                       H
```

is penta-1,3-diene.

28.2.1 The shape of alkenes

The carbon atoms which form the double bond are hybridized sp^2. Th molecule is flat, all the atoms being in the same plane. No rotation possible about the double bond. The H—C—H bond angles are 117 less than 120°, as the four electrons in the C=C double bond repel mo than the pairs of electrons in the C—H single bonds. An orbital pictu of the bonding is shown in **Figure 28.1**.

Figure 28.1 The bonding in ethene

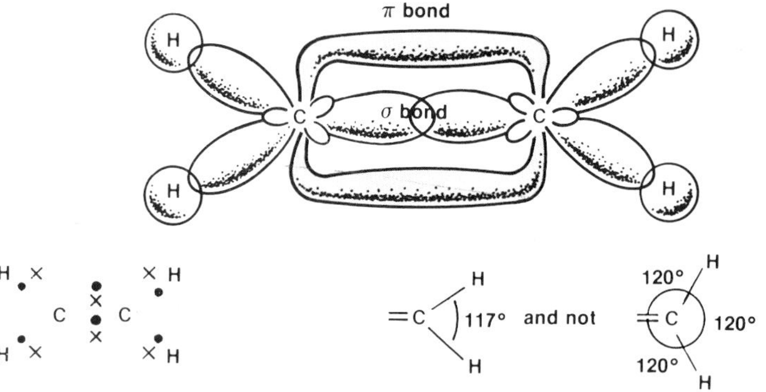

H•× C ×•×• C ×•H
H•× ×•H

=C(H)(H) 117° and not =C 120° 120° 120°

28.3 Physical properties of alkenes

Table 28.1 Melting and boiling temperatures of some alkenes

Alkene	T_m/K	T_b/K
Ethene	104	169
Propene	88	225
But-l-ene	88	267
Cyclohexene	170	357

The double bond has little effect on the intermolecular forces which a *between* the molecules, so their physical properties are very similar those of the alkanes in terms of melting and boiling temperatures ar solubility in water. **Table 28.1** shows the melting and boiling temperatur of some alkenes.

28.4 Infra-red spectra of alkenes

igure 28.2 The infra-red spectrum of ct-l-ene

The infra-red spectrum of oct-1-ene is shown in **Figure 28.2.** The main peak characteristic of an alkene is that at 1650 cm^{-1} due to a C=C stretching vibration.

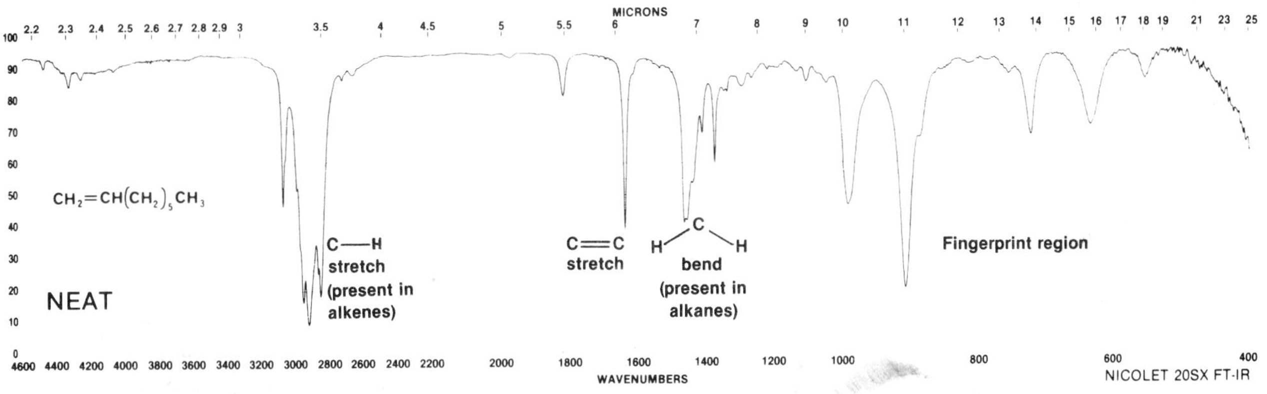

28.5 Reactivity of alkenes

The difference between alkenes and alkanes is that alkenes have a carbon–carbon double bond. The bond energies for C—C and C=C are respectively 347 kJ mol^{-1} and 612 kJ mol^{-1} so that at first sight we might expect alkenes to be less reactive than alkanes. This is not so for three reasons.

Firstly, a double bond consists of a σ bond plus a π bond. The σ bond is the same as in C—C so the π bond must have an energy of $612 - 347 = 265$ kJ mol^{-1}. So it is relatively easy to break the π part of carbon–carbon double bonds, while the σ part remains intact. This happens in most reactions of the C=C bond and it is rare for the carbon chain to actually break in two when alkenes react.

Secondly, the C=C bond forms an electron-rich area in a molecule, which can easily be attacked by positively charged reagents (**electrophiles,** see **section 26.9.5**).

Thirdly, if the π part of the double bond is broken, each of the carbon atoms becomes capable of forming a new bond. Thus addition reactions can take place. So the dominant feature of the chemistry of alkenes is **electrophilic addition reactions.**

able 28.2 Ethene bond nergies/kJ mol^{-1}

σ	π
347	265

28.6 Reactions of alkenes

28.6.1 Combustion

All alkenes will burn in air. Alkenes with more than one double bond produce a noticeably sooty flame. This is because they have a greater percentage of carbon than alkanes and there is often insufficient air to burn all the carbon. Unburnt carbon is left behind as soot. Combustion of alkenes is less important than that of alkanes as the more reactive alkene molecules are too useful for other purposes to be wasted by use as fuels. You should be able to construct balanced equations for complete combustion of any of the alkenes.

$$H_2C{=}CH_2(g) + 3O_2(g) \longrightarrow 2CO_2(g) + 2H_2O(l)$$

ethene — oxygen — carbon dioxide — water

28.6.2 Reaction with halogens

Alkenes react rapidly with chlorine gas or with solutions of bromine c iodine, in an organic solvent like tetrachloromethane, in the dark to giv 1, 2-dihalogenoalkanes:

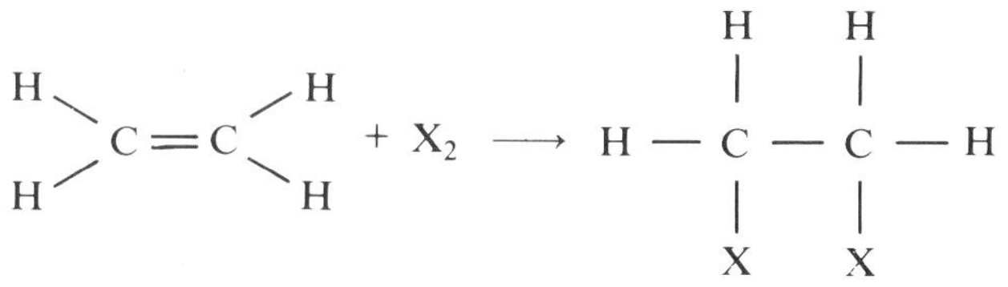

$$H_2C{=}CH_2 + X_2 \longrightarrow H{-}CHX{-}CHX{-}H$$

This is an electrophilic addition reaction. The decolorizing of a bromin solution is often used as a test for the presence of carbon–carbon multipl bonds. Both alkenes and alkynes (which have carbon–carbon triple bond: will decolorize bromine solutions.

The fact that the reaction occurs in the dark rules out a radica mechanism such as occurs in the reaction of halogens with alkanes. Furthe clues to the mechanism are that if bromine dissolved in water is reacte with ethene, some 2-bromoethan-1-ol is formed, and if the bromine dissolved in sodium chloride solution, some 1-bromo-2-chloroethane produced.

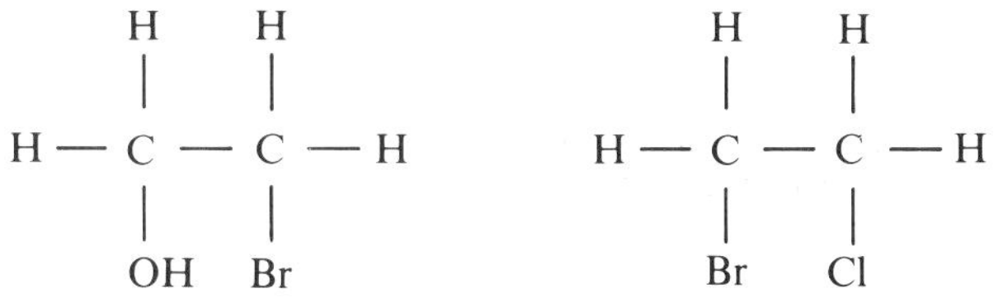

$$H{-}CH(OH){-}CH(Br){-}H$$

2-bromoethan-1-ol

$$H{-}CH(Br){-}CH(Cl){-}H$$

1-bromo-2-chloroethane

The mechanism is as follows. At any instant, a bromine molecule is likel to have an instantaneous dipole: $Br^{\delta+}{-}Br^{\delta-}$

This is because the electrons are in constant motion and at any momer are unlikely to be distributed exactly symmetrically. An instant later, th dipole could be reversed.

The δ^+ end of this dipole is attracted to the electron-rich π bond i the alkene. The π bond attracts the $Br^{\delta+}$ and repels the electrons share by the two bromine atoms, thus strengthening the dipole:

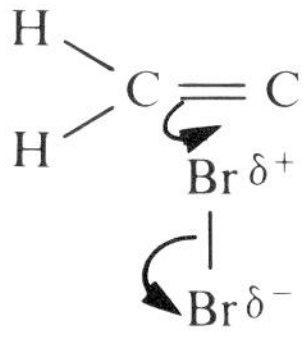

$$H_2C{=}CH_2 \quad \cdots \quad Br^{\delta+}{-}Br^{\delta-}$$

Eventually, two of the electrons from the π bond form a bond with th $Br^{\delta+}$ and the electrons shared by the bromine atoms are repelled alon with the $Br^{\delta-}$ to form a Br^- ion. This leaves a positively charged specie which is called a **carbocation**. The positive charge on the carbocation i localized on the carbon atom which is not bonded to the bromine, as thi is forming only three bonds (has six electrons in its outer shell) and i therefore electron deficient.

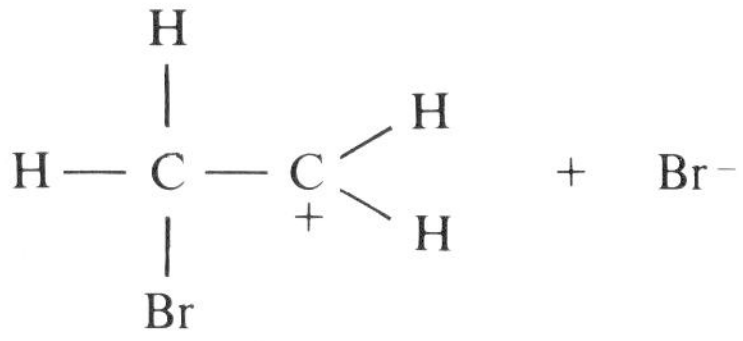

$$H{-}CH_2(Br){-}\overset{+}{C}H_2 \quad + \quad Br^-$$

a carbocation

The carbocation is rapidly attacked by any negative ion. The onl negative ion present in a non-aqueous solution will be the Br^- io but in aqueous solution OH^- will be present, as will Cl^- in a solution of sodium chloride.

$$\underset{\text{}}{H{-}\overset{H}{\underset{Br}{C}}{-}\overset{H}{\underset{Br}{C}}{-}H} \xleftarrow{Br^-} H{-}\overset{H}{\underset{Br}{C}}{-}\overset{+}{C}H_2 \xrightarrow{Cl^-} H{-}\overset{H}{\underset{Br}{C}}{-}\overset{H}{\underset{Cl}{C}}{-}H$$

$$H{-}\overset{H}{\underset{Br}{C}}{-}\overset{+}{C}H_2 \xrightarrow{OH^-\text{ (from water)}} H{-}\overset{H}{\underset{Br}{C}}{-}\overset{H}{\underset{OH}{C}}{-}H$$

So the addition takes place in two steps:

1. formation of the carbocation, followed by
2. rapid reaction with a negative ion.

As the $Br^{\delta+}{-}Br^{\delta-}$ attacks the electron-rich π bond, it is acting as an electrophile. So the whole reaction is an **electrophilic addition reaction**. **Figure 28.3** shows the same mechanism in terms of electronic orbitals.

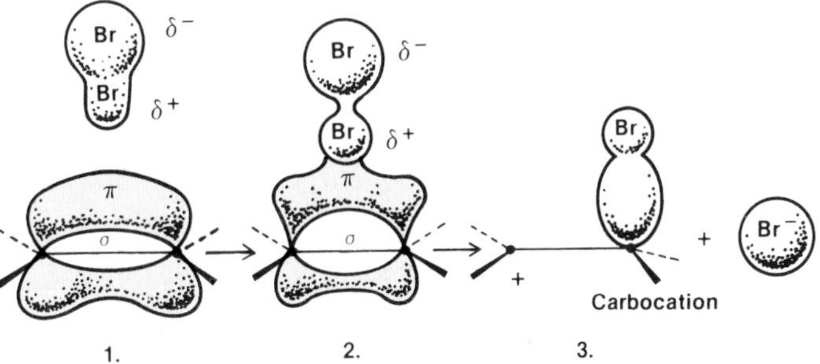

Figure 28.3 Orbitals involved in the reaction of ethene with bromine

28.6.3 Reaction with hydrogen halides

Hydrogen halides, HCl, HBr and HI add on across the double bond to form a halogenoalkane, for example:

$$\underset{\text{ethene}}{H_2C{=}CH_2} + \underset{\text{hydrogen bromide}}{HBr} \longrightarrow \underset{\text{bromoethane}}{H{-}\overset{H}{\underset{H}{C}}{-}\overset{H}{\underset{Br}{C}}{-}H}$$

The reaction is an electrophilic addition and the mechanism is similar to the addition of halogens. In this case the hydrogen halide has a permanent dipole:

$$H_2C{=}CH_2 + H^{\delta+}{-}Br^{\delta-} \longrightarrow H{-}\overset{H}{\underset{H}{C}}{-}\overset{+}{C}H_2 + Br^- \longrightarrow H{-}\overset{H}{\underset{H}{C}}{-}\overset{H}{\underset{Br}{C}}{-}H$$

Markovnikov's rule

If hydrogen bromide adds on to ethene, then only one product is possible bromoethane. With, say, propene there are two possible product depending on which of the two carbon atoms of the double bond th bromine bonds to. In fact the product is almost entirely 2-bromopropane

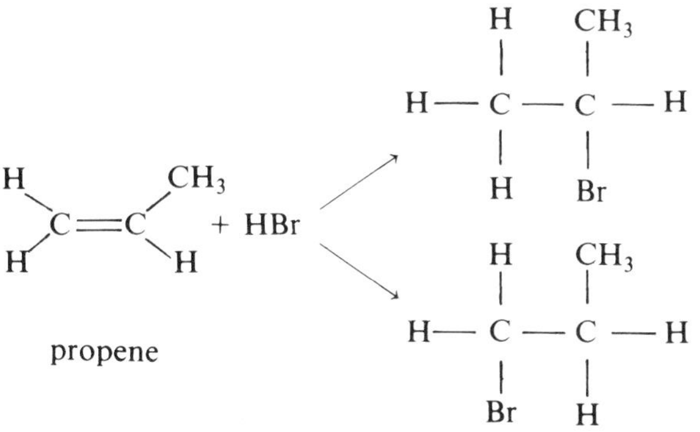

$$CH_2{=}CH{-}CH_3 \text{ (propene)} + HBr \rightarrow CH_3{-}CHBr{-}CH_3 \text{ (2-bromopropane)} \quad \text{or} \quad CH_2Br{-}CH_2{-}CH_3 \text{ (1-bromopropane)}$$

propene

2-bromopropane

1-bromopropan

This is an example of **Markovnikov's rule** which predicts that wher hydrogen halides add on to alkenes, the hydrogen adds on to the carbor atom which already has most hydrogens. The reason for this is found ir the stability of the intermediate carbocation. Two are possible:

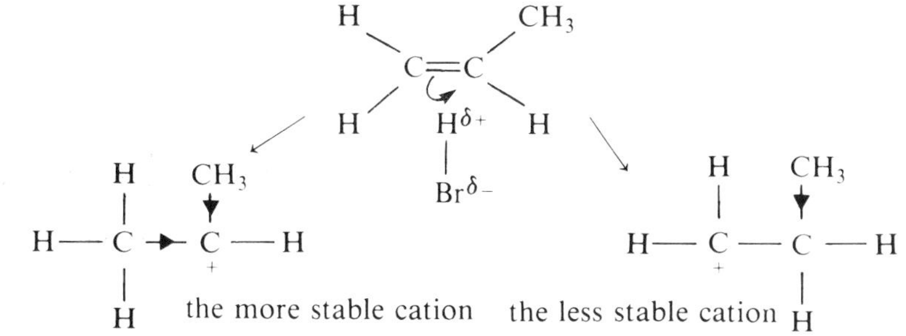

$$H_2C{=}CH(CH_3) + H^{\delta+}{-}Br^{\delta-} \rightarrow CH_3{-}\overset{+}{C}H{-}CH_3 \quad \text{or} \quad \overset{+}{C}H_2{-}CH_2{-}CH_3$$

the more stable cation the less stable cation

In the first of these, the positive charge is stabilized by the tendency o CH_3—to release electrons (see **section 26.9.3**) so this carbocation i favoured and the product we get is the one formed from this, i.e 2-bromopropane. Other alkyl groups like ethyl-, propyl-, etc. also release electrons.

28.6.4 Reaction with concentrated sulphuric acid

Concentrated sulphuric acid also adds on across the double bond. The reaction occurs at room temperature and is exothermic:

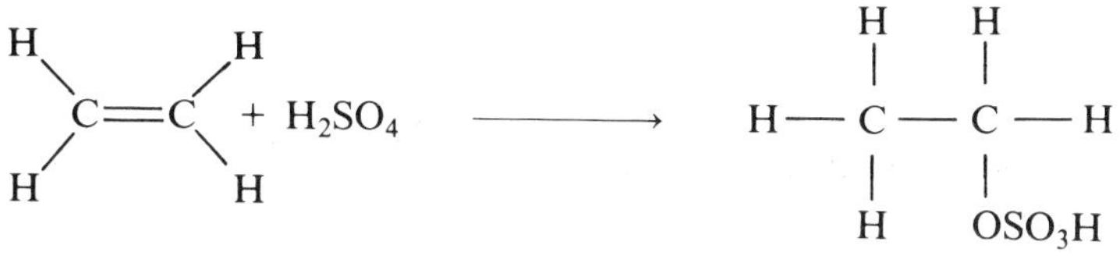

$$H_2C{=}CH_2 + H_2SO_4 \longrightarrow H{-}CH_2{-}CH_2{-}OSO_3H$$

ethyl hydrogensulphate

On addition of water to the product, an alcohol is formed:

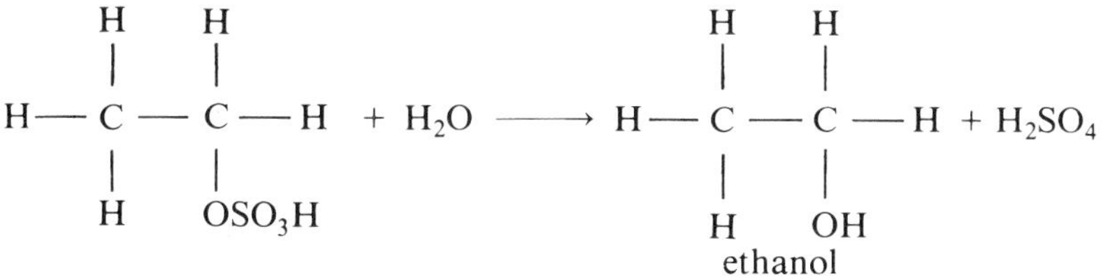

$$H{-}CH_2{-}CH_2{-}OSO_3H + H_2O \longrightarrow H{-}CH_2{-}CH_2{-}OH + H_2SO_4$$

ethanol

so that the overall effect is to add water across the double bond.

The original reaction with sulphuric acid is another electrophilic addition, the electrophile being the H^+ ion:

$$H_2C{=}CH_2 + H^+ \longrightarrow H{-}CH_2{-}\overset{+}{C}H_2{-}H + HSO_4^- \longrightarrow H{-}CH_2{-}CH_2(OSO_3H)$$

The carbocation formed then reacts rapidly with the negatively charged hydrogensulphate ion.

With an unsymmetrical alkene such as propene, the carbocation is exactly the same as that found in the reaction with hydrogen bromide, and in the same way, the hydrogen will bond to the carbon atom which already has more hydrogens.

In general, Markovnikov's rule states that when an unsymmetrical reagent HZ adds on across a carbon–carbon double bond, the hydrogen bonds to the carbon atom which already has more hydrogens.

So:

$$RHC{=}CH_2 \xrightarrow[\text{followed by } H_2O]{H_2SO_4} H{-}CH(OH)R{-}CH_2{-}H \text{ not } H{-}CHR{-}CH(OH){-}H$$

28.6.5 Oxidation of alkenes

By alkaline potassium manganate(VII)

A diol is produced rapidly and the purple colour of the MnO_4^- ion disappears and a precipitate of brown manganese(IV) oxide appears. This reaction is used as an alternative to the bromine test for multiple bonds. The reaction is believed to proceed through the intermediate shown:

$$\underset{\text{ethene}}{H_2C{=}CH_2} \xrightarrow{MnO_4^-/OH^-} \left[H{-}CH_2(O){-}CH_2(O){-}H \;\; (\text{O–Mn(=O)O}^- \text{ bridging})\right] \longrightarrow \underset{\text{ethane-1,2-diol}}{H{-}CH(OH){-}CH(OH){-}H}$$

Ethane-1, 2-diol, also known as ethylene glycol, is used as antifreeze in car cooling systems. It is also used in the manufacture of terylene.

By oxygen

Reaction of ethene with oxygen using a silver catalyst produces epoxyethane:

$$\underset{\text{ethene}}{H_2C{=}CH_2} \xrightarrow[450\ K]{O_2,\ Ag\ catalyst} \underset{\text{epoxyethane}}{H{-}CH_2{-}CH_2{-}H \ (\text{C–O–C ring})}$$

The product is unstable due to **ring strain**. The C—C—O bond angle and the C—O—C bond angle must both be approximately 60°, whereas the normal bond angle would be 109.5°. Epoxyethane reacts with water to form ethane-1, 2-diol and this is the industrial method of preparation of this compound.

Ball and spring model of epoxyethane showing bond strain

A

EPOXY RESINS

Epoxy resins have the general formula:

$$\underset{\text{(epoxide ring: O bridging } CH_2 \text{ and } CH)}{CH_2-CH-CH_2}\left[-O-C_6H_4-C(CH_3)_2-C_6H_4-OCH_2-CH(OH)-CH_2-\right]_n O-C_6H_4-C(CH_3)_2-C_6H_4-OCH_2-\underset{\text{(epoxide ring)}}{CH-CH_2}$$

Epoxy resins are used in two-part adhesives such as Araldite®. One of their major advantages is their ability to bond well to a large variety of surfaces including plastics, glass, wood and concrete. Other plus factors include low shrinkage, moisture resistance and the fact that they set without giving off any water or solvent, unlike many other glues. The strength of epoxy resins is such that the bond is often stronger than the actual materials being glued. Liquid resins have $n = 2$ or less.

The resins set when cured by mixing with a hardener which cross-links the resin chains. The active ingredient of the hardener is often a diamine—a molecule containing two $—NH_2$ groups such as:

$$H_2N-C_6H_4-CH_2-C_6H_4-NH_2$$

This can link two epoxy groups on different chains. The epoxy groups are very reactive due to their highly strained rings.

Epoxy resin adhesive

The cross-linking takes place like this:

$$\begin{array}{c} \underset{\text{(epoxide)}}{CH_2-CH}-CH_2[\sim\sim\sim]_n CH_2-\underset{\text{(epoxide)}}{CH-CH_2} \\ H_2N\sim\sim\sim NH_2 \qquad\qquad H_2N\sim\sim\sim NH_2 \\ \underset{\text{(epoxide)}}{CH_2-CH}-CH_2[\sim\sim\sim]_n CH_2-\underset{\text{(epoxide)}}{CH-CH_2} \end{array}$$

$$\downarrow$$

$$\begin{array}{c} CH_2(NH\sim)-CH(OH)-CH_2[\sim\sim\sim]_n CH_2-CH(OH)-CH_2(NH\sim) \\ \wr \qquad\qquad\qquad\qquad \wr \\ CH_2(NH\sim)-CH(OH)-CH_2[\sim\sim\sim]_n CH_2-CH(OH)-CH_2(NH\sim) \end{array}$$

The resulting cross-linked polymer is solid. Notice how many polar N—H and O—H groups it has. These are responsible for the adhesion of the glue to many surfaces. No small molecules are eliminated during the cross-linking process. If they were, these could interfere with the setting process.

By ozone (trioxygen)

If ozone-enriched oxygen is bubbled through a solution of an alkene, an unstable ozonide is first produced. This then reacts with water to give two carbonyl compounds (see Chapter 32):

$$\underset{H}{\overset{R}{>}}C=C\underset{H}{\overset{R'}{<}} + O_3 \longrightarrow \underset{\text{an ozonide}}{H-\overset{R}{C}-O-\overset{R'}{C}-H \ (\text{bridged by } O-O)} \xrightarrow{H_2O} R-C\underset{H}{\overset{O}{\lessgtr}} + R'-C\underset{H}{\overset{O}{\lessgtr}}$$

If the two carbonyl compounds are identified (by formation of derivatives with Brady's reagent, see **section 32.6**), then the position of the double

A

POLYUNSATURATED FATS

Fats and oils constitute around 40% of total food intake in Western diets. Recent evidence points to a link between the consumption of saturated fats and heart disease. Dieticians are now recommending replacement of saturated fats by polyunsaturates in the diet. The degree of unsaturation of fats or oils can be determined by reacting them with excess iodine which adds on across the double bonds. The remaining iodine is titrated with sodium thiosulphate solution. Since most fats and oils are mixtures, the degree of unsaturation is reported as the **iodine value**, which is the number of grams of iodine needed to react with 100 g of oil. Typical iodine values include:

butter	25–30
olive oil	80–90
soya oil	80–140

In practice, iodine monochloride, ICl, is often used as it is more reactive than iodine.

bond in the carbon chain of the original alkene can be deduced.

This reaction, called **ozonolysis**, is unusual in that the C=C bond is completely broken. In the majority of reactions of alkenes, only the π bond is broken and the σ bond remains intact.

28.6.6 Addition of hydrogen

Hydrogen adds across double bonds at room conditions using a highly porous nickel catalyst called Raney nickel. Alkanes are produced:

$$H_2C{=}CH_2 + H_2 \xrightarrow[\text{Ni cat.}]{\text{Raney}} H{-}CH_2{-}CH_2{-}H$$

Industrially this reaction is used in the manufacture of margarines, **Figure 28.4.** Vegetable oils are 'hardened' by reducing the number of double bonds. Saturated oils have somewhat higher melting points than unsaturated ones. The reaction is carried out at around 420 K and 5×10^2 kPa pressure with a powdered nickel catalyst. Under these conditions not all the double bonds are hydrogenated, only enough to harden the oil to the required extent.

Figure 28.4 **A typical unsaturated oil before reaction with hydrogen**

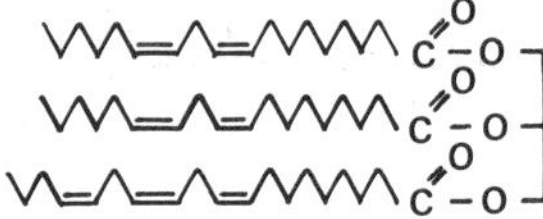

28.6.7 Polymerization

Alkenes can polymerize, joining together to form long chains of very high relative molecular mass (up to 10^6):

$$H_2C{=}CH_2 + H_2C{=}CH_2 + H_2C{=}CH_2 \longrightarrow -CH_2-CH_2-CH_2-CH_2-CH_2-CH_2-$$

ethene   poly(ethene)

This type of reaction is called **addition polymerization** as no molecule is eliminated. The polymer is named poly(ethene) even though it is actually an alkane and is unreactive, as we would expect of an alkane. The reaction is often represented:

$$H_2C{=}CH_2 \quad H_2C{=}CH_2 \quad H_2C{=}CH_2$$

although the reaction is either radical or ionic, depending on th conditions.

Poly(ethene), usually called polythene, was originally made by a high temperature (around 600 K) and high-pressure (around 3×10^5 kPa process with traces of peroxide. These initiate a radical polymerizatio which produces a product with side chains and a relatively low molecula mass (around 10^5). This is called low density polythene. Nowadays hig density polythene with no chain branching and a higher molecular mas is produced, by using Ziegler–Natta catalysts (mixtures of triethy aluminium and titanium(IV) chloride) see **section 35.3.1.** These catalyst also allow for milder conditions—around 5×10^3 kPa and 350 K. Th polymerization proceeds via an ionic mechanism. High density polythen has both a higher density and a higher softening temperature than th low density form—properties which result from the closer packing of th polymer chains.

A

ADDITION POLYMERS

Many other addition polymers are in everyday use as well as poly(ethene). They are summarized in the table.

Monomer	Polymer	Systematic chemical name	Common name or trade name (in capitals)	Typical uses
$CH_2{=}CH_2$	$-[CH_2-CH_2]_n-$	Poly(ethene)	Polythene ALKATHENE	Washing-up bowls
CH_3–$CH{=}CH_2$	$-[CH(CH_3)-CH_2]_n-$	poly(propene)	Polypropylene	Rope
Cl–$CH{=}CH_2$	$-[CH(Cl)-CH_2]_n-$	Poly(chloroethene)	Polyvinylchloride PVC	Records
CN–$CH{=}CH_2$	$-[CH(CN)-CH_2]_n-$	Poly(propenenitrile)	Polyacrylonitrile Acrylic fibre COURTELLE	Clothing
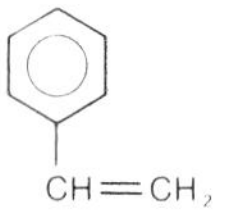 C_6H_5–$CH{=}CH_2$	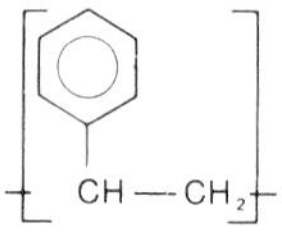 $-[CH(C_6H_5)-CH_2]_n-$	Poly(phenylethene)	Polystyrene	Plastic models Ceiling tiles
$CH_2{=}C(CH_3)CO_2CH_3$	$-[C(CH_3)(CO_2CH_3)-CH_2]_n-$	Poly(methyl 2-methyl propenoate)	Polymethylmethacrylate Acrylic PERSPEX	Cockpit, canopies in aircraft False teeth

28.7 Preparation of alkenes

Industrially alkenes are obtained by cracking alkane fractions from crud oil. In the laboratory there are two main methods of preparing alkenes.

28.7.1 Dehydration of alcohols

This may be done by passing alcohol vapour over a heated aluminiu oxide catalyst as shown in **Figure 28.5.**

R

REFLUXING

Many organic reactions take some time to go to completion and also need fairly high temperatures. Under these conditions, it is probable that one or more of the reactants, products or solvents would boil away. An added hazard is that many organic vapours are flammable and/or toxic. All these problems can be solved by using the technique of **refluxing**. A water-cooled Liebig condenser is fitted vertically to the reaction flask. Vapours which escape condense on the walls of the condenser and drip back into the reaction flask. Thus no volatile liquid should be lost. If required, a drying tube can be fitted to the top of the condenser. This tube contains granules of a suitable drying agent, e.g. calcium chloride, and prevents water vapour entering the flask.

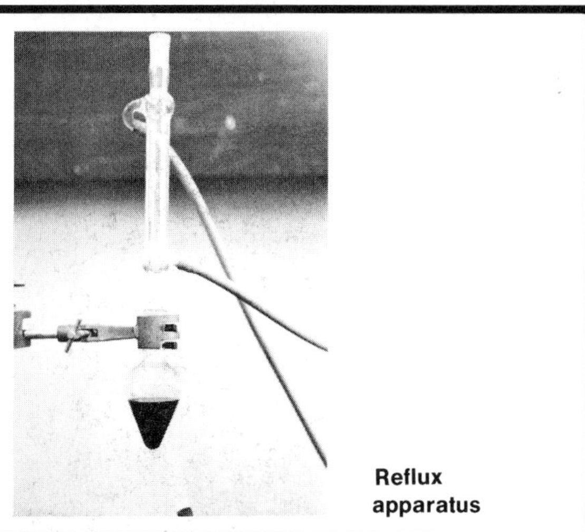

Reflux apparatus

igure 28.5 **Preparation of ethene by ehydration of ethanol**

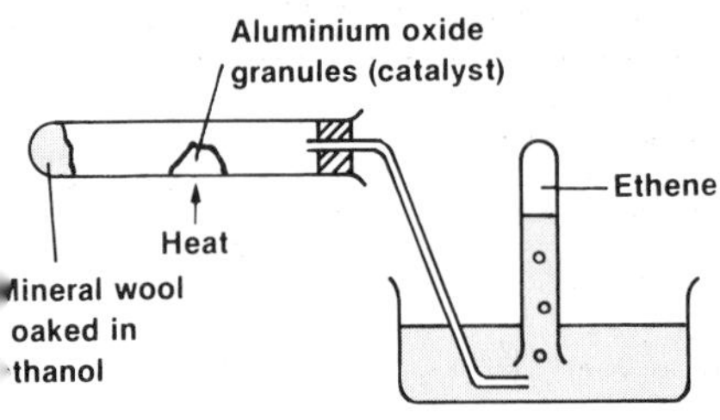

$$CH_3CH_2OH \xrightarrow[Al_2O_3\ \text{catalyst}]{\text{heat}} H_2C{=}CH_2 + H_2O$$

(The H and OH on adjacent carbons are circled and lost as H_2O.)

Alternatively, excess concentrated sulphuric acid can be used as a dehydrating agent. Excess alcohol leads to the formation of an ether—a compound with the functional group R—O—R.

$$CH_3CH_2OH \xrightarrow[(\text{conc. } H_2SO_4)\ 440\,K]{-H_2O} H_2C{=}CH_2$$

28.7.2 From halogenoalkanes

Halogenoalkanes refluxed with a strong base dissolved in alcohol lose a hydrogen halide to form alkenes:

$$CH_3CHBrCH_3 + KOH \longrightarrow CH_3CH{=}CH_2 + KBr + H_2O$$

(The H and Br on adjacent carbons are circled.)

The mechanism is discussed in **section 30.7.4.**

Ethene cannot be made by this reaction. A reaction with the solvent occurs and an ether forms.

1-bromobutane gives but-1-ene:

$$CH_3CH_2CH_2CH_2Br \xrightarrow{-HBr} CH_3CH_2CH{=}CH_2$$

but 2-bromobutane gives a mixture of but-1-ene and but-2-ene (bot *cis*- and *trans*-isomers):

```
                                      H   H   H
                                      |   |   |       H
                                  H — C — C — C = C /
                                      |   |         \
 H    H    H    H                     H   H           H
 |    |    |    |               ↗
 H — C — C — C — C — H
 |    |    |    |               ↘
 H   (H   (Br)  H)                    H               H
                                      |               |
                                  H — C — C = C — C — H
                                      |   |   |   |
                                      H   H   H   H
```

28.8 Economic importance of alkenes

Alkenes are very important industrially. The chemistry of the two mos important, ethene and propene, is summarized in **Figures 28.6** and **28.7** Both alkenes are obtained from naphtha (obtained from crude oi distillation) by cracking. Note the variety of materials which are produced Can you imagine the world without them?

Supplies of crude oil, and hence naphtha, will be exhausted in th foreseeable future, so chemists are already considering alternative route to these products. One possibility is that the chemical industry will retur to coal as a source of organic compounds, as coal reserves are of the orde of ten times greater than those of oil. Coal can provide organic compounds

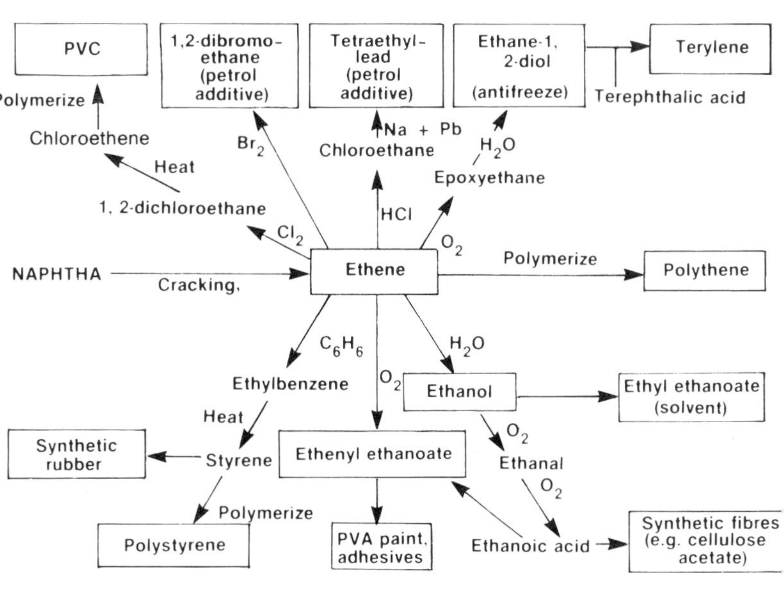

Figure 28.6 Industrial uses of ethene. Note its importance in the manufacture of plastics and synthetic fibres

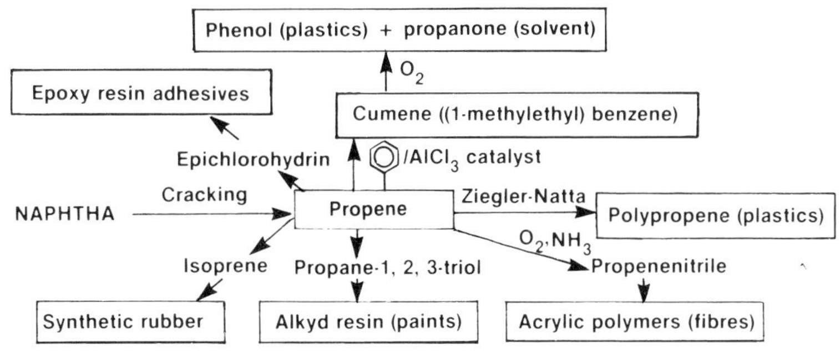

Figure 28.7 Industrial uses of propene

A

RUBBER

Natural rubber is obtained from the milky sap or **latex** of the rubber tree grown in Asia and South America. Its empirical formula, C_5H_8, was determined as long ago as the time of Michael Faraday but its structure as a polymer of relative molecular mass approximately 10^6 was not established until the 1920s. This delay was largely due to the reluctance of the chemists of the time to accept that such large molecules existed.

Tapping a rubber tree

Rubber is, in fact, a polymer of methyl-1,3-butadiene (isoprene). The polymer has one double bond per isoprene unit and in natural rubber all the double bonds are *cis*:

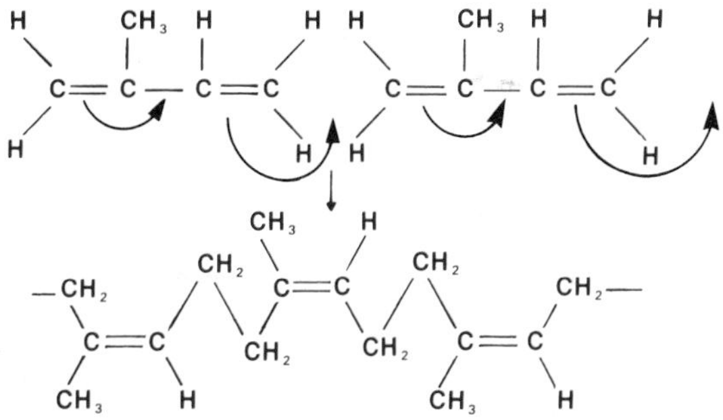

Natural rubber

The isomer in which all the double bonds are *trans* is a substance called **gutta percha** which is denser than rubber and is hard and inelastic. It is used to make the outer cover of golf balls.

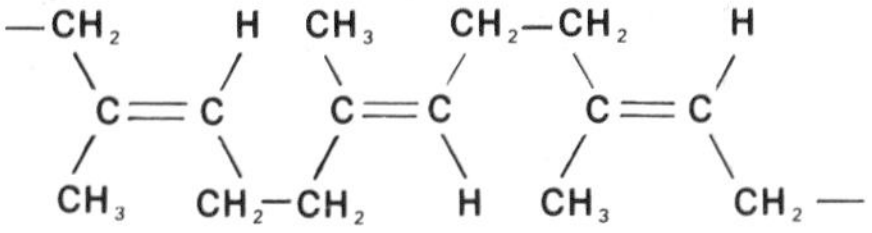

Gutta percha

Pure natural rubber is not very hard-wearing and becomes soft and sticky when slightly warmed. It would be quite useless for, say, car tyres. Its properties are improved by **vulcanizing**—heating with sulphur compounds which **cross-link** the polyisoprene chains. This prevents the molecules sliding over one another when the rubber is stretched, making it more elastic. Synthetic rubbers are made by polymerizing derivatives of buta-1,3-diene with Ziegler–Natta catalysts which ensure that the geometry of the chains is all *cis*. One such product is neoprene—poly(2-chloro-1,3-butadiene):

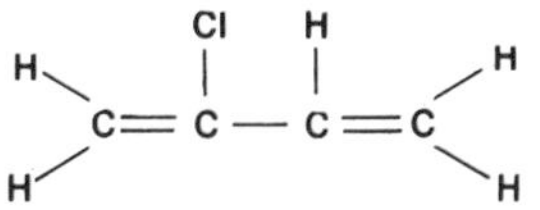

2-chloro-1,3-butadiene

It is more resistant to chemical attack than natural rubber and is used, for example, to make petrol pump hoses.

Coal tar, which is rich in benzene and other aromatic compounds, can be obtained from coal by distillation. The residue from this distillation is coke, an impure form of carbon, which can be reacted with steam to give a mixture of carbon monoxide and hydrogen:

$$C(s) + H_2O(g) \longrightarrow CO(g) + H_2(g)$$

This mixture can be used to synthesize methanol and thus other organic compounds. Another likely source of organic compounds is from vegetable material, often called **biomass**. Starch and cellulose in plant material can be fermented to ethanol, from which many organic products can be made. This is somewhat ironic: at present ethanol is made from ethene; in a few decades time, the reverse may well be the case.

28.9 Summary

- Alkenes are hydrocarbons with one or more double bonds. They are referred to as unsaturated compounds.
- The general formula of a chain alkene with one double bond is C_nH_{2n}.
- No rotation is possible at double bonds so the ethene molecule is flat.
- The physical properties of alkenes are similar to those of alkanes. van der Waals forces are the main intermolecular interaction.
- Alkenes have a peak at about $1650\,cm^{-1}$ in their IR spectra. This corresponds to a C=C stretching vibration.
- The C=C bond provides an electron-rich area in the molecule so alkenes are readily attacked by electrophiles.
- Typically the reactions of alkenes are electrophilic additions.
- The addition of unsymmetrical reagents, HZ, follows Markovnikov's rule—that the hydrogen atom adds on to the carbon atom which already has more hydrogens.
- Markovnikov's rule can be explained in terms of stabilization of the intermediate carbocation by the electron releasing properties of alkyl groups.
- Alkenes can be prepared in the laboratory by dehydration of alcohols or by elimination of hydrogen halides from halogenoalkanes.
- Industrially alkenes are obtained from the cracking of naphtha, see **section 27.9.2.**
- The reactions of alkenes are summarized in **Figure 28.8.**

Figure 28.8 Summary of the reactions of alkenes using propene as an example

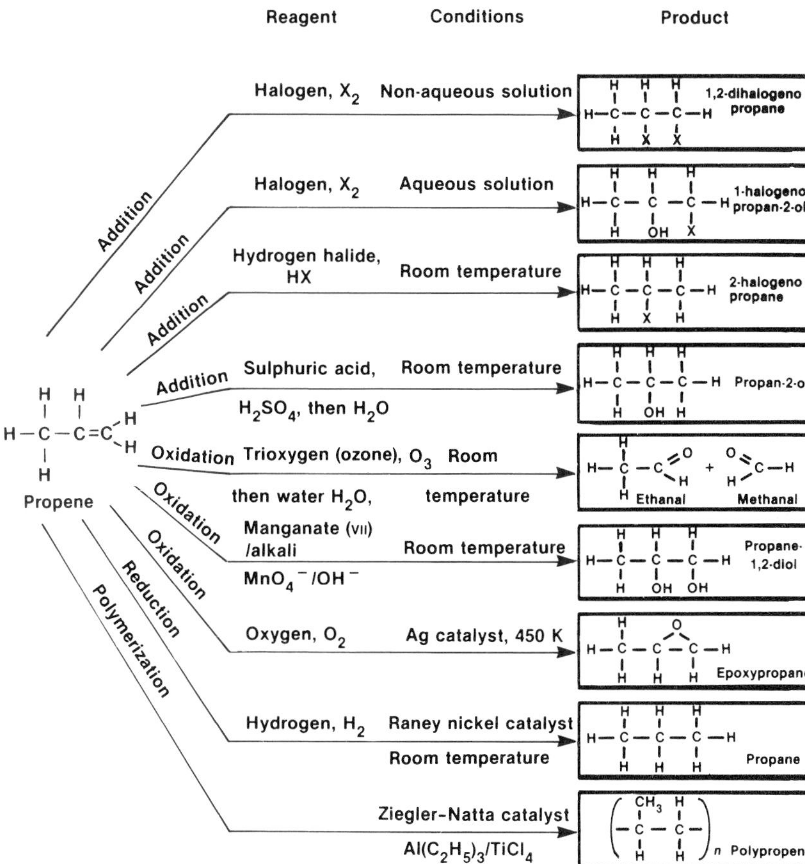

28.10 Questions

1 Name the following:

A)
```
            H
            |
        H — C — H
    H       |
    |       |        H
H — C   —   C  =  C<
    |                H
    H
```

B)
```
        H     H
        |     |
H           
 >C  =  C  —  C  =  C<  H
H                      H
```

C)
```
    H     Cl
    |     |          H
H — C  —  C  =  C<
    |                H
    H
```

D)
```
    H     H
    |     |          Cl
H — C  —  C  =  C<
    |                H
    H
```

E)
```
    Cl    H
    |     |          H
H — C  —  C  =  C<
    |                H
    H
```

F)
```
Cl         H
  >C  =  C<
H          H
```

2 Write displayed formulae for:

A) but-1-ene;
B) *cis*-but-2-ene;
C) 2-bromobuta-1,3-diene;
D) cyclohexene.

3 Write structural formulae for the main products of the following reactions:

A)
```
    H     H
    |     |          H
H — C  —  C  =  C<            +   Cl₂
    |                H
    H
```

B)
```
          H     H     H
          |     |     |
H
 >C  =  C  —  C  —  C  — H  +  HBr
H               |     |
                H     H
```

C)
```
H                      H
 >C  =  C  —  C  =  C<          + HBr
H       |     |        H
        H     H
```

D)
```
H          H
 >C  =  C<        + H2SO4 followed by H2O
R          H
```

What minor product(s) is/are possible in (B), (C) and (D)?

4 Starting from propan-1-ol, how would you produce propan-2-ol in two steps?

5 The compound
```
    H     H     H     H
    |     |     |     |
H — C  —  C  —  C  —  C — H
    |     |     |     |
    H     Br    H     H
```
is refluxed with potassium hydroxide in ethanol. What three alkenes will be formed? Each of the products is reacted separately with hydrogen bromide. One will produce two isomeric products, the other two will produce only one. Give structural formulae for all the products and explain.

29 Arenes

29.1 Introduction

Arenes, often called aromatic compounds, are organic compounds which contain a benzene ring, C_6H_6. The simplest arene is benzene. The benzene ring is particularly stable because of its unique structure and bonding, and benzene and related compounds have characteristic properties. Compounds with two or more benzene-type rings which share a pair of carbons also share the characteristic properties of benzene and are also classed as arenes. Naphthalene, anthracene and phenanthrene are examples, but these are rarely met at A level.

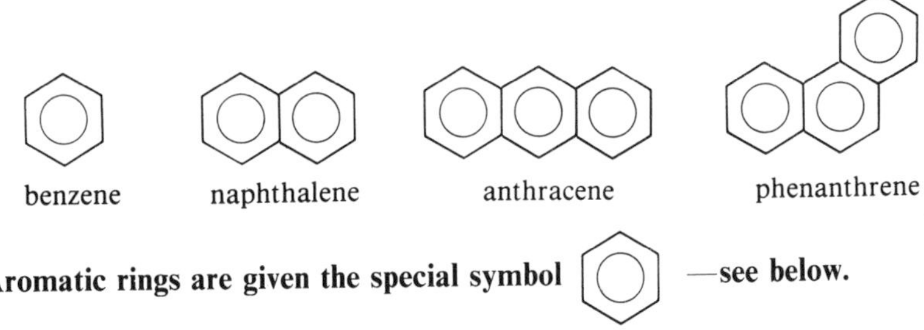

Aromatic rings are given the special symbol [symbol] —see below.

29.2 Bonding and structure of benzene

igure 29.1 The bonding in benzene

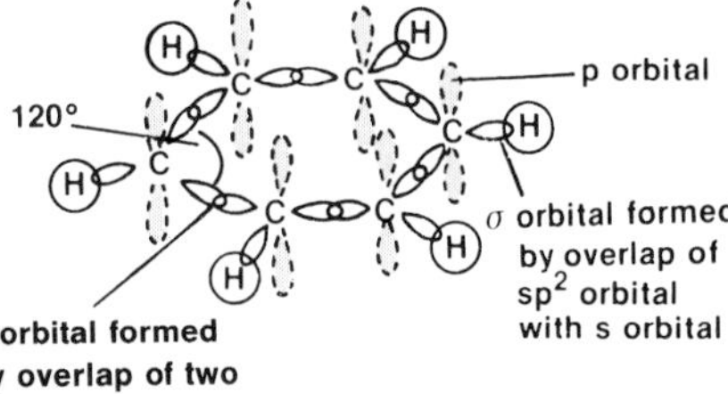

igure 29.2 The angles of a regular exagon are 120°

igure 29.3 Kekulé-type structure for enzene

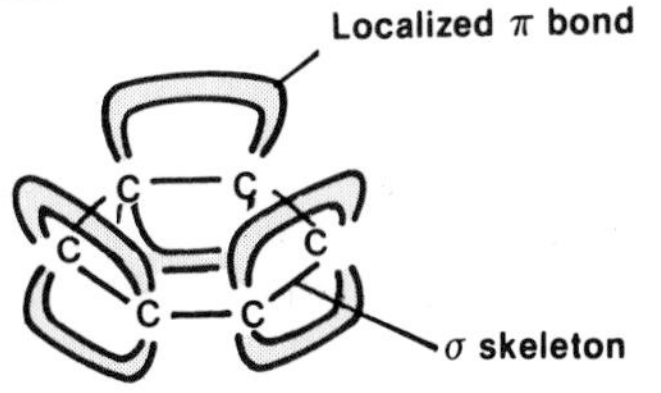

The bonding and structure of benzene and related compounds was a long standing puzzle to the early organic chemists (see box overleaf). Such compounds were first isolated from sweet-smelling oils such as balsam, hence the name 'aromatic' which now refers to their structures rather than their aromas.

The benzene molecule consists of a flat regular hexagon with a carbon atom at each 'corner'. Each carbon is hybridized so that it has three sp^2 orbitals and one p orbital. A hexagonal skeleton of six σ bonds is formed by the overlap of sp^2 orbitals. The other sp^2 orbitals form σ bonds with hydrogen, **Figure 29.1.** Notice the similarity to the bonding in graphite, see **section 10.3.3.**

As the angle between sp^2 orbitals is 120° and this is the angle of a regular hexagon, **Figure 29.2**, the benzene molecule has no ring strain. The 'spare' p orbitals could be imagined to overlap to form three localized π bonds, **Figure 29.3.** We shall call this a 'Kekulé-type' structure, see box

Figure 29.4 The delocalized π orbital in benzene

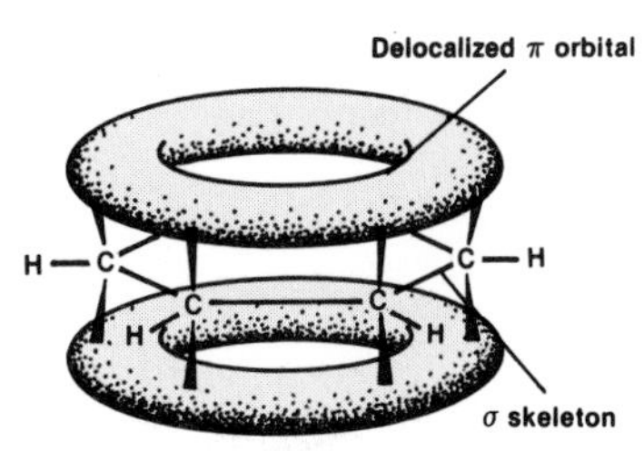

THE STRUCTURE OF BENZENE

Benzene's formula, C_6H_6, implies a good deal of unsaturation (the chain alkane of length C_6 has the formula C_6H_{14}). Formulae like:

$$CH_2{=}C{=}\overset{H}{\overset{|}{C}}{-}\overset{H}{\overset{|}{C}}{=}C{=}CH_2$$

were proposed, but benzene rarely undergoes the addition reactions which would be expected of a compound of this structure. In 1865 Friedrich August Kekulé proposed a ring structure following a dream about atoms 'in snake-like motion' where 'one of the snakes had seized hold of its own tail':

or ... in skeletal notation

However, there were two objections to this. Firstly, a cyclic triene should show addition reactions which benzene rarely does. Secondly, this structure should give rise to two isomeric disubstituted compounds:

and

Kekulé himself suggested a solution to the second dilemma by proposing that benzene consisted of structures in rapid equilibrium:

Later, this rapid alternation between two structures evolved into the idea of **resonance** between two structures, both of which contribute to the actual structure. The actual structure was thought to be a 'hybrid' (sort of average) of the two. Such resonance hybrids were believed to be more stable than either of the separate structures. Current ideas about the structure of benzene are described in the main text.

Friedrich August Kekulé

Model of benzene

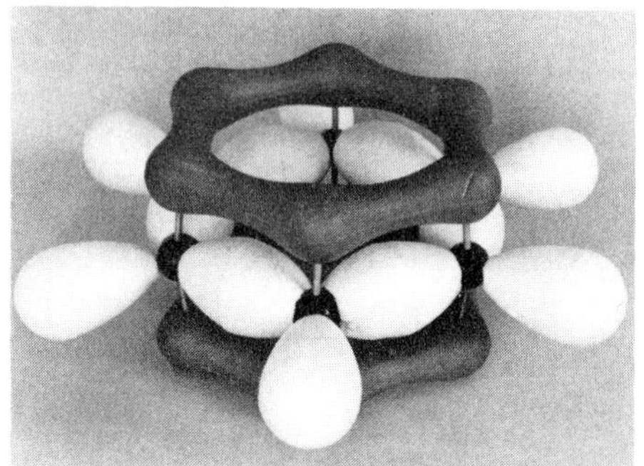

above. However, it is now believed that the p orbitals overlap to give delocalized orbital all the way round the hexagon. This results in an orbita consisting of a doughnut shaped cloud of electron density above an below the plane of the hexagon, **Figure 29.4.** Wherever a delocalize structure occurs, the resulting molecule is more stable than woul otherwise be expected.

29.2.1 Evidence for the delocalized structure and extra stability of benzene

Bond lengths

In **section 10.2.2** we saw that there is a relationship between bond lengt and bond order (single, double, triple, etc.) for bonds between the sam pair of atoms. X-ray diffraction shows that all the C—C bond lengths i benzene are the same, 0.140 nm, see **Figure 29.5.**

A C—C single bond in ethane is 0.154 nm and a C=C double bond i ethene is 0.134 nm in length. The C—C bond in benzene is midway i length between a single and a double bond. This is consistent with the p orbitals overlapping the whole molecule rather than just three of the si C—C σ bonds.

Electron density map

The electron density contour map of benzene (**Figure 29.5**) clearly show the high electron density equally spread around the whole hexagon. Agai this is consistent with a delocalized π orbital rather than three localize ones.

igure 29.5 **An electron density ɔntour map of benzene obtained by ·ray diffraction**

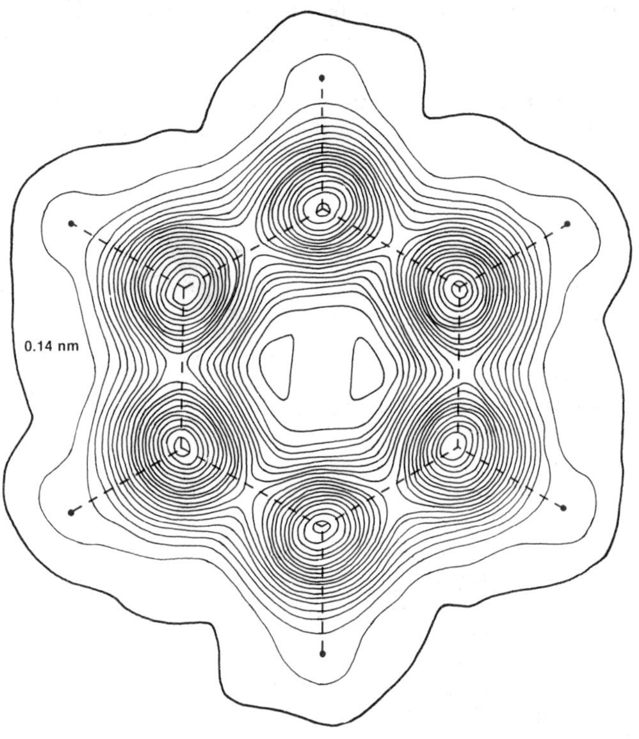

Thermochemical evidence

Thermochemical cycles can be used to compare the actual benzene molecule with the hypothetical Kekulé-type structure (which, incidentally, would be called 1, 3, 5-cyclohexatriene). In every case, the real benzene molecule turns out to be much more stable than the Kekulé-type structure.

ENTHALPIES OF HYDROGENATION

The enthalpy change on hydrogenation of cyclohexene is:

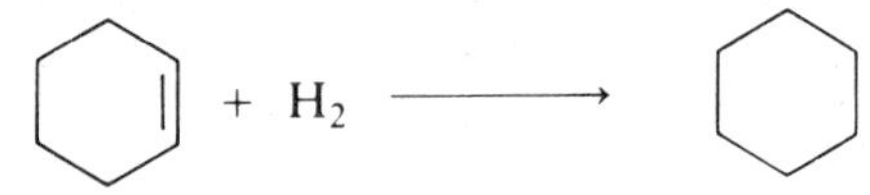

cyclohexene + H_2 ⟶ cyclohexane $\Delta H^\ominus = -120\,\text{kJ mol}^{-1}$

Therefore the enthalpy change of hydrogenation of hypothetical Kekulé-type benzene should be three times this:

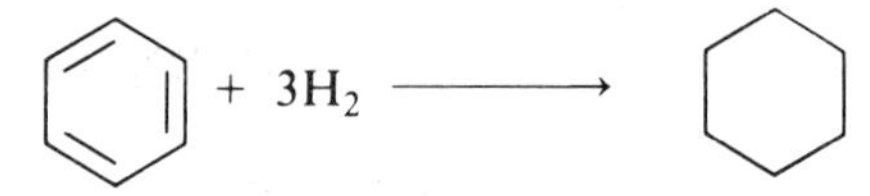

'Kekulé-type' benzene + $3H_2$ ⟶ cyclohexane $\Delta H^\ominus = -360\,\text{kJ mol}^{-1}$

The enthalpy change of hydrogenation of 'real' benzene is:

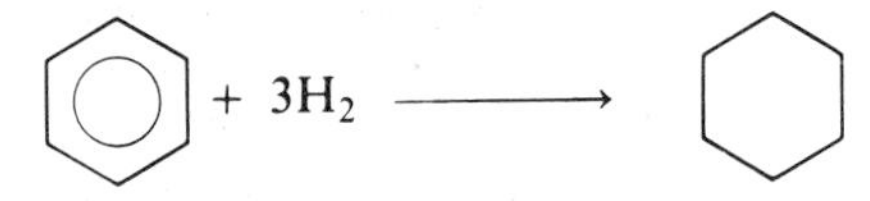

benzene + $3H_2$ ⟶ cyclohexane $\Delta H^\ominus = -208\,\text{kJ mol}^{-1}$

We can put both real and Kekulé-type benzene on an enthalpy diagram with cyclohexane as the common point, **Figure 29.6.** 'Real' benzene is $152\,\text{kJ mol}^{-1}$ more stable than Kekulé-type benzene.

igure 29.6 **Enthalpy diagram to how enthalpies of hydrogenation of enzene and hypothetical Kekulé-type enzene**

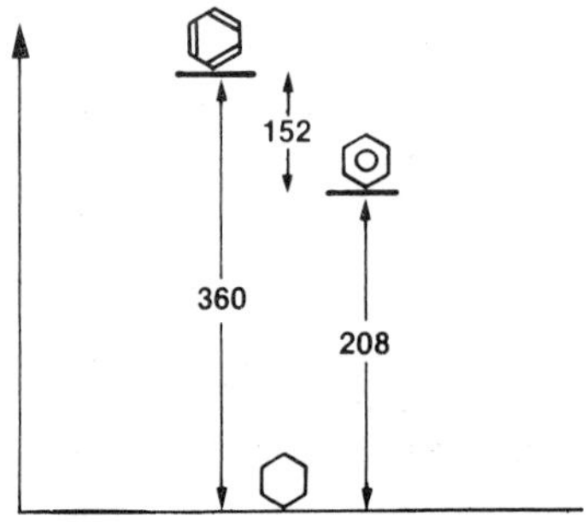

Enthalpy/kJ mol^{-1}

152

360

208

ENTHALPY CHANGE OF ATOMIZATION

We can calculate the enthalpy change of atomization of 'real' benzene from the thermochemical cycle:

$\Delta_{at}H^{\ominus}$ (benzene) → 6C (g) + 6H (g)

$\Delta_f H^{\ominus}$ (benzene) = 49 kJ mol^{-1}

$6 \times \Delta_{at}H^{\ominus}$(C)
$6 \times (+717)$ kJ mol^{-1} = 4302 kJ mol^{-1}

$6 \times \Delta_{at}H^{\ominus}$(H)
$6 \times (+218)$ kJ mol^{-1}
= 1308 kJ mol^{-1}

−49 kJ mol^{-1} → 6 C(gr) + 3H_2 (g) → +5610 kJ mol^{-}

$$\Delta_{at}H^{\ominus}\ (\text{benzene}) = -49 + 4302 + 1308\ \text{kJ mol}^{-1}$$
$$= +5561\ \text{kJ mol}^{-1}$$

The enthalpy change of atomization of Kekulé-type benzene can b calculated from bond energies:

$$\Delta_{at}H^{\ominus}\ (\text{Kekulé benzene}) = (3 \times \text{BE C}=\text{C}) + (3 \times \text{BE C}-\text{C}) + (6 \times \text{BE C}-\text{H})$$

Using tabulated bond energies:

$$\Delta_{at}H^{\ominus}\ (\text{Kekulé benzene}) = (3 \times 612) + (3 \times 347) + (6 \times 413)$$
$$= 1836 + 1041 + 2478$$
$$= +5355\ \text{kJ mol}^{-1}$$

Again, both values can be put on an enthalpy diagram with the separat atoms as the common factor, **Figure 29.7.** This calculation shows tha 'real' benzene is 206 kJ mol^{-1} more stable than Kekulé-type benzene. Th discrepancy between this calculation and the previous one can be explaine by the fact that tabulated bond energies are only averages.

Figure 29.7 Enthalpy diagram to show the enthalpies of atomization of benzene and hypothetical Kekulé-type benzene

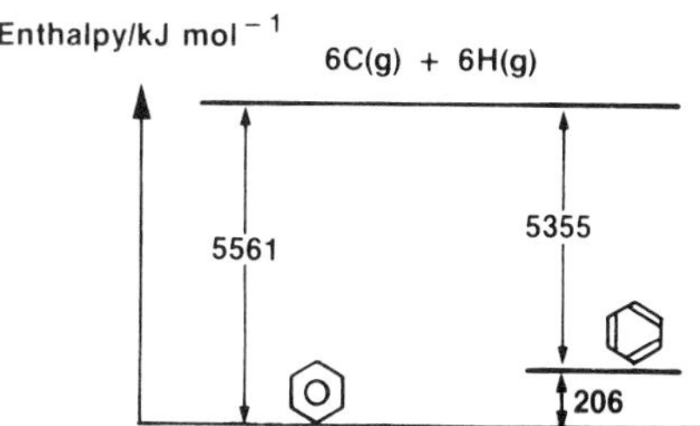

29.2.2 Summary

It is clear that benzene is more stable than expected by something lik 150 to 200 kJ mol^{-1}. This is often called the **aromatic stabilization energy** The stability is due to the bonding which consists of σ bonds plus delocalized π bond which extends over the whole ring. This produces ring of high electron density around the molecule. The resulting bond may be thought of as having a bond order of around 1.5—intermediat between double and single. Both the high electron density and the unusua stability, sometimes called the **aromatic stability**, are important in th chemistry of benzene and its derivatives.

29.3 Naming aromatic compounds

Substituted arenes are generally named as derivatives of benzene, so tha benzene forms the root of the name:

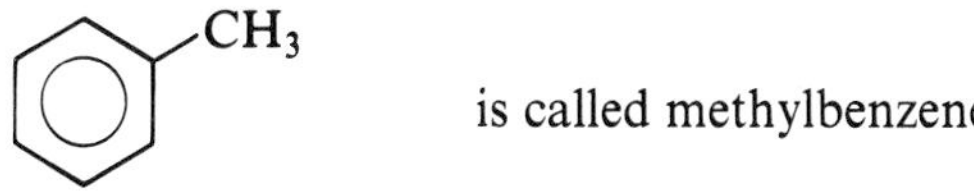

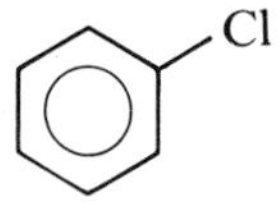

is called chlorobenzene, and so on

With more than one substituent, locants are used to show where they are attached. The ring is numbered from the carbon where the principal group (see **section 26.5**) is attached. So:

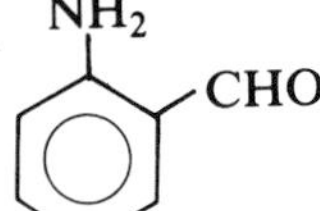

is 2-aminobenzenecarbaldehyde

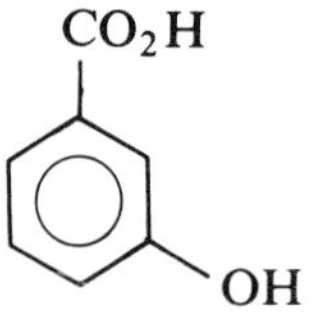

is 3-hydroxybenzenecarboxylic acid

However:

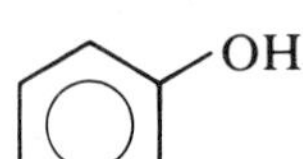

is still called phenol

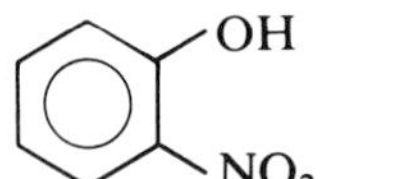

is called 2-nitrophenol

Benzene rings with two —OH groups are named systematically:

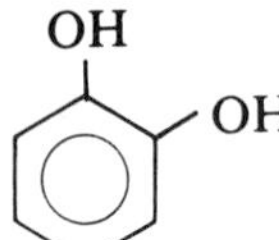

is benzene-1, 2-diol.

When a benzene ring is regarded as a side chain, 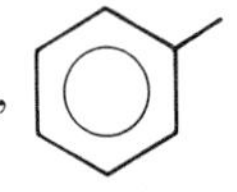 is called the phenyl group. Sometimes a conflict arises.

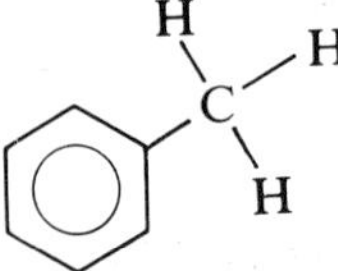

could be named phenylmethane or methylbenzene. It is usually called methylbenzene because most of its reactions are those of benzene rather than those of methane. The name draws attention of this.

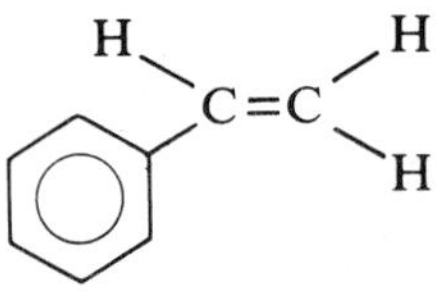

could be named phenylethene or ethenylbenzene.

In this case the main reactions are those of ethene rather than those of benzene so phenylethene is usually chosen.

As always, it is more important to be able to work from the name to the correct structure than to correctly name a compound whose formula you know.

Examples

Ethylbenzene:

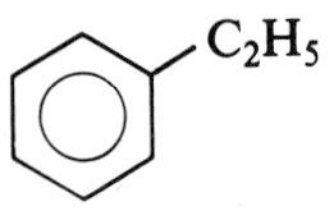

Nitrobenzene: NO_2

Benzenecarboxylic acid: CO_2H

1, 2-dimethylbenzene: CH_3 CH_3

Non-systematic names are still used for many derivatives of benzene.

29.4 Physical properties of arenes

Benzene is a liquid at room temperature, it boils at 353 K and freezes a 279 K. The boiling temperature is comparable with that of hexane (342 K but benzene's melting temperature is much higher than hexane's (178 K This is because benzene's flat hexagonal molecules pack together very wel in the solid state, making them harder to separate on melting.

Side chains on the benzene ring increase the boiling temperature b about 30 K per carbon atom of the side chain, as the larger molecule have larger van der Waals forces. However, side chains on the benzen ring *lower* the melting temperature considerably as the side chain reduce the molecules' symmetry and prevents them packing together efficiently See **Table 29.1.**

Table 29.1 Melting and boiling temperatures of hexane, plus some aromatic compounds

Compound	T_m/K	T_b/K
Hexane	178	342
Benzene	279	353
Methylbenzene	178	384
Ethylbenzene	178	409
Propylbenzene	174	432
1, 2-dimethylbenzene	248	418
1, 3-dimethylbenzene	225	412
1, 4-dimethylbenzene	286	411

Napthalene, , is a volatile solid at room temperatur ($T_m = 354$ K, $T_b = 491$ K). It is used in mothballs—the smell demonstratin its volatility. The increased melting and boiling temperatures are wha would be expected from its relative molecular mass.

Like other hydrocarbons which are non-polar, arenes do not mix wit water, but mix with other hydrocarbons and other non-polar solvents.

29.5 Infra-red spectra of arenes

The infra-red spectrum of benzene is shown in **Figure 29.8.** The importan peaks are at approximately 3030 cm^{-1} due to a C—H stretch and a approximately 1500 cm^{-1} due to a C—C stretch in the aromatic ring Substituted benzene derivatives would, of course, also show peaks characteristic of their substituents.

Figure 29.8 The infra-red spectrum of benzene

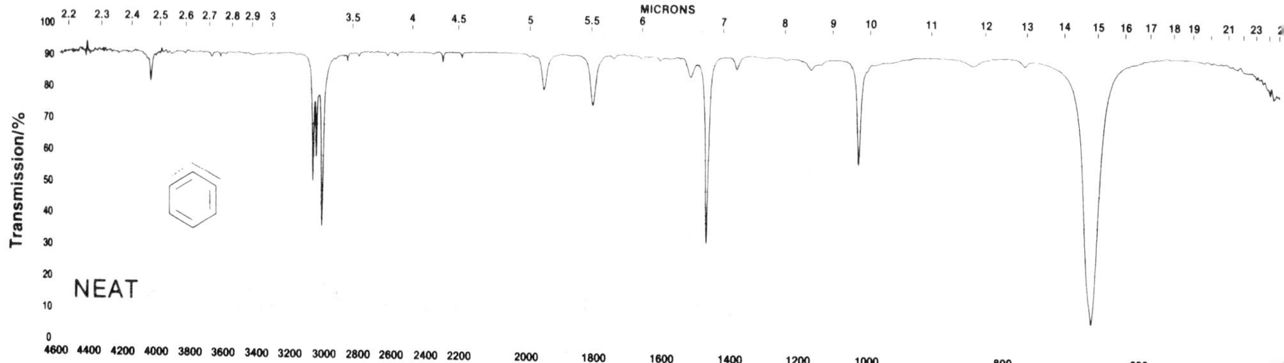

29.6 Mass spectra of arenes

The mass spectrum of phenylethanone is given in **Figure 29.9**.

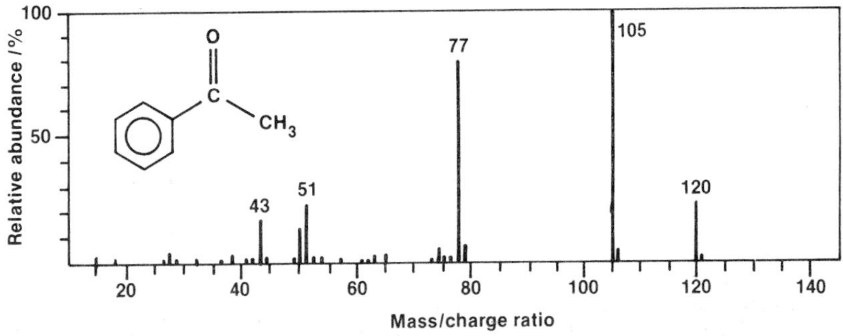

Figure 29.9 The mass spectrum of phenylethanone

Figure 29.10 Fragments of phenylethanone

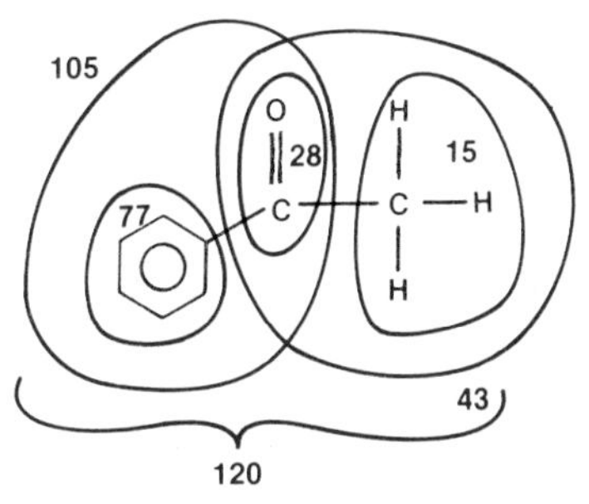

The parent ion peak is at mass number 120 (the small peak at mass number 121 is an isotope peak—a molecule containing either ^{13}C or ^{2}H). The peak at mass number 105 is the parent ion which has lost a $—CH_3$ group. That at mass number 77 is the phenyl group, $C_6H_5^+$, formed by the loss of CH_3CO. The unit CH_3CO^+ forms the peak at mass number 43. The peak at mass number 51 is less easily assigned—it is $C_4H_3^+$, a fragment of the benzene ring. The $C_6H_5^+$ peak of mass number 77 is found in the mass spectrum of many aromatic compounds. See **Figure 29.10.**

29.7 Reactivity of arenes

Two important factors govern the reactivity of arenes.

Firstly, the ring has an area of high electron density and is therefore susceptible to attack by electrophiles.

Secondly, although the ring is unsaturated, and addition reactions might be expected, the stability of the aromatic ring is such that the system is rarely destroyed. Substitution reactions are far more common than addition reactions. Any reaction which destroys the aromatic system will require an amount of energy equal to the aromatic stabilization energy to be put in, as well as any other energy change involved in the reaction.

The above two points mean that the majority of the reactions of aromatic systems are **electrophilic substitution reactions**. You should bear in mind that aromatic compounds which have substituents will show the reactions of the substituents as well as those of the aromatic system.

29.8 Reactions of arenes

29.8.1 Combustion

Arenes will burn. In a limited supply of oxygen, as in air, they burn with a noticeably smoky flame because of their high carbon:hydrogen ratio, which means there is often unburned carbon remaining, giving rise to soot.

Benzene, C_6H_6, has a C:H ratio of 1:1 compared with, say, hexane, C_6H_{14}, whose C:H ratio is 6:14 (1:2.3). A smoky flame is a useful pointer to the presence of an aromatic system in a compound.

29.8.2 Electrophilic substitution reactions

These involve attack by an electrophile, El^+, which is attracted by the high electron density of the aromatic ring. A bond forms with the cloud of π electrons which are attracted towards the electrophile. The species formed is called a π complex:

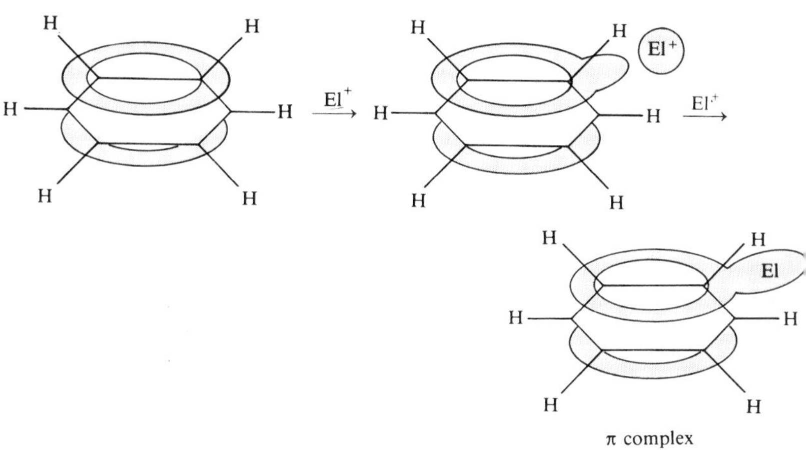

π complex

Eventually a bond is formed between one of the carbon atoms and the electrophile. To do this, the carbon must use the p orbital which was formerly part of the delocalized system and in doing so the aromatic system is destroyed. To regain the stability of the aromatic system, the carbon ejects a hydrogen ion:

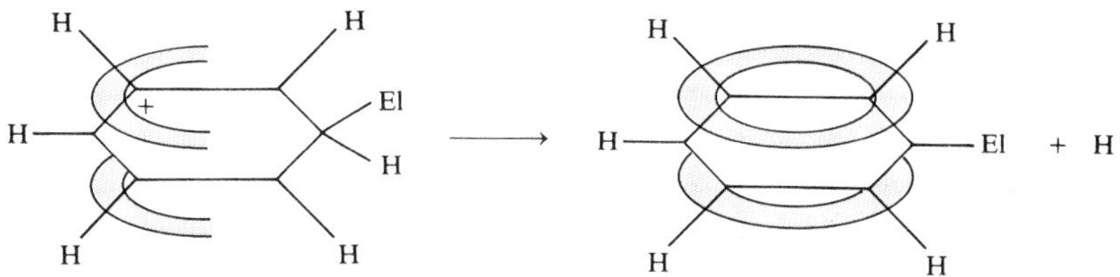

The sum total of this is the replacement or substitution of H^+ by El^+.

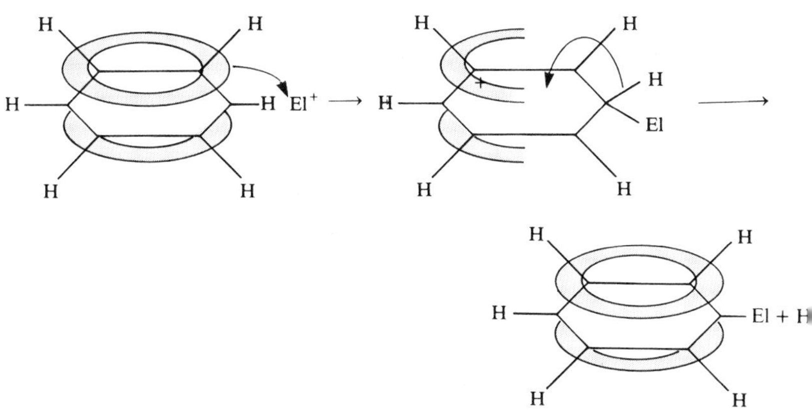

Nitration

A mixture of concentrated nitric and sulphuric acids generates the electrophile NO_2^+ by the reaction:

$$HNO_3 + 2H_2SO_4 \longrightarrow NO_2^+ + H_3O^+ + 2HSO_4^-$$

This is because H_2SO_4 is a stronger acid than HNO_3 and donates a proton (H^+) to HNO_3 which then loses a molecule of water to give NO_2^+.

The overall product of the reaction of the NO_2^+ ion with benzene is nitrobenzene:

$$C_6H_6 + NO_2^+ \longrightarrow C_6H_5NO_2 + H^+$$

nitrobenzene

A

TNT

Methylbenzene, ($C_6H_5CH_3$), formerly called toluene, is more susceptible to attack by electrophiles than benzene, as the methyl group releases electrons into the ring. Electron-releasing groups direct further substitution to the 2, 4 and 6 positions. If nitration of methylbenzene is carried out under sufficiently vigorous conditions (refluxing with a mixture of fuming nitric acid and sulphuric acid) the ring can be made to accept three nitro-substituents. 1-methyl-2, 4, 6-trinitrobenzene (trinitrotoluene, TNT) is formed. However, there is little danger of forming this product by accident. Nitro groups are electron-withdrawing and each additional one makes the ring less reactive and less likely to react with another NO_2^+ ion. So nitration of methylbenzene under mild conditions is safe enough to be used as a school practical.

TNT is an important high explosive with both military and peaceful applications. It is a solid of low melting temperature (less than 373 K) and can therefore be melted easily. This property is used both in filling shells, etc. and by bomb disposal teams who can 'steam' TNT out of unexploded bombs.

The explosion of TNT is shown in the following equation:

$$2\ CH_3C_6H_2(NO_2)_3 + 10\tfrac{1}{2}O_2 \longrightarrow 14CO_2 + 3N_2 + 5H_2O$$

The reaction is strongly exothermic. The rapid formation of a lot of gas as well as heat produces the destructive effect. The water will, of course, be steam at the temperature of an explosion.

Many other compounds with several nitrogen atoms in the molecule are explosive. This is because unreactive nitrogen gas (N_2) is readily formed from them, and in doing so, the high N≡N bond energy (945 kJ mol^{-1}) is given out. Another example is 1, 3, 5-trinitrophenol (picric acid). This can explode on impact and is therefore useful as a detonator to set off other explosives.

Making TNT shells, 1940

A little dinitrobenzene is also formed by attack of NO_2^+ on already formed nitrobenzene. The nitro group withdraws electrons from the ring and so nitrobenzene is less susceptible to attack by electrophiles than benzene itself. So only a small percentage of dinitrobenzene is formed. Like most electron-withdrawing groups, the nitro group directs further substituents to the 3 position (see below), so the di-substituted isomer produced is 1, 3-dinitrobenzene:

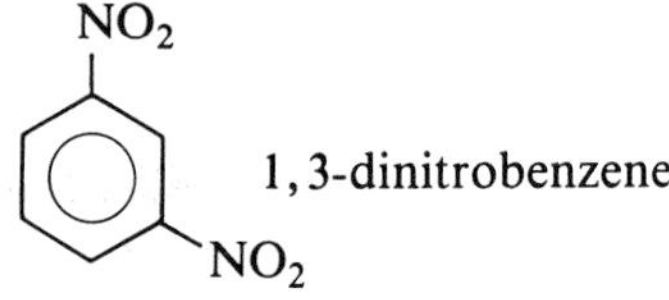

To produce nitrobenzene, benzene is refluxed with a mixture of nitric and sulphuric acids at 330 K. The dinitrobenzenes are solids and can be separated by cooling the mixture. Reduction of nitro groups produces the $—NH_2$ or amino group, see **section 34.2.6.**

Halogenation

In **section 28.6.2** we saw that halogens could act as electrophiles by virtue of their instantaneous dipoles $X^{\delta+}—X^{\delta-}$. However, they are not good enough electrophiles to react with benzene, except in the presence of a catalyst. Suitable catalysts are aluminium halides or iron filings which react with the halogen to form an iron(III) halide. The catalysts work as shown using bromine and iron filings as an example:

$$Br^{\delta+}—Br^{\delta-}: + Fe^{\delta+}(—Br^{\delta-})_3 \rightleftharpoons Br^+ + FeBr_4^-$$

A small concentration of the electrophile Br^+ is formed which then attack the benzene ring. The mechanism is as for nitration. First a π complex i formed, followed by the formation of a bond between Br^+ and one of th carbon atoms:

$$C_6H_6 + Br^+ \longrightarrow [C_6H_6Br]^+ \longrightarrow C_6H_5Br + H^+$$

An H^+ ion is then lost.

The overall reaction is:

$$C_6H_6(l) + Br_2(l) \xrightarrow[\text{room temperature}]{\text{Fe catalyst}} C_6H_5Br(l) + HBr(g)$$

The hydrogen bromide can be detected, as it forms white fumes on contac with moist air. Its presence confirms that hydrogen has been remove from the benzene and so the reaction must be *substitution* rather tha addition. The equivalent reaction works with chlorine, and the AlCl catalyst used works in the same way as $FeBr_3$. Neither reaction need UV light (although halogens will *add on* to arenes in UV light, see below Iodine reacts less readily, the equilibrium below being set up:

$$C_6H_6 + I_2 \rightleftharpoons C_6H_5I + HI$$

Addition of a base, to remove the hydrogen iodide as it forms, shifts th equilibrium to the right to give a reasonable yield of iodobenzene.

All the halogens, being electronegative, withdraw electrons from th benzene ring, making it less susceptible to attack by electrophiles, so littl disubstituted product is formed. Halogens are exceptions to the genera rule that electron-withdrawing substituents direct further substitutio to the 3 position. Halogens direct to the 2 and 4 positions so some 1, 2 dihalogenobenzene and 1, 4-dihalogenobenzene may be formed.

Interhalogen compounds such as iodine monochloride, ICl, also attac benzene rings. In this case, only iodobenzene is formed. This is becaus iodine monochloride has a permanent dipole, $I^{\delta+}—Cl^{\delta-}$, due to th greater electronegativity of chlorine than iodine. The benzene ring is thu attacked only by the $I^{\delta+}$.

Friedel–Crafts reactions

These reactions also involve the use of aluminium chloride as a catalyst This was first discovered by C Friedel and J Crafts.

Both halogenoalkanes, R—Cl, and acid chlorides, R—C(=O)Cl, reac with $AlCl_3$ (which is an electron-deficient compound as the aluminiu atom has only six electrons) to form $AlCl_4^-$ and R^+ or $R—C^+{=}O$ respectively:

$$R—Cl + AlCl_3 \rightleftharpoons R^+ + AlCl_4^-$$

$$R—C(=O)Cl + AlCl_3 \longrightarrow R—C^+{=}O + AlCl_4^-$$

Both R^+ and $R—C^+{=}O$ are good electrophiles which attack the benzen ring to form substitution products in the same way as other electrophiles:

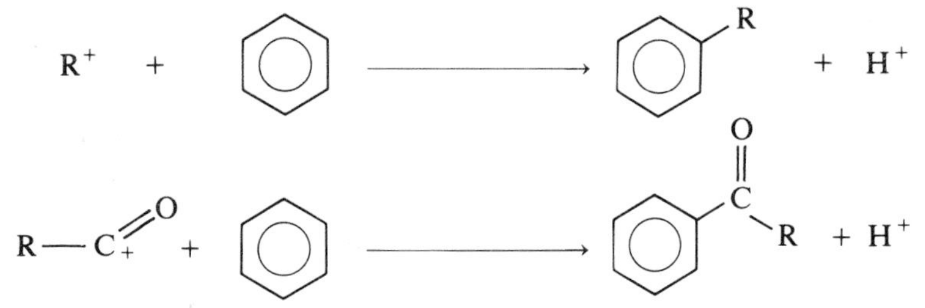

$$R^+ + C_6H_6 \longrightarrow C_6H_5R + H^+$$

$$R{-}C^+{=}O + C_6H_6 \longrightarrow C_6H_5{-}CO{-}R + H^+$$

The products are alkyl substituted arenes or aromatic ketones (see Chapter 32) respectively. The overall reactions are:

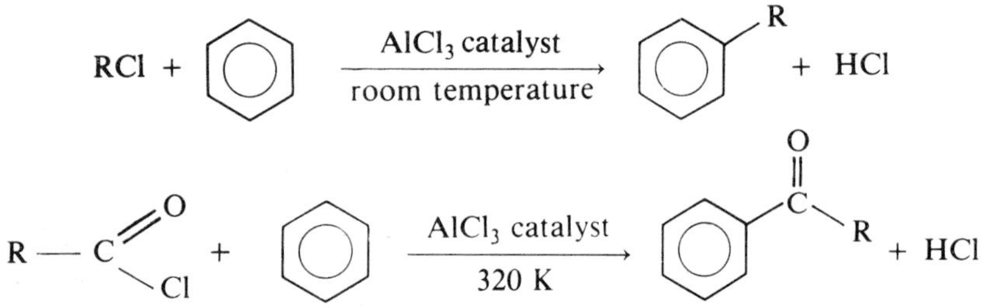

$$RCl + C_6H_6 \xrightarrow[\text{room temperature}]{AlCl_3\ \text{catalyst}} C_6H_5R + HCl$$

$$R{-}COCl + C_6H_6 \xrightarrow[320\ K]{AlCl_3\ \text{catalyst}} C_6H_5{-}CO{-}R + HCl$$

Notice the similarity between the way the aluminium chloride acts as a catalyst and the action of the iron(III) bromide on the halogenation reaction. Both act by generating a good electrophile.

Alkylation can also be achieved by reacting benzene with an alkene in the presence of an acid and an aluminium chloride catalyst. Here the electrophile is produced by the protonation of the alkene by the acid. An electrophile such as $H{-}CH_2{-}C^+H{-}H$ ($CH_3CH_2^+$) is the attacking species:

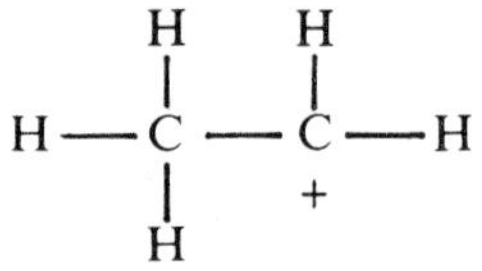

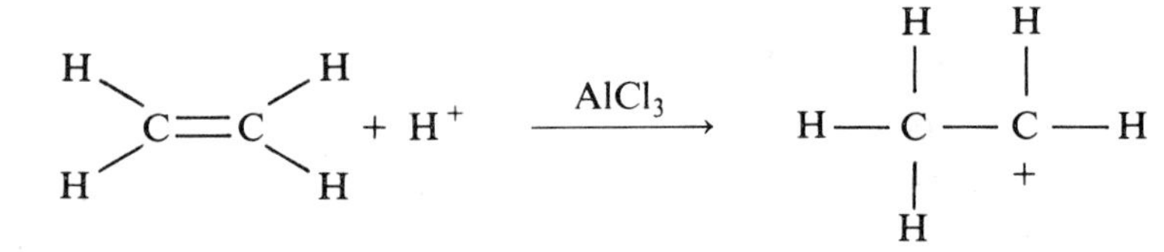

$$H_2C{=}CH_2 + H^+ \xrightarrow{AlCl_3} H{-}CH_2{-}C^+H_2$$

Then:

$$CH_3{-}C^+H_2 + C_6H_6 \longrightarrow C_6H_5{-}CH_2{-}CH_2{-}H + H^+$$

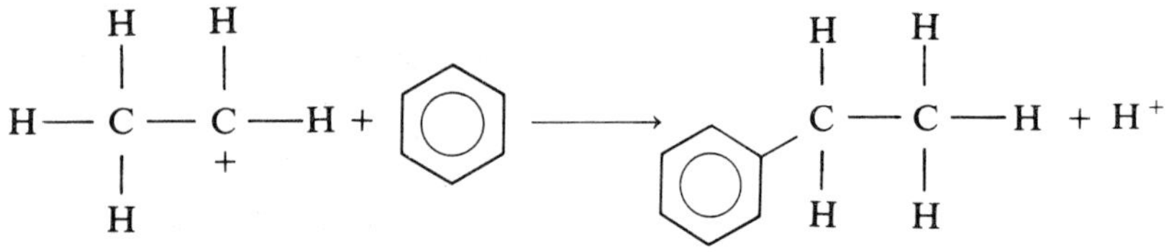

Sulphonation

When benzene is refluxed with concentrated sulphuric acid or fuming sulphuric acid (a solution of sulphur trioxide in sulphuric acid) benzenesulphonic acid is formed. The electrophile is thought to be the $HSO_3{}^+$ ion:

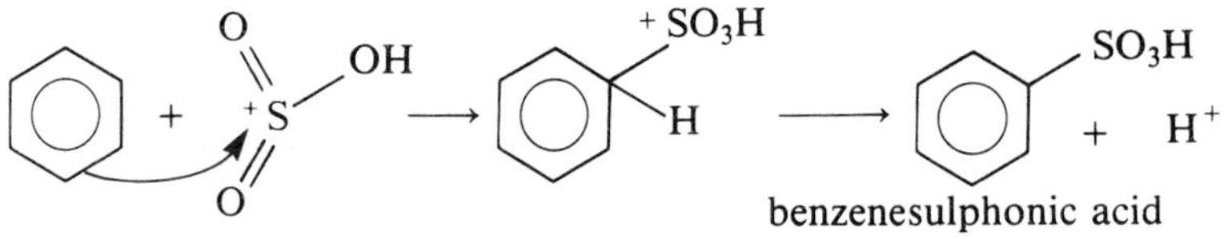

$$C_6H_6 + {}^+S(=O)_2OH \longrightarrow C_6H_6^+(SO_3H)(H) \longrightarrow C_6H_5SO_3H + H^+$$

benzenesulphonic acid

Notice that the product has a C—S bond unlike that in ethyl hydrogensulphate, $C_2H_5SO_4H$, see **section 28.6.4**, which has a C—O bond.

29.8.3 The effect of substituents

Substituents on the benzene ring affect electrophilic substitution reactions in two ways.

MANUFACTURE OF DETERGENTS

We can now see how some familiar reactions can be strung together to manufacture a useful product. A long chain alkene such as dodec-1-ene can be obtained from the cracking of crude oil derivatives. This can be made to add on to the benzene ring in the presence of hydrogen chloride and an aluminium chloride catalyst.

$$CH_3(CH_2)_9CH{=}CH_2 \ (\text{dodecene}) + C_6H_6 \xrightarrow[HCl]{AlCl_3} CH_3(CH_2)_{11}C_6H_5$$

dodecylbenzene

This is a variation of a Friedel–Crafts reaction. The product dodecylbenzene is then reacted with sulphur trioxide to form 4-dodecylbenzenesulphonic acid.

$$CH_3(CH_2)_{11}C_6H_5 + SO_3 \longrightarrow CH_3(CH_2)_{11}C_6H_4SO_3H$$

4-dodecylbenzenesulphonic acid

This is then neutralized by sodium hydroxide to form the sodium salt which acts as the detergent.

$$CH_3(CH_2)_{11}C_6H_4SO_3H + NaOH \longrightarrow CH_3(CH_2)_{11}C_6H_4SO_3^-Na^+ + H_2O$$

sodium 4-dodecylbenzenesulphonate (a detergent)

Note how the reactions used are essentially those which can be carried out in the laboratory although the conditions vary slightly. For example, sulphur trioxide gas (hard to handle in the laboratory) is used, rather than sulphuric acid, for the sulphonation step. This detergent makes up around 10% by weight of most commercial washing powders.

Figure 29.11 Electron-withdrawing and releasing substituents

C_6H_5–NO_2	Increasing electron withdrawing effect	means	Slower reaction with electrophiles
C_6H_5–COCl			
C_6H_5–Cl			
C_6H_6			
C_6H_5–CH_3	Increasing electron releasing effect	means	Faster reaction with electrophiles
C_6H_5–OCH_3			
C_6H_5–OH			
C_6H_5–NH_2			

Firstly, they may withdraw electrons from the ring making it les reactive, or release electrons onto the ring making it more reactive.

Secondly, they tend to direct further substituents on to certain positions on the ring. Some common substituents are listed in **Figure 29.11.** In general, substituents which activate the ring (by releasing electrons and making the ring more easily attacked by electrophiles) lead to further substitution at the 2, 4 and 6 positions. This is because any increased electron density tends to occur mainly at these positions. Deactivating substituents tend to *withdraw* electrons from the 2, 4 and 6 positions so when further substitution does occur, it tends to be at the 3 and 5 positions. Halogens are an exception; they deactivate the ring but direct further substituents to the 2 and 4 positions.

29.8.4 Addition reactions of arenes

While the typical reactions of arenes are substitutions in which the aromatic system remains intact, addition reactions are possible under suitable conditions.

Addition of hydrogen

Hydrogen adds on to benzene rings in the presence of a nickel catalyst at 420 K to give cyclohexane:

$$C_6H_6 + 3H_2 \xrightarrow[420\ K]{\text{Ni catalyst}} C_6H_{12}$$

Remember in skeletal notation, the benzene molecule has one hydrogen at each 'corner' and cyclohexane has two. Note that addition of hydrogen to an arene needs somewhat more vigorous conditions than addition to an alkene (see **section 28.6.6**).

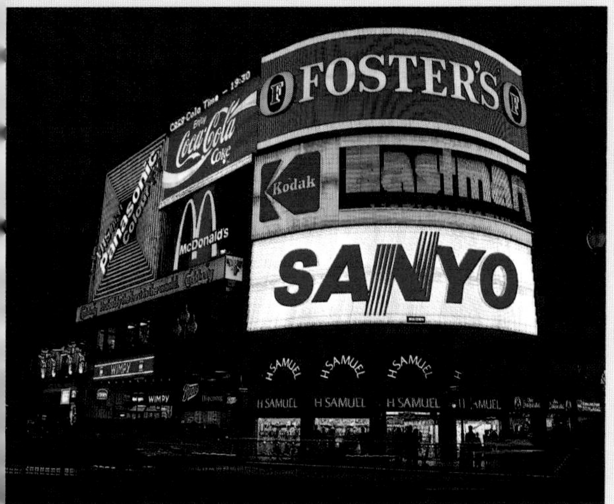

Plate 15 Piccadilly Circus lights at night. Many advertising signs are lit by passing electricity through tubes filled with inert gases. Neon gives red, argon blue and other colours are produced by mixtures of gases in tubes of coloured glass. See page 283

Plate 16 Silicon is used in microchips – see page 317

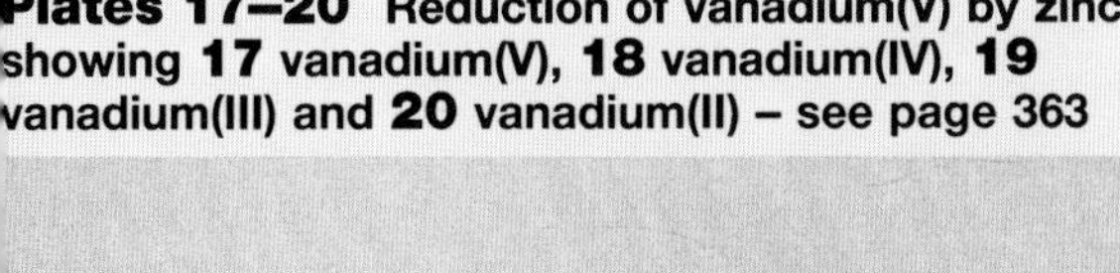

Plates 17–20 Reduction of vanadium(V) by zinc showing **17** vanadium(V), **18** vanadium(IV), **19** vanadium(III) and **20** vanadium(II) – see page 363

Plate 17 Vanadium(V)

Plate 18 Vanadium(IV)

Plate 19 Vanadium(III)

Plate 20 Vanadium(II)

Plate 21 Computer graphic image of hen egg white lysozyme

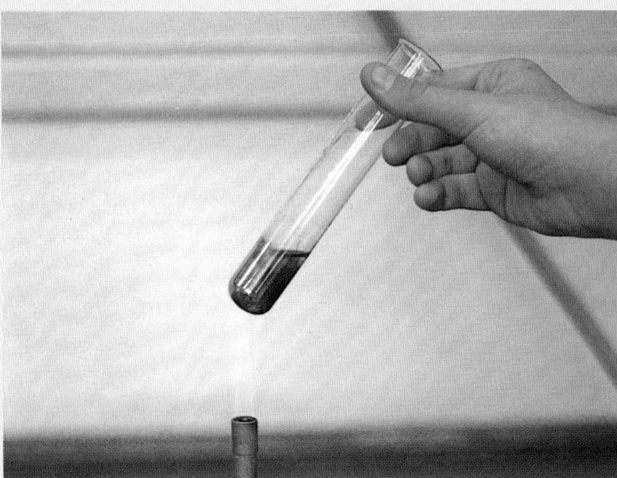

Plate 22 The silver mirror produced by aldehydes and silver ions in aqueous ammonia solution – see page 500

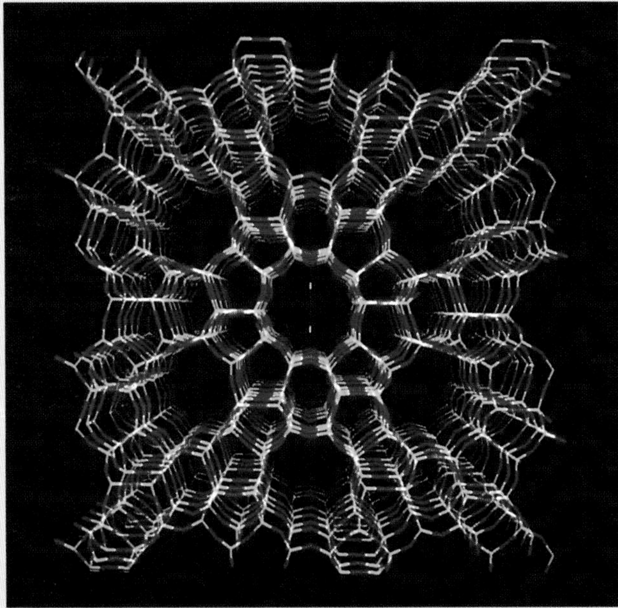

Plate 23 Computer graphic image of a zeolite – see page 317

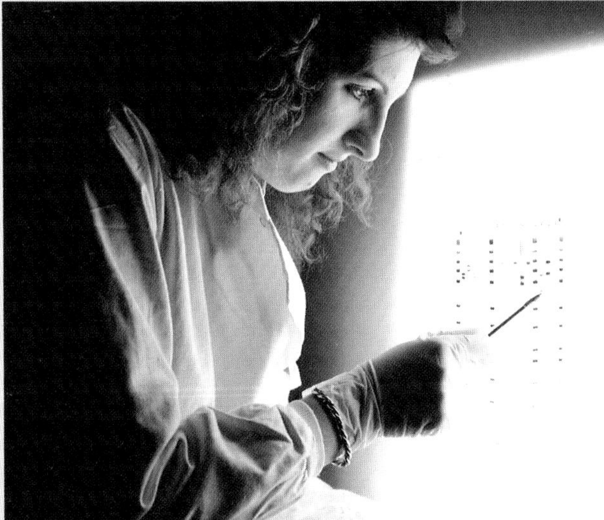

Plate 24 Electrophoresis is used to separate fragments of DNA. This can be done to identify the person from whom a small sample of tissue or blood has come. It is developing into a valuable forensic technique. It can also be used to determine paternity. See page 559

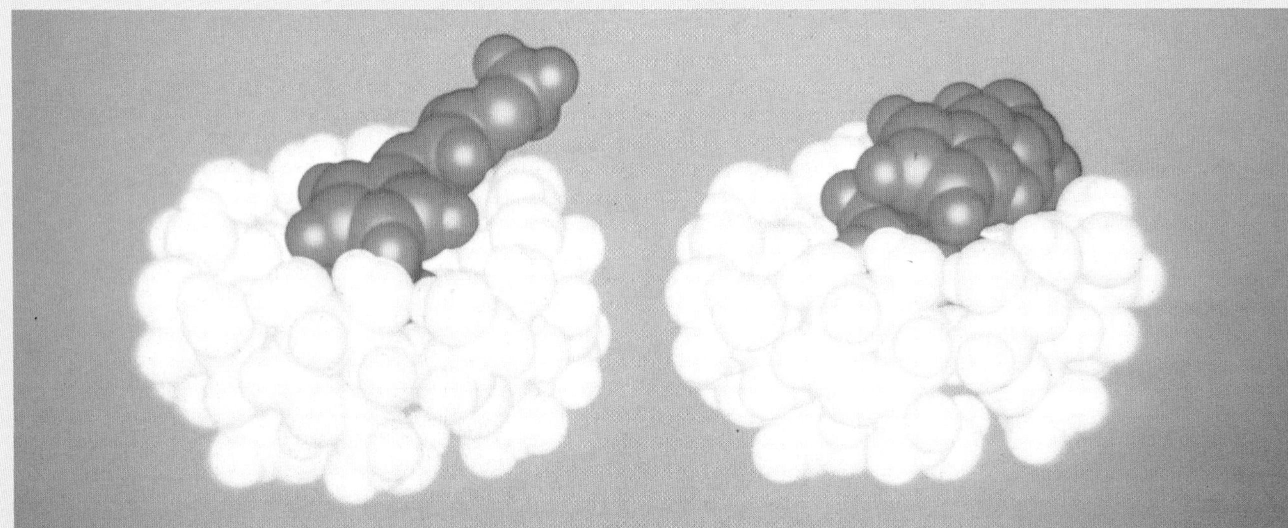

Plate 25 Computer graphic image of cyclodextrine molecules – see page 562

Addition of halogens

Both chlorine and bromine add on to benzene at room temperature when exposed to UV light (strong sunlight is sufficient). Both reactions go via a radical mechanism, for example:

$$C_6H_6 + 3Cl_2 \xrightarrow{\text{UV light}} C_6H_6Cl_6$$

1,2,3,4,5,6-hexachlorocyclohexane

Note that this reaction is quite different from the substitution reactions with halogens which take place in the dark, require a catalyst, and produce hydrogen halides.

The product of chlorine addition, 1, 2, 3, 4, 5, 6-hexachlorocyclohexane, has a number of geometrical isomers, one of which is the insecticide Gammexane.

29.9 Chemical evidence for the bonding and structure of benzene

The reactions we have looked at so far provide confirming evidence that benzene does not have a Kekulé-type structure and does have aromatic stability. The majority of reactions of benzene are **substitutions**, rather than additions. Kekulé-type benzene would be expected to undergo addition reactions easily, as it is an alkene. For example, benzene does not decolorize bromine water as would an alkene. It reacts only with *liquid* bromine (much more concentrated) and only with the help of a catalyst. Even then the reaction is substitution which leaves the aromatic system unchanged, rather than addition which would destroy the aromatic system and require input of the aromatic stabilization energy.

Addition reactions of benzene occur only under rather vigorous conditions—with radicals which react fairly indiscriminately, and with hydrogen at high temperatures with a catalyst.

29.10 Reactions of side chains

Groups attached to aromatic rings will undergo their own characteristic reactions, sometimes modified by being attached to the ring. We shall deal here only with alkane and alkene side chains; other functional groups will be considered in the relevant chapters.

29.10.1 Oxidation

Alkyl side chains attached to benzene rings are more easily oxidized than benzene itself. Strong oxidizing agents such as acidified manganate(VII) ions or acidified dichromate(VI) ions oxidize alkyl side chains of any lengths to a carboxylic acid group, for example:

$$C_6H_5C_2H_5 \xrightarrow{MnO_4^-/H^+} C_6H_5CO_2H$$

ethylbenzene → benzenecarboxylic acid

Weaker oxidizing agents, like manganese(IV) oxide, oxidize the side chains to an aldehyde:

methylbenzene $\xrightarrow{MnO_2}$ benzenecarbaldehyde

Benzene itself does not react with any of these oxidizing agents.

29.10.2 Radical halogenation of the side chain

Radical halogenation of alkyl groups occurs more readily than that of aromatic rings so when chlorine is bubbled into methylbenzene and exposed to UV light, substitution of the side chain occurs:

$$C_6H_5CH_3 \xrightarrow[\text{UV light}]{Cl_2} C_6H_5CH_2Cl \xrightarrow[\text{UV light}]{Cl_2} C_6H_5CHCl_2 \xrightarrow[\text{UV light}]{Cl_2} C_6H_5CCl_3$$

methyl benzene (chloromethyl) benzene (dichloromethyl) benzene (trichloromethyl) benzene

Hydrogen chloride is also formed.

29.10.3 Phenylethene

Phenylethene, $C_6H_5CH=CH_2$, has an alkene side chain attache

to the benzene ring. The alkene group is more reactive than benzene an so it is more sensible to refer to it as phenylethene than ethenylbenzene. It i more commonly known by its non-systematic name of styrene. Th compound undergoes all the reactions typical of alkenes includin polymerization to poly(phenylethene) or polystyrene.

A

EXPANDED POLYSTYRENE

You are probably familiar with expanded polystyrene. It is the very light plastic used as an insulator in ceiling tiles, disposable coffee cups and fast-food trays, and as a packaging material for delicate goods from eggs to hi-fi components.

It is called a **foamed plastic** and its lightness is due to its structure consisting of a vast number of gas-filled bubbles. It can be manufactured in two ways. One is by blowing gas into the liquid plastic to form a foam which then solidifies. The other is to manufacture granules of polystyrene, each of which incorporates a 'blowing agent'—a substance which produces a gas when heated. A quantity of granules is heated in a mould where the granules melt and the gas expands, forming bubbles, to fill the mould. If you look closely at, for example, hi-fi packaging,

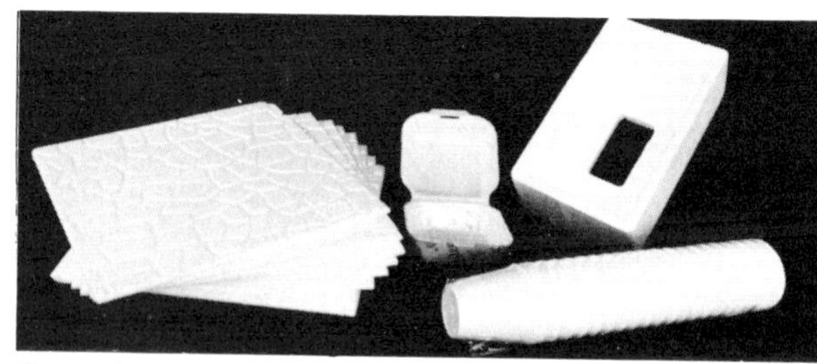

Expanded polystyrene

you will be able to pick out the individual granules. Blowing agents include pentane, a volatile liquid, and sodium hydrogencarbonate, which decomposes to carbon dioxide on heating. Gaseous foaming agents include chlorofluorocarbons which are damaging the ozone layer (see box *Atmospheric chemistry*, **section 15.3**) and most manufacturers have introduced alternatives.

29.11 Preparation of arenes

Industrially, benzene is obtained by the fractional distillation of coal tar and crude oil. Only small amounts are present in crude oil and more is made by the process called **reforming**. Here alkanes are heated under pressure with a suitable catalyst. Some of the unbranched chain alkanes form rings and then lose hydrogen to form aromatic compounds.

For example, hexane gives benzene, and heptane gives methylbenzene. In the laboratory, alkyl substituted benzenes can be prepared by the Friedel–Crafts reaction discussed on page 440.

29.12 Economic importance of arenes

Benzene itself has been used as a solvent but it is a mild carcinogen and has largely been replaced by the less toxic methylbenzene. Benzene and other arenes are added to petrol to improve the octane rating, see box *Octane rating of petrol*, **section 27.9.** However, because of the toxicity of these compounds, it is debatable whether they are preferable to lead-based additives.

Benzene is the starting material for the manufacture of a number of other important compounds which lead to a wide variety of finished products, especially plastics and fibres. Some of the most important processes and reactions are summarized in **Figure 29.12.**

igure 29.12 **Some industrial uses of enzene and its derivatives**

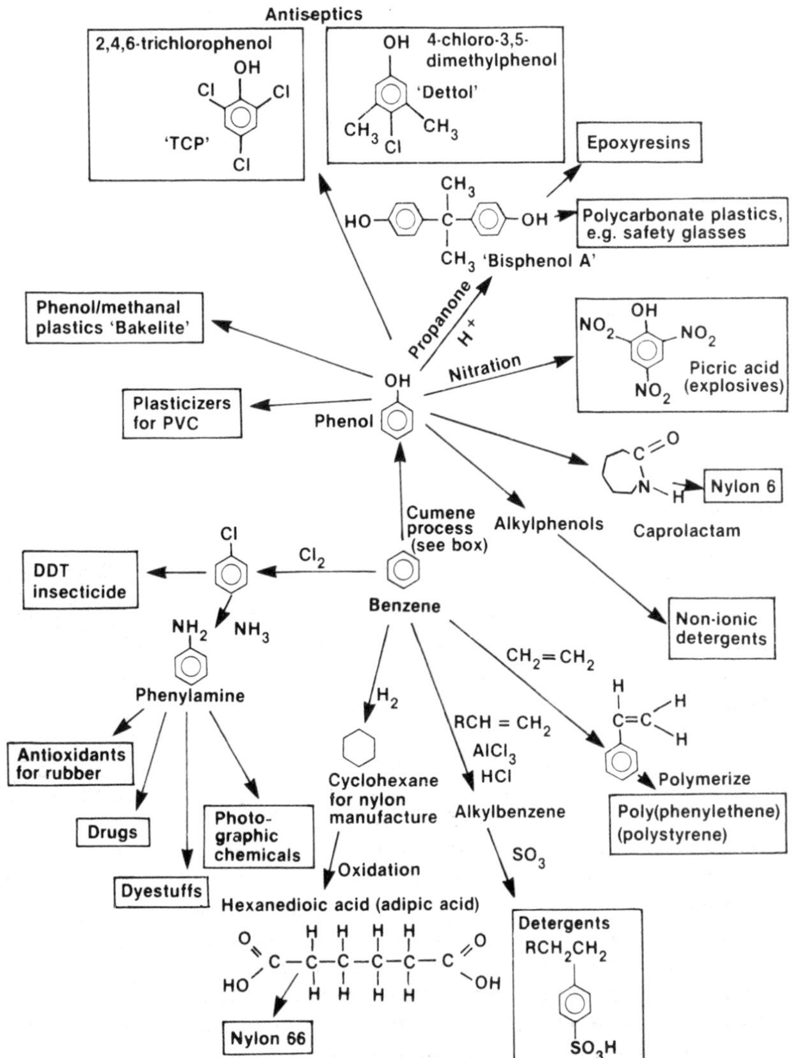

A

THE CUMENE PROCESS

Phenol is an important intermediate for the manufacture of many products, as **Figure 29.12** shows. Almost all the phenol produced today is made by the **cumene process**.

Benzene and propene are heated together with an aluminium chloride catalyst to produce (1-methylethyl)-benzene which is called cumene:

$$C_6H_6 + CH_3CH{=}CH_2 \xrightarrow{AlCl_3} C_6H_5CH(CH_3)_2$$

Cumene

This is then oxidized by air to give a compound called a hydroperoxide which is decomposed in the presence of acid to phenol and propanone:

$$C_6H_5CH(CH_3)_2 \xrightarrow{O_2} C_6H_5C(CH_3)_2OOH \longrightarrow C_6H_5OH + CH_3COCH_3$$

Phenol Propanone

The economics of this process are very favourable. The raw materials are air, benzene and propene—the last two are plentiful products of the petroleum industry. Two saleable products are produced, as propanone is in demand as a solvent and in the manufacture of bisphenol A for the manufacture of epoxy and polycarbonate resins (see box, **section 32.9.3**). In the manufacture of bisphenol A, phenol is reacted with propanone. The cumene process produces one mole of propanone to every mole of phenol—approximately 6 tonnes of propanone for 10 tonnes of phenol.

29.13 Summary

- Arenes (or aromatic compounds) are derivatives of benzene, C_6H_6.
- Benzene has a flat regular hexagon of carbon atoms linked by σ bonds. The p orbitals overlap to form a delocalized π system:

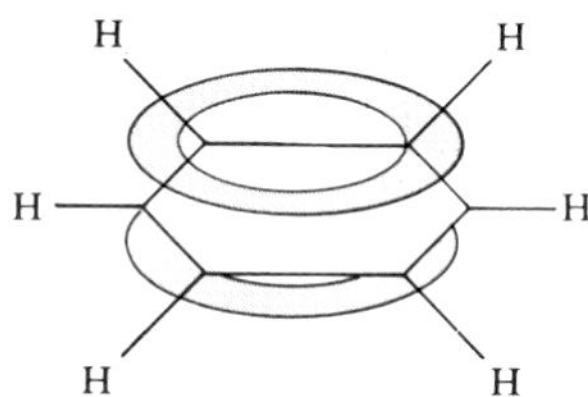

- All the C—C bond lengths in benzene are the same.
- The bonds are intermediate between single and double bonds.
- The delocalized π bonds give the ring a high electron density.
- Thermochemical cycles indicate that benzene is approximately 150 kJ mol^{-1} more stable than expected. This is called its aromatic stabilization energy.
- Aromatic hydrocarbons are typical of non-polar molecules in their boiling temperatures and solubility.
- The mass spectra of arenes frequently contain a peak at mass number 77 due to the $C_6H_5^+$ fragment.
- The reactions of benzene are summarized in **Figure 29.13**.
- Typically arenes react by electrophilic substitution reactions.
- Substituents on a benzene ring may withdraw electrons and make the ring less reactive or release electrons and make it more reactive to attack by electrophiles.
- Electron-releasing groups direct further substituents to the 2, 4 and 6 positions. Electron-withdrawing groups direct further substituents to the 3 and 5 positions.
- Halogens are exceptions. They withdraw electrons but direct to the 2, 4 and 6 positions.
- Alkyl side chains are more susceptible to radical substitution reactions than the benzene ring.
- Oxidation of alkyl side chains results in the side chain becoming an aldehyde or carboxylic acid group, depending on the strength of the oxidizing agent.
- Benzene is the precursor of many economically important compounds, notably plastics and fibres.

igure 29.13 Summary of the eactions of benzene

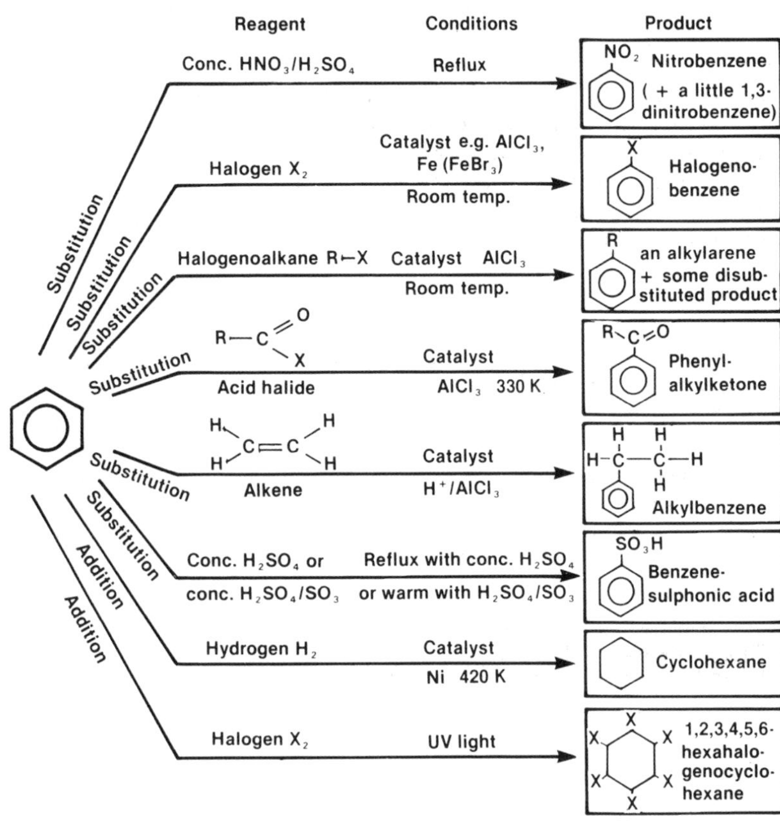

29.14 Questions

1 Name the following:

A) Br, Br

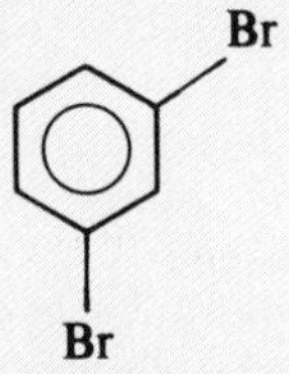

B) $CH_2CH_2CH_3$

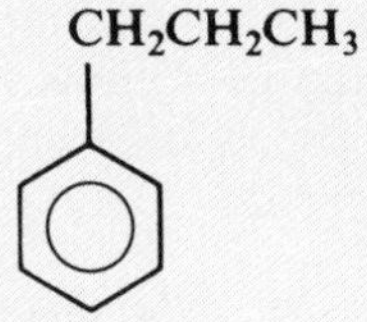

C) O=C—H, Cl

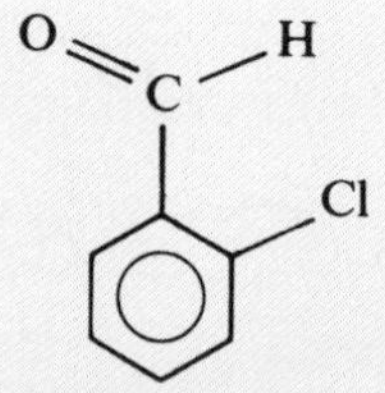

D) H, H, H, C=C—C—H, H

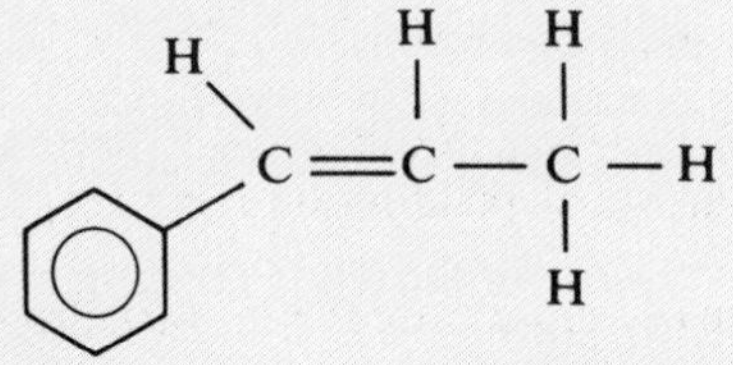

2 Write displayed structural formulae for:

A) 2-chloromethylbenzene;

B) 4-chloromethylbenzene;

C) benzene-1, 4-dicarboxylic acid;

D) 2-phenylprop-1-ene.

3 Three hydrocarbons, X, Y, Z had the following reactions:

X burned with a smoky flame, reacted with concentrated sulphuric acid and with bromine on exposure to UV light, producing hydrogen bromide and an organic product.

Y burned with a smoky flame, reacted with concentrated sulphuric acid and reacted with bromine solution in the dark.

Z burned with a relatively smokeless flame, had no reaction with concentrated sulphuric acid and reacted with bromine only on exposure to UV light producing hydrogen bromide as well as an organic product.

A) Classify X, Y, Z as an alkane, alkene or arene. Explain your reasoning.

B) Identify the type of organic compound produced in the reaction with sulphuric acid in each case.

4 Give the formula of the product of each of the following reactions. Bear in mind the type of product and the position of any substituents on the ring. Assume only monosubstitution takes place.

A) 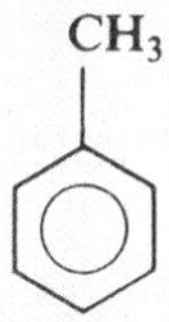 CH_3

$+HNO_3/H_2SO_4$;

B) 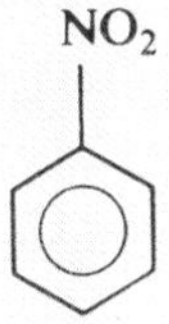 NO_2

$+HNO_3/H_2SO_4$;

C) 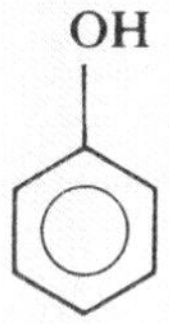 OH

$+Br_2/Fe$;

D) 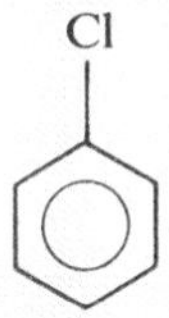 Cl

$+H_2SO_4/SO_3$.

5 'Benzene has a delocalized system of electrons. This makes it more stable than expected.'

A) What is meant by this statement?

B) Summarize the evidence which supports the statement. Use the following headings:

i chemical reactivity;
ii thermochemical;
iii X-ray diffraction.

6 Br

reaction A ← → fuming H_2SO_4 compound C

concentrated HNO_3 | concentrated H_2SO_4

compound B

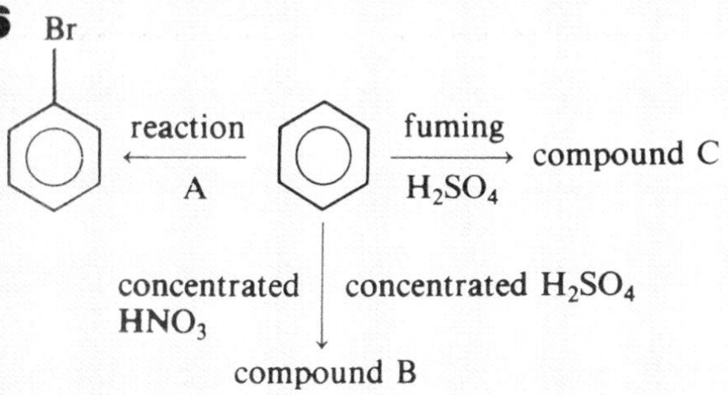

A) State the reagent and conditions required for reaction A.

B) Give the name and structural formula for compound B.

C) Give the name and structural formula for compound C.

(Nuffield 1983)

7

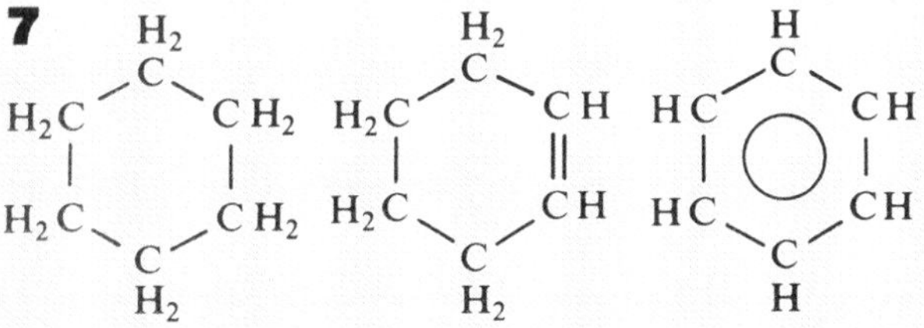

cyclohexane cyclohexene benzene

Write an account of the chemical reactions of the three hydrocarbons: cyclohexane, cyclohexene and benzene. Give equations wherever possible, and point out important similarities and differences in their reactions. How may the differences be accounted for in terms of the bonding in these molecules?

(Nuffield 1986)

8 Methyl benzene, 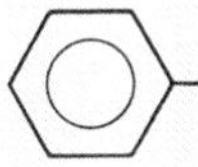 $-CH_3$, can be nitrated using a mixture of nitric acid and sulphuric acid. In an attempt to find out what nitration products result, an experiment was conducted as follows:

'10 drops of concentrated sulphuric acid were carefully added to 10 drops of concentrated nitric acid in a test tube, while shaking the mixture and cooling the test tube under a stream of cold water. The mixture of acids was then added to 5 drops of methylbenzene in another test tube, again shaking this mixture under a stream of cold water. The contents of the beaker were transferred to a separating funnel and the organic layer separated off.'

A) When methylbenzene is nitrated:

i What type of reaction (addition, elimination, etc.) is said to take place?

ii What class of reagent (electrophile, nucleophile, etc.) attacks methylbenzene?

iii What nitrogen-containing ion is believed to be involved in the actual nitration stage of the reaction?

iv Why is the mixture cooled?

B) Suggest briefly how the organic layer resulting from the nitration might be

i treated to remove residual acids; **ii** dried.

C) The methyl group in methylbenzene is said to direct incoming nitro groups to the 2 and 4 positions of the benzene ring. Using this information, draw structural formulae to show *three* of the possible products of nitration of methylbenzene.

D) Give the name of a technique suitable for the separation of the possible products of nitration of methylbenzene which could also serve to identify them.

E) How do the conditions for the nitration of phenol differ, if at all, from the conditions for the nitration of methylbenzene in respect of:

i temperature;
ii concentration of acids?

(Nuffield 1986)

9 Benzene undergoes a substitution reaction with chlorine according to the following equation:

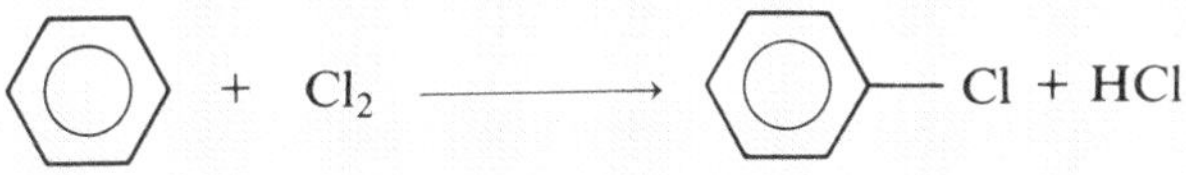

$$C_6H_6 + Cl_2 \longrightarrow C_6H_5Cl + HCl$$

This reaction occurs only in the presence of a substance such as iron filings. Iodine does not react directly with benzene under these conditions.

In contrast, benzene readily undergoes a substitution reaction with the interhalogen, iodine monochloride, ICl.

A) Suggest a reason for the ready reaction of iodine monochloride with benzene.

B) Write an equation for the reaction of benzene with iodine monochloride.

(Nuffield 1981, part question)

10 This question is about three hydrocarbons whose formulae are:

A)

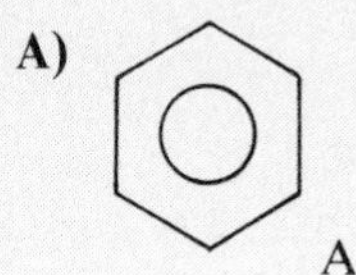

A

B)

CH_2

H_2C CH

| ‖

H_2C CH

CH_2 B

C)

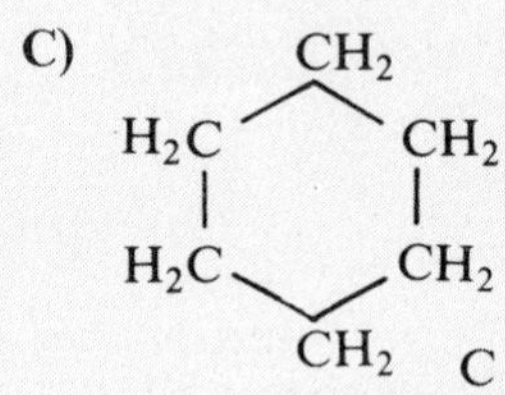

A) Give the names of the hydrocarbons B and C;

B) Each of the compounds will react with bromine in the ratio 1 mol compound: 1 mol Br_2. Copy and complete the following table:

Compound	Conditions for reacting with Br_2	Formula of organic product	Type of attack on organic compound (nucleophilic/ electrophilic/ free radical)
A			
B			
C			

(London 1980, part question)

11 A) Draw the full structural formula of E, 4-phenylbut-1-ene. Indicate on your diagram:

i one carbon–carbon bond which has length 0.154 nm;

ii one carbon–carbon bond which has length 0.139 nm;

iii one carbon–carbon bond which has length 0.133 nm.

B) Using your knowledge of simple alkenes and arenes, predict equations and give the structural formulae of the products obtained when E reacts with:

i hydrogen bromide;

ii bromine water;

iii hydrogen with a palladium or platinum catalyst at room temperature and pressure;

iv hydrogen with a nickel catalyst at high temperature and pressure.

C) Illustrate the meaning of the term optical isomerism, using one of the products of the reactions in part (**B**) above.

(Oxford 1987)

30 Halogenoalkanes and halogenoarenes

30.1 Introduction

Figure 30.1 PVC

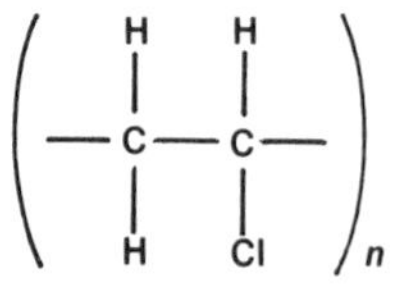

Figure 30.2 Chloroethene – VCM

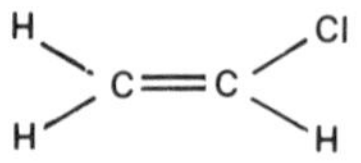

Figure 30.3 'Jenclean'

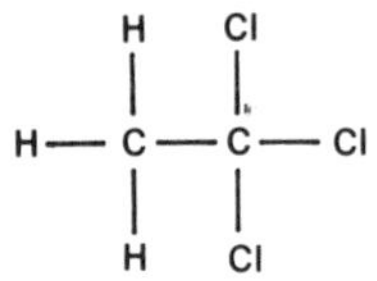

Figure 30.4 ptfe

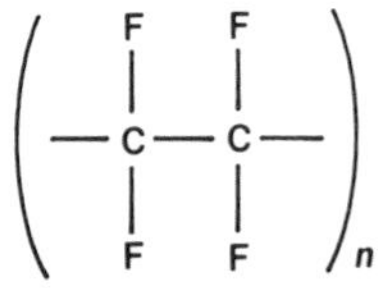

These two groups of compounds have one or more halogen atoms replacing hydrogens in an alkane or arene skeleton respectively. There are not many naturally occurring organohalogen compounds—one of the few is the hormone thyroxine (see box overleaf)—but they are among the most important groups of synthetic compounds. You will certainly be familiar with PVC (polyvinylchloride, systematic name poly(chloroethene), see Figure 30.1), the plastic used for making records, electrical cable insulation, floor tiles, etc. This plastic is made from the monomer chloroethene (industrially known as vinyl chloride monomer, VCM, see Figure 30.2), which has recently made the news as a suspected carcinogen after being used and considered safe for years. Another 'everyday' organohalogen compound is 1, 1, 1-trichloroethane, trade name 'Jenclean', see Figure 30.3, which is used as a dry cleaning solvent. The smell of recently dry-cleaned clothes is typical of halogenoalkanes. Halogenoalkane vapours also act as anaesthetics, so you should not sleep in a recently dry-cleaned sleeping bag or take home dry-cleaning in an unventilated car. Poly(tetrafluoroethene), ptfe, see Figure 30.4, is the non-stick coating on frying pans and is also used for low-friction bearings. Chemical inertness is vital for these applications, as it is for freons, which are chlorofluorocarbons used as cooling liquids in fridges and as aerosol propellants. However, in the last few years it has become clear that these volatile compounds are taking part in photochemical reactions in the atmosphere which are depleting the ozone layer and thus causing too much UV light to reach the earth's surface.

30.2 Formulae, naming and structure

Halogenoalkanes have one or more halogen (F, Cl, Br or I) atoms replacing hydrogen atoms while halogenoarenes have a halogen atom or atoms bonded *directly* to a benzene or other aromatic ring, **Figure 30.5.** This distinction is important as the halogenoalkanes and the halogenoarenes have significant differences in reactivity.

Figure 30.5

A halogeno*arene*

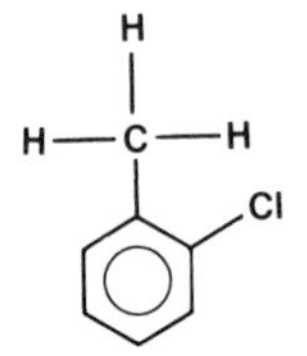

Chlorine attached directly to benzene ring

A halogeno*alkane*

Chlorine not attached directly to benzene ring

30.2.1 Naming

The prefixes fluoro-, chloro-, bromo- and iodo- are used, together with locants where necessary, to indicate the position of the halogen atom or atoms on the chain or ring.

The prefixes di-, tri-, tetra-, etc. are used as usual to indicate how many atoms of a particular halogen are present.

The prefix per- is used to indicate that all the hydrogen atoms of the parent compound have been replaced by a halogen.

A

THYROXINE

Thyroxine

Thyroxine is a hormone which controls the metabolic rate. It is secreted by the thyroid gland in the throat. Lack of it causes reduced physical and mental development (cretinism) while overproduction of thyroxine produces swelling of the thyroid gland (goitre) and hyperactivity.

The thyroid gland concentrates iodine, and uses it to produce thyroxine. This may include radioactive iodine isotopes if there is nuclear fallout. Radioactive iodine can be displaced if large quantities of non-radioactive iodine isotopes are taken. For this reason the government has stockpiled potassium iodide (KI) and potassium iodate (KIO_3) tablets for use in case of a radioactive fallout incident. These were available, but not needed in this country, at the time of the Chernobyl accident.

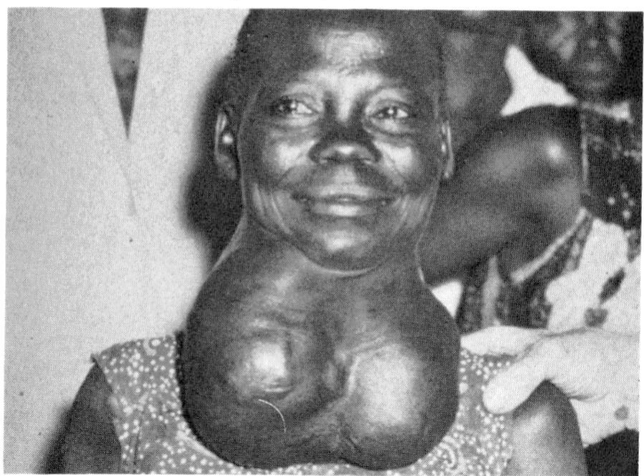

This condition is called hyperthyroidism and is caused by overproduction of thyroxine

Examples

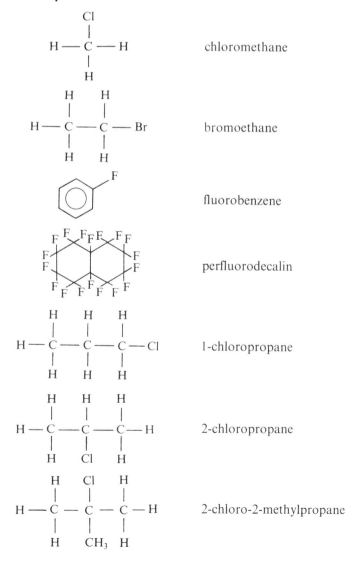

CH_3Cl	chloromethane
CH_3CH_2Br	bromoethane
C_6H_5F	fluorobenzene
$C_{10}F_{18}$	perfluorodecalin
$CH_3CH_2CH_2Cl$	1-chloropropane
$CH_3CHClCH_3$	2-chloropropane
$(CH_3)_3CCl$	2-chloro-2-methylpropane

A

ARTIFICIAL BLOOD

Recently a number of fluoro compounds such as perfluorodecalin, perfluorotripropylamine and perfluorobromooctane have been tried out as 'artificial blood' for use in transfusions, operations and accidents. These substitutes dissolve oxygen well (though not as well as real blood). They can be used irrespective of blood group and are used as a temporary measure only, the patient's own body eventually making sufficient real blood.

```
    Cl  Cl
    |   |
H — C — C — H        1,2-dichloroethane
    |   |
    H   H

    Cl  H
    |   |
Cl— C — C — H        1,1-dichloroethane
    |   |
    H   H
```

1,3,5-triiodobenzene

Primary halogenoalkane

Secondary halogenoalkane

Tertiary halogenoalkane

Halogenoalkanes may be:

primary, with a halogen at the end of the chain;

```
    H   H   H   H
    |   |   |   |
H — C — C — C — C — Br
    |   |   |   |
    H   H   H   H
```

secondary, with a halogen in the middle of the chain;

```
    H   H   Br  H
    |   |   |   |
H — C — C — C — C — H
    |   |   |   |
    H   H   H   H
```

tertiary, with a halogen at a branch in the chain.

```
        H
        |
    H — C — H
    H   |   H
    |   |   |
H — C — C — C — H
    |   |   |
    H   Br  H
```

All have the molecular formula C_4H_9Br, so they are isomers.

Try naming the compounds shown above and in the margin.

When a compound contains two or more different halogens the substituents are given in alphabetical order, so:

```
    H   I   Cl  H   H   H
    |   |   |   |   |   |
H — C — C — C — C — C — C — H
    |   |   |   |   |   |
    H   H   H   H   H   H
```

is 3-chloro-2-iodohexane *not* 2-iodo-3-chlorohexane.

When writing the formulae of halogen compounds, the symbol X i often used to represent any halogen.

30.3 Physical properties of organohalogen compounds

Table 30.1 Boiling temperatures of 1-halogenoalkanes/K

Parent alkane \ Halogen	F	Cl	Br	I
Methane	195*	249*	273*	315
Ethane	135*	285*	312	345
Propane	276*	320	332	363
Butane	306	351	374	403

Increasing T_b (across the halogens); Increasing T_b (down the alkanes)

* = gas at room temperature (298 K)

Although C—X bonds are polar $C^{\delta+}$—$X^{\delta-}$, they are not polar enoug to make the halogenoalkanes or halogenoarenes soluble in water. Eve those with the shortest hydrocarbon chains are immiscible. The mai intermolecular interaction is dipole–dipole bonding. They will, howeve dissolve readily in alcohols and ethers. They mix well with hydrocarbon hence their use as degreasing agents and dry-cleaning fluids. Boilin temperatures depend on the number and type of halogen atoms in th molecule. They have higher boiling temperatures than the correspondin alkanes due to their higher relative molecular masses and polarities, se **Table 30.1.** Note the trends:

increasing T_b with increased chain length;

increasing T_b as we go down the halogen group.

30.4 Infra-red spectra of organohalogen compounds

Table 30.2 Infra-red peaks for carbon-halogen bonds

Bond	Wave number/cm^{-1}
C—F	1400–1000
C—Cl	800–600
C—Br	600–500
C—I	500

The main feature, other than the peaks observed in the alkanes, is th C—X stretch. This is seen in the ranges shown in **Table 30.2** for differen halogens.

Since most infra-red instruments operate in the range 600–4000 cm^{-1} only the C—Cl and C—F bonds are usually observed. The infra-re spectrum of 1-chlorobutane is shown in **Figure 30.6.**

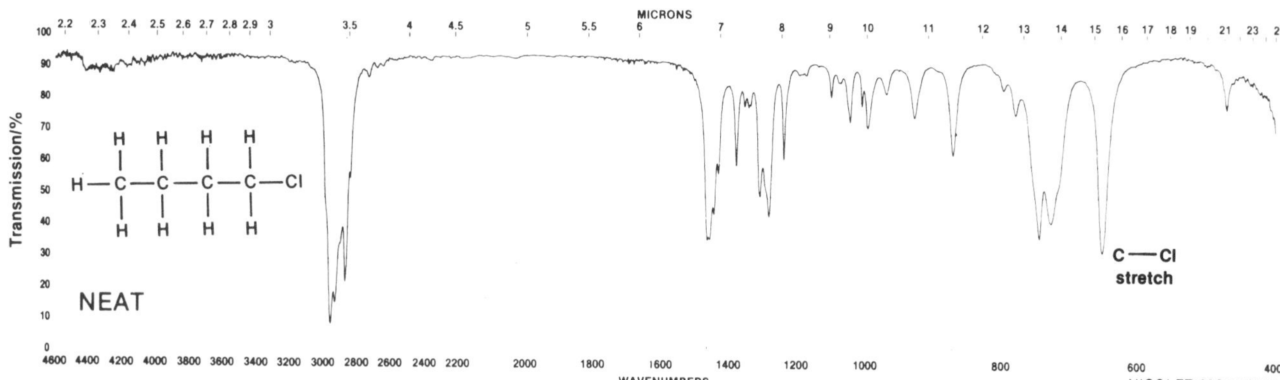

Figure 30.6 Infra-red spectrum of 1-chlorobutane

30.5 Mass spectra of organohalogen compounds

Part of the mass spectrum of 1-chlorobutane is shown in **Figure 30.7.** Remember that the mass spectrometer separates different isotopes. So the mass of the parent ion is not 92.5 (M_r for 1-chlorobutane, C_4H_9Cl). There are *two* parent ion peaks. One represents $C_4H_9{}^{35}Cl$ of mass 92 and the other

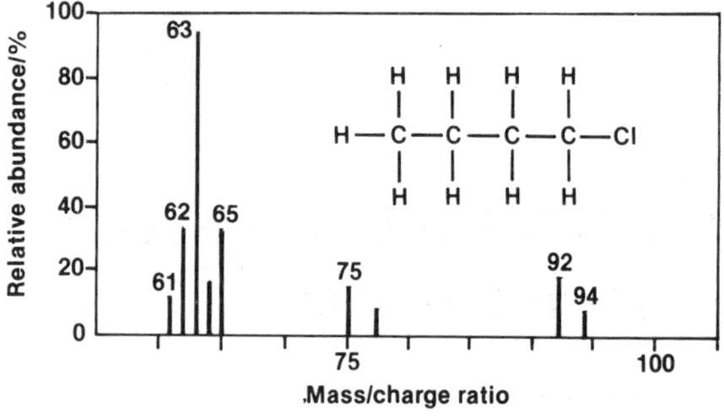

igure 30.7 Mass spectrum of ·chlorobutane. (Only the part of the pectrum above mass/charge ratio = 55 shown)

$C_4H_9{}^{37}Cl$ of mass 94. Notice that these peak heights are in the approximate ratio 3:1, the abundance ratio of ^{35}Cl:^{37}Cl. This feature of twin peaks separated by two mass units with heights in the ratio 3:1 is characteristic of the mass spectra of all chlorine compounds.

30.6 Reactivity of halogenoalkanes

Two important factors which govern the reactivity of the C—X bond are the C—X bond energy and the C—X bond polarity.

30.6.1 Bond energies

able 30.3 **Carbon-halogen bond nergies**

ond	Tabulated BE/kJ mol^{-1}
—F	467 ↑
—H	413]
—Cl	346 stronger
—Br	290
—I	228

These are listed in **Table 30.3.** The trend is due to the shared electrons being closer to the nucleus in fluorine than in iodine.

So bond energies alone would predict that iodo-compounds, with the weakest bonds, would be the most reactive, and fluoro-compounds, with the strongest, would be the least reactive.

30.6.2 Bond polarities

able 30.4 **The electronegativities of e halogens**

alogen	Electronegativity
	4.0
l	3.0
r	2.8
	2.5
	2.1]
	2.5]

Electronegativities are listed in **Table 30.4.** All the halogens are more electronegative than carbon, so the polarity will be $C^{\delta+}$—$X^{\delta-}$. The greater the difference between C and the halogen, the greater the δ^+ character of C. The $C^{\delta+}$ will be open to attack by **nucleophiles.**
Electronegativity would predict that fluoro-compounds, which have the greatest polarity, would be the most reactive, and iodo-compounds, with the least polarity, would be the least reactive.

So the two factors predict different outcomes. Which is the more important has been decided by experiment: it is the bond strength which governs the reaction rate. Iodoalkanes are more reactive than fluoroalkanes. But the polarity of the bonds will still predict the *nature* of the reagent that will attack. Halogenoalkanes are likely to be attacked by **nucleophiles** at the $C^{\delta+}$. As they are saturated, addition reactions are not possible. So we would expect **nucleophilic substitution** or **elimination** reactions.

30.7 Reactions of halogenoalkanes

30.7.1 Nucleophilic substitution

$:Nu^-$ is used to represent any negatively charged nucleophile. This will be the overall reaction:

$$R-\overset{\displaystyle H}{\underset{\displaystyle H}{|}}{\!\!\!C}^{\delta+}-X^{\delta-} \quad (:Nu^-) \longrightarrow R-\overset{\displaystyle H}{\underset{\displaystyle H}{|}}{\!\!\!C}-Nu \quad + \quad :X^-$$

However there are two ways this can happen, called S_N2 and S_N1.

S_N2

The nucleophile attaches itself to the $C^{\delta+}$ atom *at the same time* as th halogen leaves. There is a charged species which we can imagine as a intermediate step called an **activated complex**:

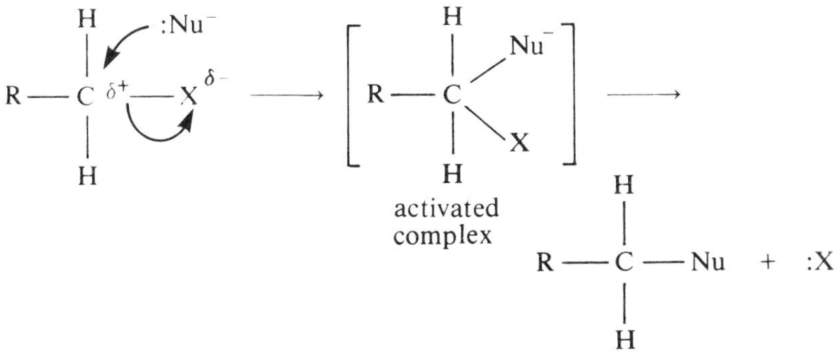

$$\mathrm{R{-}CH_2^{\delta+}{-}X^{\delta-}} + :\mathrm{Nu^-} \longrightarrow \left[\mathrm{R{-}CH_2(Nu^-)(X)}\right] \longrightarrow \mathrm{R{-}CH_2{-}Nu} + :\mathrm{X}$$

activated complex

The halogen leaves as a halide ion and is often called the **leaving group**

With this mechanism, the rate of reaction depends on the concentratio of the nucleophile *and* the concentration of the halogenoalkane.

It is a **substitution** by a **nucleophile** whose rate is determined by *tw* molecules—in short, S_N2.

S_N1

The halogen first leaves the halogenoalkane, leaving a positive ion (carbocation) which is then attacked by the nucleophile:

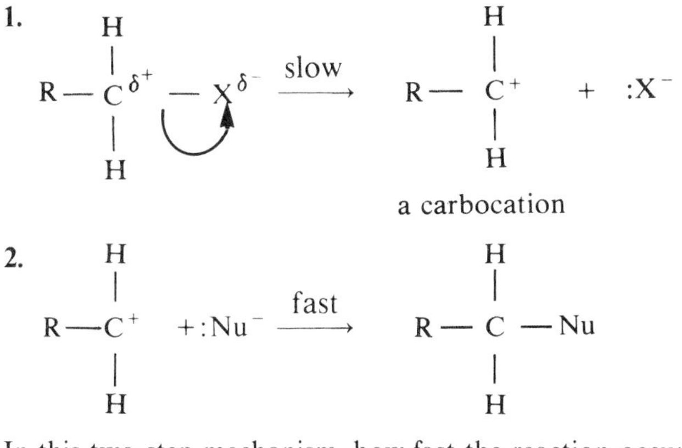

1. $$\mathrm{R{-}CH_2^{\delta+}{-}X^{\delta-}} \xrightarrow{\text{slow}} \mathrm{R{-}CH_2^{+}} + :\mathrm{X^-}$$

a carbocation

2. $$\mathrm{R{-}CH_2^{+}} + :\mathrm{Nu^-} \xrightarrow{\text{fast}} \mathrm{R{-}CH_2{-}Nu}$$

In this two-step mechanism, how fast the reaction occurs will depend o the rate of the slowest step, which is the first. We have **substitution** by **nucleophile** whose rate is determined by *one* molecule—in short, S_N1.

Increasing the concentration of the nucleophile will not increase th rate of reaction, but increasing the concentration of the halegenoalkan will.

Can we predict which sorts of halogenoalkanes will react by S_N1, S_N or perhaps by both?

For the S_N1 mechanism to be the most important, the carbocatio has to be relatively stable.

Look at the different halogenoalkanes:

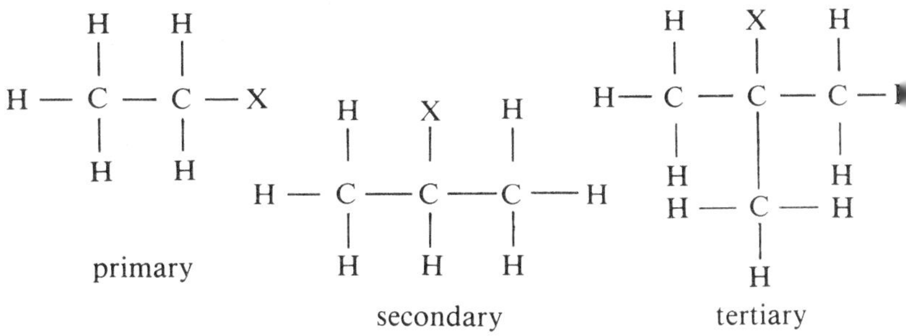

$\mathrm{H_3C{-}CH_2{-}X}$ primary

$\mathrm{H_3C{-}CHX{-}CH_3}$ secondary

$\mathrm{(CH_3)_3C{-}X}$ tertiary

E

GRIGNARD REAGENTS

Grignard reagents, RMgX, are formed by adding magnesium turnings to a solution of halogenoalkane in dry ether:

$$RX + Mg \longrightarrow RMgX$$

They are quite unstable but can be used to carry out a number of synthetically useful reactions as summarized below, using CH_3CH_2MgX as an example:

The $C^{\delta-}$—$Mg^{\delta+}$ bond in Grignard reagents is very polar (the electronegativities are C = 2.5, Mg = 1.2) and you may find it easier to remember their reactions if you think of them as 'ionic compounds', $R^- + MgX^+$, R^- acting as an excellent nucleophile.

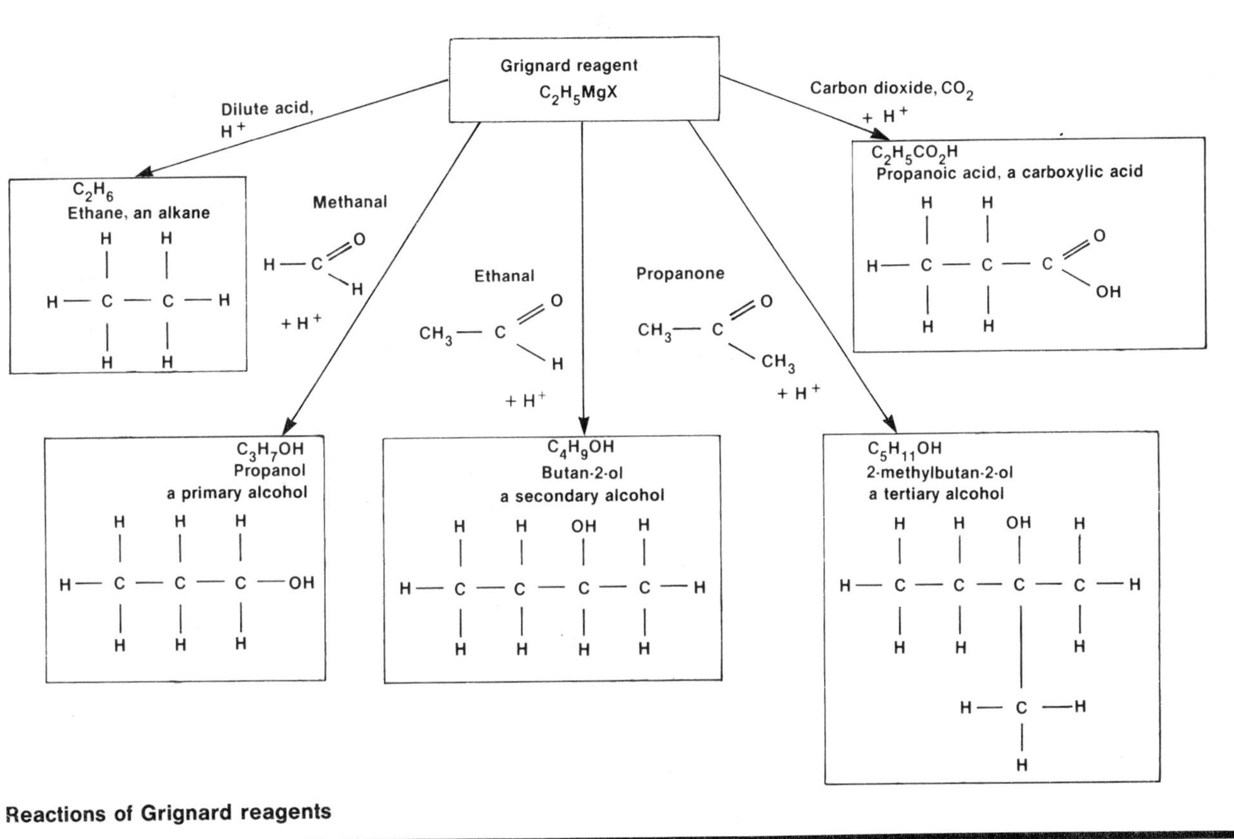

Reactions of Grignard reagents

Remember that alkyl groups release electrons (inductive effect, see **section 26.9.3**). This means that tertiary halogenoalkanes, which have three alkyl groups, give the most stable ion due to the inductive effect of the three R groups stabilizing the positive charge, and allowing the ion time to be attacked by the nucleophile. The next most stable would be the secondary, then the primary (not very stable at all).

This is in fact the case. Tertiary halogenoalkanes react via the S_N1, secondary by both the S_N1 and the S_N2, and primary by the S_N2 mechanism.

30.7.2 Experimental evidence for S_N1 and S_N2 mechanisms

The reaction of halogenobutanes with aqueous sodium hydroxide

If a few drops of 20% sodium hydroxide are added to a few drops of halogenobutane dissolved in ethanol (ethanol is a co-solvent in which both mix), and after a minute, nitric acid is added to neutralize any remaining sodium hydroxide, then a precipitate is formed on testing the resulting neutral solution with silver nitrate.

The results can be interpreted as follows. The OH^- ion is a goo nucleophile. The precipitate shows that halide ions are present:

$$AgNO_3(aq) + X^-(aq) \longrightarrow AgX(s) + NO_3^-(aq)$$

The halogen in the halogenobutane is covalently bonded to the carbo so the C—X bond must have broken. The carbon has bonded with th OH^-, so a substitution reaction has occurred:

$$R{-}X + OH^- \longrightarrow ROH + X^-$$

An alcohol is produced.

The reaction may go via S_N1, S_N2 or both, depending on the halogeno alkane:

primary secondary tertiary

S_N1 ⟶

⟵ S_N2

Also, the rate will depend on the halogen. Iodine is faster than bromir which is faster than chlorine because of the C—X bond strength Fluoro-compounds are unreactive due to the great strength of the C— bond.

Water is a weaker nucleophile than the OH^- ion but it will still rea slowly:

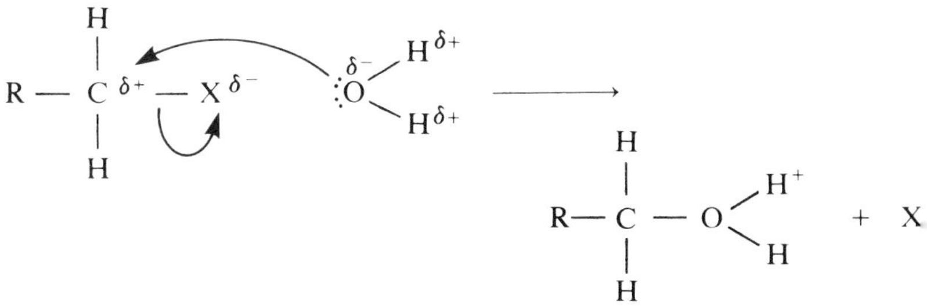

$$R{-}CH_2^{\delta+}{-}X^{\delta-} + \ddot{O}^{\delta-}(H^{\delta+})_2 \longrightarrow R{-}CH_2{-}O^{+}H_2 + X^-$$

The resulting ion must lose a proton (H^+) to form an alcohol:

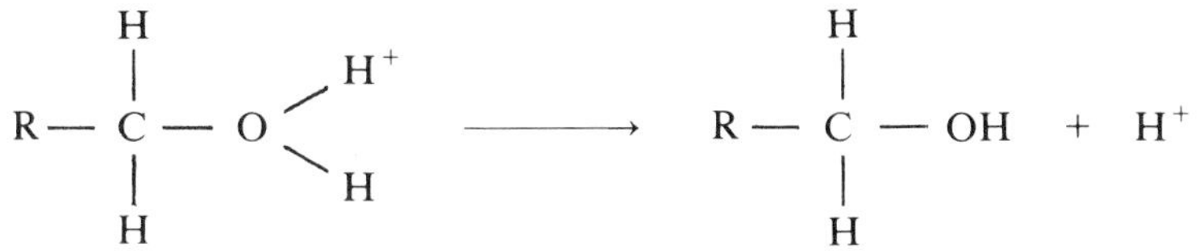

$$R{-}CH_2{-}O^{+}H_2 \longrightarrow R{-}CH_2{-}OH + H^+$$

30.7.3 Other nucleophilic substitution reactions

Halogenoalkanes can react with a variety of nucleophiles, e.g. H_2O, OH^- NH_3, RO^-, ^-CN, RCO_2^-.

Water (ethanol as a co-solvent)

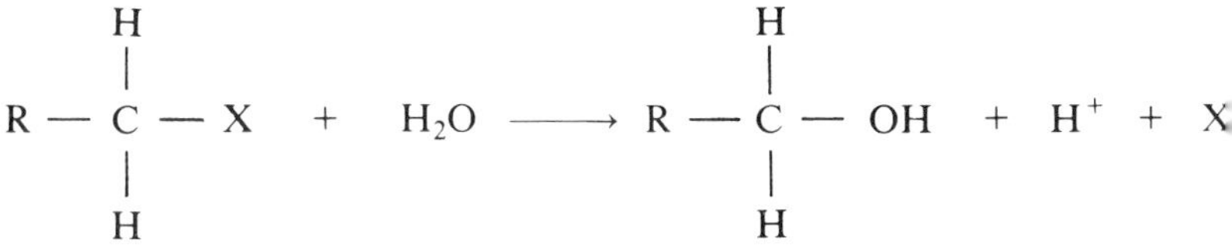

$$R{-}CH_2{-}X + H_2O \longrightarrow R{-}CH_2{-}OH + H^+ + X^-$$

an alcohol

Hydroxide (in aqueous solution)

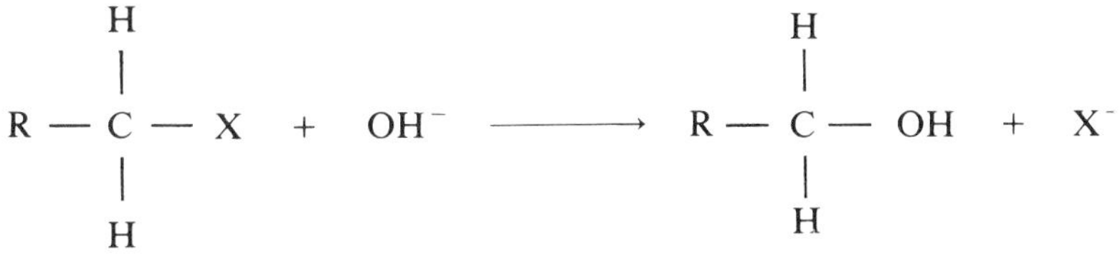

$$R{-}CH_2{-}X + OH^- \longrightarrow R{-}CH_2{-}OH + X^-$$

an alcohol

Ammonia (concentrated alcoholic solution, under pressure)

$$R-CH_2^{\delta+}-X^{\delta-} + :NH_3 \longrightarrow R-CH_2-\ddot{N}H_2 + X^- + H^+$$

a primary amine (also a nucleophile)

$$R-CH_2^{\delta+}-X^{\delta-} + H_2\ddot{N}-CH_2-R \longrightarrow R-CH_2-\ddot{N}H-CH_2-R + X^- + H^+$$

a secondary amine (also a nucleophile)

$$R-CH_2^{\delta+}-X^{\delta-} + H-\ddot{N}(CH_2R)-CH_2-R \longrightarrow R-CH_2-\ddot{N}(CH_2R)-CH_2-R + X^- + H^+$$

a tertiary amine (also a nucleophile)

$$R-CH_2^{\delta+}-X^{\delta-} \quad R-CH_2-\ddot{N}(CH_2R)-CH_2-R \longrightarrow R-CH_2-N(CH_2R)_2-CH_2-R + X^-$$

a quaternary ammonium salt:
no lone pair so not a nucleophile

Ammonia is a nucleophile due to its lone pair of electrons. Note t similarity of the first reaction with the one with water above. The prima amine which is produced also has a lone pair and is also a nucleophi If the halogenoalkane is in excess, secondary and tertiary amines can formed and also quaternary ammonium salts (see **section 34.2.5**).

Because of this, the reaction of halogenoalkanes with ammonia is n a very efficient way of preparing specific amines.

Cyanide ions (heat under reflux with potassium cyanide in ethanol)

$$\mathrm{R{-}\overset{\displaystyle H}{\underset{\displaystyle H}{|}}\!\!C^{\delta+}{-}X^{\delta-} + :\bar{C}{\equiv}N \longrightarrow R{-}\overset{\displaystyle H}{\underset{\displaystyle H}{|}}\!\!C{-}C{\equiv}N + X}$$

Note that the length of the carbon chain increases by one. For examp if we had started with bromo*ethane* we would have produced *propan* nitrile. For this reason this is an important reaction in synthesis, s Chapter 36. Nitriles can easily be converted into other compounds, f example they can be hydrolysed to carboxylic acids or reduced to amin (see **section 34.3.5**). If you are ever confronted with a problem in synthes where the target molecule has one more carbon atom than the starti molecule, this reaction is almost certain to be involved.

Ethoxide ion, $CH_3CH_2O^-$ (solution of sodium ethoxide in ethanol

$$\mathrm{R{-}\overset{\displaystyle H}{\underset{\displaystyle H}{|}}\!\!C^{\delta+}{-}X^{\delta-} + \ddot{:}OC_2H_5 \longrightarrow R{-}\overset{\displaystyle H}{\underset{\displaystyle H}{|}}\!\!C{-}OC_2H_5 + X}$$

an ether

The same reaction takes place with ethanol instead of sodium ethoxi but more slowly as ethanol is a poorer nucleophile than ethoxide as has no negative charge, in the same way as water is a poorer nucleoph than hydroxide.

All the above reactions are nucleophilic substitutions and occur v one or both of the S_N1 or S_N2 mechanisms.

Neutral nucleophiles like H_2O, NH_3 or CH_3CH_2OH will need to lo a proton (H^+) to form a neutral product.

All these reactions are essentially the same and you will almost certain find it easier to remember the basic pattern and work out the produ with a particular nucleophile rather than trying to recall a list of separa reactions.

30.7.4 Elimination reactions

2-chloro-2-methylpropane, $\mathrm{H{-}\overset{\displaystyle H}{\underset{\displaystyle H}{|}}\!\!C{-}\overset{\displaystyle Cl}{\underset{\displaystyle H{-}C{-}H \atop \displaystyle H}{|}}\!\!C{-}\overset{\displaystyle H}{\underset{\displaystyle H}{|}}\!\!C{-}H}$,

E

BASES AND NUCLEOPHILES

The term 'nucleophile' is used in organic chemistry for a species which attacks and forms a bond with an electron-deficient carbon atom $C^{\delta+}$. Thus a nucleophile must have a lone pair of electrons. A base, by the Lowry–Brönsted theory (see **section 12.2**) reacts with a proton, H^+, so it, too, needs a lone pair of electrons and indeed the Lewis definition of a base (see **section 12.2**) is that it is a lone pair donor. So really nucleophiles and bases are very similar. Both are electron pair donors; they differ only in what they donate their lone pair to. Not surprisingly, good bases are generally good nucleophiles. In inorganic chemistry the idea of a ligand is used. This is a species which donates a lone pair to form a bond with a transition metal ion. Good ligands tend also to be good bases and good nucleophiles. Ammonia is an example.

gure 30.8 **Elimination of hydrogen loride from 2-chloro-2-methylpropane**

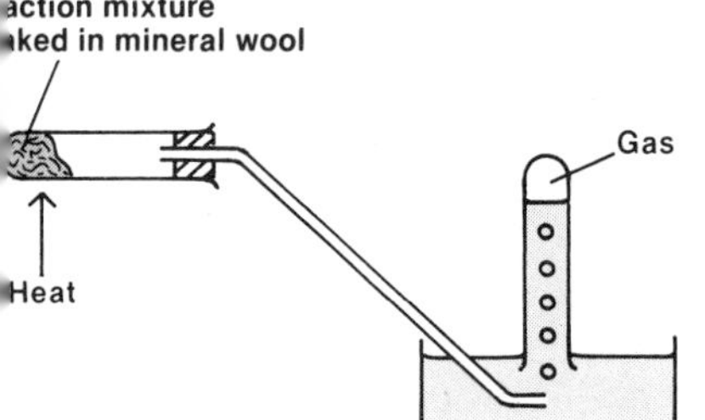

reacts with potassium hydroxide in ethanol. A mixture of 2-chloro-2-methylpropane and a 20% solution of potassium hydroxide in ethanol is soaked in mineral wool and heated as shown in **Figure 30.8.** The gaseous product is found to burn and also to decolorize bromine water, suggesting it is an alkene. The results can be interpreted as follows. The only likely alkene is 2-methylpropene,

```
      H
      |             H
      |           /
H  —  C  —  C  =  C
      |     |      \
      |     |        H
      H  H—C—H
            |
            H
```

indicating that HCl has been eliminated:

```
                         HCl
      H    (Cl    H)
      |      |    |
                                 -HCl
H  —  C  —  C  —  C  — H + KOH  ------>
      |     |     |
      H     |     H
        H — C — H
            |
            H

                            H
                            |          H
                            |        /
                      H  —  C  —  C = C         + KCl + H2O
                            |     |    \
                            |     |      H
                            H     |
                             H —  C — H
                                  |
                                  H
```

Earlier we saw the OH^- ion acting as a nucleophile (the lone pair forming a new bond with $C^{\delta+}$). Here it acts as a *base*, the lone pair forming a new bond with H^+. The reagent (OH^-) is the same as in the substitution reaction but the conditions are different (pure ethanol rather than aqueous ethanol and high temperature rather than room temperature).

```
      H    Cl     H
      |     |     |
H  —  C  —  C  —  C  —  H   +   :OH⁻  ------>
      |     |     |
      H  H—C—H    H
            |
            H

                          H
                          |            H
                          |          /
                    H  —  C  —  C = C         + H2O + Cl⁻
                          |     |    \
                          |     |      H
                          H  H—C—H
                                |
                                H
```

This reaction is a useful way of introducing double bonds into a molecule.

Since the reagents are the same, there is competition betwe substitution and elimination:

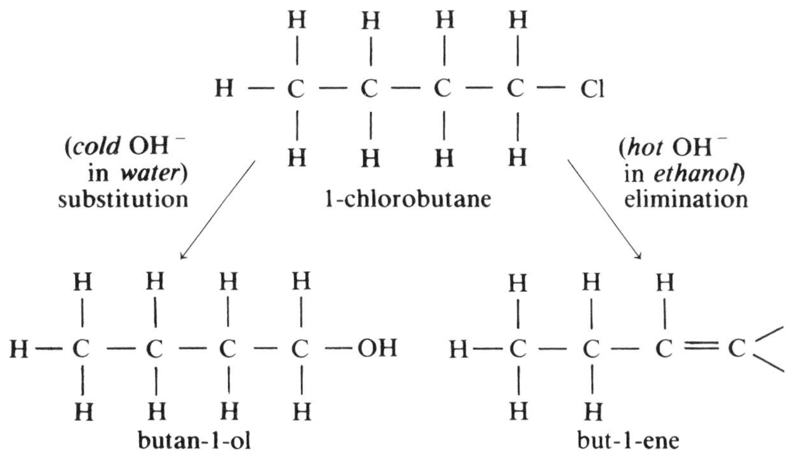

Which type of reaction predominates depends on the type of haloge oalkane (as well as the conditions) as shown below.

primary — secondary — tertiary

elimination →

← substitution

In some cases a mixture of isomeric elimination products is possible:

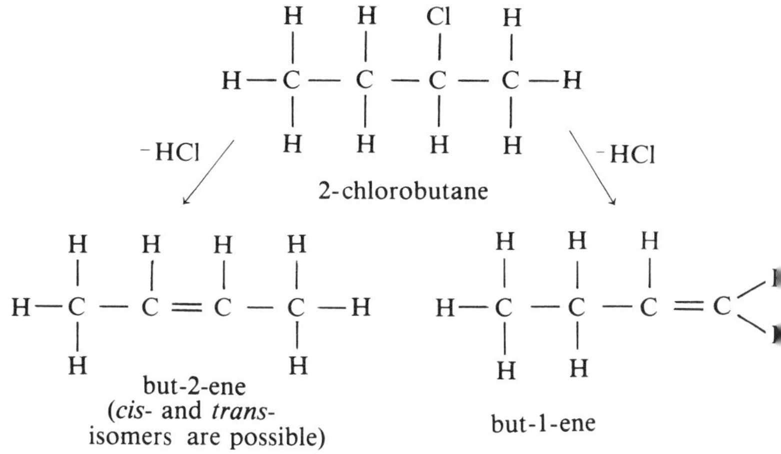

2-chlorobutane

–HCl ↙ ↘ –HCl

but-2-ene (*cis*- and *trans*- isomers are possible)

but-1-ene

30.8 Preparation of halogenoalkanes

30.8.1 By addition of H—X across a carbon–carbon double bond

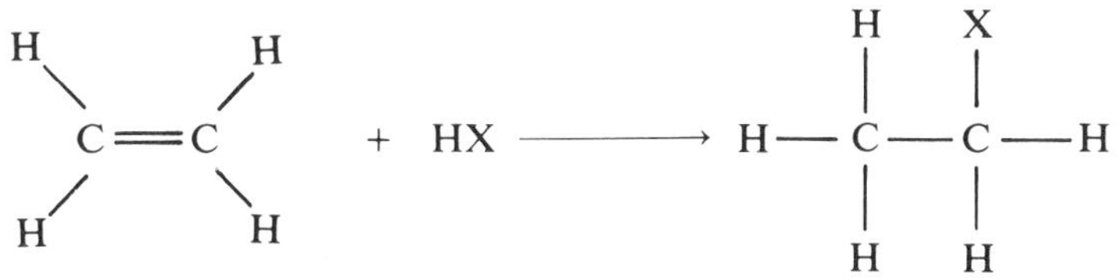

$$H_2C{=}CH_2 + HX \longrightarrow H{-}CH_2{-}CH_2X$$

Addition of X—X produces dihalogenoalkanes:

$$CH_2{=}CH_2 + X_2 \longrightarrow H-\underset{H}{\overset{X}{C}}-\underset{H}{\overset{X}{C}}-H$$

See **section 28.6.2.**

30.8.2 From alcohols

We can use (1) a hydrogen halide, H—X, (2) a phosphorus halide (PX_3 or PX_5), or (3) sulphur dichloride oxide (thionyl chloride, $SOCl_2$), see **section 31.4.4.**

1. $$R{-}OH + \begin{Bmatrix} HCl \\ HBr \end{Bmatrix} \longrightarrow \begin{Bmatrix} R{-}Cl \\ R{-}Br \end{Bmatrix} + H_2O$$

HCl or HBr is usually generated in the reaction vessel by the reaction of H_2SO_4 with NaCl or NaBr:

$$H_2SO_4(l) + 2NaBr(s) \longrightarrow Na_2SO_4(s) + 2HBr(g)$$

So the alcohol is refluxed with $NaBr/H_2SO_4$, and the halogenoalkane separated by distillation.

2. $$ROH + PCl_5 \longrightarrow RCl + POCl_3 + HCl$$

or

$$3ROH + PI_3 \longrightarrow 3RI + H_3PO_3$$

(Phosphorus triiodide is usually made in the reaction vessel by mixing iodine and red phosphorus.)

3. $$ROH(l) + SOCl_2(l) \longrightarrow RCl(l) + SO_2(g) + HCl(g)$$

Here, as both the other products are gases, the problem of separating the RCl is reduced.

30.9 Reactivity of halogenoarenes

gure 30.9 **Bonding in chlorobenzene**

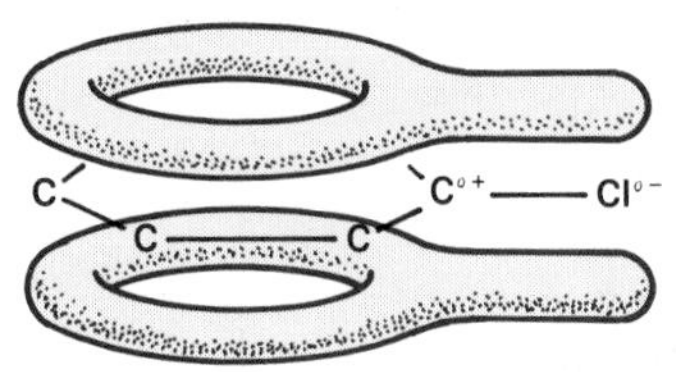

Remember that arenes are benzene type rings which have an aromatic π system. In the halogenoarenes, the C—X bond is shorter and thus stronger than in the halogenoalkanes (e.g. C—Cl in chloromethane is 0.177 nm, and C—Cl in chlorobenzene is 0.169 nm). This is due to the overlap of p orbitals on the Cl atom with the aromatic system in benzene providing extra bonding of π type as well as the C—Cl σ bond, **Figure 30.9.** The same thing happens in aryl amines and phenols (see **sections 34.2.4** and **31.6**). This overlap also reduces the δ^+ character of the carbon to which the halogen is bonded. So there are two effects:

1. a stronger bond;

2. less positive charge on the carbon.

Both effects suggest that halogenoarenes will be less reactive than halogenoalkanes. This is what is found in practice. For example, chlorobenzene will undergo nucleophilic substitution with aqueous sodium hydroxide only at 600 K and 2×10^4 kPa pressure:

$$C_6H_5Cl + OH^- \xrightarrow[\text{high pressure}]{\text{high temp.}} C_6H_5OH + Cl^-$$

chlorobenzene   phenol

Compare this with chloroalkanes which will react readily at room temperature and pressure.

Elimination reactions do not occur. For these to occur would involve disruption of the π system of the aromatic ring and this is most unlikel The aromatic ring will react in the same way as in benzene itself—typical by electrophilic substitution reactions. However, the attachment of electronegative halogen atom draws electrons away from the ring a makes it less reactive to electrophiles.

Unusually for deactivating substituents, the halogens are 2- a 4-directing. For example:

Cl — conc. HNO_3/ H_2SO_4 → Cl, NO_2 and Cl, NO_2

chlorobenzene — 1-chloro-2-nitrobenzene — 1-chloro-4-nitrobenzen

Note that compounds like (chloromethyl)benzene,

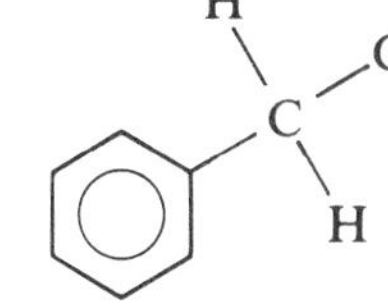

behave like halogeno*alkanes*, not halogenoarenes as the halogen is n bonded directly to the arene ring. So there is no overlap of electrons fro the halogen with the π system of the ring.

30.10 Preparation of halogenoarenes

1. Directly from the arene plus halogen with a catalyst (see **section 29.8.**: for example:

$+ Br_2$ —Fe catalyst→ Br $+$ HBr

2. From a diazonium salt (see **section 34.2.5**), warm with copper(I) halid

$N_2^+Cl^-$ —CuCl, HCl (conc.)→ Cl $+ N_2$

or warm with potassium iodide:

$N_2^+Cl^-$ $+$ KI → I $+$ KCl $+$ N_2

30.11 Halogenoalkenes

These contain both a carbon–carbon double bond and a halogen ato There are two possibilities:

1. The halogen may be bonded to one of the carbon atoms involved the double bond:

```
      H   H
      |   |        Cl
                  /
H — C — C = C              1-chloroprop-1-ene
      |           \
                   H
      H
```

2. The halogen may be bonded to a carbon not involved in the double bond:

```
      Cl  H
      |   |        H
                  /
H — C — C = C              1-chloroprop-2-ene
      |           \
                   H
      H
```

igure 30.10 **Bonding in chloroethene**

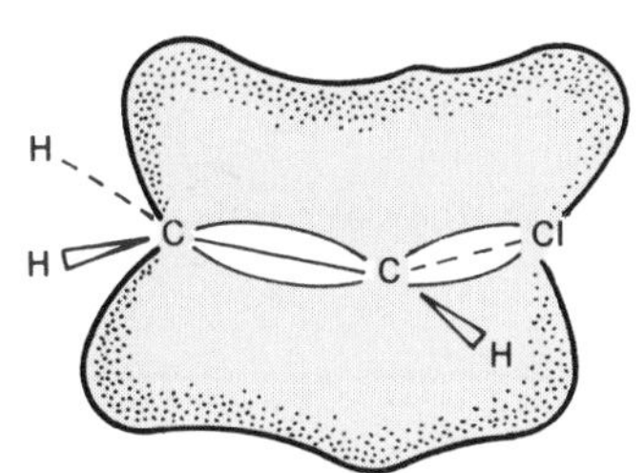

If the halogen atom is bonded to a carbon atom not involved in the double bond, the reactions are similar to those of halogenoalkanes plus those of alkenes.

If the halogen is attached to one of the carbon atoms in the double bond, the C—X bond is less reactive. This is due to overlap of full p orbitals on the halogen with the π orbitals of the double bond, see **Figure 30.10.** This is very similar to the situation in halogenoarenes (see above). It results in increased C—X bond strength and reduced δ^+ character of the carbon. Both tend to reduce reactivity.

30.12 Economic importance of organohalogen compounds

roducts made of PVC

Organohalogen compounds have very considerable economic, social and environmental implications. A number of these are mentioned in the introduction, including PVC plastics, dry-cleaning solvents, ptfe non-stick coatings and chlorofluorocarbons used for aerosol propellants, refrigerants and 'blowing agents' used to produce foam packaging for take-away food cartons, etc. (see box *Expanded polystyrene,* **section 29.10.3**). In all these uses the chlorofluorocarbons, or 'freons' as they are known, inevitably end up in the atmosphere when the aerosol is sprayed, the fridge scrapped or the food container incinerated. In the atmosphere the freons eventually decompose to chlorine atoms under the action of UV light. These act as catalysts in a series of complex reactions which ultimately lead to destruction of ozone, see box *Atmospheric chemistry*, **section 15.4.** This gas is normally present in the atmosphere and absorbs UV light. Its absence would cause a host of problems ranging from death of plankton in the sea to a greater incidence of skin cancer. Substitutes have now been developed to allow freons to be gradually phased out.

igure 30.11 **Fluothane. Can you see ow the systematic name fits the rules?**

```
    F   Cl
    |   |
F — C — C — Br
    |   |
    F   H
```

Another important halogenoalkane is fluothane, 1-bromo-1-chloro-2, 2, 2-trifluoroethane, **Figure 30.11**, which is an anaesthetic gas which has been used in over 500 million operations, see box *Anaesthetics,* **section 26.1.**

igure 30.12 **DDT – dichlorophenyl-ichloroethane – both a halogenoalkane nd a halogenoarene**

```
              Cl
              |
         Cl — C — Cl
              |
Cl —⟨◯⟩— C —⟨◯⟩— Cl
              |
              H
```

The insecticide DDT, **Figure 30.12**, is both a halogenoalkane and a halogenoarene. It initially brought massive benefits by the control of insect-borne diseases such as malaria. However, it has also caused great problems by getting into the food chain, causing the death of birds and small mammals. It is another example of the pros and cons of using chemicals in the environment.

PVC is one of the most widely used plastics and also one of the most versatile, in that its properties can be modified by various additives, so that it can be made flexible enough for rainwear and rigid enough for drainpipes. The plastic is made by polymerizing chloroethene which made from ethene, a product of cracking petroleum, as follows:

$$\underset{\text{ethene}}{H_2C{=}CH_2} + \underset{\text{chlorine}}{Cl_2} \longrightarrow \underset{\text{1,2-dichlorethane}}{H{-}CHCl{-}CHCl{-}H}$$

$$\underset{\text{1,2-dichloroethane}}{H{-}CHCl{-}CHCl{-}H} \xrightarrow{\text{heat}} \underset{\text{chloroethene}}{H_2C{=}CHCl} + \underset{\text{hydrogen chloride}}{HCl}$$

Halogen-containing organic compounds (usually chloro-compounds a chlorine is the cheapest halogen) are used industrially as solvents and als as intermediates in reactions. This is because they are both easily mad and easily converted into other materials.

30.13 Summary

- Halogenoalkanes have a halogen atom attached to an alkyl group.
- Halogenoarenes have a halogen atom bonded *directly* to an aromati ring.
- Halogenoalkanes may be classified as primary, secondary or tertiary if they have respectively one, two or three R groups attached to the carbon to which the halogen is bonded.
- Halogen-containing compounds have higher boiling temperatures than their parent alkane or arene due to their increased relative molecular mass and polarity.
- Halogenoalkanes and halogenoarenes do not dissolve in water, but mix with alkanes and other non-polar solvents.
- The $C^{\delta+}{-}X^{\delta-}$ bond is polarized as shown and therefore nucleophiles attack the $C^{\delta+}$.
- The typical reactions of halogenoalkanes are nucleophilic substitutions

$$R{-}X + {:}Nu^- \longrightarrow R{-}Nu + X^-$$

- These may occur by S_N1 or S_N2 mechanisms:

primary — secondary — tertiary
S_N1 increases →
← S_N2 increases

- Elimination reactions can also occur with loss of HX and formation of an alkene:

$$R{-}CH_2{-}CH_2X \longrightarrow RHC{=}CH_2 + HX$$

- Substitution and elimination reactions are often in competition:

primary — secondary — tertiary
elimination →
← substitution

- Halogenoarenes are less reactive than halogenoalkanes as the C—X bond has some double bond character and the carbon atom has less C^{δ} character.
- The halogen atom of a halogenoarene deactivates the aromatic ring

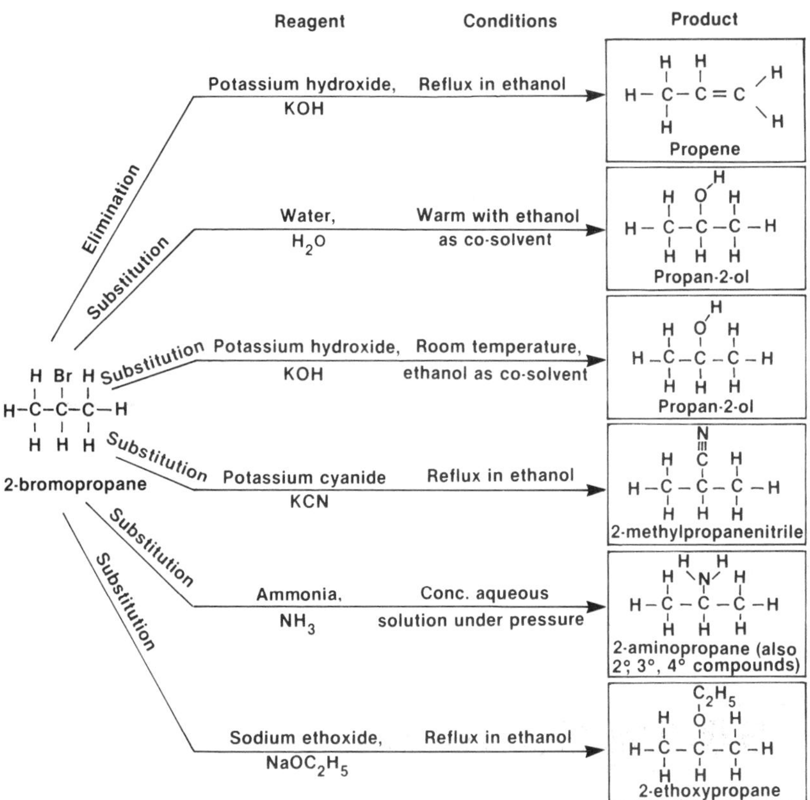

igure 30.13 Summary of the eactions of halogenoalkanes using -bromopropane as an example

but directs further substituents to the 2 and 4 positions.

- Elimination reactions do not occur with halogenoarenes.
- The reactions of halogenoalkanes are summarized in **Figure 30.13.**

30.14 Questions

1 Give the systematic names of the following compounds:

A)

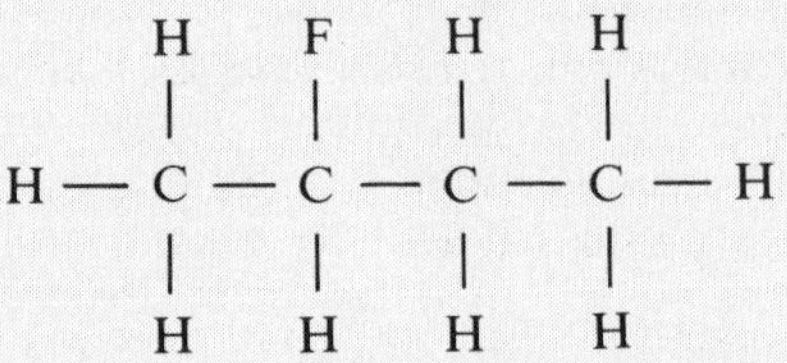

```
      H   F   H   H
      |   |   |   |
  H — C — C — C — C — H
      |   |   |   |
      H   H   H   H
```

B) (benzene ring with Br)

C) (benzene ring with Br on adjacent carbons: Br, Br)

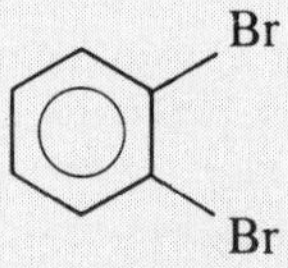

D)

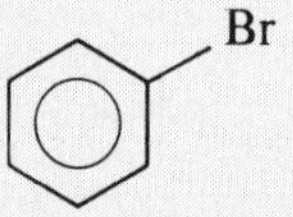

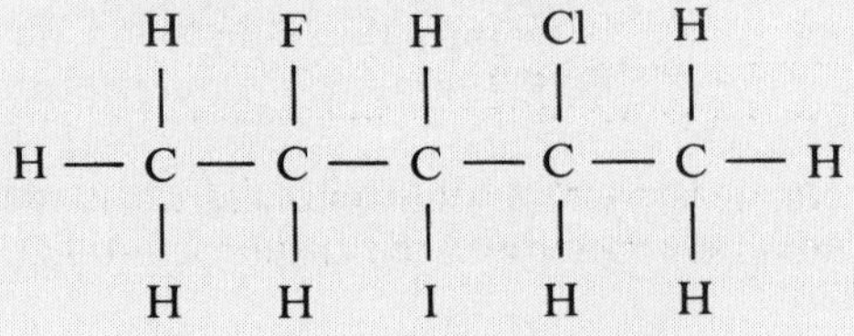

```
      H   F   H   Cl  H
      |   |   |   |   |
  H — C — C — C — C — C — H
      |   |   |   |   |
      H   H   I   H   H
```

E)

```
      H   Cl
      |   |      H
  H — C — C = C <
      |          H
      H
```

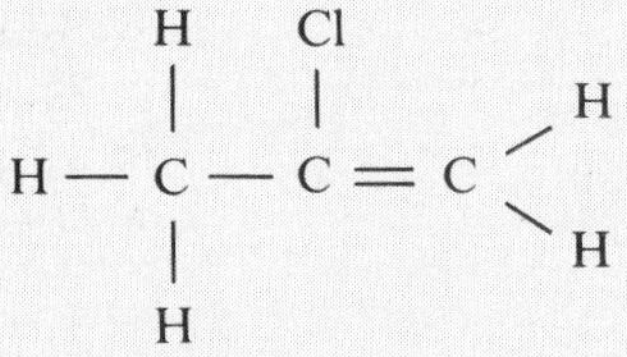

2 Write displayed structural formulae for:

A) 1, 1-dichloropropane
B) 2-bromo-1-chlorobutane
C) 1, 3, 5-trifluorobenzene
D) (chloromethyl)benzene
E) 1, 4-dichlorocyclohexane
F) 1-bromo-2-methylhexane

3 A) Classify the following reactions as addition, elimination or substitution:

i

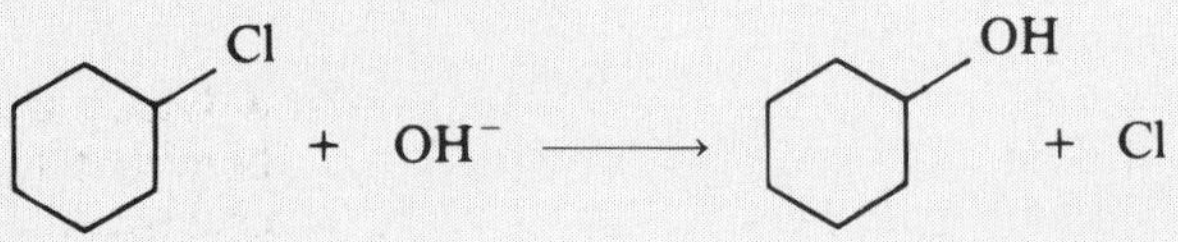

(cyclohexane)–Cl + OH^- ⟶ (cyclohexane)–OH + Cl

ii (cyclohexane ring)—Cl $+ OH^- \longrightarrow$ (cyclohexene ring) $+ H_2O + Cl^-$

iii

$$\underset{H}{\overset{H}{\diagdown}}C=C\underset{H}{\overset{Cl}{\diagup}} + Br_2 \longrightarrow H-\underset{Br}{\overset{H}{\underset{|}{\overset{|}{C}}}}-\underset{H}{\overset{Cl}{\underset{|}{\overset{|}{C}}}}-Br$$

B) (i) and (ii) use the same reagents. Describe the conditions required for each.

4 List the following compounds in order of their rate of reaction with aqueous OH^-. Try to explain the factors which governed your choice:

A) (benzene ring)—Cl **C)** (cyclohexane ring)—I

B) (cyclohexane ring)—Cl **D)** (cyclohexane ring)—Br

5 Mark the polarity δ^+ and δ^- on the following molecules:

A)
$$Cl-\underset{H}{\overset{H}{\underset{|}{\overset{|}{C}}}}-Cl$$

C)
$$H-\underset{H}{\overset{H}{\underset{|}{\overset{|}{C}}}}-Br$$

B) (benzene ring)—Cl

D)
$$H-\underset{H}{\overset{H}{\underset{|}{\overset{|}{C}}}}-Cl$$

Which carbon atom would have the greatest δ^+ character? Explain.

6 What substitution products are produced when 1-bromobutane reacts with the following nucleophiles?

A) OH^- **C)** C_2H_5OH
B) NH_3 **D)** $C_2H_5O^-$

Which reaction, with (C) or with (D), would go faster? Explain.

Would you expect a different rate if the isomer 2-bromo-2-methylpropane reacted with $C_2H_5O^-$? Explain.

7 The reaction of 1-bromobutane with aqueous sodium hydroxide is said to be a <u>bimolecular</u> <u>nucleophilic</u> <u>substitution</u>. Explain each of the underlined words.

8 When 2-chlorohexane is reacted with a strong base dissolved in ethanol at high temperature, a mixture of products containing different alkenes is formed. Chlorocyclohexane under the same conditions produces only one. Draw structures for the products of each reaction and explain why the first reaction gives a mixture of products and the second only one. What products would 3-chloropentane produce under the same conditions?

9 If you were given a sample of bromoethane, how would you convert it to 1,2-dibromoethane in two steps?

10 You are asked to prepare propanoic acid from ethanol. What reaction is likely to be involved at some stage? Try to outline the compounds which might be involved. (Do not give reagents or conditions.)

11

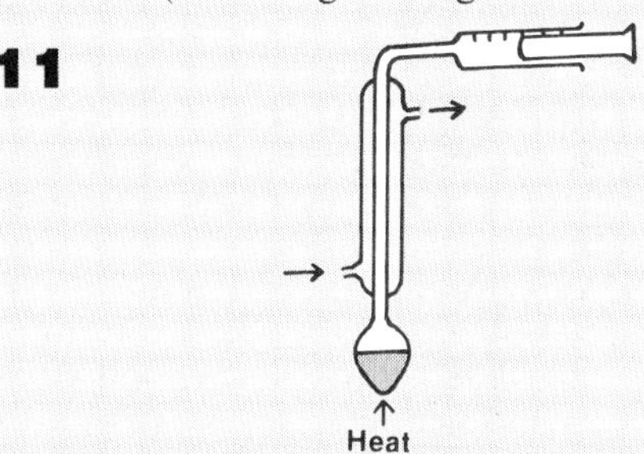

The apparatus above was used to study the reaction of 2-bromo-2-methylpropane with sodium hydroxide in ethanol solution. 0.27 g of 2-bromo-2-methylpropane was refluxed with excess alkali until no further volume change was seen on the syringe. The syringe reading was 45 cm^3. What two products are possible? Which of these would be collected in the syringe? How many moles of 2-bromo-2-methylpropane were used? How many moles of gaseous product were formed? What figure does this give for percentage elimination and percentage substitution under these conditions? What important precaution should be taken to ensure accuracy of the volume readings in this experiment? (Take the volume of 1 mole of gas to be 24 000 cm^3.)

12 Research/debate. Freons are being implicated in the destruction of the ozone layer. What is the importance of the ozone layer and what consequences will its destruction have for human beings? What other 'culprits' have been suggested? If you were in a position of power, what strategies would you propose to deal with this problem? Consider carefully the pros and cons of any suggestions you make. How could you put them over to a non-scientific person? Are there any uses of freons important enough for them to continue in use? How long would your reforms take to work? The exact part that freons play in the chemistry of the upper atmosphere is still uncertain. What difficulties are there in carrying out this sort of research?

31 Alcohols and phenols

31.1 Introduction

Alcohol is familiar to most people

Carbolic acid sprays used to be used as antiseptics

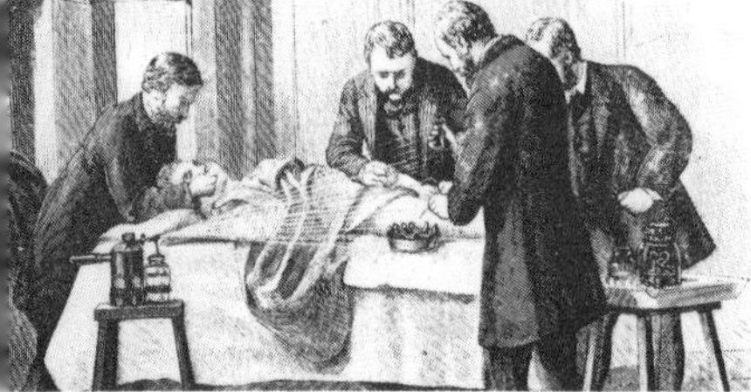

Both alcohols and phenols have the functional group —OH attached to a carbon skeleton. With alcohols the carbon skeleton is an alkyl group and with phenols the —OH is attached directly to an arene ring. We shall see that this has a considerable effect on the reactions of the —OH group.

Both alcohols and phenols are familiar in everyday life. Most people have drunk alcoholic drinks at some time and experienced the pleasurable and sometimes less than pleasurable effects of this particular drug. The brewing industry is certainly big business.

Phenols are familiar as the active ingredients in many disinfectants and antiseptics. They have been used in this way since the 1860s when Joseph Lister first used phenol itself (then called carbolic acid) to sterilize both wounds and instruments during surgery. He is believed to have discovered the value of phenol when he noticed that sailors who had amputations at sea suffered far fewer postoperative infections than amputees in hospitals. The reason was the nautical practice of dipping the stump into molten tar which contained phenol. Nowadays phenol is more important for the manufacture of phenolic plastics with trade names such as Bakelite.

A

PHENOL-METHANAL PLASTICS

Phenol and methanal (formaldehyde) react together with either an acid or an alkaline catalyst to form a polymer. Water is eliminated in the reaction which is therefore called **condensation polymerization**:

Further reaction produces **cross-links** between the chains as shown above right.

This produces a polymer called Bakelite, after its inventor, Baekeland. The plastic sets hard when it forms and cannot be melted again. It is therefore called a **thermosetting** polymer.

Bakelite is hard and rather brittle. However, it is cheap and is used for making electrical sockets, motor engine distributor caps and saucepan handles, where resistance to heat is important.

Can you spot the Bakelite?

31.2 Formulae, naming and structure

Both alcohols and phenols have one (or more) —OH group. In the case of alcohols this group is attached to an alkyl group and in the case of phenols it is attached directly to an aromatic ring. Phenylmethanol is an alcohol rather than a phenol as the —OH group is attached to an alkyl group (which happens to have an aromatic substituent), not *directly* to an aromatic ring.

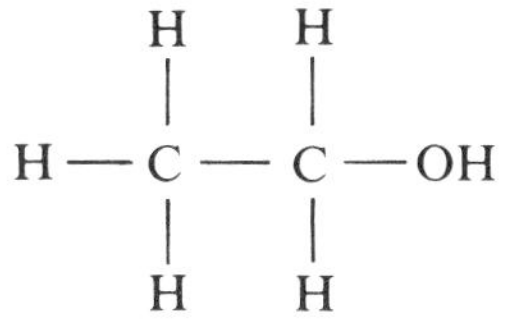

```
     H   H
     |   |
 H — C — C — OH
     |   |
     H   H
```

ethanol, an alcohol

```
    OH
    |
  (ring)
    |
   CH3
```

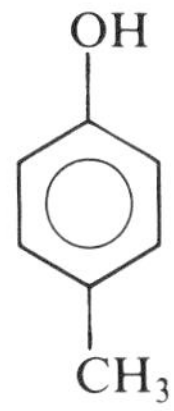

4-methylphenol, a phenol

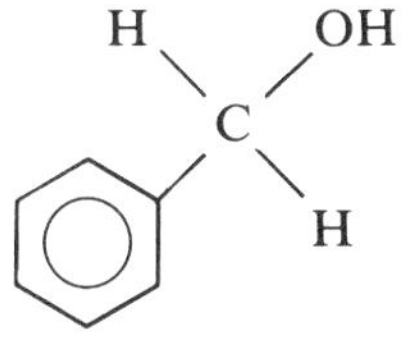

```
      H     OH
        \  /
         C
        / \
 (ring)     H
```

phenylmethanol

31.2.1 Naming

Alcohols are named using the suffix -ol if the —OH group is the principal group (or, of course, the only one). See **Table 26.4, section 26.5**. Otherwise the prefix hydroxy- is used. Locants are used if necessary to indicate where the —OH group is and di-, tri-, etc. to indicate how many of them are present in the molecule:

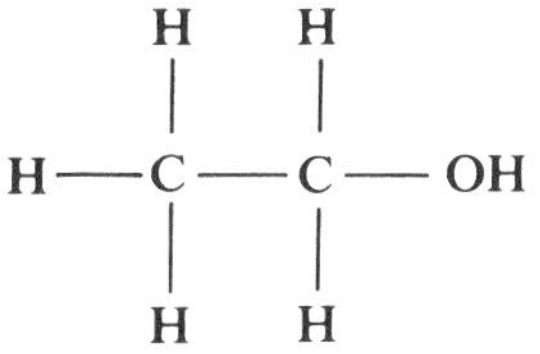

```
     H   H
     |   |
 H — C — C — OH
     |   |
     H   H
```

ethanol - no locant is needed

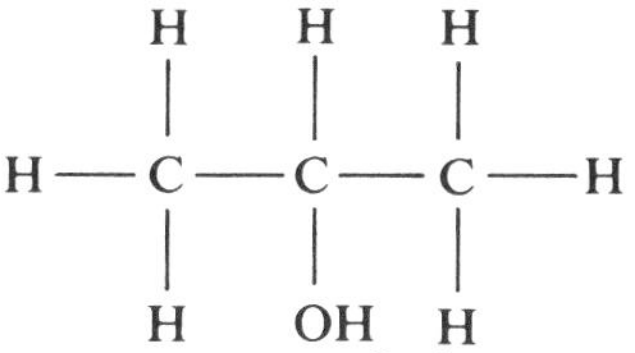

```
     H   H   H                  H   H   H
     |   |   |                  |   |   |
 H — C — C — C — H          H — C — C — C — OH
     |   |   |                  |   |   |
     H   OH  H                  H   H   H
```

propan-2-ol 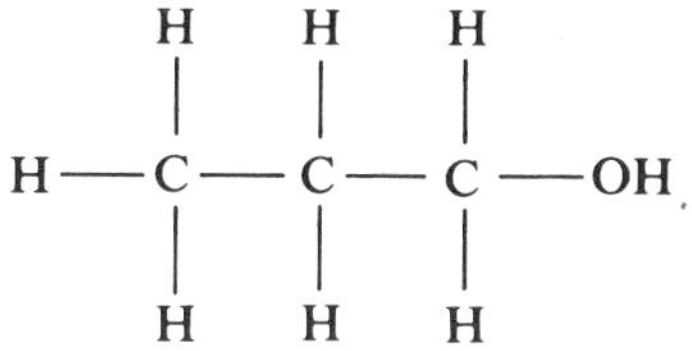 propan-1-ol

The last two are isomers.

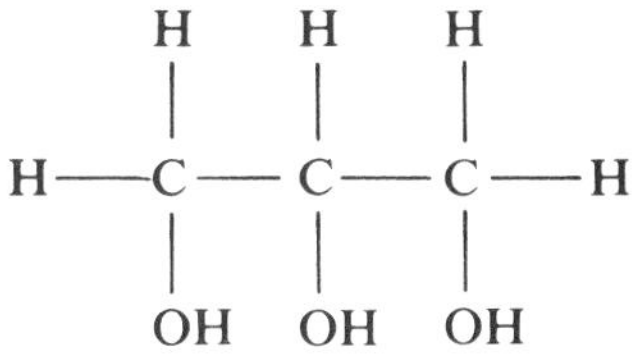

```
     H   H   H
     |   |   |
 H — C — C — C — H
     |   |   |
     OH  OH  OH
```

propane-1,2,3-triol

```
     H   OH
     |   |      O
 H — C — C — C //
     |   |     \
     H   H      OH
```

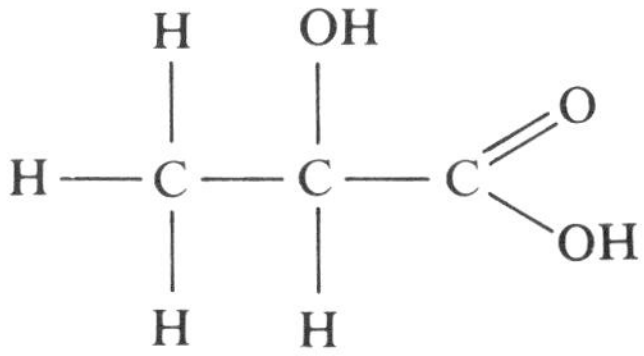

2-hydroxypropanoic acid

This compound (trivial name lactic acid) is named using the prefix hydroxy- rather than the suffix -ol as the —OH is not the principal group. The —OH group in brown is part of the carboxylic acid functional group.

Phenols are named as derivatives of the parent compound, phenol if the —OH group is the principal group, but the prefix hydroxy- is used otherwise. Aromatic compounds with two or more —OH groups are named using the suffix -ol:

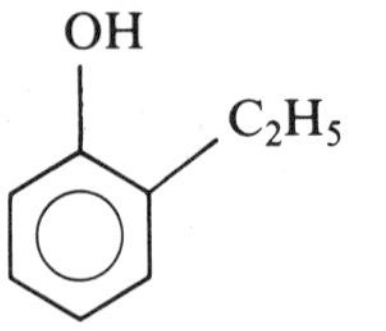

OH

C_2H_5

2-ethylphenol - the —OH is the principal group

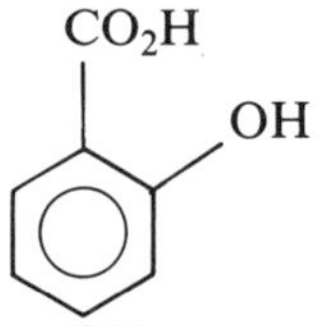

CO_2H

OH

2-hydroxybenzenecarboxylic acid - the —OH is not the principal group

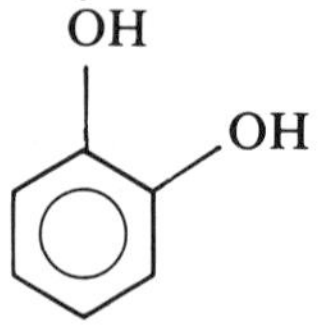

OH

OH

benzene-1,2-diol

31.2.2 Classification of alcohols

Alcohols are classified as primary (1°), secondary (2°) or tertiary (3°) according to how many other groups (R—) are bonded to the carbon which has the —OH group.

In a primary alcohol, this carbon has one R— group:

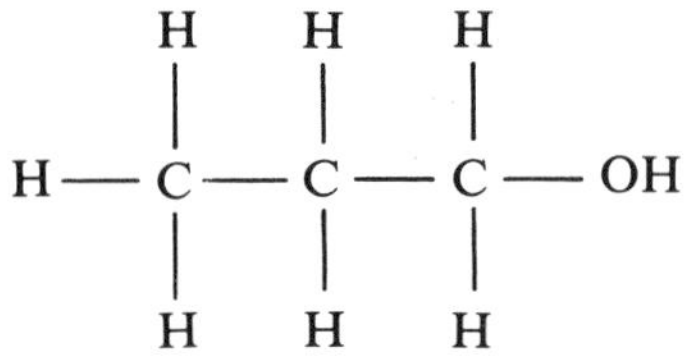

$$\begin{array}{ccccccc} & & H & & H & & H & & \\ & & | & & | & & | & & \\ H & — & C & — & C & — & C & — & OH \\ & & | & & | & & | & & \\ & & H & & H & & H & & \end{array}$$

propan-1-ol, a primary alcohol

$$\text{Methanol,}\quad \begin{array}{ccccc} & & H & & \\ & & | & & \\ H & — & C & — & OH \\ & & | & & \\ & & H & & \end{array}$$

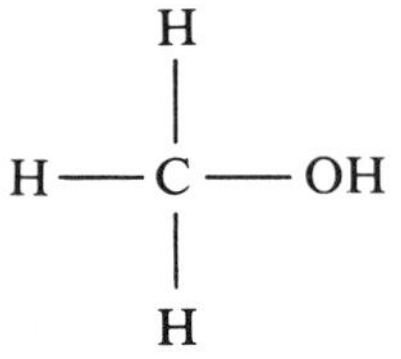

, where the carbon has no R— groups, is also considered to be a primary alcohol.

In a secondary alcohol, the —OH group is attached to a carbon with two R— groups:

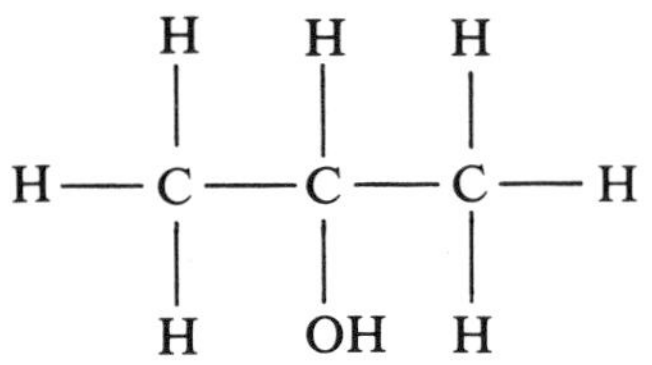

$$\begin{array}{ccccccccc} & & H & & H & & H & & \\ & & | & & | & & | & & \\ H & — & C & — & C & — & C & — & H \\ & & | & & | & & | & & \\ & & H & & OH & & H & & \end{array}$$

propan-2-ol, a secondary alcohol

Tertiary alcohols have three R— groups attached to the carbon which is bonded to the —OH:

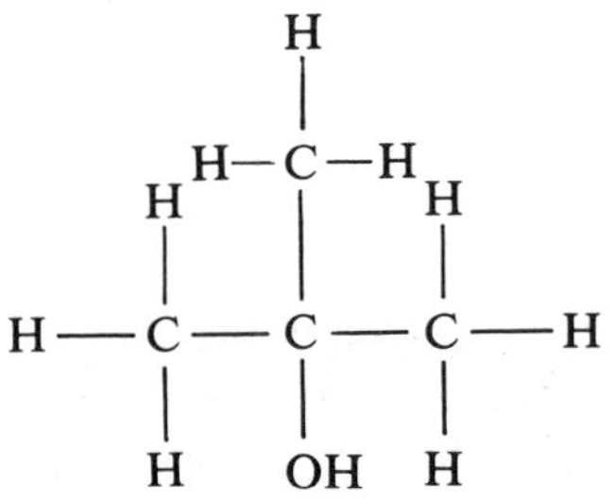

$$\begin{array}{ccccccccc} & & & & H & & & & \\ & & & & | & & & & \\ & & & H & -C- & H & & & \\ & & H & & | & & H & & \\ & & | & & | & & | & & \\ H & — & C & — & C & — & C & — & H \\ & & | & & | & & | & & \\ & & H & & OH & & H & & \end{array}$$

2-methylpropan-2-ol, a tertiary alcohol

31.2.3 Shape

In both alcohols and phenols the C—O—H angle is approximately 109°. It is based on the tetrahedral angle reduced slightly by the two lone pair on the oxygen atom.

Figure 31.1 Hydrogen bonding in methanol

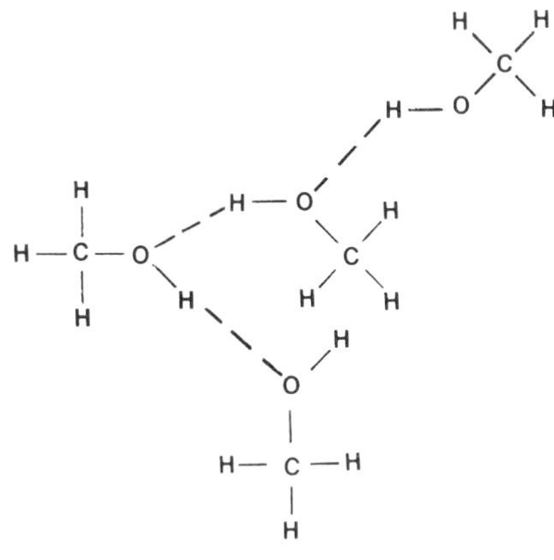

31.2.4 Physical properties

The presence of the —OH group means that hydrogen bonding occur between the molecules of both alcohols and phenols, see **Figure 31.1.** Th hydrogen bonding has two consequences. Firstly it leads to boilin temperatures (and melting temperatures) being much higher than thos of alkanes of comparable relative molecular mass. For example, butan-1-c ($M_r = 74$) boils at 390 K compared with pentane ($M_r = 72$) which boils a 309 K. The shorter-chain alcohols are all liquids at room temperatur decanol is the first solid. Some data are given in **Table 31.1**.

Secondly, alcohols with relatively short carbon chains dissolve well i water. All the alcohols up to and including the propanols mix completel with water. From this point, the solubility in water gradually decrease with increasing chain length.

Phenol is a solid, $T_m = 316$ K, with a distinctly antiseptic smell. As yo might expect, it is only sparingly soluble in cold water. A second —OH group in the molecule increases the solubility and benzene-1,2-diol i considerably more soluble than phenol.

Alcohols mix well with less polar solvents such as hydrocarbons.

Table 31.1 Melting and boiling temperatures of some alcohols and phenol

Compound	T_m/K	T_b/K
Methanol, CH_3OH	179	338
Ethanol, CH_3CH_2OH	156	352
Propan-1-ol, $CH_3CH_2CH_2OH$	147	371
Propan-2-ol, $CH_3CHOHCH_3$	184	356
Phenol	316	455

31.3 Infra-red spectra of alcohols

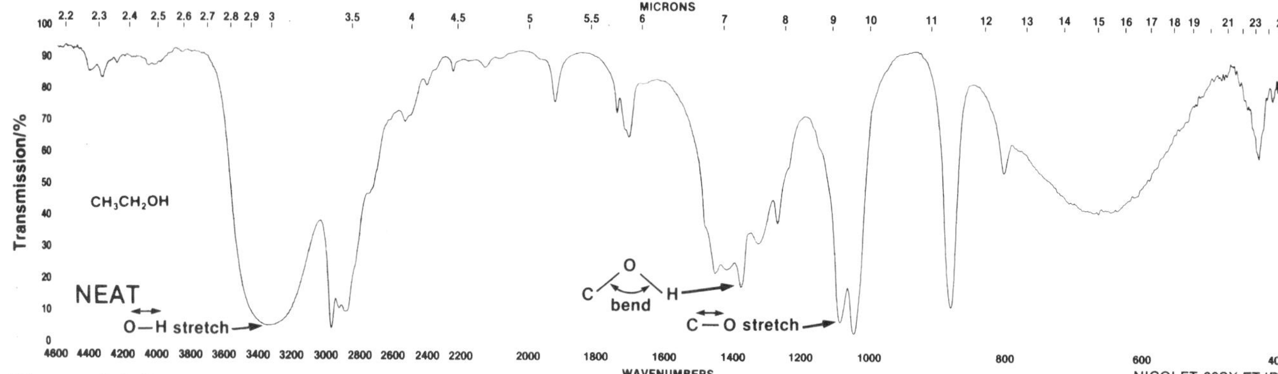

Figure 31.2 Infra-red spectrum of ethanol

Figure 31.3 The effect of hydrogen bonding on the vibrational frequency of the O—H bond

—O—H

No hydrogen bonding

—O—H- - - O

If the hydrogen is bonded to another molecule by hydrogen bonding, this effectively increases the mass of the hydrogen atom and reduces the frequency of vibration

The infra-red spectra of alcohols show a peak caused by an O—H stretching vibration at between 3300 cm^{-1} and 3600 cm^{-1}, see **Figure 31.2** The large range is caused by hydrogen bonding. If the hydrogen is bonde to another molecule by hydrogen bonding this effectively increases th mass of the hydrogen atom and reduces the frequency of the vibration.

The peak is often broad, also because of hydrogen bonding. Becaus they are relatively weak, hydrogen bonds are constantly breaking an reforming, so that different alcohol molecules may be hydrogen bonde (directly or indirectly) to different numbers of other molecules. This mean that in any sample of alcohol molecules, there will be a spread of molecule with different numbers of others bound to them and consequently a spread of vibrational frequencies, **Figure 31.3** (opposite).

A C—O—H bending vibration is seen at around 1350 cm^{-1} and a C—O stretching vibration at between 1050 and 1200 cm^{-1}.

31.4 Reactivity of alcohols

able 31.2 **Tabulated bond energies r the bonds in alcohols**

ond	BE/kJ mol^{-1}
—H	413
—C	347
—O	336
—H	464

31.4.1 Bond energies

The tabulated bond energies for the bonds in an alcohol are given in **Table 31.2.** These suggest that the C—O bond is the weakest and might be the most likely to break.

31.4.2 Polarity

The electronegativities of the carbon, hydrogen and oxygen atoms are:

$$C = 2.5; \quad H = 2.1; \quad O = 3.5$$

$$\begin{array}{ccccccc} & & H & & H & & \\ & & | & & | & & \\ H & — & C & — & C^{\delta+} & — O^{\delta-} — & H^{\delta+} \\ & & | & & | & & \\ & & H & & H & & \end{array}$$

This means that an alcohol molecule is polarized as shown above for ethanol. The δ^+ charge on the carbon means that it can be attacked by nucleophiles, while the δ^- oxygen might be attacked by positively charged reagents like H^+. Alternatively, the $O^{\delta-}$ atom can be thought of as being able to react as a nucleophile.

The hydrocarbon skeleton is relatively non-polar and will not be attacked by either type of reagent. Because the oxygen is so electronegative, if either the C—O or O—H bond breaks, it will tend to do so heterolytically so that the electrons in the bond go to the oxygen atom:

either

$$\begin{array}{ccccccc} & H & & H & & & \\ & | & & | & & & \\ H— & C & — & C & —O—H & \longrightarrow & H—\overset{H}{\underset{H}{C}}—\overset{H}{\underset{H}{C^+}} + {}^-OH \\ & | & & | & & & \\ & H & & H & & & \end{array}$$

or

$$\begin{array}{ccccccc} & H & & H & & & \\ & | & & | & & & \\ H— & C & — & C & —O—H & \longrightarrow & H—\overset{H}{\underset{H}{C}}—\overset{H}{\underset{H}{C}}—O^- + H^+ \\ & | & & | & & & \\ & H & & H & & & \end{array}$$

31.5 Reactions of alcohols

31.5.1 Combustion

In a plentiful supply of oxygen alcohols burn readily to give carbon dioxide and water:

$$C_2H_5OH(l) + 3O_2(g) \longrightarrow 2CO_2(g) + 3H_2O(l)$$

Ethanol is often used as a fuel, for example in picnic stoves which burn methylated spirits. This is ethanol with a small percentage of poisonous methanol added to make it unfit to drink so that it can be sold without

Alcohol-burning stove

the tax which is levied on alcoholic drinks. A blue dye is also added t show that it cannot be drunk.

31.5.2 Selective oxidation

Combustion is, of course, oxidation but alcohols can be oxidized mo selectively by oxidizing agents such as sodium dichromate(VI), potassiu dichromate(VI) or potassium manganate(VI) in acid solution.

Primary alcohols are oxidized to **aldehydes** which can themselves b further oxidized to **carboxylic acids**. For example:

oxidizing agent in excess - reflux

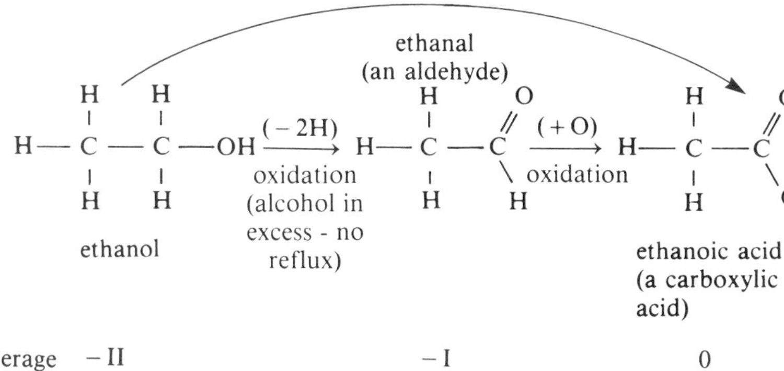

Average Ox(C)	−II	−I	0

Secondary alcohols are oxidized to **ketones** which are not readily oxidiz further:

propan-2-ol $\xrightarrow{(-2H)}$ propanone **(a ketone)**

Tertiary alcohols are not easily oxidized at all. Vigorous oxidation resul in breaking of the carbon chain and carboxylic acids are produce This difference in behaviour can be used to distinguish between 1°, 2° an 3° alcohols. The reason for the failure of ketones to oxidize further, c tertiary alcohols to oxidize at all, is because in each case, oxidation wou require the breaking of a relatively strong carbon–carbon bond rath than a C—H bond which must break for further oxidation of an aldehyd

Acidified dichromate(VI) ions are the most commonly used oxidizin agent. In the reaction the orange dichromate(VI) ions are reduced t green chromium(III) ions:

$$\overset{+VI}{Cr_2O_7^{2-}} + 14H^+ + 6e^- \longrightarrow \overset{+III}{2Cr^{3+}} + 7H_2O$$

down 3 × 2 = 6

Bearing in mind the average oxidation number changes:

$$\overset{-II}{C_2H_5OH} \longrightarrow \overset{-I}{C_2H_4O}$$

up 1 × 2 = 2

we can work out a balanced equation for the reaction of ethanol an acidified dichromate(VI). Ethanol and dichromate(VI) must react togeth in a 3:1 ratio to give ethanal:

$$Cr_2O_7^{2-}(aq) + \underset{\text{ethanol}}{3C_2H_5OH(l)} + 8H^+(aq)$$

$$\longrightarrow \underset{\text{ethanal}}{3CH_3CHO(g)} + 2Cr^{3+}(aq) + 7H_2O(l)$$

The breathalyser – now being replaced by more sophisticated devices

Similarly for the overall reaction to ethanoic acid; then a 3:2 ethanol: $Cr_2O_7^{2-}$ ratio is required:

$$3C_2H_5OH(l) + 2Cr_2O_7^{2-}(aq) + 16H^+(aq) \longrightarrow 3CH_3CO_2H(aq) + 4Cr^{3+}(aq) + 11H_2O(l)$$

igure 31.4 Apparatus for the xidation of ethanol to ethanal

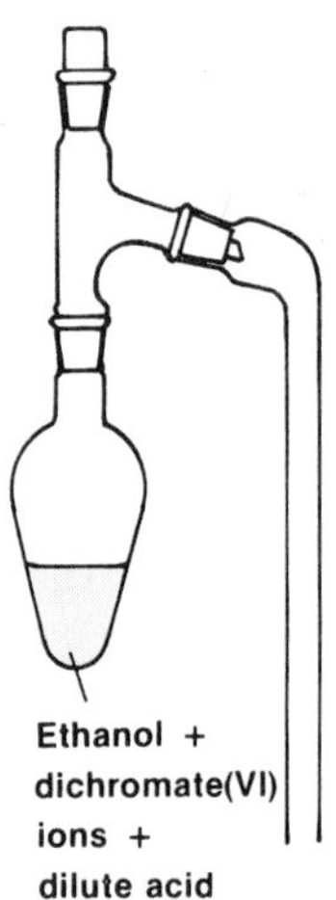

The oxidation of ethanol is the basis of the breathalyser used by police forces to rapidly estimate the ethanol content of the breath of suspected drunken drivers (see photo on opposite page). As the driver blows into the bag, ethanol molecules reduce orange $Cr_2O_7^{2-}$ ions to green Cr^{3+} ions. If more than a certain amount of the crystals changes colour the driver is likely to be 'over the limit' and must give a blood or urine sample for more sophisticated analysis by GLC (see **section 11.3.3**).

To prepare ethanal from ethanol, we use fairly gentle oxidizing conditions. A mixture containing less than the required ratio of dichromate(VI) to ethanol is used with dilute acid and heated gently in the apparatus shown in **Figure 31.4**. Ethanal ($T_b = 294$ K) vaporizes as soon as it is formed and distils over, thus avoiding further oxidation to ethanoic acid. Ethanol, T_b 352 K, remains in the flask. To prepare ethanoic acid, an excess of dichromate(VI) is used with concentrated acid and the mixture refluxed (**Figure 31.5**). Any ethanol or ethanal which boils will condense and drip back into the flask until it is eventually oxidized to the acid. After refluxing for around 20 minutes, the ethanoic acid (T_b = 391 K) can be distilled off along with the water by rearranging the apparatus as in **Figure 31.6**.

31.5.3 Alcohols as acids and bases

igure 31.5 Reflux apparatus for xidation of ethanol to ethanoic acid

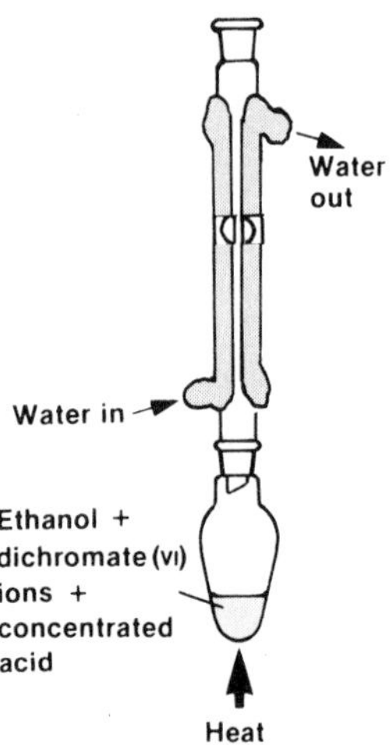

Acids

$$R{-}CH_2{-}\ddot{O}{-}H \rightleftharpoons R{-}CH_2{-}\ddot{O}{:}^- + H^+$$

$$R{-}CH_2{-}\ddot{O}{-}H \underset{}{\overset{+H^+}{\rightleftharpoons}} R{-}CH_2{-}\dot{O}^+H_2$$

An alcohol, like water, is capable of donating a proton or accepting one (see Chapter 12). If it donates a proton to some other species B it is acting as an acid. To do this, the strong O—H bond must break heterolytically leaving a carbon-containing negative ion called an **alkoxide ion**:

$$R{-}CH_2{-}O{-}H + B \rightleftharpoons R{-}CH_2{-}O^- + BH^+$$

an alkoxide ion

igure 31.6 Apparatus for distilling f ethanoic acid from the reaction mixture

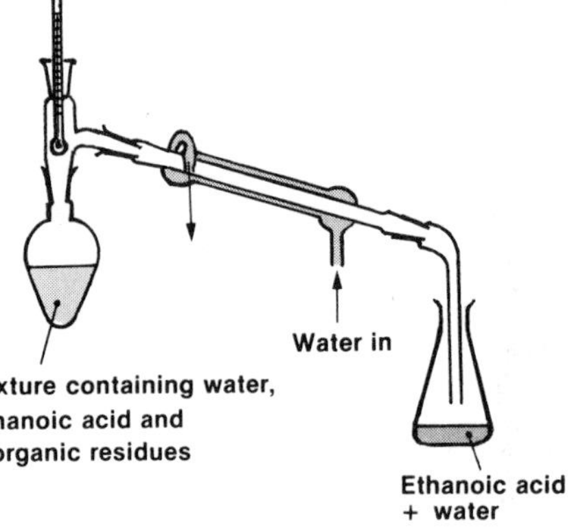

Alcohols are weaker acids than water. The only reactions which they undergo which show evidence of acidity are those with alkali metals. These are less vigorous than the reactions of the same metals with water as alcohols are weaker acids than water, for example:

$$Na(s) + C_2H_5OH(l) \longrightarrow C_2H_5O^-Na^+(s) + \tfrac{1}{2}H_2(g)$$

sodium ethoxide

The evolution of hydrogen with sodium is a useful test for the presence of an —OH group in an organic compound. Sodium ethoxide is a white, ionic solid which can be isolated by evaporating excess ethanol. Alkoxide ions are strong bases and good nucleophiles because of the negative charge on the oxygen atom.

Bases

Because of the lone pair of electrons on the oxygen atom alcohols ca accept a proton from an acid HA, thus acting as bases:

$$R-CH_2-\ddot{O}H + HA \longrightarrow R-CH_2-\dot{O}^+H_2 + A^-$$

The alcohol is said to be **protonated**. Protonation is an important fir step in a number of reactions of alcohols.

31.5.4 Formation of halogenoalkanes

Alcohols react with hydrogen halides (HCl, HBr, HI) to give th corresponding halogenoalkane. The rate of reaction is in the ord $HI > HBr > HCl$. Strong acids catalyse the reaction so the hydrogen halid is often generated in the reaction flask by the reaction of an alkali met halide with concentrated sulphuric acid (not in the case of hydrogen iodid as sulphuric acid oxidizes hydrogen iodide to iodine).

$$2KBr(s) + H_2SO_4(l) \longrightarrow K_2SO_4(s) + 2HBr(g)$$

For example:

$$C_2H_5OH(l) + HBr(g) \xrightarrow{H^+ \text{ catalyst}} C_2H_5Br(l) + H_2O(l)$$

The first step in the reaction is the rapid protonation of the —OH grou

$$H-CH_2-CH_2-\dot{O}H + H^+ \xrightarrow{\text{fast}} H-CH_2-CH_2-\overset{+}{O}H_2$$

Some of the positive charge resides on the carbon to which the $-\overset{+}{O}H_2$ group is attached.

The reaction then continues in one of two ways: either the $C^{\delta+}$ i attacked by the nucleophile X^- and loses H_2O—an S_N2 mechanism a the rate of this step depends on the concentration of both $C_2H_5OH_2$ and X^-:

$$H-CH_2-\overset{+}{C}H_2-OH_2 \;\; (\ddot{X}^-) \xrightarrow{\text{slow}} H-CH_2-CH_2-X + H_2O$$

or the protonated alcohol loses H_2O leaving a carbocation which is rapidl attacked by the nucleophile X^-. This is an S_N1 mechanism as the rate the slow step depends only on the concentration of $C_2H_5OH_2{}^+$:

$$H-CH_2-CH_2-\overset{+}{O}H_2 \xrightarrow{\text{slow}} \underset{\text{carbocation}}{H-CH_2-\overset{+}{C}H_2} + H_2O \xrightarrow[X^-]{\text{fast}} H-CH_2-CH_2-X$$

The S_N1 mechanism involves a carbocation. This is more stable the mor alkyl groups which are attached to it, as they release electrons and stabiliz the positive charge. So tertiary alcohols tend to react by S_N1 and primar by S_N2 mechanisms. This is the same pattern we found for the reactio of halogenoalkanes to alcohols—the reverse of this reaction (see **sectio 30.7.2**).

Alternatively, the conversion of alcohols to halogenoalkanes can be brought about by using phosphorus halides—PCl_5, PBr_3 or PI_3 (for the last two the phosphorus trihalide is made in the reaction flask by mixing red phosphorus and the halogen). The equations for these reactions are:

$$3ROH + PX_3 \longrightarrow 3RX + H_3PO_3$$
phosphonic acid

$$ROH + PCl_5 \longrightarrow RCl + POCl_3 + HCl$$
phosphorus trichloride oxide

For preparing chloroalkanes, a further reagent can be used: sulphur dichloride oxide, $SOCl_2$. This reagent has the advantage that all the products other than the halogenoalkane are gases and so there is no problem separating them from the halogenoalkane, for example:

$$CH_3CH_2OH(l) + SOCl_2(l) \longrightarrow CH_3CH_2Cl(l) + SO_2(g) + HCl(g)$$

31.5.5 Dehydration of alcohols

Alcohols can be dehydrated by passing their vapours over heated aluminium oxide, for example:

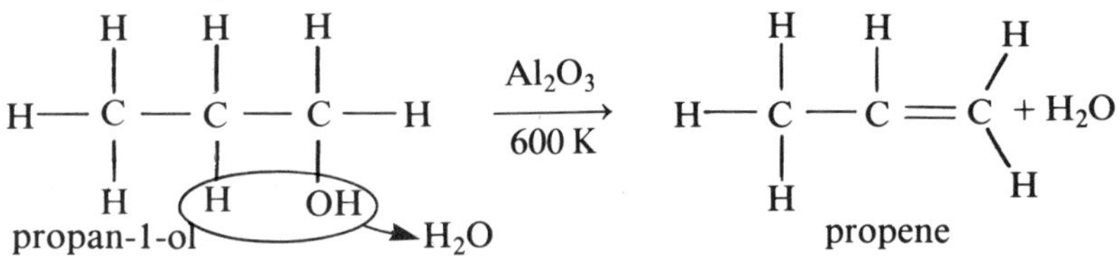

$$H-CH_2-CH(H)-CH_2(OH) \xrightarrow[600\ K]{Al_2O_3} H-CH_2-CH=CH_2 + H_2O$$

propan-1-ol (H and OH circled → H_2O) propene

gure 31.7 **Dehydration of an alcohol**

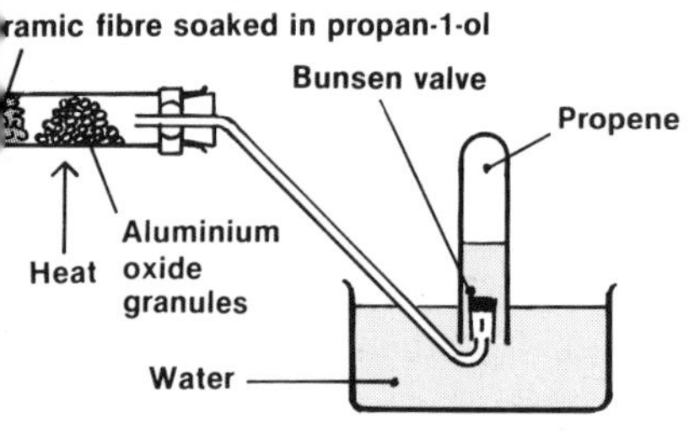

The apparatus used is shown in **Figure 31.7.**

With longer-chain or branched alcohols, there may be more than one possible product:

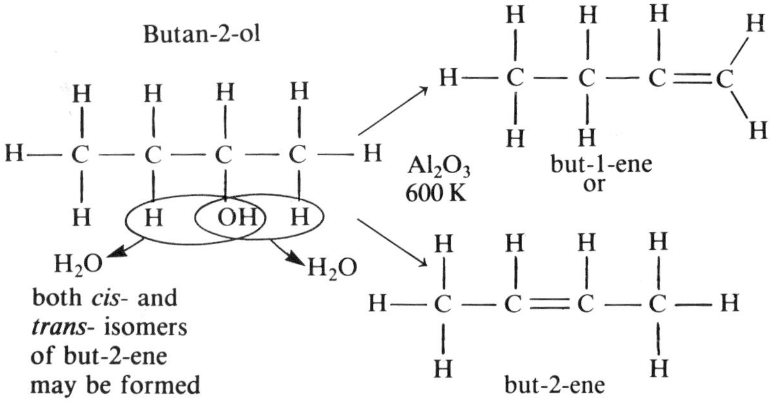

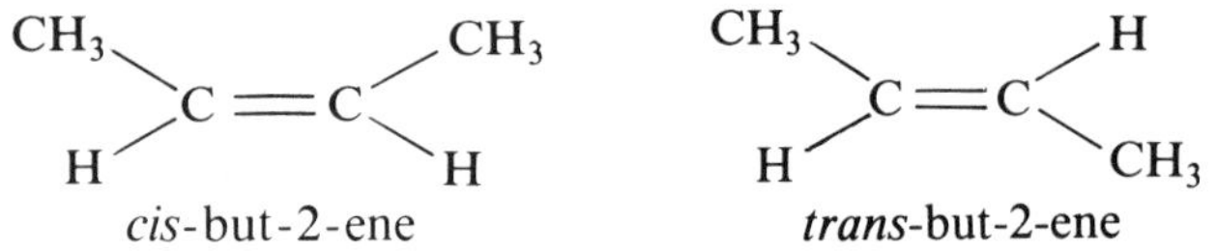

Butan-2-ol

$$H-CH_2-CH(OH)-CH_2-CH_3 \xrightarrow[600\ K]{Al_2O_3} H-CH_2-CH_2-CH=CH_2 \text{ (but-1-ene)}$$

or

$$H-CH_2-CH=CH-CH_2-H \text{ (but-2-ene)}$$

H_2O / H_2O

both *cis*- and *trans*- isomers of but-2-ene may be formed

$$\underset{cis\text{-but-2-ene}}{(CH_3)(H)C{=}C(CH_3)(H)} \qquad \underset{trans\text{-but-2-ene}}{(CH_3)(H)C{=}C(H)(CH_3)}$$

Saytzeff's rule predicts that the predominant product is the isomer in which the alkene has the most alkyl groups attached, so but-2-ene (two alkyl groups) is the main product in the example above, because in but-1-ene, the alkene has only one alkyl group.

Phosphoric(V) acid is an alternative dehydrating agent. Sulphuric acid can also be used to dehydrate alcohols if used in excess, for example:

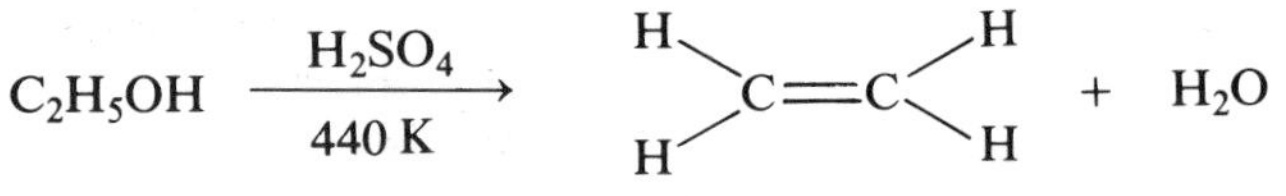

$$C_2H_5OH \xrightarrow[440\ K]{H_2SO_4} H_2C{=}CH_2 + H_2O$$

However, if an excess of alcohol is used the result is the formation of a **ether**:

$$\mathrm{CH_3CH_2{-}O{-}H \;+\; H{-}O{-}CH_2CH_3} \xrightarrow[400\ \mathrm{K}]{\text{excess alcohol}} \mathrm{CH_3CH_2{-}O{-}CH_2CH_3} + H_2O$$

ethanol → ethoxyethane (an ether)

This reaction is known as **Williamson's continuous ether synthesis**.

31.5.6 Formation of esters

Alcohols react with carboxylic acids. A molecule of water is eliminated a an ester is produced, for example:

$$\mathrm{CH_3OH} + \mathrm{CH_3COOH} \underset{\text{catalyst}}{\overset{H^+}{\rightleftharpoons}} \mathrm{CH_3{-}O{-}CO{-}CH_3} + H_2O$$

methanol + ethanoic acid ⇌ methyl ethanoate

The reaction is catalysed by a strong acid such as sulphuric acid and dc not go to completion. An equilibrium mixture containing a good deal all four components is formec. The overall reaction:

acid + alcohol ⇌ ester + water

is rather similar to:

acid + base ⟶ salt + water

and suggests that the alcohol is behaving in some sense as a base.

The mechanism of esterification is discussed in **section 14.6.2.** T esterification takes place more readily with acid chlorides, see Chapter : than with carboxylic acids:

$$\mathrm{R{-}\overset{\delta+}{C}(={O}^{\delta-}){-}Cl^{\delta-}} + \mathrm{H\ddot{O}{-}R'} \longrightarrow \mathrm{R{-}C(=O){-}\overset{+}{O}H{-}R'} + Cl^-$$

$$\mathrm{R{-}C(=O){-}\overset{+}{O}H{-}R'} \longrightarrow \mathrm{R{-}C(=O){-}O{-}R'} + H^+$$

The overall reaction is as follows:

E

THE MECHANISM OF ESTERIFICATION

Tracer experiments with alcohols containing the ^{18}O isotope show that all the ^{18}O is found in the ester and none in the water:

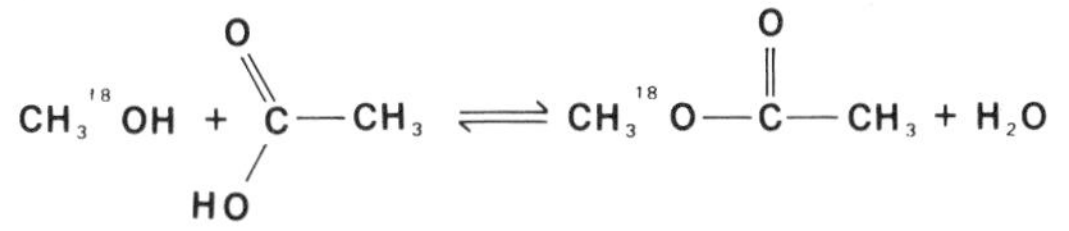

$$CH_3{}^{18}OH + HO{-}C(=O){-}CH_3 \rightleftharpoons CH_3{}^{18}O{-}C(=O){-}CH_3 + H_2O$$

A mass spectrometer can easily detect the difference in mass between methyl ethanoate containing ^{16}O, $M_r = 74$, and that containing ^{18}O, $M_r = 76$.

This indicates that a new bond is formed between the oxygen of the alcohol and the carbon of the carboxylic acid followed by loss of a molecule of water containing the oxygen from the —OH group of the acid. The mechanism is discussed in **section 14.6.2.**

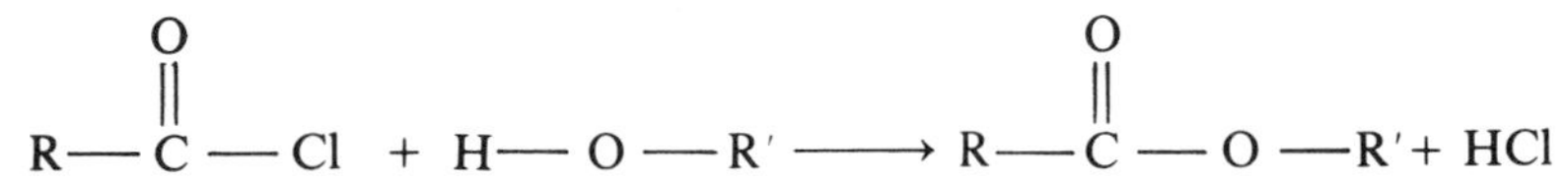

$$R{-}C(=O){-}Cl + H{-}O{-}R' \longrightarrow \underset{\text{an ester}}{R{-}C(=O){-}O{-}R'} + HCl$$

The chlorine atom makes the carbon of the $-C^{\delta+}(=O^{\delta-})Cl^{\delta-}$ group strongly δ^+ as it is attached to *two* electronegative atoms. This allows even a weak nucleophile like an alcohol to attack it.

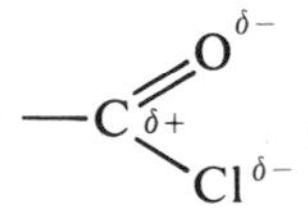

The esterification reaction with an acid chloride has two advantages over that with an acid: it occurs more quickly and it goes to completion.

31.5.7 The triiodomethane (iodoform) reaction

The reaction occurs only with alcohols containing the group:

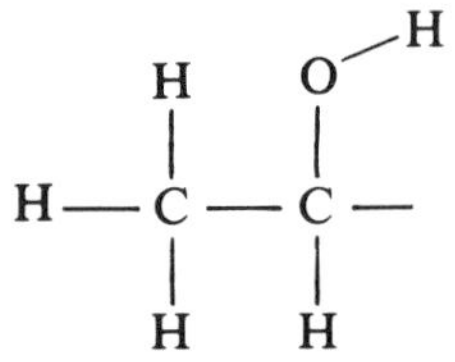

$$H{-}C(H)(H){-}C(H)(O{-}H){-}$$

so it will work with ethanol and propan-2-ol but not methanol or propan-1-ol. The alcohol is warmed with a solution of sodium hydroxide and iodine. Alcohols with the group above will form a yellow precipitate of triiodomethane by the reaction:

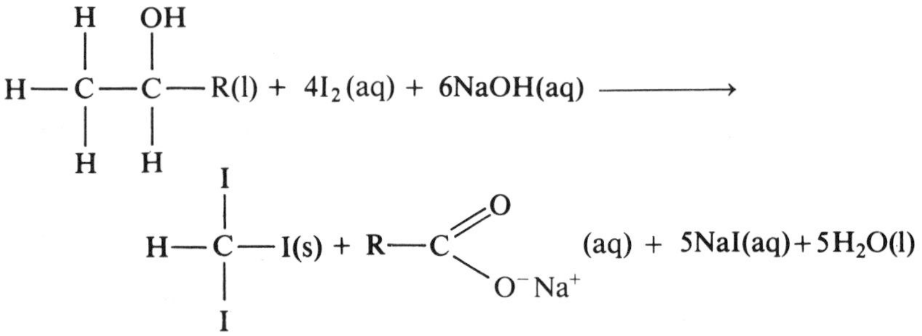

$$H{-}C(H)(H){-}C(H)(OH){-}R(l) + 4I_2(aq) + 6NaOH(aq) \longrightarrow$$

$$H{-}C(I)(I){-}I(s) + R{-}C(=O){-}O^-Na^+(aq) + 5NaI(aq) + 5H_2O(l)$$

31.6 Reactivity of phenols

At first sight phenols seem to be very similar to alcohols. However, in phenols, the bonding in the benzene ring interacts with that in the —OH group, so that in many respects their reactivity is quite different from

Figure 31.8 **Interaction of the oxygen lone pair with the aromatic system in phenol**

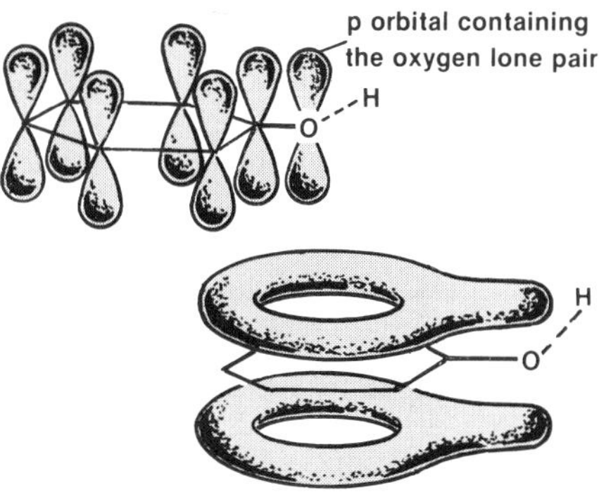

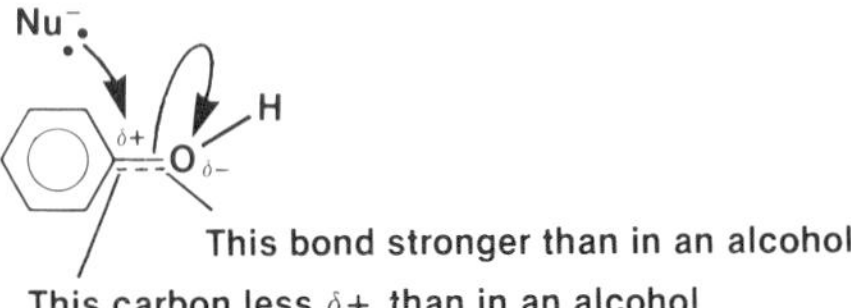

Figure 31.9 **Phenol is less susceptible than alcohols to attack by nucleophiles**

Nu^- H $\delta+$ O $\delta-$

This bond stronger than in an alcohol

This carbon less $\delta+$ than in an alcohol

alcohols. **Figure 31.8** shows how one of the lone pairs on the oxygen atc in phenol can overlap with the delocalized π system on the benzene ri to form an extended delocalized π orbital. This means that the C— bond in phenol has some double bond character making it stronger th the C—O bond in an alcohol. This is confirmed by measurement of bo lengths. The C—O bond in ethanol has a length of 0.143 nm while th in phenol is 0.136 nm long. Compare this with a typical C=O length 0.122 nm. The extra strength of the C—O bond in phenol (which cou be represented C⋯O to stress its double bond character) means that will be harder to break the C—O bond in phenols than in alcohols.

A second factor caused by this overlapping of p orbitals is that the charge of the carbon bonded to the oxygen can be spread over the wh delocalized system, **Figure 31.9**, thus making this carbon less susceptil to attack by nucleophiles than is the corresponding carbon in alcoho These two factors combine to make nucleophilic substitution reactio less likely in phenols than in alcohols, i.e. phenols are less reactive nucleophiles than are alcohols. In the phenoxide ion, $C_6H_5O^-$, whi results from loss of a proton by phenol, the p–π overlap means that t negative charge can be delocalized over the whole molecule, leading extra stability of the ion. No such delocalization (and therefore stabilit is available in alkoxide ions.

The fact that the —OH group is attached to the benzene ring has tw further effects on the reactivity of phenol. Firstly, elimination reactio are unlikely as they would involve disruption of the aromatic syste Secondly, oxidation is not likely either except under vigorous conditio The carbon to which the —OH group is attached has no hydrogen bond to it, so phenol is effectively a tertiary alcohol. In other words, oxidati of phenol would involve breaking C—C bonds *and* destruction of t aromatic system.

As well as reactions of the —OH group, phenol's benzene ring can al react. The —OH group releases electrons on to the benzene ring maki it more reactive than benzene itself. This may seem odd as oxygen is mo electronegative than carbon and would be expected to attract electro towards itself. It does in fact attract electrons through the C—O σ bon but this is outweighed by the release of electrons from oxygen's lone pai on to the benzene ring through the p–π overlap. The net effect is an over release of electrons on to the ring.

31.7 Reactions of phenol

31.7.1 Combustion

Phenol burns in air with a smoky flame characteristic of aromat compounds.

31.7.2 Redox reactions

Phenol resists oxidation under the conditions which oxidize alcohols. Th is because phenol has no hydrogen on the carbon to which the —O group is attached. So loss of hydrogen, the first step of oxidation, cann occur. Phenol can be reduced by heating molten phenol with powder zinc. Benzene is produced:

$$C_6H_5OH(l) + Zn(s) \xrightarrow{700\ K} C_6H_6(l) + ZnO(s)$$

31.7.3 Reactions of the—OH group in phenol

Acidity

Phenol is considerably more acidic than alcohols, pK_a for phenol = 9.9, compared with pK_a for ethanol = 16, and for a typical weak acid, ethanoic acid, $pK_a = 4.8$. This is due to the delocalization of the negative charge in the phenoxide ion which stabilizes it:

$$C_6H_5\text{—}OH + H_2O \rightleftharpoons C_6H_5\text{—}O^- + H_3O^+$$

phenoxide ion

Phenol is only slightly soluble in cold water but dissolves in warm water and in aqueous sodium hydroxide solution:

$$C_6H_5\text{—}OH(s) + NaOH(aq) \longrightarrow (C_6H_5\text{—}O^- + Na^+)(aq) + H_2O(l)$$

sodium phenoxide

The equilibrium can be displaced to the left by adding concentrated hydrochloric acid, the phenol precipitating out of solution.

Note that *alkoxides* cannot be made by reaction of an alcohol with an alkali metal hydroxide, only by reaction with an alkali metal itself. This is a further illustration of the greater acidity of phenols compared with alcohols. Alkali metal phenoxides can also be made by direct reaction of molten phenol with the alkali metal, for example:

arbolic soap

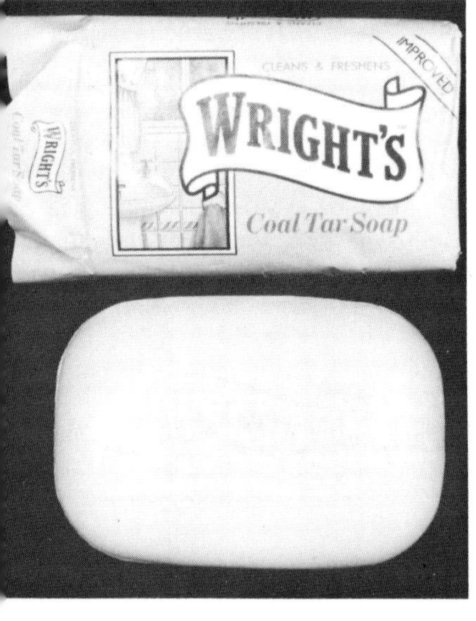

$$K(s) + C_6H_5\text{—}OH(l) \longrightarrow (C_6H_5\text{—}O^- + K^+)(s) + \tfrac{1}{2}H_2(g)$$

The phenoxide ion generated is a strong nucleophile.

Phenol is not a strong enough acid to displace carbon dioxide from sodium carbonate solution. This test enables it to be distinguished from carboxylic acids, which produce carbon dioxide when added to a carbonate.

Electron-withdrawing substituents like —NO_2 increase the acid strength of phenol by further delocalizing the negative charge remaining after the molecule has lost a proton. Electron-releasing substituents like —CH_3 make phenol less acidic.

Phenol was once known as **carbolic acid**. The name is still used in carbolic soaps, which contain small amounts of phenol and derivatives as antiseptics. (Pure phenol is an unpleasantly corrosive substance causing blistering of the skin and it should be treated with care.) Phenols, being stronger acids and therefore weaker bases than alcohols, are less easily protonated than alcohols.

Substitution of the —OH group by halogens

This reaction takes place much less easily than in alcohols due to the increased strength of the C—OH bond and the reduced δ^+ character of the carbon which is less easily attacked by nucleophiles. Hydrogen halides do not react at all and even phosphorus pentachloride produces only a poor yield of chlorobenzene:

$$C_6H_5OH + PCl_5 \longrightarrow C_6H_5Cl + POCl_3 + HCl$$

Dehydration

This does not occur as it would involve disruption of the aromatic system.

COAL TAR

Coal is a fossilized plant material which is a mixture of compounds containing nitrogen, oxygen and sulphur as well as carbon and hydrogen. The composition varies with its source. Heating coal in the absence of air decomposes it and four main products are formed—coal gas, ammoniacal liquor, coal tar and coke. Coal gas is a mixture of hydrogen, methane and carbon monoxide, and until around 1970 was the gas supplied to homes as fuel. Ammoniacal liquor is a solution containing ammonia, and coal tar is a complex mixture of hundreds of organic compounds. Coke is an impure form of carbon used as a fuel and a chemical reducing agent in, for example, the blast furnace.

Coal tar is a viscous black liquid and is a valuable source of organic compounds, especially aromatics, including phenol. Although oil has overtaken coal as the major source of organic compounds, production from coal tar has remained steady over the past three decades and may be expected to increase in importance as crude oil reserves are used up.

Petroleum can be produced from coal and this is done on a large scale in South Africa which has little oil. A mixture of carbon monoxide and hydrogen, called **synthesis gas**, is first made by heating coal with steam and oxygen. The synthesis gas is then heated over an iron catalyst where the following reactions occur:

$$2nH_2 + nCO \longrightarrow (CH_2)_n + nH_2O$$

$$nH_2 + 2nCO \longrightarrow (CH_2)_n + nCO_2$$

The process is called the Fischer–Tropsch synthesis. The cost of petrol made by this process is comparable with that of petrol derived from crude oil.

Ether formation

Reaction of phenoxide ions with halogenoalkanes produces ethers by a nucleophilic substitution reaction, for example:

$$C_6H_5\bar{O}: + \overset{\delta-}{Br}-\overset{\delta+}{C}H_2-CH_3 \longrightarrow C_6H_5-O-CH_2-CH_3 + Br^-$$

ethoxybenzene (an ether)

This method is called **Williamson's synthesis**, as distinct from Williamson's *continuous* synthesis, see **section 31.5.5.**

Formation of esters

Phenols are weaker nucleophiles than alcohols as their lone pair electrons are partially delocalized on to the benzene ring, thus they do not form esters by reaction with carboxylic acids directly. However, the phenoxide ion, $C_6H_5O^-$, is a better nucleophile than phenol and it will react with acid derivatives such as acid chlorides (or acid anhydrides, see Chapter 33) which are themselves more reactive than the parent acid:

$$C_6H_5\bar{O}: + CH_3-\overset{\delta+}{C}(=\overset{\delta-}{O})-\overset{\delta-}{Cl} \xrightarrow[\text{temperature}]{\text{room}} C_6H_5-O-\overset{O}{\overset{\|}{C}}-CH_3 + Cl^-$$

phenyl ethanoate (an ester)

The reaction is carried out by dissolving the phenol in sodium hydroxide solution to generate the phenoxide ion before adding the acid chloride.

Reaction with iron(III) chloride

Phenol produces a purple complex when mixed with iron(III) chloride, in which the phenol is acting as a ligand (see **section 25.3.2**). This is often used as a test for a phenol or, in fact, any —OH group bonded to an unsaturated system.

Reactions of the benzene ring in phenol

Because of the net electron-releasing effect of the —OH group the aromatic

OVER-THE-COUNTER PAINKILLERS—ASPIRIN AND PARACETAMOL

There is an enormous market for non-prescription pain-killers. A large number of brands is available but all are formulated from a choice of four active ingredients—aspirin, paracetamol, codeine and ibuprofen. Of these the most used are aspirin and paracetamol.

Aspirin (structure: benzene ring bearing $O-C(=O)-CH_3$ and $C(=O)-O-H$ on adjacent carbons)
2-ethanoyloxybenzene-carboxylic acid

Paracetamol (structure: $H_3C-C(=O)-N(H)-C_6H_4-OH$)
4-(N-methanoylamino) phenol

The structures above show that both compounds are related to phenol.

The two drugs are comparable in their ability to relieve mild pain and both also have the effect of reducing fever. Aspirin also has anti-inflammatory properties. Neither drug is free from side effects. Aspirin may cause irritation and bleeding of the stomach while long-term use of paracetamol may cause liver damage. Indeed, it has been suggested that, were aspirin a new drug, it would not be approved for over-the-counter sale without prescription. A number of simple derivatives of aspirin are also useful. The calcium salt is sold as soluble aspirin, the ionic salt being more soluble in water than the covalently bonded acid, while the sodium salt is mixed with a hydrogencarbonate to give effervescent aspirin which fizzes when added to water. The carbon dioxide gas evolved helps the tablet to dissolve.

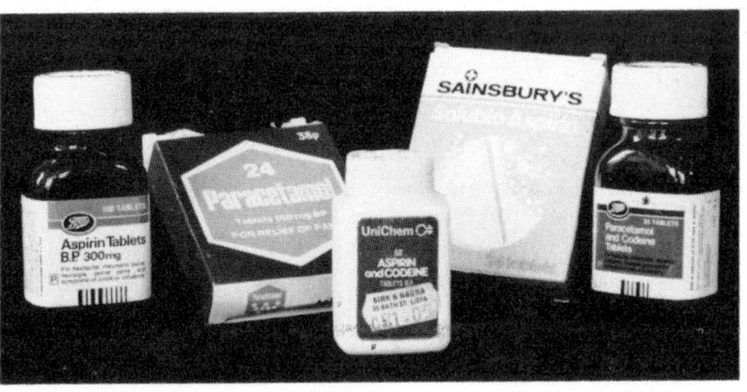

All non-prescription painkillers are formulated from a choice of four main ingredients

ring in phenol is much more reactive towards electrophilic substitution than the benzene ring. This may be shown in various ways:

1. the same reaction occuring more rapidly;
2. more substitution taking place under the same conditions;
3. less vigorous conditions being required for the same reaction.

Like most ring activators, the —OH group directs further substituents to the 2, 4 and 6 positions.

Electrophilic substitution reactions

NITRATION

Phenol can be nitrated by dilute nitric acid, at room temperature:

$$\text{phenol} \xrightarrow[\text{room temperature}]{HNO_3(aq)} \text{2-nitrophenol and 4-nitrophenol}$$

Compare this with benzene, where heating with a mixture of concentrated nitric and sulphuric acids is required for nitration. Nitration of phenol under more vigorous conditions yields 2,4,6-trinitrophenol—the explosive picric acid:

$$\text{phenol} \xrightarrow[\text{reflux}]{\text{conc. } HNO_3/H_2SO_4} \text{2,4,6-trinitrophenol}$$

HALOGENATION

Phenol will react at room temperature with aqueous bromine or chlori
to give a trisubstituted product. No catalyst is needed:

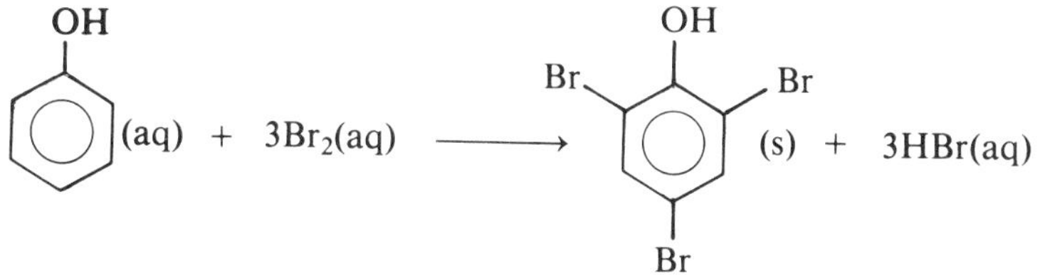

$$\text{(phenol, OH)(aq)} + 3Br_2(aq) \longrightarrow \text{(OH, Br, Br, Br)(s)} + 3HBr(aq)$$

Compare this with benzene where liquid bromine and a catalyst a
required yet produce only a monosubstituted product.

These two examples show the greatly increased reactivity of the rin
towards electrophiles.

Addition reactions

Phenol can be hydrogenated under similar conditions to those require
by benzene. Cyclohexanol is produced:

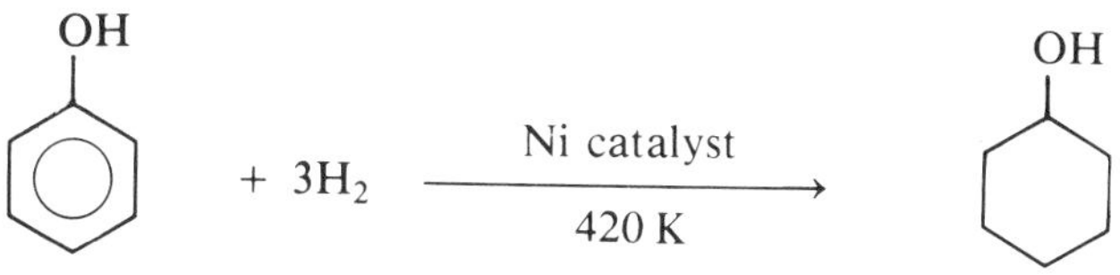

$$\text{(phenol, OH)} + 3H_2 \xrightarrow[\text{420 K}]{\text{Ni catalyst}} \text{(OH)}$$

cyclohexanol

31.8 Preparation of alcohols and phenols

31.8.1 Alcohols

The main methods of preparing alcohols in the laboratory are summarize
below. The reactions are discussed in more detail in the chapters on th
relevant starting materials.

From alkenes

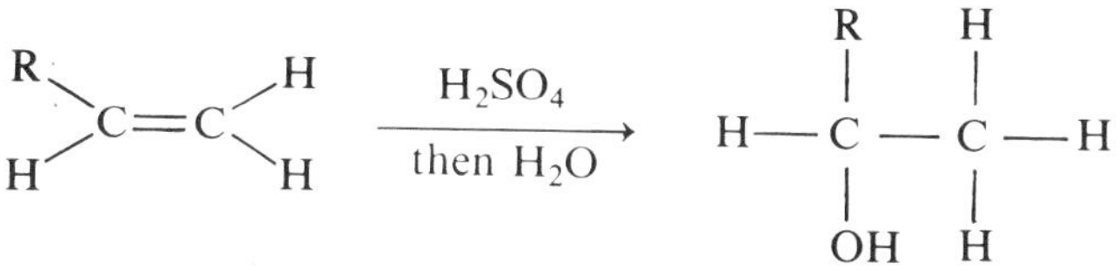

$$\underset{H}{\overset{R}{}}C{=}C\underset{H}{\overset{H}{}} \xrightarrow[\text{then } H_2O]{H_2SO_4} H-\underset{OH}{\overset{R}{C}}-\underset{H}{\overset{H}{C}}-H$$

With unsymmetrical alkenes (those with more alkyl groups on one carbon
than the other), the —OH group becomes attached to the *more substitute*
carbon atoms, see **section 28.6**.

From halogenoalkanes

Boiling halogenoalkanes with aqueous alkali produces alcohols:

$$R{-}X + OH^- \xrightarrow[\text{aqueous solution}]{\text{reflux}} R{-}OH + X^-$$

Reduction of aldehydes, ketones, carboxylic acids and esters

On reduction, aldehydes, carboxylic acids and esters yield primary
alcohols, while ketones give secondary alcohols. (This is the reverse of the
oxidation of alcohols to aldehydes to carboxylic acids.) A number of
reducing agents can be used as shown in the summary overleaf.

$$\left.\begin{array}{l}\text{aldehyde}\\ \\ \text{ketone}\end{array}\right\} \begin{array}{c}\xrightarrow{Na/C_2H_5OH}\\ \text{or}\\ \xrightarrow{NaBH_4(aq)}\\ \text{or}\\ \xrightarrow[\text{LiAlH}_4\text{ in ethoxyethene}]{}\end{array} \left\{\begin{array}{l}1^\circ\text{ alcohol}\\ \\ 2^\circ\text{ alcohol}\end{array}\right.$$

$$\left.\begin{array}{l}\text{acid}\\ \\ \text{acid chloride}\\ \text{ester}\end{array}\right\} \begin{array}{c}\xrightarrow{\text{LiAlH}_4\text{ in ethoxyethane}}\\ \text{or}\\ \xrightarrow[Na/C_2H_5OH]{}\\ \text{or}\\ \xrightarrow[H_2/\text{catalyst}]{}\end{array} \begin{array}{l}\text{alcohol}\\ \\ \text{(esters, } R{-}CO_2R'\text{, give}\\ \textit{two}\text{ alcohols } RCH_2OH \text{ and } R'OH)\end{array}$$

Hydrolysis of esters

Esters can be hydrolysed in either acidic or alkaline conditions to produce alcohols and carboxylic acids (salts in alkaline conditions):

$$R{-}\overset{\overset{O}{\|}}{C}{-}OR' \xrightarrow{H_2O,\ H^+ \text{ or } OH^-} R{-}\overset{\overset{O}{\|}}{C}{-}OH + R'OH$$

31.8.2 Phenol

Phenol can be prepared from benzenesulphonic acid by heating it with solid sodium hydroxide and acidifying the sodium phenoxide produced:

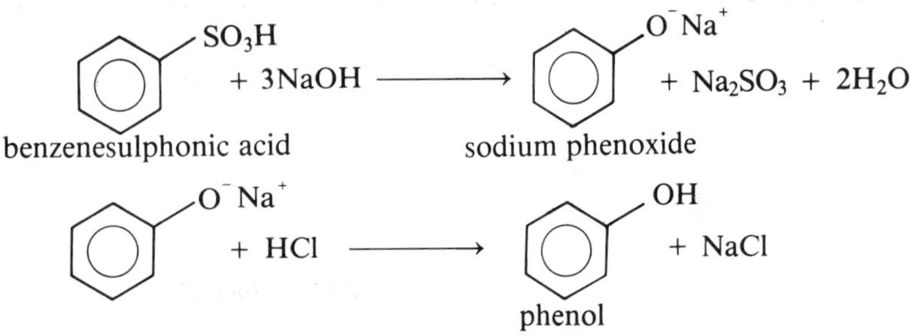

$$C_6H_5SO_3H + 3NaOH \longrightarrow C_6H_5O^-Na^+ + Na_2SO_3 + 2H_2O$$

benzenesulphonic acid → sodium phenoxide

$$C_6H_5O^-Na^+ + HCl \longrightarrow C_6H_5OH + NaCl$$

phenol

31.9 Economic importance of alcohols and phenols

31.9.1 Alcohols

Fermentation

Ethanol is the alcohol present in all alcoholic drinks. It is made by **fermentation** in which carbohydrates like starch are first broken down into sugars and then into ethanol, by the action of enzymes obtained from yeast. The key step is:

$$C_6H_{12}O_6(aq) \xrightarrow{\text{enzymes}} 2C_2H_5OH(aq) + 2CO_2(g)$$

glucose (a sugar) → ethanol + carbon dioxide

ALCOHOLIC DRINKS

Alcoholic drinks are produced by fermentation. As well as a source of carbohydrate, which may vary from barley in the case of beer, to grapes in the case of wine, other ingredients are added for flavour, for example hops in the brewing of beer. The carbon dioxide produced during fermentation is retained in some drinks such as beer and sparkling wines where it causes fizziness.

As well as ethanol, alcoholic drinks contain small quantities of many other organic compounds including carboxylic acids, unfermented sugars, aldehydes and esters—the exact blend in any particular drink being responsible for its flavour.

Alcoholic drinks are usually divided into beers, of alcohol content under 5%, wines, with around 10% alcohol, and spirits, with over 30% alcohol. The maximum alcohol content which can be produced by fermentation is about 15%, depending on the type of yeast used. Stronger drinks must be made by adding extra alcohol, to produce fortified wines like sherry, or by distilling off the alcohol to produce spirits like brandy, rum, whisky and vodka.

THE USE OF ALCOHOLS IN COSMETICS

Ethanol is an ingredient in many cosmetics. It comprises around 70% of a typical after-shave lotion. Ethanol is quite volatile and its function is to evaporate, absorbing its ethalpy change of vaporization and cooling the skin. This closes the pores in the skin which have been opened by the hot water used in shaving. The ethanol also acts as a mild antiseptic preventing infection in any shaving nicks.

Ethanol is also an ingredient in many types of nail varnish. It acts as a solvent for the lacquer. When applied to the nails the ethanol evaporates leaving a layer of lacquer on the nail.

Nail varnish remover may also contain ethanol, but more frequently contains propanone or ethyl ethanoate to dissolve the lacquer.

Polyols (alcohols with more than one —OH group) are found in moisturizing creams where they act as humectants (they attract water vapour from the air). Water vapour molecules from the air are attracted by the —OH groups of the alcohols in the moisturizing cream and are 'trapped' by hydrogen bonding. Alcohols used in this way are propane-1,2,3-triol (glycerine), propane-1,2-diol and hexane-1,3,4,6-tetraol (sorbitol).

```
    H    H    H
    |    |    |
H — C —— C —— C — H        Glycerine
    |    |    |
    OH   OH   OH

    H    H    H
    |    |    |
H — C —— C —— C — H        Propane-1,2-diol
    |    |    |
    OH   OH   H

    H    H    H    H    H    H
    |    |    |    |    |    |
H — C —— C —— C —— C —— C —— C — H    Sorbitol
    |    |    |    |    |    |
    OH   H    OH   OH   H    OH
```

THE STRENGTH OF ALCOHOLIC DRINKS

Some alcoholic drinks contain a high enough percentage of alcohol to be flammable. This was the basis of an early method of measuring their concentration or strength. 100° proof spirit was the weakest mixture of alcohol and water which, when poured onto gunpowder, would not stop it igniting. Nowadays 100° proof is defined as having a density of 0.92 g cm^{-3} at 51 °F (approximately 282 K). This corresponds to about 60% alcohol.

Brandy has a high percentage of alcohol

A

THE EFFECT OF ALCOHOL ON THE BODY

Alcohol, along with caffeine, is probably the most widely used drug of all. There are few adults who have not experienced its effects. Although it is often regarded as a stimulant, it is in fact a sedative. Its apparently stimulating effects, notable at parties, result from alcohol's ability to reduce normal inhibitions. Alcohol tends to heighten the mood of the drinker, making the extrovert party-goer more extroverted, the depressive more depressed, and so on.

Alcohol is absorbed into the bloodstream through the walls of the stomach, small intestine and colon at a rate which depends on a number of factors, including the presence of food which slows down absorption. Once in the bloodstream it is transported to all parts of the body. Thus the same dose of alcohol will have a less marked effect on a large person than on a smaller one. One effect is to open up small blood vessels in the skin, allowing a greater blood flow to the skin and cooling the core of the body.

High doses of alcohol produce disorientation, confusion, blurred vision, slurred speech, poor muscle control, nausea and vomiting—no conditions under which to drive! The maximum level of alcohol in the blood at which driving is permitted in the UK is 80 mg of alcohol per 100 cm³ of blood, which corresponds to a level in the breath of 35 micrograms per 100 cm³ of breath. However, this is in no sense a 'safe' level—it is obviously safer not to drink and drive at all.

The body can dispose of about 10 cm³ of alcohol per hour, mainly by oxidation to ethanal, although some is excreted in sweat, urine and the breath. Since a pint of beer contains about 20 cm³ of alcohol (approximately 3% of approximately 600 cm³), it takes the body two hours to remove all the alcohol, so the body would take ten hours to disperse all the alcohol in five pints of beer. These are all average figures, but it would be quite possible for a drinker to be over the legal limit for driving the morning after a heavy night's drinking. It seems likely that the breakdown products such as ethanal are responsible for many of the symptoms of a hangover.

Alcohol is *not* recommended for hypothermia victims as it lowers the core temperature of the body

Fermentation was once a major source of ethanol for industrial use. Now ethanol is mostly made by catalytic hydration of ethene obtained from cracking:

$$H_2C{=}CH_2 + H_2O \xrightarrow[\text{catalyst}]{\text{phosphoric acid}} H{-}\underset{H}{\overset{H}{C}}{-}\underset{H}{\overset{H}{C}}{-}OH$$

igure 31.10 **Industrial chemistry of nethanol**

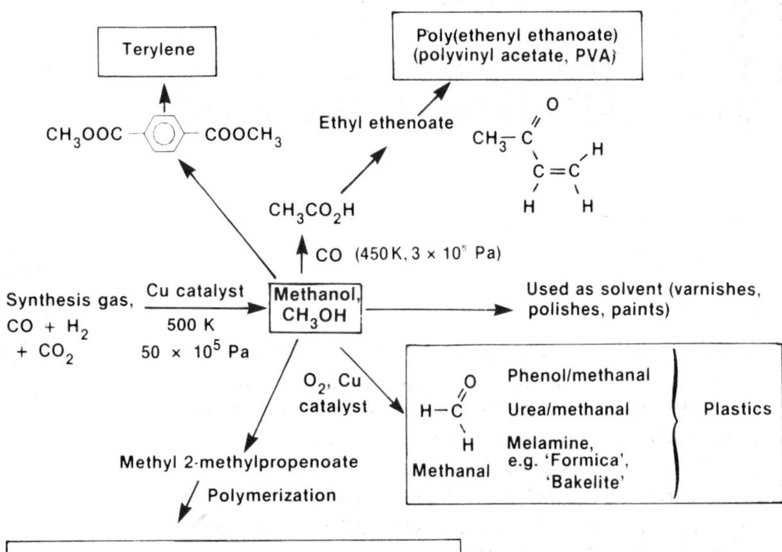

Figure 31.11 Industrial chemistry of ethanol

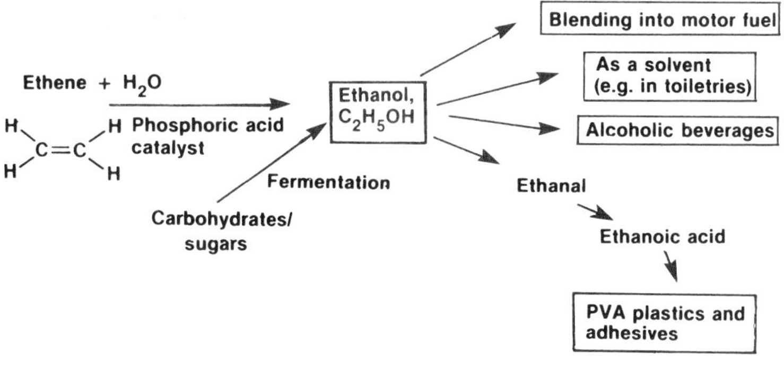

Figure 31.12 Industrial chemistry of ethane-1, 2-diol

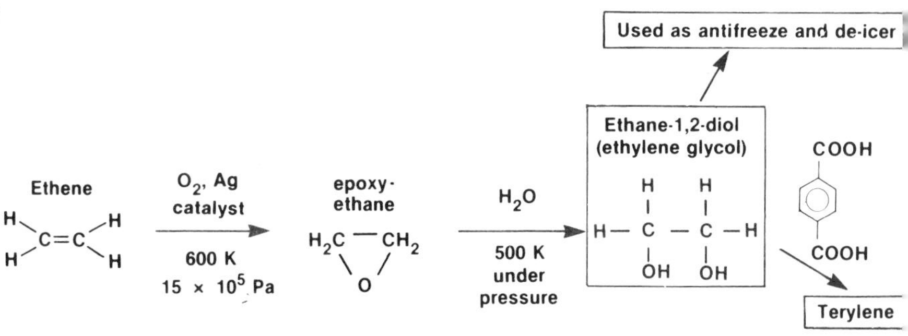

However, in the future, fermentation may again become a significant source of ethanol as stocks of crude oil are depleted. Already Brazil produces alcohol by fermentation for use as a motor fuel (see box *Alternative fuels* **section 9.3.3**).

Alcohols have many uses in finished products, for example, many toiletries contain alcohols and ethanol is the liquid (dyed red) in many thermometers. However, their major uses are as intermediates in the manufacture of other organic compounds. This is because of their reactivity—they can be easily made and easily converted into other compounds.

The industrial chemistry of some important alcohols is summarized in **Figures 31.10, 31.11** and **31.12**.

A

DISINFECTANTS AND ANTISEPTICS

Antiseptics and disinfectants both kill germs. Antiseptics can be applied to the skin while disinfectants are normally applied only to surfaces as they may irritate the skin. Phenol is a disinfectant but causes blistering of the skin, while its derivatives have both better germicidal properties and are safer to use as antiseptics.

The structures of phenol and some other germicides are given opposite with their germ-killing power compared with that of phenol. Household disinfectants are aqueous solutions of phenol derivatives. As phenols are not particularly soluble in water, a soap is first made which dissolves the active ingredients. It is usual to add colouring and a perfume—often one smelling of pine.

Once a compound is known to have a particular medical effect such as germ killing, it is usual to synthesize close derivatives of it and test them for the required effect. This process often leads to the discovery of compounds which have fewer side effects.

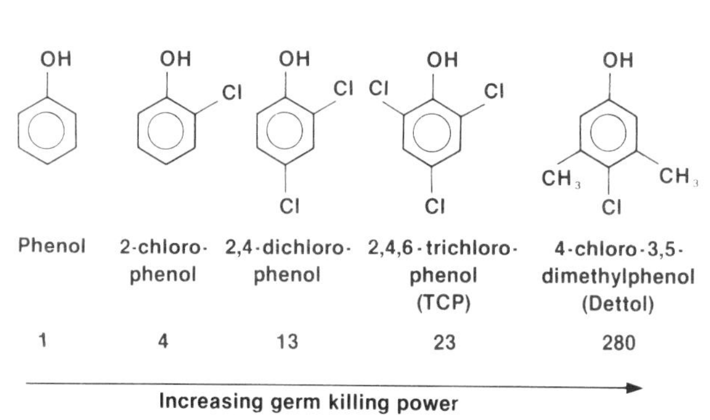

31.9.2 Phenol

The industrial chemistry of phenol is shown in **Figure 31.13.**

gure 31.13 Industrial chemistry of enol

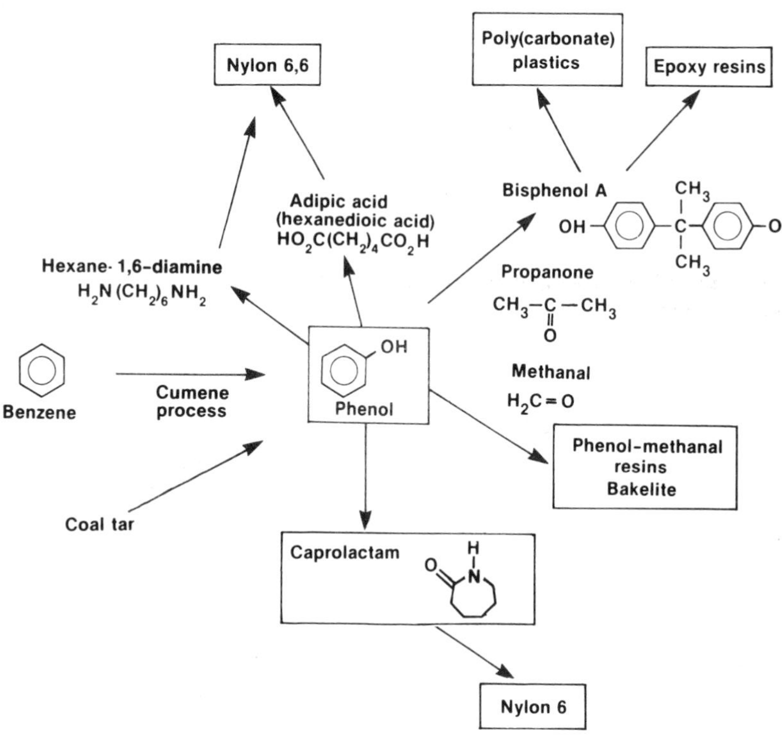

31.10 Summary

- Alcohols and phenols both contain the —OH functional group—bonded to an alkyl group in alcohols and directly to a benzene ring in phenols.
- Alcohols are classified as primary, secondary or tertiary if they have respectively 1, 2 or 3 R— groups attached to the carbon bearing the —OH group.
- Alcohols and phenols have significantly higher melting and boiling temperatures than alkanes or arenes of comparable relative molecular mass because of hydrogen bonding between the molecules.
- Alcohols with chains of four carbons or less mix well with water. All the alcohols mix well with non-polar solvents.
- The infra-red spectra of alcohols show peaks at 3300–3600 cm^{-1} due to the O—H stretch, a C—O—H bending vibration at around 1350 cm^{-1} and the C—O stretch at between 1050 and 1200 cm^{-1}.
- The $C^{\delta+}$—$O^{\delta-}$—$H^{\delta+}$ unit is polar. The $C^{\delta+}$ can be attacked by nucleophiles and the $O^{\delta-}$ can act as a weak nucleophile.
- The reactions of alcohols are summarized in **Figure 31.14** (overleaf).
- Phenols show reactions of the benzene ring and of the —OH group.
- The —OH group releases electrons on to the ring making it more reactive to electrophilic substitution reactions, especially at the 2, 4 and 6 positions.
- The C—OH bond in phenol is strengthened because of overlap between the π orbital on the ring and the lone pairs of the oxygen. This also reduces the δ^+ character of the carbon. Both factors make phenols less reactive towards nucleophiles.
- The phenoxide ion is stabilized by delocalization of the negative charge. This makes phenols stronger acids than alcohols.
- The reactions of phenol are summarized in **Figure 31.15**.

Figure 31.14 Reactions of alcohols using ethanol as an example

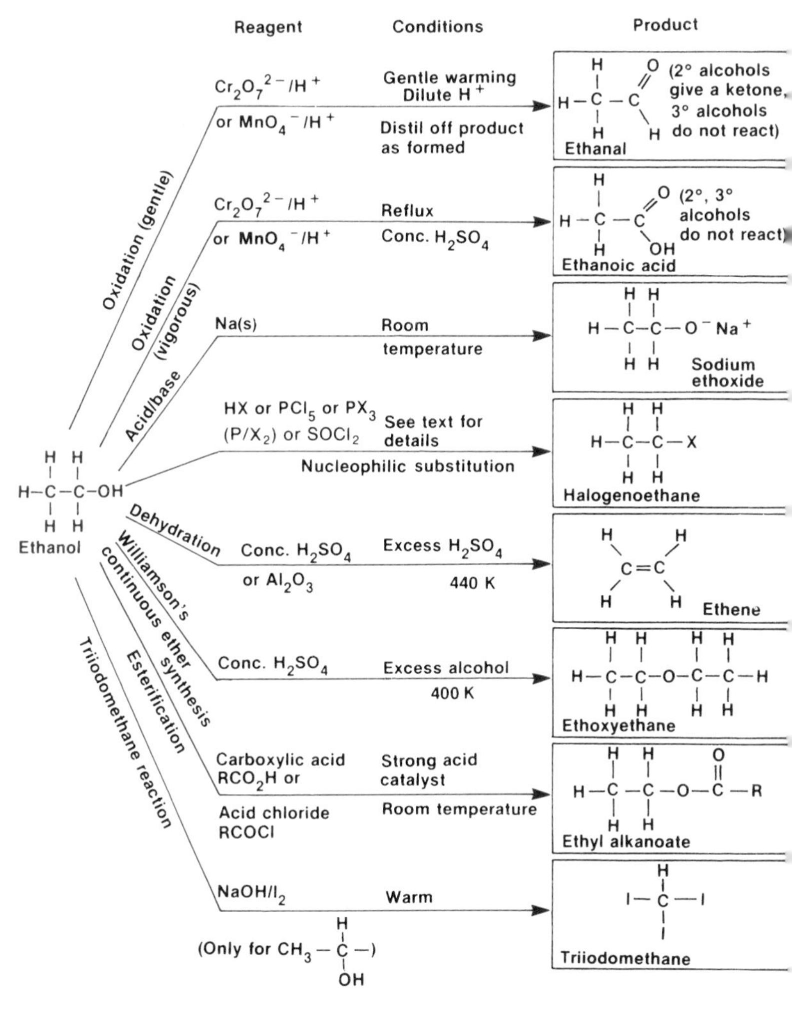

Figure 31.15 Reactions of phenol

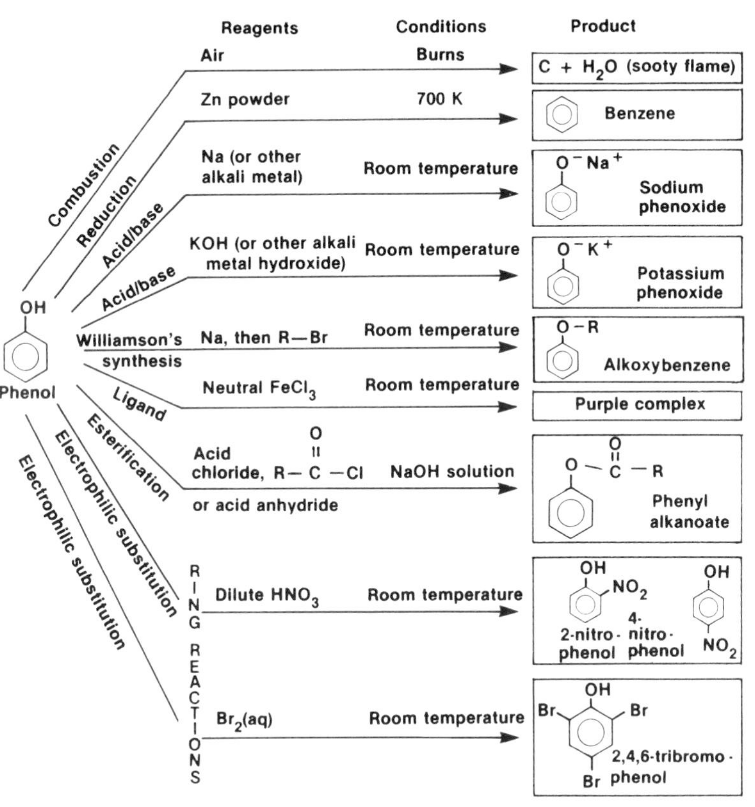

• Alcohols can be prepared by the following reactions:

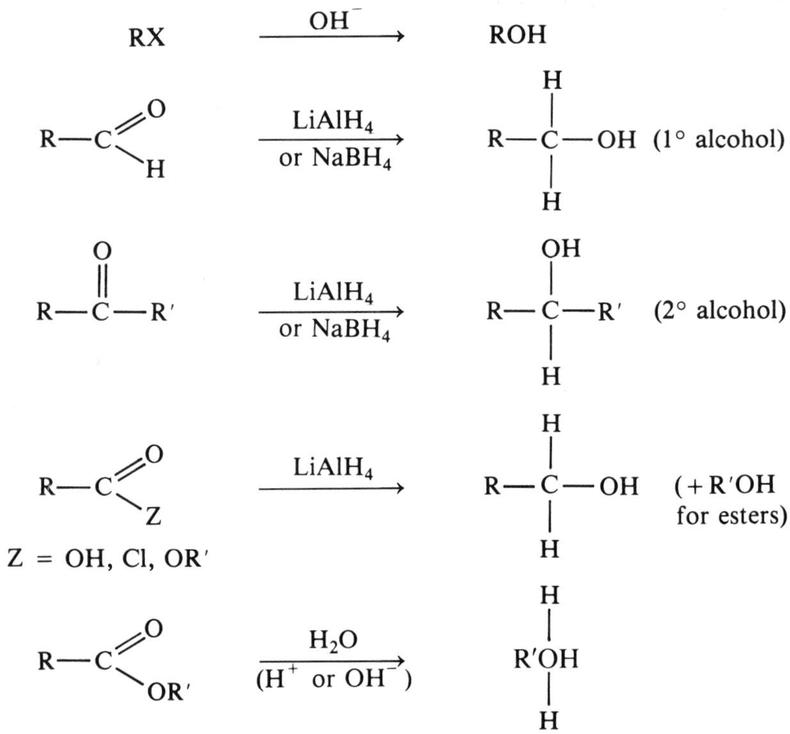

$$\mathrm{RX} \xrightarrow{\mathrm{OH^-}} \mathrm{ROH}$$

$$\mathrm{R{-}C({=}O)H} \xrightarrow[\text{or } \mathrm{NaBH_4}]{\mathrm{LiAlH_4}} \mathrm{R{-}CH_2{-}OH} \quad (1^\circ \text{ alcohol})$$

$$\mathrm{R{-}C({=}O){-}R'} \xrightarrow[\text{or } \mathrm{NaBH_4}]{\mathrm{LiAlH_4}} \mathrm{R{-}CH(OH){-}R'} \quad (2^\circ \text{ alcohol})$$

$$\mathrm{R{-}C({=}O)Z} \xrightarrow{\mathrm{LiAlH_4}} \mathrm{R{-}CH_2{-}OH} \quad (+\mathrm{R'OH} \text{ for esters})$$

$Z = \mathrm{OH, Cl, OR'}$

$$\mathrm{R{-}C({=}O)OR'} \xrightarrow[(\mathrm{H^+} \text{ or } \mathrm{OH^-})]{\mathrm{H_2O}} \mathrm{H{-}R'OH{-}H}$$

• Phenol is prepared by the action of sodium hydroxide on benzene-sulphonic acid.

• Both alcohols and phenols are important industrial intermediates because of their reactivity. Many of the final products are plastics or fibres.

31.11 Questions

1 Name the following:

A)

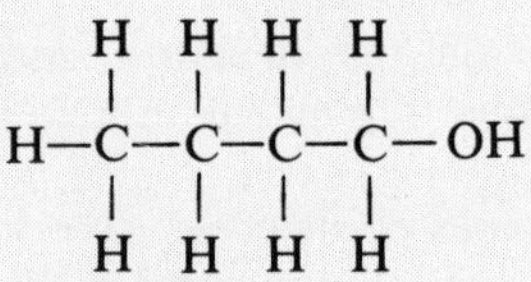

H H H H
H–C–C–C–C–OH
H H H H

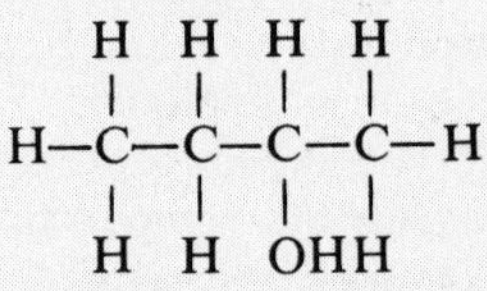

B)

H H H H
H–C–C–C–C–H
H H OH H

C) (benzene ring with OH and CH_3 on adjacent carbons)

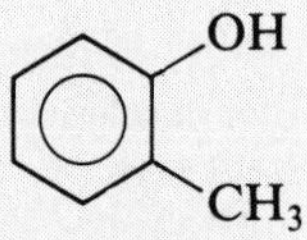

D)

OH
H–C–H
(benzene ring)

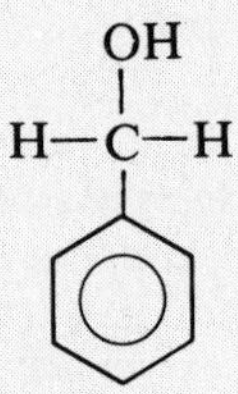

2 Give the structural formula of:
A) 2-chloropropan-1-ol;
B) 2-methylpropan-2-ol;
C) 4-nitrophenol;
D) benzene-1, 4-diol.

3 Classify the following as 1°, 2° or 3° alcohols or phenols:

A)

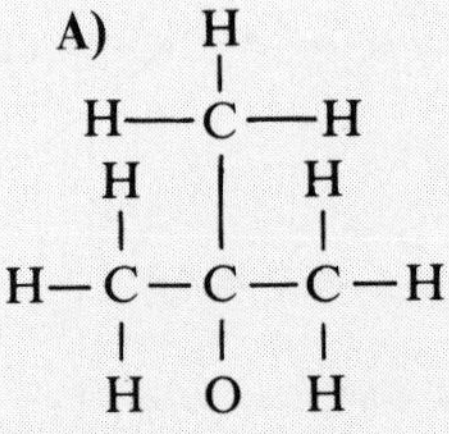

H
H—C—H
H H
H–C–C–C–H
H O H

B)

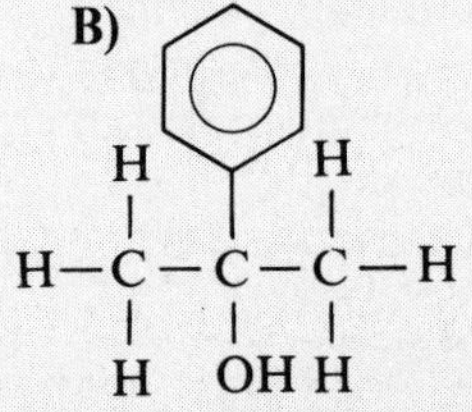

H H
H–C–C–C–H
H OH H

C)

H H H H H
H–C–C–C–C–C–H
H H OH H H

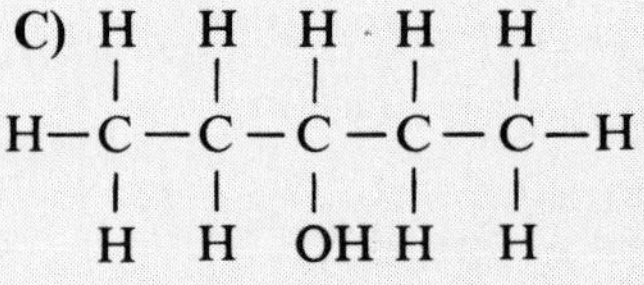

D) 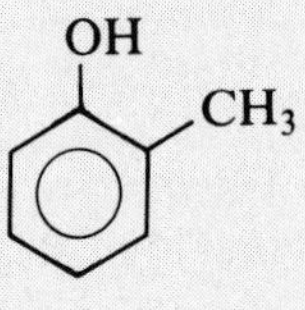 (benzene ring with OH and CH_3 on adjacent carbons)

4 Compare and contrast the reactions of phenol with those of ethanol. Explain any difference in terms of the bonding and structure of the molecules. Give reaction mechanisms where appropriate.

5 How would the reactions of the following alcohols differ?

A)
$$H_3C-\underset{OH}{\overset{CH_3}{\overset{|}{\underset{|}{C}}}}-CH_3$$

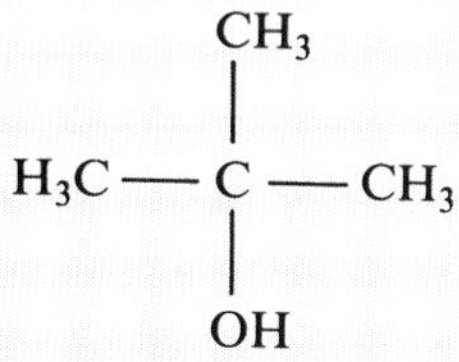

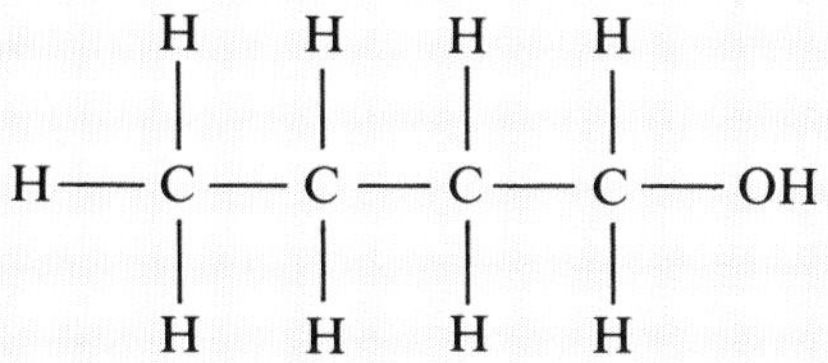

B)
$$H-\underset{H}{\overset{H}{C}}-\underset{H}{\overset{H}{C}}-\underset{H}{\overset{H}{C}}-\underset{H}{\overset{H}{C}}-OH$$

C)
$$H-\underset{H}{\overset{H}{C}}-\underset{H}{\overset{H}{C}}-\underset{OH}{\overset{H}{C}}-\underset{H}{\overset{H}{C}}-H$$

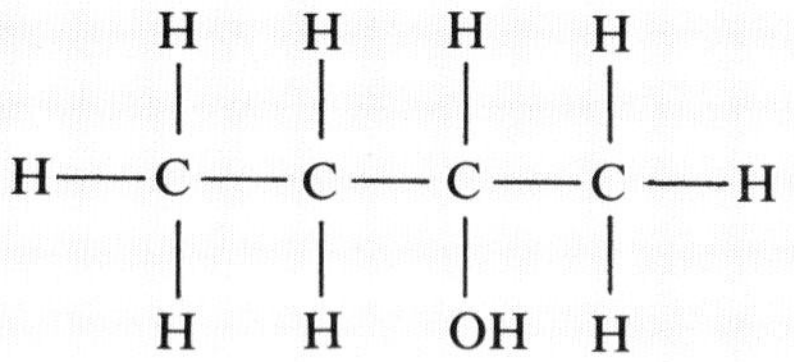

6 The nitro group, $—NO_2$, is an electron-withdrawing group, while the methyl group, $—CH_3$, releases electrons. Put the following in order of acid strength. Explain your answer.

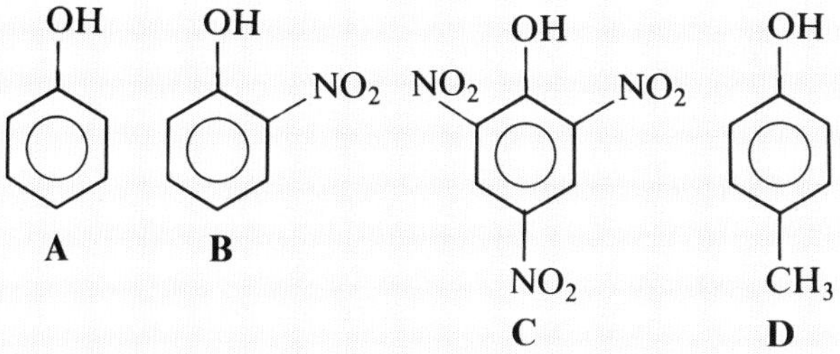

A: phenol (C_6H_5OH); B: 2-nitrophenol (OH, NO_2 ortho); C: 2,4,6-trinitrophenol (OH; NO_2, NO_2, NO_2); D: 4-methylphenol (OH, CH_3 para)

7 The following diagram shows the molecular formulae of a number of related compounds.

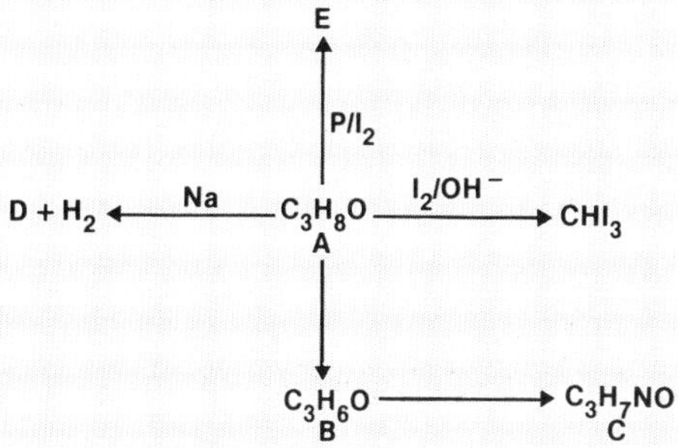

$$D + H_2 \xleftarrow{Na} \underset{A}{C_3H_8O} \xrightarrow{I_2/OH^-} CHI_3$$

$$\underset{A}{C_3H_8O} \xrightarrow{P/I_2} E$$

$$\underset{A}{C_3H_8O} \longrightarrow \underset{B}{C_3H_6O} \longrightarrow \underset{C}{C_3H_7NO}$$

A) i Give the names and structural formulae of *three* compounds of molecular formula C_3H_8O.

ii Which *one* of these compounds corresponds to compound *A* in the diagram? Give your reasons.

B) What reagent(s) would be needed to:

i prepare compound *B* from compound *A*?

ii convert compound *B* into compound *C*?

C) i Identify the compounds *D* and *E*.

ii Give the name and formula of the organic product of the reaction between compounds *D* and *E*.

D) In an experiment a student obtained 39 g of solid CHI_3 from 10 g of *A*. What was his percentage yield?

(London 1979)

8 A) Describe briefly, giving essential conditions and equations, the laboratory conversion of propene into:

i propan-2-ol;

ii propane-1, 2-diol.

B) Sodium metal is added carefully to propan-2-ol and then the product is warmed with iodoethane. Compound E, $C_5H_{12}O$, distills over.

i Deduce the structural formula of *E*.

ii Name and write down the structure of a functional group isomer of *E*.

C) State and explain the type of stereoisomerism shown by propane-1,2-diol. Illustrate your answer with clear diagrams.

(Oxford 1986)

9 A) Arrange the following compounds A, B and C in order of increasing acid strength and give reasons for your order.

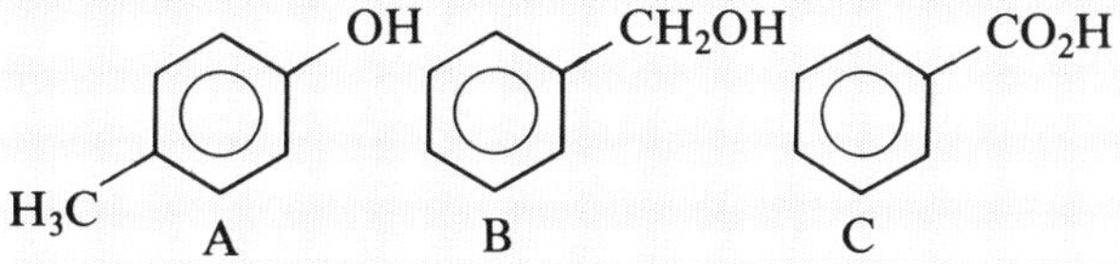

A: 4-methylphenol (H_3C–C_6H_4–OH); B: phenylmethanol ($C_6H_5CH_2OH$); C: benzoic acid ($C_6H_5CO_2H$)

B) Describe a test-tube experiment which would distinguish between the isomers A and B above, and describe what you would see for each isomer.

(Oxford 1987)

10 Linalool, boiling point 198 °C, is a naturally occurring compound that is optically active. The (−) form occurs in rose oil and the (+) form occurs in orange oil.

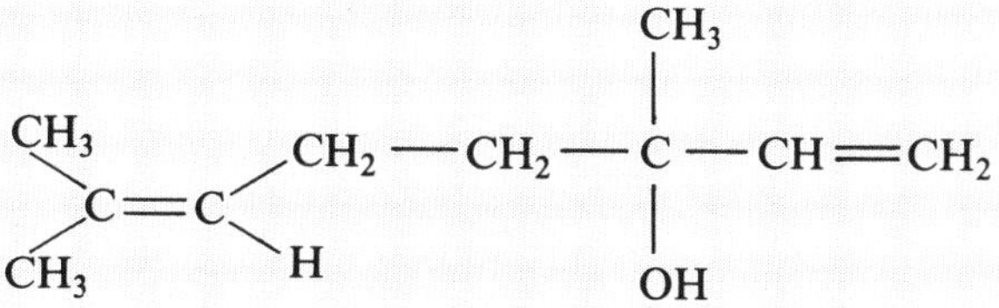

$$(CH_3)_2C{=}CH-CH_2-CH_2-\underset{OH}{\overset{CH_3}{\overset{|}{\underset{|}{C}}}}-CH{=}CH_2$$

A) Copy the structural formula drawn above. Indicate, by means of a * on the structural formula, the chiral centre (asymmetric carbon atom) in the structure.

B) What difference could be detected experimentally between the (−) form and the (+) form of linalool?

C) What specific structural features could be deduced from the infra-red spectrum of linalool?

D) What reaction would you expect between linalool and the following reagents, drawing structrural formulae if it helps you to explain your answer:

i sodium metal;

ii bromine;

iii ethanoyl chloride?

E) What reagent(s) would you use to reform linalool from the product of the reaction with ethanoyl chloride?

F) Linalool can be extracted from orange peel. What type of procedure would you use to attempt to obtain a sample of linalool (impure)?

(Nuffield 1984)

32 Carbonyl compounds

32.1 Introduction

The *carbonyl* group is $\rangle C{=}O$. The term carbonyl compound includes two functional groups—aldehydes and ketones.

In aldehydes, the carbon bonded to the oxygen (the carbonyl carbon) has at least one hydrogen atom bonded to it, so the general formula of an aldehyde is:

$$\begin{array}{l} R \\ \quad \rangle C{=}O \\ H \end{array}$$ **, sometimes written as RCHO.**

In ketones, the carbonyl carbon has two R groups so the formula of a ketone is:

$$\begin{array}{l} R \\ \quad \rangle C{=}O \\ R' \end{array}$$

The R groups in both aldehydes and ketones may be alkyl or aryl.

Although aldehydes and ketones are not met often in everyday life, they have been described as the backbone of organic chemistry because of their importance in the synthesis of other compounds.

32.2 Formulae, naming and structure

Aldehydes have the general formula $\begin{array}{l} R \\ \quad \rangle C{=}O \\ H \end{array}$ where R may be H.

Ketones have the general formula $\begin{array}{l} R \\ \quad \rangle C{=}O \\ R' \end{array}$. The two R groups may be the same or different, but neither can be H.

32.2.1 Naming

Aldehydes are named using the suffix -al when the aldehyde is the principal group and the prefix oxo- when it is not. The carbon of the aldehyde functional group is counted as part of the carbon chain of the root. So:

The bonding in ethanal

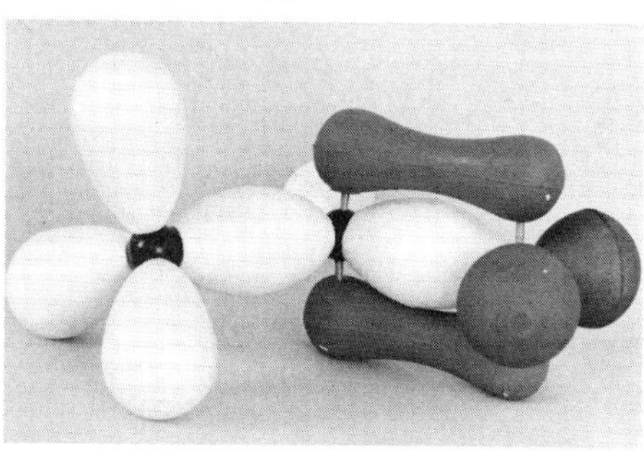

```
     H
     |      O
     |    //
H —  C — C
     |    \
     H     H
```

is ethanal

```
O          O
 \\       //
   C — C
  /      \
H         OH
```

is oxoethanoic acid as the aldehyde is not the pri cipal group (see **section 26.5**).

The aldehyde group can only occur at the end of a chain so locants a not needed. When an aldehyde is a substituent on, for example, a benze ring, the suffix -carbaldehyde is used and the carbon is not counted part of the root name.

```
            O
            ||
So:         C
           / \
   (C6H5)     H
```

is counted as a derivative of benzene (not as derivative of methylbenzene) and called benzenecarbaldehyde. It is som times still called benzaldehyde.

Ketones are named using the suffix -one when the ketone is the princip group. When it is not the principal group, it shares with aldehydes t prefix oxo-. No confusion should occur as in ketones the carbonyl grou must come within the chain whereas in aldehydes it must come at t end. As with aldehydes, the carbon atom of the ketone functional grou is counted as part of the root. So the simplest ketone:

```
    H
    |
H — C — H
     \
      C = O
     /
H — C — H
    |
    H
```

is called propanone

```
    H   O      O
    |   ||   //
H — C — C — C
    |         \
    H          OH
```

is 2-oxopropanoic acid as the ketone is not t principal group (see **section 26.5**).

No ketone with fewer than three carbon atoms is possible. However:

```
    H
    |
H — C — H
    |
    C = O
    |
  (C6H5)
```

is correctly named phenylethanone.

Locants are not needed in propanone or in butanone:

```
    H   H   O   H
    |   |   ||  |
H — C — C — C — C — H
    |   |       |
    H   H       H
```

as there is no possible ambiguity abo the position of the carbonyl group.

gure 32.1 **Bonding in the carbonyl oup**

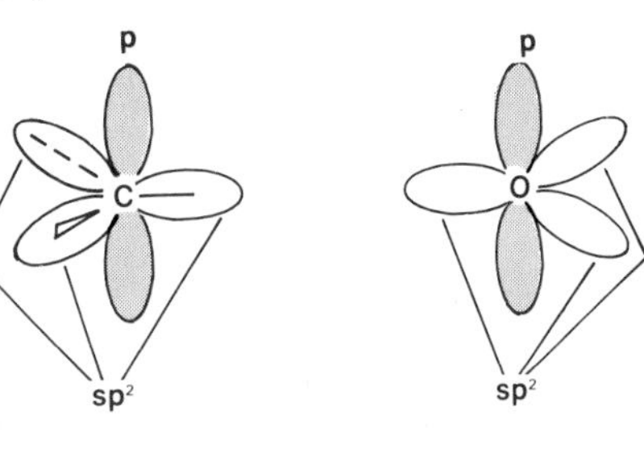

However there are two possible isomeric pentanones:

```
      H   H   O   H   H
      |   |   ‖   |   |
  H—C—C—C—C—C—H
      |   |       |   |
      H   H       H   H
```

pentan-3-one

```
      H   O   H   H   H
      |   ‖   |   |   |
  H—C—C—C—C—C—H
      |       |   |   |
      H       H   H   H
```

pentan-2-one

32.2.2 Shapes

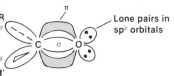

gure 32.2 **Approximate bond angles in ethanal**

CH_3 124° 118° C=O H 118°

The carbon atom of the carbonyl group is hybridized sp^2. The C—O σ bond is formed by overlap of an sp^2 orbital on the carbon with an sp^2 orbital on the oxygen and the π bond by overlap of p orbitals, **Figure 32.1.** The molecule is therefore flat and has a trigonal shape (R, R′, C and O are all in the same plane) with the R—C—R′ angle (or R—C—H angle) a little less than 120° as the four electrons in the C=O bond repel more than the two electrons in the C—R and C—R′ (or C—H) bonds, **Figure 32.2.**

32.3 Physical properties of carbonyl compounds

rmalin is used as a preservative

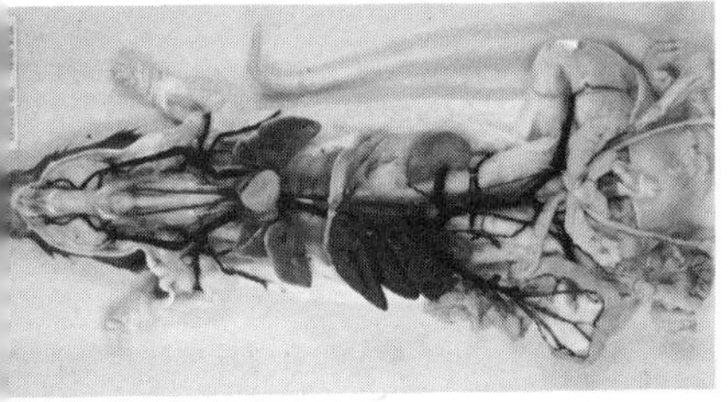

The carbonyl group is strongly polar $C^{\delta+}=O^{\delta-}$ so dipole–dipole forces between the molecules are quite significant. This leads to boiling temperatures higher than those of alkanes of comparable relative molecular mass but not as high as those of alcohols, where hydrogen bonding can occur between the molecules:

butane,	$CH_3CH_2CH_2CH_3$	$M_r = 58$	$T_b = 273\,K$
propanone,	CH_3COCH_3	$M_r = 58$	$T_b = 329\,K$
propan-1-ol,	$CH_3CH_2CH_2OH$	$M_r = 60$	$T_b = 370\,K$

The shorter-chain carbonyl compounds are completely miscible with water due to the formation of hydrogen bonds, **Figure 32.3.**

gure 32.3 **Hydrogen bonding tween propanone and water**

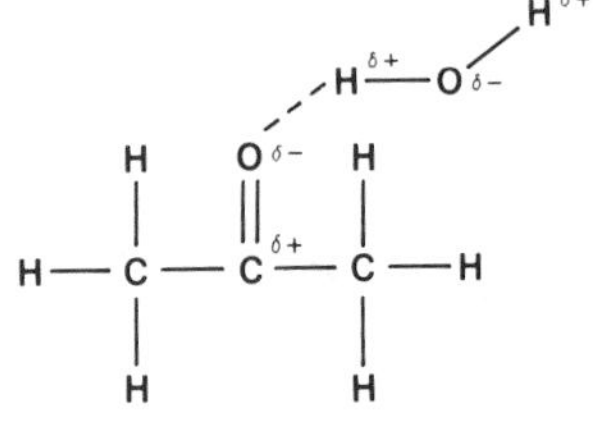

Solubility decreases with increasing length of the carbon chain. Propanone is sometimes used for quickly drying laboratory glassware. The wet glassware is rinsed with propanone which then evaporates rapidly as it is much more volatile than water.

Methanal, HCHO, is a gas at room temperature, often encountered as a 40% aqueous solution called 'formalin' used for preserving biological specimens, but other short-chain aldehydes and ketones are liquids. They have characteristic, fairly pleasant smells (propanone is found in many brands of nail-varnish remover). Benzenecarbaldehyde smells of almonds and is used to scent soaps and flavour food.

32.4 Infra-red spectra of carbonyl compounds

The infra-red spectrum of propanone is shown in **Figure 32.4.** The carbonyl group shows a prominent peak at around 1700 cm^{-1} due to a C=O stretching vibration. The peak is usually strong and sharp. In aldehydes it is at between 1740 and 1720 cm^{-1} while in ketones it is between 1700 and 1680 cm^{-1}.

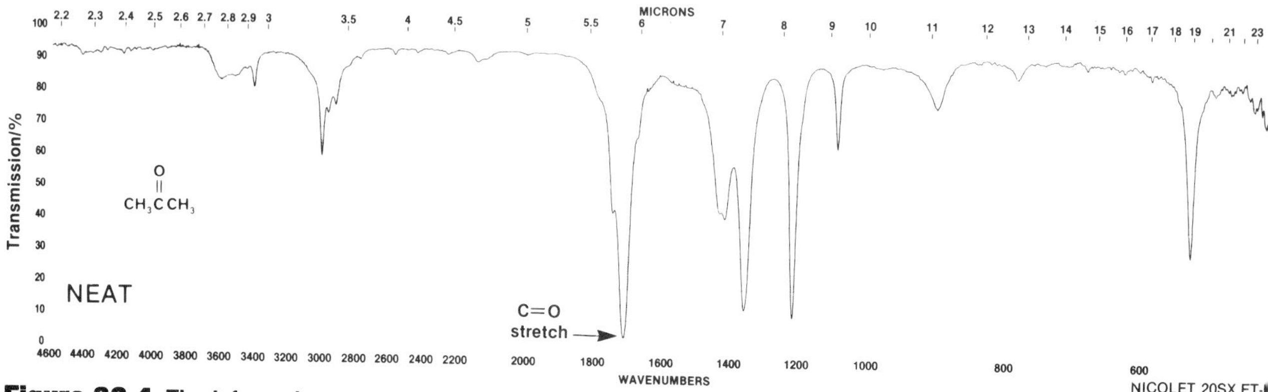

Figure 32.4 **The infra-red spectrum of propanone**

32.5 Reactivity of carbonyl compounds

The C=O bond in carbonyl compounds is strong (approximate 740 kJ mol^{-1}). This is stronger than C=C (612 kJ mol^{-1}) and more tha twice as strong as C—O (358 kJ mol^{-1}) while C=C is less than twice strong as C—C (347 kJ mol^{-1}). Despite this, most reactions of carbon compounds involve the C=O bond. The reason is the large difference electronegativity between carbon and oxygen which makes the C=O bon strongly polar.

This means that nucleophilic reagents can attack the $C^{\delta+}$ whi electrophiles like H^+ can attack the $O^{\delta-}$. Since there is a double bon carbonyl compounds are unsaturated and therefore addition reactions a possible. The most typical reactions of the carbonyl group are **nucleophili additions**.

Reactions of the rest of the carbon skeleton can take place, especial those of aromatic rings. The carbonyl group is a deactivating substitue as, indeed, are most unsaturated groups.

32.6 Reactions of carbonyl compounds

Carbonyl compounds react in a wide variety of ways.

32.6.1 Nucleophilic addition reactions

The general pattern for this reaction is as follows:

$$RR'C^{\delta+}{=}O^{\delta-} \; + \; :Nu^- \longrightarrow RR'C(Nu)O^- \xrightarrow[\text{(from solvent)}]{H^+} RR'C(Nu)O{-}H$$

Not all nucleophiles are negatively charged, so the fine details of th reaction may differ with neutral nucleophiles.

Addition of hydrogensulphite ions

The nucleophile is the hydrogensulphite ion (hydrogensulphate(IV) ion) SO_3H^-, generated by adding a solution of sodium hydrogensulphite:

$$RR'C^{\delta+}{=}O^{\delta-} \; + \; :SO_3H^- \longrightarrow RR'C(SO_3H)O^- \xrightarrow{Na^+} RR'C(SO_3^-Na^+)OH$$

The overall effect is the addition of $NaHSO_3$ across the C=O double bond.

Addition of hydrogen cyanide

Here the nucleophile is ^{-}CN:

$$R(R')C^{\delta+}{=}O^{\delta-} + {:}CN^{-} \longrightarrow R(R')C(CN)O^{-} \xrightarrow{H^+} R(R')C(CN)OH$$

a 2-hydroxynitrile

A mixture of sodium cyanide and acid is used to avoid using the poisonous gas hydrogen cyanide. The ^{-}CN group is called **nitrile** and it reacts with acids to give a carboxylic acid, so reaction of 2-hydroxynitriles gives 2-hydroxycarboxylic acids.

Addition of alcohols

These also add on across the double bond of aldehydes in the same way, $RO^{\delta-}{-}H^{\delta+}$ being the nucleophile:

$$R(H)C{=}O \xrightarrow[\text{HCl catalyst}]{R'O{-}H} R(H)C(OR')OH$$

This product is called a **hemiacetal**.

Further reaction can take place to give an **acetal**:

$$\underset{\text{a hemiacetal}}{R(H)C(OR')OH} + R''OH \longrightarrow \underset{\text{an acetal}}{R(H)C(OR')OR''} + H_2O$$

Hemiacetals and acetals cannot usually be isolated, as they are stable only in solution. **Ketals**, analogous compounds made from ketones, are more difficult to prepare.

Addition of water

Water can also add on across the C=O bond, for example:

$$H_2C{=}O + H_2O \rightleftharpoons H_2C(OH)_2$$

This equilibrium is significantly over to the right only for methanal and ethanal. The product is stable only in solution.

All these reactions involve addition of $H^{\delta+}Z^{\delta-}$ across the double bond $C^{\delta+}{=}O^{\delta-}$ in such a way that the $H^{\delta+}$ adds on to the $O^{\delta-}$ and the $Z^{\delta-}$ to the $C^{\delta+}$.

32.6.2 Addition–elimination (condensation) reactions

As the name implies, these reactions involve first addition to form an unstable intermediate which then rapidly loses (**eliminates**) a molecule of water.

Reaction with hydrazine ($H_2\ddot{N}—\ddot{N}H_2$)

The reaction occurs as follows:

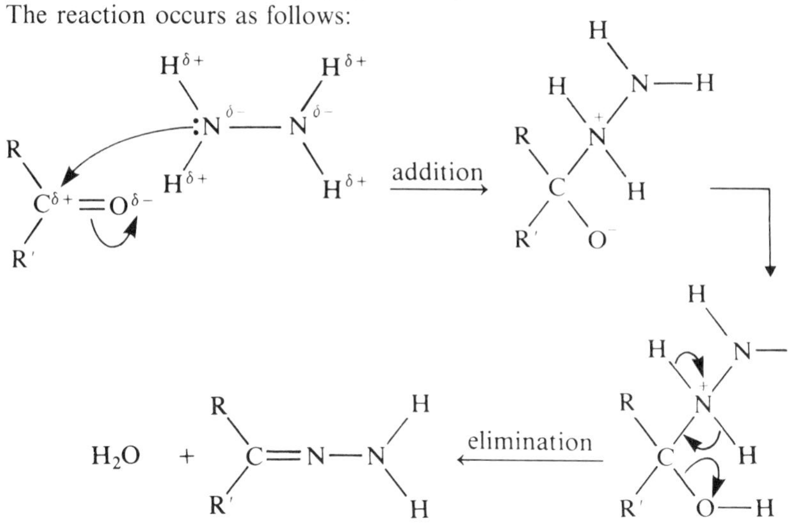

addition

elimination

H_2O + $RR'C=N—NH_2$

The product is called a **hydrazone**. You may well find it easier to rememb the product by concentrating on the elimination step:

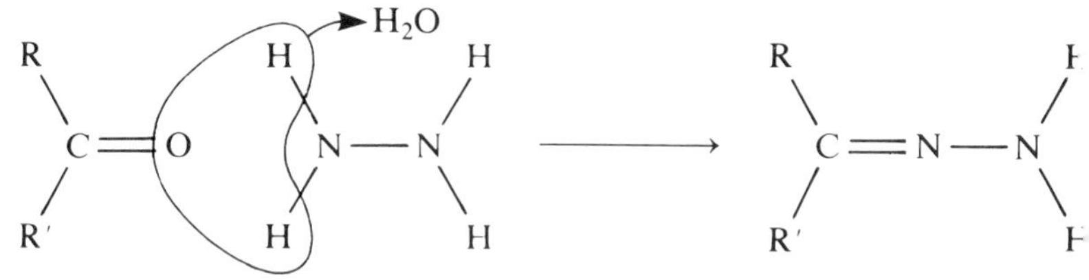

hydrazine a hydrazone

Other compounds related to ammonia will undergo similar reactions:

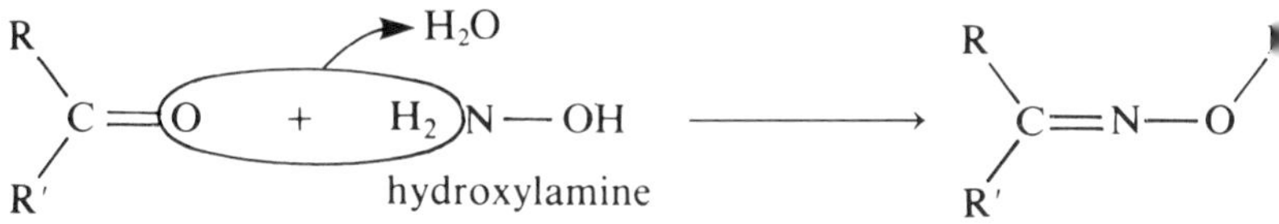

$RR'C=O$ + $H_2N—OH$ ⟶ $RR'C=N—OH$ + H_2O

hydroxylamine an oxime

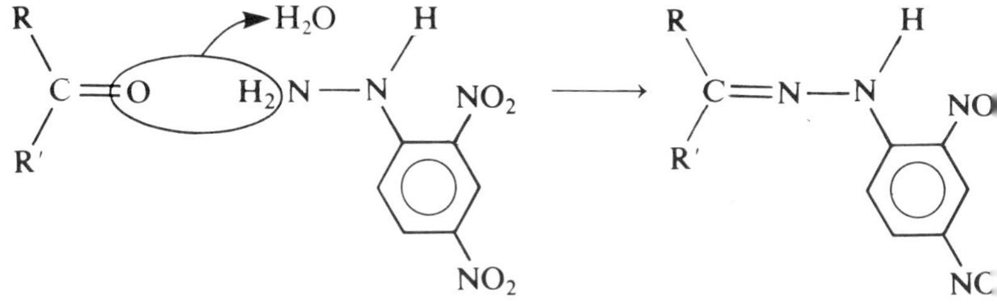

2,4-dinitrophenylhydrazine (Brady's reagent) a 2,4-dinitrophenylhydrazo

The derivatives are usually named by reference to the starting materi e.g. 'ethanal oxime' or 'propanone 2,4-dinitrophenylhydrazone'.

Identification of carbonyl compounds

The derivatives made using Brady's reagent are orange coloured crystalli solids. They can be prepared easily by mixing the carbonyl compou with an acid solution of Brady's reagent in methanol. The orange cryst of the 2,4-dinitrophenylhydrazone can be filtered off and purified recrystallization (see box, opposite). Their melting temperatures can easily measured. This allows identification of the original carbon compound by reference to tables of melting points of 2,4-dinitrophen hydrazine derivatives.

E

RECRYSTALLIZATION

This is a technique used for purifying solids, for instance for removing traces of unreacted reactants from the product of a reaction. A solvent is needed in which the product to be purified is much more soluble at its boiling temperature than at, say, 273 K, the temperature of ice. Boiling solvent is added to the impure product until just enough has been added to dissolve it all. If there are any insoluble impurities they can be removed by filtering the hot solution. The solution is then cooled to 273 K with ice. The product is less soluble at this temperature. It will precipitate out and can be removed by filtration. Any impurities, which should be present only in fairly small quantities, will remain dissolved in the solvent. The whole process can be repeated if necessary. The effectiveness of the method depends on the selection of a suitable solvent in which the product is much more soluble at high temperatures than low temperatures.

E

MEASUREMENT OF MELTING TEMPERATURES

The accurate measurement of melting temperatures is a useful indicator of how pure a specimen of a compound is. The melting temperature of a compound can be found from reference books. A pure specimen of the compound should melt over a very small temperature range. Impure compounds will melt gradually over a range of a few degrees. One apparatus for measuring melting temperatures is the Thiele tube, see below.

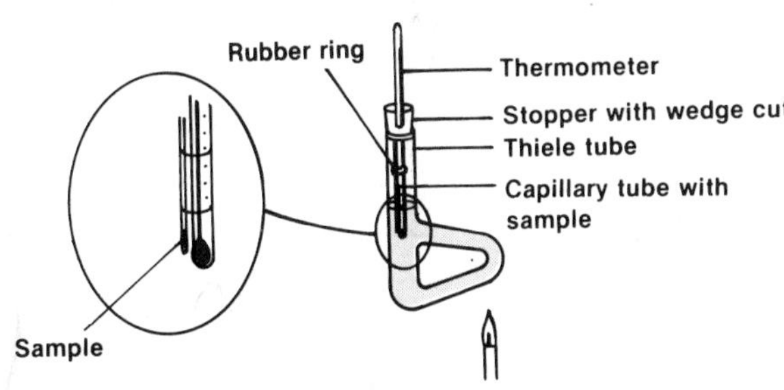

A small sample of the solid is placed in a capillary tube attached to the bulb of a thermometer inserted into a Thiele tube filled with an oil of high boiling temperature. The apparatus is gently heated as shown. Convection currents in the oil heat the capillary tube and thermometer. The melting temperature is noted and the temperature at which the sample solidifies on cooling can be used to check the first reading.

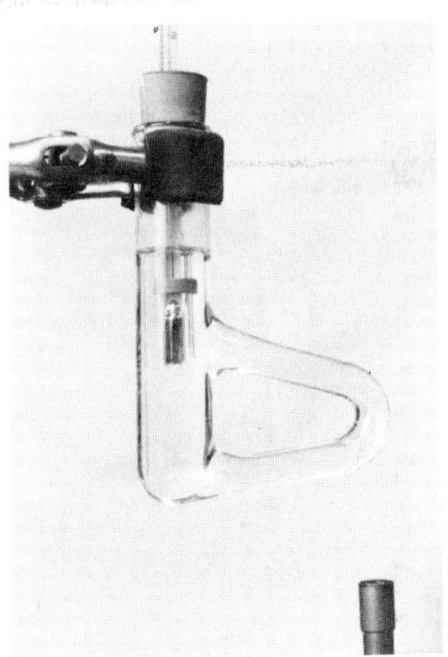

Thiele tube in use

Comparison of reactivity of aldehydes and ketones with nucleophiles

Both nucleophilic addition reactions and addition–elimination reactions involve nucleophilic attack on the carbonyl carbon, $C^{\delta+}$. Aldehydes react more readily than ketones for two reasons. Firstly, in a ketone, the electron-releasing effect of *two* alkyl groups reduces the positive charge on the carbonyl carbon making it less easily attacked by nucleophiles:

$$\mathrm{R} \rightarrow \mathrm{C}^{\delta+}{=}\mathrm{O}^{\delta-} \leftarrow \mathrm{R}$$

positive charge on $C^{\delta+}$ reduced by two R→ groups in a ketone

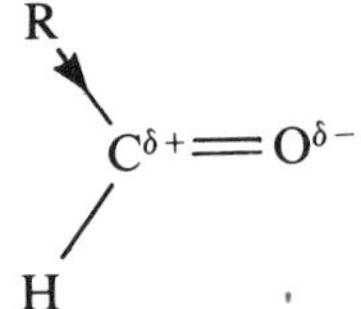

positive charge on $C^{\delta+}$ reduced by only one R→group in an aldehyde (none at all in methanal)

Secondly, the two alkyl groups of a ketone tend to get in the way of attacking nucleophiles. This is sometimes called **steric hindrance**.

32.6.3 Redox reactions

Aldehydes can be oxidized to carboxylic acids or reduced to prima alcohols:

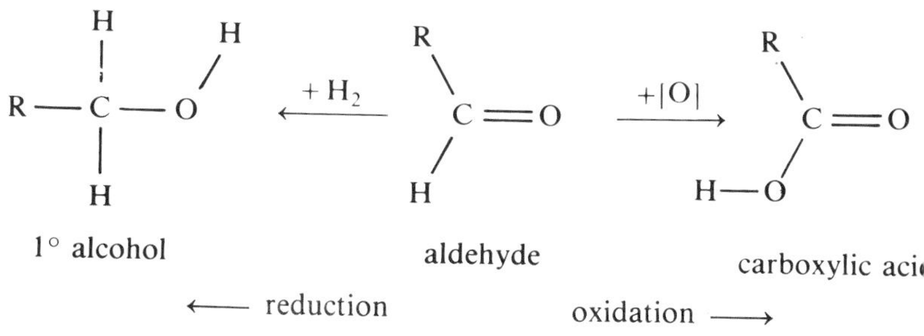

$$R-CH_2-OH \xleftarrow{+H_2} R-CHO \xrightarrow{+[O]} R-COOH$$

1° alcohol — aldehyde — carboxylic aci

⟵ reduction oxidation ⟶

Ketones *cannot* be oxidized easily but they can be reduced to secondar alcohols:

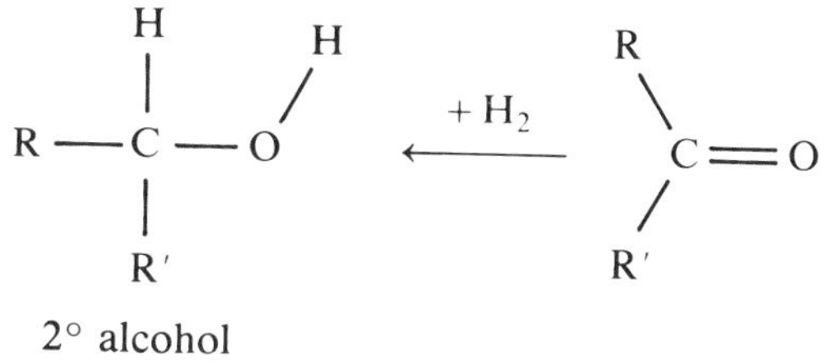

$$R-CH(R')-OH \xleftarrow{+H_2} R-CO-R'$$

2° alcohol

Oxidation

The usual oxidizing agents for conversion of aldehydes to carboxylic acid are acidified dichromate(VI) ions or acidified manganate(VII) ions. Ver powerful oxidizing agents are required to oxidize ketones. Such reaction result in breaking the carbon chain.

DISTINGUISHING ALDEHYDES FROM KETONES

Weak oxidizing agents can oxidize aldehydes but not ketones. This is th basis of two tests to distinguish the two.

1. Benedict's test

Benedict's solution is a mixture of copper(II) ions in an alkaline solutior with a complexing agent. An alternative way of making this solution is to mix Fehling's A solution (containing the Cu^{2+}) and Fehling's B solution (containing the alkali and a complexing agent). Warming an aldehyde with Benedict's solution (or a mixture of Fehling's A and B) produces a brick-red precipitate of copper(I) oxide, as the copper(II) oxidizes the aldehyde to a carboxylic acid and is itself reduced to copper(I). Ketones give no reaction to this test.

2. The silver mirror test

Aldehydes (but not ketones) are oxidized by Ag^+ ions in aqueous ammonia solution. The Ag^+ is reduced to metallic silver. When an aldehyde is warmed with Tollen's reagent (a solution of Ag^+ ions in aqueous ammonia), metallic silver is formed. In a spotlessly clean test tube a silver mirror will be formed on the inside of the test tube, see **Plate 22.** Ketones give no reaction to this test. The reaction of aldehydes with Ag^+ ions was once used as a method of silvering mirrors.

Reduction

Both aldehydes and ketones can be reduced to alcohols by a variety of reducing agents:

- sodium in ethanol which generates hydrogen
- hydrogen with a catalyst of nickel or platinum or palladium
- lithium tetrahydridoaluminate(III), $LiAlH_4$, in ethoxyethane solution followed by addition of water

- sodium tetrahydridoborate(III), $NaBH_4$, in aqueous solution.

Both of the last two reducing agents generate the nucleophile H^-, the hydride ion:

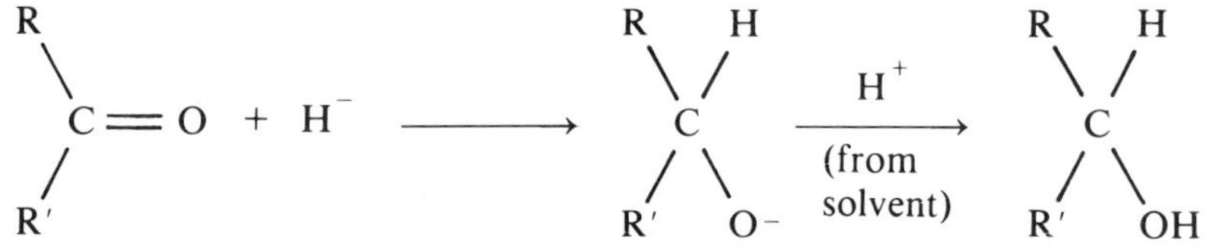

$$\begin{matrix}R \\ \quad\diagdown \\ \quad\quad C{=}O \\ \quad\diagup \\ R'\end{matrix} + H^- \longrightarrow \begin{matrix}R \quad H \\ \diagdown\,\diagup \\ C \\ \diagup\,\diagdown \\ R' \quad O^-\end{matrix} \xrightarrow[\text{(from solvent)}]{H^+} \begin{matrix}R \quad H \\ \diagdown\,\diagup \\ C \\ \diagup\,\diagdown \\ R' \quad OH\end{matrix}$$

This reaction could therefore also be classified as a nucleophilic addition reaction.

32.6.4 Reactions of the alkyl groups of carbonyl compounds

Halogens react with carbonyl compounds. The reaction is acid catalysed and substitution of hydrogen atoms in the alkyl group by halogen atoms occurs, for example:

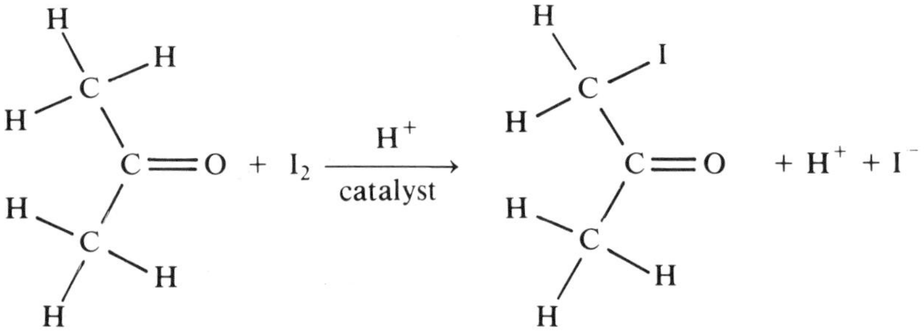

$$CH_3COCH_3 + I_2 \xrightarrow[\text{catalyst}]{H^+} CH_2ICOCH_3 + H^+ + I^-$$

propanone iodopropanone

Further substitution can occur, firstly of the same methyl group and then, when that is fully substituted, of the second. The mechanism is discussed in Chapter 14.

32.6.5 The triiodomethane reaction

Aldehydes and ketones which contain the group:

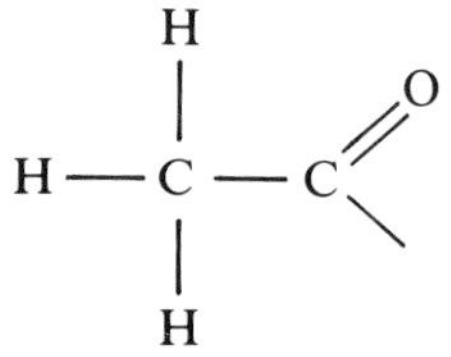

$$\begin{matrix} & H & & \\ & | & & \diagup\!\!\!\diagup O \\ H - & C & - & C \\ & | & & \diagdown \\ & H & & \end{matrix}$$

give a positive **triiodomethane reaction**. Reaction with iodine and sodium hydroxide produces yellow crystals of triiodomethane, for example:

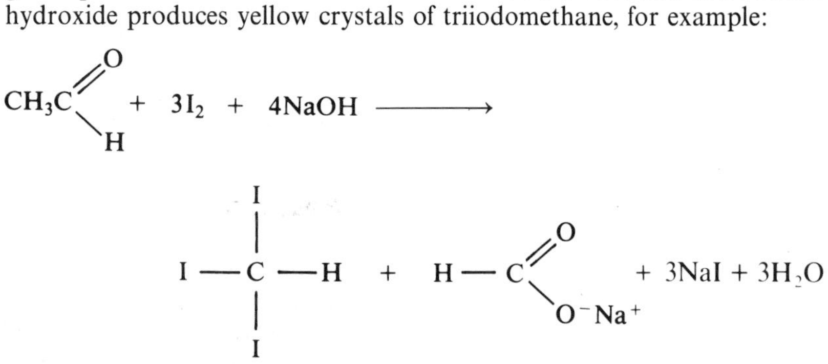

$$CH_3CHO + 3I_2 + 4NaOH \longrightarrow$$

$$CHI_3 + HCOO^-Na^+ + 3NaI + 3H_2O$$

triiodomethane

A

SWEETENERS—THE TRIANGLE OF SWEETNESS

The sweetness of sugars has been known for thousands of years. The average person in the UK consumes around 750 g of sucrose weekly which provides 12 500 kJ (or about 3000 Calories) of energy. Many people try to reduce this energy intake without sacrificing the sweetness by using artificial sweeteners. One of these, lead ethanoate ('sugar of lead') has been known since Roman times, but is no longer used due to the toxicity of lead compounds. The table lists some sweet compounds and compares their sweetness with that of sucrose (table sugar). The values are approximate as they can only be found by using human tasters to find the concentration of a solution which is as sweet as a standard sucrose solution.

Substance	Sweetness	Comments
Sucrose	1	Natural sugars
Glucose	0.5	
Fructose	1.5	
Cyclamate	30	Some evidence suggests a possible link with cancer
Saccharin	350	Bitter aftertaste
Aspartame	200	Trade name Nutrasweet. Unstable, therefore short shelf-life in products
Sucralose	650	Still being tested for safety

Most artificial sweeteners have been discovered by accident, as it is not normal practice to taste newly synthesized compounds until they have been extensively tested for toxicity—a long and expensive process—and there is no way to measure sweetness without tasting. There does, however, seem to be some link between chemical structure and sweetness. Most sweet compounds seem to have a system AH, B where A and B are electronegative atoms. These are thought to form hydrogen bonds with a protein in the taste buds. To do this effectively, A and B must be around 0.3 nm apart.

The structures of some sweet compounds are shown with the AH, B atoms in blue.

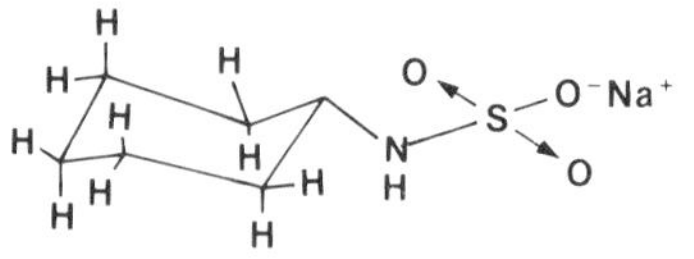

Sodium cyclamate

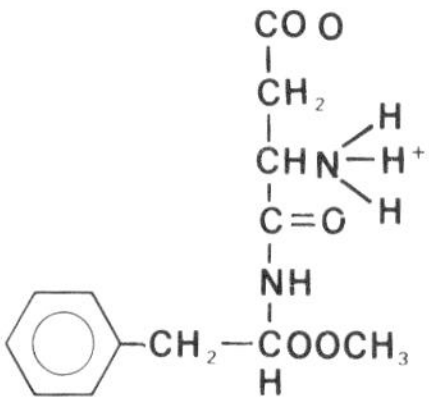

Aspartame

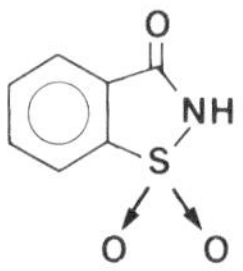

Saccharin

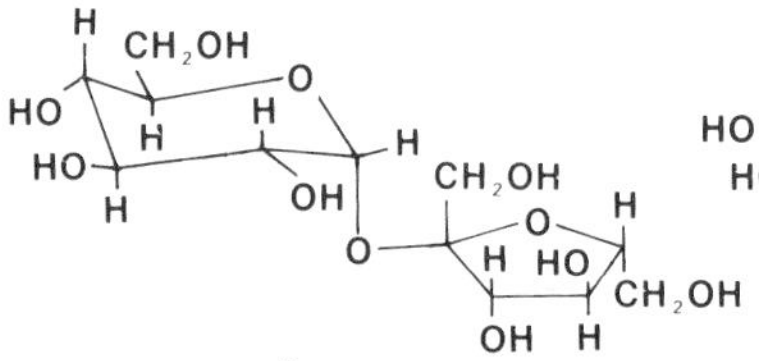

Sucrose

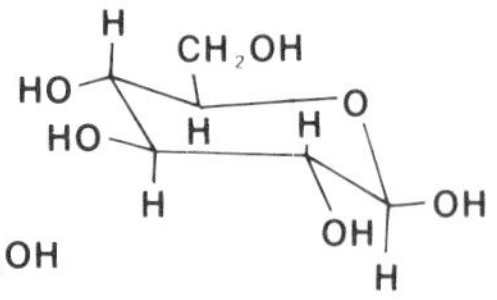

Glucose

More recently, it has been suggested that a third group X is involved which is non-polar and thus repellent to hydrogen bonds. This leads to the idea of a 'triangle of sweetness' with the dimensions shown:

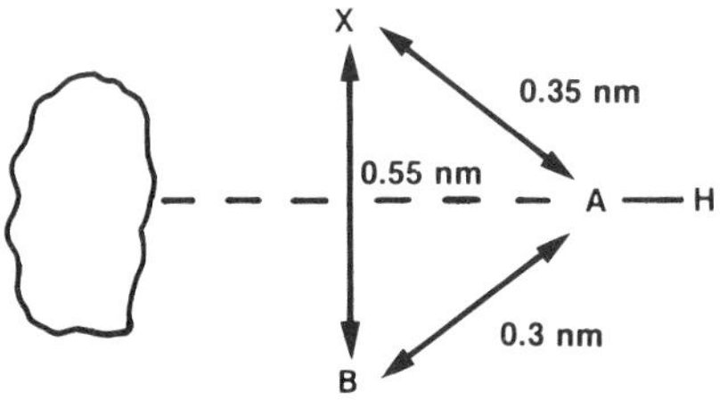

These theories lead to the possibility of synthesizing new sweeteners by design rather than by trial and error.

32.7 Preparation of carbonyl compounds

32.7.1 From alcohols

Primary alcohols are oxidized by acidified dichromate(VI) ions t aldehydes and secondary alcohols to ketones. Aldehydes are susceptibl to further oxidation to carboxylic acids. However, aldehydes have lowe boiling points than the corresponding alcohols (as alcohols form hydroge bonds with one another and aldehydes do not). This means that aldehyde can be distilled off the reaction mixture before they can be further oxidized.

32.8 Carbohydrates

Carbohydrates are a group of compounds of general formula $C_n\ (H_2O)_m$ which explains the name. They occur naturally in all plants where the are produced by **photosynthesis**—the endothermic reaction of water and

igure 32.5 Glucose

carbon dioxide which is fuelled by light energy:

$$6CO_2 + 6H_2O \xrightarrow[\text{photosynthesis}]{\text{light}} \underset{\text{glucose}}{C_6H_{12}O_6} + 6O_2$$

One of the simplest carbohydrates is the sugar glucose, a sweet-tasting compound used in many sweets.

One way of representing the formula of glucose is as in **Figure 32.5**, which shows that it is an aldehyde. However, most glucose molecules exist in a cyclic form, in which the —OH group on carbon number five has reacted with the aldehyde group. This is just the same as the reaction of an alcohol with an aldehyde to form a hermiacetal (see page 497).

In this case both the functional groups are part of the same molecule so we have an **intramolecular** (within the molecule) reaction:

the chain form of glucose

the ring form of glucose

igure 32.6 The ring form of glucose

lodel of glucose

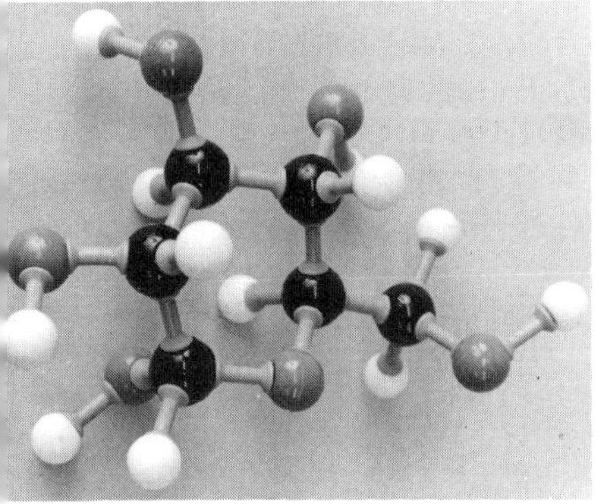

lodel of fructose

Any of the other —OH groups could take part in the same reaction, but a six-membered ring is the most stable as there is least ring strain and it allows the C—C—C bond angles in the ring to closely approach 109.5°.

The ring is not flat, as shown in **Figure 32.6.** In aqueous solution the ring form is in equilibrium with the chain form.

Fructose is an isomer of glucose. It is also a sugar, sweeter than glucose. It is found in fruits and honey. It is now made on a large scale by the action of enzymes on corn—hence it is called corn syrup—and it is used extensively in the food industry.

The functional group of fructose is a ketone:

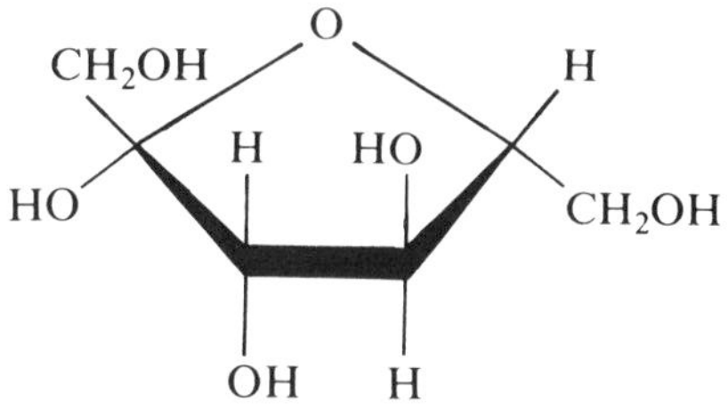

Fructose can also form a ring in a similar way to glucose. This time the ring is five membered (see left).

The ring form is in equilibrium with the chain form, as in glucose.

Both glucose and fructose (as well as other sugars) are called **monosaccharides** ('single sugars'). Two sugars can link together with the elimination of a molecule of water to form a **disaccharide** ('double sugar'). The best known example is sucrose in which a glucose molecule is linked to a fructose molecule. Sucrose is the sugar we buy in the shops.

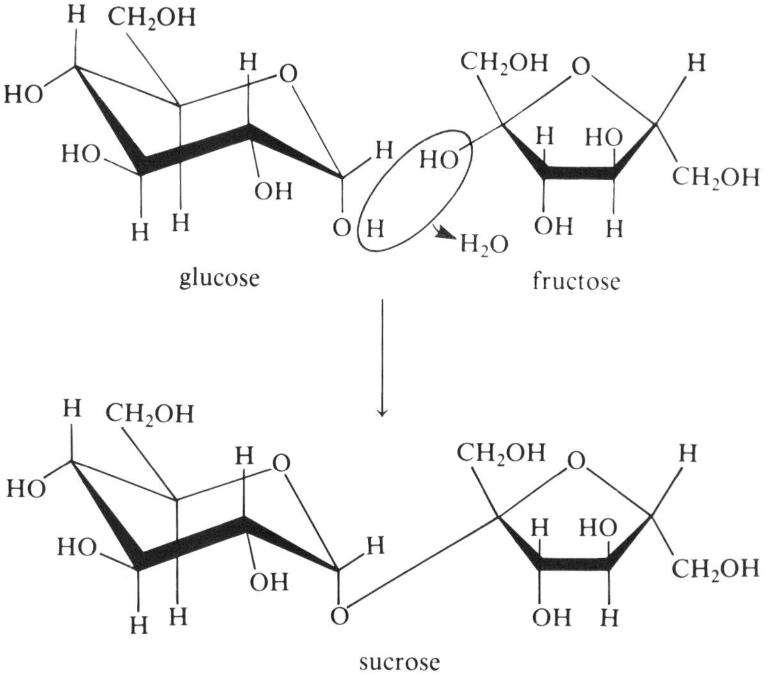

glucose fructose

sucrose

Figure 32.7 The chiral carbons of glucose

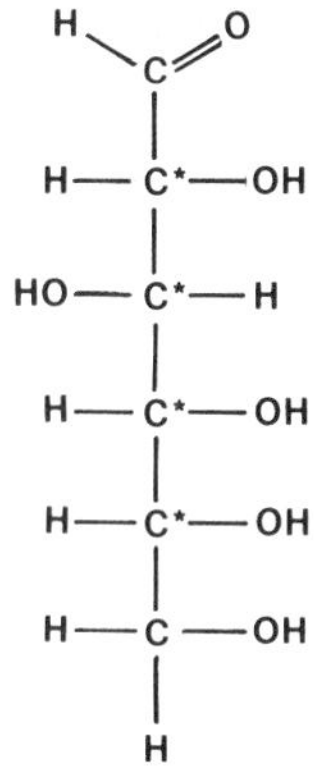

Both glucose and fructose have a number of isomers. It is also worth noting that glucose has four chiral carbon atoms and so exists in optically active forms (see **section 26.7.2**). The four chiral carbon atoms are marked in **Figure 32.7.** Each of these carbons is bonded to four different groups. Notice that you must consider the whole group to which the carbon atom is bonded, not just the first atom.

32.8.1 Polysaccharides

Both starch and cellulose are **polysaccharides**—polymers made up of a few thousand monosaccharides each linked together by elimination of a water molecule. The links in starch can be broken down either by the action of enzymes, such as salivary amylase which breaks down starch in the digestive system, or by boiling with dilute acid. This reaction, where water is added, is called a **hydrolysis** reaction. Starch is synthesized in plants and forms their main food reserve. Bread and potatoes are foods with an especially high starch content. Cellulose is responsible for the structure of plant material. We cannot digest cellulose though it is useful in our diets as roughage.

Foods containing starch

32.8.2 Tests for carbohydrates

Reducing sugars

Sugars like glucose which have an aldehyde functional group, sometimes called **aldoses**, are oxidized by Benedict's (Fehling's) solution and produce a brick-red precipitate of copper(I) oxide by reducing the copper(II) to copper(I). They are therefore called **reducing sugars**. Benedict's test can therefore be used as a test for reducing sugars.

Although it is a ketone, fructose will also produce a positive result with Benedict's test. This is because of the following equilibrium:

$$
\begin{array}{ccc}
\begin{array}{c}
\mathrm{H} \\
| \\
\mathrm{H-C-OH} \\
| \\
\mathrm{C=O} \\
| \\
\mathrm{HO-C-H} \\
| \\
\mathrm{H-C-OH} \\
| \\
\mathrm{H-C-OH} \\
| \\
\mathrm{H-C-OH} \\
| \\
\mathrm{H}
\end{array}
& \rightleftharpoons &
\begin{array}{c}
\mathrm{H\diagdown\ \ \diagup\!\!\!/O} \\
\mathrm{C} \\
| \\
\mathrm{H-C-OH} \\
| \\
\mathrm{HO-C-H} \\
| \\
\mathrm{H-C-OH} \\
| \\
\mathrm{H-C-OH} \\
| \\
\mathrm{H-C-OH} \\
| \\
\mathrm{H}
\end{array}
\end{array}
$$

Sucrose does not react with Benedict's solution and is therefore not a reducing sugar.

odine and starch give a blue–black colour

Starch

Starch produces a deep blue–black colour when treated with a solution of iodine in potassium iodide solution. This is used as a test for starch. Cellulose does not give a positive result with this test.

32.8.3 Reactions of carbohydrates

Dehydration

Addition of concentrated sulphuric acid dehydrates carbohydrates, the main product being carbon:

$$C_n(H_2O)_m \xrightarrow{H_2SO_4} nC + mH_2O$$

Hydrolysis

Polysaccharides can be broken down into monosaccharides by boiling with dilute acid or by the use of a suitable enzyme:

$$\underset{\text{polysaccharide}}{(C_6(H_2O)_5)_n + nH_2O} \xrightarrow[\text{or enzyme catalyst}]{H^+(aq)} \underset{\text{monosaccharide}}{nC_6(H_2O)_6}$$

A molecule of water is required to break each link in the polysaccharide, hence the term hydrolysis. For example:

$$\underset{\text{sucrose}}{\text{(G)}\diagdown_{O}\diagup\text{(F)}} + H_2O \longrightarrow \underset{\text{glucose}}{\text{(G)}\diagdown OH} + \underset{\text{fructose}}{HO\diagup\text{(F)}}$$

Oxidation

Reducing sugars, including glucose and fructose, are oxidized easily and therefore give a positive result with Benedict's test. Di- and polysaccharides do not. Hydrolysis of di- and polysaccharides produces reducing sugars which can be oxidized.

32.9 Economic importance of carbonyl compounds

32.9.1 Methanal

The major use of methanal is in the manufacture of thermosetting polymers. Bakelite, a copolymer of phenol and methanal, has been

H_2O H_2O

$H{-}N(H){-}C({=}O){-}N(H){-}H \quad O{=}CH_2 \quad H{-}N(H){-}C({=}O){-}N(H){-}H \quad O{=}CH_2 \quad H{-}N(H){-}C({=}O){-}N(H){-}H$ e

H^+ catalyst

$\sim N(H){-}C(O){-}N(H){-}CH_2{-}N(H){-}C(O){-}N(H){-}CH_2{-}N(H){-}C(O){-}N(H)\sim$

described (see box, **section 31.1**). A similar polymer can be made fro methanal and urea, NH_2CONH_2. It, too, is a condensation polymer, molecule of water being eliminated with each urea–methanal link forme as shown above. New bonds formed are in brown.

Cross-links can form:

cross-link

to form a rigid thermosetting polymer, which is used as an adhesive stick the 'chips' together in chipboard and also for moulding electric sockets. A similar polymer is made from methanal and melamir **Figure 32.8.** —NH_2 groups on melamine are linked by methanal in t same way as the —NH_2 groups in urea. 'Melamine' is used as the nar for the resulting polymer which is used for making picnic crockery as w as kitchen worktops. Melamine has a low density, is colourless (but c be coloured if required), electrically insulating and unaffected by lig moderate heat and water.

Methanal is manufactured by the oxidation of methanol by air usi a copper catalyst at 800 K:

$$\underset{\text{methanol}}{CH_3OH} + \tfrac{1}{2}O_2 \xrightarrow[\text{catalyst}]{\text{Cu}} \underset{\text{methanal}}{HCHO} + H_2O$$

Figure 32.8 **Melamine**

H_2N NH_2 NH_2

Melamine is used for kitchen worktops

igure 32.9 Industrial chemistry of hanal

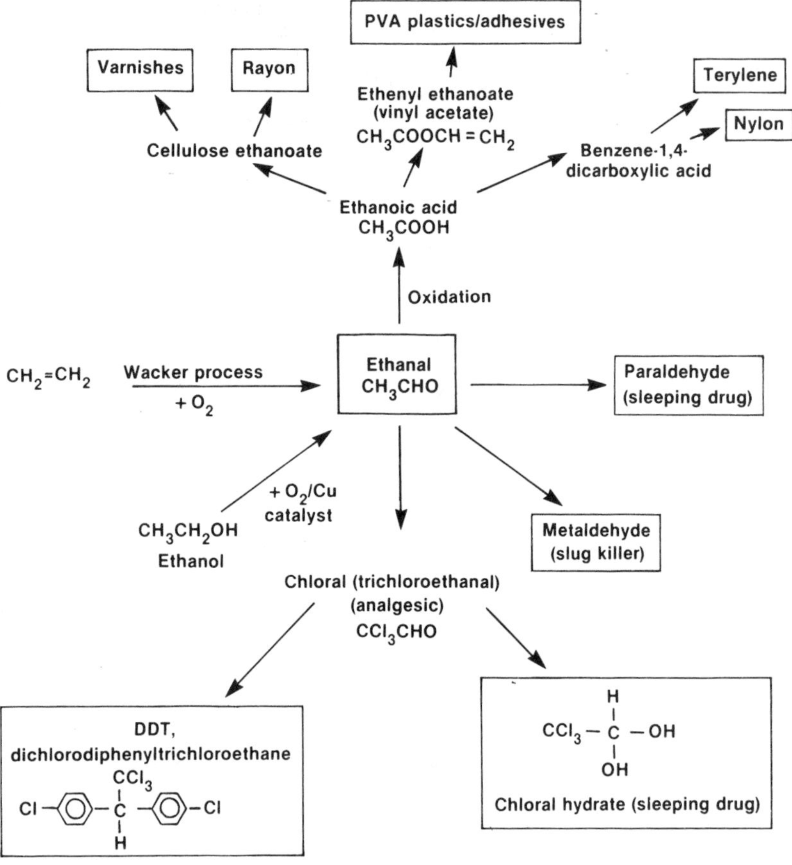

32.9.2 Ethanal

Figure 32.9 summarizes some of the industrial uses of ethanal. Ethanal is manufactured by the oxidation of ethanol in a similar way to methanal opposite, or by the Wacker process in which ethene reacts with oxygen with a catalyst system of aqueous palladium(II) chloride and copper(II) chloride at 300–330 K.

32.9.3 Propanone

The majority of propanone is made as a co-product of the cumene process by which phenol is made (see box **section 29.12**). The rest is made by the oxidation of propan-2-ol by air at 770 K and 3×10^2 kPa with a copper catalyst. Hydrogen peroxide is a co-product:

$$\underset{\text{propan-2-ol}}{CH_3CH(OH)CH_3} + O_2 \xrightarrow[\text{Cu catalyst}]{770\text{ K. } 3\times10^2\text{ kPa}} \underset{\text{propanone}}{CH_3COCH_3} + H_2O_2$$

The uses of propanone are summarized in **Figure 32.10.**

Figure 32.10 **Industrial chemistry of propanone**

Poly(methyl 2-methylpropenoate) (Perspex) $\text{+CH}_2\text{–C(CH}_3\text{)(CO}_2\text{CH}_3\text{)+}_n$

Methyl 2-methylpropenoate $CH_2=C(CH_3)CO_2CH_3$

Propene $CH_2=CHCH_3$ —Cumene process→ Propanone $CH_3–CO–CH_3$

Propan-2-ol —O_2, 3×10^2 kPa, 770 K, Cu catalyst→ Propanone

Propanone → Solvent

Propanone → Bisphenol A $HO–C_6H_4–C(CH_3)_2–C_6H_4–OH$

Bisphenol A → Epoxy resins

Bisphenol A → Poly(carbonate) plastics

Cellulose sheets being washed after emerging from sulphuric acid bath

32.9.4 Other uses of carbonyl compounds

Butanone is used industrially as a solvent. A number of carbony compounds are used as perfumes and flavourings, for example benzene carbaldehyde which smells of almonds. Sugars and starches are of enormou importance in the food industry.

Cellulose is obtained from wood by dissolving away the lignin in a aqueous solution of calcium hydrogensulphite and filtering off th insoluble cellulose. Cellulose is used to make the fibre rayon by treatin it with a mixture of carbon disulphide and sodium hydroxide, when i forms a viscous solution called a xanthate. This is forced through fin holes into a bath of sulphuric acid which precipitates the cellulose as fin threads. Forcing the viscous solution (called viscose) through a narro slit into the acid bath produces sheets of cellophane:

$$\underset{\text{cellulose}}{(\text{rest of molecule})\text{–C–OH}} + CS_2 + NaOH \longrightarrow$$

$$\underset{\text{cellulose xanthate}}{(\text{rest of molecule})\text{–C–O–C(=S)–S}^-} + Na^+ + H_2O$$

formation of the xanthate

$$\underset{\text{cellulose xanthate}}{(\text{rest of molecule})\text{–C–O–C(=S)–S}^-} + H^+ \longrightarrow \underset{\text{cellulose}}{(\text{rest of molecule})\text{–C–OH}} + C$$

The explosive cellulose trinitrate (guncotton) is made by nitrating thre of the —OH groups per glucose unit of cellulose with a mixture of nitri and sulphuric acids.

A

BISPHENOL A

Bisphenol A is an important industrial intermediate compound used in the manufacture of epoxy resins and poly(carbonate) plastics. Epoxy resins are used in adhesives such as Araldite, see box, **section 28.6.5.** Poly(carbonate)s are strong, transparent and heat resistant. They are used for the lenses of safety glasses, riot shields and as an almost unbreakable substitute for glass.

Poly(carbonate)s are used for police riot shields

Bisphenol A is made from propanone and phenol:

$$(CH_3)_2C{=}O + 2\,HO{-}C_6H_4{-}H \xrightarrow[H^+/\text{thiol cat.}]{320\ K} HO{-}C_6H_4{-}C(CH_3)_2{-}C_6H_4{-}OH + H_2O$$

Bisphenol A

Poly(carbonate)s are made by reacting bisphenol A with carbonyl chloride, $COCl_2$, in dichloromethane solution. Hydrogen chloride is eliminated in this condensation polymerization:

$$HO{-}C_6H_4{-}C(CH_3)_2{-}C_6H_4{-}OH + Cl{-}CO{-}Cl + HO{-}C_6H_4{-}C(CH_3)_2{-}C_6H_4{-}OH + Cl{-}CO{-}Cl \xrightarrow{-HCl}$$

$$\sim O{-}C_6H_4{-}C(CH_3)_2{-}C_6H_4{-}O{-}\overset{O}{\overset{\|}{C}}{-}O{-}C_6H_4{-}C(CH_3)_2{-}C_6H_4{-}O{-}\overset{O}{\overset{\|}{C}}\sim \text{ etc.}$$

A poly(carbonate)

Epoxy resins are made by the reaction of bisphenol A with epichlorohydrin:

$$H{-}\underset{H}{C}H(O){-}\underset{H}{C}(O){-}\underset{H}{\overset{H}{C}}{-}Cl \quad \text{(epoxide ring: O bridging the first two C)}$$

The reaction is as follows:

$$\underset{\text{Epichlorohydrin}}{CH_2(O)CH{-}CH_2Cl} + \underset{\text{Bisphenol A}}{HO{-}C_6H_4{-}C(CH_3)_2{-}C_6H_4{-}OH} + NaOH$$

$$\downarrow$$

$$CH_2(O)CH{-}CH_2{-}\left[O{-}C_6H_4{-}C(CH_3)_2{-}C_6H_4{-}OCH_2{-}\overset{OH}{CH}{-}CH_2\right]_n{-}O{-}C_6H_4{-}C(CH_3)_2{-}C_6H_4{-}OCH_2{-}CH(O)CH_2 + NaCl$$

An epoxy resin

The value of n and hence whether the resins are liquid ($n \leqslant 2$) or solid ($n > 2$) depends on the ratio of bisphenol A to epichlorohydrin used.

32.10 Summary

- Aldehydes and ketones both contain the carbonyl functional group, $>C{=}O$. Aldehydes are $\underset{H}{\overset{R}{>}}C{=}O$ where R may be H. Ketones are

$$\begin{matrix} R \\ \quad \diagdown \\ \quad C{=}O \\ \quad \diagup \\ R' \end{matrix}$$ where neither R nor R′ is H.

- The carbonyl group is flat and trigonal shaped with the bond angl approximately 120°.
- The carbonyl group is strongly polarized, $C^{\delta+}{=}O^{\delta-}$.
- Dipole–dipole forces operate between the molecules and hydrogen bonds can be formed with water.
- The $C^{\delta+}$ may be attacked by nucleophiles or the $O^{\delta-}$ by electrophiles, say, H^+.
- The reactions of carbonyl compounds are summarized in **Figure 32.**

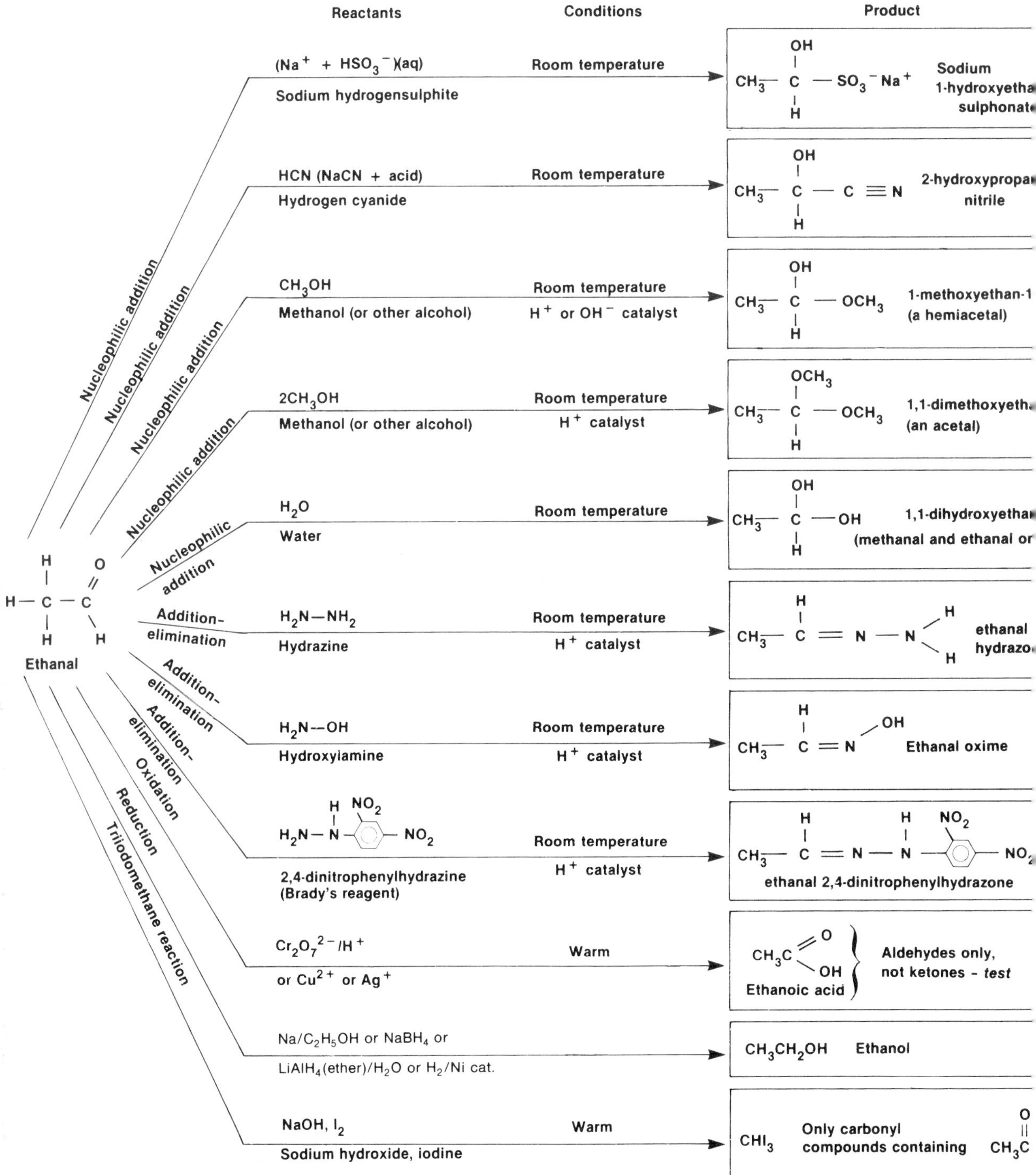

Figure 32.11 Reactions of carbonyl compounds using ethanal as an example

- Aldehydes can be distinguished from ketones by Benedict's or the silver mirror test. Aldehydes produce a brick-red precipitate when warmed with Benedict's solution (or a mixture of Fehling's A and B solutions). Aldehydes produce a silver mirror when boiled with Tollen's reagent, a mixture of silver nitrate and aqueous ammonia. Ketones do not react with either of these solutions.
- An aldehyde or ketone can be identified by making a derivative with 2,4-dinitrophenylhydrazine (Brady's reagent), recrystallizing it and measuring its melting point.
- Aldehydes and ketones can be prepared in the laboratory by oxidation of primary alcohols (aldehydes) or secondary alcohols (ketones) with acidified dichromate(VI) ions.
- Carbohydrates have the general formula $C_n(H_2O)_m$.
- Glucose is a sugar (sweet-tasting carbohydrate) of formula $C_6H_{12}O_6$. It has five —OH groups and one aldehyde group. Most glucose molecules exist as six-membered rings.
- Fructose is a sugar isomeric with glucose but the carbonyl group is a ketone. Most fructose molecules exist as five-membered rings.
- Simple sugars (monosaccharides) can link together with the elimination of water to form di-, tri-, polysaccharides, etc.
- Sucrose is a disaccharide consisting of a molecule of glucose and one of fructose.
- Starch and cellulose are both polysaccharides.
- Glucose and fructose are reducing sugars and give a positive reaction with Benedict's test. Sucrose does not.
- Starch gives a blue–black complex with iodine.
- Methanal is important industrially for the manufacture of thermosetting plastics.
- Ethanal is an important intermediate for the manufacture of plastics, fibres and drugs.
- Propanone is used as a solvent and is also used in the manufacture of polymers.

32.11 Questions

1 Name the following compounds:

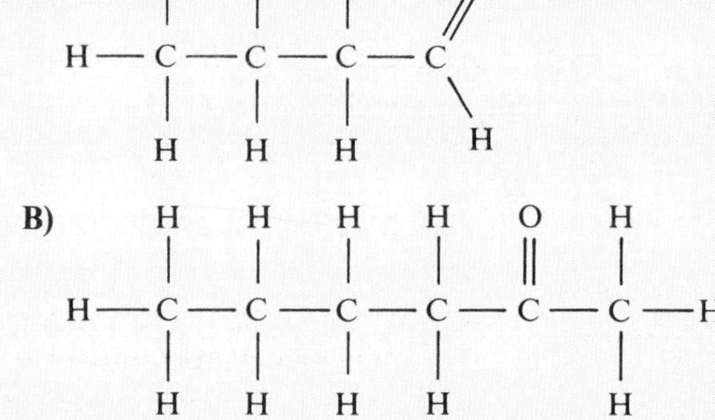

D) $CH_3COCH_2COCH_3$ (H—C(H)(H)—C(=O)—C(H)(H)—C(=O)—C(H)(H)—H)

2 Write structural formulae for:
A) pentanal; **B)** chloroethanal; **C)** butanone; **D)** cyclopentanone.

3 Give the structural formula of the organic product of the following reactions:

A) C_2H_5CHO + HCN

B) CH_3COCH_3 + Br_2

C) CH_3COCH_3 + $NaBH_4$

D) HCHO + $K_2Cr_2O_7/H^+$

E) $CH_3-\overset{\overset{\displaystyle O}{\|}}{C}-CH_3 + I_2$

F) $CH_3-\overset{\overset{\displaystyle O}{\|}}{C}-C_2H_5 + H_2N-NH_2$

4 A) Sketch and label the apparatus needed for the conversion of ethanol into aqueous ethanal. Give the full equation for the reaction and explain any precautions taken to ensure that the aqueous solution of the product contains no organic impurity.

B) Give the equation and the full structural formula of the product formed by the reaction of ethanal with aqueous $NaHSO_3$. Classify this type of reaction.

C) The triiodomethane test is positive for ethanol and ethanal but negative for methanol and methanal. Briefly explain this difference.

(Oxford 1987)

5 A) Give one example of the use of each of the following reagents in organic chemistry. For each example, name the type of reaction and give the full structural formula of the organic product.

i 2, 4-dinitrophenylhydrazine;
ii HCN in the presence of KCN;
iii Ammoniacal silver nitrate solution.

B) Suggest a mechanism for reaction (ii) above.

(Oxford 1988, part question)

6 The sex hormone, oestrone, may be represented by either of the structures shown in **Figures 1** and **2**. All the atoms are shown in **Figure 1** and **Figure 2** is an abbreviated formula.

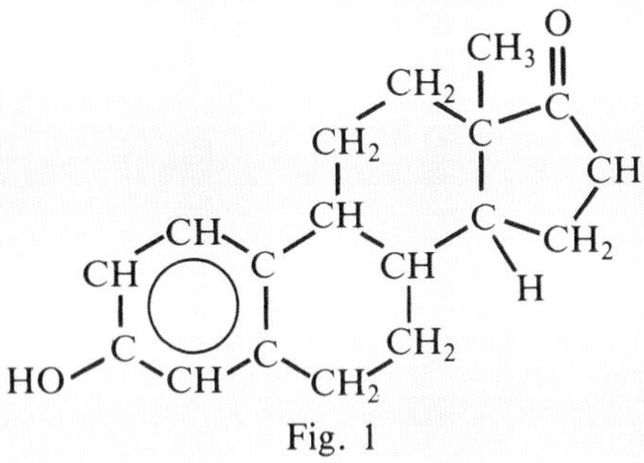

Fig. 1

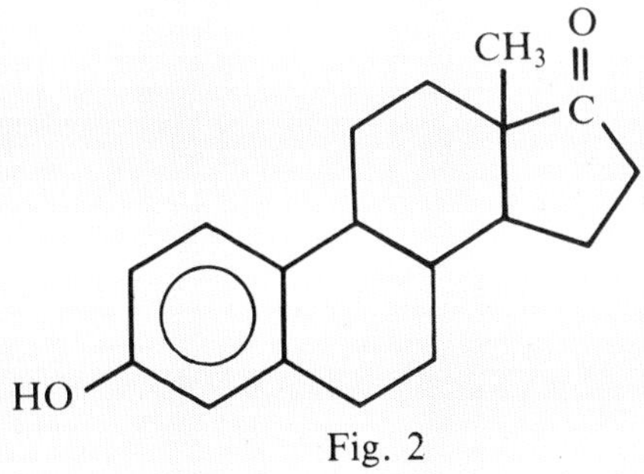

Fig. 2

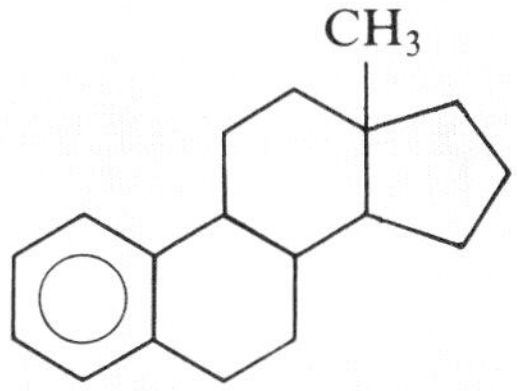

A) On a copy of the full structural formula, **Figure 1**, place a circle around the four carbon atoms which are chiral centres.

B) Copy and complete the outline structure thereby showing the structural formulae of substances which will be formed when oestrone reacts with each of the reagents named.

i aqueous sodium hydroxide;
ii aqueous bromine;
iii hydrogen cyanide;
iv 2, 4-dinitrophenylhydrazine.

C) What physical method might be usefully applied to the determination of the structure of a molecule as large as that of oestrone, and what sort of information would the method give?

(Cambridge 1987)

7 Describe a procedure by which you would attempt to prepare in the laboratory a pure sample of propanal (CH_3CH_2CHO) from propan-1-ol, stating the reagents involved.

Give at least *four* of the typical reactions of aldehydes and ketones.

State briefly how you would attempt to identify an unknown aldehyde or ketone.

(Nuffield 1984)

8 Compare and contrast the reactions of an alkene (containing a carbon–carbon double bond) with those of an aldehyde or ketone (containing a carbon–oxygen double bond). Give the mechanism of at least one reaction of each functional group (C=C and C=O). Account for any differences or similarities.

9 Suggest a method of preparing iodopropanone, $CH_2I-\overset{\overset{\displaystyle O}{\|}}{C}-CH_3$, from propene, $CH_2=CH-CH_3$. Three steps are required.

10 Compound X has the formula C_4H_9Br. On boiling with aqueous sodium hydroxide, compound Y, formula $C_4H_{10}O$, is produced. This can be oxidized with $K_2Cr_2O_7/H^+$ to Z, C_4H_8O and produces a yellow precipitate with Brady's reagent. Z does not react with Benedict's solution. Give the structures of X, Y and Z and that of the product of the reaction of Z with Brady's reagent.

33 Carboxylic acids and their derivatives

33.1 Introduction

This chapter deals with a number of closely related groups of compounds.

Carboxylic acids have the functional group: $-\mathrm{C}\overset{\displaystyle /\!\!/ \mathrm{O}}{\underset{\displaystyle \backslash \mathrm{OH}}{}}$

This is interesting because it includes two arrangements we have seen before—the carbonyl group, C=O, and the hydroxy group, —OH. We shall see how the presence of both groups on the same carbon atom leads to considerable changes in the properties of each group. The most obvious is that carboxylic acids are (not surprisingly) significantly acidic, while alcohols are very weak acids indeed. In the acid derivatives the —OH group of the acid is replaced by different groups as shown:

ester $-\mathrm{C}(=\mathrm{O})-\mathrm{O}-\mathrm{R}'$

amide $-\mathrm{C}(=\mathrm{O})-\mathrm{NH_2}$

acid chloride $-\mathrm{C}(=\mathrm{O})-\mathrm{Cl}$

anhydride $-\mathrm{C}(=\mathrm{O})-\mathrm{O}-\mathrm{C}(=\mathrm{O})-\mathrm{R}'$

The easiest way to understand the chemistry of the derivatives is by comparison with that of the parent acids.

The most familiar carboxylic acid is ethanoic acid (acetic acid) which is responsible for the sour taste of vinegar. Short-chain *esters* are fairly volatile and have pleasant fruity smells, one example being 3-methylbutyl ethanoate which smells of pear drops:

$$
\begin{array}{ccccccccccccccc}
 & & & & & & & & & & H & & & & \\
 & & & & & & & & & & | & & & & \\
 & & & & & & & & & H- & C & -H & & & \\
 & & & & & & & & & & | & & & & \\
 & & O & & & & H & & H & & | & & H & & \\
 & & \| & & & & | & & | & & | & & | & & \\
CH_3 & — & C & — & O & — & C & — & C & — & C & — & C & — & H \\
 & & & & & & | & & | & & | & & | & & \\
 & & & & & & H & & H & & H & & H & &
\end{array}
$$

3-methylbutyl ethanoate – 'pear drops'

Fats **and** ***oils*** **are esters with longer carbon chains.** ***Anhydrides*** **and** ***aci chlorides*** **are more reactive than their 'parent' carboxylic acids and hav important uses as industrial intermediates for this reason.** ***Amides*** **are les reactive.** ***Proteins*** **are poly(amides); the amide group is closely linked wit the chemistry of life. Nylon is also a poly(amide) while Terylene is poly(ester).**

33.2 Formulae, naming and structure

The carboxylic acid functional group is $—C\begin{array}{l}\diagup\!\!\diagup O\\ \diagdown OH\end{array}$, sometimes writte as —COOH or —CO_2H. The carbon of the functional group has only on 'spare' bond so the functional group can only occur at the end of a carbo chain, not in the middle.

33.2.1 Naming

Carboxylic acids are named using the suffix -oic acid. The carbon ator of the functional group is counted as part of the carbon chain of the root. S

$$H—C\begin{array}{l}\diagup\!\!\diagup O\\ \diagdown OH\end{array}$$ is methanoic acid

$$
\begin{array}{ccccc}
 & & H & & \\
 & & | & & \\
H & — & C & — & C\begin{array}{l}\diagup\!\!\diagup O\\ \diagdown OH\end{array} \\
 & & | & & \\
 & & H & &
\end{array}
$$
is ethanoic acid, and so on

Where there are substituents or side chains on the carbon chain, they ar numbered using locants, counting from the carbon of the carboxylic aci as carbon number one. So:

$$
\begin{array}{ccccccc}
 & & H & & Br & & \\
 & & | & & | & & \\
H & — & C & — & C & — & C\begin{array}{l}\diagup\!\!\diagup O\\ \diagdown OH\end{array} \\
 & & | & & | & & \\
 & & H & & H & &
\end{array}
$$
is 2-bromopropanoic acid

$$
\begin{array}{ccccccccc}
 & & H & & CH_3 & & H & & \\
 & & | & & | & & | & & \\
H & — & C & — & C & — & C & — & C\begin{array}{l}\diagup\!\!\diagup O\\ \diagdown OH\end{array} \\
 & & | & & | & & | & & \\
 & & H & & H & & H & &
\end{array}
$$
is 3-methylbutanoic acid

When the functional group is attached to, for example, a benzene ring the suffix -carboxylic acid is used and the carbon of the functional grou is *not* counted as part of the root. So:

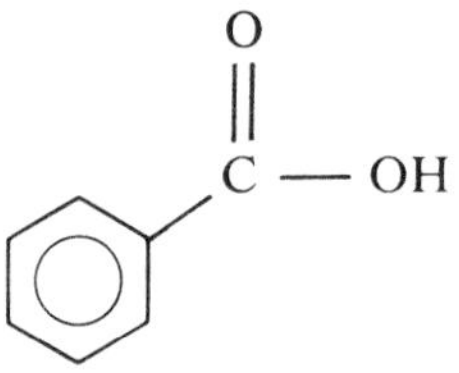

C_6H_5—C(=O)—OH is benzenecarboxylic acid (still sometimes called benzoic acid)

Acid derivatives

The acid derivatives all have the formula: R—C(=O)—Z

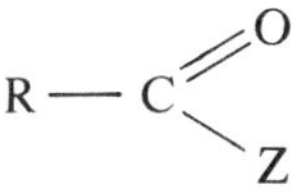

If R is an alkyl group, R—C(=O)— is called the **acyl** group.

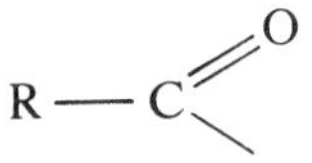

Acid chlorides (or **acyl chlorides**) have the formula: R—C(=O)—Cl and are named using the suffix -oyl chloride. So:

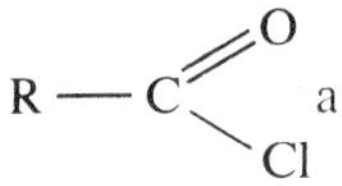

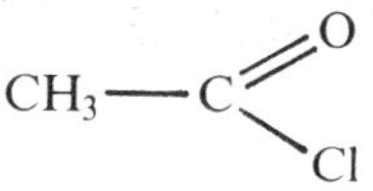

CH_3—C(=O)—Cl is ethanoyl chloride.

Anhydrides have the formula: R—C(=O)—O—C(=O)—R′

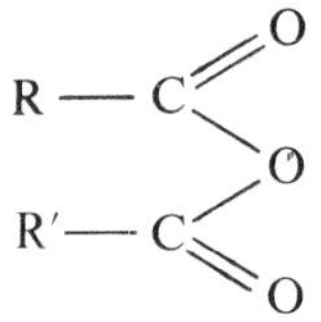

They can be thought of as two molecules of carboxylic acid from which a molecule of water has been eliminated, for example:

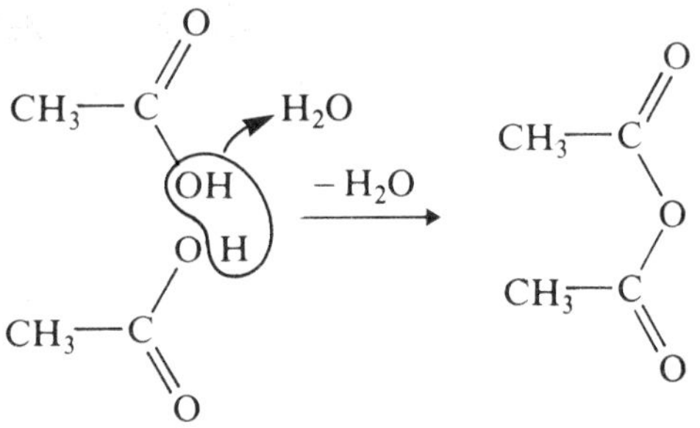

CH_3—C(=O)—OH + HO—C(=O)—CH_3 $\xrightarrow{-H_2O}$ CH_3—C(=O)—O—C(=O)—CH_3 + H_2O

Symmetrical anhydrides, derived from two molecules of the same acid, are simply named as the anhydride of the parent acid so the above compound is called ethanoic anhydride. Mixed anhydrides, derived from two different acids, are named by listing the parent acids in alphabetical order. So:

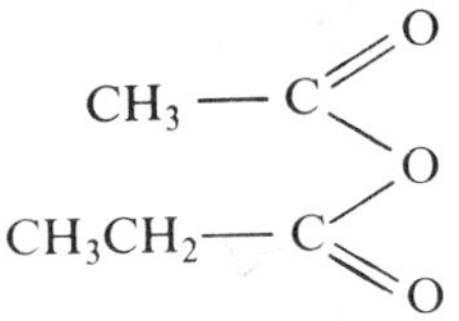

CH_3—C(=O)—O—C(=O)—CH_3CH_2 is named ethanoic propanoic anhydride

Esters have the general formula: R—C(=O)—O—R′

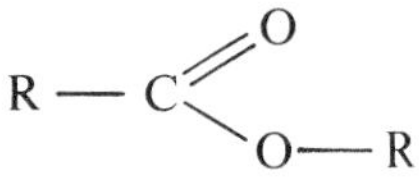

and they are named as alkyl or aryl derivatives of the parent acid, so

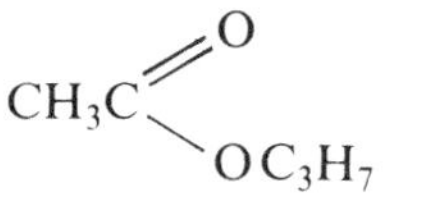

CH_3C(=O)OC_3H_7 is propyl ethanoate

and

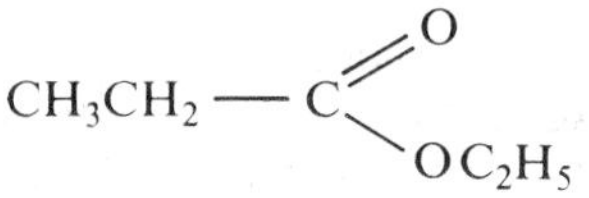

CH_3CH_2—C(=O)OC_2H_5 is ethyl propanoate

gure 33.1 Dot cross diagrams of the OH and —NH_2 groups (see overleaf)

Figure 33.2 Bonding in the carboxylic acid functional group

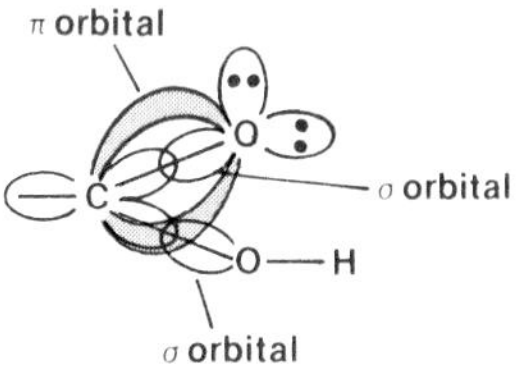

Figure 33.3 The shape of carboxylic acids

119.5°
0.125 nm
122°
0.131 nm
116°
0.095 nm

The molecule is not quite flat so the angles do not add up to 360°

Take care with the names of esters; it is easy to get them 'the wrong w round'.

Amides have the —OH group of the parent carboxylic acid replac by a $—NH_2$ group (which has the same number of electrons as the —C group **Figure 33.1**). They are named using the suffix -amide, so:

$CH_3—C(=O)NH_2$ is ethanamide

33.2.2 Shape

The carbon atom of a carboxylic acid is hybridized sp^2. The C=O dou bond consists of a σ bond formed by overlap of sp^2 orbitals on both carbon and the oxygen atom plus a π orbital formed from overlap o orbitals on each atom, **Figure 33.2.**

The carboxylic acid group is virtually flat and trigonal, with all bond angles approximately 120° as we would expect from electron p repulsion theory, **Figure 33.3.**

33.3 Physical properties of carboxylic acids

Figure 33.4 A molecule of a carboxylic acid forming hydrogen bonds with water

Figure 33.5 A hydrogen bonded dimer of two carboxylic acid molecules

The carboxylic acid group can form hydrogen bonds with water molecul **Figure 33.4**, or with other molecules of carboxylic acid. This results carboxylic acids dissolving well in water, provided that their carbon cha are fairly short. The carboxylic acids up to and including C_4 (butan acid) are completely soluble in water. The acids also have much hig melting temperatures than the alkanes of similar relative molecular ma Pure ethanoic acid ($M_r = 60$) melts at 290 K while butane ($M_r = 58$) me at 135 K. Ethanoic acid's high melting point means that it will freeze a cool laboratory forming ice-like (glacial) ethanoic acid.

A carboxylic acid molecule may also form hydrogen bonds w another similar molecule to form a dimer, **Figure 33.5.** These are fou when there are no other molecules with which to form hydrogen bon They occur in pure liquid acids, in the vapour phase and in solutions carboxylic acids in non-polar solvents like hexane. The dimers, o formed, cannot hydrogen bond to other molecules, therefore boili temperatures of the acids are relatively low.

Carboxylic acids will dissolve increasingly well in non-polar solve as their carbon chains get longer.

The acids have characteristic smells. You will recognize the smell ethanoic acid as that of vinegar, while butanoic acid causes the smell rancid butter.

33.4 Infra-red spectra of carboxylic acids

Figure 33.6 The infra-red spectrum of ethanoic acid

The infra-red spectrum of ethanoic acid is shown in **Figure 33.6.** The m important peaks in the spectrum are those at 3100 cm^{-1} due to an O—

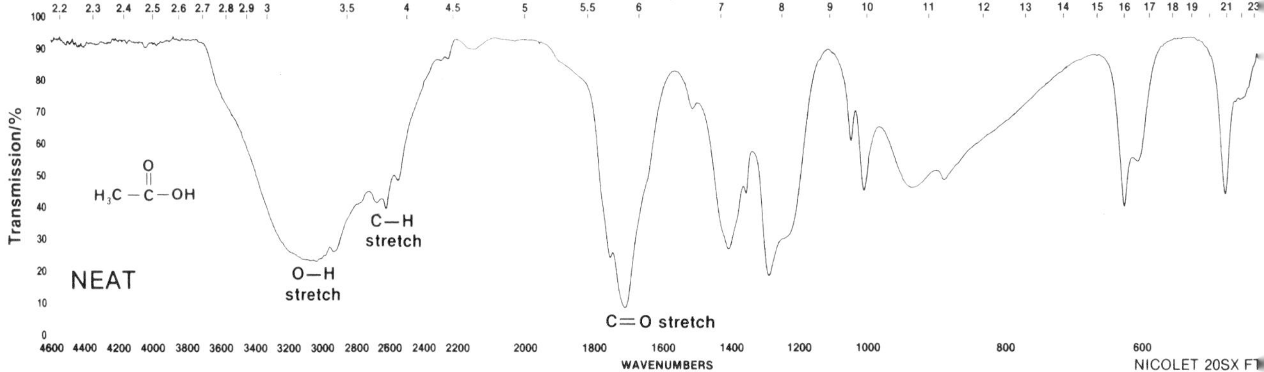

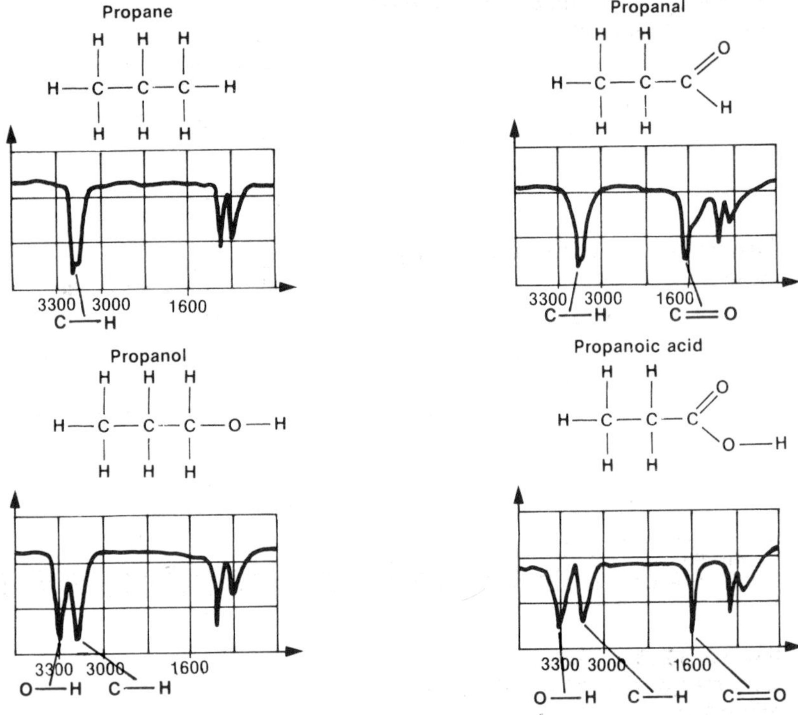

Figure 33.7 The infra-red spectra of propane, propanol, propanal and propanoic acid. Note how the spectra are additive

stretch and 1700 cm^{-1} due to a C═O stretch. We have met both these before, the former in alcohols and the latter in aldehydes and ketones. The O—H peak is broadened due to hydrogen bonding as in alcohols (see **section 31.3**). The additive nature of infra-red spectra is shown in **Figure 33.7**.

33.5 Reactivity of carboxylic acids

The carboxylic acid group is polarized as shown: R—$C^{\delta+}$(═$O^{\delta-}$)—O—H

The $C^{\delta+}$ is therefore likely to be attacked by nucleophiles while the $O^{\delta-}$ may be attacked by positively charged species like H^+. However, the $C^{\delta+}$ tends to attract electrons from the C—OH bond which reduces the δ^+ character of this carbon atom and makes it less easily attacked by nucleophiles than the carbonyl carbon in aldehydes and ketones. This is similar to the electron-releasing effect of the —OH group in phenol. The oxygen of the —OH group in turn attracts electrons from the O—H bond, weakening this bond and allowing the loss of the hydrogen as a H^+ ion.

So there are two important features of the chemistry of carboxylic acids: nucleophilic attack and loss of H^+, i.e. acidity.

33.5.1 Nucleophilic attack

Nucleophiles attack the $C^{\delta+}$:

R—$C^{\delta+}$(═$O^{\delta-}$)—O—H ← $:Nu^-$ ⟶ R—C(O^-)(Nu)—O—H

This is normally followed by loss of the —OH group as an OH^- ion:

$$R\text{—}C(O^-)(OH)\text{—}Nu \longrightarrow R\text{—}C(=O)\text{—}Nu + OH^-$$

33.5.2 Loss of a proton—acidity

If the hydrogen of the —OH group is lost, a negative ion—a **carboxylate ion**—is left. The negative charge is shared over the carbonyl group, making the resulting ion more stable:

$$R\text{—}C(=O)\text{—}O\text{—}H \rightleftharpoons R\text{—}C(=O)\text{—}O^- + H^+$$

a carboxylate ion

In this ion, the negative charge can be delocalized over the atoms shown in brown:

$$R\text{—}C(=O)\text{—}O^- \longleftrightarrow R\text{—}C(\text{—}O^-)=O$$

The double-headed arrow indicates that both forms contribute to the actual structure which is often represented as:

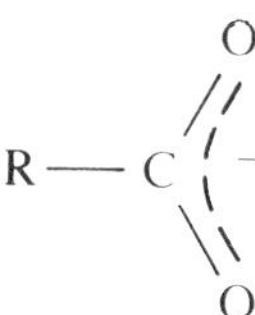

Figure 33.8 Bonding in the carbonyl group

The bonding can be described as overlap between the π bond of the C=O group and a lone pair in a p orbital on the other oxygen, **Figure 33.8**. There is evidence for this delocalization in the bond lengths in the ethanoate ion in, for example, sodium ethanoate. Both C—O bonds are identical in length, 0.127 nm, which is about half-way between C=O (0.123 nm) and C—O (0.136 nm), **Figure 33.9.**

Figure 33.9 Bond lengths in the ethanoate ion

R—C (0.127 nm, 0.127 nm)

Carboxylic acids are weak acids, so the equilibrium:

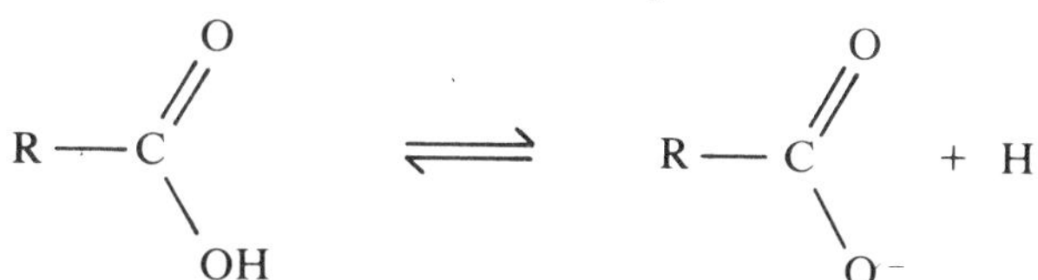

is well over the left as shown by the pK_a value of 4.8. They are, however, generally stronger acids than phenols and will displace carbon dioxide from solutions of carbonates—a test which can be used to distinguish unsubstituted carboxylic acids from unsubstituted phenols.

The effect of substituents on acidity

The acidity of a carboxylic acid is affected by the electron-releasing or electron-attracting power of the group attached to the $—CO_2H$ group.

Electron-attracting groups increase the acidity by spreading out even further the negative charge on the carboxylate ion formed when H^+ is lost. Conversely, electron-releasing groups such as $CH_3—$ reduce the acid

able 33.1 pK_a **s of some carboxylic cids. Remember—larger values of pK_a epresent *weaker* acids**

cid	Formula	pK_a
ethanoic	HCO_2H	3.8
hanoic	CH_3CO_2H	4.8
opanoic	$CH_3CH_2CO_2H$	4.9
utanoic	$CH_3CH_2CH_2CO_2H$	4.8
hloroethanoic	CH_2ClCO_2H	2.86
chloroethanoic	$CHCl_2CO_2H$	1.3
ichloroethanoic	CCl_3CO_2H	0.7
chloropropanoic	$CH_3CHClCO_2H$	2.8
chloropropanoic	$CH_2ClCH_2CO_2H$	4.1
omoethanoic	CH_2BrCO_2H	2.9
uoroethanoic	CH_2FCO_2H	2.6
enzenecarboxylic	$C_6H_5CO_2H$	4.2

igure 33.10 **The benzenecarboxylate n showing interaction between e —CO_2^- group and the benzene ng**

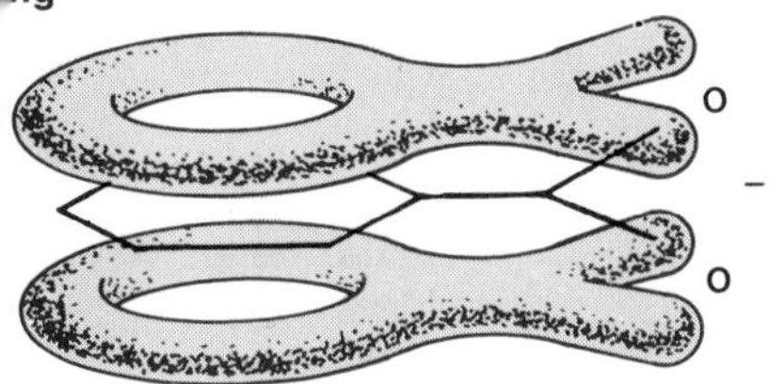

strength. You can see evidence for this in **Table 33.1.** Compared with methanoic acid, ethanoic acid is a weaker acid as it has an electron-releasing CH_3— substituent. Compared with ethanoic acid, the chloro-, dichloro- and trichloro-derivatives get successively stronger as they have more electron-withdrawing chlorine atoms.

You can also see the effect of changing the halogen atom:

$$CH_2BrCO_2H > CH_2ClCO_2H > CH_2FCO_2H$$

pK_a	2.9	2.86	2.6

——— stronger acid ———→

The halogen atom has less effect the further it is from the —CO_2H group:

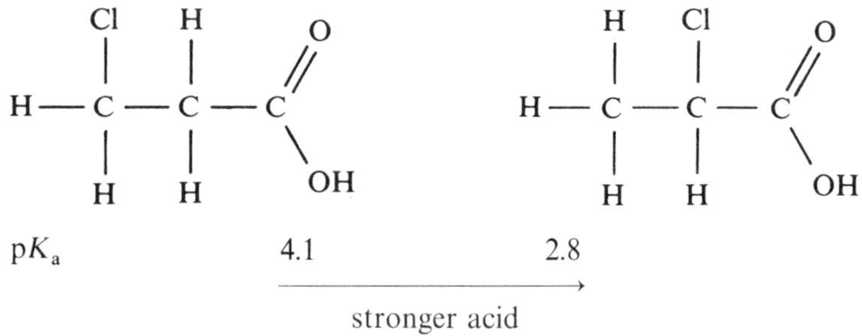

Benzenecarboxylic acid is stronger than ethanoic acid because the negative charge on the carboxylate ion can be delocalized by interaction with the benzene ring as shown in **Figure 33.10**.

33.6 Reactions of carboxylic acids

33.6.1 As acids

Carboxylic acids form ionic salts by reaction with the more reactive metals, alkalis or metal carbonates. For example:

$$2CH_3—C(=O)OH\ (aq) + Mg(s) \longrightarrow$$

ethanoic acid   magnesium

$$2CH_3—C(=O)O^-\ (aq) + Mg^{2+}(aq) + H_2(g)$$

magnesium ethanoate   hydrogen

$$CH_3CH_2C(=O)OH\ (aq) + NaOH\ (aq) \longrightarrow$$

propanoic acid   sodium hydroxide

$$CH_3CH_2C(=O)O^-\ (aq) + Na^+(aq) + H_2O(l)$$

sodium propanoate   water

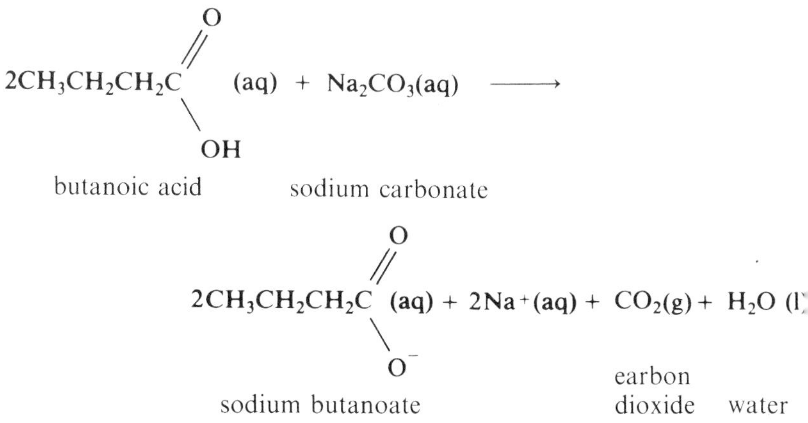

$$2CH_3CH_2CH_2C(=O)OH\,(aq) + Na_2CO_3(aq) \longrightarrow$$

butanoic acid sodium carbonate

$$2CH_3CH_2CH_2C(=O)O^-\,(aq) + 2Na^+(aq) + CO_2(g) + H_2O\,(l)$$

sodium butanoate carbon dioxide water

Reaction of carboxylic acids with ammonia gives ammonium salts which can then be dehydrated by strong heating to give **amides**, for example:

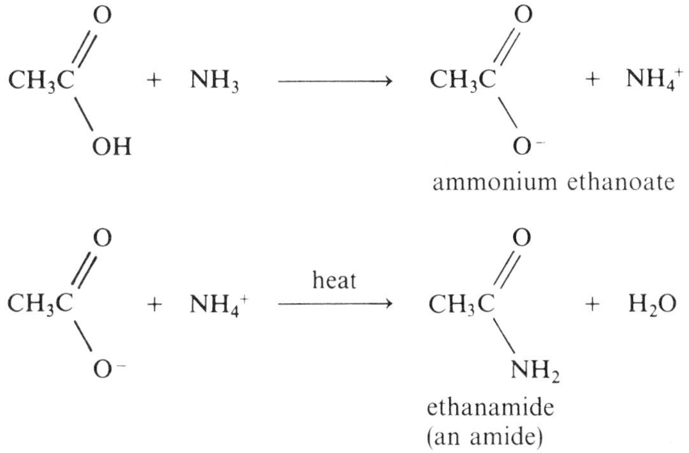

$$CH_3C(=O)OH + NH_3 \longrightarrow CH_3C(=O)O^- + NH_4^+$$

ammonium ethanoate

$$CH_3C(=O)O^- + NH_4^+ \xrightarrow{\text{heat}} CH_3C(=O)NH_2 + H_2O$$

ethanamide (an amide)

33.6.2 Nucleophilic substitution reactions

Reaction with phosphorus pentachloride

Phosphorus pentachloride and sulphur dichloride oxide (thionyl chloride) both generate the nucleophile $Cl^{\delta-}$ which replaces the —OH group of the acid to produce an **acid chloride**:

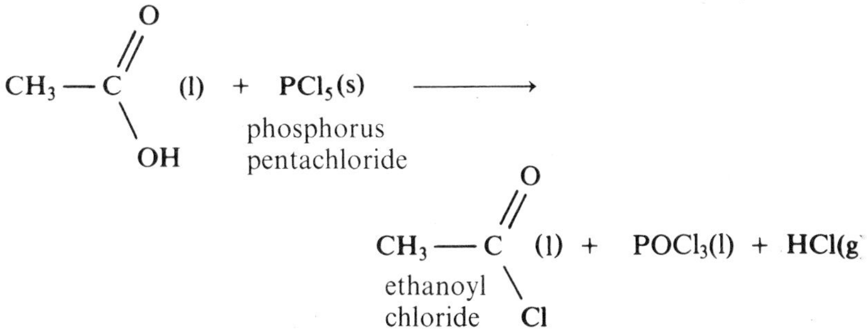

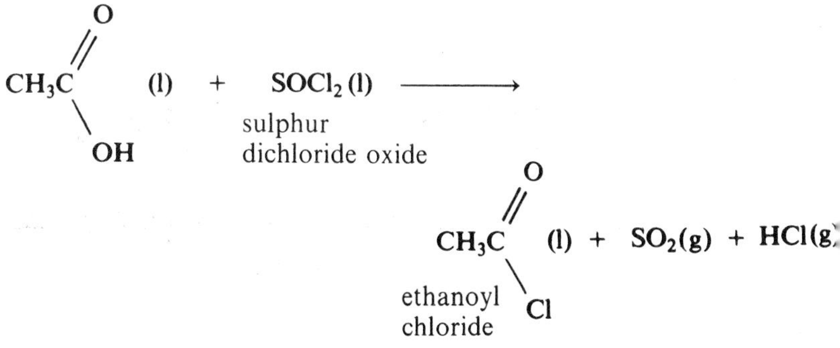

$$CH_3-C(=O)OH\,(l) + PCl_5(s) \longrightarrow$$

phosphorus pentachloride

$$CH_3-C(=O)Cl\,(l) + POCl_3(l) + HCl(g)$$

ethanoyl chloride

$$CH_3C(=O)OH\,(l) + SOCl_2\,(l) \longrightarrow$$

sulphur dichloride oxide

$$CH_3C(=O)Cl\,(l) + SO_2(g) + HCl(g)$$

ethanoyl chloride

Formation of esters

Alcohols react reversibly with carboxylic acids in the presence of a strong acid catalyst to form esters:

$$\underset{\text{ethanoic acid}}{CH_3C(=O)OH} + \underset{\text{ethanol}}{C_2H_5OH} \xrightarrow[\text{catalyst}]{H^+} \underset{\text{ethyl ethanoate}}{CH_3C(=O)O{-}C_2H_5} + \underset{\text{water}}{H_2O}$$

Alcohols are rather weak nucleophiles and will only attack the $C^{\delta+}$ of the acid when the acid has been protonated by the strong acid catalyst, thus increasing the positive charge. The mechanism is discussed in more detail in **section 14.6.2**.

33.6.3 Dehydration

Phosphorus pentoxide will remove a molecule of water from two molecules of carboxylic acid to yield an anhydride:

$$CH_3{-}C(=O)OH + HO{-}C(=O){-}CH_3 \xrightarrow[P_2O_5]{\text{heat}} \underset{\text{ethanoic anhydride}}{CH_3{-}C(=O){-}O{-}C(=O){-}CH_3} \quad (-H_2O)$$

33.6.4 Reduction

Lithium tetrahydridoaluminate(III) will reduce carboxylic acids to primary alcohols:

$$CH_3{-}C(=O)OH \xrightarrow[\text{in ethoxyethane followed by addition of water}]{LiAlH_4} \underset{\text{ethanol}}{CH_3{-}CH_2{-}OH}$$

Less powerful reducing agents will not reduce carboxylic acids. As aldehydes are more easily reduced to alcohols than are carboxylic acids, it is not possible to reduce carboxylic acids to aldehydes.

33.7 Preparation of carboxylic acids

33.7.1 Oxidation of alcohols or aldehydes

Both primary alcohols and aldehydes can be oxidized to carboxylic acids, by refluxing with excess potassium dichromate(VI) and acid (see **section 31.5.2**):

$$R{-}CH_2{-}OH \xrightarrow[\text{reflux}]{Cr_2O_7^{2-}/H^+} R{-}C({=}O){-}OH$$

an alcohol → a carboxylic acid

$$R{-}C({=}O){-}H \xrightarrow[\text{reflux}]{Cr_2O_7^{2-}/H^+} R{-}C({=}O){-}OH$$

an aldehyde → a carboxylic acid

33.7.2 From acid derivatives

All the acid derivatives can be hydrolysed (reacted with water) to give parent acid. The basic reaction can be represented:

$$R{-}C({=}O){-}Z + H_2O \longrightarrow R{-}C({=}O){-}OH + H$$

The conditions vary with the acid derivative used—acid chlorides a anhydrides react rapidly even with moist air. Amides must be refluxed some time with an acid or alkali. Details can be found in **section 33.8**

33.7.3 From nitriles

Nitriles, R—C≡N, are hydrolysed to carboxylic acids by boiling w aqueous solutions of mineral acids:

$$R{-}C{\equiv}N + 2H_2O + H^+ \xrightarrow{\text{heat}} R{-}C({=}O){-}OH + NH$$

33.8 Acid derivatives

33.8.1 Physical properties

Table 33.2 Lists some physical properties of derivatives of ethanoic a with those of the parent acid for comparison.

Table 33.2 Properties of derivatives of ethanoic acid

Property	Ethanoyl chloride, CH_3COCl	Ethanoic anhydride $CH_3CO_2COCH_3$	Ethanoic acid, CH_3CO_2H	Methyl ethanoate, $CH_3CO_2CH_3$	Ethanami CH₃CONH
T_m/K	161	200	290	175	355
T_b/K	324	413	391	330	494
Solubility in water	← React	→	Very soluble	Soluble	Very soluble
M_r	78.5	102	60	74	59

Although it is not easy to make comparisons because of the differe in relative molecular masses, some patterns can be seen. The es anhydride and acid chloride have lower melting temperatures than parent acid while that of the amide is greater. This is because the ami with two N—H bonds, can form more hydrogen bonds than the acid, wh the other three can form none as they have no hydrogen bonded to an electronegative atom.

The same sort of pattern is seen in the boiling temperatures and for the same reason (except that the anhydride has a slightly higher boiling temperature than the acid as it has a considerably larger relative molecular mass). The amide is very soluble in water, as is the parent acid, due to the formation of hydrogen bonds with water. The ester is less soluble as there is less hydrogen bonding. The other two derivatives react rapidly with water so it is not possible to measure their solubilities.

33.8.2 Infra-red spectra of acid derivatives

All the acid derivatives contain a carbonyl group and therefore show a strong C=O stretching vibration at around $1750\,cm^{-1}$, although this may be shifted up or down by as much as $50\,cm^{-1}$. The O—H stretch present in carboxylic acids is, of course, absent although amides have an N—H stretching vibration at almost the same frequency. This peak is also broadened by hydrogen bonding.

33.8.3 Reactions of acid derivatives

Acid–base properties

Esters, anhydrides and acid chlorides have no acidic hydrogens. Amides can lose a proton but they are much weaker acids than carboxylic acids. As they have a lone pair, they can also accept a proton and act as bases. However, this lone pair can become involved in delocalization, see **Figure 33.11**, and is less available to accept a proton than it would be in, for example, ammonia. So amides are weak bases.

gure 33.11 Delocalization of the ne pair on the —NH_2 group

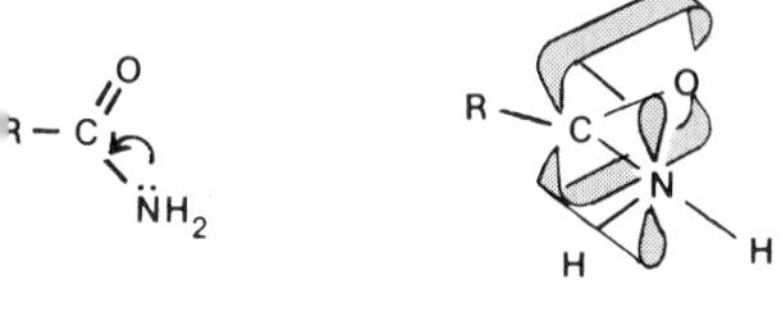

Nucleophilic substitution reactions

In all the acid derivatives, the carbonyl carbon atom has a δ^+ charge and is therefore attacked by nucleophiles. The general reaction is:

$$R{-}C(=O){-}Z \quad + \quad :Nu^- \longrightarrow R{-}C(=O){-}Nu \quad + \quad Z^-$$

How readily the reaction occurs depends on three factors:

1. how good the nucleophile is;
2. the magnitude of the δ^+ charge on the carbonyl carbon, which in turn depends on the electron-releasing or attracting power of Z;
3. the ease of expulsion of Z^- (the **leaving group**).

Factors 2 and 3 tend to be linked—groups which strongly attract electrons generally form stable negative ions, Z^-.

Acid derivatives can be listed in order of their reactivity towards nucleophiles:

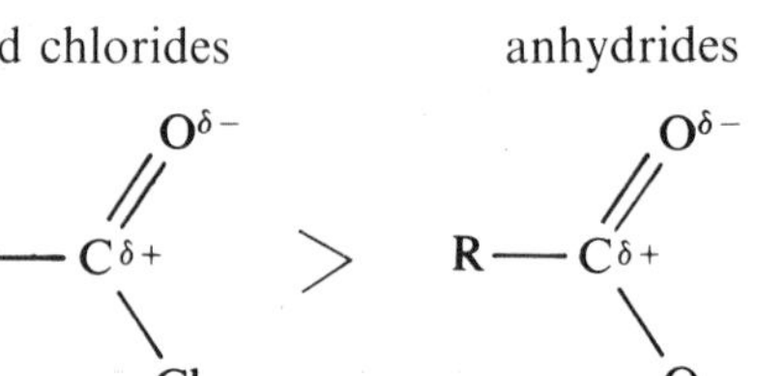

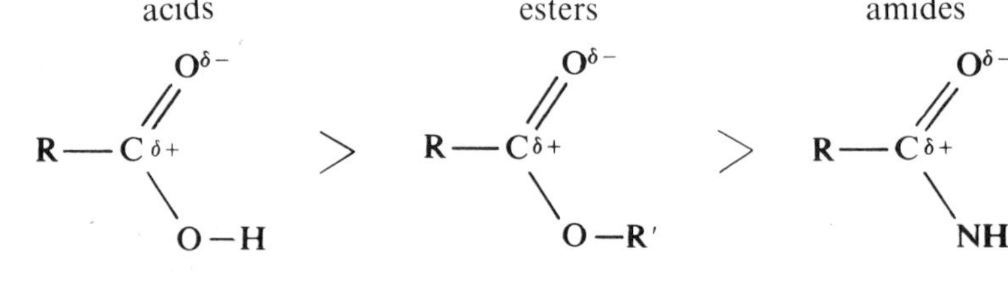

Figure 33.12 Comparison between an acid chloride and a chloroalkane

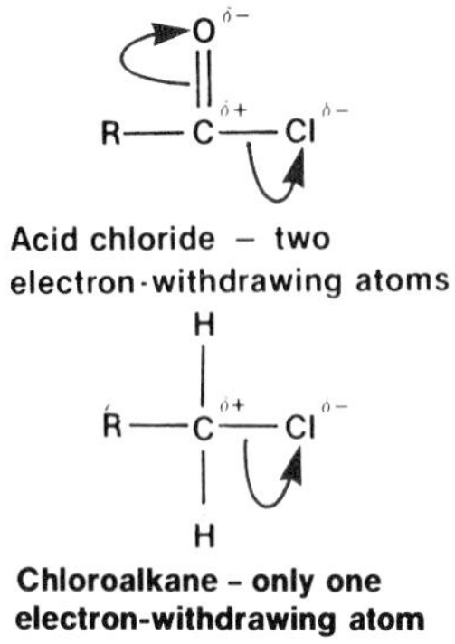

R—C(=O$^{\delta-}$)—$^{\delta+}$Cl$^{\delta-}$

Acid chloride – two electron-withdrawing atoms

R—CH$_2$$^{\delta+}$—Cl$^{\delta-}$

Chloroalkane – only one electron-withdrawing atom

The Z groups of acid chlorides and anhydrides withdraw electrons fro the carbonyl carbon making them more reactive than carboxylic acids. I esters, acids and amides, the Z group releases electrons on to the carbon carbon making them less reactive. Acids and esters are very similar i reactivity.

Acid chlorides are much more reactive towards nucleophiles than a chloroalkanes. For example, even a weak nucleophile like water will rea vigorously with an acid chloride but only very slowly indeed with chloroalkane. This is because of the presence of the oxygen atom in th acid chloride. Both this *and* the chlorine atom withdraw electrons fro the carbon to which the chlorine is attached, so this carbon has a muc greater δ^+ charge in an acid chloride than in a chloroalkane. See **Figu 33.12**. The products of the reactions of some nucleophiles with some ac derivatives are shown in **Table 33.3.**

Acid Derivative / Nucleophile	Acid Chloride	Anhydride	Carboxylic acid	Ester	Amide
	R—C(=O)Cl	R—C(=O)—O—C(=O)—R	R—C(=O)OH	R—C(=O)OR′	R—C(=O)NH$_2$
	← increasing reactivity				
Ammonia, NH_3	R—C(=O)NH$_2$ Amide	R—C(=O)NH$_2$ Amide	R—C(=O)O$^-$ NH_4^+ Ammonium salt ↓ $-H_2O$ R—C(=O)NH$_2$ Amide	R—C(=O)NH$_2$ Amide	—
Amine, R″ NH_2	R—C(=O)NHR″	R—C(=O)NHR″	R—C(=O)NHR″	R—C(=O)NHR″	—
	← N-substituted amides →				
Alcohol, R″OH	R—C(=O)O—R″ Ester	R—C(=O)O—R″ Ester	R—C(=O)O—R″ Ester (reflux with H^+ catalyst)	R—C(=O)O—R″ Different ester (reflux with H^+ catalyst)	—
Water, H_2O	R—C(=O)OH Carboxylic acid	R—C(=O)OH Carboxylic acid	—	R—C(=O)O—H R′—OH Carboxylic acid + alcohol (reflux with H^+ catalyst)	R—C(=O)OH Carboxylic acid (reflux with H^+catalyst)

Table 33.3 Reactions of some acid derivatives with electrophiles. All reactions take place at room temperature unless otherwise stated

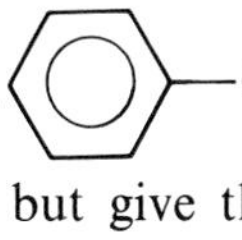

Note that these reactions involve conversion of one acid derivative in another. Carboxylic acids, for example, can be prepared by hydrolys (reaction with water) of any derivative under suitable conditions. Oth nucleophiles may be used and give different products. Alcohols a relatively poor nucleophiles and phenols are even worse so alkoxid RO^- or phenoxide, C_6H_5—O^- ions may be used instead, as these a better nucleophiles but give the same products. Acid chlorides can b converted into anhydrides by treatment with the sodium salt of a carboxylic acid which produces the nucleophile:

$$R{-}C(=O){-}O^- \qquad R{-}\overset{\delta+}{C}(=O){-}Cl^{\delta-} + R'{-}C(=O){-}O^- \longrightarrow R{-}C(=O){-}O{-}C(=O){-}R' + Cl^-$$

This enables mixed anhydrides to be made. Not all these reactions go to completion. In particular, the hydrolysis of esters produces an equilibrim mixture containing the ester, water, acid and alcohol. The acid catalyst, of course, affects only the rate at which equilibrium is reached, not the composition of the equilibrium mixture.

Bases also catalyse hydrolysis of esters. In this case the salt of the acid is produced. This removes the acid from the reaction mixture and moves the equilibrium over to the right, for example:

$$CH_3{-}C(=O){-}O{-}CH_3 + H_2O \underset{\text{catalyst}}{\overset{\text{NaOH}}{\rightleftharpoons}} CH_3{-}C(=O){-}OH + CH_3OH$$

$$CH_3{-}C(=O){-}OH \xrightarrow{\text{NaOH}} \underset{\text{sodium ethanoate}}{CH_3C(=O){-}O^-} + Na^+ + H_2O$$

Reduction of acid derivatives

ACID CHLORIDES

Acid chlorides can be reduced to aldehydes by hydrogen using a poisoned palladium catalyst to prevent further reduction to alcohols as follows:

$$R{-}C(=O){-}Cl + H_2 \xrightarrow[\text{catalyst}]{\text{poisoned Pd}} \underset{\text{aldehyde}}{R{-}C(=O){-}H} + HCl$$

With an unpoisoned catalyst, hydrogen will reduce acid chlorides to primary alcohols:

$$R{-}C(=O){-}Cl + 2H_2 \xrightarrow[\text{Pd catalyst}]{} \underset{1^\circ\text{ alcohol}}{R{-}CH_2{-}OH} + HCl$$

A

SOAP MANUFACTURE

$$\begin{array}{l} H-\overset{\displaystyle H}{\overset{|}{C}}-O-\overset{\displaystyle O}{\overset{\|}{C}}-R \\ H-\overset{|}{C}-O-\overset{\displaystyle O}{\overset{\|}{C}}-R \\ H-\underset{\displaystyle H}{\underset{|}{\overset{|}{C}}}-O-\overset{\displaystyle O}{\overset{\|}{C}}-R \end{array} + 3NaOH \longrightarrow \begin{array}{l} H-\overset{\displaystyle H}{\overset{|}{C}}-OH \\ H-\overset{|}{C}-OH \\ H-\underset{\displaystyle H}{\underset{|}{\overset{|}{C}}}-OH \end{array} + 3(Na^+ + R-\overset{\displaystyle O}{\overset{\|}{C}}-O^-)$$

Fat or oil
R is a $C_{12}-C_{18}$ chain

Propane-1,2,3-triol

Soap

Soaps are sodium or potassium salts of fatty acids (the name given to naturally occurring carboxylic acids with long chains). They are made by boiling fats or oils with sodium (or potassium) hydroxide. This yields the salt of the fatty acid and propane-1,2,3-triol (glycerol, sometimes called glycerine). This is an example of alkaline hydrolysis of an ester and is called **saponification** (see above).

Potassium soaps tend to be liquids while those containing sodium are usually solid. The box *Detergents and soaps*, **section 16.3**, describes how soaps work. They are inefficient in acidic solutions because of the formation of free fatty acid which is less soluble in water than the salts:

$$R-\overset{\displaystyle O}{\overset{\|}{C}}-O^-(aq) + H^+(aq) \rightleftharpoons R-\overset{\displaystyle O}{\overset{\|}{C}}-OH(s)$$

This equlibrium is well over to the right.

Soaps are also inefficient in hard water (which contains dissolved calcium or magnesium salts) because of the formation of insoluble calcium or magnesium salts—scum:

$$\underset{\textbf{Soap}}{2R-\overset{\displaystyle O}{\overset{\|}{C}}-O^-} + Ca^{2+}(aq) \longrightarrow \underset{\textbf{Scum}}{\left[2R-\overset{\displaystyle O}{\overset{\|}{C}}-O^- + Ca^{2+}\right](s)}$$

The reaction can also be brought about by using lithium tetrahydrido-aluminate(III) or sodium tetrahydridoborate(III).

ANHYDRIDES

These can be reduced in a similar way and with similar reagents to those used for acid chlorides, forming either aldehydes or primary alcohols:

$$(RCO)_2O \xrightarrow[\text{poisoned Pd catalyst}]{H_2} 2R-\overset{\displaystyle O}{\overset{\|}{C}}-H \quad \text{aldehyde}$$

$$(RCO)_2O \xrightarrow{H_2/Pd \text{ or } LiAlH_4} 2R-\underset{\displaystyle H}{\underset{|}{\overset{\displaystyle H}{\overset{|}{C}}}}-OH \quad 1^\circ \text{ alcohol}$$

ESTERS

These can be reduced by catalytic hydrogenation, lithium tetrahydrido-aluminate(III) or sodium in ethanol to give a mixture of two alcohols:

$$R-\overset{\displaystyle O}{\overset{\|}{C}}-O-R' \xrightarrow{\text{reduction}} \underset{1^\circ \text{ alcohol}}{R-\underset{\displaystyle H}{\underset{|}{\overset{\displaystyle H}{\overset{|}{C}}}}-OH} + R'OH$$

may be 1°, 2° or 3° alcohol depending on R

AMIDES

Amides are reduced by catalytic hydrogenation or lithium tetrahydridoaluminate(III) to yield primary amines:

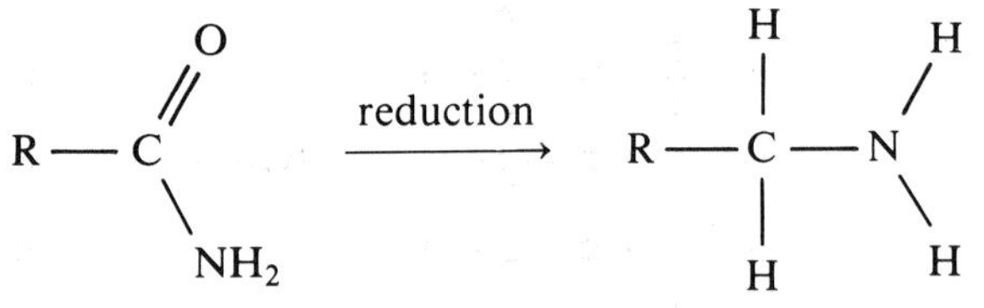

$$R{-}C({=}O){-}NH_2 \xrightarrow{\text{reduction}} R{-}CH_2{-}NH_2$$

1° amine

Other reactions of acid derivatives

DEHYDRATION OF AMIDES

Amides can be dehydrated with phosphorus pentoxide to produce nitriles (compounds containing the —C≡N group):

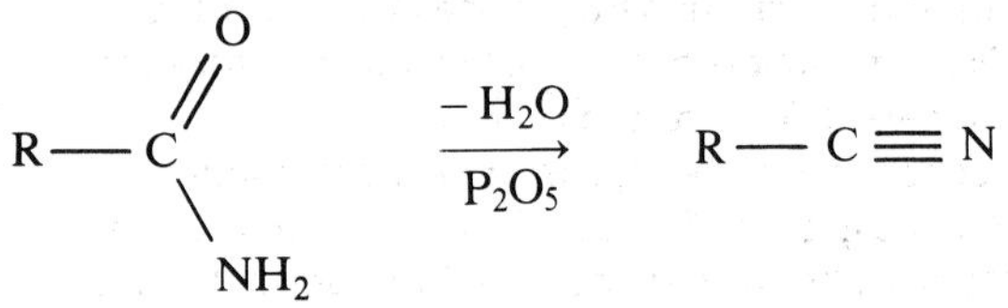

$$R{-}C({=}O){-}NH_2 \xrightarrow[P_2O_5]{-H_2O} R{-}C{\equiv}N$$

FRIEDEL–CRAFTS ACYLATION

Acid chlorides react with aromatic hydrocarbons in the presence of a catalyst of aluminum chloride. This generates the electrophile

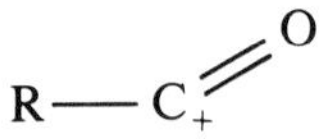

$$R{-}C^{+}{=}O$$

which attacks the benzene ring. An aromatic ketone is produced:

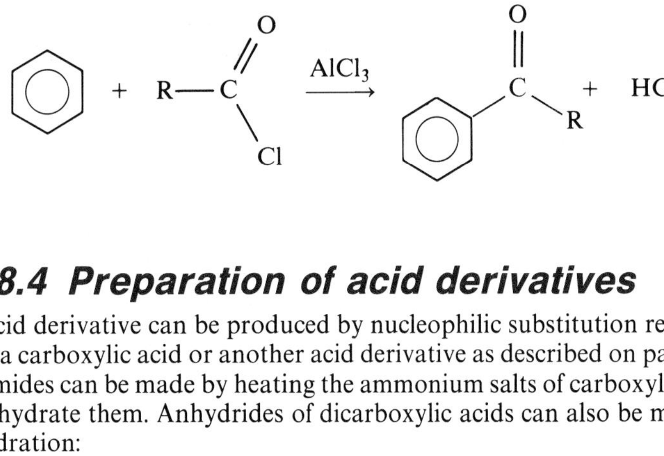

$$C_6H_6 + R{-}C({=}O){-}Cl \xrightarrow{AlCl_3} C_6H_5{-}C({=}O){-}R + HCl$$

33.8.4 Preparation of acid derivatives

An acid derivative can be produced by nucleophilic substitution reactions from a carboxylic acid or another acid derivative as described on page 524.

Amides can be made by heating the ammonium salts of carboxylic acids to dehydrate them. Anhydrides of dicarboxylic acids can also be made by dehydration:

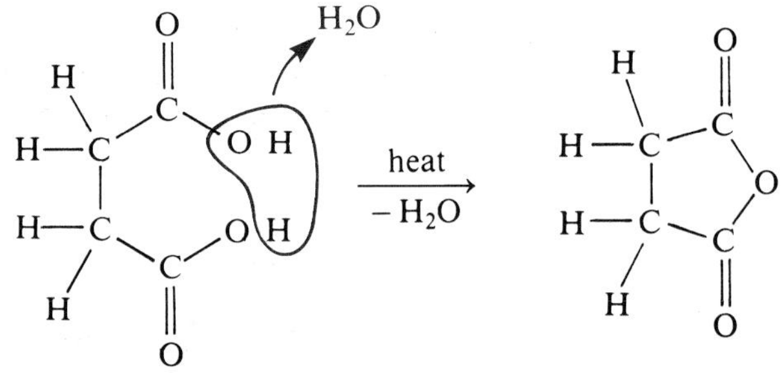

$$HOOC{-}CH_2{-}CH_2{-}COOH \xrightarrow[-H_2O]{\text{heat}} \text{(cyclic anhydride)} + H_2O$$

butane-1,4-dioic acid butane-1,4-dioic anhydride

Butene-1,4-dioic acid exists in *cis-* and *trans-* forms as follows:

$$\begin{array}{cc} \text{H(C=O)... } & \\ \end{array}$$

cis-butene-1,4-dioic acid *trans*-butene-1,4-dioic acid

Only the *cis* form forms an anhydride. In the *trans* form the —OH grou are too far apart and no rotation is possible.

33.9 Economic importance of carboxylic acids and their derivatives

Ethanoic acid is an important industrial chemical. It is made by fo different processes:

1. Naphtha is oxidized by air at a pressure of 50×10^2 kPa and 450 K Significant quantities of methanoic and propanoic acids are also forme These are not in such demand as ethanoic acid and present dispos problems.
2. Butane can be oxidized in similar conditions.
3. Methanol is reacted with carbon monoxide at 450 K and 30×10^2 kP to give ethanoic acid in 99% yield.
4. Ethanal (produced from ethene by the Wacker process) can be oxidize to ethanoic acid.

The industrial chemistry of ethanoic acid is summarized in **Figure 33.1** Dicarboxylic acids are important in the manufacture of poly(esters) an poly(amides).

Animal and vegetable oils and fats are esters of the alcohol propane 1,2,3-triol (glycerol). Do not confuse these with the oils derived from crud

Figure 33.13 The industrial chemistry of ethanoic acid

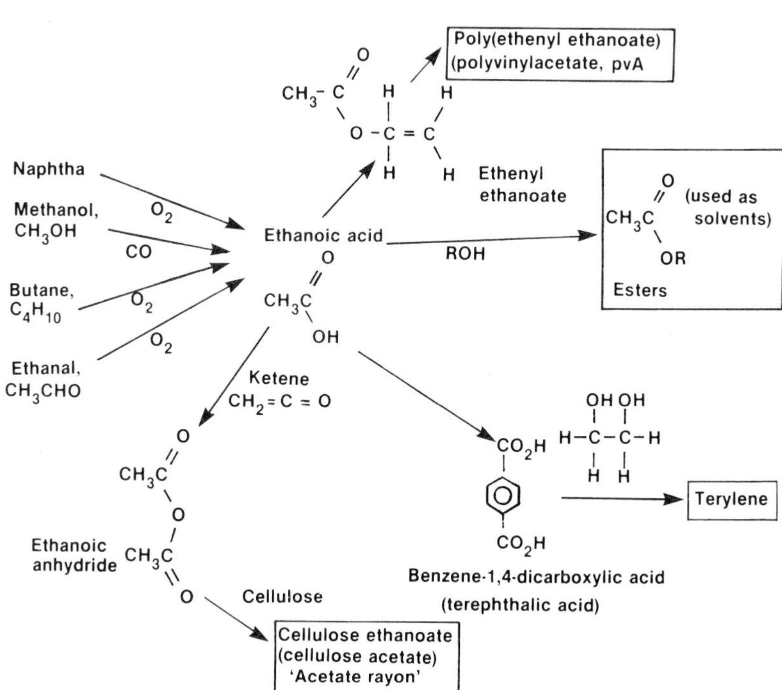

A

POLYESTERS AND POLYAMIDES

Polyesters are made by the reaction of a diol with a dicarboxylic acid. The acid and diol are joined by an ester linkage and a molecule of water is eliminated for each ester linkage formed. This is the basis of the name **condensation polymerization**.

For example, benzene-1,4-dicarboxylic acid and ethane-1,2-diol react to form a polyester called Terylene after the non-systematic name for benzene-1,4-dicarboxylic acid—terephthalic acid. The polymer is melted and forced through find holes to make the fibre Terylene.

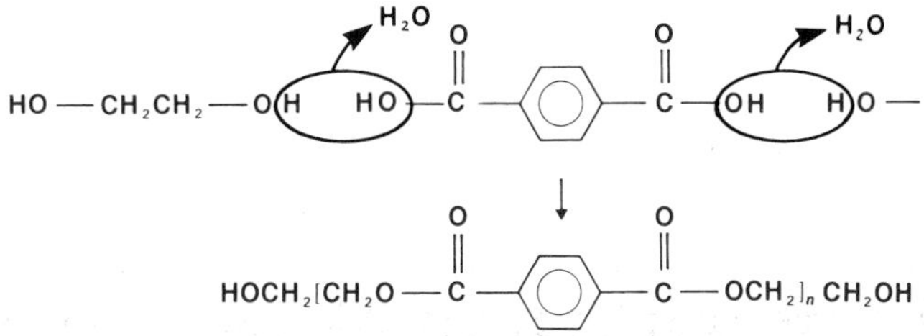

HO—CH₂CH₂—O(H HO)—C(=O)—C₆H₄—C(=O)—(OH H)O—CH₂—CH₂—OH (H_2O eliminated at each linkage)

↓

HOCH₂[CH₂O—C(=O)—C₆H₄—C(=O)—OCH₂]ₙ CH₂OH

Nylons are polyamides. Some forms are made by the copolymerization (polymerization of two different monomers) of a diamine with a dicarboxylic acid or a diacid dichloride. Different types of nylon with slightly different properties are formed depending on the lengths of the carbon chains of the two monomers. The reaction between hexane-1,6-diamine and hexanedioyldichloride gives nylon-6,6 while nylon-6,10 is produced from decanedioylodichloride and hexane-1,6-diamine:

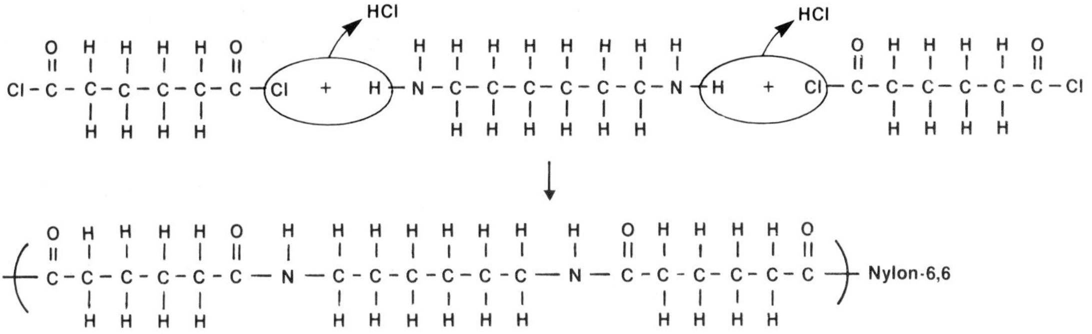

$$Cl-CO-(CH_2)_4-CO-Cl + H-NH-(CH_2)_6-NH-H + Cl-CO-(CH_2)_4-CO-Cl \xrightarrow{-HCl} \left[-CO-(CH_2)_4-CO-NH-(CH_2)_6-NH-CO-(CH_2)_4-CO-\right] \text{ Nylon-6,6}$$

Nylon-6 is made from a single monomer, a cyclic amide called caprolactam:

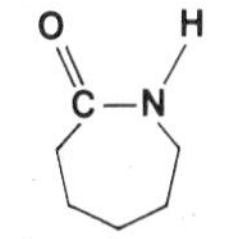

Caprolactam

The C—N bond in the ring breaks and reforms to make a linear polymer:

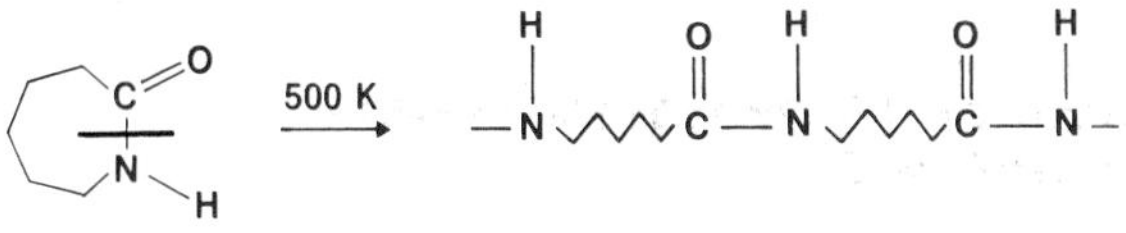

$$n\ \text{caprolactam} \xrightarrow{500\text{ K}} -NH\sim\sim C(=O)-NH\sim\sim C(=O)-NH-$$

oils which are hydrocarbons. Edible oils and fats contain three molecules of long-chain (around C_{12}–C_{18}) carboxylic acids called fatty acids, combined with propane-1,2,3-triol. As the non-systematic name of this triol is glycerol, fats and oils are referred to as **triglycerides**:

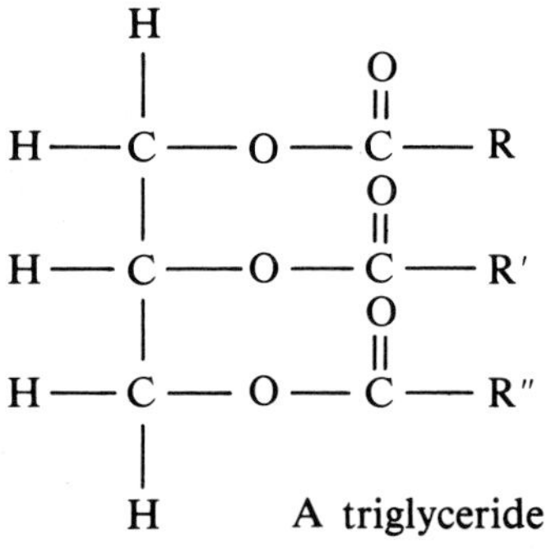

$$\begin{array}{l} H_2C-O-CO-R \\ HC-O-CO-R' \\ H_2C-O-CO-R'' \end{array}$$

A triglyceride

A typical triglyceride may contain two or three different fatty acids. Common fatty acids include stearic and palmitic acids which are saturated and oleic and linoleic acids which are unsaturated. The only difference

tearic acid
$H_3(CH_2)_{16}CO_2H$
almitic acid
$H_3(CH_2)_{14}CO_2H$
leic acid
$H_3(CH_2)_7CH{=}CH(CH_2)_7CO_2H$
inoleic acid
$H_3(CH_2)_4(CH{=}CHCH_2)_2(CH_2)_6CO_2H$

between fats and oils is that fats are solid at room temperature while o are liquids. When eaten, fats and oils are broken down by hydroly to give uncombined fatty acids. Some of these cannot be produc by the body and have to be present in the diet. These are called **essenti fatty acids**.

33.10 Summary

- The functional groups of acid derivatives are:

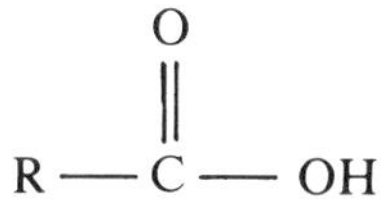

$R-\overset{\overset{\displaystyle O}{\|}}{C}-OH$ carboxylic acid

$R-\overset{\overset{\displaystyle O}{\|}}{C}-Cl$ acid chloride

$R-\overset{\overset{\displaystyle O}{\|}}{C}-O-R'$ ester

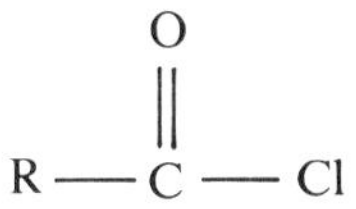

$R-C(=O)-O-C(=O)-R'$ anhydride

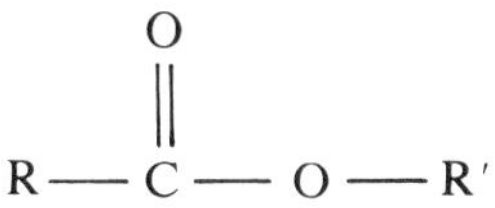

$R-\overset{\overset{\displaystyle O}{\|}}{C}-NH_2$ amide

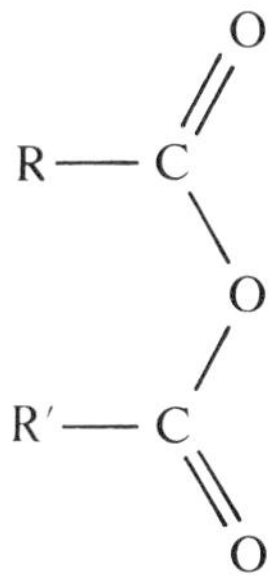

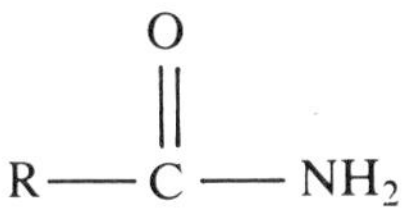

A

ETHANOIC ANHYDRIDE

Ethanoic anhydride is used in the manufacture of aspirin and cellulose ethanoate (cellulose acetate or acetate rayon) fibres. In each case it is used to form an ester by reaction with —OH groups.

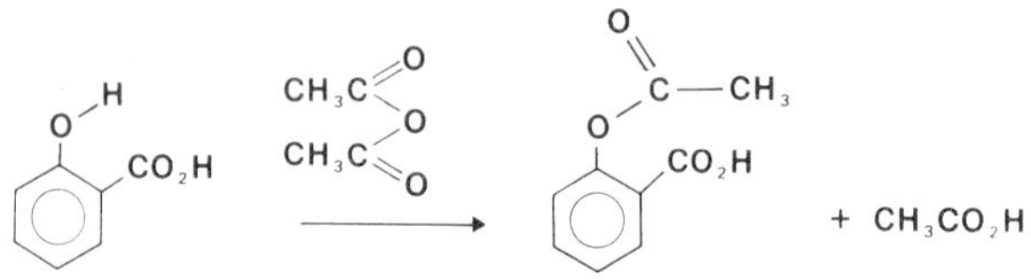

2-hydroxybenzene-carboxylic acid (made by the reaction of phenol and carbon dioxide) + $(CH_3CO)_2O$ → 2-ethanoyloxybenzene-carboxylic acid (aspirin) + CH_3CO_2H

Cellulose is obtained from wood pulp. Cellulose ethanoate will dissolve in non-aqueous solvents and the solution is forced through fine holes. Fibres are formed as the solvent evaporates.

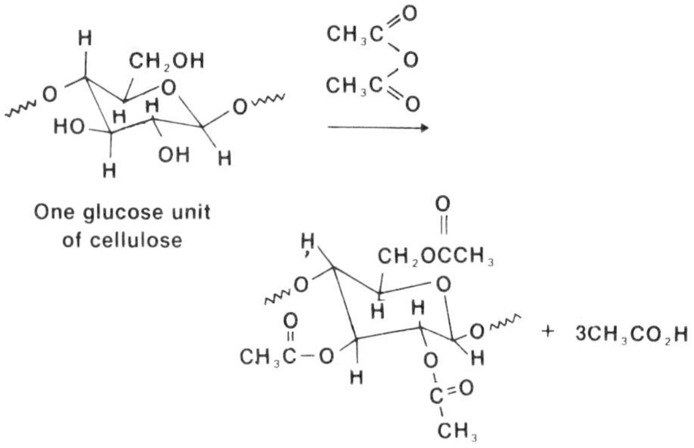

One glucose unit of cellulose + $(CH_3CO)_2O$ → One unit of cellulose ethanoate + $3CH_3CO_2H$

- The group $R—C(=O)—Z$ is flat and trigonal, with all bond angles approximately 120°.
- The group $\rangle C^{\delta+}=O^{\delta-}$ (with Z attached to C) is polarized as shown. The δ^+ character of the carbon is determined by the electron-attracting or releasing power of Z.
- The intermolecular forces which operate are:

Carboxylic acids, amides	hydrogen bonding
esters, acid chlorides, anhydrides	dipole–dipole forces

- The reactions of carboxylic acids are summarized in **Figure 33.14**.
- Carboxylic acids, RCO_2H, where R is an alkyl group, are weak acids.
- Electron-withdrawing substituents such as Cl— or C_6H_5— make the acids stronger as they delocalize the negative charge of the $R—CO_2^-$ ion.

igure 33.14 **Reactions of carboxylic ids using ethanoic acid as an example**

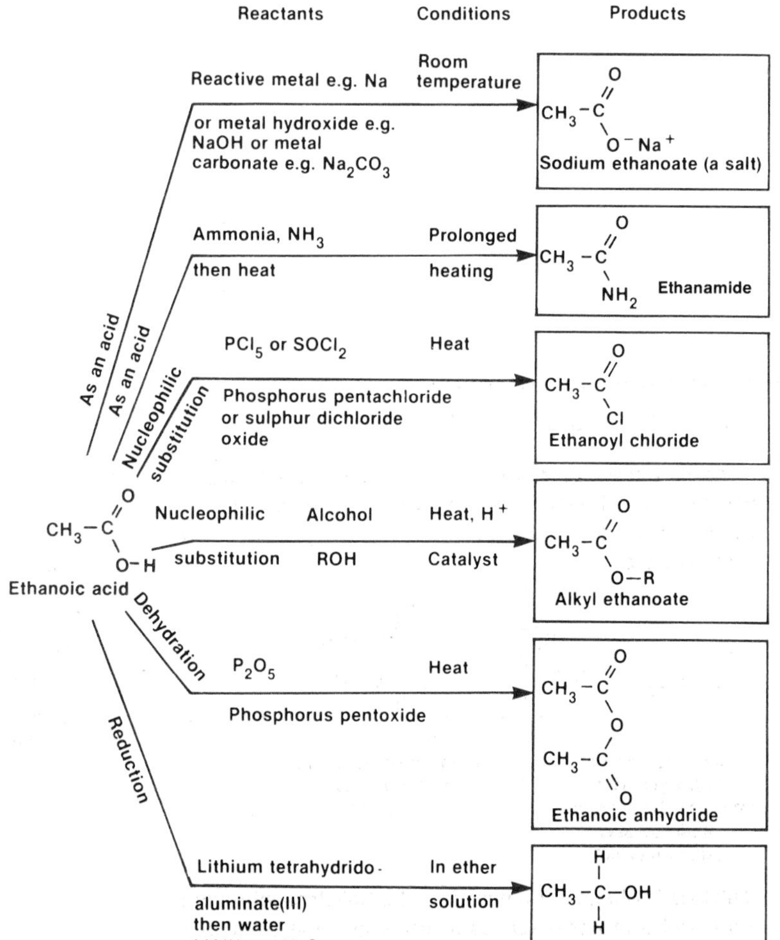

• The $R-C(=O)O^-$ ion has two identical bonds of length between that c C—O and that of C=O due to delocalization.
• Acid derivatives can be converted into one another or the parent aci by nucleophilic substitution reactions using an appropriate nucleophil

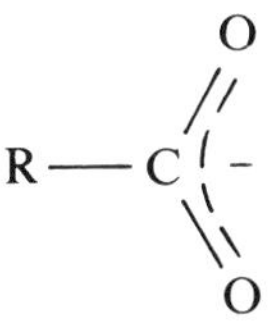

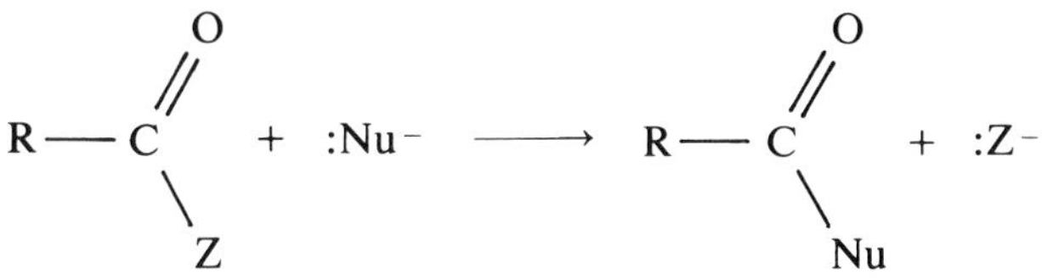

$$R-C(=O)-Z + :Nu^- \longrightarrow R-C(=O)-Nu + :Z^-$$

• The order of reactivity of acid derivatives is:

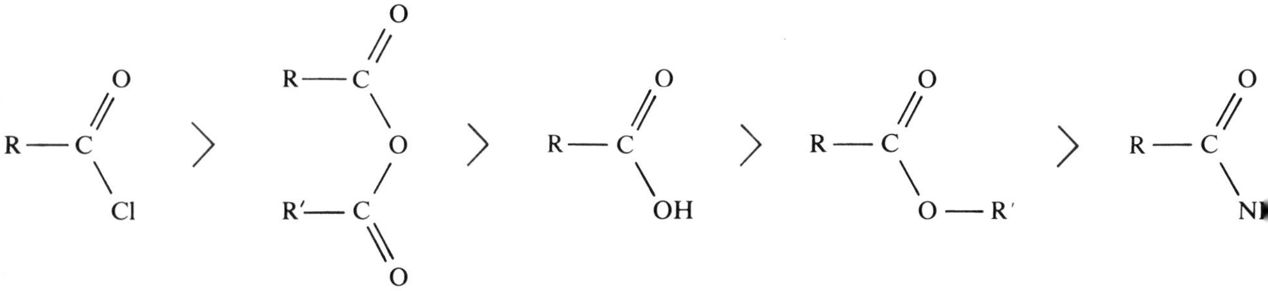

$$R-C(=O)Cl > (R-CO)(R'-CO)O > R-C(=O)OH > R-C(=O)O-R' > R-C(=O)N$$

It is easy to prepare less reactive derivatives from more reactive ones.
• Acid chlorides, anhydrides and esters can be reduced to 1° alcohols. With acid chlorides and anhydrides, the reduction can be stopped at ar aldehyde by using a suitable reducing agent.
• Amides can be dehydrated to nitriles.
• Carboxylic acids can be prepared by oxidation of alcohols or aldehydes or by hydrolysis of other acid derivatives.

33.11 Questions

1 Name the following:

A) 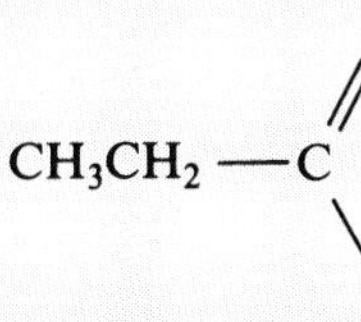 $CH_3CH_2-C(=O)-OH$

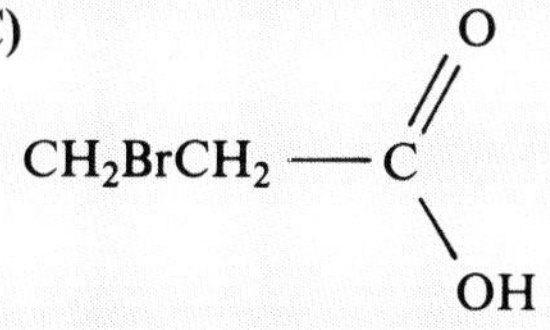

B) 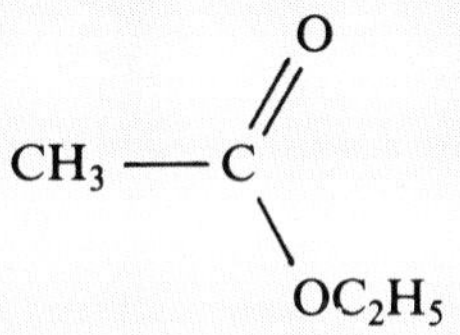 $CH_3-C(=O)-OC_2H_5$

C) $CH_2BrCH_2-C(=O)-OH$

D) 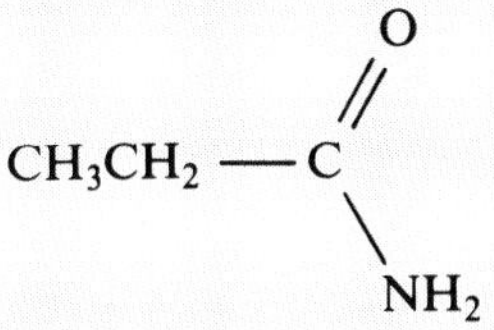 $CH_3CH_2-C(=O)-NH_2$

2 Give the structural formula of:
A) 2-chloropropanamide;
B) methyl propanoate;
C) ethanoic propanoic anhydride;
D) propanoyl chloride.

3 Give the products of the following reactions:
A) propanoic acid and ammonia, followed by heating;
B) ethanoyl chloride plus methanol;
C) butanoic anhydride plus water;
D) butanoic acid plus phosphorus pentachloride.

4 List the following in order of acid strength. Explain your reasoning.
A) CH_2BrCO_2H; **C)** CH_2ClCO_2H;
B) CH_2FCO_2H; **D)** CHF_2CO_2H.

5 **A)** Give reagents, essential conditions and equations for the conversion of ethanoic acid into:
i ethanoic anhydride;
ii ethanamide.
B) Ethanoic anhydride is a liquid at room temperature but ethanamide is a solid. Comment briefly on this.
C) Ethyl ethanoate was warmed briefly with sodium hydroxide in water enriched in $H_2{}^{18}O$. After neutralization, a sample of the organic components was investigated by mass spectrometry. It was shown that compounds of relative molecular mass 46, 60, 62 and 88 were present.
Explain these results and give the mechanism of the reaction.
(Oxford 1987)

6 Describe, and attempt to classify, the reactions of the —OH group in a variety of organic compounds. Refer to specific reactions in your answer, giving structural formulae where possible.

(Nuffield 1979)

7 Deduce the structures of the compounds A to G, giving the reason for selecting each structure.

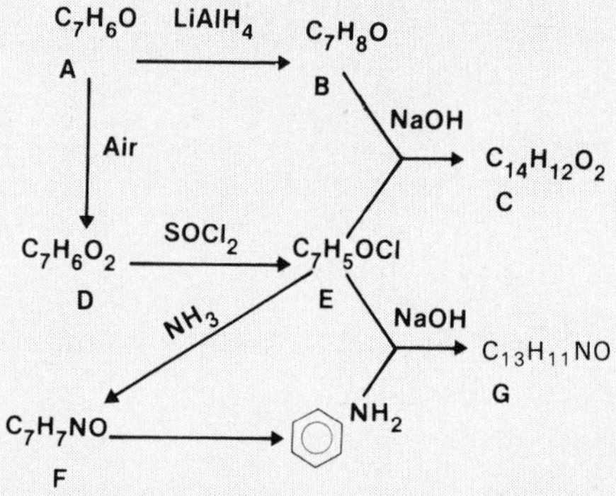

How would you show that the compound G was identical with one of the same formula from the laboratory stock-room?

(London 1979)

8 The following is a description of an experiment in which ethyl benzoate is hydrolysed. 5 cm^3 (0.03 mol) of ethyl benzoate are mixed with 30 cm^3 of sodium hydroxide solution and heated under reflux for about fifteen minutes, until all the oily drops have disappeared. The flask is then fitted for fractional distillation and the mixture distilled until the boiling point reaches 78 °C to give a distillate A. The contents of the flask are allowed to cool and then slowly acidified with concentrated hydrochloric acid, when a precipitate B separates. This precipitate is purified by recrystallization from water.

A) Write the equation for the reaction between ethyl benzoate and sodium hydroxide solution.

B) Sketch the apparatus required for heating the mixture under reflux.

C) During the distillation, where would you place the bulb of the thermometer and why?

D) i What is the liquid A?

ii Give one *chemical* test and its result, which would support your answer to (i).

E) i What is the white precipitate B?

ii Write the equation for its formation.

F) i Describe how you would separate and recrystallize the precipitate B. Indicate the apparatus you would use.

ii State how you would check that your recrystallized specimen was pure.

(London 1978)

9 Comment on the pK_a values for the series of acids:

Structural formula of acid	**pK_a**
$CH_3CH_2CH_2CO_2H$	4.82
$CH_3CH_2CHClCO_2H$	2.84
$CH_3CHClCH_2CO_2H$	4.06
$CH_2ClCH_2CH_2CO_2H$	4.52

(Nuffield 1980, Special paper, part question)

34 Organic nitrogen compounds

34.1 Introduction

This chapter deals with two main groups of compounds—*amines* and *nitriles*. Amines are best regarded as derivatives of ammonia in which one or more hydrogen atoms in the ammonia molecule have been replaced by organic groups. Nitriles contain the functional group —C≡N. Both amines and nitriles are very reactive groups of compounds and are therefore useful as intermediates in synthesis.

Other nitrogen-containing functional groups are amides, dealt with in Chapter 33, and hydrazones and oximes, mentioned in Chapter 32.

34.2 Amines

34.2.1 Formulae, naming and structure

Naming

Primary amines have the general formula R—NH_2, where the R— can be an alkyl or aryl group. Such amines are named as alkyl or aryl derivatives of ammonia using the suffix -amine, for example:

CH_3—NH_2 methylamine

C_2H_5—NH_2 ethylamine

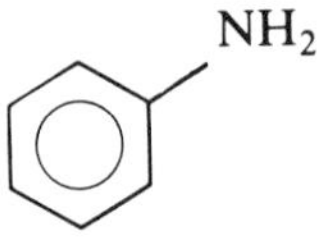

C_6H_5—NH_2 phenylamine (often still called aniline and sometimes aminobenzene)

If the amine is not the principal group, the prefix amino- is used, for example:

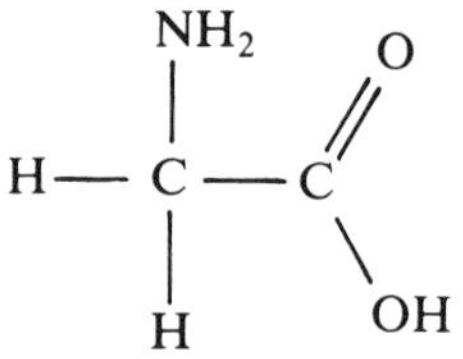

$$\mathrm{H{-}C(NH_2)(H){-}C({=}O)OH}$$ aminoethanoic acid

This notation may also be used to distinguish between isomers, for example:

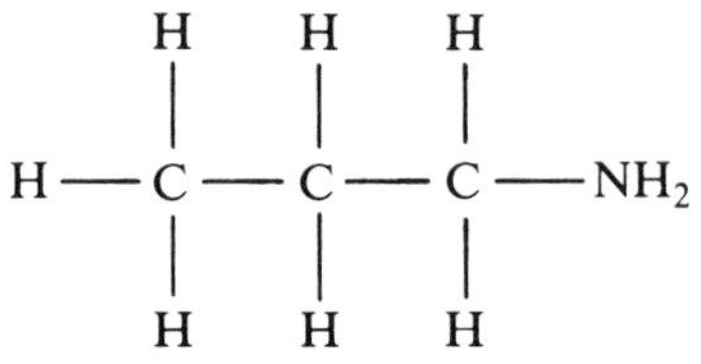

$$\mathrm{H{-}CH_2{-}CH_2{-}CH_2{-}NH_2}$$ is called 1-aminopropane

and

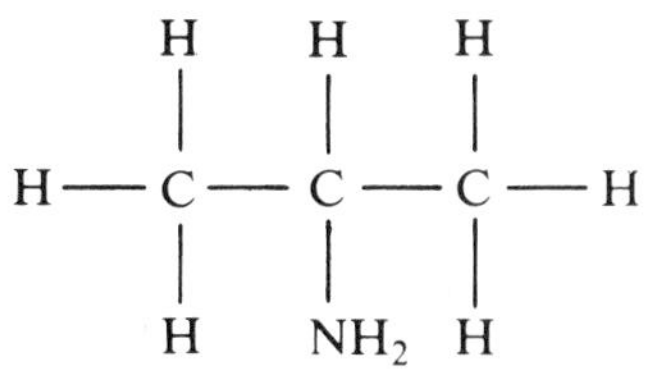

$$\mathrm{H{-}\underset{H}{\overset{H}{C}}{-}\underset{NH_2}{\overset{H}{C}}{-}\underset{H}{\overset{H}{C}}{-}H}$$ is called 2-aminopropane

Secondary and tertiary amines have the general formulae RR′N—H an RR′R″N respectively. They are named as di- and trisubstituted derivative of ammonia respectively, for example:

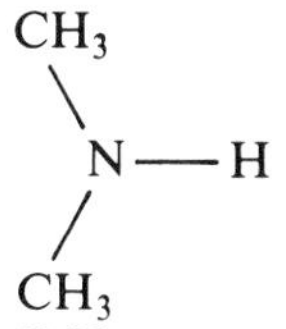

$(CH_3)_2N{-}H$ dimethylamine

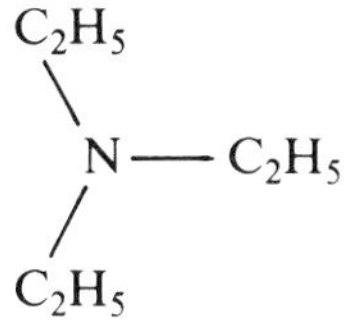

$(C_2H_5)_2N{-}C_2H_5$ triethylamine

$C_6H_5{-}N(CH_3){-}H$ N-methylphenylamine—the substituents ar listed in alphabetical order, methyl befor phenyl. N indicates that the methyl group is bonde to the nitrogen, rather than to the benzene rin

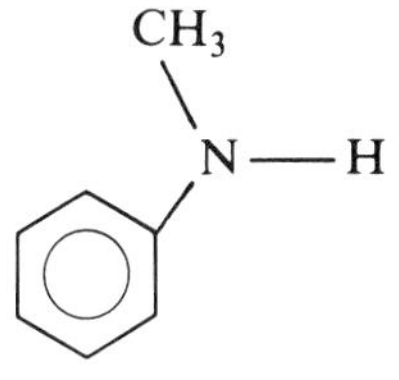

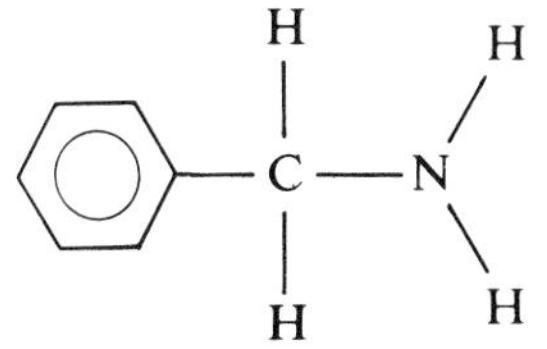

$C_6H_5{-}CH_2{-}NH_2$ (phenylmethyl)amine—an alkyl amine, not a aromatic one as the nitrogen is not attache directly to the aromatic ring. The brackets hel to make it clear that the phenyl group is no bonded directly to the nitrogen.

Notice the rather different way in which the terms primary, secondary an tertiary are used with amines compared to alcohols. In amines, 1°, 2° an 3° refer to the number of substituents (R groups) on the nitrogen atom In alcohols, 1°, 2° and 3° refer to the number of substituents on the carbo atom bonded to the —OH group:

amines alcohols

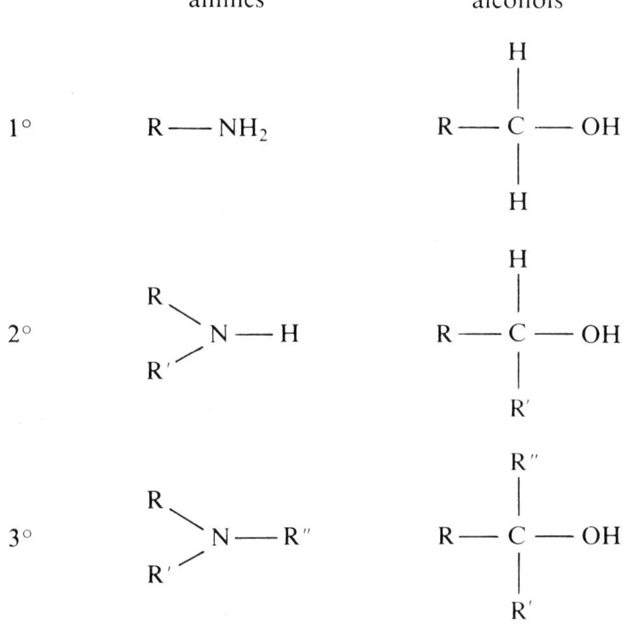

gure 34.1 The shape of the nethylamine molecule

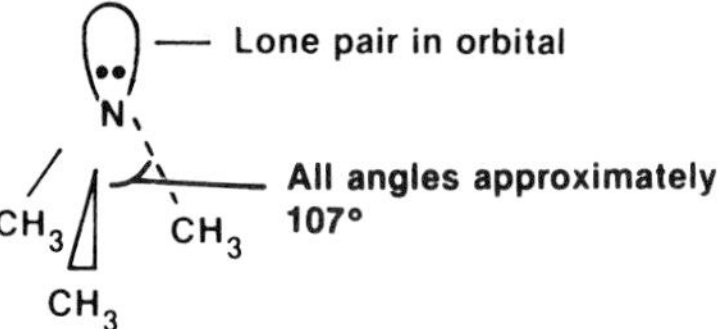

Ammonia is pyramidal with bond angles of approximately 107° (see **section 10.3.3**). Amines retain this basic shape, the nitrogen atom being hybridized sp^3 with the lone pair of electrons in one of the sp^3 orbitals, **Figure 34.1.**

34.2.2 Physical properties of amines

Boiling temperatures

Amines are polar:

$$R-\underset{H}{\overset{H}{C}}{}^{\delta+}-N^{\delta-}\begin{matrix}H\\H\end{matrix}$$

Primary and secondary amines can hydrogen bond to one another as well as to molecules of solvents such as water or alcohols. However, as nitrogen is less electronegative than oxygen (electronegativities: O = 3.5, N = 3.0), the hydrogen bonds are not as strong as those in alcohols.

So the boiling temperatures of amines are lower than those of comparable alcohols:

methylamine	CH_3-NH_2	$T_b = 267\,K$	$M_r = 31$
methanol	CH_3-OH	$T_b = 338\,K$	$M_r = 32$

Shorter-chain amines—methylamine, ethylamine, dimethylamine and trimethylamine—are gases at room temperature and those with slightly longer chains are volatile liquids. They have characteristic fishy smells, indeed rotting fish and animal flesh smell of di- and triamines produced by the decomposition of proteins.

Disubstituted amines have only one hydrogen bonded to nitrogen and trisubstituted ones none so there is less hydrogen bonding and the boiling temperatures get lower, illustrated by the three isomeric amines of formula C_3H_9N in **Table 34.1.**

ble 34.1 Boiling temperatures of me isomeric amines compared with alcohol

Name	Formula	T_b/K		M_r
1-aminopropane	$C_3H_7-NH_2$	321	lower	59
Ethylmethylamine	$C_2H_5-NHCH_3$	310	T_b	59
Trimethylamine	$CH_3-N(CH_3)_2$	276	↓	59
Propan-1-ol	C_3H_7OH	370		60

Propan-1-ol, shown for comparison, has the highest boiling temperature as oxygen is more electronegative than nitrogen.

Solubility

Primary amines with chain lengths up to about C_4 are very soluble in water. Di- and tri-substituted amines are less water soluble as there is less hydrogen bonding. Most amines are soluble in less polar solvents.

Phenylamine is not very soluble in water due to the benzene ring.

34.2.3 Infra-red spectra of amines

Figure 34.2 shows the infra-red spectrum of diethylamine. The main peak of interest is the N—H stretch which occurs in the range 3300–3500 cm^{-1}. Like the O—H stretch in alcohols, it is somewhat broadened by hydrogen bonding. This peak is similar in frequency to the O—H stretch in alcohols because the masses of nitrogen and oxygen atoms are very similar as are the strengths of the O—H and N—H bonds.

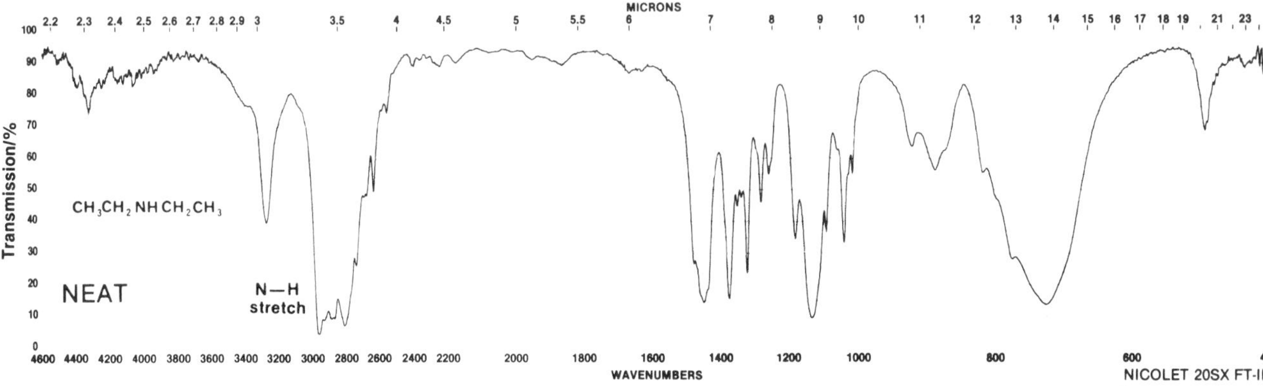

Figure 34.2 Infra-red spectrum of diethylamine

34.2.4 Reactivity of amines

One of the main features of the chemistry of amines is the fact that th have a lone pair of electrons. This may be used to form a bond with a H^+ ion, an electron-deficient carbon atom or with a transition metal io The amine is said to be acting as a **base**, a **nucleophile** or a **ligand** respective but the three types of behaviour are essentially similar. As nitrogen is le electronegative than oxygen the lone pair is less strongly held than in, fo example, alcohols so that amines are better bases, nucleophiles and ligan than alcohols.

Amines as bases

Amines can accept a proton (H^+ ion):

$$\underset{\text{amine}}{R\text{—}\ddot{N}H_2} + H^+ \rightleftharpoons \underset{\text{alkylammonium ion}}{RNH_3^+}$$

An amine's strength as a base (its ability to accept a proton) is governed b the availability of the lone pair of electrons. This is increased by electro releasing substituents such as alkyl groups, so secondary alkylamines a stronger bases than primary ones (see **Table 34.2**; a *smaller* value of p means the compound is a *stronger* base).

$R \rightarrow \ddot{N}H_2$ — an alkylamine

$R \rightarrow \ddot{N}\text{—}H$ (with a second R bonded to N by an arrow) — a dialkylamine

Table 34.2 Base strengths of some amines

Name	Formula	pK_b	
Ammonia	$\ddot{N}H_3$	4.74	
Methylamine	$CH_3\ddot{N}H_2$	3.36	$R\ddot{N}H_2$
Ethylamine	$C_2H_5\ddot{N}H_2$	3.28	1° amines
1-aminopropane	$C_3H_7\ddot{N}H_2$	3.23	
Dimethylamine	$(CH_3)_2\ddot{N}H$	3.23	$R_2\ddot{N}H$
Diethylamine	$(C_2H_5)_2\ddot{N}H$	3.07	2° amines
Trimethylamine	$(CH_3)_3\ddot{N}$	4.20	$R_3\ddot{N}$
Triethylamine	$(C_2H_5)_3\ddot{N}$	3.36	3° amines
Phenylamine	$C_6H_5\text{—}NH_2$	9.30	Aryl amine

However, surprisingly, trialkylamines are weaker bases than dialkylamin in aqueous solution. This is because the R_3NH^+ ion, formed when trialkylamine accepts a proton, only has one hydrogen. Thus it cann form as many hydrogen bonds with the oxygens of water molecules a can the $R_2NH_2^+$ (with two hydrogens) formed by a dialkylamine. Th makes the R_3NH^+ ions less soluble than the $R_2NH_2^+$ ion and counterac the increased availability of the lone pair in R_3N:. In non-hydroge bonding solvents the expected order of base strength:

$$R_3N: > R_2NH > RNH_2 > NH_3$$

is observed.

Aryl groups *withdraw* electrons by overlap of the lone pair electrons wi the delocalized π system on the benzene ring, so arylamines are weak bases than ammonia:

$C_6H_5\text{—}\ddot{N}H_2$ (curved arrow from the N lone pair to the benzene ring)

A similar effect is seen with phenol (a stronger acid, hence a weaker bas than alcohols).

To allow more efficient overlap of the lone pair-containing orbitals wi the π system, phenylamine is a shallower pyramidal shape than alkylamin and has an H—N—H angle of 114°, **Figure 34.3.** The C—N bond h

Figure 34.3 The shape of the phenylamine molecule

Extended π system — N—H, 114°, H

Compare

N, 107°, CH_3, CH_3, CH_3

ble 34.3 **C—N bond lengths**

d	Length
-N	0.147 nm
=N	0.130 nm
≈N	0.135 nm
henylamine	

ble 34.4. **Bond polarities of amines mpared with alcohols**

$\overset{2.5}{\delta^+C}-\overset{3.5}{O^{\delta-}}$

$\overset{2.5}{\delta^+C}-\overset{3.0}{N^{\delta-}}$

some double bond character as indicated by the bond lengths in **Table 34.3**.

Bond energies and polarity

The C—N bond in amines is relatively weak and might therefore be expected to break easily, more easily than, say, the C—O bond in alcohols. The bond energies are:

$$C—N\ 286\,kJ\,mol^{-1} \quad C—O\ 358\,kJ\,mol^{-1} \quad C—C\ 347\,kJ\,mol^{-1}$$

However, the C—N bond is less polar than a C—O bond because nitrogen is less electronegative than oxygen, see **Table 34.4**. This means that there is less tendency for the C—N bond to break heterolytically, i.e. to give $—C^+$ and $:\bar{N}$. So it is unusual for amines to undergo elimination reactions or substitution reactions which involve the C—N bond breaking.

34.2.5 Reactions of amines

Combustion

Amines will burn in air to form carbon dioxide, water and nitrogen. Arylamines give the characteristic smoky flame of aromatic compounds.

As bases

Amines will react with acids to form salts, for example:

$$C_2H_5\ddot{N}H_2 + H^+ + Cl^- \longrightarrow C_2H_5NH_3^+ + Cl^-$$

ethylamine → ethylammonium chloride (ethylamine hydrochloride)

The products are ionic compounds which can be crystallized.

In the case of a relatively insoluble amine like phenylamine, the amine will dissolve in excess hydrochloric acid due to the formation of the soluble ionic salt. Addition of a strong base removes the proton from the salt and regenerates the insoluble amine:

$$C_6H_5—NH_2 \xrightarrow{+H^+ + Cl^-} C_6H_5—NH_3^+\,Cl^- \xrightarrow{+OH^-} C_6H_5—NH_2 + H_2O + Cl^-$$

phenylamine → **phenylammonium chloride (phenylamine hydrochloride) water-soluble ionic salt** → **phenylamine**

The smell of a solution of an amine is lost on addition of acid due to the formation of the ionic, and therefore involatile, salt. The formation of a soluble salt derivative of an amine is a common way of making long-chain amines water soluble. It is often used to make amines soluble for use as drugs. The antidepressant amitriptylene is an example. The free amine would be insoluble in water but amitriptylene hydrochloride, being ionic, is water soluble, **Figure 34.4.** Since an equilibrium exists between the amine and its hydrochloride, the action of the drug is unaffected:

$$drug + H^+ + Cl^- \rightleftharpoons drug\ H^+ + Cl^-$$

gure 34.4 **Amitriptylene and its drochloride**

Formation of amides

Amines react with acid chlorides or anhydrides to form N-substituted amides. This is a nucleophilic substitution reaction in which the amine acts as the nucleophile, for example:

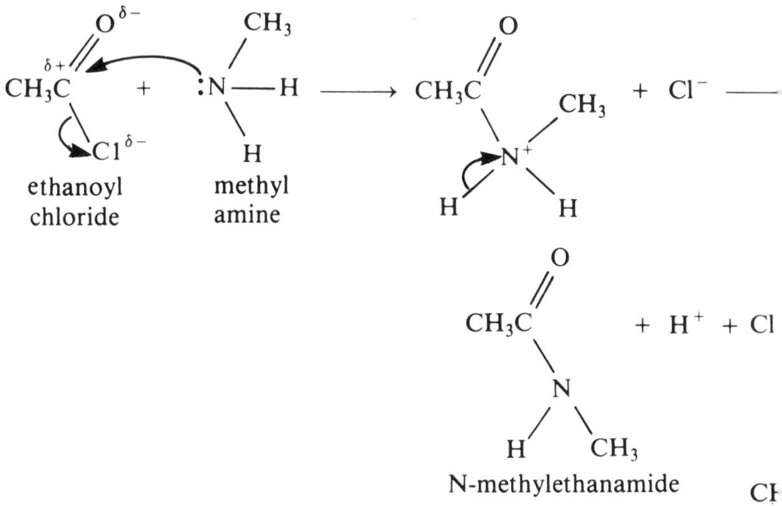

$$CH_3C(=O^{\delta-})^{\delta+}Cl^{\delta-} + :N(CH_3)(H)H \longrightarrow CH_3C(=O)N^+(CH_3)(H)H + Cl^-$$

ethanoyl chloride; methyl amine

$$\longrightarrow CH_3C(=O)N(H)CH_3 + H^+ + Cl$$

N-methylethanamide

Nylon is a polyamide

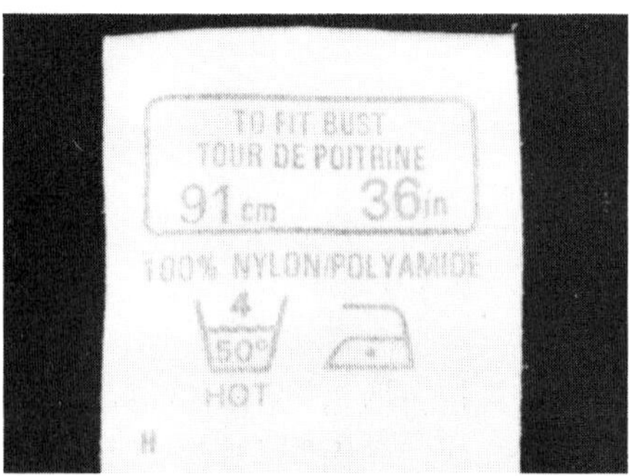

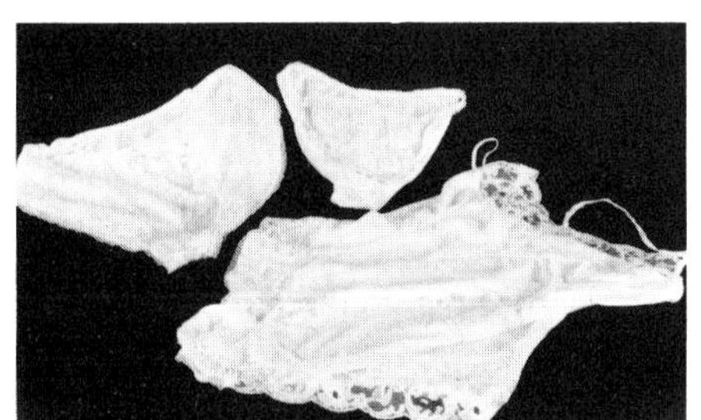

Notice that the final step requires the loss of a H^+ ion from the $—N^+(CH_3)—H$ group.

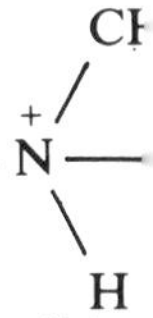

Starting with a tertiary amine would result in a species which did n have a proton to lose so this reaction occurs only with primary a secondary amines. The reaction is important in the formation poly(amides) like nylon.

Reaction of amines with benzoyl chloride, $C_6H_5C(=O)Cl$, produc solid N-substituted amides which can be purified by recrystallizatic The melting point of these derivatives can be used to identify the origin amine.

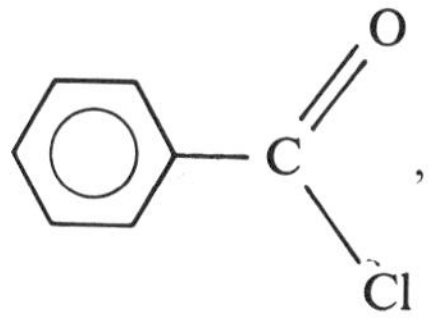

Reaction with halogenoalkanes

Reaction occurs when an amine is heated with excess halogenoalkane alcohol. This is also a nucleophilic substitution reaction:

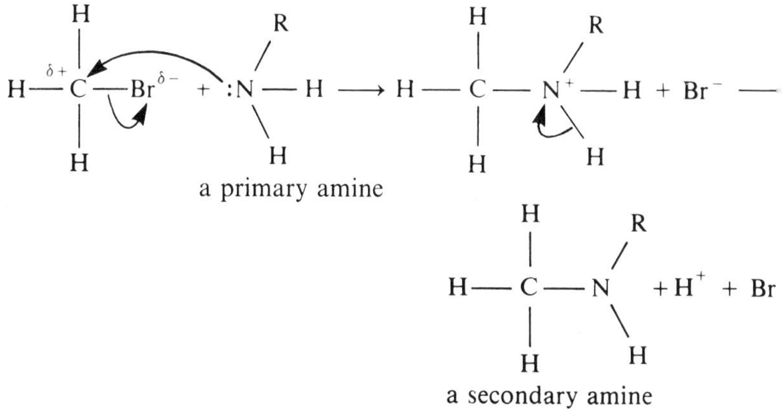

$$H_3C^{\delta+}—Br^{\delta-} + :N(R)(H)—H \longrightarrow H_3C—N^+(R)(H)—H + Br^-$$

a primary amine

$$\longrightarrow H_3C—N(R)H + H^+ + Br$$

a secondary amine

A secondary amine is produced first, but if the halogenoalkane is in exce the secondary amine (a better nucleophile than a primary amine) al reacts forming a tertiary amine:

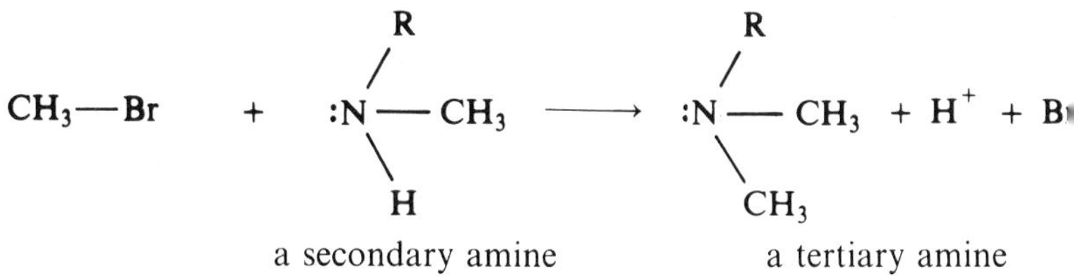

$$CH_3—Br + :N(R)(H)—CH_3 \longrightarrow :N(R)(CH_3)—CH_3 + H^+ + Br$$

a secondary amine; a tertiary amine

Finally, the tertiary amine can also react producing a **quaternary ammonium salt**:

$$CH_3{-}Br + :N(R)(CH_3){-}CH_3 \longrightarrow CH_3{-}\overset{+}{N}(R)(CH_3){-}CH_3 + Br^-$$

a quaternary ammonium salt

Reaction with nitrous acid (nitric(III) acid)

Nitrous acid is unstable and is produced in the reaction vessel by the reaction of sodium nitrite (sodium nitrate(III)) with dilute hydrochloric acid:

$$NaNO_2(aq) + HCl(aq) \longrightarrow NaCl(aq) + \underset{\text{nitrous acid}}{HNO_2(aq)}$$

Nitrous acid reacts with amines to produce the ion $R{-}\overset{+}{N}{\equiv}N$ called either an alkyl or an aryldiazonium ion depending on the nature of R.

$$R{-}NH_2 \xrightarrow{HNO_2/HCl} R{-}\overset{+}{N}{\equiv}N + Cl^-$$

Alkyldiazonium ions are unstable and rapidly decompose to give a stable nitrogen molecule and a carbocation, R^+. R^+ readily reacts with any available species in the solution to give a mixture of products. Some possibilities are shown in **Figure 34.5**.

Figure 34.5 Possible reactions of R^+

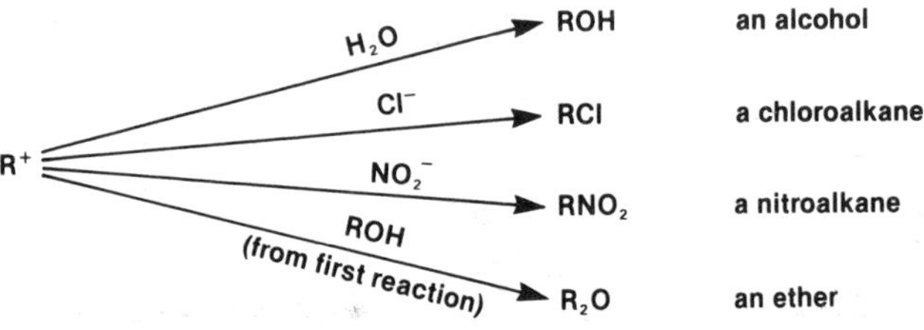

An alkene can also be formed by loss of H^+, for example:

$$C_2H_5^+ \longrightarrow H_2C{=}CH_2 + H^+$$

Aryldiazonium ions are more stable, providing the solution is kept cold (below about 278 K, 5 °C). They are, however, explosively unstable in the solid state so that they are always used in solution. Their stability in solution is due to delocalization of the charge on to the aromatic ring caused by overlap of p orbitals in the $-\overset{+}{N}{\equiv}N$ group with the π system of the benzene ring, see **Figure 34.6**.

Figure 34.6 Bonding in the benzenediazonium ion

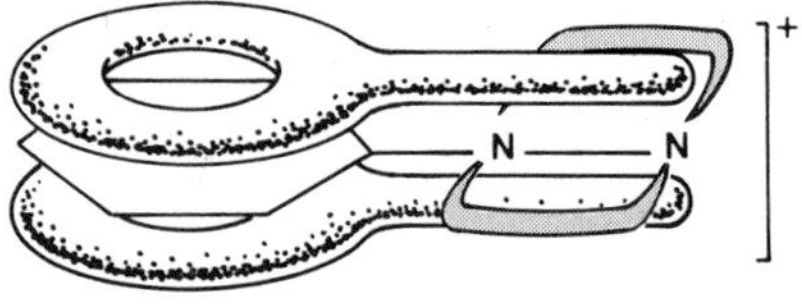

Aromatic diazonium compounds undergo a number of useful reactions. We shall use benzenediazonium chloride ($C_6H_5{-}\overset{+}{N}{\equiv}N + Cl^-$) produced by the reaction of phenylamine with nitrous acid as an example.

Allowing an aqueous solution to warm up to room temperature produces phenol and nitrogen. Heating to 373 K with a source of Cl^-, Br^-, I^- or CN^- and a suitable catalyst produces a halogen-substituted or cyano-substituted benzene. The catalysts are copper(I) compounds, **Figure 34.7** (overleaf).

COUPLING REACTIONS

Aryldiazonium ions react with phenols or aromatic amines in electrophilic

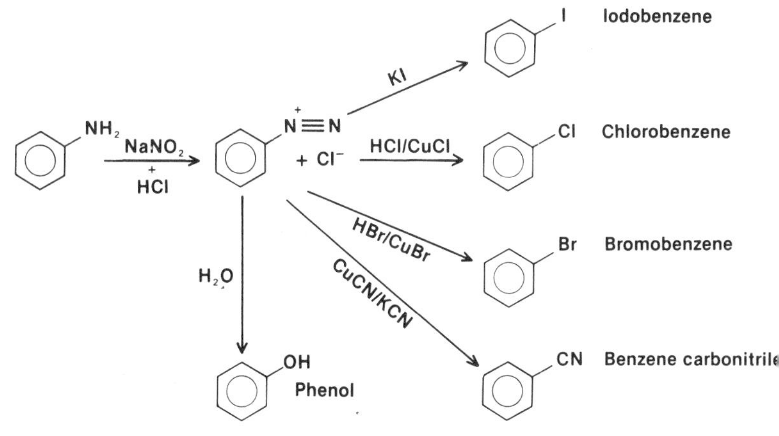

Figure 34.7 Reactions of benzenediazonium chloride

substitution reactions which result in the joining of the diazoniu compounds with the phenol or amine. The reaction is carried out in alkali conditions.

$$C_6H_5-\overset{+}{N}{\equiv}N + H-C_6H_4-OH \longrightarrow$$

phenol

$$C_6H_5-N{=}N-C_6H_4-OH + H$$

4-hydroxyazobenzene
(yellow)

$$C_6H_5-\overset{+}{N}{\equiv}N + H-C_6H_4-NH_2 \longrightarrow$$

phenylamine

$$C_6H_5-N{=}N-C_6H_4-NH_2 + H$$

4-phenylazophenylamine
(yellow)

These compounds all contain the unit —N=N—. Remember that t nitrogen atoms have a lone pair so the bond angle around the N ato will be approximately 120° and the —N=N— unit will be flat. A bett representation of the shape of such molecules is shown in **Figure 34.** The compounds with an —N=N— linkage can exist as a pair of *cis*/*tra* isomers, in just the same way that alkenes can be *cis* or *trans*:

Figure 34.8 The structure of diazonium compounds

R R R
120°
N=N N=N
120°
R
cis- *trans*-

H H C=C R R	H R C=C R H
cis	*trans*

The products of coupling reactions have intense colours—the tw examples given are yellow. Many such compounds are used as dyes. T —N=N— links the π systems of the two aromatic rings to give extended delocalized system, **Figure 34.9.** Such extended delocaliz systems are described as **conjugated**. Molecules with extended systems conjugation are frequently coloured. Some of these coloured compoun can act as indicators—methyl orange is an example.

Figure 34.9 The conjugated system in 4-phenylazophenylamine

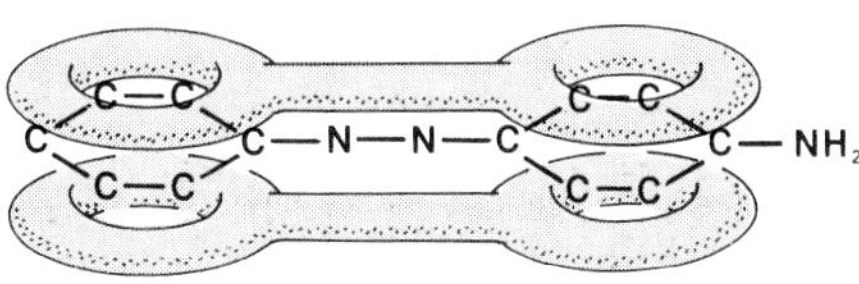

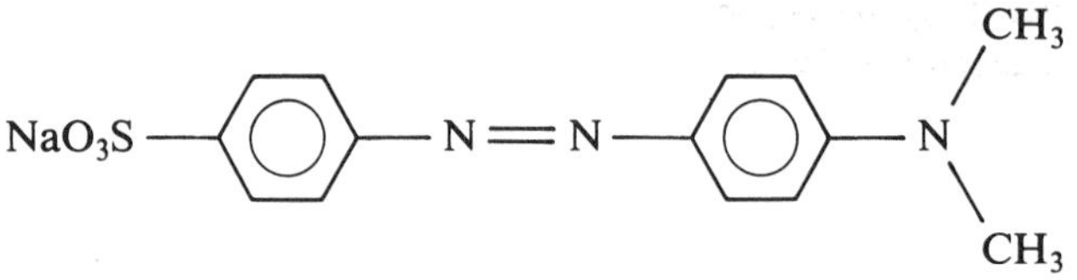

$$NaO_3S-C_6H_4-N{=}N-C_6H_4-N(CH_3)_2$$

methyl orange

Reactions of the benzene ring in aromatic amines

The substituents $—NH_2$, —NHR and $—NR_2$ release electrons on to the aromatic ring and so make it more reactive towards electrophilic substitution. The 2 and 4 positions are most activated.

Amines as ligands

The lone pairs of amines makes them good ligands (compounds which form dative covalent bonds with transition metal ions—see **section 25.3.2**). A good example is the reaction of amines with copper(II) ions where the blue coloured complexes of formula $Cu(H_2O)_2(RNH_2)_2^{2+}$ are formed.

The compound ethane-1,2-diamine, often called by its non-systematic name ethylene diamine, is often encountered as a bidentate ligand:

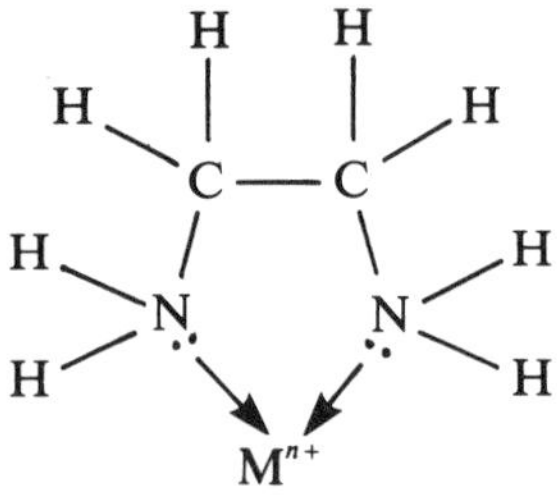

$$H_2\ddot{N}-CH_2-CH_2-\ddot{N}H_2 \rightarrow M^{n+}$$

34.2.6 Preparation of amines

Reduction of nitriles

Nitriles are reduced to primary amines by lithium tetrahydridoaluminate(III):

$$R—C{\equiv}N \xrightarrow[\text{ether}]{LiAlH_4} R—CH_2—NH_2$$

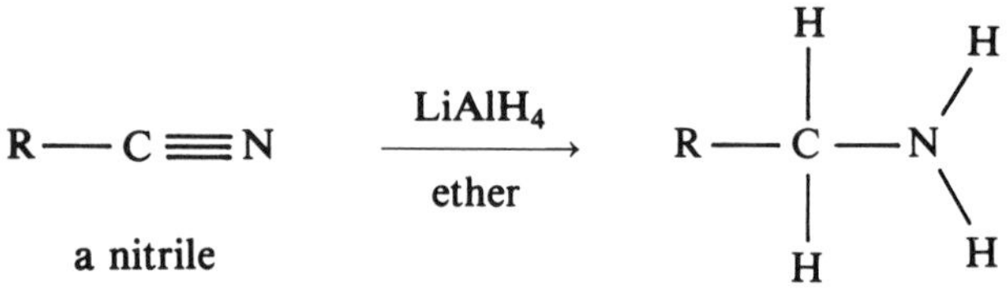

a nitrile

Other reducing agents such as sodium/ethanol or hydrogen with a catalyst can be used.

Reduction of amides

Lithium tetrahydridoaluminate(III) will also reduce amides to amines. N-substituted amides will produce 2° or 3° amines, for example:

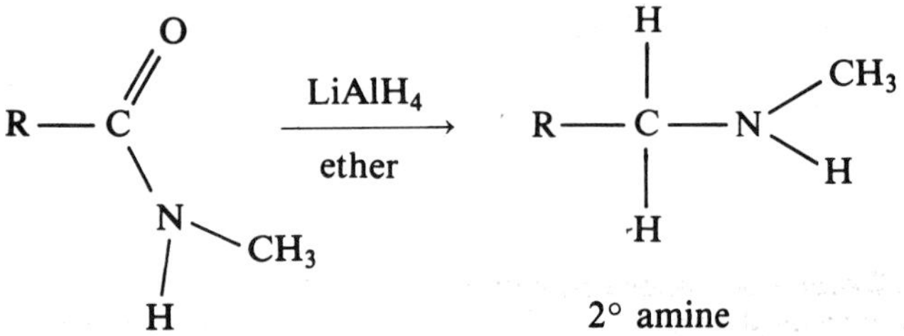

$$R—C(=O)—NH—CH_3 \xrightarrow[\text{ether}]{LiAlH_4} R—CH_2—NH—CH_3$$

N-substituted amide — 2° amine

Reduction of nitro compounds

This method is useful for producing aromatic amines from aromatic nitro compounds. Only 1° amines can be made. A variety of reducing agents can

A

DYES

The most obvious requirement of a dye is that it is coloured. Only slightly less important are that it must bind efficiently to the cloth so that it does not wash out and that it does not fade on exposure to light and to air.

Different dyes bond to fabrics in different ways. **Direct dyes** bond to the fabric only by weak intermolecular forces—van der Waals, dipole–dipole and hydrogen bonding. These are not particularly strong and direct dyes have a tendency to wash out. The structure of the dye 'direct brown 138' is shown. Notice that it has three diazo groups —N=N— and it has a number of polar $—NH_2$ groups which can form hydrogen bonds with the —OH and —NH groups which are present in fabrics such as cotton (cellulose has —OH groups), nylon and wool (both poly(amides) with —NH groups).

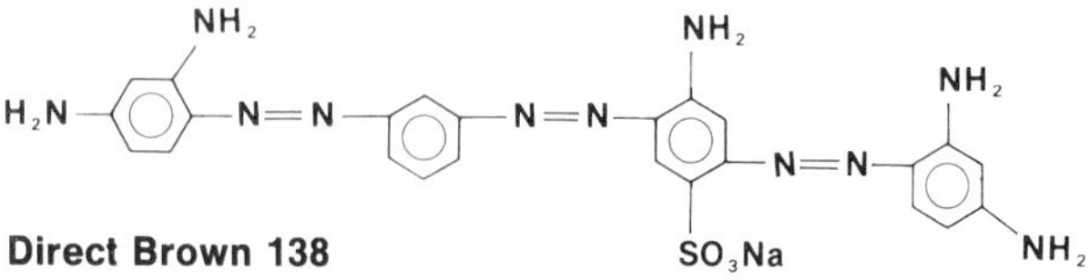

Direct Brown 138

Vat dyes contain ketone groups, C=O, and are insoluble. These are reduced before dyeing to C—OH groups and converted to their soluble sodium salts, $C—O^- \ Na^+$. The dye molecules become trapped in the structure of the fibre and are then oxidized back to the insoluble ketone form to prevent them being washed out. Indigo, the dye with which blue jeans are dyed, is such a dye.

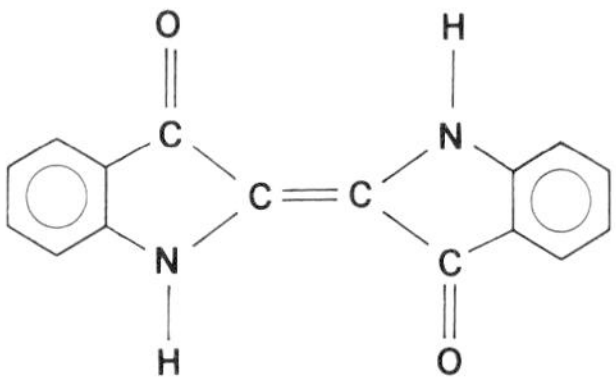

Indigo

Azoic dyes make use of the coupling reaction of diazonium salts. The fibre is first impregnated with a phenol or naphthol, applied as the soluble salt:

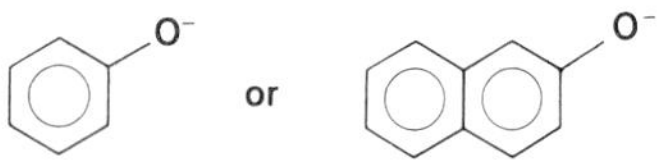

or

The cloth is then treated with a diazonium salt and the coupling reaction takes place on the material. The dye is thus formed actually on the cloth, for example:

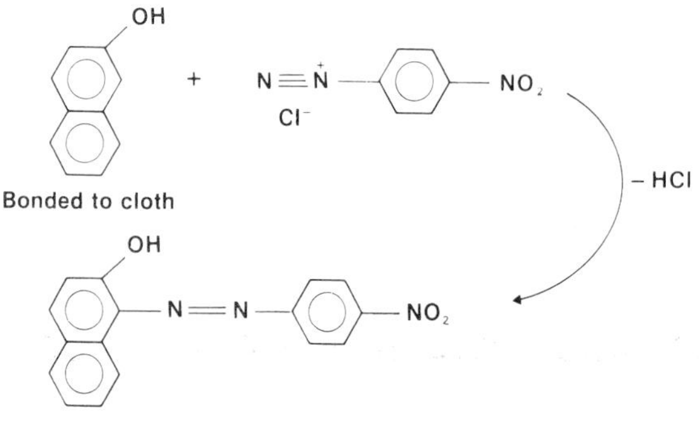

Bonded to cloth

– HCl

'Pau red' still bonded to cloth

Reactive dyes actually form covalent bonds with fibre molecules and are therefore extremely colour fast. A dye molecule is reacted with the molecule trichlorotriazine:

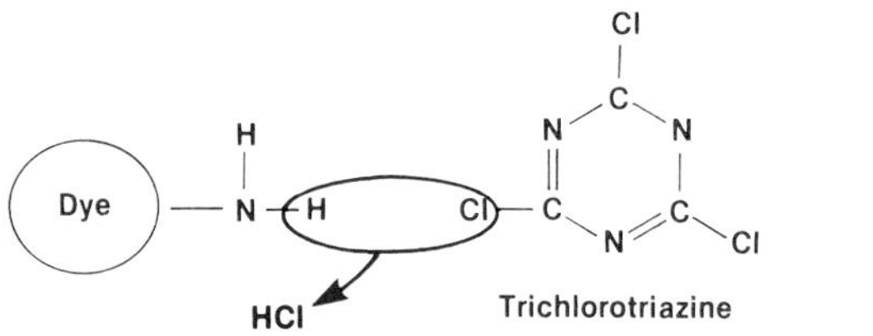

Trichlorotriazine

Trichlorotriazine can react with either —OH groups (present in cotton) or —NH groups (present in wool and nylons), thus effectively bonding the dye to the fabric.

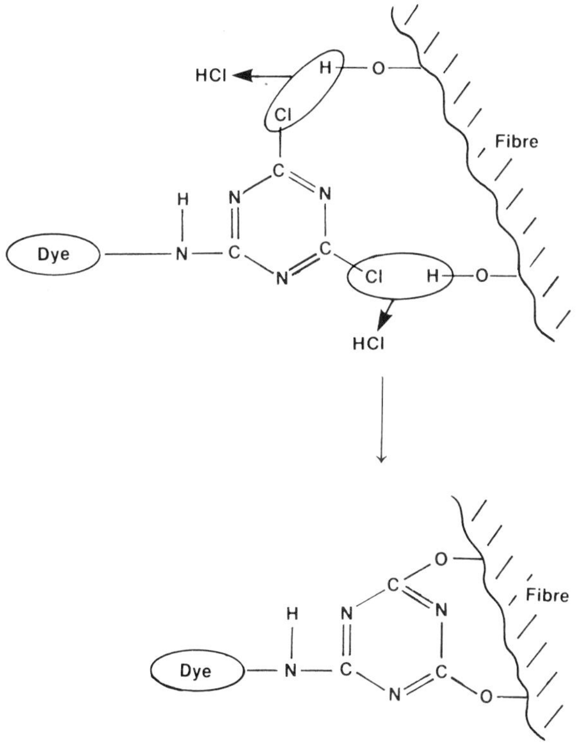

Bonding the dye to the fabric

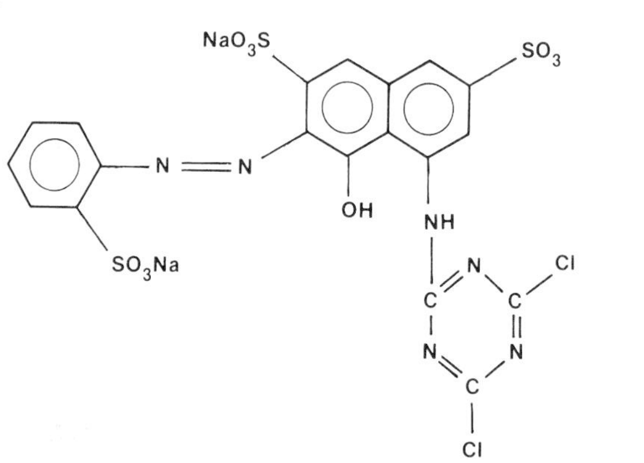

Procion Brilliant Red 2BS—the first reactive dye, first synthesized by ICI in the 1950s

be used including lithium tetrahydridoaluminate(III), tin and hydrochloric acid, and catalytic hydrogenation, for example:

$$C_6H_5NO_2 \xrightarrow{\text{reduction}} C_6H_5NH_2$$

nitrobenzene → phenylamine

Other nitrogen-containing compounds such as oximes and hydrazones can also be reduced to amines.

Reaction of ammonia with halogenoalkanes

The disadvantage of this method is that a mixture of primary, secondary and tertiary amines is produced as well as quaternary ammonium salts. The products must be separated by fractional distillation. The reaction takes place on heating a mixture of ammonia and the halogenoalkane dissolved in alcohol in a sealed tube.

$$R—Br + NH_3 \longrightarrow RNH_2 + HBr$$ (1° amine)

$$\longrightarrow R_2NH$$ (2° amine)

$$\longrightarrow R_3N$$ (3° amine)

$$\longrightarrow R_4N^+Br^-$$ (quaternary ammonium salt)

The reaction of ammonia with alcohols

A similar reaction takes place between ammonia and alcohols (but not phenols) when they are heated together under pressure with a suitable catalyst such as aluminium oxide. Again, a mixture of primary, secondary and tertiary amines is formed:

$$R—OH + NH_3 \longrightarrow RNH_2 + H_2O$$ (1° amine)

$$\longrightarrow R_2NH$$ (2° amine)

$$\longrightarrow R_3N$$ (3° amine)

This is a nucleophilic substitution reaction. Phenols are less susceptible to nucleophilic attack than alcohols (see **section 31.6**).

Hofmann degradation of amides

This reaction produces a primary amine with one less carbon atom than the original amide (hence the term 'degradation'). The amide is reacted with bromine and alkali:

$$R—C(=O)NH_2 + Br_2 + 4NaOH \longrightarrow R—NH_2 + 2NaBr + Na_2CO_3 + 2H_2O$$

So, ethanamide would be converted into methylamine by this reaction but into ethylamine by reduction. The Hofmann degradation is useful in planning syntheses because of this shortening of the carbon chain.

34.2.7 Economic importance of amines

Dyestuffs

Aromatic amines are important in the manufacture of azo dyes—those containing an —N=N— group linking two aromatic rings. These are

Figure 34.10 Structures of some azo dyes

OH, N=N, SO_3Na

'Acid orange 7'

NH_2, H_2N, N=N, N=N, NH_2, N=N, NH_2, SO_3Na, NH_2

'Direct brown 138'

produced from diazonium salts by coupling reactions. Two examples a shown in **Figure 34.10.**

Diamines such as hexane-1,6-diamine are used in the manufacture nylon.

Many drugs of different types have amine functional groups. Son examples are shown in **Table 34.5.** The amine group is marked in brow

Table 34.5 Some drugs with amine functional groups

Name	Formula	Type
Librium	H, N, N, CH_3, Cl, N, O^-	Tranquillizer
Sulphanilamide	H, N, H, SO_2N, H, H	Antibacterial
Amphetamine	CH_2—CH—N(H)H, CH_3	Stimulant
Polaramine	N, CH_3, N, CH_3	Antihistamine
Amitriptylene	CH_3, N, CH_3	Antidepressant
Pethidine	N, CH_3, C=O, O	Pain killer

Amines are also industrially important in the manufacture of cation detergents. These are quaternary ammonium compounds

$$\sim\!\sim\!\sim\!\sim\!\sim \overset{\displaystyle CH_3}{\underset{\displaystyle CH_3}{\overset{|}{\underset{|}{N^+}}}} \!-\! CH_3 \quad Br^-$$

with a long hydrocarbon chain and a positively charged organic grou Such detergents are not particularly good at cleaning, but have germicid properties and are often an ingredient in nappy washing solutions. The are also used in hair and fabric conditioners. Both wet fabric and wet ha

A

ANTIHISTAMINE DRUGS

Allergies are abnormal bodily reactions to a foreign substance called an allergen, usually a protein. Common allergens are pollen, dust and housemites, but a number of people are allergic to substances in food such as the yellow dye tartrazine. The symptoms of an allergy, which include skin rashes, runny nose, sore throat, swelling, headache, etc. are caused by the release of a chemical called histamine.

Ethylamine group

NCH_2CH_2

Histamine

Histamine may be released locally, for example at the site of an insect bite, or into the whole bloodstream. Over 50 types of drug are available to counteract the effects of histamine. Histamine contains an ethylamine group. Many antihistamine drugs also contain an ethylamine or similar group. Such drugs compete with histamine for receptor sites and thus block the action of the real histamine. Other drugs which happen to have an ethylamine or similar group also have antihistamine properties as well as their intended mode of action. Examples include some of the so-called tricyclic antidepressants such as imipramine and desipramine. The reason for the label 'tricyclic' is obvious.

Imipramine R = CH_3
Desipramine R = H

pick up negative charges on their surfaces. These attract positively charged cationic detergent molecules which coat the surface of the fabric or hair, lubricating the surfaces and preventing build-up of static electricity.

Aromatic amines are used to manufacture antioxidants which are added to, for example, the rubber in tyres. One such compound is N-phenylnaphthalene-1-amine:

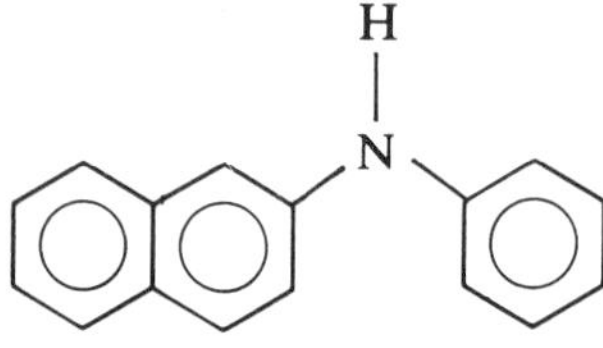

lyurethane dashboard

lyurethanes give off toxic fumes when rned

Antioxidants of different sorts are added to a wide variety of products, including food. They are compounds which are more easily oxidized than the product they protect so that they, rather than the product, are attacked by oxygen in the air.

Finally, amines are used in the manufacture of polyurethanes. These are versatile polymers, whose properties can be tailored to meet a great many uses, from foams used in cavity wall insulations and car dashboard paddings to elastomeric fibres such as Lycra used for making stretchy fabrics for swimwear and underwear. Polyurethanes are made by a polymerization process from a diol (or polyol, to provide cross-linking) and a diisocyanate:

$$\underset{\text{a diol}}{HO-R-OH} + \underset{\text{a diisocyanate}}{O=C=N-R'-N=C=O}$$

$$\downarrow$$

$$\underbrace{HO-R-O-\overset{\overset{\displaystyle O}{\|}}{C}-\overset{\overset{\displaystyle H}{|}}{N}-R'-N=C=O}$$

the urethane linkage

Amines are used in the manufacture of the isocyanates. One problem with polyurethanes is that, when burned, they give off toxic gases including nitric acid, nitrogen dioxide and hydrogen cyanide. Their use as a furniture foam has been increasingly curtailed for this reason.

34.3 Nitriles

34.3.1 Formulae, naming and structure

Naming

Nitriles contain the functional group —C≡N. As with carboxylic acid this group can only occur at the end of a carbon chain. The —C≡ group is named using the suffix -nitrile when it is the principal group ar the prefix cyano- when it is not. When the suffix -nitrile is used, the carbc of the functional group is counted as part of the root so:

$CH_3C{\equiv}N$ is ethanenitrile

and

$CH_3CH_2C{\equiv}N$ is propanenitrile

Where the —C≡N is attached to, say, a benzene ring, the carbon of tl functional group is not counted as part of the root and the suff -carbonitrile is used, for example:

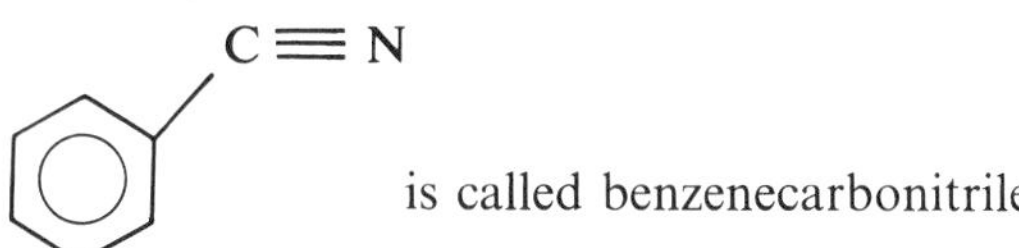

is called benzenecarbonitrile

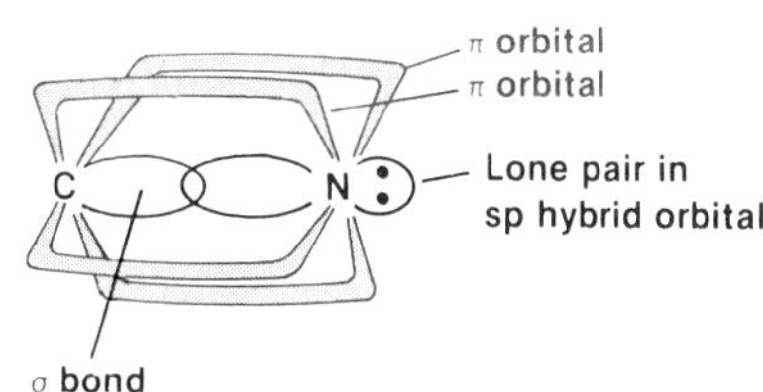

Figure 34.11 The bonding in the nitrile group

The —C≡N group is formed by a σ bond and two π bonds, **Figure 34.1** Both the carbon and nitrogen atoms are sp hybridized. The σ bond formed by overlap of an sp orbital from each atom and the two π bon at right angles to one another are formed by overlap of p orbitals. Tl nitrogen atom has a lone pair in the remaining sp orbital. The ur R—C≡N is linear. The geometry and bonding of the —C≡N: unit very similar to the —C≡C— unit found in alkynes.

*Iso*nitriles have the functional group —N≡C.

34.3.2 Physical properties of nitriles

The —$\overset{\delta+}{C}{\equiv}\overset{\delta-}{N}$ unit is moderately polar. The electronegativity valu are C = 2.5, N = 3.0. Thus the short-chain nitriles are liquids at roo temperature and are fairly soluble in both water and less polar solvent

34.3.3 Infra-red spectra

The infra-red spectrum of benzenecarbonitrile is shown in **Figure 34.1** Nitriles have a peak at between 2200 cm^{-1} and 2300 cm^{-1} due to tl —C≡N stretching vibration.

34.3.4 Reactivity of nitriles

The most important feature of the reactivity of the nitrile group is i unsaturation which means that addition reactions are possible.

Figure 34.12 Infra-red spectrum of benzenecarbonitrile

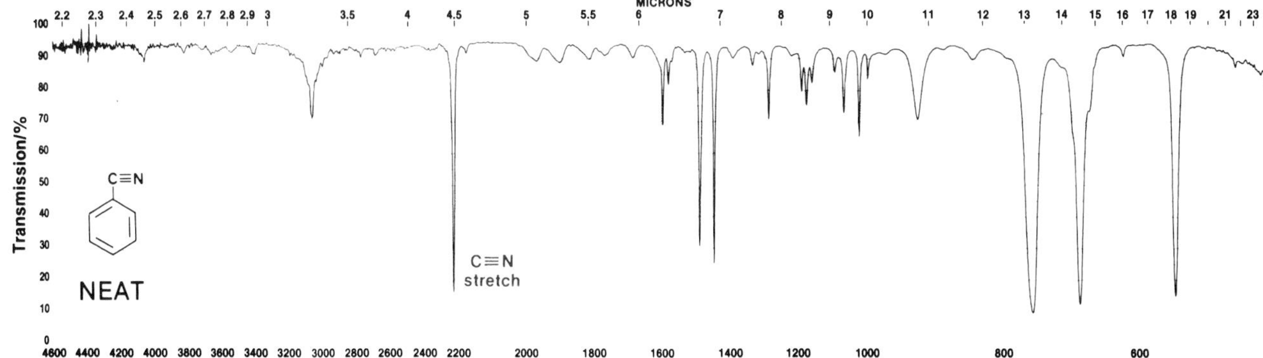

34.3.5 Reactions of nitriles

Reduction

Lithium tetrahydridoaluminate(III) or sodium in ethanol will reduce nitriles to primary amines:

$$R—C\equiv N \xrightarrow{[+4H]} R—CH_2—NH_2$$

The use of lithium tetrahydridoaluminate(III) involves reaction in ether solution followed by addition of water as is usual with this reducing agent.

Hydrolysis

Nitriles are hydrolysed by boiling with either acids or alkalis to form first the amide and then the carboxylic acid (or its salt if alkali has been used):

$$R—C\equiv N \xrightarrow[H^+]{H_2O} R—C(=O)NH_2 \xrightarrow{H_2O} R—C(=O)OH + NH_4^+$$

$$R—C\equiv N \xrightarrow{H_2O,\ NaOH} R—C(=O)O^-\ Na^+ + NH_3$$

34.3.6 Preparation of nitriles

From halogenoalkanes

Potassium cyanide dissolved in alcohol reacts with halogenoalkanes to yield nitriles by a nucleophilic substitution reaction:

$$CH_3CH_2^{\delta+}—I^{\delta-} + :CN^- \longrightarrow CH_3CH_2—C\equiv N + I^-$$

iodoethane → propanenitrile

The order of reactivity of different halogens is I > Br > Cl.

This reaction is very important in planning organic syntheses as it enables the length of the carbon chain to be increased by one. In the above example, iodoethane produces propanenitrile. Halogenoarenes are less reactive than halogenoalkanes and this is not a feasible method for producing aromatic nitriles. They are made from diazonium salts as described below.

From diazonium salts

Diazonium salts are made from aromatic amines by reaction with nitrous acid (nitric(III) acid) below 5 °C. Treatment of these salts with aqueous potassium cyanide in the presence of copper(I) cyanide produces an aromatic nitrile:

$$C_6H_5—\overset{+}{N}\equiv N + Cl^- + KCN \xrightarrow{CuCN} C_6H_5—C\equiv N + KCl + N_2$$

benzenediazonium chloride → benzenecarbonitrile

By dehydration of amides

Heating with phosphorus pentoxide dehydrates amides to nitriles:

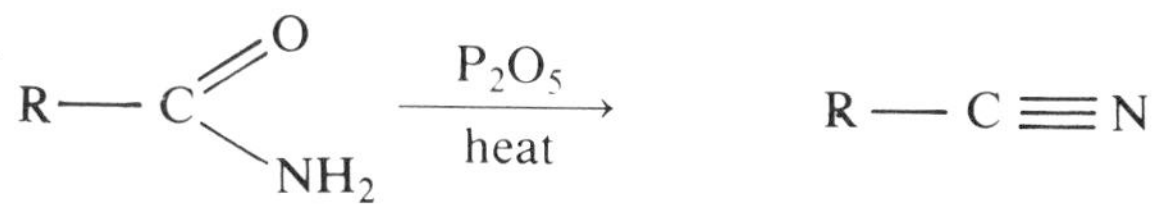

$$R{-}C(=O)NH_2 \xrightarrow[\text{heat}]{P_2O_5} R{-}C{\equiv}N$$

Nitriles may also be made by the action of CN^- ions on aldehydes or ketones. In this case, a hydroxynitrile is produced (see **section 32.6.1**), for example:

$$CH_3{-}CHO \xrightarrow{CN^-/H^+} CH_3{-}C(OH)(H){-}C{\equiv}N$$

ethanal → 2-hydroxypropanenitrile

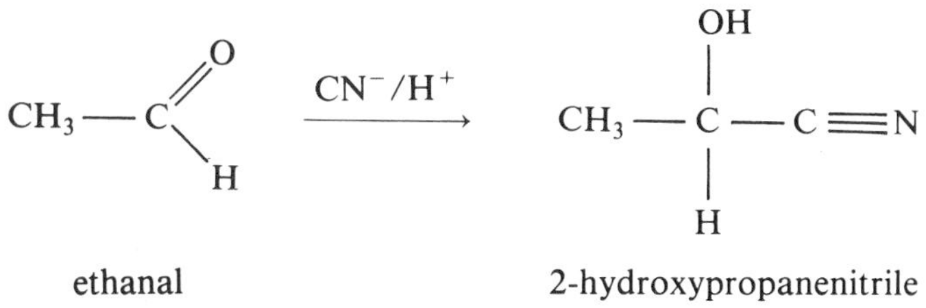

34.3.7 Economic importance of nitriles

Probably the most important nitrile industrially is propenenitrile, which has both an alkene and a nitrile functional group:

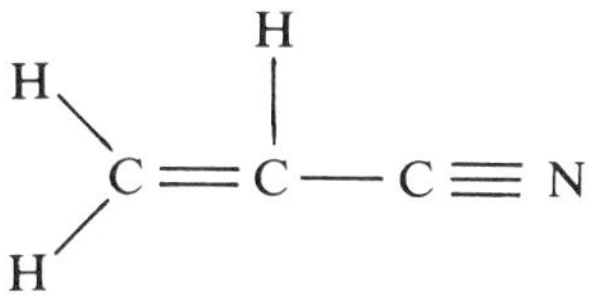

$$H_2C{=}CH{-}C{\equiv}N$$

It can be polymerized to poly(propenenitrile) by an addition polymerization process which starts with an aqueous solution of the monomer from which the insoluble polymer precipitates out:

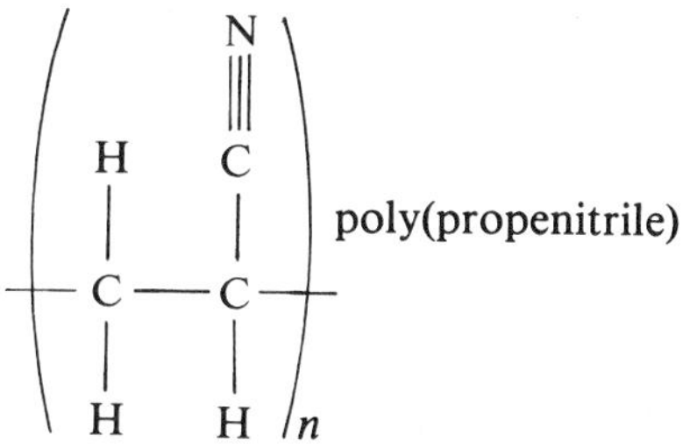

$$\left({-}CH_2{-}CH(C{\equiv}N){-} \right)_n \quad \text{poly(propenitrile)}$$

Figure 34.13 Reaction scheme for the manufacture of nylon-66

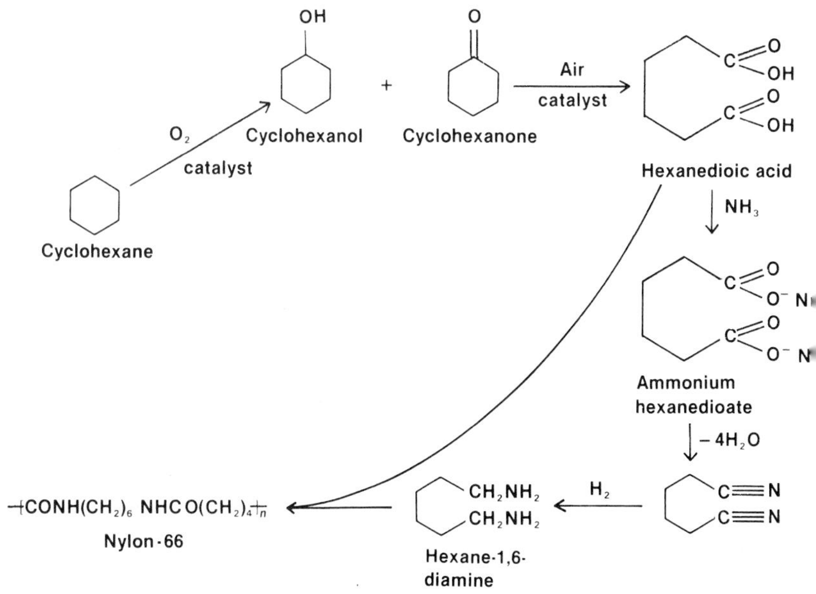

The non-systematic name of propenenitrile is acrylonitrile which is the origin of the name of the polymer 'acrylic' and one of its trade names Acrilan. Other brand names include Orlon and Courtelle. The main use is as a fibre similar in many ways to wool. The fabrics are stronger and lighter than wool but garments are flammable and do not retain their shape so well. In some acrylic fibres, a small percentage of chloroethene is included in the polymer to make it easier to dye.

Nitriles are also involved in the manufacture of monomers for the production of Perspex and nylon. The latter is a good illustration of an industrial application of the preparation and reactions of nitriles, see **Figure 34.13** (opposite).

34.4 Summary

34.4.1 Amines

- Amines are derived from ammonia, NH_3, by replacement of one or more hydrogen atoms by alkyl or aryl groups.
- Primary amines have the general formula: $R—N(H)(H)$
- Secondary amines have the general formula: $R—N(R')(H)$
- Tertiary amines have the general formula: $R—N(R')(R'')$
- Amines have a pyramidal shape based on that of ammonia.
- Amines are polar, and primary and secondary amines can hydrogen bond so short-chain amines are soluble in water.
- Short-chain amines are gases or volatile liquids.
- Amines have a lone pair of electrons making them able to accept H^+ ions and act as bases.

Figure 34.14 The reactions of amines using ethylamine as an example

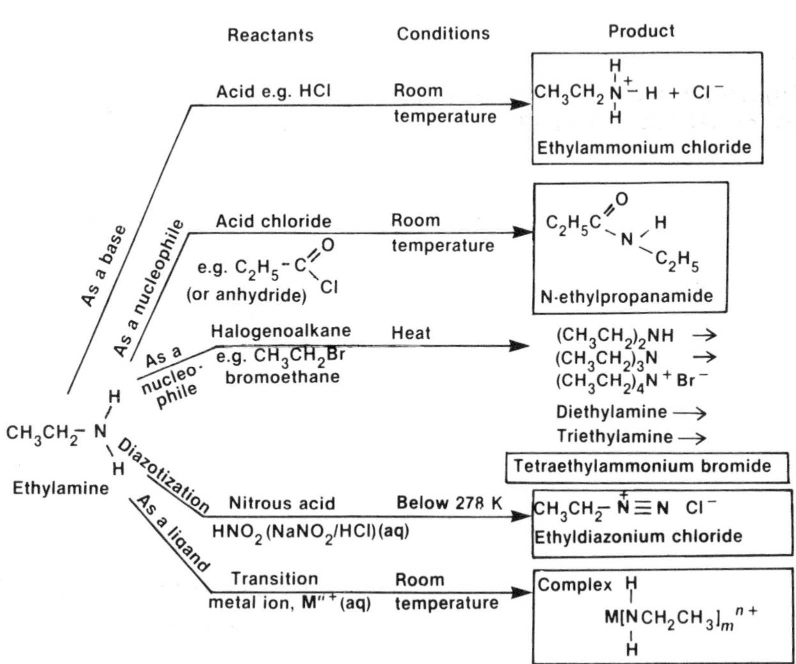

- Base strengths run in the order:

$$2^\circ \text{ amine} > 1^\circ \text{ amine} > 3^\circ \text{ amine} > NH_3 \gg \text{arylamine}$$

in aqueous solution.
- The reactions of amines are summarized in **Figure 34.14.**
- Alkyldiazonium ions decompose to $R^+ + N_2$, followed by reaction of R^+ with solvent to give ROH and other products. Aryldiazonium ions are stable below 278 K but can be made to undergo the reactions shown in **Figure 34.7** as well as coupling reactions.
- The $-NH_2$ group activates the benzene ring, especially at the 2 and 4 positions.
- Amines can be prepared by reduction of nitriles, amides or nitro compounds, or by the reaction of ammonia with halogenoalkanes or alcohols.
- The Hofmann degradation of amides produces an amine with one carbon atom less than the starting amide.
- Amines are industrially important in the manufacture of dyes, polyamides, polyurethanes and a number of types of drugs.

34.4.2 Nitriles

- Nitriles contain the functional group $-C{\equiv}N$.
- Nitriles are moderately polar and those with short chains are liquids which dissolve quite well in water.
- The reactions of nitriles are summarized in **Figure 34.15.**

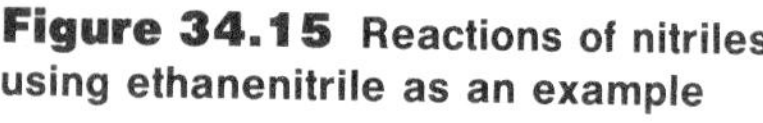

Figure 34.15 Reactions of nitriles using ethanenitrile as an example

$CH_3-C{\equiv}N$ Ethanenitrile

	Reactants	Conditions	Product
Reduction	**Lithium tetrahydrido-** aluminate(III) $LiAlH_4$ or sodium in ethanol	In ether solution followed by addition of H_2O at room temperature	$CH_3-CH_2-NH_2$ Ethylamine
Hydrolysis	H_2O, H^+ or H_2O, OH^-	Heat	CH_3COOH or CH_3COO^- Ethanoic acid or ethanoate ion

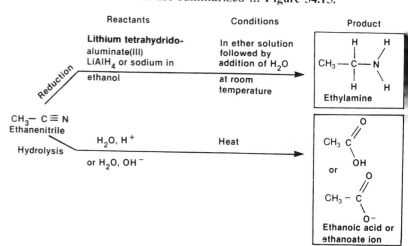

- Nitriles can be made by dehydration of amides.
- Alkylnitriles can be made by the reaction of halogenoalkanes with sodium cyanide and arylnitriles by treating a diazonium salt with potassium cyanide.
- A nitrile prepared from a halogenoalkane has one more carbon than the alkane, which is useful in syntheses.

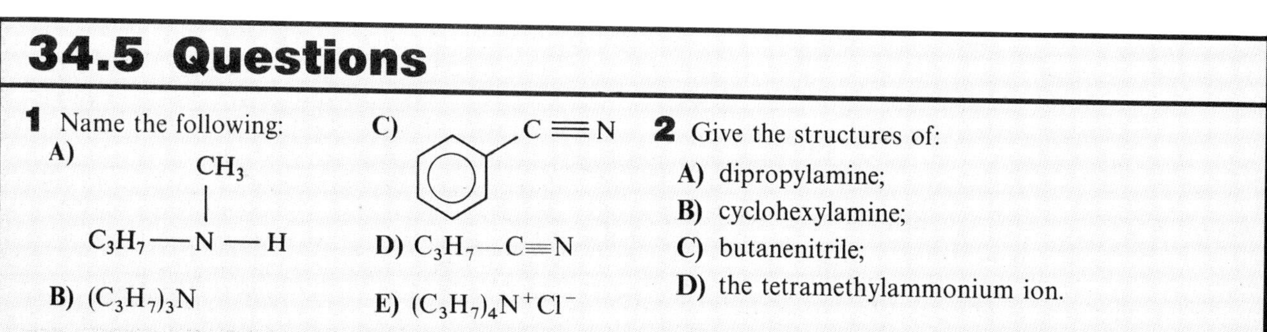

34.5 Questions

1 Name the following:

A) $C_3H_7-N(CH_3)-H$

B) $(C_3H_7)_3N$

C) $C_6H_5-C{\equiv}N$

D) $C_3H_7-C{\equiv}N$

E) $(C_3H_7)_4N^+Cl^-$

2 Give the structures of:

A) dipropylamine;

B) cyclohexylamine;

C) butanenitrile;

D) the tetramethylammonium ion.

3 Arrange the following in order of their strengths as bases, weakest first:

$CH_3—NH_2$,

$(CH_3)_2NH$,

[phenylamine: benzene ring with NH_2]

[2,4,6-trimethylphenylamine: benzene ring with NH_2 and three CH_3 groups]

Hint: $CH_3—$ groups release electrons.

4 Give the products of the following reactions:
A) ethylmethylamine with dilute acid;
B) propylamine plus ethanoyl chloride;
C) methylamine plus bromoethane;
D) ethanenitrile plus lithium tetrahydridoaluminate(III) in ether followed by addition of water;
E) Propanamide plus phosphorus pentoxide.

5 Propenenitrile is an important compound which is polymerized to form a fibre known as Orlon which is used for making clothes. The propenenitrile monomer is manufactured from propene, ammonia and oxygen in the presence of a catalyst.

$$CH_3CHCH_2 + NH_3 + \tfrac{3}{2}O_2 \xrightarrow[450\,^\circ C]{Mo} CH_2{=}CHCN$$

A) Give a diagram to show *all* the bonds in propenenitrile and indicate its three-dimensional structure.
B) Give a diagram to represent the structure of Orlon, showing at least two of the repeating units.
C) Describe the mechanism by which the polymerization takes place.
D) Indicate the likely sources of the raw materials for the process:
i propene;
ii ammonia;
iii oxygen.

In the laboratory, propenenitrile could be made by the action of hydrogen cyanide on an aldehyde.
E) Give equations to describe this process, indicating the mechanism by which it occurs.

In industry, aldehydes are made from petroleum feedstock by reaction with carbon monoxide. By reference to availability of raw materials, transfer of materials, separation of products and any technological, economic or environmental issues you may wish to consider, present a reasoned argument for not using the laboratory process industrially.

(London, specimen)

6 A) Give the name and structural formula of one example in each case of:
i A, a primary aliphatic amine;
ii B, a primary aromatic amine;
iii C, an amide.
iv D, a nitrile.
B) Describe, stating reagents and reaction conditions, how you would prepare:
i compound A, from a nitrile;
ii compound B, from the parent hydrocarbon;
iii compound C, from the parent acid;
iv) compound D, from an amide.

(London 1980)

7 A compound has the formula:

$$H_2N—C_6H_4—CH{=}CHCO_2C_3H_7$$

How would it react with:
A) bromine;
B) cold dilute hydrochloric acid;
C) boiling, dilute hydrochloric acid;
D) potassium manganate(VII) (potassium permanganate);
E) nitrous acid, i.e. dilute sulphuric acid and sodium nitrite?

Comment on the stereochemistry of the compound.

(London 1979)

8 Write an essay on the amines, the amides and the alkyl cyanides (alkyl nitriles), including a description of their methods of preparation and their reactions.

Starting from pentan-1-ol, how would you attempt to obtain samples of butyl cyanide (pentanenitrile), pentylamine and hexanamide? For each reaction step, name the reagents and conditions needed.

(Nuffield 1985, Special paper)

9 A) Write down the reagents and the equations for the preparation of a solution of benzenediazonium chloride from phenylamine.
B) Explain why this reaction should be kept cold.
C) i Cold benzendiazonium chloride solution is added to sodium phenoxide solution. Write down the formula of the organic product D.
ii Draw the full structural formulae of the two stereoisomers of D and state the type of stereoisomerism they show.

(Oxford 1987)

10 A substance containing carbon, hydrogen and oxygen only, when heated under reflux with aqueous acid for several hours, could be split into two components.

The lower-boiling component was a liquid at room temperature. 1.0 g, on complete combustion, produced 2.2 g of carbon dioxide and 1.2 g of water. On volatilization in a gas syringe at 100 °C and one atmosphere pressure, 0.10 g of the liquid produced 51 cm^3 of vapour. When the liquid was oxidized, the product gave a crystalline derivative with 2,4-dinitrophenylhydrazine, but would not react with Fehling's solution. The higher-boiling component from the heating under reflux was neutralized with ammonia and the product heated with phosphorus(V) oxide, P_2O_5. This yielded a product of relative molecular mass 97, whose composition by mass was 74.2% carbon, 11.3% hydrogen and 14.4% nitrogen.

Using the information provided, suggest a possible identity for each of the substances involved, and explain the reaction of the original substance with aqueous acid when heated under reflux.

(Nuffield 1981)

35 Polymers

35.1 Introduction

Polymers are compounds consisting of giant molecules made up of many smaller molecules (*monomers*) linked together. Many substances of importance in everyday life are polymers including plastics like PVC, Perspex and Bakelite and fibres such as nylon, Terylene and acrylics. Polymers are also closely linked with the chemistry of life.

Proteins are polymers which occur in all living things. They appear in a wide variety of places—in structural materials like muscle fibres, hair and cartilage, in hormones like insulin which control particular aspects of metabolism and enzymes which catalyse innumerable chemical reactions in the body. DNA, deoxyribonucleic acid, the molecule which stores genetic information and ensures that we resemble our parents, is also a polymer.

35.2 Proteins

Proteins are polymers built up from monomers called **amino acids**.

35.2.1 Amino acids

These molecules have two functional groups—a carboxylic acid group and a primary amine. All the important naturally occurring amino acids, of which there are about 20, are 2-amino acids, that is, the amino group is on the carbon next to the $—CO_2H$ group, **Figure 35.1.**

igure 35.1 2-aminopropanoic id – a 2-amino acid, also called alanine

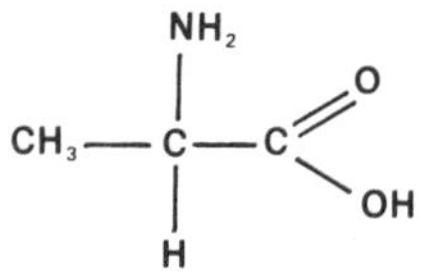

$$CH_3—\overset{\displaystyle NH_2}{\underset{\displaystyle H}{C}}—C\begin{matrix}\nearrow O \\ \searrow OH\end{matrix}$$

The test for proteins is the **biuret test** in which an alkaline solution of the suspected protein is warmed with copper sulphate solution. A violet colour indicates the presence of protein.

2-amino acids have the general formula:

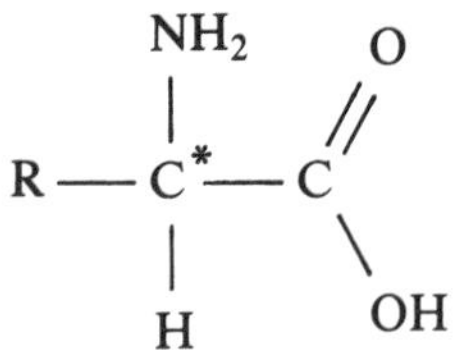

$$R—\overset{\displaystyle NH_2}{\underset{\displaystyle H}{C^*}}—C\begin{matrix}\nearrow O \\ \searrow OH\end{matrix}$$

Optical activity

R can be a whole variety of groups (including ones with functional groups). Except for the case when $R{=}H$, $—NH_2$ or $—CO_2H$, the carbon marked * has four different groups bonded to it and is therefore **asymmetric**.

Amino acids can therefore exist in two mirror image forms which are optically active. See **section 26.7.2** for more detail of optical activity.

Acid and base properties

Amino acids have both an acidic and a basic functional group. The carboxylic acid group has a tendency to lose a proton (act as an acid) as shown (see page 558).

R

TWENTY NATURALLY OCCURRING AMINO ACIDS

Formula	Name	Abbreviation
H_2NCHCO_2H | H	Glycine	gly
H_2NCHCO_2H | CH_3	Alanine	ala
H_2NCHCO_2H | $CHCH_3$ | CH_3	Valine	val
H_2NCHCO_2H | CH_2 | $CH(CH_3)_2$	Leucine	leu
H_2NCHCO_2H | CHC_2H_5 | CH_3	Isoleucine	ile
HN—CH—CO_2H CH_2 CH_2 CH_2	Proline (Note: Proline is a *secondary amine*)	pro
H_2N—CH—CO_2H | CH_2 | C CH NH	Tryptophan	try
H_2NCHCO_2H | CH_2 | CH_2SCH_3	Methionine	met
H_2NCHCO_2H | CH_2—C_6H_5	Phenylalanine	phe
H_2NCHCO_2H | CH_2OH	Serine	ser
H_2NCHCO_2H | CHOH | CH_3	Threonine	thr
H_2NCHCO_2H | CH_2SH	Cysteine	cys
H_2NCHCO_2H | CH_2 | $CONH_2$	Asparagine	asn
H_2NCHCO_2H | CH_2 | CH_2CONH_2	Glutamine	gln
H_2NCHCO_2H | CH_2—C_6H_4—OH	Tyrosine	tyr
H_2NCHCO_2H | CH_2 | C=CH | | HN N \ // CH	Histidine	his
H_2NCHCO_2H | $(CH_2)_3$ | NH | HN=C—NH_2	Arginine	arg
H_2NCHCO_2H | $(CH_2)_3$ | CH_2NH_2	Lysine	lys
H_2NCHCO_2H | CH_2CO_2H	Aspartic acid	asp
H_2NCHCO_2H | CH_2 | CH_2CO_2H	Glutamic acid	glu

E

OPTICAL ACTIVITY OF AMINO ACIDS— THE 'CORN' LAW

Since all 2-amino acids (except glycine) have an asymmetric carbon (one with four different groups attached) they all have two mirror image forms called a pair of **enantiomers**. Solutions of the same concentration of each enantiomer will rotate the plane of polarization of polarized light through the same angle but in opposite directions. The symbols + and − are used to indicate this. + denotes a clockwise rotation as seen by an observer looking towards the light source and − indicates an anti-clockwise rotation.

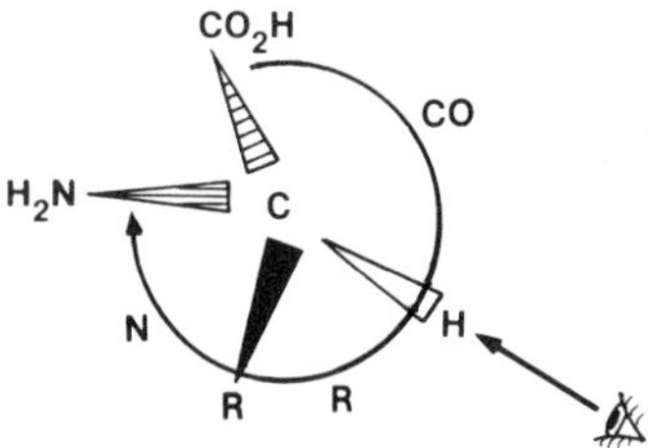

The **absolute configuration** of an amino acid (the position of all the groups in space) can be found by X-ray diffraction. Amino acids whose absolute configuration is as shown are called L-amino acids. You can remember this by the so-called CORN law. Look along the H—C bond. If the other three groups, going clockwise, are in the order:

$$-CO_2H, -R, -NH_2$$

then the compound is an L-amino acid, otherwise it is the mirror image which is given the symbol D. All the naturally occurring amino acids are of the L-configuration.

Perhaps surprisingly, there is no simple relationship between absolute configuration (shown by D or L) and the direction of rotation of polarized light (shown by + or −). For example, L-glutamic acid is + but L-cysteine is −. What is certain, however, is that if the L-form of a given substance is + then the D-form will be −. So the two enantiomers of glutamic acid are L(+) glutamic acid and D(−) glutamic acid. L(−) and D(+) glutamic acids are impossible.

A

THE ISOELECTRIC POINT OF AMINO ACIDS AND PROTEINS— ELECTROPHORESIS

Amino acids can both accept and donate protons. They therefore act as buffers.

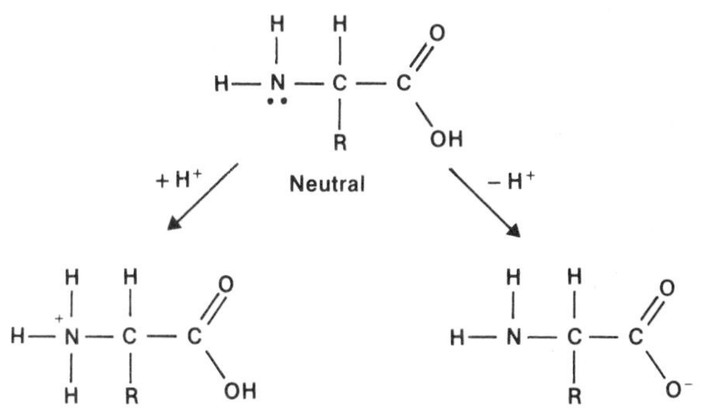

In strongly acidic solutions they will exist as cations and in strongly basic ones as anions. At some intermediate pH called the **isoelectric point**, they will be neutral. Exactly what this pH is will depend on the relative strengths of the $-NH_2$ group as a base and the $-CO_2H$ group as an acid, both of which depend on the nature of the —R group. Proteins, too, will have an isoelectric point. Mixtures of amino acids or proteins can be separated by **electrophoresis**. The apparatus is shown schematically in the diagram. Proteins or amino acids whose isoelectric point is lower than the pH of the buffer will exist as anions and move towards the anode; those whose isoelectric pH is greater than the pH of the buffer will exist as cations and migrate to the cathode.

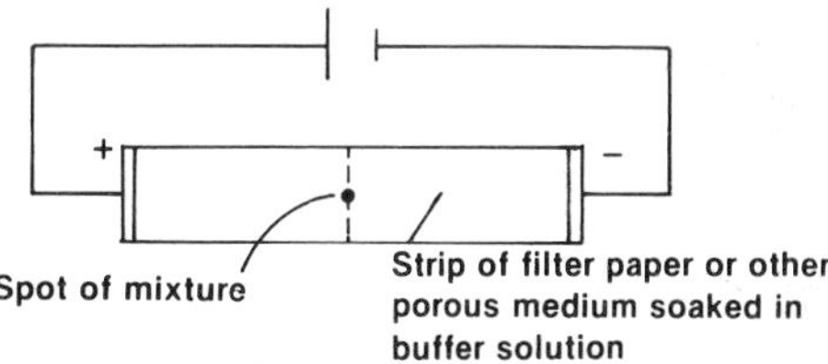

Any species whose isoelectric point is the same as the buffer will not move. The rate of migration will depend on the amount of charge on the amino acid or protein and its mobility as well as the voltage. The rate of migration at a given pH will enable a particular protein or amino acid to be identified. The technique is a quick method of identifying complex compounds. It is regularly used in hospital laboratories where an analysis of proteins present in blood samples can be an important pointer in the diagnosis of a number of diseases.

$$-\mathrm{C}(=\mathrm{O})\mathrm{OH} \rightleftharpoons -\mathrm{C}(=\mathrm{O})\mathrm{O^-} + \mathrm{H^+}$$

The amine group has a tendency to accept a proton (act as a base):

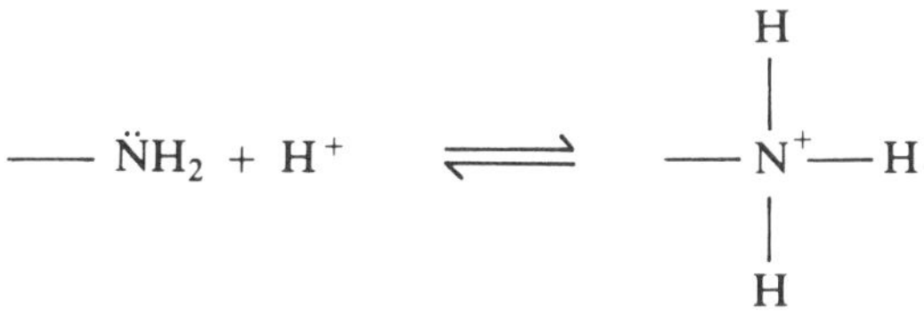

$$-\ddot{\mathrm{N}}\mathrm{H_2} + \mathrm{H^+} \rightleftharpoons -\mathrm{N^+H_3}$$

Amino acids exist largely as **zwitterions** in which the carboxylic acid
deprotonated and the amino group protonated, the molecule remainin
neutral overall:

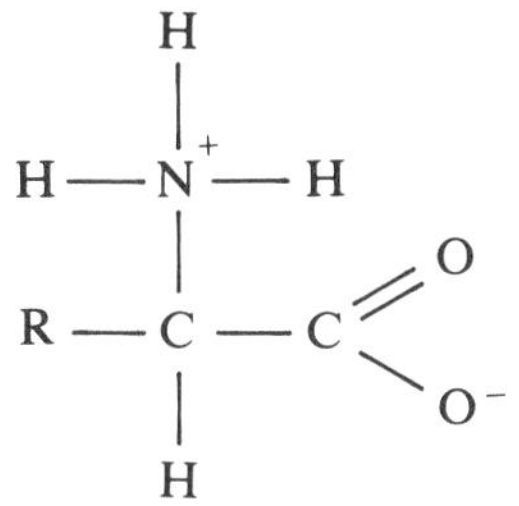

$$\mathrm{R-CH(N^+H_3)-C(=O)O^-}$$

a zwitterion

This ionic nature explains the high melting temperatures, and the fact tha
amino acids dissolve well in water and poorly in non-polar solvent
Typically, amino acids are white solids at room temperature resemblin
ionic salts in many of their properties.

35.2.2 Peptides, polypeptides and proteins

The amino group of one amino acid can react with the carboxylic aci
group of another to form an amide (see Chapter 33). A molecule of wate
is eliminated:

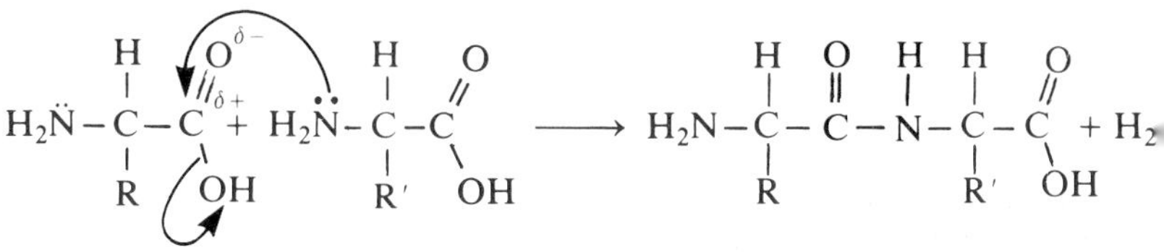

$$\mathrm{H_2\ddot{N}-CH(R)-C^{\delta+}(=O^{\delta-})OH + H_2\ddot{N}-CH(R')-C(=O)OH \longrightarrow H_2N-CH(R)-C(=O)-NH-CH(R')-C(=O)OH + H_2}$$

a dipeptide

The amide linkage is shown in brown. The resulting molecule is a dime
containing two amino acid units. Compounds formed by the linkage c
two or more amino acids are called **peptides**; one with two amino acid
is called a dipeptide and in this context the amide linkage is often calle
a peptide linkage. The dipeptide still retains $-NH_2$ and $-CO_2H$ group
and can react further to give tri- and tetrapeptides, and so on. Molecule
containing up to about 50 amino acids are called **polypeptides** whil
molecules with more amino acid units than this are called **proteins**.

35.2.3 The structure of proteins

Proteins have complex structures. Three levels of structure can b
distinguished in every protein.

Firstly there is the **primary structure**. This refers to the sequence c
different amino acids along the protein chain. As there are around twent
common amino acids which can be in any order, the number of possibilitie
is enormous. It is rather like making a necklace with supplies of beads c

twenty different shades. The primary structure can be written down using the three letter names of the proteins given in the box *Twenty naturally occurring amino acids*. For example, just one short sequence of the protein insulin (the hormone controlling sugar metabolism) runs:

—ala—glu—ala—leu—tyr—

One end of any protein or peptide has a free $—NH_2$ group and the other a free $—CO_2H$ group. These are often referred to as the N-terminal end and the C-terminal end respectively. It is usual to write the N-terminal end on the left.

For example, the sequence gly—ala would represent:

$$\begin{array}{ccccccccc} & & H & & O & & H & & CH_3 \\ & & | & & \| & & | & & | \\ H_2N & — & C & — & C & — & N & — & C & — & CO_2H \\ & & | & & & & & & | \\ & & H & & & & & & H \\ & & \underbrace{\quad}_{\text{gly}} & & & & & & \underbrace{\quad}_{\text{ala}} \end{array}$$

while ala—gly would represent:

$$\begin{array}{ccccccccc} & & CH_3 & & O & & H & & H \\ & & | & & \| & & | & & | \\ H_2N & — & C & — & C & — & N & — & C & — & CO_2H \\ & & | & & & & & & | \\ & & H & & & & & & H \\ & & \underbrace{\quad}_{\text{ala}} & & & & & & \underbrace{\quad}_{\text{gly}} \end{array}$$

X-ray diffraction studies of protein molecules have shown that they have a helical or spiral structure—an idea first proposed by the US chemist Linus Pauling. A typical protein helix has 18 amino acids for every five turns of the helix, see **Figure 35.2.**

The helical structure of proteins is called the **secondary structure**. The chain is held in its helical shape by hydrogen bonding between N—H and O=C groups. See **Figures 35.3** and **35.4.** Finally, the protein helix may be bent, twisted or folded into a particular shape. This is the **tertiary structure**.

The easiest way to visualize the primary, secondary and tertiary structures of a protein is to imagine a length of wire representing the protein chain. The wire is painted in strips of different colours, each strip representing an amino acid. The primary structure is the pattern of coloured strips on the wire. If the wire is wound into a spiral, the tightness

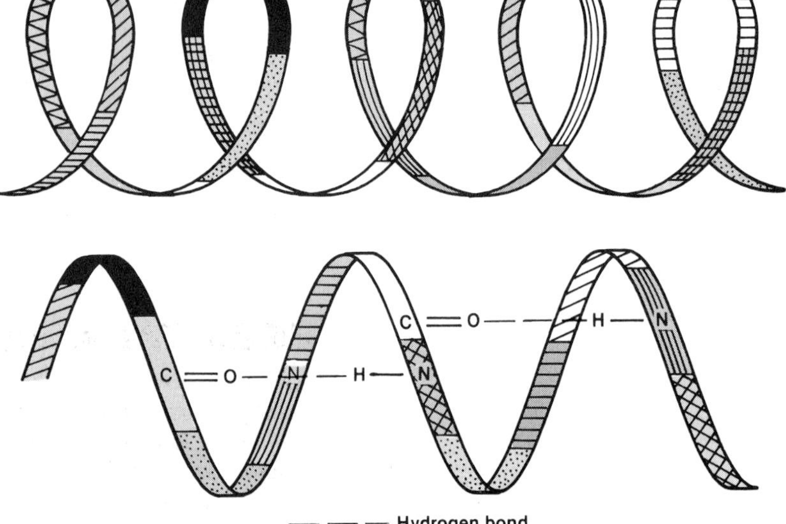

igure 35.2 The helical structure of a rotein. The shaded strips represent mino acid links. 18 amino acids occupy ve turns of the helix

igure 35.3 The helix of a protein is eld by hydrogen bonds

Figure 35.4 The helical structure of a protein (keratin). Hydrogen bonds which maintain the shape are shown in brown

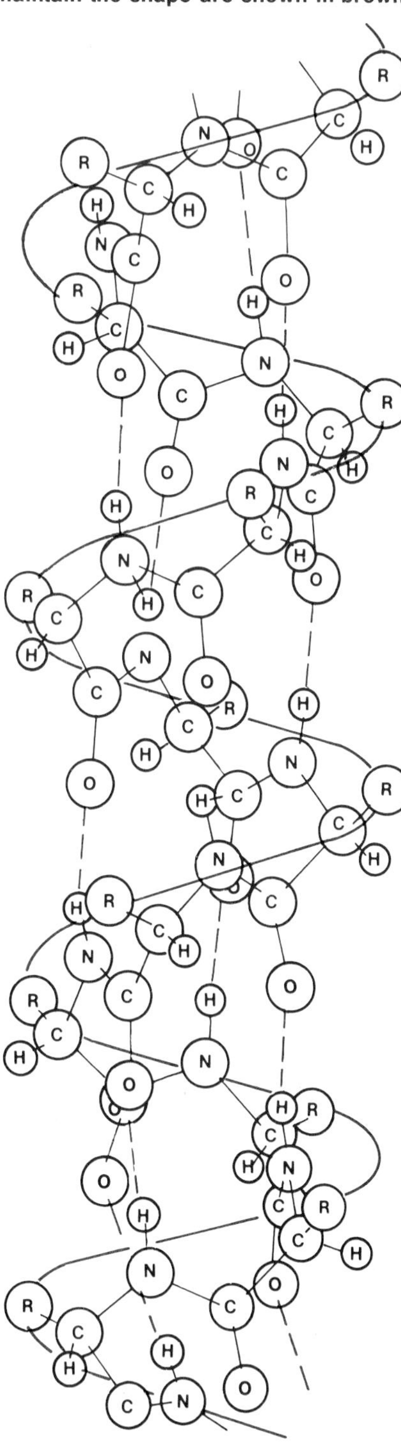

of the spiral represents the secondary structure. The resulting spiral coul then be folded like a twisted telephone wire to represent the tertiar structure.

35.2.4 Determination of protein structure

The secondary and tertiary structures of proteins can be investigated onl by X-ray diffraction techniques. The primary structure (the amino aci sequence) can be investigated by normal chemical techniques.

Boiling a protein with hydrochloric acid of concentration 6 mol dm^{-} for 24 hours hydrolyses all the amide (peptide) linkages and produces a mixture of all the amino acids in the original protein. The mixture ca be separated, and the amino acids identified, by two-way **paper chromato graphy** (see **section 11.3.3**). Amino acids, being colourless, do not giv visible spots but can be located by spraying the paper with **ninhydri** which reacts with amino acids to give a purple colour. This allows th amino acids present in the original protein to be found by comparing th *two* R_f values obtained for each spot with R_f values obtained for know amino acids in the same solvents.

A quantitative separation by, for example, column chromatograph (see **section 11.3.3**) will give the relative numbers of each amino acid. Fo example insulin, one of the simplest proteins, has a total of 51 amino acid as shown in **Table 35.1**

Determining the **sequence** is more difficult and a number of method are used. Only an outline is possible here.

N-terminal and C-terminal analysis

Reagents are available which will react with either the N-terminal end o the C-terminal end of a protein or peptide. If the protein is then hydrolysed the amino acids which were at the ends of the chain can be identified a these are attached to the reagent.

Partial hydrolysis

Hydrolysis under milder conditions than those used for complet hydrolysis can result in the protein being broken into peptides, each o three or four amino acids. These can be identified by chromatography a described above. The amino acid sequence of the original protein can the be deduced from the overlapping peptide fragments. The principle can b seen from the following simple example.

Partial hydrolysis of a tetrapeptide yielded three fragments:

$$\text{N-terminal end} \begin{cases} \text{gly—val} \\ \text{ala—gly} \\ \text{val—gly} \end{cases}$$

The original peptide must have had the sequence below to yield th fragments shown:

N-terminal end ala—gly—val—gly

Enzyme controlled hydrolysis

This is a complementary technique to partial hydrolysis. Certain enzyme will catalyse the hydrolysis of specific peptide bonds only. For example trypsin will break only the peptide bonds formed by the C-terminal end of the amino acids lys and arg. So partial hydrolysis with this enzym will produce peptides in which these amino acids are at the C-terminal end.

Frederick Sanger and his co-workers at Cambridge took ten years i the late 1940s and early 1950s to determine the amino acid sequence o the simple protein insulin. Sanger won the 1958 Nobel prize for this work Nowadays, many of the procedures can be automated and the process i much faster.

e coiling of the coils form a tertiary ucture. The structure of proteins is ıch more complex

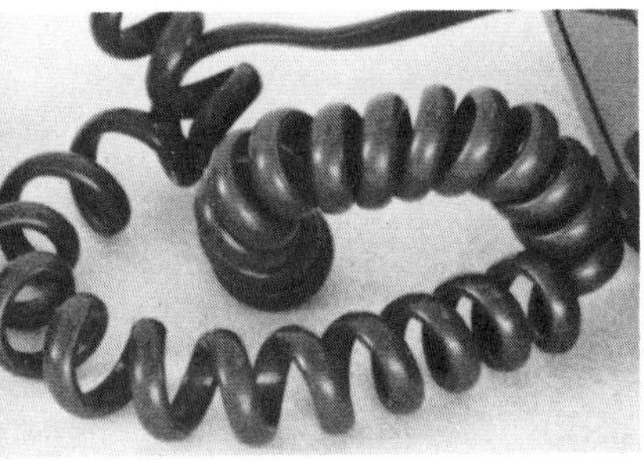

able 35.1 **The 51 amino acids which ake up insulin**

e	3
	5
n	3
	3
	2
	6
	6
	4
	3
	3
	4
	4
	1
	1
	1
	1
	1
	—
	51

gure 35.5 **Typical graph of reaction te against temperature for an zyme-catalysed reaction**

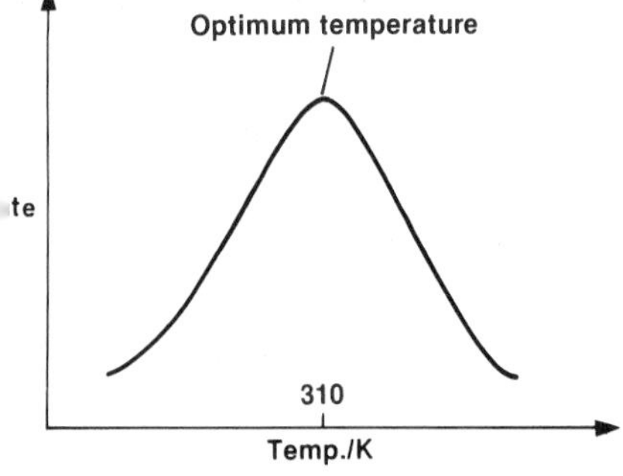

35.2.5 *The structure of proteins and their properties*

Following X-ray diffraction studies, it has been possible to build scale models of the structures of proteins and to see how their structure influences their properties. We shall look at just a few examples.

Enzymes

Enzymes are immensely efficient protein-based catalysts which occur in living systems. They can accelerate reactions by factors of up to 10^{10}. (This means that a reaction which is complete in *1 second* with the enzyme would take *300 years* without it.) Enzymes are very specific, usually catalysing just one reaction of one compound, called the **substrate**. Enzymes bond very efficiently to the substrate and the molecule with which it is reacting, and temporarily hold the two molecules in the correct orientation for them to react. The products are then released by the enzyme. The shape of the enzyme is thus very important. **Plate 21** shows a model of the shape of the enzyme lysozyme, a naturally occurring antibiotic found in chicken egg white. It catalyses the breakdown of poly-sugars found in the cell walls of bacteria thus killing the bacteria. The substrate fits into a deep cavity in the enzyme called the **active site** of the enzyme.

Because the catalytic activity of enzymes is so dependent on their shape, enzymes are very sensitive to changes in temperature and pH which can disrupt the hydrogen bonds responsible for secondary and tertiary structures. Most enzymes will rapidly **denature** (change their shape) at temperatures above about 320 K. This is why enzymes have an optimum temperature as catalysts. Below the optimum temperature, the reaction is slow for the usual reasons. Above this temperature, the enzyme denatures and loses its catalytic effect, **Figure 35.5.**

The stretchiness of wool

Wool is a protein fibre where the helix is, as usual, held together by hydrogen bonds, **Figure 35.6.** When wool is gently stretched, the hydrogen bonds stretch, **Figure 35.7**, and the fibre extends. Releasing the tension allows the hydrogen bonds to return to their normal length and the fibre returns to its original shape. However, washing at high temperatures can break the hydrogen bonds and a garment may permanently lose its shape.

Hair perming and permanent creasing of clothes

As well as hydrogen bonding, at least two other types of bond are involved in the secondary and tertiary structures of proteins.

One is bonding between the side chains or R groups of the amino acids. An acidic R group such as that of aspartic acid can react with a basic

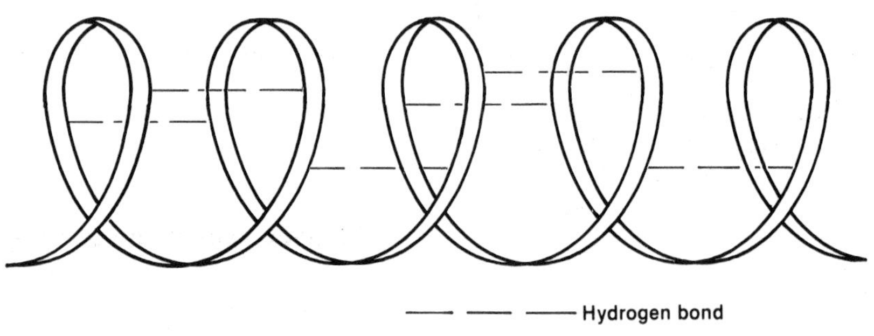

gure 35.6 **Hydrogen bonds in wool**

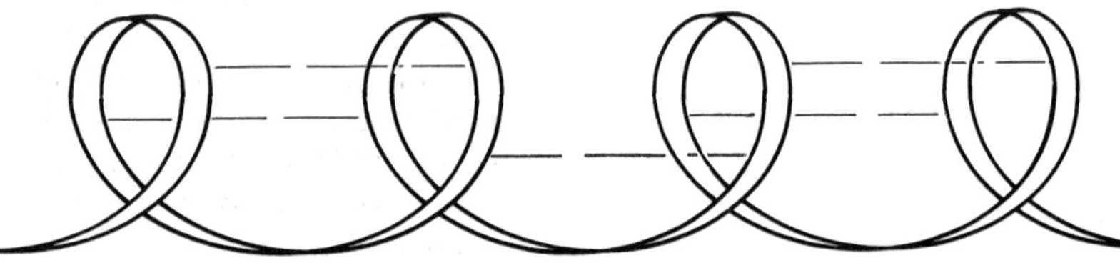

igure 35.7 **The wool is gently retched**

Figure 35.8 **Aspartic acid and lysine**

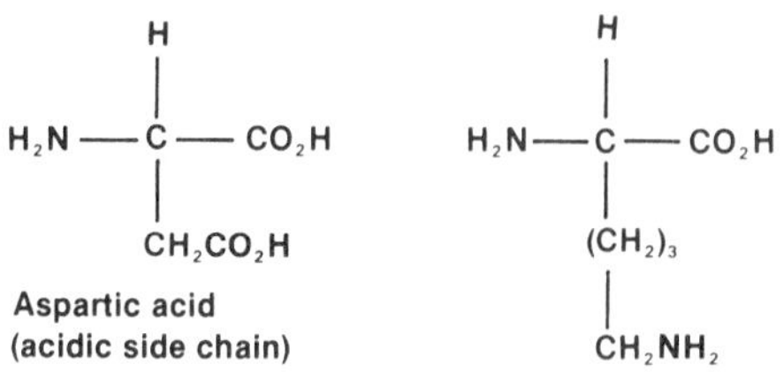

$H_2N—C(H)(CH_2CO_2H)—CO_2H$

Aspartic acid (acidic side chain)

$H_2N—C(H)((CH_2)_3CH_2NH_2)—CO_2H$

Lysine (basic side chain)

Figure 35.9 **Cysteine**

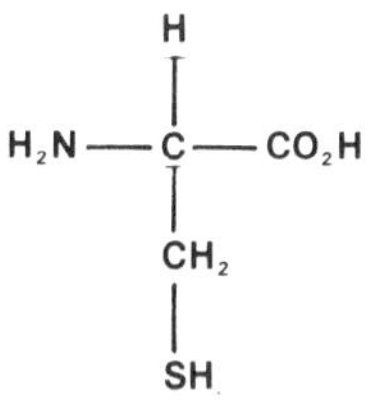

$H_2N—C(H)(CH_2SH)—CO_2H$

Cysteine

Figure 35.10 (right) **Two cysteine molecules can link via a weak S—S bond to form the 'double' amino acid cystine**

Figure 35.11 **Sulphur–sulphur bridges in the insulin molecule**

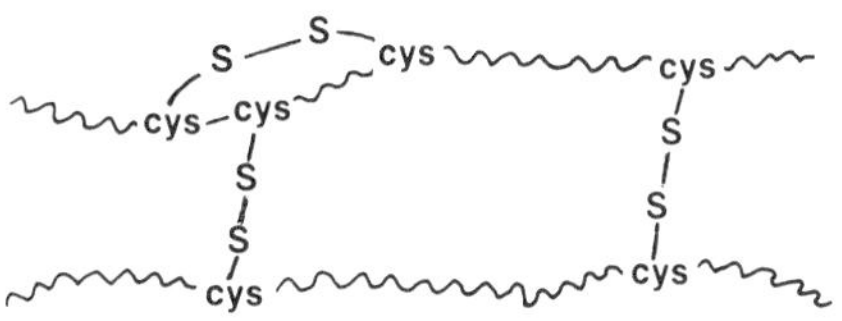

cys—S—S—cys, cys—S—S—cys, cys—S—S—cys

side chain on another amino acid, such as lysine, **Figure 35.8.**

The second is bonding between two molecules of the amino acid cystei which has the side chain —CH_2SH, **Figure 35.9.** Two cysteine molecul can be oxidized to form a S—S bond. The resulting 'double' amino ac is called cystine and can link two protein chains, as shown in **Figure 35.1** Insulin has three S—S bridges, **Figure 35.11.** The S—S bonds in cysti are easily broken by mild reducing agents and reformed by mild oxidizi agents:

$$\text{cys—S—S—cys} \xrightarrow[+2H]{\text{reduction}} \text{cys—SH + HS—cys}$$

$$\text{cys—SH + HS—cys} \xrightarrow[-2H]{\text{oxidation}} \text{cys—S—S—cys}$$

This is the basis of both permanent waving of hair and permanent creasi of woollen garments. Both hair and wool contain the protein keratin whic

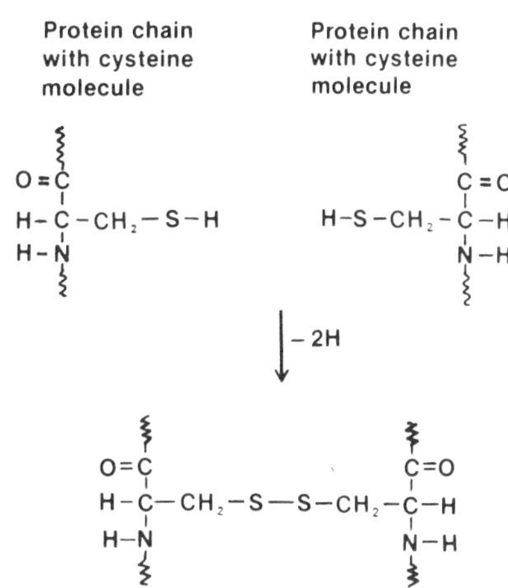

Cystine molecule linking two protein chains

A

THE USE OF ENZYMES

Enzymes are such efficient catalysts that chemists have always been challenged by the possibility of getting them to catalyse reactions of economic or social importance. A large part of the 'new' science of biotechnology is concerned with doing just this.

People have long used enzymes to catalyse the reactions that make alcohol from carbohydrates and yoghurt from milk. More recently biological washing powders have used protease enzymes to dissolve protein-based stains such as blood and egg. Much work in the last decade which has proved successful has been involved with efforts to bind enzymes to solid supports so that reactants can be passed over them without the enzyme being washed away.

A recently developed medical use of enzymes is in the treatment of slipped discs. A slipped disc occurs when a section of tissue from between the vertebrae in the back or the neck bulges out causing painful pressure on a nerve and, in some cases, paralysis of the limb served by the nerve. Conventional treatment requires surgery to remove the bulging portion of the disc. As an alternative, the enzyme chymopapain can be injected into the bulging disc to denature the protein of the disc causing it to shrink and relieve the pressure on the nerve. No incision is required and the procedure is much quicker and less painful than surgery. A similar enzyme is sometimes used to 'tenderize' tough joints of meat by dissolving tough fibres.

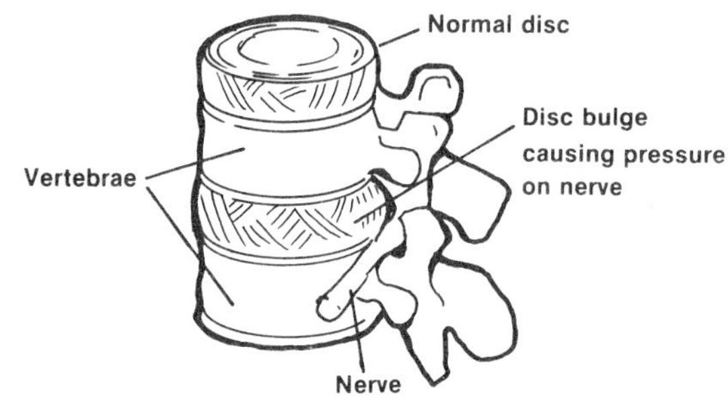

Another approach to the use of enzymic catalysts in industrial processes is to synthesize simple molecules that mimic the behaviour of enzymes. Considerable progress has been made with molecules called cyclodextrines which consist of rings of several glucose molecules. Such research is likely to improve our understanding of how enzymes work as well as opening up the possibility of 'designer catalysts'—molecules tailored specifically to catalyse a particular reaction.

e protein in the egg white (albumin) is natured by high temperature. The ange is irreversible. Once the drogen bonds have been broken it is mensely improbable that they will form in exactly the same way on oling

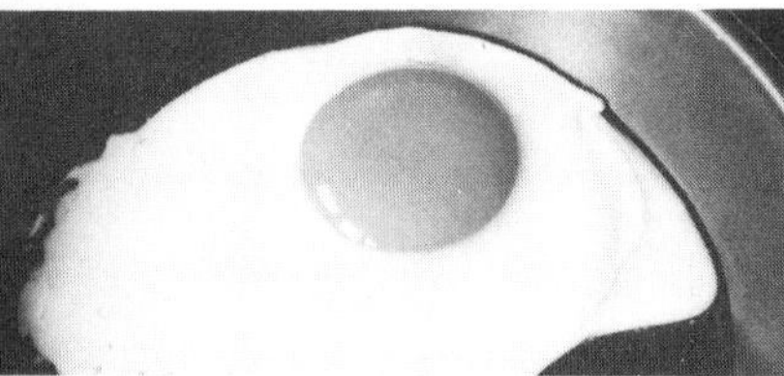

contains a large proportion of cystine with —S—S— bridges.

In perming, a mild reducing agent is used to break the S—S bonds. The hair is then curled using rollers and the S—S bonds reformed by treating the hair with a gentle oxidizing agent (usually hydrogen peroxide). This 'locks' the hair in its curled arrangement.

A similar sequence of applying a reducing agent, creasing and then oxidizing is used in permanent creasing. Both processes are illustrated schematically in **Figure 35.12.**

gure 35.12 (right) Breaking and orming S—S bridges can produce rmanent changes in the shapes of otein molecules

35.3 Synthetic polymers

This is a large and varied group of materials, many of which have been discussed in earlier chapters under the relevant functional group.

The first completely synthetic polymer was Bakelite which was patented in 1907. Since then, a large number of different types of polymer have been developed with an enormous range of properties to suit them for a vast number of applications from cling film to canoes, climbing ropes to non-stick coatings, tights to motor cycle helmets. **Figure 35.13** shows the enormous increase in production of polymers since 1860 along with some landmarks in their development.

gure 35.13 130 years of synthetic lymers

allace Carothers

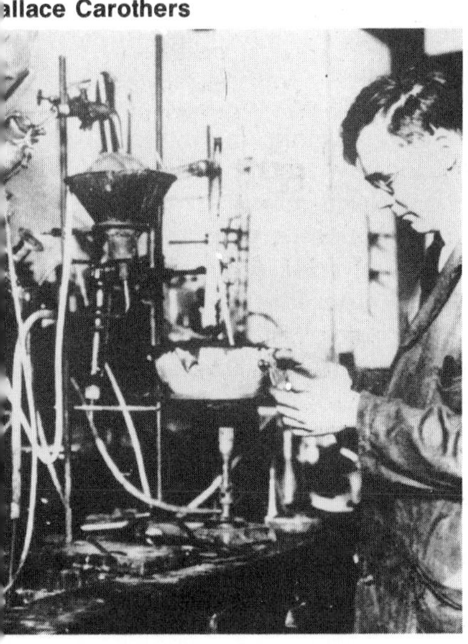

One way of classifying polymers is by the type of reaction by which they are made.

Addition polymers are made from unsaturated monomers and the empirical formula of the polymer is the same as that of the monomer. Only one type of monomer is needed—this is called **homopolymerization.**

Condensation polymers are made from two different monomers, both of which have two functional groups. For each pair of monomers joined together, a small molecule such as water or hydrogen chloride is expelled. Polymerization reactions which involve two (or more) different types of monomer are called **co-polymerization** processes.

35.3.1 Addition polymers

Addition polymers are commonly made from monomers which a derivatives of ethene of general formula:

$$\mathrm{H_2C{=}CHR}$$

Some examples are given in **Table 35.2**.

Table 35.2 Addition polymers

R	Name of polymer	Common or trade name
—H	Poly(ethene)	Polythene(Alkathene)
$—CH_3$	Poly(propene)	Polypropylene
—Cl	Poly(chloroethene)	PVC (polyvinylchloride)
—C≡N	Poly(propenenitrile)	Acrylic (Acrilan, Courtelle)
—C_6H_5	Poly(phenylethene)	Polystyrene

A number of methods of polymerization are used. One of the most comm is **radical polymerization** which requires an **initiator** to supply the initi radicals and continues via the **propagation** and **termination** steps typic of a **chain reaction**, see **section 27.7.3.**

Ziegler–Natta catalysts

Another common method is by the use of catalysts which include transition metal. One such catalyst system is aluminium trieth $Al(C_2H_5)_3$, and titanium(IV) chloride, $TiCl_4$, and is called a **Ziegler–Nat catalyst**. Ziegler–Natta catalysts have two important attributes. Firstl they allow polymerization to take place under relatively mild conditio of temperature and pressure. Secondly, they produce polymers of ve regular geometry which has an important effect on their properties.

The geometry of polymer chains

In an addition polymer derived from a substituted ethene:

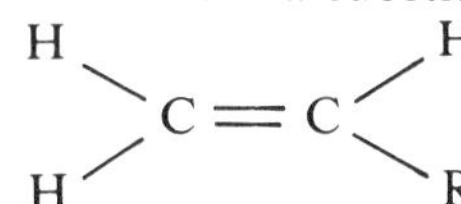

many geometrical arrangements of the product may occur. If we imagi the main carbon chain to be in a horizontal plane, then the substitue may either stick up above or be down below the plane. If all the substituen are above the plane, the arrangement is called **isotactic**; if substituents a alternately above and below the plane, it is called **syndiotactic**; and if t substituents are randomly arranged above and below the plane, t configuration is described as **atactic**, see **Figure 35.14.**

Polymers made using Ziegler–Natta type catalysts are more regul than those made by radical polymerization or other mechanisms. T regularity or otherwise of polymer chains has considerable bearing on t properties of the finished plastic. For example, in poly(chloroether (PVC), there is weak bonding, similar to hydrogen bonding between t chlorine atoms on one chain and hydrogen atoms on the next, whi prevents the chains sliding over one another and makes the polymer tou and rigid, see **Figure 35.15.** The effect of these forces is greater if t polymer is regular, so that a hydrogen atom on one chain always lin up with a chlorine atom on the other. An irregular arrangement mig allow two chlorines to line up, which would lead to repulsion betwe the chlorines.

Polar substituent groups can lead to increased attraction between t chains while bulky side chains can reduce attraction by forcing the chai apart.

Figure 35.14 Geometries of addition polymers

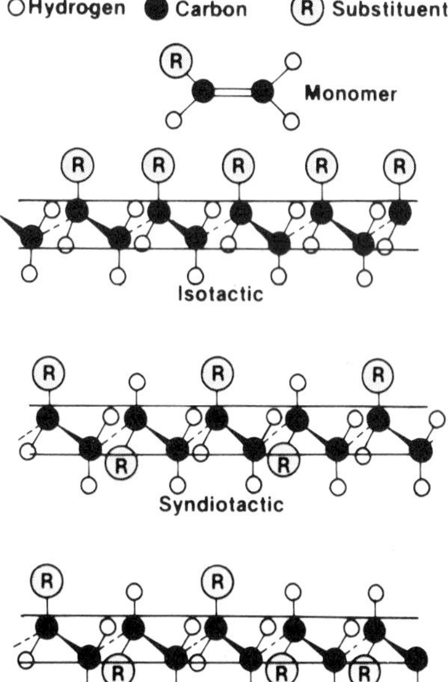

Modification of the properties of polymers

The properties of polymers can be 'tailored' by a number of methods meet the requirements of a particular application.

gure 35.15 Weak bonding between ains in poly(chloroethene)

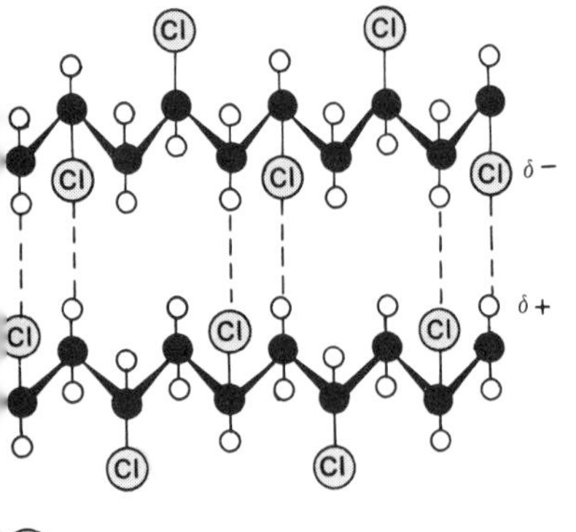

Cl Chlorine ● Carbon ○ Hydrogen

– – – Weak hydrogen bond

gure 35.16 Cross-linking

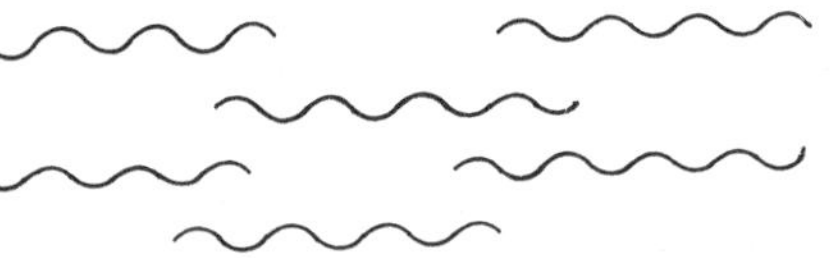

Polymer without cross-links

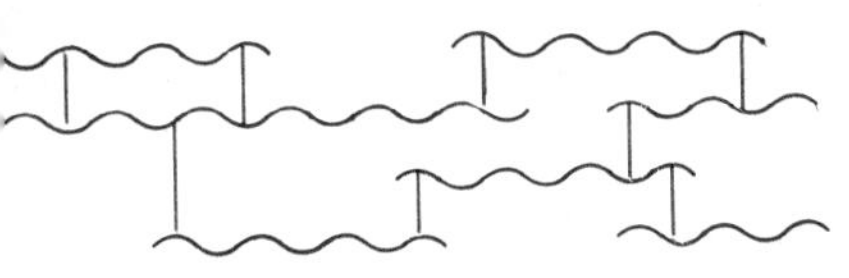

Polymer with cross-links

A plasticizer can be added. This is a small molecule which gets in between the polymer chains and reduces the attraction between them making the plastic more flexible. Its effect is rather like a lubricant.

Fillers may be added to make the plastic cheaper and other additives may include pigments, flame retardants and stabilizers to prevent decomposition by ultraviolet light.

Another way of modifying the properties of polymers is by co-polymerizing—adding a small quantity of another monomer at the polymerization stage. This also modifies the attraction between the chains.

Polymer properties are also affected by the relative molecular mass of the polymer, which can be controlled at the manufacturing stage. Polymers of low relative molecular mass will have a relatively low viscosity when molten, which may make them easier to mould. It is also possible to introduce some **cross-linking** (with covalent bonds) into polymers. Cross-links join polymer chains and make the resulting polymer less elastic and more rigid, see **Figure 35.16.**

In general, polymers without cross-links soften on heating, as molecular vibration pushes the chains apart and reduces the bonding between them. Such plastics are described as **thermoplastic** or **thermosoftening**. They can therefore be moulded by heating them, shaping them and allowing them to cool.

Cross-linked polymers do not usually soften on heating and are described as **thermosetting**. They have to be moulded into shape when manufactured. They eventually decompose on strong heating. Thermo-setting plastics are useful where resistance to heat is important, for example saucepan handles.

35.3.2 Condensation polymers

Condensation polymers are normally made from two different monomers, each of which has two functional groups. For example, a diol and a dicarboxylic acid would react together to give a polyester by eliminating molecules of water, **Figure 35.17.** The same polymer could have been

Dicarboxylic acid

Diol

H)O – A – O(H HO)– C(=O) – B – C(=O) –(OH H)O – A – O(H HO)– C(=O) – B – C(=O) –(OH

H_2O H_2O H_2O

↓

··· O – A – O – C(=O) – B – C(=O) – O – A – O – C(=O) – B – C(=O) – ···

A polyester

igure 35.17 (right) Making a olyester

produced from a diacid dichloride rather than a dicarboxylic acid in which case hydrogen chloride would have been eliminated instead of water.

Other common synthetic condensation polymers are given in **Table 35.3** (overleaf). Manufacture of condensation polymers can usually be achieved simply by mixing the monomers if they are reactive enough.

Cross-linking can be brought about by adding to the reaction mixture a small amount of a compound with *three* functional groups. For example, polyesters can be cross-linked by addition of some propane-1, 2, 3-triol. The cross-linked resins obtained in this case are called alkyd resins and are used in paints.

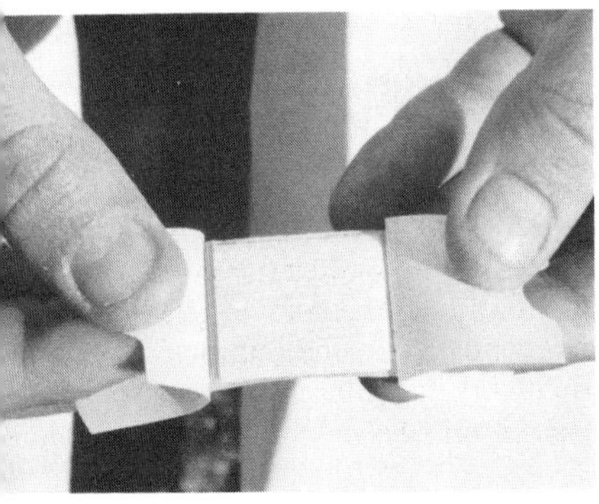

Silicones

Silicones are the odd ones out in this chapter in that they are not wholly organic polymers, the backbone of the molecule being a Si—O—Si chain (see left). Silicones have a wide range of useful properties including repelling water—they are used for waterproofing fabrics; lubricating—they are added to polishes; and non-stick ability—they are used to coat the backing of self-adhesive stickers (see photo, above left). See also **section 24.3.**

— Si(R)(R) — O — Si(R)(R) — O — Si(R)(R) —

Table 35.3 Condensation polymers

Monomer A	Monomer B	Linkage	Name	Molecule eliminated
$HO-\overset{O}{\overset{\|}{C}}-A-\overset{O}{\overset{\|}{C}}-OH$ Dicarboxylic acid	$HO-B-OH$ Diol	$-A-\overset{O}{\overset{\|}{C}}-O-B-$	Polyester, e.g. Terylene	H_2O
$Cl-\overset{O}{\overset{\|}{C}}-A-\overset{O}{\overset{\|}{C}}-Cl$ Diacid dichloride	$HO-B-OH$ Diol	$-A-\overset{O}{\overset{\|}{C}}-O-B-$	Polyester	HCl
$HO-\overset{O}{\overset{\|}{C}}-A-\overset{O}{\overset{\|}{C}}-OH$ Dicarboxylic acid	$H-\overset{H}{\overset{\|}{N}}-B-\overset{H}{\overset{\|}{N}}-H$ Diamine	$-A-\overset{O}{\overset{\|}{C}}-\overset{H}{\overset{\|}{N}}-B-$	Polyamide, e.g. Nylon	H_2O
$Cl-\overset{O}{\overset{\|}{C}}-A-\overset{O}{\overset{\|}{C}}-Cl$ Diacid dichloride	$H-\overset{H}{\overset{\|}{N}}-B-\overset{H}{\overset{\|}{N}}-H$ Diamine	$-A-\overset{O}{\overset{\|}{C}}-\overset{H}{\overset{\|}{N}}-B-$	Polyamide	HCl
$H-\overset{H}{\overset{\|}{N}}-\overset{O}{\overset{\|}{C}}-\overset{H}{\overset{\|}{N}}-H$ Urea	$H_2C=O$ Methanal	$-\underset{H}{\overset{H}{C}}-\overset{H}{\overset{\|}{N}}-\overset{O}{\overset{\|}{C}}-\overset{H}{\overset{\|}{N}}-\underset{H}{\overset{H}{C}}-$	Urea–methanal	H_2O
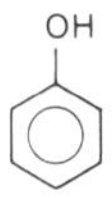 OH-substituted benzene ring, Phenol	$H_2C=O$ Methanal	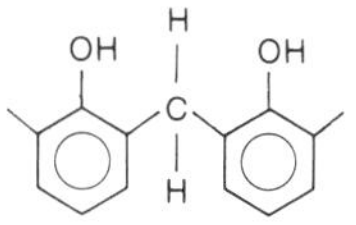 two OH-substituted benzene rings linked by $-\underset{H}{\overset{H}{C}}-$	Phenol–methanal, Bakelite	H_2O
$O=C=N-A-N=C=O$ Diisocyanate	$HO-B-OH$ Diol	$-A-\overset{H}{\overset{\|}{N}}-\underset{O}{\underset{\|}{C}}-O-B-$	Polyurethane	None, so strictly polyurethanes are n[ot] condensation polym[ers] although they have different monomers
$HO-\underset{R}{\overset{R}{Si}}-OH$	$HO-\underset{R}{\overset{R}{Si}}-OH$ Dialkylsilanediol	$-\underset{R}{\overset{R}{Si}}-O-\underset{R}{\overset{R}{Si}}-$	Silicone	H_2O

35.4 Summary

- Polymers are large molecules made of many small molecules (monomers) linked together.
- Amino acids are bifunctional compounds with both $-NH_2$ and $-CO_2H$ groups. Their general formula is:

$$H_2N-\underset{H}{\overset{R}{C}}-CO_2H$$

- Peptides are poly(amino acids) with less than 50 amino acids. Proteins are polypeptides with more than 50 amino acids.
- The primary structure of a protein refers to its amino acid sequenc[e]

A

PAINT

Paint consists of two main components, a **pigment** which provides the colour and a **vehicle**—the liquid mixture by which the pigment is applied to the surface. The vehicle itself is in two parts, a **binder**, which binds the pigment on to the surface, and a **solvent**, a fairly volatile liquid which evaporates after the paint has been applied leaving the binder and pigment behind on the surface. The volatility of the solvent is important—if it is too great, the paint will dry too quickly and be difficult to apply; if it is not volatile enough, the paint will dry too slowly and tend to form runs on the painted surface. In emulsion paint, the solvent is water; in gloss it is a mixture of hydrocarbons.

The most important white pigment is titanium(IV) oxide (coated with silica and alumina) which scatters light well and is chemically inert so the paint does not 'yellow'. Colours are provided by organic dyes such as azo dyes (see box, **section 34.2.5**) or transition metal compounds. Perhaps the most important component is the binder. In gloss paints, this is usually an **alkyd resin**—a polyester which is cross-linked by triol groups. As the paint dries the alkyd resin reacts with oxygen in the air to form a polymer which is insoluble in the solvent. The number of cross-links is critical—too many and the film would be brittle. Some flexibility is required to cope with expansion of the surface caused by temperature changes or absorption of water. Paints also contain additives to control viscosity and to prevent the paint separating in the can during storage.

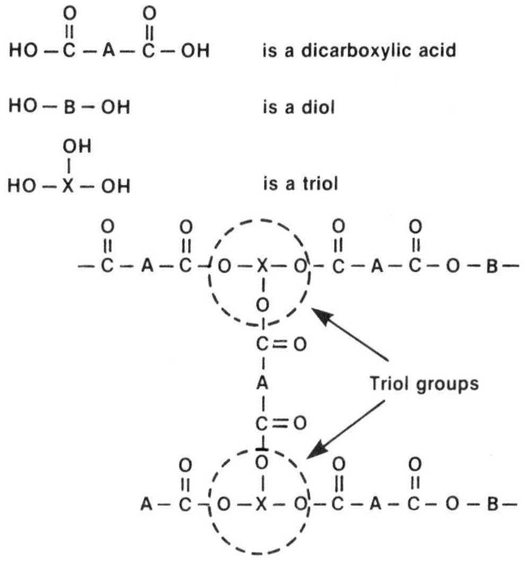

$HO-\overset{O}{\overset{\|}{C}}-A-\overset{O}{\overset{\|}{C}}-OH$ **is a dicarboxylic acid**

$HO-B-OH$ **is a diol**

$HO-\overset{OH}{\overset{|}{X}}-OH$ **is a triol**

$-\overset{O}{\overset{\|}{C}}-A-\overset{O}{\overset{\|}{C}}-O-X-O-\overset{O}{\overset{\|}{C}}-A-\overset{O}{\overset{\|}{C}}-O-B-$

C=O

A

C=O

Triol groups

$A-\overset{O}{\overset{\|}{C}}-O-X-O-\overset{O}{\overset{\|}{C}}-A-\overset{O}{\overset{\|}{C}}-O-B-$

- The secondary structure refers to its helical shape maintained largely by hydrogen bonding.
- The tertiary structure refers to folding of the helix.
- The primary structure of proteins may be investigated by hydrolysis to the separate amino acids or to short peptide chains, followed by two-way paper chromatography to identify the fragments.
- Enzymes are protein-based catalysts found in biological systems. They are both very efficient and very specific catalysts.
- Enzymes are easily denatured by moderately high temperatures and changes in acidity or alkalinity. They have an optimum temperature and optimum pH.
- Addition polymers are produced by linking unsaturated monomers. The polymer has the same empirical formula as the monomer.
- Condensation polymers are usually formed from two different monomers, each of which has two functional groups. Small molecules are eliminated in the polymerization reaction.
- Ziegler–Natta type catalysts can produce geometrically regular addition polymers.
- Polymers can be described as thermoplastic (which soften on heating) or thermosetting (which do not soften on heating).
- Thermosetting polymers generally have many cross-links between their chains.

35.5 Questions

1 Amino acids are important compounds which, when linked together, form proteins.

A) Glycine, $H_2N—CH_2—CO_2H$, is represented as having covalent bonding, yet its melting point (235 °C) is more characteristic of an ionic compound. Explain this observation.

B) The tertiary structure of a protein is often maintained by hydrogen bonding between the amino acid units in a protein chain. Show, by means of a diagram, how a chain containing an alanine unit might be hydrogen bonded to one containing valine.

$$H_2N—CH(CH_3)—CO_2H$$
alanine

$$CO_2H—CH(NH_2)—CH(CH_3)—CH_3$$
valine

(London, specimen A/S, part question)

2 Write an essay on 'synthetic polymers'. Aspects which you might consider include:

A) the nature of the chemical reactions used to make the polymers;

B) the methods used for their fabrication;

C) some specific applications of polymers; and

D) the effect of structure on the properties of a polymer.

(Nuffield 1982)

3 Study the formulae below and answer the questions that follow:

$ClOC—C_6H_4—COCl$ benzene-1,4 dioyl chloride

$CH_2OH—CH_2OH$ ethane-1,2-diol

A) Draw a diagram to illustrate the formula of the polymer formed by reacting these two compounds together.

B) Name the type of polymerization taking place in this reaction.

C) Outline a practical procedure by which you would attempt to produce polymer fibres using this reaction.

D) Account for the fact that this polymer produces excellent fibres, but that the polymer formed from benzene-1,2-dioyl chloride and ethane-1,2-diol does not.

(Nuffield 1983, part question)

4 This question concerns the nature and properties of amino acids and proteins. The simplest amino acid is glycine, usually represented as:

$$H_2NCH_2CO_2H$$

A) Give the principal structure of glycine:

i in aqueous acid solution;

ii in aqueous alkaline solution.

B) i How may amino acids be obtained from a protein?

ii Give a balanced equation for the conversion of the dipeptide glycylglycine to glycine.

C) In what capacity does protein material play a role in the control of metabolism?

D) Why, in this role, are the proteins so very sensitive to:

i changes in pH;

ii increases in temperature?

(COSSEC, specimen A/S, part question)

5 Two polymerization reactions may be represented by the following equations:

$$nCH_2{=}CH_2 \longrightarrow -\!\!\!+ CH_2—CH_2 +\!\!\!-_n$$
polymer A

$$n\,C_6H_5OH + nCH_2O \longrightarrow [—C_6H_3(OH)—CH_2—]_n + nH_2O$$
polymer B

A) Give the names of

i polymer A,

ii polymer B.

B) Name the type of polymerization involved in the formation of

i polymer A,

ii polymer B.

(COSSEC, specimen A/S, part question)

36 Synthetic routes

36.1 Introduction

This chapter deals with the problem of devising a series of reactions for making (*synthesizing*) a given molecule, often called the *target molecule.* Sometimes (often, in exam questions) you will be told the molecule from which you must start. In real life, the starting molecule will be chosen on the grounds of availability, expense and reactivity, along with other considerations like ease of storage, toxicity and so on.

Synthesis of a target molecule is a common problem in industries involved in drug or pesticide manufacture. Frequently, a molecule is found to have a particular pharmaceutical effect, for example, as an antibiotic. Drug companies may then synthesize, on a small scale, a number of compounds of similar structure and screen them for possible antibiotic properties. Any promising compounds may then be made in larger quantities for thorough investigation of their effectiveness, side-effects and so on. A large-scale method of manufacture must then be devised for any commercially viable products.

36.2 Basic ideas

gure 36.1 Devising a synthesis of mpound X from compound A

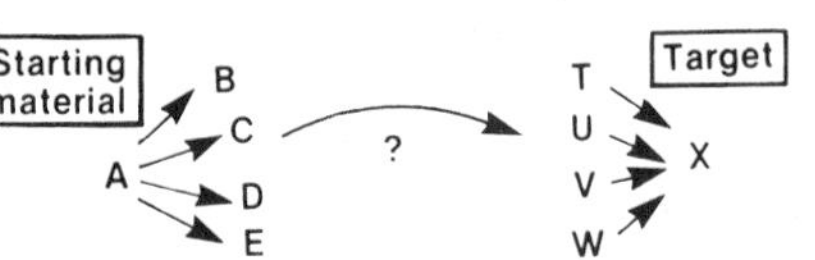

If you are given the problem of devising a scheme for making target molecule X from starting material A, one way of proceeding is to write down all the compounds which can be made from A and all the ways in which X can be prepared, **Figure 36.1.** A way may then suggest itself whereby B, C, D or E can be converted, in one or more steps, to T, U, V or W. It is important to keep the number of steps as small as possible to maximize the yield of the target. Remember that no reaction will ever have a 100% yield and that a three-step synthesis in which each step has an 80% yield will produce an overall yield of 80 × 80 × 80% = 51.2%.

Sometimes you will be able to see straight away that a particular reaction will be needed. If, for example, the target molecule has one more carbon atom than the starting material, it is almost certain that the reaction of cyanide ions with a halogenoalkane will be needed at some stage as this reaction increases the length of the carbon chain by one, for example:

$$\underset{\text{bromomethane}}{CH_3Br} + CN^- \longrightarrow \underset{\text{ethanenitrile}}{CH_3C{\equiv}N} + Br^-$$

Similarly the Hofmann degradation of amides reduces the carbon chain length by one, so this may well be required if the target molecule has one less carbon atom than the starting material, for example:

$$\underset{\text{ethanamide}}{CH_3C(=O)NH_2} + Br_2 + 4NaOH \longrightarrow \underset{\text{methylamine}}{CH_3NH_2} + Na_2CO_3 + 2NaBr + 2H_2O$$

36.3 Methods of preparation of functional groups

These have been given in the chapter about each functional group, b are listed here as a summary, **Table 36.1.** Remember that some compoun are readily available commercially, for example, alkanes, alkenes and number of aromatic compounds which are available from crude oil coal tar. These rarely need to be made in the laboratory and are not list here. Only the basic reaction has been given here. Fuller details of reagen conditions, etc. will be found in the appropriate chapter.

Table 36.1 **Summary of methods of synthesizing functional groups. The table refers to non-aromatic compounds, i.e. R is an alkyl or substituted alkyl group**

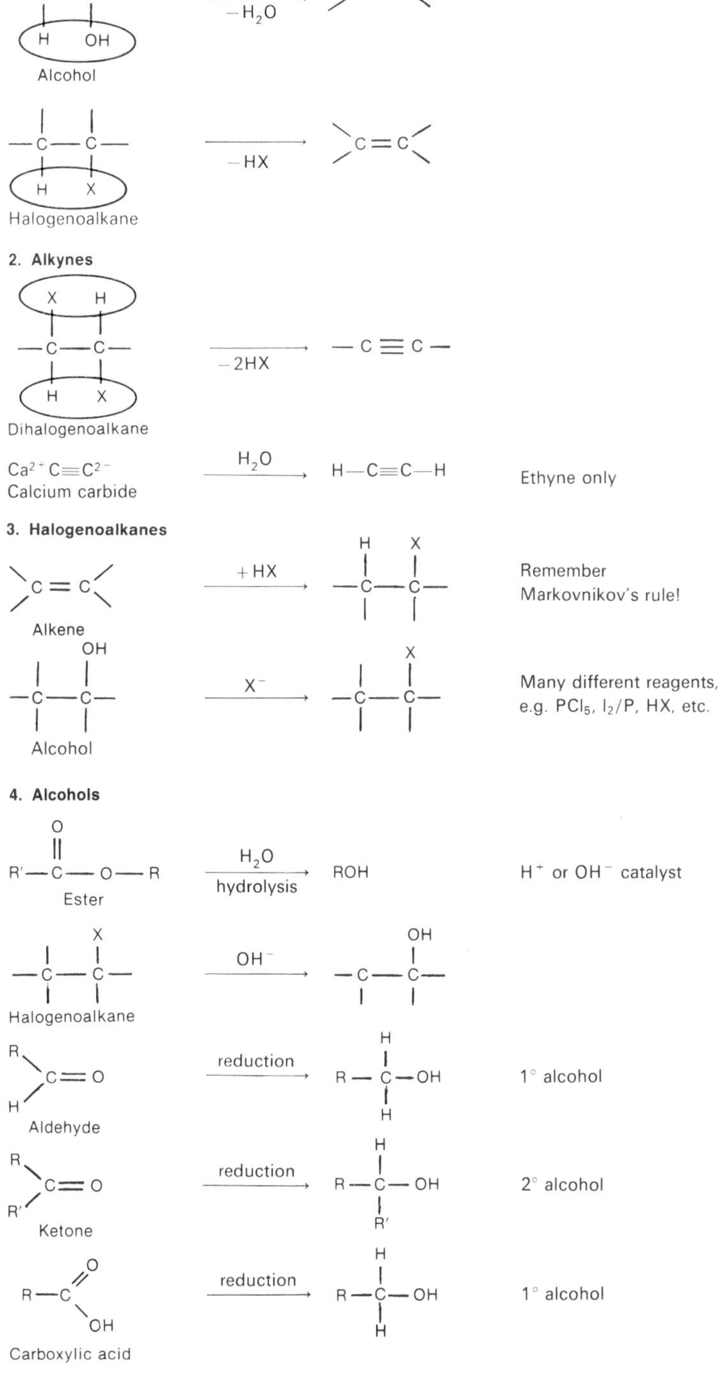

Notes

$$\underset{\text{Amine}}{R{-}NH_2} \xrightarrow[(NaNO_2/HCl)]{HNO_2} R{-}OH$$

5. Aldehydes

$$\underset{1^\circ\ \text{alcohol}}{R{-}CH_2{-}OH} \xrightarrow{\text{gentle oxidation}} R{-}CHO$$

Distil off product to prevent further oxidation

$$\underset{\text{Acid chloride}}{R{-}COCl} \xrightarrow[\text{Pd catalyst}]{H_2,\ \text{poisoned}} R{-}CHO$$

6. Ketones

$$\underset{2^\circ\ \text{alcohol}}{RR'CH{-}OH} \xrightarrow{\text{oxidation}} RR'C{=}O$$

7. Carboxylic acids

$$\underset{1^\circ\ \text{alcohol}}{R{-}CH_2{-}OH} \xrightarrow{\text{oxidation}} R{-}COOH$$

$$\underset{\text{Nitrile}}{R{-}C{\equiv}N} \xrightarrow{H^+/H_2O} R{-}COOH$$

$$\underset{\text{Amide}}{R{-}CONH_2} \xrightarrow{H^+/H_2O} R{-}COOH$$

$$\underset{\text{Other acid derivative}}{R{-}COZ} \xrightarrow{H_2O} R{-}COOH$$

8. Acid derivatives

These can be interconverted by use of a suitable nucleophile

$$\underset{\text{Acid derivative}}{R{-}COZ} + W^- \longrightarrow R{-}COW + Z^-$$

9. Amines

$$\underset{\text{Nitroalkane}}{R{-}NO_2} \xrightarrow{\text{reduction}} R{-}NH_2$$

$$\underset{\text{Amide}}{R{-}CONH_2} \xrightarrow{\text{reduction}} R{-}CH_2{-}NH_2$$

$$\underset{\text{Nitrile}}{R{-}C{\equiv}N} \xrightarrow{\text{reduction}} R{-}CH_2{-}NH_2$$

$$\underset{\text{Halogenoalkane}}{R{-}X} \xrightarrow{NH_3} R{-}NH_2$$

Also 2°, 3°, 4° compounds

$$\underset{\text{Alcohol}}{R{-}OH} \xrightarrow{NH_3} R{-}NH_2$$

Also 2°, 3° compounds

$$\underset{\text{Amide}}{R{-}CONH_2} \xrightarrow{\text{Hofmann degradation}} R{-}NH_2$$

The product has 1 carbon less than the starting material

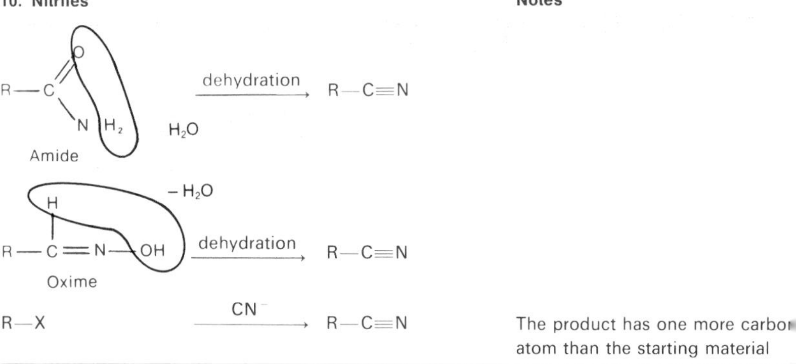

The interrelationships between these functional groups are shown Figure 36.2.

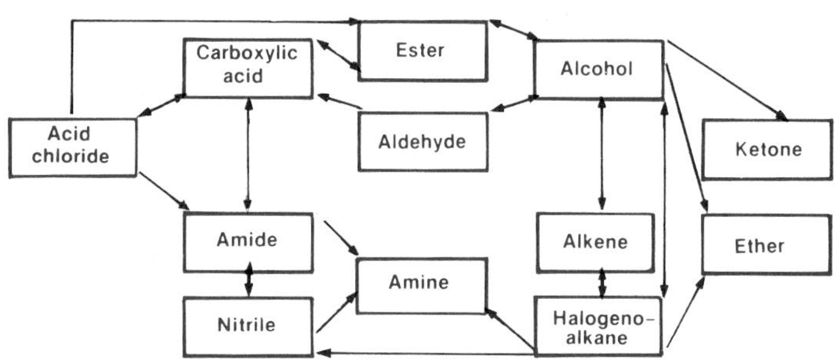

Figure 36.2 Interrelationships between functional groups

36.4 Reagents used in organic chemistry

This section considers a number of reagents commonly used in organ chemistry.

36.4.1 Oxidizing agents

The most commonly used oxidizing agent is acidified dichromate(VI) ion $Cr_2O_7^{2-}/H^+$. This will oxidize primary alcohols to aldehydes an aldehydes to carboxylic acids. Secondary alcohols are oxidized to ketone A similar oxidizing agent is a solution of manganate(VII) ions, which ma be used in acid, neutral or alkaline solution.

36.4.2 Reducing agents

A number of these are used.

In general, lithium tetrahydridoaluminate(III), $LiAlH_4$, will redu C=O or C≡N, but not C=C. It is used in ether solution followed b the addition of water.

Sodium tetrahydridoborate(III) is a similar but milder reducing agen It also reduces C=O, but not C=C. It can be used in aqueous solutio To reduce C=C but not C=O, hydrogen with a nickel catalyst is used.

Reducing agents such as iron or tin and hydrochloric acid may be use to reduce $R—NO_2$ to $R—NH_2$.

36.4.3 Dehydrating agents

Phosphorus pentoxide is frequently used to remove a molecule water from an organic compound, for example, to convert an amid

$$R—C\begin{matrix}{=}O\\ \backslash NH_2\end{matrix}$$

, into a nitrile, R—C≡N, or alcohol to an alkene.

Sulphuric acid is also a dehydrating agent but may react in other wa (for example, it reacts with alkenes to give alkylhydrogensulphates) an so phosphorus pentoxide is preferred.

36.5 Examples of reaction schemes

36.5.1 How can we convert 1-bromopropane into propanoic acid?

Firstly, note that both the starting material and the target have the same number of carbon atoms so no alteration to the carbon skeleton is needed. Secondly, write down all the compounds which can be made in one step from the starting material (1-bromopropane) and all those from which the target (propanoic acid) can be made in one step, as shown in **Figure 36.3.** In this case, as you can see, two of the compounds are the same. 1-bromopropane can be converted into propan-1-ol which can be converted into propanoic acid. So the conversion we require can be done in two steps:

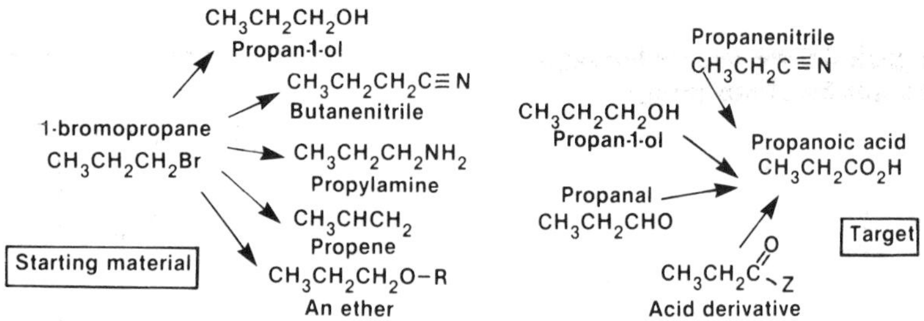

gure 36.3 Devising a synthesis of opanoic acid from 1-bromopropane

1. $CH_3CH_2CH_2Br$ (1-bromopropane) $\xrightarrow{\text{reflux with } OH^-(aq)}$ $CH_3CH_2CH_2OH$ (propan-1-ol)

2. $CH_3CH_2CH_2OH$ (propan-1-ol) $\xrightarrow{\text{reflux with } Cr_2O_7^{2-}/H^+}$ $CH_3CH_2CO_2H$ (propanoic acid)

Both these reactions have a good yield.

36.5.2 How can we convert ethanol into propylamine?

Firstly, note that the target molecule (propylamine) has one more carbon atom than the starting material (ethanol). This indicates that the formation of a nitrile is involved at some stage.

Secondly, write down all the compounds which can be made from the starting material and all the compounds from which the target can be made, **Figure 36.4.** Here, no compound which can be made in one step from the starting material can be converted into the product, so more than two steps must be required. But we already know that the formation

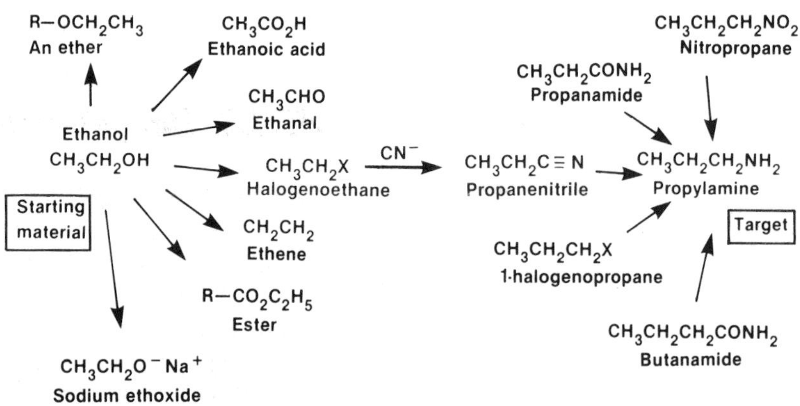

gure 36.4 Devising a synthesis of opylamine from ethanol

of a nitrile is required. A halogenoethane can be converted in propanenitrile so the synthesis can be completed in three steps:

1. CH_3CH_2OH (ethanol) $\xrightarrow[H_2SO_4/NaBr]{\text{reflux with}}$ CH_3CH_2Br (bromoethane)

2. CH_3CH_2Br (bromoethane) $\xrightarrow{CN^-}$ $CH_3CH_2C{\equiv}N$ (propanenitrile)

3. $CH_3CH_2C{\equiv}N$ (propanenitrile) $\xrightarrow{LiAlH_4}$ $CH_3CH_2CH_2NH_2$ (propylamine)

Another halogenoalkane could have been chosen, i.e. the chloro- iodo-compounds.

36.5.3 How can we convert propanoic acid into ethanol?

Firstly, note that the target molecule has one carbon atom less than th starting material, which suggests that one stage could involve the Hofman degradation of an amide. This immediately suggests that the first ste should be the conversion of propanoic acid to propanamide and th second should be the Hofmann degradation of propanamide to ethylamin Ethylamine can be converted to ethanol by treatment with nitrous aci

1. $CH_3CH_2CO_2H$ (propanoic acid) $\xrightarrow[\text{heat}]{NH_3\text{, then}}$ $CH_3CH_2CONH_2$ (propanamide)

2. $CH_3CH_2CONH_2$ (propanamide) $\xrightarrow{Br_2/NaOH}$ $CH_3CH_2NH_2$ (ethylamine)

3. $CH_3CH_2NH_2$ (ethylamine) $\xrightarrow[(HCl/NaNO_2)]{HNO_2}$ CH_3CH_2OH (ethanol)

36.5.4 How can we convert propan-1-ol into bromopropanone?

Firstly, both starting material and target have the same number of carbc atoms. Write down all the compounds that can be made from the startin material and those from which the target can be made, **Figure 36.5.** Clearl no two are the same, so a two-step synthesis is not possible. Notice that a the compounds which can be converted to the target have an oxyge atom bonded to the second carbon atom. This is the key to this synthesi How can we get a functional group onto carbon number two when virtual all the substances we can make from the starting material have the functional groups on carbon number one? The exception is propen $CH_3CH{=}CH_2$. Addition of H—Z to this obeys Markovnikov's rule (s **section 28.6**) so that the functional group will be attached to carbc number two. Addition of sulphuric acid followed by water yiel propan-2-ol which can be converted by oxidation into propanone whic

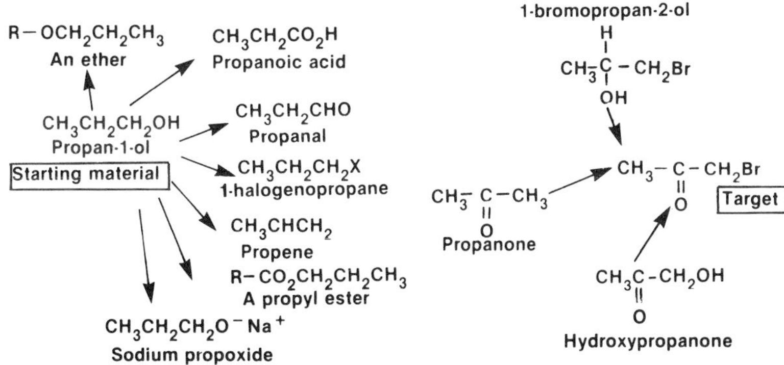

Figure 36.5 Devising a synthesis of bromopropanone from propan-1-ol

from **Figure 36.5**, can be converted into the target. So the conversion can be achieved by:

1. $CH_3CH_2CH_2OH$ propan-1-ol $\xrightarrow[Al_2O_3]{\text{hot vapour}}$ $CH_3{-}CH{=}CH_2$ propene
2. $CH_3{-}CH{=}CH_2$ propene $\xrightarrow[H_2O]{H_2SO_4 \text{ then}}$ $CH_3{-}CH(OH){-}CH_3$ propan-2-ol
3. $CH_3{-}CH(OH){-}CH_3$ propan-2-ol $\xrightarrow{Cr_2O_7^{2-}/H^+}$ $CH_3{-}CO{-}CH_3$ propanone
4. $CH_3{-}CO{-}CH_3$ propanone $\xrightarrow{Br_2/H^+}$ $CH_3{-}CO{-}CH_2Br$ bromopropanone

As an alternative to step 2 we could have added, say, HBr to give 2-bromopropane and then refluxed with aqueous alkali to give propan-2-ol. However it is always better to use the minimum number of steps to get the maximum yield.

Dehydration followed by addition of sulphuric acid or HX is a useful way of converting a primary alcohol to a secondary alcohol.

A

INDUSTRIAL MANUFACTURE COMPARED WITH LABORATORY SYNTHESIS

It is often the case that the industrial process used for manufacturing a particular chemical is different from that used in the laboratory. Some of the factors to be considered include:

- In the laboratory it is difficult to obtain very high temperatures and pressures.
- Raw material costs are much more important in an industrial process.
- Industrially, it is very helpful if a product can be made by a continuous process rather than in small batches.
- The question of by-products is important. In a small-scale preparation they can be thrown away. Industrially this may present a large-scale disposal problem, particularly if any by-products are toxic or otherwise environmentally unacceptable. Ideally, by-products should be saleable too, in which case they are known as co-products. Failing this they should be innocuous and easily disposed of.
- The **energy balance** of an industrial process is important. The energy generated by exothermic processes can be used elsewhere in the plant or to offset energy requirements of endothermic processes. This is rarely a concern in laboratory preparations.
- Generally, it is desirable to avoid elaborate purification procedures in large-scale processes.
- Ideally, very toxic reactants such as cyanides would be avoided in industrial processes, while these might be acceptable in small quantities and under careful control in the laboratory.

Some of these points are illustrated by the laboratory and industrial preparations of propanone. The obvious starting material for making propanone is propene. Propene is readily available from the cracking of petroleum fractions and is reactive because of its double bond.

The obvious laboratory synthesis would be:

1. $CH_3{-}C(H){=}CH_2$ $\xrightarrow[\text{then } H_2O]{H_2SO_4}$ $CH_3{-}C(H)(OH){-}CH_3$

Propene (Hydration) Propan-2-ol

2. $CH_3{-}C(H)(OH){-}CH_3$ $\xrightarrow[\text{Oxidation}]{Cr_2O_7^{2-}/H^+}$ $CH_3{-}C(=O){-}CH_3$

Propan-2-ol Propanone

In industry, some propanone is made by a similar route but the hydration is done by the direct reaction of steam and propene at 600 K with a catalyst of phosphoric(V) acid and silicon(IV) oxide. The oxidation is carried out using oxygen (obtained from air) as a cheap oxidizing agent. At around 600 K and 20×10^2 kPa pressure with a copper catalyst, propanone and hydrogen peroxide (a saleable co-product) are produced.

An alternative industrial process which produces propanone is the **cumene process** (see box, section 29.12), which involves reacting propene with benzene followed by oxidation and cleavage of the hydroperoxide which is produced. The advantage of this apparently roundabout process is that phenol is also produced and, indeed, is the main product. Again, air is the oxidizing agent which is much cheaper than dichromate ions. Yields of over 85% are obtained.

36.6 Questions

1 Give reactions and conditions for the conversion in one step of:
A) 1-bromobutane to pentanenitrile;
B) ethanol to ethanal;
C) propanoic acid to propanol;
D) butene to butan-2-ol;
E) cyclohexanol to cyclohexene.

2 How would you bring about the following conversions in two steps?
A) ethene to ethanoic acid;
B) ethanol to propanenitrile;
C) propanoic acid to ethylamine;
D) phenylamine to phenol;
E) propanone to 2-bromopropane;
F) benzene to phenylamine.

3 How would you convert in three steps:
A) propan-1-ol to propanone;
B) benzene to benzenecarbonitrile;
C) ethene to ethane-1, 2-dicarboxylic acid?

4 **A)** For each of the substances given below, draw the full structural formula and write the name of the class of compounds to which it belongs:
i $C_2H_5OC_2H_5$;
ii CH_3COCl;
iii $C_6H_5NO_2$.
B) For each of the three compounds given in (A), write a balanced equation in which the compound appears as a product and state the conditions under which each preparation takes place.
(Oxford 1986)

5 The diagram below shows two synthetic pathways by which the compound W might be prepared:
A) Identify the compounds Q, R, T and V by systematic names or structural formulae;
B) State the reagents and/or conditions for the conversion steps p, r, s, t and v;
C) Suggest a structure for the single organic product Z which is formed by the elimination of water from W. (London, specimen A/S, part question)

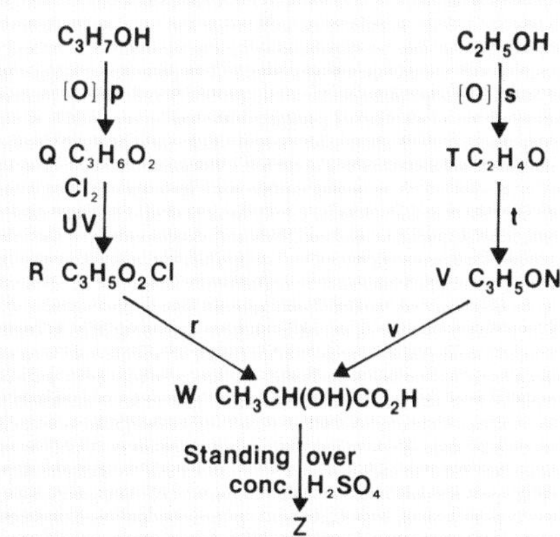

6 Describe how you would carry out any *four* of the following conversions. Balanced equations are not required but you should give reagents, structures of intermediate compounds and, as far as possible, conditions of reactions. You should try to accomplish each conversion in no more than four stages.
A) $CH_2{=}CH_2$ to CH_3CO_2H;
B) $CH_3CHBrCH_3$ to $CH_3CHBrCH_2Br$;
C) C_6H_6 to $C_6H_5CO_2H$;
D) CH_3CN to C_2H_5OH;
E) $C_2H_5CO_2H$ to $C_2H_5NH_2$;
F) C_6H_6 to $C_6H_5NHCOCH_3$.

7 Describe an organic synthesis consisting of two or more steps which you have carried out in the laboratory. Your answer should start with a clear outline of the reactions involved. (You may find a simple flow diagram is a convenient way of presenting this information.) In your description you should consider carefully each of the operations involved in the synthesis emphasizing, for example, reaction conditions, apparatus used, method(s) of purification and safety considerations. (Nuffield 1982)

37 Identifying organic compounds

37.1 Introduction

This chapter deals with the problem of how we can determine the structure of an organic compound, that is to say, what atoms are present, what bonds hold them together and how they are arranged in space. This is clearly a tall order when presented with a few grams of solid or a few cubic centimetres of a liquid. The solution lies in detective work, piecing together evidence of different types from different sources.

In practice it is fairly unusual to have no idea whatever of the identity of a compound. More often encountered is the situation where a chemist believes that he or she has synthesized or extracted a particular compound and all that is needed is confirmation of its identity—a much easier problem.

37.2 Purity

Firstly we need to know whether our unknown compound is pure. For a solid, we can measure its melting temperature. A pure solid should have a sharp melting temperature while a mixture will melt gradually over a range of few degrees. If we merely need to confirm a suspected identity, a **mixed melting point** determination can be done by mixing our unknown with a specimen of the compound we suspect it is. If the mixture melts at the same temperature as the pure compound then the identity is confirmed. If two different compounds have been mixed the melting temperature will be lower and less sharp. As a further check on purity, chromatography (see **section 11.3.3**) can be used. A pure substance will show no evidence of separation with any type of chromatography—thin layer, paper or gas–liquid as appropriate.

37.3 Qualitative analysis

The aim here is to determine what elements are present in the sample. Instrumental methods are normally used, for example, mass spectrometry, but the 'classical' method is **sodium fusion.** Other than carbon and hydrogen, present in all organic compounds, the only elements likely to be present in an organic compound are oxygen, nitrogen, halogens, sulphur and, occasionally, phosphorus. Sodium fusion tests for nitrogen, halogens and sulphur. The unknown compound is heated with sodium and plunged into water. This treatment converts nitrogen into sodium cyanide, sulphur into sodium sulphide and halogens into sodium halides, which can be tested for as shown in **Table 37.1** (overleaf).

Table 37.1 Tests for elements following sodium fusion

Element	Test	Positive result
Nitrogen, as CN^-	Add a mixture of iron(II) sulphate and iron(III) sulphate solutions and acidify	Blue–green colour (Prussian blue)
Sulphur, as S^{2-}	Add sodium pentacyanonitrosylferrate (II)	Purple colour
Halogens, as:	Acidify and boil to remove any CN^- or S^{2-}, then add silver nitrate	
Cl^-		White precipitate, soluble in excess NH_3(aq)
Br^-		Off-white precipitate, sparingly soluble in excess NH_3(aq)
I^-		Yellow precipitate, insoluble in excess NH_3(a

The presence of carbon and hydrogen can be shown by burning th substance and testing for carbon dioxide and water.

37.4 Quantitative analysis

This determines the masses of each element present and thus the empiric (simplest) formula. Mass spectrometry can help here but otherwise th basic method is **combustion analysis** which involves burning the unknow compound in excess oxygen in an apparatus like that in **Figure 37.1** an

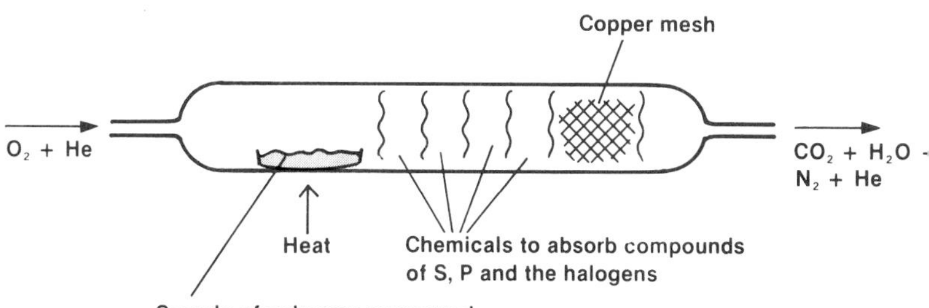

Figure 37.1 Combustion analysis

weighing the water and carbon dioxide produced. The method describe determines carbon, hydrogen and nitrogen only. Halogens and sulphu must be determined separately. Oxygen is assumed to represent th difference after the other three elements have been measured. The sampl is placed in a platinum boat and burned completely in a stream of oxyge (diluted with helium as a carrier gas, to flush through unused oxygen an combustion products). The combustion products then pass through a serie of substances which absorb any compounds of sulphur, phosphorus an the halogens leaving a gas stream containing the helium and unused oxyge plus the carbon dioxide, water and nitrogen oxides from combustion c the sample. The stream then passes over a heated copper mesh whic combines with the excess oxygen and reduces oxides of nitrogen to nitroge gas:

$$2Cu(s) + O_2(g) \longrightarrow 2CuO(s)$$
$$xCu(s) + NO_x(g) \longrightarrow xCuO(s) + \tfrac{1}{2}N_2(g)$$

The emerging gas stream thus contains the carbon dioxide, water an nitrogen from the sample. The amount of water is determined by passin the gas through magnesium chlorate(VII), which absorbs water, and th carbon dioxide is determined by absorbing it in soda lime, leaving jus nitrogen and the helium carrier gas. Traditionally, the amounts of wate and carbon dioxide were measured by weighing the absorbants, but no the thermal conductivity of the gas stream is measured before and afte absorption, which is also the method used to determine the nitrogen.

From the masses of the combustion products the empirical (simplest formula can be found as shown in the example below.

Example: Compound X

0.23 g of a compound X containing carbon, hydrogen and oxygen onl yielded 0.44 g of carbon dioxide and 0.27 g of water on complete combustion in oxygen. What is its formula?

Carbon

$$0.44\,\text{g of } CO_2 (M_r = 44) \text{ is } 0.44/44 = 0.01\,\text{mol } CO_2$$

As each mol of CO_2 has 1 mol of C, the sample contained 0.01 mol of C atoms.

Hydrogen

$$0.27\,\text{g of } H_2O\ (M_r = 18) \text{ is } 0.27/18 = 0.015\,\text{mol } H_2O$$

As each mol of H_2O has 2 mol of H, the sample contained 0.03 mol of H atoms.

Oxygen

0.01 mol of carbon atoms ($A_r = 12$) is 0.12 g
0.03 mol of hydrogen atoms ($A_r = 1$) is 0.03 g
total of C plus H is 0.15 g

The rest (0.23 – 0.15 g) must be oxygen, so the sample contained 0.08 g of oxygen.

$$0.08\,\text{g of oxygen } (A_r = 16) \text{ is } 0.08/16 = 0.005\,\text{mol oxygen atoms}$$

Formula
So the sample contains:

0.01 mol C
0.03 mol H
0.005 mol O

Dividing through by the smallest (0.005) gives the ratio:

C 2
H 6
O 1

so the **empirical formula** is C_2H_6O.

37.5 Relative molecular mass

The empirical formula of compound X in the example could represent a range of **molecular formulae**—C_2H_6O, $C_4H_{12}O_2$, etc., represented by $(C_2H_6O)_n$. To find the value of n we need to know the relative molecular mass, M_r. A variety of methods is available.

Gas volume measurements (see **section 15.3**) can be used for a gas or a volatile liquid:

In the above example, 0.23 g of the substance (empirical formula C_2H_6O) had a volume of 155 cm³ at 373 K and 10^5 Pa pressure. Using the ideal gas equation:

$$PV = nRT$$

$$n = \frac{PV}{RT}$$

$$n = \frac{10^5 \times 155 \times 10^{-6}}{8.3 \times 373}$$

$$= 5 \times 10^{-3}\,\text{mol}$$

P = pressure in Pa (= 10^5)
V = volume in m³ (= 155×10^{-6})
n = number of moles of gas = ?
R = gas constant (= $8.3\,\text{J K}^{-1}\,\text{mol}^{-1}$)
T = temperature in K (= 373)

So: 5×10^{-3} mol has a mass of 0.23 g
1 mol has a mass of $0.23/5 \times 10^{-3}$ g
1 mol has a mass of 46 g
so M_r is 46

The empirical formula is C_2H_6O. The molecular formula is (C_2H_6O) What is n?

The M_r of the empirical formula is $(2 \times 12) + (6 \times 1) + (1 \times 16) = 4$ The M_r of the sample is 46, so $n = 1$; the molecular formula (in this cas is the same as the empirical formula.

Other methods for determining M_r include depression of meltir temperature or elevation of boiling temperature (see **section 16.3.5**) f solids, rate of diffusion (see **section 15.4.2**) for a gas, and mass spectromet (see **section 19.3.4**). The latter is the preferred technique as it can give oth information, as well as just the relative molecular mass.

37.6 Mass spectrometry

The principle of this technique has been described in **section 8.5** and son of its applications can be found in **section 19.3.4**. For organic structu determination it can offer four types of information:

1. Relative molecular mass can be determined from the mass/charge rat of the parent or molecular ion.
2. The presence of certain elements can be deduced from isotope peak For example, chlorine-containing compounds will produce pairs peaks differing by two mass units, whose abundances are in the rat 3:1, due to the isotopes ^{35}Cl and ^{37}Cl.
3. High resolution mass spectrometry, which measures relative molecul masses to four or five decimal places, enables us to distinguish betwe compounds containing different elements but which have the san relative molecular mass to the nearest whole number.
4. Fragmentation of the ions as they fly through the mass spectromet can also give clues to structure, as we shall see in the examples.

The mass spectrum of compound X is given in **Figure 37.2.** The pare ion confirms the value of $M_r = 46$.

Figure 37.2 The mass spectrum of compound X

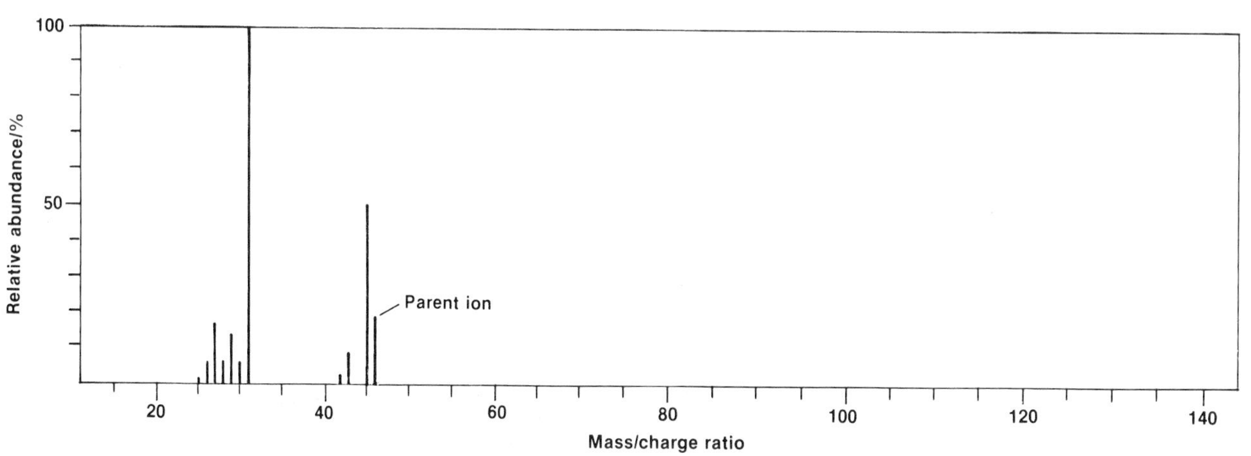

37.7 Functional groups

Once we know the molecular formula we need to know the function groups present. This is done by a combination of study of the chemic reactions of the compound and infra-red spectroscopy.

37.7.1 Chemical reactions

Some tests are very straightforward. Is the compound acidic (suggests carboxylic acid or a phenol) or basic (suggests an amine)? Is the compoun solid (suggests long carbon chain or ionic bonding), liquid (suggests medium length carbon chain or polar or hydrogen bonding) or gas (suggests short carbon chain, little or no polarity)? Does it dissolve in water (suggests polar groups) or not (suggests no polar groups)?

Does the compound burn with a smoky flame (suggests high C:H ratio, possibly aromatic) or non-smoky flame (suggests low C:H ratio, probably non-aromatic)? Some specific chemical tests are listed in **Table 37.2.**

able 37.2 Chemical tests for nctional groups

Test	Result	Inference	Notes
Add Br_2 dissolved in CH_3CCl_3	Decolorizes	C=C present	Br_2 will react with alkanes in strong light so test must be done in dark
Add Na	H_2 given off	—OH Present	Could be carboxylic acid, phenol or alcohol
If —OH present, add $Na_2CO_3(aq)$	CO_2 produced	$—CO_2H$ present	But could be a phenol with strongly electron-withdrawing groups
Add neutral $FeCl_3(aq)$	Purple colour	C_6H_5—OH present	
Add Brady's reagent	Orange precipitate	Aldehyde or ketone present	
Heat with: Benedict's solution or Tollen's reagent	Brick-red precipitate / Silver mirror	Aldehyde present	
Heat with I_2/NaOH(aq)	Yellow precipitate	CH_3—CH(OH)— or CH_3—C(=O)— present	Triiodomethane reaction
Add $CuSO_4(aq)$ and NaOH(aq) and warm	Lilac colour	Peptide link present, —C(=O)—N(H)—	Biuret test
Add ninhydrin solution	Blue colour	Amino acid or peptide present	

This may be sufficient to identify the compound. In the example of compound X, the molecular formula was C_2H_6O. If this compound gave off hydrogen on adding sodium, the only possibility would be ethanol, CH_3CH_2OH. This could be confirmed by a positive triiodomethane test.

37.7.2 Infra-red (IR) spectra

This method has been described in **section 19.3.3.** Particular functional groups produce peaks in different areas of the spectrum, as summarized in **Figure 37.3.**

igure 37.3 Infra-red absorptions of me functional groups

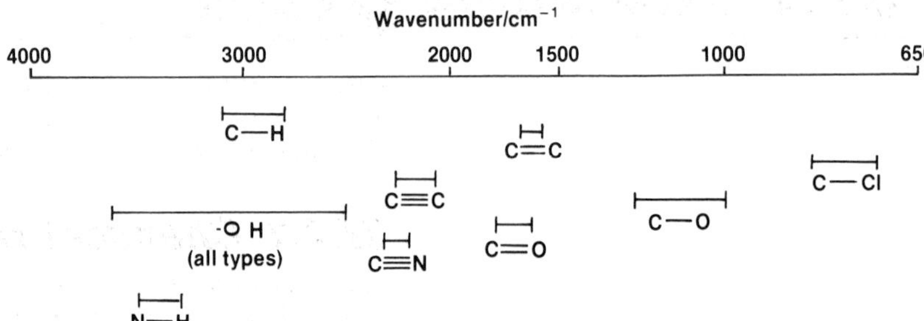

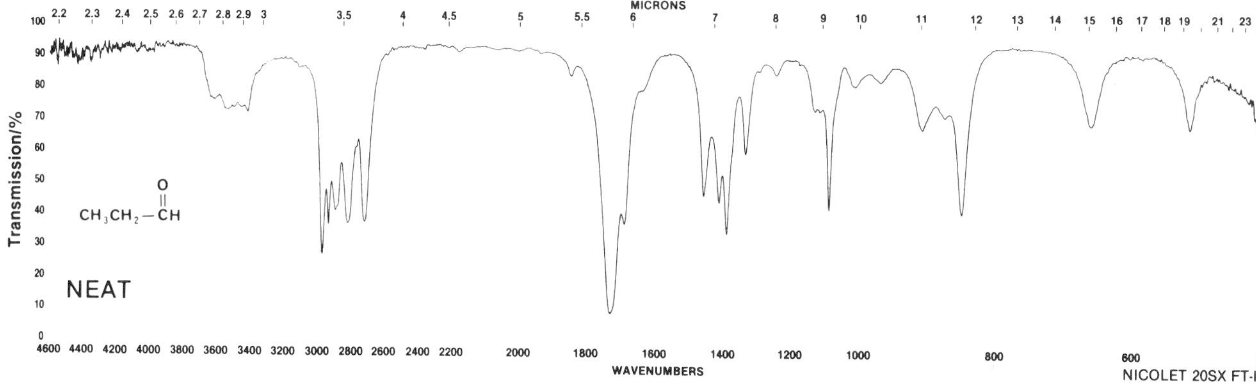

Figure 37.4 **The infra-red spectrum of propanal. Remember 'peaks' point downwards**

A typical infra-red spectrum is given in **Figure 37.4.**

Variations in the rest of the molecule can shift the peak slightly, shown in **Table 37.3** for the C=O group, which appears in sever functional groups.

You must, however, use common sense. There is no point in decidi that a compound is a carboxylic acid on the basis of a peak, at, sa $1720\,cm^{-1}$, if the compound is not acidic.

Table 37.3 **Position of carbonyl (C=O) peaks in the I.R. spectra of different compounds**

Molecule	Wave number/cm^{-1}
Ketone (R_2CO)	1725–1700
Aldehyde (RCHO)	1740–1720
Aromatic ketone (R_2CO)	1700–1680
Aromatic aldehyde (RCHO)	1715–1695
Carboxylic acid (RCO_2H)	1725–1700
Ester (RCO_2R)	1750–1730
Amide ($RCONH_2$)	1680–1640
Acid chloride (RCOCl)	1815–1790

37.7.3 Nuclear magnetic resonance (NMR) spectra

The method has been described in **section 19.3.5.** It gives information abo the number and environment of hydrogen atoms in the molecu Resonance occurs when the applied magnetic field is of just the rig magnitude to allow a hydrogen nucleus to 'flip' to a higher energy sta However, the hydrogen nucleus is protected from the external field some extent by the electrons. This is called **shielding**. Hydrogen ator bonded to electronegative atoms like oxygen, are shielded less than on bonded to, for example carbon, as the oxygen pulls the electrons awa The amount of shielding is measured in NMR spectra by the **chemic shift** (δ) from a reference compound tetramethylsilane (TMS), **Figure 37.** The *greater* the chemical shift, the *less* the shielding. Most hydrogen ator in organic molecules have values of δ between 0 and 10, see **Figure 37.**

Figure 37.5 **Tetramethylsilane – TMS**

$$H_3C-Si(CH_3)_2-CH_3$$

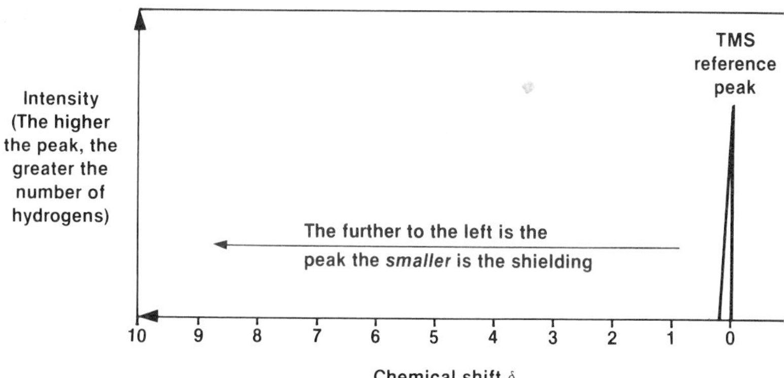

Figure 37.6 **NMR spectra use the compound TMS as a reference**

In ethanol, there are three types of hydrogen atom:

$$H_a-O-C(H_b)_2-C(H_c)_2-H_c$$

The three-dimensional representation makes it clear that all three H_cs a identical. One atom, H_a, attached to the oxygen atom, has little shieldin

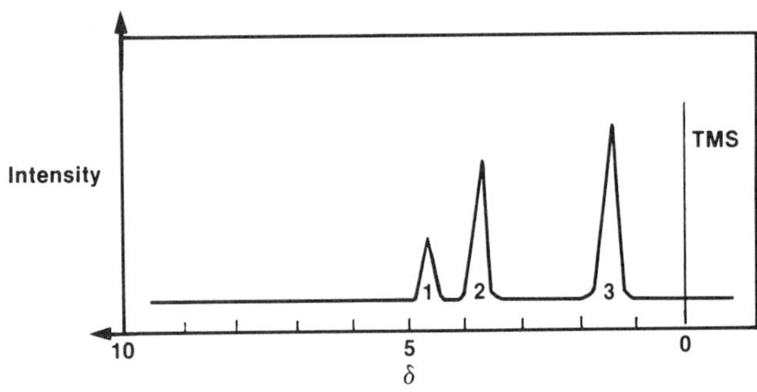

'gure 37.7 The NMR spectrum of hanol

able 37.4 Typical hydrogen chemical ift values (δ)

pe of hydrogen atom	Chemical shift
$-CH_3$	0.9
$-CH_2-R$	1.3
CH	2.0
$_3-C(=O)OR$	2.0
$CH_3-C(=O)-$	2.1
)$-CH_3$	2.3
$-C\equiv C-H$	2.6
$-CH_2-Hal$	3.2–3.7
$-O-CH_3$	3.8
$-O-H$	4.5
$HC=CH_2$	4.9
$HC=C$	5.9
)$-OH$	7
)$-H$	7.3
$-C(=O)H$	9.7
$-C(=O)O-H$	11.5

as the oxygen pulls the electrons away from it. Two atoms, H_bs, attached to carbon 1, are more shielded as the effect of the oxygen is less, while three atoms, H_cs, bonded to carbon 2, are the most shielded as the oxygen is now two atoms away. So the NMR spectrum is as shown in **Figure 37.7.** The peak areas are in the ratio 1:2:3. So NMR tells us how many types of hydrogen atoms are in the molecule and something about their environments. For example, ethanol, CH_3CH_2OH has three types of hydrogen as shown above. Its isomer, methoxymethane, CH_3-O-CH_3, has only one type of hydrogen (there are six of them) with a shielding approximately the same as the H_b hydrogens in ethanol, as they are bonded to a carbon adjacent to an oxygen. Its NMR spectrum is shown in **Figure 37.8.** Some values of chemical shift, δ, are given in **Table 37.4.**

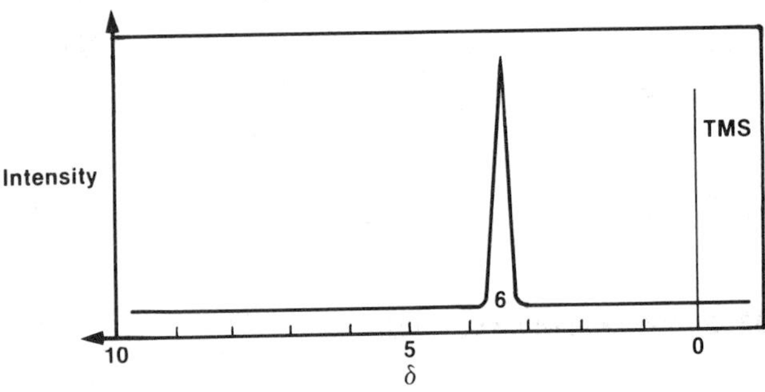

'igure 37.8 The NMR spectrum of nethoxymethane

The following examples work through the above techniques to identify two compounds, Y and Z.

37.7.4 Compound Y

Compound Y contains C, H and O only.

0.44 g, on combustion, gave 0.88 g CO_2 and 0.36 g H_2O. The infra-red spectrum, NMR spectrum and mass spectrum are given in **Figure 37.9.** Compound Y gave an orange precipitate with Brady's reagent and a brick-red precipitate with Benedict's test.

Empirical formula:

$$0.44\,g \text{ gave } 0.88/44 = 0.02\,mol\ CO_2, \text{ i.e. } 0.02\,mol\ C$$
$$\text{and } 0.36/18 = 0.02\,mol\ H_2O, \text{ i.e. } 0.04\,mol\ H$$

Figure 37.9 IR and overleaf, NMR and mass spectra for compound Y

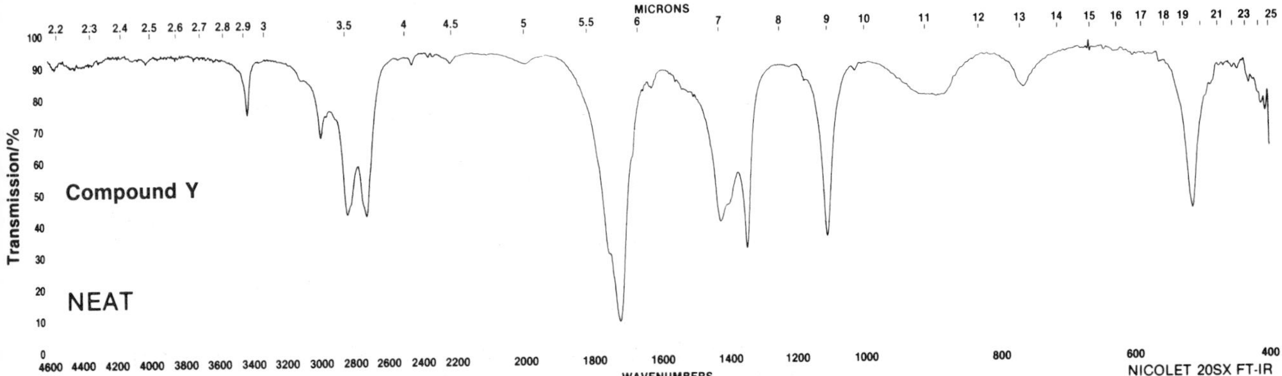

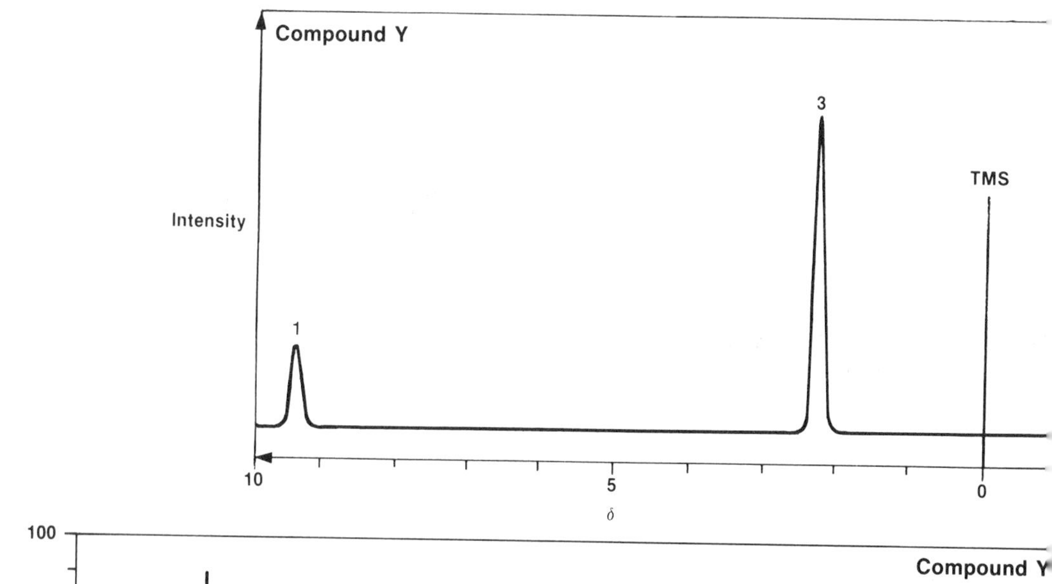

$$0.02 \text{ mol C} = 12 \times 0.02 = 0.24 \text{ g C}$$
$$0.04 \text{ mol H} = 1 \times 0.04 = 0.04 \text{ g H}$$
$$\text{total C} + \text{H} = 0.28 \text{ g}$$
$$\text{rest: } 0.44 - 0.28 = 0.16 \text{ g is oxygen}$$
$$0.16/16 = 0.01 \text{ mol O}$$
$$\text{empirical formula} = C_2H_4O$$
$$M_r \text{ of empirical formula} = (2 \times 12) + (4 \times 1) + (1 \times 16) = 44$$

mass spectrum gives a molecular ion peak at $M_r = 44$, see **Figure 37.9**

so molecular formula $= C_2H_4O$

A positive Benedict's test indicates an aldehyde so the only possibility is ethanal:

CH_3—C(=O)—H

From **Figure 37.9** this is confirmed by:

1. a peak at approximately 1750 cm^{-1} in the infra-red spectrum due to C=O
2. the NMR shows two types of hydrogen atoms in the ratio 3:1. The single proton is very unshielded—this is the —C(=O)—H—the δ value fits

with the value in **Table 37.4** for such a hydrogen. The three hydrogens of $\delta \simeq 2$ are those of the CH_3 group.

3. the mass spectrum shows fragments of $M_r = 29$ and $M_r = 15$ corresponding to the C—C bond breaking (see margin, left).

15 29

37.7.5 Compound Z

Compound Z contains C, H and O only.

0.74 g on combustion gave 1.32 g CO_2 and 0.54 g H_2O. The infrared spectrum, NMR spectrum and mass spectrum of Z are shown in **Figure 37.10.** The compound is acidic, effervescing on addition of sodium carbonate solution.

gure 37.10 IR, NMR and mass ectra for compound Z

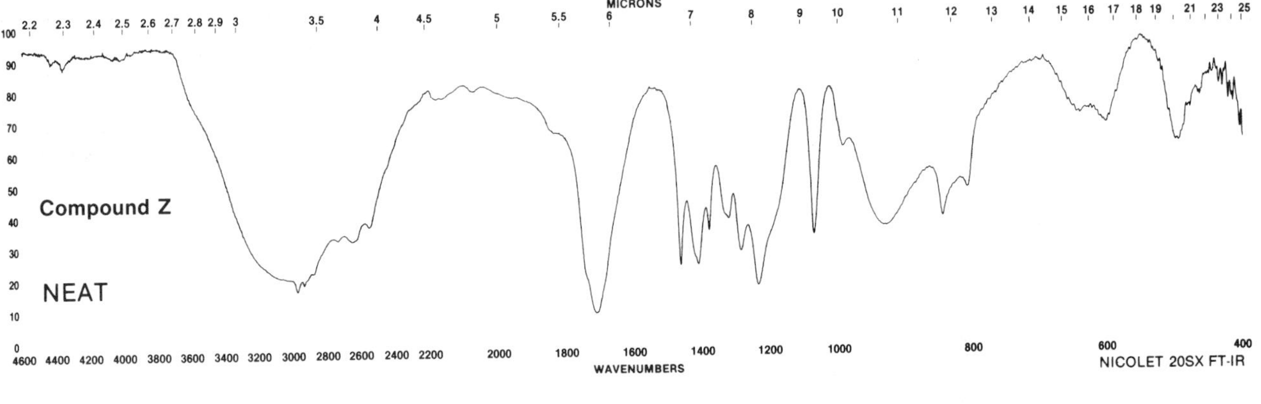

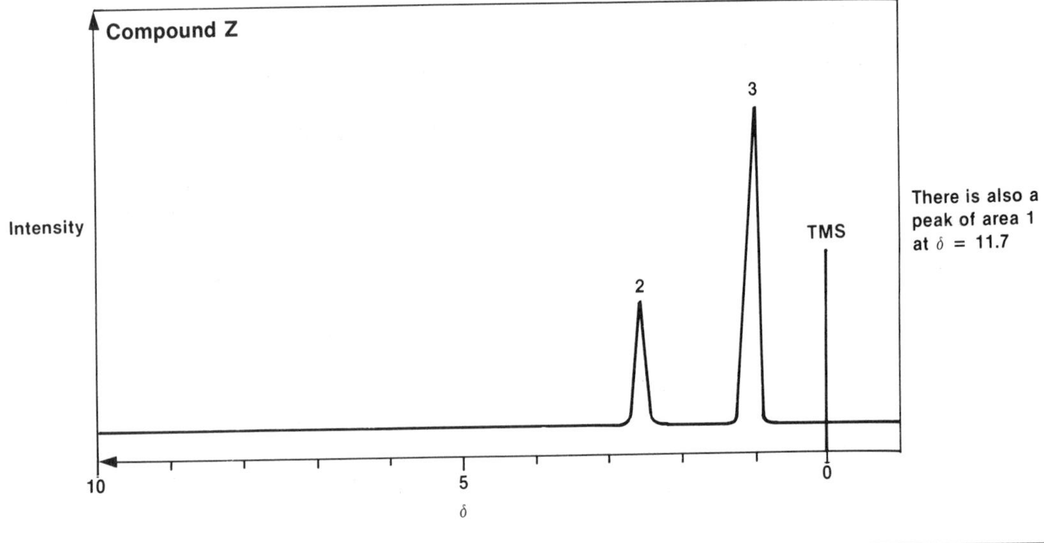

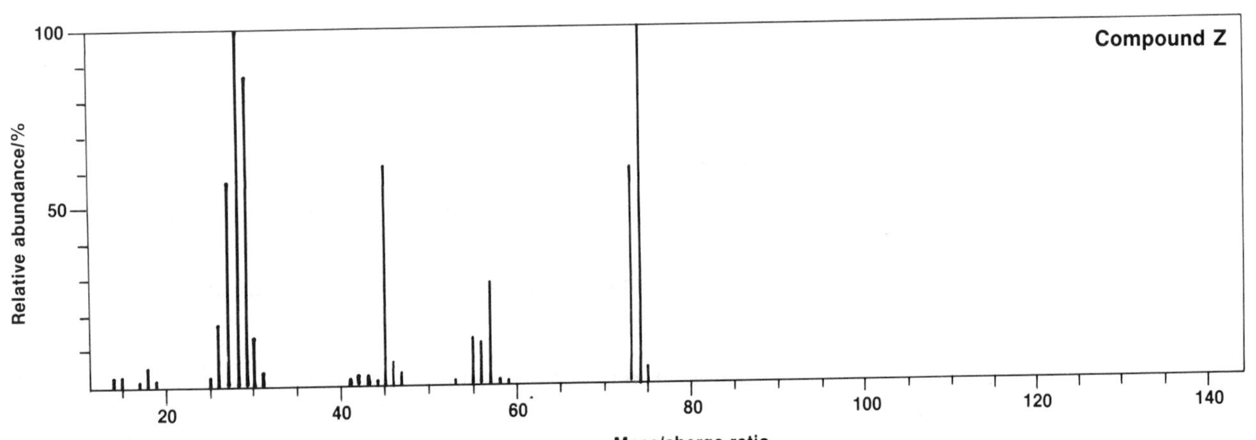

Empirical formula:

$$0.74\,\text{g gave } 1.32/44 = 0.03\,\text{mol } CO_2\text{, i.e. } 0.03\,\text{mol C}$$
$$0.54/18 = 0.03\,\text{mol } H_2O\text{, i.e. } 0.06\,\text{mol H}$$
$$0.03\,\text{mol C} = 12 \times 0.03 = 0.36\,\text{g C}$$
$$0.06\,\text{mol H} = 1 \times 0.06 = 0.06\,\text{g H}$$
$$\text{total C} + \text{H} = 0.42\,\text{g}$$
$$\text{rest: } 0.74 - 0.42 = 0.32\,\text{g is oxygen}$$
$$0.32/16 = 0.02\,\text{mol O}$$
$$\text{empirical formula} = C_3H_6O_2$$
$$M_r \text{ of empirical formula} = 74$$

mass spectrum gives a molecular ion at $M_r = 74$, see **Figure 37.10**
so molecular formula = $C_3H_6O_2$

Effervescence with sodium carbonate suggests a carboxylic acid, so th only possibility is propanoic acid:

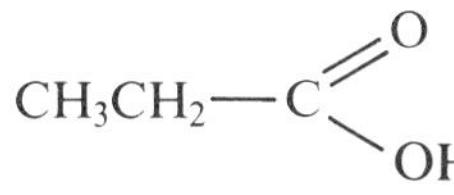

$CH_3CH_2—C(=O)OH$

From **Figure 37.10** this is confirmed by:

1. peaks in the infra-red spectrum at about 1720 cm^{-1} caused by C=O and a very broad peak at about 3000 cm^{-1} due to O—H (broadened due to hydrogen bonding)
2. the NMR shows three types of hydrogen atoms in the ratio 1:2:3 a the formula for propanoic acid suggests. The single hydrogen has a value expected for a —C(=O)—O—H hydrogen
3. the mass spectrum shows peaks at 45, corresponding to $C^+(=O)OH$ and at 29, corresponding to $CH_3CH_2{}^+$.

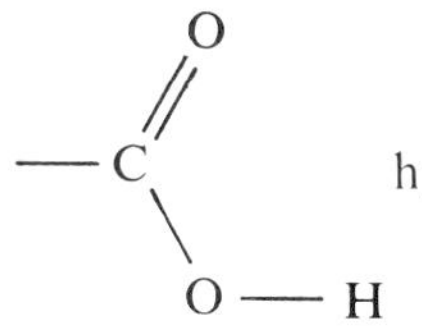

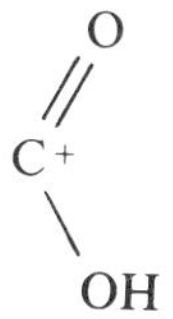

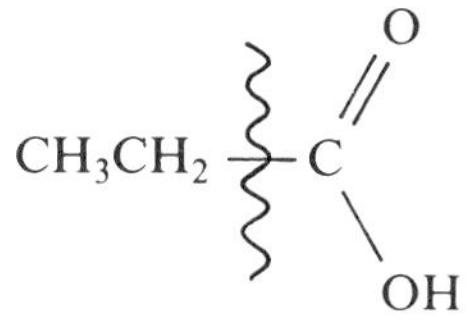

CH_3CH_2 ⌇ $C(=O)OH$

29 45

37.8 Summary

- Organic structure determination involves the following stages:
  A check on the purity of the substance;
  Qualitative analysis to determine the elements present;
  Quantitative analysis by combustion to determine the empirical formula;
  Determination of the relative molecular mass to find the molecular formula;
  Chemical tests to determine which functional groups are present.
- Mass spectrometry can be used to find:
  Relative molecular mass from the mass of the parent ion;
  The presence of certain elements from isotopic abundance ratios;
  Molecular formula (by high resolution mass spectrometry);
  Clues to structure from fragmentation patterns.
- Infra-red spectra can suggest which functional groups are present.
- NMR spectra can tell how many hydrogen atoms of different types are

present and something about their environment.

- A complete structure determination usually requires information from a number of techniques.

37.9 Questions

1 Compound A is a liquid containing C, H and O only. On combustion, 0.36 g gave 0.88 g CO_2 and 0.36 g H_2O. The liquid was neutral and mixed with water but did not decolorize bromine water. It formed an orange precipitate with Brady's reagent but did not react with Benedict's solution. The IR spectrum, NMR spectrum and mass spectrum are shown.

A) Calculate the empirical formula and molecular formula.

B) What is the structure of the compound? Explain how the evidence leads to your conclusion and give as much confirming evidence as possible.

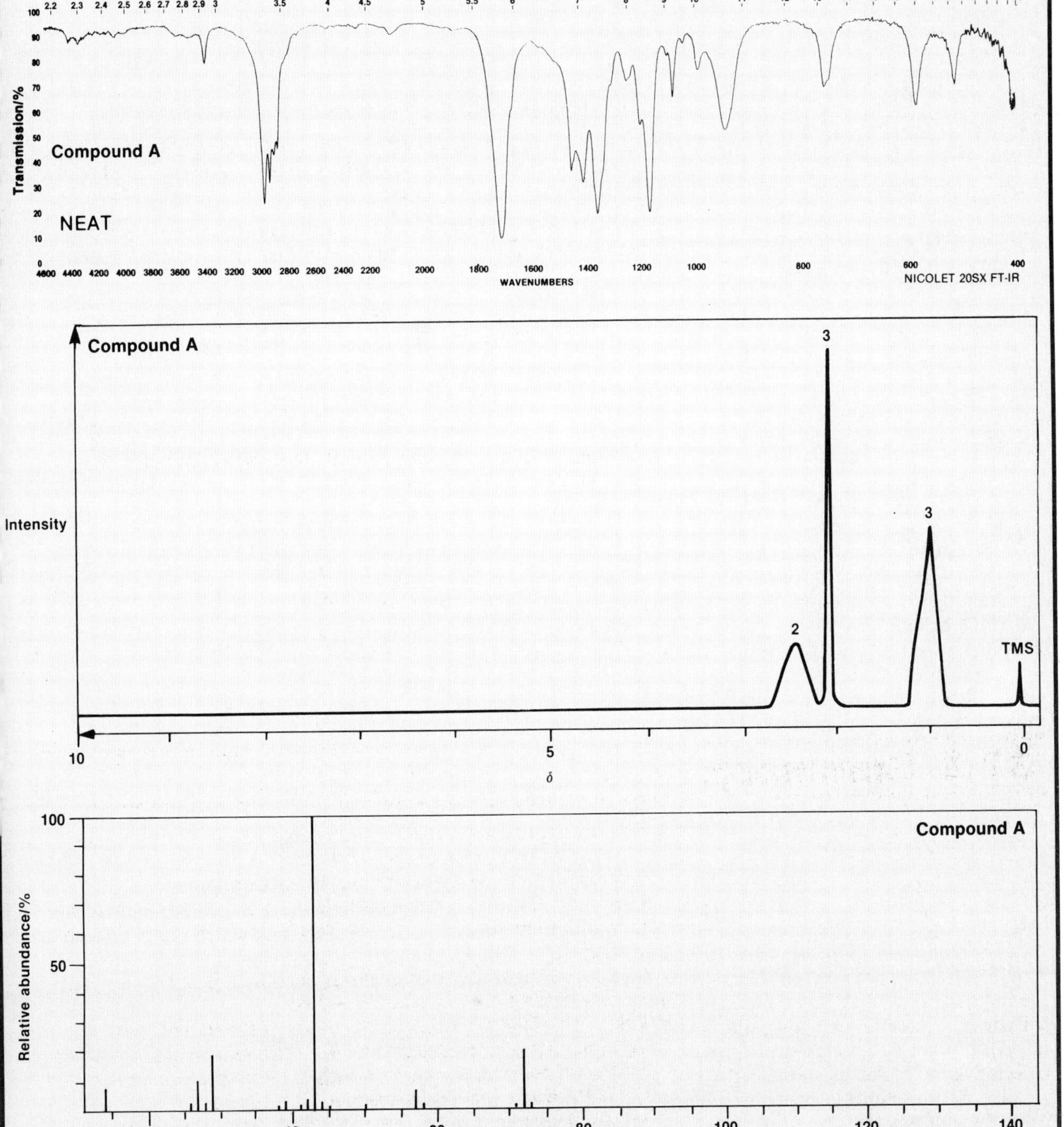

2 Chlorine ($A_r = 35.457$) exists as two isotopes ^{35}Cl and ^{37}Cl. Interpret the following mass spectra, identifying all the peaks as far as you can and giving reasons for your answers.

(a) Molecular chlorine, Cl_2

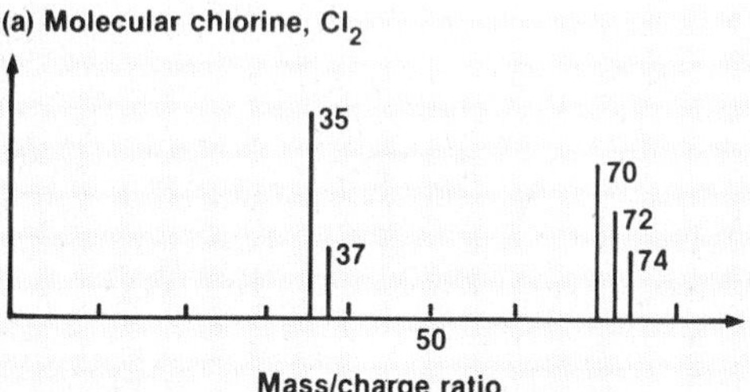

(b) Chloroethane

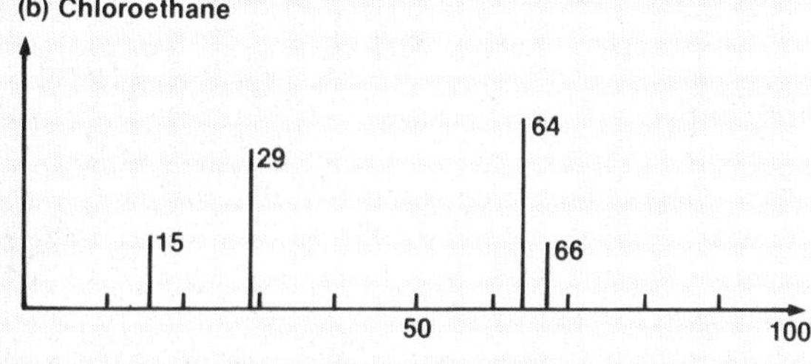

(c) A product $C_2H_4Cl_2$ formed when ethene reacts with molecular chlorine

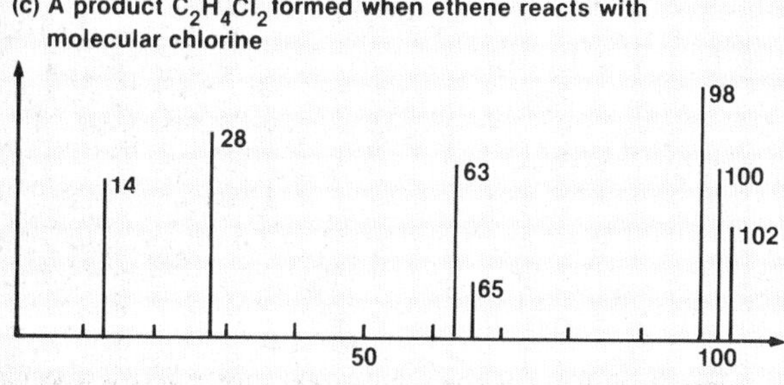

(London, specimen A/S, part question)

3 Give a brief account of how a compound gives rise to a characteristic infra-red absorption spectrum.

An organic compound, A, of molecular formula $C_4H_5O_2N$ and $M_r = 99$, has the following properties:

A) A is fairly soluble in water, yielding a weakly acidic solution;

B) When refluxed with dilute acid, A forms a compound B of composition C = 40.7%, H = 5.1%, O = 54.2%, ($M_r = 118$) which gives a more strongly acidic solution than A;

C) A is optically active.

Suggest structures for A and B, explaining how you reach your conclusions.

(Nuffield 1986)

4 Two organic compounds, A and B, are isomers with the composition by mass of carbon 70.5%, hydrogen 5.9% and oxygen 23.6%. A is moderately soluble in water and B is a pleasant-smelling liquid. Their mass spectra are shown above right.

A) i What is the empirical formula of A and B? (Relative molecular masses: C = 12, O = 16, H = 1)

ii What is the molecular formula of A and B? Justify your answer.

B) Give the formulae of the molecular fragments corresponding to the following peaks:

Mass/charge ratio
136
105
91
77

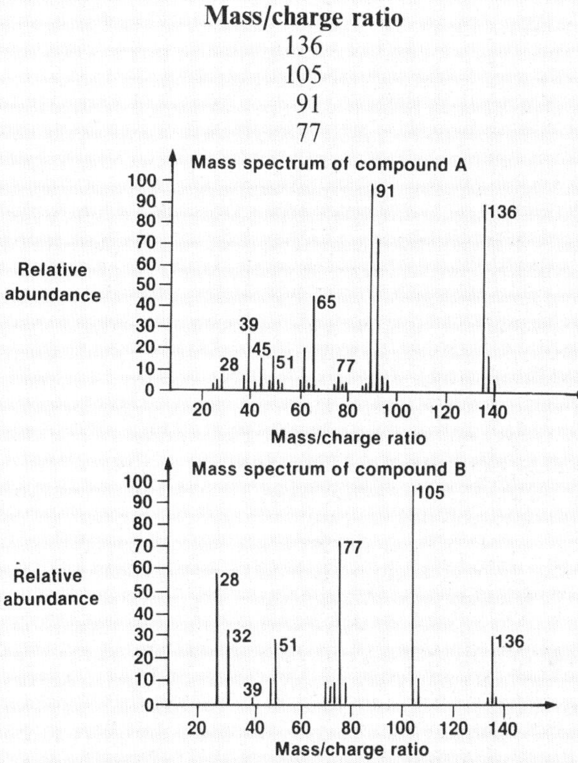

C) What structural formulae would you predict for A and B?

D) Describe *two* tests or chemical reactions in which the behaviour of A and B would differ.

(Nuffield 1984)

5 A pleasant-smelling liquid, A, was hydrolysed in acidic conditions and gave two organic products.

One product, B, was readily soluble in water and was a monocarboxylic acid. Its mass spectrum is shown below.

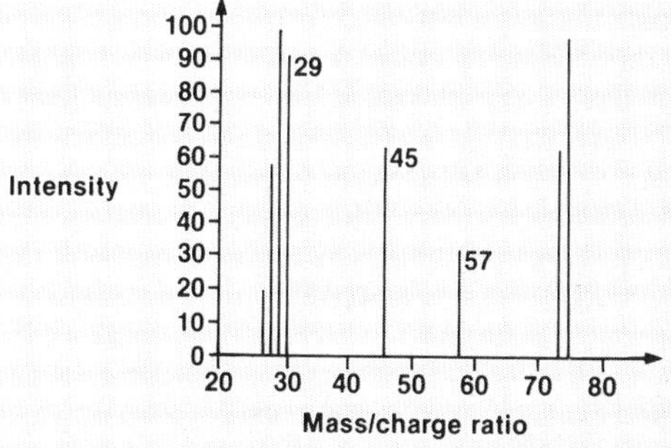

A) i What is the relative molecular mass of B?

ii Suggest fragment ions which could be responsible for the peaks listed below. (Relative atomic masses: H = 1, C = 12, O = 16)

29
45
57

iii Hence deduce the displayed formula and name of B.

The second product, C, was also readily soluble in water and on analysis by mass was found to contain 60.0% carbon, 13.3% hydrogen and 26.7% oxygen.

The relative molecular mass of C was found to be 60. (Relative atomic masses: H = 1, C = 12, O = 16)

B) i Calculate the molecular formula of C.

ii Draw three possible displayed formulae for isomers having this molecular formula.

C was oxidized to D by refluxing with acidified sodium dichromate(VI). D was neither acidic, nor did it react on warming with Fehling's solution but it contained the same number of carbon atoms per molecule as C.

C) i To what class of compound does D belong?

ii Hence deduce the correct formula for C and name it.

D) From your answers to parts (A), (B) and (C) draw the displayed formula for A.

E) Outline two experiments you could do in your school laboratory to confirm the identity of D.

(Nuffield 1988, part question)

6 A compound containing an amide group was hydrolysed and gave two organic compounds as products.

One compound was sparingly soluble in water, burnt with a smoky flame and was a monocarboxylic acid; 0.21 g of this monobasic acid when titrated with 0.10 M sodium hydroxide gave an end-point of 17.2 cm^3. Calculate the relative molecular mass of the organic acid and suggest a structural formula.

The second compound was readily soluble in water and gave a purple coloration when warmed with ninhydrin; but the solution was not optically active when tested in a polarimeter. Analysis by mass of a sample produced the result C 32.0%, H 6.7%, O 42.7%, N 18.7%. Calculate the empirical formula of the second compound and suggest a possible structural formula.

Draw a graphical formula of the original compound (a structural formula showing all atoms and bonds).

Give *two* examples of chemical reactions by which simple amides can be formed.

(Nuffield 1983)

7 A substance containing carbon, hydrogen and oxygen only, when heated under reflux with aqueous acid for several hours, could be split into two components.

The lower boiling component was a liquid at room temperature. 1.0 g, on complete combustion, produced 2.2 g of carbon dioxide and 1.2 g of water. On volatilization in a gas syringe at 100 °C and one atmosphere pressure, 0.10 g of the liquid produced 51 cm^3 of vapour. When the liquid was oxidized, the product gave a crystalline derivative with 2,4-dinitrophenylhydrazine, but would not react with Fehling's solution. The higher boiling component from the heating under reflux was neutralized with ammonia and the product heated with phosphorus(V) oxide, P_2O_5. This yielded a product of relative molecular mass 97, whose composition by mass was 74.2% carbon, 11.3% hydrogen and 14.4% nitrogen. Using the information provided, suggest a possible identity for each of the substances involved, and explain the reaction of the original substance with aqueous acid when heated under reflux.

(Nuffield, 1981)

Numerical answers

Chapter 8

3. 63.6
4. 118.8 kJ mol^{-1}
3960 kJ mol^{-1}
39.6 kJ mol^{-1}
1.188×10^{10} kJ mol^{-1}
5. $v = 8.05 \times 10^{14}$ s^{-1}
A) 5.313×10^{-19} J per atom
B) 318.7 kJ mol^{-1}
7. 2.88×10^{-11} J per atom
1.73×10^{13} J mol^{-1}
4.08×10^{-4} g
8. A) 9922 kJ mol^{-1}
11. $v = 1.59 \times 10^{14}$ s^{-1}
12. $t_{\frac{1}{2}} = 2.95$ days
8860 count min^{-1}
13. 1.697×10^{-39} m^3
2.2×10^{14} kg m^{-3}
14. A) 2000 counts min^{-1}
B) 2 hours
16. A) 1.5×10^{13} s^{-1}
B) 2×10^{-5} m
17. 11 200 years

Chapter 9

1. A) 46.2 kJ mol^{-1}
C) 46.2 kJ mol^{-1}
2. C) 2160 J
D) -216 kJ mol^{-1}
3. A) 6.5×10^{-3} mol
B) 10.302 kJ K^{-1}
C) i) -1999.0785 kJ mol^{-1}
ii) -2649.615 kJ mol^{-1}
iii) -3310.275 kJ mol^{-1}
4. B) 12.4 K
C) 129.0 g
5. $+2061$ kJ mol^{-1}
6. A) -70 kJ mol^{-1}
B) -217 kJ mol^{-1}
C) -97 kJ mol^{-1}
D) -195 kJ mol^{-1}
E) $+301$ kJ mol^{-1}
8. -62 kJ mol^{-1}
9. A) -85 kJ mol^{-1}
B) -86 kJ mol^{-1}
11. B) i) 84.5 kJ
ii) -1690 kJ mol^{-1}
C) $+3918.0$ kJ mol^{-1}
D) $+3919.0$ kJ mol^{-1}

Chapter 10

3. -733.7 kJ mol^{-1}
-2254.8 kJ mol^{-1}
-3844.5 kJ mol^{-1}
-2862.1 kJ mol^{-1}
7. $+174$ kJ mol^{-1}

Chapter 11

4. 56.25
5. C) 4.12
7. 25.15 kPa
9. CO 0.458 mol
H_2O 0.458 mol
H_2 0.541 mol
CO_2 0.541 mol
CO 0.88 mol
H_2O 0.88 mol
H_2 0.12 mol
CO_2 9.11 mol
10. A) 0.909 g
B) 0.972 g
11. 0.45
12. 0.522, 0.488
13. A 14%, B 82%, C 4%
15. 10^3 Pa
16. 48.4

Chapter 12

1. A) i) 0.699
ii) 3
iii) 0.3
B) i) 1.69
ii) 2.69
iii) 0
C) i) 12
ii) 11
iii) 13.69
D) i) 2.1
ii) 2.6
iii) 3.1
E) i) 7.8
ii) 8.3
2. 2.5×10^{-5} mol dm^{-3}
3. 4.6
4. 0.63
7. B) 2.87
8. B) 11.5
9. B) i) 1.31×10^{-3} mol dm^{-3}
ii) 1.7×10^{-5} mol dm^{-3}

Chapter 13

4. A) $+0.93$ V
B) -0.55 V
C) $+0.97$ V
D) $+0.13$ V
7. A) -1.66 V
B) 9.9×10^{-3} mol dm^{-3}
8. B) -0.25 V
C) 0.95 V
9. C) iii) 0.24 mol dm^{-3}

Chapter 14

2. A) -5×10^{-3} mol dm^{-3} s^{-1}
B) $+3 \times 10^{-3}$ mol dm^{-3} s^{-1}
3. C) 8.3×10^4 dm^6 mol^{-2} s^{-1}
9. D) 6.75 mol dm^{-3} s^{-1}
10. B) 10.18 kJ mol^{-1}
11. 109 kJ mol^{-1}
13. A) 103.75 kJ mol^{-1}

Chapter 15

1. A) 94.19 cm^3
B) 52.72 cm^3
C) 1.365 dm^3
D) 91.84 m^3
E) 1.69 pints
3. 154.4
4. 99.96
5. A) 235.8
7. A) i) 83.3 s
ii) 17.7 s
C) 64

Chapter 16

7. B) 70 mm Hg
C) 88 mm Hg
8. B) 328.9
9. A) 28% propanone
B) 50% propanone
10. 0.52×10^{-3} K
1.86×10^{-3} K
2.47 kPa
11. 1.04 K, 3.72 K, 4946.8 kPa
12. 120
60

Chapter 17

2. Ni^{2+}
4. 3.55×10^6 amps
5. 3328.125 kg

apter 18

. **A)** 16
B) 1/16
C) 87.5%

. **B)** **i)** $-7.8\ J\ K^{-1}\ mol^{-1}$
ii) $+876.4\ J\ K^{-1}\ mol^{-1}$
iii) $-174.8\ J\ K^{-1}\ mol^{-1}$
iv) $+119\ J\ K^{-1}\ mol^{-1}$

. **B)** **i)** $+493\ kJ\ mol^{-1}$
ii) $-52\ kJ\ mol^{-1}$
C) 5522.9 K

. **A)** **a)** $198.8\ J\ K^{-1}\ mol^{-1}$
b) **i)** $308.7\ J\ K^{-1}\ mol^{-1}$
ii) $46\ J\ K^{-1}\ mol^{-1}$
c) **i)** $109.9\ J\ K^{-1}\ mol^{-1}$
ii) $-152.8\ J\ K^{-1}\ mol^{-1}$
B) **a)** $360.8\ J\ K^{-1}\ mol^{-1}$
b) **i)** $-1154.3\ J\ K^{-1}\ mol^{-1}$
ii) $-172\ J\ K^{-1}\ mol^{-1}$
c) **i)** $-793.5\ J\ K^{-1}\ mol^{-1}$
ii) $188.8\ J\ K^{-1}\ mol^{-1}$

. $-521\ kJ\ mol^{-1}$
$-523\ kJ\ mol^{-1}$

. $1.74 \times 10^3\ mol\ dm^{-3}$

7. **B)** $G = 0$ when $T = 900\ K$
D) 2.79×10^{-3}

9. **A)** $-1327\ kJ\ mol^{-1}$
$-94.9\ J\ K^{-1}\ mol^{-1}$
B) $-1298.7\ kJ\ mol^{-1}$
C) $-758.4\ kJ\ mol^{-1}$

10. **A)** 1 in 1.2×10^{24}
B) 3.8×10^{10} years
C) **i)** $2^{6 \times 10^{23}}$
ii) $5.7\ J\ K^{-1}\ mol^{-1}$
iii) $160.7\ J\ K^{-1}\ mol^{-1}$

Chapter 19

1. 0.3202 nm
10.5°

4. **A)** C_2H_4O
B) 44

Chapter 22

1. $-102\ kJ\ mol^{-1}$
$-306\ kJ\ mol^{-1}$

3. $+73\ kJ\ mol^{-1}$

Chapter 23

9. **D)** **ii)** 0.19 g

Chapter 24

(Group III) **2.** **E)** **i)** 175.5
(Group V) **1.** **B)** $+942\ kJ\ mol^{-1}$
6. **E)** 83.4%
H) **i)** 236
(Group VI) **1.** **C)** $-394.5\ kJ\ mol^{-1}$
(Group VII) **3.** **A)** **i)** 0.015 mol of I_2
ii) $7.5\ cm^3$
4. **A)** 50.8 g
C) **i)** 14.3 g
ii) 55.3 g
D) 60 °C
E) 100% KIO_3
6. **F)** $-511\ kJ\ mol^{-1}$

Chapter 25

4. **B)** **ii)** 0.79 V
iii) $-457\ kJ\ mol^{-1}$

5. **D)** **i)** 0.31 V
iii) $-90\ kJ\ mol^{-1}$ of Cr^{3+}

9. **B)** **i)** $4 \times 10^{-3}\ mol$
iii) $2 \times 10^{-3}\ mol$

Chapter 27

5. $6 \times 10^{14}\ s^{-1}$

6. 50%

Chapter 30

11. $1.875 \times 10^{-3}\ mol$
95% elimination
5% substitution

Chapter 31

7. **D)** 59.4%

Index

The Periodic Table

Key: Relative atomic mass / Symbol / Name / Atomic number

Group; s-block (I–II), d-block, p-block (III–0)

Period	I	II											III	IV	V	VI	VII	0
1	1.0 H Hydrogen 1																	4.0 He Helium 2
2	6.9 Li Lithium 3	9.0 Be Beryllium 4											10.8 B Boron 5	12.0 C Carbon 6	14.0 N Nitrogen 7	16.0 O Oxygen 8	19.0 F Fluorine 9	20.2 Ne Neon 10
3	23.0 Na Sodium 11	24.3 Mg Magnesium 12											27.0 Al Aluminium 13	28.1 Si Silicon 14	31.0 P Phosphorus 15	32.1 S Sulphur 16	35.5 Cl Chlorine 17	39.9 Ar Argon 18
4	39.1 K Potassium 19	40.1 Ca Calcium 20	45.0 Sc Scandium 21	47.9 Ti Titanium 22	50.9 V Vanadium 23	52.0 Cr Chromium 24	54.9 Mn Manganese 25	55.9 Fe Iron 26	58.9 Co Cobalt 27	58.7 Ni Nickel 28	63.5 Cu Copper 29	65.4 Zn Zinc 30	69.7 Ga Gallium 31	72.6 Ge Germanium 32	74.9 As Arsenic 33	79.0 Se Selenium 34	79.9 Br Bromine 35	83.8 Kr Krypton 36
5	85.5 Rb Rubidium 37	87.6 Sr Strontium 38	88.9 Y Yttrium 39	91.2 Zr Zirconium 40	92.9 Nb Niobium 41	95.9 Mo Molybdenum 42	(99) Tc Technetium 43	101.1 Ru Ruthenium 44	102.9 Rh Rhodium 45	106.4 Pd Palladium 46	107.9 Ag Silver 47	112.4 Cd Cadmium 48	114.8 In Indium 49	118.7 Sn Tin 50	121.8 Sb Antimony 51	127.6 Te Tellurium 52	126.9 I Iodine 53	131.3 Xe Xenon 54
6	132.9 Cs Caesium 55	137.3 Ba Barium 56	138.9 * La Lanthanum 57	178.5 Hf Hafnium 72	181.0 Ta Tantalum 73	183.9 W Tungsten 74	186.2 Re Rhenium 75	190.2 Os Osmium 76	192.2 Ir Iridium 77	195.1 Pt Platinum 78	197.0 Au Gold 79	200.6 Hg Mercury 80	204.4 Tl Thallium 81	207.2 Pb Lead 82	209.0 Bi Bismuth 83	(210) Po Polonium 84	(210) At Astatine 85	(222) Rn Radon 86
7	(223) Fr Francium 87	(226) Ra Radium 88	(227) † Ac Actinium 89	(261) Unq Unnil-quadium 104	(262) Unp Unnil-pentium 105	(263) Unh Unnil-hexium 106												

f-block

*Lanthanides	140.1 Ce Cerium 58	140.9 Pr Praseo-dymium 59	144.2 Nd Neodymium 60	(147) Pm Promethium 61	150.4 Sm Samarium 62	152.0 Eu Europium 63	157.3 Gd Gadolinium 64	158.9 Tb Terbium 65	162.5 Dy Dysprosium 66	164.9 Ho Holmium 67	167.3 Er Erbium 68	168.9 Tm Thulium 69	173.0 Yb Ytterbium 70	175.0 Lu Lutetium 71
†Actinides	232.0 Th Thorium 90	(231) Pa Protactinium 91	238.1 U Uranium 92	(237) Np Neptunium 93	(242) Pu Plutonium 94	(243) Am Americium 95	(247) Cm Curium 96	(245) Bk Berkelium	(251) Cf Californium	(254) Es Einsteinium	(253) Fm Fermium	(256) Md Mendele-vium	(254) No Nobelium	(257) Lr Lawrencium